ENGINEERING MECHANICS
(Statics & Dynamics)

5th Revised Edition

Dr. D.S. Kumar

"For Namrata, Ritvik, Arshia and Ankit who have added a lot of joy to my life."

ENGINEERING MECHANICS

(Statics & Dynamics)

by

DR. D.S. KUMAR

B.Sc. Engg. (Hons.), M.Sc. Engg, Ph.D, FIE

Formerly :

Professor Mech. Engg, Punjab Engg. College, Chandigarh,
Director Academic Affairs, Punjab Technical Univ., Jalandhar,
Principal S.U.S. College of Engg. & Technology, Tangori (Mohali) and
Advisor, Manav Rachna Educational Institutions, Faridabad

S.K. KATARIA & SONS®

Publisher of Engineering Books

4885/109, Prakash Mahal, Dr. Subhash Bhargav Lane,
Opposite Delhi Medical Association, Daryaganj, New Delhi-110002 (INDIA)
Phones: +91-11-23243489, 43551243; Mobile: +91-9871775858
e-mail: katariabook@gmail.com; katariabooks@yahoo.com
Website : www.skkatariaandsons.com

(e-books, subscriptions and bulk buying also available on our website)

Published by:
S.K. Kataria & Sons®
4885/109, Prakash Mahal, Dr. Subhash Bhargav Lane,
Opposite Delhi Medical Association, Daryaganj, New Delhi-110002
Phones: +91-11-23243489, 43551243; Mobile: +91-9871775858
e-mail: katariabook@gmail.com; katariabooks@yahoo.com
Website: www.skkatariaandsons.com

Head Office:
Opp. Clock Tower, Ludhiana (Pb)
Ph. : 2726401

First Edition: January, 2003; Second Edition: 2004; Third Edition: July 2006; Fourth Edition: July, 2012; Fifth Edition: January, 2016 Reprint : 2017 Reprint : 2018 Reprint: 2019 Reprint: 2021 Reprint: 2022 Reprint: 2023

Price: ₹ 795/-

Laser Typesetting:
Plus Computers, Shahdara, Delhi

Printed at:
AL-AZIZ BOOKS, SAHIBABAD

Preface to the Fifth Edition

The author is grateful to the teachers and students of various engineering institutions for their appreciation and healthy criticism of the earlier editions of the book.

To strengthen the text, the book has now been thoroughly revised and edited by incorporating the suggestions made both by teachers and students. The salient features of this revised edition are :

- Considerable changes and additions have been made in the entire text to clear the points that were not apparent or needed additional information for better understanding. Some chapters have been completely overhauled and rewritten with the inclusion of additional articles.
- Topics on Shear Force and Bending Moment, and Power Transmission : Methods and Devices have been added to complete the curriculum requirements of many examining bodies.
- Several of the drawings of earlier editions have been replaced by new ones to improve presentation and clarity. Mistakes and misprints too have been rectified.
- A wide variety of conventional problems set by various engineering boards have been selected, solved and added in this text at the appropriate places.
- Learning objectives have been added to the introduction of each chapter to set the purpose and goal of the topics described therein.
- Each chapter ends with a review and summary of the material covered in that chapter; this is to help the students to comprehend the subject matter prior to sitting for any test or examination.
- There are also sections on objective questions keeping in view the recent trend of such questions being asked in the various university and competitive examinations.

The author hopes that these features will greatly enhance the utility of the book for the student community and teaching faculty. Any suggestions from the readers for further updating the book will be greatly appreciated and gratefully acknowledged.

D.S. Kumar

World Spa, Gurgaon
January, 2016

Preface to the First Edition

Engineering Mechanics forms a core subject which is taught to students of all disciplines of engineering in all the technical institutions. The study of this subject is aimed at developing a thorough understanding of the basic concepts and principles of mechanics and their application to solve engineering problems.

The author had taught this subject for several years at Punjab Engg. college, Chandigarh and this monograph is essentially based on the lectures delivered by him. From the experience gained through useful class discussions and feed back, these class notes were revised to improve their clarity and necessary explanatory notes were added during each teaching semester. The subject matter has thus been thoroughly tested in the class room and found suitable.

The text presents an amalgam of numerous ideas and influences and is the outcome of the distillation of the available text books written by distinguished authors. Obviously then a question arises as to 'Why another book on Engineering Mechanics' The answer to this question is reflected by the following salient features incorporated in this book:

- Precise and logical presentation of the basic concepts and principles
- Text written in simple, lucid and easily understandable language
- Full use of line diagrams made to supplement the text and to explain a particular phenomenon as clearly as possible
- Solutions provided to a wide variety of problems of standard comparable to those set for engineering degree, AMIE, GATE and Engineering Service Examinations
- Consistent use of SI units and notations throughout the text

The author strongly believes that this treatise on Engineering Mechanics will contribute effectively to the existing literature in the field, and will be a useful text for the students of engineering curriculum.

Acknowledgements: On completion of the book **'Engineering Mechanics'** I wish to acknowledgement my thanks and gratitude to all those who have been associated with this work directly/indirectly. Acknowledgement is due and is hereby made to all the authors whose works have been freely consulted in the preparation of this text. I owe an enormous debt of gratitude to my colleagues and students for much valued assistance in the form of discussion and feedback. A word of thanks is due to the publishers too for their considerable patience and cooperation throughout. Above all, I with to place on record my sincerest thanks and apologies to my near and dear ones who willingly endured certain hardships as a result of my per-occupation with this work. A special mention in this respect is reserved for my grand-children **Tanu** and **Ritvik.**

My effort in writing this book will by amply rewarded if the readers send their constructive suggestions and objective criticism with a view to enhance the usefulness of the book. Further, I would gratefully acknowledge the contribution of the readers by way of bringing to my attention the errors and printing mistakes that might have inadvertently crept into the book.

CITM, Faridabad
January, 2003

D.S. Kumar

Contents

Dynamics : Kinematics and Kinetics

1 Basic Concepts

Learning objectives : Attention has been focussed in this chapter to introduce the reader to the following basic concepts of mechanics

- *matter, particle and body; space, time, motion and trajectory*
- *fundamental laws of motion*
- *scalar and vector quantities, force and its representation, system of forces*
- *principle of transmissibility*
- *system of units; dimensional homogenity*

Engineering mechanics is a basic subject which describes and predicts the effects of forces on rigid bodies. It essentially deals with the study of bodies at rest or in motion when subjected to external mechanical disturbances. A knowledge of its basic concepts and principles is a must for an engineer engaged in the design and construction of various types of machines and structures.

Refer the chart given below for tree diagram depicting the various branches of the physical science of mechanics and their subdivisons.

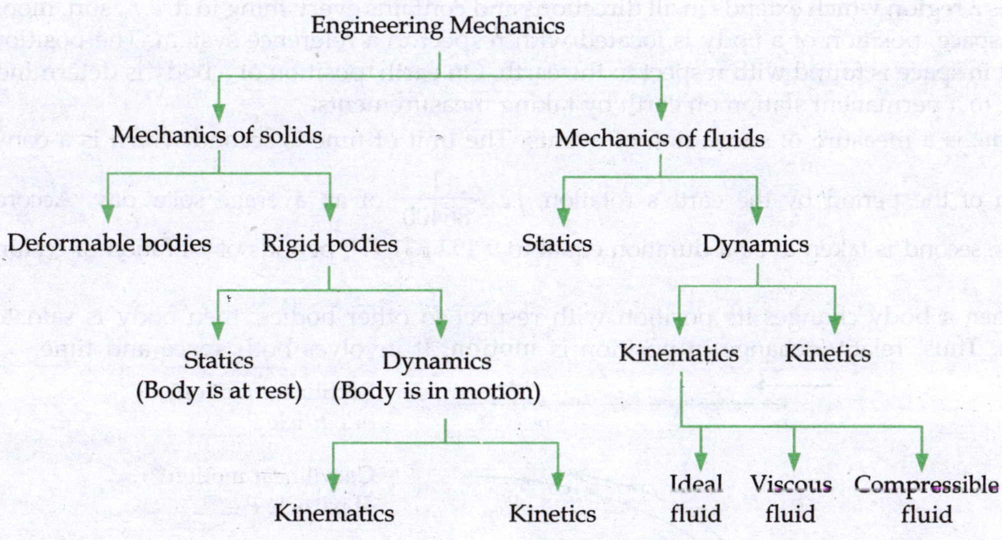

Classification of Engineering Mechanics

The studies and concepts in the domain of engineering mechanics are :

- *Mechanics of solids and mechanics of fluids* : Mechanics of solids is the study of solids at rest or in motion, where as mechanics of fluids deals with the study of liquids and gases at rest or in motion.
- *Deformable bodies and rigid bodies* : The deformable bodies deform when acted upon by forces where as no deformation occurs in a rigid body. The rigid body may change its position or orientation under the action of applied force. However the relative position of the particles constituting the rigid body remains unchanged.
- *Statics and dynamics* : Statics deals with forces in terms of their distribution and effect on a body in equilibrium, *i.e.*, at absolute or relative rest. *Dynamics* deals with the study of bodies in motion.
- *Kinematics and kinetics* : Kinematics is concerned with the description of motion of objects independent of causes of motion. Here study is made of motion interrelationship among position, velocity, acceleration and time without taking into account of the forces causing motion. In *Kinetics*, both the motion and its causes are considered. Kinetics relates the action of forces and the resulting motion.

Both kinematics and kinetics relate to the domain of the dynamics of a rigid body.

1.1. MATTER, PARTICLE AND BODY

Matter is anything that occupies space, possesses mass and offers resistance to any external force. Iron, stone, wood salt, water and air etc represent matter.

A *particle* is an object that has infinitely small volume (occupies negligible space) but has a mass which can be considered to be concentrated at a point.

A *body* has a definite shape and consists of numbers of particles.

A body in which the distance between any two particles remains constant, *i.e.*, the size and shape of the body do not change, is called a *rigid* body. A rigid body exhibits no relative deformation between two parts when it is acted upon by a force system.

An *elastic* body undergoes deformation but regains its original shape after removal of the external forces.

1.2. SPACE, TIME, MOTION AND TRAJECTORY

Space is a region which extends in all directions and contains every thing in it, *e.g.*, sun, moon, stars etc. In space, position of a body is located with respect to a reference system. The position of an aircraft in space is found with respect to the earth. On earth, position of a body is determined with respect to a permanent station on earth by taking measurements.

Time is a measure of succession of events. The unit of time is second which is a convenient fraction of the period by the earth's rotation, *i.e.*, $\frac{1}{86400}$ of an average solar day. According to SI units, second is taken to be a duration equal to 9 192 631 770 periods of vibration of cesium – 133 atom.

When a body changes its position with respect to other bodies, then body is said to be in motion. Thus, relative change in position is **motion**. It involves both space and time.

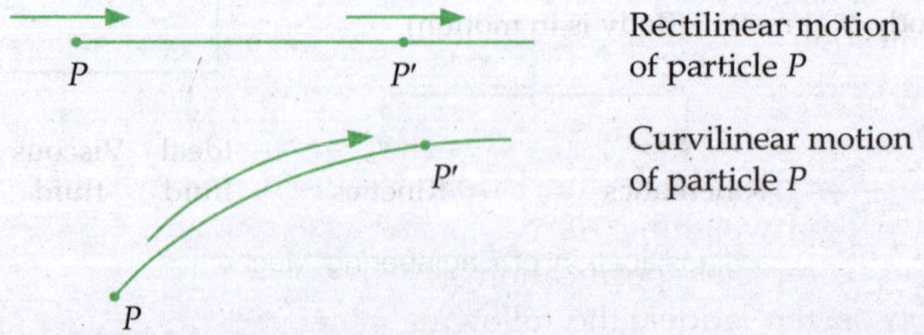

Fig. 1.1

Trajectory is path followed by a body during its motion. It may be a straight line or a curve. Straight line path is observed in *rectilinear motion*, whereas curved path is observed in *curvilinear motion*.

1.3. SCALAR AND VECTOR QUANTITIES

A quantity is said to be *scalar* if it is completely defined by its magnitude alone. Mass, length, volume, temperature and time etc., are scalar quantities.

Any quantity which possesses magnitude as well as direction is called a *vector* quantity. A vector quantity needs both magnitude and direction for its complete specification. Some examples of vector quantities are displacement, velocity, acceleration, momentum and force.

A vector quantity is represented by a directed segment of a straight line. The length of the line represents the magnitude of the vector on some scale. The direction of the line indicates the direction of the vector, and arrowhead indicates the sense of direction. For example, a force of 50 N acting in the *x-y* plane and making 30° with *x*-axis is represented by segment *OA* as shown in Fig. 1.2.

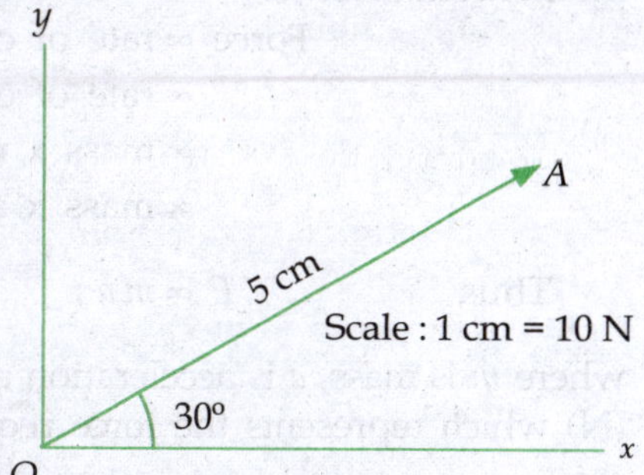

Fig. 1.2

A vector is denoted by a letter of English alphabet with an arrow at its top such as $\vec{a}$, $\vec{A}$, $\vec{X}$ etc. The magnitude of any vector $\vec{A}$ is denoted by $|\vec{A}|$.

In Bow's method of force representation, a force is designated by writing two capital letters one on either side of it. With reference to Fig. 1.3, a force of P_1 = 10 kN is represented by *AB* and P_2 = 15 kN is represented by *CD*.

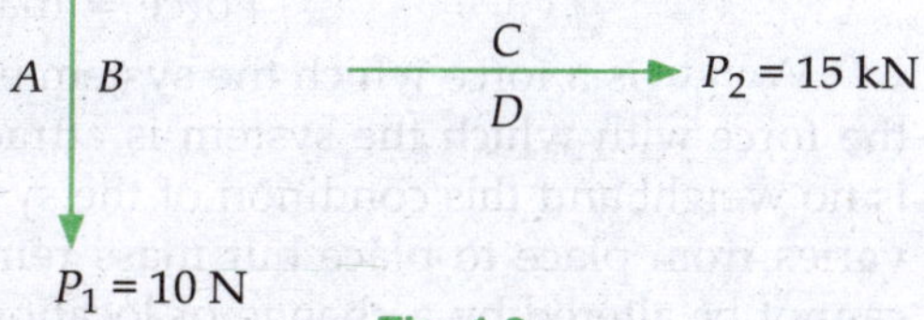

Fig. 1.3

1.4. MASS, FORCE AND WEIGHT

Mass is an indication of the quantity of matter present within a system. The more mass means more matter.

Force is an external agent which tends to change the speed or direction of a system. A force is applied whenever the system needs to be accelerated or decelerated. A force on a rigid body may produce one or both of the following effects:

— change its state of rest or motion
— accelerate or retard its motion
— change its shape or size
— turn or rotate it, and
— keep it in equilibrium.

A force is a vector quantity determined completely by its

— magnitude
— point of application
— line of action, and
— direction.

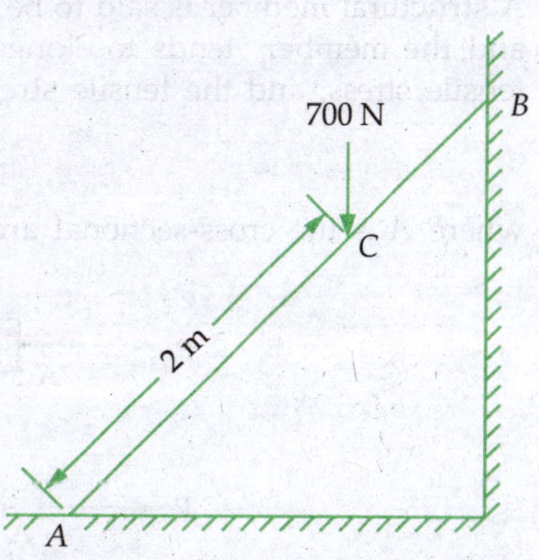

Fig. 1.4

With reference to Fig. 1.4, *AB* is the ladder that rests against a wall and a person weighing 700 N stands at point *C* on the ladder. For complete specification of the force exerted by the person on the ladder, the following characteristics need to be specified:

— magnitude is 700 N
— the point of application is at a distance of 2 m from the floor along the ladder
— the line of action of force is vertical, and
— the direction is downward.

Newton's second law of motion states that force is proportional to rate of change of momentum. That is

$$\begin{aligned}\text{Force} &\propto \text{rate of change of momentum}\\ &\propto \text{rate of change of (mass x velocity)}\\ &\propto \text{mass x rate of change of velocity}\\ &\propto \text{mass x acceleration}\end{aligned}$$

Thus $\quad F \propto m\,a\,; \qquad F = \dfrac{m\,a}{g_c}$

where m is mass, a is acceleration and g_c is proportionality constant. The SI unit of force is newton (N) which represents the force required to accelerate 1 kg mass with an acceleration of 1 m/s^2

Thus $\quad 1\text{ N} = \dfrac{1\text{ kg} \times 1\text{ m/s}^2}{g_c}$ and $g_c = \dfrac{1\text{ kg m}}{\text{Ns}^2}$

The constant of proportionality becomes unity for defining unit of force as newton and accordingly dropped. Such a practice corresponds to writing Newton's second law as

$$\text{Force} = \text{mass x acceleration}; \quad F = m\,a$$

Weight is a force which the system exerts due to gravitational acceleration. Weight is actually the force with which the system is attracted towards the earth. Where there is no gravity, there is no weight and this condition of the system is known as weightlessness. The weight of a system varies from place to place but mass remains constant. The mass of a given collection of matter cannot be altered by a change of location, a deformation of shape, a change of temperature, and through chemical reactions which the matter may undergo.

The weight (W) of system equals the product of mass (m) and local gravitational acceleration. The mass of a system remains constant whereas weight varies with change in value of gravitational acceleration from one place to another. The value of g at sea level is 9.8066 m/s^2 and is generally taken as 9.81 m/s^2.

1.5. TENSION AND COMPRESSION

A structural member is said to be in tension when it is subjected to two equal and opposite pulls and the member tends to elongate/increase in length. The stress induced in the bar is called tensile stress, and the tensile stress σ_t at any cross-section X-X is given by

$$\sigma_t = \frac{P}{A}$$

where A is the cross-sectional area of the bar.

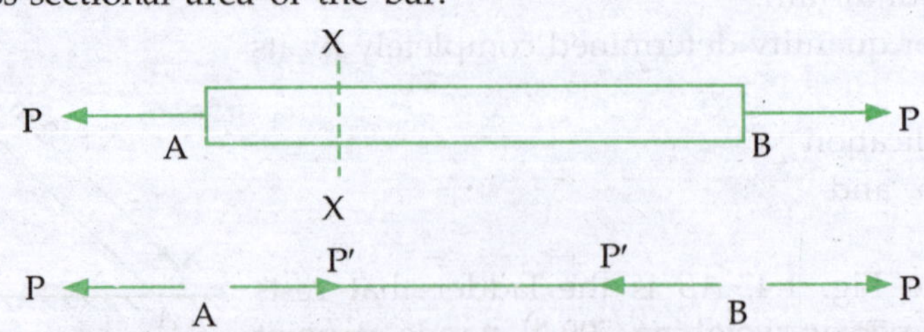

Fig. 1.5. Bar under tension

At every point of bar AB, equal and opposite forces P' (where $P' = P$) act. The force P' acting at point A balances the external force P; hence when considering the equilibrium of A, the arrow showing P' must be directed toward B. Likewise, when considering the equilibrium of B, the arrow points towards A.

The hoisting ropes used in cranes and passenger elevators are the elements subjected to tensile loading. An element under tension is called a *tie*.

If the structural member is subjected to two equal and opposite pushes, and the member tends to shorten/decrease in length, the member is said to be in compression. The stress so produced is called compressive stress, and the compressive stress σ_c at any cross-section X-X is given by

$$\sigma_c = \frac{P}{A}$$

Under equilibrium conditions of the bar, the resistance force P' (where $P' = P$) will be directed as shown in Fig. 1.6. A bar under compression is called a *strut*.

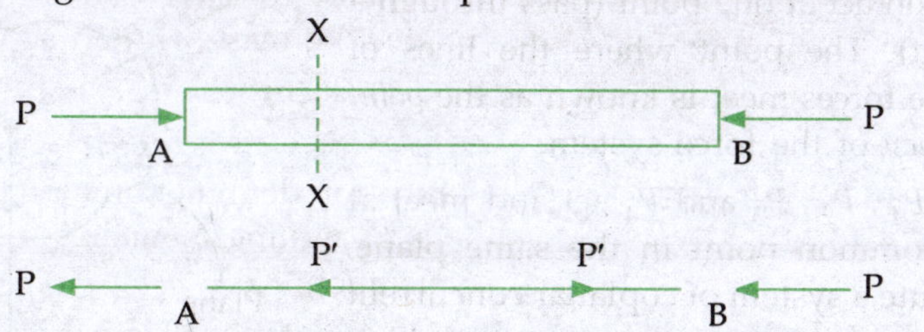

Fig. 1.6. Bar under compression

1.6. SYSTEM OF FORCES

When several forces of different magnitude and direction act upon a body, they constitute a force system.

Considering the plane in which forces are applied and depending upon the position of line of action, forces may be classified as :

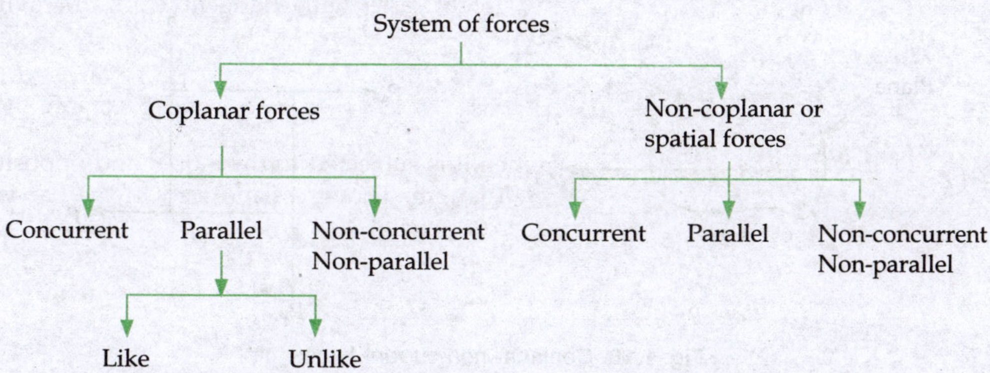

- *Collinear forces:* The lines of action of all forces lie along the same straight line and in the same plane.

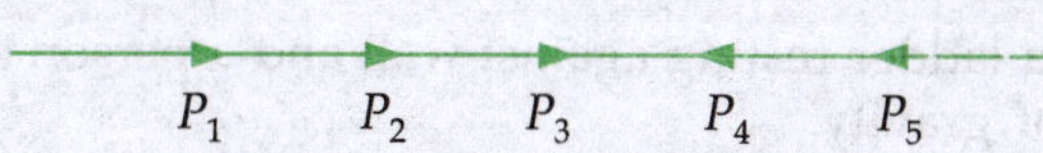

Fig. 1.7. Collinear force system

The forces P_1, P_2, P_3, P_4 and P_5 are acting along the same straight line, and constitute a collinear force system.

Example: Force on a rope in a tug of war.

- *Coplanar parallel forces* : The lines of action of all forces are parallel to each other and lie in a single plane.

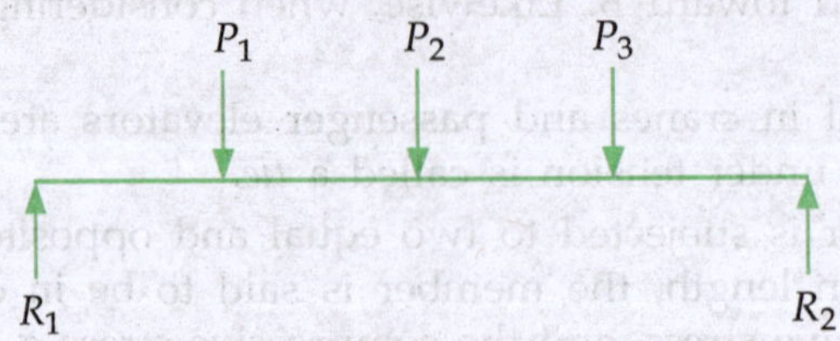

Fig. 1.8. Coplanar parallel forces

Example: System of vertical loads (including reactions) acting on a beam.

- *Coplanar concurrent forces*: All forces lie in the same plane, have different directions but their lines of action act at one point (pass through a single point). The point where the lines of action of the forces meet is known as the *point of concurrency* of the force system.

  The forces P_1, P_2, P_3 and P_4 act and meet at the same common point in the same plane, and constitute a system of coplanar concurrent forces.

  Example: Forces on a rod resting against a wall.

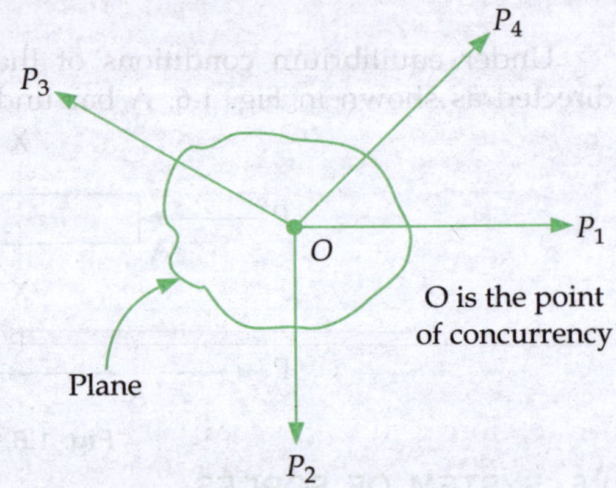

Fig. 1.9. Coplanar concurrent forces

- *Coplanar non-concurrent forces*: All forces lie in the same plane but their lines of action do not pass through a single point.

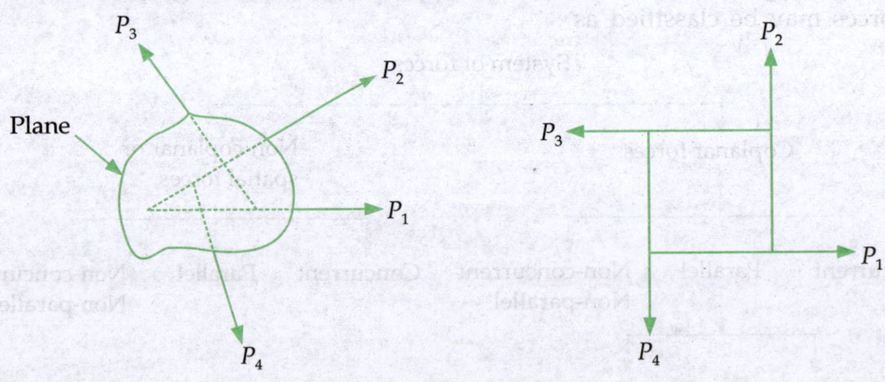

Fig. 1.10. Coplanar non-current forces

The forces P_1, P_2, P_3 and P_4 lie in the same place but act at different points, and constitute a non-concurrent force system.

Example: Forces on a ladder resting against wall and a person standing on a rung which is not at its centre of gravity.

- *Non-coplanar concurrent forces*: All forces do not lie in the same plane but their lines of action pass through a single point.

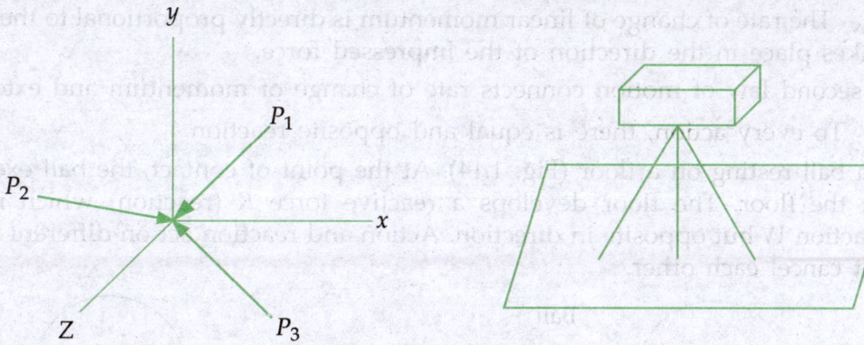

Fig. 1.11

Example: forces on a tripod carrying a camera.

- *Non-coplanar and non-concurrent forces:* All forces do not lie in the same plane, and their lines of action do not meet at a single point.
  The forces P_1, P_2, P_3, P_4 and P_5 lie in different planes Further, their lines of action donot pass through a single common point.
  Example: Forces acting on a moving bus.

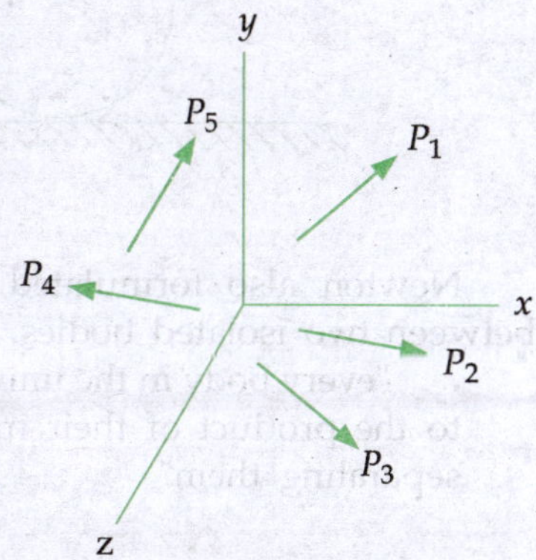

Fig. 1.12.

- *Coplanar like and unlike parallel force systems* : The different forces in a coplanar parallel force system lie in the same plane and have their lines of action parallel to each other. Further
  (*i*) the system is called the coplanar like parallel force system if the parallel forces point in the same direction [Fig. 1.13 (*a*)]

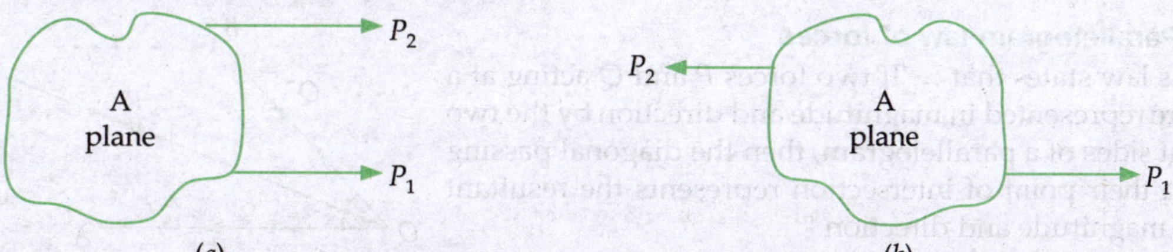

Fig. 1.13. Parallel force systems

  (*ii*) the system is called the coplanar unlike parallel force system if the forces point in opposite directions [Fig. 1.13(*b*)].

1.7. LAWS OF MECHANICS

The following are the basic laws of mechanics that form the foundation for making an analysis of the mechanics related problems of machines and structure.

1.7.1. Newtons laws of motion

First law: Every body continues in its state of rest or of uniform motion in a straight line if there is no unbalanced force acting upon it.

Newton's first law of motion gives the concept of inertia. Inertia is the property of matter by which a resistance is offered for change in the state of rest or uniform motion.

Second law: The rate of change of linear momentum is directly proportional to the impressed force and it takes place in the direction of the impressed force.

Newton's second law of motion connects rate of change of momentum and external force.

Third law: To every action, there is equal and opposite reaction.

Consider a ball resting on a floor (Fig. 1.14). At the point of contact, the ball exerts a force *W* (action) on the floor. The floor develops a reactive force *R* (reaction) which is equal in magnitude to action *W* but opposite in direction. Action and reaction act on different bodies and as such do not cancel each other.

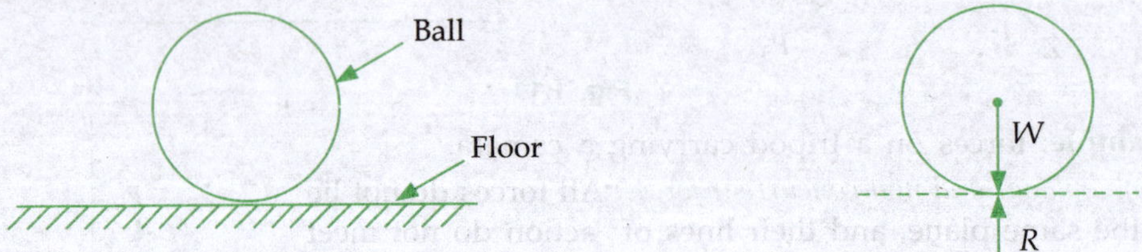

Fig. 1.14

Newton also formulated the **Law of gravitation**, which governs the mutual attraction between two isolated bodies. Newton's law of gravitation states that

"every body in the universe attracts every other body with a force directly proportional to the product of their mass and inversely proportional to the square of the distance separating them"

$$F = G\frac{m_1 m_2}{r^2}$$

where m_1 and m_2 are the masses of the two bodies, *r* is the distance separating them and *G* is the universal gravitation constant ($G = 6.67 \times 10^{-11}\ \text{Nm}^2/\text{kg}^2$). The line joining the centres of bodies is then the line of action of the force of attraction.

1.7.2. Parallelogram law of forces

This law states that : "If two forces *P* and *Q* acting at a point are represented in magnitude and direction by the two adjacent sides of a parallelogram, then the diagonal passing through their point of intersection represents the resultant both in magnitude and direction."

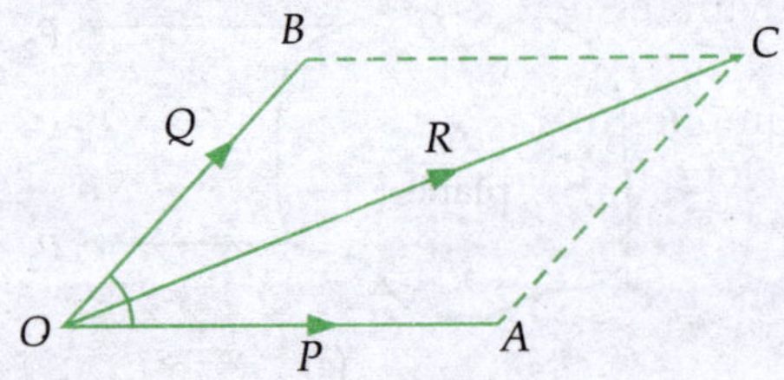

Fig. 1.15. Parallelogram law of forces

With reference to Fig. 1.15, *P* and *Q* are the two forces acting on the body at point *O*. The force *P* is represented in magnitude and direction by *OA* whereas the force *Q* is represented in magnitude and direction by *OB*. The resultant of these two forces is obtained by the diagonal *OC* of the parallelogram.

1.7.3. Sine law and cosine law

Refer to the triangle *ABC* with sides *a*, *b*, *c* and included angles α, β and γ as shown in Fig. 1.16.

Then as per sine law :

$$\frac{a}{\sin\alpha} = \frac{b}{\sin\beta} = \frac{c}{\sin\gamma}$$

and as per cosine law :

$$a^2 = b^2 + c^2 - 2\,bc\cos\alpha$$
$$b^2 = c^2 + a^2 - 2\,ca\cos\beta$$
$$c^2 = a^2 + b^2 - 2\,ab\cos\gamma$$

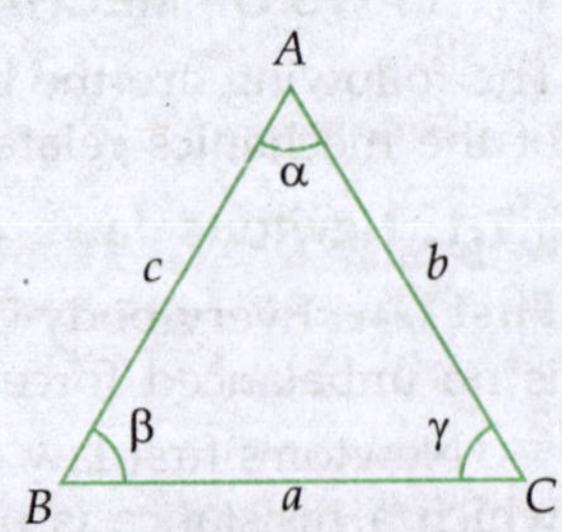

Fig. 1.16. Sine law and cosine law

1.7.4. Lami's theorem

This theorem states that :

"If three forces acting at a point are in equilibrium, then each force is proportional to the sine of the angle between the other two forces."

With reference to Fig. 1.17, P, Q and R are the three forces acting at the point O. Then according to Lami's theorem,

$$\frac{P}{\sin\alpha} = \frac{Q}{\sin\beta} = \frac{R}{\sin\gamma}$$

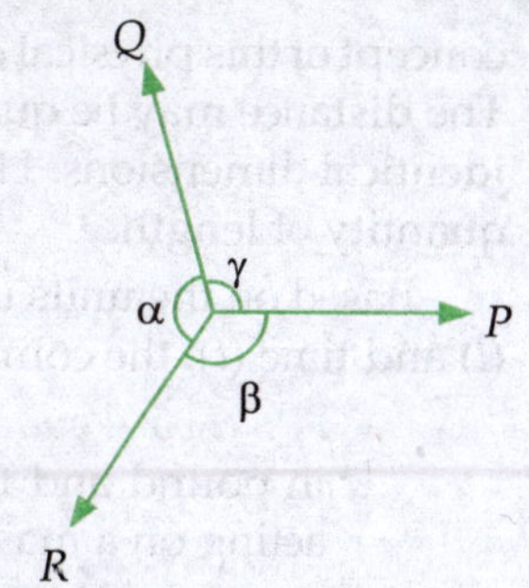

Fig. 1.17. Lami's theorem

1.7.5. Principle of transmissibility

When the point of application of a force acting on a body is shifted to any other point on the line of action of the force without changing its direction, there occurs no change in the equilibrium state of the body. This implies that a force acting at any point on a body may also be considered to act at any other point along its line of action without changing its effect on the body.

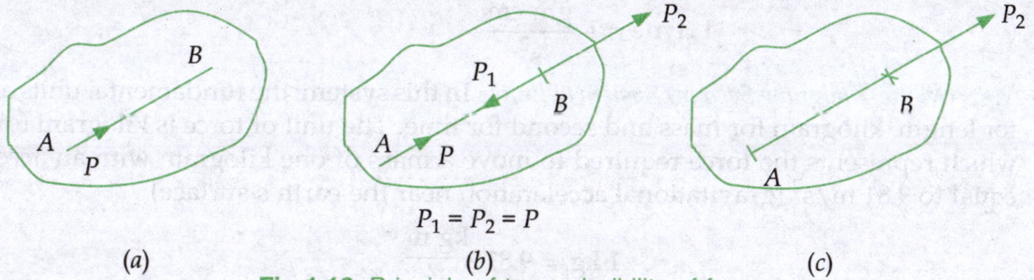

Fig 1.18. Principle of transmissibility of forces

Consider a force P acting at point A on rigid body (Fig 1.17). B is another point on the line of action of force P. At point B, apply two oppositely directed forces (P_1 and P_2) equal to and collinear with P. Such an application will in no way alter the action of given force P. At point A, forces P and P_1 are equal but opposite and accordingly cancel each other. That leaves a force $P_2 = P$ at B. This implies that a force acting at any point on a body may also be considered to act at any other point along its line of action without changing its effect on the body.

These laws of mechanics and their applications have been dealt with at appropriate places in the text.

1.8. EQUILIBRIUM, RESULTANT AND EQUILIBRANT

When two or more than two forces act on a body in such a way that the body remains in a state of rest or of uniform motion (no acceleration), then the system of forces is said to be in *equilibrium*

When a body is acted upon by a system of forces, then vectorial sum of all the forces is known as *resultant*. Hence resultant refers to the single force which produces the same effect as is done by the combined effect of several forces.

A number of forces may act on a body in such a manner that the body is not in equilibrium. The resultant of several forces may cause a change of state of rest or of uniform motion. A single force may have to be applied to the body to bring it in equilibrium state. That single force is known as *equilibrant*. Equilibrant is equal and opposite to the resultant of several forces acting on the body.

1.9. DIMENSIONS AND UNITS

Dimensions refer to the qualitative characteristics for physical quantities, while units are standards of comparison for the quantitative measure of dimensions. For elaborating the difference, let us consider the distance between two points. The term length when applied to it gives the qualitative concept of this physical quantity. The term unit would, however, indicate the magnitude of distance. The distance may be quantitatively expressed as a metre or a mile. The metre and mile will then be identical dimensions. However, they will be different units as each contains a different amount of quantity of length.

Based on the units used to represent the three fundamental quantities namely, mass (m), length (l) and time (t), the commonly used system of units are :

- *FPS (Foot-Pound-Second) System of units:* In this system the length is expressed in foot, mass in pound and time in second. The unit of force is poundal which is defined as the force acting on a mass of one pound and producing an acceleration of 32.2 ft/s^2

$$1 \text{ poundal} = 32.2 \frac{\text{pound ft}}{\text{s}^2}$$

- *CGS (Centimetre-Gram-Second) System of units:* In this system the length is expressed in centimetre, mass in gram and time in second. The unit of force is dyne which is defined as the force acting on a mass of one gram and producing an acceleration of 1 cm/s^2.

$$1 \text{ dyne} = 1 \frac{\text{gm cm}}{\text{s}^2}$$

- *MKS (Metre-Kilogram-Second) System of units*: In this system, the fundamental units are metre for length, kilogram for mass and second for time. The unit of force is kilogram force (Kg_f) which represents the force required to move a mass of one kilogram with an acceleration equal to 9.81 m/s^2 (gravitational acceleration near the earth's surface)

$$1 \text{ kg}_f = 9.81 \frac{\text{kg m}}{\text{s}^2}$$

Almost all major countries of the world use SI units (System International d' Unit or International system of units) in which the fundamental units are metre (m) for measuring length, kilogram (kg) for measuring mass, and second (s) for measuring time. The force is expressed in newton (N); Newton is the amount of force which causes one kilogram mass to move with an acceleration of 1 m/s^2

$$1 \text{ N} = 1 \frac{\text{kg m}}{\text{s}^2}$$

The relation between different units of force are:

$$1 \text{ kg}_f = 9.81 \text{ N} = 9.81 \times 10^5 \text{ dyne}$$

Some derived units pertinent to mechanics are:

Quantity	Unit	Symbol
Area	square metre	m^2
Velocity	metre per second	m/s
Acceleration	metre per second square	m/s^2
Density	kilogram per cubic metre	kg/m^3
Force	newton	N
Moment	newton-metre	Nm
Stress	newton per square meter	N/m^2 = pascal
Work and energy	joule	J = Nm
Power	watt	W = J/s = Nm/s

- **Joule** is the work done when the point of application of a force of one newton is displaced through one meter along its line of action.

$$\text{Work} = \text{force} \times \text{distance moved}$$
$$= \text{newton} \times \text{metre} = \text{Nm} = \text{J or joule}$$

- **Watt** represents a work equivalent of one joule done per second

$$\text{Power} = \text{work done per unit time}$$
$$= \text{J/s} = \text{W or watt} ; \text{W} = 1 \text{ J/s}$$

- **Stress** is force per unit area. A stress equivalent to one newton per square metre is called Pascal

$$1 \text{ Pa} = 1 \text{ N/m}^2$$

To facilitate working with magnitudes which may be multiples or submultiples of SI units, we use a prefix before the unit to suit a particular situation. The commonly used prefixes are listed in Table 1.1.

Table 1.1. SI System : Standard Multipliers

Factor	Prefix	Factor	Prefix
10^{12}	tera, T	10^{-3}	milli, m
10^{9}	giga, G	10^{-6}	micro, μ
10^{6}	mega, M	10^{-9}	nano, n
10^{3}	kilo, k	10^{-12}	pico, p
10^{-1}	deci, d	10^{-15}	femto, f
10^{-2}	centi, c	10^{-18}	atto, a

The use of tera, mega, kilo, milli, micro and pico is preferred.

The SI system of units is comprehensive, coherent, rational, brief and the easiest for the numerical and practical purposes, when used all by themselves.

Some of the additional comments on SI units are :

— While writing plurals, do not add s to the symbols. Thus write kg and not kgs.

— Units with name of scientists are not to be written with capital letters. Thus write newton and not Newton as unit of force.

— There are to be no dots and no dashes. For example, work is in Nm, not N.m, not N-m etc.

— After every three digits, there is to be a gap but no comma, *e.g.*, 2 000 and not 2,000

— For practical purposes, use of units like tonne, degree celcius, litre, speed in km/hr is permitted but should be restricted as far as possible.

1.10. DIMENSIONAL HOMOGENEITY

The fundamental theory of dimensional analysis is based on the following axiom :

"Equations describing a physical phenomenon must be dimensionally homogeneous and the units therein must be consistent."

An equation is said to be dimensionally homogeneous if the dimensions of various terms on the two sides of the equation are identical. A dimensionally homogeneous equation is independent of the fundamental units of measurement if the units therein are consistent. This implies that a length dimension can be added to or subtracted from only a length dimension. Addition of area to volume is meaningless since area and volume are dimensionally incompatible. Consistency of units demands that if one term of an equation is measured in a particular unit of velocity say m/s, then all other terms in the equation must be measured in the same units of velocity.

The concept of dimensional homogeneity can be elaborated by considering the time period of oscillation T of a simple pendulum of length l and mass m :

$$T = 2\pi\sqrt{l/g}$$

Substituting the dimensions on both sides of the equation :

$$[T] = [1]\,[L/LT^{-2}]^{1/2} = [T]$$

The dimension of 2π, being constant, is taken as unity. The equation is thus dimensionally homogeneous because dimensions on both sides of the equation are same. Substituting the value of gravitational acceleration g = 9.807 m/s^2, equation is modified to

$$T = 2\pi\sqrt{l/9.807} = 2.006\sqrt{l}$$

This modified equation is not dimensionally homogeneous because the right hand side does not have the dimensions of time. Further, it is valid only when the time is measured in seconds and the length is measured in metres.

Some well known examples of dimensionally homogeneous relationships used in mechanics are:

- Force = rate of change of momentum

Now : force = mass × acceleration $\equiv M \times LT^{-2} = MLT^{-2}$

rate of change of momentum $= \dfrac{\text{mass} \times \text{velocity}}{\text{time}} = \dfrac{M \times (LT^{-1})}{T} = MLT^{-2}$

Obviously the dimensions of force are the same as that of the rate of change of momentum.

- Work done = change in kinetic energy

Now : work done = force × displacement $\equiv MLT^{-2} \times L = ML^2T^{-2}$

change in kinetic energy $= \dfrac{1}{2}$ mass × (velocity)$^2 \equiv \dfrac{1}{2} M \times (LT^{-1})^2 \equiv \dfrac{1}{2} ML^2T^{-2}$

Obviously the dimensions of work done are the same as those of change in kinetic energy.

It is worthwhile to point out that two dimensionally homogeneous equations can be multiplied or divided without affecting the homogeneity. However, an equation resulting from the addition or subtraction of two dimensionally homogeneous equations may not be dimensionally homogeneous. The students may verify this fact by considering the following two equations which are dimensionally homogeneous :

$$v = u + at \,; \qquad s = ut + \frac{1}{2} at^2$$

where s is the distance travelled in time t by a particle having velocity u and acceleration a; v represents the final velocity.

Dimensional homogeneity helps to convert the units from one system to another. Let it be required to convert pressure from kg$_f$/cm^2 to N/m^2. Recalling that : 1 kg$_f$ = 9.807 N and 1 cm = 0.01 m, the conversion factor becomes

$$= \frac{\text{kg}_f/\text{cm}^2}{\text{N/m}^2} = \left[\frac{\text{kg}_f}{\text{N}}\right] \times \left[\frac{\text{m}}{\text{cm}}\right]^2 = 9.807 \times \left(\frac{1}{0.01}\right)^2 = 9.807 \times 10^4$$

Thus any pressure in kg$_f$/cm^2 can be converted to N/m^2 by multiplying it with 9.807×10^4.

It also helps to determine the dimensions of a physical quantity. Consider the hydrostatic equation $p = wh$, where p is the pressure intensity, h is the height of fluid column and w is the specific weight of the fluid.

Dimensionally, $p = [ML^{-2} T^{-2}] [L] = ML^{-1} T^{-2}$

Further, if w is taken in N/m^3 and h in m of fluid column, then units of pressure intensity would be

$$p = \left(\frac{N}{m^3}\right) \times m = \frac{N}{m^2}$$

REVIEW AND SUMMARY

1. Mechanics is the science that describes the conditions when a body is said to be at rest or in motion under the effect of forces.
2. Statics deals with bodies at rest, while dynamics concerns the bodies which are in motion. Dynamics is further divided into kinematics and kinetics.
   Kinematics deals with the relationship between displacement, velocity, acceleration and time of a given motion, without considering the forces that cause the motion.
   Kinetics is the study of relationship between forces acting on the body, and the mass and motion of the body.
3. A rigid body has definite amount of matter and the distance between the particles forming it remains constant before and after the application of force.
4. A quantity is said to be scalar when it is completely defined by its magnitude alone. Mass, volume, temperature and energy etc., are scalar quantities.
   Any physical quantity that possesses magnitude as well as direction is called a vector quantity. Displacement, velocity, acceleration and momentum etc., are vector quantities.
5. Force is an external agency that tends to change the state of rest or of uniform motion of a body. Force is a vector quantity, and is completely defined by its magnitude, direction and point of application. The application of force on a body may :
   - change its state of rest or motion
   - accelerate or retard its motion
   - change its shape and size
   - turn or rotate it, and
   - keep it in equilibrium
6. Depending upon the position of line of action, forces may be collinear (along the same line), non-coplanar (not lying in the same plane), non-concurrent (not meeting at a point) and coplnar concurrent (lying in the same plane and meeting at a point), etc.
7. According to the principle of transmissibility, the effect of force on a body is same at every point in its line of action.
8. Law of dimensional homogeneity states that a basic equation representing a phenomenon must be valid for all system of units.
9. The resultant force and the equilibrant force are collinear and equal in magnitude but opposite in directions.

REVIEW QUESTIONS

A. Conceptual and conventional questions :

1. Differentiate between statics, kinetics and kinematics.
2. Define matter, particle and body. How does a rigid body differ from an elastic body?
3. Define space, time, motion and trajectory.
4. When is the motion of a particle called rectilinear, and when curvilinear ?

5. State Newton's law of motion.
6. What is Newton's law of gravitation?
7. Differentiate between scalar and vector quantities. How a vector quantity is represented?
8. Given below is a list of some of the quantities used in mechanics

   mass, displacement, momentum, area, volume, velocity, weight, speed, temperature and force.

   Classify these into scalar and vector quantities.
9. Make a clear distinction between mass, force and weight.
10. What are the two ways for the representation of forces ?
11. Represent graphically a force of 50 N acting upwards at 60º to the horizontal and towards the right.
12. State the effects which a force may produce when it acts on a body.
13. Enumerate the characteristics of a force.
14. What is a force system ? What is its point of concurrency ?
15. How forces are specified depending upon the position of their line of action?
16. Sketch a non-concurrent and non-parallel coplanar force system.
17. State the difference between equilibrium, resultant and equilibrant.
18. Explain the principle of transmissibility of forces.
19. How does the term dimension differ from the term unit?
20. State the unit of force into CGS, MKS and SI system of units. How are they related to each other ?
21. Name various systems of units. List the salient features of the system which is being used these days.
22. What is meant by dimensional homogeneity ? Point out its utility.

B. Complete the following statements with most appropriate word/words :

1. Mechanics is a science which describes and predicts the state of .... or .... under the action of forces.
2. .... relates to bodies in motion without any reference to forces.
3. A body which does not deform under the action of applied forces is called the ....
4. A .... is a body of infinitely small volume and is considered to be concentrated at a point.
5. A .... quantity is one that has magnitude as well as direction.
6. Magnitude, direction, sense and .... are the characteristics of a force.
7. According to .... of forces, the effect of a force upon a body is the same at every point in its line of action.
8. Fundamental principles of mechanics are ....
9. Forces whose lines of action pass through a common point are called .... forces.
10. A .... is a single force which can replace two or more forces and produce the same effect as the forces.
11. Equlibrant is equal and opposite to the .... of several forces acting or a body.
12. .... is a method of designating a force by writing two capital letters one or either side of the force.
13. The units which are used for the measurement of fundamental quantities are called .... units.

Answers :

1. rest, motion 2. kinematics 3. rigid body 4. particle 5. vector 6. point of application 7. principle of transmissibility 8. Newton's laws of motion and gravitation, parallelogram law, principle of transmissibility etc. 9. concurrent 10. resultant 11. resultant 12. Bow's notation 13. basic

C. Multiple choice questions :

1. Identify the wrong statement about matter, particle and body
   (*a*) matter is any thing that occupies space and possesses mass
   (*b*) a body has definite shape and consists of number of particles
   (*c*) a particle has infinitely small volume and its mass is considered to be concentrated at a point
   (*d*) a body exhibits no relative deformation between two parts when it is acted upon by a force system
2. For analysis purposes, a body can be modelled as rigid if
   (*a*) it has continuous distribution of mass
   (*b*) it is characterized by some preferred configuration of its own
   (*c*) relative motion between its parts is negligible
   (*d*) its dimensions are small compared with the co-ordinates describing its motion
3. Which of the following quantities is not a vector quantity?
   (*a*) energy (*b*) momentum (*c*) acceleration (*d*) force
4. Newtons first law of motion gives the concept of
   (*a*) work (*b*) force (*c*) inertia (*d*) energy
5. Newton's second law of motion connects
   (*a*) change of momentum and velocity (*b*) momentum and acceleration
   (*c*) momentum and rate of change of force
   (*d*) rate of change of momentum and external force
6. A force which combines with two or more forces to produce equilibrium is called
   (*a*) resultant (*b*) equilibrant (*c*) couple (*d*) moment
7. Which of the following products is not equal to one ?
   (*a*) micro × mega (*b*) nano × giga (*c*) pico × tera (*d*) hecto × deci
8. According to the principle of transmissibility of forces, the effect of force on a body is
   (*a*) same at every point in its line of action
   (*b*) different at different points in its line of action
   (*c*) minimum when it acts at the centre of gravity of the body
   (*d*) maximum when its acts at the centre of gravity of the body
9. The unit of k in the relation $U = \dfrac{ky}{y^2 + a^2}$ is
   (*a*) m/s (*b*) ms (*c*) Jm (*d*) J/s
   In the given expression U represents the potential energy, y represents the displacement and a represents the maximum displacement, *i.e.*, amplitude.

Answers :

1. (*d*) 2. (*c*) 3. (*a*) 4. (*c*) 5. (*d*) 6. (*b*)
7. (*d*) 8. (*a*) 9. (*c*)

2 Co-planar Forces

Learning objectives : The subject matter included in this chapter will enable the readers to

- *understand the concept of resultant and equilibrant, composition and resolution of forces*
- *state and explain the parallelogram law, triangle law and polygon law for addition of forces*
- *deal with the analytical and graphical methods for determining the resultant of coplanar force system*
- *draw the free body diagram; consideration of internal and external forces*
- *state the law of sines and Lami's theorem*

When a number of concurrent forces act on a body to keep the body in equilibrium, then resultant of these forces will be the vector sum of all these forces. The process of determination of resultant in known as *composition* of forces. Also from a given force, two forces in two directions can be obtained, which will produce the same effect as produced by single given force. This process is known as *resolution* of a force.

2.1. PARALLELOGRAM LAW OF FORCES

The parallelogram law of forces is used to determine the resultant of two forces acting at a point in a plane and inclined to each other at an angle.

It states that:

"If two forces, acting at a point be represented in magnitude and direction by the two adjacent sides of a parallelogram, then their resultant is represented in magnitude and direction by the diagonal of the parallelogram passing through that point."

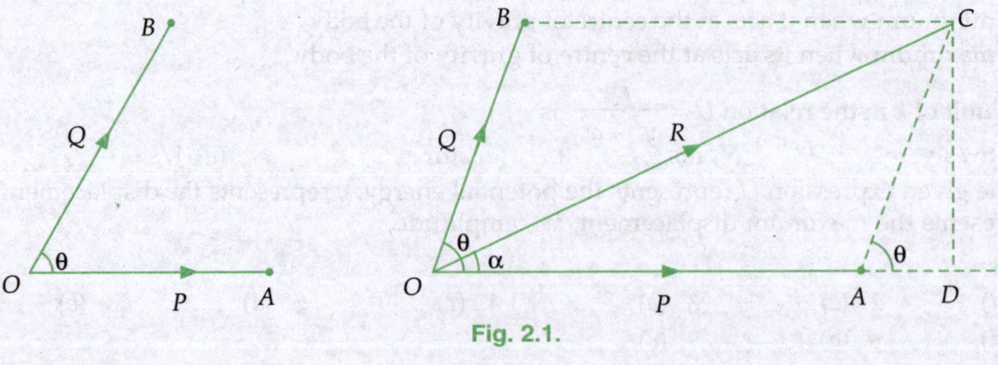

Fig. 2.1.

Consider two forces P and Q acting on a body at O as shown in Fig 2.1. The force P is represented in magnitude and direction by $\overrightarrow{OA}$ whereas the force Q is represented in magnitude and direction by $\overrightarrow{OB}$. Let the angle between the two forces be θ. The resultant of these two forces is obtained by the diagonal $\overrightarrow{OC}$ of the parallelogram $OACB$, as shown in Fig. 2.1.

The relationship between P, Q and R can be derived as follows:

Drop perpendicular from C and let it meet OA extend at point D. In $\Delta\, CAD$, side CA is parallel and equal to OB, *i.e.*, it represents force Q.

The resultant R of P and Q is given by

$$R = OC = \sqrt{OD^2 + CD^2} = \sqrt{(OA + AD)^2 + CD^2}$$

But

$$OA = P$$

$$AD = AC\cos\theta = Q\cos\theta$$

$$CD = AC\sin\theta = Q\sin\theta$$

$$\therefore \quad R = \sqrt{(P + Q\cos\theta)^2 + (Q\sin\theta)^2} = \sqrt{P^2 + Q^2\cos^2\theta + 2PQ\cos\theta + Q^2\sin^2\theta}$$

$$= \sqrt{P^2 + Q^2(\sin^2\theta + \cos^2\theta) + 2PQ\cos\theta}$$

$$= \sqrt{P^2 + Q^2 + 2PQ\cos\theta} \qquad \text{...(2.1)}$$

The inclination of the resultant R to the direction of force P is given by

$$\tan\alpha = \frac{CD}{OD} = \frac{CD}{OA + AD} = \frac{Q\sin\theta}{P + Q\cos\theta}$$

$$\alpha = \tan^{-}\left[\frac{Q\sin\theta}{P + Q\cos\theta}\right] \qquad \text{...(2.2)}$$

Note: It is not necessary that for the law of parallelogram for forces to be valid, one of the two forces should be along the x-axis. The forces P and Q may be in any direction as shown in Fig. 2.2.

If the angle between the forces is θ, then

$$R = \sqrt{P^2 + Q^2 + 2PQ\cos\theta}$$

The direction of resultant will be

$$\alpha = \tan^{-1}\frac{Q\sin\theta}{P + Q\cos\theta}$$

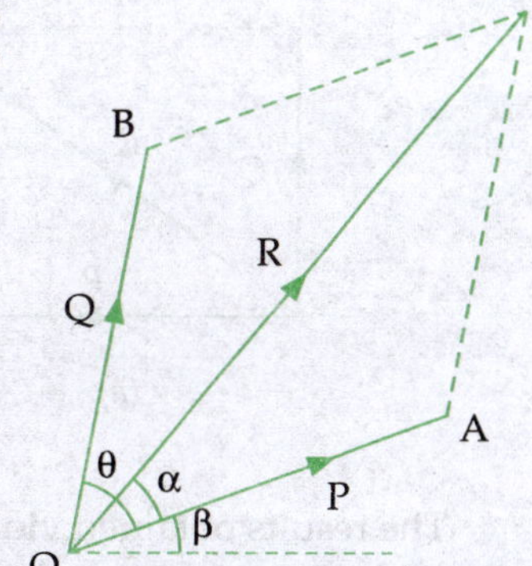

Fig. 2.2

α is the angle which the resultant makes with the direction of P. If β is the angle which the force P makes with the x-axis, the angle made by the resultant with x-axis will be $(\alpha + \beta)$.

Special cases:

(*i*) When the two forces are equal and θ is the angle between them:

$$R = \sqrt{P^2 + Q^2 + 2PQ\cos\theta} = \sqrt{P^2 + P^2 + 2P \times P\cos\theta}$$

$$= \sqrt{2P^2(1 + \cos\theta)} = \sqrt{2P^2 \times 2\cos^2\frac{\theta}{2}} = 2P\cos\frac{\theta}{2} \qquad \text{...(2.3)}$$

and $\alpha = \tan^{-1}\left(\dfrac{Q\cos\theta}{P+Q\cos\theta}\right) = \tan^{-1}\left(\dfrac{P\sin\theta}{P+P\cos\theta}\right) = \tan^{-1}\left(\dfrac{\sin\theta}{1+\cos\theta}\right)$

$$= \tan^{-1}\left[\frac{2\sin\frac{\theta}{2}\cos\frac{\theta}{2}}{2\cos^2\frac{\theta}{2}}\right] = \tan^{-1}\left(\tan\frac{\theta}{2}\right)$$

$$\therefore \quad \alpha = \frac{\theta}{2} \qquad \text{...(2.4)}$$

i.e., the resultant bisects the angle between the forces.

(*ii*) When the two forces act at right angles; *i.e.*, $\theta = 90°$

$$R = \sqrt{P^2+Q^2+2PQ\cos 90°} = \sqrt{P^2+Q^2} \quad (\because \cos 90° = 0) \qquad \text{...(2.5)}$$

and $\alpha = \tan^{-1}\dfrac{Q\sin 90°}{P+Q\cos 90°} = \tan^{-1}\left(\dfrac{Q}{P}\right) \qquad (\because \sin 90° = 1)$

(*iii*) When the two forces act in the same line and same sense, *i.e.*, $\theta = 0°$

$$R = \sqrt{P^2+Q^2+2PQ\cos 0°} = \sqrt{P^2+Q^2+2PQ} \qquad (\because \cos 0° = 1)$$

$$= P + Q \qquad \text{...(2.6)}$$

Apparently the resultant is maximum when the forces are collinear and act in the same direction.

(*iv*) When the two forces have the same line of action but opposite senses, *i.e.*, $\theta = 180°$

$$R = \sqrt{P^2+Q^2+2PQ\cos 180°} = \sqrt{P^2+Q^2-2PQ} \qquad (\because \cos 180° = -1)$$

$$= P - Q \qquad \text{...(2.7)}$$

Obviously the resultant is minimum when the two forces are collinear but act in opposite direction.

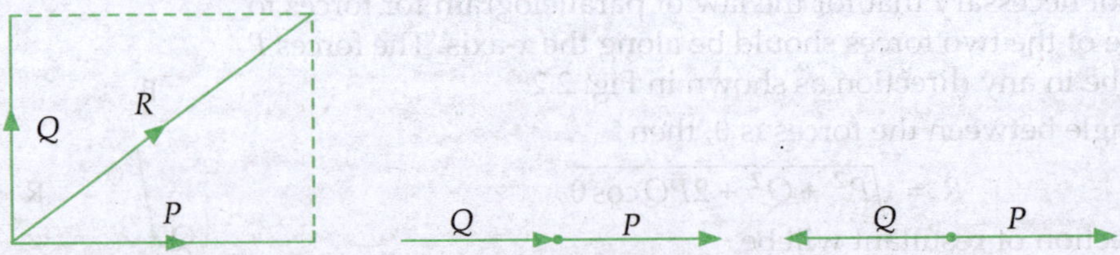

Fig. 2.3

The results obtained vide equations (2.5), (2.6) and (2.7) have been represented in Fig 2.3. These results lead us to conclude that when the forces acting on a body are collinear, their resultant is equal to the algebraic sum of the forces.

EXAMPLE 2.1

(*i*) *Two forces of equal mangnitude P act at an angle θ to each other. What will be their resultant ?*

(*ii*) *Resultant of two equal forces is equal to either of them. Determine the angle between the forces.*

(*iii*) *The resultant of two forces (P + Q) and (P – Q) equals* $\sqrt{3P^2+Q^2}$ *. Show that the forces are then inclined to each other at an angle of 60°.*

(iv) *The resultant of two forces P and Q is R. If one of the forces is reversed in direction, The resultant is S. Then for the identity*

$$R^2 + S^2 = 2(P^2 + Q^2)$$

to hold good, show that the forces can have any angle of inclination between them.

Solution : Solution to these forces stems from the parallelogram law of forces.

(i) $R^2 = P^2 + Q^2 + 2PQ\cos\theta = P^2 + P^2 + 2P^2\cos\theta$

$$= 4P^2\left(\frac{1+\cos\theta}{2}\right) = 4P^2\cos^2\frac{\theta}{2}$$

$$\therefore \quad R = 2P\cos\frac{\theta}{2}$$

(ii) $R = 2P\cos\frac{\theta}{2}$ as outlined above.

Since the resultant equals either of the two equal forces,

$$P = 2P\cos\frac{\theta}{2}; \quad \cos\frac{\theta}{2} = \frac{1}{2}$$

$$\frac{\theta}{2} = 60^\circ$$

$$\therefore \quad \theta = 120^\circ \text{ or } \frac{2\pi}{3}$$

(iii) $3P^2 + Q^2 = (P+Q)^2 + (P-Q)^2 + 2(P+Q)(P-Q)\cos\theta$

$$= 2P^2 + 2Q^2 + 2(P^2 - Q^2)\cos\theta$$

or $\quad (P^2 - Q^2) = 2(P^2 - Q^2)\cos\theta$

or $\quad \cos\theta = \frac{1}{2}; \quad \theta = 60^\circ \text{ or } \frac{\pi}{3}$

(iv) $R^2 = P^2 + Q^2 + 2PQ\cos\theta$

$$S^2 = P^2 + (-Q)^2 + 2P(-Q)\cos\theta = P^2 + Q^2 - 2PQ\cos\theta$$

upon summation

$$R^2 + S^2 = 2(P^2 + Q^2)$$

Obviously for the given identify to hold good, the forces can have any angle of inclination between them.

EXAMPLE 2.2

Two coplanar forces with magnitude 20 kN and 15 kN are oriented with angle of inclination equal to 60 degree. What would be their resultant in magnitude and direction if

(a) both the forces act away from their point of intersection O,

(b) first force acts away from point O and the second force acts towards the point.

Solution : **Case (*a*) :** $P = 20$ kN, $Q = 15$ kN and $\theta = 60$ (Fig. 2.4.)

Invoking parallelogram law of forces

$$\text{Resultant } R = \sqrt{P^2 + Q^2 + 2PQ\cos\theta}$$

$$= \sqrt{20^2 + 15^2 + 2\times 20\times 15\times\cos 60}$$

$$= \sqrt{400 + 225 + 300}$$

$$= \sqrt{925} = \mathbf{30.41\ kN}$$

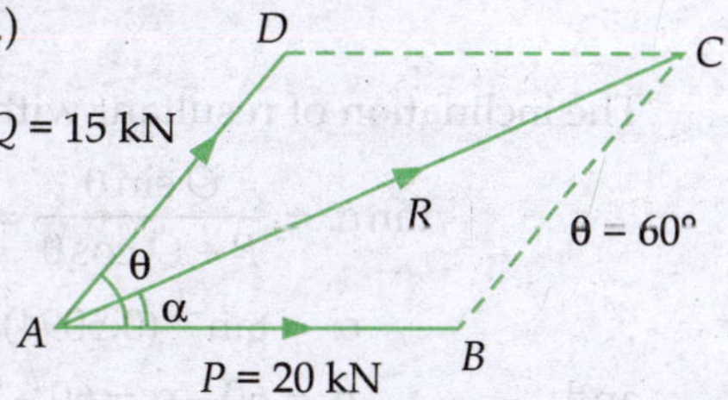

Fig. 2.4. (a)

Inclination α of resultant with the direction of force P,

$$\tan\alpha = \frac{Q\sin\theta}{P+Q\cos\theta} = \frac{15\sin 60^\circ}{20+15\cos 60^\circ} = \frac{15\times 0.866}{20+15\times 0.2} = 0.4723$$

$\therefore \quad \alpha = \tan^{-1}(0.4723) = \mathbf{25.28^\circ}$ (clockwise)

Case (*b*) :

$$R = \sqrt{P^2+Q^2+2PQ\cos\theta}$$
$$= \sqrt{20^2+15^2+2\times 20\times 15\cos 120^\circ}$$
$$= \sqrt{400+225-300}$$
$$= \sqrt{325} = \mathbf{18.03\ N}$$

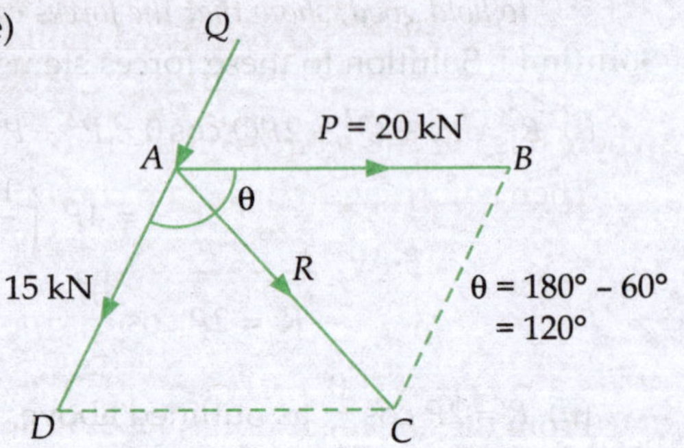

Fig. 2.4. (b)

$$\tan\alpha = \frac{Q\sin\theta}{P+Q\cos\theta} = \frac{15\sin 120^\circ}{20+15\cos 120^\circ}$$
$$= \frac{15\times 0.866}{20+15\times(-0.5)} = 1.0392$$

$\therefore \quad \alpha = \tan^{-1}(1.0392) = \mathbf{46.1^\circ}$ (anti-clockwise)

EXAMPLE 2.3

Two locomotives on opposite banks of a canal pull a vessel moving parallel to the banks by means of two horizontal ropes. The tensions in these ropes have been measured to be 20 kN and 24 kN while the angle between them is 60°. Find the resultant pull on the vessel and the angle between each of the ropes and the sides of the canal.

Solution: The vessel is acted upon at point O by the force system as depicted in the figure given below:

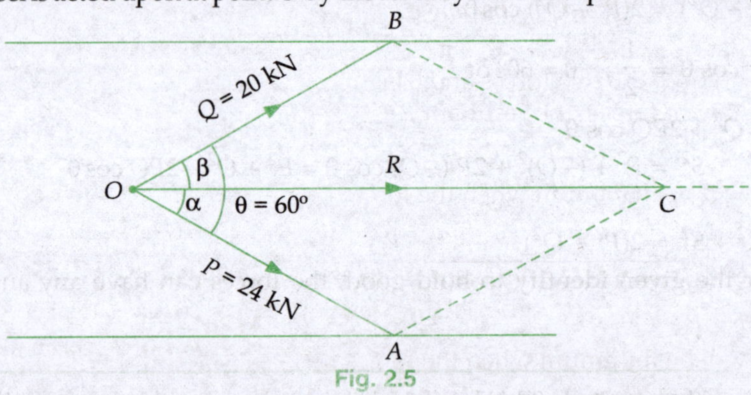

Fig. 2.5

The resultant force exerted on the vessel can be worked out by applying the parallelogram law of forces. That is

$$R = \sqrt{P^2+Q^2+2PQ\cos\theta} = \sqrt{24^2+20^2+2\times 24\times 20\times\cos 60^\circ}$$
$$= \sqrt{576+400+480} = \sqrt{1456} = \mathbf{38.16\ N}$$

The inclination of resultant with direction of force P is

$$\tan\alpha = \frac{Q\sin\theta}{P+Q\cos\theta} = \frac{20\sin 60^\circ}{24+20\cos 60^\circ} = \frac{20\times 0.866}{24+20\times 0.5} = 0.5094$$

$\therefore \quad \alpha = \tan^{-1}(0.5094) = 27^\circ$

and $\quad \beta = 60 - \alpha = 60 - 27 = 33^\circ$

EXAMPLE 2.4

The magnitude of two forces is such that when acting at right angles produce a resultant force of $\sqrt{20}$ and when acting at 60 degree produce a resultant equal to $\sqrt{28}$. Work out the magnitude of the two forces.

Solution : Let P and Q be the magnitude of the two forces. From parallelogram law of forces,

$$R^2 = P^2 + Q^2 + 2PQ\cos\theta$$

where θ is the angle between the two forces.

Then according to the given conditions,

$$20 = P^2 + Q^2 + 2PQ\cos 90^\circ = P^2 + Q^2 \qquad (\because \cos 90^\circ = 0)$$

and $$28 = P^2 + Q^2 + 2PQ\cos 60^\circ = P^2 + Q^2 + PQ \qquad \left(\because \cos 60^\circ = \frac{1}{2}\right)$$

From the above identities $PQ = 8$ and $Q^2 = \dfrac{64}{P^2}$

That gives: $$20 = P^2 + \frac{64}{P^2}$$

or $$P^4 - 20P^2 + 64 = 0$$

or $$(P^2 - 16)(P^2 - 4) = 0$$

$\therefore$ $P = 2$ or 4 units

The corresponding values of force Q will be 4 or 2 units.

EXAMPLE 2.5

The greatest and least resultant of two forces acting on a body are 35 kN and 5 kN respectively. Determine the magnitude of the forces. What would be the angle between these forces if the magnitude of the resultant is stated to be 25 kN?

Solution : Let P and Q be the two forces and θ be the angle of inclination between them. Then from the parallelogram law of forces, the resultant R is

$$R^2 = P^2 + Q^2 + 2PQ\cos\theta$$

The resultant will be maximum when the forces are collinear and act in the same direction, *i.e.*, θ = 0°. That gives

$$R^2 = \sqrt{P^2 + Q^2 + 2PQ\cos 0^\circ} = \sqrt{P^2 + Q^2 + 2PQ} = P + Q$$

$$\therefore \quad 35 = P + Q \qquad \text{...}(i)$$

The resultant will be minimum when the forces are collinear and act in the opposite direction, *i.e.*, θ = 180°. That gives

$$R = \sqrt{P^2 + Q^2 + 2PQ\cos 180^\circ} = \sqrt{P^2 + Q^2 - 2PQ} = P - Q$$

$$\therefore \quad 5 = P - Q \qquad \text{...}(ii)$$

From identities (*i*) and (*ii*)

$$P = \mathbf{20\ kN} \text{ and } Q = \mathbf{15\ kN}$$

(*b*) Let θ be the angle between the force P = 20 kN and Q = 15 kN when their resultant is 25 kN. Then

$$25^2 = 20^2 + 15^2 + 2 \times 20 \times 15\cos\theta$$

$$625 = 400 + 225 + 600\cos\theta$$

$$\cos\theta = 0 \text{ and this implies that } \theta = 90^\circ$$

Thus the given system of forces are at right angles to each other when the resultant is 25 kN.

EXAMPLE 2.6

Two forces, one of which is double the other has resultant of 260 N. If the direction of the larger fo reversed and the other remains unaltered, the resultant reduces to 180 N. Determine the magnitude of the force and the angle between the forces.

Solution : Let P and Q (= $2P$) be the two forces and θ be the angle of inclination between them. From parallelogram law of forces

$$R^2 = P^2 + Q^2 + 2PQ \cos\theta$$

Then according to the given conditions,

$$260^2 = P^2 + (2P)^2 + 2P\,(2P) \cos\theta$$

$$= P^2 + 4P^2 + 4P^2 \cos\theta = 5P^2 + 4P^2 \cos\theta \qquad ...(i)$$

and

$$180^2 = P^2 + (-2P)^2 + 2P\,(-2P) \cos\theta$$

$$= P^2 + 4P^2 - 4P^2 \cos\theta = 5P^2 - 4P^2 \cos\theta \qquad ...(ii)$$

Adding identities (i) and (ii) we get

$$10P^2 = 260^2 + 180^2 = 67600 + 32400 = 100000$$

$$\therefore \quad P = \sqrt{\frac{100000}{10}} = \mathbf{100\ N}$$

$$Q = 2P = 2 \times 100 = \mathbf{200\ N}$$

Substituting $P = 100$ N in identity (*i*), we get

$$260^2 = 5 \times 100^2 + 4 \times 100^2 \cos\theta$$

$$\cos\theta = \frac{260^2 - 5 \times 100^2}{4 \times 100^2} = 0.44$$

$$\therefore \quad \theta = \cos^{-1}(0.44) = \mathbf{63.89°}$$

EXAMPLE 2.7

The resultant of two forces P and Q acting at a point is R. The resultant R gets doubled when Q is either doubled or its direction is reversed. Show that P, Q and R conform to the ratio

$$P : Q : R = \sqrt{2} : \sqrt{3} : \sqrt{2}$$

Solution : From parallelogram law of forces, the resultant of forces P and Q with θ angle of inclination between them is given by

$$R^2 = P^2 + Q^2 + 2PQ \cos\theta \qquad ...(i)$$

When Q is doubled

$$(2R)^2 = P^2 + (2Q)^2 + 2P \times (2Q) \cos\theta$$

$$4R^2 = P^2 + 4Q^2 + 4\,PQ \cos\theta \qquad ...(ii)$$

When Q is reversed in direction,

$$(2R)^2 = P^2 + (-Q)^2 + 2P \times (-Q) \cos\theta$$

$$4R^2 = P^2 + Q^2 - 2PQ \cos\theta \qquad ...(iii)$$

Adding identities (*i*) and (*iii*)

$$5R^2 = 2P^2 + 2Q^2 \qquad ...(iv)$$

Multiplying identity (*iii*) by 2 and adding the result to identity (*ii*)

$$12\,R^2 = 3P^2 + 6Q^2$$

or

$$4R^2 = P^2 + 2Q^2 \qquad ...(v)$$

From identities (*iv*) and (*v*)

$$5R^2 = 2P^2 + (4R^2 - P^2)$$

$$\text{or} \quad R^2 = P^2 \quad \text{or} \quad R = P$$

Then from identity (*v*)

$$4P^2 = P^2 + 2Q^2$$

$$3P^2 = 2Q^2 \; ; \; Q = \sqrt{3/2}\,P$$

Therefore, $P : Q : R \equiv P : \sqrt{3/2}\,P : P \equiv \sqrt{2} : \sqrt{3} : \sqrt{2}$

EXAMPLE 2.8

Two forces equal to P and 2 P respectively act on a particle. When the first force is increased by 120 N and the second force is doubled, the direction of the resultant remains the same. Determine the value of force P.

Solution : Let α be the inclination which the resultant makes with force P. Then for the Ist case

$$\tan\alpha = \frac{2P\sin\theta}{P + 2P\cos\theta} \quad \text{...(i)}$$

where θ is the angle between the two given forces.
Subsequently when the forces get changed

$$\tan\alpha = \frac{4P\sin\theta}{(P + 120) + 4P\cos\theta} \quad \text{...(ii)}$$

From identities (*i*) and (*ii*)

$$\frac{2P\sin\theta}{P + 2P\cos\theta} = \frac{4P\sin\theta}{(P + 120) + 4P\cos\theta)}$$

$$\text{or} \quad \frac{1}{P + 2P\cos\theta} = \frac{2}{(P + 120) + 4P\cos\theta}$$

$$\text{or} \quad (P + 120) + 4P\cos\theta = 2P + 4P\cos\theta$$

That gives: $\boldsymbol{P = 120}$ **N**

EXAMPLE 2.9

The resultant of two forces P and Q is at right angles to P. Show that the angle between the forces is $\cos^{-1}(-P/Q)$.

Solution : Let θ be the angle between the two given forces P and Q. Then the inclination α which the resultant makes with force P is

$$\tan\alpha = \frac{Q\sin\theta}{P + Q\cos\theta}$$

Since $\alpha = 90^\circ$ (given), we get

$$\tan 90^\circ = \frac{Q\sin\theta}{P + Q\cos\theta}$$

$$\infty = \frac{Q\sin\theta}{P + Q\cos\theta}$$

This is possible only when the denominator is zero

$$P + Q\cos\theta = 0 \quad \text{or} \quad \cos\theta = (-P/Q)$$

$$\therefore \quad \theta = \cos^{-1}(-P/Q)$$

EXAMPLE 2.10

Find the magnitude of two concurrent forces which conform to the following data:

angle between the forces = 75°

angle between the resultant and one of forces = 45°

magnitude of the resultant = 2000 N

Solution : Let P and Q be the two forces and θ be the angle between them. Further, let α be the inclination which the resultant makes with force P. Then from parallelogram law of forces

$$(i)\quad \tan\alpha = \frac{Q\sin\theta}{P + Q\cos\theta}; \qquad \tan 45^\circ = \frac{Q\sin 75^\circ}{P + Q\cos 75^\circ}$$

$$1 = \frac{0.996Q}{P + 0.259Q} \qquad \therefore \quad P = 0.707Q$$

$$(ii)\quad R^2 = P^2 + Q^2 + 2PQ\cos\theta$$

$$(2000)^2 = (0.707Q)^2 + Q^2 + 2 \times (0.707Q) \times Q\cos 75^\circ$$

$$= 0.5Q^2 + Q^2 + 0.366Q^2 = 1.866Q^2$$

$$\therefore \qquad Q = \sqrt{\frac{(2000)^2}{1.866}} = \mathbf{1464\ N}$$

and $\qquad P = 0.707Q = 0.707 \times 1464 = \mathbf{1035\ N}$

EXAMPLE 2.11

Two forces P_1 and P_2 are inclined to each other at an angle θ and their resultant is at right angle to force P_1. There is another set of forces P_1 and P_3 with the angle of inclination θ and with their resultant at right angles to force P_3. Show that the force P_1 is goemetric mean between P_2 and P_3.

Solution : Refer Fig. 2.6 (*a*) for the first force system.

Inclination of resultant with force P_1,

$$\tan\alpha = \frac{P_2\sin\theta}{P_1 + P_2\cos\theta}$$

or $$\frac{P_2\sin\theta}{P_1 + P_2\cos\theta} = \tan 90^\circ = \infty$$

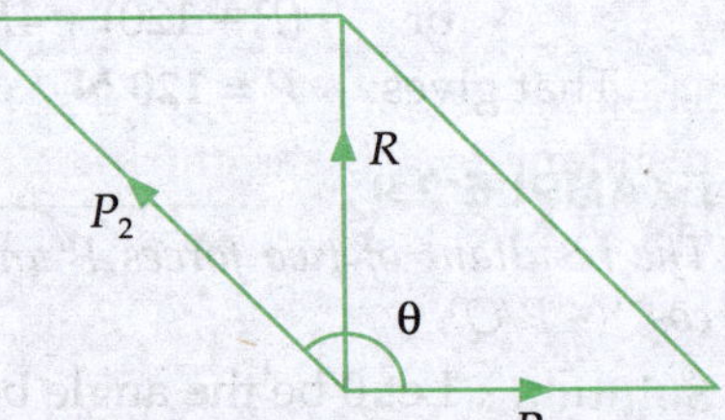

Fig. 2.6. (a)

That gives : $P_1 + P_2\cos\theta = 0$ or $P_1 = -P_2\cos\theta$ ...(*i*)

For the second force system indicated in Fig. 2.6 *b*

$$\frac{P_1\sin\theta}{P_3 + P_1\cos\theta} = \tan 90^\circ = \infty$$

That is : $P_3 + P_1\cos\theta = 0$ or $P_3 = -P_1\cos\theta$ ...(*ii*)

From equation (*i*) and (*ii*),

$$\frac{P_1}{P_3} = \frac{P_2}{P_1} \quad \text{or} \quad P_1 = \sqrt{P_2 P_3}$$

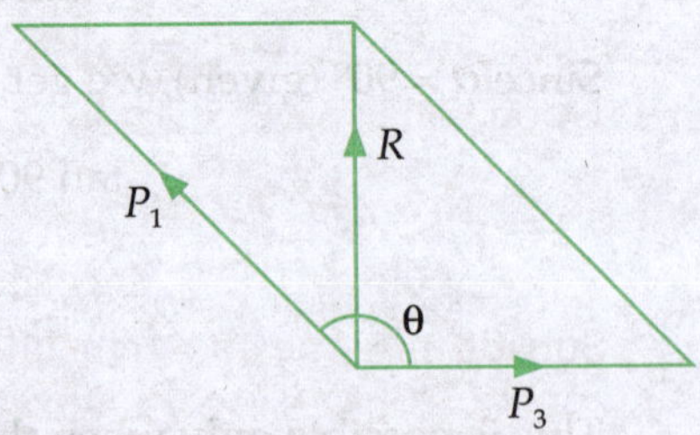

Fig. 2.6. (b)

That is the force P_1 is geometric mean between forces P_2 and P_3.

EXAMPLE 2.12

A vehicle is being pulled by two forces P and Q as shown in the adjoining Fig. 2.7 (a). Presuming that the resultant of these forces lies along the axis of the vehicle, make calculations for the pull P and the magnitude of the resultant pull.

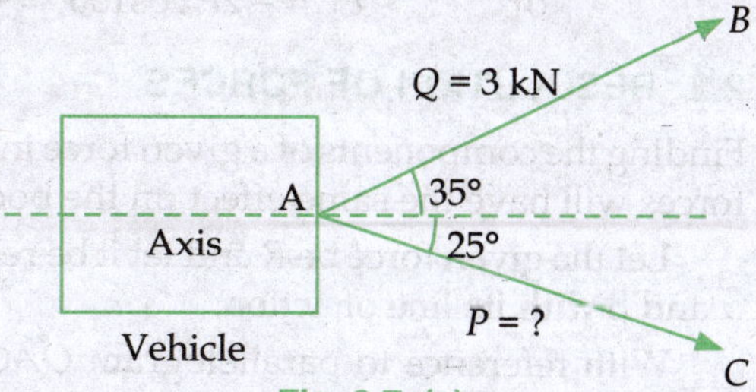

Fig. 2.7. (a)

Solution : The parallelogram law of forces applied to the given arrangement and depicted in Fig. 2.7 *b* shows that :

(*i*) the included angle between the pulls P and Q is

$$\theta = 35 + 25 = 60°$$

(*ii*) the inclination of resultant with the pull P is

$$\alpha = 25°$$

Then :
$$\tan\alpha = \frac{Q\sin\theta}{P + Q\cos\theta}$$

$$\tan 25° = \frac{3\sin 60°}{P + 3\cos 60}$$

or
$$P\tan 25° = 3\sin 60° - 3\tan 25° \text{ cost } 60°$$

or
$$0.466\,P = 2.598 - 0.699 = 1.899$$

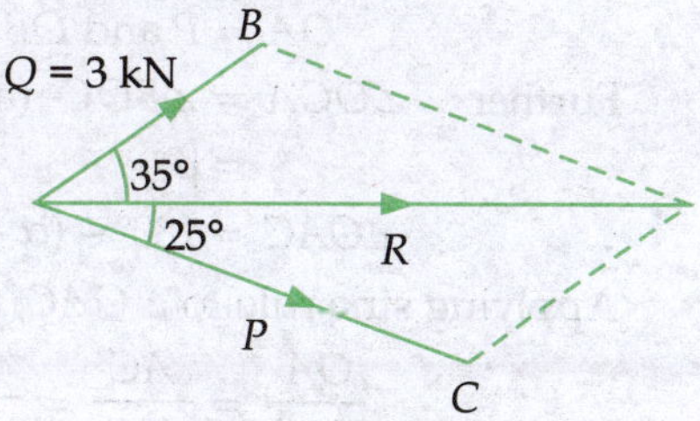

Fig. 2.7. (b)

$\therefore$
$$\text{Pull } P = \frac{1.899}{0.466} = \mathbf{4.075\ kN}$$

$$\text{Resultant } R = \sqrt{P^2 + Q^2 + 2PQ\cos\theta}$$

$$= \sqrt{4.075^2 + 3^2 + 2\times 4.075\times 3\times\cos 60°} = \mathbf{6.15\ kN}$$

EXAMPLE 2.13

Determine the horizontal force and a force inclined at an angle of 60° with the vertical. The resultant of these forces is stated to be a vertical force F.

Solution : Let horizontal force be F_1 and the inclined force be F_2. The angle of inclination θ between these forces is 150° and the resultant $R = F$ is at right angle to force F_1. Then from parallelogram law of forces

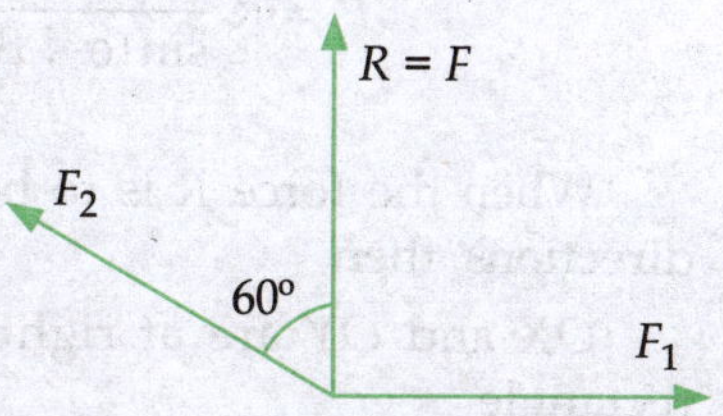

Fig. 2.8

$$R^2 = F^2 = F_1^2 + F_2^2 + 2F_1F_2\cos 150° \qquad \text{...(i)}$$

$$\tan\alpha = \tan 90° = \frac{F_2\sin 150°}{F_1 + F_2\cos 150°}$$

Since $\tan 90° = \infty$, the above identity will be true if

$$F_1 + F_2\cos 150° = 0 \qquad \text{...(ii)}$$

Upon squaring:

$$F_1^2 + F_2^2\cos^2 150° + 2F_1F_2\cos 150° = 0$$

or
$$F_1^2 + 2F_1F_2\cos 150° = -F_2^2\cos^2 150°$$

Substituting the above result in identity (i), we get

$$F^2 = F_2^2 - F_2^2\cos^2 150° = F_2^2(1 - \cos^2 150°) = F_2^2\sin^2 150° = 0.25F_2^2$$

That gives:
$$F_2 = \sqrt{\frac{F^2}{0.25}} = \mathbf{2\,F}$$

Then from identity (*ii*),

$$F_1 + 2F \cos 150^\circ = 0$$

or $$F_1 = -2F \cos 150^\circ = -2F\left(-\sqrt{3}/2\right) = \sqrt{3}\,F$$

2.2. RESOLUTION OF FORCES

Finding the components of a given force in two given directions is called *resolution*. These component forces will have the same effect on the body as the given single force.

Let the given force be R and let it be required to find its components in directions making angles α and β with its line of action.

With reference to parallelogram $OACB$, the sides OA and OB represent the components of the given force R along OX and OY respectively. That is

$$OA = P \text{ and } OB = Q$$

Further $\angle OCA = \angle BOC$ (alternate angles) $= \beta$

$\therefore$ $\angle OAC = 180^\circ - (\alpha + \beta)$

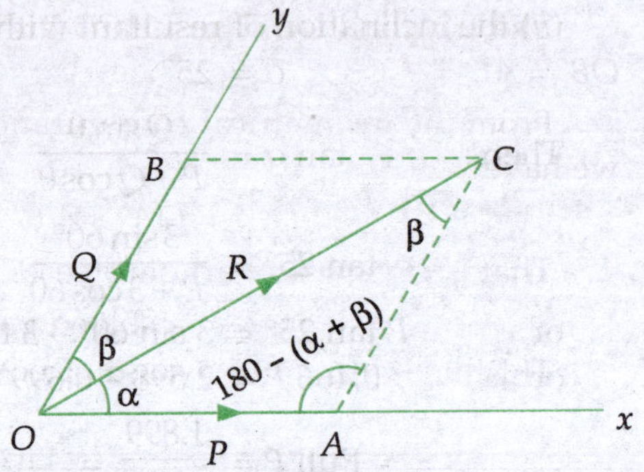

Fig. 2.9

Applying sine rule to $\Delta\, OAC$, we get

$$\frac{OA}{\sin\beta} = \frac{AC}{\sin\alpha} = \frac{OC}{\sin\{180 - (\alpha+\beta)\}}$$

AC is parallel and equal to OB which represents Q

$$\therefore \quad \frac{P}{\sin\beta} = \frac{Q}{\sin\alpha} = \frac{R}{\sin(\alpha+\beta)} \qquad ...(2.8)$$

Thus the resolved parts of the given force R are

$$P = R\,\frac{\sin\beta}{\sin(\alpha+\beta)} \quad \text{and} \quad Q = R\,\frac{\sin\alpha}{\sin(\alpha+\beta)} \qquad ...(2.9)$$

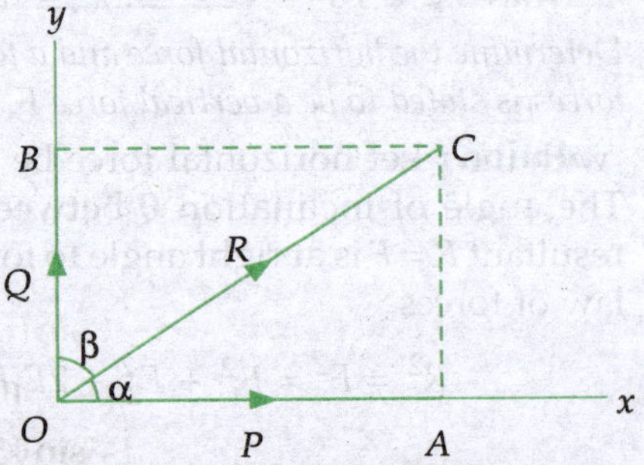

Fig. 2.10

When the force R is to be resolved along perpendicular directions, then

OX and OY are at right angles and $OACB$ becomes a rectangle

$$\alpha + \beta = 90^\circ \quad \text{or} \quad \beta = 90^\circ - \alpha$$

$$\therefore \quad P = R\,\frac{\sin\beta}{\sin(\alpha+\beta)} = R\,\frac{\sin(90^\circ - \alpha)}{\sin 90^\circ} = R\cos\alpha \qquad ...(2.10)$$

i.e., the projection of OC (representing R) on OX is a measure of resolved component along the direction OX

$$Q = R\,\frac{\sin\alpha}{\sin(\alpha+\beta)} = R\,\frac{\sin\alpha}{\sin 90^\circ} = R\sin\alpha \qquad ...(2.11)$$

i.e., the projection of OC (representing R) on OY is a measure of resolved component along the direction of OY

When the components P and Q are at right angles to each other, they are called the rectangular components of force R.

2.3. THEOREM OF RESOLVED PARTS

"The algebraic sum of the resolved parts of two forces in a given direction is equal to the resolved part of their resultant in the same direction."

Let OC represent the resultant of two forces OA and OB acting at point O as shown in Fig. 2.11. Further, let OX be the direction along which resolved parts of OA, OB and OC are to be worked out.

Drop perpendiculars form B, A and C to meet OX at B', A' and C' respectively.

The triangles OBB' and ACD are congruant and that gives $OB' = AD = A'C'$.

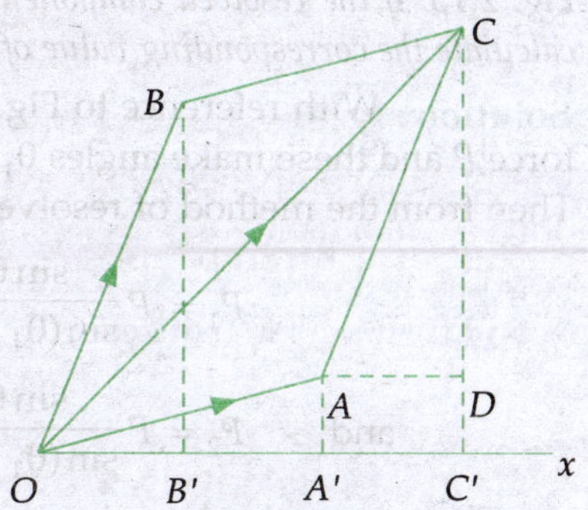

Fig. 2.11

From the geometrical configuration of the system of forces, we have

$$OC' = OA' + A'C' = OA' + AD = OA' + OB'$$

That is: Resolved part of force OC on OX

= resolved part of force OA on OX + resolved part of force OB on OX

This theorem can be extended to any number of coplanar-concurrent forces.

EXAMPLE 2.14

The resultant of two forces is 10 N and it is inclined at 60° to one of the forces whose magnitude is 5 N. Determine the magnitude and direction of the other force.

Solution : The given force system may be depicted as shown in Fig. 2.12.

$$P = OA = 5 \text{ N}$$
$$R = OC = 10 \text{ N}$$
$$\alpha = 60^\circ$$

Let the other force Q be oriented along OB making an angle β with the resultant.

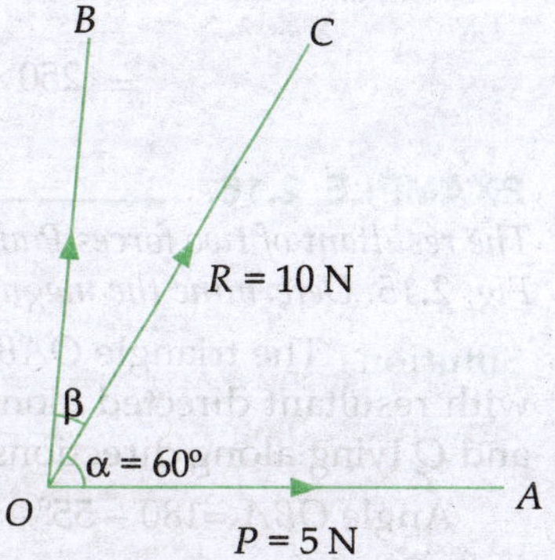

Fig. 2.12

From the method of resolved parts,

$$P = R\,\frac{\sin\beta}{\sin(\alpha+\beta)} \quad \ldots(i)$$

$$Q = R\,\frac{\sin\alpha}{\sin(\alpha+\beta)} \quad \ldots(ii)$$

Substituting $P = 5$ N and $R = 10$ N in expression (i), we get

$$5 = 10\,\frac{\sin\beta}{\sin(\alpha+\beta)}\,; \qquad 2\sin\beta = \sin(\alpha+\beta)$$

or $\quad 2\sin\beta = \sin\alpha\cos\beta + \cos\alpha\sin\beta = \sin 60^\circ\cos\beta + \cos 60^\circ\sin\beta$

$$= 0.866\cos\beta + 0.5\sin\beta$$

That gives:

$$1.5\sin\beta = 0.866\cos\beta$$

$$\tan\beta = \frac{0.866}{1.5} = 0.5773 \quad \text{and} \quad \beta = \tan^{-1}(0.5773) = 30^\circ$$

Now, considering identity (ii) and substituting $\beta = 30^\circ$,

$$Q = 10\,\frac{\sin 60^\circ}{\sin 90^\circ} = 10 \times \frac{0.866}{1} = 8.66 \text{ N}$$

EXAMPLE 2.15

A force 250 is to resolved into components along the lines A-A′ and B-B′ oriented as shown in the adjoining Fig. 2.13. If the resolved component along BB′ equals 125 N, what would be the value of angle α ? Also calculate the corresponding value of force component along A-A′.

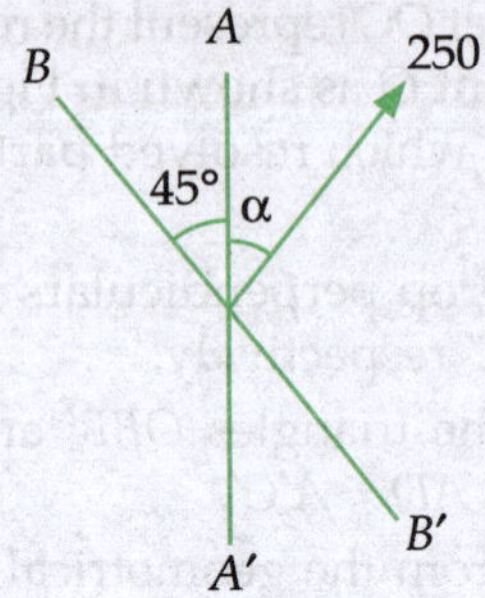

Fig. 2.13

Solution : With reference to Fig. 2.14, P_1 and P_2 are the components of force P and these make angles θ_1 and θ_2 respectively on either side of P. Then from the method of resolved parts

$$P_1 = P\frac{\sin\theta_2}{\sin(\theta_1+\theta_2)}$$

and
$$P_2 = P\frac{\sin\theta_1}{\sin(\theta_1+\theta_2)}$$

Then as per the data given :

$$125 = 250\,\frac{\sin\alpha}{\sin 135^\circ}$$

or
$$\sin\alpha = \frac{125}{250}\times\sin 135^\circ = 0.3536$$

∴ Required angle of inclination

$$\alpha = \sin^{-1}(0.3536) = \mathbf{20.71^\circ}$$

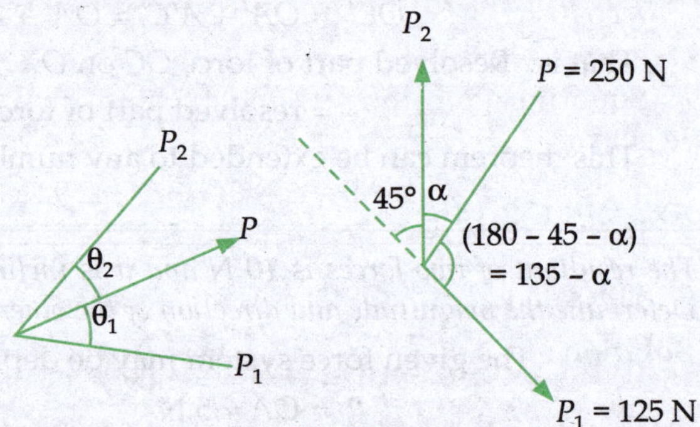

Fig. 2.14

$$\text{Component } P_2 = 250\frac{\sin(135-20.71)^\circ}{\sin 135^\circ}$$

$$= 250\times\frac{0.9115}{0.7071} = \mathbf{322.27\ N}$$

EXAMPLE 2.16.

The resultant of two forces P and Q is 200 N and it makes an angle of 30° with the horizontal as shown in Fig. 2.15. Determine the magnitude of components P and Q.

Solution : The triangle OAB constitutes the force triangle with resultant directed along OA, and the components P and Q lying along directions OB and BA respectively.

Angle $OBA = 180 - 55^\circ - 40^\circ = 85^\circ$

Applying sine rule to ΔOAB, we get

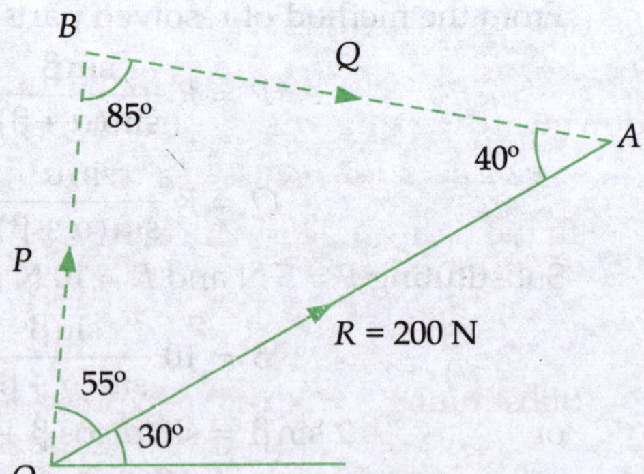

Fig. 2.15

$$\frac{OA}{\sin 85^\circ} = \frac{OB}{\sin 40^\circ} = \frac{BA}{\sin 55^\circ}$$

or
$$\frac{200}{\sin 85^\circ} = \frac{P}{\sin 40^\circ} = \frac{Q}{\sin 55^\circ}$$

$$\therefore\quad P = 200\times\frac{\sin 40^\circ}{\sin 85^\circ} = 200\times\frac{0.643}{0.996} = \mathbf{129.1\ N} \text{ from } O \text{ to } B$$

$$Q = 200\times\frac{\sin 55^\circ}{\sin 85^\circ} = 200\times\frac{0.819}{0.996} = \mathbf{164.4\ N} \text{ from } B \text{ to } A$$

EXAMPLE 2.17.

The force system shown in the adjoining Fig. 2.16 has a resultant of 2 kN acting up along the Y-axis. Make calculations for the force F both in magnitude and direction.

Solution : The resultant force $R = 2$ kN acts upwards along O-Y axis. Resolving all the forces in vertical direction, we get

$$F \sin\theta - 2.4 \sin 30^\circ = 2$$

or $$F \sin\theta = 2 + 2.4 \sin 30^\circ = 2 + 2.4 \times \frac{1}{2}$$

$$= 3.2 \text{ kN} \quad ...(i)$$

Resolving all the forces in horizontal direction, we get

$$F \cos\theta + 2.4 \cos 30^\circ - 5 = 0$$

$$F \cos\theta = 5 - 2.4 \cos 30^\circ = 5 - 2.4 \times 0.866 = 2.922 \quad ...(ii)$$

Squaring identities (*i*) and (*ii*) and then adding, we get

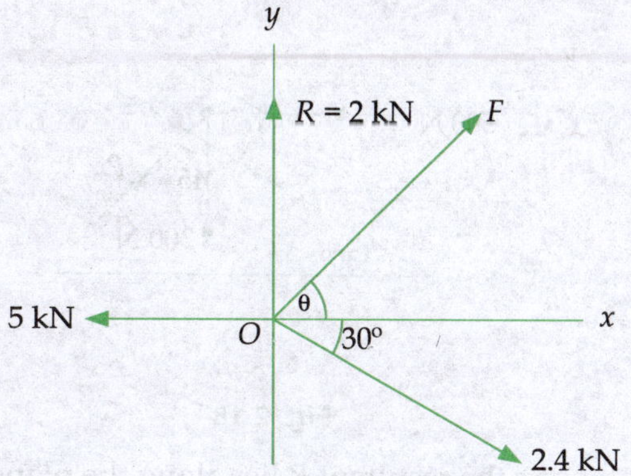

Fig. 2.16

$$F^2 (\sin^2\theta + \cos^2\theta) = 3.2^2 + 2.922^2 = 18.778$$

$$\therefore \quad F = \sqrt{18.778} = \textbf{4.33 kN}$$

The orientation of force can be obtained by dividing identity (*i*) by identity (*ii*). That gives

$$\tan\theta = \frac{3.2}{2.922} = 1.095$$

$$\therefore \quad \theta = 2.18 \tan^{-1} (1.095) = \textbf{47.6°}$$

EXAMPLE 2.18

Two cables which have known tensions of 40 N and 60 N are attached to the top of a tower AB as shown in Fig. 2.17. What tension will be induced in the guy wire AC if the resultant of the forces exerted at the top A by the cables acts vertically downwards?

Solution : Since the resultant of the forces acts vertically downwards, we have $\Sigma F_x = 0$ and that gives

$$40 \cos 15° - 60 \cos 30° + T \cos\theta = 0 \quad ...(i)$$

In the right angled triangle ABC

$$\theta = \tan^{-1}\left(\frac{AB}{BC}\right) = \tan^{-1}\left(\frac{15}{10}\right) = 56.31°$$

Substituting $\theta = 56.31°$ in expression (*i*), we get

$$40 \cos 15° - 60 \cos 30° + T \cos 56.31° = 0$$

or $$38.64 - 51.96 + 0.555\, T = 0$$

$$\therefore \quad \text{Tension in the guy wire } T = \frac{51.96 - 38.64}{0.555} = \textbf{24 N}$$

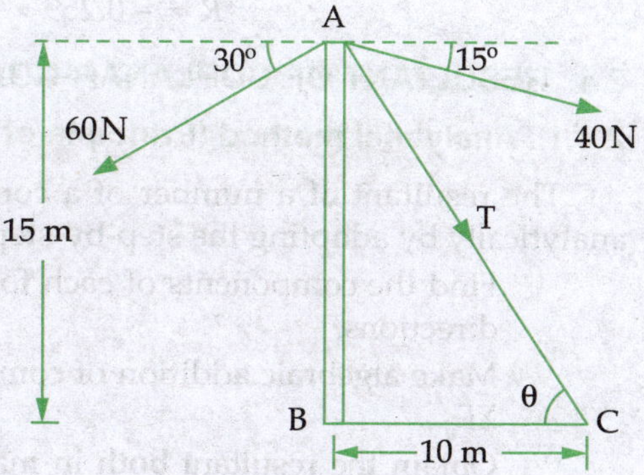

Fig. 2.17

EXAMPLE 2.19

A block of weight 300 N is acted upon by a horizontal force F = 500 N and a pressure P exerted by the inclined as shown in Fig. 2.18. If the resultant of the force system lies parallel to the plane, work out the magnitude of pressure P and the resultant force.

Solution : Let the co-ordinates be chosen parallel and perpendicular to the plane.

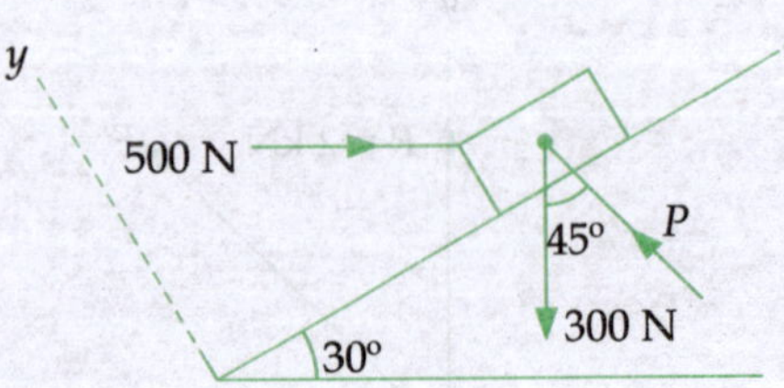

Fig. 2.18

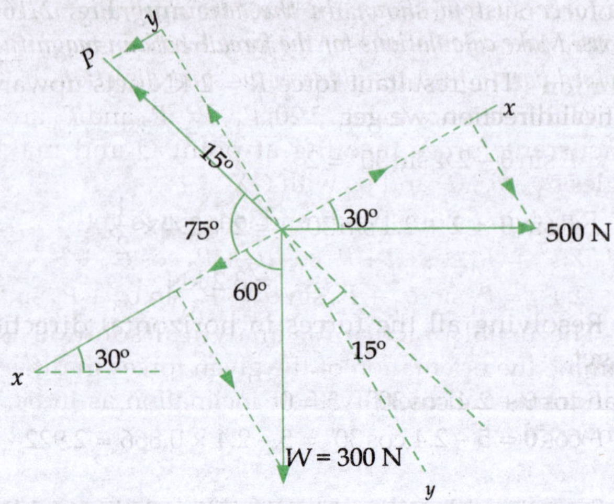

Fig. 2.19

As the resultant R lies along the plane, its component perpendicular to the plane is zero. Then with reference to Fig. 2.19.

$$R = -P\sin 15^\circ + 500\cos 30^\circ - 300\cos 60^\circ$$
$$= -0.259\,P + 433 - 150$$
$$= -0.259\,P + 283$$

Also
$$P\cos 15^\circ - 500\sin 30^\circ - 300\sin 60^\circ = 0$$
$$P\cos 15^\circ = 500\sin 30^\circ + 300\sin 60^\circ$$
$$P \times 0.966 = 250 + 259.8 = 509.8$$
$$P = \frac{509.8}{0.966} = 527.74 \text{ N}$$
$$\therefore \quad R = -0.259 \times 527.74 + 283 = \textbf{146.31 N}$$

2.4. RESULTANT OF COPLANAR-CONCURRENT FORCES

2.4.1. Analytical method (Principle of resolved parts)

The resultant of a number of a coplanar-concurrent forces acting on a body is worked out analytically by adopting the step-by-step procedure outlined below:

(i) Find the components of each force in the system in two mutually perpendicular X and Y directions.

(ii) Make algebraic addition of components in each direction to get two components ΣF_x and ΣF_y.

(iii) Obtain the resultant both in magnitude and direction by combining the two component forces ΣF_x and ΣF_y which are mutually prependicular.

$$\text{Resultant, } R = \sqrt{(\Sigma F_x)^2 + (\Sigma F_y)^2}$$

and its inclination θ to X-axis is given by

$$\tan\theta = \frac{\Sigma F_y}{\Sigma F_x}$$

This analytical method is based on theorem of resolved parts which states that "*the algebraic sum of the resolved parts of two forces in a given direction is equal to the resolved part of their resultant in the same direction*".

With reference to Fig. 2.20; P_1, P_2, P_3 and P_4 are the concurrent forces meeting at point O and making angles α_1, α_2, α_3 and α_4 with OX.

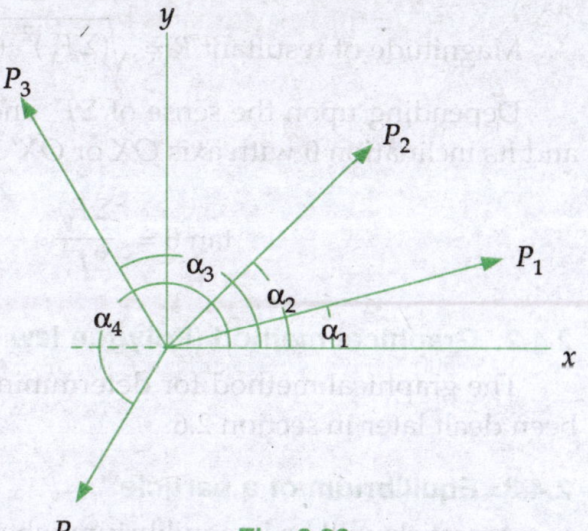

Fig. 2.20

Resolving along x-axis and y-axis, we get

$$\Sigma F_x = P_1 \cos \alpha_1 + P_2 \cos \alpha_2 + P_3 \cos \alpha_3 + P_4 \cos \alpha_4$$
$$\Sigma F_y = P_1 \sin \alpha_1 + P_2 \sin \alpha_2 + P_3 \sin \alpha_3 + P_4 \sin \alpha_4$$

The usual practice for analytical solution is to examine the orientation of the given forces in different quandrants and identify their inclination as indicated below :

1.	1st quadrant $F_{x1} = P_1 \cos \alpha_1$ $F_{y1} = P_1 \sin \alpha_1$	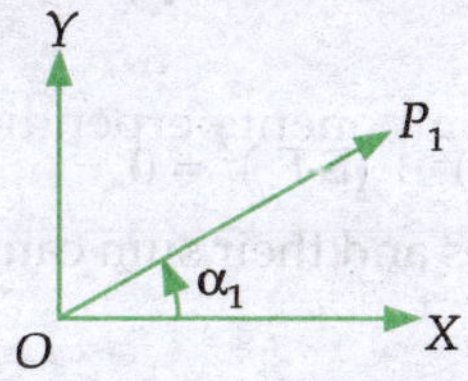	Inclination α_1 with axis OX measured in anti-clockwise direction.
2.	2nd quadrant $F_{x2} = -P_2 \cos \alpha_2$ $F_{y2} = P_2 \sin \alpha_2$	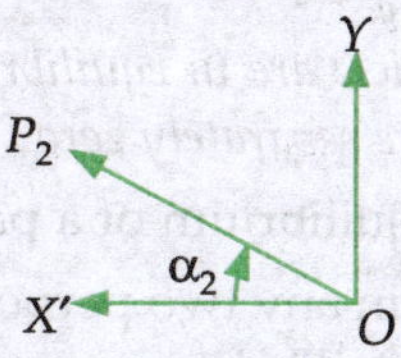	Inclination α_2 with axis OX' measured in clockwise direction.
3.	3rd quadrant $F_{x3} = -P_3 \cos \alpha_3$ $F_{y3} = -P_3 \sin \alpha_3$	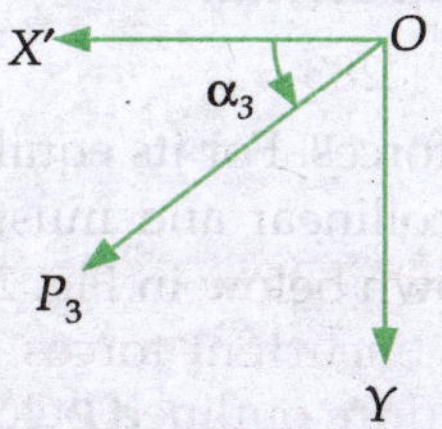	Inclination α_3 with axis OX' measured in anti-clockwise direction.
4.	4th quadrant $F_{x4} = P_4 \cos \alpha_4$ $F_{y4} = -P_4 \sin \alpha_4$	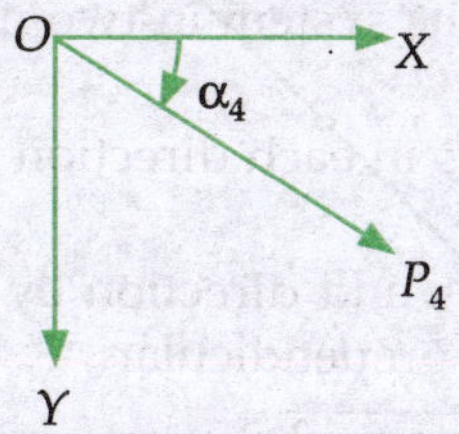	Inclination α_4 with axis OX measured in clockwise direction.

Each one of the indicated forces is thus split up into respective x and y components, and that gives:

$$\Sigma F_x = P_1 \cos \alpha_1 - P_2 \cos \alpha_2 - P_3 \cos \alpha_3 + P_4 \cos \alpha_4$$
$$\Sigma F_y = P_1 \sin \alpha_1 - P_2 \sin \alpha_2 - P_3 \sin \alpha_3 + P_4 \sin \alpha_4$$

Magnitude of resultant $R = \sqrt{(\Sigma F_x)^2 + (\Sigma F_y)^2}$

Depending upon the sense of ΣF_x and ΣF_y, the resultant may lie in any one of the quadrants and its inclination θ with axis OX or OX' is given by

$$\tan\theta = \frac{\Sigma F_y}{\Sigma F_x}$$

2.4.2. Graphical method (Polygon law of forces)

The graphical method for determining the resultant of coplanar concurrent force systems has been dealt later in section 2.6.

2.4.3. Equilibrium of a particle

A particle will be in equilibrium when resultant of all the forces acting on it is zero

$$\text{Resultant } R = \sqrt{(\Sigma F_x)^2 + (\Sigma F_y)^2}$$

For a particle to be in equilibrium:

$$R = 0; \quad (\Sigma F_x)^2 + (\Sigma F_y)^2 = 0$$

Now $(\Sigma F_x)^2$ and $(\Sigma F_y)^2$ are positive quantities and their sum cannot be zero unless each of then is zero.

$$\therefore \qquad \Sigma F_x = 0 \quad \text{and} \quad \Sigma F_y = 0 \qquad ...(2.11)$$

Hence, if any number of forces acting at a particle are in equilibrium, then the algebraic sum of their resolved parts in any two perpendicular directions are separately zero.

Equations 2.11 are called the equations of equilibrium of a particle in plane.

Conversely, if the sum of the resolved parts in any two perpendicular directions are separately zero, then the forces acting at any point are in equilibrium.

Equilibrium of a two-force and three-force body.

A **two-force body** is subjected to only two forces. For its equilibrium, the resultant force must be zero. Accordingly the two forces must be collinear and must have the same magnitude but opposite direction. These aspects have been shown below in Fig. 2.21 (*a*) and (*b*).

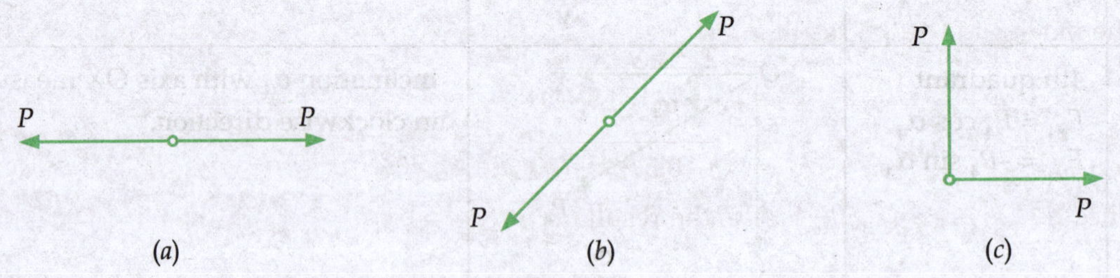

Fig. 2.21

The particle subjected to two forces of equal magnitude (Fig. 2.21*c*) shall not be in equilibrium because the forces have different lines of action.

A three-force body is subjected to only three forces. For the particle to be in equilibrium, the acting forces must be concurrent and the resultant of any two of them must be equal, opposite and collinear with the third force. The three forces would form a closed triangle. This aspect has been illustrated in Fig. 2.22.

Law of sines can then be used to obtain solution to the problem. That is, each force is proportional to the sine of the angle between the other two forces.

$$\frac{F_1}{\sin\alpha} = \frac{F_2}{\sin\beta} = \frac{F_3}{\sin\gamma}$$

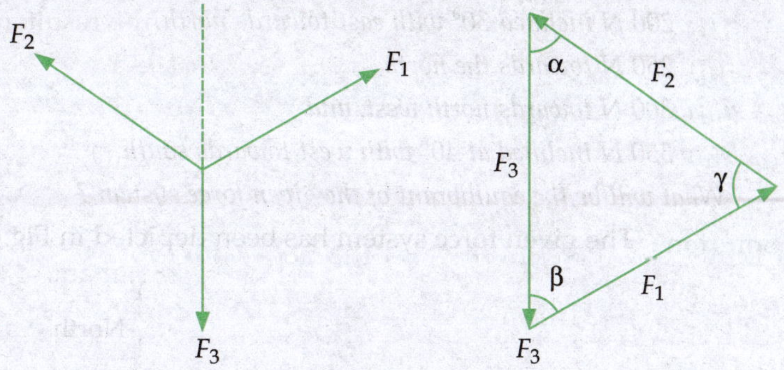

Fig. 2.22

EXAMPLE 2.20

Forces P, 2P, $3\sqrt{3}P$ *and 4P act at a point O in directions making angles of 60°, 150° and 300° respectively with the first force. Find the magnitude and direction of the resultant of the given force system.*

Solution : Refer Fig. 2.23 for the orientation of the given forces with the first force which has been drawn horizontally.

Resolving the given forces along the horizontal and vertical directions, we get

$$\Sigma F_x = P + 2P\cos 60° + 3\sqrt{3}P\cos 150° + 4P\cos 300°$$

$$= P + 2P \times 0.5 + 3\sqrt{3}P \times \left(-\frac{\sqrt{3}}{2}\right) + 4P \times (0.5)$$

$$= P + P - 4.5\,P + 2P = -\,0.5P$$

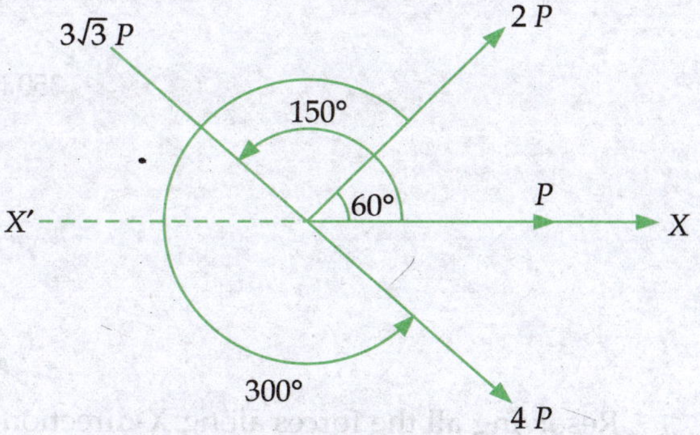

Fig. 2.23

and $\Sigma F_y = 0 + 2P\sin 60° + 3\sqrt{3}P\sin 150° + 4P\sin 300°$

$$= 0 + 1.732\,P + 2.598\,P - 3.464\,P = 0.866\,P$$

The magnitude of the resultant is

$$R = \sqrt{(\Sigma F_x)^2 + (\Sigma F_y)^2}$$

$$= \sqrt{(-0.5P)^2 + (0.866P)^2} \simeq P$$

Since ΣF_x is *–ve* and ΣF_y is *+ve*, the resultant lies in the second quadrant and its inclination α with *OX′* is

$$\tan\alpha = \frac{\Sigma F_y}{\Sigma F_x} = \frac{0.866\,P}{0.5\,P} = 1.732 \qquad \therefore \alpha = 60°$$

Then the inclination of the resultant with the direction of first force is (180 – 60) = 120°

EXAMPLE 2.21

Determine the magnitude and direction of the resultant of the following set of forces acting on a body

(i) *200 N inclined 30° with east towards north,*

(ii) *250 N towards the north,*

(iii) *300 N towards north west, and*

(iv) *350 N inclined at 40° with west towards south.*

What will be the equilibrant of the given force system ?

Solution : The given force system has been depicted in Fig. 2.24.

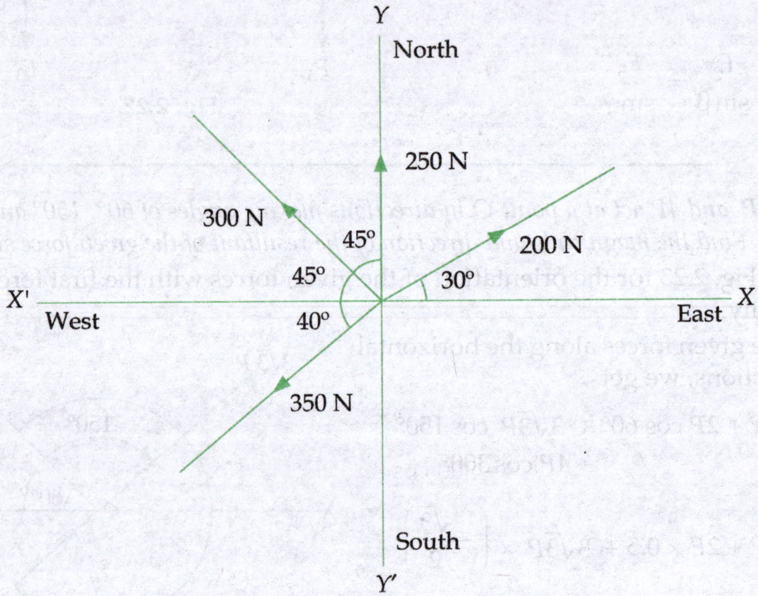

Fig. 2.24

Resolving all the forces along X-direction

$$\Sigma F_x = 200 \cos 30° + 250 \cos 90° + 300 \cos 135° + 350 \cos 220°$$
$$= 200 \times 0.866 + 250 \times 0 + 300 \times (-0.707) + 350 \times (-0.766)$$
$$= 173.2 + 0 - 212.1 - 268.1 = -307 \text{ N (along } OX')$$

Resolving all the forces along Y-direction

$$\Sigma F_y = 200 \sin 30° + 250 \sin 90° + 300 \sin 135° + 350 \sin 220°$$
$$= 200 \times 0.5 + 250 \times 1 + 300 \times 0.707 + 350 \times (-0.642)$$
$$= 100 + 250 + 212.1 - 224.7 = 337.4 \text{ N (along OY)}$$

Alternatively

Refer table given below which shows the location of the given forces in different quandrants and the direction of their resolved parts (horizontal and vertical components). As per the normal sign convention, the x-components acting to the right and y-components acting upward are taken as positive.

S.No.	Force and its location	Horizontal component F_x	Vertical component F_y
1.	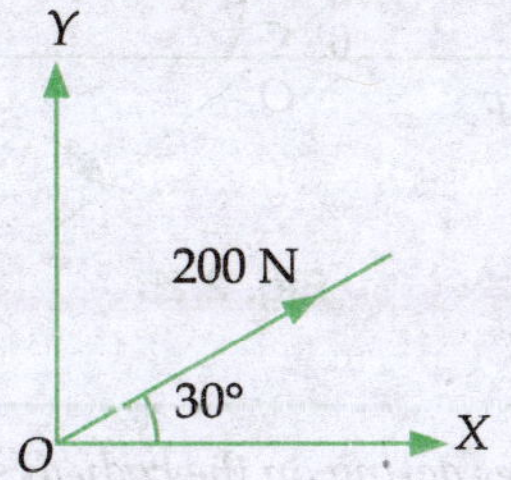	200 cos 30° = 173.2 N	200 sin 30° = 100 N
2.	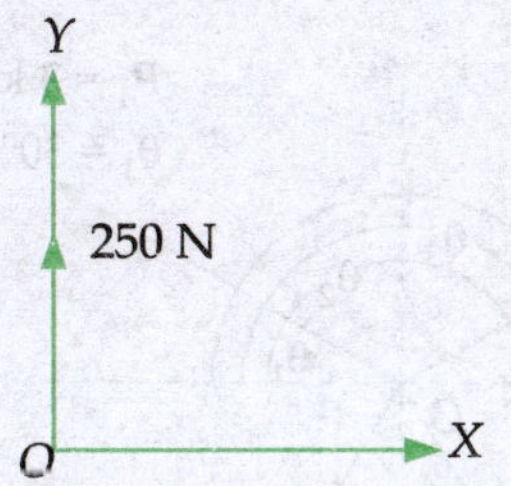	0	250 N
3.	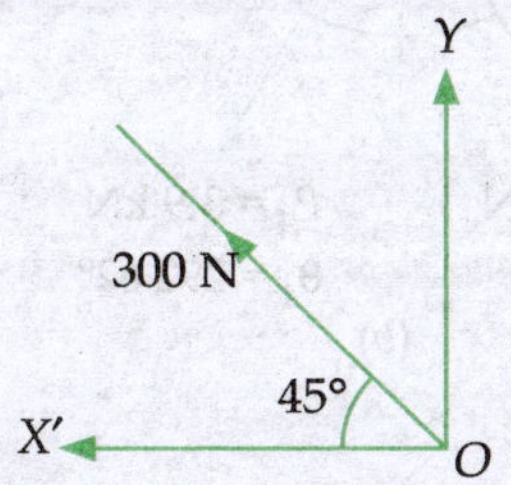	– 300 cos 45° = – 212.1 N	300 sin 45° = 212.1 N
4.	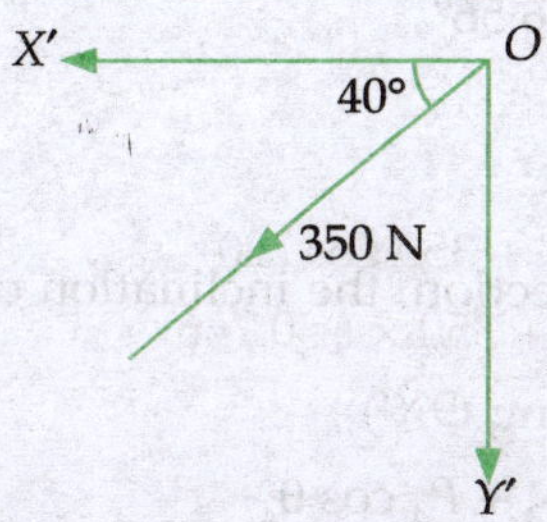	– 350 cos 40° = – 268.1 N	– 350 sin 40° = – 224.97 N

Then : $\Sigma F_x = 173.2 + 0 - 212.1 - 268.1 = -307$ N (along OX')

$\Sigma F_y = 100 + 250 + 212.1 - 224.97 = 337.1$ N (along OY')

The magnitude of resultant force is

$$R = \sqrt{(\Sigma F_x)^2 + (\Sigma F_y)^2} = \sqrt{(-307)^2 + (337.4)^2} = \mathbf{456\ N}$$

Since ΣF_x is negative and ΣF_y is positive, the resultant lies in the second quadrant and its inclination θ' with OX' is

$$\theta' = \tan^{-1}\left(\frac{\Sigma F_y}{\Sigma F_x}\right) = \tan^{-1}\left(\frac{337.4}{307}\right) = 47.7°$$

Obviously the inclination of the resultant with OX (positive direction) will be

$$\theta = 180 - 47.7 = 132.3°$$

An equal and opposite force will be required to hold the body in position. The equilibrant will thus have a magnitude of 456 N, inclined at 47.7° to the x-axis (fourth quadrant) and is opposite to the resultant. Both the resultant and equilibrant are shown in Fig. 2.25.

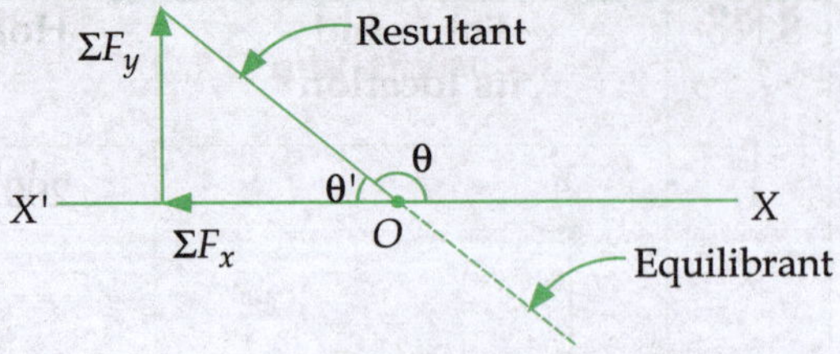

Fig. 2.25

EXAMPLE 2.22

Determine the resultant, both in magnitude and direction, of the four forces acting on the body as shown in the figure given below Fig. 2.26 a.

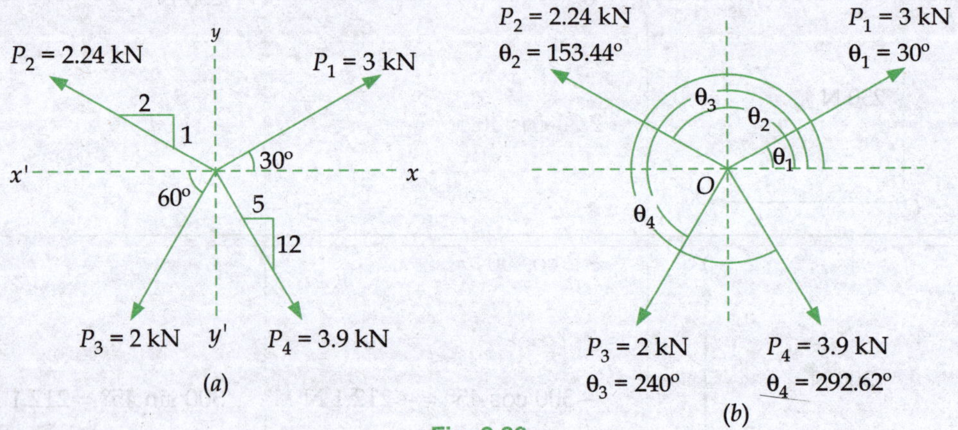

Fig. 2.26

Solution : Inclination of force 2.24 kN with OX′ $= \tan^{-1}\left(\frac{1}{2}\right) = 26.56°$

Inclination of force 3.9 kN with OY′ $= \tan^{-1}\left(\frac{5}{12}\right) = 22.62°$

Measuring all angles from axis O-X in the anticlockwise direction, the inclination of different forces will be as depicted in Fig. 2.26 *b*.

Resolving all the forces in horizontal direction,

$$\begin{aligned}\Sigma F_x &= P_1 \cos\theta_1 + P_2 \cos\theta_2 + P_3 \cos\theta_3 + P_4 \cos\theta_4 \\ &= 3 \cos 30° + 2.24 \cos 153.44° + 2 \cos 240° + 3.9 \cos 292.62° \\ &= 2.598 - 2.004 - 1.0 + 1.50 = 1.094 \text{ kN}\end{aligned}$$

Resolving all the forces in vertical direction,

$$\begin{aligned}\Sigma F_y &= P_1 \sin\theta_1 + P_2 \sin\theta_2 + P_3 \sin\theta_3 + P_4 \sin\theta_4 \\ &= 3 \sin 30° + 2.24 \sin 153.44° + 2 \sin 240° + 3.9 \sin 292.62° \\ &= 1.50 + 1.00 - 1.73 - 3.60 = -2.83 \text{ kN}\end{aligned}$$

Alternatively

Refer table given below which shows the location of the given system of forces in different quadrants and the direction of their resolved parts (horizontal and vertical components. As per the normal sign convention, the horizontal (x) components acting to the right and vertical (y) components acting upward are taken as positive.

S.No.	Force and its orientation	Horizontal component F_x	Vertical component F_y
1.	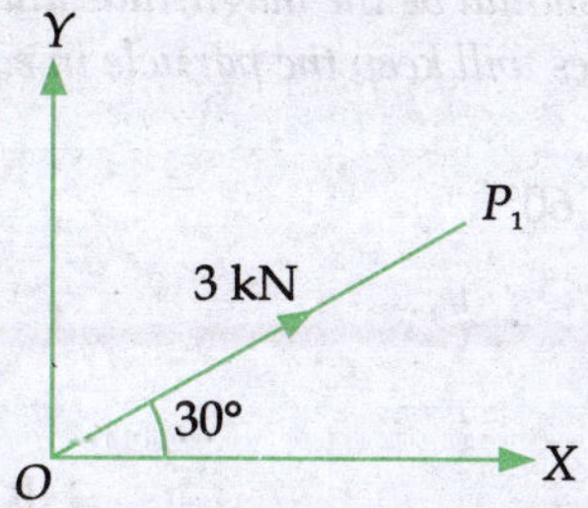	3 cos 30° = 2.598 kN	3 sin 30° = 1.50 kN
2.	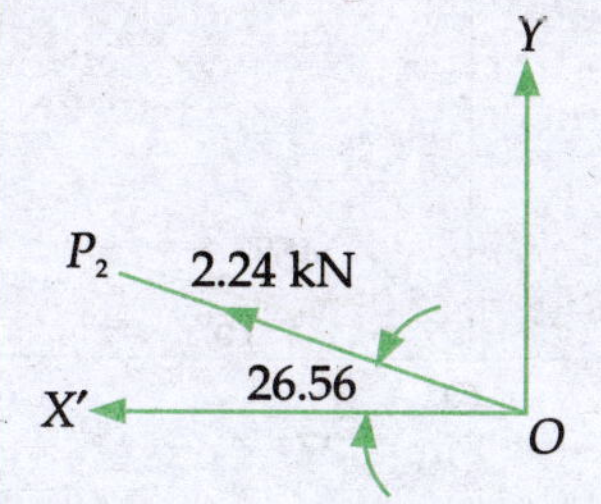	– 2.24 cos 26.56° = – 2.004 kN	2.24 sin 26.56° = 1.001 kN
3.	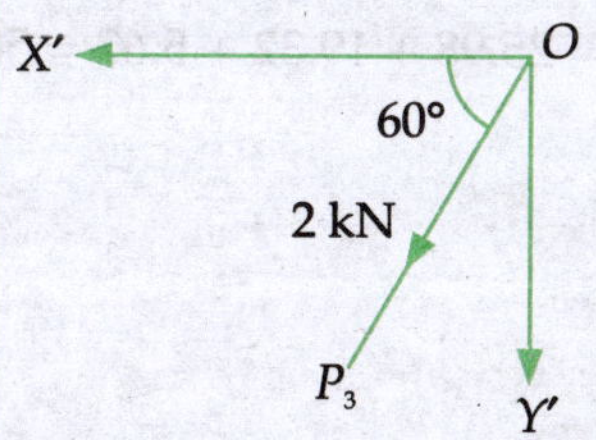	– 2 cos 60° = – 1 kN	– 2 sin 60° = – 1.732 kN
4.	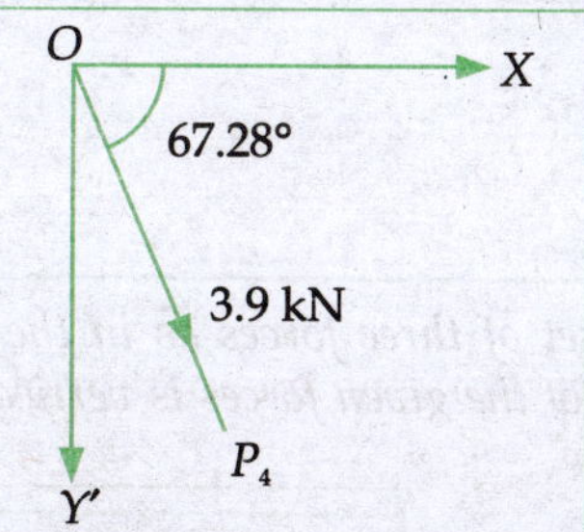	3.9 cos 67.28° = 1.506 kN	– 3.9 sin 67.28° = – 3.597kN

The magnitude of the resultant force is

$$R = \sqrt{(\Sigma F_x)^2 + (\Sigma F_y)^2} = \sqrt{(1.094)^2 + (-2.83)^2}$$

$$= \sqrt{1.197 + 8.009} = 3.034 \text{ kN}$$

and the inclination of resultant with the horizontal is

$$\alpha = \tan^{-1}\left(\frac{\Sigma F_y}{\Sigma F_x}\right)$$

$$= \tan^{-1}\left(\frac{-2.83}{1.094}\right) = 68.86°$$

EXAMPLE 2.23

A particle positioned at point O is acted upon by three forces as shown in Fig. 2.27. Determine the value of force P such that the resultant of these three forces is horizontal. What should be the magnitude and direction of a fourth force which acting at point O along with the given three forces will keep the particle in equilibrium state?

Solution : Angle between force P and the axis 0 - x is (15 + 45) = 60°

Resolving all the forces along y-direction,

$\Sigma F_y = -30 \sin 30° + 20 \sin 15° + P \sin 50°$

The negative sign with force of 30 N is due to the fact that its y-component is acting downward.

Since the resultant is horizontal, $\Sigma F_y = 0$

$$\therefore \quad 0 = -30 \times 0.5 + 20 \times 0.2588 + P \times 0.866$$
$$= -15 + 5.176 + 0.866\,P$$

That gives :

$$P = \frac{15 - 5.176}{0.866} = 11.344 \text{ N}$$

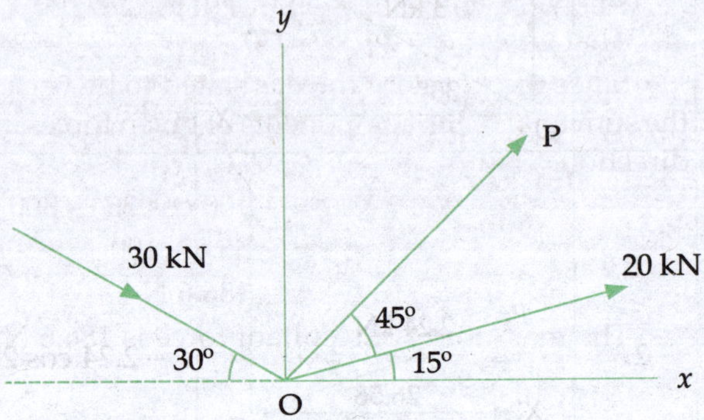

Fig. 2.27

Resolving all forces along X-direction,

$$\Sigma F_x = 30 \cos 30° + 20 \cos 15° + 11.344 \cos 60° = 25.98 + 19.32 + 5.67 = 50.97 \text{ kN}$$

The equilibrant would be – 50.97 kN as shown below.

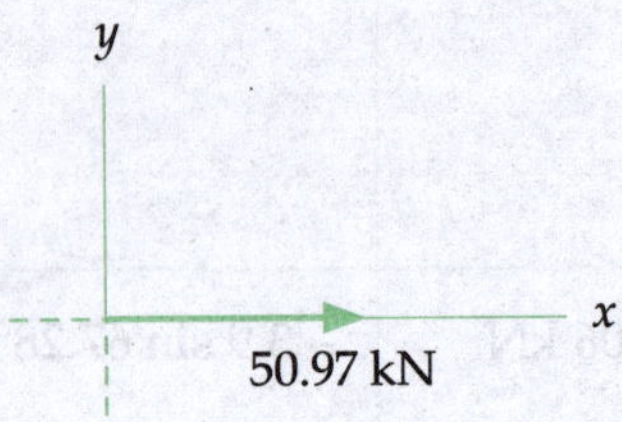

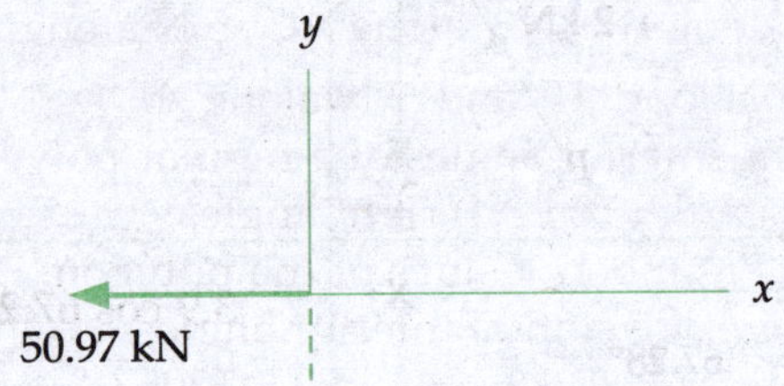

EXAMPLE 2.24

A hoist trolley is in a state of equilibrium when acted upon by a set of three forces as in the adjoining (Fig. 2.28 a). Make calculations for the inclination α if the resultant of the given forces is vertical. Proceed to determine the corresponding magnitude of the resultant.

Solution : Refer Fig. 2.28 b which shows the horizontal and vertical components of the given set of forces. Since the resultant is vertical, the sum of forces acting in the horizontal direction is zero.

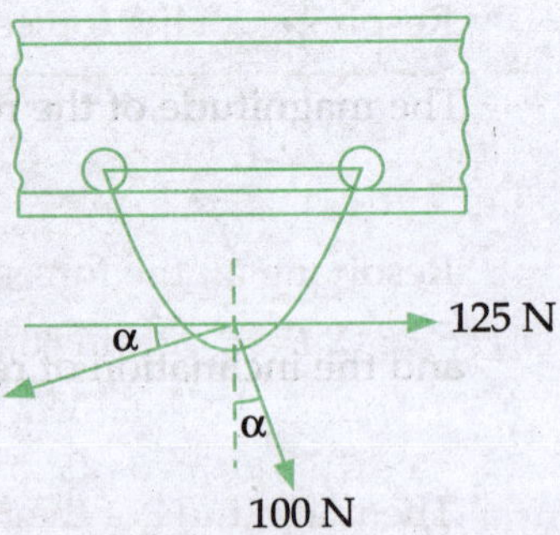

Fig. 2.28 a

That is $\quad \Sigma F_x = 125 + 100 \sin \alpha - 200 \cos \alpha = 0$

or $\quad 125 + 100 \sin \alpha = 200 \cos \alpha$

or $\quad 1 + 0.8 \sin = 1.6 \cos \alpha$

Squaring both sides, one obtains

$$1 + 0.64 \sin^2 \alpha + 1.6 \sin \alpha = 2.56 \cos^2$$
$$= 2.56\,(1 - \sin^2 \alpha)$$

or $\quad 3.2 \sin^2 \alpha + 1.6 \sin \alpha - 1.56 = 0$

Solution of this quadratic equation gives

$$\sin\alpha = \frac{-1.6 \pm \sqrt{1.6^2 - 4 \times 3.2 \times (-1.56)}}{2 \times 3.2}$$

$$= \frac{-1.6 \pm 4.74}{6.4}$$

Considering positive sign : $\sin\alpha = 0.4906$

That gives : $\alpha = 29.38° \simeq 30°$

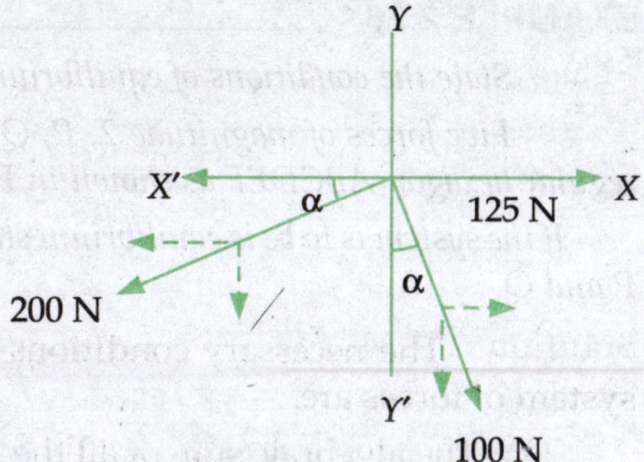

Fig. 2.28 b

Since the resultant force is stated to be vertical, it equals the sum total of the components of given forces in the vertical direction. That is

$$R = \Sigma F_y = -100\cos 30° - 300\sin 30°$$
$$= -86.6 - 100$$
$$= \mathbf{-186.6\ N}$$

The magnitude of resultant force is 186.6 N and it acts vertically in the downward direction.

EXAMPLE 2.25

Find the resultant of forces 2, $\sqrt{3}$, 5, $\sqrt{3}$ and 2 N that act at an angular point of a regular hexagon towards the other angular points taken in order.

Solution : Let *ABCDEF* be the regular hexagon with A as the angular point where forces are acting. The forces are:

2 N along *AB* ; $\sqrt{3}$ along *AC* ; 5 N along *AD* ;

$\sqrt{3}$ N along *AE* and 2 N along *AF*

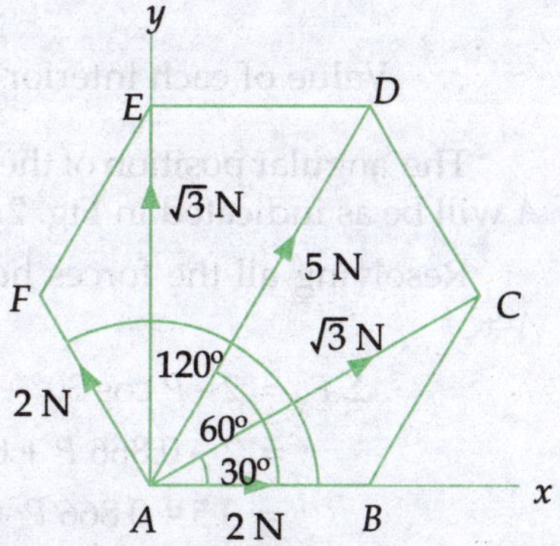

Fig. 2.29

Sum of the interior angles of a regular polygon

= (2*n* – 4) right angles

where *n* is the number of sides of the polygon. For a hexagon, *n* = 6 and therefore sum of interior angles

= (2 × 6 – 4) × 90 = 720°

$\therefore$ Value of each interior angle $= \dfrac{720}{6} = 120°$

Since included angle of a regular hexagon is 120°, the angular position of these forces will be as indicated in Fig. 2.29.

Resolving all the forces horizontally, *i.e.*, along *AX*

$$\Sigma F_x = 2 + \sqrt{3}\cos 30° + 5\cos 60° + \sqrt{3}\cos 90° + 2\cos 120°$$

$$= 2 + \sqrt{3} \times \frac{\sqrt{3}}{2} + 5 \times \frac{1}{2} + \sqrt{3} \times 0 + 2\left(-\frac{1}{2}\right) = 5\text{ N}$$

Resolving all the forces vertically, i.e., along *AY*

$$\Sigma F_y = \sqrt{3}\sin 30° + 5\sin 60° + \sqrt{3}\sin 90° + 2\sin 120°$$

$$= \sqrt{3} \times \frac{1}{2} + 5 \times \frac{\sqrt{3}}{2} + \sqrt{3} \times 1 + 2 \times \frac{\sqrt{3}}{2} = 5\sqrt{3}\text{ N}$$

The magnitude of the resultant force is

$$R = \sqrt{(\Sigma F_x)^2 + (\Sigma F_y)^2} = \sqrt{5^2 + (5\sqrt{3})^2} = \sqrt{100} = \mathbf{10\ N}$$

and the inclination of the resultant with horizontal is

$$\alpha = \tan^{-1}\left(\frac{\Sigma F_y}{\Sigma F_x}\right) = \tan^{-1}\frac{5\sqrt{3}}{5} = \tan^{-1}(\sqrt{3}) = \mathbf{60°}$$

EXAMPLE 2.26

(*a*) *State the conditions of equilibrium of a coplanar system of forces.*

(*b*) *Five forces of magnitude 2, P, Q, 4 and 3 kN act along AB, CA, AD, AE and FA respectively in a regular hexagon ABCDEF as shown in Fig. 2.30 a.*

If the system is to be is equilibrium state, determine the value of forces P and Q.

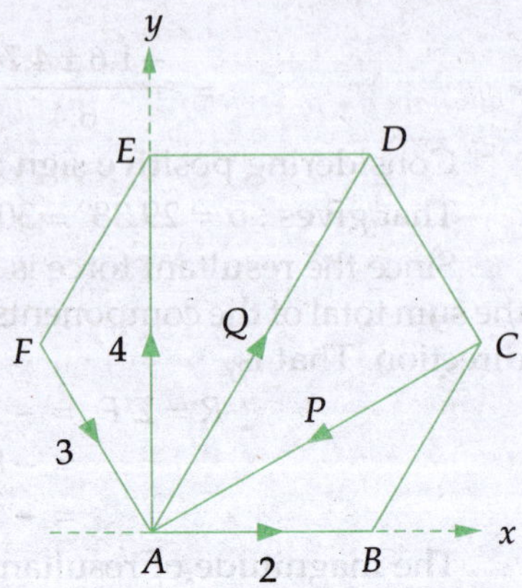

Fig. 2.30 a

Solution : The necessary conditions of equilibrium of a coplanar system of forces are:

(*i*) The algebraic sum of all the forces in a horizontal plane is equal to zero: $\Sigma F_x = 0$

(*ii*) The algebraic sum of all the forces in a vertical plane is equal to zero: $\Sigma F_y = 0$

(*iii*) The algebraic sum of all the moments about any point is equal to zero: $\Sigma M = 0$.

(*b*) Sum of the interior angles of a regular polygon

$$= (2n - 4) \text{ right angles}$$

where *n* is the number of sides of the polygon. For a hexagon, *n* = 6 and therefore sum of interior angles

$$= (2 \times 6 - 4) \times 90 = 720°$$

$$\therefore \quad \text{Value of each interior angle} = \frac{720}{6} = 120°$$

The angular position of the given forces at corner *A* will be as indicated in Fig. 2.30 *b*.

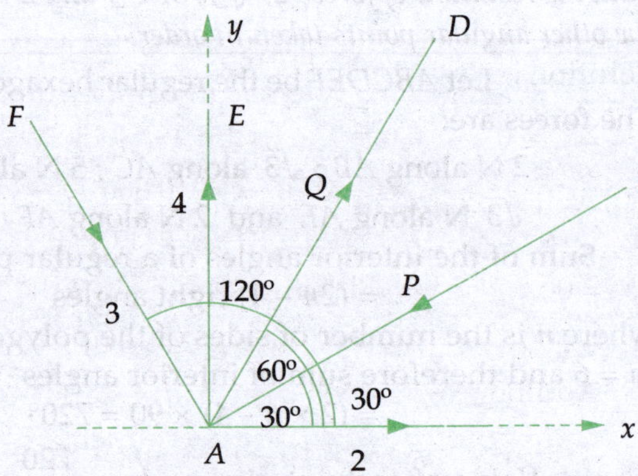

Fig. 2.30 b

Resolving all the forces horizontally, *i.e.*, along *A-x*

$$\Sigma F_x = 2 - P\cos 30° + Q\cos 60° - 3\cos 120°$$
$$= 2 - 0.866\,P + 0.5\,Q + 1.5$$
$$= 3.5 - 0.866\,P + 0.5\,Q$$

Resolving all the forces vertically, *i.e.*, along *A-y*

$$\Sigma F_y = -P\sin 30° + Q\sin 60° + 4 - 3\sin 120°$$
$$= -0.5\,P + 0.866\,Q + 4 - 3 \times 0.866$$
$$= 1.402 - 0.5\,P + 0.866\,Q$$

Since the forces are in equilibrium: $\Sigma F_x = 0$ and $\Sigma F_y = 0$

$$\therefore \quad 3.5 - 0.866\,P + 0.5\,Q = 0 \quad \ldots(a)$$

$$\text{and} \quad 1.402 - 0.5\,P + 0.866\,Q = 0 \quad \ldots(b)$$

Multiplying identity (*a*) by 0.866 and identity (*b*) by 0.5, we get:

$$3.031 - 0.75\,P + 0.433\,Q = 0 \quad \ldots(c)$$

$$0.701 - 0.25\,P + 0.433\,Q = 0 \quad \ldots(d)$$

Subtracting identity (*d*) from identify (*c*) and solving for *P*, we get

$$\text{Force } P = \frac{3.031 - 0.701}{0.75 - 0.25} = \mathbf{4.66\ kN}$$

Substituting this value of P in identify (c), we get

$$\text{Force } Q = \frac{0.75 \times 4.66 - 3.031}{0.433} = 1.071 \text{ kN}$$

EXAMPLE 2.27

Calculate the tensile force in the cables AB and BC as shown in Fig. 2.31 a. Presume the pulleys to be frictionless.

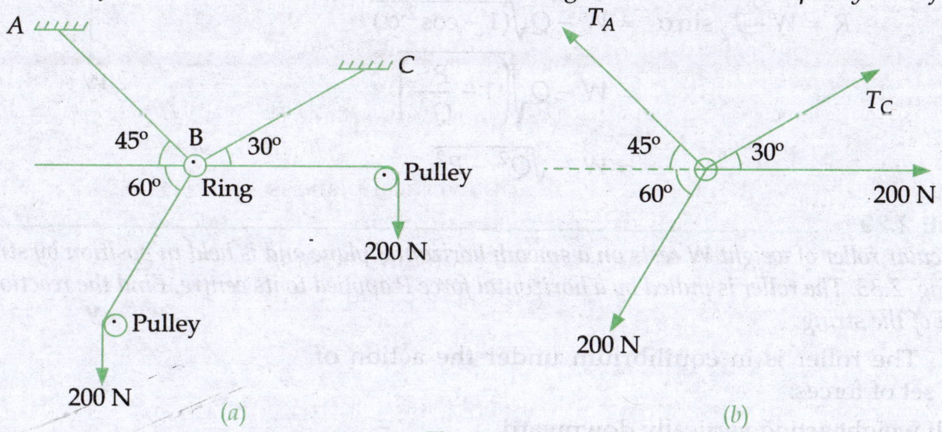

Fig. 2.31

Solution : Since the system is in equilibrium,

$\Sigma F_x = 0$; $\quad 0 = - T_A \cos 45° + T_C \cos 30° + 200 - 200 \cos 60°$

$\quad = - 0.707\, T_A + 0.866\, T_C + 200 - 100$

or $\quad - T_A + 1.22\, T_C + 141.0 = 0 \quad$ ...(*i*)

$\Sigma F_y = 0$; $\quad 0 = T_A \sin 45° + T_C \sin 30° - 200 \sin 60° = 0.707\, T_A + 0.5\, T_C - 173$

or $\quad T_A + 0.707\, T_C - 244 = 0 \quad$ ...(*ii*)

Adding expressions (*i*) and (*ii*),

$$1.927\, T_C - 103 = 0 \,; \quad T_C = \frac{103}{1.927} = 53.45 \text{ N}$$

Then from expression (*ii*)

$$T_A + 0.707 \times 53.45 - 244 = 0$$

$$\therefore \quad T_A = 244 - 0.707 \times 53.45 = 244 - 37.79 = 206.21 \text{ N}$$

EXAMPLE 2.28

A ball of weight W rests upon a smooth horizontal plane and has attached to its centre two strings AB and AC which pass over frictionless pulleys at B and C and carry loads P and Q as shown in Fig. 2.32. If the string AB is horizontal, find the angle α that the string AC makes with the horizontal when the ball is in a position of equilibrium. Also determine the pressure between the ball and the plane.

Solution : Let T_1 and T_2 be the tensions in the strings AB and AC respectively. Since the pulleys are frictionless

$$T_1 = P \quad \text{and} \quad T_2 = Q$$

Resolving the forces horizontally we get

$$T_1 = T_2 \cos \alpha$$

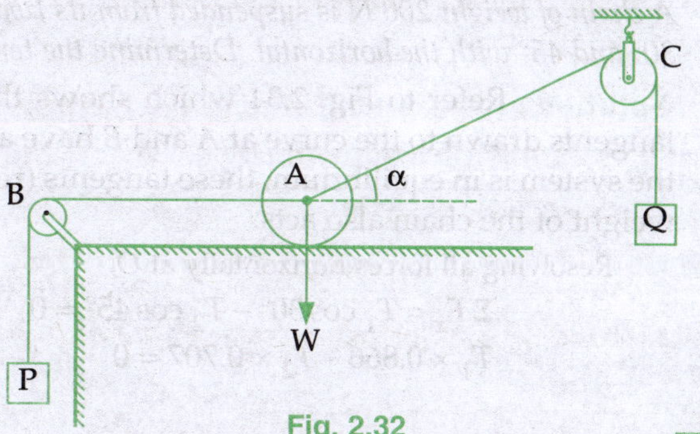

Fig. 2.32

or $P = Q\cos\alpha$

$$\therefore \quad \alpha = \cos^{-1}\left(\frac{P}{Q}\right)$$

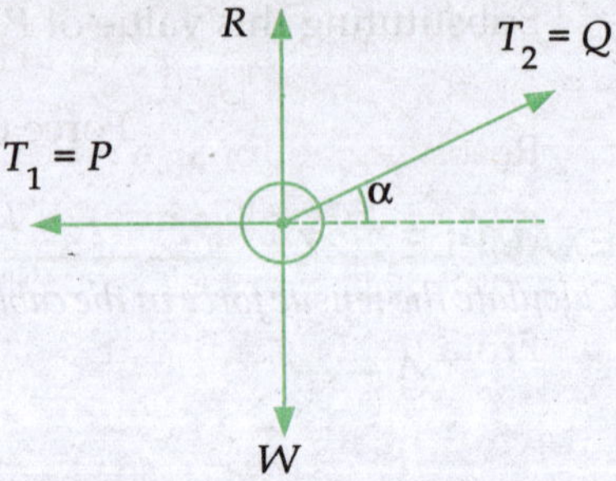

Resolving the forces vertically, the pressure R between the ball and the plane can be written as

$$R = W - T_2 \sin\alpha = W - Q\sqrt{(1-\cos^2\alpha)}$$

$$= W - Q\sqrt{\left(1 - \frac{P^2}{Q^2}\right)}$$

$$= W - \sqrt{Q^2 - P^2}$$

EXAMPLE 2.29

A right circular roller of weight W rests on a smooth horizontal plane and is held in position by string AC as shown in Fig. 2.33. The roller is pulled by a horizontal force P applied to its centre. Find the reaction at B and the tension of the string.

Solution : The roller is in equilibrium under the action of following set of forces:

(*i*) Self weight acting vertically downward

(*ii*) Reaction R_b acting vertically upward

(*iii*) Force P acting horizontally towards right

(*iv*) Tension T of the string acting in the direction from C towards A

For these concurrent forces, the conditions of equilibrium are:

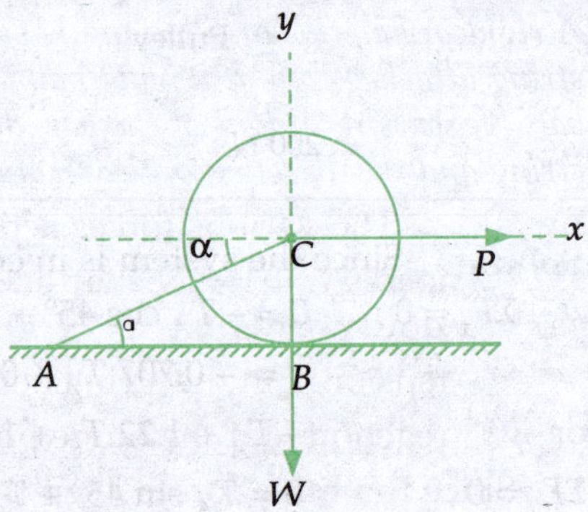

Fig. 2.33

$\Sigma F_x = 0;$ $\quad P - T\cos\alpha = 0$

$$\therefore \quad T = \frac{P}{\cos\alpha}$$

$\Sigma F_y = 0;$ $\quad R_b = T\sin\alpha - W = 0$

or $\quad R_b = T\sin\alpha + W$

$$= \frac{P}{\cos\alpha}\sin\alpha + W = P\tan\alpha + W$$

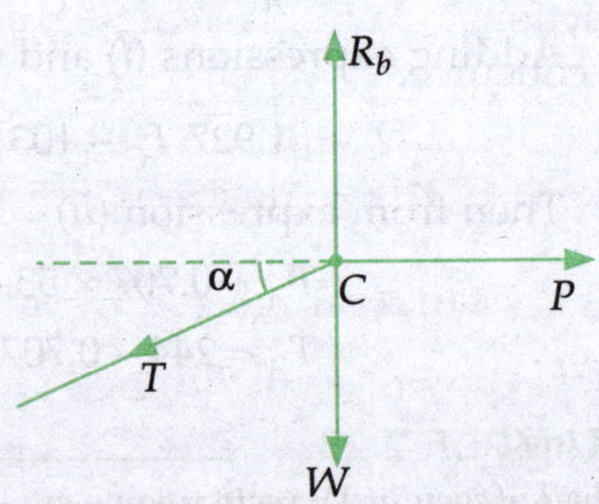

EXAMPLE 2.30

A chain of weight 200 N is suspended from its two ends on a wall in such a way that its ends make angles of 30° and 45° with the horizontal. Determine the tensions produced in the ends of the chain.

Solution : Refer to Fig. 2.34 which shows the chain AB suspended from its ends A and B. The tangents drawn to the curve at A and B have an inclination of 30° and 45° with the horizontal. Since the system is in equilibrium, these tangents (representing tensions T_1 and T_2) meet at point O where weight of the chain also acts.

Resolving all forces horizontally at O,

$$\Sigma F_x = T_1\cos 30^\circ - T_2\cos 45^\circ = 0$$

$$T_1 \times 0.866 - T_2 \times 0.707 = 0$$

$$\therefore \quad T_2 = T_1 \times \frac{0.866}{0.707} = 1.225\,T_1 \qquad \ldots(i)$$

Resolving all forces vertically at *O*,

$$\Sigma F_y = T_1 \sin 30^\circ + T_2 \sin 45^\circ - 200 = 0$$

$$T_1 \times 0.5 + T_2 \times 0.707 = 200 \qquad \ldots(ii)$$

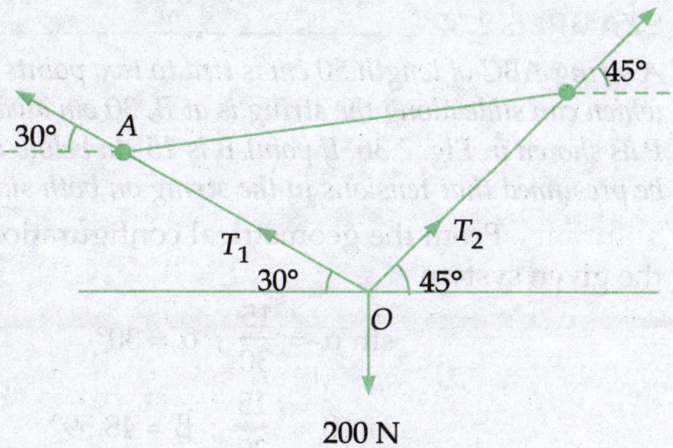

Fig. 2.34

From (*i*) and (*ii*),

$$0.5\,T_1 + 0.707 \times 1.225\,T_1 = 200$$

$$0.5\,T_1 + 0.866\,T_1 = 200$$

$$\therefore \quad T_1 = \frac{200}{0.5 + 0.866} = 146.41 \text{ N}$$

$$T_2 = 1.225\,T_1 = 1.225 \times 146.41 = 179.35 \text{ N}$$

EXAMPLE 2.31

A rigid prismatic bar 5 m long is supported in a vertical plane by hinge at one end and by a horizontal string attached to it at its mid point (Fig. 2.35) A 150 N weight hangs vertically from the free end of the bar. Neglecting weight of the bar, determine tension in the string and the direction of the reaction at the hinge.

Solution : The various forces acting on the bar are:

(*i*) load W = 150 N that acts at end B vertically downwards

(*ii*) tension T in the string acting horizontally

(*iii*) reaction R at the hinge point A.

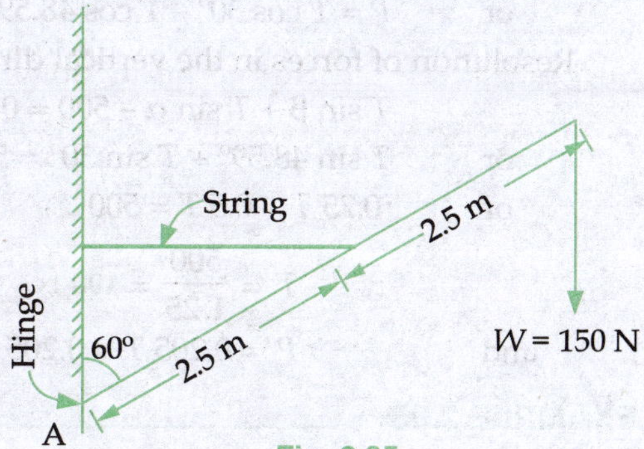

Fig. 2.35

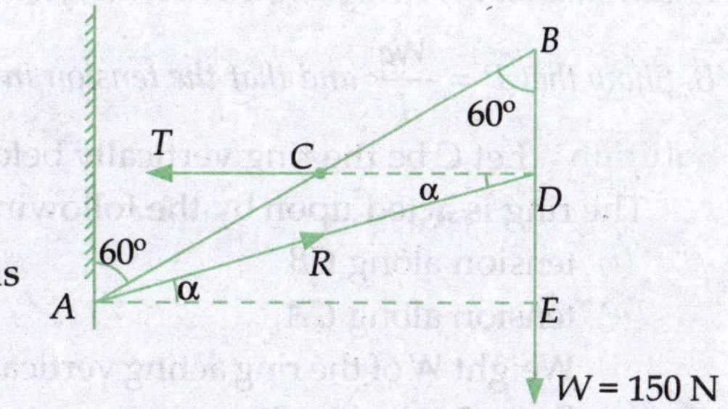

Since the three forces keep the bar in equilibrium, they must be concurrent, i.e., meet at a point. Accordingly reaction R would act along AD and D would be the point of concurrency. Then

$$BD = BC \cos 60^\circ = 2.5 \cos 60^\circ = 1.25 \text{ m} = DE$$

$$AE = AB \sin 60^\circ = 5 \sin 60^\circ = 4.33 \text{ m}$$

$$\tan\alpha = \frac{DE}{AE} = \frac{1.25}{4.33} = 0.2887$$

$$\alpha = \tan^{-1}(0.2887) = 16.1^\circ$$

The direction of the reaction at the hinge is thus inclined at 16.1° with the horizontal.

Applying the equilibrium conditions

$$\Sigma F_x = 0; \qquad R\cos\alpha - T = 0$$

$$\text{or} \qquad R\cos\alpha = T \qquad \ldots(i)$$

$$\Sigma F_y = 0; \qquad R\sin\alpha - 150 = 0$$

$$\text{or} \qquad R\sin\alpha = 150 \qquad \ldots(ii)$$

From expressions (*i*) and (*ii*),

$$\tan\alpha = \frac{150}{T}; \quad T = \frac{150}{\tan\alpha} = \frac{150}{\tan 16.1^\circ} = 519.57 \text{ N}$$

EXAMPLE 2.32

A string ABC of length 50 cm is tied to two points A and C at the same level. A smooth ring of weight 500 N which can slide along the string is at B, 30 cm away from A along the string and pulled by a horizontal force P as shown in Fig. 2.36. If point B is 15 cm below the level of AC, determine the magnitude of force P. It may be presumed that tensions in the string on both sides of B are same.

Solution : From the geometrical configuration of the given system

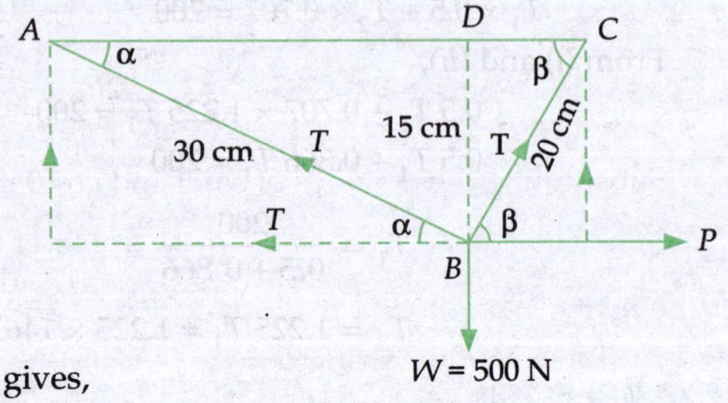

Fig. 2.36

$$\sin\alpha = \frac{15}{30}\ ;\ \alpha = 30^\circ$$

$$\sin\beta = \frac{15}{20}\ ;\ \beta = 48.59^\circ$$

Since the system of forces is in equilibrium,

$$\Sigma F_x = 0 \text{ and } \Sigma F_y = 0$$

Resolution of forces in the horizontal direction gives,

$$P + T\cos\beta - T\cos\alpha = 0$$

or $P + T\cos 48.59^\circ - T\cos 30^\circ = 0$

or $P = T\cos 30^\circ - T\cos 48.59^\circ = 0.866\,T - 0.661\,T = 0.205\,T$

Resolution of forces in the vertical direction gives,

$$T\sin\beta + T\sin\alpha - 500 = 0$$

or $T\sin 48.59^\circ + T\sin 30^\circ = 500$

or $0.75\,T + 0.5\,T = 500$

$$\therefore \quad T = \frac{500}{1.25} = 400 \text{ N}$$

and $P = 0.205\,T = 0.205 \times 400 = 82$ N

EXAMPLE 2.33

A string of length l is fastened to two points A and B at the same level at a distance a apart. A ring of weight W can slide on the string and a horizontal force P is applied to it such that it is in equilibrium vertically below B. Show that $P = \dfrac{Wa}{l}$ *and that the tension in the string is* $\dfrac{W(l^2 + a^2)}{2l}$.

Solution : Let C be the ring vertically below point B. Further, if $BC = x$ then $AB = (l - x)$.

The ring is acted upon by the following set of forces,

(i) tension along CB

(ii) tension along CA

(iii) Weight W of the ring acting vertically downwards

(iv) horizontal force P

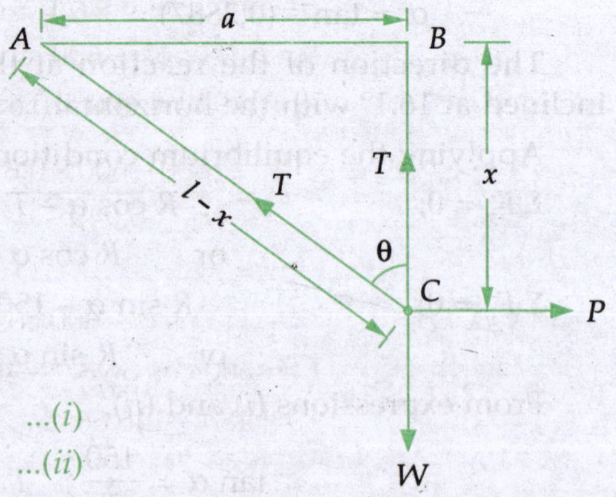

Fig. 2.37

Since the ring slides along the string

tension along CB = tension along CA = T (say)

For these concurrent forces, the conditions of equilibrium are:

$\Sigma F_y = 0;\quad T + T\cos\theta = W \quad \text{or} \quad T(1 + \cos\theta) = W \qquad ...(i)$

$\Sigma F_x = 0;\quad P = T\sin\theta \qquad ...(ii)$

From the geometry of the arrangement

$$(l - x)^2 = x^2 + a^2$$

$$l^2 + x^2 - 2lx = x^2 + a^2; \quad x = \frac{l^2 - a^2}{2l}$$

$$l - x = l - \frac{l^2 - a^2}{2l} = \frac{l^2 + a^2}{2l}$$

$$\cos\theta = \frac{BC}{AC} = \frac{x}{l - x} = \frac{l^2 - a^2}{2l} \times \frac{2l}{l^2 + a^2} = \frac{l^2 - a^2}{l^2 + a^2}$$

$$\sin\theta = \frac{AB}{l - x} = a \times \frac{2l}{l^2 + a^2} = \frac{2al}{l^2 + a^2}$$

Substituting these values of sin θ and cos θ in expressions (*i*) and (*ii*), we get

$$T\left(1 + \frac{l^2 - a^2}{l^2 + a^2}\right) = W; \quad T\left(\frac{2l^2}{l^2 + a^2}\right) = W$$

∴ Tension of the string $T = \frac{W(l^2 + a^2)}{2l^2}$

Horizontal force $P = T\sin\theta = \frac{W(l^2 + a^2)}{2l^2} \times \frac{2al}{l^2 + a^2} = \frac{Wa}{l}$

EXAMPLE 2.34

Determine tension in the cables AB and AC which are tied together at A, and are loaded as shown in Fig. 2.38 a. Consider the system to be in equilibrium.

Solution : Refer (Fig. 2.38 *b*) which shows the different forces acting at point *A*.

T_1 and T_2 denote the tensions in the cables *AB* and *AC* respectively, and these tensions act away from point *A* which is under consideration.

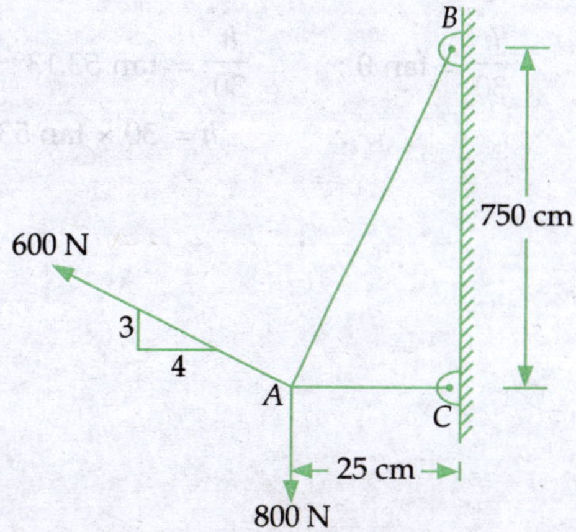

Fig. 2.38 a

From the geometry of the given arrangement :

$\tan\alpha = \frac{750}{25} = 3$ ∴ $\alpha = 71.56°$

$\tan\beta = \frac{3}{4} = 0.75$ ∴ $\beta = 36.87°$

Since the given system is in equilibrium; we have

(*i*) $\Sigma F_y = 0$: $T_1 \sin\alpha + 600\sin\beta - 800 = 0$

$T_1 \sin 71.56° + 600 \sin 36.87° - 800 = 0$

$T_1 \times 0.949 + 600 \times 0.6 - 800 = 0$

∴ $T_1 = \frac{800 - 600 \times 0.6}{0.949}$

$= $ **463.65 N**

(*ii*) $\Sigma F_x = 0$: $T_2 + T_1 \cos\alpha - 600\cos\beta = 0$

$T_2 = 600 \cos 36.87° - 463.65 \cos 71.56°$

$= 480 - 146.66$

$= $ **333.34 N**

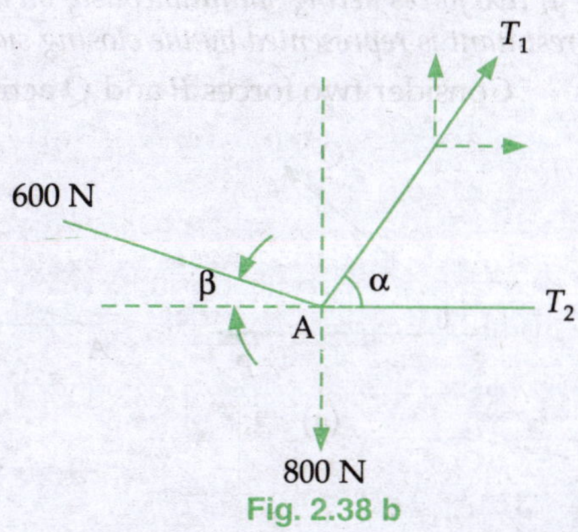

Fig. 2.38 b

EXAMPLE 2.35

A collar of 200 N weight slides on a frictionless vertical rod and is connected to a 250 N counterweight as shown in Fig. 2.39. Calculate the value of distance h for which the system will remain in equilibrium. Assume pulley to be frictionless.

Solution : The collar is in equilibrium under the action of following set of forces :

(*i*) weight of the collar,

(*ii*) tension in the string. Since the pulley is frictionless, this tension equals the counterweight.

(*iii*) reaction; it is horizontal due the vertical rod being frictionless.

Resolving , the forces vertically and applying the condition for equilibrium : $\Sigma F_y = 0$, we get

$$250 \sin\theta - 200 = 0$$

or $$\sin\theta = \frac{200}{250} = 0.8$$

$$\therefore \quad \theta = \sin^{-1}(0.8) = 53.13°$$

Then from the geometrical configuration of the arrangement,

$$\frac{h}{30} = \tan\theta \; ; \quad \frac{h}{30} = \tan 53.13°$$

$$\therefore \quad h = 30 \times \tan 53.3 = \mathbf{40.25\ cm}$$

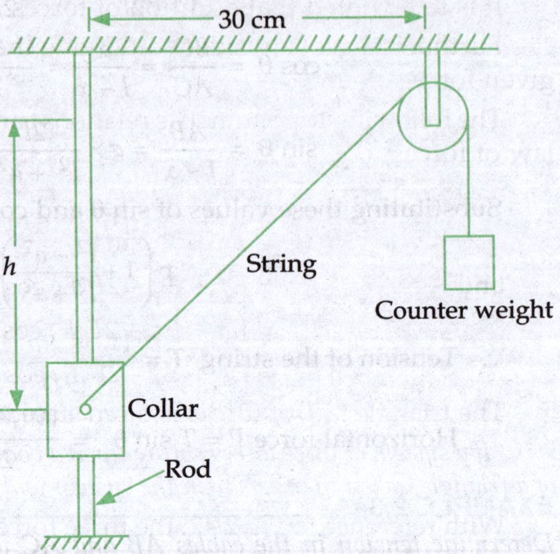

Fig. 2.39

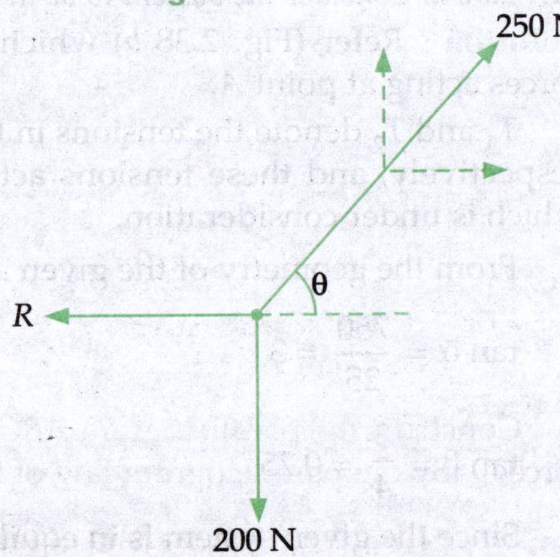

2.5. TRIANGLE LAW OF FORCES

"If two forces acting simultaneously on a body are represented by the sides of a triangle taken in order, their resultant is represented by the closing side of the triangle taken in the opposite order."

Consider two forces *P* and *Q* acting on the body as shown in Fig 2.40 *a*.

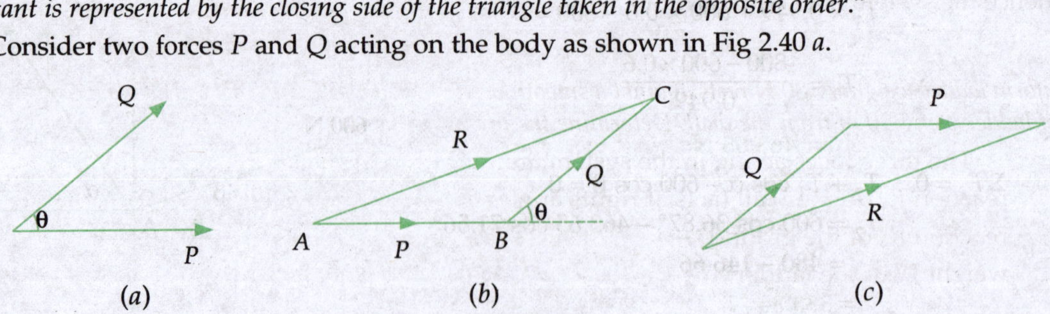

Fig. 2.40

Line AB be drawn to represent force P and BC to represent Q. The triangle ABC is completed by drawing the closing line AC. Line AC represents the resultant in magnitude, line of action and direction.

It is to be noted that addition of forces P and Q in any order gives the same resultant R.

Further, the closing side would give the equivalent if it is taken in the same order as the two given forces.

The following trigonometric relations can be applied while working out solutions by the triangle law of forces:

$$\frac{F_1}{\sin\alpha} = \frac{F_2}{\sin\beta} = \frac{F_3}{\sin\gamma}$$

and

$$F_1^2 = F_2^2 + F_3^2 - 2F_2F_3\cos\alpha$$
$$F_2^2 = F_1^2 + F_3^2 - 2F_1F_3\cos\beta$$
$$F_3^2 = F_1^2 + F_2^2 - 2F_1F_2\cos\gamma$$

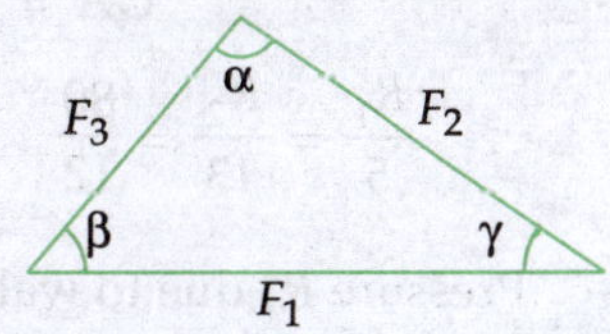

Fig. 2.41

The triangle law of forces is often stated as follows :

"If a system of three forces acting upon a body can be represented in magnitude and direction by the sides of a triangle, taken in order, then the system will be in equilibrium."

With reference to Fig 2.42, the three forces P, Q and R acting at point O have been expressed by the sides AB, BC and CA of the triangle ABC.

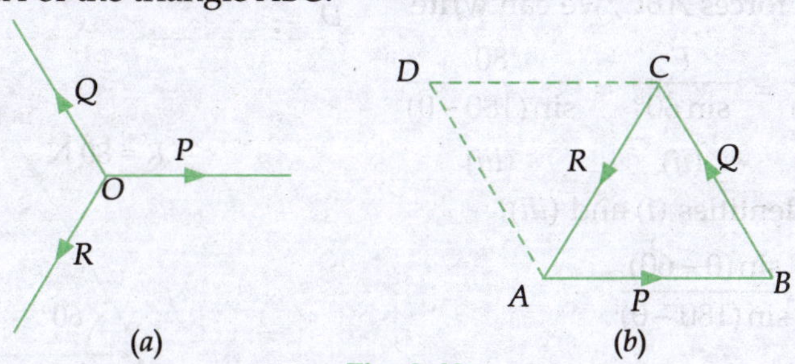

Fig. 2.42

Construct the parallelogram $ABCD$. The side AD, being equal and parallel to BC represents the force Q. By the parallelogram law of forces,

$$\vec{AB} + \vec{AD} = \vec{AC}$$
$$P + Q = -\vec{CA} = -R$$

$\therefore$ Resultant of P, Q and $R = -R + R = 0$

and hence the system is in equilibrium.

EXAMPLE 2.36

A uniform ladder weighing 80 N rests against a smooth vertical wall at a height of 12 m above the ground; the foot of ladder being 10 m from the wall. Determine the pressure due to wall.

Solution : The three foces acting in the system are:

(*i*) reaction R_1 due to wall (acts at rights angles to wall as the wall is smooth)

(*ii*) reaction R_2 of the ground

(*iii*) weight W of the ladder

Under equilibrium conditions, the reaction R_2 passes through the point of intersection of R_1 and W. The three forces R_1, R_2 and W are then represented respectively by the sides AN, OA and ON of the triangle OAN.

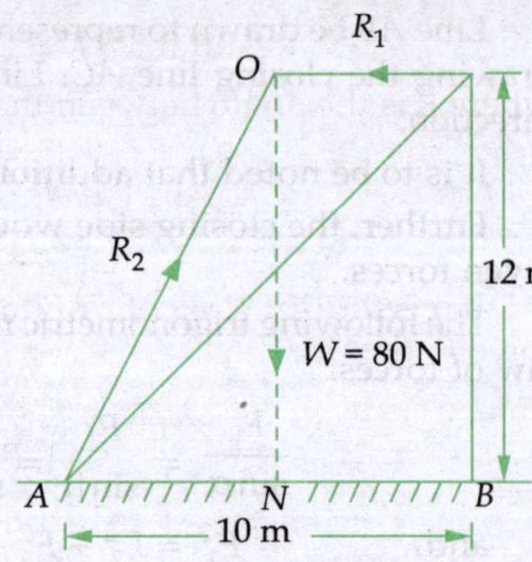

Fig. 2.43

$$\frac{R_1}{AN} = \frac{R_2}{OA} = \frac{W}{ON};$$

$$AN = 5 \text{ m}$$

$$OA = \sqrt{5^2 + 12^2} = 13 \text{ m}$$

$$\therefore \quad \frac{R_1}{5} = \frac{R_2}{13} = \frac{80}{12}$$

Pressure R_1 due to wall $= \dfrac{80 \times 5}{12} = \mathbf{33.3\ N}$

EXAMPLE 2.37

The resultant of two forces 60 N and F acting at a point is stated to be of 80 N magnitude with inclination at 60° to the 60 N force. Determine the magnitude of force F and its inclination with the 60 N force.

Solution : Refer Fig. 2.44 for the vector diagram for the given set of forces and their resultant. The angle of inclination between the force F and the 60 N force has been taken as θ. Applying the law of sines to this triangle of forces ABC, we can write

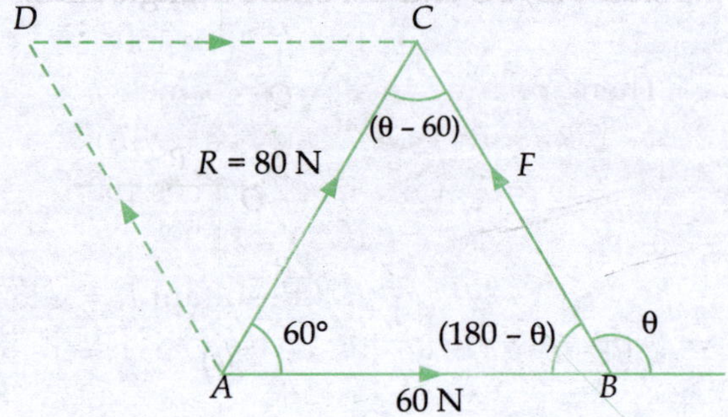

Fig. 2.44

$$\frac{60}{\sin(\theta - 60)} = \frac{F}{\sin 60°} = \frac{80}{\sin(180 - \theta)}$$

$$(i) \qquad (ii) \qquad (iii)$$

From equality of identities (*i*) and (*iii*),

$$\frac{60}{80} = \frac{\sin(\theta - 60)}{\sin(180 - \theta)}$$

or $$\frac{3}{4} = \frac{\sin\theta\cos 60 - \cos\theta \sin 60}{\sin\theta}$$

or $3 \sin\theta = 4 \sin\theta \times 0.5 - 4$

$\cos\theta \times 0.866$

$$= 2\sin\theta - 3.464 \cos\theta$$

That gives : $\theta = \tan^{-1}(-3.464) = 106.1°$

From equality of identities (*ii*) and (*iii*),

$$F = \frac{80 \sin 60°}{\sin(180 - \theta)} = \frac{80 \sin 60°}{\sin(180 - 106.1)}$$

$$= \frac{80 \times 0.866}{0.961} = 72.09 \text{ N}$$

EXAMPLE 2.38

A stud is acted upon by two forces P = 2 kN and Q = 3 kN as shown in Fig. 2.45. Determine the resultant of these concurrent forces by using

(*a*) *Parallelogram law of forces*

(*b*) *Triangle law of forces*

Solution : (*a*) Resultant $R=\sqrt{P^2+Q^2+2PQ\cos\theta}$

where θ is the angle between the forces *P* and *Q*; θ = 30°

$$R=\sqrt{2^2+3^2+2\times2\times3\cos 30^\circ}$$

$$=\sqrt{4+9+10.39}=4.836\text{ N}$$

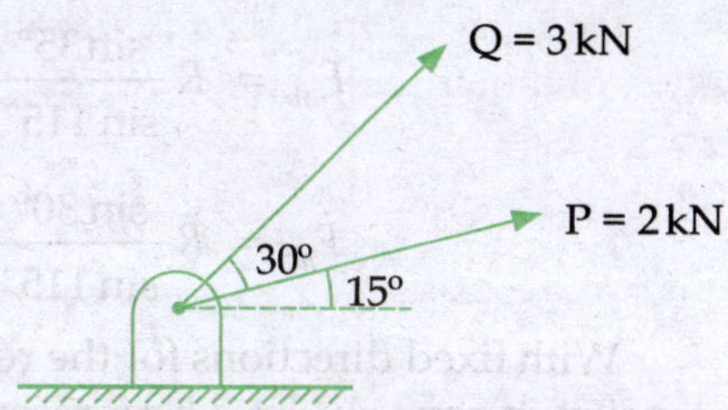

Fig. 2.45

If α is the direction of resultant with respect to direction of force *P*, then

$$\tan\alpha=\frac{Q\sin\theta}{P+Q\cos\theta}$$

$$=\frac{3\sin 30^\circ}{2+3\cos 30^\circ}=\frac{3\times0.5}{2+3\times0.866}=0.326$$

$$\alpha=\tan^{-1}0.326=18.06^\circ$$

∴ Angle made by resultant with *x*-axis = 18.06 + 15 = 33.06°

(*b*) The force diagram of these concurrent forces is as shown below:

$$P=OA=2\text{ kN}$$

$$Q=AB=3\text{ kN}$$

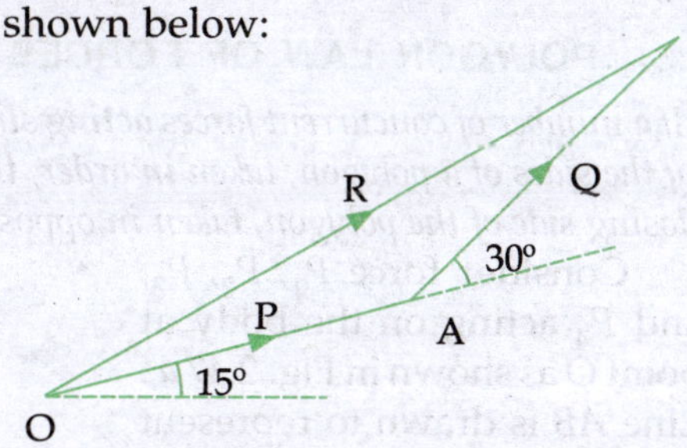

and the included angle between these forces is

$$\angle OAB=180^\circ-30^\circ=150^\circ$$

Then applying the law of cosines

$$R^2=P^2+Q^2-2PQ\cos(OAB)$$

$$=2^2+3^2-2\times2\times3\cos 150^\circ$$

$$=4+9+10.39=23.39$$

∴ Resultant $R=\sqrt{23.39}=4.836$ N

Applying the law of sines

$$\frac{R}{\sin OAB}=\frac{Q}{\sin AOB}$$

$$\sin AOB=\frac{Q}{R}\sin OAB=\frac{3}{4.836}\sin 150=0.310$$

$$\therefore\ \angle AOB=\sin^{-1}(0.31)=18.06^\circ$$

This is the angle which the resultant makes with direction of force *P*.

∴ Angle of inclination of the resultant with the horizontal = 15 + 18.06 = 33.06°

EXAMPLE 2.39

The resultant of two ropes pulling a cart, as shown in Fig. 2.46, acts along the axis of the cart and has a magnitude of 6 kN. What would be the tension in the ropes when θ = 30°? Proceed to workout the value of θ so that the tension in the rope AB is minimum. Also find out this minimum tension.

Solution : With reference to the force diagram shown below, the law of sines gives

$$\frac{F_{ab}}{\sin 35^\circ}=\frac{F_{ac}}{\sin 30^\circ}=\frac{R}{\sin(180-30-35)}$$

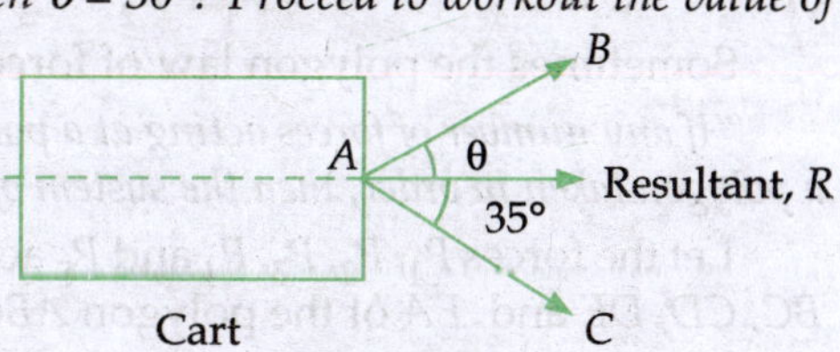

Fig. 2.46

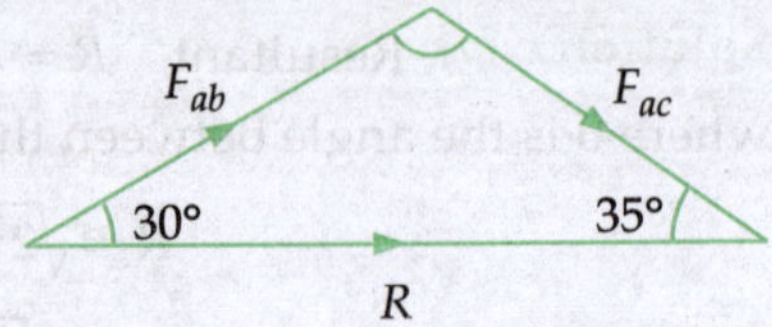

$$\therefore \quad F_{ab} = R\frac{\sin 35^\circ}{\sin 115^\circ} = 6 \times \frac{0.573}{0.906} = 3.795 \text{ kN}$$

$$F_{ac} = R\frac{\sin 30^\circ}{\sin 115^\circ} = 6 \times \frac{0.5}{0.906} = 3.311 \text{ kN}$$

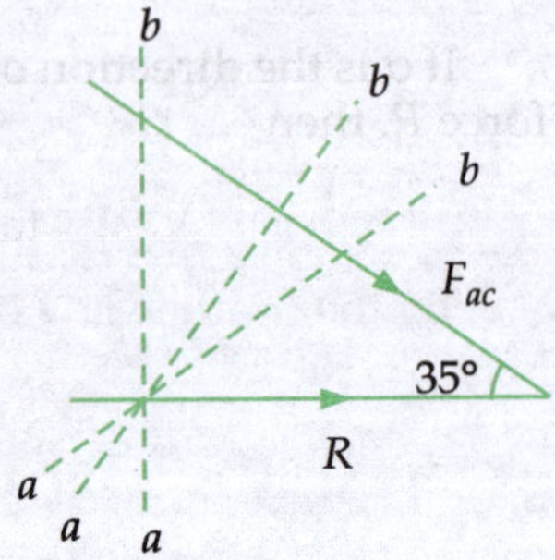

With fixed directions for the resultant and the line of action of tension in rope *AC* let *a-b* represent the line of action of tension in rope *AB*. It is observed that tension in rope *AB* will be minimum when the line of action of tensions in *AB* and *AC* are perpendicular. Hence for minimum tension in rope *AB*,

$$\theta = 90 - 35 = 55^\circ$$

Minimum value of tension in rope *AB* will be then given by

$$F_{ab} = R\frac{\sin 35^\circ}{\sin(180 - 35 - 55)} = 6 \times \frac{0.573}{1} = 3.438 \text{ N}$$

2.6. POLYGON LAW OF FORCES

"If a number of concurrent forces acting simultaneously on a body are represented in magnitude and direction by the sides of a polygon, taken in order, then the resultant is represented in magnitude and direction by the closing side of the polygon, taken in opposite order."

Consider force P_1, P_2, P_3, and P_4 acting on the body at point *O* as shown in Fig. 2.47 *a*. Line *AB* is drawn to represent force P_1, line *BC* to represent P_2, line *CD* to represent P_3 and line *DE* to represent P_4. The polygon is completed by drawing the closing line *AE*. This closing line *AE* represents the resultant of the given system in magnitude, line of action and direction.

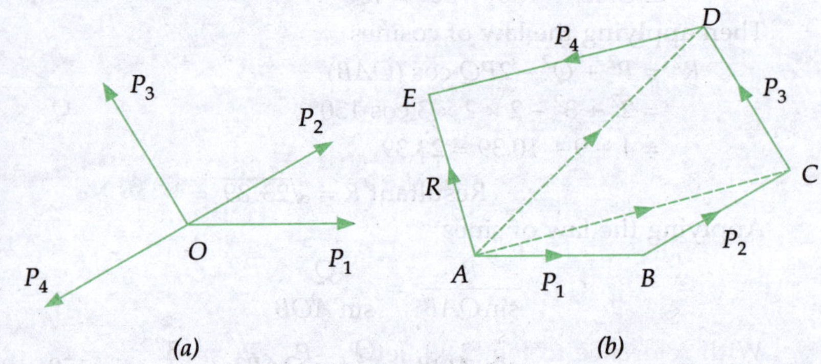

Fig. 2.47

The polygon law of forces follows from the triangle law of forces as is evident from the steps outlined below:

$$\vec{AC} = \vec{AB} + \vec{BC}$$

$$\vec{AD} = \vec{AC} + \vec{CD} = \vec{AB} + \vec{BC} + \vec{CD}$$

$$\vec{AE} = \vec{AD} + \vec{DE} = \vec{AB} + \vec{BC} + \vec{CD} + \vec{DE}$$

or $$R = P_1 + P_2 + P_3 + P_4$$

Sometimes the polygon law of forces is stated as follows:

"If any number of forces acting at a point can be re-presented in magnitude and direction by the sides of a polygon taken in order, then the system of forces is in equilibrium."

Let the forces P_1, P_2, P_3, P_4 and P_5 acting at point *O* be represented in magnitude by the sides *AB*, *BC*, *CD*, *DE* and *EA* of the polygon *ABCDE*. Join *AC* and *BD*.

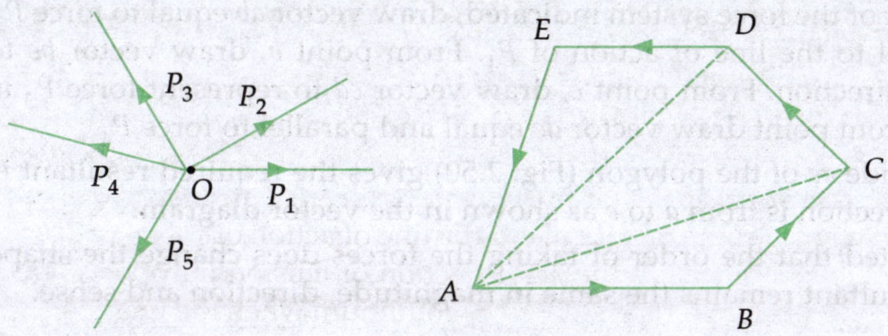

Fig. 2.48

By the triangle law of forces.

$$\vec{AC} = \vec{AB} + \vec{BC}$$

$$\vec{AD} = \vec{AC} + \vec{CD} = \vec{AB} + \vec{BC} + \vec{CD}$$

$$\vec{AE} = \vec{AD} + \vec{DE} = \vec{AB} + \vec{BC} + \vec{CD} + \vec{DE}$$

i.e., $\vec{AE}$ represents the resultant of forces P_1, P_2, P_3, and P_4

$\therefore$ Vector sum of all the forces $= \vec{AE} + \vec{EA} = 0$

and hence the system is in equilibrium.

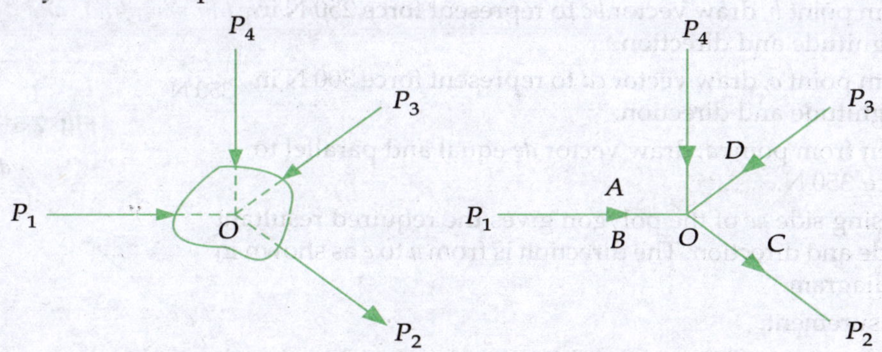

Fig. 2.49.

With reference to Fig. 2.49, let P_1, P_2, P_3 and P_4 be the coplanar concurrent forces acting simultaneously on a body, and O be the point of concurrency of three forces.

The polygon law of forces provides a convenient method for determining graphically the resultant of these coplanar concurrent forces. The steps involved in the method are outlined below:

(*a*) Construct the position (space) diagram showing the various forces (or loads) along with their magnitude and line of action (Fig. 2.49).

(*b*) Name the forces in the space diagram by using Bow's notation by which each force is named by two capital letters placed on its either side.

(*c*) Construct the force diagram by starting from a convenient point. Go on adding all the forces vectorially one by one to some suitable scale.

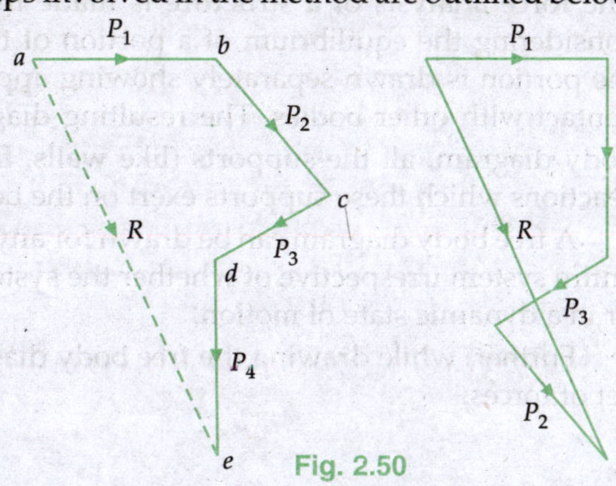

Fig. 2.50

In the context of the force system indicated, draw vector *ab* equal to force P_1 to some suitable scale and parallel to the line of action of P_1. From point *b*, draw vector *bc* to represent P_2 in magnitude and direction. From point *c*, draw vector *cd* to represent force P_3 in magnitude and direction. Then from point draw vector *de* equal and parallel to force P_4.

The closing side *ae* of the polygon (Fig. 2.50) gives the required resultant in magnitude and direction. The direction is from *a* to *e* as shown in the vector diagram.

It is to be noted that the order of taking the forces does change the shape of the polygon. However, the resultant remains the same in magnitude, direction and sense.

EXAMPLE 2.40

Find graphically the resultant of the concurrent force system shown in Fig. 2.51.

Solution : Refer to Fig. 2.51 for the space diagram and name the various forces by Bow's notation.

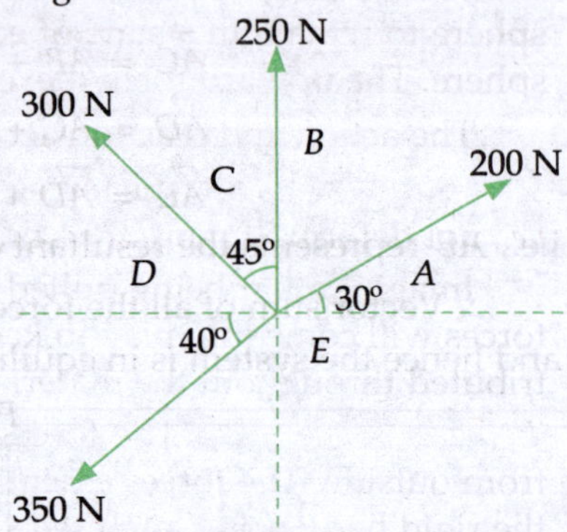

Fig. 2.51

Proceed to draw the vector diagram by adopting the following steps:

(*i*) Select some suitable point *a*, draw vector *ab* equal to force 200 N to some suitable scale and parallel to the line of action of this force. (Let 1 cm = 50 N).

(*ii*) From point *b*, draw vector *bc* to represent force 250 N in magnitude and direction.

(*iii*) From point *c*, draw vector *cd* to represent force 300 N in magnitude and direction.

(*iv*) Then from point *d*, draw vector *de* equal and parallel to force 350 N.

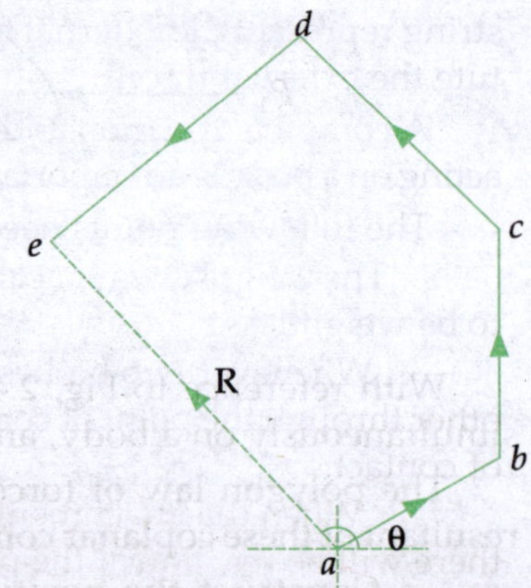

The closing side *ae* of the polygon gives the required resultant in magnitude and direction. The direction is from *a* to *e* as shown in the vector diagram.

By measurement:

$$ae = 9.1 \text{ cm}$$

$\therefore$ Magnitude of resultant = 9.1 × 50 = 455 N

Angle θ (with the horizontal) = 133°

2.7. FREE BODY DIAGRAM

The force analysis of a structure is made in a simplified way by considering the equilibrium of a portion of the structure. For that, the portion is drawn separately showing applied forces, self weight and reactions at the point of contact with other bodies. The resulting diagram is known as *free body diagram* (FBD). In a free body diagram, all the supports (like walls, floors, hinges etc.) are removed and replaced by the reactions which these supports exert on the body.

A free body diagram can be drawn for any single body of a system, for any subsystem or for the entire system irrespective of whether the system is in equilibrium, *i.e.,* at rest or in uniform motion or in a dynamic state of motion.

Further, while drawing the free body diagrams, one must have consideration of the following set of forces:

Body force: The body force is the force that is exerted by one body on another body that lies at a distance. There is no physical contact between the bodies under consideration. Gravitation force and the magnetic force are the body forces.

Contact force: The contact force or the surface force represents the force that is exerted by one body on another body through the point of contact. Friction force is the contact force.

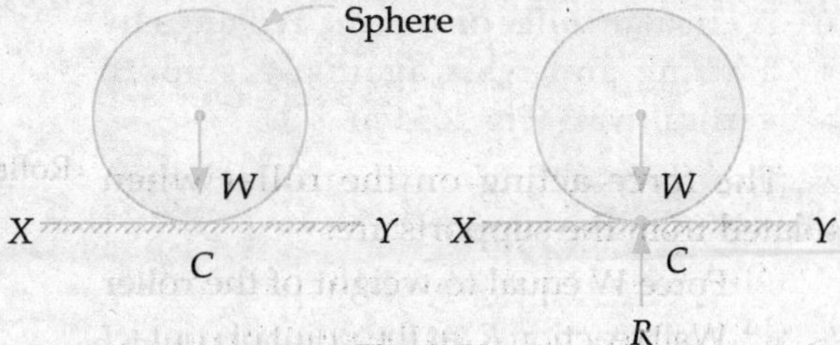

Fig. 2.52

With reference to the figure shown above, a sphere of weight W rests on a horizontal plane $X - Y$ and exerts a downward force equivalent to its weight at the point of contact C. For the sphere to remain in a state of equilibrium, the plane too exerts an equal and opposite force on the sphere. The upward force R exerted by the plane is called the reactive force or simply the reaction.

The action and reaction act at the same point but act on different bodies.

Internal forces: The forces which hold togather the particles of the body and help it to be rigid, *i.e.*, not deform. If more than one body is involved, internal forces hold the bodies together.

Imagine a bar being pulled by two equal and opposite forces applied at the ends. The internal forces will come into play to keep the bar undeformed. These forces cause stresses and strains distributed throughout the material of the body.

External forces: The forces which act on a body or system of bodies externally, *i.e.*, applied from outside. The forces essentially denote the action of other bodies (walls, floors, hinges etc.) on the rigid body being analysed.

Consider a sphere suspended by a string. The weight of the sphere and tension induced in the string represent the external forces. The reactions developed at the contact points, if any, also constitute the external forces.

A complete isolation and systematic representation of all external (applied and reactive) forces acting on a body is an important and effective tool for the solution of problems in mechanical systems.

The following points need to be kept in mind while drawing the free body diagram :

(*i*) The weight always acts on a body vertically downwards at its centre of gravity unless specified to be weightless.

(*ii*) Whenever two bodies are in contact with one another, each exerts a reaction force upon the other through the point of contact. If the contact is smooth, the reaction will be normal to the surface of contact.

(*iii*) When the attachment of a body with another body is by means of string, wire or cable etc., there will be reaction at the point of attachment.

(*iv*) Only those forces need to be considered which act on the body itself.

Given below are a few examples of systems and mechanisms together with their free body diagrams:

(*i*) A sphere resting on a frictionless plane surface (Fig. 2.53)

The forces acting on the sphere when isolated from the surface are:

(*a*) Force W equal to the weight of the sphere. This weight acts downward through the centroid of the sphere.

(*b*) Reaction R at the point of contact with the surface. This reaction acts upwards normal to the surface as it is frictionless.

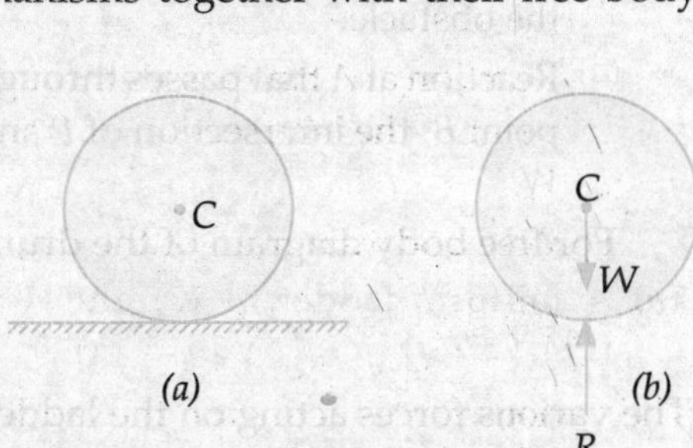

Fig. 2.53

As the sphere is in equilibrium, *R* and *W* are equal and collinear, and the free body diagram be as shown in Fig. 2.53

(ii) A circular roller of weight *W* hangs by a string and rests against a smooth vertical wall (Fig. 2.54 *a*)

The force acting on the roller when isolated from the supports are:

(*a*) Force *W* equal to weight of the roller

(*b*) Wall reaction R_c at the point of contact C with the wall. The reaction will be normal to the wall as it is smooth

(*c*) Tension *T* in the string along *BA*.

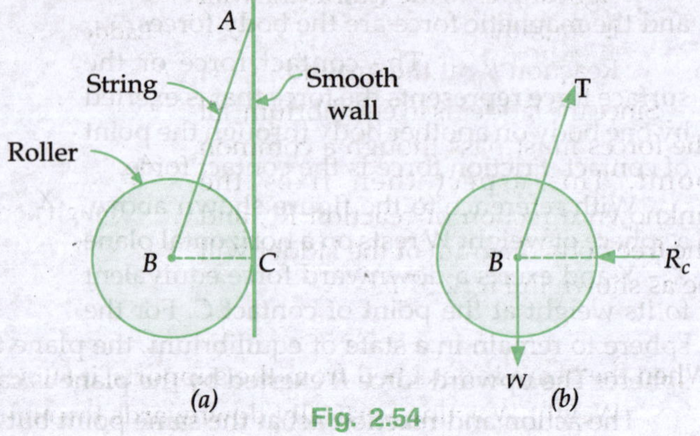

Fig. 2.54

As the roller is in equilibrium, all the forces will be concurrent and the free body diagram will be as shown in Fig. 2.54 *b*.

(iii) A sphere resting in a V-shaped groove (Fig. 2.55 *a*)

The sphere is isolated from the inclined surface forming the groove, and is considered to be acted upon by the following set of forces:

(*a*) Weight of the sphere acting vertically downward through its centre *C*.

(*b*) Reaction R_a acting normal to the inclined plane at the contact point *A*.

(*c*) Reaction R_b acting normal to the inclined plane at the contact point *B*.

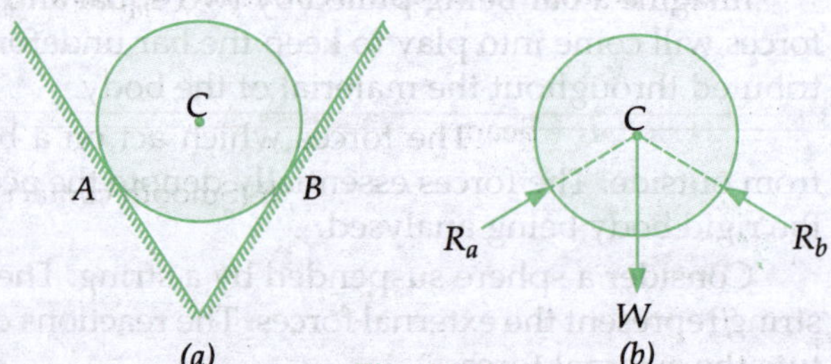

Fig. 2.55

Since the sphere is in equilibrium, all the forces meet at point C and the free body diagram will be as shown in Fig. 2.55 *b*.

(iv) A drum being rolled along the horizontal comes across a stepped obstacle (Fig. 2.56 *a*)

At the instant of obstacle being overcome, the forces acting on the drum are:

(*a*) Weight *W* of the drum acting vertically downwards through its centre *C*.

(*b*) Pull *P* required to be applied at the position shown to overcome the obstacle.

(*c*) Reaction at *A* that passes through point *B*, the intersection of *P* and *W*.

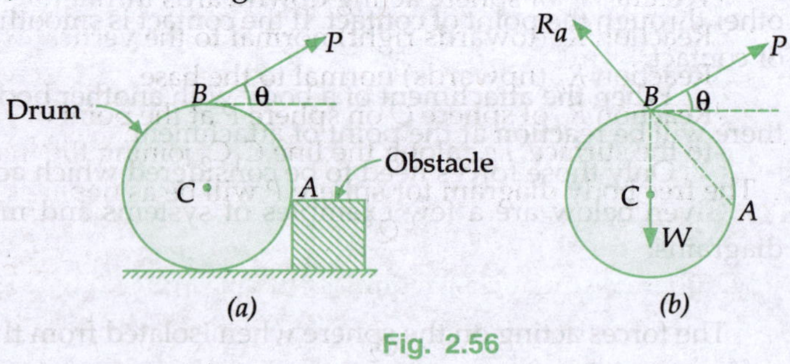

Fig. 2.56

For free body diagram of the drum, refer to Fig. 2.56 *b*.

(v) A uniform ladder of weight *W* leans a against a smooth wall and rests on a rough floor (Fig. 2.57 *a*)

The various forces acting on the ladder when it is isolated from the wall and floor are:

(*a*) Force *W* equal to the weight of the ladder and acting vertically downward from the mid of the ladder.

(*b*) Reaction R_b of the wall. This acts right angles to the wall as the wall is smooth.

(*c*) Reaction R_a at the ground.

Since the ladder is in equilibrium, all the forces must pass though a common point. This aspect then fixes the unknown direction of reaction R_a, and the free body diagram of the ladder will be as shown in Fig. 2.57 *b*.

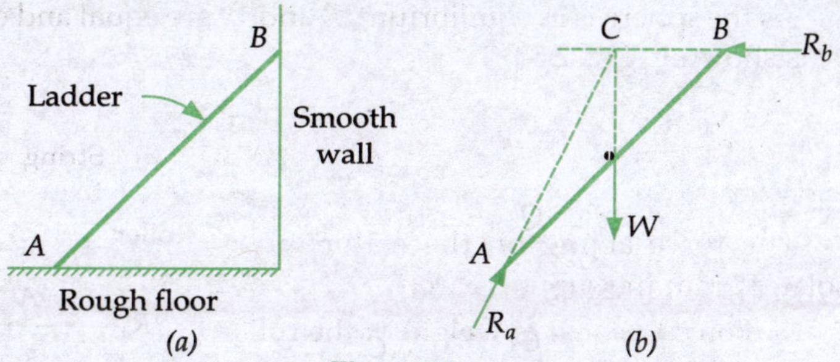

Fig. 2.57.

(*vi*) A beam loaded and supported (Fig. 2.58 *a*)

When the beam is detached from the supports, it subjected to following set of forces:

(*a*) Weight *W* acting vertically downwards through mass centre of the beam
(*b*) Reaction R_b normal to the beam at its smooth contact with the corner
(*c*) Horizontal applied force *P* and the couple *M*
(*d*) vertical and horizontal reactions (R_{av} and R_{ah}) exerted at the pin connection at *A*.

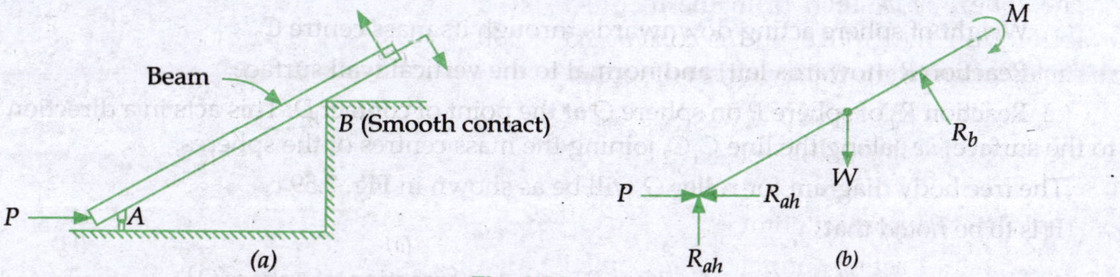

Fig. 2.58

The beam will be in equilibrium under the action of five forces and one couple, and its free body diagram will be as shown in Fig. 2.58 *b*.

(*vii*) Two spheres *P* and *Q* placed in a vessel (Fig. 2.59 *a*)

Forces acting on sphere P

(*a*) Weight W_1 of sphere acting downwards through its mass centre C_1
(*b*) Reaction R_a (towards right) normal to the vertical wall surface
(*c*) Reaction R_b (upwards) normal to the base
(*d*) Reaction R_d of sphere *Q* on sphere *P* at the point of contact *D*. This acts in a direction normal to the surface, *i.e.*, along the line C_1C_2 joining the mass centres of the spheres

The free body diagram for sphere *P* will be as depicted in Fig. 2.59 *b*.

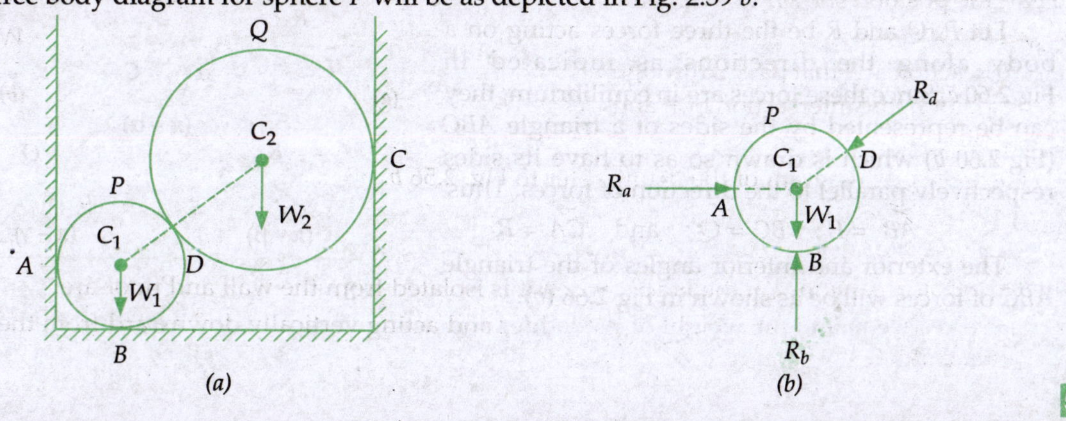

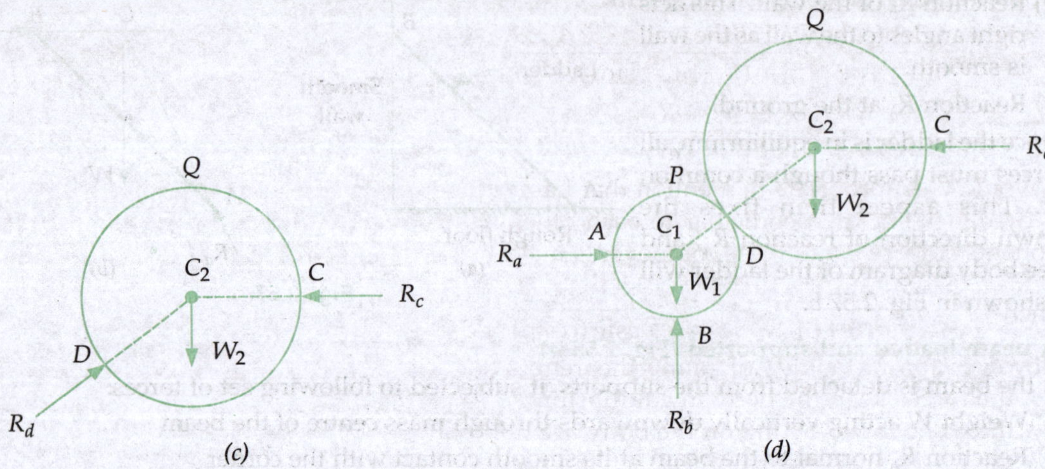

Fig. 2.59

Forces acting on sphere Q

(*a*) Weight of sphere acting downwards through its mass centre C_2

(*b*) Reaction R_c (towards left) and normal to the vertical wall surface.

(*c*) Reaction R_d of sphere *P* on sphere *Q* at the point of contact *D*. This acts in a direction normal to the surface, *i.e.*, along the line C_1C_2 joining the mass centres of the spheres.

The free body diagram for roller *Q* will be as shown in Fig. 2.59 *c*.

It is to be noted that

$$\left\{\begin{matrix}\text{reaction of sphere } P\\ \text{on sphere } Q\end{matrix}\right\} \quad \text{and} \quad \left\{\begin{matrix}\text{reaction of sphere } Q\\ \text{on sphere } P\end{matrix}\right\}$$

are equal in magnitude, opposite in direction and are collinear. Obviously when both the spheres are taken together, these reactions cancel and do not appear in the free body diagram (Fig. 2.59 *d*).

2.8. LAMI'S THEOREM

"If a body is in equilibrium under the action of three forces, then each force is proportional to the sine of the angle between the other two forces."

Let *P*, *Q* and *R* be the three forces acting on a body along the directions as indicated in Fig 2.60 *a*. Since these forces are in equilibrium, they can be represented by the sides of a triangle *ABC* (Fig 2.60 *b*) which is drawn so as to have its sides respectively parallel to the direction of forces. Thus

$$\vec{AB} = P; \quad \vec{BC} = Q; \quad \text{and} \quad \vec{CA} = R$$

The exterior and interior angles of the triangle *ABC* of forces will be as shown in Fig 2.66 (*b*).

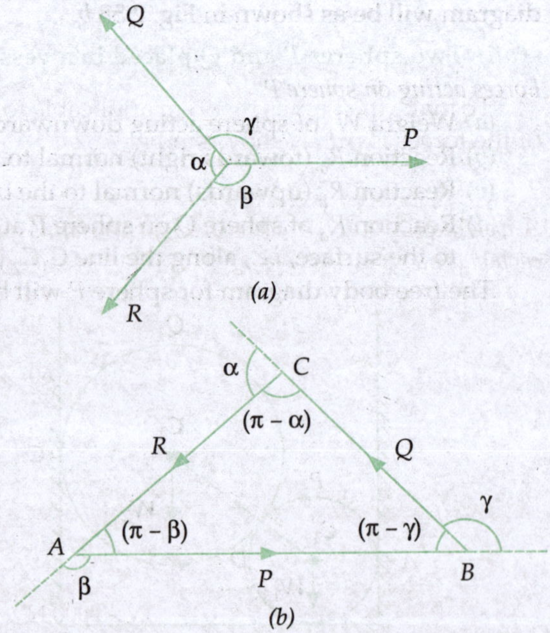

Fig. 2.60

Applying sine rule for the triangle ABC

$$\frac{AB}{\sin(\pi-\alpha)} = \frac{BC}{\sin(\pi-\beta)} = \frac{CA}{\sin(\pi-\gamma)}$$

$$\therefore \quad \frac{P}{\sin\alpha} = \frac{Q}{\sin\beta} = \frac{R}{\sin\gamma}$$

It would be appropriate to mention that Lami's theorem can be used to resolve a given force along two lines not perpendicular to each other. With reference to Fig. 2.61. P_1 and P_2 are the components of force P and these components make angles α and β respectively on either side of P.

Then $P_1 = \dfrac{P\sin\beta}{\sin(\alpha+\beta)}$ and $P_2 = \dfrac{P\sin\alpha}{\sin(\alpha+\beta)}$

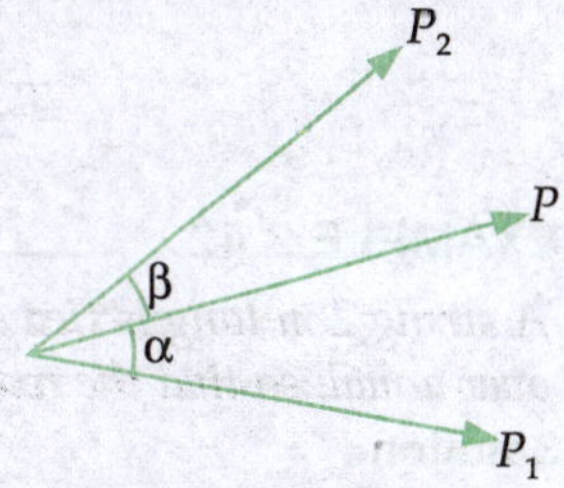

Fig. 2.61

However, for the validity of Lami's theorem to hold good,

- the forces keep the body in equilibrium
- the three forces acting on the body are non-parallel
- the forces are concurrent, *i.e.*, act at a point on the body
- the forces are either directed towards or away from the point of concurrence Fig. 2.62
- the angle between any two forces is less than 180°.

It needs to be noted that the Lami's theorem is not meant to find the resultant of the three concurrent forces. It is only to provide the relation between a force and the angle between the other two forces.

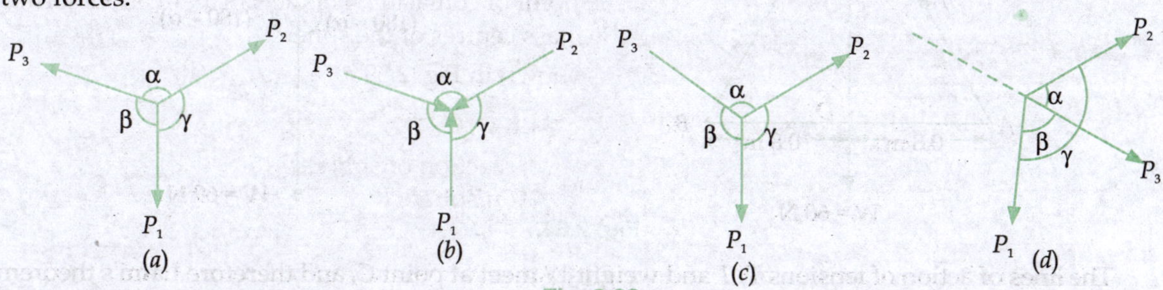

Fig. 2.62

To make the Lami's theorem applicable to the force system depicted in Fig. 2.62 *c*, the orientation of the forces is put as shown in Fig 2.62 *d*.

EXAMPLE 2.41

A weight of 2000 N is supported by two chains AC and BC as shown in Fig. 2.63. Determine the tension in each chain.

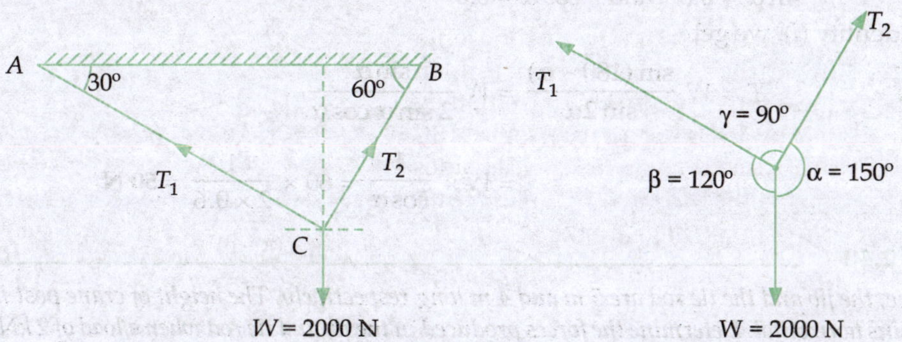

Fig. 2.63

Solution : Let T_1 and T_2 be the tensions in chains AC and BC respectively. Since the lines of action of these tensions and the weight W meet at a point, Lami's theorem can be applied. That is

$$\frac{T_1}{\sin\alpha} = \frac{T_2}{\sin\beta} = \frac{W}{\sin\gamma}$$

or

$$\frac{T_1}{\sin 150^\circ} = \frac{T_2}{\sin 120^\circ} = \frac{2000}{\sin 90^\circ}$$

$$\therefore \quad T_1 = 2000 \times \frac{\sin 150^\circ}{\sin 90^\circ} = 2000 \times \frac{0.5}{1} = 1000 \text{ N}$$

$$T_2 = 2000 \times \frac{\sin 120^\circ}{\sin 90^\circ} = 2000 \times \frac{0.866}{1} = 1732 \text{ N}$$

EXAMPLE 2.42

A string 2 m long is tied to the ends of a uniform rod that weighs 60 N and is 1.6 m long. The string passes over a nail, so that the rod hangs horizontally. Make calculations for the tension in the string.

Solution :

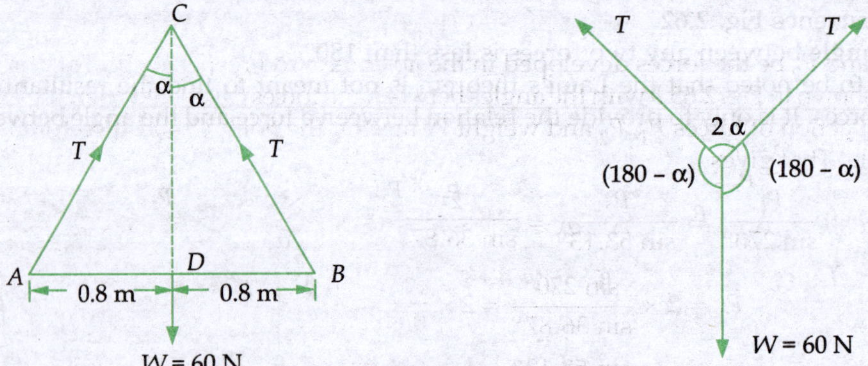

Fig. 2.64.

The lines of action of tensions T, T and weight W meet at point C, and therefore Lami's theorem is applicable. That is

$$\frac{W}{\sin 2\alpha} = \frac{T}{\sin(180 - \alpha)} = \frac{T}{\sin(180 - \alpha)} \qquad ...(i)$$

From the geometry of the arrangement

$$AC = BC = 1 \text{ m}; \quad CD = \sqrt{1^2 - 0.8^2} = 0.6 \text{ m}$$

$$\sin\alpha = 0.8 \quad \text{and} \quad \cos\alpha = 0.6$$

From identity (*i*), we get

$$T = W\frac{\sin(180 - \alpha)}{\sin 2\alpha} = W\frac{\sin\alpha}{2\sin\alpha\cos\alpha}$$

$$= W \times \frac{1}{\cos\alpha} = 60 \times \frac{1}{2 \times 0.6} = \mathbf{50\ N}$$

EXAMPLE 2.43

In a jib crane, the jib and the tie rod are 5 m and 4 m long respectively. The height of crane post is 3 m and the tie rod remains horizontal. Determine the forces produced in the jib and tie rod when a load of 2 kN is suspended at the crane head.

Solution : Refer Fig. 2.65 *a* for the arrangement of the system.

$$\sin\theta = \frac{3}{5} = 0.6\,; \quad \theta = 36.87^\circ$$

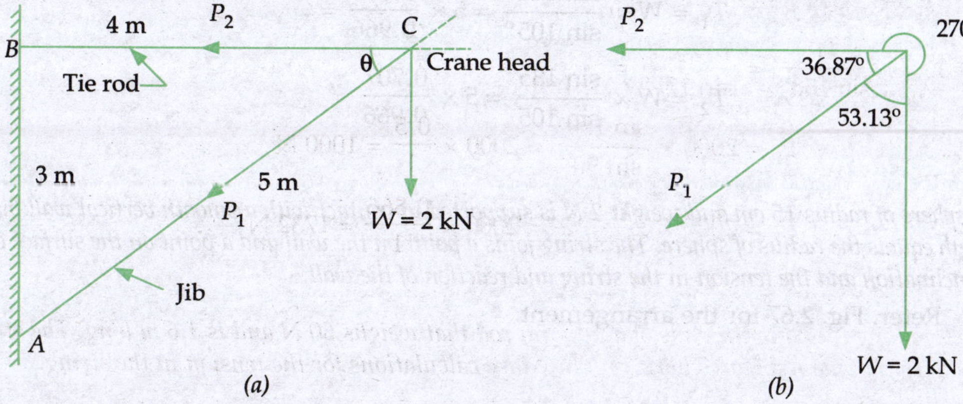

Fig. 2.65

Let P_1 and P_2 be the forces developed in the jib and tie rod respectively. The three forces P_1, P_2 and W are shown in Fig. 2.65 *b* with the angles between the forces calculated from the given directions. The lines of action of forces P_1, P_2 and weight W meet at the point C, and therefore Lami's theorem is applicable. That gives

$$\frac{P_1}{\sin 270^\circ} = \frac{P_2}{\sin 53.13^\circ} = \frac{2}{\sin 36.87^\circ}$$

$$\therefore \quad P_1 = 2 \times \frac{\sin 270^\circ}{\sin 36.87^\circ} = 2 \times \frac{1}{0.6} = -3.33 \text{ kN}$$

$$P_2 = 2 \times \frac{\sin 53.13^\circ}{\sin 36.87^\circ} = 2 \times \frac{0.8}{0.6} = 2.667 \text{ kN}$$

The – ve sign indicates that the direction of force P_1 is opposite to that shown in Fig. 2.71 (*a*). Obviously the tie rod will be under tension and the jib will in compression.

EXAMPLE 2.44

A machine weighing 5 kN is supported by two chains attached to some point on the machine. One chain goes to the hook in the ceiling and has an indication of 45° with the horizontal. The other chain goes to the eye bolt in the wall and is inclined at 30° to the horizontal. Make calculations for the tensions induced in the chain.

Solution : Refer Fig. 2.66 for the arrangement. The machine is acted upon by the following set of forces:

(*i*) weight of machine $W = 5$ kN acting vertically downwards,

(*ii*) tension T_1 in the chain OA which goes to hook in the ceiling,

(*iii*) tension T_2 in the chain OB which goes to the eye bolt.

These forces are concurrent and meet at point O. Applying Lami's theorem

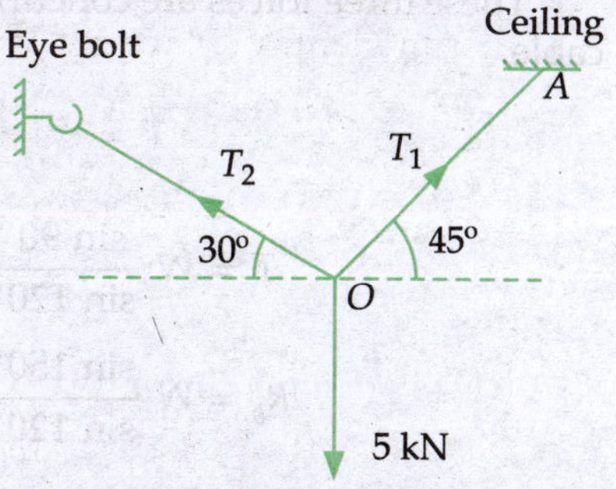

Fig. 2.66.

$$\frac{T_1}{\sin(90+30)} = \frac{T_2}{\sin(90+45)} = \frac{W}{\sin(180-30-45)}$$

or $$\frac{T_1}{\sin 120°} = \frac{T_2}{\sin 135°} = \frac{W}{\sin 105°}$$

$$\therefore \quad T_1 = W \times \frac{\sin 120°}{\sin 105°} = 5 \times \frac{0.866}{0.966} = 4.48 \text{ kN}$$

$$T_2 = W \times \frac{\sin 135}{\sin 105} = 5 \times \frac{0.707}{0.966} = 3.66 \text{ N}$$

EXAMPLE 2.45

A smooth sphere of radius 15 cm and weight 2 N is supported in contact with a smooth vertical wall by a string whose length equals the radius of sphere. The string joins a point on the wall and a point on the surface of sphere. Workout inclination and the tension in the string and reaction of the wall.

Solution : Refer. Fig. 2.67 for the arrangement.

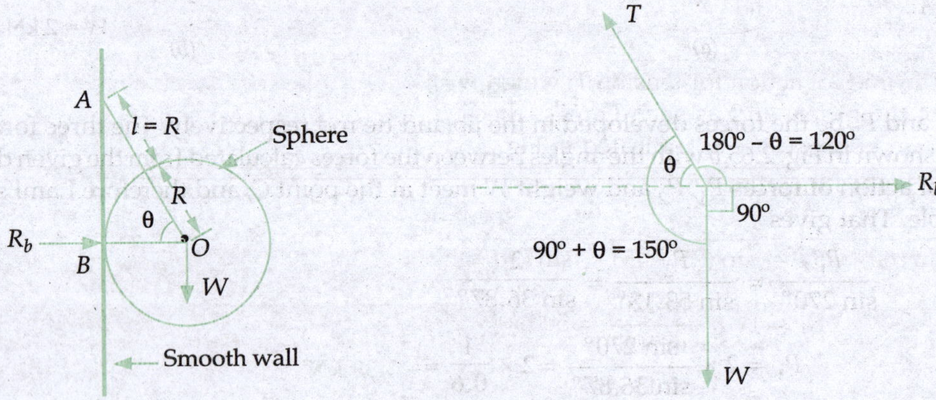

Fig. 2.67

In triangle AOB,

$$\cos\theta = \frac{OB}{OA} = \frac{R}{2R} = \frac{1}{2}\,; \quad \theta = 60°$$

The sphere is in equilibrium under the action of following forces:

(i) Weight $W = 2$ N of the sphere which acts vertically downwards through the centre O.

(ii) Tension T in the string

(iii) Reaction R_b of the wall at the point of contact B. Since the wall is smooth, this reaction acts perpendicular to the wall.

These three forces are concurrent, *i.e.*, meet at point O and as such the Lami's theorem is applicable.

$$\frac{T}{\sin 90°} = \frac{W}{\sin 120°} = \frac{R_b}{\sin 150°}$$

$$\therefore \quad T = W\frac{\sin 90°}{\sin 120°} = 2 \times \frac{1}{0.866} = 2.31 \text{ N}$$

$$R_b = W\frac{\sin 150°}{\sin 120°} = 2 \times \frac{0.5}{0.866} = 1.15 \text{ N}$$

EXAMPLE 2.46

A roller of weight 500 N rests on a smooth inclined plane and is kept free from rolling down by a string as shown in Fig. 2.68 a. Work out tension in the string and reaction at the point of contact B.

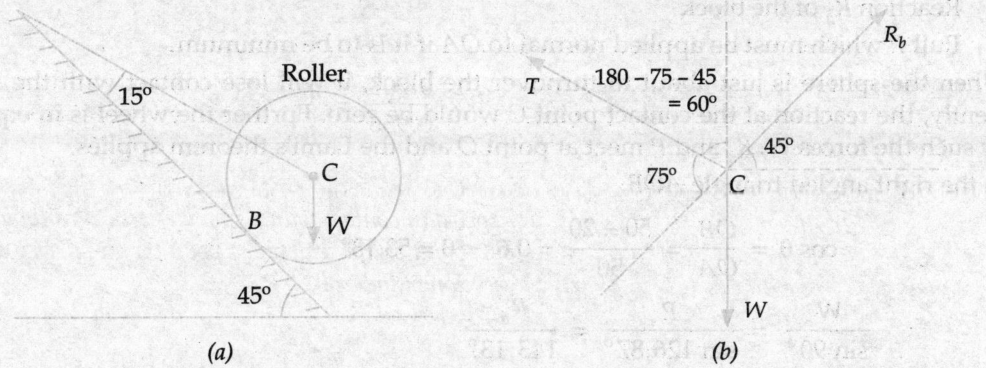

Fig. 2.68

Solution : The lines of action for tension T, weight W and reaction R_b at the contact point meet at C (the centre of the roller) and as such Lami's theorem is applicable. The angles between various segments around point C are as indicated in Fig. 2.68 *b*.

Invoking Lami's theorem,

$$\frac{W}{\sin(60+45)} = \frac{R_b}{\sin(75+45)} = \frac{T}{\sin(90+45)}$$

$$\therefore \quad R_b = W \times \frac{\sin 120^\circ}{\sin 105^\circ} = 500 \times \frac{0.866}{0.966} = 448.24 \text{ N}$$

$$T = W \times \frac{\sin 135^\circ}{\sin 105^\circ} = 500 \times \frac{0.707}{0.966} = 365.94 \text{ N}$$

EXAMPLE 2.47

A uniform wheel of 50 cm diameter and 1 kN weight rests against a rigid rectangular block of thickness 20 cm (Fig. 2.69 a). Considering all surfaces smooth, determine :

(a) *least pull to be applied through the centre of wheel to just turn it over the corner of the block,*

(b) *reaction of the block*

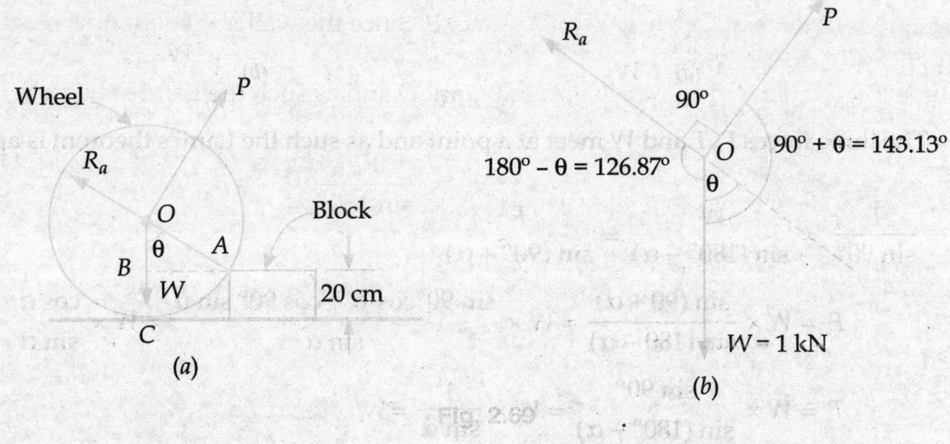

Fig. 2.69

Solution : The wheel is acted upon by the following forces when it is just about to turnover the block.

(*i*) Weight $W = 1$ kN of the wheel acting vertically downwards through centre of the sphere

(*ii*) Reaction R_a of the block

(*iii*) Pull P which must be applied normal to OA if it is to be minimum.

When the sphere is just about to turnover the block, it will lose contact with the floor and apparently, the reaction at the contact point C would be zero. Further the wheel is in equilibrium and as such the forces W, R_a and P meet at point O and the Lami's theorem applies.

In the right angled triangle AOB,

$$\cos\theta = \frac{OB}{OA} = \frac{50-20}{50} = 0.6\,; \quad \theta = 53.13^\circ$$

$$\frac{W}{\sin 90^\circ} = \frac{P}{\sin 126.87^\circ} = \frac{R_a}{143.13^\circ}$$

$$\therefore \quad P = W \times \frac{\sin 126.87^\circ}{\sin 90^\circ} = 1 \times \frac{0.8}{1} = \mathbf{0.8\ kN}$$

$$R_a = W \times \frac{\sin 143.13^\circ}{\sin 90^\circ} = 1 \times \frac{0.6}{1} = \mathbf{0.6\ kN}$$

EXAMPLE 2.48

Figure 2.70 a shows a weight W tied to the end of a cord of length l. Determine the magnitude of force F to pull the weight at an angle α as indicated in the figure. Proceed to find the tension in the cord.

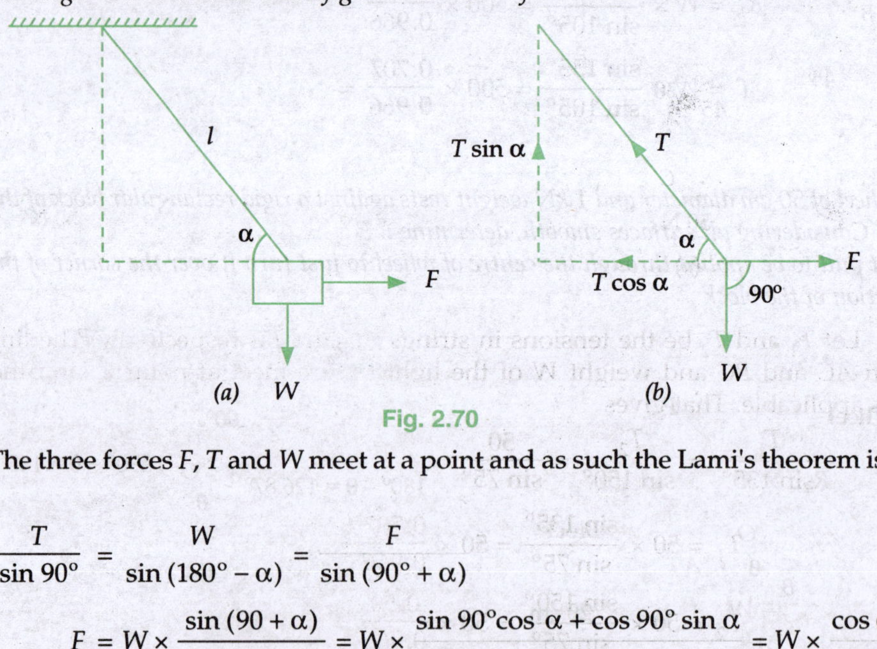

Fig. 2.70

Solution : The three forces F, T and W meet at a point and as such the Lami's theorem is applicable. Therefore

$$\frac{T}{\sin 90^\circ} = \frac{W}{\sin(180^\circ - \alpha)} = \frac{F}{\sin(90^\circ + \alpha)}$$

$$\therefore \quad F = W \times \frac{\sin(90+\alpha)}{\sin(180-\alpha)} = W \times \frac{\sin 90^\circ \cos\alpha + \cos 90^\circ \sin\alpha}{\sin\alpha} = W \times \frac{\cos\alpha}{\sin\alpha} = \mathbf{W \cot\alpha}$$

$$T = W \times \frac{\sin 90^\circ}{\sin(180^\circ - \alpha)} = W \times \frac{1}{\sin\alpha} = \mathbf{W\ cosec\ \alpha}$$

Alternatively :

Since the system is in equilibrium

$$\Sigma F_x = 0; \quad F = T\cos\alpha \qquad ...(i)$$

$$\Sigma F_y = 0; \quad W = T\sin\alpha \qquad ...(ii)$$

Upon division: $\dfrac{F}{W} = \dfrac{\cos\alpha}{\sin\alpha}$ or $F = W\cot\alpha$

Substituting $F = W\cot\alpha = W\dfrac{\cos\alpha}{\sin\alpha}$ in identity (*i*), we get

$$W\frac{\cos\alpha}{\sin\alpha} = T\cos\alpha$$

or $\quad T = \dfrac{W}{\sin\alpha} = W\,\text{cosec}\,\alpha$

EXAMPLE 2.49

An electric light fixture weighing 50 N hangs from point C by two strings AC and BC as shown in Fig. 2.71 a. The string AC is inclined at 60° to the horizontal and string BC is 45° to the vertical. Using Lami's theorem or otherwise determine the forces in the strings AC and BC.

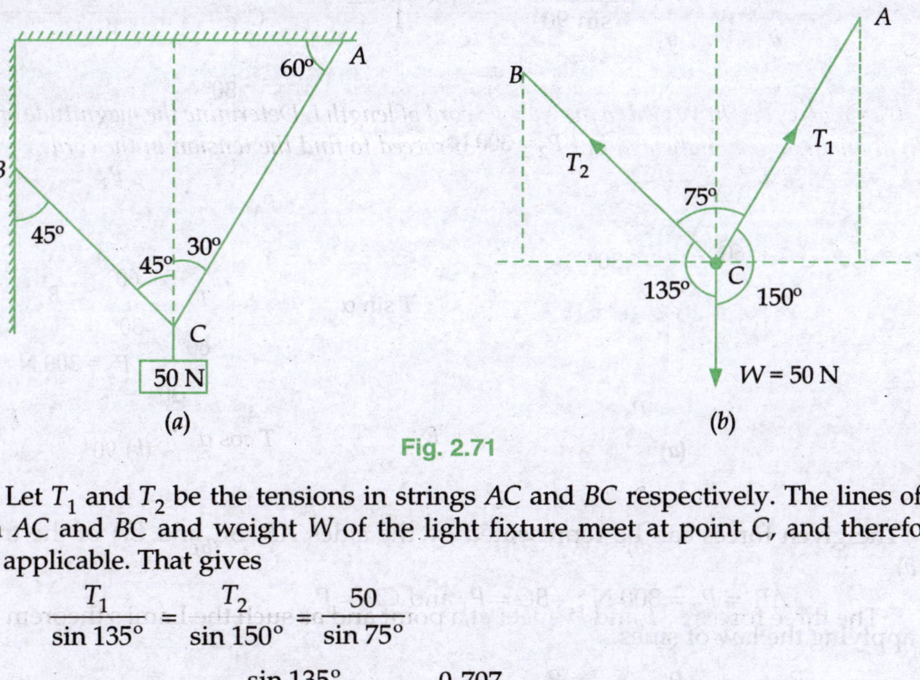

Fig. 2.71

Solution : Let T_1 and T_2 be the tensions in strings *AC* and *BC* respectively. The lines of action of tensions in *AC* and *BC* and weight *W* of the light fixture meet at point *C*, and therefore Lami's theorem is applicable. That gives

$$\frac{T_1}{\sin 135^\circ} = \frac{T_2}{\sin 150^\circ} = \frac{50}{\sin 75^\circ}$$

$$\therefore \quad T_1 = 50 \times \frac{\sin 135^\circ}{\sin 75^\circ} = 50 \times \frac{0.707}{0.966} = 36.594 \text{ N}$$

$$T_2 = 50 \times \frac{\sin 150^\circ}{\sin 75^\circ} = 50 \times \frac{0.50}{0.966} = 25.88 \text{ N}$$

Alternatively :

Since the light fixture is in equilibrium, $\Sigma F_x = 0$ and $\Sigma F_y = 0$. Resolving all the forces horizontally, we get

$$T_1 \sin 30^\circ - T_2 \sin 45^\circ = 0$$
$$T_1 \times 0.5 - T_2 \times 0.707 = 0$$
$$T_2 = \frac{0.5}{0.707} T_1 = 0.707\, T_1$$

Resolving all the forces vertically, we get

$$T_1 \cos 30^\circ + T_2 \cos 45^\circ - 50 = 0$$
$$T_1 \times 0.866 + T_2 \times 0.707 = 50$$

Substituting $T_2 = 0.707\, T_1$ in the above identity, we get

$$0.866\, T_1 + 0.707 \times 0.707\, T_1 = 50$$

or $$0.866\, T_1 + 0.5\, T_1 = 50$$

$\therefore$ $$T_1 = \frac{50}{1.366} = \mathbf{36.6\ N}$$
$$T_2 = 0.707\, T_1 = 0.707 \times 36.6 = \mathbf{25.88\ N}$$

EXAMPLE 2.50

A body acted upon by three forces P_1, P_2 and P_3, as shown in Fig. 2.72 a, is in equilibrium. If P_2 = 300 N, make calculations for the forces P_1 and P_3.

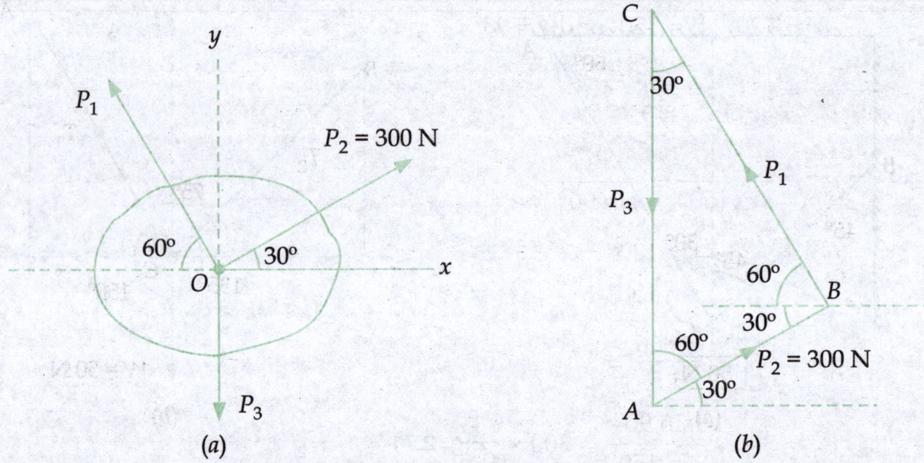

Fig. 2.72

Solution : The given forces can be represented by the sides *AB*, *BC* and *CA* of the triangle *ABC* (Fig. 2.72 *b*)

$$AB = P_2 = 300\text{ N};\quad BC = P_1 \text{ and } CA = P_3$$

Then applying the law of sines,

$$\frac{P_1}{\sin 60^\circ} = \frac{P_2}{\sin 30^\circ} = \frac{P_3}{\sin 90^\circ}$$

That gives: $$P_1 = P_2 \frac{\sin 60^\circ}{\sin 30^\circ} = 300 \times \frac{0.866}{0.5} = \mathbf{519.6\ N}$$

$$P_3 = P_2 \frac{\sin 90^\circ}{\sin 30^\circ} = 300 \times \frac{1}{0.5} = \mathbf{600\ N}$$

Alternatively :

Since the system of forces is in equilibrium

$$\Sigma F_x = 0 \quad \text{and} \quad \Sigma F_y = 0$$

Resolution of forces along X-axis gives :

$$P_2 \cos 30^\circ + P_1 \cos 120^\circ + P_3 \cos 270^\circ = 0$$

All angles have been measured anticlockwise from axis O-X

$$0.866\, P_2 - 0.5\, P_1 + 0 = 0$$

$$\therefore \quad P_1 = \frac{0.866}{0.5} P_2 = \frac{0.866}{0.5} \times 300 = 519.6 \text{ N}$$

Resolution of all forces along Y-axis gives :

$$P_2 \sin 30^\circ + P_1 \sin 120^\circ + P_3 \sin 270^\circ = 0$$

$$0.5\, P_2 + 0.866\, P_1 - P_3 = 0$$

$$\therefore \quad P_3 = 0.5\, P_2 + 0.866\, P_1$$

$$= 0.5 \times 300 + 0.866 \times 519.6 = 150 + 450 = 600 \text{ N}$$

Alternatively :

Since the three forces meet at a point, solution to this problem can be obtained by invoking the Lami's theorem.

$$\alpha = 120^\circ \,;\ \beta = 150^\circ \text{ and } \gamma = 90^\circ$$

$$\frac{P_1}{\sin \alpha} = \frac{P_2}{\sin \beta} = \frac{P_3}{\sin \gamma}$$

or

$$\frac{P_1}{\sin 120^\circ} = \frac{300}{\sin 150^\circ} = \frac{P_3}{\sin 90^\circ}$$

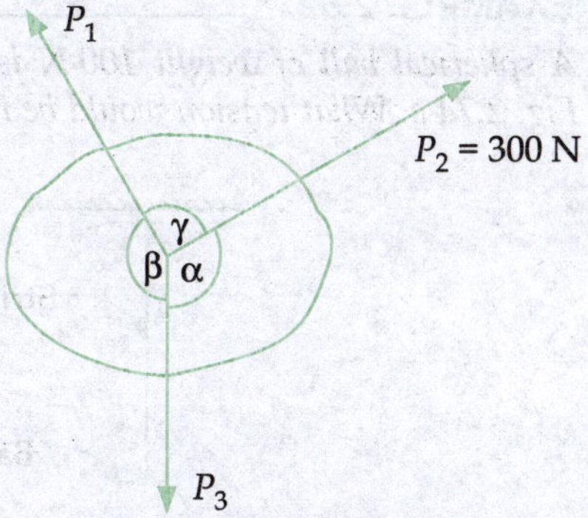

$$\therefore \quad P_1 = 300 \times \frac{\sin 120^\circ}{\sin 150^\circ}$$

$$= 300 \times \frac{0.866}{0.5} = 519.6 \text{ N}$$

$$P_3 = 300 \times \frac{\sin 90^\circ}{\sin 150^\circ} = 300 \times \frac{1}{0.5} = 600 \text{ N}$$

EXAMPLE 2.51

A string ABCDE whose extremity A is fixed has weights W_1 and W_2 attached to it at B and C, and passes round a smooth peg at D carrying a weight of 800 N at the free end E (Fig. 2.73). If in a state of equilibrium, BC is horizontal and AB and CD make angles of 150° and 120° respectively with BC, make calculations for

(a) *the tensions in portions AB, BC, CD and DE of the string*

(b) *the value of weights W_1 and W_2*

(c) *the pressure on the peg D.*

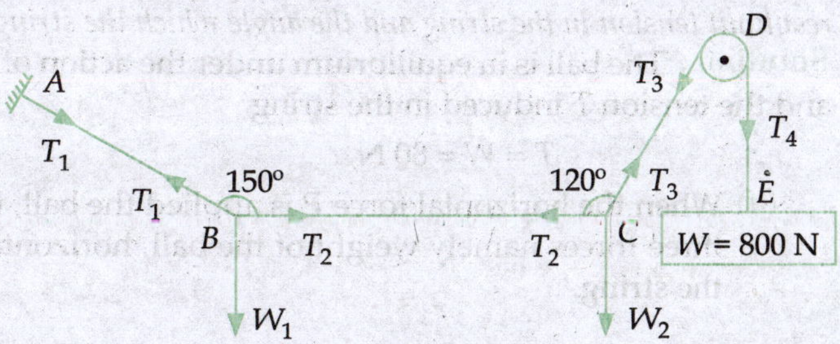

Fig. 2.73

Solution : Let T_1, T_2, T_3, T_4 be the tensions in segments AB, BC, CD and DE of the string.

Under equilibrium conditions,

$$T_3 = T_4 = 800 \text{ N}$$

Applying Lami's theorem at point C,

$$\frac{T_2}{\sin 150^\circ} = \frac{T_3}{\sin 90^\circ} = \frac{W_2}{\sin 120^\circ}$$

$$\therefore \quad T_2 = T_3 \frac{\sin 150^\circ}{\sin 90^\circ} = 800 \times \frac{0.5}{1} = 400 \text{ N}$$

$$W_2 = T_3 \frac{\sin 120^\circ}{\sin 90^\circ} = 800 \times \frac{0.866}{1} = 692.8 \text{ N}$$

Applying Lami's theorem at point B,

$$\frac{T_1}{\sin 90^\circ} = \frac{T_2}{\sin 120^\circ} = \frac{W_1}{\sin 150^\circ}$$

$$\therefore \quad T_1 = T_2 \frac{\sin 90^\circ}{\sin 120^\circ} = 400 \times \frac{1}{0.866} = 461.89 \text{ N}$$

$$W_1 = T_2 \frac{\sin 150^\circ}{\sin 120^\circ} = 400 \times \frac{0.5}{0.866} = 230.95 \text{ N}$$

(*c*) Pressure on the peg at D $= T_3 \sin 60^\circ + W$

$= 800 \sin 60^\circ + 800 = 692.82 + 800 = 1492.82$ N

EXAMPLE 2.52

A spherical ball of weight 100 N is attached to a string and is suspended from the ceiling as shown in Fig. 2.74 a. What tension would be induced in the string ?

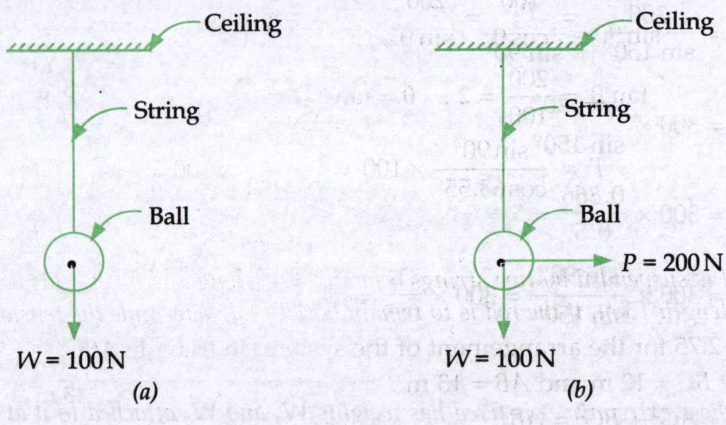

Fig. 2.74

Subsequently a horizontal force of 200 N is applied to the ball as shown in Fig. 2.74 b. Calculate the resultant tension in the string and the angle which the string makes with the vertical.

Solution : The ball is in equilibrium under the action of two forces namely, the weight W of the ball and the tension T induced in the string

$$\therefore \quad T = W = 80 \text{ N}$$

(*ii*) When the horizontal force P is applied the ball, it will be in equilibrium under the action of three forces namely weight of the ball, horizontal force applied and the tension induced in the string.

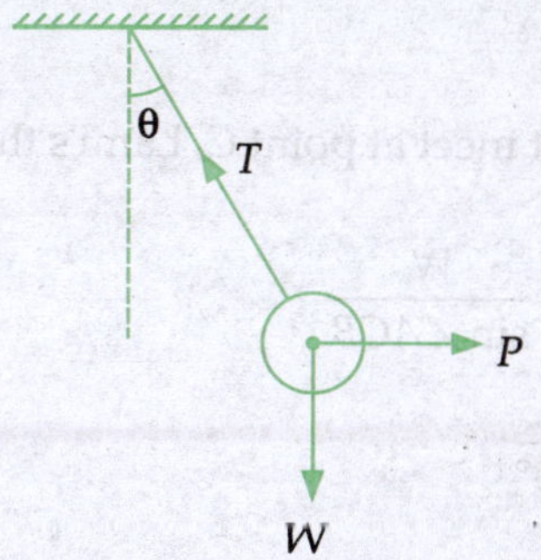

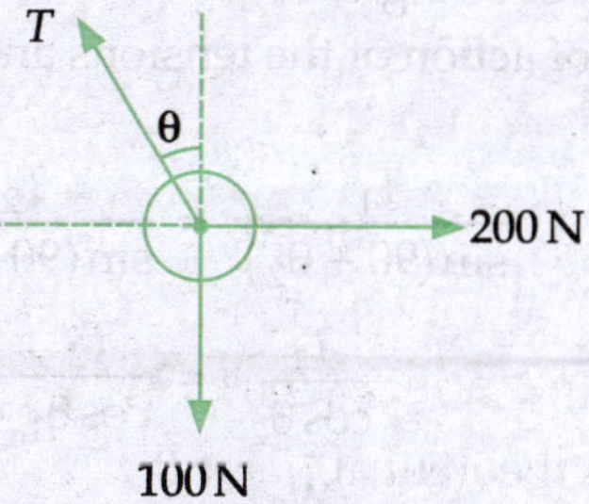

With reference to free body diagram, we have under equilibrium state

$$\Sigma F_x = 0: \quad -T \sin\theta + 200 = 0 \quad \ldots(i)$$

$$\Sigma F_y = 0: \quad T \cos\theta - 100 = 0 \quad \ldots(ii)$$

From expressions (*i*) and (*ii*)

$$\tan\theta = \frac{200}{100} = 2; \quad \theta = 63.55°$$

$$T = \frac{200}{\sin 63.55} = 223.46 \text{ N}$$

Alternatively

Invoking Lami's theorem

$$\frac{T}{\sin 90°} = \frac{100}{\sin(90+\theta)} = \frac{200}{\sin(180-\theta)}$$

or $$\frac{T}{\sin 90} = \frac{100}{\cos\theta} = \frac{200}{\sin\theta}$$

and hence $$\tan\theta = \frac{200}{100} = 2; \quad \theta = \tan^{-1}(2) = 63.55°$$

$$T = \frac{\sin 90°}{\cos 63.55°} \times 100 = \frac{1}{0.4454} \times 100 = 224.52 \text{ N}$$

EXAMPLE 2.53

A body of weight 20 N is suspended by two strings 5 m and 12 m long and other ends being fastened to the extremities of a rod of length 13 m. If the rod is to remain horizontal, determine the tensions in the strings.

Solution: Refer Fig. 2.75 for the arrangement of the system. In triangle *ABC*,

$AC = 5$ m; $BC = 12$ m and $AB = 13$ m

$5^2 + 12^2 = 13^2; \quad AC^2 + BC^2 = AB^2$

and accordingly ΔABC is a right angled triangle with $\angle ACB = 90°$

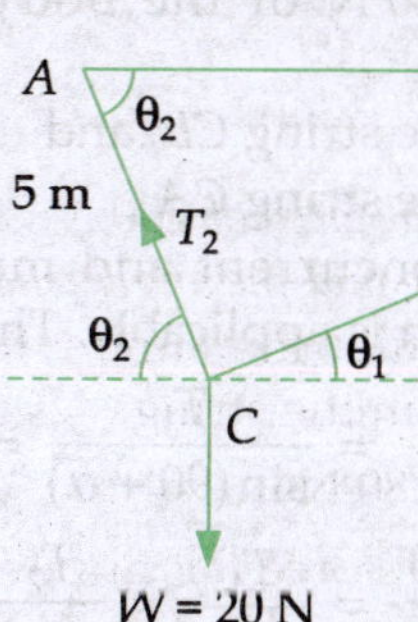

Fig. 2.75

Further: $\sin\theta_1 = \frac{5}{13}$ and $\cos\theta_1 = \frac{12}{13}$

$\sin\theta_2 = \frac{12}{13}$ and $\cos\theta_2 = \frac{5}{13}$

The weight is acted upon by the following set of forces :

(*i*) weight of body $W = 20$ N acting vertically downwards,

(ii) tension T_1 in string CB, and

(iii) tension T_2 in string CA

Since the lines of action of the tensions and weight meet at point C, Lami's theorem holds good. That is

$$\frac{T_1}{\sin(90+\theta_2)} = \frac{T_2}{\sin(90+\theta_1)} = \frac{W}{\sin \angle ACB}$$

or
$$\frac{T_1}{\cos\theta_2} = \frac{T_2}{\cos\theta_1} = \frac{W}{\sin 90°}$$

$$\therefore \quad T_1 = \frac{W}{\sin 90°} \times \cos\theta_2 = \frac{20}{1} \times \frac{5}{12} = 7.69 \text{ N}$$

$$T_2 = \frac{W}{\sin 90°} \times \cos\theta_1 = \frac{20}{1} \times \frac{12}{13} = 18.46 \text{ N}$$

EXAMPLE 2.54

A body of weight 20 N is suspended by two strings 5 m and 12 m long and other ends being fastened to the extremities of a rod of length 13 m. If the rod be so held that the body hangs immediately below the middle point, find out the tensions in the strings.

Solution: Refer Fig. 2.76 for the arrangement of the system. In triangle ABC,

$$AC = 5 \text{ m}; BC = 12 \text{ m and } AB = 13 \text{ m}$$

$$5^2 + 12^2 = 13^2; \quad (AC)^2 + (BC)^2 = AB^2$$

and accordingly ΔABC is a right angled triangle with $\angle ACB = 90°$

Applying sine rule to triangles ACD and BCD,

$$\frac{6.5}{\sin(90-\alpha)} = \frac{CD}{\sin(90-\beta)}; \quad \frac{6.5}{\cos\alpha} = \frac{CD}{\cos\beta} \quad \ldots(i)$$

and
$$\frac{6.5}{\sin\alpha} = \frac{CD}{\sin\beta} \quad \ldots(ii)$$

From expressions (i) and (ii),

$$\frac{6.5}{\cos\alpha} \times \cos\beta = \frac{6.5}{\sin\alpha} \times \sin\beta$$

or $\quad \tan\alpha = \tan\beta \quad \therefore \quad \alpha = \beta$

The system is acted upon by the following set of forces :

(i) weight W = 20 N of the body acting vertically downwards

(ii) tension T_1 in the string CB, and

(iii) tension T_2 in the string CA

These forces are concurrent and meet at point C, and therefore Lami's theorem is applicable. That gives

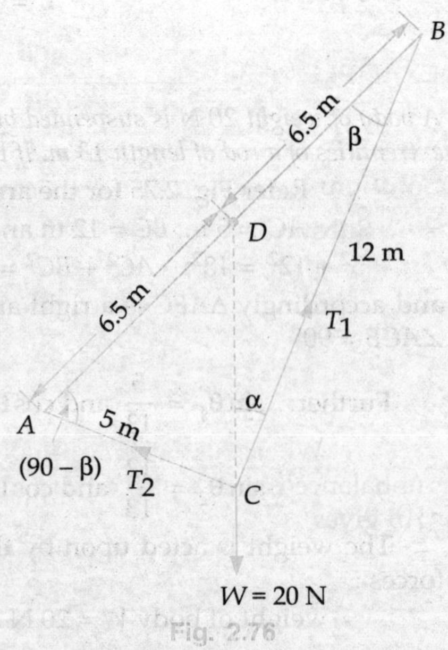

Fig. 2.76

$$\frac{W}{\sin 90°} = \frac{T_1}{\sin(90+\alpha)} = \frac{T_2}{\sin(180-\alpha)}$$

or
$$\frac{20}{1} = \frac{T_1}{\cos\alpha} = \frac{T_2}{\sin\alpha}$$

Since $\alpha = \beta$, we can write

$$20 = \frac{T_1}{\cos\beta} = \frac{T_2}{\sin\beta}$$

From the given geometry $\cos\beta = \dfrac{12}{13}$ and $\sin\beta = \dfrac{5}{13}$. Therefore

$$T_1 = 20\cos\beta = 20\times\frac{12}{13} = \mathbf{18.46\ N}$$

and
$$T_2 = 20\sin\beta = 20\times\frac{5}{13} = \mathbf{7.7\ N}$$

EXAMPLE 2.55

A small ball of weight W is held by two wires as shown in Fig. 2.77 a. Estimate the tension induced in the wire AB (i) before the wire BC is cut, and (ii) after the wire BC is cut.

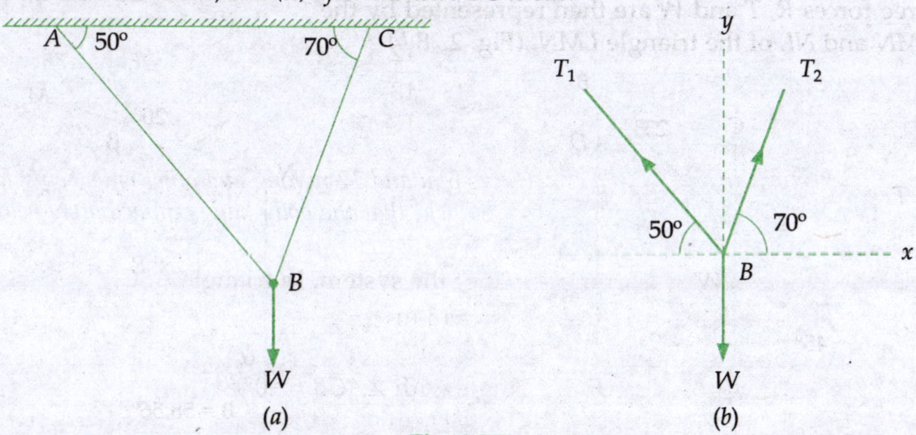

Fig. 2.77

Solution: Before the wire CB is cut, the ball is in static equilibrium.

Let T_1 and T_2 be the tensions in wires BA and BC respectively. Since the system is in equilibrium,

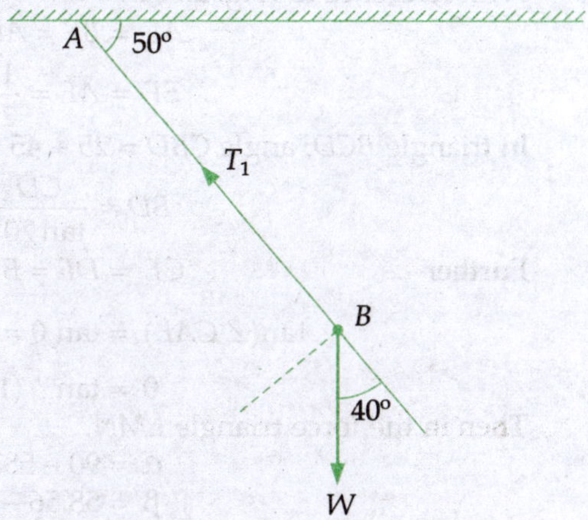

$\Sigma F_x = 0;\quad T_2\cos 70° - T_1\cos 50° = 0$

$$T_2 = \frac{T_1\cos 50°}{\cos 70°} = \frac{0.643T_1}{0.342} = 1.88\,T_1$$

$\Sigma F_y = 0;\quad T_2\sin 70° + T_1\sin 50° - W = 0$

or $\quad 0.94\,T_2 + 0.766\,T_1 - W = 0$

Substituting for $T_2 = 1.88T$, we get

$0.94\times 1.88\,T_1 + 0.766\,T_1 = W$

$\therefore$ Tension in wire AB, $T_1 = \dfrac{W}{0.94\times 1.88 + 0.766}$

$= \mathbf{0.395\ W}$

(*ii*) When the string BC is cut, the ball is no longer in static equilibrium. It is acted upon by an unbalanced force that acts normal to AB and accelerates the ball. Resolution of forces along the wire AB gives

$T_1 - W\cos 40° = 0$

or $\quad T_1 = W\cos 40° = \mathbf{0.766\ W}$

EXAMPLE 2.56

A joist of length 4 m and weighing 200 N is raised by pulling a rope as shown in Fig. 2.78. Determi... tension T induced in the rope and reaction at end A of the joist

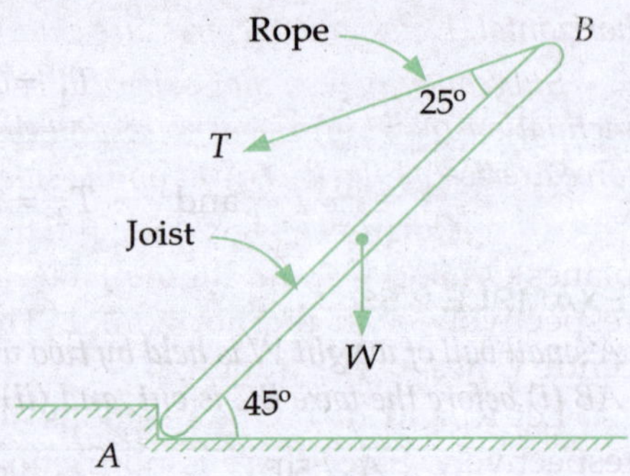

Fig. 2.78

Solution : The three forces acting in the system are:

(i) weight W of the joist

(ii) force T exerted by the rope

(iii) reaction R of the ground at A

Under equilibrium conditions, the reaction R passes through the point of intersection of W and T. (Fig. 2.78 *a*)

The three forces R, T and W are then represented by the sides LM, MN and NL of the triangle LMN. (Fig. 2.78 *b*)

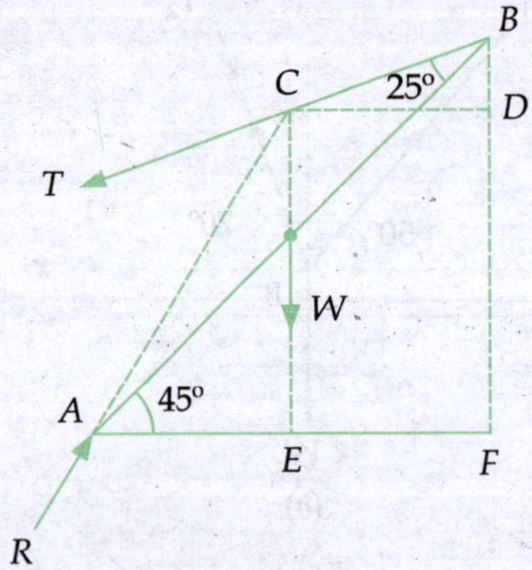

Fig. 2.78 a

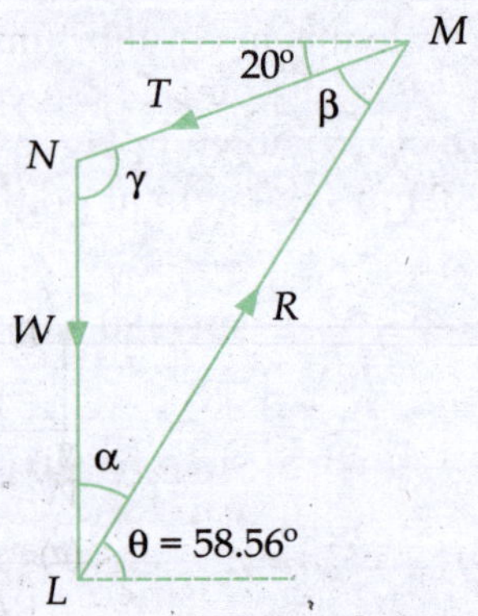

Fig. 2.78 b

With reference to (Fig. 2.78 *a*)

$$AF = BF = AB\cos 45° = 4 \times 0.707 = 2.828 \text{ m}$$

$$EF = AE = \frac{1}{2} \times AF = 1.414 \text{ m}$$

In triangle BCD, angle $CBD = 25 + 45 = 70°$

$$BD = \frac{CD}{\tan 70°} = \frac{EF}{\tan 70°} = \frac{1.414}{2.447} = 0.515 \text{ m}$$

Further $$CE = DF = BF - BD = 2.828 - 0.515 = 2.313 \text{ m}$$

$$\tan(\angle CAE) = \tan\theta = \frac{CE}{AE} = \frac{2.313}{1.414} = 1.6358$$

$$\theta = \tan^{-1}(1.6358) = 58.56°$$

Then in the force triangle LMN,

$$\alpha = 90 - 58.56° = 31.44°$$

$$\beta = 58.56 - 20 = 38.56°$$

$$\gamma = 180 - 31.44 - 38.56 = 110°$$

From the law of sines

$$\frac{T}{\sin\alpha} = \frac{W}{\sin\beta} = \frac{R}{\sin\gamma}$$

$$\therefore \quad T = W\frac{\sin\alpha}{\sin\beta} = 200\frac{\sin 31.44°}{\sin 38.56°} = \mathbf{167.45\ N}$$

$$R = W\frac{\sin\gamma}{\sin\beta} = 200\frac{\sin 110°}{\sin 38.56°} = \mathbf{301.67\ N}$$

EXAMPLE 2.57

A heavy spherical ball of weight W rests in a V-shaped trough whose sides are inclined at α and β to the horizontal. Determine the pressure exterted on each side. Neglect friction.

Subsequently a similar spherical ball is placed on the side of inclination α and it is made to rest on the first ball. Work out the force exerted by the lower ball on the side inclined at β.

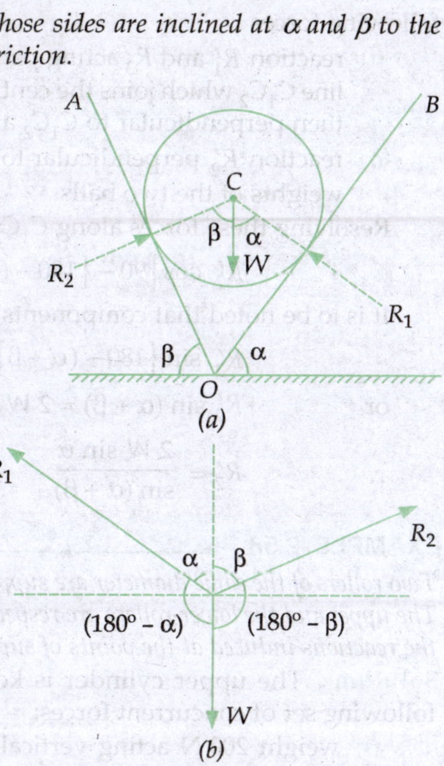

Fig. 2.79

Solution : Reference Fig. 2.79 *a*, the trough is formed by two planes *OA* and *OB* and these planes are inclined at α and β respectively with the horizontal. The spherical ball with centre *C* lies in this trough.

Let R_1 and R_2 be the reactions at surfaces *OA* and *OB* respectively. Since there is no friction, these reactions will be normal to the respective surfaces.

Since the system is in equilibrium condition, the forces R_1, R_2 and *W* meet at point *C*, the centre of the ball. The system of forces is as shown in Fig. 2.79 *b*.

Applying Lami's theorem at point *C*,

$$\frac{R_1}{\sin(180-\beta)} = \frac{R_2}{\sin(180-\alpha)} = \frac{W}{\sin\{180-(\alpha+\beta)\}}$$

or

$$\frac{R_1}{\sin\beta} = \frac{R_2}{\sin\alpha} = \frac{W}{\sin(\alpha+\beta)}$$

$$\therefore \qquad R_1 = W\frac{\sin\beta}{\sin(\alpha+\beta)}$$

and

$$R_2 = W\frac{\sin\alpha}{\sin(\alpha+\beta)}$$

(*b*) Refer Fig. 2.80 which shows the two balls placed in the trough.

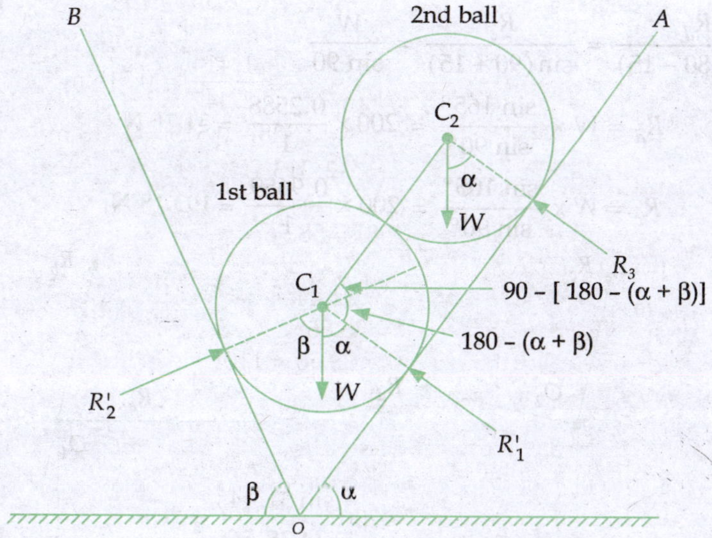

Fig. 2.80

The two balls may be assumed to form a single body which is in equilibrium under the following forces :

(*i*) reaction R_1' and R_3 acting perpendicular to surface *OA*. Since the balls are of equal size, the line C_1C_2 which joins the centres of the two balls is parallel to *OA*. These reactions would be then perpendicular to C_1C_2 also.

(*ii*) reaction R_2' perpendicular to surface OB.

(*iii*) weights of the two balls.

Resolving these forces along C_1C_2, we get

$$R_2' \cos\left[90 - \{180 - (\alpha + \beta\}\right] - W \sin\alpha - W \sin\alpha = 0$$

It is to be noted that components of R_1' and R_3 along C_1C_2 are zero.

$$R_2' \sin\{180 - (\alpha + \beta\} = 2W\sin\alpha$$

or $$R_2' \sin(\alpha + \beta) = 2W \sin\alpha$$

$\therefore$ $$R_2' = \frac{2W \sin\alpha}{\sin(\alpha + \beta)}$$

EXAMPLE 2.58

Two rollers of the same diameter are supported by an inclined plane and a vertical wall as shown in Fig. 2.81. The upper and the lower rollers are respectively 200 N and 250 N in weight. Assuming smooth surfaces, find the reactions induced at the points of support A, B, C and D.

Solution : The upper cylinder is kept in equilibrium by the following set of concurrent forces:

(*i*) weight 200 N acting vertically downward through its centre O_2

(*ii*) reaction R_c acting perpendicular to the inclined plane

(*iii*) pressure R_d from the lower cylinder in the direction O_1O_2

Since these three forces are concurrent, the Lami's theorem applies. That gives

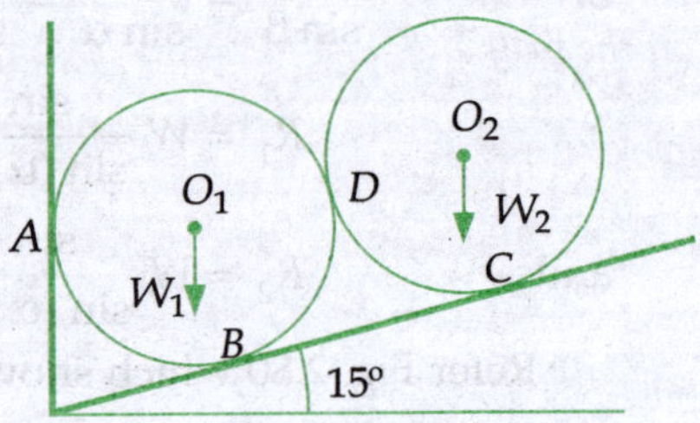

Fig. 2.81

$$\frac{R_d}{\sin(180 - 15)} = \frac{R_c}{\sin(90 + 15)} = \frac{W}{\sin 90}$$

$$\therefore \quad R_d = W \times \frac{\sin 165^\circ}{\sin 90^\circ} = 200 \times \frac{0.2588}{1} = \textbf{51.76 N}$$

$$R_c = W \times \frac{\sin 105^\circ}{\sin 90^\circ} = 200 \times \frac{0.9659}{1} = \textbf{193.18 N}$$

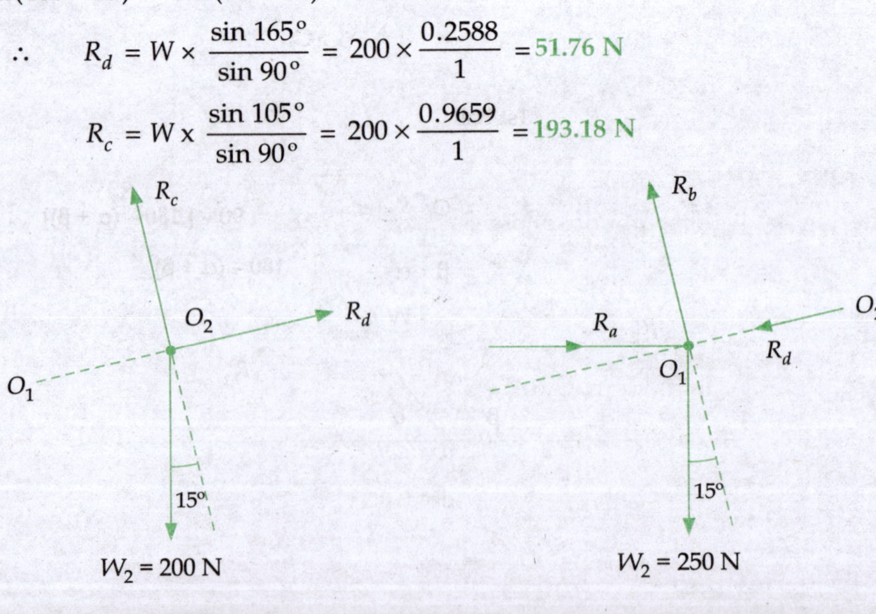

The following set of forces keep the lower cylinder in equilibrium :

(i) weight 250 N acting vertically downward through its centre O_1

(ii) reaction R_a acting perpendicular to the vertical wall

(iii) reaction R_b acting perpendicular to the inclined plane

(iv) pressure R_d from the upper cylinder in the direction O_2O_1

Resolving the forces along O_1O_2, we get

$$R_a \cos 15^\circ - W_1 \sin 15^\circ - R_d = 0$$

$$R_a \times 0.966 - 250 \times 0.259 - 51.76 = 0$$

$$\therefore \quad R_a = \frac{(250 \times 0.259) + 51.76}{0.966} = \frac{116.51}{0.966} = 120.61 \text{ N}$$

Resolving the forces perpendicular to O_1O_2, we get

$$R_b - R_a \sin 15^\circ - W_1 \cos 15^\circ = 0$$

$$R_b = R_a \sin 15^\circ + W_1 \cos 15^\circ$$

$$= (120.61 \times 0.2588) + (250 \times 0.9659)$$

$$= 31.21 + 241.47 = 272.68 \text{ N}$$

EXAMPLE 2.59

Two smooth spheres P, Q each of radius 25 cm and weighing 500 N, rest in a horizontal channel having vertical walls (Fig. 2.82). If the distance between the walls is 90 cm, make calculations for the pressure exerted on the wall and floor at points of contact A, B and C.

Solution : The following points need consideration

(i) the spheres are smooth and as such the pressures at various points of contact would be normal to the surface.

(ii) at the point of contact between the two spheres, the reactions would act along the line joining their centres.

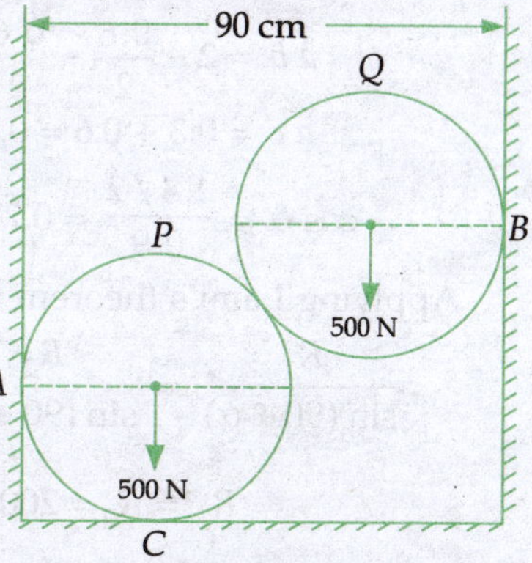

Fig. 2.82

With reference to the adjoining figure the line $C_1 C_2$ makes an angle α with the horizontal line passing through centre C_1 of sphere P.

$$\cos \alpha = \frac{b - r - r}{2r}$$

$$= \frac{90 - 25 - 25}{50} = \frac{40}{50}$$

$$\therefore \quad \alpha = \cos^{-1}(0.8) = 36.87^\circ$$

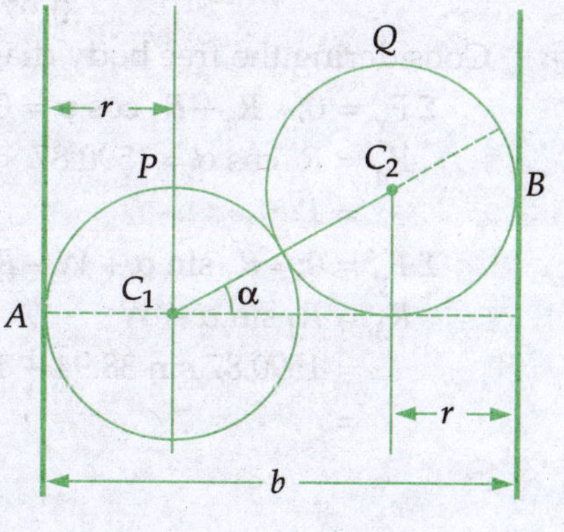

Considering the equilibrium of sphere Q

$$\Sigma F_x = 0; \quad R_b - R \cos \alpha = 0$$

$$\Sigma F_y = 0; \quad R \sin \alpha - 500 = 0$$

$$\therefore \quad R = \frac{500}{\sin \alpha} = \frac{500}{\sin 36.87} = 833.33 \text{ N}$$

$$R_b = R \cos \alpha = 833.33 \times \cos 36.87 = 666.66 \text{ N}$$

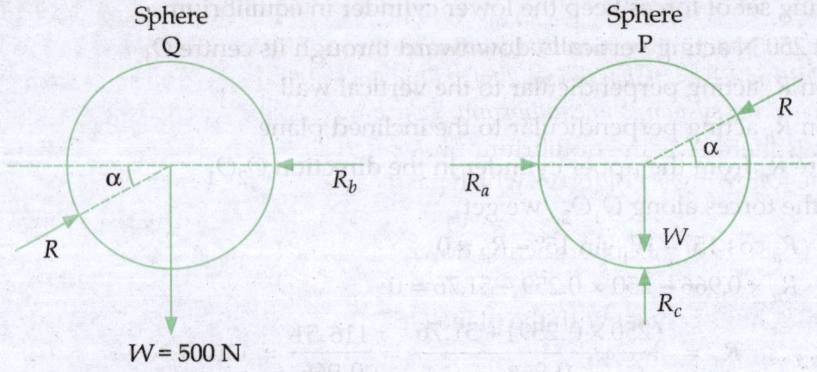

Considering equilibrium of sphere P,

$$\Sigma F_x = 0; \quad R_a - R\cos\alpha = 0$$

$$R_a = R\cos\alpha = 833.33 \times \cos 36.87 = \mathbf{666.67\ N}$$

$$\Sigma F_y = 0; \quad R_c - W - R\sin\alpha = 0$$

$$R_c = W + R\sin\alpha = 500 + 833.33 \times \sin 36.87 = \mathbf{1000\ N}$$

EXAMPLE 2.60

Refer to the system of cylinders arranged as depicted in Fig. 2.83. The cylinders A and B weigh 1000 N each and the weight of cylinder C is 2000 N. Determine the forces exerted at the contact points.

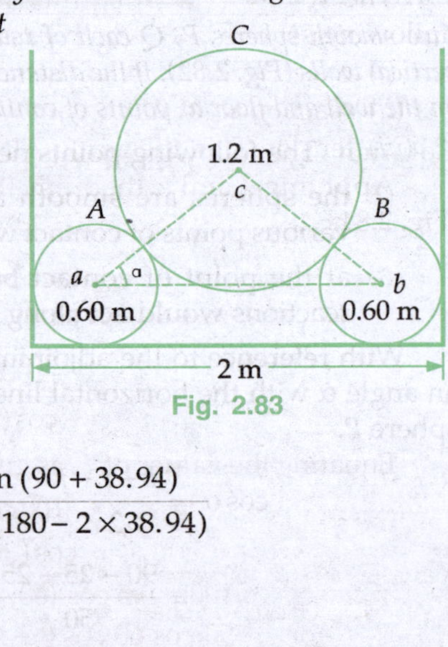

Fig. 2.83

Solution : a, b and c are centres of spheres.

$$ab = 2 - \frac{0.6}{2} - \frac{0.6}{2} = 1.4 \text{ m}$$

$$ac = 0.3 + 0.6 = 0.9 \text{ m}$$

$$\cos\alpha = \frac{1.4/2}{0.9} = 0.7777; \quad \alpha = 38.94°$$

Applying Lami's theorem to the forces acting on sphere C,

$$\frac{R_1}{\sin(90+\alpha)} = \frac{R_2}{\sin(90+\alpha)} = \frac{2000}{\sin(180-2\alpha)}$$

$$R_1 = R_2 = 2000 \times \frac{\sin(90+\alpha)}{\sin(180-2\alpha)} = 2000 \times \frac{\sin(90+38.94)}{\sin(180-2\times 38.94)}$$

$$= 2000 \times \frac{0.7777}{0.9777} = \mathbf{1590.87\ N}$$

Considering the free body diagram of cylinder A,

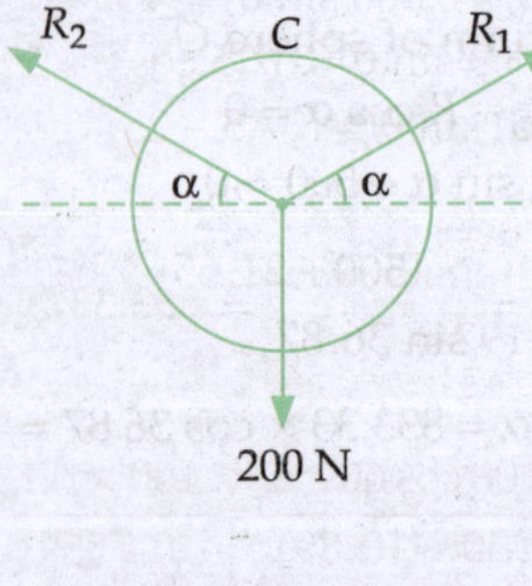

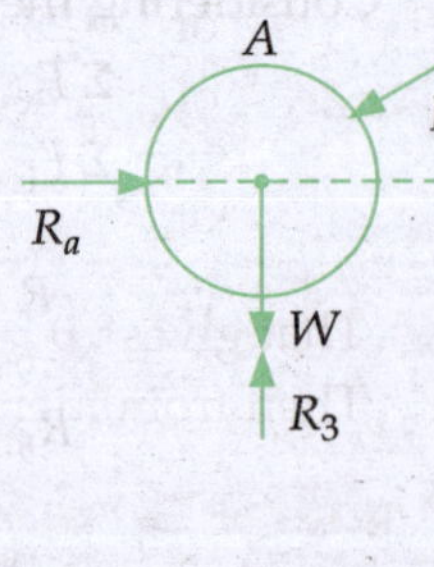

$$\Sigma F_x = 0; \quad R_a - R_1\cos\alpha = 0$$

$$R_a = R_1\cos\alpha = 1590.87 \times \cos 38.94$$

$$= \mathbf{1237.38\ N}$$

$$\Sigma F_y = 0; \quad R_1\sin\alpha + W - R_3 = 0$$

$$R_3 = R_1\sin\alpha + W$$

$$= 1590.87 \sin 38.94 + 1000$$

$$= \mathbf{1999.87\ N}$$

EXAMPLE 2.61

Two spheres of equal radius are connected by a flexible string AB. The spheres rest on two mutually perpendicular planes DE and EF as shown in Fig. 2.84 a. Find the tension in the string and the angle θ it makes with the horizontal when the system is in equilibrium. Take P = 50 N and Q = 100 N.

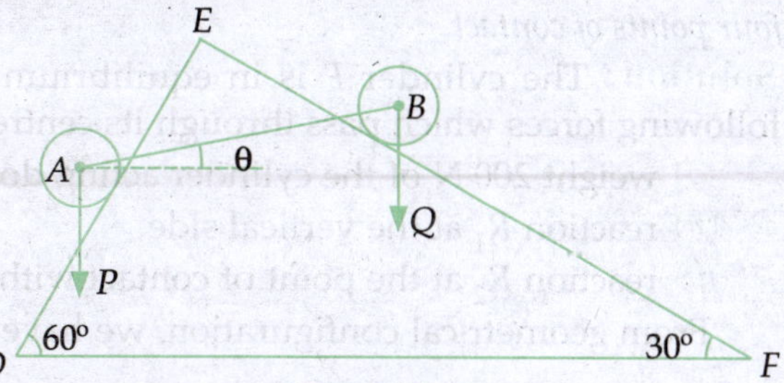

Fig. 2.84 a

Solution : The sphere A is in equilibrium under the action of the following set of forces which pass through its centre.

(*i*) weight P = 50 N of the sphere acting downwards

(*ii*) reaction R_1 at its contact with the plane *DE*

(*iii*) tension T_s in the string *AB*.

Applying Lami's theorem to these forces shown acting in Fig. 2.84 *b*.

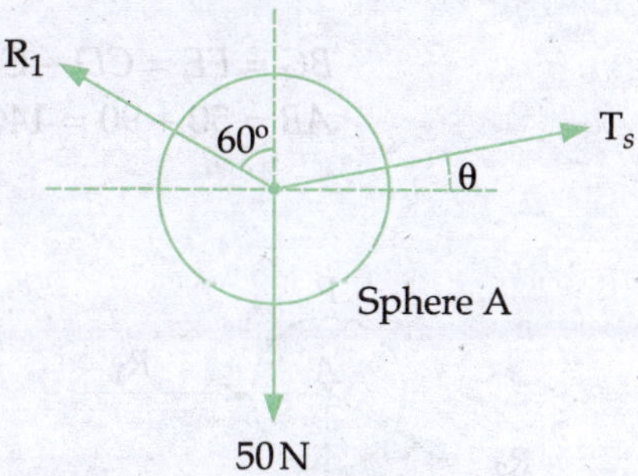

Fig. 2.84 b

$$\frac{T_s}{\sin(90+30)} = \frac{R_1}{\sin(90+\theta)} = \frac{50}{\sin(60+90-\theta)}$$

$$T_s = \frac{-50\cos 30}{-\cos(60-\theta)} = \frac{50\cos 30}{\cos(60-\theta)} \quad ...(i)$$

The sphere *B* is in equilibrium under the action of the following set of forces which pass through its centre.

(*i*) weight Q = 100 N of the sphere acting downwards

(*ii*) reaction R_2 at its contact with plane *EF*

(*iii*) tension T_s in the string *AB*

Applying Lami's theorem to these forces shown acting in Fig. 2.84 *c*

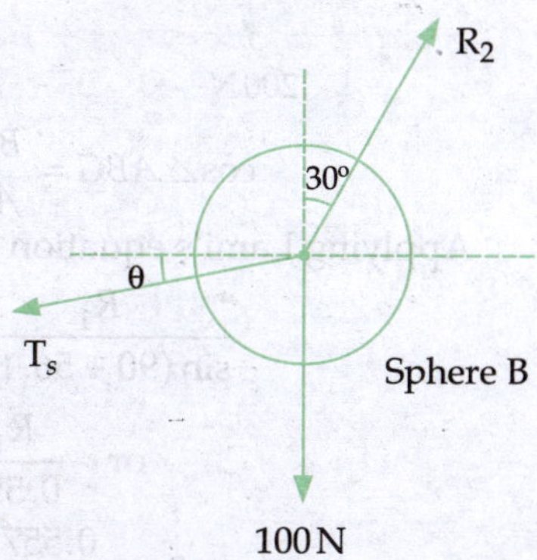

Fig. 2.84 c

$$\frac{T_s}{\sin(90+60)} = \frac{R_2}{\sin(90-\theta)} = \frac{100}{\sin(90+30+\theta)}$$

$$T_s = \frac{-100\cos 60}{-\cos(30+\theta)} = \frac{100\cos 60}{\cos(30+\theta)} \quad ...(ii)$$

Equating the values of T_s as given by identities (*i*) and (*ii*) above

$$\frac{100\cos 60}{\cos(30+\theta)} = \frac{50\cos 30}{\cos(60-\theta)}$$

or $100 \times 0.5 \cos(60-\theta) = 50 \times 0.866 \cos(30+\theta)$

or $50(\cos 60 \cos\theta + \sin 60 \sin\theta) = 43.3 (\cos 30 \cos\theta - \sin 30 \sin\theta)$

or $50 (0.5 \cos\theta + 0.866 \sin\theta) = 43.3 (0.866 \cos\theta - 0.5 \sin\theta)$

or $25 \cos\theta + 43.3 \sin\theta = 37.5 \cos\theta - 21.65 \sin\theta$

or $(43.3 + 21.65) \sin\theta = (37.5 - 25) \cos\theta$

or $\frac{\sin\theta}{\cos\theta} = \tan\theta = \frac{37.5-25}{43.3+21.65} = \frac{12.5}{64.95} = 0.1924$

That gives : $\theta = \tan^{-1}(0.1924) =$ **10.89°**

Then from identity (*ii*)

$$T_s = \frac{100\cos 60}{\cos(30+10.89)} = \frac{100 \times 0.5}{0.7559} = \mathbf{66.15\ N}$$

EXAMPLE 2.62

Two cylinders P and Q rest in a channel as shown in Fig. 2.85. The cylinder P has diameter of 100 mm and weighs 200 N, whereas the cylinder Q has diameter of 180 mm and weighs 500 N. If the bottom width of the channel is 180 mm and with one side vertical and the other inclined at 60°, determine the pressures at all the four points of contact.

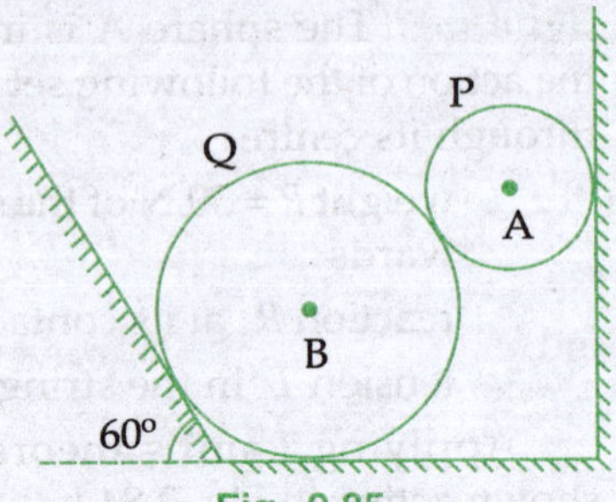

Fig. 2.85

Solution : The cylinder *P* is in equilibrium under the action of following forces which pass through its centre *A*

(*i*) weight 200 N of the cylinder acting downwards

(*ii*) reaction R_1 at the vertical side

(*iii*) reaction R_2 at the point of contact with cylinder *Q*

From geometrical configuration, we have

$$\angle BCF = 60°; CF = BF \cot 60° = \frac{180}{2} \times 0.577 = 52 \text{ mm}$$

$$BG = FE = CD - CF - ED = 180 - 52 - 50 = 78 \text{ mm}$$

$$AB = 50 + 90 = 140 \text{ mm}$$

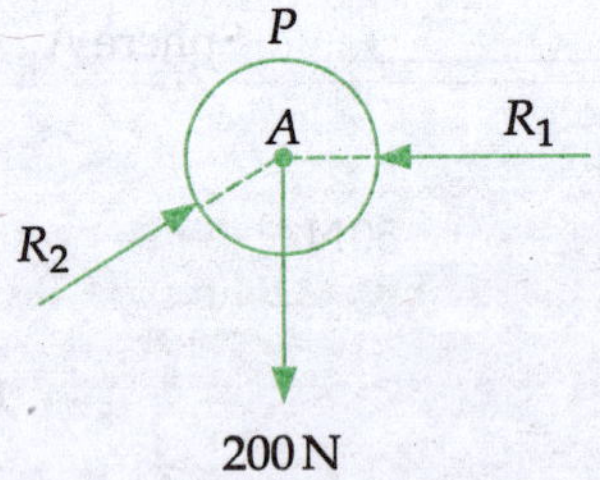

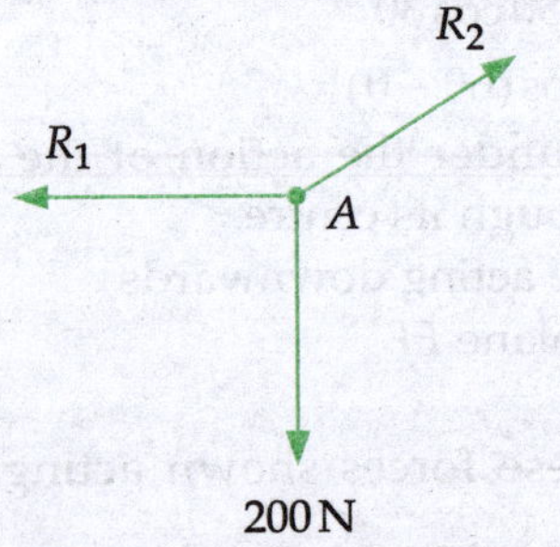

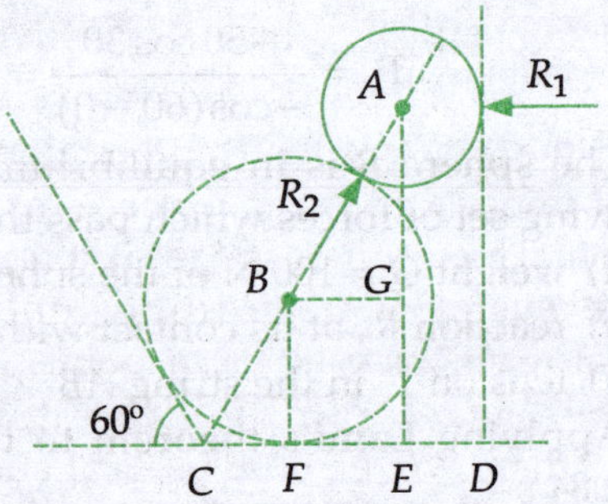

$$\cos\angle ABG = \frac{BG}{AB} = \frac{78}{140} = 0.557; \angle ABG = 56.15°$$

Applying Lami's equation to the system of forces acting at point *A*

$$\frac{R_1}{\sin(90 + 56.15)} = \frac{R_2}{\sin 90°} = \frac{200}{\sin(180 - 56.15)}$$

$$\text{or} \quad \frac{R_1}{0.557} = \frac{R_2}{1} = \frac{200}{0.830}$$

$$\therefore \quad R_1 = \frac{0.557 \times 200}{0.830} = 134.22 \text{ N}$$

$$\text{and} \quad R_2 = \frac{1 \times 200}{0.830} = 240.96 \text{ N}$$

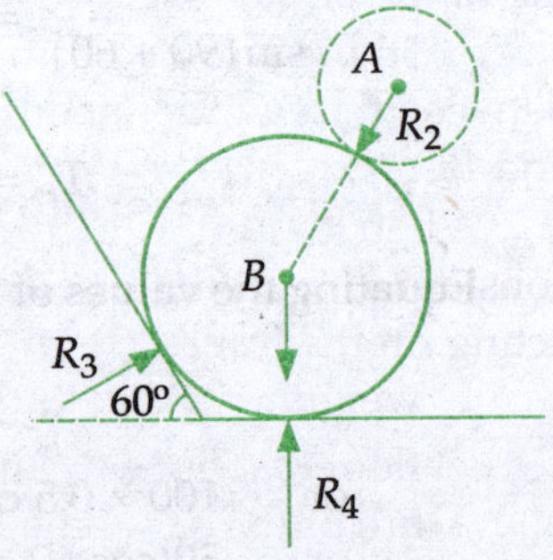

(*a*)

(*b*) The cylinder *Q* is in equilibrium under the action of following forces which pass through its centre *B*.

(*i*) weight 500 N of the cylinder acting downwards

(*ii*) reaction R_2 = 240.96 N of cylinder *P*

(*iii*) reaction R_3 at the inclined surface

(*iv*) reaction R_4 at the base of channel

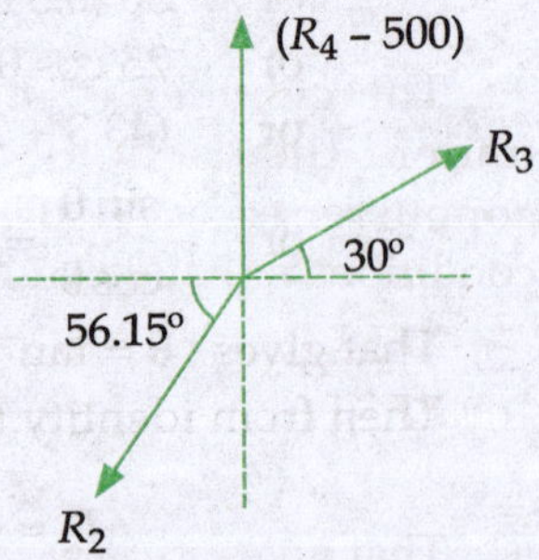

(*b*)

The lines of action of reaction R_4 (acting upwards) and weight of the cylinder (acting downwards) coincide with each other.

$$\therefore \quad \text{Net upward force} = (R_4 - 500)$$

Applying Lami's equation to the system of forces acting at point B

$$\frac{R_2}{\sin 60} = \frac{R_3}{\sin(90 + 56.15)} = \frac{R_4 - 500}{\sin(180 + 30 - 56.15)}$$

or $$\frac{R_2}{0.866} = \frac{R_3}{0.557} = \frac{R_4 - 500}{0.4407}$$

$$\therefore \quad R_3 = \frac{R_2 \times 0.557}{0.866} = \frac{240.96 \times 0.557}{0.866} = \mathbf{154.98\ N}$$

and $$R_4 - 500 = \frac{R_2 \times 0.4407}{0.866} = \frac{240.96 \times 0.4407}{0.866} = 112.62$$

$$\therefore \quad R_4 = 500 + 122.62 = \mathbf{622.62\ N}$$

EXAMPLE 2.63

A steel cylinder 1 m in diameter and weighing 500 N is placed between two cross arms which make an angle of 60° with each other and are pinned at P as shown in Fig. 2.97. A rope attached to the arms near their upper ends prevents them from slipping. Assuming the floor to be smooth, determine the tension in the rope.

Solution : It is to be noted that

(i) the floor is smooth and frictionless. It implies that even a very small horizontal force will result into motion of the system. Accordingly only vertical reactions will act at points A and B.

(ii) there will be both vertical and horizontal components of reactions at the hinge point.

(iii) there will be normal reactions between the cylinder and the arms AD and BC.

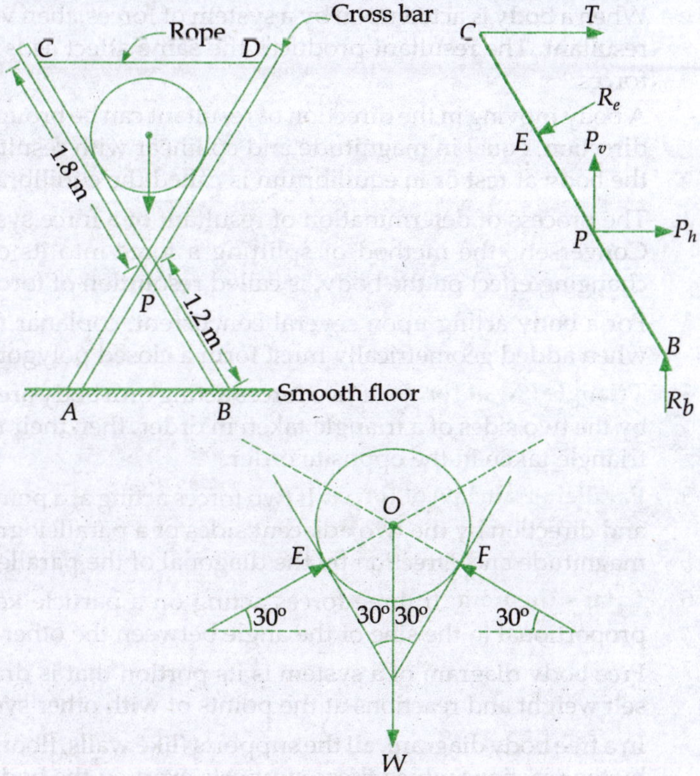

Fig. 2.86

When the entire arrangement is considered as a single body, the forces acting on the frame are :

(i) weight of the cylinder, and

(ii) floor reactions R_a and R_b which are of the same magnitude.

$$\therefore \quad R_a = R_b = \frac{500}{2} = 250\ \text{N}$$

The cylinder is in equilibrium under the action of coplanar concurrent forces. Then from the conditions of equilibrium:

$\Sigma F_x = 0$: $\quad R_e \cos 30° - R_f \cos 30°$; $\quad R_e = R_f$

$\Sigma F_y = 0$: $\quad R_e \sin 30° + R_f \sin 30° - 500 = 0$

or $\quad 0.5\, R_e + 0.5\, R_e - 500 = 0 \qquad (\because\ R_e = R_f)$

That gives: $R_e = R_f = 500$ N

Alternatively: The three forces R_e, R_f and W meet at a point and as such the Lami's theorem is applicable. Therefore

$$\frac{R_e}{\sin 120^\circ} = \frac{R_f}{\sin 120^\circ} = \frac{W}{\sin 120^\circ}$$

$$\therefore \quad R_e = R_f = W = 500 \text{ N}$$

Considering the equilibrium of arm BC and taking moments about the hinge point P, we get

$\Sigma M_p = 0$: $\quad T \times 1.8 \cos 30^\circ - R_e \times PE - R_b \times 1.2 \sin 30^\circ$

In triangle OEP : $\dfrac{OE}{PE} = \tan 30^\circ$; $\qquad PE = \dfrac{OE}{\tan 30^\circ} = \dfrac{0.5}{0.577} = 0.866$ m

$$\therefore \quad T \times 1.8 \cos 30^\circ = R_e \times 0.866 + R_b \times 1.2 \sin 30^\circ$$

$$1.559\, T = 500 \times 0.866 + 250 \times 1.2 \times 0.5 = 5.83$$

$$\text{Tension in the rope } T = \frac{583}{1.559} = \mathbf{373.95 \text{ N}}$$

REVIEW AND SUMMARY

1. When a body is acted upon by a system of forces, then vectorial sum of all the forces is known as resultant. The resultant produces the same affect as is done by the combined effect of several forces.

   A body moving in the direction of resultant can be brought to rest by applying a force in opposite direction, equal in magnitude and collinear with resultant force. This single force which keeps the body at rest or in equilibrium is called the equilibrant force.
2. The process of determination of resultant of a force system is known as composition of forces. Conversely, the method of splitting a force into its components in two directions, without changing effect on the body, is called resolution of force.
3. For a body acting upon several concurrent, coplanar forces to be in equilibrium, these forces when added geometrically must form a closed polygon.
4. **Triangle law of forces:** If two forces acting on a body are represented in magnitude and direction by the two sides of a triangle taken in order, then their resultant is given by the third side of the triangle taken in the opposite order.
5. **Parallelogram law of forces:** If two forces acting at a point of a body are represented in magnitude and direction by the two adjacent sides of a parallelogram, then their resultant is given both in magnitude and direction by the diagonal of the parallelogram passing through that point.
6. **Lami's theorem:** If three forces acting on a particle keep it in equilibrium, then each force is proportional to the sine of the angle between the other two forces.
7. Free body diagram of a system is its portion that is drawn separately showing applied forces, self weight and reactions at the points of with other systems.

   In a free body diagram, all the supports (like walls, floors, hinges, etc.) are removed and replaced by the reactions which these supports exert on the body.
8. Internal forces are those forces which hold the particles of the body together and help it to be rigid. These forces cause internal stresses and strains distributed throughout the material of the body.

   External forces represent those forces which act on a body or a system of bodies externally, *i.e.*, applied from outside. These forces essentially denote the action of other bodies (walls, floors, hinges, etc.) on the rigid body being analysed.

REVIEW QUESTIONS

A. Conceptual and conventional questions :

1. Define the term resultant of a force system.
2. Differentiate between resolution and composition of forces.
3. Write the statement for triangle law of forces and parallelogram law of forces.
4. State and prove: (*a*) parallelogram law of forces, (*b*) theorem of resolution parts, and (*c*) Lami's theorem.
5. Three coplanar forces acting at a point are stated to be in equilibrium. Show that each force is proportional to the sine of the angle between the other two forces.
6. Two forces act at a point. State the range of the values of their resultant.
7. Is it possible for a force to have a component larger in magnitude than the force itself?
8. Comment on the validity of following statements:
   (*i*) Rectangular components of a force P are forces acting at right angles to the force P.
   (*ii*) Lines of action of any two components of a force intersect on the line of action of the force.
   (*iii*) There are many possible solutions to the determination of the components of a given force.
9. Outline the steps involved in the analytical method for the determination of resultant of coplanar force system.
10. What is a force polygon? How a force polygon is drawn to determine the magnitude of the resultant of a concurrent force system?
11. Define the term free body diagram and state the importance of drawing such a diagram.
12. Show that if a body is in equilibrium under the action of three forces (concurrent), those forces will always lie along the sides of a closed triangle.
13. The resultant of two equal forces acting at a point also equals to P. Determine the angle between the two forces.
14. A force of 200 N is resolved into two components. If one of the components is equal to 120 N and makes an angle of 30° with the 200 N force, find the other component and the angle between the components.
15. Refer Fig. 2.87 for a system of concurrent coplanar forces. If the resultant is zero, find the magnitude and direction of force P.
16. A 1 kN force has been resolved into components along AB and AC directions in the x-y plane specified by the angles α and β as shown in Fig. 2.88. If the component along AC is 2 kN and the component along AB is 1.6 N, determine the angles α and β.

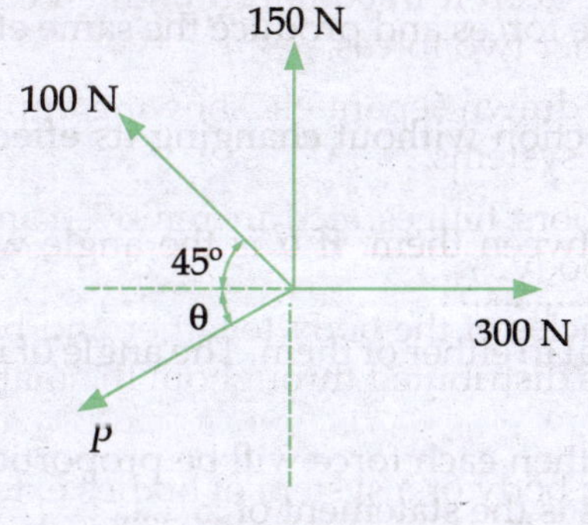

Fig. 2.87

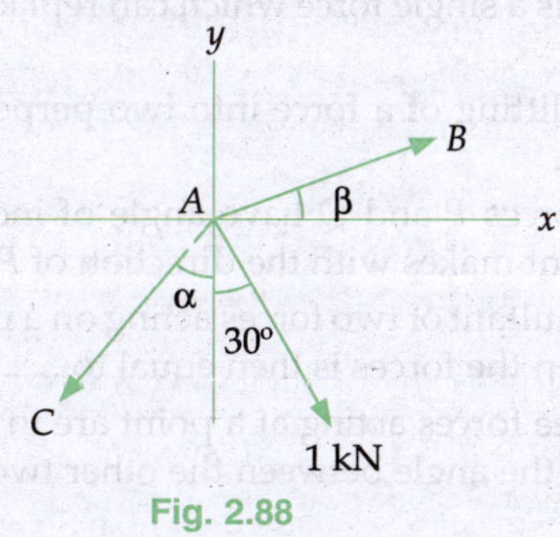

Fig. 2.88

17. A body is free to slide on a smooth vertical circular wire and is connected by a string, equal in length to the radius of the circle, to the highest point of the circle. Find the tension developed in the string, and the reaction of the circle.

(Each is equal to weight of the body)

18. A particle is acted upon by the force system shown in Fig. 2.89 Determine the magnitude and direction of the resultant.

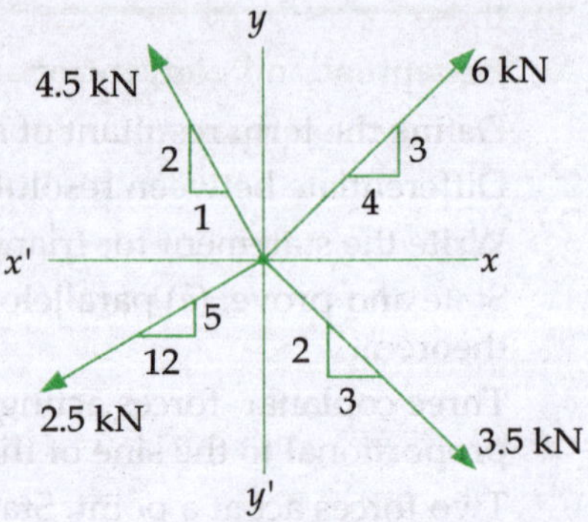

Fig. 2.89

19. Three forces 2 *P*, 3 *P* and 4 *P* act along three sides of an equilateral triangle taken in order. Find the magnitude and line of action of the resultant force.

(1.732 *P*, 210°)

20. A small block of weight 100 N is placed on an inclined plane which makes an angle θ = 30° with the horizontal. What is the component of this weight : (*i*) parallel to the inclined plane, and (*ii*) perpendicular to the inclined plane.

(8.66 N, 50 N)

21. A lamp weighing 5 N is suspended from the ceiling by a chain. It is pulled aside by a horizontal chord until the chain makes an angle of 60° with the ceiling. Determine the tensions in the chain and the chord.

(5.77 N, 2.89 N)

22. A weight *W* is supported by the two strings at right angles to one another and attached to two points in the same horizontal line. Prove that the tensions induced in the strings are inversely proportional to their lengths.

23. The top of an electric pole is connected with a stay wire which makes an angle of 60° with the horizontal. If a horizontal force of 100 N is essential for keeping the pole vertical, determine tension induced in the stay wire. Also workout vertical component of this tension and state the purpose served by it.

(200 N, 173 N, resistance against lifting)

24. A body of 50 N weight is suspended by two strings 5 m and 12 m long. The other ends of the strings are fastened to the extremities of a rod which is 13 m long. The rod is so held that the body hangs vertically below its mid point. Workout the tensions induced in the strings for the given arrangement.

(46.15 N, 19.23 N)

B. Complete the following statements with most appropriate word/words :

1. Forces whose lines of action pass through a common point are called ...... forces
2. A .... is a single force which can replace two or more forces and produce the same effect as the forces.
3. The splitting of a force into two perpendicular direction without changing its effect is called ....
4. Two forces *P* and *Q* have angle of inclination θ between them. If α is the angle which their resultant makes with the direction of *P*, then tan α = ...
5. The resultant of two forces acting on a particle is equal to either of them. The angle of inclination between the forces is then equal to ....
6. "If three forces acting at a point are in equilibrium, then each force will be proportional to the sine of the angle between the other two forces". This is the statement of ....

7. If a system of coplanar concurrent forces is in equilibrium then the vector diagram drawn with these forces must be a .... figure.
8. Equations of equilibrium of a particle subjected to coplanar force system are ....
9. A body isolated from all other members which are connected to it is called the .... body.
10. .... is equal and opposite to the resultant of several forces acting on the body.

Answers :

1. concurrent 2. resultant 3. resolution of forces 4. $\dfrac{Q\sin\theta}{P+Q\cos\theta}$ 5. 120° 6. Lami's theorem 7. closed 8. $\Sigma F_x = 0$ and $\Sigma F_y = 0$ 9. free body 10. equilibrant

C. Multiple choice questions :

1. Resultant of two equal forces is equal to either of them. The angle between the forces is
   (*a*) 0° (*b*) 60° (*c*) 90° (*d*) 120°
2. Four forces $P, 2P, 3P$ and $4P$ act along the sides, taken in order, of a square. The resultant force is
   (*a*) zero (*b*) $12\sqrt{2}\,P$ (*c*) $2P$ (*d*) $\sqrt{5}\,P$
3. Given that $\vec{P}+\vec{Q}=\vec{R}$ and also $P = Q = R$. Then the angle between forces $\vec{p}$ and $\vec{Q}$ are
   (*a*) 0 (*b*) $\dfrac{\pi}{3}$ (*c*) $\dfrac{2\pi}{3}$ (*d*) π
4. The sum of two forces acting at a point is 16 N. If the resultant force is 8 N and its direction is perpendicular to the minimum force, then the forces are :
   (*a*) 6 N and 10 N (*b*) 8 N and 8 N
   (*c*) 4 N and 12 N (*d*) 2 N and 14 N
5. A system of three forces acts on a body and keeps it in equilibrium. The forces need to be
   (*a*) coplanar only (*b*) concurrent only
   (*c*) coplanar as well as concurrent (*d*) coplanat but may or may not be concurrent
6. Two forces can be in equilibrium if they are
   1. equal in magnitude 2. opposite in direction 3. collinear in action
   Which of these are the essential conditions for equilibrium?
   (*a*) 1 and 3 (*b*) 2 and 3 (*c*) 1 and 3 (*d*) 1, 2 and 3
7. A number of forces acting at a point will lie in equilibrium if
   (*a*) sum of the forces is zero
   (*b*) sum of the resolved parts is zero
   (*c*) algebric sum of the forces is zero
   (*d*) sum of the resolved parts in any two mutually perependicular direction is zero
8. A circular roller of weight W hangs by a tie rod and rests against a smooth vertical wall as shown in Fig. 2.90. The tension in the tie rod AB will be
   (*a*) less than weight W
   (*b*) greater then W
   (*c*) equal to W
   (*d*) data insufficient for making force analysis

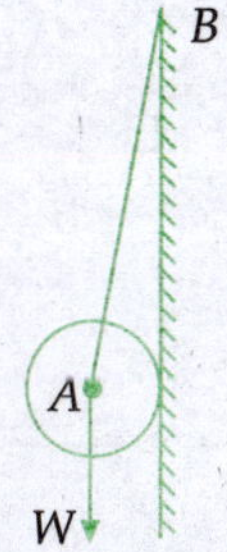

Fig. 2.90

9. The free body diagram of a body shows the body
   (*a*) with its surroundings and external forces acting on it
   (*b*) isolated from all external forces
   (*c*) isolated from all surroundings
   (*d*) isolated from its surroundings and all external effects acting on it

10. The truss shown in Fig. 2.91 supports a load of 100 N
    The load on member AB is
    (*a*) 50 N (*b*) 57.7 N
    (*c*) 86.6 N (*d*) 100 N

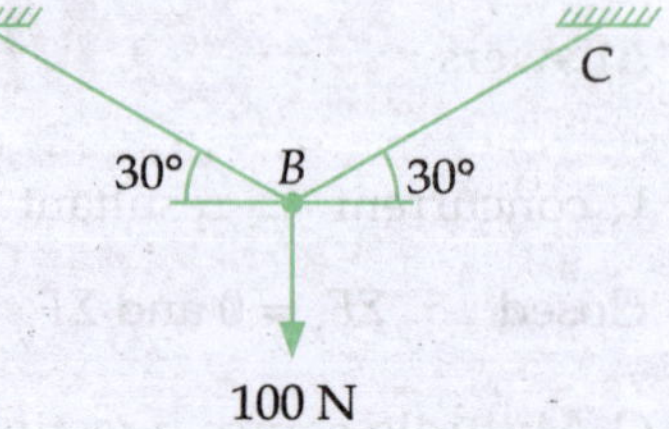

Fig. 2.91

11. A weight of 500 N is supported by two ropes as shown in Fig. 2.92
    The values of tensions T_1 and T_2 are respectively
    (*a*) 433 N and 250 N
    (*b*) 250 N and 433 N
    (*c*) 353.5 N and 250 N
    (*d*) 250 N and 353.5 N

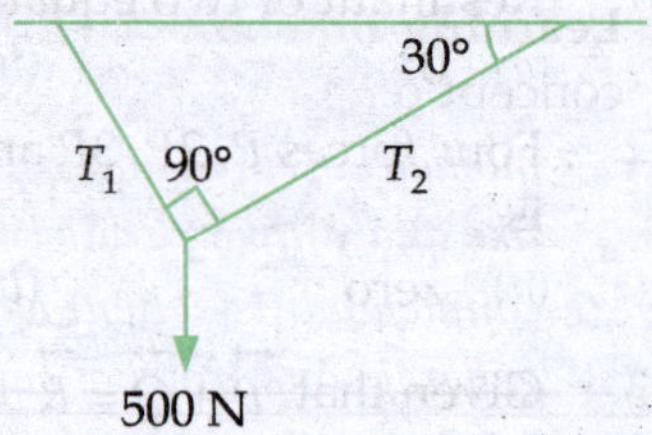

Fig. 2.92

12. The point *A* is in equilibrium under the action of an applied force of 600 N as shown in Fig. 2.93. The values of tensions T_{AB} and T_{AC} are respectively
    (*a*) 520 N and 300 N
    (*b*) 300 N and 520 N
    (*c*) 450 N and 150 N
    (*d*) 150 N and 450 N

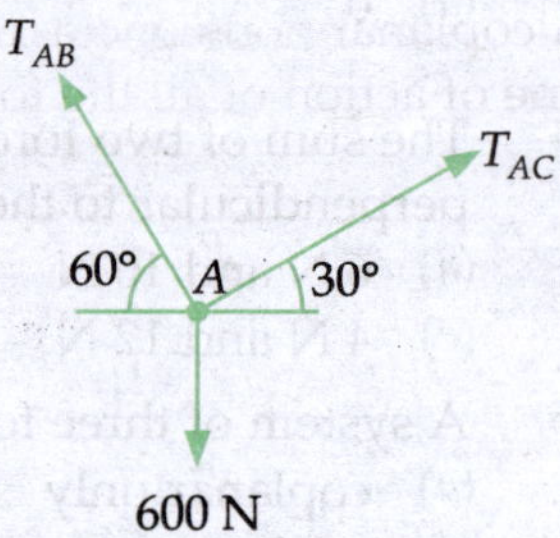

Fig. 2.93

Answers :

1. (*d*)	2. (*b*)	3. (*c*)	4. (*a*)	5. (*c*)	6. (*d*)
7. (*d*)	8. (*b*)	9. (*d*)	10. (*d*)	11. (*b*)	12. (*a*)

Moment of Force and Parallel Forces 3

Learning objectives : After having made a study of this chapter, the reader will develop a concept of

- *moment of force and principle of moments; Varignon's theorem*
- *like and unlike parallel forces and their resultant*
- *couple and its salient aspects ; force-couple system*
- *general conditions of equilibrium*

A coplanar non-concurrent force system consists of a set of forces that lie in the same plane but the line of action of all the forces do not meet at a single point.

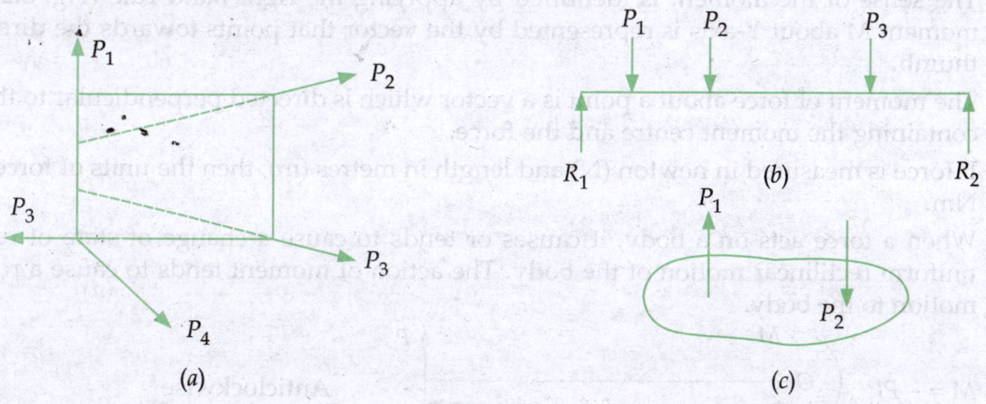

Fig. 3.1

Attention has been focussed in this chapter to study the effect of the resultant of these forces at a point which is away from the lines of action of the system of forces. Reference has also been made to the parallel forces and the effect they produce on a body. The general conditions for the equilibrium of a body have also been stated.

3.1. MOMENT OF A FORCE

Moment of force about a point is defined as the turning or rotational effect of a force about that point. It is measured by the product of force and the perpendicular distance of the line of action of the force from that point.

With reference to Fig. 3.2,

P = force acting on a body

l = perpendicular distance between the point O and line of action of force P

Then moment of force P about point O

$$= P \times l$$

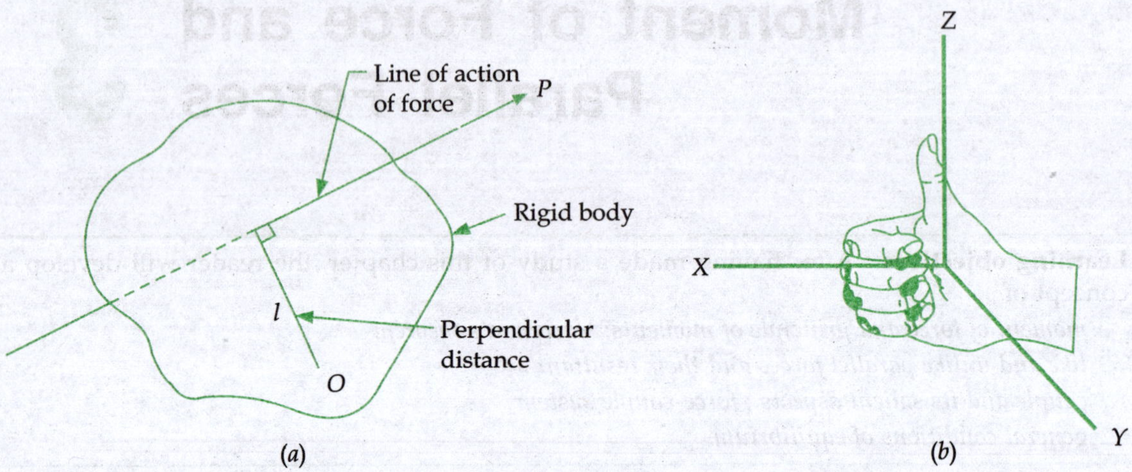

Fig. 3.2. Concept of moment of force and direction of a moment

The point O is called the *moment centre* and the distance l is the *moment arm* or *arm of the force*.

- The sense of the moment is identified by applying the right hand rule (Fig. 3.2 *b*). The moment M about Y-axis is represented by the vector that points towards the direction of thumb.
- The moment of force about a point is a vector which is directed perpendicular to the plane containing the moment centre and the force.
- If force is measured in newton (N) and length in metres (m), then the units of force will be Nm.
- When a force acts on a body, it causes or tends to cause a change of state of rest or of uniform rectilinear motion of the body. The action of moment tends to cause a rotational motion to the body.

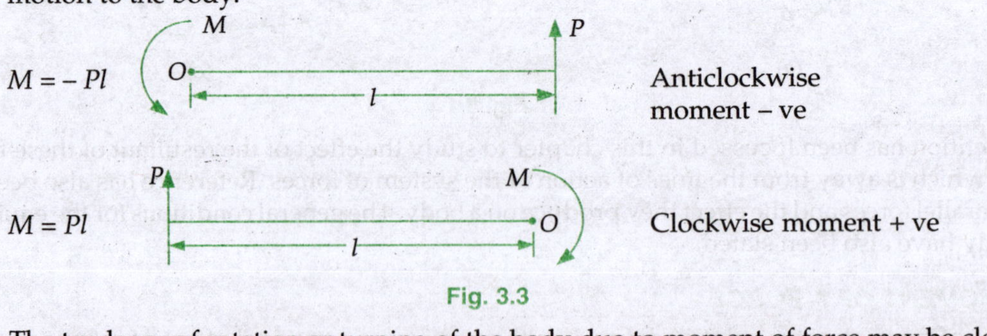

Fig. 3.3

- The tendency of rotation or turning of the body due to moment of force may be clockwise (in the same direction in which hands of the clock move) or anticlockwise (in a direction opposite to the movement of hands of a clock). The corresponding moments are referred to as clockwise moment and anticlockwise moment.

The general convention is to take the clockwise moments as positive and anticlockwise moments as negative (Fig. 3.3).

- If a number of moments act on a body (Fig. 3.4), then the resultant moment will be the algebraic sum of these moments, and the effect of resultant moment on the body will be governed by its turning tendency.

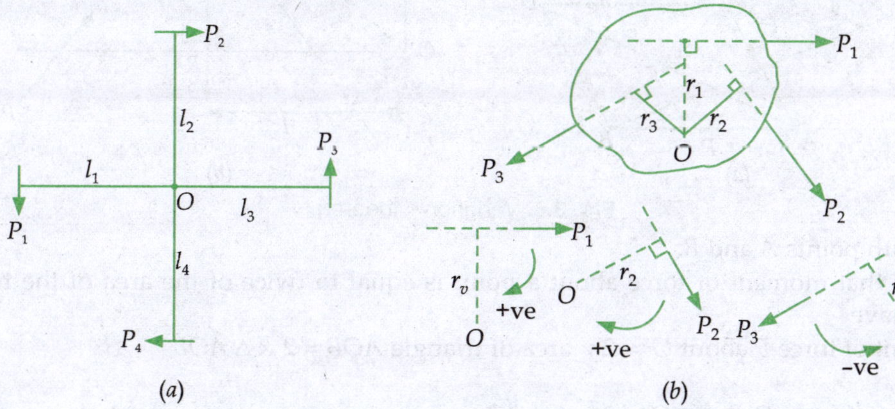

Fig. 3.4. Number of moments acting on a body

With reference to (Fig. 3.4 *a*), the resultant moment *M* about the point *O* is given by

$$M = -P_1l_1 + P_2l_2 - P_3l_3 + P_4l_4$$

Likewise, with reference to (Fig. 3.4 *b*), the resultant moment of forces P_1, P_2 and P_3 about point *O* is

$$M = r_1P_1 + r_2P_2 - r_3P_3$$

3.2. GRAPHICAL REPRESENTATION OF MOMENT

Let a force *P* be represented in magnitude and direction by the vector *AB*. Further, let *O* be the point about which the moment is to be determined.

Moment of force *P* about point *O*,

$$= P \times OM$$

where *OM* is the perpendicular dropped from point *O* on line *AB*

$$= AB \times OM = 2\left[\frac{1}{2} \times AB \times OM\right]$$

$$= 2 \text{ (area of triangle } OAB).$$

Hence the moment of force about any point is geometrically equal to twice the area of the triangle whose base is the line that represents the force and whose vertex is the point about which the moment is required to be found out.

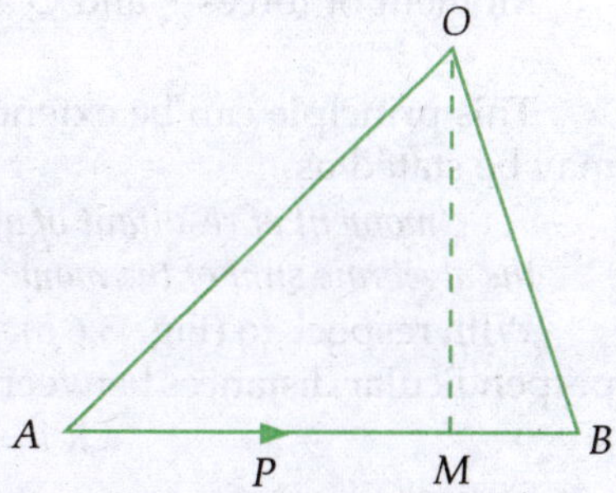

Fig. 3.5. Representation of a moment

3.3. VARIGNON'S THEOREM : LAW OF MOMENTS

"Moment of a resultant of two forces, about a point lying in the plane of the forces, is equal to the algebraic sum of moments of these two forces about the same point."

Consider two concurrent forces *P* and *Q* represented in magnitude and direction by *AB* and *AC* respectively. Let *O* be the point about which moment is to be taken. Through *O*, draw a line parallel to the direction of force *P* and let this line meet the line of action of force *Q* at point C. With *AB* and *AC* as adjacent sides, complete the parallelogram *ABCD*. The diagram *AD* of this parallelogram represents in magnitude and direction the resultant of forces *P* and *Q*.

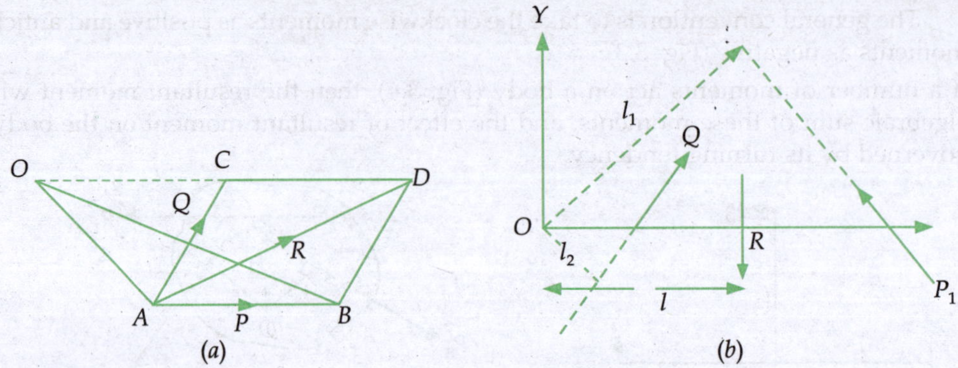

Fig. 3.6. Varignon's theorem

Join O with points A and B.

Recalling that moment of force about a point is equal to twice of the area of the triangle so formed, we have

Moment of force P about O $= 2 \times$ area of triangle $AOB = 2 \times \Delta\, AOB$ ...(i)

Likewise,

moment of force Q about O $= 2 \times \Delta\, AOC$ ...(ii)

moment of force R about O $= 2 \times \Delta\, AOD$

From geometrical configuration of Fig. 3.6,

$$\Delta\, AOD = \Delta\, AOC + \Delta\, ACD = \Delta\, AOC + \Delta\, ABD$$

Further, the Δ's AOB and ABD are on the same base AB and between the same lines and as such are equal in area. Then

$$\Delta\, AOD = \Delta\, AOC + \Delta\, AOB$$

The moment of force R about O may then be re-written as

$$= 2 \times (\Delta\, AOC + \Delta\, AOB) \quad \text{...(iii)}$$

From the identities (i), (ii) and (iii), it is evident that

Moment of forces P and Q about point O

= moment of resultant R about O

This principle can be extended for any number of forces and the generalised law of moments may be stated as

"moment of resultant of a number of forces about a point lying in the plane of forces is equal to the algebraic sum of the moments of these forces about the some point."

With respect to (Fig. 3.6 b), R is the resultant of forces P and Q; l and l_1 and l_2 are the respective perpendicular distances between R, P and Q from any point O. Then according to Varignon's theorem

$$R\,l = P l_1 + Q l_2$$

EXAMPLE 3.1

A force of 200 N is acting at a point B as shown in the adjoining figure. Determine the moment of this force about O.

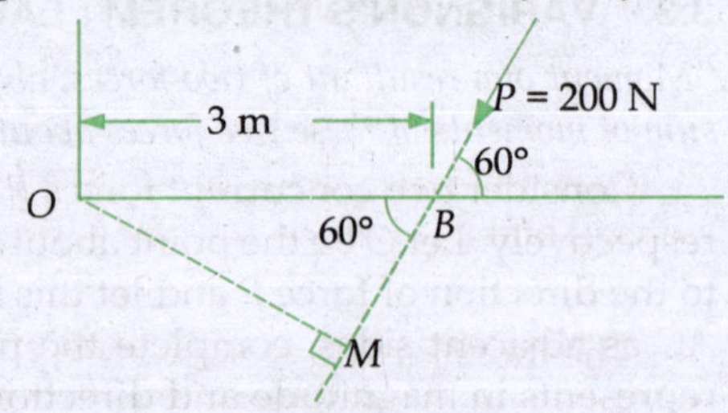

Fig. 3.7

Solution : Moment of force about point O

= force × (perpendicular distance between point O and the line of action of force)

$= P \times OM = P \times OB \cos 60°$

$= 200 \times (3 \times 0.866)$

= **519.6 Nm** (clockwise)

EXAMPLE 3.2

A force of 200 N is acting at point B of a lever AB which is hinged at its lower end as shown in Fig. 3.8. Find the moment of force about the hinged end.

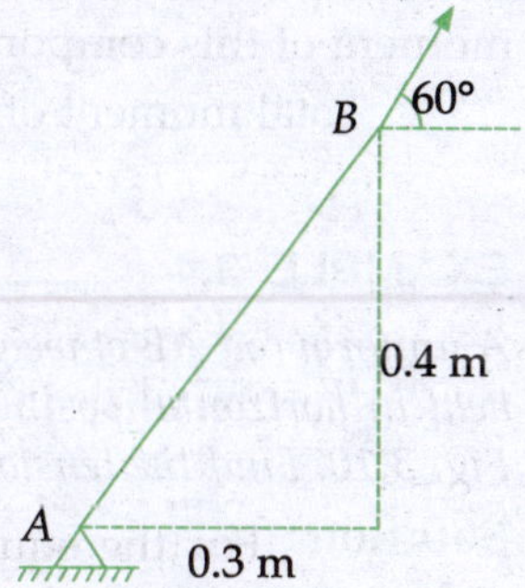

Fig. 3.8

Solution : Moment of force about point A

= force × (perpendicular distance between point A and the line of action of force)

$= P \times AM$

Now, $AB = \sqrt{0.4^2 + 0.3^2} = 0.5$ m

$$\alpha = \tan^{-1}\left(\frac{0.4}{0.3}\right) = 53.13^\circ$$

$$\gamma = 90 - 53.13 = 36.87^\circ$$

$$\beta = 36.87 - 30 = 6.87^\circ$$

$$AM = AB \sin 6.87^\circ = 0.5 \times 0.1196 = 0.0598 \text{ M}$$

$$\therefore \quad M_a = 200 \times 0.0598 = \textbf{11.96 Nm (anticlockwise)}$$

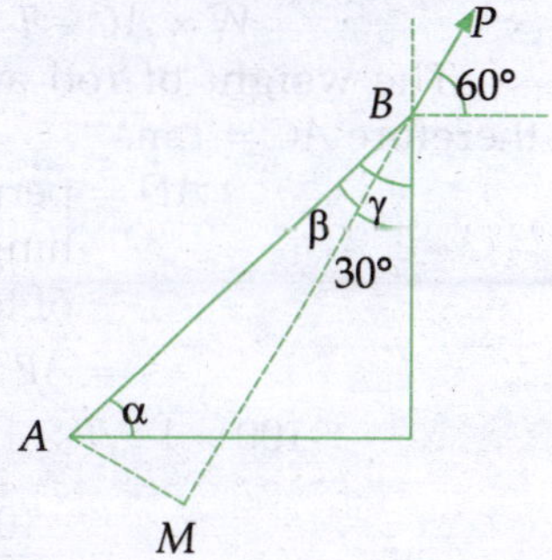

Alternatively :

The horizontal and vertical components of the given force are:

$$P_x = 200 \cos 60^\circ = 100 \text{ N}$$

$$P_y = 200 \sin 60^\circ = 173.2 \text{ N}$$

From Varignon's theorem, moment of a given force about a specified point is equal to the sum of the moments of its components about that point.

$$\therefore \quad M_a = 100 \times 0.4 - 173.2 \times 0.3$$

$$= 40 - 51.96 = \textbf{– 11.96 Nm (anticlockwise)}$$

EXAMPLE 3.3

A force of 600 N acts on a bracket as shown in Fig. 3.9. Determine the moment of force about points B and C.

Solution : The force of 600 N acting on the bracket can be resolved into horizontal (x-direction) and vertical (y-direction) components.

$F_x = 600 \cos 60° = 300$ N and the perpendicular distance of its line of action from point B is 12 cm. The corresponding moment is

$$= 300 \times 0.12 \; 36 \text{ Nm (clockwise)}$$

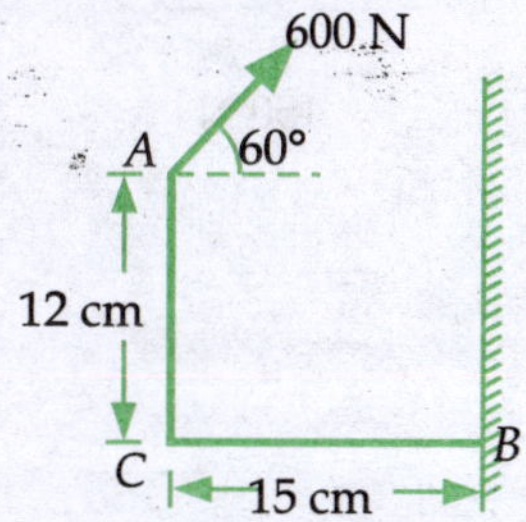

Fig. 3.9

$F_y = 600 \sin 60° = 519.6$ N and the perpendicular distance of its line of action from point B is 15 cm. The corresponding moment is

$$= 519.6 \times 0.15 = 77.94 \text{ Nm (clockwise).}$$

∴ Total moment of the applied force about point B is

$$M_B = 36 + 77.94 = \textbf{113.94 Nm}$$

The moment is a vector perpendicular to the plane of the figure and pointing into the paper.

(*b*) Moment of F_x component of force about point C,

$$= 300 \times 0.12 = 36 \text{ Nm}$$

The point C lies on the line of action of vertical component of the force, and accordingly the moment of this component about point C is zero.

∴ Total moment of the applied force about point is

$$M_c = 36 \text{ Nm}$$

EXAMPLE 3.4

A uniform rod AB of weight 100 N is hinged at A so as to be able to rotate about A in a vertical plane. It is held in horizontal position by a string attached to B and inclined at 60° to the horizontal as shown in Fig. 3.10. Find the tension induced in the string.

Solution : For the equilibrium of the system, the algebraic sum of the moments of all forces about a point lying in the same plane in zero. Then taking moments about hinge point A and equating it to zero, we get

$$W \times AC - T \times AD = 0$$

The weight of rod acts at its mid point C and therefore $AC = 1$ m.

AD = perpendicular distance between hinge point A and line of action of tension T

$= AB \sin 60° = 2 \times 0.866 = 1.732$ m

$$\therefore \quad 100 \times 1 - T \times 1.732 = 0$$

$$T = \frac{100}{1.732} = \mathbf{57.74\ N}$$

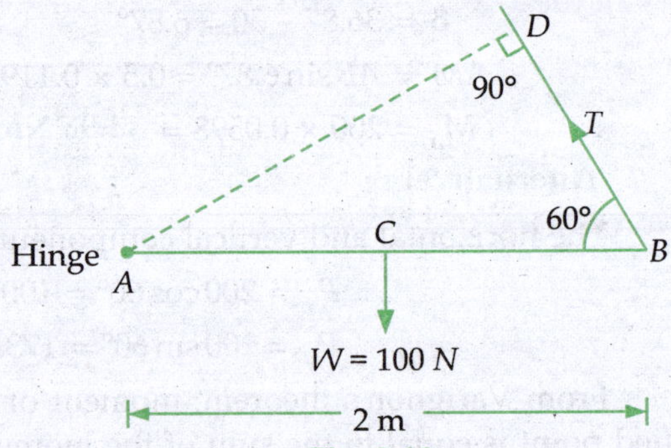

Fig. 3.10

EXAMPLE 3.5

A rigid bar ABCD of a particular geometry is acted upon by a set of three forces as shown in Fig. 3.11. Replace this force system to a force and a moment acting at the fixed end A.

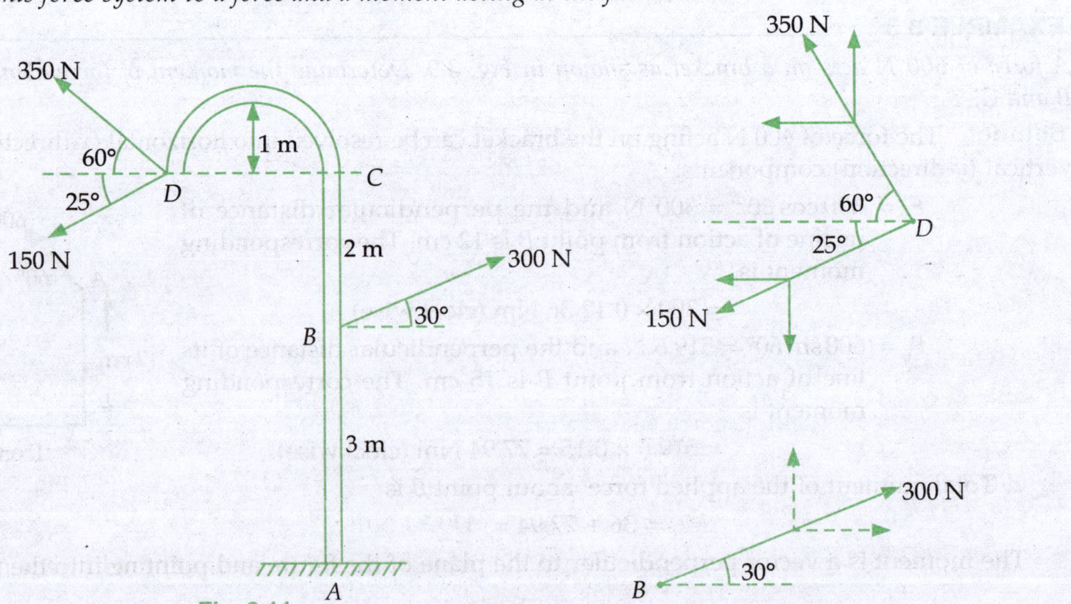

Fig. 3.11

Solution :

Resolving the given system of forces horizontally, we get

$$\Sigma F_x = 300\cos 30° - 350\cos 60° - 150\cos 25°$$
$$= 259.81 - 175 - 135.95 = -51.14 \text{ N}$$

Resolving the given system of forces vertically, we get

$$\Sigma F_y = 300\sin 30° + 350\sin 60° - 150\sin 25°$$
$$= 150 + 303.11 - 63.39 = 389.72 \text{ N}$$

Magnitude of resultant of the given force system,

$$R = \sqrt{(\Sigma F_x)^2 + (\Sigma F_y)^2} = \sqrt{(-51.14)^2 + (389.72)^2}$$
$$= \sqrt{2615.3 + 151881.68} = \mathbf{393.06} \text{ N}$$

Since ΣF_x is –ve and ΣF_y is +ve, the resultant lies in the second quadrant and its inclination with the x-axis is

$$\alpha = \tan^{-1}\left(\frac{\Sigma F_y}{\Sigma F_x}\right) = \tan^{-1}\left(\frac{389.72}{51.14}\right) = 82.52°$$

The resultant moment at the fixed end A can be determined by taking moments of the given forces about this point. That gives :

$$\Sigma M_A = 300\cos 30° \times 3 + 350\sin 60° \times 2 - 350\cos 60° \times 5$$
$$- 150\cos 25° \times 5 - 150\sin 25° \times 2$$
$$= 779.42 + 606.22 - 875 - 679.37 - 126.78$$
$$= \mathbf{-295.51 \text{ Nm}} \quad \text{(anti-clockwise)}$$

EXAMPLE 3.6

A uniform rod of 10 m length has a self weight of 5 N. The rod carries a weight of 30 N hung from one of its ends. From what point the rod be suspended so that it remains horizontal ?

Solution : The self weight of the rod 5 N acts at the mid-point C of the rod.

Let the rod be suspended at point F which is at distance x from its end B where a load of 30 N is hung.

Taking moments about point F, we get

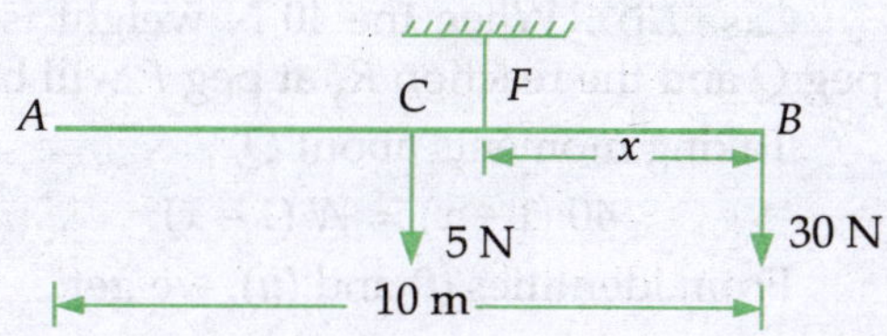

Fig. 3.12

$$5 \times CF - 30x = 0$$
$$5(BC - x) = 30x$$
or $$5(5 - x) = 30x$$
or $$25 - 5x = 30x$$
$$\therefore \quad x = \frac{25}{35} = \mathbf{0.7143 \text{ m}}$$

Hence for the rod to remain horizontal, it needs to be suspended from point F such that $BF = 0.7143$ m.

EXAMPLE 3.7

A uniform rod ABCD weighs 8 N and is supported at B and C as shown in Fig. 3.13. Determine the minimum force at A which will just overturn the rod.

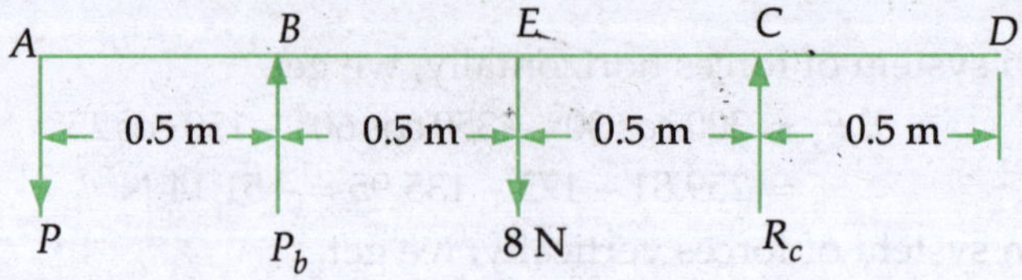

Fig. 3.13

Solution : Let P be the minimum force to be applied at end A which will just overturn the rod. At that instant reaction R_c will be zero.

Taking moments about point B,

$$8 \times 0.5 - P \times 0.5 = 0$$

$$\therefore \quad P = \frac{8 \times 0.5}{0.5} = 8 \text{ N}$$

EXAMPLE 3.8

A heavy uniform rod AB of 4 m length becomes in a limiting position of being tilted when either

(i) 100 N weight is suspended from A, or

(ii) 40 N weight is suspended from B

The rod rests horizontally on two pegs which are set one metre apart on either sides of the middle point of rod. Determine the weight of rod and the distance of the pegs from the mid-point of rod.

Solution : Case (i) : P and Q are the two pegs and let peg P be at distance x from mid-point C. Then the location of peg Q is at distance $(1 - x)$ from the mid-point C.

Let R_1 = reaction at peg P

R_2 = reaction at peg Q

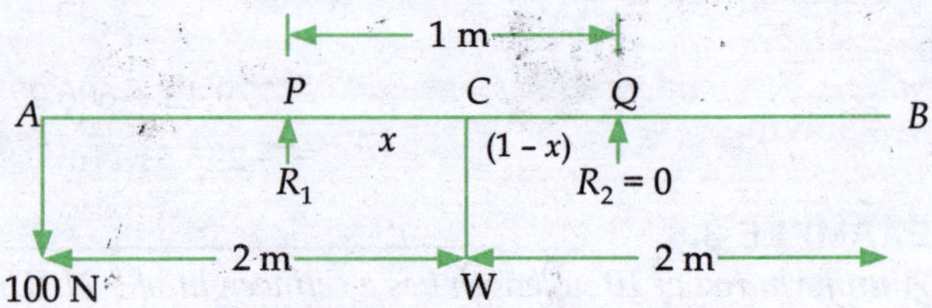

Fig. 3.14 a

The rod will tilt and turn about peg P and the reaction R_2 at peg Q will be zero. Then taking moments about P,

$$100\,(2 - x) = W x \qquad \dots(i)$$

Case (ii) : When the 40 N weight is suspended from end B, the rod will tilt and turn about peg Q and the reaction R_1 at peg P will be zero.

Taking moments about Q,

$$40\,(1 + x) = W\,(1 - x) \qquad \dots(ii)$$

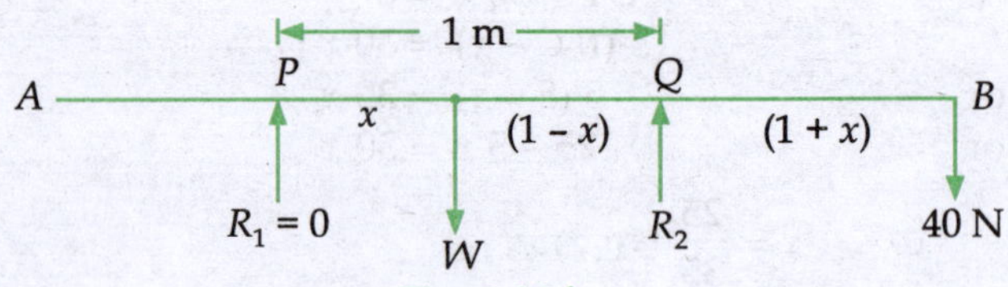

Fig. 3.14 b

From identities (i) and (ii), we get

$$\frac{100}{40}\,\frac{2-x}{1+x} = \frac{x}{1-x}$$

Upon simplification

$$5(2 - x)\,(1 - x) = 2\,x(1 + x) \quad \text{or} \quad 3x^2 - 17x + 10 = 0$$

Solution of this quadratic equation gives :

$$x = \frac{-(-17) \pm \sqrt{(-17)^2 - 4 \times 3 \times 10}}{2 \times 3} = \frac{17 \pm 13}{6} = 5 \text{ or } \frac{2}{3}$$

$$\therefore \quad x = \frac{2}{3} \text{ m as it cannot be 5 m}$$

Then from identity (*i*), the weight of rod is

$$W = \frac{100\left(2-\frac{2}{3}\right)}{\frac{2}{3}} = \mathbf{200\ N}$$

EXAMPLE 3.9

A 30 cm long rod rests on two pegs whose distance apart is 15 cm. Weights of 3 W and 5 W respectively are suspended from its ends. Determine the position of pegs if its reactions are to be equal.

Solution : Let the pegs be so located at points *E* and *F* such that

$AE = x$ cm

and $BF = (15 - x)$ cm

Since the reactions at ends *E* and *F* are equal, we have

$$R_e = R_f = \frac{1}{2}(3\ W + 5\ W) = 4\ W$$

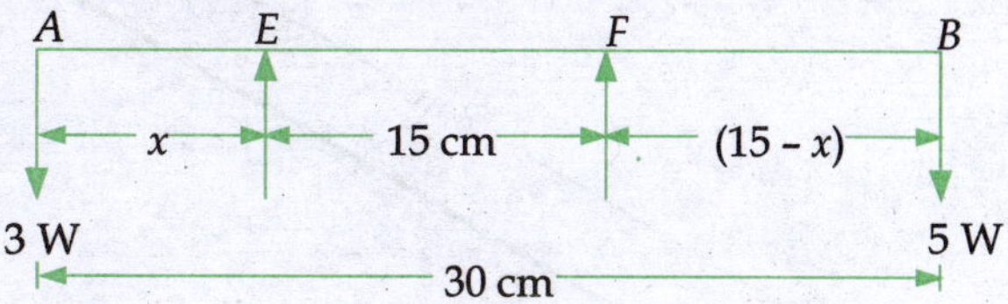

Fig. 3.15

For the rod to be in a state of equilibrium, the algebraic sum of the moments of the forces about *E* must be zero. That gives:

$$-3\ W x - 4\ W \times 15 + 5\ W(30 - x) = 0$$

The clockwise moments have been taken positive and anticlock moments negative

$$-3\ W x - 60\ W + 150\ W - 5\ W x = 0$$

Solving: $x = \frac{90}{8} = \mathbf{11.25\ cm}$

$\therefore$ $AE = 11.25$ cm and $BF = 15 - 11.25 = \mathbf{3.75\ cm}$

EXAMPLE 3.10

The lever of a lever safety valve is 100 cm long, weighs 4 N and this weight acts at its CG which is 12 cm away from the valve. The valve weighs 3 N, is 10 cm in diameter and is located at a distance of 8 cm from the fulcrum. Make calculations for the weight to be suspended at the end of the lever which will just release the steam at a pressure of 1×10^5 N/m².

Solution : Refer Fig. 3.16

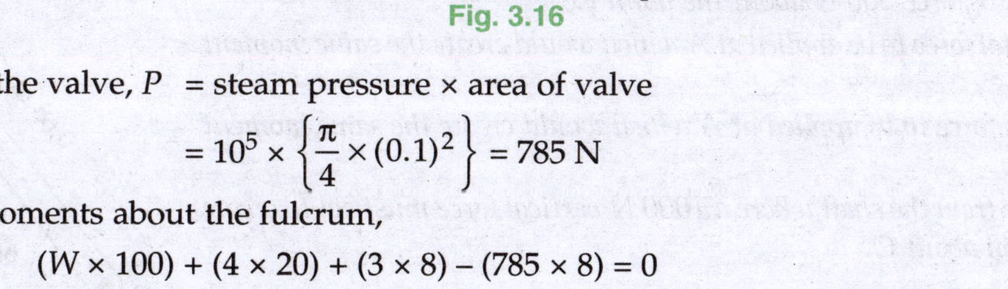

F = fulcrum; *V* = valve; *G* is position of CG

Fig. 3.16

Force on the valve, P = steam pressure × area of valve

$$= 10^5 \times \left\{\frac{\pi}{4} \times (0.1)^2\right\} = 785\ N$$

Taking moments about the fulcrum,

$$(W \times 100) + (4 \times 20) + (3 \times 8) - (785 \times 8) = 0$$

or $\quad 100\,W = (785 \times 8) - (4 \times 20) - (3 \times 8)$

$= 6280 - 80 - 24 = 6176$

$\therefore \quad W = \dfrac{6176}{100} = \mathbf{61.76\ N}$

EXAMPLE 3.11

A heavy uniform bar AB rests in equilibrium with end A in contact with a smooth vertical wall and a point D in contact with another smooth vertical as shown in Fig. 3.17. Determine the inclination of the bar to the horizontal.

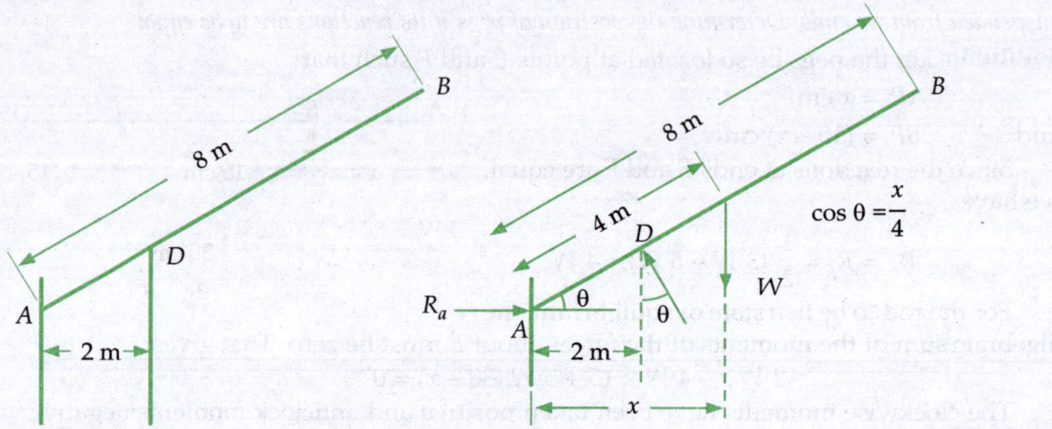

Fig. 3.17

Solution : Let the bar make an angle θ with the horizontal. At D, a reaction R_d acts perpendicular to the bar

Considering equilibrium of the bar,

$$\Sigma F_x = 0; \quad R_a - R_d \sin\theta = 0; \quad R_a = R_d \sin\theta \qquad ...(i)$$

$$\Sigma F_y = 0; \quad R_d \cos\theta - W = 0; \quad W = R_d \cos\theta \qquad ...(ii)$$

$\Sigma M = 0$, Taking moments about point A

$$W\,x = R_d \times AD = \frac{W}{\cos\theta} \times \frac{2}{\cos\theta} = \frac{2\,W}{\cos^2\theta} = \frac{2W \times 16}{x^2}$$

$$\therefore \quad x^3 = 32 \quad \text{and} \quad x = (32)^{1/3} = 3.174$$

$$\cos\theta = \frac{3.174}{4} = 0.7935$$

$$\theta = \cos^{-1}(0.7935) = \mathbf{37^\circ\ 28'}$$

EXAMPLE 3.12.

A vertical force 500 N is applied to the end A of a lever which is attached to a shaft at point O (Fig. 3.18). Make calculations for :

(*a*) *moment of force 500 N about the point O;*

(*b*) *horizontal force to be applied at A which would create the same moment about O;*

(*c*) *smallest force to be applied at A which would create the same moment about O;*

(*d*) *distance from the shaft where a 1000 N vertical force must act to create the same moment about O.*

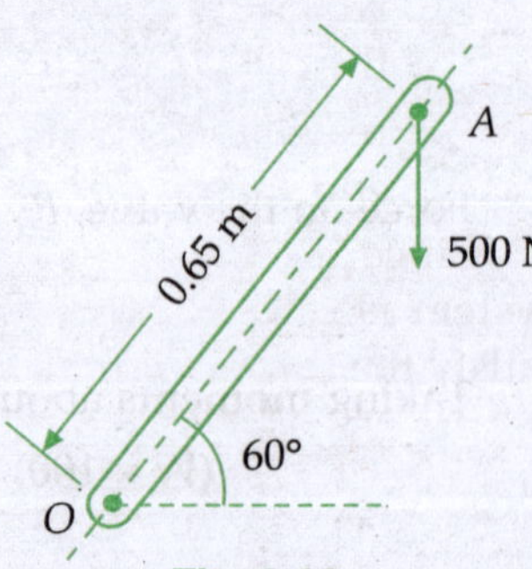

Fig. 3.18

Solution : Recall that moment of force about a point is measured by the product of force and the perpendicular distance of the line of action of the force from that point, and refer to the figures given below :

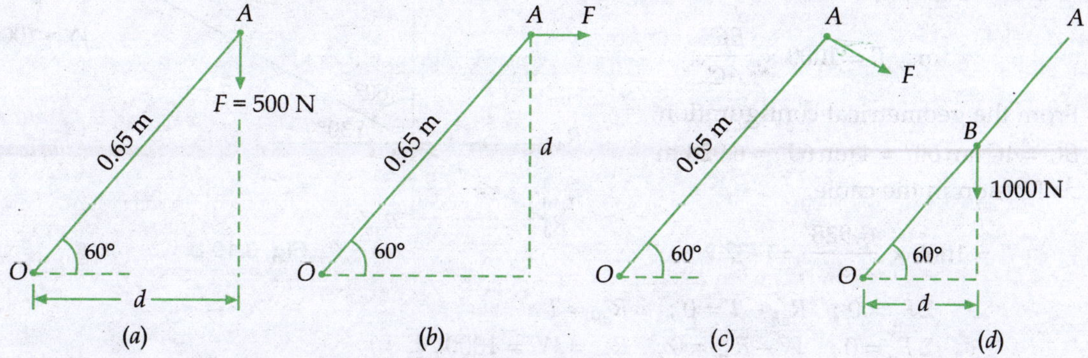

(*a*) Moment of vertical force F = 500 N about O is

$$M_o = F \times d = 500 \times (0.65 \cos 60°) = \mathbf{162.5\ Nm}$$

(*b*) Moment of horizontal force F about O is

$$M_o = F \times d = F \times (0.65 \sin 60°) = 0.563\ F$$

Then according to the given condition,

$$0.563\ F = 162.5 \text{ and that gives : } F = \frac{162.5}{0.563} = \mathbf{287.74\ N}$$

(*c*) The smallest force would occur when d is maximum and maximum value of d is 0.65 m. Accordingly

$$F \times 0.65 = 162.5 \quad \text{or} \quad F = \frac{162.5}{0.65} = \mathbf{250\ N}$$

(*d*) As per the given condition :

$$1000 \times OB \cos 60° = 162.5$$

where OB is the distance from the shaft where 1000 N vertical force has to be applied

$$OB = \frac{162.5}{1000 \times \cos 60°} = \frac{162.5}{1000 \times 0.5} = \mathbf{0.325\ m}$$

EXAMPLE 3.13

The cable and the rigid bar shown in Fig. 3.19 a support a load of 1000 N. Determine the tensions in the cable and the reaction at the hinge A.

Solution : Refer (Fig. 3.19 *b*) for the free body diagram. The various forces acting in the rigid bar are:

(*i*) weight W = 1000 N and the tension in the cable at B.

(*ii*) reactive forces R_{ah} and R_{av} at the hinged end A.

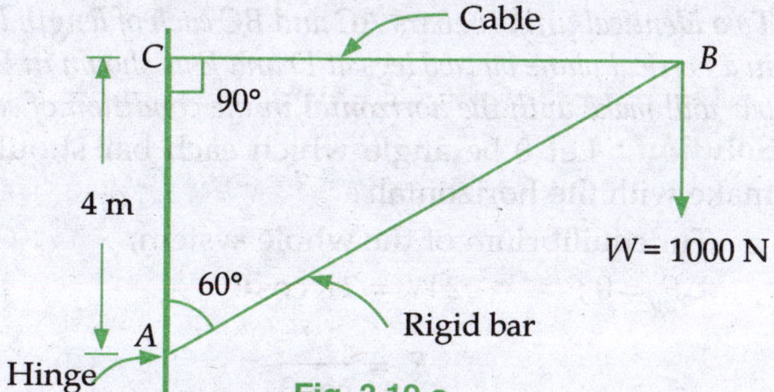

Fig. 3.19 a

Since the bar is in equilibrium, the force system (T, W, R_{ah} and R_{av}) acting on the bar satisfy the three conditions of equilibrium. That is

$$\Sigma M = 0 ; \quad \Sigma F_x = 0 \quad \text{and} \quad \Sigma F_y = 0$$

Taking moment about the hinge A and equating to zero,

$$1000 \times BC - T \times AC = 0$$

$$\text{or} \quad T = 1000 \times \frac{BC}{AC}$$

From the geometrical configuration,

$BC = AC \tan 60° = 4 \tan 60° = 6.928$ m

∴ Tension in the cable,

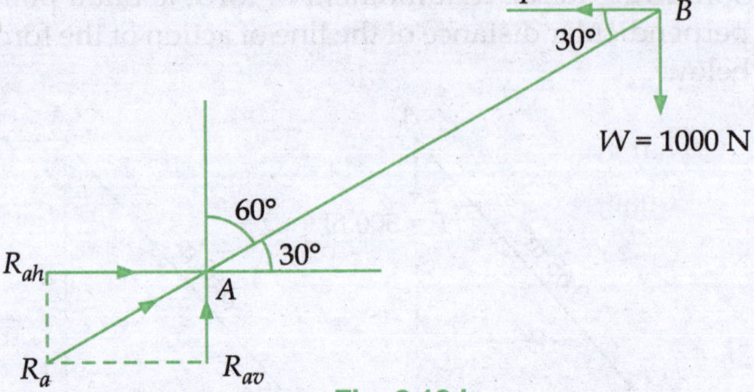

Fig. 3.19 b

$$T = 1000 \times \frac{6.928}{4} = 1732 \text{ N}$$

$$\Sigma F_x = 0; \quad R_{ah} - T = 0; \quad R_{ah} = T$$

$$\Sigma F_y = 0; \quad W - R_{av} = 0; \quad R_{av} = W = 1000 \text{ N}$$

∴ Total reaction at the hinge,

$$R_a = \sqrt{R_{ah}^2 + R_{av}^2} = \sqrt{1732^2 + 1000^2} = 2000 \text{ N}$$

$$\theta = \tan^{-1}\frac{R_{av}}{R_{ah}} = \tan^{-1}\frac{1000}{1732} = 30°$$

where θ is the angle which the reaction at A makes with the x-axis.

Alternatively :

Since the bar is in equilibrium, the three forces W, T and R_a would be concurrent and Lami's theorem would apply.

That gives

$$\frac{W}{\sin 150°} = \frac{R_a}{\sin 90°} = \frac{T}{\sin 120°}$$

$$\therefore \quad R_a = W \times \frac{\sin 90°}{\sin 150°} = 1000 \times \frac{1}{0.5} = 2000 \text{ N}$$

$$\text{and} \quad T = W \times \frac{\sin 120°}{\sin 150°} = 1000 \times \frac{0.866}{0.5} = 1732 \text{ N}$$

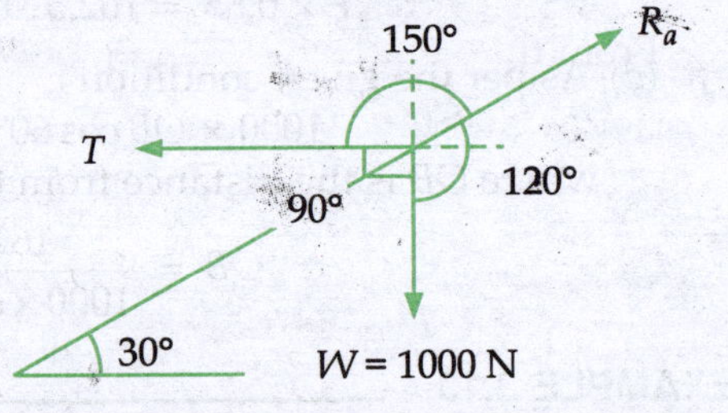

EXAMPLE 3.14

Two identical uniform bars AC and BC each of length l and weight W are hinged together at C and supported in a vertical plane by two legs at D and E as shown in Fig. 3.20. Neglecting friction, find the angle θ that each bar will make with the horizontal in the condition of equlibrium.

Solution : Let θ be angle which each bar should make with the horizontal.

For equilibrium of the whole system,

$$\Sigma F_y = 0; \qquad 2W = 2R\cos\theta$$

$$R = \frac{W}{\cos\theta} \qquad \text{...(i)}$$

For equilibrium of bar BC, $\Sigma M = 0$

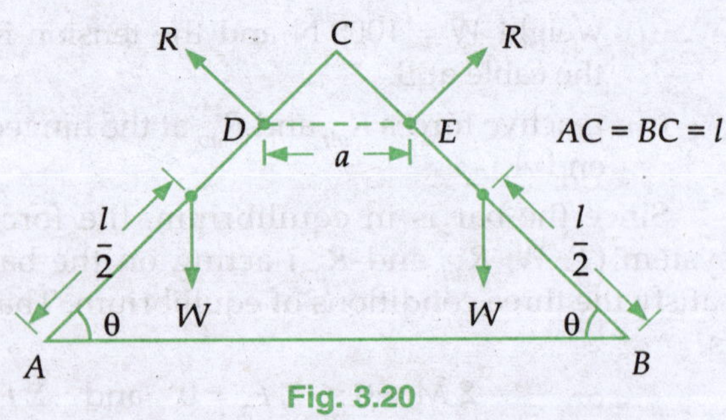

Fig. 3.20

Taking moments about point C,

$$R \times EC = W \times \frac{l}{2}\cos\theta \quad \text{or} \quad R \times \frac{a}{2\cos\theta} = W \times \frac{l}{2}\cos\theta$$

That gives : $R = \frac{Wl}{a}\cos^2\theta$ ...(*ii*)

Equating identities (*i*) and (*ii*), we get

$$\frac{Wl}{a}\cos^2\theta = \frac{W}{\cos\theta} \quad \text{or} \quad \cos^3\theta = \frac{a}{l};$$

$$\therefore \quad \theta = \cos^{-1}\left(\frac{a}{l}\right)^{\frac{1}{3}}$$

EXAMPLE 3.15

Two rods AB and BC of equal length, are freely jointed at B, are placed in a vertical plane with A and C joined by an inextensible string and rest on a smooth horizontal plane as shown in Fig. 3.21. The weight of rod AB is 2 W and that of BC is 4 W. If the inclination of rods at B is 90°, make calculations for tension in the string.

Solution : The weight of rod *AB* acts at its mid-point *D* and that of rod *BC* acts through its mid point *E*.

Considering the equilibrium of the entire system

$\Sigma F_y = 0;\ R_a + R_c = 2\,W + 4\,W = 6\,W$

$\Sigma M = 0;$ Taking moments of all forces about point *A*,

$$R_c \times AC = 2\,W \times AF + 4\,W \times AH$$

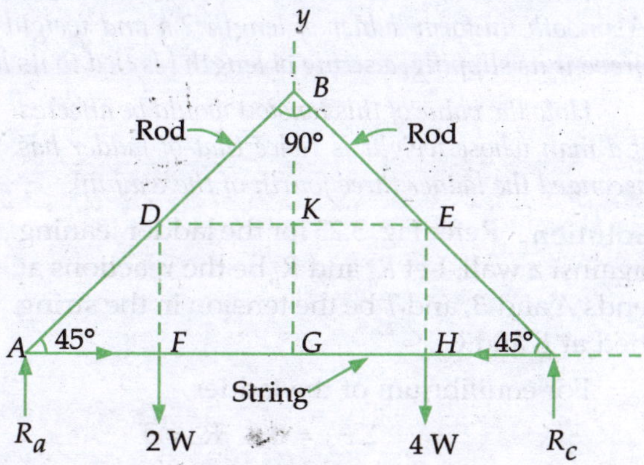

Fig. 3.21

From the geometry of the arrangement

$$AC = 2\,AG = 2 \times l\cos 45^\circ = 2\,l \times 0.707 = 1.414\,l$$

$$AF = AD\cos 45^\circ = \frac{l}{2} \times 0.707 = 0.3535\,l$$

$$AH = AC - HC = AC - AF = 1.414\,l - 0.3535\,l = 1.0605\,l$$

Substituting the relevant values in the expression for moment of all the forces about point *A*, we get

$$R_c \times 1.414\,l = 2\,W \times 0.3535\,l + 4\,W \times 1.0605\,l$$

or $\quad 1.414\,R_c = 0.707\,W + 4.242\,W$

$\therefore \quad R_c = \frac{(0.707 + 4.242)\,W}{1.414} = 3.5\,W$

and $\quad R_a = 6\,W - 3.5\,W = 2.5\,W$

(*b*) Considering equilibrium of rod *BC*, we take moments about end *B*

$$T \times BG + 4\,W \times EK - R_c \times GC = 0$$

or $\quad T \times AB\sin 45^\circ + 4W \times BE\sin 45^\circ - R_c \times BC\sin 45^\circ = 0$

or $\quad T \times AB + 4\,W \times BE - R_c \times BC = 0$

or $\quad T \times l + 4\,W \times \frac{l}{2} - 3.5\,W \times l = 0$

Solution gives : $T = 3.5\,W - 2\,W = 1.5\,W$

EXAMPLE 3.16

A weight of 800 N is hanging from a wooden beam which is carried by two persons as indicated in the adjoining figure. Neglecting weight of the beam, determine the load shared by each person.

Solution : Considering equilibrium of the beam,

$$\Sigma F_y = 0 ; \quad P + Q - 800 = 0 \quad \text{or} \quad P + Q = 800$$

Taking moments about lower end A of the beam,

$$Q \times AE - W \times AD = 0$$

or $\quad Q \times AB \cos 60^\circ - W \times AC \cos 60^\circ = 0$

or $\quad Q \times (1 \times 0.5) - 800 \times (0.4 \times 0.5) = 0$

or $\quad 0.5\,Q = 160 ; \; Q = \dfrac{160}{0.5} = \mathbf{320\ N}$

and $\quad P = 800 - Q = 800 - 320 = \mathbf{480\ N}$

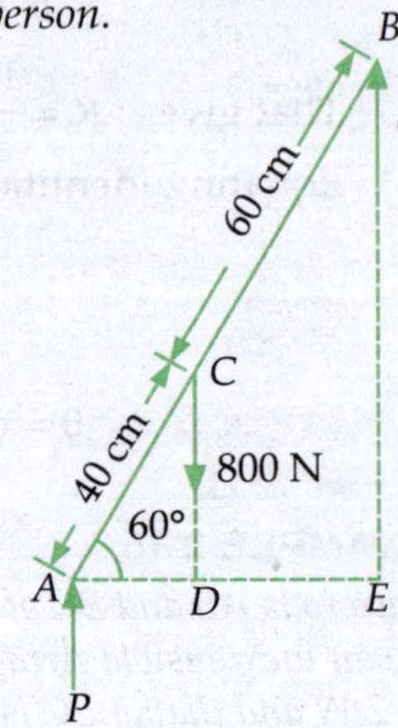

Fig. 3.22

EXAMPLE 3.17

A smooth uniform ladder of length 2 a and weight W rests on the ground and against a vertical wall. To prevent its slipping, a string of length l is tied to its lower end and to the junction of the wall with the ground.

How the value of this tension would be affected if a man whose weight is twice that of ladder has ascended the ladder three fourth of the way up.

Solution : Refer Fig. 3.23 for the ladder leaning against a wall. Let R_a and R_b be the reactions at ends A and B, and T be the tension in the string tied at B and C.

For equilibrium of the ladder,

$$\Sigma F_x = 0 ; \quad R_a = T$$
$$\Sigma F_y = 0 ; \quad R_b = W$$
$$\Sigma M = 0$$

Taking moments about point B,

$$R_a \times AC = W \times BD$$

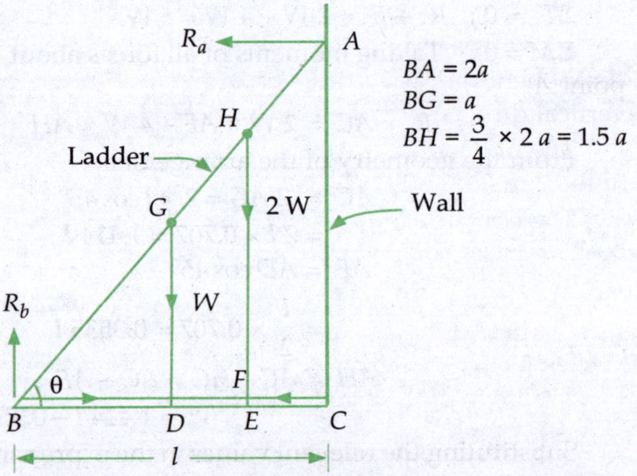

Fig. 3.23

or $\quad R_a \times 2\,a \sin\theta = W \times a \cos\theta$

or $\quad T \times 2\,a \times \dfrac{\sqrt{4a^2 - l^2}}{2a} = W\,a\,\dfrac{l}{2a}$

$\therefore \quad T = \dfrac{W\,l}{2\sqrt{4a^2 - l^2}}$

$$\cos\theta = \frac{l}{2a}$$

$$\sin\theta = \sqrt{1 - \cos^2\theta}$$

$$= \sqrt{1 - \frac{l^2}{4a^2}} = \frac{\sqrt{4a^2 - l^2}}{2a}$$

(*b*) When a man with weight 2 W stands at H located at distance

$$AH = \frac{3}{4} \times 2\,a = 1.5\,a$$

For equilibrium of the system

$$\Sigma F_x = 0; \quad R_a = T$$
$$\Sigma F_y = 0; \quad R_b = W + 2\,W = 3\,W$$
$$\Sigma M = 0; \quad \text{Taking moments about point } B$$
$$R_a \times AC = W \times BD + 2\,W \times BE$$

or $$R_a \times 2\,a \sin\theta = W \times a\cos\theta + 2\,W \times 1.5\,a\cos\theta = 4\,W a\cos\theta$$

or $$T \times 2\,a\,\frac{\sqrt{4a^2 - l^2}}{2\,a} = 4\,W a \times \frac{l}{2\,a} = 2\,W l$$

$$\therefore \quad T = \frac{2\,W l}{\sqrt{4a^2 \;\; l^2}}$$

EXAMPLE 3.18

Determine reactions at A and B (Fig. 3.24)

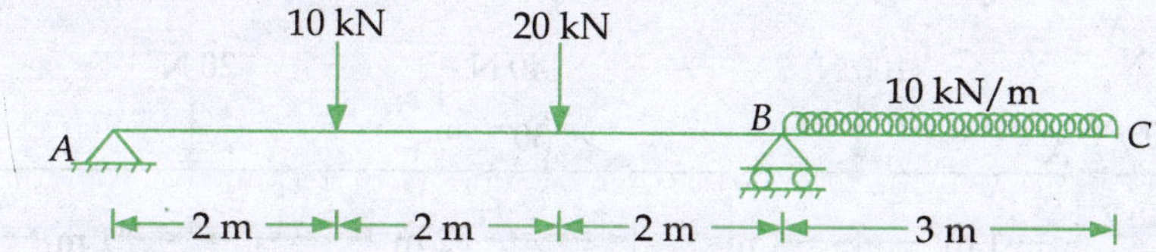

Fig. 3.24

Solution : At the hinged support, the reaction R_a can be in any direction. R_{av} and R_{ah} are its components in the vertical and horizontal directions. At the roller support, the reaction R_b will be in vertical direction only.

The uniformly distributed load has a total magnitude of 10 × 3 = 30 kN, and acts at the middle of the 3 m length.

The equivalent free body diagram for the beam will then be as shown below:

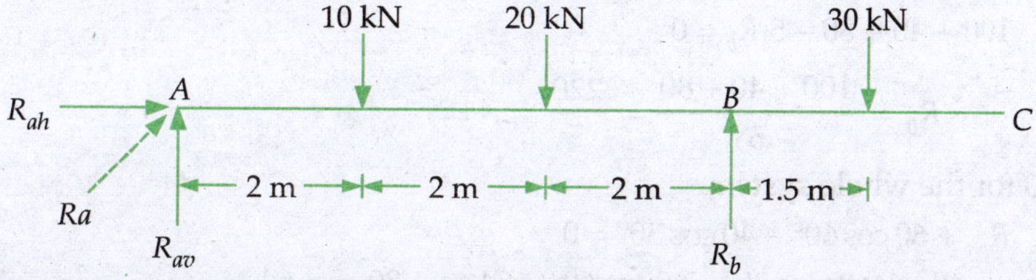

The reactions R_{ah}, R_{av} and R_b can be evaluated by applying the conditions of equilibrium. That is

$$\Sigma M = 0; \quad \Sigma F_x = 0; \quad \Sigma F_y = 0$$

(*i*) Taking moments about A and equating it to zero;

$$(10 \times 2) + (20 \times 4) - (R_b \times 6) + (30 \times 7.5) = 0$$
$$20 + 80 - 6\,R_b + 225 = 0$$
$$\therefore \quad R_b = \frac{225 + 20 + 80}{6} = \frac{325}{6} = \mathbf{54.17\ kN}$$

(*ii*) $\Sigma F_x = 0$ for the whole system.

$$\therefore \quad R_{ah} = 0$$

This is because no other horizontal force, active or reactive, is acting on the beam.

Further, $\Sigma F_y = 0$. That gives $R_{av} - 10 - 20 + R_b - 30 = 0$

$$\therefore \quad R_{av} = 10 + 20 - R_b + 30 = 10 + 20 - 54.17 + 30 = \mathbf{5.83\ kN}$$

EXAMPLE 3.19

Determine reactions at A and B for the beam loaded as shown in Fig. 3.25.

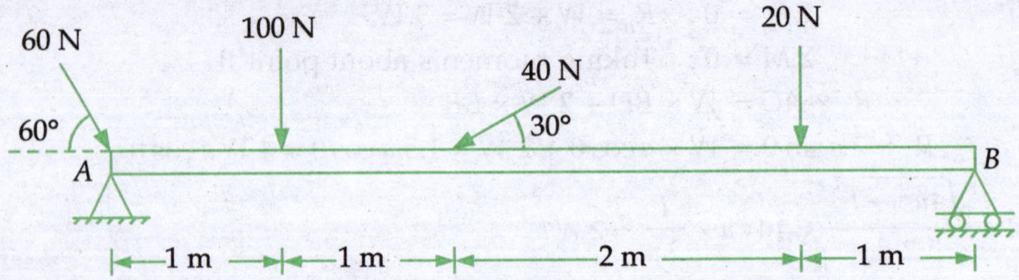

Fig. 3.25

Solution : At the hinged support the reaction R_a can be in any direction. R_{av} and R_{ah} are its components in the vertical and horizontal directions. At the roller support, the reaction R_b will be in vertical direction only.

The equivalent free body diagram for the beam will be then as shown below.

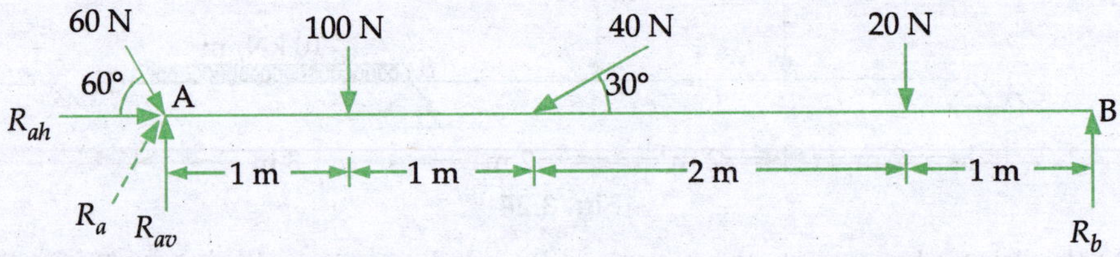

The reactions R_{ah}, R_{av} and R_b can be evaluated by applying the conditions of equilibrium. That is

$$\Sigma M = 0\,;\quad \Sigma F_x = 0\,;\quad \Sigma F_y = 0$$

(*i*) Taking moments about *A* and equating it to zero,

$$(100 \times 1) + (40 \sin 30° \times 2) + (20 \times 4) - (R_b \times 5) = 0$$

$$100 + 40 + 80 - 5\,R_b = 0$$

$$\therefore \quad R_b = \frac{100 + 40 + 80}{5} = \frac{220}{5} = 44 \text{ N}$$

(*ii*) $\Sigma F_x = 0$ for the whole system.

$$R_{ah} + 60 \cos 60° - 40 \cos 30° = 0$$

$$\therefore \quad R_{ah} = 40 \cos 30° - 60 \cos 60° = 34.64 - 30 = 4.64 \text{ N}$$

(*iii*) Further, $\Sigma F_y = 0$. That gives

$$R_{av} - 60 \sin 60° - 100 - 40 \sin 30° - 20 + Rb = 0$$

$$R_{av} = 60 \sin 60° + 100 + 40 \sin 30° + 20 - R_b$$

$$= 51.96 + 100 + 20 + 20 - 44 = 147.96 \text{ N}$$

Total reaction at hinge support,

$$R_a = \sqrt{R_{ah}^2 + R_{av}^2} = \sqrt{4.64^2 + 147.96^2} = 148 \text{ N}$$

$$\theta = \tan^{-1}\frac{R_{av}}{R_{ah}} = \tan^{-1}\frac{147.96}{4.64} = 88.20°$$

where θ is the angle which the reaction at *A* makes with the *x*-axis.

EXAMPLE 3.20

Figure 3.26 shows a system of levers supporting a load of 100 N. Determine the reactions at A and B on the lever.

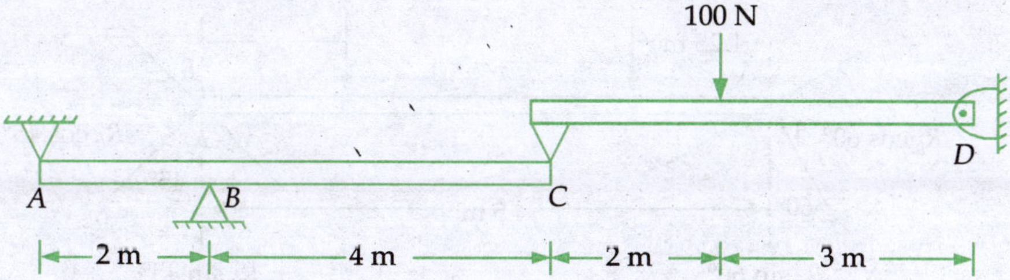

Fig. 3.26

Solution : Refer to the free body diagrams for the upper and lower beams.

Taking moments about end C and equating it to zero,

$$100 \times 2 - R_d \times 5 = 0\,; \quad R_d = \frac{100 \times 2}{5} = 40 \text{ N}$$

$$\Sigma F_y = 0\,; \quad R_c - 100 + R_d = 0$$

$$R_c = 100 - R_d = 100 - 40 = 60 \text{ N}$$

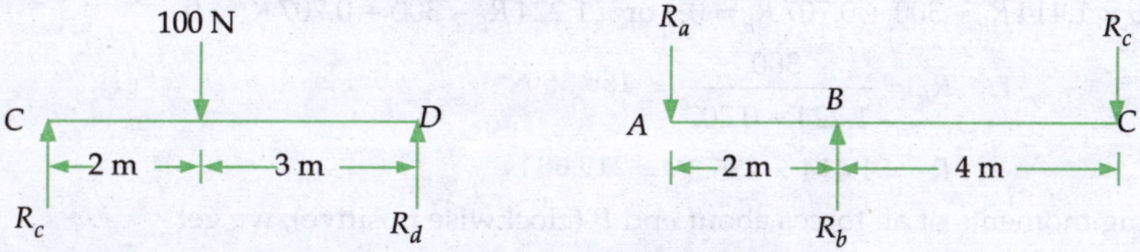

Taking moments about end A and equating it to zero,

$$R_c \times 6 - R_b \times 2 = 0\,; \quad R_b = \frac{6R_c}{2} = \frac{6 \times 60}{2} = 180 \text{ N}$$

$$\Sigma F_y = 0\,; \quad R_a - R_b + R_c = 0$$

$$R_a = R_b - R_c = 180 - 60 = 120 \text{ N}$$

Thus the reactions at points A and B on the lever are:

R_b = 180 N (upwards) and R_a = 120 N (downwards)

EXAMPLE 3.21

A weightless bar AB is placed in horizontal position on smooth inclines as shown in Fig. 3.27. Calculate the distance x at which the 100 N load be placed from end B so that the bar remains horizontal.

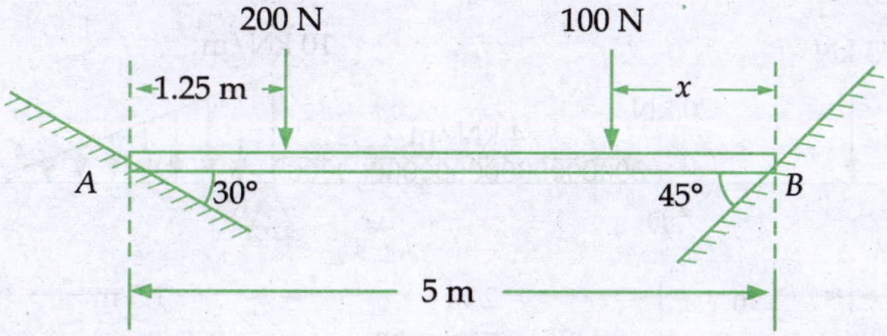

Fig. 3.27

Solution : Reactions at ends A and B of the bar will be normal to the smooth inclines and can be resolved horizontally and vertically as shown below

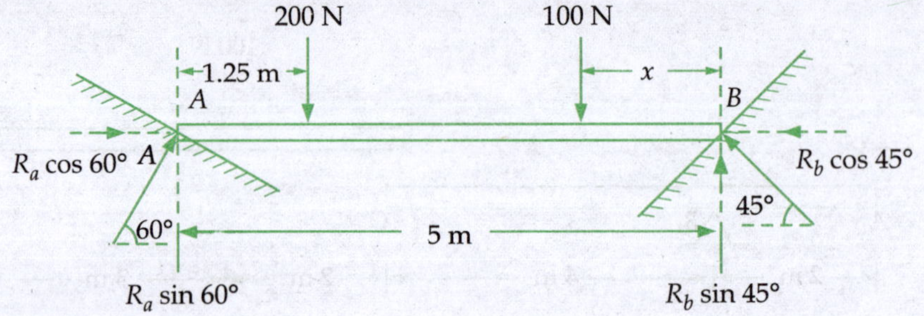

Under equilibrium conditions : $\Sigma F_x = 0$, $\Sigma F_y = 0$ and $\Sigma M = 0$

$\Sigma F_x = 0$: $R_a \cos 60° - R_b \cos 45° = 0$

$$R_a = \frac{R_b \cos 45°}{\cos 60°} = 1.414\, R_b \quad ...(i)$$

$\Sigma F_y = 0$: $R_a \sin 60° - 200 - 100 + R_b \sin 45° = 0$

$$0.866\, R_a - 300 + 0.707\, R_b = 0 \quad ...(ii)$$

From expressions (*i*) and (*ii*),

$0.866 \times 1.414 R_b - 300 + 0.707 R_b = 0$ or $1.224 R_b - 300 + 0.707 R_b = 0$

$$\therefore \quad R_b = \frac{300}{1.224 + 0.707} = \mathbf{155.36\ N}$$

$$R_a = 1.414 \times 155.36 = \mathbf{219.68\ N}$$

Taking moments of all forces about end B (clockwise positive), we get

$$-100x - 200 \times 3.75 + R_a \sin 60° \times 5 = 0$$

It is to be noted that component $R_a \cos 60°$ of R_a passes through B, and hence its moment about B is zero

$$-100x - 750 + 219.68 \sin 60° \times 5 = 0 \qquad (\because R_a = 219.68\ \text{N})$$

or
$$-100x - 750 + 951.24 = 0$$

$$\therefore \quad x = \frac{951.24 - 750}{100} = \mathbf{2.01\ m}$$

EXAMPLE 3.22

A beam has been loaded and supported as shown in Fig. 3.28 given below. Determine the reactions at the support points A and B.

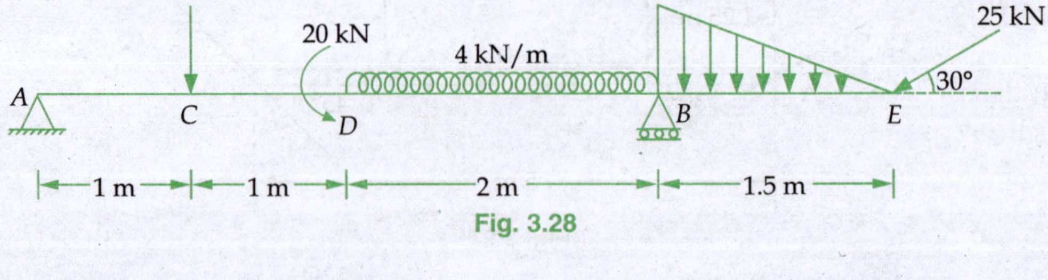

Fig. 3.28

Solution : Equivalent loads and their position of acting are as shown below :

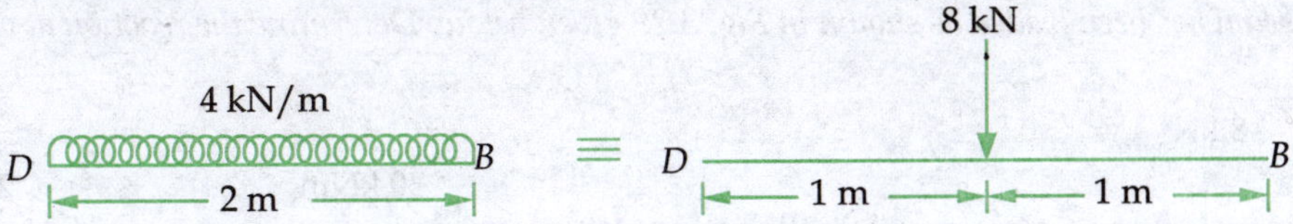

The total load is equal to 4 × 2 = 8 kN and it acts at the centre.

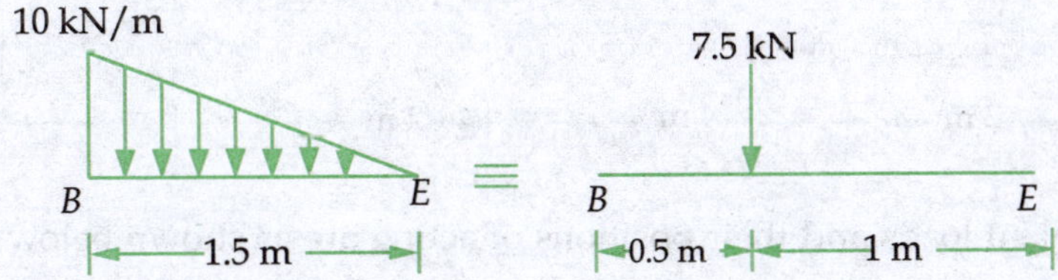

The total load is $\frac{1}{2} \times 1.5 \times 10 = 7.5$ kN and it acts at $\frac{1}{3} \times 1.5 = 0.5$ m from the base, *i.e.*, from point *B*.

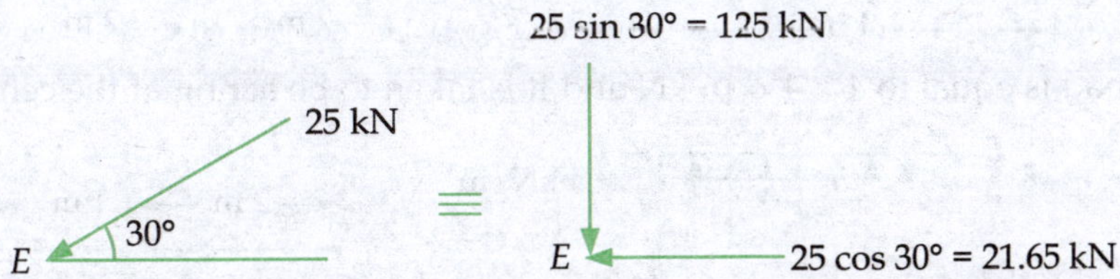

These equivalent loads have been indicated in the diagram of the beam drawn below.

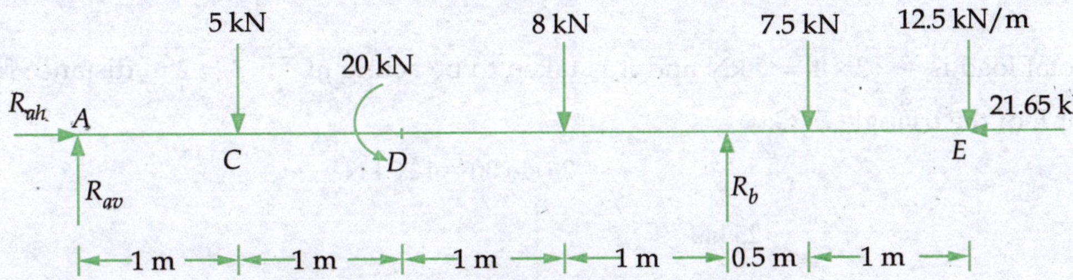

For equilibrium of the beam : $\Sigma F_x = 0$; $\Sigma F_y = 0$ and $\Sigma M = 0$

$\Sigma F_x = 0$; $R_{ah} - 21.65 = 0$; $R_{ah} = 21.65$ kN

$\Sigma F_y = 0$; $R_{av} - 5 - 8 + R_b - 7.5 - 12.5 = 0$ $\quad\therefore\quad R_{av} + R_b = 33$

Taking moments about the end *A* (clockwise positive)

$\Sigma M_a = 0$; $\quad 5 \times 1 - 20 + 8 \times 3 - R_b \times 4 + 7.5 \times 4.5 + 12.5 \times 5.5 = 0$

$\quad 5 - 20 + 24 - 4R_b + 33.75 + 68.75 = 0$; $R_b =$ **27.875 kN**

Then $\quad R_{av} = 33 - R_b = 33 - 27.875 =$ **5.125 kN**

Thus the reactions at support points *A* and *B* are

$$R_{ah} = 21.65 \text{ kN} \quad ; \quad R_{av} = 5.125 \text{ and } R_b = 27.875 \text{ kN}$$

EXAMPLE 3.23

A cantilever beam has been loaded as shown in Fig. 3.29 given below. Determine the reaction at cantilever end A.

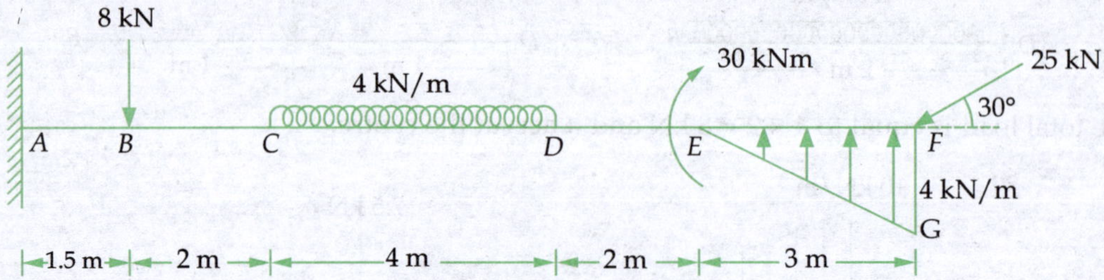

Fig. 3.29

Solution : Equivalent loads and their positions of acting are as shown below :

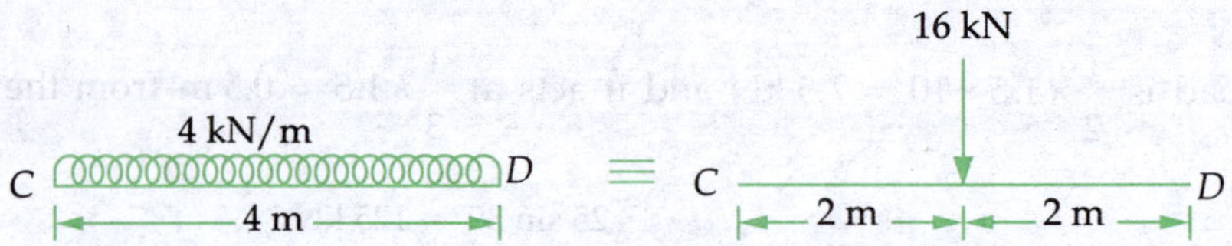

The total load is equal to $4 \times 4 = 16$ kN and it is taken to be acting at the centre.

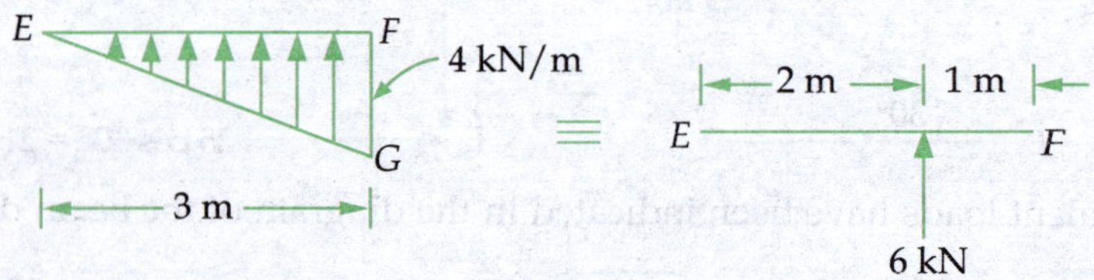

The total load is $\frac{1}{2} \times 3 \times 4 = 6$ kN and it is taken to be acting at $\frac{2}{3} \times 3 = 2$ m distance from the apex point *E* of the triangle *EFG*.

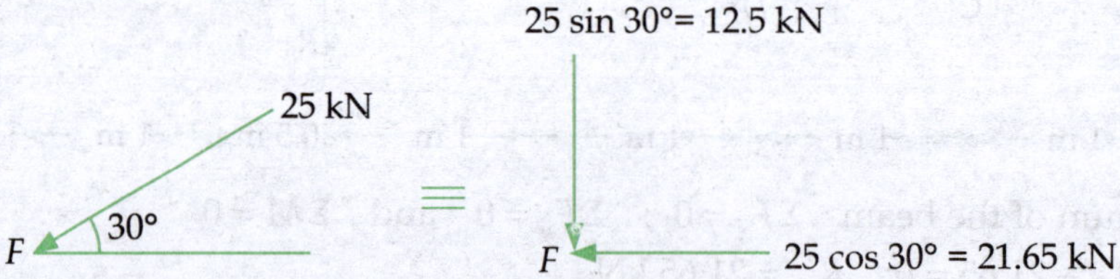

These equivalent loads have been indicated in the diagram of the beam drawn below.

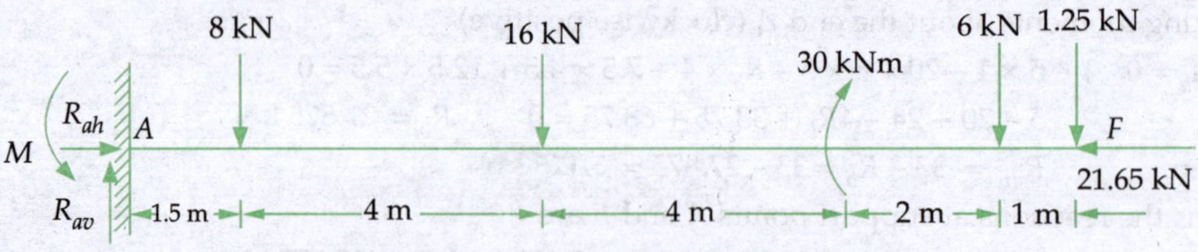

For equilibrium of the beam : $\Sigma F_x = 0$; $\Sigma F_y = 0$ and $\Sigma M = 0$

$$\Sigma F_x = 0; \quad R_{ah} - 21.65 = 0 \quad ; \quad R_{ah} = \mathbf{21.65\ kN}$$

$$\Sigma F_y = 0; \quad R_{av} - 8 - 16 + 6 - 12.5 = 0 \quad ; \quad R_{av} = \mathbf{30.5\ kN}$$

Since the end A of the beam is fixed and end F is free, the beam AF is a cantilever and hence a moment of couple M is produced at A as shown in Fig. 3.40.

Taking moments about the cantilever end A,

$$\Sigma M_a = 0 \;\; ; \;\; -M + 8 \times 1.5 + 16 \times 5.5 + 30 - 6 \times 11.5 + 12.5 \times 12.5 = 0$$

$$-M + 12 + 88 + 30 - 69 + 156.25 = 0$$

$$\therefore \qquad M = \textbf{217.25 kNm (anticlockwise).}$$

Thus the reactions at the fixed end are :

$$R_{uh} = 21.65 \text{ kN} \; ; \quad R_{uv} = 30.5 \text{ kN} \quad \text{and} \quad M = 217.25 \text{ kNm}$$

EXAMPLE 3.24

Compute the simplest resultant force for the loads acting on a cantilever beam as shown in the figure given below. What force and moment is transmitted by this force to the supporting wall at A?

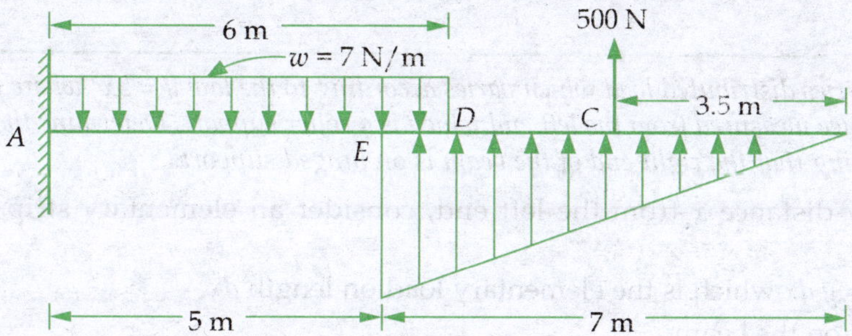

Fig. 3.30

Solution : Equivalent loads and their position of acting are as shown below:

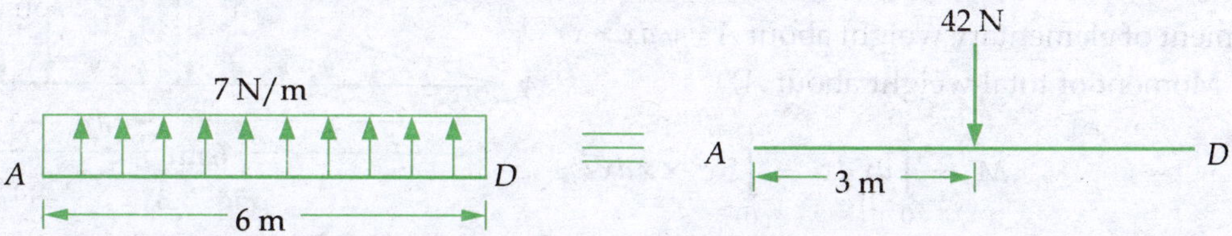

The load is equal to 7 × 6 = 42 N and its acts at the centre of portion AD

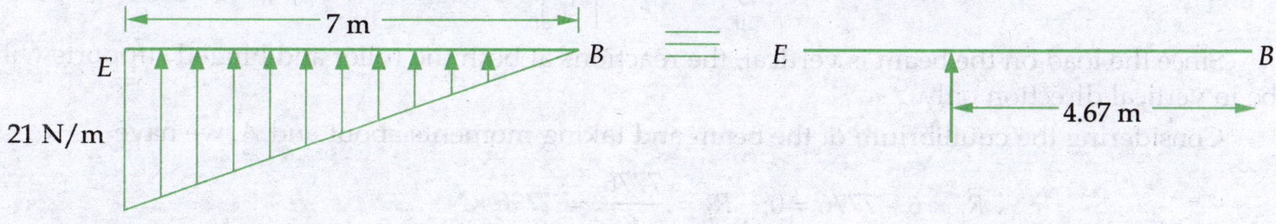

The total load is $\frac{1}{2} \times 7 \times 21 = 73.5$ N and it acts at $\frac{2}{3} \times 7 = 4.67$ from the apex, *i.e.*, from point B.

These equivalent loads have been indicated in the diagram of the beam drawn below:

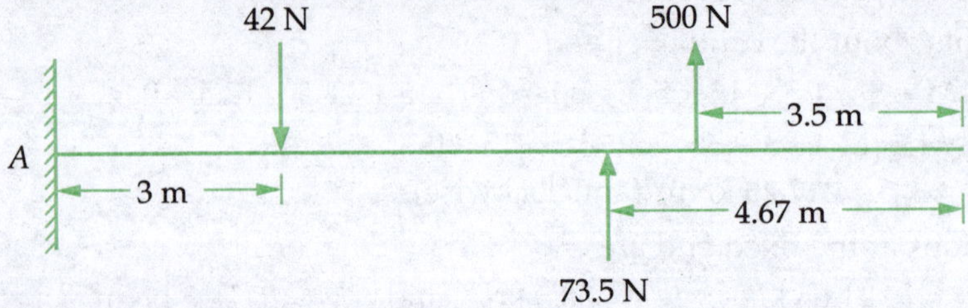

Resultant force acting at the wall = 42 – 73.5 – 500 = **– 531.5 N** (upward)

Moment acting at the wall,

$$M_A = 42 \times 3 - 500\,(12 - 3.5) - 73.5\,(12 - 4.67)$$
$$= 126 - 4250 - 538.75 = -4662.75 \text{ N} \quad \text{(anti clockwise)}$$

EXAMPLE 3.25

A beam 6 m long carries distributed load which varies according to the law $y = 5x^3$ where y is in kN/m and x in meter is the distance measured from the left end which is a roller support. Determine the support reactions on the beam presuming that the right end of the beam is on hinged support.

Solution : At any distance x from the left end, consider an elementary strip of height h and thickness dx.

Area of strip = $y\,dx$ which is the elementary load on length dx

∴ Total load on the beam,

$$W = \int_0^l y\,dx = \int_0^6 5x^3\,dx = 5\left|\frac{x^4}{4}\right|_0^6 = 1620 \text{ kN}$$

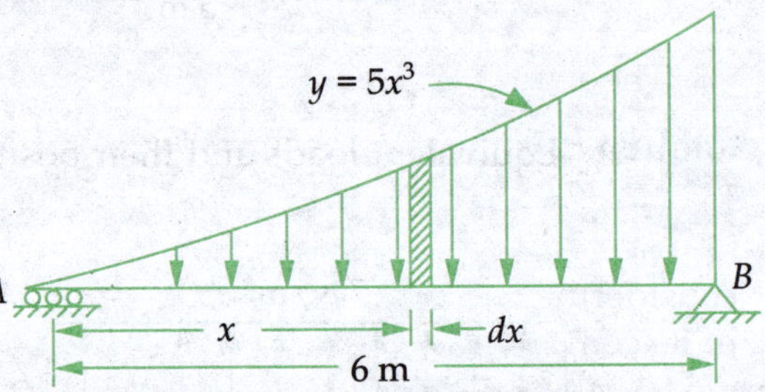

Fig. 3.31

Moment of elementary weight about $A = y\,dx \times x$

∴ Moment of total weight about A,

$$M_a = \int_0^l yx\,dx = \int_0^l 5x^3 \times x\,dx$$

$$= 5\int_0^6 x^4\,dx = 5\left|\frac{x^5}{5}\right|_0^6 = 7776 \text{ kNm} \quad \text{(clockwise)}$$

Since the load on the beam is vertical, the reactions at both the roller and hinged supports will be in vertical direction only.

Considering the equilibrium of the beam and taking moments about and A, we have

$$R_b \times 6 - 7776 = 0; \quad R_b = \frac{7776}{6} = 1296 \text{ kN}$$

and $R_a = W - R_b = 1620 - 1296 = 324 \text{ kN}$

3.4. RESULTANT OF COPLANAR, NON-CONCURRENT FORCE SYSTEM

Below are outlined the different steps which are followed to determine the resultant of coplanar, non-current force system in magnitude, direction and position :

1. Resolve the given forces horizontally and vertically and determine

ΣF_x = algebraic sum of the component of all forces in x-direction, and

ΣF_y = algebraic sum of the component of all forces in y-direction

2. Determine the magnitude of the resultant force by using the relation

$$R = \sqrt{(\Sigma F_x)^2 + (\Sigma F_y)^2}$$

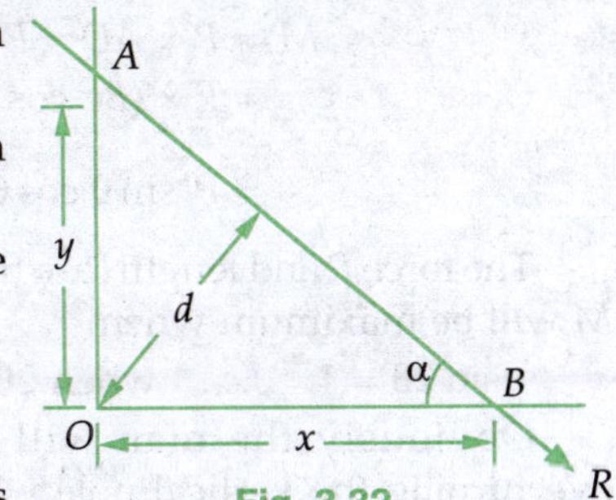

Fig. 3.32

3. Find the direction of the resultant force with respect to x-axis by using the relation

$$\alpha = \tan^{-1}\left(\frac{\Sigma F_y}{\Sigma F_x}\right)$$

Depending upon the sign of ΣF_x and ΣF_y, decide in which of the quadrant the resultant lies.

4. Obtain the algebraic sum of the moments of all forces about any given point say O.

5. Mark the position of resultant such that it produces the same direction of moment about point O.

6. Apply Varignon's theorem of moment to find the exact position of resultant. That is

$$\Sigma M_O = R \times d$$

where ΣM_O = moment of given forces about point O

d = perpendicular distance of line of action of resultant R from point O.

The x and y intercepts of the line of action of resultant can then be worked out as

$$x = \frac{d}{\sin\alpha} \quad \text{and } y = \frac{d}{\cos\alpha}$$

Alternatively, the intercepts can be determined by considering the moment of R about O as the sum of moments of its components about O (Varignon's theorem)

$$R \times d = \Sigma M_O$$

$$\Sigma F_x \times 0 + \Sigma F_y \times x = \Sigma M_O$$

$$\therefore \quad x = \frac{\Sigma M_O}{\Sigma F_y}$$

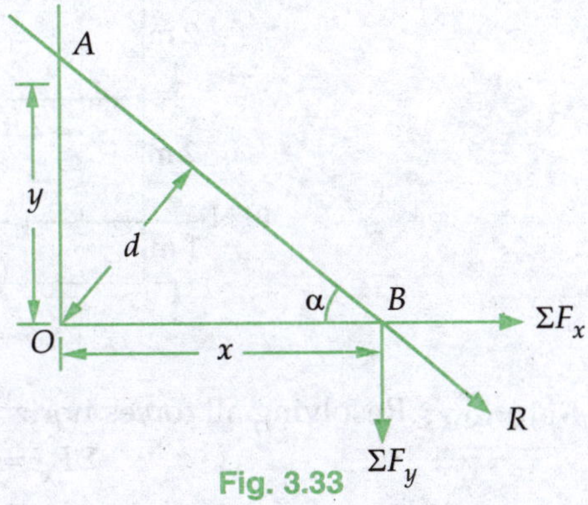

Fig. 3.33

Similarly by resolving the resultant into its components at A, it can be calculated that

$$y = \frac{\Sigma M_O}{\Sigma F_x}$$

EXAMPLE 3.26

One end of a rope of length l is fixed to a vertical telegraph pole. A man standing on the ground pulls at the other end of the rope with a given force. Find the point at which the rope should be tied to the pole in order that the man will have the best chance of over turning the pole about its end fixed to the ground.

Solution : With reference to Fig. 3.34, AB is the vertical pole grounded at point A.

Let P be the force applied through rope BC (length l) towards C where the man is standing.

Moment of force P about point A,

$$M = P \times AD = P \times AC \sin\theta$$
$$= P \times (BC\cos\theta)\sin\theta$$
$$= Pl\sin\theta\cos\theta = \frac{1}{2}Pl\sin 2\theta$$

The force P and length l are fixed and as such the moment M will be maximum when

$\sin 2\theta = 1$, *i.e.*, when $2\theta = 90°$ and $\theta = 45°$

Obviously the man will have the best chance of overturning the vertical pole when he pulls it through the rope at an angle of 45° with the horizontal.

$$AB = BC\sin\theta = l\sin 45° = \frac{l}{\sqrt{2}}$$

Hence to achieve the objective, the rope should be tied to the pole at point B which is at height $l/\sqrt{2}$ above the point A.

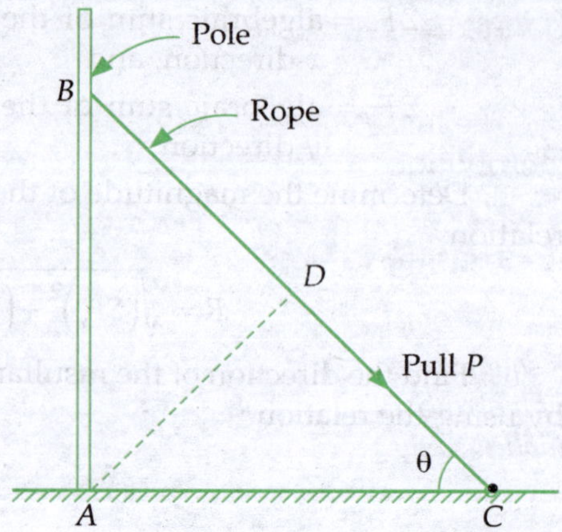

Fig. 3.34

EXAMPLE 3.27

Determine the resultant of four forces and one couple that act on the plate as shown in the Fig. 3.35 given below:

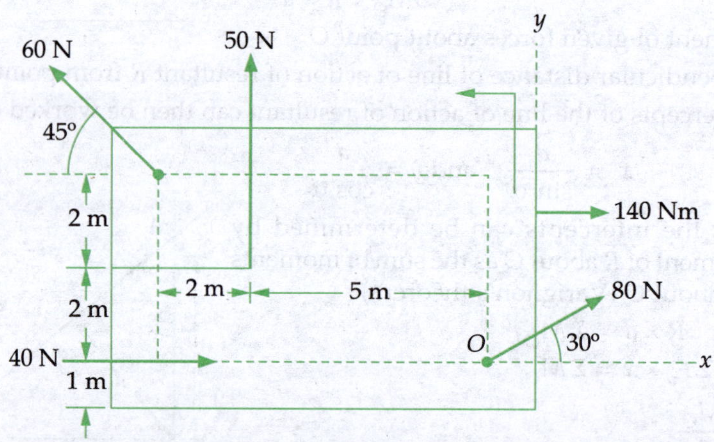

Fig. 3.35

Solution : Resolving all forces horizontally,

$$\Sigma F_x = 80\cos 30° - 60\cos 45° + 40$$
$$= 69.28 - 42.43 + 40 = 66.85 \text{ N}$$

Resolving all forces vertically,

$$\Sigma F_y = 80\sin 30° + 50° + 60\sin 45° = 132.43 \text{ N}$$

Magnitude of resultant force,

$$F = \sqrt{(\Sigma F_x)^2 + (\Sigma F_y)^2} = \mathbf{148.34\ N}$$

Let α be the angle which the resultant force makes with the horizontal direction, Then

$$\tan\alpha = \frac{\Sigma F_y}{\Sigma F_x} = \frac{132.43}{66.85} = 1.981$$

That gives: α = 63.21°

Taking moments about point O (considering clockwise moments positive),

$$M_0 = -140 + 50 \times 5 + 60 \sin 45° \times 7 - 60 \cos 45° \times 4$$

The lines of application of 40 N and 80 N forces pass through O and as such they will have no contribution towards moments.

$$M_0 = -140 + 250 + 296.98 - 169.70 = 237.28 \text{ Nm}$$

Thus the system will have a clockwise moment equivalent to 237.28 Nm at point O.

EXAMPLE 3.28

A square block of 25 cm side and weighing 20 N is hinged at A and rests on rollers at B (Fig. 3.36) It is pulled by a string attached at C and inclined at 30° with the horizontal. Make calculations for the force P to be applied so that the block gets just lifted off the roller.

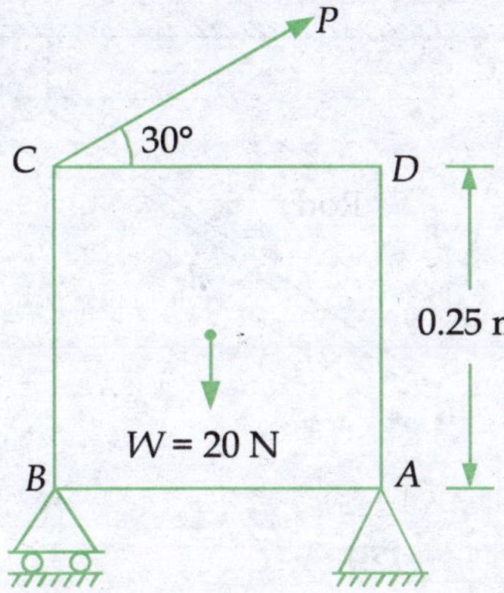

Fig. 3.36

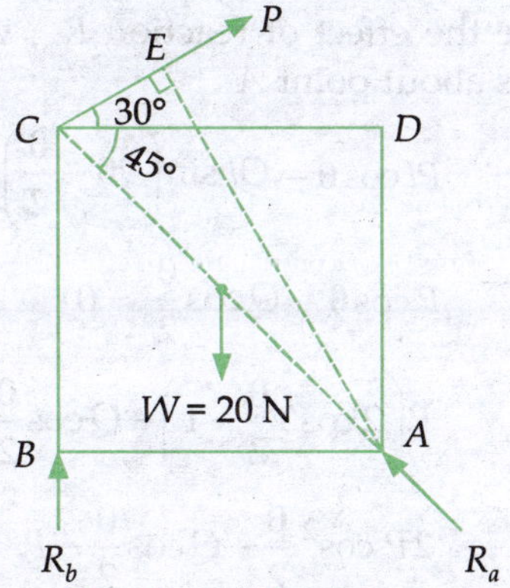

Free body diagram

Solution : With reference to free body diagram, the block is subjected to the following set of forces :

(*i*) force P,

(*ii*) weight of block W = 20 N acting vertically downward,

(*iii*) reaction R_a at the hinge point A

(*iv*) when the block is at the state of just being lifted off the roller, reaction $R_b = 0$.

Taking moments about point A,

$$M_a = 20 \times \frac{0.25}{2} - P \times AE$$

Now, $$AE = AC \sin 45° = \sqrt{0.25^2 + 0.25^2} \times \sin 45° = 0.3414 \text{ m}$$

$$\therefore \quad M_a = \frac{20 \times 0.25}{2} - P \times 0.3414$$

Under equilibrium state, this moment is zero. Therefore

$$0 = 20 \times \frac{0.25}{2} - P \times 0.3414$$

That gives : $$P = \frac{20}{0.3414} \times \frac{0.25}{2} = 7.323 \text{ N}$$

EXAMPLE 3.29.

A uniform rod AB of negligible weight and length l is hinged at the end A and supported by a string as shown in Fig. 3.37. What should be the inclination θ of the rod if it is in a state of equilibrium? Take P = 2Q.

Solution: The various forces acting on the rod are :

(*i*) Weight P acting downwards at end B

(*ii*) Reaction R_a acting at the hinge point A in a direction which is unknown

(*iii*) Tension T in the string. Presuming the pulley to be smooth and frictionless, $T = Q$

From the given configuration; $AB = AC$ and as such the triangle ABC is isosceles. Accordingly

$$\angle ABC = \angle ACB = \frac{180 - \theta}{2} = 90 - \frac{\theta}{2}$$

Under equilibrium state, the moment of all the forces about any point is zero. To eliminate the effect of reaction R_a, we take moments about point A

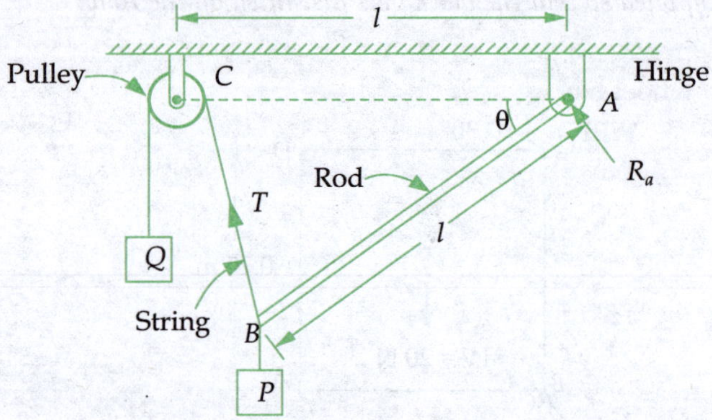

Fig. 3.37

$$Pl\cos\theta - Ql\sin\left(90 - \frac{\theta}{2}\right) = 0$$

or $$P\cos\theta - Q\cos\frac{\theta}{2} = 0$$

or $$P\left(2\cos^2\frac{\theta}{2} - 1\right) - Q\cos\frac{\theta}{2} = 0$$

or $$2P\cos^2\frac{\theta}{2} - Q\cos\frac{\theta}{2} - P = 0$$

Solution of this quadratic equation in $\cos\frac{\theta}{2}$ gives

$$\cos\frac{\theta}{2} = \frac{Q \pm \sqrt{Q^2 + 8P^2}}{4P} = \frac{1}{4}\left[\frac{Q}{P} \pm \sqrt{\frac{Q^2}{P^2} + 8}\right]$$

Substituting $\frac{Q}{P} = \frac{1}{2}$ (given) and taking the positive value

$$\cos\frac{\theta}{2} = \frac{1}{4}\left[\frac{1}{2} + \sqrt{\frac{1}{4} + 8}\right] = 0.843$$

$$\frac{\theta}{2} = \cos^{-1}(0.843) = 32.5°$$

∴ Inclination of the rod θ = **65°**

EXAMPLE 3.30

A uniform wheel weighing 20 kN and of 600 mm diameter rests against 150 mm thick rigid block as shown in Fig. 3.38 a.

Considering all surfaces to be smooth, determine : (a) the least pull through the centre of wheel to just turn the wheel over the corner of the block, (b) the reaction of the block.

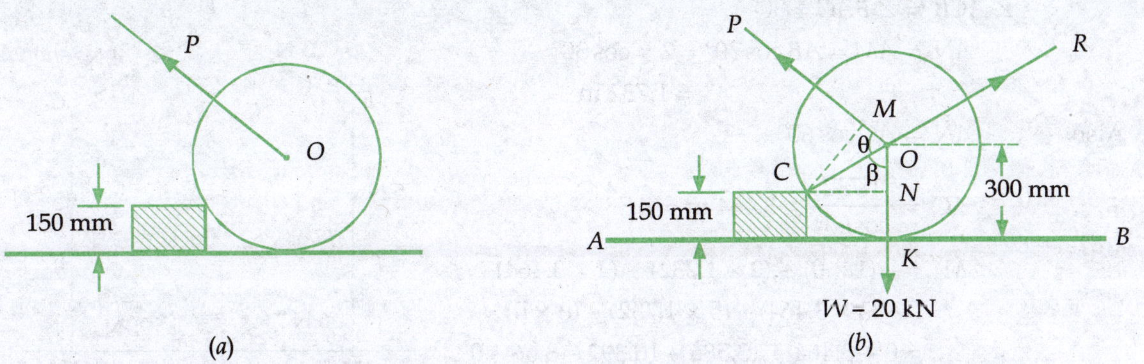

Fig. 3.38

Solution : When the wheel is about to turn, its contact with the ground will be lost. Hence the wheel has to be in equilibrium under the action of its weight W and the force P.

With reference to Fig. 3.38 *b*, CM and CN are the perpendiculars dropped on the lines of action of forces P and W respectively.

$$OC = OK = 300 \text{ mm} \; ; \quad ON = OK - NK = 300 - 150 = 150 \text{ mm}$$

$$CN = \sqrt{OC^2 - ON^2} = \sqrt{300^2 - 150^2} = 259.81 \text{ mm}$$

Taking moments about point C,

$$P \times CM - W \times CN = 0$$

$$P \times OC \sin\theta - 20 \times 259.81 = 0$$

$$\therefore \quad P = \frac{20 \times 259.81}{OC \sin\theta} = \frac{20 \times 259.81}{300 \sin\theta} = \frac{17.32}{\sin\theta}$$

The force P will be minimum when $\sin\theta$ is maximum. For that

$$\sin\theta = 1 \quad \text{or} \quad \theta = 90^\circ$$

Hence, P_{min} = **17.32 kN** when pull is applied perpendicular to OC.

(*b*) The reaction R can be determined by resolving W along OC

$$R = W \cos\beta \text{ where } \beta = \angle CON$$

$$\cos\beta = \frac{ON}{OC} = \frac{150}{300} = \frac{1}{2}$$

$$\therefore \quad R = 20 \times \frac{1}{2} = \mathbf{10 \text{ kN}}$$

The force P acts perpendicular to OC and as such its resolved part along OC is zero.

EXAMPLE 3.31

A regular hexagon ABCDEF of side 2 m in length is subjected to forces 1, 2, 3, 4, 5 and 6 N along sides AB, CB, DC, DE, EF and FA respectively as shown in the Fig. 3.39 given below. Make calculations for the sum of moments at point A.

Solution : Starting from force represented by side AB, we take moments about point A and get

$$M_a = -1 \times 0 + 2 \times AN + 4 \times AC - 3 \times AE - 5 \times AM - 6 \times 0$$

The clockwise moments have been taken positive and anticlockwise moments as negative.

From the geometrical configuration of the system,

$$\angle BAN = \angle FAM = 30^\circ$$

$$\angle ACB = \angle BAC = 30^\circ$$

$$AN = AM = AB\cos 30^\circ = 2 \times \cos 30^\circ$$

$$= 1.732 \text{ m}$$

Also $\quad AN = AC\cos 60^\circ$

$$AC = \frac{1.732}{\cos 60^\circ} = 3.464 \text{ m} = AE$$

$\therefore$
$$M_a = -(1 \times 0) + (2 \times 1.732) + (4 \times 3.464) - (3 \times 3.464) - (5 \times 1.732) - (6 \times 0)$$

$$= 0 + 3.464 + 13.586 - 10.392 - 8.66 - 0$$

$$= \mathbf{-1.728 \text{ Nm}}$$

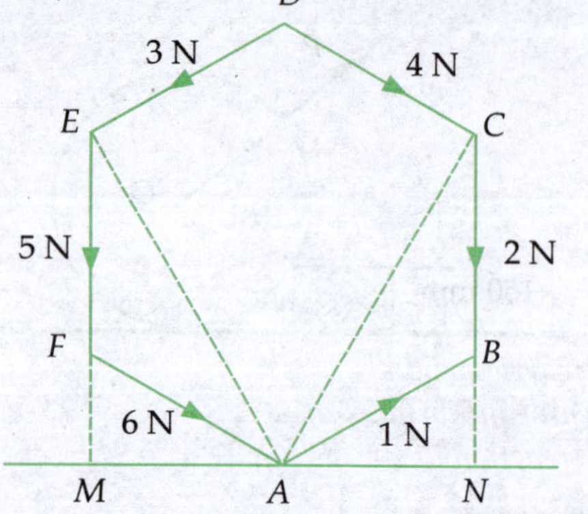

Fig. 3.39

The negative sign suggests that the resultant moment is anticlockwise.

EXAMPLE 3.32

Forces of 20, 30, 40, 50, 60 and 70 N act along the sides AB, CB, CD, ED, EF and FA respectively of a regular hexagon ABCDEF (Fig. 3.40) whose each side measures 10 cm. Make calculations for the algebraic sum of the moments of the forces about the centre of the hexagon and one of the vertices.

Solution : Let O be the centre of the hexagon and OP be the perpendicular drawn from the centre onto the side AB.

The triange OAB is an equilateral triangle and that gives

$$OP = \sqrt{(OA)^2 - (AP)^2} = \sqrt{10^2 - 5^2} = 8.66 \text{ cm}$$

Obviously the distance of each side from the centre O is 8.66 cm.

Algebraic sum of the moments about O,

$$M_O = -20 \times 8.66 + 30 \times 8.66 - 40 \times 8.66 + 50 \times 8.66 - 60 \times 8.66 - 70 \times 8.66$$

The clockwise moments have been taken as positive and anticlockwise moments negative.

$$M_O = 8.66(-20 + 30 - 40 + 50 - 60 - 70)$$

$$= -952.60 \text{ N cm}$$

The negative sign suggests that the resultant moment about the centre of the hexagon is anticlockwise.

(*b*) From the geometrical configuration of the system

$$\angle BAN = \angle FAM = 30^\circ$$

$$\angle ACB = \angle BAC = 30^\circ$$

$$AN = AM = AB\cos 30^\circ$$

$$= 10 \times 0.866 = 8.66 \text{ cm}$$

Also $\quad AN = AC\cos 60^\circ$

$$AC = \frac{8.66}{\cos 60^\circ} = 7.32 \text{ cm} = AE$$

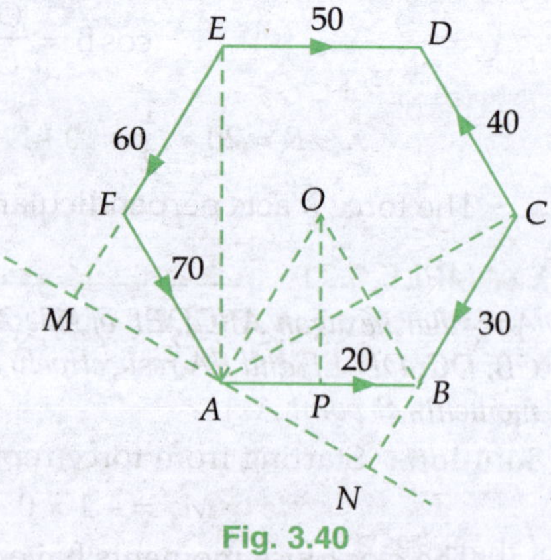

Fig. 3.40

$\therefore$ Moments about vertex A,

$$M_a = -(20 \times 0) + (30 \times 8.66) - (40 \times 17.32) + (50 \times 17.32) - (60 \times 8.66) - (70 \times 0)$$
$$= 8.66\,(30 - 80 + 100 - 60) = 8.66 \times (-10)$$
$$= -86.60 \text{ N cm}$$

The negative sign suggests that the resultant moment is anticlockwise about the vertex A.

EXAMPLE 3.33

A beam AB of length 2 m, hinged at A and supported at B by a cord which passes over two frictionless pulleys, carries a 50 kN load as shown in Fig. 3.41. Determine the position of 100 kN load on the beam if it is to remain in equilibrium in horizontal position. Also determine the reaction at the hinged end.

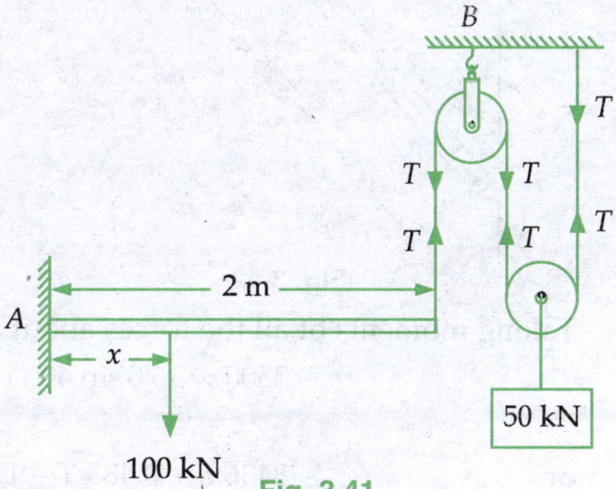

Fig. 3.41

Solution : Since the pulleys are weightless and frictionless, the tension T in the entire length of cord will be same.

For the equilibrium of pulley supporting 50 kN load,

$$\Sigma F_y = 0\,; \quad T + T - 50 = 0\,; \quad T = 25 \text{ kN}$$

The free body diagram of the beam is as shown and for its equilibrium

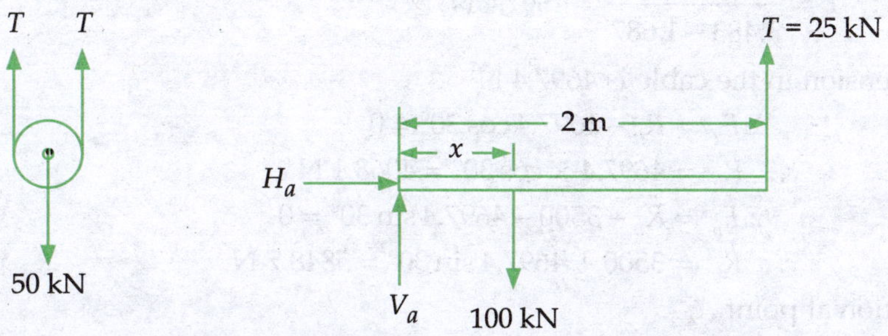

$$\Sigma F_x = 0\,; \quad H_a = 0$$
$$\Sigma F_y = 0\,; \quad V_a - 100 + 25 = 0\,; \quad V_a = 75 \text{ N}$$

Therefore the reaction at the hinged end is 75 N in the vertical direction.

$$\Sigma M_a = 0\,; \quad 100 \times x - 25 \times 2 = 0, \quad x = 0.5 \text{ m}$$

EXAMPLE 3.34

A load P of 3500 N is acting on the boom which is held by a cable BC as shown in Fig. 3.42. Neglecting weight of the boom :

(a) *draw the free body diagram of the boom,*

(b) *find the tension in cable BC, and*

(c) *determine the reaction at A*

Solution : *(a, b)* The free body diagram of the beam will be as shown and the equations of equilibrium of the coplanar system of forces are :

$$\Sigma F_x = 0, \Sigma F_y = 0 \text{ and } \Sigma M = 0$$

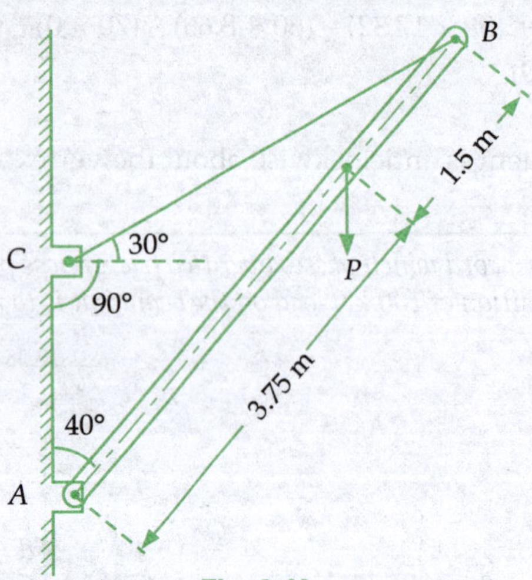

Fig. 3.42

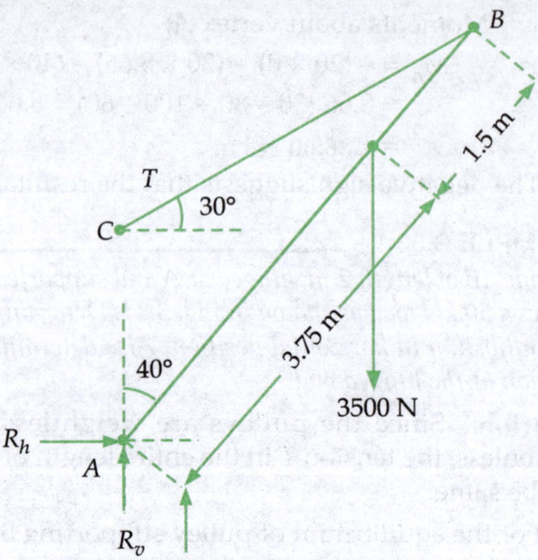

Free body diagram

Taking moments of all the forces about point A, we get

$$-3500 \times 3.75 \sin 40^\circ + T \cos 30^\circ \times (3.75 + 1.5) \cos 40^\circ - T \sin 30^\circ \times (3.75 + 1.5) \sin 40^\circ = 0$$

or $$-8436.6 + 3.483\,T - 1.687\,T = 0$$

That gives : $$T = \frac{8436.6}{3.483 - 1.687} = \mathbf{4697.4\ N}$$

Thus the tension in the cable is 4697.4 N

(c) $$\Sigma F_x = R_h - 4697.4 \cos 30^\circ = 0$$

$$\therefore \quad R_h = 4697.4 \times \cos 30^\circ = 4068.1 \text{ N}$$

$$\Sigma F_y = R_v - 3500 - 4697.4 \sin 30^\circ = 0$$

$$\therefore \quad R_v = 3500 + 4697.4 \sin 30^\circ = 5848.7 \text{ N}$$

Total reaction at point A,

$$R = \sqrt{R_h^2 + R_v^2} = \sqrt{(4068.1)^2 + (5848.7)^2} = \mathbf{7124.5\ N}$$

If θ is the angle which the reaction at A makes with the x-axis, then

$$\theta = \tan^{-1}\left(\frac{R_v}{R_h}\right) = \tan^{-1}\left(\frac{5848.7}{4068.1}\right) = \mathbf{55.2^\circ}$$

EXAMPLE 3.35

A rigid link ABC, supported by cylinders at A and C, is loaded by a force of 25 kN as shown in Fig. 3.43. Each cylinder has a diameter of 20 cm and AB = BC = 30 cm. Neglecting friction and self weights of cylinders and link, estimate the support reactions at D, E and F.

Solution : The different forces acting on the link are :

(*i*) Reactions R_d and R_e passing through A

(*ii*) reaction R_f passing through C

(*iii*) 25 kN force at B.

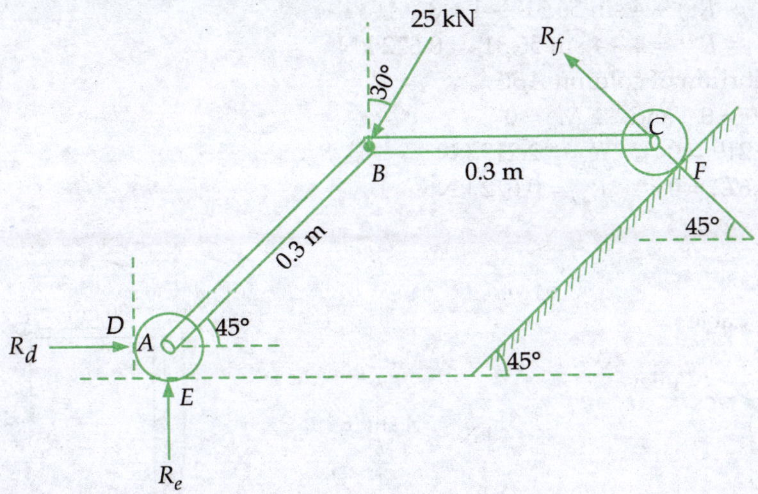

Fig. 3.43.

Under equilibrium conditions: $\Sigma F_x = 0$; $\Sigma F_y = 0$ and $\Sigma M = 9$

Taking moments of all forces about point A (clockwise positive), we get

$$\Sigma M_a = (25 \cos 30°) \times 0.3 \cos 45° - (25 \sin 30°) \times 0.3 \sin 45°$$
$$- R_f \sin 45° \times 0.3 \sin 45° - R_f \cos 45° \times (0.3 \cos 45° + 0.3)$$

$$\therefore \quad 0 = 4.592 - 2.651 - 0.15\, R_f - 0.362\, R_f$$

That gives: $$R_f = \frac{4.592 - 2.651}{0.15 + 0.362} = \mathbf{3.789\ kN}$$

Now, $$\Sigma F_x = R_d - 25 \sin 30° - R_f \sin 45° = 0$$

$$\therefore \quad R_d = 25 \sin 30 + R_f \sin 45°$$
$$= 25 \times 0.5 + 3.789 \times 0.707 = \mathbf{15.179\ kN}$$

$$\Sigma F_y = R_e - 25 \cos 30° + R_f \cos 45° = 0$$

$$\therefore \quad R_e = 25 \cos 30° - R_f \cos 45°$$
$$= 25 \times 0.866 - 3.789 \times 0.707 = \mathbf{18.971\ kN}$$

EXAMPLE 3.36

A column and cable-pulley arrangement supports a load of 4 kN as shown in Fig. 3.44. The weight is suspended by the cable and pulley attached at pin C. The pulley has a radius of 15 cm and the weights of column, pulley and cable are negligibly small. Make calculations for the support reactions at joint A of the column.

Solution : From the given slope of the cable,

$$\alpha = \tan\left(\frac{3}{2}\right) = 56.31°$$

For the equilibrium of pulley and cable arrangements,

$$\Sigma F_x = 0 \quad \text{and} \quad \Sigma F_y = 0$$

$\Sigma F_x = 0$: $$R_{cx} - 4 \cos 56.31° = 0$$

$$\therefore \quad R_{cx} = 4 \cos 56.31° = 2.219\ \text{kN}$$

$\Sigma F_y = 0$: $R_{cy} + 4 \sin 56.31° - 4 = 0$

$\therefore$ $R_{cy} = 4 - 4 \sin 56.31° = 0.672$ kN

For the equilibrium of column *ABC*,

$\Sigma F_x = 0$, $\Sigma F_y = 0$ and $\Sigma M_o = 0$

$\Sigma F_x = R_{ax} - 2.219 = 0$; $R_{ax} = 2.219$ kN

$\Sigma F_y = R_{ay} - 0.672 = 0$; $R_{ay} = 0.672$ kN

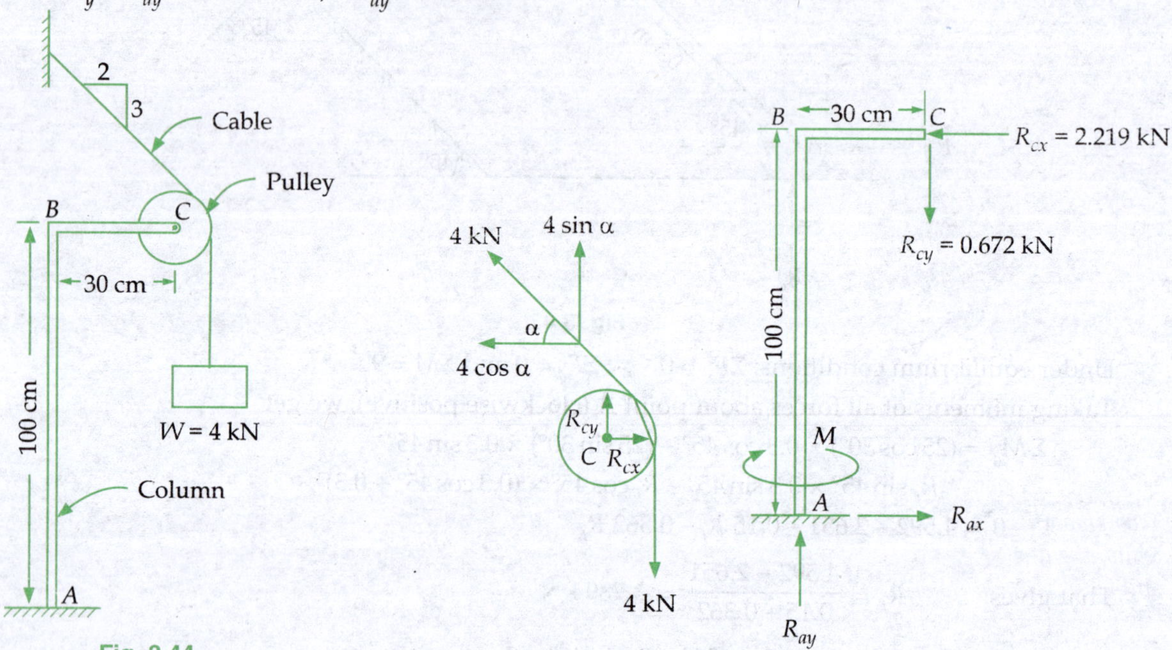

Fig. 3.44

If *M* is the clockwise moment developed at *A*, then taking moments about end *A* (clockwise + *ve*), we get

$$M - 2.219 \times \frac{100}{100} + 0.672 \times \frac{30}{100} = 0$$

$$\therefore \quad M = 2.219 \times 1 - 0.672 \times 0.3 = 2.0174 \text{ kNm} \quad \text{(clockwise).}$$

EXAMPLE 3.37

A beam of 3 m span is loaded and supported as shown in Fig. 3.45. Neglecting weight of the beam and assuming pulley to be frictionless, make calculations for the tension in the cable and the reaction at pin support.

Solution : Since the pulley is weightless and frictionless, the tension *T* in the entire length of the cable will be same. From the given configuration,

$$\tan \alpha = \frac{4}{2} = 2, \quad \alpha = 63.43°$$

With reference to freebody diagram of the beam as shown let the components of reaction at the pin support lie along the directions indicated.

Under conditions of equilibrium : $\Sigma F_x = 0$; $\Sigma F_y = 0$ and $\Sigma M = 0$. Taking moment of the given system of forces about pin support *A*,

$$\Sigma M_a = 10 \times 1 - T \sin 63.43° \times 2 + 5 \times 3 = 0$$

$$10 - 1.789\, T + 15 = 0; \quad T = 13.97 \text{ kN}$$

$$\Sigma F_x = R_{ah} - T\cos 63.43^\circ = 0$$

$$\therefore \quad R_{ah} = 13.97 \times \cos 63.43 = \mathbf{6.248\ kN}$$

$$\Sigma F_y = R_{av} - 10 + T\sin 63.43^\circ - 5 = 0$$

$$\therefore \quad R_{av} = 10 - 13.97\sin 63.43^\circ + 5$$

$$= \mathbf{2.50\ kN}$$

Resultant reaction at point A,

$$R_a = \sqrt{(R_{ah})^2 + (R_{av})^2}$$

$$= \sqrt{(6.248)^2 + (2.50)^2} = \mathbf{6.73\ kN}$$

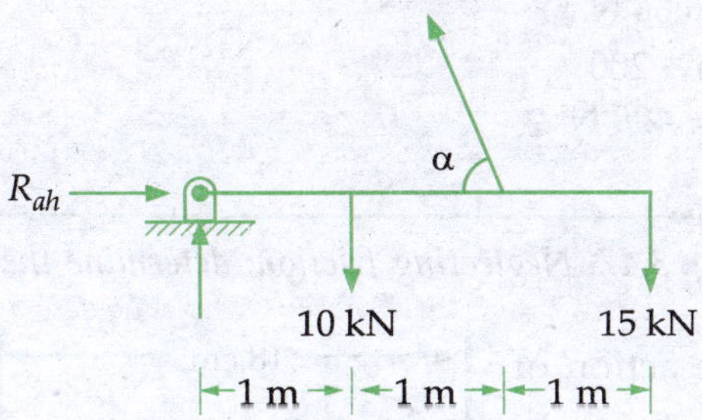

Free body diagram

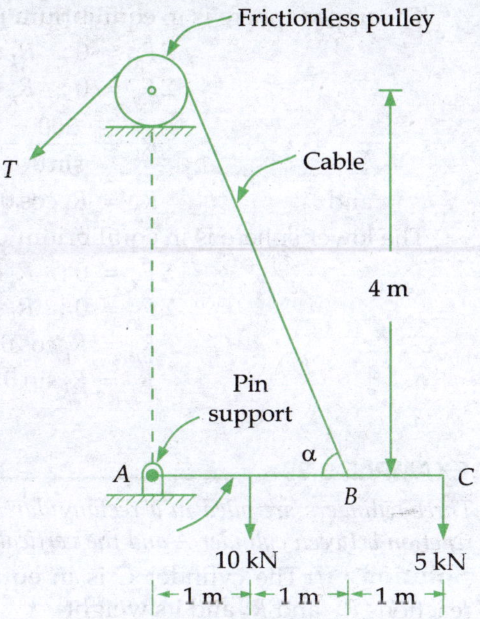

Fig. 3.45

EXAMPLE 3.38

Two smooth spheres, each of radius 20 cm and weight 200 N, rest in a horizontal channel having vertical walls, the distance between which is 72 cm as shown in Fig. 3.46. Find the pressures at contact points E, F and G.

Solution : Let R_e, R_f and R_g be the reactions at surface points E, F and G respectively.

From the geometrical configuration

$$OO' = 20 + 20 = 40 \text{ cm}$$

$$OM = 32 \text{ cm}$$

$$\cos\theta = \frac{32}{40} = 0.8\ ;\ \theta = 36.87^\circ\ ;\ \sin\theta = 0.6$$

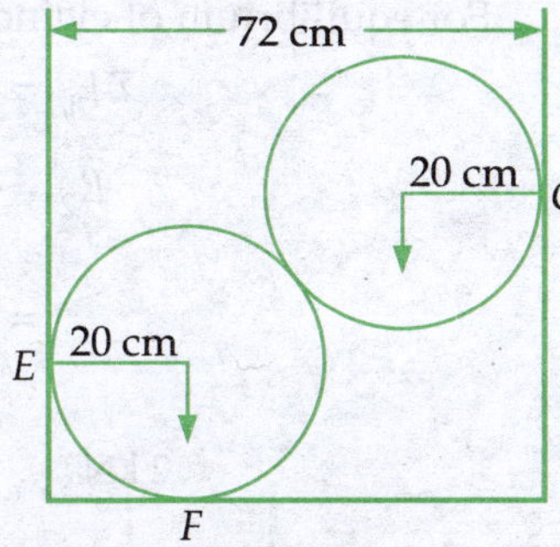

Fig. 3.46

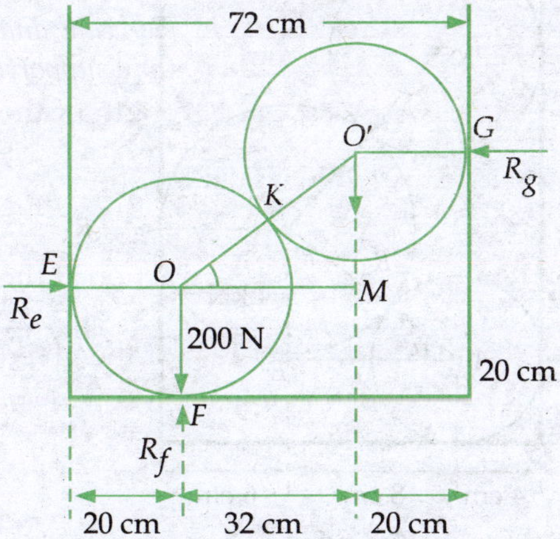

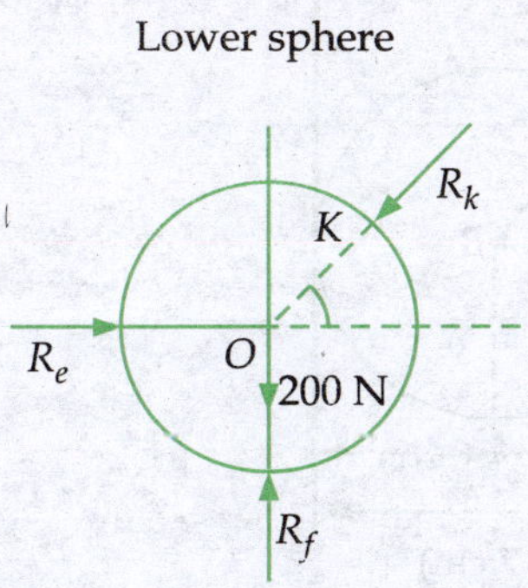

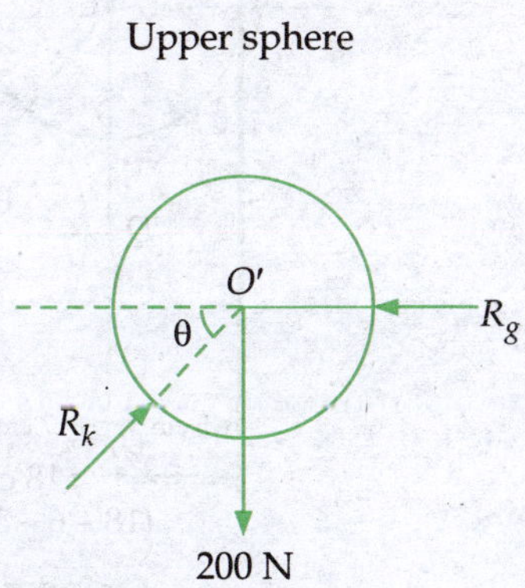

The upper sphere is in equilibrium under the action of forces R_g, R_k and its weight. For that

$$\Sigma F_x = 0; \quad R_k \cos\theta - R_g = 0$$

$$\Sigma F_y = 0; \quad R_k \sin\theta - 200 = 0$$

$$\therefore \quad R_k = \frac{200}{\sin\theta} = \frac{200}{0.6} = \mathbf{333.33\ N}$$

and $$R_g = R_k \cos\theta = 333.33 \times 0.8 = \mathbf{266.66\ N}$$

The lower sphere is in equilibrium under the action of R_e, R_f, R_k and its weight. For that

$$\Sigma F_x = 0; \quad R_k \cos\theta - R_e = 0$$

$$\Sigma F_y = 0; \quad R_k \sin\theta + 200 - R_f = 0$$

$$\therefore \quad R_e = R_k \cos\theta = 333.33 \times 0.8 = \mathbf{266.66\ N}$$

$$R_f = R_k \sin\theta + 200 = 333.33 \times 0.6 + 200$$

$$= 200 + 200 = \mathbf{400\ N}$$

EXAMPLE 3.39

Three cylinders are piled in a rectangular ditch as shown in Fig. 3.47. Neglecting friction, determine the reaction between cylinder A and the vertical wall.

Solution : (*i*) The cylinder C is in equilibrium under the action of reactions R_1 and R_2 and its weight

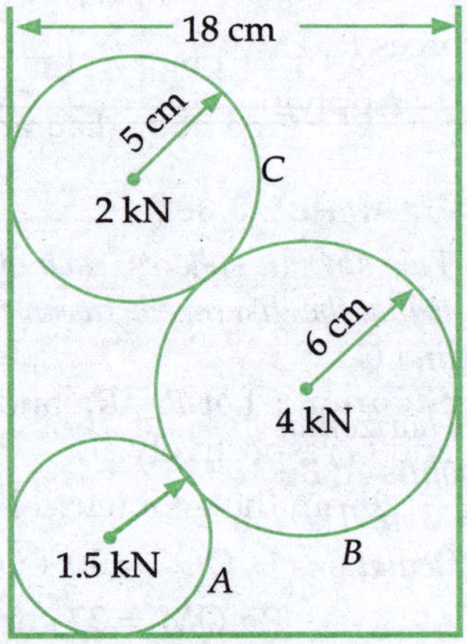

Fig. 3.47

In $\Delta\,CBD$: $BC = 6 + 5 = 11$ cm

$BD = 18 - 5 - 6 = 7$ cm

$$\cos\theta = \frac{BD}{BC} = \frac{7}{11} = 0.636; \ \theta = 50.48^\circ$$

For equilibrium of cylinder C,

$$\Sigma F_y = 0; \quad R_2 \sin\theta - 2 = 0$$

$$\therefore \quad R_2 = \frac{2}{\sin\theta} = \frac{2}{\sin 50.48^\circ}$$

$$= \frac{2}{0.7714} = \mathbf{2.593\ kN}$$

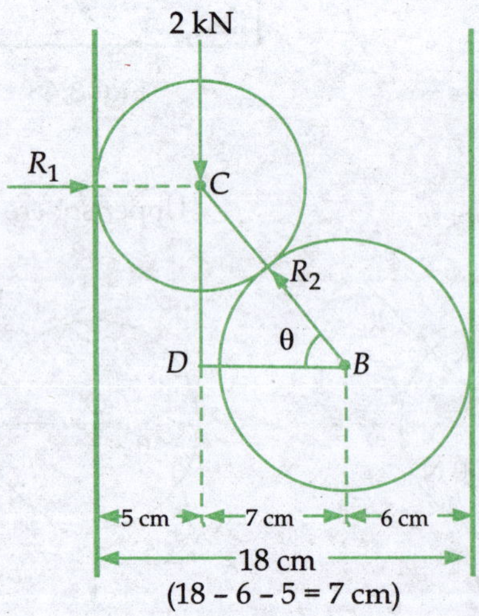

Cylinder C

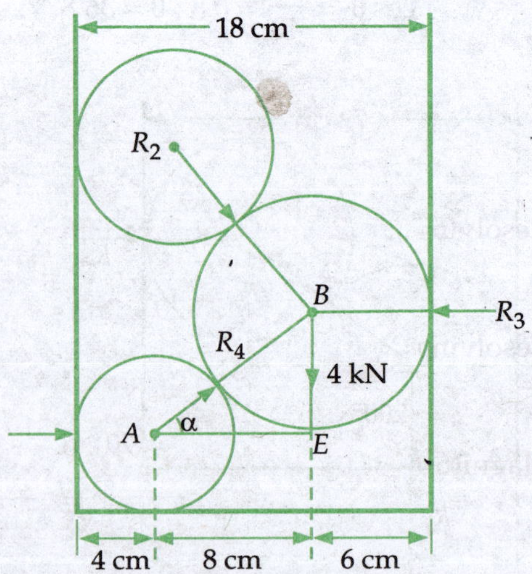

Cylinder B

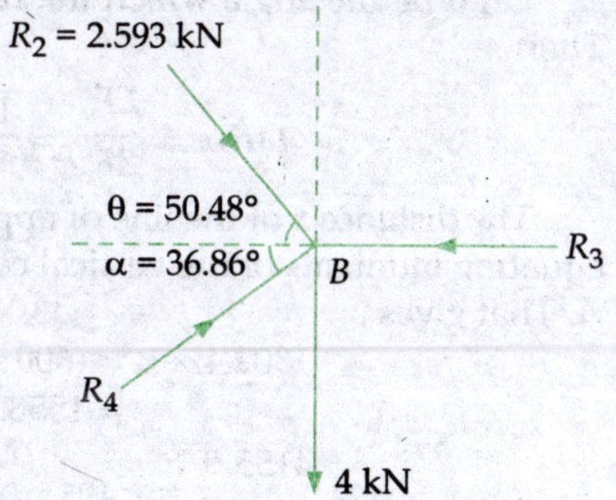

(*ii*) The cylinder *B* is in equilibrium under the action of forces R_2, R_3, R_4 and its weight

In ΔABE : $AB = 6 + 4 = 10$ cm

$$AE = 18 - 6 - 4 = 8 \text{ cm}$$

$$\cos\alpha = \frac{AE}{AB} = \frac{8}{10} = 0.8 \; ; \alpha = 36.86°$$

For equlibrium of cylinder *B*,

$$\Sigma F_y = 0; \; R_2 \sin\theta + 4 - R_4 \sin\alpha = 0$$

or $$2.593 \sin 50.48° + 4 - R_4 \sin 36.86° = 0$$

or $$2.593 \times 0.7714 + 4 - R_4 \times 0.6 = 0$$

$\therefore$ $$R_4 = \frac{2.593 \times 0.7714 + 4}{0.6}$$

$$= \frac{2+4}{0.6} = 10 \text{ kN}$$

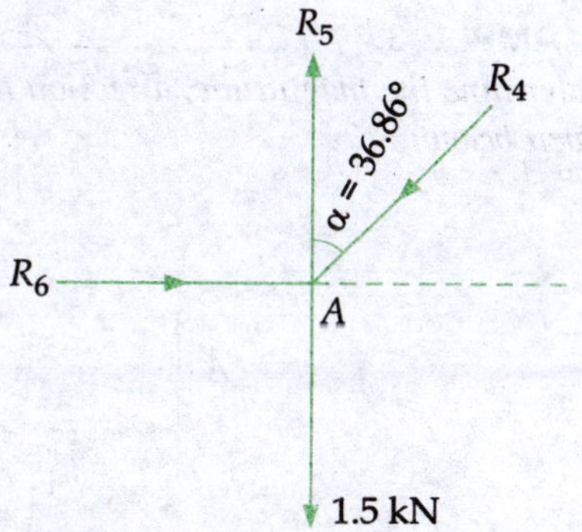

(*iii*) The cylinder *A* is in equilibrium under the action of forces R_4, R_5, R_6 and its weight

Applying the condition $\Sigma F_x = 0$, we have

$$R_6 - R_4 \cos\alpha = 0$$

$$R_6 = R_4 \cos\alpha = 10 \times \cos 36.86 = 8 \text{ kN}$$

EXAMPLE 3.40

A horizontal line ABCD measuring 9 m is acted upon by forces of magnitude 400, 600, 400 and 200 N at points A, B, C and D respectively with downward direction. These points are so located that AB = BC = CD = 3 m. The lines of action of the forces are inclined at 90°, 60°, 45° and 30° respectively with AB. Make calculations for the magnitude, position and direction of the resultant.

Solution : Refer to Fig. 3.48 for the given force system.

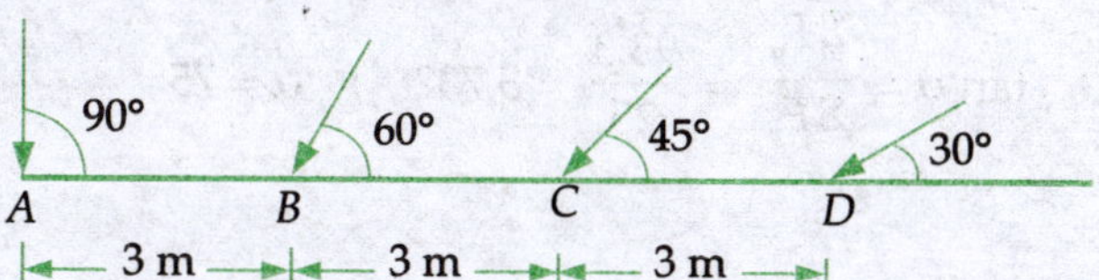

Fig. 3.48

Resolving all the forces horizontally,

$$\Sigma F_x = 200\cos 30° + 400\cos 45° + 600\cos 60° + 400\cos 90°$$

$$= 173.20 + 282.8 + 300 + 0 = 756 \text{ N}$$

Resolving all the forces vertically,

$$\Sigma F_y = 200\sin 30° + 400\sin 45° + 600\sin 60° + 400\sin 90°$$

$$= 100 + 282.8 + 519.6 + 400 = 1302.4 \text{ N}$$

Magnitude of the resultant force,

$$F = \sqrt{(\Sigma F_x)^2 + (\Sigma F_y)^2} = \sqrt{756^2 + 1302.4^2} = 1505.9 \text{ N}$$

Let α be the angle which the resultant makes with the horizontal direction, *i.e.*, with the
Then

$$\tan\alpha = \frac{\Sigma F_y}{\Sigma F_x} = \frac{1302.4}{756} = 1.723 \quad \therefore \quad \alpha = \tan^{-1}(1.723) = 59.87°$$

The distance x of the line of application of the resultant force from end A can be worked out by equating moments of the vertical components of the given forces and the resultant force about end A. That gives

$$1302.4 \times x = (600 \sin 60°) \times 3 + (400 \sin 45° \times 6 + (200 \sin 30° \times 9)$$
$$= 1558.84 + 1697.06 + 900 = 4155.9$$

$$\therefore \quad x = \frac{4155.9}{1302.4} = 3.191 \text{ m}$$

EXAMPLE 3.41

Determine the magnitude, direction and position of the resultant of the system of forces shown in Fig. 3.49 given below.

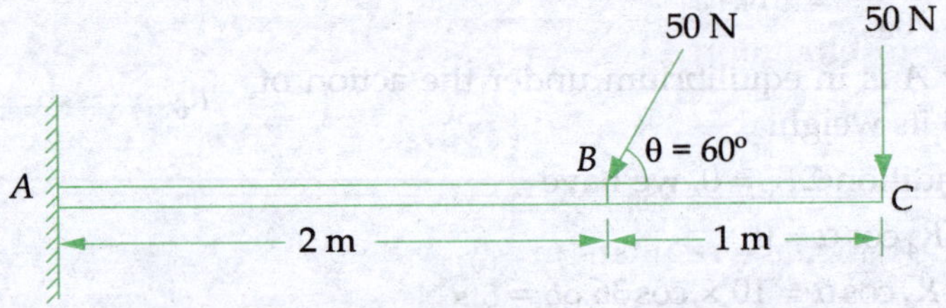

Fig. 3.49

Solution: Resolving the given forces horizontally and vertically, we get

$$\Sigma F_x = -50 \cos 60° = -25 \text{ N}$$
$$\Sigma F_y = -50 \sin 60° - 50 = -93.3 \text{ N}$$

$$\text{Magnitude of resultant } R = \sqrt{(\Sigma F_x)^2 + (\Sigma F_y)^2}$$
$$= \sqrt{(-25)^2 + (-93.3)^2} = 96.59 \text{ N}$$

If α be the angle which the resultant makes with the horizontal direction, then

$$\tan\alpha = \frac{\Sigma F_y}{\Sigma F_x} = \frac{93.3}{25} = 3.732 \quad ; \quad \alpha = 75°$$

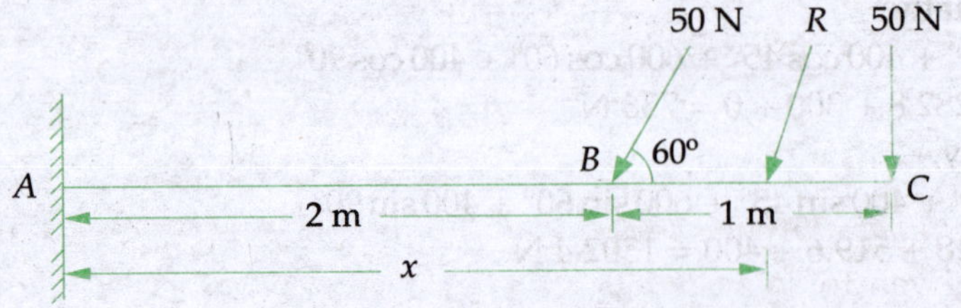

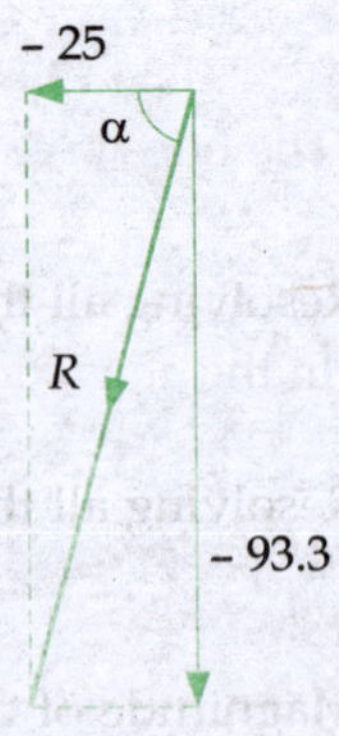

Let the resultant R be acting at distance x from the fixed end. Then under equilibrium conditions, moment of vertical component of resultant about A

= sum of the moments of vertical components of given forces about A

That gives : $93.3\,x = 50 \sin 60° \times 2 + 50 \times 3 = 86.60 + 150 = 236.6$

$$\therefore \quad x = \frac{236.6}{93.3} = \textbf{2.536 m}$$

EXAMPLE 3.42

Forces equal to P, 2P, 3P and 4P act along the sides AB, BC, CD and DA of a square ABCD. Find the magnitude, direction and line of action of the resultant.

Solution : Refer to Fig. 3.50 for the given force system

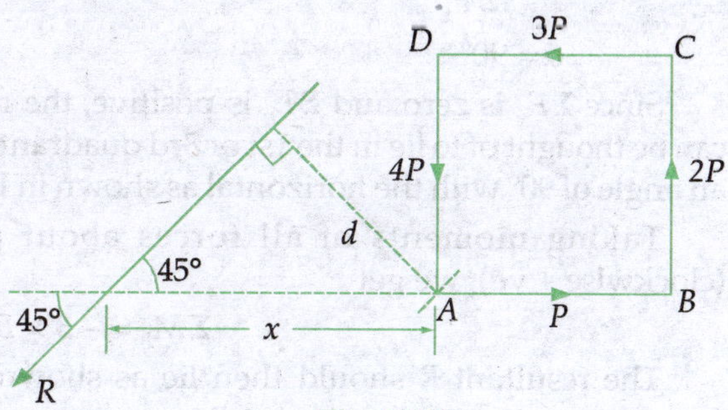

Fig. 3.50

Resolving all the forces horizontally and vertically,

$$\Sigma F_x = P - 3P = -2P$$
$$\Sigma F_y = 2P - 4P = -2P$$

Magnitude of the resultant force,

$$R = \sqrt{(\Sigma F_x)^2 + (\Sigma F_y)^2}$$
$$= \sqrt{(-2P)^2 + (-2P)^2} = \mathbf{2\sqrt{2}\,P}$$

Let α be the angle which the resultant makes with the horizontal. Then

$$\tan \alpha = \frac{\Sigma F_y}{\Sigma F_x} = \frac{-2P}{-2P} = 1; \qquad \alpha = 45°$$

Since both ΣF_x and ΣF_y are negative, resultant lies in the 3rd quadrant making an angle of 45° with horizontal as shown in Fig. 3.50.

Taking moments of all forces about point A (clockwise + ve), we get

$$\Sigma M_a = -2P \times a - 3P \times a = -5Pa$$

The resultant R should then lie as shown in Fig. 3.71 so that it can produce anti-clockwise (negative) moment about point A.

If d is the perpendicular distance of resultant from A, then

$$R \times d = \Sigma M_a$$
$$\therefore \quad d = \frac{\Sigma M_a}{R} = \frac{5Pa}{2\sqrt{2}\,P} = \frac{5a}{2\sqrt{2}}$$

The horizontal distance x of the line of action of resultant from point A is

$$x = \frac{d}{\sin \alpha} = \frac{5a}{2\sqrt{2}} \div \sin 45° = 2.5\,a$$

In the above identities a is a measure of the side of each side of the square block.

EXAMPLE 3.43

Three forces equal to 1 kN, 2 kN and 3 kN are respectively acting in order along the three sides of an equilateral triangle. Make calculations for the magnitude, direction and position of their resultant.

Solution : Refer Fig. 3.51 for the given force system. Resolving all the forces horizontally and vertically.

$$\Sigma F_x = -1 \cos 60° + 2 - 3 \cos 60° = -0.5 + 2 - 1.5 = 0$$
$$\Sigma F_y = -1 \sin 60° + 3 \sin 60° = -0.866 + 2.598 = 1.732 \text{ kN}$$

Magnitude of the resultant force,

$$R = \sqrt{(\Sigma F_x)^2 + (\Sigma F_y)^2}$$

$$= \sqrt{0^2 + 1.732^2} = 1.732 \text{ N}$$

Let α be the angle which the resultant makes with the horizontal.

$$\tan \alpha = \frac{\Sigma F_y}{\Sigma F_x} = \frac{1.732}{0} = \infty$$

$$\therefore \quad \alpha = 90^\circ$$

Since ΣF_x is zero and ΣF_y is positive, the resultant can be thought of to lie in the 1st or 3rd quadrant making an angle of 90° with the horizontal as shown in Fig. 3.44.

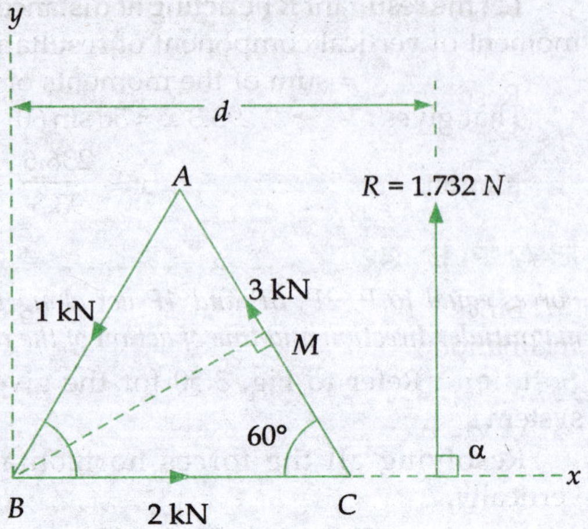

Fig. 3.51

Taking moments of all forces about point B (clockwise + ve), we get

$$\Sigma M_b = -3 \times BM = -3 \times a \sin 60^\circ$$

The resultant R should then lie as shown in Fig. 3.50 so that it gives rise to anti-clockwise (negative) moment about point B.

If d is the perpendicular distance of resultant from B, then

$$R \times d = \Sigma M_b$$

$$\therefore d = \frac{\Sigma M_b}{R} = \frac{3a \sin 60^\circ}{1.732} = 3a$$

where a is a measure of each side of the equilateral triangle.

EXAMPLE 3.44

A force system comprises of three forces of 3 kN, 4 kN and – 5 kN and these forces act along the sides AB, BC and CA of an equilateral triangle with each side 12 cm. A is the apex and side BC forms the base of the triangle. Determine the resultant both in magnitude and direction with line BC.

Solution : Refer Fig. 3.52 for the given force system. The origin O is at the point B and the x-axis is taken along the line BC.

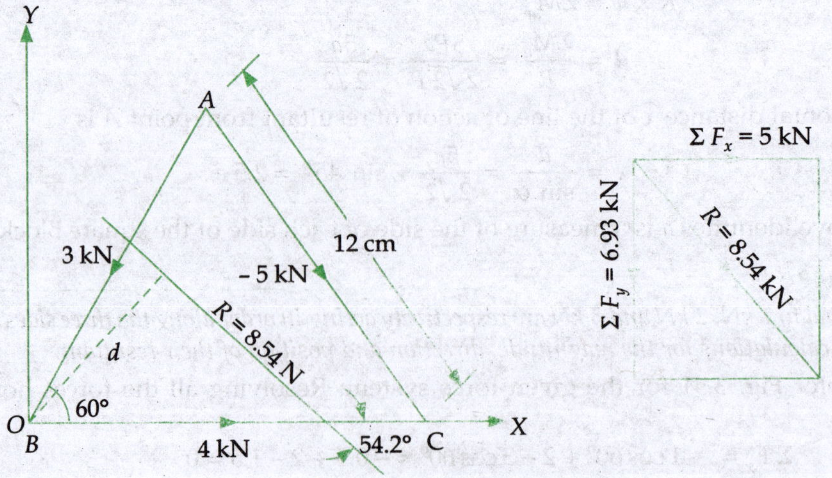

Fig. 3.52

Resolving at the forces horizontally (along X-axis)

$$\Sigma F_x = -3\cos 60° + 4 + 5\cos 60° = -3 \times 0.5 + 4 + 5 \times 0.5 = 5 \text{ kN}$$

Resolving all the forces vertically (along Y-axis)

$$\Sigma F_y = -3\sin 60° - 5\sin 60° = -3 \times 0.866 - 5 \times 0.866 = -6.93 \text{ kN}$$

Magnitude of the resultant force

$$R = \sqrt{(\Sigma F_x)^2 + (\Sigma F_y)^2} = \sqrt{5^2 + (-6.93)^2} = 8.54 \text{ kN}$$

Since ΣF_x is positive and ΣF_y is negative, the resultant lies in the fourth quadrant, and its inclination with the horizontal direction line *BC* is

$$\tan\alpha = \frac{\Sigma F_y}{\Sigma F_x} = \frac{6.93}{5} = 1.386 \qquad \therefore \quad \alpha = 54.2°$$

(*b*) The perpendicular distance *d* of resultant *R* from origin (end *B*) can be worked out by equating the moment of the given forces and the resultant force about end *B*. That gives :

$$8.54 \times d = 5 \times 12 \sin 60°$$

It needs to be noted that the lines of action of 3 kN and 4 kN forces pass through point *B* and as such these force make no contribution towards moments

$$\therefore \quad d = \frac{5 \times 12 \times \sin 60°}{8.54} = 6.08 \text{ cm}$$

EXAMPLE 3.45

A horizontal beam AD of length 12 m is acted upon by a set of forces as shown in Fig. 3.53

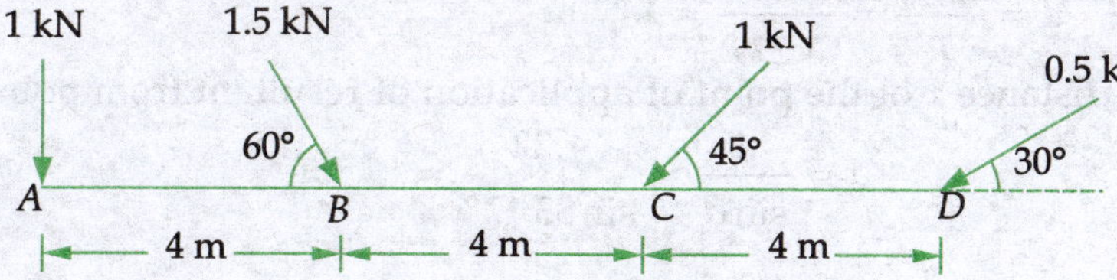

Fig. 3.53

Determine the magnitude, direction and position of the resultant.

Solution : Resolving the forces along *x*-and *y*-directions, we get

$$\begin{aligned}\Sigma F_x &= 0 + 1.5\cos 60° - 1\cos 45° - 0.5\cos 30° \\ &= 0.75 - 0.707 - 0.433 = -0.39 \text{ kN} \\ \Sigma F_y &= -1 - 1.5\sin 60° - 1\sin 45° - 0.5\sin 30° \\ &= -1 - 1.299 - 0.707 - 0.25 = -3.256 \text{ kN}\end{aligned}$$

Resultant force

$$R = \sqrt{(\Sigma F_x)^2 + (\Sigma F_y)^2} = \sqrt{(-0.39)^2 + (3.256)^2} = \sqrt{10.7536} = 3.279 \text{ kN}$$

Inclination of resultant with the horizontal,

$$\tan\alpha = \frac{\Sigma F_y}{\Sigma F_x} = \frac{3.256}{0.39} = 8.333; \qquad \therefore \quad \alpha = 83.17°$$

Since both ΣF_x and ΣF_y are negative, the resultant lies in the 3rd quadrant making an angle of 83.17° with horizontal as shown in Fig. 3.54

Taking moments of all forces about point A (clockwise $+ve$), we get

$$\Sigma M_a = (1.5 \sin 60°) \times 4 + (1 \sin 45°) \times 8 + (0.5 \sin 30°) \times 12$$
$$= 5.196 + 5.657 + 3 = 13.853 \text{ kNm}$$

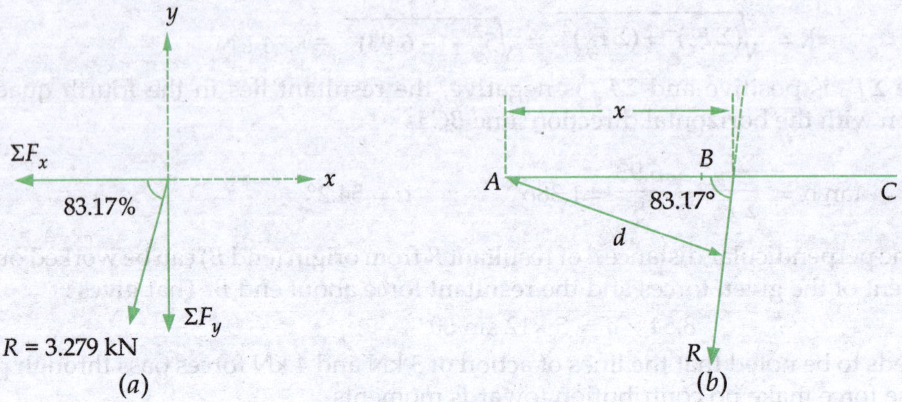

Fig. 3.54

The resultant R should then lie as shown in Fig. 3.54 so that it can produce clockwise (positive) moment about point A.

If d is the perpendicular distance of resultant from A, then

$$R \times d = \Sigma M_a$$

$$\therefore \quad d = \frac{\Sigma M_a}{R} = \frac{13.853}{3.279} = 4.22 \text{ m}$$

The horizontal distance x of the point of application of resultant from point A is

$$x = \frac{d}{\sin \alpha} = \frac{4.22}{\sin 83.17°} = 4.25 \text{ m}$$

EXAMPLE 3.46

A plate measuring 6 m × 4 m is acted upon a set of forces in its plane as depicted in Fig. 3.55. Determine the magnitude, direction and position of the resultant force.

Solution : Resolving the forces in the horizontal and vertical directions, we get

$$\Sigma F_x = 80 - 100 \cos 30° + 120 \cos 45° - 60 \cos 30°$$
$$= 80 - 86.6 + 84.85 - 51.96 = +26.29 \text{ N} \rightarrow$$

The force 100 N acting at point E has no component in the x-direction.

$$\Sigma F_y = -100 \sin 30° + 120 \sin 45° + 100 + 60 \sin 30°$$
$$= -50 + 84.85 + 100 + 30 = 164.85 \text{ N} \uparrow$$

The force 80 N acting at point A has no component in the y-direction.

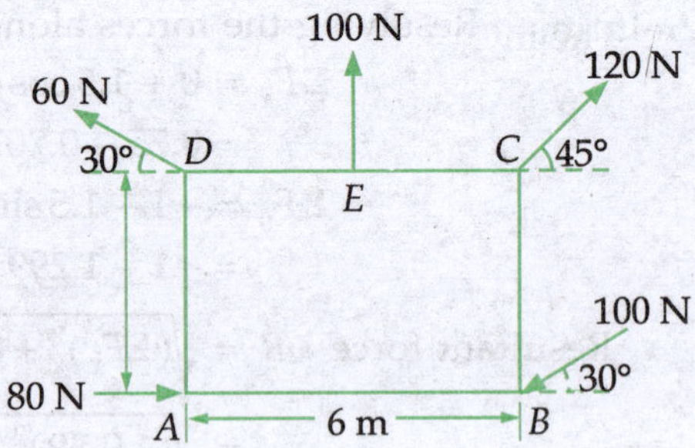

Fig. 3.55

$$\text{Resultant force } R = \sqrt{(\Sigma F_x)^2 + (\Sigma F_y)^2}$$
$$= \sqrt{(26.29)^2 + (164.85)^2} = \sqrt{27866.68} = 166.93 \text{ N}$$

Inclination α of the resultant with the horizontal,

$$\tan\alpha = \frac{\Sigma F_y}{\Sigma F_x} = \frac{164.85}{26.29} = 6.27 \quad \therefore \quad \alpha = 80.94^\circ$$

Since both ΣF_x and ΣF_y are positive, the resultant lies in first quadrant at an angle 80.94° with horizontal, *i.e.*, with x-axis as shown in (Fig. 3.56 *a*).

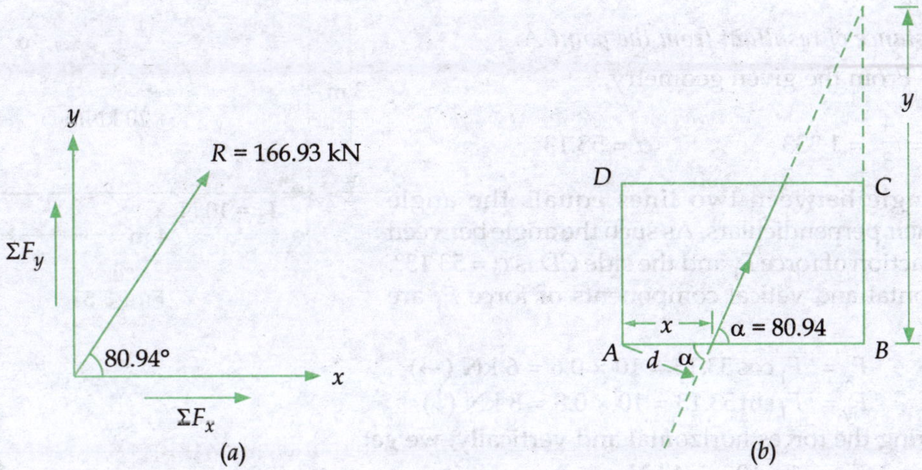

Fig. 3.56

(*b*) **Position of resultant force:**

Taking moments about point A (clockwise + *ve*), we get

$$\Sigma M_a = (100\sin 30^\circ) \times 6 - (120\sin 45^\circ) \times 6 + (120\cos 45^\circ) \times 4 - 100 \times 3 - (60\cos 30^\circ) \times 4$$

$$= 300 - 509.1 + 339.4 - 300 - 207.8 = -377.5 \text{ Nm (anticlockwise)}$$

The resultant R should then lie as shown in (Fig. 3.56 *b*) so that it can produce anticlockwise (negative) moment about point A.

If d is the perpendicular distance of the resultant from A, then

$$R \times d = \Sigma M_a \quad \therefore \quad d = \frac{\Sigma M_a}{R} = \frac{377.5}{166.93} = 2.261 \text{ m}$$

The intercepts x on x-axis and y on y-axis are then given by

$$x = \frac{d}{\sin\alpha} = \frac{2.261}{\sin 80.94} = 2.29 \text{ m}$$

$$y = \frac{d}{\cos\alpha} = \frac{2.261}{\cos 80.94} = 14.36 \text{ m}$$

The intercepts on x-axis and y-axis can also be worked out as

$$x = \frac{\Sigma M_a}{\Sigma F_y} = \frac{377.5}{164.85} = 2.29 \text{ m}$$

$$y = \frac{\Sigma M_a}{\Sigma F_x} = \frac{377.5}{26.29} = 14.36 \text{ m}$$

Since intercept y is greater than BC, the resultant meets the arm when produced.

EXAMPLE 3.47

A rectangular body ABCD is acted upon by a set of forces as indicated in the Fig. 3.57. The force F_1 is oriented perpendicular to the diagonal AC of the body. Determine the :

(a) resultant of the force system,

(b) points where the resultant cuts the X and Y axis, and

(c) distance of resultant from the point A.

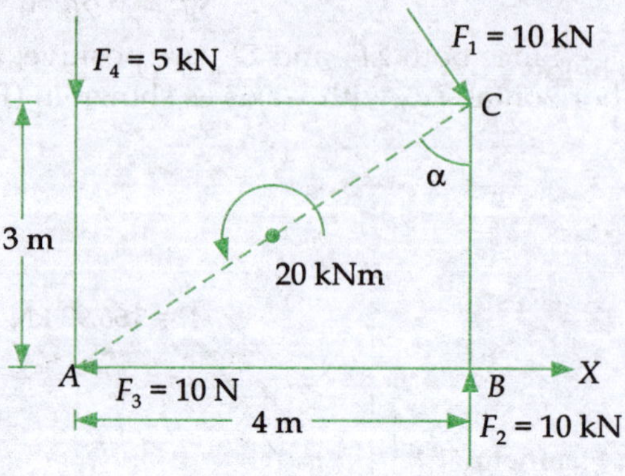

Fig. 3.57

Solution : From the given geometry,

$$\tan \alpha = \frac{4}{3} = 1.333 \quad \therefore \quad \alpha = 53.13°$$

The angle between two lines equals the angle between their perpendiculars. As such the angle between the line of action of force F_1 and the side CD is $\alpha = 53.13°$. The horizontal and vetical components of force F_1 are then :

$$F_x = F_1 \cos 53.13 = 10 \times 0.6 = 6 \text{ kN } (\rightarrow)$$

$$F_y = F_1 \sin 53.13 = 10 \times 0.8 = 8 \text{ kN } (\downarrow)$$

Resolving the forces horizontal and vertically, we get

$$\Sigma F_x = 6 - 10 = -4 \text{ kN}$$

$$\Sigma F_y = -8 + 10 - 5 = -3 \text{ kN}$$

Magnitude of resultant

$$R = \sqrt{(\Sigma F_x)^2 + (\Sigma F_y)^2} = \sqrt{(-4)^2 + (-3)^2} = 5 \text{ kN}$$

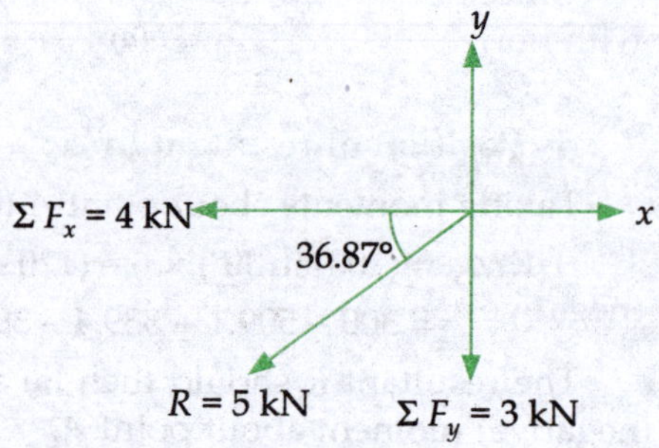

Since both ΣF_x and ΣF_y are *–ve*, the resultant R lies in third quadrant and its inclination with horizontal or *x*-axis is

$$\tan \alpha = \frac{\Sigma F_y}{\Sigma F_x} = \frac{3}{4} = 0.75 \quad \therefore \alpha = 36.87°$$

(*b*) The distance *x* between point of intersection of resultant with *x*-axis and the origin *A* can be worked out by equating the moment of the given force system and the resultant force about origin *A*. That gives :

$$5 \sin 36.87° \times x = 10 \cos 53.13° \times 3 + 10 \sin 53.3° \times 4 - 10 \times 4 - 20$$

$$= 18 + 32 - 40 - 20 = -10 \text{ kNm}$$

$$\therefore \quad x = -\frac{10}{5 \sin 36.87°} = -3.33 \text{ m}$$

The negative sign indicates that the resultant R intersects the X-axis to the left of origin.

Further, $\quad y = x \tan 36.87° = 3.33 \times 0.75$

$$= 2.5 \text{ m}$$

That is the resultant R intersects the Y-axis at 2.5 m above the origin.

The perpendicular distance, d = AM, of the resultant from the origin is

$$d = x \sin 36.87° = 3.33 \times 0.6 \simeq 2 \text{ m}$$

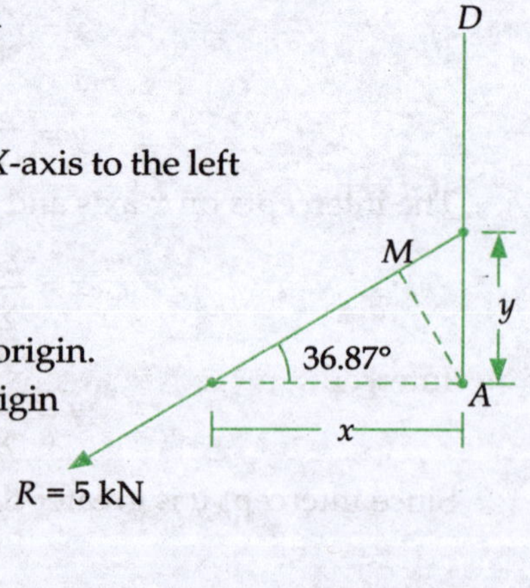

EXAMPLE 3.48

For the force system applied to an angle bracket as shown in Fig. 3.58, determine the magnitude and direction of the resultant. What would be the distance of line of action of the resultant from point O.

Solution : The component of forces in the horizontal and vertical directions are determined as below :

$$\Sigma F_x = 150\cos 35^\circ - 80$$
$$= 122.87 - 80 = 42.87 \text{ kN}$$
$$\Sigma F_y = 150\sin 35^\circ + 20 - 40 = 66 \text{ kN}$$

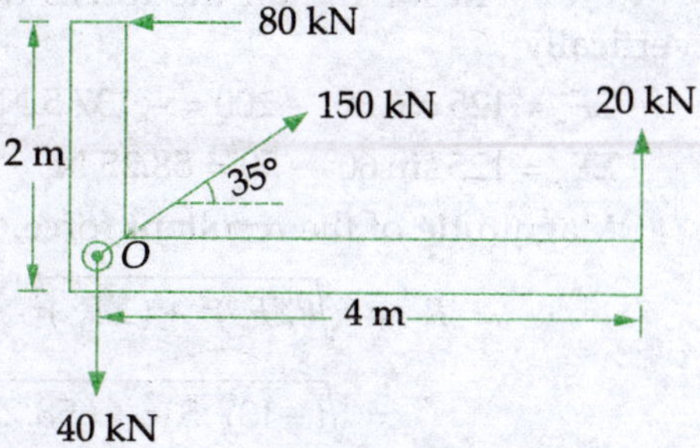

Fig. 3.58

Resultant $R = \sqrt{(\Sigma F_x)^2 + (\Sigma F_y)^2}$

$$= \sqrt{42.87^2 + 66^2} = \sqrt{6193.84}$$
$$= 78.7 \text{ kN}$$

Inclination α (Fig. 3.59 *a*) of the resultant with horizontal,

$$\alpha = \tan^{-1}\left(\frac{\Sigma F_y}{\Sigma F_x}\right) = \tan^{-1}\left(\frac{66}{42.87}\right) =$$

$\tan^{-1}(1.5395) = 57°$

Since both ΣF_x and ΣF_y are positive, the resultant lies in the first quadrant making an angle 57° with horizontal, *i.e.*, with *x*-axis as shown in (Fig. 3.59 *b*).

Position of resultant force

Taking moments of all forces about point *A* (clockwise + *ve*) we get

$$\Sigma M_o = -80 \times 2 - 20 \times 4 = -240 \text{ kN} \quad \text{(anticlockwise)}$$

The resultant *R* should then lie as shown in Fig. 3.59 so that it can produce anticlockwise (negative) moment about point *O*.

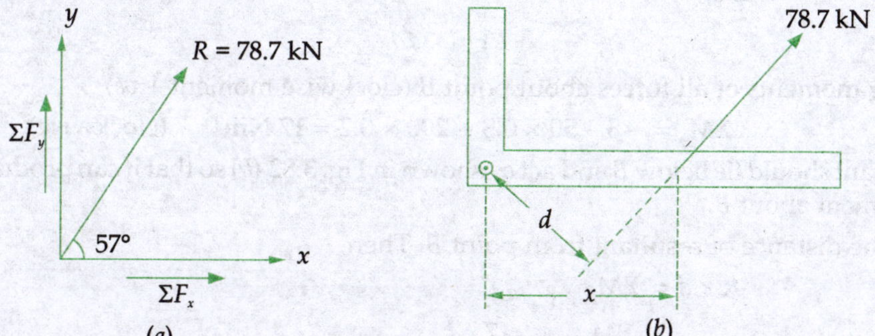

Fig. 3.59

If *d* is the perpendicular distance of resultant from *O*, then

$$R \times d = \Sigma M_o$$

$$\therefore \quad d = \frac{\Sigma M_o}{R} = \frac{240}{78.7} = 3.049 \text{ m}$$

Intercept on *x*-axis is given by

$$x = \frac{d}{\sin\alpha} = \frac{3.049}{\sin 57} = 3.625 \text{ m}$$

EXAMPLE 3.49

An angle bracket has been subjected to three forces and a couple as shown in Fig. 3.60. Determine the resultant of this system of forces. Proceed to locate the points where the line of action of the resultant intersects line AB and the line BC.

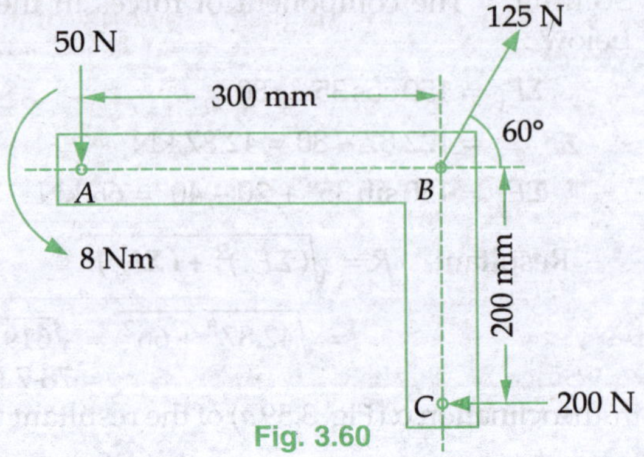

Fig. 3.60

Solution : Resolving all the forces horizontally and vertically

$$\Sigma F_x = 125\cos 60^\circ - 200 = -137.5 \text{ N}$$

$$\Sigma F_y = 125\sin 60^\circ - 50 = 58.25 \text{ N}$$

Magnitude of the resultant force,

$$R = \sqrt{(\Sigma F_x)^2 + (\Sigma F_y)^2}$$

$$= \sqrt{(-137.5)^2 + (58.25)^2} = 149.33 \text{ N}$$

If α be the angle which the resultant makes with the horizontal, then

$$\tan\alpha = \frac{\Sigma F_y}{\Sigma F_x} = \frac{58.25}{137.5} = 0.4236 \quad \therefore \quad \alpha = 22.96^\circ$$

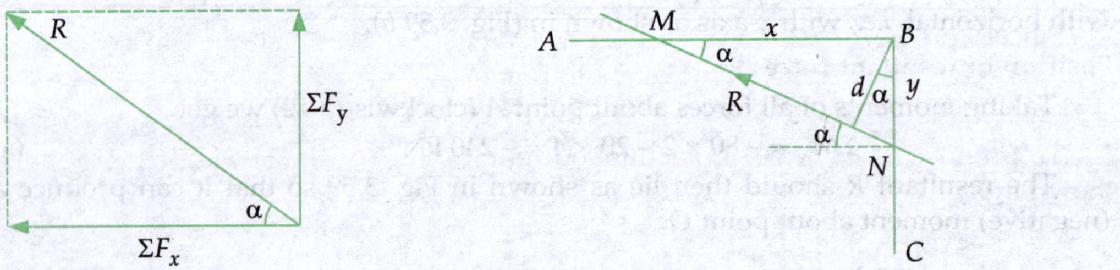

Fig. 3.61

(*b*) Taking moments of all forces about point *B* (clockwise moment + *ve*)

$$\Sigma M_b = -8 - 50 \times 0.3 + 200 \times 0.2 = 17 \text{ Nm} \qquad \text{(clockwise)}$$

The resultant should lie below *B* and act as shown in Fig. 3.82 (*b*) so that it can produce a clockwise (positive) moment about *B*.

Let *d* be the distance of resultant from point *B*. Then,

$$R \times d = \Sigma M_b$$

$$d = \frac{\Sigma M_b}{R} = \frac{17}{149.33} = 0.1138 \text{ m}$$

The intercepts on *x*-axis and *y*-axis are then worked out as

$$x = BM = \frac{d}{\sin\alpha} = \frac{0.1138}{\sin 22.96^\circ} = 0.292 \text{ m}$$

$$y = BN = \frac{d}{\cos\alpha} = \frac{0.1138}{\cos 22.96^\circ} = 0.124 \text{ m}$$

The resultant thus intersects *AB* at 0.292 m from *B* and *BC* at 0.124 m from *B*.

EXAMPLE 3.50

Determine the resultant of the system of forces and a couple acting on a lamina shown in Fig. 3.62. Also locate the points of intersection of the resultant with the arms AB, BC and CD of the lamina.

Solution : Resolving all the forces horizontally and vertically, we get

$\Sigma F_x = 50 + 75 \sin 30° = 87.5$ N (→)

$\Sigma F_y = -100 + 75 \cos 30° = -35.05$ N (↓)

Magnitude of the resultant force,

$$R = \sqrt{(\Sigma F_x)^2 + (\Sigma F_y)^2}$$

$$= \sqrt{(87.5)^2 + (-35.05)^2} = 94.26 \text{ N}$$

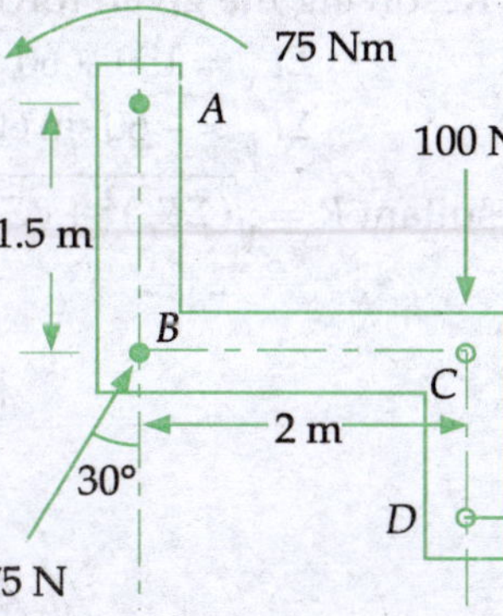

Fig. 3.62

Since ΣF_x is positive and ΣF_y is negative the resultant lies in the fourth quadrant.

Let α be the angle which the resultant makes with the horizontal. Then

$$\alpha = \tan^{-1}\left(\frac{\Sigma F_y}{\Sigma F_x}\right) = \tan^{-1}\left(\frac{35.05}{87.5}\right) = 21.83°$$

Since ΣF_x is negative and ΣF_y is positive, the resultant lies in the second quadrant.

(*b*) Taking moments of all forces about point *B* (clockwise + *ve*)

$$\Sigma M_b = 100 \times 2 - 50 \times 1 - 75 = 75 \text{ Nm (clockwise)}$$

To produce a clockwise moment about point *B*, the orientation of the resultant should be as shown in Fig. 3.63.

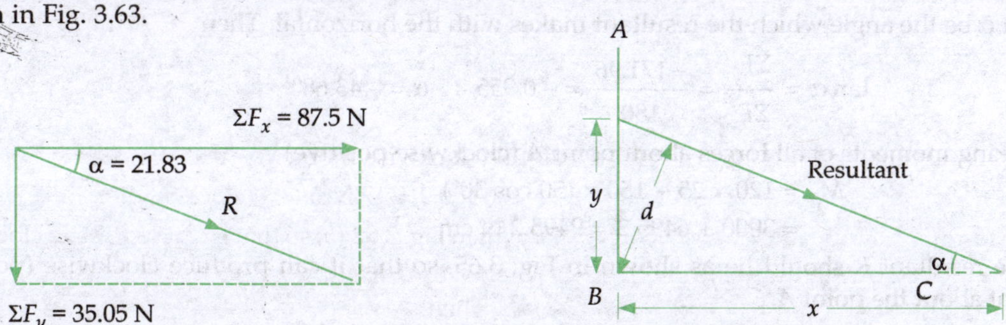

Fig. 3.63

If *d* is the perpendicular distance of line of action of resultant from *B*, then

$$R \times d = \Sigma M_b \quad ; \quad d = \frac{\Sigma M_b}{R} = \frac{75}{94.26} = \mathbf{0.796 \text{ m}}$$

The intercepts on *x*-axis and *y*-axis are then worked out as

$$x = \frac{d}{\sin \alpha} = \frac{0.796}{\sin 21.83} = \mathbf{2.141 \text{ m}}$$

Since the intercept *x* is greater than *BC*, the resultant meets the arm *BC* when produced.

$$y = \frac{d}{\cos \alpha} = \frac{0.796}{\cos 21.83} = \mathbf{0.857 \text{ m}}$$

EXAMPLE 3.51

Determine the magnitude and position of the resultant of set of forces acting on the bell crank lever as shown in Fig. 3.64.

Solution : Resolving the given forces horizontally and vertically, we get

$$\Sigma F_x = 150 + 60\cos 60^\circ = 180 \text{ N}$$

$$\Sigma F_y = -60\sin 60^\circ - 120 = -171.96 \text{ N}$$

$$\therefore \quad \text{Resultant } R = \sqrt{(\Sigma F_x)^2 + (\Sigma F_y)^2} = \sqrt{180^2 + (-171.96)^2} = 248.94 \text{ N}$$

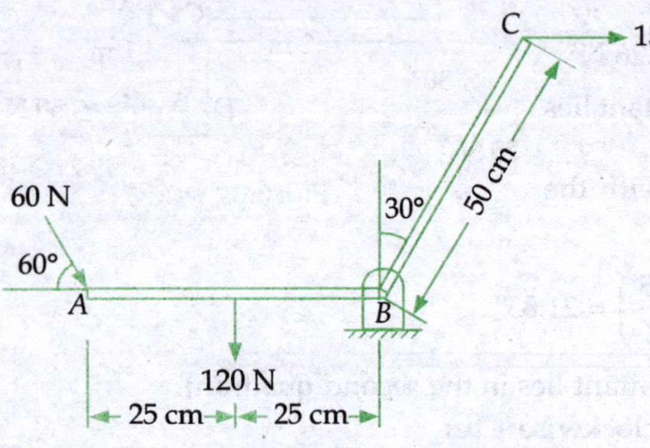

Fig. 3.64

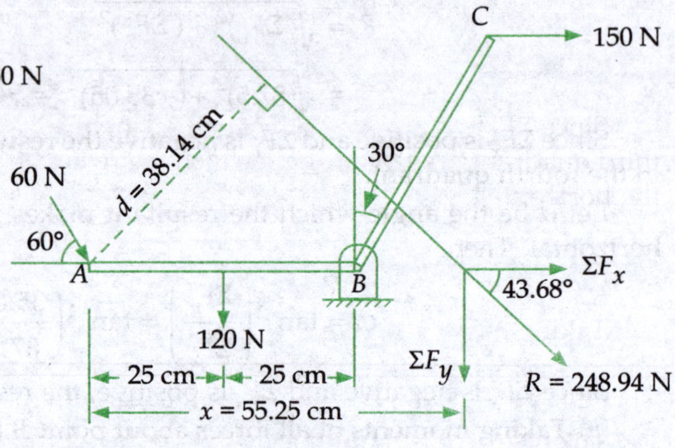

Fig. 3.65

Since ΣF_x is positive and ΣF_y is negative, the resultant lies in the fourth quadrant.

Let α be the angle which the resultant makes with the horizontal. Then

$$\tan\alpha = \frac{\Sigma F_y}{\Sigma F_x} = \frac{-171.96}{180} = -0.955 \;; \quad \alpha = -43.68^\circ$$

Taking moments of all forces about point A (clockwise positive)

$$M_a = 120 \times 25 + 150 \times (50\cos 30^\circ)$$

$$= 3000 + 6495.2 = 9495.2 \text{ N cm}$$

The resultant R should lie as shown in Fig. 3.65, so that it can produce clockwise (positive) moment about the point A.

If d is the perpendicular distance of resultant from point A, then

$$R \times d = \Sigma M_a$$

$$\therefore \quad d = \frac{\Sigma M_a}{R} = \frac{9495.2}{248.94} = 38.14 \text{ cm}$$

If x is the horizontal distance of the line of action from point A, then from the geometry of Fig. 3.69,

$$x = \frac{d}{\sin\alpha} = \frac{38.14}{\sin 43.68} = \mathbf{55.23 \text{ cm}}$$

EXAMPLE 3.52

For the force system acting on a body OABC as shown in Fig. 3.66, make calculations for the magnitude and direction of resultant force. Proceed to determine (i) the distance of resultant from point O, and (ii) the points where the resultant meets the x-axis and y-axis.

Solution : From the geometry of the given force system

$$\theta = \tan^{-1}\left(\frac{4}{3}\right) = 53.12^\circ$$

Resolving the given forces horizontally and vertically, we get

$$\Sigma F_x = 20\cos 53.12^\circ - 20 = -8 \text{ kN}$$
$$\Sigma F_y = -10 - 20\sin 53.12^\circ + 20 = -6 \text{ kN}$$

$$\text{Resultant } R = \sqrt{(\Sigma F_x)^2 + (\Sigma F_y)^2}$$
$$= \sqrt{(-8)^2 + (-6)^2} = 10 \text{ kN}$$

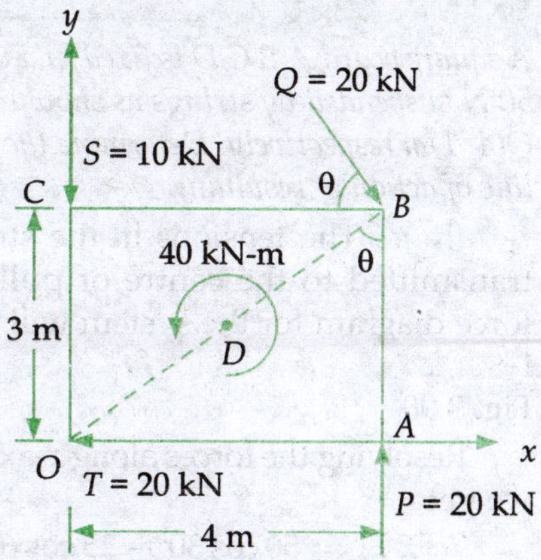

Fig. 3.66

Since ΣF_x and ΣF_y are both negative, the resultant lies in the third quadrant. If α is the angle which the resultant makes with the horizontal, then

$$\tan\alpha = \frac{\Sigma F_y}{\Sigma F_x} = \frac{6}{8} = 0.75 \qquad \therefore \quad \alpha = 36.87^\circ$$

Taking moment of the given system of forces about point O (clockwise positive) we have,

$$\Sigma M_o = (20\cos 53.12^\circ) \times 3 + (20\sin 53.12^\circ) \times 4 - 20 \times 4 - 40$$
$$= 36 + 64 - 80 - 40 = -20 \text{ kN m}$$

The resultant R should lie as shown in Fig. 3.88, so that it can produce anticlockwise (negative) moment about the point O.

If d is the perpendicular distance of line of action of resultant from point O, then

$$R \times d = \Sigma M_o\,; \quad d = \frac{\Sigma M_o}{R} = \frac{20}{10} = 2 \text{ m}$$

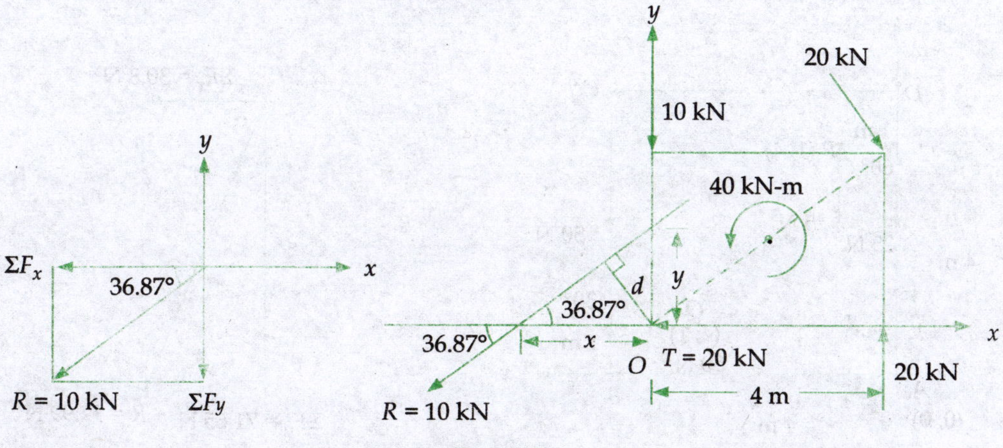

Fig. 3.67

The intercepts x on x-axis and y on y-axis are then given by (from the geometry of Fig. 3.88),

$$x = \frac{d}{\sin\alpha} = \frac{2}{\sin 36.87} = \frac{2}{0.6} = 3.33 \text{ m}$$

$$y = \frac{d}{\cos\alpha} = \frac{2}{\cos 36.87} = \frac{2}{0.8} = 2.5 \text{ m}$$

EXAMPLE 3.53

A square board A B C D is fixed in a vertical plane. Two pulleys P and Q on the board carry loads of 25 N and 50 N suspended by strings as shown in Fig. 3.68. The co-ordinates of the centre of pulleys are P (1, 4)m and Q(4, 1)m respectively. Determine the magnitude and direction of the resultant and the distance from A to the line of action of resultant.

Solution : The tensions in the strings will be transmitted to the centre of pulleys and the force diagram for the system will be as shown in Fig. 3.90

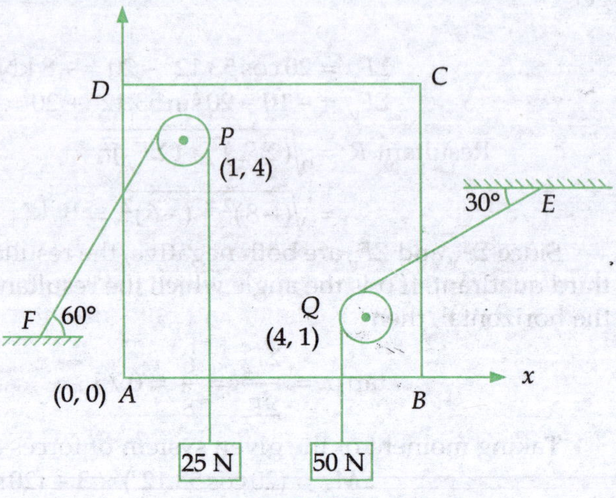

Fig. 3.68

Resolving the forces along x-axis and y-axis, we get

$$\Sigma F_x = 50 \cos 30° - 25 \cos 60 = 30.8 \text{ N}$$

$$\Sigma F_y = 50 \sin 30° - 25 \sin 60 - 50 - 25 = -71.65 \text{ N}$$

$$\text{Resultant } R = \sqrt{(\Sigma F_x)^2 + (\Sigma F_y)^2} = \sqrt{(30.8)^2 + (-71.65)^2} = 77.99 \text{ N}$$

Since ΣF_x is positive and ΣF_y is negative, the resultant lies in the fourth quadrant, and its inclination with the x-axis is,

$$\alpha = \tan^{-1}\left(\frac{\Sigma F_y}{\Sigma F_x}\right) = \tan^{-1}\left(\frac{71.65}{30.8}\right) = 66.74° \text{ (clockwise)}$$

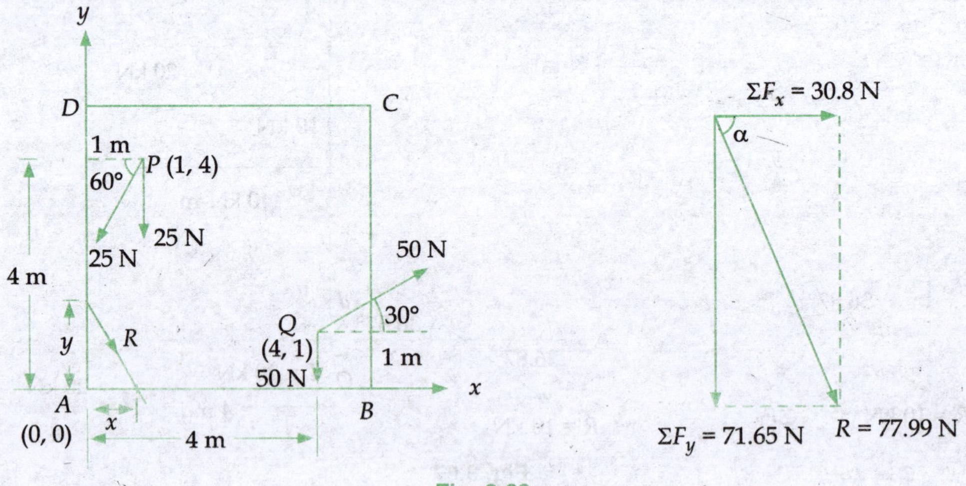

Fig. 3.69

Taking moment about point A (clockwise positive)

$$\Sigma M_a = 25 \times 1 + 25 \sin 60° \times 1 - 25 \cos 60° \times 4 + 50 \times 4 + 50 \cos 30° \times 1 - 50 \sin 30° \times 4$$

$$= 25 + 21.65 - 50 + 200 + 43.3 - 100 = 139.95 \text{ Nm}$$

The resultant is placed in such a fashion that it gives clockwise moment about point A

If d is the perpendicular distance of theline of action of resultatn from point A.

$$d = \frac{\Sigma M_a}{R} = \frac{139.95}{77.99} = \mathbf{1.794\ m}$$

The intercepts of the resulant on x-axis and y-axis are:

$$x = \frac{\Sigma M_a}{\Sigma F_y} = \frac{139.95}{71.65} = \mathbf{1.953\ m}$$

$$y = \frac{\Sigma M_a}{\Sigma F_x} = \frac{139.95}{30.8} = \mathbf{4.544\ m}$$

EXAMPLE 3.54

Determine the resultant of the four forces acting tangentially to a circle of radius 3 m as shown in Fig. 3.70. What will be the location of the resultant with respect to centre of the circle?

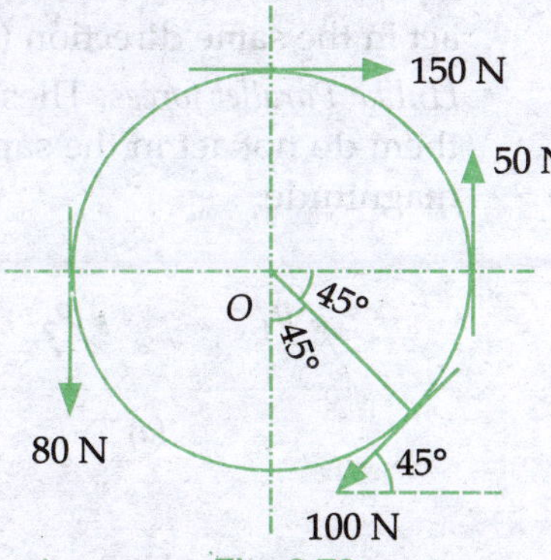

Fig. 3.70

Solution : For these coplanar and non-current forces

$$\Sigma F_x = 150 - 100 \cos 45^\circ = 79.29 \text{ N}$$

$$\Sigma F_y = 50 - 80 - 100 \sin 45^\circ = -100.71 \text{ N}$$

$$\text{Resultant } R = \sqrt{(\Sigma F_x)^2 + (\Sigma F_y)^2}$$

$$= \sqrt{(79.29)^2 + (-100.71)^2}$$

$$= 128.18 \text{ N}$$

Since ΣF_x is positive and ΣF_y is negative, the resultant lies in the fourth quadrant.

Let α be the angle which the resultant makes with the horizontal direction as shown in (Fig. 3.71 *a*). Then

$$\alpha = \tan^{-1}\left(\frac{\Sigma F_y}{\Sigma F_x}\right) = \tan^{-1}\left(\frac{100.71}{79.29}\right) = 51.8^\circ$$

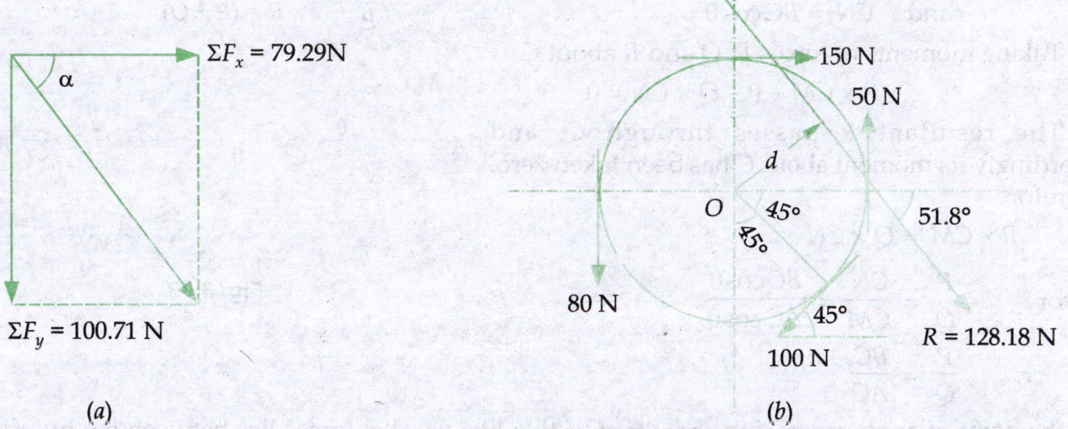

Fig. 3.71

Taking moments of all the forces about the centre (clockwise positive)

$$M_o = 150 \times 3 - 50 \times 3 + 100 \times 3 - 80 \times 3 = 360 \text{ (clockwise)}$$

The resultant R should then lie as shown in (Fig. 3.71 *b*) so that it can produce clockwise (positive) moment about the centre.

If d is the perpendicular distance of resultant from the centre, then

$$R \times d = \Sigma M_o$$

$$\therefore \quad d = \frac{\Sigma M_o}{R} = \frac{360}{128.18} = 2.795 \text{ m}$$

The resultant will act at a perpendicular distance of 2.795 m from the centre.

3.5. PARALLEL FORCES

Forces which donot meet at a point but lie in the same plane are called coplanar non-concurrent forces. The non-concurrent forces may be parallel or non-parallel. The lines of action of parallel forces are parallel to each other and those of non-parallel forces are not parallel. The parallel forces are classified as

- *Like parallel forces.* These forces have their lines of action parallel to each other and all of them act in the same direction (Fig. 3.72 *a*). These forces may be equal or unequal in magnitude
- *Unlike Parallel forces.* These forces have their lines of action parallel to each other and all of them do not act in the same direction (Fig. 3.91 *b*). These forces may be equal or unequal in magnitude.

Fig. 3.72

3.5.1. Resultant of like parallel forces

With reference to Fig. 3.73, two like parallel forces act at points A and B, and their resultant R meets AB at C. The resultant R will be equal to $(P + Q)$ and parallel to P and Q.

Through C, draw MN perpendicular to P and Q. Then

$$CM = AC \cos\theta$$

$$\text{and} \quad CN = BC \cos\theta$$

Taking moments of forces P, Q and R about C,

$$P \times CM + 0 - Q \times CN = 0$$

The resultant R passes through C, and accordingly its moment about C has been taken zero. Therefore

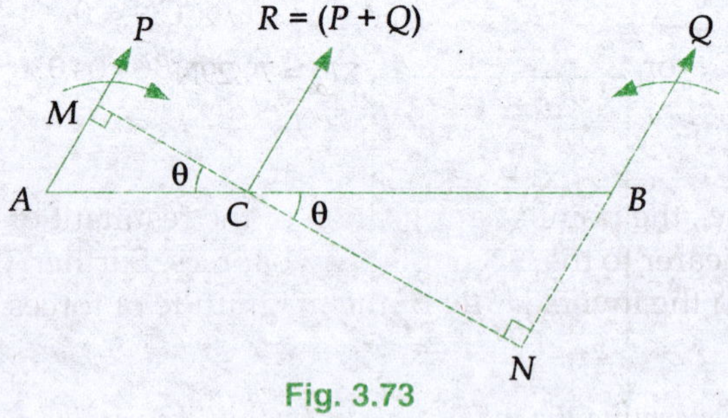

Fig. 3.73

$$P \times CM = Q \times CN$$

$$\text{or} \quad \frac{P}{Q} = \frac{CN}{CM} = \frac{BC\cos\theta}{AC\cos\theta}$$

$$\therefore \quad \frac{P}{Q} = \frac{BC}{AC}$$

i.e., the point of application of the resultant of two like parallel forces lies between the two forces. Further, it divides the distance between these forces internally in the inverse ratio of the magnitude of forces.

EXAMPLE 3.55

Two like parallel forces are acting at a distance of 200 mm from each other. These forces are equivalent to a single force of 250 N and its line of action is at a distance of 50 mm from one of the forces. Make calculations for the magnitude of the two forces.

Solution : With reference to Fig. 3.73, let P and Q be the two like parallel forces which act at points A and B, and let the resultant meet AB at C. Then

$$AB = 200 \text{ mm} \; ; \quad AC = 50 \text{ mm} \quad \text{and} \quad BC = 200 - 50 = 150 \text{ mm}$$

$$P + Q = R = 250 \text{ N} \qquad ...(i)$$

Applying the principle of moments and taking moments about point C,

$$P \times AC = Q \times BC$$

$$P = Q \times \frac{BC}{AC} = Q \times \frac{150}{50} = 3\,Q \qquad ...(ii)$$

From expressions (*i*) and (*ii*),

$$3Q + Q = 250 \; ; \quad Q = \frac{250}{4} = \mathbf{62.5 \text{ N}}$$

$$P = 3 \times 62.5 = \mathbf{187.5 \text{ N}}$$

3.5.2. Resultant of unlike parallel forces

With reference to Fig. 3.74, two unlike parallel forces act at points A and B and their resultant R meets AB at C. The point C will lie outside of B and on the same side as the larger force P. The resultant R will be equal to $(P - Q)$ and parallel to P and Q.

The moment of resultant R about C is zero, and as such the algebraic sum of the moments of P and Q about C must be zero. That implies that these moments must be equal and opposite and that is possible only if C lies outside AB.

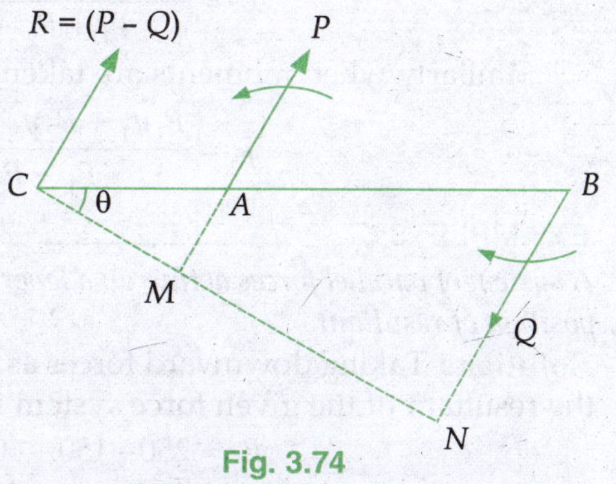

Fig. 3.74

Through C, draw MN perpendicular to P and Q. Then

$$CM = AC\cos\theta \quad \text{and} \quad CN = BC\cos\theta$$

Taking moments of forces P, Q and R about C

$$P \times CM - Q \times CN = 0$$

or

$$P \times AC\cos\theta = Q \times BC\cos\theta$$

$$\therefore \quad \frac{P}{Q} = \frac{BC}{AC}$$

i.e., the point of application of the resultant of two unlike parallel forces lies outside the two forces nearer to the larger of the two forces. Further, it divides the distance between these forces externally in the inverse ratio of the magnitude of forces.

EXAMPLE 3.56

Determine the magnitude of two unlike parallel forces acting at a distance of 400 mm from each other. The resultant of these forces is 80 N and it acts at a distance of 160 mm from the greater of the two forces.

Solution : With reference to Fig. 3.74, let P and Q be the two unlike parallel forces which act at A and B, and let the resultant R meet AB externally at C. Then

$$AB = 400 \text{ mm} \; ; \quad AC = 160 \text{ mm} \quad \text{and} \quad BC = 560 \text{ mm}$$

$$P - Q = R = 80 \text{ N} \qquad ...(i)$$

Applying the principle of moments and taking moments about point C,

$$P \times AC = Q \times BC$$

$$P = Q \times \frac{BC}{AC} = Q \times \frac{560}{160} = 3.5\,Q \quad \text{...(ii)}$$

From expressions (*i*) and (*ii*),

$$3.5\,Q - Q = 80; \quad Q = \frac{80}{2.5} = 32 \text{ N}$$

$$P = 3.5\,Q = 3.5 \times 32 = 112 \text{ N}$$

3.5.3. Centre of parallel forces

When a number of parallel forces act at different points in a body, the magnitude of their resultant is obtained by the algebraic sum of all these parallel forces. The centre of the forces is worked out by taking moments.

Let $A(x_1, y_1)$, $B(x_2, y_2)$ and $C(x_3, y_3)$ be the points on a body which are acted upon by forces P_1, P_2 and P_3 respectively.

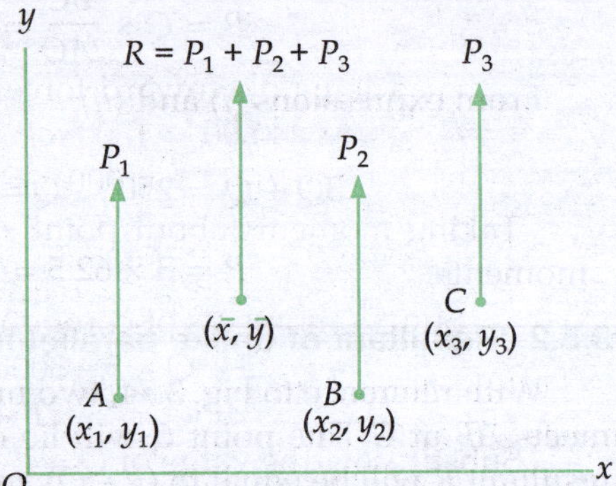

Fig. 3.75.

The resultant of these forces is

$$R = P_1 + P_2 + P_3 + \ldots$$

and its point of application $(\bar{x}, \bar{y})$ is worked out by taking moments about the two axes.

Talking moments about y-axis, we get

$$(P_1 + P_2 + P_3 + \ldots)\,\bar{x} = P_1 x_1 + P_2 x_2 + P_3 x_3 + \ldots$$

$$\therefore \quad \bar{x} = \frac{P_1 x_1 + P_2 x_2 + P_3 x_3 + \ldots}{P_1 + P_2 + P_3 + \ldots} = \frac{\Sigma Px}{\Sigma P}$$

Similarly when moments are taken about x-axis, we get

$$\bar{y} = \frac{P_1 y_1 + P_2 y_2 + P_3 y_3 + \ldots}{P_1 + P_2 + P_3} = \frac{\Sigma Py}{\Sigma P}$$

EXAMPLE 3.57

A system of parallel forces acting on a lever is as shown in Fig. 3.76. Determine the magnitude, direction and position of resultant.

Solution : Taking downward forces as positive, the resultant of the given force system is

$$R = 250 - 150 + 60 = 160 \text{ N}$$

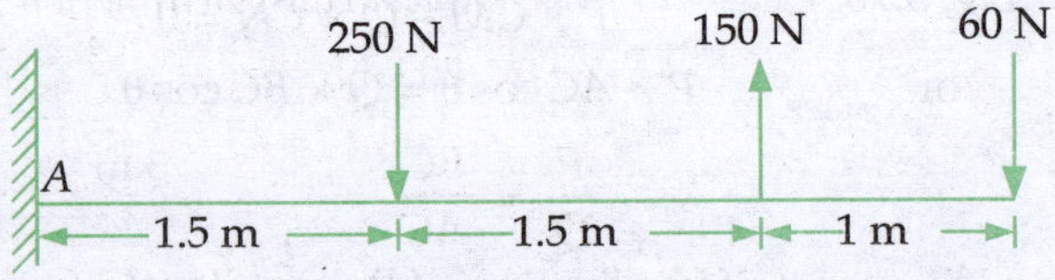

Fig. 3.76

Thus the resultant has a magnitude of 160 N and it acts vertically downwards.

Let this resultant act at a distance x from the fixed end A. Taking moments about point A (clockwise moments positive) and applying the principle of moments

$$\text{moment of resultant} = \text{sum of the moments of its components}$$

$$160\,x = (250 \times 1.5) - (150 \times 3) + (60 \times 4)$$

$$160\,x = 375 - 450 + 240 = 165$$

$$\therefore \quad x = \frac{165}{160} = 1.031 \text{ m}$$

Thus the resultant lies at a distance of 1.031 m from the fixed end A.

EXAMPLE 3.58

For the system of parallel forces depicted in Fig. 3.77, the resultant has a magnitude of 600 N vertically upwards and it acts through a point which is 4 m to the right of force 1500 N. Workout the values of forces P and Q.

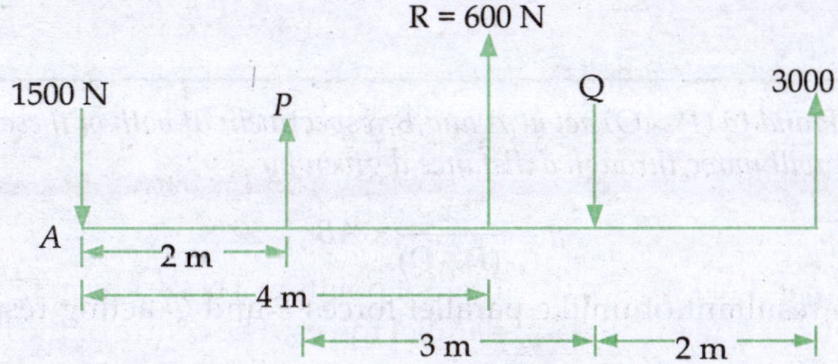

Fig. 3.77

Solution : Taking downward forces as positive, we have

$$-600 = 1500 - P + Q - 3000 = -P + Q - 1500$$

$$P - Q = -900 \quad ...(i)$$

Taking moments about point A (clockwise moments positive) and applying the principle of moments,

Sum of moments of components = moment of resultant

$$-P \times 2 + Q \times 5 - 3000 \times 7 = -600 \times 4$$

$$-2P + 5Q = 21000 - 2400 = 18600 \quad ...(ii)$$

Substituting the value of $Q = P + 900$ from expression (i)

$$-2P + 5\,(P + 9800) = 18600$$

$$3P = 18600 - 4500 = 14100$$

$$\therefore \quad P = \frac{14100}{3} = \mathbf{4700\ N}$$

and $\quad Q = P + 900 = 4700 + 900 = \mathbf{5600\ N}$

EXAMPLE 3.59

A coplanar parallel force system consisting of three forces acts on a rigid bar AB as shown in Fig. 3.78. Determine the simplest equivalent action for the force system.

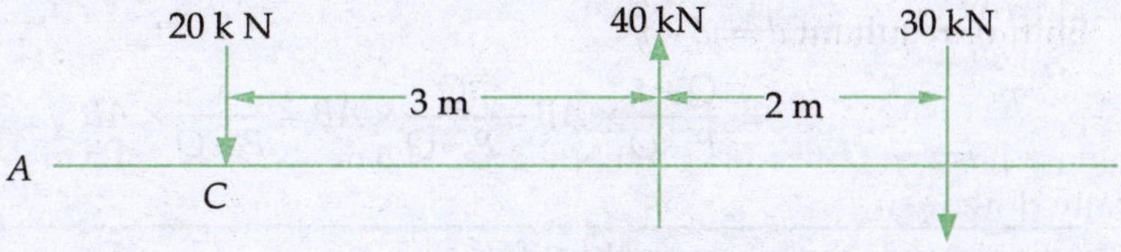

Fig. 3.78

If an additional force of 10 kN acts along the bar from A to B, what would be the simplest equivalent action?

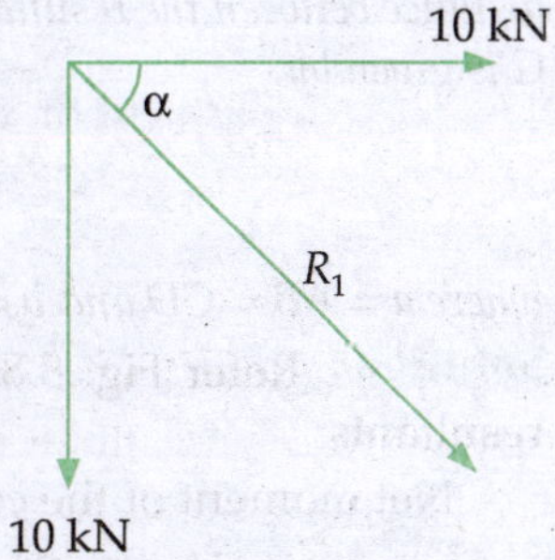

Solution : Let R be the simplest equivalent action. Then

$$R = \Sigma F = -20 + 40 - 30 = -10 \text{ kN} \quad (\downarrow)$$

The location of R can be worked out by taking moments about point C. That gives

$$10 \times x - 40 \times 3 + 30 \times 5 = 0; \quad x = 3 \text{ m}$$

Thus the equivalent action R acts downwards along the line of action of 40 kN force.

(*b*) The resultant of 10 kN acting downwards and 10 kN force acting horizontally from A is

$$R_1 = \sqrt{10^2 + 10^2} = 14.14 \text{ kN}$$

This resultant acts at $\alpha = \tan^{-1}(10/10) = 45^\circ$ below x-axis as shown below.

EXAMPLE 3.60

Two unlike parallel force P and Q (P > Q) act at A and B respectively. If both of these forces are increased by x, show that the resultant will move through a distance d given by

$$d = \frac{x}{(P-Q)} \times AB$$

Solution : Let R be the resultant of unlike parallel forces P and Q acting respectively at points A and B.

$$R = (P - Q)$$

Let this resultant act at distance y from point A.

Taking moments about point A, we get

$$R \times y = Q \times AB \quad ; \quad y = \frac{Q}{R} \times AB = \frac{Q}{P-Q} \times AB$$

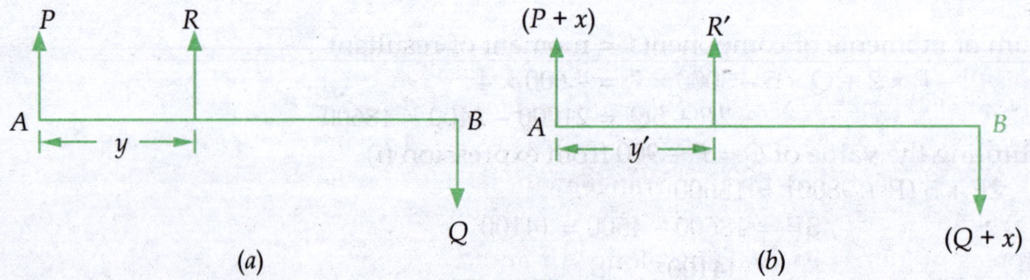

Fig. 3.79

When forces P and Q are increased by amount x, then

$$R' = (P + x) - (Q + x) = P - Q$$

and $\quad R' \times y' = (Q + x) \times AB$

or $\quad y' = \frac{Q+x}{R'} \times AB = \frac{Q+x}{P-Q} \times AB$

$\therefore \quad$ Shift of resultant $d = y' - y$

$$= \frac{Q+x}{P-Q} \times AB - \frac{Q}{P-Q} \times AB = \frac{x}{P-Q} \times AB$$

EXAMPLE 3.61

A rectangle ABCD has forces P and Q acting along its sides as shown in Fig. 3.80 a. Show that the perpendicular distance between the resultants of P and Q at A and that of P and Q at C is given by

$$= \frac{Pa - Qb}{\sqrt{P^2 + Q^2}}$$

where a = AB = CD and b = BC = DA

Solution : Refer Fig. 3.80 *b* for the system of forces and their resultants.

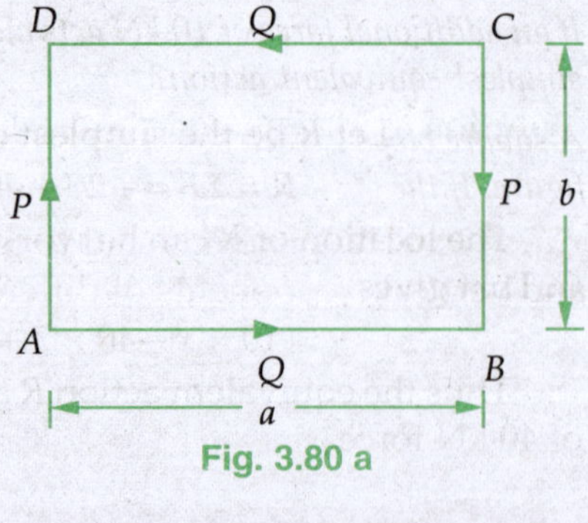

Fig. 3.80 a

Net moment of the couples formed by forces P and Q

$= P \times a - Q \times b = Pa - Qb \quad \ldots(i)$

Resultant of the forces P and Q at point A

$$R_1 = \sqrt{P^2 + Q^2}$$

Resultant of the forces P and Q at point C

$$R_2 = \sqrt{P^2 + Q^2}$$

The resultants R_1 and R_2 are equal and constitute a couple with perpendicular distance x between them.

Moment of the couple formed by the resultants

$$= \sqrt{P^2 + Q^2} \times x \quad \text{...}(ii)$$

Equating the moments given by expressions (*i*) and (*ii*),

$$\sqrt{P^2 + Q^2} \times x = Pa - Qb$$

$$\therefore \quad x = \frac{Pa - Qb}{\sqrt{P^2 + Q^2}}$$

which is the required expression.

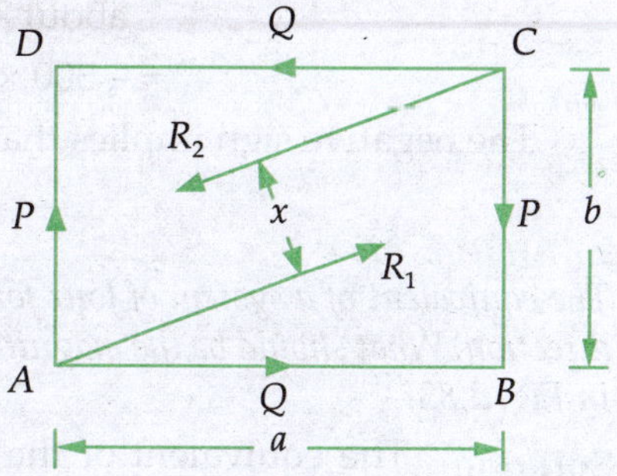

Fig. 3.80 b

EXAMPLE 3.62

A rectangle ABCD has sides AB = CD = 80 mm and BC = DA = 60 mm. Forces of 150 N each act along AB and CD, and forces of 100 N each act along BC and DA. Make calculations for the resultant moment of the force system.

Solution : Refer to Fig. 3.81 for the arrangement of the force system.

The forces of 150 N each acting along AB and CD generate a couple of moment

$$= 150 \times 60 = 9000 \text{ Nmm}$$

The forces of 100 N each acting along BC and DA produce a couple of moment,

$$= 100 \times 80 = 8000 \text{ Nm}$$

$$\therefore \text{ Resultant moment} = 9000 + 8000 = 17000 \text{ Nmm}$$

(anticlockwise).

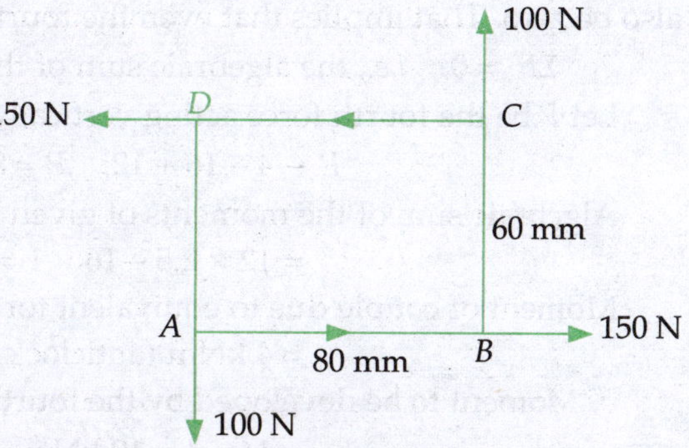

Fig. 3.81

Alternatively, the above result can be obtained by taking moments of all forces about any point, say A. That gives

$$M = (100 \times 80) + (150 \times 60) + (100 \times 0) - (150 \times 0)$$

$$= 8000 + 9000 = 17000 \text{ Nmm (anticlockwise)}$$

EXAMPLE 3.63

A square block of each side 1.5 m is acted upon by a system of forces along its sides as shown in the adjoining figure. If the system reduces to a couple, determine the magnitude of the forces P and Q, and the couple.

Solution : The system is said to reduce to a couple and as such the algebraic sum of the horizontal and vertical components of different forces is zero.

That gives :

$$\Sigma F_x = 150 - 150 \cos 45° - P = 0$$

$$\therefore \quad P = 150 - 150 \times 0.707 = 43.95 \text{ N}$$

Also, $\Sigma F_y = 300 - 150 \sin 45° - Q = 0$

$$\therefore \quad Q = 300 - 150 \times 0.707 = 193.95 \text{ N}$$

Moment of couple = algebraic sum of the moments of forces about any corner, say A

$$= -300 \times 1.5 - 43.95 \times 1.5 = -515.925 \text{ Nm}$$

The negative sign implies that the moment is anticlockwise.

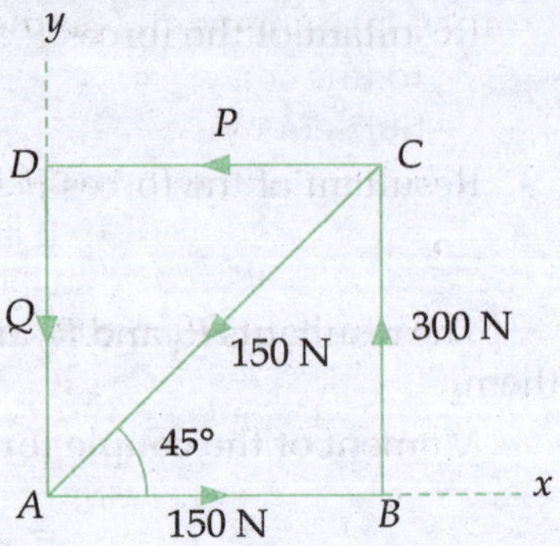

Fig. 3.82

EXAMPLE 3.64

The equivalent of a system of four forces is a couple of moment of magnitude 4 kNm acting in anticlockwise direction. What should be the magnitude and location of the fourth force if three of the four forces are as shown in Fig. 3.83?

Solution : The equivalent of the given system of forces is a couple and accordingly

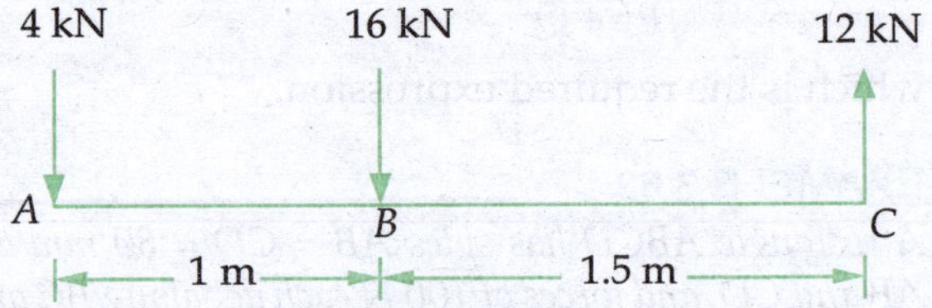

Fig. 3.83

(i) $\Sigma F_h = 0$, *i.e.*, the algebraic sum of the horizontal components of different forces is zero.

Since the given three forces are all vertical, they do not have any horizontal component. Obviously the horizontal component of the fourth force would also be zero. That implies that even the fourth force would be a vertical force.

(ii) $\Sigma F_v = 0$, *i.e.*, the algebraic sum of the vertical components of different forces is zero.

Let P be the fourth force acting vertically upwards. Then

$$P - 4 - 16 + 12; \quad P = 8 \text{ kN } (\uparrow)$$

Algebraic sum of the moments of given three forces about point A

$$= 12 \times 2.5 - 16 \times 1 = 14 \text{ kNm (anticlockwise)}$$

Moment of couple due to equivalent force system

$$= 4 \text{ kNm (anticlockwise)}$$

$\therefore$ Moment to be developed by the fourth force

$$= 14 - 4 = 10 \text{ kNm}$$

This moment of 10 kNm has to be clockwise so that net moment about point A is equal to 4 kNm (anticlockwise). For providing clockwise moment, the vertical force $P = 8$ kN should lie on left of point A at distance

$$x = \frac{10}{8} = 1.25 \text{ m}$$

3.5.4. Graphical method for the resultant of parallel forces

Consider a body acted upon by like parallel forces P_1, P_2, P_3. Draw the relevant space diagram and let the forces be named according to Bow's notation.

The vector diagram for the force system is then drawn by following the steps outlined below:

1. Take any suitable point a, draw ab parallel and equal to force P_1 as per some suitable scale.
2. Proceed to draw bc and cd parallel and equal to force P_2 and P_3 respectively.
3. Taking some convenient point o, join oa, ob, oc and od.

4. Mark any point l on the line of action of force P_1 and through this point draw line Rl parallel to line ao.
   Subsequently,
   lm is drawn || to ob meeting line of action of force P_2,
   mn is drawn || to oc meeting line of action of force P_3
   ns is drawn || to od
5. Extend Rl and Sn to meet at point k.

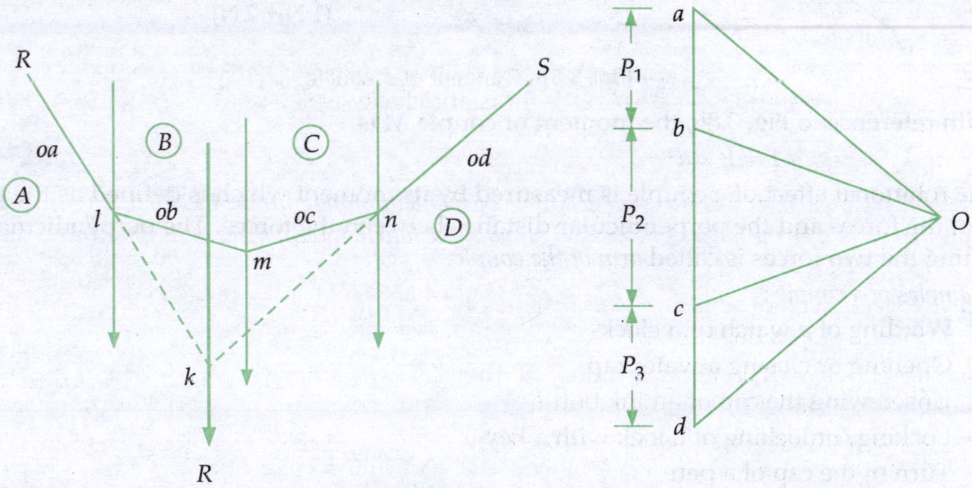

Fig. 3.84

Result : The magnitude of the resultant force is given by ad (as per the chosen scale) and its position is indicated by line drawn through k and parallel to ad.

While finding the resultant of unlike parallel forces, the same approach is adopted but due consideration is given to the direction of the forces while drawing the vector diagram. This is illustrated in the figure given below :

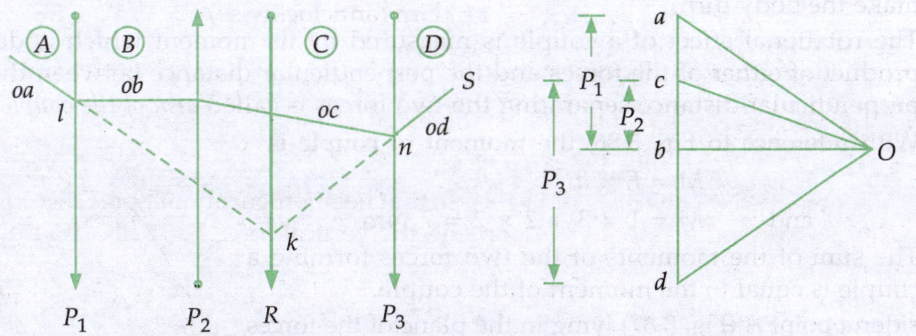

$ab = P_1$ (drawn downward), $bc = P_2$ (drawn upward)
$cd = P_3$ (drawn downward)

Fig. 3.85

3.6. COUPLE

Two parallel forces equal in magnitude but opposite in direction, and separated by a finite distance are said to form a couple. If F and F' are two such forces, then the couple is denoted by (F, F'). A body acted upn by a couple spins round but remains at the same spot.

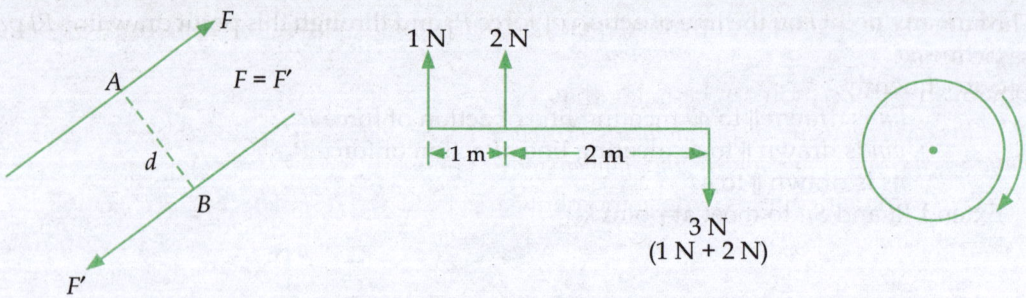

Fig. 3.86. Concept of a couple

With reference to Fig. 3.86, the moment of couple M is

$$M = F \times d$$

The rotational effect of a couple is measured by its moment which is defined as the product of either of the forces and the perpendicular distance between the forces. The perpendicular distance separating the two forces is called *arm of the couple*.

Examples of a couple :

(*i*) Winding of a watch or a clock
(*ii*) Opening or closing a water tap
(*iii*) Unscrewing the cap of an ink bottle
(*iv*) Locking/unlocking of a lock with a key
(*v*) Turn of the cap of a pen
(*vi*) Forces applied to the handle of a screw

The salient aspects of a couple are :

(*i*) The algebraic sum of the vertical and horizontal components of the forces (F, F') forming a couple is zero, *i.e.*, the resultant of the forces (F, F') is zero.

(*ii*) A zero resultant of the forces forming the couple implies that a couple cannot produce translation. The couple causes only rotation and provides a turning effect, *i.e.*, tends to make the body turn.

(*iii*) The rotational effect of a couple is measured by its moment which is defined as the product of either of the forces and the perpendicular distance between the forces. The perpendicular distance separating the two forces is called *arm of the couple*.

With reference to Fig. 3.86, the moment of couple is

$$M = F \times d$$

and $$M = 1 \times 3 + 2 \times 2 = 7 \text{ Nm}$$

(*iv*) The sum of the moments of the two forces forming a couple is equal to the moment of the couple.

Consider a point A (Fig. 3.87) lying in the plane of the forces F and F' that form a couple. Then the sum of the moments of the two forces about point A is

$$M = F' d_2 - F d_1$$
$$= F d_2 - F d_1 \quad (\because F' = F)$$
$$= F(d_2 - d_1) = F d$$
= moment of the couple

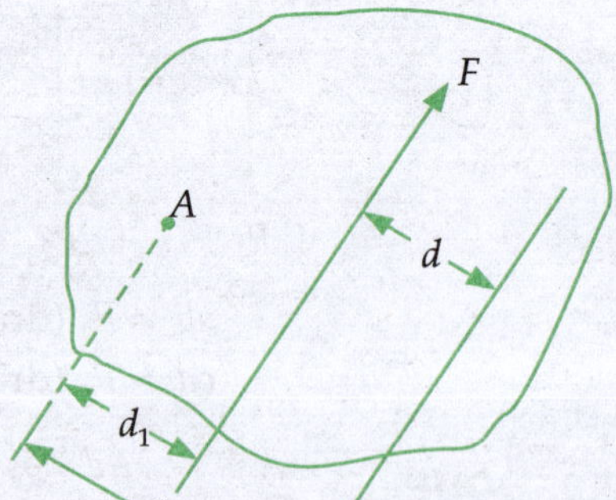

Fig. 3.87

The sum of the moments does not depend on the location of point A; it will have the same value and same sense.

Obviously a deduction can be made that

"the algebraic sum of the moments of two forces forming a couple about any point in their plane is constant and equal to the moment of the couple."

(*v*) Two caplanar couples, whose moments are equal and opposite, balance each other.

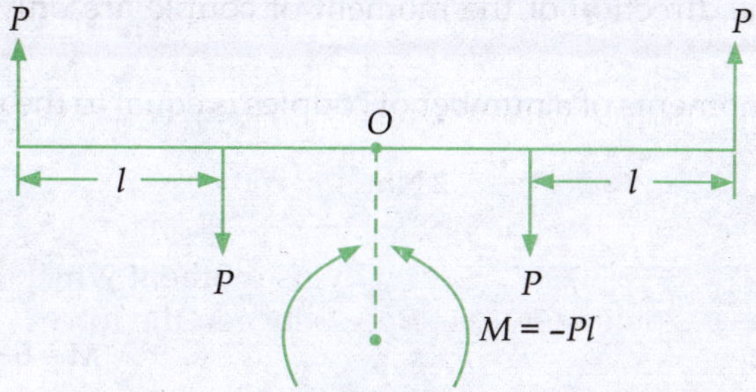

Fig. 3.88

With reference to Fig. 3.88, there act two couples whose moments are equal and opposite. These couples will balance each other and accordingly there will be no turning effect at point O.

(*vi*) Any two couples will be equivalent if their moments are equal, both in magnitude and direction.

Consider the system of forces depicted in Fig. 3.89

In Fig. 3.89 (*a*) : $M = P \times l = P\,l$ (anticlockwise)

In Fig. 3.89 (*b*) : $M = \frac{P}{2} \times 2\,l = P\,l$ (anticlockwise)

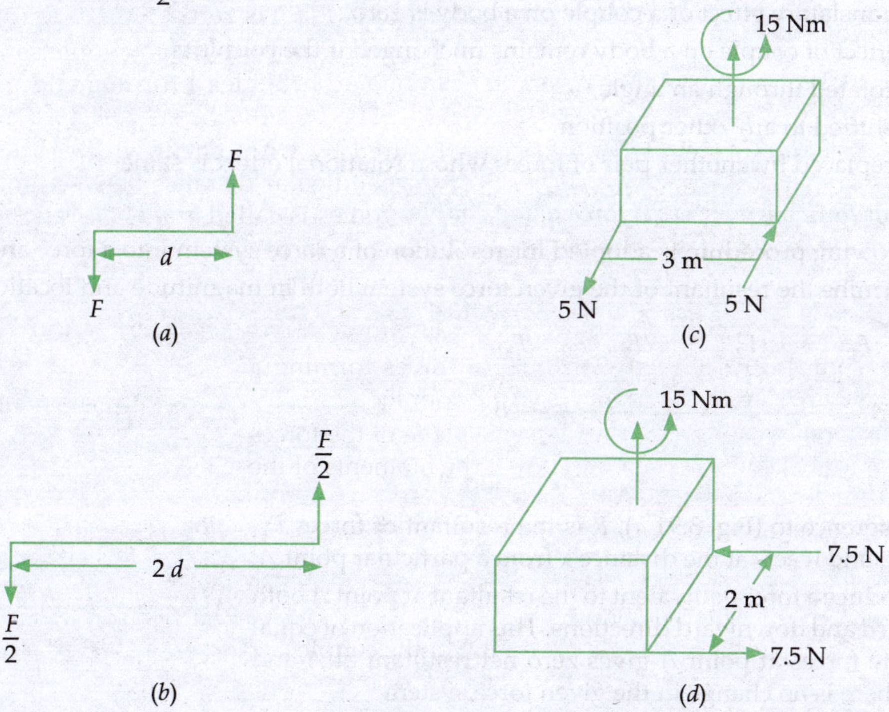

Fig. 3.89

Further, the two couples depicted in Fig. 3.89 *c* and *d* have a different force magnitude and direction but they have same moment magnitude and direction. They two represent an equivalent couple.

Two equivalent couples have the same effect on the body, and it does not matter

— where the two forces form the couple

— what magnitudes and directions the forces have

The magnitude and the direction of the moment of couple are only factors which decide the equivalence of couples.

(*vii*) Algebraic sum of moments of a number of couples is equal to the moment of a single couple.

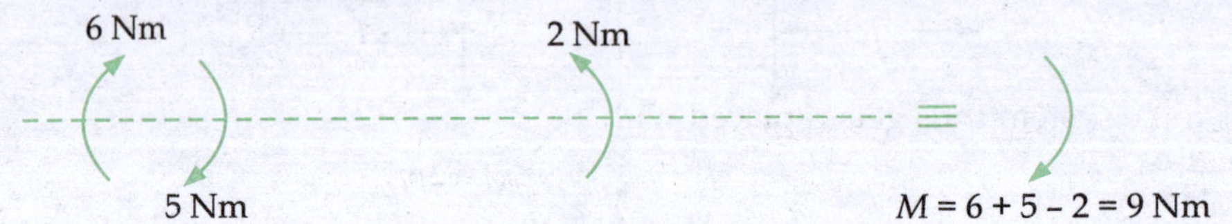

Fig. 3.90

(*viii*) A single force P and a couple M acting in the same plane on a body cannot balance each other. However, they are together equivalent to a single force at a distance $e = M/P$ from its original line of action.

Further :

- a couple cannot be balanced by a single force, but can be balanced only by a couple of opposite sense.
- any number of coplanar couples can be reduced to a single couple of moment equal to the algebraic sum of the moments of all the couples.
- the translatory effect of a couple on a body is zero.
- the effect of couple on a body remains unchanged if the couple is
  - (*a*) rotated through an angle,
  - (*b*) shifted to any other position,
  - (*c*) replaced by another pair of forces whose rotational effect is same

Resolution of force system into a force and a couple

The following procedure is adopted for resolution of a force system into a force and a couple :

(*i*) Determine the resultant of the given force system both in magnitude and location.

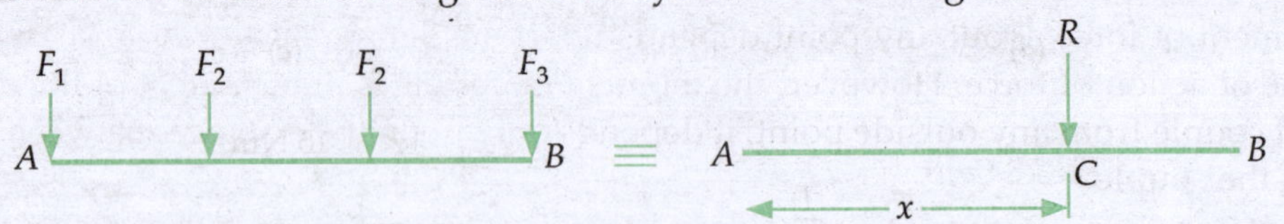

Fig. 3.91 a

With reference to (Fig. 3.91 *a*), R is the resultant of forces F_1, F_2, F_3 and F_4 and it acts at the distance x from a particular point A.

(*ii*) Introduce a force equivalent to the resultant at point A both in the upward and downward directions. This application of equal and opposite forces at point A gives zero net resultant at A and obviously there is no change in the given force system.

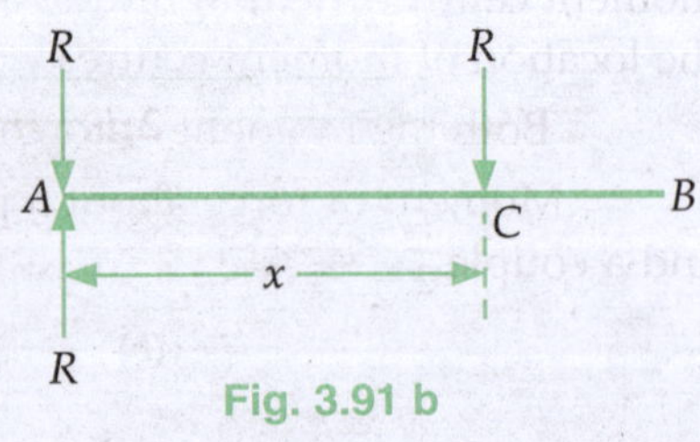

Fig. 3.91 b

(iii) The downward resultant R and the upward force R applied at A constitute a couple. The moment of this couple is in anti-clockwise direction and equals the product of R and the distance x between te forces. That is $M_A = R \times x$

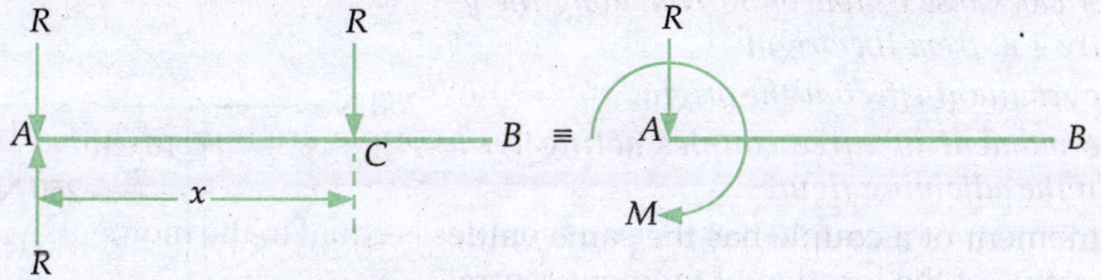

Fig. 3.91 c

The given force system thus gets replaced with a force and a couple at point A.

Moment and couple

The differences between a moment and a couple may be stated as :

(i) Two parallel forces having the same magnitude but acting in opposite direction form a couple. The moment of couple is the product of either of the forces and the perpendicular distance between them.

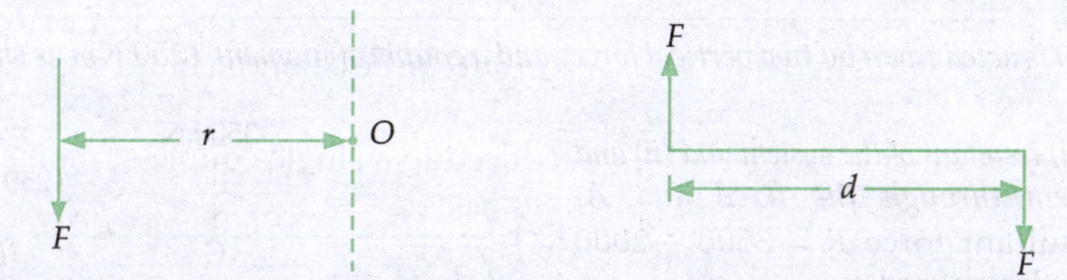

Moment about an axis : $F \times r$ Couple with moment : $F \times d$

Fig. 3.92

The product of a force and the perpendicular distance of the line of action of the force from a point or axis is referred to as the moment of force about that point or axis.

(ii) Both the moment and couple have sense of direction (clockwise or anti-clockwise) and are expressed in the units Nm.

(iii) Moment of a force varies where as the moment of a couple is always constant.

The moment of force about any point depends upon the perpendicular distance of that point from the line of action of force. However, the moment of couple is independent of the distance of the centre of couple from any outside point; it depends only upon the distances between the forces constituting the couple.

Further, the moment of couple has the same value and same sense irrespective of the location of moment centre. The magnitude and the sense of the moment of force does change with change in the location of moment centre.

(iv) Both the moment of force and couple tend to cause a rotational effect on the body.

(v) Moment of force about a point can be translated into moment of force about another point and a couple.

EXAMPLE 3.65

(a) A body is acted upon by the following two couples:

(i) a clockwise couple of 30 Nm on the x-axis with its centre 4 m from the origin, and

(ii) a counter clockwise couple of 50 Nm along the y-axis with its centre 2 m from the origin.

Calculate the resultant effect at the origin.

(b) Find the moment of three couples acting on a L-bar as shown in the adjoining figure.

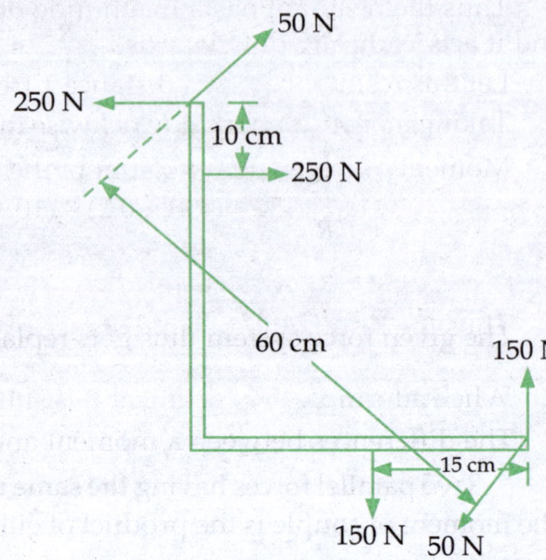

Fig. 3.93

Solution : The moment of a couple has the same value and sense irrespective of the location of moment centre.

∴ Resultant moment

$$= 30 - 50 = -20 \text{ Nm (anti-clockwise)}$$

(*b*) Resultant moment on the bar due to the three given couples is

$= -(250 \times 0.1) + (50 \times 0.6) - (150 \times 0.15)$

$= -25 + 30 - 22.5$

$= -17.5$ Nm (anticlockwise)

EXAMPLE 3.66.

A cantilever beam is acted upon by two vertical forces and a couple of moment 1250 Nm as shown in the figure given below:

Determine: (i) resultant of the system and (ii) and equivalent system through the fixed and A.

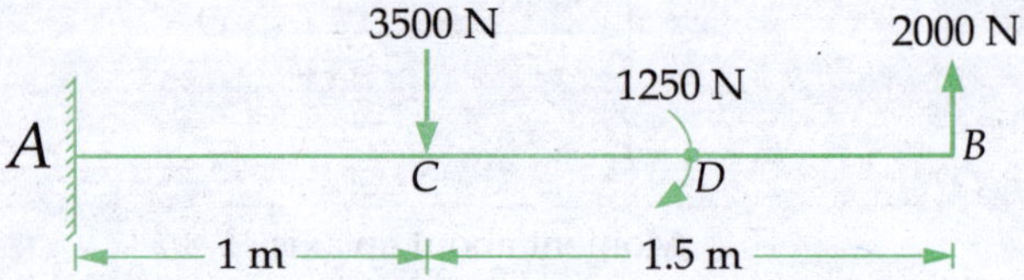

Fig. 3.94

Solution : Resultant force $R = 3500 - 2000 = 1500$ N acting downward

Resultant moment about *A* (clockwise moment positive)

$= 3500 \times 1 + 1250 - 2000 \times 2.5$

$= 3500 + 1250 - 5000 = -250$ Nm (anticlockwise)

The resultant force acts downward and it will give the anti-clockwise moment if it acts towards left of *A*. Let *x* be the distance of resultant force (1500 N) from *A*. Then

$$1500\,x = 250 \qquad \therefore \quad x = 0.166 \text{ m}$$

Hence the resultant of the system is 1500 N downward and it acts at a distance of 0.166 m left of fixed and *A*.

(*ii*) This means that we have to determine a single resultant force and a single moment through *A*.

Single resultant force = 1500 N (downward)

Single resultant moment = 250 N (anticlockwise)

EXAMPLE 3.67

A rigid bar is subjected to a system of parallel forces as shown in Fig. 3.95.

Reduce this system to

(a) a single force

(b) a single force-moment system at A

(c) a single force moment system at B

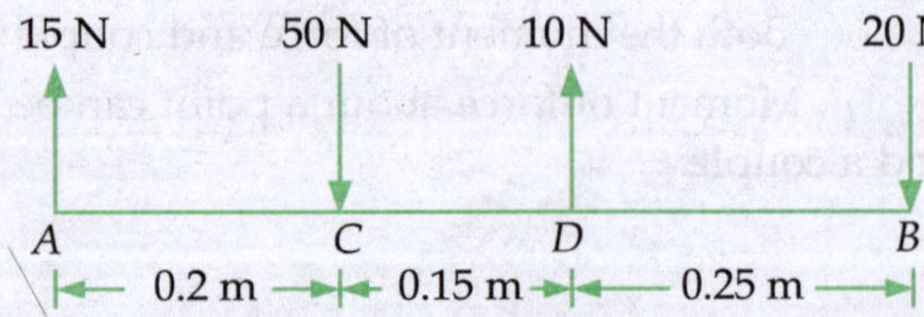

Fig. 3.95

Solution : (*a*) A single force or resultant

Taking downward forces as positive, the resultant of the given force system is

$$R = -15 + 50 - 10 + 20 = \mathbf{45\ N}$$

Thus the resultant has a magnitude of 45 N, its line of action is parallel to that of given forces and it acts vertically downwards.

Let this resultant act at a distance x from the end A.

Taking moments about A (clockwise moments positive) and applying the principle of moments,

Moment of the resultant = sum of the moments of its components

$$45x = (50 \times 0.2) - (10 \times 0.35) + (20 \times 0.6) = 10 - 3.5 + 12 = 18.5$$

$$\therefore \quad x = \frac{18.5}{45} = \mathbf{0.411\ m}$$

Thus the resultant lies at a distance of 0.411 from A.

(*b*) Single force-moment at *A*

When a force of 45 N acting at E is shifted to A, it is a accompanied by a moment

$$M_a = (45 \times 0.411) = \mathbf{18.5\ Nm}$$

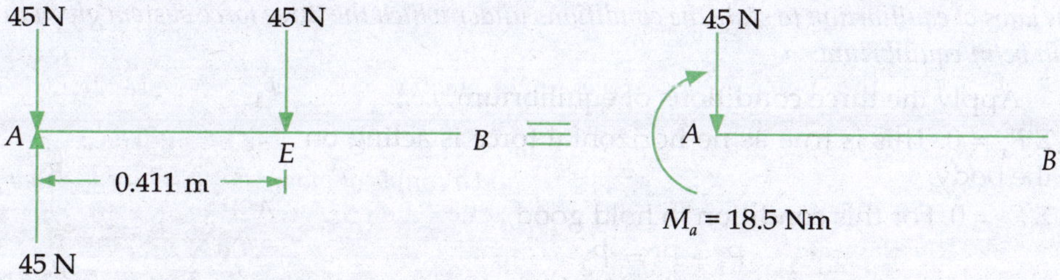

(*c*) Single force - moment at *B*

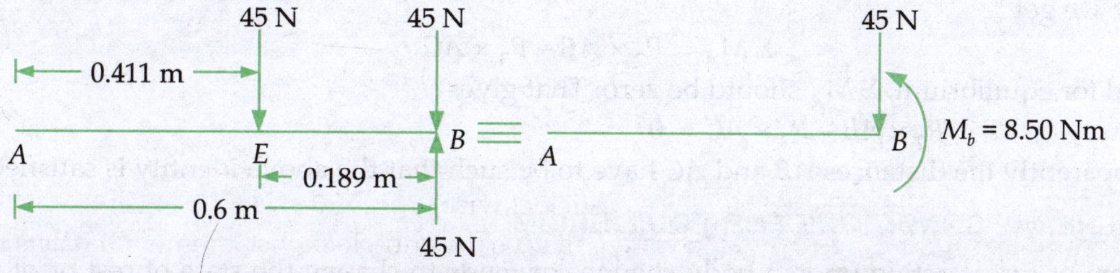

When the force of 45 N acting at E is moved to B, it is accompanied by a anticlock wise moment

$$M_b = -(45 \times 0.189) = \mathbf{-8.50\ Nm}$$

EXAMPLE 3.68

Check whether the parallel force system depicted in Fig. 3.96 is in equilibrium. If not, can this system be brought to equilibrium by applying a single force.

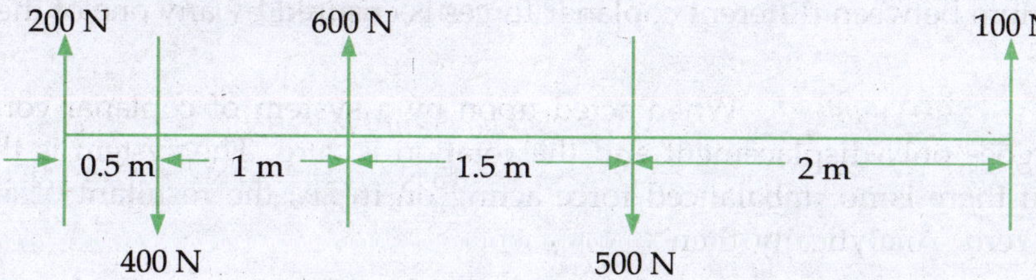

Fig. 3.96

Solution: Considering downward forces as positive, the resultant of the given force system is

$$R = -200 + 400 - 600 + 500 - 100 = 0$$

Thus the resultant force is zero.

Taking moments (clockwise moments positive) about a point on the line of action of force 200 N at the left end,

$$M = (400 \times 0.5) - (600 \times 1.5) + (500 \times 3) - (100 \times 5)$$
$$= 200 - 900 + 1500 - 500 = 300 \text{ Nm } (\circlearrowright)$$

Obviously the given force system of parallel forces reduces to a clockwise couple of 300 Nm. It may be recalled that the conditions of equilibrium of coplanar parallel system are

$$(a)\ \Sigma F = 0 \quad \text{and} \quad (b)\ \Sigma M = 0$$

The given system is thus not in equilibrium.

It may be further recalled that a couple cannot be balanced by a single force. Therefore, the given force system cannot be brought to equilibrium by applying a single force.

EXAMPLE 3.69

Apply the laws of equilibrium to state the conditions under which the three force system given in the adjoining figure will be in equilibrium.

Solution: Apply the three conditions of equilibrium, *i.e.*,

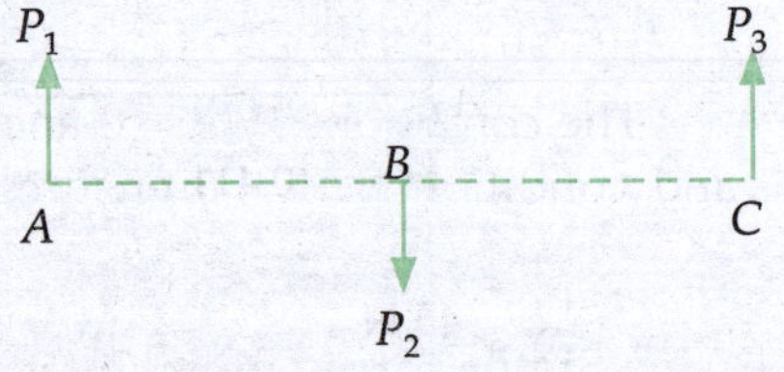

Fig. 3.97

(*i*) $\Sigma F_x = 0$. This is true as no horizontal force is acting on the body.

(*ii*) $\Sigma F_y = 0$. For this condition to hold good,

$$P_1 + P_3 = P_2$$

(*iii*) $\Sigma M = 0$. Taking moments of all the forces about point A, we get

$$\Sigma M_A = P_2 \times AB - P_3 \times AC$$

and for equilibrium, ΣM_A should be zero. That gives

$$P_2 \times AB - P_3 \times AC = 0$$

Apparently the distances AB and AC have to be such that the above identity is satisfied.

3.7. GENERAL CONDITIONS FOR EQUILIBRIUM

A system of forces acting upon a body changes or tends to change the state of rest or of uniform motion of the body. The body may also undergo rotation if a couple also acts on the body.

Obviously, the body is likely to experience

- displacement parallel to x-axis due to component F_x of the resultant force
- displacement parallel to y-axis due to component F_y of the resultant force
- rotation due to resultant couple.

The equilibrium between different coplanar forces is ensured by any one of the following sets of equations :

A : *Concurrent force system :* When acted upon by a system of coplanar concurrent forces, the body undergoes only displacement and the rotation is zero. The system is then said to be in equilibrium if there is no unbalanced force acting on it, *i.e.*, the resultant of all forces acting on the body is zero. Analytically then

(*i*) $\Sigma F_x = 0$; $\Sigma F_y = 0$ (*ii*) $\Sigma F_x = 0$; $\Sigma M_A = 0$ (*iii*) $\Sigma M_A = 0$; $\Sigma M_B = 0$

where

- ΣF_x and ΣF_y are the algebraic sum of forces in the x and y directions respectively
- ΣM_A and ΣM_B are the algebraic sums of moments of forces about points A and B.
- A and B are the points in the plane so chosen that they donot lie on the x-axis or y-axis.

B : *Coplanar non-concurrent forces :* When acted upon by a system of normal coplanar non-concurrent forces, the body undergoes displacement (due to resultant force) as well as rotation due to resultant couple. The system is then said to in equilibrium if the algebraic sum of all the forces and their moments about any point in their plane is zero. Mathematically,

(i) $\Sigma F_x = 0 \; ; \Sigma F_y = 0 \; ; \Sigma M_A = 0$

The parameter ΣM_A is the algebraic sum of the moments of forces about any point in the plane.

(ii) $\Sigma F_x = 0 \; ; \Sigma M_A = 0 \; ; \Sigma M_B = 0$

The points A and B lie in the plane and are so chosen that $\Sigma M_B = 0$ and the line joining A and B is not perpendicular to x-axis.

(iii) $\Sigma M_A = 0 \; ; \Sigma M_B = 0 \; ; \Sigma M_C = 0$

The points A, B and C lie in the plane and are so chosen that they donot lie in the same straight line.

The conditions $\Sigma F_x = 0$ and $\Sigma F_y = 0$ ensure that the system does not reduce to a single force and condition $\Sigma M = 0$ ensures that it does not reduce to a couple.

C : *Parallel force system :*

(i) $\Sigma F = 0 \; ; \Sigma M_A = 0$ (ii) $\Sigma M_A = 0 \; ; \Sigma M_B = 0$

where :

— ΣF is the algebraic sum of forces parallel to the line of action of forces

— ΣM_A and ΣM_B are the algebraic sum of moments about points A and B

— A and B are the points in the plane so chosen that the line joining A and B is not parallel to the forces.

REVIEW AND SUMMARY

1. The moment of force about a point is defined as the product of force and the perpendicular distance of the line of action of the force from that point. The moment of force measures the tendency of the force to rotate the rigid body about an axis.
2. **Varignon's theorem :** The moment of the resultant of a number of forces about any point is equal to the algebraic sum of the moments of all the forces of the system about the same point.
3. The lines of action of parallel forces are parallel to each other.
   (a) When the two parallel forces act in the same direction, their resultant equals their sum and its point of application lies between the two forces. Further, it divides the distance between the two forces internally in the inverse ratio of the magnitude of forces.
   (b) When the two parallel forces act in opposite directions, their resultant equals their difference and its point of application lies outside the two forces nearer to the larger of the two forces. Further, it divides the distance between the forces externally in the inverse ratio of the magnitude of forces.
4. Two parallel forces equal in magnitude but opposite in direction, and separated by a finite distance are said to form a couple.

The rotational effect of a couple is measured by its moment which equals the product of either of the two forces and the perpendicular distance between the forces.

5. The salient aspects of a couple are :
   - a couple consists of a pair of equal and opposite forces separated by a finite distance
   - the algebraic sum of the vertical and horizontal components of forces forming a couple is zero
   - a couple can not be balanced by a single force, but can be balanced only by a couple of opposite sense
   - a system of forces acting on a rigid body can be replaced by a single force acting at a given point and a couple; this force and couple is known as equivalent force-couple system
   - two system of forces are said to be equivalent when they can be replaced by the same force-couple system at a given point.
   - the translatory effect of a couple on a body is zero.
   - the effect of couple on a body remains unchanged if the couple is
     (*a*) rotated through an angle
     (*b*) shifted to any other position
     (*c*) replaced by another pair of forces whose rotational effect is same
6. When acted upon by a system of coplanar non-current forces, the body undergoes displacement (due to resultant force) as well as rotation due to resultant couple. The system would then be in equilibrium if

$$\Sigma F_x = 0\,; \quad \Sigma F_y = 0 \quad \text{and} \quad \Sigma M = 0$$

The conditions $\Sigma F_x = 0$ and $\Sigma F_y = 0$ ensure that the system does not reduce to a single force, and the condition $\Sigma M = 0$ ensures that it does not reduce to a couple.

REVIEW QUESTIONS

A. Conceptual and conventional questions :

1. Define moment and moment of force.
2. List some engineering applications of moment.
3. State the possible displacements of a body subjected to a general force system.
4. State and prove Varignon's theorem.
5. What is a couple? Enumerate its various characteristics.
6. Can a couple be balanced by a single force applied to a body?
7. Is it true that the line of action of the resultant of the parallel forces always passes between the lines of action of these forces?
8. Outline the steps involved in the graphical method for the determination of the resultant of parallel forces.
9. State the equilibrium conditions for bodies acted upon by a system of : (*i*) coplanar concurrent forces, (*ii*) coplanar non-concurrent forces.
10. Find the magnitude of the two like parallel forces acting at a distance of 24 cm, whose resultant is 200 N and its line of action is at a distance of 6 cm from one of the forces. **(50 N, 150 N)**
11. Three like parallel force 20 N, 40 N and 60 N are acting at points *A*, *B* and *C* respectively on a straight line *ABC*. The distances are *AB* = 3 m, *BC* = 4 m. Find the resultant and also the distance of the resultant from point *A* on line *ABC*. **(120 N, 4.5 m)**

12. Four parallel forces of magnitudes 100 N, 200 N, 50 N and 400 N are shown in Fig. 3.98. Determine the magnitude of the resultant and also the distance of the resultant from point A.

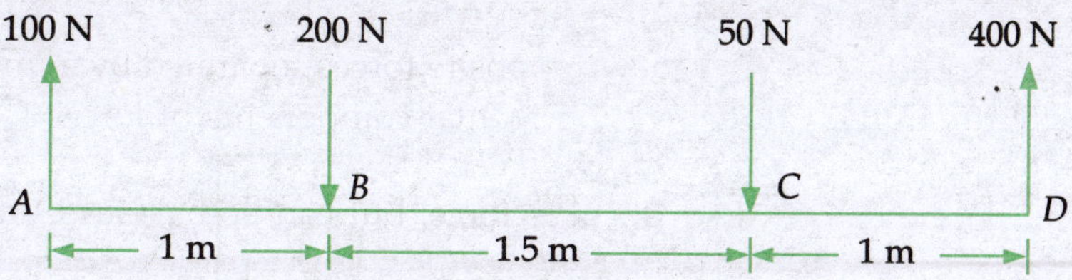

Fig. 3.98

(R = 250 N & 4.3 m)

13. Two like parallel forces P and Q act on a rigid body at A and B respectively. If P and Q be inter changed in position, show that the point of application of the resultant will be displaced through a distance d along AB, where

$$d = \frac{P - Q}{P + Q} \times AB$$

14. $ABCD$ is a rectangle in which $AB = CD = 100$ mm and $BC = DA = 80$ mm. Forces of 100 N each act along AB and CD and forces of 50 N each act along BC and DA. Find the resultant moment of two couples. **(13 Nmm)**

15. Three forces, acting on a rigid body, are represented in magnitude, direction and line of action by the three sides of a triangle taken in order. Prove that the forces are equivalent to a couple whose moment is equal to twice the area of the triangles.

16. A beam AB of span 8 m is hinged at A and is on rollers at B. It carries a uniformly distributed load, a concentrated load and an externally applied moment as shown in Fig 3.99. Determine the reactions at A and B for the loading shown.

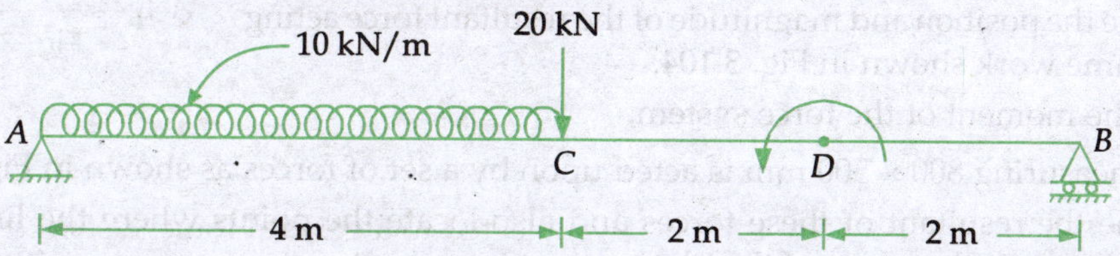

Fig. 3.99

(R_b = 10 kN down and, R_a = 70 kN)

17. For the lever system depicted in Fig. 3.100, determine the force P required to keep it in equilibrium.

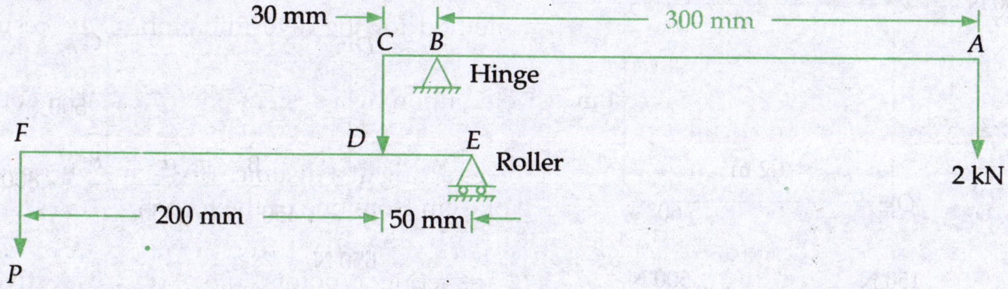

Fig. 3.100

18. Figure 3.101 shows a horizontal beam AD of a jib-crane supported horizontally at the p and by a tie rod BC. Calculate the tension in the tie rod when a load of 1 kN is acting at poi... on the beam.

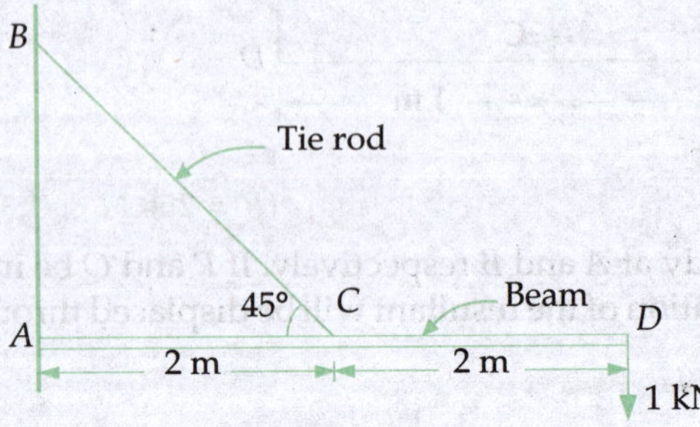

Fig. 3.101

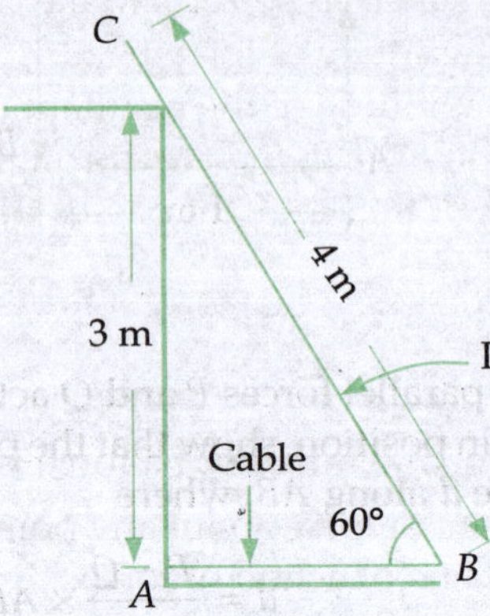

Fig. 3.102

19. A post BC of weight 18 N is prevented from sliding with the help of cable AB as shown in Fig. 3.102. Assuming all surfaces to be smooth, work out the tension developed in the cable.

(5.19 N)

20. A uniform rod AB of 1 m length and 1 kg mass is suspended from its ends as shown in Fig. 3.103. What amount of tensions would be induced in the strings when a mass of 5 kg is suspended from a point C on the rod?

(T_1 = 16.85 N, T_2 = 41.85 N)

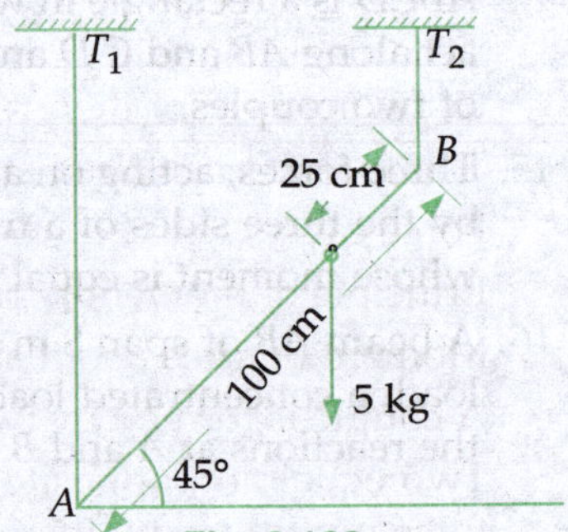

Fig. 3.103

21. Determine the position and magnitude of the resultant force acting on the frame work shown in Fig. 3.104.

What is the moment of the force system.

22. A plate measuring 800 × 500 mm is acted upon by a set of forces as shown in Fig. 3.105. Determine the resultant of these forces and also locate the points where the line of action of resultant intersets the edges of the plate.

(1706.76 N, 54,1°, 193.19 mm, 238.48 mm, 329.48 mm)

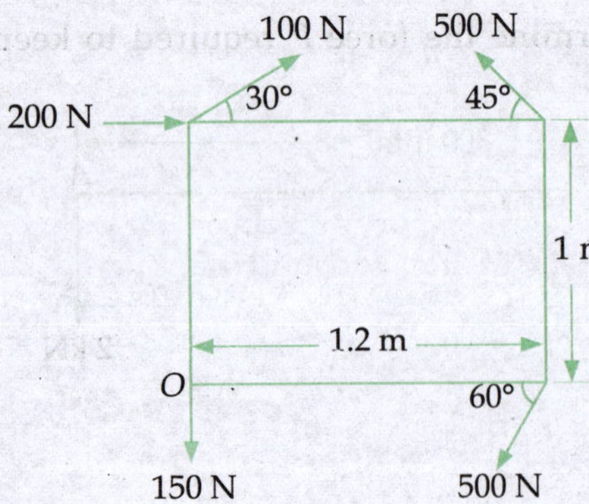

Fig. 3.104

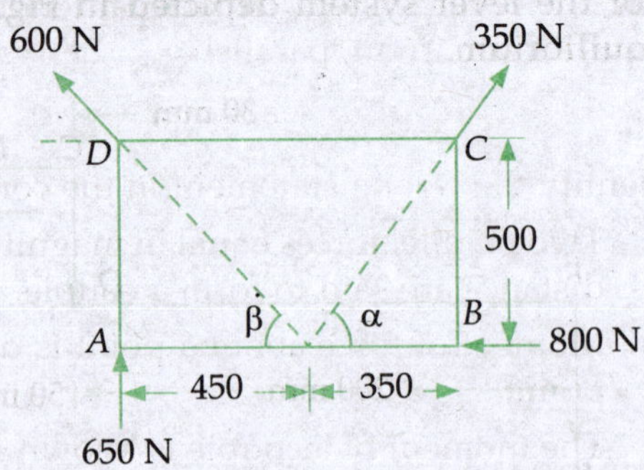

Fig. 3.105

23. A circular disc of 8 cm radius is acted upon by set of two forces as shown in Fig. 3.106.
The resultant moment of these forces about a point A on the disc is stated to be zero. Make calculations for :
(a) the magnitude of force P
(b) the magnitude of the resultant of two forces
(c) the point on the y-axis through which the line of action of the resultant passes through

(61.6 N, 141.25 N, 2.20 cm)

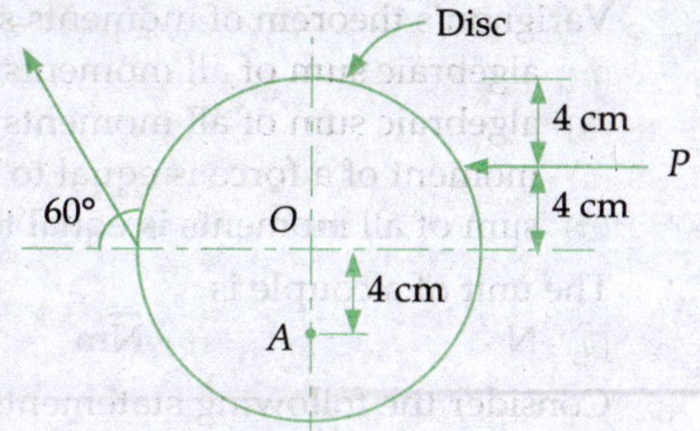

Fig. 3.106

B. Complete the following statements with most appropriate word/words :

1. The rotational tendency of a force is called ....
2. The moment of force about any point is the .... of force and .... distance between the point and line of action of force.
3. The moment of force about any pivot point is a .... quantity and it varies .... with its distance from the point.
4. "The algebraic sum of the moments of a system of forces about any point in their plane is equal to the moment of their resultant about that point."
This is the generalised statement of ....
5. Force causes .... displacement while moment causes .... displacement.
6. Principle of moment is based on .... theorem.
7. Moment of force about an axis indicates .... of the body about that axis.
8. A couple is formed by two .... and .... parallel forces.
9. Two equal and opposite parallel forces produce a .... whose moment is equal to either force multiplied by their .... distance.
10. The moment of a couple is known as ....
11. If a body is in equilibrium under the action of three forces, then the forces must be .... or ....
12. The conditions of equilibrium for coplanar and non-concurrent forces are ....

Answers :

1. moment 2. product, perpendicular 3. vector, directly 4. Varignon's theorem 5. linear, angular 6. Varignon's 7. rotational tendency 8. equal, opposite 9. couple, perpendicular 10. torque 11. concurrent, parallel 12. $\Sigma F_x = 0$, $\Sigma F_y = 0$ and $\Sigma M = 0$

C. Multiple choice questions :

1. Identify the wrong statement in the context of moment, couple and torque
(a) Two parallel forces equal in magnitude but opposite in direction, and separated by a finite distance are said to form a couple.
(b) Moment of force about a point is defined as the turning tendency of the force about that point.
(c) The moment of a couple is known as torque
(d) The sum of forces forming a couple in any direction has a definite non-zero value.

2. Varignon's theorem of moments states that the
   (*a*) algebraic sum of all moments is zero
   (*b*) algebraic sum of all moments is equal to the moment of the resultant about the origin
   (*c*) moment of a force is equal to the moment sum of its components
   (*d*) sum of all moments is equal to the resultant couple
3. The unit of a couple is
   (*a*) N (*b*) Nm (*c*) N/m (*d*) Nm^2
4. Consider the following statements :
   The effect of couple in a body remains unchanged if the couple is
   1. rotated through an angle
   2. shifted to any other position
   3. replaced by another pair of forces whose rotational effect is same
   Which of these statements are correct ?
   (*a*) 1 and 2 (*b*) 1 and 3 (*c*) 2 and 3 (*d*) 1, 2, and 3
5. Consider the following statements
   1. The translatory effect of a couple on a body is zero.
   2. A couple can be balanced by a couple of opposite sense
   3. The moment of a couple about any point is the same.
   Which of these statements is/are correct?
   (*a*) 2 and 3 (*b*) 3 only (*c*) 1 and 2 (*d*) 1, 2, and 3
6. A rectangle *ABCD* has sides *AB* = *CD* = 80 mm and *BC* = *DA* = 60 mm. Forces of 150 N each act along *AB* and *CD*, and forces of 100 N each act along *BC* and *DA*. The resultant moment of the force system is
   (*a*) 8000 Nmm (*b*) 9000 Nmm (*c*) 1000 Nmm (*d*) 17000 Nmm
7. All the forces acting at a point
   (*a*) may not be replaced by a single force
   (*b*) may be replaced by a single force
   (*c*) may be replaced by an equilibrant
   (*d*) may be replaced by a couple
8. Consider the following statements :
   1. A single force *F* acting at certain distance *d* from a point gives rise to a moment *Fd* about that point.
   2. Two equal and opposite forces each of a certain magnitude *F* and separated by a distance *d* constitute a couple whose magnitude is *Fd*.
   3. Both moment and couple have the same units and a sense of direction that may be clockwise or counter clockwise.
   Which of these statements is/are correct.
   (*a*) 1 and 2 (*b*) 1, 2 and 3 (*c*) 1 only (*d*) 2 only.

Answers :

1. (*d*) 2. (*c*) 3. (*b*) 4. (*d*) 5. (*d*) 6. (*d*)
7. (*b*)

4 Concurrent Forces in Space

Learning objectives : The contents of this chapter will make the reader familar with

- *vectors and their associated algebra*
- *concept of concurrent forces in a space system*
- *components and moment of a force*
- *conditions for the equilibrium of a particle in space*

In a space force system, the lines of action of different forces lie in different planes. Attention has been directed in this chapter to understand the concept of space system, to find the resultant of space forces and to set up the conditions for the equilibrium of such forces when they are concurrent, i.e., they meet at a point. The force system on a tripod carrying a camera is a classical example of non-coplanar concurrent forces.

4.1. VECTOR ALGEBRA

Vector approach is generally considered to be easier and more suitable while dealing with space forces. For that knowledge about basic concepts of vectors is essential. A brief review of vectors and their associated algebra has been made in the following paragraphs.

4.1.1. Vector and its representation

Vectors are mathematical expressions possessing magnitude and direction.

A vector quantity may be written either as thick bold letter (**A**) or a capital letter with arrow over it ($\vec{A}$). The plane capital letter (A) denotes the magnitude of that vector. Thus

$$\vec{A} = |\vec{A}| = A \quad ...(4.1)$$

where $\vec{A}$ is a vector quantity and A is its magnitude.

4.1.2. Components of a vector

The vector $\vec{A}$ represented by OP can be resolved into three components A_x, A_y and A_z along the rectangular axes.

$$\vec{A} = A_x \boldsymbol{i} + A_y \boldsymbol{j} + A_z \boldsymbol{k} \quad ...(4.2)$$

where $\boldsymbol{i}, \boldsymbol{j}$ and $\boldsymbol{k}$ are referred to as unit vectors along the positive x, y and z directions respectively. A unit vector is a vector having a unit magnitude or unit length.

Unit vector corresponding to vector $\vec{A}$

$$= \frac{\vec{A}}{|\vec{A}|} = \frac{\vec{A}}{\sqrt{A_x^2 + A_y^2 + A_z^2}} \quad \text{...(4.3)}$$

The components A_x, A_y and A_z are called scalar compo-nents of the vector $\vec{A}$.

If the vector $\vec{A}$ makes angles α, β and γ with the x, y and z axes respectively, then

$$\vec{A} = (A\cos\alpha)\,\boldsymbol{i} + (A\cos\beta)\,\boldsymbol{j} + (A\cos\gamma)\,\boldsymbol{k} \quad \text{...(4.4)}$$

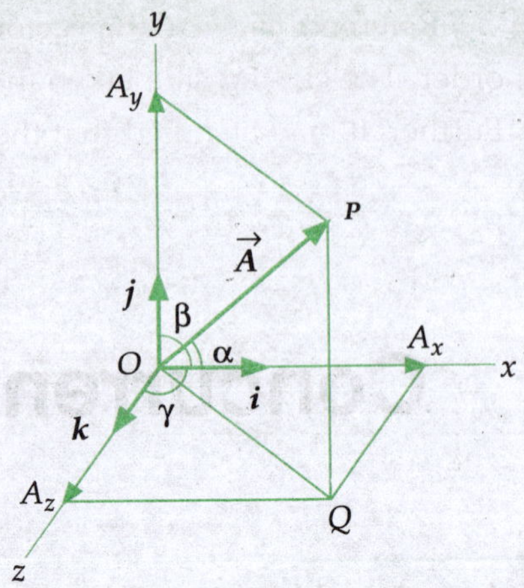

Fig. 4.1. Components of a vector

EXAMPLE 4.1.

A force of 25 N makes an angle of 30, 45 and 75 degree with the x, y and z-axis. What should be the corresponding force vector?

Solution :

$$\vec{F} = F_x i + F_y j + F_z k$$
$$= (F\cos\alpha)\,i + (F\cos\beta)\,j + (F\cos\gamma)\,k$$
$$= (25\cos 30)\,i + (25\cos 45)\,j + (25\cos 75)\,k$$
$$= 21.65\,i + 17.68\,j + 6.47\,k$$

4.1.3. Position vector of a given point

The position vector of a point $P(x, y, z)$ with respect to point $O(0, 0, 0)$ is the vector OP. It is denoted by $\vec{r}$ and is used to specify the position of point P relative to O. (Fig. 4.2)

$$\vec{r} = (x-0)\,\boldsymbol{i} + (y-0)\,\boldsymbol{j} + (z-0)\,\boldsymbol{k}$$
$$= x\boldsymbol{i} + y\boldsymbol{j} + z\boldsymbol{k} \quad \text{...(4.5)}$$

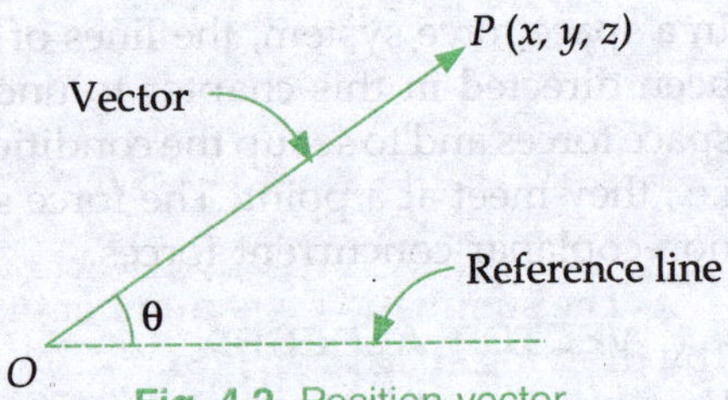

Fig. 4.2. Position vector

A vector is generally represented by a line segment with an arrow at one end.

(*i*) the length of the line represents the magnitude of the vector quantity

(*ii*) the arrow mark specifies the sense of direction, *i.e.*, to or from a point

(*iii*) the orientation of the line shows the direction of the vector, *i.e.*, the angle which the vector makes with a reference line.

4.1.4. Addition and subtraction of vectors

The resultant of two vectors may be obtained according to parallelogram law or triangle law.

Reference Fig. 4.3, the resultant $\vec{R}$ of two vectors $\vec{A}$ and $\vec{B}$ is the diagonal of the parallelogram for which $\vec{A}$ and $\vec{B}$ are the adjacent sides.

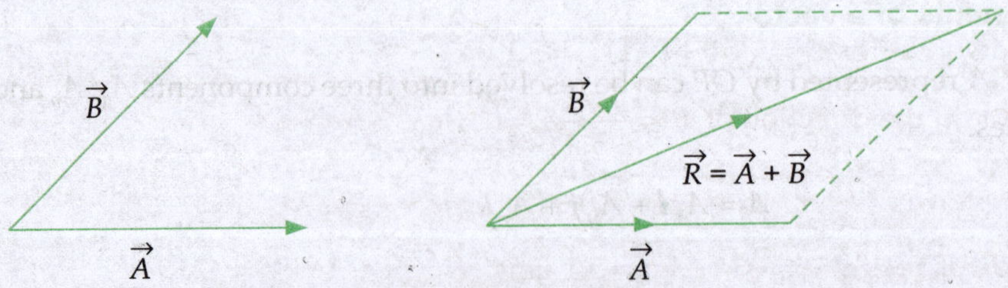

Fig. 4.3. Addition of vectors by parallelogram law

Reference Fig. 4.4, the vectors $\vec{A}$ and $\vec{B}$ are represented by the two sides of a triangle taken in order. The closing side taken in opposite order represents the resultant $\vec{R}$ of the vector $\vec{A}$ and $\vec{B}$. Further, it may be noted that the vector addition is commutative. That is

$$\vec{R} = \vec{A} + \vec{B} = \vec{B} + \vec{A} \quad \text{...(4.6)}$$

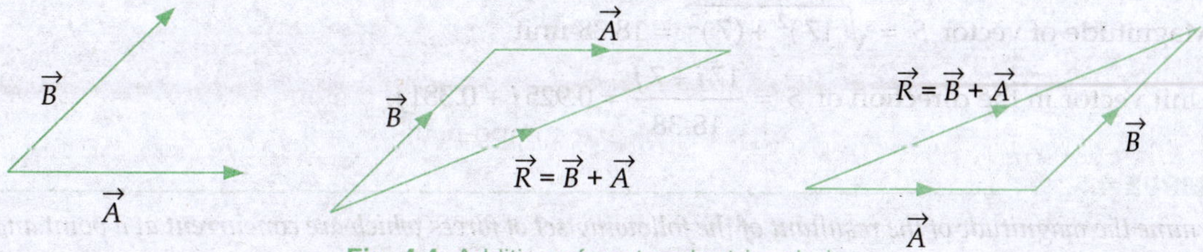

Fig. 4.4. Addition of vectors by triangle law

The difference between vector $\vec{A}$ and $\vec{B}$ can be obtained by addition of vectors $\vec{A}$ and $-\vec{B}$ (Fig. 4.5).

$$\vec{A} - \vec{B} = \vec{A} + (-\vec{B})$$

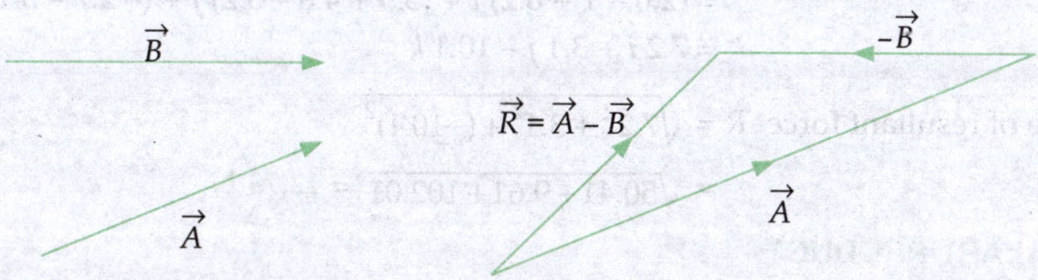

Fig. 4.5. Subtraction of vectors

The summation of a number of vectors can be obtained by the repeated use of parallelogram or by drawing space polygon.

EXAMPLE 4.2

Two vectors are prescribed as

$$\vec{A} = 3i + 2j \text{ and } \vec{B} = 2i - 3j$$

If $\vec{P} = \vec{A} + \vec{B}$ *and* $\vec{Q} = \vec{A} - \vec{B}$, *then find*

(a) *magnitude of vectors* $\vec{P}$ *and* $\vec{Q}$

(b) *unit vectors along P and Q, and*

(c) *magnitude and unit vector along* $\vec{S} = 3\vec{P} + 2\vec{Q}$

Solution : $\vec{P} = \vec{A} + \vec{B} = (3i + 2j) + (2i - 3j) = 5i - j$

Magnitude of vector $\vec{P} = \sqrt{5^2 + (-1)^2} = 5.1$ unit

Unit vector in the direction of $\vec{P} = \dfrac{5i - j}{5.1} = 0.98\, i - 0.196\, j$

$$\vec{Q} = \vec{A} - \vec{B} = (3i + 2j) - (2i - 3j) = i + 5j$$

Magnitude of vector $Q = \sqrt{1^2 + 5^2} = 5.1$ unit

Unit vector in the direction of $\vec{Q} = \dfrac{i + 5j}{5.1} = 0.196i + 0.98j$

(*b*) $\vec{S} = 3\vec{P} + 2\vec{Q} = 3(5i - j) + 2(i + 5j)$

$= (15i - 3j) + (2i + 10j) = 17i + 7j$

Magnitude of vector $\vec{S} = \sqrt{(17)^2 + (7)^2} = 18.38$ unit

Unit vector in the direction of $\vec{S} = \dfrac{17i + 7j}{18.38} = 0.925i + 0.381j$

EXAMPLE 4.3.

Determine the magnitude of the resultant of the following set of forces which are concurrent at a point and are expressed in newtons

$$\vec{F_1} = 2.0\,i + 3.5\,j - 2.7\,k\ ;\quad \vec{F_2} = -i + 4.8\,j - 3.1\,k\ ;\quad \vec{F_3} = 6.2\,i + 5.2\,j - 4.3\,k$$

Solution : Resultant force $\vec{F} = \vec{F_1} + \vec{F_2} + \vec{F_3}$

$= (2.0 - 1 + 6.2)\,i + (3.5 + 4.8 - 5.2)\,j + (-2.7 - 3.1 - 4.3)\,k$

$= 7.2\,i + 3.1\,j - 10.1\,k$

Magnitude of resultant force $R = \sqrt{7.2^2 + 3.1^2 + (-10.1)^2}$

$= \sqrt{50.41 + 9.61 + 102.01} = 12.73$ **N**

4.2. DOT (SCALAR) PRODUCT

The dot product of vectors $\vec{A}$ and $\vec{B}$ is a scalar quantity, and is defined as the product of the magnitude of the vectors and cosine of their included angle θ

$$\vec{A} \cdot \vec{B} = |\vec{A}|\,|\vec{B}|\cos\theta = AB\cos\theta$$

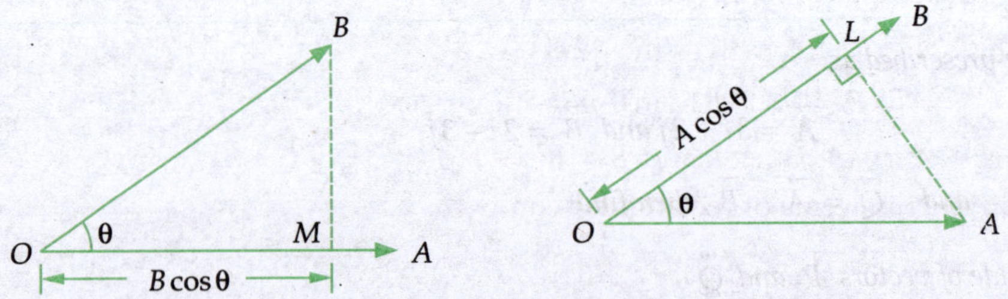

Fig. 4.6. Dot or scalar product

It is to be noted that :

(*i*) When $\theta = 0°$, i.e., when the vectors $\vec{A}$ and $\vec{B}$ are along the same direction

$$\vec{A} \cdot \vec{B} = AB\cos 0° = AB$$

(*ii*) When $\theta = 90°$, i.e., when the vectors $\vec{A}$ and $\vec{B}$ are perpendicular to each other

$$\vec{A} \cdot \vec{B} = AB\cos 90° = 0$$

(*iii*) For $0° \le \theta \le 90°$, the product is positive

For $90° \le \theta \le 360°$, the product is negative

(iv) $\vec{A} \cdot \vec{B}$ = A times projection of $\vec{B}$ on $\vec{A}$

= B times projection of $\vec{A}$ on $\vec{B}$

(v) In terms of components,

$$\vec{A} \cdot \vec{B} = (A_x \boldsymbol{i} + A_y \boldsymbol{j} + A_z \boldsymbol{k}) \times (B_x \boldsymbol{i} + B_y \boldsymbol{j} + B_z \boldsymbol{k})$$
$$= A_x B_x + A_y B_y + A_z B_z$$

i, ***j*** and ***k*** are the unit vectors along x, y and z directions respectively, and

$$\boldsymbol{i} \cdot \boldsymbol{i} = \boldsymbol{j} \cdot \boldsymbol{j} = \boldsymbol{k} \cdot \boldsymbol{k} = 1 \quad (\because \theta = 0^\circ \text{ and } \cos 0^\circ = 1)$$
$$\boldsymbol{i} \cdot \boldsymbol{j} = \boldsymbol{j} \cdot \boldsymbol{k} = \boldsymbol{k} \cdot \boldsymbol{i} = 0 \quad (\because \theta = 90^\circ \text{ and } \cos 90^\circ = 0)$$

(vi) The angle between the vectors $\vec{A}$ and $\vec{B}$ is

$$\cos\theta = \frac{\vec{A} \cdot \vec{B}}{|A||B|} = \frac{\vec{A} \cdot \vec{B}}{AB}$$

(vii) The dot product of force and displacement defines the term work in the field of dynamics.

(viii) The following laws hold good for the dot product of vectors

$$\vec{A} \cdot \vec{B} = \vec{B} \cdot \vec{A} \quad \text{(Dot product is commutative)}$$
$$\vec{A} \cdot (\vec{B} + \vec{C}) = \vec{A} \cdot \vec{B} + \vec{A} \cdot \vec{C}$$
$$m(\vec{A} \cdot \vec{B}) = (m\vec{A}) \cdot \vec{B} = \vec{A}(m \cdot \vec{B}) \quad \text{where } m \text{ is a scalar quantity.}$$

4.3. CROSS (VECTOR) PRODUCT

The cross product of vectors $\vec{A}$ and $\vec{B}$ is a vector quantity and is defined as the product of the magnitude of the vectors and the sine of the smaller angle between the vectors. Its direction is perpendicular to the plane containing the vectors.

If $\vec{n}$ is the unit vector which gives the direction of the resultant vector, then

$$\vec{R} = \vec{A} \times \vec{B} = |A||B| \sin\theta\, \vec{n} = AB \sin\theta\, \vec{n}$$

If $\vec{A} = A_x \boldsymbol{i} + A_y \boldsymbol{j} + A_z \boldsymbol{k}$ and $\vec{B} = B_x \boldsymbol{i} + B_y \boldsymbol{j} + B_z \boldsymbol{k}$

then the cross product in terms of rectangular components is given by

$$\vec{A} \times \vec{B} = \begin{vmatrix} \boldsymbol{i} & \boldsymbol{j} & \boldsymbol{k} \\ A_x & A_y & A_z \\ B_x & B_y & B_z \end{vmatrix}$$

$$= (A_y B_z - B_y A_z)\, \boldsymbol{i} - (A_x B_z - B_x A_z)\, \boldsymbol{j} + (A_x B_y - B_x A_y)\, \boldsymbol{k}$$

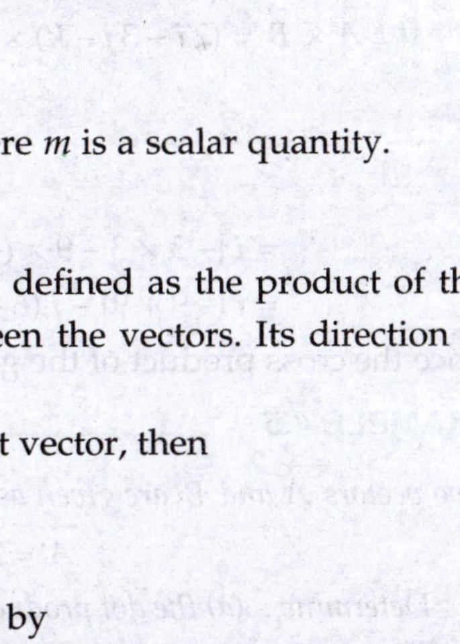

Fig. 4.7. Cross or vector product

Further it is to be noted that :

(i) $\vec{A} \times \vec{B} = 0$ if $\theta = \theta$ or π

(ii) The cross product of a unit vector with itself, such as $i \times i$, is equal to zero because both vectors have the same magnitude. Like wise : $j \times j = 0$ and $k \times k = 0$.

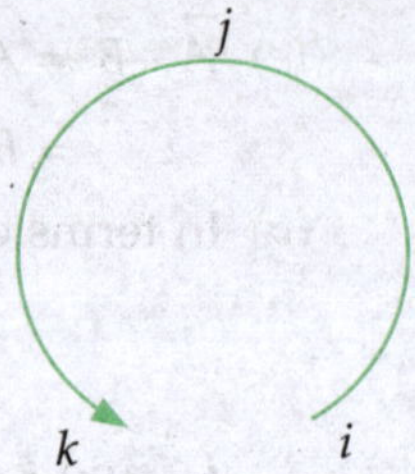

Fig. 4.8

Further for determination of the sign of the vector product of two unit vectors, the three letters representing the unit vectors are arranged in a circle and in counter clockwise order. The product of two unit vectors will be positive if they follow each other in counter clockwise order and will be negative if they follow each other in clockwise order. That is

$$i \times j = k \;\; ; \;\; j \times k = i \;\; ; \;\; k \times i = j$$

and

$$j \times i = -k \;\; ; \;\; k \times j = -i \;\; ; \;\; i \times k = -j$$

(iii) The cross product of position vector and force vector defines the term moment of force.

EXAMPLE 4.4

(a) Check whether the following set of vectors are perpendicular to each other or not

$$\vec{A} = 6i + 10j - 5k \quad \text{and} \quad \vec{B} = 5i + 2j + 10k$$

(b) Check whether the following set of vectors are parallel to each other or not

$$\vec{A} = 2i - 3j - k \quad \text{and} \quad \vec{B} = -6i + 9j + 3k$$

Solution : $\vec{A} \cdot \vec{B} = (6\boldsymbol{j} + 10\boldsymbol{j} - 5\boldsymbol{k}) \cdot (5\boldsymbol{i} + 2\boldsymbol{j} + 10\boldsymbol{k})$

$$= 6 \times 5 + 10 \times 2 + (-5 \times 10) = 0$$

Since the dot product of the given vectors is zero, the vectors are perpendicular to each other.

(b) $\vec{A} \times \vec{B} = (2\boldsymbol{i} - 3\boldsymbol{j} - \boldsymbol{k}) \times (-6\boldsymbol{i} + 9\boldsymbol{j} + 3\boldsymbol{k})$

$$= \begin{vmatrix} i & j & k \\ 2 & -3 & -1 \\ -6 & 9 & 3 \end{vmatrix}$$

$$= \boldsymbol{i}\,[-3 \times 3 - 9 \times (-1)] - \boldsymbol{j}\,[2 \times 3 - (-6) \times (-1)] + \boldsymbol{k}\,[2 \times 9 - (-6 \times -3)]$$

$$= \boldsymbol{i}\,(-9 + 9) - \boldsymbol{j}\,(6 - 6) + \boldsymbol{k}\,(18 - 18) = 0$$

Since the cross product of the given vectors is zero, the vectors are parallel to each other.

EXAMPLE 4.5

Two vectors $\vec{A}$ and $\vec{B}$ are given as

$$\vec{A} = 2i + 3j \quad \text{and} \quad \vec{B} = 3i - j$$

Determine : (a) the dot product and cross product of the vectors, (b) the angle between vectors $\vec{A}$ and $\vec{B}$ (c) the included angle between vector $\vec{A}$ and the vector resulting from the cross product.

Solution : The dot product is given as

$$\vec{A} \cdot \vec{B} = A_x B_x + A_y B_y$$

$$= 2 \times 3 + 3 \times (-1) = 6 - 3 = 3 \text{ units}$$

The cross product is expressed as

$$\vec{A} \times \vec{B} = \begin{vmatrix} \boldsymbol{i} & \boldsymbol{j} & \boldsymbol{k} \\ A_x & A_y & A_z \\ B_x & B_y & B_z \end{vmatrix} = \begin{vmatrix} \boldsymbol{i} & \boldsymbol{j} & \boldsymbol{k} \\ 2 & 3 & 0 \\ 3 & -1 & 0 \end{vmatrix}$$

$$= [3 \times 0 - (-1) \times 0]\,\boldsymbol{i} - (2 \times 0 - 3 \times 0)\,\boldsymbol{j} + [2 \times (-1) - 3 \times 3]\,\boldsymbol{k}$$

$$= 0 - 0 - 11\,\boldsymbol{k} = -11\,\boldsymbol{k}$$

(b) From the definition of dot product : $\vec{A} \cdot \vec{B} = AB\cos\theta$

$$\cos\theta = \frac{\vec{A}\cdot\vec{B}}{AB} = \frac{3}{\sqrt{2^2+3^2}\times\sqrt{3^2+(-1)^2}} = \frac{3}{\sqrt{13}\times\sqrt{10}} = 0.263$$

$$\theta = \cos^{-1}(0.263) = 74.75^\circ$$

Check : From the definition of cross product : $|\vec{A}\times\vec{B}| = AB\sin\theta$

$$\sin\theta = \frac{\vec{A}\times\vec{B}}{AB} = \frac{11}{\sqrt{13}\times\sqrt{10}} = 0.965$$

$$\theta = \sin^{-1}(0.965) = 74.79^\circ$$

(c) $$\vec{A}\times(\vec{A}\times\vec{B}) = \begin{vmatrix} i & j & k \\ 2 & 3 & 0 \\ 0 & 0 & -11 \end{vmatrix} = -33\,i + 22\,j$$

The above vector has a magnitude of $\sqrt{(-33)^2+(-22)^2} = 39.66$

$$\sin\theta = \frac{|\vec{A}\times(\vec{A}\times\vec{B})|}{|\vec{A}|\times|\vec{A}\times\vec{B}|} = \frac{39.66}{\sqrt{13}\times 11} = 1.00$$

Thus the angle between $\vec{A}$ and $(\vec{A}\times\vec{B})$ is 90°. The included angle between vector $\vec{B}$ and the vector $(\vec{A}\times\vec{B})$ would also workout to be 90°.

Remarks : The included angle between the vector resulting from the cross product and either of the constituent vectors must be 90°.

EXAMPLE 4.6

Determine the work done when a force $\vec{F} = 2i + 3j - k$ moves form point P (3, 1, – 1) to point Q (2, – 1, 1).

Solution : Distance moved = position vector of point P – position vector of point Q

$$\vec{S} = (3i + j - k) - (2i - j + k) = i + 2j - 2k$$

Work done by the force is given by the dot product of force applied and distance moved

$$\text{Work done} = \vec{F}\cdot\vec{S} = (2i + 3j - k)\cdot(i + 2j - 2k)$$

$$= 2\times 1 + 3\times 2 + (-1)\times(-2) = 10 \text{ units}$$

4.4. COMPONENT OF A FORCE

Force is a vector quantity having both magnitude and direction

Let a force $\vec{F}$ be designated by vector OP. If through point P, planes are drawn parallel to the coordinate planes, then a rectangular box or parallelopiped gets formed (Fig. 4.9). The force $\vec{F}$ is then represented by the diagonal OP of the box, and the three components F_x, F_y and F_z are represented by the edges of the box.

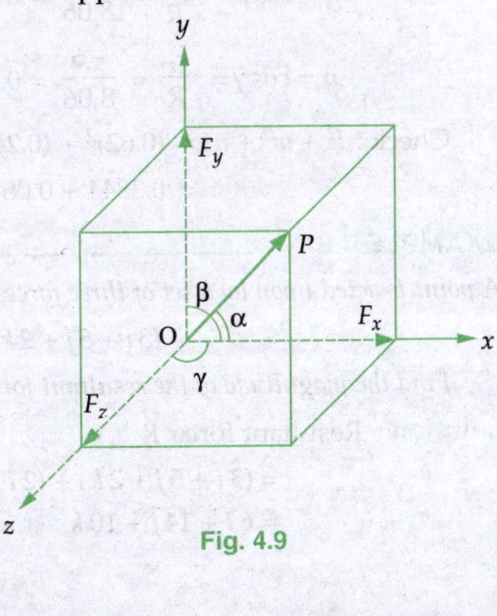

Fig. 4.9

Salient aspects :

(i) If $\boldsymbol{i}, \boldsymbol{j}$ and $\boldsymbol{k}$ are the unit vectors in the positive x, y and z directions respectively, then force $\vec{F}$ is designated as

$$\vec{F} = F_x \boldsymbol{i} + F_y \boldsymbol{j} + F_z \boldsymbol{k}$$

$$\text{Magnitude } F = |\vec{F}| = \sqrt{F_x^2 + F_y^2 + F_z^2}$$

The components F_x, F_y and F_z are called the scalar components of the force vector $\vec{F}$.

(ii) If α, β and γ are the angles which the force vector $\vec{F}$ makes with the coordinate axes x, y and z axes respectively, then

$$F_x = F\cos\alpha\,; \quad F_y = F\cos\beta \quad \text{and} \quad F_z = F\cos\gamma$$

That gives

$$\vec{F} = (F\cos\alpha)\,\boldsymbol{i} + (F\cos\beta)\,\boldsymbol{j} + (F\cos\gamma)\,\boldsymbol{k}$$

(iii) The cosines of angles α, β and γ are called direction cosines and are denoted as

$$l = \cos\alpha\,; \quad m = \cos\beta \quad \text{and} \quad n = \cos\gamma$$

The direction cosines are not independent and are related by the expression

$$l^2 + m^2 + n^2 = 1$$

EXAMPLE 4.7

Consider the $\vec{F} = (5i + 2j - 6k)$ kN that passes through the origin. Determine the magnitude of the force and the angles it makes with the x, y and z-axis.

Solution : Resultant of force, $R = \sqrt{5^2 + 2^2 + (-6)^2}$

$$= \sqrt{25 + 4 + 36} = 8.06 \text{ kN}$$

The direction cosines of the resultant are :

$$l = \cos\alpha = \frac{F_x}{R} = \frac{5}{8.06} = 0.62 \quad ; \quad \alpha = 51.68°$$

$$m = \cos\beta = \frac{F_y}{R} = \frac{2}{8.06} = 0.248 \quad ; \quad \beta = 75.64°$$

$$n = \cos\gamma = \frac{F_z}{R} = \frac{-6}{8.06} = 0.744 \quad ; \quad \gamma = -138.07°$$

$$\text{Check : } l^2 + m^2 + n^2 = (0.62)^2 + (0.248)^2 + (0.744)^2$$

$$= 0.3844 + 0.0615 + 0.5535 = 1.0$$

EXAMPLE 4.8

A point is acted upon by a set of three forces given by

$$(3i + 5j + 2k),\ (2i + 7j + 3k) \text{ and } (i + 2j + 5k)$$

Find the magnitude of the resultant force and its direction cosines.

Solution : Resultant force $\vec{R}$

$$= (3\boldsymbol{i} + 5\boldsymbol{j} + 2\boldsymbol{k}) + (2\boldsymbol{i} + 7\boldsymbol{j} + 3\boldsymbol{k}) + (\boldsymbol{i} + 2\boldsymbol{j} + 5\boldsymbol{k})$$

$$= 6\boldsymbol{i} + 14\boldsymbol{j} + 10\boldsymbol{k}$$

Magnitude of resultant = $\sqrt{6^2 + 14^2 + 10^2}$ = **18.22 units**

Direction cosines of the resultant are :

$$l = \cos\alpha = \frac{F_x}{F} = \frac{6}{18.22} = 0.329\,; \qquad \alpha = 70.79^\circ$$

$$m = \cos\beta = \frac{F_y}{F} = \frac{14}{18.22} = 0.768; \qquad \beta = 39.82^\circ$$

$$n = \cos\gamma = \frac{F_z}{F} = \frac{10}{18.22} = 0.5488; \qquad \gamma = 56.71^\circ$$

Check : $l^2 + m^2 + n^2 = (0.329)^2 + (0.768)^2 + (0.5488)^2 \simeq 1$

EXAMPLE 4.9

A force 25 kN starts from a point P(1, 3, – 2) and passes through the point Q (– 8, 5, 3). Express the force in terms of unit vectors i, j and k.

Solution : The vector $\vec{r} = \overrightarrow{PQ}$ represents the position of point Q with respect to point P. That is

$$\overrightarrow{PQ} = \vec{r} = (-8-1)\,\boldsymbol{i} + (5-3)\,\boldsymbol{j} + [3-(-2)]\,\boldsymbol{k}$$

$$= -9\boldsymbol{i} + 2\boldsymbol{j} + 5\boldsymbol{k}$$

$$\text{Unit vector along } PQ = \frac{-9i+2j+5k}{\sqrt{(-9)^2+(2)^2+(5)^2}} = \frac{-9i+2j+5k}{\sqrt{110}}$$

$$= -0.858\,\boldsymbol{i} + 0.1907\,\boldsymbol{j} + 0.4767\,\boldsymbol{k}$$

Force vector $\vec{F} = |\vec{F}| \times \vec{n}$

= magnitude of force × unit vector in the direction of force

$= 25\,(-0.858\,\boldsymbol{i} + 0.1907\,\boldsymbol{j} + 0.4767\,\boldsymbol{k})$

$= \mathbf{-21.45\,i + 4.767\,j + 11.917\,k}$

EXAMPLE 4.10

A space force has components of 10 N, 20 N and – 50 N along the x, y and z axes respectively. Write the vector equation for the force and calculate : (a) the magnitude of the force, (b) the direction cosines of the force and (c) the angles made by the force along the coordinate axes.

Solution : Refer Fig. 4.10 for the coordinate axes and the orientation of the given forces.

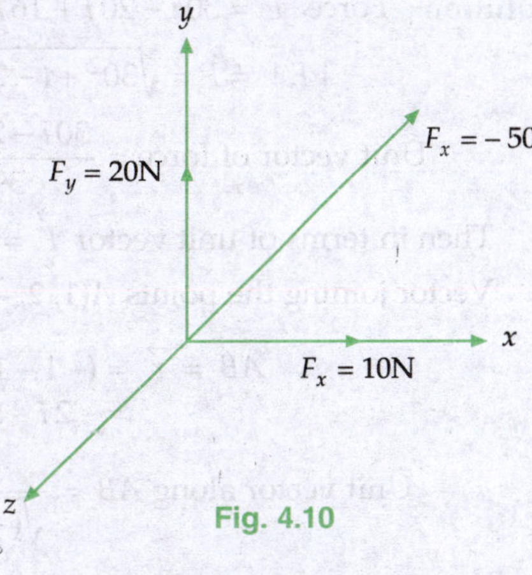

Fig. 4.10

Vector equation of force

$$\vec{F} = F_x\boldsymbol{i} + F_y\boldsymbol{j} + F_z\boldsymbol{k} = 10\,\boldsymbol{i} + 20\,\boldsymbol{j} - 50\,\boldsymbol{k}$$

(*a*) Magnitude of force

$$F = |\vec{F}| = \sqrt{F_x^2 + F_y^2 + F_z^2}$$

$$= \sqrt{10^2 + 20^2 + (-50)^2} = \mathbf{54.77\ N}$$

(*b*) Direction cosines of the force

$$l = \cos\alpha = \frac{F_x}{F} = \frac{10}{54.77} = 0.1826$$

$$m = \cos\beta = \frac{F_y}{F} = \frac{20}{54.77} = 0.3652$$

$$n = \cos\gamma = \frac{F_z}{F} = \frac{-50}{54.77} = -0.9129$$

Check : $l^2 + m^2 + n^2 = (0.1826)^2 + (0.3652)^2 + (-0.9129)^2$

$= 0.0333 + 0.1334 + 0.8334 \simeq 1.0$

(c) Angles made by the forces along the coordinate axes

$\alpha = \cos^{-1}(0.1826) = 79.48°$ $\beta = \cos^{-1}(0.3652) = 68.58°$ $\gamma = \cos^{-1}(-0.9129) = 155.91°$

EXAMPLE 4.11

A space force of magnitude 125 N makes angles of 30° and 65° with the x and y axes respectively. Make calculations for the scalar components of the force along the x, y and z axes and write the vector equation for the force.

Solution : $l = \cos\alpha = \cos 30° = 0.866$

$m = \cos\beta = \cos 65° = 0.423$

From the relation; $l^2 + m^2 + n^2 = 1$, we have

$$n = \cos\gamma = \sqrt{1 - l^2 - m^2} = \sqrt{1 - (0.866)^2 - (0.423)^2}$$

$$= \sqrt{1 - 0.75 - 0.179} = 0.266$$

That gives

$$\gamma = \cos^{-1}(0.266) = 74.57°$$

Scalar components of the force are :

$F_x = F\cos\alpha = 125\cos 30° = 108.25$ N

$F_y = F\cos\beta = 125\cos 65° = 52.83$ N

$F_z = F\cos\gamma = 125\cos 74.57° = 33.26$ N

Vector equation of the force

$$= F_x\boldsymbol{i} + F_y\boldsymbol{j} + F_z\boldsymbol{k} = 108.25\boldsymbol{i} + 52.83\boldsymbol{j} + 33.26\boldsymbol{k}$$

EXAMPLE 4.12

The x, y, z components of a force are 30 kN, – 20 kN and 16 kN respectively. Determine the component of this force along the line joining A(1, 2, – 3) and B(– 1, – 3, 4).

Solution : Force $\vec{F} = 30\boldsymbol{i} - 20\boldsymbol{j} + 16\boldsymbol{k}$

$$|\vec{F}| = F = \sqrt{30^2 + (-20)^2 + 16^2} = \sqrt{1556} = 39.446 \text{ kN}$$

$$\text{Unit vector of force} = \frac{30i - 20j + 16k}{39.446} = 0.760\boldsymbol{i} - 0.507\boldsymbol{j} + 0.406\boldsymbol{k}$$

Then in terms of unit vector $\vec{F} = 39.446\,(0.760\boldsymbol{i} - 0.507\boldsymbol{j} + 0.406\boldsymbol{k})$.

Vector joining the points $A(1, 2, -3)$ and $B(-1, -3, 4)$ is

$$\overrightarrow{AB} = \vec{r} = (-1-1)\boldsymbol{i} + (-3-2)\boldsymbol{j} + [4-(-3)]\boldsymbol{k}$$

$$= -2\boldsymbol{i} - 5\boldsymbol{j} + 7\boldsymbol{k}$$

$$\text{Unit vector along } AB = \frac{-2i - 5j + 7k}{\sqrt{(-2)^2 + (-5)^2 + (7)^2}} = \frac{-2i - 5j + 7k}{\sqrt{78}}$$

$$= -0.226\boldsymbol{i} - 0.566\boldsymbol{j} + 0.792\boldsymbol{k}$$

The component of force F along AB is

$= (\vec{F}$ in terms of unit vectors$) \cdot ($unit vector along $AB)$
$= 39.446\,(0.760\,\boldsymbol{i} - 0.507\,\boldsymbol{j} + 0.406\,\boldsymbol{k}) \cdot (-0.226\,\boldsymbol{i} - 0.566\,\boldsymbol{j} - 0.792\,\boldsymbol{k})$
$= 39.446\,[0.760 \times (-0.226) - 0.507 \times (-0.566) + 0.406 \times 0.792]$
$= 39.446\,(-0.1718 + 0.2869 + 0.3215) = 17.224$ kN

EXAMPLE 4.13

A force $\vec{F}$ acts at the origin of a coordinate system in a direction defined by the angles $\theta_x = 68^o$ and $\theta_z = 55^o$. The component of force $\vec{F}$ along y-direction is – 125 N. Determine :

(a) *angle θ_y* (b) *magnitude of force $\vec{F}$*

(c) *components of force along x and z directions, and*

(d) *component of force on a line that passes through the origin and point (1, 1, 1)*

Solution : The direction cosines along the x and z directions are :

$$l = \cos\theta_x = \cos 68^o = 0.3746$$
$$n = \cos\theta_z = \cos 55^o = 0.5736$$

From the relation : $l^2 + m^2 + n^2 = 1$, we have

$$m = \cos\theta_y = \sqrt{1 - l^2 - n^2} = \sqrt{1 - (0.3746)^2 - (0.5736)^2} = \pm\, 0.7332$$

That gives : $\theta_y = 42.85^o$ and 137.15^o

(b) From the relation : $F_y = F\cos\theta_y$ we have

$$\text{Magnitude of force } F = \frac{F_y}{\cos\theta_y} = \frac{-125}{\cos 137.15} = 170.53 \text{ N}$$

(c) Then scalar components of force in the x and z directions are :

$$F_x = F\cos\theta_x = 170.53 \times \cos 68^o = 63.88 \text{ N}$$
$$F_z = F\cos\theta_z = 170.53 \times \cos 55^o = 97.81 \text{ N}$$

Vector equation of force

$$\vec{F} = F_x\boldsymbol{i} + F_y\boldsymbol{j} + F_z\boldsymbol{k} = 63.88\,\boldsymbol{i} - 125\,\boldsymbol{j} + 97.81\,\boldsymbol{k}$$

(d) The coordinates of origin O and the given point A : $O(0, 0, 0)$ and $A(1, 1, 1)$

$$\text{Vector } \overrightarrow{OA} = (1-0)\,\boldsymbol{i} + (1-0)\,\boldsymbol{j} + (1-0)\,\boldsymbol{k} = \boldsymbol{i} + \boldsymbol{j} + \boldsymbol{k}$$

$$\text{Unit vector along } OA = \frac{\boldsymbol{i} + \boldsymbol{j} + \boldsymbol{k}}{\sqrt{1^2 + 1^2 + 1^2}} = \frac{\boldsymbol{i} + \boldsymbol{j} + \boldsymbol{k}}{1.732}$$

Force vector $\vec{F}$ = magnitude of force × unit vector along the direction OA

$$= 170.53 \times \frac{\boldsymbol{i} + \boldsymbol{j} + \boldsymbol{k}}{1.732} = 98.46\,(\boldsymbol{i} + \boldsymbol{j} + \boldsymbol{k})$$

Thus the components of force on a line that passes through the origin and point (1, 1, 1) are

$$F_x = 98.46 \text{ N}; \quad F_y = 98.46 \text{ N} \quad \text{and} \quad F_z = 98.96 \text{ N}$$

EXAMPLE 4.14

Forces 30 kN, 20 kN, 25 kN and 100 kN are respectively directed through the points whose coordinates are A(2, 1, 5), B(3, – 1, 4), C(– 3, – 2, 1) and D(4, 1, – 2). If these forces are concurrent at the origin O, make calculations for the resultant of the system.

Solution : Distance $OA = \sqrt{(2-0)^2+(1-0)^2+(5-0)^2} = 5.478$

$$OB = \sqrt{(3)^2+(-1)^2+(4)^2} = 5.1$$

$$OC = \sqrt{(-3)^2+(-2)^2+(1)^2} = 3.742$$

$$OD = \sqrt{(4)^2+(1)^2+(-2)^2} = 4.582$$

Unit vectors along these directions are :

$$n_1 = \frac{1}{5.478}(2\boldsymbol{i}+\boldsymbol{j}+5\boldsymbol{k}) \quad ; \quad n_2 = \frac{1}{5.1}(3\boldsymbol{i}-\boldsymbol{j}+4\boldsymbol{k})$$

$$n_3 = \frac{1}{3.742}(-3\boldsymbol{i}-2\boldsymbol{j}+\boldsymbol{k}) \quad ; \quad n_4 = \frac{1}{4.582}(4\boldsymbol{i}+\boldsymbol{j}-2\boldsymbol{k})$$

Now force vector $\vec{F} = |\vec{F}| \times \vec{n}$

= magnitude of force × unit vector in the of direction force

$$\vec{F_1} = 30 \times \frac{(2\boldsymbol{i}+\boldsymbol{j}+5\boldsymbol{k})}{5.478} = 10.95\boldsymbol{i} + 5.48\boldsymbol{j} + 27.38\boldsymbol{k}$$

$$\vec{F_2} = 20 \times \frac{(3\boldsymbol{i}-\boldsymbol{j}+4\boldsymbol{k})}{5.1} = 11.76\boldsymbol{i} - 3.92\boldsymbol{j} + 15.69\boldsymbol{k}$$

$$\vec{F_3} = 25 \times \frac{(-3\boldsymbol{i}-2\boldsymbol{j}+\boldsymbol{k})}{3.742} = -20.04\boldsymbol{i} - 13.36\boldsymbol{j} + 6.68\boldsymbol{k}$$

$$\vec{F_4} = 100 \times \frac{(4\boldsymbol{i}+\boldsymbol{j}-2\boldsymbol{k})}{4.582} = 87.30\boldsymbol{i} + 21.82\boldsymbol{j} - 43.65\boldsymbol{k}$$

$$\text{Resultant } \vec{R} = \vec{F_1} + \vec{F_2} + \vec{F_3} + \vec{F_4} = 89.97\boldsymbol{i} + 10.02\boldsymbol{j} + 6.1\boldsymbol{k}$$

Magnitude of resultant force

$$|\vec{R}| = R = \sqrt{(89.97)^2+(10.02)^2+(6.1)^2} = 90.73 \text{ N}$$

The scalar components of the resultant along the x, y and z axes are 89.97 N, 10.02 N and 6.1 N respectively.

Direction cosines of the resultant are,

$$l = \cos\alpha = \frac{89.97}{90.73} = 0.9916 \; ; \quad \alpha = 7.43^\circ$$

$$m = \cos\beta = \frac{10.02}{90.73} = 0.1104 \; ; \quad \beta = 83.66^\circ$$

$$n = \cos\gamma = \frac{6.1}{90.73} = 0.0672 \; ; \quad \gamma = 86.15^\circ$$

Check : $l^2 + m^2 + n^2 = (0.9916)^2 + (0.1104)^2 + (0.067)^2 = 1.0$

EXAMPLE 4.15

Express the resultant in vectorial form of the forces acting on a bolt as shown in Fig. 4.11.

Proceed to determine the resultant both in magnitude and direction.

Solution : Let the orientation of each force be measured from the x-axis in the anti-clockwise direction. Then the x and y components of each force work out as :

$$F_{x_1} = F_1 \cos 270° = 110 \cos 270° = 0$$
$$F_{x_2} = F_2 \cos 345° = 100 \cos 345° = 96.59 \text{ N}$$
$$F_{x_3} = F_3 \cos 30° = 150 \cos 30° = 129.9 \text{ N}$$
$$F_{x_4} = F_4 \cos 110° = 80 \cos 110° = -27.36 \text{ N}$$

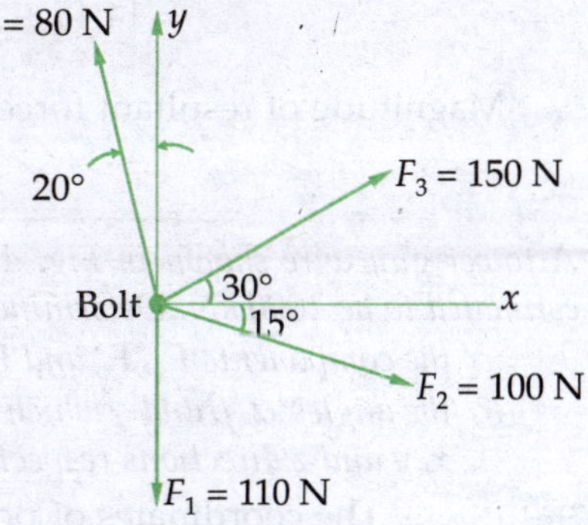

Fig. 4.11

and likewise

$$F_{y_1} = 110 \sin 270° = -110 \text{ N} \quad ; \quad F_{y_2} = 100 \sin 345° = -25.88 \text{ N}$$
$$F_{y_3} = 150 \sin 30° = 75 \text{ N} \quad ; \quad F_{y_4} = 80 \sin 110° = 75.17 \text{ N}$$

Then resultant of the given four forces in vectorial form is

$$\vec{R} = \left(F_{x_1} + F_{x_2} + F_{x_3} + F_{x_4}\right) i + \left(F_{y_1} + F_{y_2} + F_{y_3} + F_{y_4}\right) j$$
$$= (0 + 96.59 + 129.9 - 27.36)\, i + (-110 - 25.88 + 75 + 75.17)\, j$$
$$= 199.13\, i + 14.29\, j$$

Magnitude of resultant = $\sqrt{(199.13)^2 + (14.29)^2}$ = **199.64 N**

Direction of resultant with x-axis, $\alpha = \tan^{-1}\left(\frac{14.29}{199.13}\right)$ = **4.1°**

EXAMPLE 4.16.

Determine the resultant of the concurrent force system acting at point O as shown in the Fig. 4.12 given below:

Solution : Unit vector in the direction of OA

$$= \frac{0i + 3j + 2k}{\sqrt{0^2 + 3^2 + 2^2}} = 0i + 0.832j + 0.555k$$

Unit vector in the direction of OB

$$= \frac{4i + 0j + 3k}{\sqrt{4^2 + 0^2 + 3^2}} = 0.8i + 0j + 0.6k$$

Unit vector in the direction of OC

$$= \frac{4i + 3j + 0k}{\sqrt{4^2 + 3^2 + 0^2}} = 0.8i + 0.6j + 0k$$

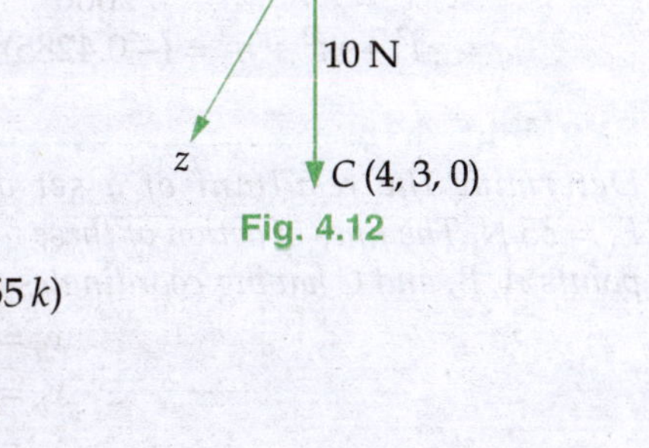

Fig. 4.12

Then:

Force F_1 in the direction of $OA = 6(0i + 0.832j + 0.555k)$
Force F_2 in the direction of $OB = 8(0.8i + 0j + 0.6k)$
Force F_3 in the direction of $OC = 10(0.8i + 0.6j + 0k)$

Resultant force $F = F_1 + F_2 + F_3$

$$= 6(0\,i + 0.832\,j + 0.555\,k) + 8(0.8\,i + 0\,j + 0.6\,k) + 10(0.8\,i + 0.6\,j + 0\,k)$$

$$= 14.4\,i + 10.99\,j + 8.13\,k$$

Magnitude of resultant force = $\sqrt{14.4^2 + 10.99^2 + 8.13^2}$ = **19.86 N**

EXAMPLE 4.17

A tower guy wire shown in Fig. 4.13 is anchored by means of a bolt at A and the tension in the wire is estimated to be 2000 kN. Determine

(a) the components F_x, F_y and F_z of the force acting on the bolt, and

(b) the angles α, β and γ which define the direction of force along the x, y and z directions respectively.

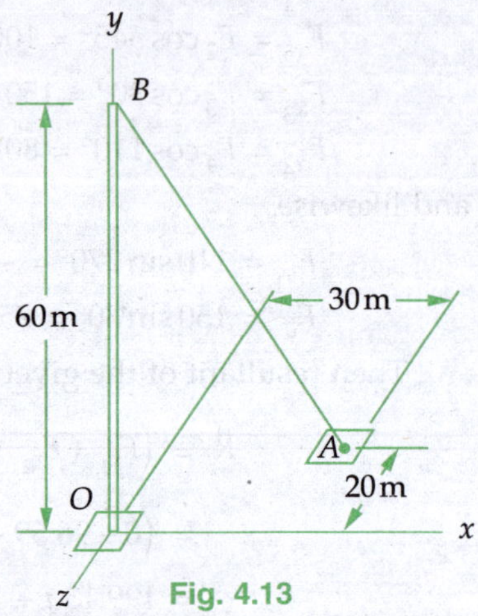

Fig. 4.13

Solution : The coordinates of points *A* and *B* are :

$$A(30, 0, -20) \text{ and } B(0, 60, 0)$$

$$\overrightarrow{AB} = \overrightarrow{OB} - \overrightarrow{OA}$$

$$= (0 - 30)\,\boldsymbol{i} + (60 - 0)\,\boldsymbol{j} + [0 - (-20)]\,\boldsymbol{k}$$

$$= -30\,\boldsymbol{i} + 60\,\boldsymbol{j} + 20\,\boldsymbol{k}$$

$$AB = |\overrightarrow{AB}| = \sqrt{(-30)^2 + (60)^2 + (20)^2} = 70 \text{ m}$$

$$\text{Unit vector along } AB = \frac{-30\,\boldsymbol{i} + 60\,\boldsymbol{j} + 20\,\boldsymbol{k}}{70}$$

$$\text{Force vector } \vec{F} = |\vec{F}| \times \vec{n}$$

= magnitude of force × unit vector in the direction of force

$$= 2000 \times \frac{-30i + 60j + 20k}{70} = -857.1\,\boldsymbol{i} + 1714.2\,\boldsymbol{j} + 571.4\,\boldsymbol{k}$$

Comparing it with the general expression $\vec{F} = F_x\boldsymbol{i} + F_y\boldsymbol{j} + F_z\boldsymbol{k}$, we have

$$F_x = -857.1 \text{ N}; \quad F_y = 1714.2 \text{ N} \quad \text{and} \quad F_z = 571.4 \text{ N}$$

These are called the scalar components of the force vector.

(*b*) The direction cosines of the force vector are :

$$l = \cos\alpha = \frac{-857.1}{2000} = -0.4285\,; \qquad \alpha = 115.37^\circ$$

$$m = \cos\beta = \frac{1714.2}{2000} = 0.8571\,; \qquad \beta = 31.0^\circ$$

$$n = \cos\gamma = \frac{571.4}{2000} = 0.2857\,; \qquad \gamma = 73.4^\circ$$

Check : $l^2 + m^2 + n^2 = (-0.4285)^2 + (0.8571)^2 + (0.2857)^2 \simeq 1$

EXAMPLE 4.18.

Determine the resultant of a set of three forces with magnitude : F_a = 50 N, F_b = 12 N and F_x = 35 N. The lines of action of these forces are concurrent at the origin O and respectively pass through the points A, B, and C having coordinates :

$$x_a = -1\,; \quad y_a = 2 \quad and \quad z_a = 4\ m$$

$$x_b = 3\,; \quad y_b = 0 \quad and \quad z_b = -3\ m$$

$$x_c = 2\,; \quad y_c = -2 \quad and \quad z_c = 4\ m$$

Solution : The coordinates of points A, B and C with O as origin are :

$$A(-1, 2, 4) ; \quad B(3, 0, -3) \quad \text{and} \quad C(2, -2, 4)$$

$$\overrightarrow{OA} = (-1-0)\, i + (2-0)\, j + (4-0)\, k = -1\, i + 2\, j - 4\, k$$

$$OA = |\overrightarrow{OA}| = \sqrt{(-1)^2 + 2^2 + 4^2} = 4.58$$

$$\text{Unit vector along } OA = \frac{-1i + 2j + 4k}{4.58} = -0.218\, i + 0.436\, j + 0.872\, k$$

$$\text{Force vector } \overrightarrow{F_a} = \text{magnitude of force } F_a \times \text{unit vector}$$

$$= 50(-0.218\, i + 0.436\, j + 0.872\, k)$$

$$= -10.9\, i + 21.8\, j + 43.6\, k$$

$$\text{Similarly } \overrightarrow{OB} = 3\, i - 3\, k$$

$$OB = |\overrightarrow{OB}| = \sqrt{3^2 + (-3)^2} = 4.24$$

$$\text{Unit vector along } OB = \frac{3i - 3k}{4.24} = 0.707\, i - 0.707\, k$$

$$\text{Force vector } \overrightarrow{F_b} = 12\, (0.707\, i - 0.707\, k) = 8.48\, i - 8.48\, k$$

$$\text{and } \overrightarrow{OC} = 2\, i - 2\, j + 4\, k$$

$$OC = |\overrightarrow{OC}| = \sqrt{2^2 + (-2)^2 + 4^2} = 4.9$$

$$\text{Unit vector along } OC = \frac{2i - 2j + 4k}{4.9} = 0.408\, i - 0.408\, j + 0.816\, k$$

$$\text{Force vector } \overrightarrow{F_c} = 35\, (0.408\, i - 0.408\, j + 0.816\, k)$$

$$= 14.28\, i - 14.28\, j + 28.56\, k$$

$$\text{Then resultant vector } \overrightarrow{R} = \overrightarrow{F_a} + \overrightarrow{F_b} + \overrightarrow{F_c}$$

$$= [-10.9\, i + 21.8\, j + 43.6\, k] + [8.48\, i - 8.48\, k] + [14.28\, i - 14.28\, j + 28.56\, k]$$

$$= (-10.9 + 8.48 + 14.28)\, i + (21.8 - 14.28)\, j + (43.6 - 8.48 + 28.56)\, k$$

$$= 11.86\, i + 7.52\, j + 63.68\, k$$

Magnitude of resultant,

$$R = |\overrightarrow{R}| = \sqrt{(11.86)^2 + (7.52)^2 + (63.68)^2}$$

$$= \sqrt{140.66 + 56.55 + 4055.14} = \mathbf{65.21\ N}$$

The scalar components of the resultant along the x, y and z axes are 11.86 N, 7.52 N and 63.68 N respectively.

Direction cosines of the resultant are :

$$l = \cos\alpha = \frac{11.86}{65.21} = 0.1818 \; ; \quad \alpha = 79.52°$$

$$m = \cos\beta = \frac{7.52}{65.21} = 0.1153 \; ; \quad \beta = 83.38°$$

$$n = \cos\gamma = \frac{63.68}{65.21} = 0.9765 \;;\quad \gamma = 12.44°$$

Check : $l^2 + m^2 + n^2 = (0.1818)^2 + (0.1153)^2 + (0.9765)^2$
$= 0.0330 + 0.0133 + 0.9535 \simeq 1.0$

4.5. MOMENT OF A FORCE

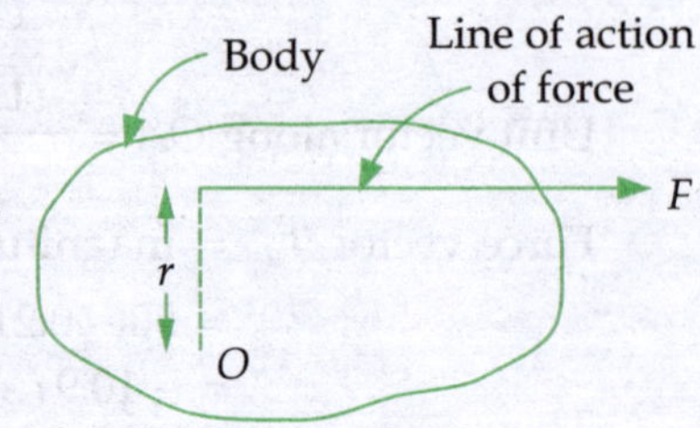

Fig. 4.14

Moment of force refers to the turning effect produced by the force $\vec{F}$ about a point. It is a vector quantity and it equals the product of force and the perpendicular distance of the line of action of force from the point.

Let $\vec{F}$ = force acting on the body

$\vec{r}$ = perpendicular distance from the point O to the line of action of force

Moment of force is then given by

$$\vec{M} = \vec{r} \times \vec{F}$$

In terms of the components of the force vector along the coordinate axes,

$$\vec{F} = F_x\boldsymbol{i} + F_y\boldsymbol{j} + F_z\boldsymbol{k}$$

and $\vec{r} = x\boldsymbol{i} + y\boldsymbol{j} + z\boldsymbol{k}$ is the position vector of any point on the line of action of forces with respect to O (the point about which turning takes place).

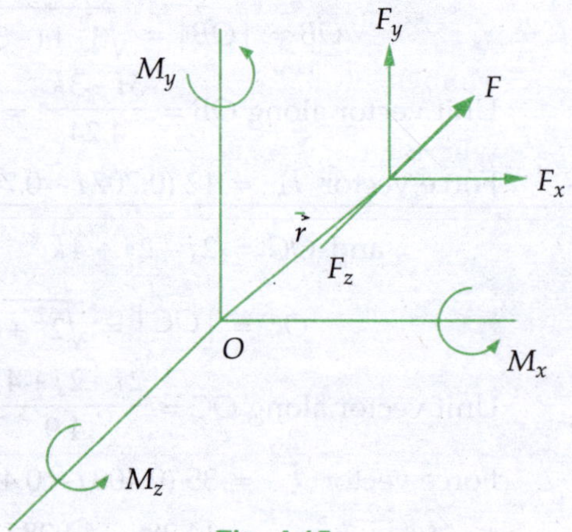

Fig. 4.15

Further $\vec{M} = \vec{r} \times \vec{F} = \begin{vmatrix} \boldsymbol{i} & \boldsymbol{j} & \boldsymbol{k} \\ x & y & z \\ F_x & F_y & F_z \end{vmatrix}$

$$= (yF_z - zF_y)\,\boldsymbol{i} - (xF_z - zF_x)\,\boldsymbol{j} + (xF_y - yF_x)\,\boldsymbol{k}$$

EXAMPLE 4.19

A force $\vec{F} = (5i - 2j - 3k)$ N acts at point P whose position vector form the origin of the coordinate axes is $(6i + 4j - 2k)$ mm. What would be the moment of the force about the origin?

Solution : Position vector $\overrightarrow{OP} = (6-0)\,i + (4-0)\,j + (-2-0)\,k$
$= 6i + 4j - 2k$

The moment of force is given by the cross product of the position vector and the force

$$\text{Moment } \vec{M} = \vec{r} \times \vec{F} = \begin{vmatrix} i & j & k \\ 6 & 4 & -2 \\ 5 & -2 & -3 \end{vmatrix}$$

$$= (-12 - 4)\,i - (-18 + 10)\,j + (-12 - 20)\,k$$
$$= (-16i + 8j - 32k) \text{ N mm}$$

Magnitude of moment of force

$$= \sqrt{(-16)^2 + 8^2 + (-32)^2} = 36.66 \text{ N mm}$$

EXAMPLE 4.20

Force of 100 N is directed along the line that is drawn from the point A(5, 2, – 1) m to the point B(3, – 4, 7)m. Make calaulations for the moment of this force about the z-axis.

Solution : Position vector $\overrightarrow{AB} = (3-5)i + (-4-2)j + (7+1)k$

$= (-2i - 6j + 8k)$ m

Unit vector along $\overrightarrow{AB} = \dfrac{-2i-6j+8k}{\sqrt{(-2)^2+(-6)^2+8^2}}$

$= \dfrac{-2i-6j+8k}{\sqrt{104}} = -0.196i - 0.294j + 0.784k$

Then: force vector along $\overrightarrow{AB} = F$ x unit vector along $\overrightarrow{AB}$

$= 100\times(-0.196i - 0.294j + 0.784k) = -19.6i - 29.4j + 78.4k$

Moment of the force about the origin O $= \overrightarrow{OA} \times \vec{F}$

where $\overrightarrow{OA} = (5-0)i + (2-0)j + (-1-0)k = 5i + 2j - k$

$$\therefore \quad M_o = \begin{vmatrix} i & j & k \\ 5 & 2 & -1 \\ -19.6 & -29.4 & 78.4 \end{vmatrix}$$

Moment about z-axis $= [5\times(-29.4) - (-19.6)\times 2]k$

$= (-147 + 39.2)k = -107.8$ k Nm

EXAMPLE 4.21.

A force vector of magnitude 500 N is directed along the line drawn from the point A(– 4, 2, 1) m to the point B (1, – 3, 6) m. Find the magnitude of the moment of this force about; x, y and z axes.

Solution : Position vector $\overrightarrow{AB}$

= position of point B (1, – 3, 6) with respect to point A (– 4, 2, 1)

$= [1-(-4)]i + (-3-2)j + (6-1)k$

$= 5i - 5j + 5k$

Then unit vector along $\overrightarrow{AB}$ (along the direction of force)

$= \dfrac{5i-5j+5k}{\sqrt{5^2+(-5)^2+5^2}} = 0.577i - 0.577j + 0.577k$

$\therefore$ Force vector $\vec{F}$ = magnitude of force × unit vector along the direction of force

$= 500\times(0.577i - 0.577j + 0.577k)$

$= 288.5i - 288.5j + 288.5k$

Position vector $\overrightarrow{OA} = (-4-0)i + (2-0)j + (1-0)k$

$= -4i + 2j + k$

Moment of the force vector about origin O,

$$\vec{M} = \begin{vmatrix} i & j & k \\ -4 & 2 & 1 \\ 288.5 & -288.5 & 288.5 \end{vmatrix}$$

$$= [(2 \times 288.5) - (-288.5 \times 1)]\, i - [(-4 \times 288.5) - (288.5 \times 1)]\, j + [-4 \times (-288.5) - 288.5 \times 2]\, k$$

$$= 865.5\, i + 1442.5\, j + 577\, j$$

Thus : Moment about x-axis = 865.5 Nm

Moment about y-axis = 1422.5 Nm

Moment about z-axis = 577 Nm

EXAMPLE 4.22

A force $\vec{F} = 3i - 4j + 12k$ acts at a point A whose coordinates are (1, – 2, 3) m. Compute:

(a) *moment of force about origin*

(b) *moment of force about the point B (2, 1, 2) m*

(c) *vector component of force F along line AB and the moment of this force about the origin.*

Solution : Position vector $\overrightarrow{OA}$ = position of point $A(1, -2, 3)$ with respect to origin $O(0, 0, 0)$.

$$= (1 - 0)\, \boldsymbol{i} + (-2 - 0)\, \boldsymbol{j} + (3 - 0)\, \boldsymbol{k} = \boldsymbol{i} - 2\boldsymbol{j} + 3\boldsymbol{k}$$

Moment of force about the origin

$$= \overrightarrow{OA} \times \vec{F} = (\boldsymbol{i} - 2\boldsymbol{j} + 3\boldsymbol{k}) \times (3\boldsymbol{i} - 4\boldsymbol{j} + 12\boldsymbol{k})$$

$$= \begin{vmatrix} \boldsymbol{i} & \boldsymbol{j} & \boldsymbol{k} \\ 1 & -2 & 3 \\ 3 & -4 & 12 \end{vmatrix}$$

$$= [(-2 \times 12) - (-4 \times 3)]\, \boldsymbol{i} - [(1 \times 12) - (3 \times 3)]\, \boldsymbol{j} + [(-4 \times 1) - (-2 \times 3)]\, \boldsymbol{k}$$

$$= -12\boldsymbol{i} - 3\boldsymbol{j} + 2\boldsymbol{k}$$

(b) Position vector $\overrightarrow{BA}$ = Position of A (1, – 2, 3) with recpect to point B (2, 1, 2)

$$= (1 - 2)\, \boldsymbol{i} + (-2 - 1)\, \boldsymbol{j} + (3 - 2)\, \boldsymbol{k} = -\boldsymbol{i} - 3\boldsymbol{j} + \boldsymbol{k}$$

Moment of force about point B

$$= \overrightarrow{BA} \times \vec{F} = (-\boldsymbol{i} - 3\boldsymbol{j} + \boldsymbol{k}) \times (3\boldsymbol{i} - 4\boldsymbol{j} + 12\boldsymbol{k})$$

$$= \begin{vmatrix} \boldsymbol{i} & \boldsymbol{j} & \boldsymbol{k} \\ 1 & -3 & 1 \\ 3 & -4 & 12 \end{vmatrix}$$

$$= [-3 \times 12 - (-4 \times 1)]\, \boldsymbol{i} - [(-1 \times 12) - (3 \times 1)]\, \boldsymbol{j} + [-1 \times (-4) - 3 \times (-3)]\, \boldsymbol{k}$$

$$= -32\boldsymbol{i} + 15\boldsymbol{j} + 13\boldsymbol{k}$$

(c) Position vector $\overrightarrow{AB} = (2-1)\,\boldsymbol{i} + [1-(-2)]\,\boldsymbol{j} + (2-3)\,\boldsymbol{k} = \boldsymbol{i} + 3\boldsymbol{j} - \boldsymbol{k}$

Unit vector along $\overrightarrow{AB}$ (along the direction of force)

$$= \frac{\boldsymbol{i}+3\boldsymbol{j}-\boldsymbol{k}}{\sqrt{1^2+3^2+(-1)^2}} = \frac{\boldsymbol{i}+3\boldsymbol{j}-\boldsymbol{k}}{\sqrt{11}} = 0.3\,\boldsymbol{i} + 0.9\,\boldsymbol{j} - 0.3\,\boldsymbol{k}$$

Projection of $\vec{F}$ on AB $= \vec{F}\cdot$ unit vector along AB

$$= (3\boldsymbol{i} - 4\boldsymbol{j} + 12\boldsymbol{k})\cdot(0.3\,\boldsymbol{i} + 0.9\,\boldsymbol{j} - 0.3\,\boldsymbol{k})$$

$$= 3\times 0.3 - 4\times 0.9 + 12\times(-0.3) = -6.3 \text{ m}$$

EXAMPLE 4.23

A force of 100 N is directed along the line drawn from point A(5, 2, 3) to the point B(2, – 5, 8). Determine the moment of this force about a point C(4, 3, 1). The distances are in metres.

Solution : Position vector $\overrightarrow{AB}$ = position of point B (2, – 5, – 8)with respect to point A (5, 2, 3)

$$= (2-5)\,\boldsymbol{i} + (-5-2)\,\boldsymbol{j} + (8-3)\,\boldsymbol{k}$$

$$= -3\boldsymbol{i} - 7\boldsymbol{j} + 5\boldsymbol{k}$$

Unit vector along AB (along the direction of force)

$$= \frac{-3i-7j+5k}{\sqrt{(-3)^2+(-7)^2+(5)^2}} = \frac{-3i-7j+5k}{9.11}$$

$\therefore$ Force vector $\vec{F}$ = magnitude of force × unit vector along the direction of force

$$= 100 \times \frac{-3i-7j+5k}{9.11}$$

$$= -32.93\,\boldsymbol{i} - 76.84\,\boldsymbol{j} + 54.88\,\boldsymbol{k}$$

Position vector $\overrightarrow{CA}$ = position of point A with respect to C

$$= (4-5)\,\boldsymbol{i} + (3-2)\,\boldsymbol{j} + (1-3)\,\boldsymbol{k} = -\boldsymbol{i} + \boldsymbol{j} - 2\boldsymbol{k}$$

$\therefore$ Moment of the force vector about point C

$$= \begin{vmatrix} i & j & k \\ -1 & 1 & -2 \\ -32.93 & -76.84 & 54.88 \end{vmatrix}$$

$$= [1\times 54.88 - (-76.84)(-2)]\,\boldsymbol{i} - [(-1)\times 54.88 - (-32.93)(-2)]\,\boldsymbol{j} + [(-1)\times(-76.84) - (-32.93)\times 1]\,\boldsymbol{k}$$

$$= [-98.80\,\boldsymbol{i} + 120.74\,\boldsymbol{j} + 109.77\,\boldsymbol{k}] \text{ Nm}$$

EXAMPLE 4.24.

A force of 600 N acts on a bracket as shown. Determine the moment of this force about point B. Use vector approach.

Solution : The amount M_B of force F is given by the vector product

$$M_B = \vec{r} \times \vec{F}$$

where $\vec{r}$ is the vector drawn from A to B.

Resolving the vector $\vec{r}$ and force $\vec{F}$ into horizontal (x-direction) and vertical (y-direction) components, we get

$$\vec{r} = \overrightarrow{AB} = 0.15\,i - 0.12\,j$$

and
$$\vec{F} = (600 \cos 60°)\,i + (600 \sin 60°)\,j$$
$$= 300\,i + 519.6\,j$$

Then the moment is determined as

$$M_B = (0.15\,i - 0.12\,j) \times (300\,i - 519.6\,j)$$
$$= (0.15) \times 300 \times (i \times i) + (0.15) \times (519.6) \times (i \times j)$$
$$- 0.12 \times 300 \times (j \times i) - 0.12 \times 519.6 \times (j \times j)$$

Recalling that : $(i \times i) = (j \times j) = 0$

$(i \times j) = k$ and $(j \times i) = -k$

We get
$$M_B = 0.15 \times 519.6\,k - 0.12 \times 300 \times (-k)$$
$$= 77.94\,k + 36\,k = 113.94\,k$$

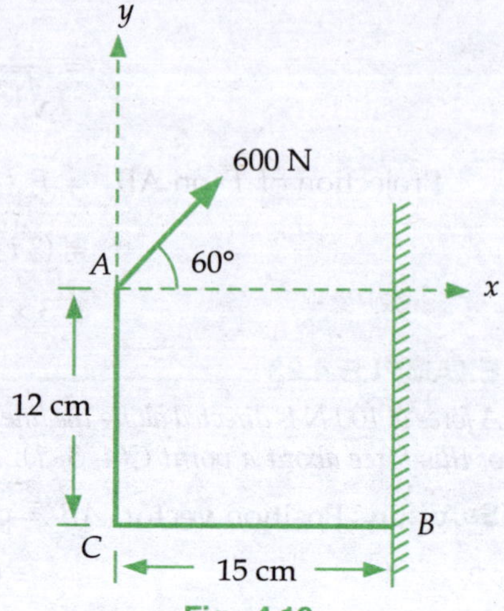

Fig. 4.16

Thus the moment has a magnitude of 113.94 Nm. The moment is a vector perpendicular to the plane of the figure and pointing into the paper.

EXAMPLE 4.25.

Determine the magnitude, direction and resultant of the system of forces shown in the Fig. 4.17. Use the vector approach.

Solution : Resolving the given forces into horizontal (x-direction) and vertical (y-direction) components, we get

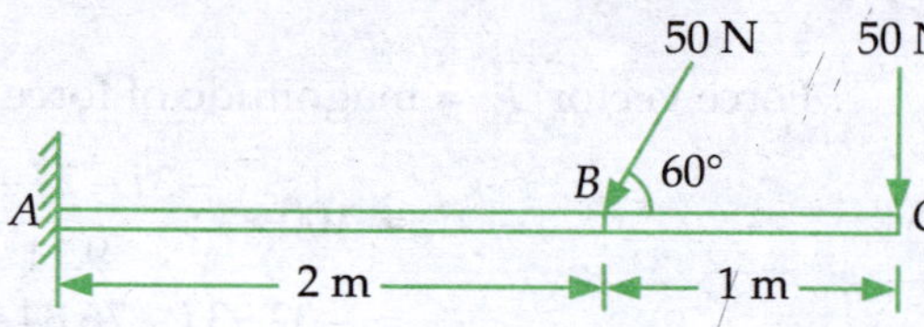

Fig. 4.17

$$\vec{R} = -50\,j + (-50 \cos 60°\,i - 50 \sin 60°\,j)$$
$$= -50\,j + (-25\,i - 43.4\,j)$$
$$= -25\,i - 93.3\,j$$

Magnitude of resultant,

$$|R| = \sqrt{(-25)^2 + (-93.3)^2} = 96.59$$

Since both the x and y-components of the resultant are negative, the resultant lies in the 3rd quadrant and its inclination α with horizontal or x-axis is

$$\alpha = \tan^{-1}\left(\frac{93.3}{25}\right) = 75°$$

Let the resultant R be acting at distance x from the fixed end. Then under equilibrium conditions,

moment of resultant about A = sum of the moments of given forces about A

$$x\,i \times (-25\,i - 93.3\,j) = 2\,i \times (-25\,i - 43.4\,j) + 3\,i\,(-50\,j)$$

Recollecting that : $i \times i = j \times j = 0$ and $(i \times j) = k$ and $(j \times i) = -k$

we get
$$-93.3\,(k)\,x = -86.6\,(k) + (-150)\,(k)$$

That gives :
$$x = \frac{86.6 + 150}{93.3} = 2.535 \text{ m}$$

EXAMPLE 4.26

A cable supporting a 20 m high vertical post is anchored to the ground as shown in Fig. 4.18. If the tensile force in the cable is 10 kN, find its moment about the z-axis passing through base of the post.

Solution : With point O as origin, the coordinates of points A and B are :

$$A\,(0, 5, 0) \quad \text{and} \quad B\,(4, 0, 3)$$

Position vector $\overrightarrow{AB}$ = position of B with respect to A

$$= (4-0)\,\boldsymbol{i} + (0-5)\,\boldsymbol{j} + (3-0)\,\boldsymbol{k}$$
$$= 4\boldsymbol{i} - 5\boldsymbol{j} + 3\boldsymbol{k}$$

Unit vector along $AB = \dfrac{4\boldsymbol{i} - 5\boldsymbol{j} + 3\boldsymbol{k}}{\sqrt{4^2 + (-5)^2 + 3^2}} = \dfrac{4\boldsymbol{i} - 5\boldsymbol{j} + 3\boldsymbol{k}}{7.07}$

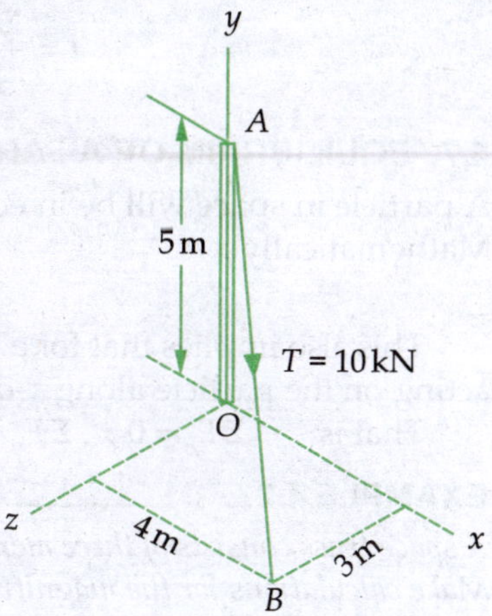

Fig. 4.18

Force vector $\vec{F}$ = magnitude of force × unit vector along the direction AB

$$= 10 \times \frac{4\boldsymbol{i} - 5\boldsymbol{j} + 3\boldsymbol{k}}{7.07} = 5.66\,\boldsymbol{i} - 7.07\,\boldsymbol{j} + 4.24\,\boldsymbol{k}$$

Moment of force $\vec{F}$ about origin $O = \vec{r} \times \vec{F}$

$$\vec{r} = \text{position vector } \overrightarrow{OA} = (0-0)\,\boldsymbol{i} + (5-0)\,\boldsymbol{j} + (0-0)\,\boldsymbol{k}$$
$$= 5\boldsymbol{j}$$

$\therefore$ Moment of force $\vec{M}_o = 5\boldsymbol{j} \times (5.66\,\boldsymbol{i} - 7.07\,\boldsymbol{j} + 4.24\,\boldsymbol{k})$

$$= -28.3\,\boldsymbol{k} + 0 + 21.2\,\boldsymbol{i} = 21.2\,\boldsymbol{i} - 28.3\,\boldsymbol{k}$$

The desired moment of $\vec{F}$ about z axis

$$\vec{M}_z = -28.3 \text{ kNm (clockwise)}$$

The minus sign is indicative of the fact that the moment is clockwise when viewed from positive z-direction.

4.6. RESULTANT OF SPACE CONCURRENT FORCES

The resultant of any number of space concurrent forces can be determined by writing the vector expression for each force and adding the vector expressions of all forces.

Consider the forces identified as

$$\vec{F_1} = F_{1x}\,\boldsymbol{i} + F_{1y}\,\boldsymbol{j} + F_{1z}\,\boldsymbol{k}$$
$$\vec{F_2} = F_{2x}\,\boldsymbol{i} + F_{2y}\,\boldsymbol{j} + F_{2z}\,\boldsymbol{k}$$
$$\vec{F_3} = F_{3x}\,\boldsymbol{i} + F_{3y}\,\boldsymbol{j} + F_{3z}\,\boldsymbol{k}$$

....

....

$$\vec{F_n} = F_{nx}\,\boldsymbol{i} + F_{ny}\,\boldsymbol{j} + F_{nz}\,\boldsymbol{k}$$

The resultant of these forces is

$$\vec{R} = \vec{F_1} + \vec{F_2} + \vec{F_3} + \ldots \vec{F_n}\,\boldsymbol{k}$$
$$= (F_{1x} + F_{2x} + F_{3x} + \ldots F_{nx})\,\boldsymbol{i} + (F_{1y} + F_{2y} + F_{3y} + \ldots F_{ny})\,\boldsymbol{j} + (F_{1z} + F_{2z} + F_{3z} + \ldots F_{nz})\,\boldsymbol{k}$$
$$= \Sigma F_x\,\boldsymbol{i} + \Sigma F_y\,\boldsymbol{j} + \Sigma F_z\,\boldsymbol{k}$$

The magnitude of the resultant force is

$$R = |\vec{R}| = \sqrt{(\Sigma F_x)^2 + (\Sigma F_y)^2 + (\Sigma F_z)^2}$$

and the direction cosines are

$$l = \frac{\Sigma F_x}{R}; \quad m = \frac{\Sigma F_y}{R}; \quad n = \frac{\Sigma F_z}{R}$$

4.7. EQUILIBRIUM OF A PARTICLE IN SPACE

A particle in space will be in equilibrium when the resultant of all the forces acting on it is zero. Mathematically :

$$\vec{R} = \Sigma F_x \boldsymbol{i} + \Sigma F_y \boldsymbol{j} + \Sigma F_z \boldsymbol{k} = 0$$

This also implies that for a particle in space to be in equilibrium, the algebraic sum of the forces acting on the particle along x-direction, y-direction and z-direction are individually equal to zero.

That is $\quad \Sigma F_x = 0 ; \quad \Sigma F_y = 0; \quad \Sigma F_z = 0$

EXAMPLE 4.27.

A space truss consists of three members AB, AC and AD which are oriented as shown in the adjoining figure. Make calculations for the magnitude and nature of forces induced in each member of the truss when a 10 kN weight is suspended at the joint A.

Solution : With O as the origin, the co-ordinates of points A, B, C and D are :

$$A(3, 0, 0) ; \quad B(0, 0, -4) ; \quad C(0, 2, 0) \quad \text{and} \quad D(0, 0, 4)$$

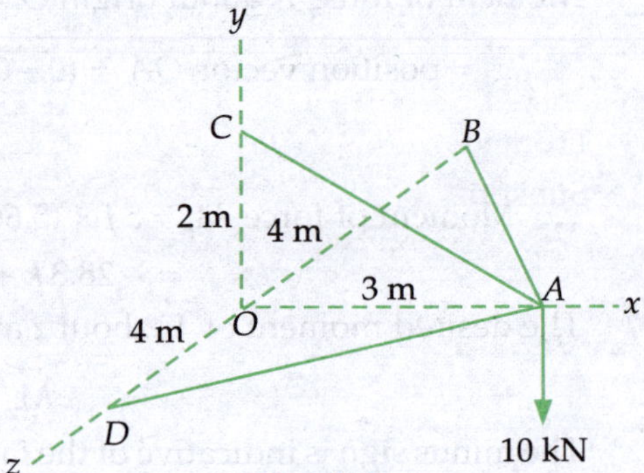

Fig. 4.19

For force in member AB :

$$\overrightarrow{AB} = (0-3)\, i + (0-0)\, j + (-4-0)\, k = -3\, i - 4\, k$$

$$AB = |\overrightarrow{AB}| = \sqrt{(-3)^2 + (-4)^2} = 5$$

$$\text{Unit vector along } AB = \frac{-3i - 4k}{5} = -0.6\, i - 0.8\, j$$

$$\text{Force vector } \overrightarrow{F_{ab}} = \text{magnitude of force } F_{ab} \times \text{unit vector}$$

$$= F_{ab} \times (-0.6\, i - 0.8\, k)$$

For force in member AC :

$$\overrightarrow{AC} = (0-3)\, i + (2-0)\, j + (0-0)\, k = -3\, i - 2\, j$$

$$AC = |\overrightarrow{AC}| = \sqrt{(-3)^2 + (-4)^2} = \sqrt{13}$$

$$\text{Unit vector along } AC = \frac{-3i + 2j}{\sqrt{13}} = -0.832\, i + 0.55\, j$$

$$\text{Force vector } \overrightarrow{F_{ac}} = F_{ac} \times (-0.832\, i + 0.555\, j)$$

For force in member AD :

$$\overrightarrow{AD} = (0-3)\, i + (0-0)\, j + (4-0)\, k = -3\, i - 4\, k$$

$$AD = |\overrightarrow{AD}| = \sqrt{(-3)^2 + (4)^2} = 5$$

Unit vector along $AD = \dfrac{-3i+4k}{5} = -0.6\,i + 0.8\,k$

Force vector $\overrightarrow{F_{ad}} = F_{ad} \times (-0.6\,i + 0.8\,j)$

Force due to 10 kN weight suspended from point A

$= -10\,j$

Resultant of all forces meeting at point A

$$\begin{aligned}\vec{R} &= F_{ab} \times (-0.6\,i - 0.8\,k) + F_{ac} \times (-0.832\,i + 0.555\,j) \\ &\qquad + F_{ad}(-0.6\,i + 0.8\,k) - 10\,j \\ &= (-0.6\,F_{ab} - 0.832\,F_{ac} - 0.6\,F_{ad})\,i \\ &\qquad + (0.555\,F_{ac} - 10)j + (-0.8\,F_{ab} + 0.8\,F_{ad})\,k\end{aligned}$$

Applying the equilibrium conditions to A : $\vec{R} = 0$

Then equating the coefficients of i, j and k to zero, we get

$$-0.6\,F_{ab} - 0.832\,F_{ac} - 0.6\,F_{ad} = 0 \quad \text{...(i)}$$
$$-0.555\,F_{ac} - 10 = 0 \quad \text{...(ii)}$$
$$-0.8\,F_{ab} + 0.8\,F_{ad} = 0 \quad \text{...(iii)}$$

From identity (*ii*) : $F_{ac} = \dfrac{10}{0.555} = 18.02$ N

From identity (*iii*) : $F_{ab} = F_{ad}$

Substituting these values in identity (*i*), we get

$$-0.6\,F_{ab} - (0.832 \times 18.02) - 0.6\,F_{ab} = 0$$

$$\therefore \quad F_{ab}\,(= F_{ad}) = -\frac{0.832 \times 18.02}{1.2} = 12.49 \text{ N}$$

EXAMPLE 4.28

A vertical pole is guyed by three cables PA, PB and PC tied at a common point P, 10 m above the ground. The base points of the cable are

A (– 3, 0, – 4), B (1, – 1, 5) and C (5, 0, – 1)

If the tensile forces in the cables are adjusted to be 15, 18 and 20 kN, make calculations for the resultant force on the pole at P.

Solution : Refer Fig. 4.20 for the arrangement of the three cables and the pole.

The pole P is 10 m above the ground at O, and as such the forces in the cables would be directed along PA, PB and PC.

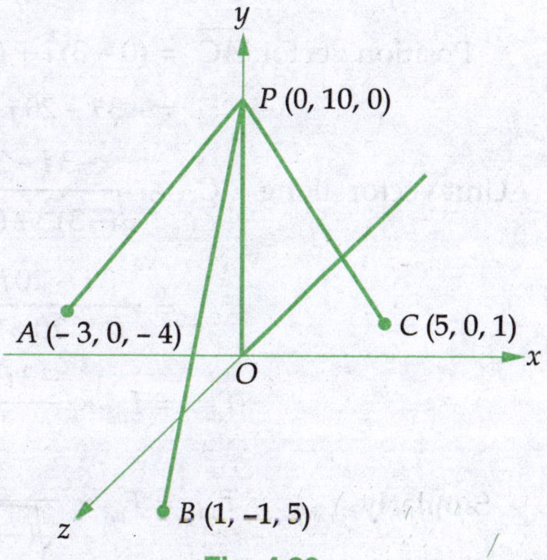

Fig. 4.20

$$\text{Distance } PA = \sqrt{(x_a - x_p)^2 + (y_a - y_b)^2 + (z_a - z_p)^2}$$
$$= \sqrt{(-3-0)^2 + (0-10)^2 + (-4-0)^2}$$
$$= \sqrt{125} = 11.18$$
$$PB - \sqrt{(1-0)^2 + (-1-10)^2 + (5-0)^2}$$
$$= \sqrt{147} = 12.12$$
$$PC = \sqrt{(5-0)^2 + (0-10)^2 + (-1-0)^2} = \sqrt{126} = 11.22$$

Unit vectors along these directions are :

$$n_1 = \frac{1}{11.18}(-3i - 10j - 4k) \qquad n_2 = \frac{1}{12.12}(i - 11j + 5k) \qquad n_3 = \frac{1}{11.22}(5i - 10j - 1k)$$

Force vector $\vec{F} = |\vec{F}| \times \vec{n}$

= magnitude of force × unit vector in the direction of force

$$\vec{F_1} = 15 \times \frac{(-3i - 10j - 4k)}{11.18} = -4.02\,i - 13.4\,j - 5.36\,k$$

$$\vec{F_2} = 18 \times \frac{(i - 11j + 5k)}{12.12} = 1.48\,i - 16.34\,j + 7.42\,k$$

$$\vec{F_3} = 20 \times \frac{(5i - 10j + 1k)}{11.22} = 8.91\,i - 17.8\,j - 1.78\,k$$

$$\text{Resultant } \vec{R} = \vec{F_1} + \vec{F_2} + \vec{F_3} = 6.37\,i - 47.54\,j + 0.28\,k$$

$$\text{Magnitude of resultant force} = \sqrt{(6.37)^2 + (-47.54)^2 + (0.28)^2} = 47.96 \text{ kN}$$

The resultant force on the pole is 47.96 kN in magnitude and it acts predominantly downwards to hold the pole in position.

EXAMPLE 4.29

A vertical tower AB shown in Fig. 4.21 is subjected to a horizontal force F = 50 kN at its top and it is anchored by two guy wires AC and AD. Compute the tensions in the guy wires and thrust in the pole.

Assume following coordinates for the points :

O (0, 0, 0) ; A (3, 20, 0) ;

B (3, 0, 0) ; C (0, 0, 4) and D (0, 0, – 4).

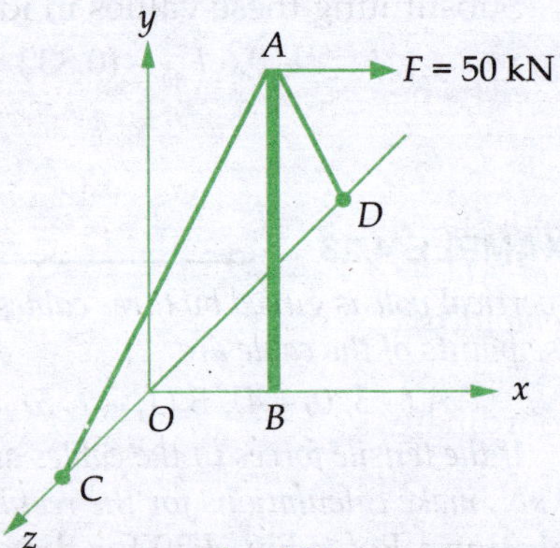

Fig. 4.21

Solution : The vector expression for tension in a guy wire is

= magnitude of tension

× unit vector along the direction of that wire

$$\text{Position vector } \overrightarrow{AC} = (0-3)\,i + (0-20)\,j + (4-0)\,k$$

$$= -3\,i - 20\,j + 4\,k$$

$$\text{Unit vector along } \overrightarrow{AC} = \frac{-3\,i - 20\,j + 4\,k}{\sqrt{(-3)^2 + (-20)^2 + 4^2}}$$

$$= \frac{-3i - 20j + 4k}{20.61}$$

$$\therefore \qquad \vec{T}_{ac} = T_{ac} \times \frac{-3i - 20j + 4k}{20.61} = T_{ac}\,[-0.146\,i - 0.97\,j + 0.194\,k]$$

$$\text{Similarly} \qquad \vec{T}_{ad} = T_{ad} \times \frac{-3i - 20j - 4k}{\sqrt{(-3)^2 + (-20)^2 + (-4)^2}} = T_{ad}\,[-0.146\,i - 0.97\,j - 0.194\,k]$$

$$\vec{T}_{ab} = T_{ab} \times \frac{0i - 20j - 0k}{\sqrt{0^2 + (-20)^2 + (0)^2}} = T_{ab}\,(-j)$$

The force $F = 50$ kN can be expressed as a force vector

$$\vec{F} = 50\,i$$

Resultant of all forces meeting at point A

$$= \vec{T}_{ac} + \vec{T}_{ad} + \vec{T}_{ab} + \vec{F}$$

$$= (-0.146\,T_{ac} - 0.146\,T_{ad} + 50)\,\boldsymbol{i} + (-0.97\,T_{ac} - 0.97\,T_{ad} - T_{ab})\,\boldsymbol{j} + (0.194\,T_{ac} - 0.194\,T_{ad})\,\boldsymbol{k}$$

Applying the equilibrium conditions to point A : $\vec{R} = 0$

Then equating the coefficients of $\boldsymbol{i}$, $\boldsymbol{j}$ and $\boldsymbol{k}$ to zero, we get

$$-0.146\,T_{ac} - 0.146\,T_{ad} + 50 = 0 \quad \text{...}(i)$$

$$-0.97\,T_{ac} - 0.97\,T_{ad} - T_{ab} = 0 \quad \text{...}(ii)$$

$$0.194\,T_{ac} - 0.194\,T_{ad} = 0\,;\; T_{ac} = T_{ad} \quad \text{...}(iii)$$

From identities (*i*) and (*iii*) we obtain

$$-0.146\,T_{ac} - 0.146\,T_{ac} + 50 = 0$$

$$\therefore \quad T_{ac} = \frac{50}{0.146 + 0.146} = 171.23 \text{ N}$$

$$T_{ad} = T_{ac} = 171.23 \text{ N}$$

Substituting these values of T_{ad} and T_{ac} in identify (*ii*), we get

$$-0.97 \times 171.23 - 0.97 \times 171.23 - T_{ab} = 0$$

$$\therefore \quad T_{ab} = -2 \times 0.97 \times 171.23 = -332.18 \text{ N}$$

The negative value for T_{ad} indicates that the tower would experience thrust rather than tension.

EXAMPLE 4.30

In the system shown in Fig. 4.22, a 5 m long beam is held in vertical position AO by three guy wires AB, AC and AD. If a tension equivalent to 600 N is induced in AD and the resultant force at A is to be vertical, compute the magnitude of forces in AC and AB.

Solution : Taking y and z axis along A and B respectively, the coordinates of various points with respect to O as origin are :

$A(0, 5, 0)$; $B(0, 0, -3)$

$C(-1.5, 0, 1)$ and $D(2, 0, 1.5)$

The vector expression for tension induced in guy wire is

= magnitude of tension × unit vector along the direction of that wire

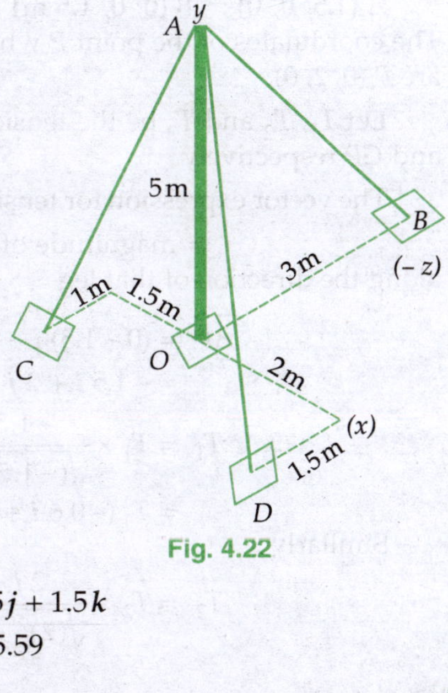

Fig. 4.22

$$\text{Position vector } \overrightarrow{AD} = (2-0)\,\boldsymbol{i} + (0-5)\,\boldsymbol{j} + (1.5-0)\,\boldsymbol{k}$$

$$= 2\boldsymbol{i} - 5\boldsymbol{j} + 1.5\boldsymbol{k}$$

$$\text{Unit vector along } \overrightarrow{AD} = \frac{2\boldsymbol{i} - 5\boldsymbol{j} + 1.5\boldsymbol{k}}{\sqrt{1^2 + (-5)^2 + 1.5^2}}$$

$$= \frac{2\boldsymbol{i} - 5\boldsymbol{j} + 1.5\boldsymbol{k}}{5.59}$$

$$\therefore \quad \vec{T}_{ad} = T_{ad} \times \frac{2\boldsymbol{i} - 5\boldsymbol{j} + 1.5\boldsymbol{k}}{5.59} = 600 \times \frac{2\boldsymbol{i} - 5\boldsymbol{j} + 1.5\boldsymbol{k}}{5.59}$$

$$= 214.67\,\boldsymbol{i} - 536.67\,\boldsymbol{j} + 161\,\boldsymbol{k}$$

Similarly : $\overrightarrow{T_{ab}} = T_{ab} \times \frac{(-5j-3k)}{\sqrt{(-5)^2+(-3)^2}} = T_{ab} \times \frac{(-5j-3k)}{5.83}$

$= -T_{ab}(0.857\,j + 0.514\,k)$

$\overrightarrow{T_{ac}} = T_{ac} \times \frac{-1.5i-5j+k}{\sqrt{(-1.5)^2+(-5)^2+1^2}} = T_{ac} \times \frac{(-1.5i-5j+k)}{5.31}$

$= -T_{ac}(0.282\,i + 0.942\,j - 0.188\,k)$

Resultant of all forces meeting at point A

$= \overrightarrow{T_{ad}} + \overrightarrow{T_{ab}} + \overrightarrow{T_{ac}}$

$= (214.67 - 0.282\,T_{ac})\,i - (536.67 + 0.857\,T_{ab} + 0.942\,T_{ac})\,j + (161 - 0.514\,T_{ab} + 0.188\,T_{ac})\,k$

As resultant force at A is to be vertical

$$\Sigma F_x = 0 \text{ and } \Sigma F_z = 0$$

That gives : $214.67 - 0.282\,T_{ac} = 0$

$\therefore$ Tension in cable AC : $T_{ac} = \frac{214.67}{0.282} = $ **761.24 N**

Also $161 - 0.514\,T_{ab} + 0.188\,T_{ac} = 0$

$\therefore$ Tension in cable AB : $T_{ab} = \frac{161 + 0.188 \times 761.24}{0.514} = $ **591.66 N**

EXAMPLE 4.31

A tripod supports a load of 2.5 kN at point P as shown in Fig. 4.23. The end points A, B and C of the three legs lie in the x-z plane. Make calculations for the force developed in each leg.

Solution : With point O as origin, the coordinates of the end points are :

A (1.5, 0, 0) ; B (0, 0, 1.5 m) and C (– 1.25, 0, – 1).

The coordinates of the point P which supports the load are P (0, 2, 0).

Let T_1, T_2 and T_3 be the tension in the legs AP, BP and CP respectively.

The vector expression for tension induced in a leg is = magnitude of tension × unit vector along the direction of that leg

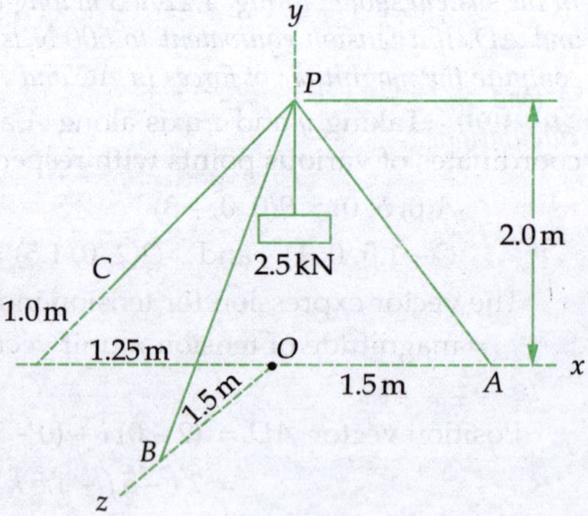

Fig. 4.23

$\overrightarrow{AP} = (0-1.5)\,i + (2-0)\,j + (0-0)\,k$

$= -1.5\,i + 2\,j$

$\therefore \vec{T_1} = T_1 \times \frac{-1.5i+2j}{\sqrt{(-1.5)^2+(2)^2}}$

$= T_1(-0.6\,i + 0.8\,j)$

Similarly :

$\vec{T_2} = T_2 \times \frac{2j-1.5k}{\sqrt{(2)^2+(-1.5)^2}} = T_2(0.8\,j - 0.6\,k)$

$$\vec{T_3} = T_3 \times \frac{1.25\boldsymbol{i} + 2\boldsymbol{j} - \boldsymbol{k}}{\sqrt{(1.25)^2 + 2^2 + 1^2}} = T_3(0.49\,\boldsymbol{i} + 0.78\,\boldsymbol{j} + 0.39\,\boldsymbol{k})$$

Also : $\vec{W} = -2.5\,\boldsymbol{j}$

Resultant of all the forces meeting at point P

$$= \vec{T_1} + \vec{T_2} + \vec{T_3} + \vec{W}$$

$$= (-0.6T_1 + 0.49T_3)\,\boldsymbol{i} + (0.8T_1 + 0.8T_2 + 0.78T_3 - 2.5)\,\boldsymbol{j} + (-0.6T_2 + 0.39T_3)\,\boldsymbol{k}$$

Applying the equilibrium conditions to point P

$$\vec{R} = \Sigma\vec{F} = 0$$

Then equating the coefficients of $\boldsymbol{i}, \boldsymbol{j}$ and $\boldsymbol{k}$ to zero, we get

$$-0.6T_1 + 0.49T_3 = 0 \quad \text{...}(i)$$

$$0.8T_1 + 0.8T_2 + 0.78T_3 - 2.5 = 0 \quad \text{...}(ii)$$

$$-0.6T_2 + 0.39T_3 = 0 \quad \text{...}(iii)$$

From identities (*i*) and (*iii*) :

$$T_1 = \frac{0.49}{0.6}T_3 = 0.816T_3 \quad \text{and} \quad T_2 = \frac{0.39}{0.6}T_3 = 0.65\,T_3$$

Substituting these values for T_1 and T_2 in identity (*ii*), we get

$$0.8 \times 0.816\,T_3 + 0.8 \times 0.65\,T_3 + 0.78T_3 = 2.5$$

or $$(0.653 + 0.52 + 0.78)\,T_3 = 2.5$$

$\therefore$ Force developed in leg CP : $T_3 = \frac{2.5}{1.953} = \mathbf{1.28\ kN}$

Force developed in leg BP : $T_2 = 0.65 \times 1.28 = \mathbf{0.832\ kN}$

Force developed in leg AP : $T_1 = 0.816 \times 1.28 = \mathbf{1.04\ kN}$

EXAMPLE 4.32

A load of 1500 N is supported at point A by three cables AB, AC and AD as shown in Fig. 4.24. Make calculations for the tensions induced in each cable.

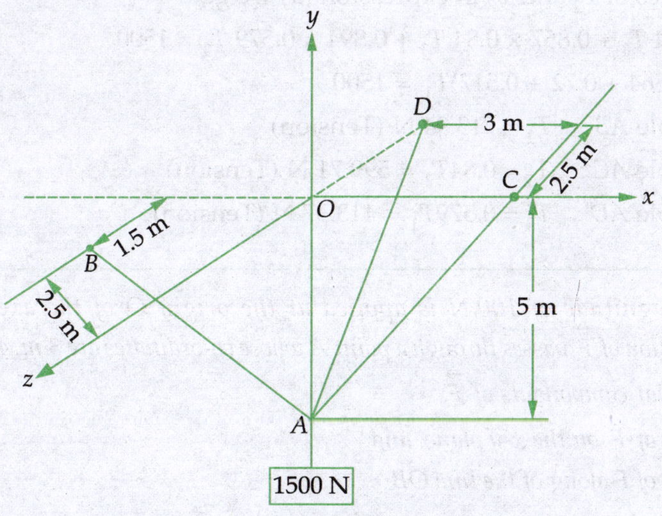

Fig. 4.24

Solution : Let T_1, T_2 and T_3 be the tensions induced in the cable *AB*, *AC* and *AD* respectively.

The co-ordinates of various points with *O* as origin are

$A\,(0, -5, 0)\,;\quad B\,(-2.5, 0, 1.5);\quad C\,(3, 0, 0)\quad \text{and}\quad D(0, 0, -2.5)$

The vector expression for tension induced in a cable is = magnitude of tension × unit vector along the direction of that cable

$$\overrightarrow{AB} = (-2.5-0)\,\boldsymbol{i} + [0-(-5)]\,\boldsymbol{j} + (1.5-0)\,\boldsymbol{k}$$

$$= -2.5\,\boldsymbol{i} + 5\,\boldsymbol{j} + 1.5\,\boldsymbol{k}$$

$$\therefore \quad \vec{T}_1 = T_1 \times \frac{-2.5\boldsymbol{i} + 5\boldsymbol{j} + 1.5\boldsymbol{k}}{\sqrt{(-2.5)^2 + (5)^2 + (1.5)^2}}$$

$$= (-0.432\,\boldsymbol{i} + 0.864\,\boldsymbol{j} + 0.259\,\boldsymbol{k})\,T_1$$

Similarly :

$$\vec{T}_2 = T_2 \times \frac{3\boldsymbol{i} + 5\boldsymbol{j}}{\sqrt{3^2 + 5^2}} = (0.514\,\boldsymbol{i} + 0.857\,\boldsymbol{j})\,T_2$$

$$\vec{T}_3 = T_3 \times \frac{5\boldsymbol{j} - 2.5\boldsymbol{k}}{\sqrt{5^2 + (-2.5)^2}} = (0.894\,\boldsymbol{j} - 0.447\,\boldsymbol{k})\,T_3$$

Also : $\quad \vec{W} = -1500\,\boldsymbol{j}$

Resultant of all the forces meeting at point *A*

$$= \vec{T}_1 + \vec{T}_2 + \vec{T}_3 + \vec{W}$$

$$= (-0.432\,T_1 + 0.514\,T_2)\,\boldsymbol{i} + (0.864\,T_1 + 0.857\,T_2 + 0.894\,T_3 - 1500)\,\boldsymbol{j} + (0.259\,T_1 + 0.447\,T_3)\,\boldsymbol{k}$$

Since the state of equilibrium exists, $\vec{R} = 0$. That is

$$-0.432\,T_1 + 0.514\,T_2 = 0\,;\quad T_2 = 0.84T_1 \qquad \text{...(i)}$$

$$-0.864\,T_1 + 0.857\,T_2 + 0.894\,T_3 - 1500 = 0 \qquad \text{...(ii)}$$

$$-0.259\,T_1 - 0.447\,T_3 = 0\,;\quad T_3 = 0.579T_1 \qquad \text{...(iii)}$$

Substituting the values of T_2 and T_3 in expression (*ii*) we get

$$0.864\,T_1 + 0.857 \times 0.84\,T_1 + 0.894 \times 0.579\,T_1 = 1500$$

or $\quad (0.864 + 0.72 + 0.517)T_1 = 1500$

$\therefore$ Tension in cable *AB* : $\quad T_1 = 713.94$ N (Tension)

Tension in cable *AC* : $\quad T_2 = 0.84T_1 = 599.71$ N (Tension)

Tension in cable *AD* : $\quad T_3 = 0.579T_1 = 413.37$ N (Tension)

EXAMPLE 4.33

A force $\vec{F}$ with a magnitude of 100 N is applied at the origin O of the axes x, y, z as shown in Fig. 4.25. The line of action of $\vec{F}$ passes through a point A whose co-ordinates are 3 m, 4 m and 5 m. Determine :

(a) *the x, y, z scalar components of $\vec{F}$*

(b) *the projection of $\vec{F}$ on the x-y plane, and*

(c) *the projection of F along of the line OB*

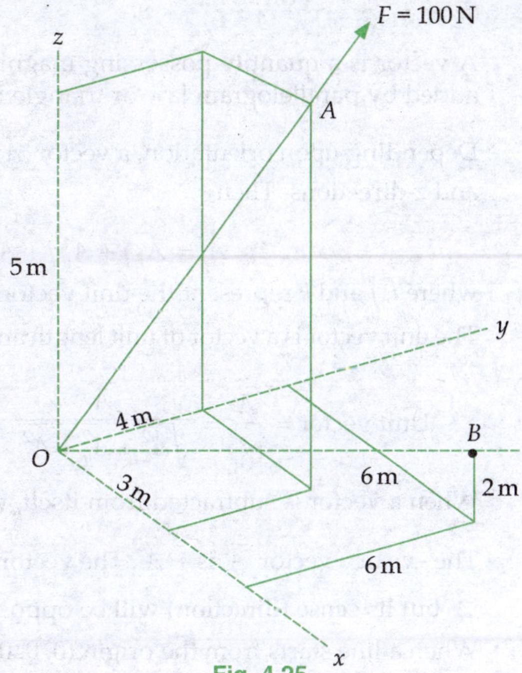

Fig. 4.25

Solution : Coordinates of point A

$= (3, 4, 5)$

Vector $\overrightarrow{OA} = (3-0)\,i + (4-0)\,j + (5-0)\,k$

$= 3\,i + 4\,j + 5\,k$

Unit vector along $\overrightarrow{OA} = \dfrac{3i + 4j + 5k}{\sqrt{3^2 + 4^2 + 5^2}}$

$= 0.424\,i + 0.566\,j + 0.707\,k$

Force vector $\vec{F}$ = magnitude of force × unit vector along the direction of force

$= 100(0.424\,i + 0.566\,j + 0.707\,k)$

$= 42.4\,i + 56.6\,j + 70.7\,k$

The scalar components of force vector $\vec{F}$ are

$F_x = 42.4$ N ; $F_y = 56.6$ N and $F_z = 70.7$ N

(*b*) Projection of $\vec{F}$ on x - y plane

Projection $= 3\,i + 4\,j$

Unit vector $= \dfrac{3i + 4j}{5} = 0.6\,i + 0.8\,j$

Then $F_{xy} = \vec{F}$ × unit vector along projection on x-y plane

$= (42.4\,i + 56.6\,j + 70.7\,k) \cdot (0.6\,i + 0.8\,j)$

$= 42.4 \times 0.6 + 56.6 \times 0.8 =$ **70.72 N**

Alternatively : The cosine of angle θ between $\vec{F}$ and x - y plane is given by

$$\cos\theta = \frac{\sqrt{3^2 + 4^2}}{\sqrt{3^2 + 4^2 + 5^2}} = 0.707\ ; \qquad \theta = 45^\circ$$

$F_{xy} = F\cos\theta = 100 \times \cos 45^\circ =$ **70.7 N**

(*c*) Vector $\overrightarrow{OB} = (6-0)\,i + (6-0)\,j + (2-0)\,k = 6\,i + 6\,j + 2\,k$

Unit vector along $\overrightarrow{OB} = \dfrac{6i + 6j + 2k}{\sqrt{6^2 + 6^2 + 2^2}} = 0.688\,i + 0.688\,j + 0.229\,k$

Then $F_{OB} = \vec{F}$ × unit vector along OB

$= (42.4\,i + 56.6\,j + 70.7\,k) \cdot (0.688\,i + 0.688\,j + 0.229\,k)$

$= 42.4 \times 0.688 + 56.6 \times 0.688 + 70.7 \times 0.229$

$= 29.17 + 38.94 + 16.19 =$ **84.3 N**

REVIEW AND SUMMARY

1. A vector is a quantity possessing magnitude and direction, and the two vector quantities are added by parallelogram law or triangle law

2. Depending upon orientation, a vector $\vec{A}$ may have component $\overrightarrow{A_x}$, $\overrightarrow{A_y}$ and $\overrightarrow{A_z}$ along the x, y and z-directions. Then

$$\vec{A} = A_x i + A_y j + A_z k$$

where i, j and k represent the unit vectors along the x, y and z-axes respectively.

3. The unit vector is a vector of unit length and it is determined by dividing a vector by its magnitude

$$\text{Unit vector} = \frac{\vec{A}}{|\vec{A}|} = \frac{\vec{A}}{\sqrt{A_x^2 + A_y^2 + A_z^2}}$$

4. When a vector is subtracted from itself, we get a zero vector or null vector.

The –ve of a vector $\vec{A}$ is $-\vec{A}$. The vector $-\vec{A}$ will have the same magnitude and inclination as $\vec{A}$ but its sense (direction) will be opposite to that of $\vec{A}$.

5. When a line starts from the origin (0, 0, 0) and ends with a point $(x, -y, z)$ in space, its position vector will be

$$= (x - 0)\,i + (-y - 0)\,j + (z - 0)\,k$$
$$= xi - yj + zk$$

6. The dot product of vectors $\vec{A}$ and $\vec{B}$ is a scalar quantity, and is defined as the product of the magnitude of the vectors and cosine of their included angle.

$$\vec{A}.\vec{B} = |\vec{A}|.|\vec{B}|\cos\theta = AB\cos\theta$$

The dot product defines the term work in the field of dynamics.

7. The cross product of vector $\vec{A}$ and $\vec{B}$ is a vector quantity and is defined as the product of the magnitude of vectors and the sine of the smaller angle between the vectors.

$$\vec{A}\times\vec{B} = AB\sin\theta\,\vec{n}$$

If $\vec{A} = A_x i + A_y j + A_z k$ and $\vec{B} = B_x i + B_y j + B_z k$, then the cross product in terms of rectangular components is given by

$$\vec{A}\times\vec{B} = \begin{vmatrix} i & j & k \\ A_x & A_y & A_z \\ B_x & B_y & B_z \end{vmatrix}.$$

$$= (A_y B_z - B_y A_z)\,i - (A_x B_z - B_x A_z)\,j + (A_x B_y - B_x A_y)\,k$$

The cross product defines the term moment of force.

REVIEW QUESTIONS

A. Conceptual and conventional questions :

1. What is meant by concurrent forces in space?
2. Define the terms unit vector and position vector.
3. Explain the concept of dot and cross products in the context of vectors.
4. Express vectorially the resultant of concurrent forces in space.
5. State the conditions of equilibrium for a particle in space.
6. Define moment of force about a point in space.
7. A line originates at the points (4, 1, – 2) and passes through the point (2, 2, 6). What will be the unit vector along this line?
8. Two vectors $\vec{A}$ and $\vec{B}$ are defined by the relations.

$$\vec{A} = 3i + 4j + 6k \quad \text{and} \quad \vec{B} = 4i + 5j - 9k$$

Determine the sum, difference and dot product of these vectors. Also find the angle between $\vec{A}$ and $\vec{B}$. ($7i + 9j - 3k$; $-i - j + 15k$; $-2z$; 109.2°)

9. Two vectors $\vec{A}$ and $\vec{B}$ with magnitude 10 units and 15 units respectively have a dot product equivalent to 50 units. Find the magnitude of their cross product. (144.4 units)
10. The coordinates of the initial and terminal points of a vector are (3, 1, – 2) and (4, – 7, 10). Specify the vector, evaluate its magnitude and direction cosines.

($i - 8j + 12k$, 14.46, 0.069, – 0.553, 0.830)

11. The line of action of a 100 N force $\vec{F}$ passes through points A (2, 5, 8) and B (7, 2, 6). The direction of force is from A to B. Express $\vec{F}$ in terms of the unit vectors i, j and k.

($81.11\,i - 48.67\,j - 32.44\,k$)

12. A constant force $4i + 3j$ moves a particle along straight line from position (6, 9, 5) to position (4, 6, 8). What work will be done by the force?
Take force in newtons and distances in metres. (– 17 Nm)
13. A force $\vec{F} = (10i + 8j - 5k)$ N acts at point A (2, 5, 6) m. What is the moment of the force about the point B (3, 1, 4) m. ($-36\,i + 15\,j - 48\,k$ Nm)
14. A point P is acted upon by the following set of forces as shown in Fig. 4.26.

F_1 = 900 N acting along line PA
F_2 = 1200 N acting along line PB
F_3 = 1500 N

If the resultant of these forces is parallel to y-axis, what should be the direction of force F_3? Proceed to calculate the magnitude of the resultant force. (– 3020 N downwards)

15. A vertical load of 1000 N is supported by three bars as shown in Fig. 4.27 Points C, O and D are in x-y plane while B is 1.5 m above this plane. The base points of the bars are :

A (0, 6, 0), B (– 3, 1.5, 0)
C (0, 0, 3) and D (3, 0, – 3)

Compute the forces in each member.

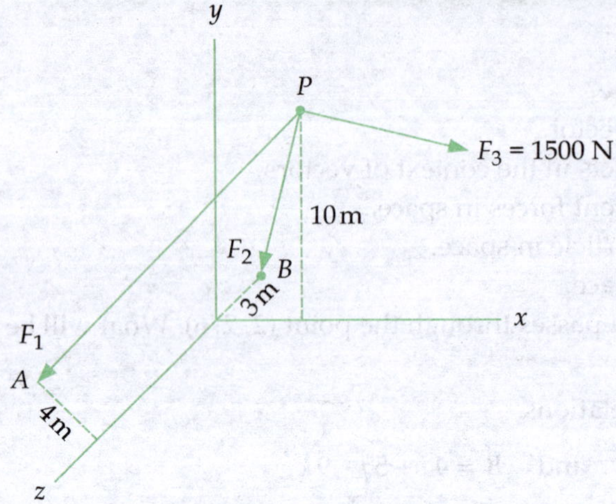

Fig. 4.26

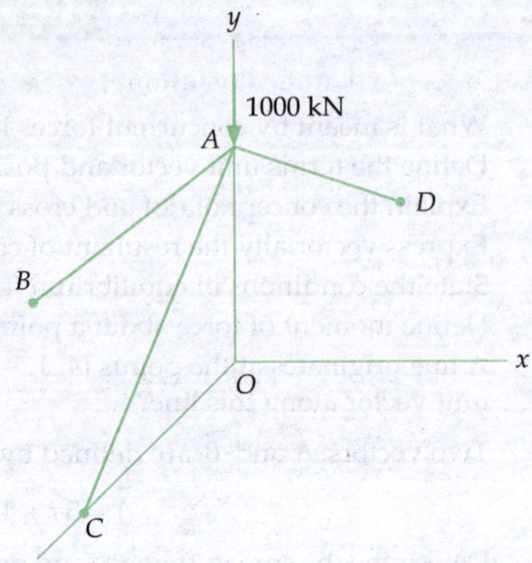

Fig. 4.27

16. A tripod carrying a load of 50 kN has its supports A, B and C which are coplanar in x-z plane. Assuming all points to be of ball and socket type, make calculations for the forces in members AD, BD and CD (Refer Fig. 4.28) **(– 13 kN, 58 kN, 50.4 kN)**

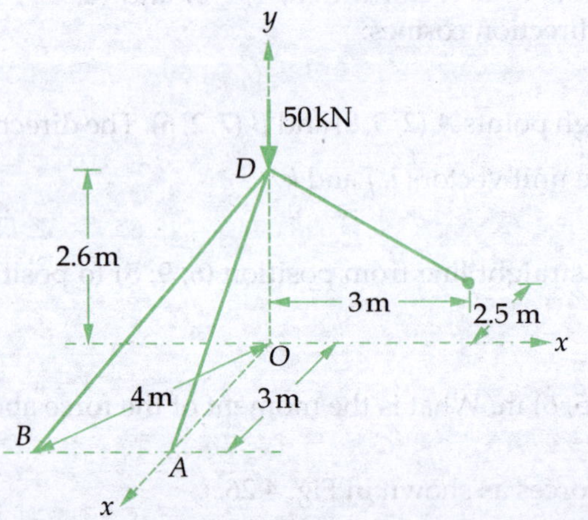

Fig. 4.28

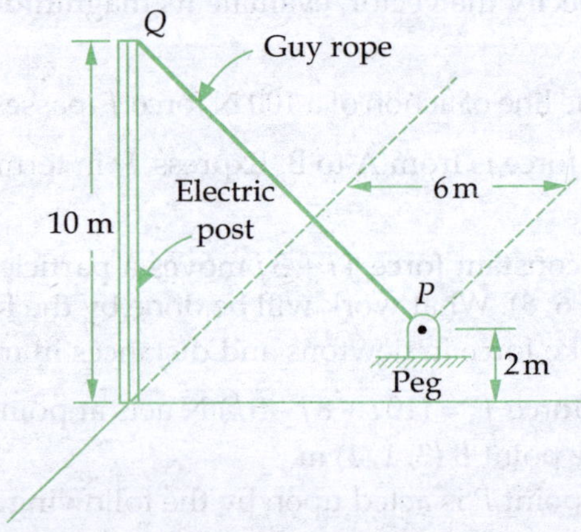

Fig. 4.29

17. An electric post is kept in equilibrium by a metal guy rope tied to a peg P as shown in Fig. 4.25. If the force in guy rope is estimated to be 1250 N, calculate the components of the force at the peg and the angles of inclination with the rectangular axes.
(– 603 N, 1060 N, 210 N, 120.3°, 31.8°, 80.3°)

18. For a concurrent force system in space depicted in Fig. 4.30, the tension in string AB is 75 kN and the resultant of three forces applied at A is vertical. Make calculations for the tensions induced in strings AC and AD, and the resultant force. **(40.5 kN, 123.7 kN, 222.6 kN)**

19. A three member space truss is formed by the member OA, OB and OC, and a weight of 10 kN is suspended at the joint O as shown in Fig. 4.31. Calculate the magnitude and nature of force induced in each member of the truss. **(– 12.5 kN, 18 kN, – 12.5 kN)**

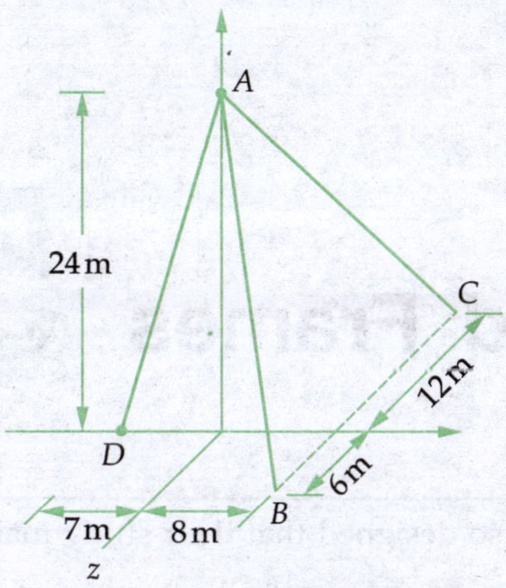

Fig. 4.30

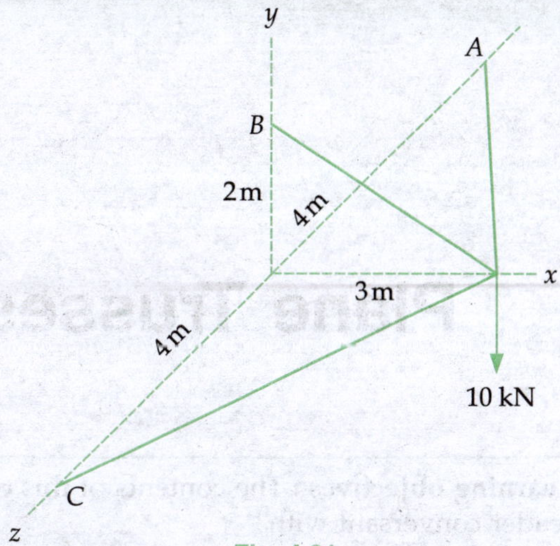

Fig. 4.31

B. Complete the following statements with most appropriate word/words :

1. The force system on a tripod carrying a camera is an example of .... forces.
2. The resultant of two force vectors can be obtained according to .... law or .... law.
3. The unit vector is obtained by dividing the vector by ....
4. A zero vector is also called a .... vector.
5. A vector $\vec{F} = 2\,i + 3\,j - 4\,k$ passes through the origin. The magnitude of its vector component along y-axis is ....
6. The dot product of two vector quantities is defined as the product of the magnitude of the vectors and .... of their included angle.
7. Two vectors will be .... to each other if their cross product is zero.
8. If l, m and n represent the direction cosines of a force system, then the sum of their squares equals ....

Answers:

1. non-coplanar concurrent 2. parallelogram, triangle 3. its magnitude 4. null 5. three
6. cosine 7. parallel 8. unity

Plane Trusses and Frames 5

Learning objectives : The contents of this chapter are so designed that their study makes the reader conversant with

- *framed structures and their application*
- *types of frames : plane frame and space frame ; statically determinate and statically indeterminate frames ; perfect, deficient and redundant beams*
- *reaction at supports of a frame*
- *analysis of a frame : method of joints, method of sections and the graphical method*

A framed structure is an assemblage of a number of bars or rods joined together in such a way to form a rigid framework; this structure is designed to resist geometrical distortion under any applied system of loading. The ends are joined by riveting, welding or by nuts and bolts. The bars used in the framed structure are slender (cross-sectional dimensions quite small compared to length) and are usually of angle section, T-section, I-section etc. The bars are called two-force members because they are either in tension or in compression. The framed structure finds application in the roofs of sheds at railway platforms, workshops and industrial buildings, and bridge trusses etc. Attention has been focussed in this chapter on the study of plane trusses, that is, the structures whose members lie in one plane and with loads applied only at the joints.

5.1. CLASSIFICATION FRAMES

The frames are generally classified as:

• **Truss and frame:** The structure is called a truss when the loads are applied only at joints and not any other point of the bar member. The truss shown in Fig. 5.1 is supported at *A* and *B*, and has loads applied at the joints *C*, *D* and *E*.

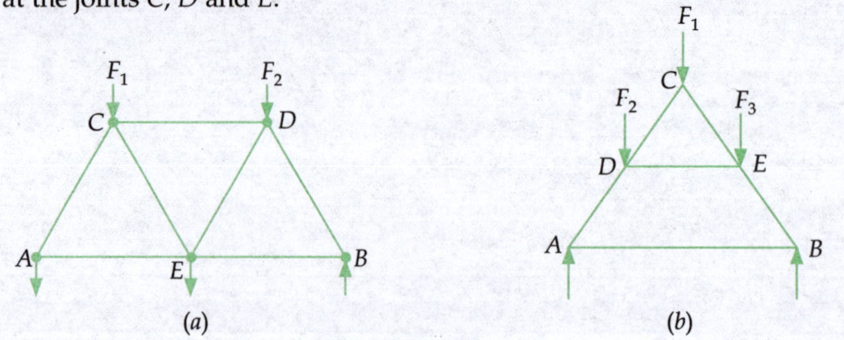

Fig. 5.1. Truss and frame

The structure is called a frame when some on its members are subjected to two or more forces. These forces are applied not only at the joints but also at some other points of the member bars. The structure shown in Fig. 5.2 represents a frame; here the member AC is subjected to loads F_1 and F_2, and the member BC is acted upon by forces F_1 and F_3.

The machines are the structures that have some movable parts and are used to transmit power.

• **Plane frame and space frame:** In a plane frame, all the members lie in a single plane and the forces act along the plane of the frame. Bridge trusses and roof trusses are the plane frames.

A frame in which all the members do not lie in the same plane is called a space frame. Tripod and suspention towers are the space frames.

• **Statically determinate and statically indeterminate frames:** The force analysis of the members of a statically determinate frame can be made by applying the equations of statics only. Equations of static equilibrium are not sufficient to determine the forces in statically indeterminate frames; there is need of considering their deformation also.

• **Perfect, deficient and redundant frames:** The structure is said to be **perfect** if the number of members is just sufficient to prevent its distortion of shape when subjected to external loads. For a perfect frame, the following correlation exists between the number of joints j and the number of members m,

$$m = 2j - 3$$

The number of joints and the number of members in the frames shown in Fig. 5.2 satisfy the identity $m = 2j - 3$ and as such these are perfect structures (frames). A perfect structure is statically determinate. Equations of static equilibrium are sufficient to determine the forces in its members without the need of considering their deformation.

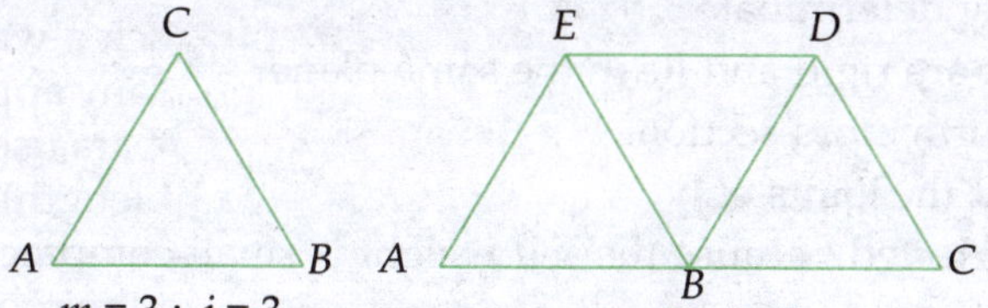

$m = 3$; $j = 3$ $m = 7$; $j = 5$

(a)   (b)

Fig. 5.2

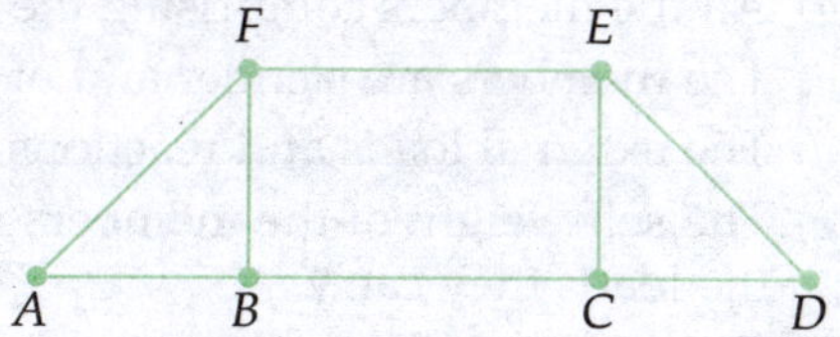

Fig. 5.3

A structure is termed **imperfect** or **deficient** frame if the number of memebrs in it is less than that required for a perfect frame. Refer Fig. 5.3 which shows a frame having 6 joints and 8 members. The number of members required for the frame to be perfect is

$$m = 2j - 3 = 2 \times 6 - 3 = 9$$

Since there are only 8 members (one less than that required for a perfect frame), the given frame is deficient or imperfect. An imperfect frame is under rigid or collapsible. Such a frame cannot prevent geometrical distortion when loaded.

A structure is termed **redundant** frame if the number of members in it is more than that required for a perfect frame. Refer Fig. 5.3 which shows a frame which has 7 joints and 12 members. The number of members required for the frame to be perfect is

$$m = 2j - 3 = 2 \times 7 - 3 = 11$$

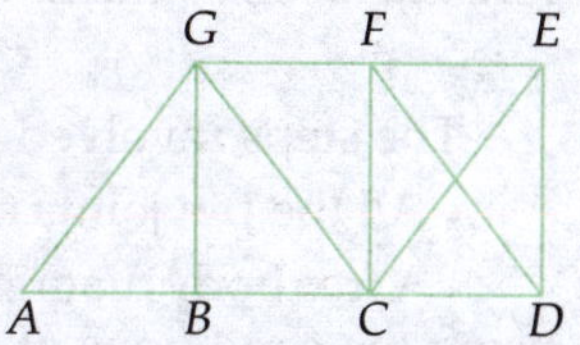

Fig. 5.4

Since the frame has 12 members (one more than that required for a perfect frame), the given frame is redundant. A redundant frame is over rigid and statically indeterminate. Further, each extra member adds one degree of indeterminancy. Thus the frame depicted in Fig. 5.4 is redundant to a single degree because one member is extra.

The frame depicted in Fig. 5.4 would remain stable even if one of the members CE or DF is removed.

5.2. REACTIONS AT SUPPORTS OF A FRAME

Statically determinate frames normally have two types of supports: one hinged and the other on rollers.

Roller supports (Fig. 5.5 *a*) are frictionless and provide a reaction at right angles to the roller base. At a hinged support (Fig. 5.5 *b*), the line of action of the reaction depends upon the load system on the frame. This reaction can be split into two components acting in the horizontal and vertical directions.

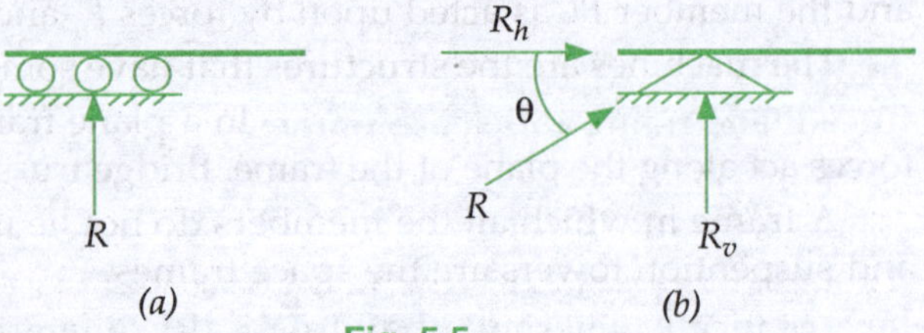

Fig. 5.5

$$R_v = R \sin \theta \text{ and } R_h = R \cos \theta$$

A roller support prevents motion only in the direction perpendicular to its seat. With pin (hinged) support, the beam has no translatory motion in any direction. In both types of supports, there is no prevention of rotation about the connections.

5.3. ANALYSIS OF A FRAME

The analysis of a truss involves the determination of :

(*i*) The reactions at the supports,

(*ii*) The internal stresses induced in the members due to external loads. That helps to workout the size and shape of the structural members forming the truss.

While making such an analysis, the following assumptions are made:

(*i*) The truss is a perfect one and statically determinate.

(*ii*) All the members comprising the truss are rigid and lie in the same plane.

(*iii*) The members are slender and of uniform cross-section.

(*iv*) The external loads and reactions act at the joints only.

(*v*) The self weight of the members is neglected because the self weight is small compared to the loads they carry.

(*vi*) The forces are transmitted from one member to another through smooth pins (no friction) fitting perfectly in the members.

The stresses in the members of a framed structure are generally worked out by adopting the following methods:

(1) Analytical method: method of joints and method of sections

(2) Graphical method

5.3.1. Method of joints

Every joint is treated separately as a free body in equilibrium, i.e., the sum of all the vertical forces as well as the horizontal forces acting on the joint is equated to zero.

$$\Sigma F_v = 0 \quad \text{and} \quad \Sigma F_h = 0$$

The steps involved are:

- All the pin joints are labelled.
- A free body diagram of the entire frame is drawn and the reaction at the supports is determined using the conditions of equilibrium.
- Each joint is treated separately as a free body. A certain direction of forces acting on the joint is assumed and the magnitude of forces is worked out by applying the conditions of equilibrium. If the magnitude of a particular force comes out positive, the assumption in respect of its direction is correct. However if the magnitude of the force comes out negative, the actual direction of force is reversed , i.e., opposite to what has been assumed.

- Start is made from a joint where there are not more than two members in which the forces are unknown, and the process is repeated from one joint to another until all the unknowns have been determined.

- Assumption is made about the nature of forces in a member and the force is determined. If the answer is positive, the assumption is correct. A negative answer would imply that the nature of force is opposite to that assumed.

Finally, the force in the member will be tensile if the member pulls the joint to which it is connected (force is directed away from the pin) whereas the force in the member will be compressive if the member pushes the joint to which it is connected (force is towards the pin). These aspects have been illustrated in Fig. 5.5. A member under tension is called a *tie*, and a member under compression is called a **strut**.

(*a*) Tension: Pulling at each end (*b*) Compression: Pushing at each end

Fig. 5.6

The following principles help to identify the members not subjected to any force when the truss is loaded:

(*i*) A single force cannot form a system in equilibrium. This implies that if there is only one force acting at a joint, then for the equilirbium of this joint, this force equals zero.
With reference to Fig. 5.7 (*a*) : $F_1 = 0$

(*ii*) When two members meeting at a joint are not collinear and there is no external force acting at the joint, then the forces in both the members are zero.
With reference to Fig. 5.7 (*b*) : $F_1 = F_2 = 0$

(*iii*) When three members are meeting at an unloaded joint and out of them two are collinear, then the force in the third member will be zero.
With reference to Fig. 5.7 (*c*) : $F_3 = 0$.

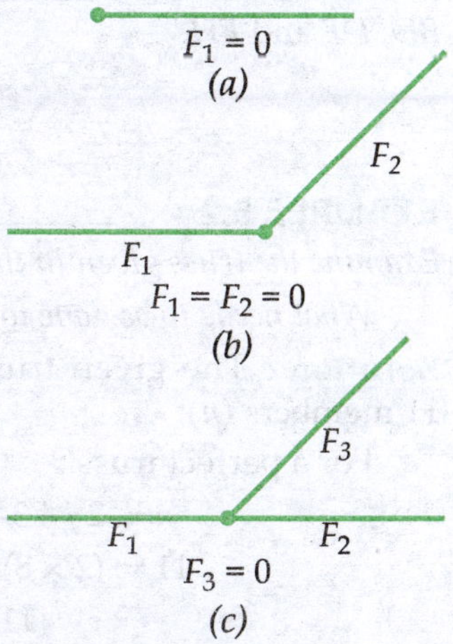

Fig. 5.7

EXAMPLE 5.1

With reference to trusses shown in Fig. 5.8, identify the members which are not subjected to any force.

Solution : (*i*) Consider joint *D* for the truss shown in Fig. 5.8 *a*. Forces at this joint are:

F_1 in member *DC* ; F_2 in member *DA* ;
F_3 in member *DF*

The joint is unloaded and the forces F_2 and F_3 are collinear

$$\therefore \quad F_1 = 0$$

Likewise forces in the members *EF* and *HG* are zero.

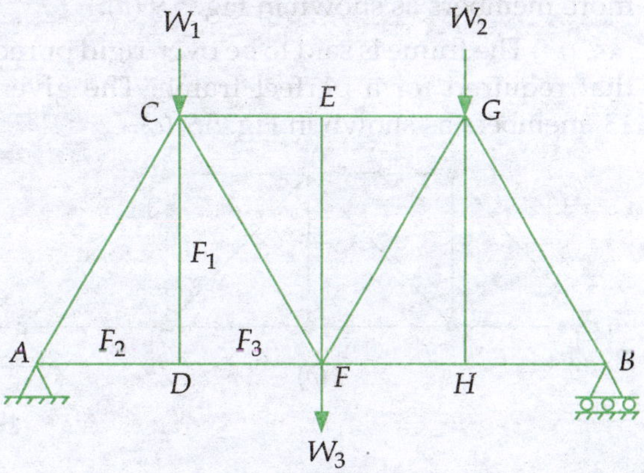

(a)

(*ii*) Considering joint D of of the cantilever truss (Fig. 5.7 *b*), there are three forces namely F_1, F_2 and F_3.

Since the joint is unloaded and the forces F_1 and F_2 are collinear

$$\therefore \quad F_3 = 0$$

Considering joint C;

Since $F_3 = 0$, the forces acting at this joint are F_4, F_5 and F_6.

The joint is unloaded and out of these three forces, two forces namely F_5 and F_6 are collinear. Accordingly, the third force F_4 is zero.

Thus the members AC and DC will not be subjected to any load.

(*iii*) Considering loading and the support system for the truss depicted in Fig. 5.7 *c*, the reaction at roller support is vertical and $R_b = W$. No reaction will be experienced at the hinged support, *i.e.*, $R = 0$.

Obviously then, there will no force in the members other than BD, DF and FH.

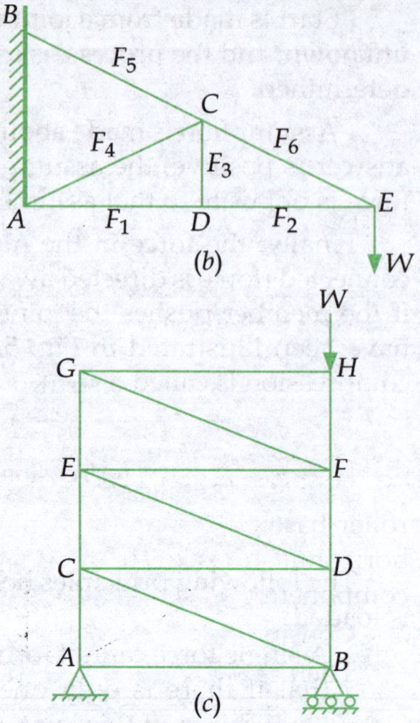

Fig. 5.8

EXAMPLE 5.2

Examine the truss given in the adjoining diagram.

What needs to be done to make it: (i) deficient, (ii) perfect and (iii) over-rigid (redundant) frame.

Solution : The given truss has 8 joints (j) and 11 members (n)

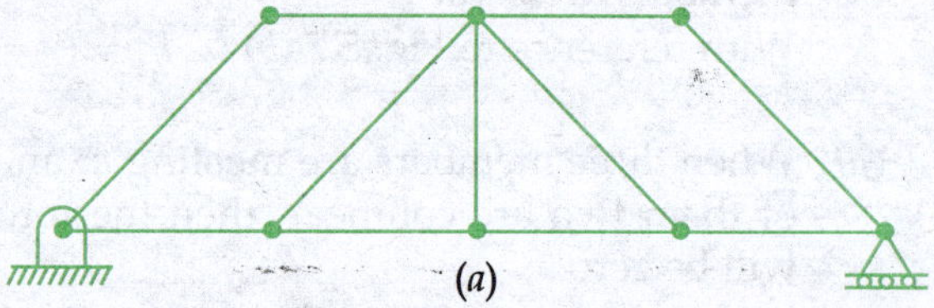

For a perfect truss :

$$n = 2j - 3$$

$$11 = (2 \times 8) - 3;$$

$$11 = 13 \text{ which is not true.}$$

Hence the given truss is already a deficient frame.

(*ii*) For the truss to be a perfect frame, it needs 13 members. That can be done by adding two more members as shown in Fig. 5.9 (*b*).

(*iii*) The frame is said to be over-rigid or redundant if the number of members in it are more than that required for a perfect frame. The given truss will become over-rigid if it has more than 13 members as shown in Fig. 5.9 (*c*).

Fig. 5.9

EXAMPLE 5.3

Determine the forces in all the members of a truss with the loading and support system shown in Fig. 5.10.

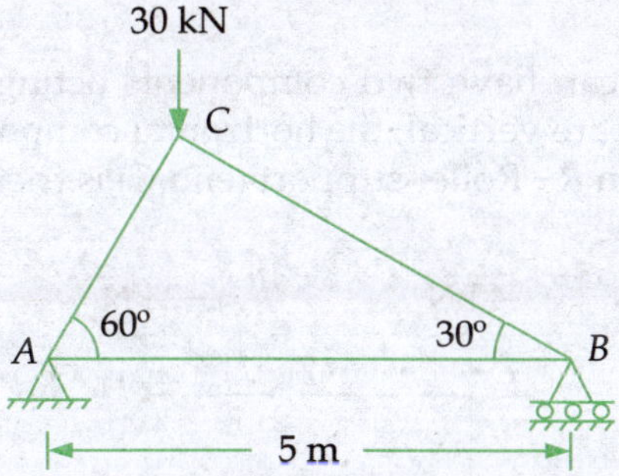

Fig. 5.10

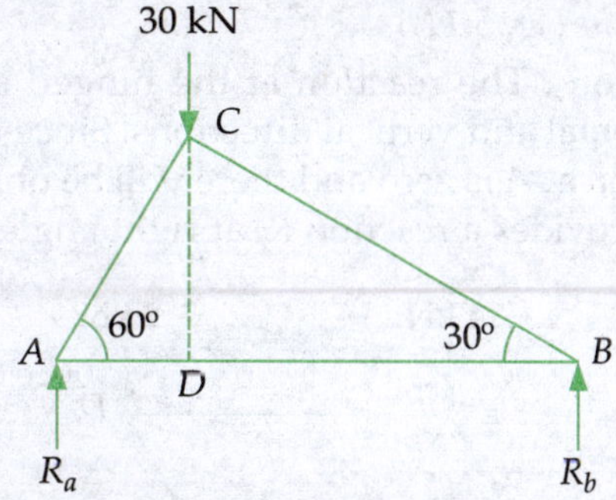

Free body diagram of the whole truss

Solution : Roller support (end B) is frictionless and provides a reaction R_b at right angles to the roller base. The reaction at the hinged support (end A) can have two components acting in the horizontal and vertical directions. Since the load of 30 kN acts vertically downward, the horizontal component of reaction at A is zero and there will be only vertical reaction R_a.

Considering the free-body diagram of the whole truss.

The triangle *ABC* is a right angled triangle with angle $ACB = 90°$

$$AC = AB \cos 60° = 5 \times 0.5 = 2.5 \text{ m}$$

Distance of line of action of 30 kN force from *A*,

$$AD = AC \cos 60° = 2.5 \times 0.5 = 1.25 \text{ m}$$

Taking moments about A, we obtain

$$R_b \times 5 = 30 \times 1.25\ ; \qquad R_b = \frac{30 \times 1.25}{5} = 7.5 \text{ kN}$$

$$R_a = 30 - 7.5 = 22.5 \text{ kN}$$

Joint A: Consider the free body diagram of joint A with the direction of forces assumed as shown. Equations of equilibrium can be written as:

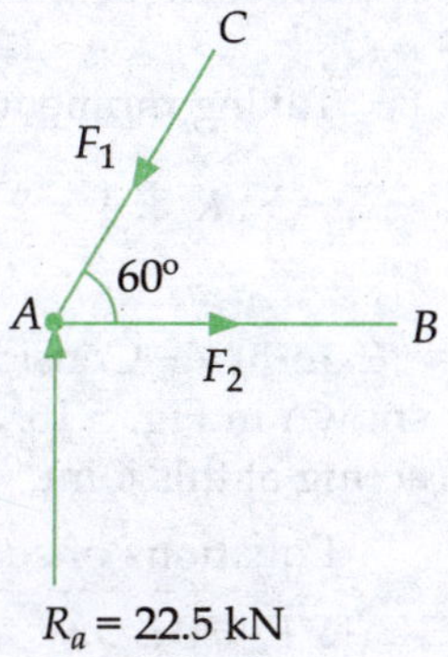

Joint A

$$\Sigma F_x = 0: \qquad F_2 - F_1 \cos 60° = 0$$

$$\Sigma F_y = 0: \qquad F_1 \sin 60° - R_a = 0$$

$$\therefore \qquad F_1 = \frac{R_a}{\sin 60°} = \frac{22.5}{0.866} = 25.97 \text{ kN (compressive)}$$

$$\text{and} \qquad F_2 = F_1 \cos 60° = 25.97 \times 0.5 = 12.99 \text{ kN (tensile)}$$

The force F_1 is acting towards the pin which means that the member *AC* is in compression.

The force F_2 is pulling the joint *A* which means that the member *AB* is in tension.

Joint *B*: Consider the free body diagram of joint *B*. The force F_2 in member *AB* has already been calculated above and found to be tensile. Hence force F_2 will pull the joint *B* and will be directed away from it as shown in Fig. 5.11.

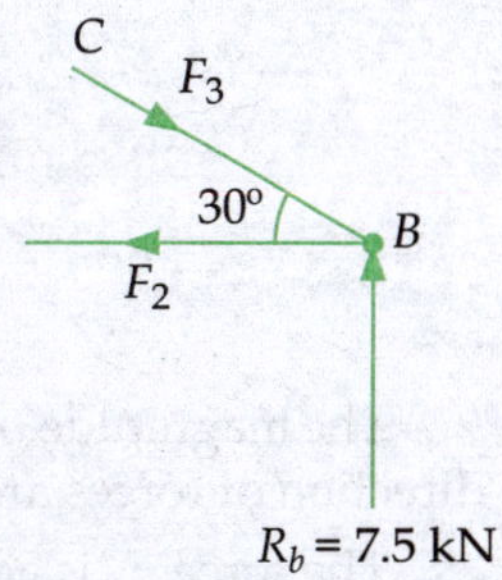

Joint B

From the equations of equilibrium,

$$\Sigma F_x = 0: \qquad F_3 \cos 30° - F_2 = 0$$

$$\therefore \qquad F_3 = \frac{F_2}{\cos 30°} = \frac{12.99}{0.866} = 15 \text{ kN (compressive)}$$

The force F_3 is acting towards the joint *B* which means that the member *BC* is in compression.

EXAMPLE 5.4

Determine the reactions and the forces in each member of a simple triangular truss supporting two loads as shown in Fig. 5.11.

Solution : The reaction at the hinged support (end *A*) can have two components acting in the horizontal and vertical directions. Since the applied loads are vertical, the horizontal component of reaction at *A* is zero and there will be only vertical reaction R_a. Roller support (end *C*) is frictionless and provides a reaction R_c at right angles to the roller base.

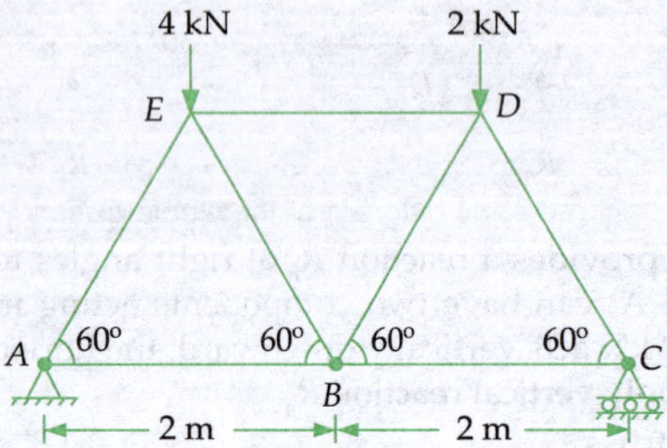

Fig. 5.11

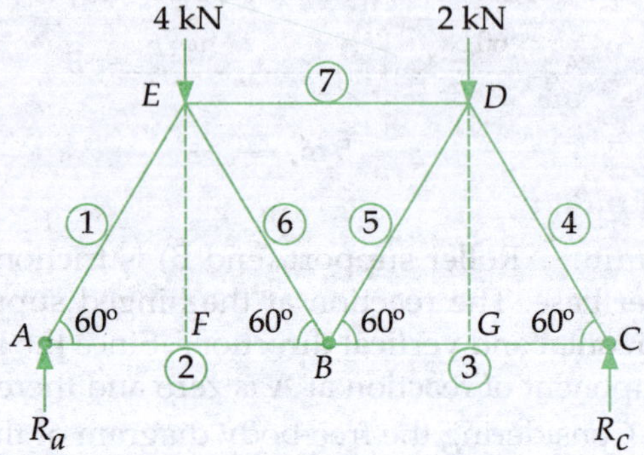

Free body diagram of the whole truss

With reference to free body diagram of the whole truss (Fig. 5.13), each of the triangular (element) constitutes an equilateral triangle of each side equal to 2 m. Then

Distance of line of action of 4 kN force from A,

$$AF = AE\cos 60^\circ = 2 \times 0.5 = 1\text{ m}$$

Distance of line of action of 2 kN force from *A*,

$$AG = AB + BG = AB + BD\cos 60^\circ = 2 + 2 \times 0.5 = 3\text{ m}$$

Taking moments about A, we get

$$R_c \times 4 = 2 \times 3 + 4 \times 1 = 10\,; \qquad R_c = \frac{10}{4} = 2.5\text{ kN}$$

$$R_a = (4 + 2) - 2.5 = 3.5\text{ kN}$$

Joint *A*: Consider the free body diagram of joint *A* with the direction of forces assumed as shown in Fig. 5.16. Start can be made with this joint because there are only two unknown forces acting at this joint.

Equations of equilibrium can be written as:

$$\Sigma F_x = 0: \qquad F_2 - F_1\cos 60^\circ = 0$$

$$\Sigma F_y = 0: \qquad R_a - F_1\sin 60^\circ = 0$$

$$\therefore \qquad F_1 = \frac{R_a}{\sin 60^\circ} = \frac{3.5}{0.866}$$

$$= 4.04\text{ kN (compressive)}$$

and

$$F_2 = F_1\cos 60^\circ = 4.04 \times 0.5$$

$$= 2.02\text{ kN (tensile)}$$

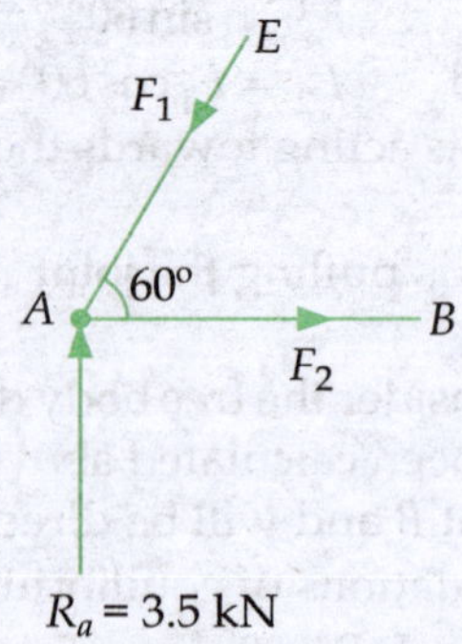

Joint A

The magnitude of forces F_1 and F_2 are both coming out to be positive and therefore the assumed direction of forces are correct.

The force F_1 is acting towards the pin *A* which means that member *AE* is in compression. The force F_2 is pulling the pin *A* which means that member *AB* is in tension.

Joint C : Consider the free body diagram of joint C with the direction of forces assumed as shown. Analysis of this joint is also possible because there are only two unknown forces F_3 and F_4 acting at this joint.

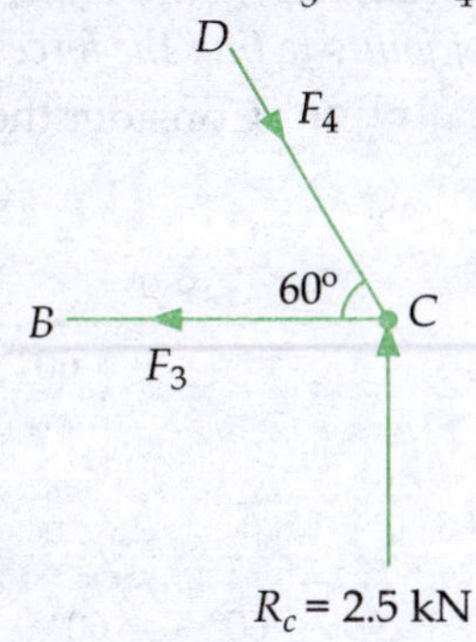

Joint C

Equations of equilibrium can be written as:

$\Sigma F_x = 0:$ $\quad F_4 \cos 60^\circ - F_3 = 0$

$\Sigma F_y = 0:$ $\quad R_c - F_4 \sin 60^\circ = 0$

$$\therefore \quad F_4 = \frac{R_c}{\sin 60^\circ} = \frac{2.5}{0.866} = 2.88 \text{ kN (compression)}$$

and $\quad F_3 = F_4 \cos 60^\circ = 2.88 \times 0.5 = 1.44$ kN (tensile)

Joint B : Consider the free body diagram of joint B with the two known values F_2 (tensile force in member AB) and F_3 (tensile force in member BC) inserted:

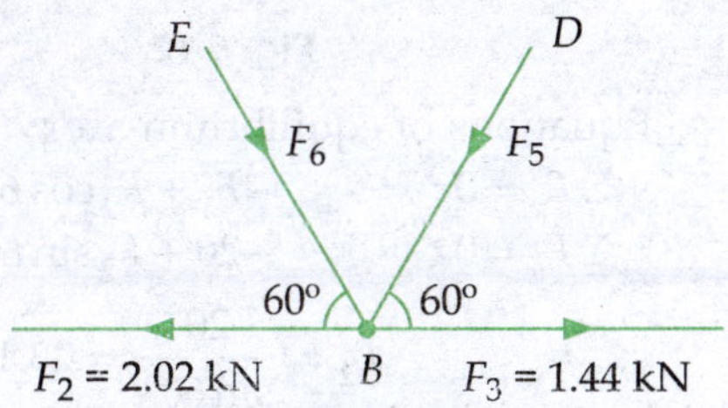

Joint B

Equations of equilibrium for this joint are:

$\Sigma F_x = 0:$ $\quad 1.44 - 2.02 + F_5 \cos 60^\circ + F_6 \cos 60^\circ$

or $\quad (F_5 + F_6) \times 0.5 = 0.58$

or $\quad F_5 + F_6 = 1.16 \qquad$ ...(*i*)

$\Sigma F_y = 0:$ $\quad F_5 \sin 60^\circ - F_6 \sin 60^\circ = 0$

or $\quad F_5 = F_6 \qquad$ ...(*ii*)

From expression (*i*) and (*ii*)

$$F_5 = F_6 = \frac{1.16}{2} = 0.58 \text{ kN}$$

Force F_5 in member BD is 0.58 kN tensile and force F_6 in member BE is 0.58 kN compressive.

Joint D : Consider the free body diagram of joint D with the known values F_4 (compressive force in member DC) and F_5 (tensile force member DB) inserted.

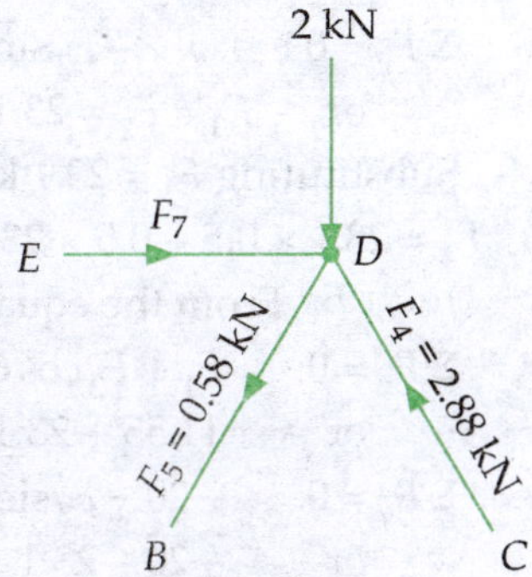

Joint D

Equations of equilibrium for this joint are:

$\Sigma F_x = 0:$ $\quad F_7 - 0.58 \cos 60^\circ - 2.88 \cos 60^\circ$

or $\quad F_7 = 0.58 \times 0.5 + 2.88 \times 0.5$

$\quad = 0.29 + 1.44 = 1.73$ kN $\quad$ (compressive)

There is no need to consider the equilibrium of joint E as the forces in all the members have been determined. The magnitude and the nature of the forces in all the members of the given truss for the applied system of loading are tabulated below :

Member	Force	Member	Force
AE	$F_1 = 4.04$ kN (C)	*BE*	$F_5 = 0.58$ kN (T)
AB	$F_2 = 2.02$ kN (T)	*BD*	$F_6 = 0.58$ kN (C)
BC	$F_3 = 1.44$ kN (T)	*DE*	$F_7 = 1.73$ kN (C)
CD	$F_4 = 2.88$ kN (C)		

EXAMPLE 5.5

A cantilever truss Wassen type is loaded as shown in the Fig. 5.12. Use the method of sections or the method of joints to find the forces in all members of the truss.

Solution : Consider the free body diagram of *joint C* with the direction of forces assumed as shown

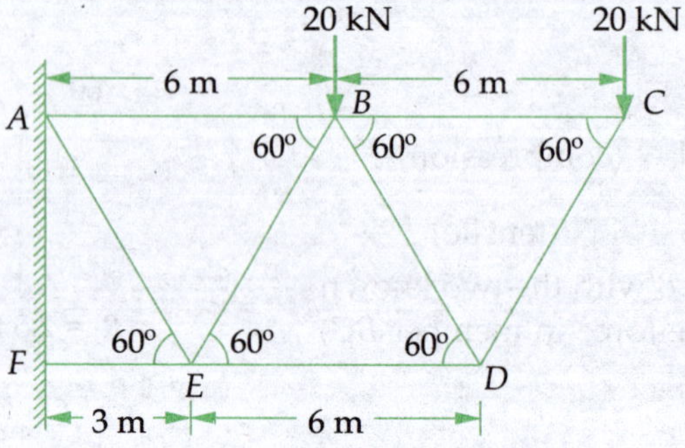

Fig. 5.12

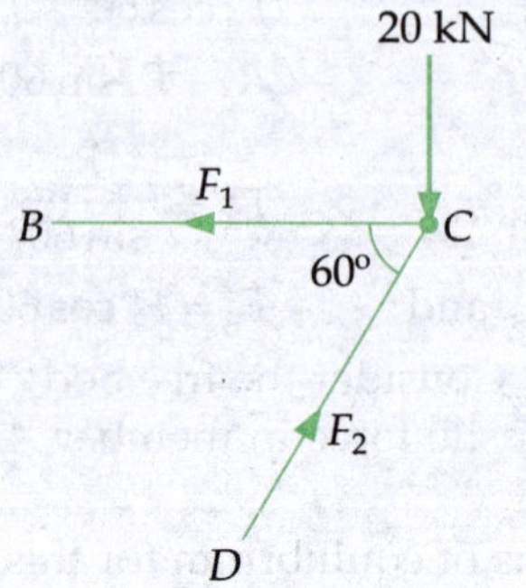

Joint C

Equations of equilibrium are :

$\Sigma F_x = 0 : \quad -F_1 + F_2 \cos 60° = 0$

$\Sigma F_y = 0 : \quad -20 + F_2 \sin 60° = 0$

$\therefore \quad F_2 = \dfrac{20}{\sin 60°} = 23.1$ kN (compression)

and $F_1 = F_2 \cos 60° = 23.1 \cos 60° = 11.55$ kN (tension)

Joint D: From the equations of equilibrium,

$\Sigma F_x = 0 : \quad -F_2 \cos 60° - F_3 \cos 60° + F_4 = 0$

or $\quad 23.1 \times 0.5 + 0.5 F_3 = F_4$

$\Sigma F_y = 0 : \quad -F_2 \sin 60° + F_3 \sin 60° = 0$

or $\quad F_3 = F_2 = 23.1$ kN see below (tension)

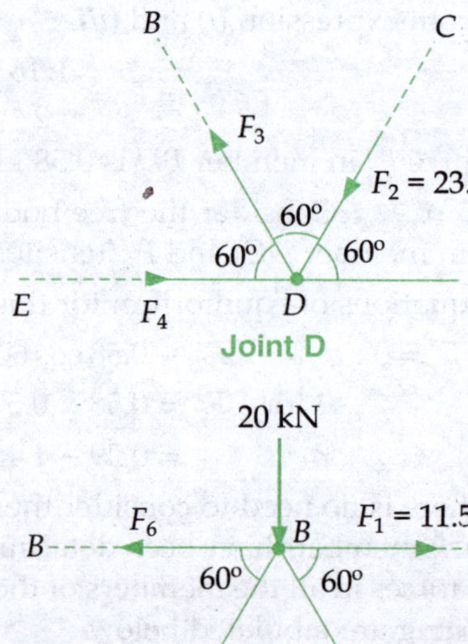

Joint D

Substituting $F_3 = 23.1$ kN in expression for F_4, we get

$F_4 = 23.1 \times 0.5 + 0.5 \times 23.1 = 23.1$ kN (compression)

Joint B: From the equations of equilibrium,

$\Sigma F_x = 0: \quad F_1 + F_3 \cos 60° - F_5 \cos 60 - F_6 = 0$

or $\quad 11.55 + 23.1 \times 0.5 - 0.5 F_5 = F_6$

$\Sigma F_y = 0: \quad -20 - F_3 \sin 60° - F_5 \sin 60° = 0$

or $\quad -20 - 23.1 \times 0.866 = 0.866 F_5$

$\therefore \quad F_5 = \dfrac{-20 - 23.1 \times 0.866}{0.866} = -46.2$ kN

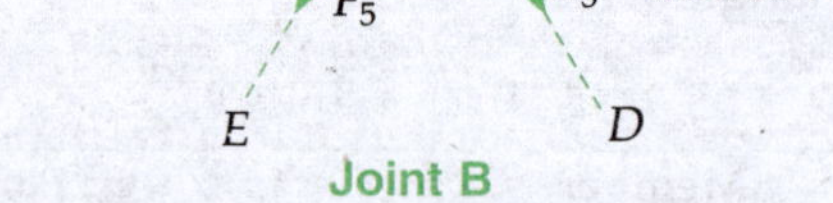

Joint B

The negative sign with the magnitude of force F_5 shows that a wrong choice has been made while assuming its direction. Obviously the assumed direction of force in member *BE* need to be reversed. Therefore $F_5 = 46.2$ kN (compression) and *BE* is a compressive member.

Substituting $F_5 = -46.2$ kN in expression for F_6, we get

$$F_6 = 11.55 + 23.1 \times 0.5 - 0.5 \times (-46.2) = 46.2 \text{ kN (tension)}$$

Joint E: From the equations of equilibrium,

$\Sigma F_x = 0; \quad -F_4 - F_5 \cos 60° - F_7 \cos 60° + F_8 = 0$

or $\quad F_8 = F_4 + F_5 \cos 60° + F_7 \cos 60°$

$\quad = 23.1 + 46.2 \times 0.5 + 0.5\, F_7$

$\Sigma F_y = 0; \quad -F_5 \sin 60° + F_7 \sin 60°$

$\therefore \quad F_7 = F_5 = 46.2$ kN (tension)

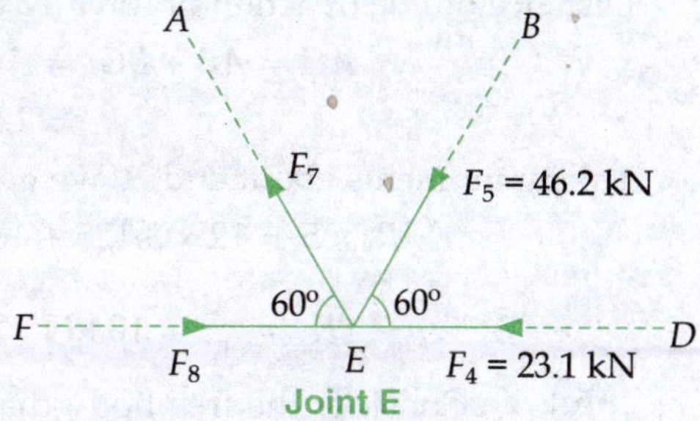

Joint E

Substituting $F_7 = 46.2$ kN in expression for F_8,
we get $F_8 = 23.1 + 46.2 \times 0.5 + 0.5 \times 46.2$

$\quad = 69.3$ kN (compression)

The forces in the various members have been tabulated below:

Member	*AB*	*BC*	*CD*	*DE*	*EF*	*AE*	*BE*	*BD*
Force (kN)	46.2	11.55	23.1	23.1	69.3	46.2	46.2	23.1
Nature	C	T	C	C	C	T	C	T

EXAMPLE 5.6

Determine the forces in all the members of the truss loaded and supported as shown in Fig. 5.13.

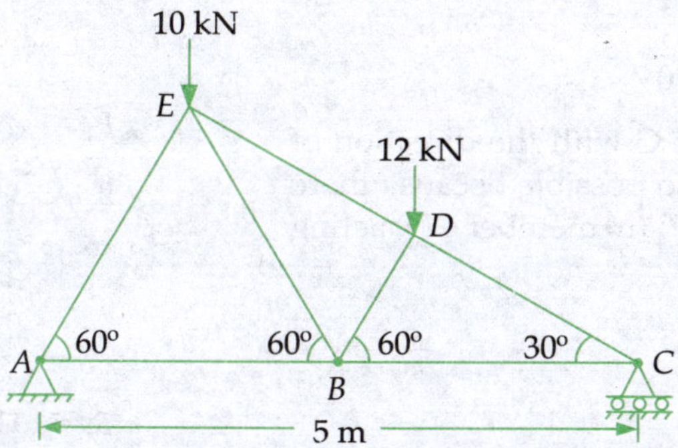

Fig. 5.13

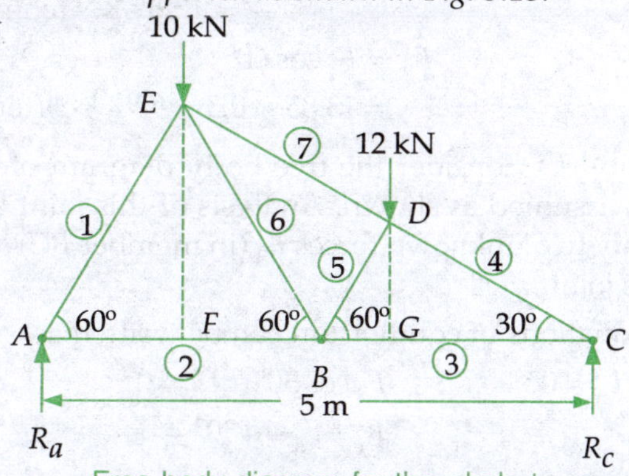

Free body diagram for the whole truss

Solution : The reactions at the supports can be determined by considering equilibrium of the entire truss. Since both the external loads are vertical, only the vertical component of the reaction at the hinged end *A* need to be considered.

With reference to free body diagram of the whole truss, the triangle *AEC* is a right angled triangle with angle *AEC* = 90°. Then

$$AE = AC \cos 60° = 5 \times 0.5 = 2.5 \text{ m}$$

$$CE = AC \sin 60° = 5 \times 0.866 = 4.33 \text{ m}$$

Further, from the give geometry, $\Delta\, ABE$ is an equilateral triangle and therefore

$$AB = BC = AE = 2.5 \text{ m}$$

Distance of line of action of force 10 kN from joint *A*,

$$AF = AE \cos 60° = 2.5 \times 0.5 = 1.25 \text{ m}$$

Again, the traingle *BDC* is a right angled triangle with angle *BDC* = 90°. Also

$$BC = AC - AB = 5 - 2.5 = 2.5 \text{ m}$$

$$BD = BC \cos 60° = 2.5 \times 0.5 = 1.25 \text{ m}$$

Distance of line of action of force 12 kN from joint A,

$$AG = AB + BG = AB + BD\cos 60^\circ$$
$$= 2.5 + 1.25 \times 0.5 = 3.125 \text{ m}$$

Taking moments about end A, we get

$$R_c \times 5 = 12 \times 3.125 + 10 \times 1.25 = 50$$

$$\therefore \quad R_c = \frac{50}{5} = 10 \text{ kN} \quad \text{and} \quad R_a = (10 + 12) - 10 = 12 \text{ kN}$$

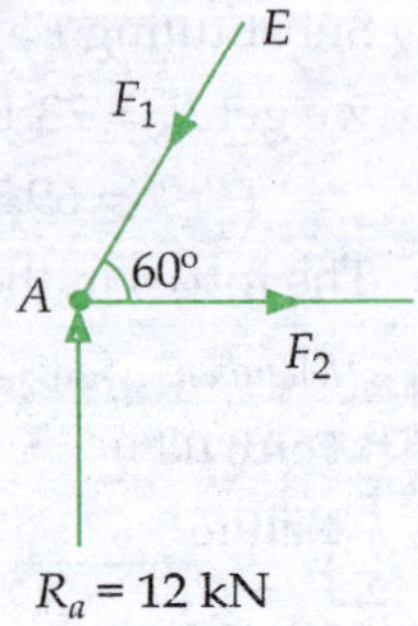

Joint A

Joint A : Consider the free body diagram of joint A with the direction of forces assumed as shown. Start can be made with this joint because there are only two unknown forces F_1 (in member AE) and F_2 (in member AB) acting at this joint.

Equations of equilibrium can be written as

$\Sigma F_x = 0:$ $\quad F_2 - F_1 \cos 60^\circ = 0$

$\Sigma F_y = 0:$ $\quad R_a - F_1 \sin 60^\circ = 0$

$$\therefore \quad F_1 = \frac{R_a}{\sin 60^\circ} = \frac{12}{0.866}$$
$$= \textbf{13.85 kN} \text{ (compression)}$$

$$F_2 = F_1 \cos 60^\circ$$
$$= 13.85 \times 0.5 = \textbf{6.92 kN} \text{ (tension)}$$

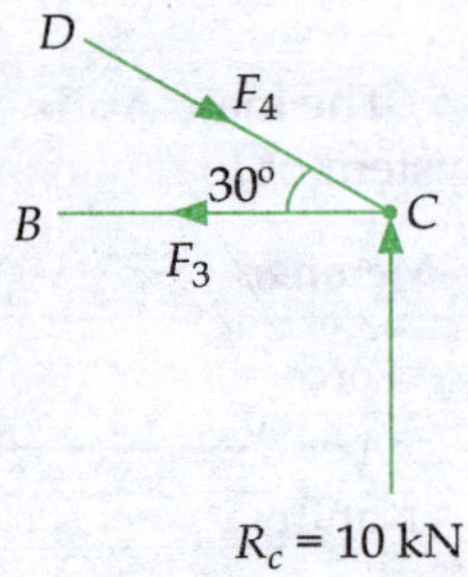

Joint C

Joint C: Consider the free body diagram of joint C with the direction of forces assumed as shown. Analysis of this joint is also possible because there are only two unknown forces F_3 (in member BC) and F_4 (in member CD) acting at this joint.

Equations of equilibrium can be written as

$\Sigma F_x = 0:$ $\quad F_4 \cos 30^\circ - F_3 = 0$

$\quad R_c - F_4 \sin 30^\circ = 0$

$$\therefore \quad F_4 = \frac{R_c}{\sin 30^\circ} = \frac{10}{0.5} = \textbf{20 kN} \text{ (compression)}$$

$$\text{and} \quad F_3 = F_4 \cos 30^\circ = 20 \times 0.866 = \textbf{17.32 kN} \text{ (tension)}$$

Joint B: Consider the free body diagram of joint B with the known values of forces F_2 (tensile force in member AB) and F_3 (tensile force in member BC) inserted.

Equations of equilibrium for this joint are:

$\Sigma F_x = 0:$ $\quad 17.32 - 6.92 - F_5 \cos 60^\circ - F_6 \cos 60^\circ = 0$

or $\quad (F_5 + F_6) \times 0.5 = 10.40$

or $\quad F_5 + F_6 = 20.80 \quad$ ...(*i*)

$\Sigma F_y = 0:$ $\quad - F_5 \sin 60^\circ + F_6 \sin 60^\circ = 0$

or $\quad F_5 = F_6 \quad$ ...(*ii*)

From expressions (*i*) and (*ii*) : $F_5 = F_6 = 10.40$ kN

Force F_5 in member BD is 10.40 kN compressive and force F_6 in member BE is 10.40 kN tensile.

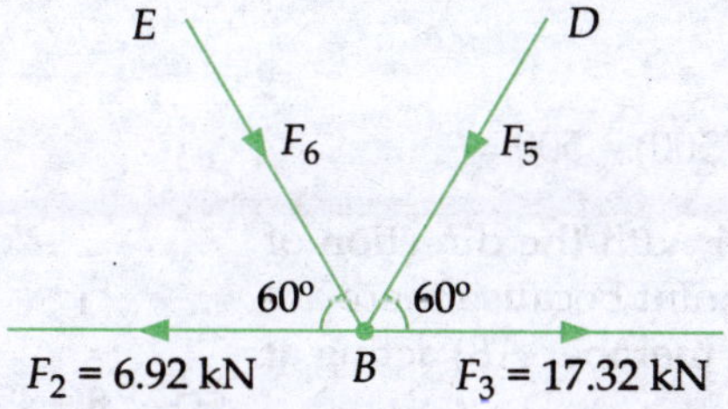

Joint B

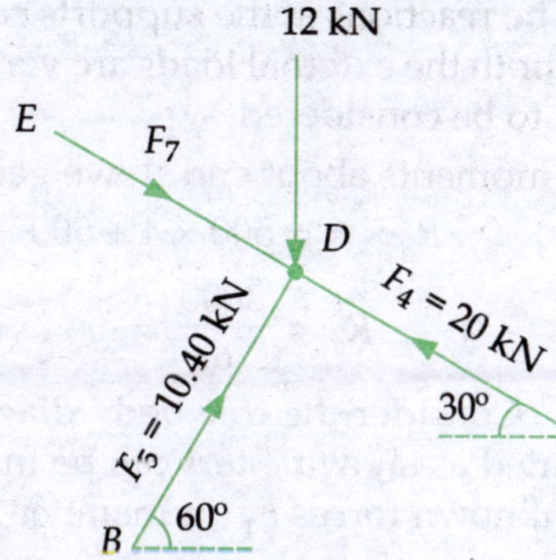

Joint D

Joint *D*: Consider the free body diagram of joint *D* with known values of forces F_4 (compressive force in member *DC*) and F_5 (compressive force in member *DB*) inserted.

Equations of equilibrium for this joint are:

$\Sigma F_x = 0$; $\quad F_7 \cos 30° + F_5 \cos 60° - F_4 \cos 30° = 0$

or $\quad F_7 \cos 30° = -F_5 \cos 60° + F_4 \cos 30°$

$$= -10.40 \times 0.5 + 20 \times 0.866 = 12.12$$

$$\therefore \quad F_7 = \frac{12.12}{\cos 30°} = 13.99 \text{ kN (compressive)}$$

The magnitude and the nature of the forces in all the members of the given truss for the applied system of loading are tabulated below:

Member	**AE**	**AB**	**BC**	**CD**	**BD**	**BE**	**DE**
Force	F_1 = 13.85 kN	F_2 = 6.92 kN	F_3 = 17.32 kN	F_4 = 20 kN	F_5 = 10.40 kN	F_6 = 10.40 kN	F_7 = 13.99 kN
Nature	C	T	T	C	C	T	C

EXAMPLE 5.6

Determine the reactions and the forces in each member of a truss supporting two loads as shown in Fig. 5.14.

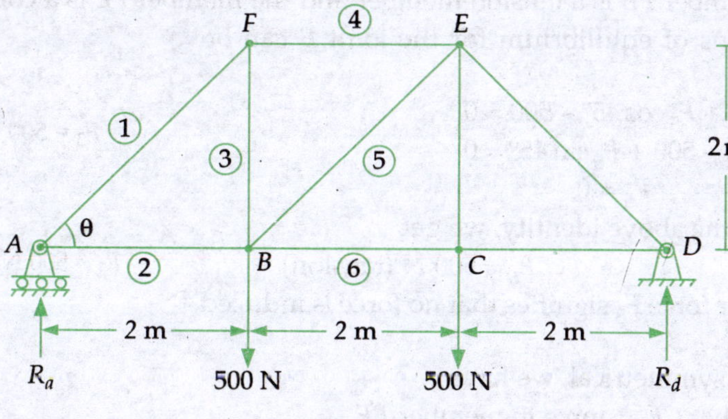

Fig. 5.14

Solution : The reactions at the supports can be determined by considering equilibrium of the truss. Since both the external loads are vertical, only the vertical component of reaction at the hinged end D need to be considered.

Taking moments about end A, we get

$$R_d \times 6 = 500 \times 4 + 500 \times 2 = 3000$$

$$\therefore \quad R_d = \frac{3000}{6} = 500 \text{ N}; \quad R_a = (500 + 500) - 500 = 500 \text{ N}$$

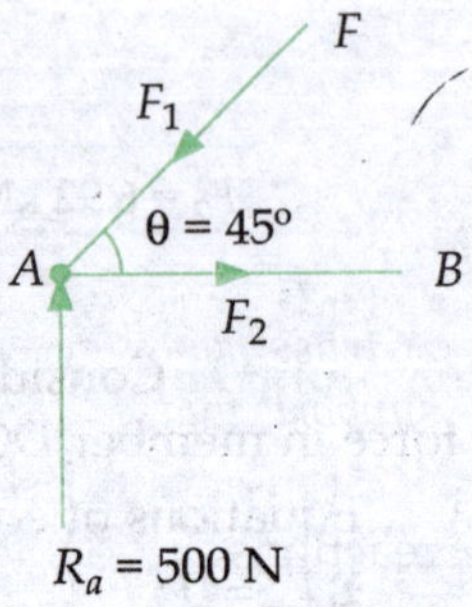

Joint A

Joint A: Consider the free body diagram of joint A with the direction of forces assumed as shown. Start can be made with this joint because there are only two unknown forces F_1 (in member AF) and F_2 (in member AB) acting at this joint.

$$\tan\theta = \frac{BF}{AB} = \frac{2}{2} = 1; \quad \theta = 45^\circ$$

Equations of equilibrium can be written as,

$$\Sigma F_x = 0: \qquad F_2 - F_1 \cos 45^\circ = 0$$

$$\Sigma F_y = 0: \qquad R_a - F_1 \sin 45^\circ = 0$$

$$\therefore \quad F_1 = \frac{R_a}{\sin 45^\circ} = \frac{500}{0.707} = 707.21 \text{ N (compression)}$$

$$F_2 = F_1 \cos 45^\circ$$

$$= 707.21 \times 0.707 = 500 \text{ N (tension)}.$$

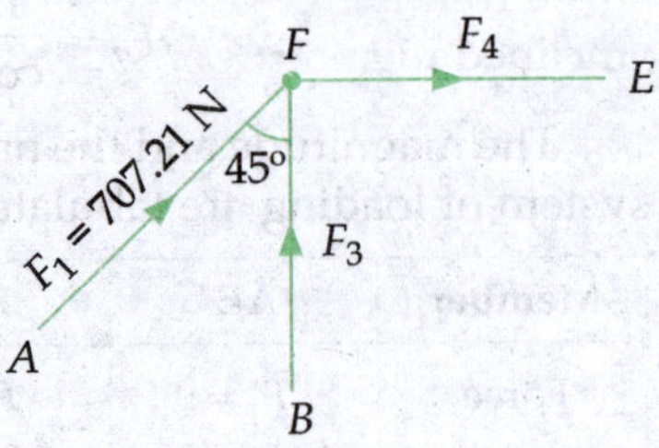

Joint F

Joint F: Equations of equilibrium for the joint F can be written as

$$\Sigma F_x = 0: \qquad F_4 + 707.21 \sin 45^\circ = 0$$

$$\therefore \quad F_4 = -707.21 \times 0.707 = -500 \text{ N}$$

$$\Sigma F_y = 0: \qquad F_3 + 707.21 \cos 45^\circ = 0$$

$$\therefore \quad F_3 = -707.21 \times 0.707 = -500 \text{ N}$$

The – ve sign with the magnitude of forces F_3 and F_4 shows that a wrong choice has been made while assuming their directions. Obviously the assumed direction of the forces in members FB and FE need to be reversed.

Therefore the member FB is a tension member and the member FE is a compression member.

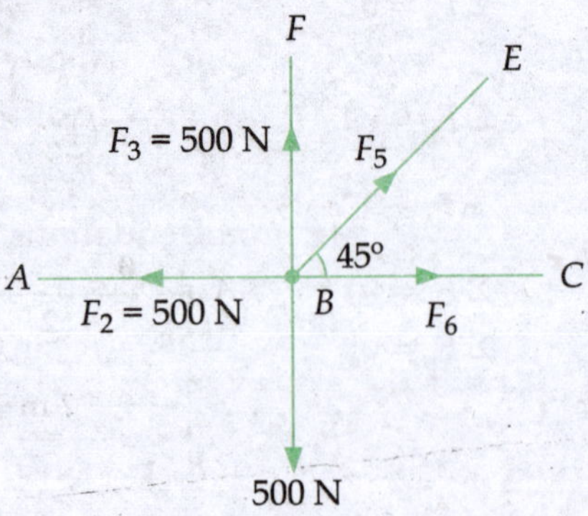

Joint B

Joint B: Equations of equilibrium for the joint B can be written as

$$\Sigma F_x = 0: \qquad F_6 + F_5 \cos 45^\circ - 500 = 0$$

$$\Sigma F_y = 0: \qquad 500 - 500 + F_5 \sin 45^\circ = 0$$

$$\therefore \qquad F_5 = 0$$

Putting $F_5 = 0$ in the above identity, we get

$$F_6 = 500 \text{ N (tension)}$$

The zero value for force F_5 signifies that no force is induced in the member BE.

Since the truss is symmetrical, we have

Force in member CE = force in member BF

= 500 N (tension)

Force in member CD = force in member AB

= 500 N (tension)

Force in member DE = force in member AF

= 707.21 (compression)

The forces in all the members of the given truss for the applied system of loading are tabulated below:

Member	AF, DE	AB, CD	BF, CE	FE	BE	BC
Force	707.21 N	500 N	500 N	500 N	0	500 N
Nature	C	T	T	C	—	T

EXAMPLE 5.7

A truss has been loaded and supported as shown in Fig. 5.15. Make calculations for the reactions at the supports and the forces in the members of the truss.

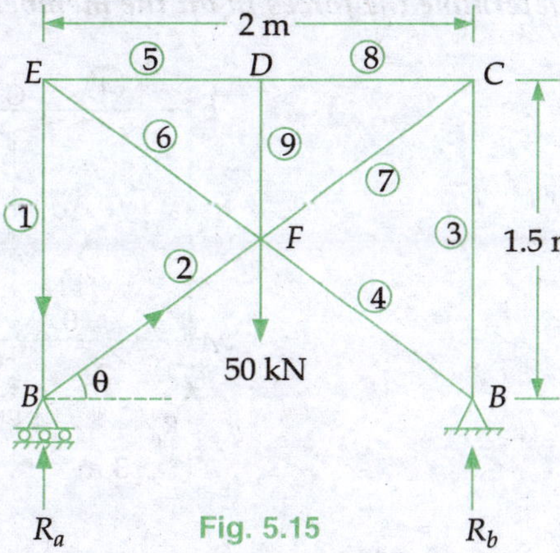

Fig. 5.15

Solution : Both the supports will have only vertical reactions as the external load of 50 kN at the joint F is vertical. Further, due to symmetry

$$R_a = R_b = \frac{50}{2} = 25 \text{ kN}$$

From the geometry of the truss, the inclination of the inclined members with the horizontal is

$$\tan\theta = \frac{0.75}{1} = \frac{3}{4}$$

That gives: $\sin\theta = \dfrac{3}{5}$ and $\cos\theta = \dfrac{4}{5}$

Joint A: From the equations of equilibrium,

$\Sigma F_x = 0:\quad F_2 \cos\theta = 0;$

$\therefore\quad F_2 = 0$ (no force)

$\Sigma F_y = 0:\quad R_a - F_1 = 0;$

$\therefore\quad F_1 = R_a = 25$ **kN** (compressive)

Joint B: From the equations of equilibrium,

$\Sigma F_x = 0:\quad F_4 \cos\theta = 0;$

$\therefore\quad F_4 = 0$ (no force)

$\Sigma F_y = 0:\quad R_b - F_3 = 0;$

$\therefore\quad F_3 = R_b = 25$ **kN** (compressive)

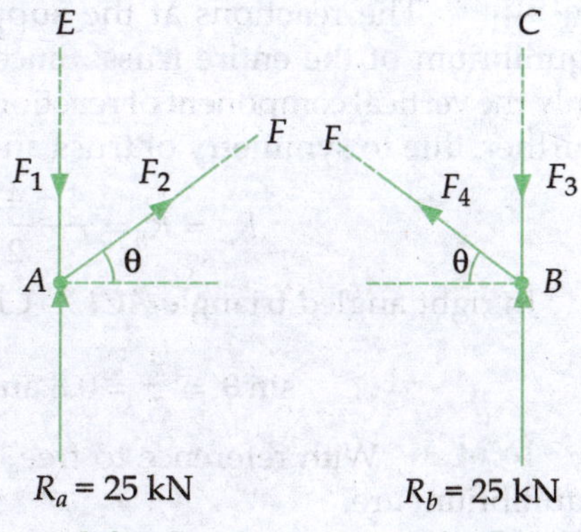

Joint A　　Joint B

Joint C: From the equations of equilibrium,

$\Sigma F_x = 0:\quad F_6 \cos\theta - F_5 = 0$

$\Sigma F_y = 0:\quad F_1 - F_6 \sin\theta = 0$

$$\therefore\quad F_6 = \frac{F_1}{\sin\theta} = \frac{25}{3/5} = \mathbf{41.67\ kN} \text{ (tensile)}$$

and $\quad F_5 = F_6 \cos\theta = 41.67 \times \dfrac{4}{5} = 33.34$ kN (compression)

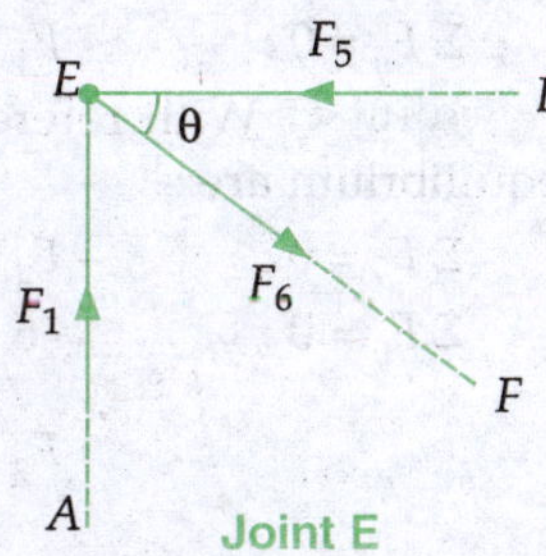

Joint E

Likewise considering equilibrium of joint C,

$$F_7 = 41.67 \text{ kN (tension)}$$

and $$F_8 = 33.34 \text{ kN (compression)}$$

Joint *D*: Considering equilibrium of forces in vertical direction,

$$F_9 = 0 \text{ (no force)}$$

The forces in various members of the truss are tabulated below:

Members	AE, BC	DE, DC	FE, FC	FD, FA, FB
Force	25 kN	33.34 kN	41.67 kN	0 (no force)
Nature	compressive	compressive	tensile	---

EXAMPLE 5.8

Determine the forces in all the members of the truss loaded and supported as shown in Fig. 5.16.

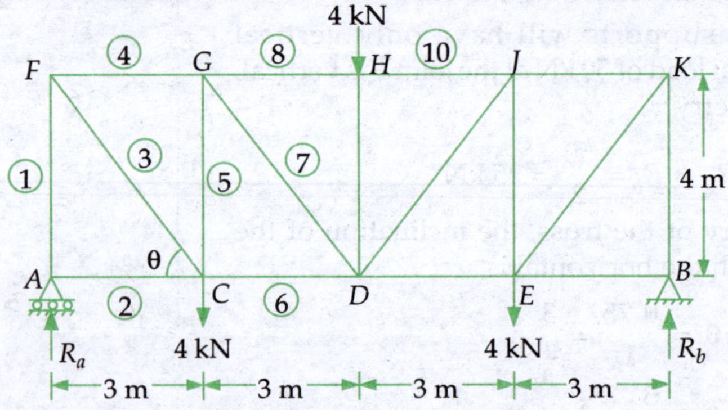

Fig. 5.16

Solution : The reactions at the supports can be determined by considering equilibrium of the entire truss. Since all the three external loads are vertical, only the vertical component of reaction at the hinged end B need to be considered. Further, due to symmetry of truss and loading

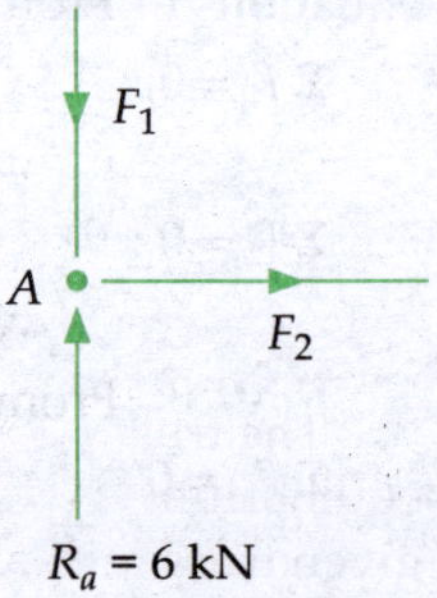

Joint A

$$R_a = R_b = \frac{4+4+4}{2} = 6 \text{ kN}$$

In right angled triangle *ACF*, $CF = \sqrt{3^2 + 4^2} = 5$ m

$$\therefore \quad \sin\theta = \frac{4}{5} = 0.8 \text{ and } \cos\theta = \frac{3}{5} = 0.6$$

Joint *A*: With reference to free body diagram of joint *A*, the equations of equilibrium are:

$$\Sigma F_x = 0: \quad F_2 = 0$$

$$\Sigma F_y = 0: \quad F_1 = 6 \text{ kN (compression)}$$

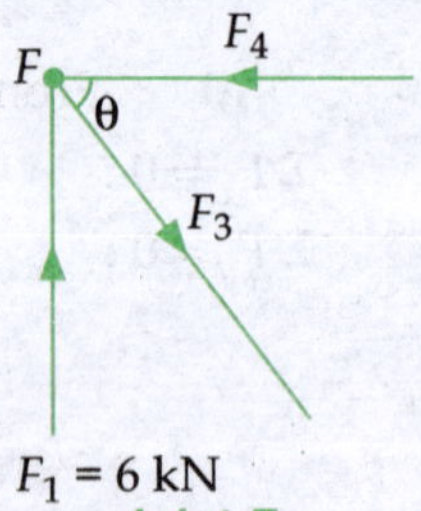

Joint F

Joint *F*: With reference to free body diagram of joint *F*, the equations of equilibrium are :

$$\Sigma F_x = 0: \quad -F_4 + F_3 \cos\theta = 0$$

$$\Sigma F_y = 0: \quad 6 - F_3 \sin\theta = 0$$

$$\therefore \quad F_3 = \frac{6}{\sin\theta} = \frac{6}{0.8} = 7.5 \text{ kN (tension)}$$

and $\quad F_4 = F_3 \cos\theta = 7.5 \times 0.6$

$= 4.5\text{ kN}$ (compression)

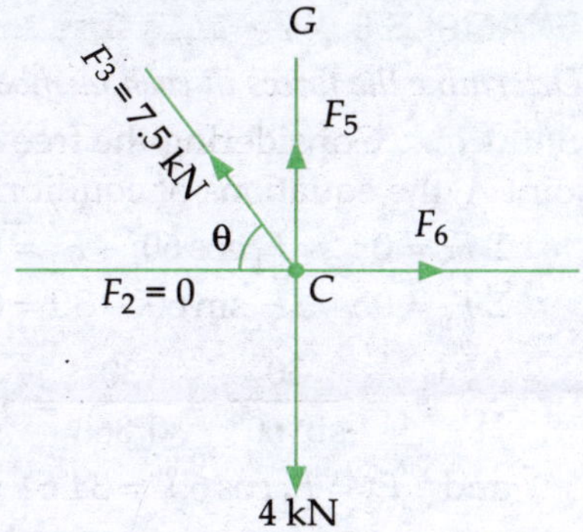

Joint C

Joint C: With reference to free body diagram of joint C, the equations of equilibrium are:

$F_x = 0: \quad F_6 - F_3 \cos\theta = 0$

$\therefore \quad F_6 = F_3 \cos\theta = 7.5 \times 0.6$

$= 4.5\text{ kN}$ (tension)

$F_y = 0: \quad F_5 + F_3 \sin\theta - 4 = 0$

$\therefore \quad F_5 = 4 - F_3 \sin\theta$

$= 4 - 7.5 \times 0.8 = -2\text{ kN}$

The negative sign with the magnitude of force F_5 shows that a wrong choice has been made while assuming its direction. Obviously the assumed direction of force in member CG need to be reversed.

Therefore, $F_5 = 2$ kN (compression) and CG is a compression member.

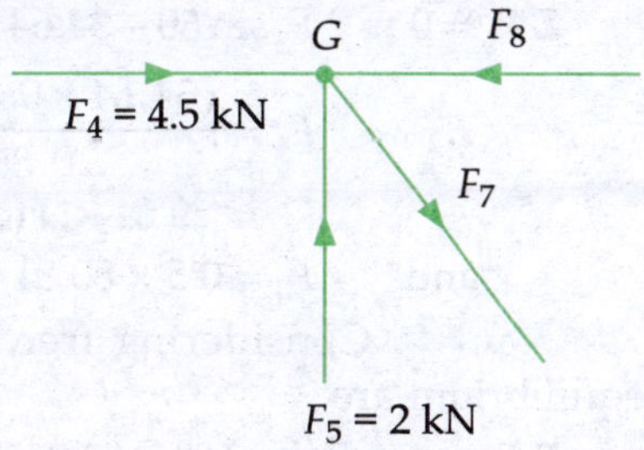

Joint G

Joint G: With reference to free body diagram of joint G, the equations of equilibrium are :

$\Sigma F_x = 0: \quad 4.5 - F_8 + F_7 \cos\theta = 0$

$\Sigma F_y = 0: \quad 2 - F_7 \sin\theta = 0$

$\therefore \quad F_7 = \dfrac{2}{\sin\theta} = \dfrac{2}{0.8} = 2.50\text{ kN}$ (tension)

and $\quad F_8 = 4.5 + F_7 \cos\theta$

$= 4.5 + 2.50 \times 0.6 = 6\text{ kN}$ (compression)

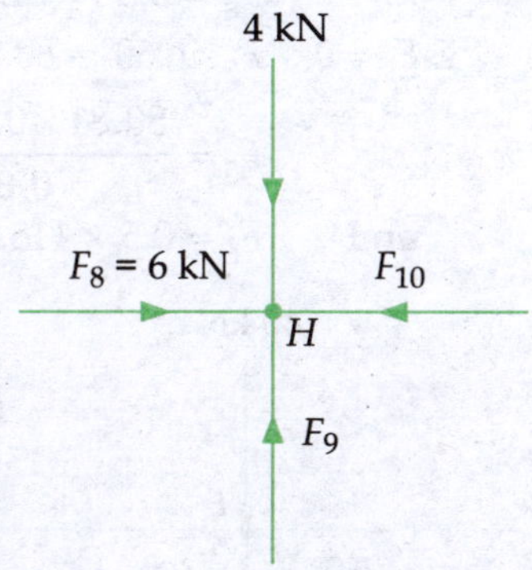

Joint H

Joint H: With reference to free body diagram of joint H, the equations of equilibrium are :

$\Sigma F_x = 0: \quad F_{10} - 6 = 0$

$\therefore \quad F_{10} = 6\text{ kN}$ (compression)

$\Sigma F_y = 0: \quad F_9 - 4 = 0$

$\therefore \quad F_9 = 4\text{ kN}$ (compression)

The truss is symmetrical and as such, there is no need to consider equilibrium of the joints on the left. Forces in all the members of the given truss for the applied system of loading are tabulated below:

Member	Force	Nature
AF, BK	6 kN	compressive
AC, BE	0	—
CF, EK	7.5 kN	tensile
FG, JK	4.5 kN	compressive
GC, JE	2 kN	compressive
CD, DE	4.5 kN	tensile
DG, DJ	2.5 kN	tensile
GH, HJ	6 kN	compressive
HD	4 kN	compressive

EXAMPLE 5.9

Determine the forces in each member of the cantilever truss loaded as shown in Fig. 5.17.

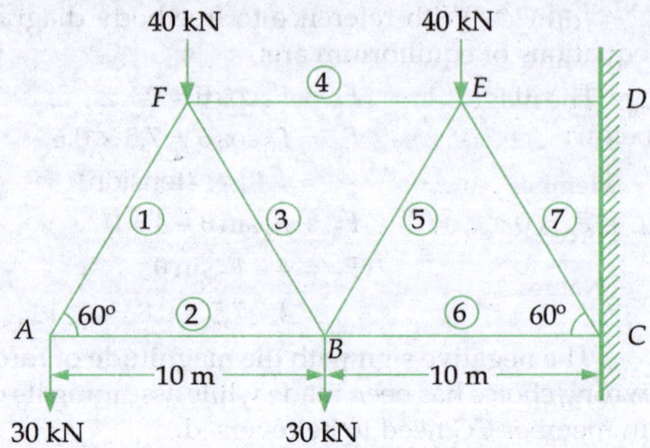

Fig. 5.17

Solution : Considering the free body diagram of joint A, the equations of equilibrium are :

$$\Sigma F_x = 0: \quad F_1 \cos 60^\circ - F_2 = 0$$

$$\Sigma F_y = 0: \quad F_1 \sin 60^\circ - 30 = 0$$

$$\therefore \quad F_1 = \frac{30}{\sin 60^\circ} = \frac{30}{0.866} = \textbf{34.64 kN} \text{ (tension)}$$

and $\quad F_2 = F_1 \cos 60^\circ = 34.64 \times 0.5$

$$= \textbf{17.32 kN} \text{ (compression)}$$

Joint F: With reference to free body diagram of joint F, the equations of equilibrium are :

$$\Sigma F_x = 0: \quad F_4 - F_3 \cos 60^\circ - 34.64 \cos 60^\circ$$

or $\quad F_4 = 0.5\, F_3 + 17.32$

$$\Sigma F_y = 0: \quad F_3 \sin 60 - 34.64 \sin 60^\circ - 40 = 0$$

$$\therefore \quad F_3 = \frac{(34.64 \times 0.866) + 40}{0.866}$$

$$= \textbf{80.81 kN} \text{ (compression)}$$

and $\quad F_4 = 0.5 \times 80.81 + 17.32 = \textbf{57.72 kN}$ (tension)

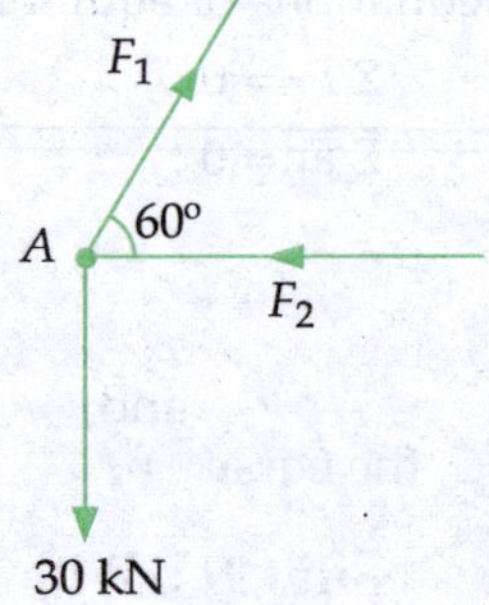

Joint A

Joint B: Considering free body diagram of joint B, the equations of equilibrium are:

$$\Sigma F_x = 0: \; F_5 \cos 60^\circ + 80.81 \cos 60^\circ + 17.32 - F_6 = 0$$

$$F_6 = 0.5\, F_5 + 80.81 \times 0.5 + 17.32 = 0.5\, F_5 + 57.72$$

$$\Sigma F_y = 0: \; F_5 \sin 60^\circ - 80.81 \sin 60^\circ - 30 = 0$$

$$\therefore \quad F_5 = \frac{80.81 \times 0.866 + 30}{0.866} = \textbf{115.45 kN} \text{ (tension)}$$

and $\quad F_6 = 0.5 \times 115.45 + 57.72 = \textbf{115.44}$ (compression)

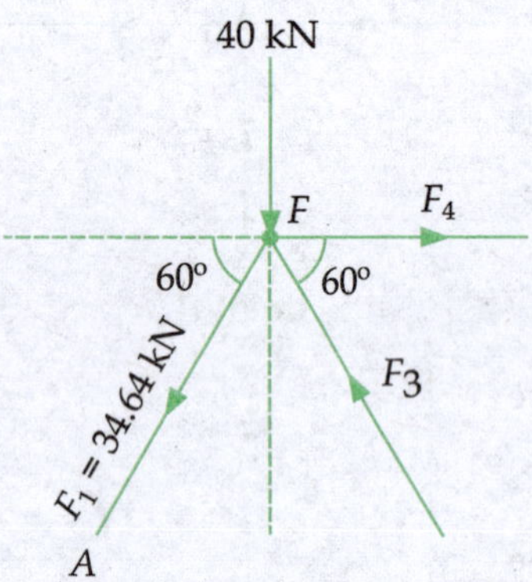

Joint F

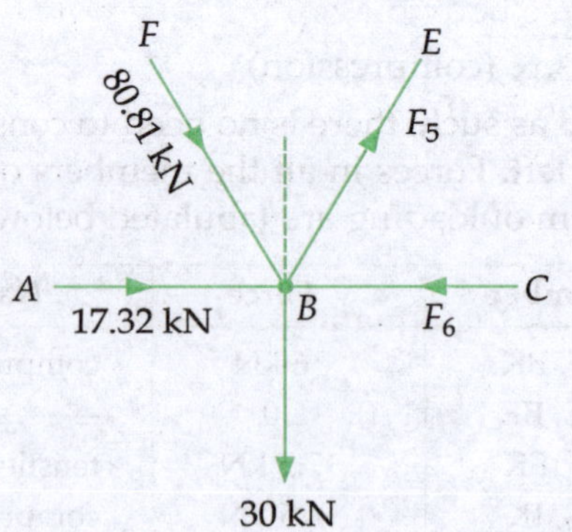

Joint B

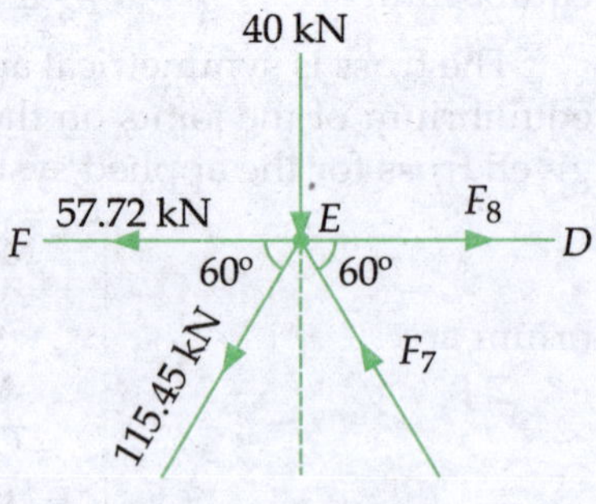

Joint E

Joint E: Considering joint E, the equations of equilibrium are:

$$\Sigma F_x = 0: \; F_8 - 57.72 - F_7 \cos 60^\circ - 115.45 \cos 60^\circ = 0$$

or $\quad F_8 = 57.72 + 0.5\, F_7 + 115.45 \times 0.5 = 0.5\, F_7 + 115.44$

$\Sigma F_y = 0: \quad F_7 \sin 60^\circ - 115.45 \sin 60^\circ - 40 = 0$

$$\therefore \quad F_7 = \frac{115.45 \times 0.866 + 40}{0.866} = 161.62 \text{ kN (compression)}$$

and $\quad F_8 = 0.5 \times 161.62 + 115.44 = 196.25$ kN (tension)

The magnitude and the nature of forces in the various members of the truss have been tabulated below:

Member	AF	AB	BF	FE	BE	BC	CE	ED
Force kN	34.64	17.32	80.81	57.72	115.45	115.44	161.62	196.25
Nature	tensile	compressive	compressive	tensile	tensile	compressive	compressive	tensile

EXAMPLE 5.10

Determine the forces in the various members of the cantilever truss loaded and supported as shown in Fig. 5.18.

Solution : From the geometrical configuration of the truss,

$$BC = \sqrt{2^2 + 1^2} = 2.236 \text{ m}$$

$$\sin\theta = \frac{1}{2.236} = 0.447$$

$$\cos\theta = \frac{2}{2.236} = 0.894$$

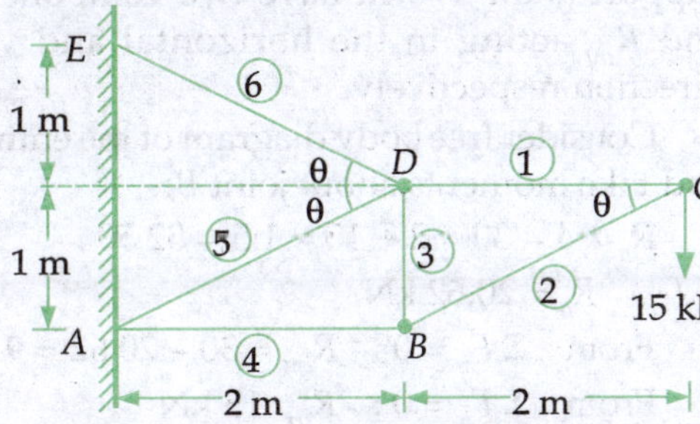

Fig. 5.18

Joint C: Considering the free body diagram of joint C, the equation of equilibrium are:

$\Sigma F_x = 0: \quad F_2 \cos\theta - F_1 = 0$

$\Sigma F_y = 0: \quad F_2 \sin\theta - 15 = 0$

$$\therefore \quad F_2 = \frac{15}{\sin\theta} = \frac{15}{0.447} = 33.56 \text{ kN (compression)}$$

and $\quad F_1 = F_2 \cos\theta = 33.56 \times 0.894 = 30.00$ kN (tension)

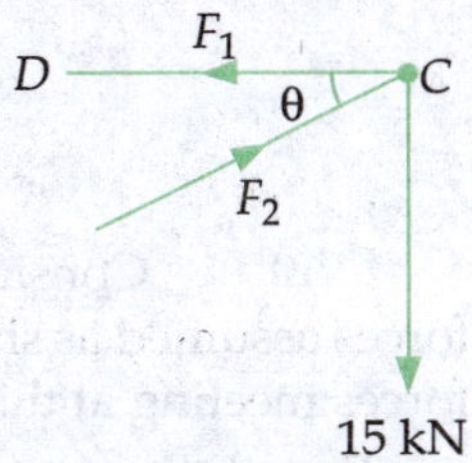

Joint C

Joint B: With reference to free body diagram of joint B, the equations of equilibrium are:

$\Sigma F_x = 0: \quad F_4 - 33.56 \cos\theta$

$\therefore \quad F_4 = 33.56 \times 0.894 = 30.00$ kN (compression)

$\Sigma F_y = 0: \quad F_3 - 33.56 \sin\theta = 0$

$\therefore \quad F_3 = 33.56 \times 0.447 = 15.00$ kN (tension)

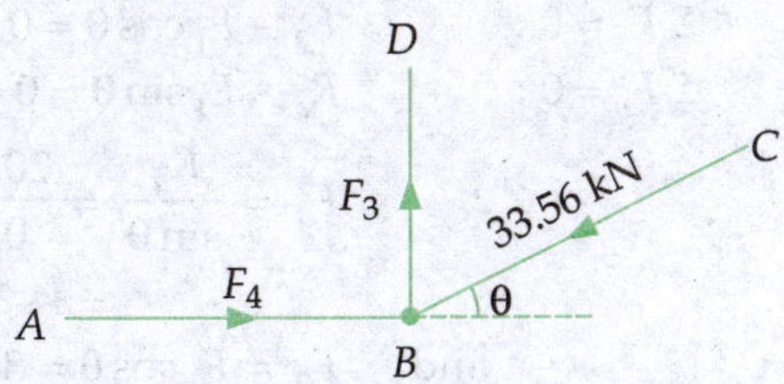

Joint B

Joint D: Considering joint D, the equations of equilibrium are:

$\Sigma F_x = 0: \quad 30 - F_5 \cos\theta - F_6 \cos\theta = 0$

$$\text{or} \quad F_5 + F_6 = \frac{30}{0.894} = 33.56 \quad \text{...(i)}$$

$\Sigma F_y = 0: \quad F_6 \sin\theta - F_5 \sin\theta - 15 = 0$

$$\text{or} \quad F_6 - F_5 = \frac{15}{0.447} = 33.56 \quad \text{...(ii)}$$

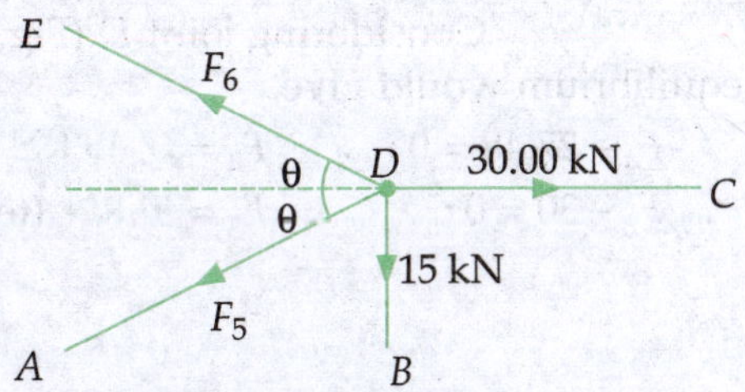

Joint D

From identities (*i*) and (*ii*),

$F_6 = 33.56$ kN (tension) and $F_5 = 0$ (no force)

The magnitude and the nature of forces in the various members of the truss have been tabulated below:

Member	CD	CB	BD	BA	AD	DE
Force	30 kN	33.56 kN	15 kN	30 kN	0	33.56 kN
Nature	tensile	compressive	tensile	compressive	—	tensile

EXAMPLE 5.11

Determine the magnitude and nature of forces in the various members of the triangular truss loaded and supported as shown in Fig. 5.19.

Solution : The roller support (joint A) is frictionless and provides a reaction R_a at right angles to the roller base. The reaction at the hinged support (joint B) will have two components R_{bx} and R_{by} acting in the horizontal and vertical direction respectively.

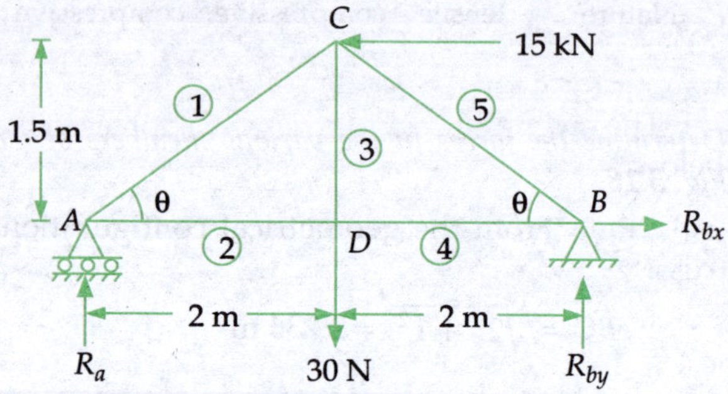

Fig. 5.19

Consider free body diagram of the entire truss and take moments about joint B;

$R_a \times 4 = 30 \times 2 + 15 \times 1.5 = 82.50$;

$R_a = 20.62$ kN

From $\Sigma F_y = 0$: $R_{by} = 30 - 20.62 = 9.38$ kN

From $\Sigma F_x = 0$: R_{bx} 15 kN

Further, from the geometrical configuration of the truss

$$AC = \sqrt{(AD)^2 + (CD)^2} = \sqrt{2^2 + 1.5^2} = 2.5 \text{ m}$$

$$\sin\theta = \frac{CD}{AC} = \frac{1.5}{2.5} = 0.6 \quad \text{and} \quad \cos\theta = \frac{AD}{AC} = \frac{2}{2.5} = 0.8$$

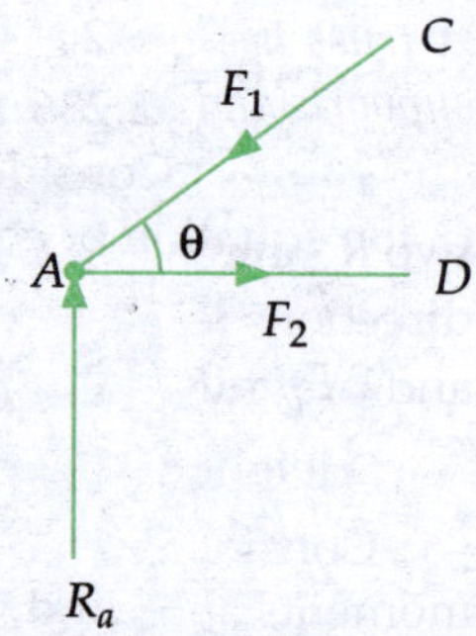

Joint A

Joint A : Consider the free body diagram of joint A with the direction of forces assumed as shown. Start can be made with this joint because of the three forces meeting at this point, only two are unknown.

Equations of equilibrium can be written as

$\Sigma F_x = 0$: $\quad F_2 - F_1 \cos\theta = 0$

$\Sigma F_y = 0$: $\quad R_a - F_1 \sin\theta = 0$

$$\therefore \quad F_1 = \frac{R_a}{\sin\theta} = \frac{20.62}{0.6}$$

$= 34.36$ kN (compression)

and $\quad F_2 = F_1 \cos\theta = 34.36 \times 0.8$

$= 27.48$ kN (tensile)

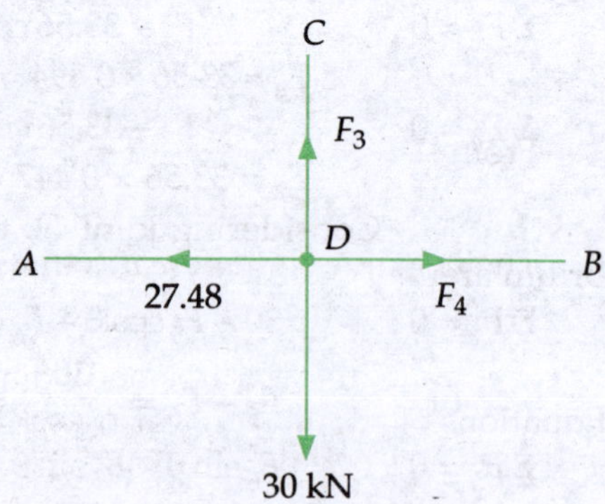

Joint D

Joint D: Considering joint D (Fig. 5.52), the equations of equilibrium would give

$F_4 - 27.48 = 0$: $\quad F_4 = 27.48$ kN (tensile)

$F_3 - 30 = 0$: $\quad F_3 = 30$ kN (tensile)

Joint B: Considering joint B, the equations of equilibrium are :

$\Sigma F_x = 0:\quad 15 - 27.48 + F_5 \cos\theta = 0$

$$\therefore \quad F_5 = \frac{27.48 - 15}{\cos\theta}$$

$$= \frac{12.48}{0.8} = \textbf{15.60 kN} \text{ (compression)}$$

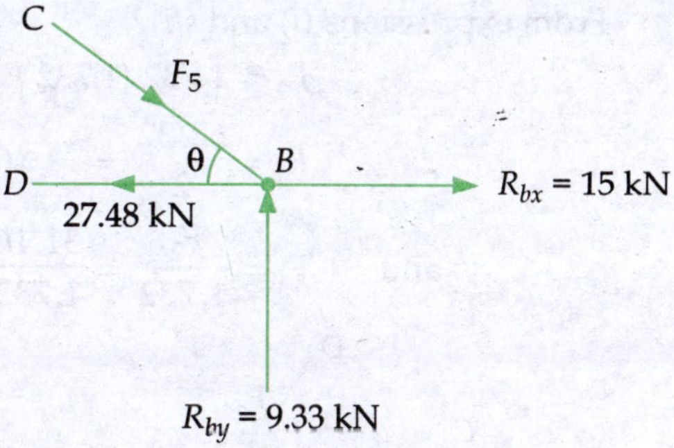

Joint B

Check:

$\Sigma F_y = 0:\quad F_5 \sin\theta - R_{by} = 0$

$$F_5 = \frac{R_{by}}{\sin\theta} = \frac{9.38}{0.6} = \textbf{15.63 kN} \text{ (compression)}$$

The magnitude and the nature of forces in the various members of the truss have been tabulated below:

Member	AD	AC	DC	DB	BC
Force	27.48 kN	34.36 kN	30 kN	27.48 kN	15.60 kN
Nature	T	C	T	T	C

EXAMPLE 5.12

A truss has been loaded and supported as shown in Fig. 5.20. Make calculations for the reactions at the supports and the forces in the members of the truss.

Solution : The reaction at the hinged support (joint A) will have two R_{ax} and R_{ay} components acting in the horizontal and vertical directions respectively. The roller support (joint B) is frictionless and provides a reaction R_b at right angles to the roller base.

Height of the truss $= 2\tan 60^\circ = 3.464$ m

Consider free body diagram of the entire truss and take moments about joint A;

$$R_b \times 4 = (10 \times 2) + (15 \times 3.464)$$

$$= 20 + 51.96 = 71.96$$

$$\therefore \quad R_b = \frac{71.96}{4} = 17.99 \text{ kN}$$

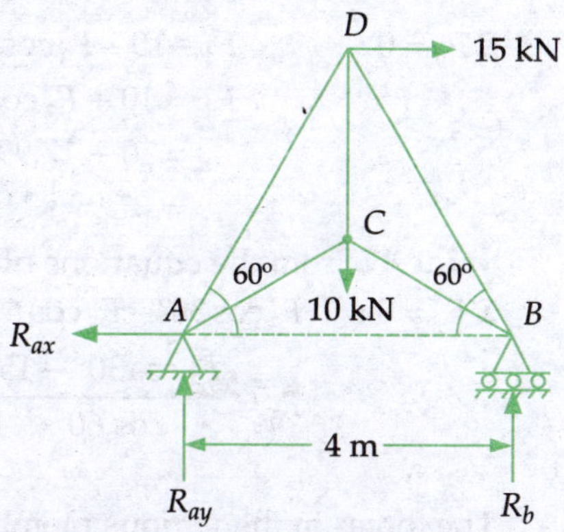

Fig. 5.20

From $\Sigma F_y = 0$;

$$R_{ay} = 10 - 17.99 = -7.99 \text{ kN}$$

The negative sign indicates that the reaction R_{ay} is acting downwards.

From $\Sigma F_x = 0$; $R_{ax} = 15$ kN

Joint B : Consider free body diagram of joint B with the direction of forces assumed as shown. Equations of equilibrium can be written as :

$\Sigma F_x = 0:\quad F_2 \cos 60^\circ - F_1 \cos 30^\circ = 0$

or $\quad 0.5\, F_2 - 0.866\, F_1;\; F_2 = 1.732\, F_1 \quad$ ...(*i*)

$\Sigma F_y = 0:\quad -F_2 \sin 60^\circ + F_1 \sin 30^\circ + R_b = 0$

or $\quad -0.866\, F_2 + 0.5\, F_1 + 17.99 = 0$

or $\quad F_1 = 1.732\, F_2 - 35.98 \quad$ ...(*ii*)

From expressions (*i*) and (*ii*)

$$F_2 = 1.732\,(1.732\,F_2 - 35.98) = 3F_2 - 62.32$$

$$\therefore \quad F_2 = \frac{62.32}{2} = \textbf{31.16 kN} \text{ (compressive)}$$

$$\text{and} \quad F_1 = \frac{F_2}{1.732} = \frac{31.16}{1.732} = \textbf{17.99 kN} \text{ (tensile)}$$

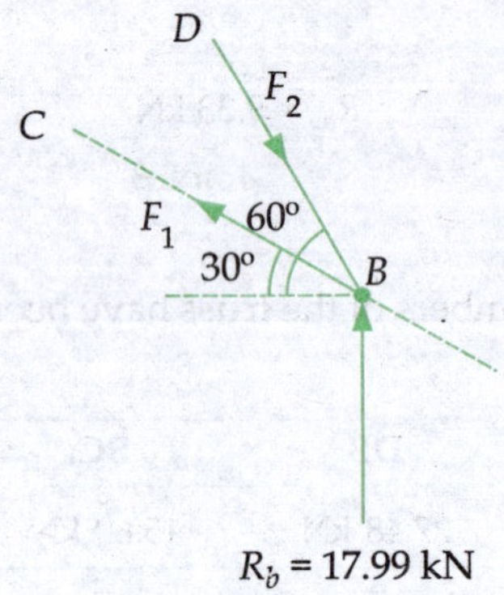

Joint B

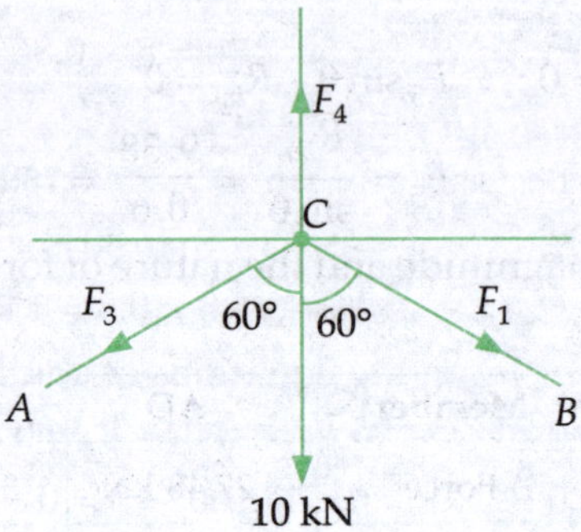

Joint C

Joint *C*: From the equations of equilibrium,

$\Sigma F_x = 0:$ $\quad F_1 \sin 60^\circ - F_3 \sin 60^\circ = 0$

$\therefore \quad F_3 = F_1 = \textbf{17.99 kN}$ (tensile)

$\Sigma F_y = 0:$ $\quad F_4 - 10 - F_3 \cos 60^\circ - F_1 \cos 60^\circ = 0$

$\therefore \quad F_4 = 10 + F_3 \cos 60^\circ + F_1 \cos 60^\circ$

$= 10 + 17.99 \times 0.5 + 17.99 \times 0.5$

$= \textbf{27.99 kN}$ (tensile)

Joint *A*: From the equations of equilibrium

$\Sigma F_x = 0:$ $\quad F_3 \cos 30^\circ - F_5 \cos 60^\circ - 15 = 0$

$$\therefore \quad F_5 = \frac{F_3 \cos 30^\circ - 15}{\cos 60^\circ} = \frac{17.99 \times 0.866 - 15}{0.5}$$

$= \textbf{1.16 kN}$ (compressive)

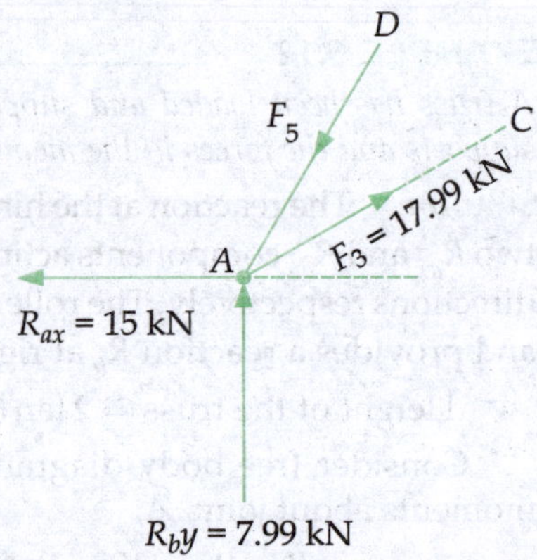

Joint A

The forces in the various members of the truss are tabulated below :

Member	AC	AD	CD	BC	BD
Force	17.99 kN	1.16 kN	27.99 kN	17.99 kN	31.16 kN
Nature	tensile	compressive	tensile	tensile	compressive

EXAMPLE 5.13

Make calculations for the forces in each member of the truss loaded and supported as shown in Fig. 5.21. All members of the truss are 2 m in length.

Solution : Let *T* be the tension in rope ; R_{bx} and R_{by} be the reactions at the hinged support as shown in the figure.

Consider equilibrium of the entire truss

$M_b = 0: \quad T \times DB - 2 \times CB - 3 \times AB = 0$

$$\therefore \quad T = \frac{2 \times CB + 3 \times AB}{DB}$$

$$= \frac{2 \times 2 + 3 \times 4}{2} = 8 \text{ kN}$$

$\Sigma F_x = 0: \quad T\cos 30 - R_{bx} = 0$

$\therefore \quad R_{bx} = T\cos 30 = 8 \times 0.866$

$= 6.93$ kN

$\Sigma F_y = 0: \quad R_{by} + T\sin 30^\circ - 2 - 3 = 0$

$\therefore \quad R_{by} = 2 + 3 - 8\sin 30^\circ$

$= 2 + 3 - 8 \times 0.5 = 1$ kN

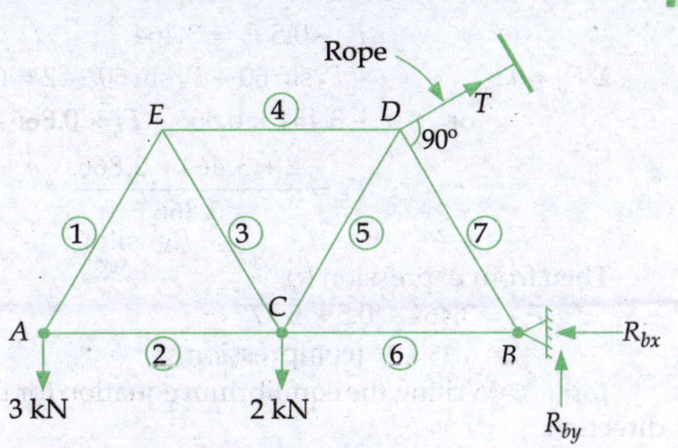

Fig. 5.21

Joint A: From the equations of equilibrium

$\Sigma F_x = 0: \quad F_1 \cos 60^\circ - F_2 = 0$

$\Sigma F_y = 0: \quad F_1 \sin 60^\circ - 3 = 0$

$$\therefore \quad F_1 = \frac{3}{\sin 60^\circ} = \textbf{3.464 kN} \text{ (tensile)}$$

and $\quad F_2 = F_1 \cos 60^\circ = 3.464 \times 0.5$

$= \textbf{1.732 kN}$ (compression)

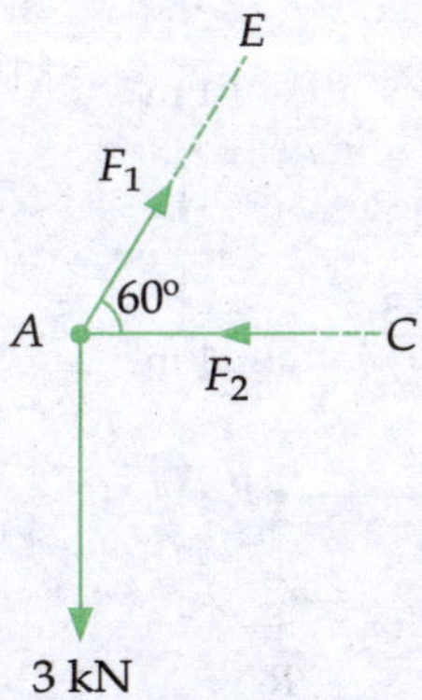

Joint A

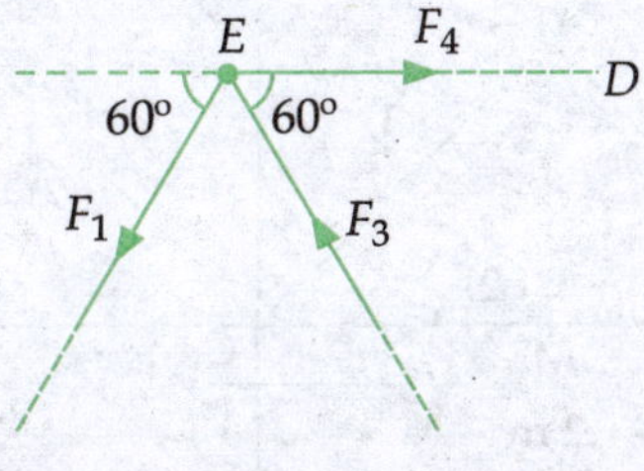

Joint E

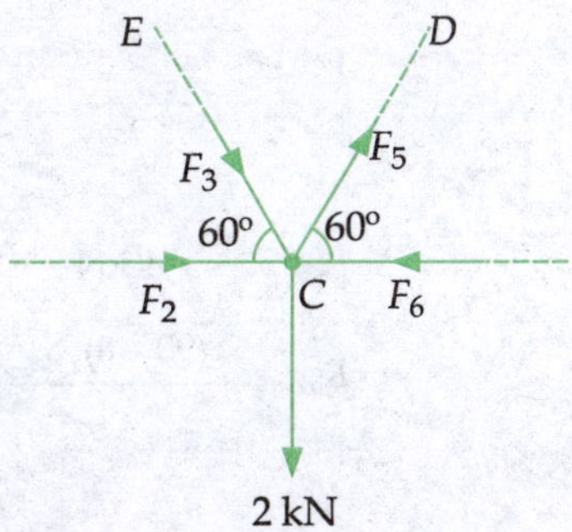

Joint C

Joint E: From the equations of equilibrium

$\Sigma F_x = 0: \quad F_4 - F_1 \cos 60^\circ - F_3 \cos 60^\circ = 0$

$\Sigma F_y = 0: \quad F_1 \sin 60^\circ + F_3 \sin 60^\circ = 0$

$\therefore \quad F_3 = F_1 = \textbf{3.464 kN}$ (compression)

and $\quad F_4 = (F_1 + F_3)\cos 60^\circ$

$= (3.464 + 3.464) \times 0.5 = \textbf{3.464 kN}$ (tensile)

Joint C: Writing the equations of equilibrium,

$\Sigma F_x = 0: \quad F_2 - F_6 + F_3 \cos 60^\circ + F_5 \cos 60^\circ = 0$

or $\quad 1.732 - F_6 + 3.464 \times 0.5 + 0.5\, F_5 = 0$

$$F_6 - 0.5\,F_5 = 3.464 \qquad \ldots(i)$$

$\Sigma F_y = 0:$ $\qquad -F_3 \sin 60 + F_5 \sin 60^\circ - 2 = 0$

or $\qquad -3.464 \times 0.866 + F_5 \times 0.866 = 2$

$$\therefore \quad F_5 = \frac{2 + 3.464 \times 0.866}{0.866}$$

$$= 5.77 \text{ kN (tensile)}$$

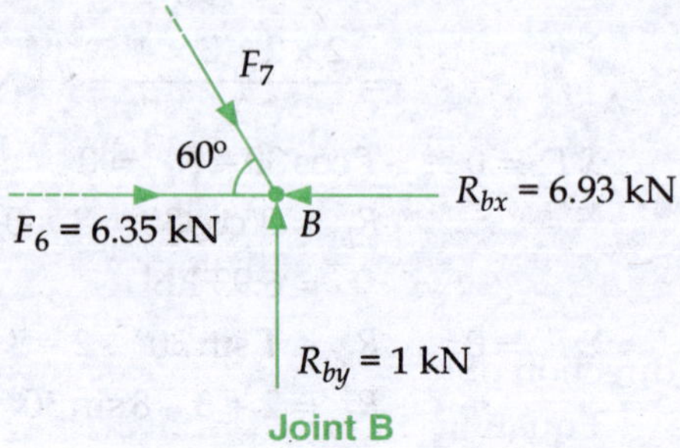

Joint B

Then from expression (*i*)

$$F_6 = 3.464 + 0.5 \times 5.77$$

$$= 6.35 \text{ kN (compression)}$$

Joint B: Writing the equilibrium equation for the vertical direction:

$$\Sigma F_y = 0: \quad F_7 \sin 60^\circ - 1 = 0$$

$$\therefore \quad F_7 = \frac{1}{0.866} = 1.155 \text{ kN (compression)}$$

EXAMPLE 5.14

For the truss loaded as shown in Fig. 5.22, make calculations for the reactions at the supports and forces set up in each member of the truss.

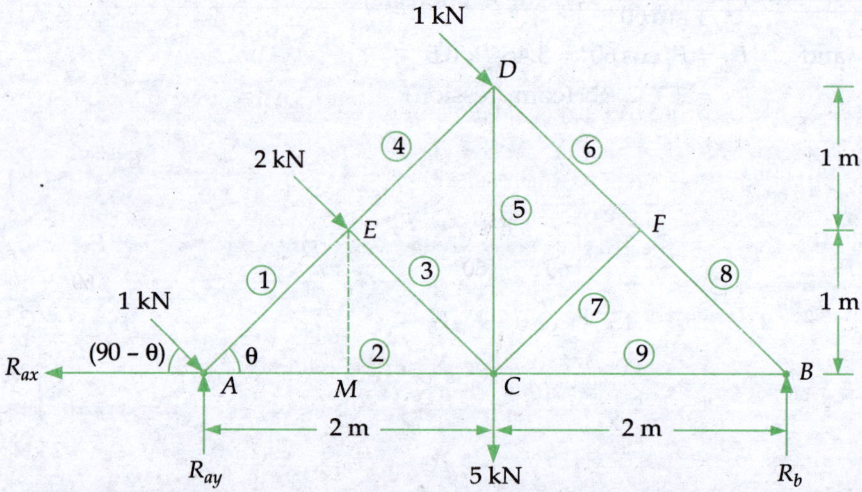

Fig. 5.22

Solution : $AD = \sqrt{(AC)^2 + (CD)^2} = \sqrt{2^2 + 2^2} = \sqrt{8} = 2.828$ m

$$AE = \frac{AD}{2} = 1.414 \text{ m}$$

$$\sin\theta = \frac{EM}{AE} = \frac{1}{1.414} = 0.707$$

$$\cos\theta = \frac{AM}{AE} = \frac{1}{1.414} = 0.707$$

Applying equilibrium conditions for the entire truss,

$\Sigma M_a = 0:$ $\qquad -2 \times AE - 1 \times AD + R_b \times AB - 5 \times AC = 0$

or $\qquad -2 \times 1.414 - 1 \times 2.828 + R_b \times 4 - 5 \times 2$

$$\therefore \quad R_b = \frac{2 \times 1.414 + 1 \times 2.828 + 10}{4} = 3.914 \text{ kN}$$

$\Sigma F_x = 0: \quad 1 \cos(90 - \theta) + 2\cos(90 - \theta) + 1\cos(90 - \theta) - R_{ax} = 0$

or $\quad R_{ax} = (1 + 2 + 1)\sin\theta = 4 \times 0.707 = 2.828$ kN

$\Sigma F_y = 0: \quad R_{ay} - 5 + R_b - 1\sin(90 - \theta)$

$\qquad - 2\sin(90 - \theta) - 1\sin(90 - \theta)$

or $\quad R_{ay} = 5 - R_b + (1 + 2 + 1)\cos\theta$

$\qquad = 5 - 3.914 + 4 \times 0.707$

$\qquad = 3.914$ kN

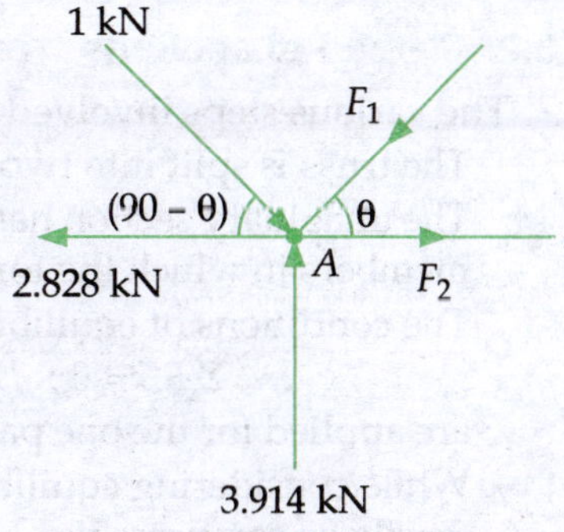

Joint A

Joint A: Consider the free body diagram of joint A with the direction of forces assumed as shown.

Equations of equilibrium are:

$\Sigma F_x = 0: \quad F_2 - F_1\cos\theta + 1\cos(90 - \theta) - 2.828 = 0$

or $\quad F_2 = F_1\cos\theta - \sin\theta + 2.828$

$\Sigma F_y = 0: \quad -F_1\sin\theta - 1\sin(90 - \theta) + 3.914$

$$\therefore \quad F_1 = \frac{3.914 - \cos\theta}{\sin\theta} = \frac{3.914 - 0.707}{0.707} = \textbf{4.536 kN} \text{ (compression)}$$

and $\quad F_2 = 4.536 \times 0.707 - 0.707 + 2.828 = 5.328$ (tension)

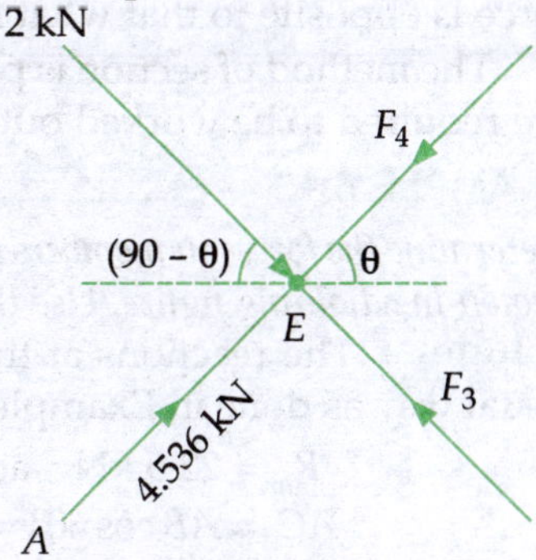

Joint E

Joint E: With reference to free body diagram of joint E, the equations of equilibrium along the direction AD and perpendicular to it are :

$F_4 - 4.536 = 0; \quad F_4 =$ **4.536 kN** (compression)

$F_3 - 2 = 0; \quad F_2 =$ **2 kN** (compression)

Joint D: With reference to free body diagram of joint D, the equations of equilibrium are:

$\Sigma F_x = 0: \quad 1\cos(90 - \theta) + 4.536\cos\theta - F_6\cos\theta$

$$\text{or} \quad F_6 = \frac{\sin\theta + 4.536\cos\theta}{\cos\theta} = \frac{0.707 + 4.536 \times 0.707}{0.707}$$

$\qquad =$ **5.536 kN** (compression)

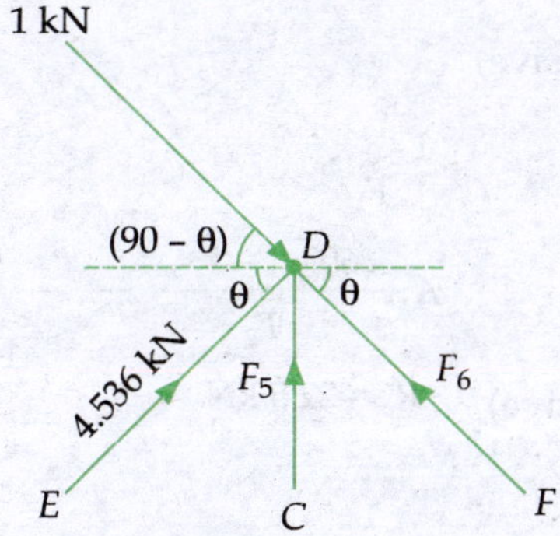

Joint D

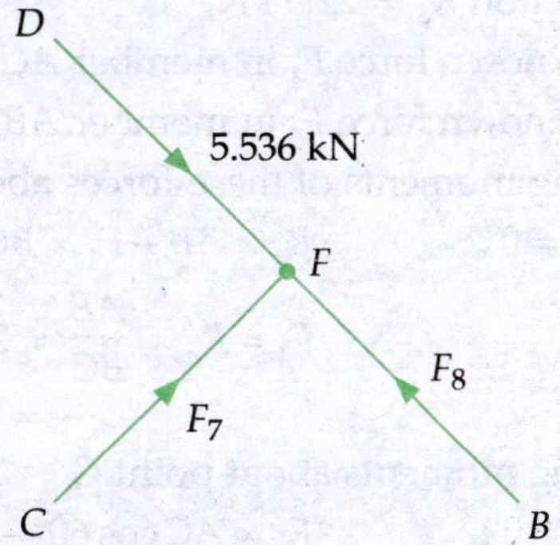

Joint F

It is apparent from free body diagram of joint F, that for its static equilibrium

$$F_8 = 5.536 \text{ kN (compression)} \quad \text{and} \quad F_7 = 0 \text{ (no force)}$$

Applying conditions of equilibrium to point B, we have

$\Sigma F_x = 0$: $\quad F_9 - 5.536 \cos\theta = 0$

$\therefore \quad F_9 = 5.536 \times 0.707 = 3.914$ kN (tension)

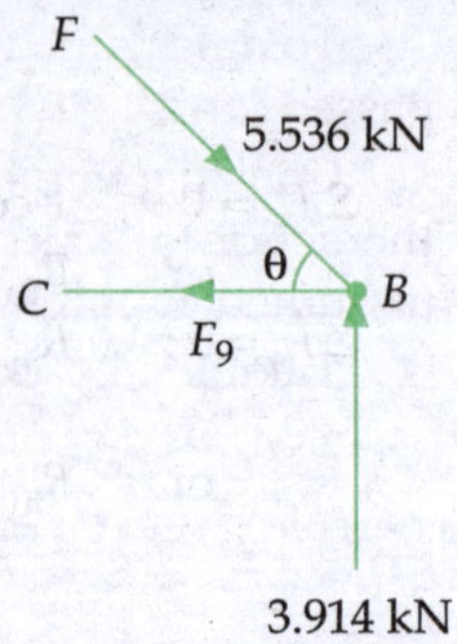

Joint B

5.3.2. Method of sections

The various steps involved are:

(*i*) The truss is split into two parts by passing an imaginary section.

(*ii*) The imaginary section has to be such that it does not cut more than three members in which the forces are to be determined.

(*iii*) The conditions of equilibrium

$$\Sigma F_x = 0; \quad \Sigma F_y = 0 \text{ and } \Sigma M = 0$$

are applied for the one part of the truss and the unknown force in the member is determined.

(*iv*) While considering equilibrium, the nature of force in any member is chosen arbitrarily to be tensile or compressive.

If the magnitude of a particular force comes out positive, the assumption in respect of its direction is correct. However, if the magnitude of the forces comes out negative, the actual direction of the force is opposite to that what has been assumed.

The method of section is particularly convenient when the forces in a few members of the frame are required to be worked out.

EXAMPLE 5.15

Determine the forces in members AC and AB of a simple triangular truss with the loading and support system shown in adjoining figure. Use the method of sections.

Solution : The reactions at the supports are determined in the usual way as done in Example 5.2.

$$R_a = 22.5 \text{ kN} \quad \text{and} \quad R_b = 7.5 \text{ kN}$$

$$AC = AB \cos 60° = 5 \times 0.5 = 2.5 \text{ m}$$

$$BC = AB \sin 60° = 5 \times 0.866 = 4.33 \text{ m}$$

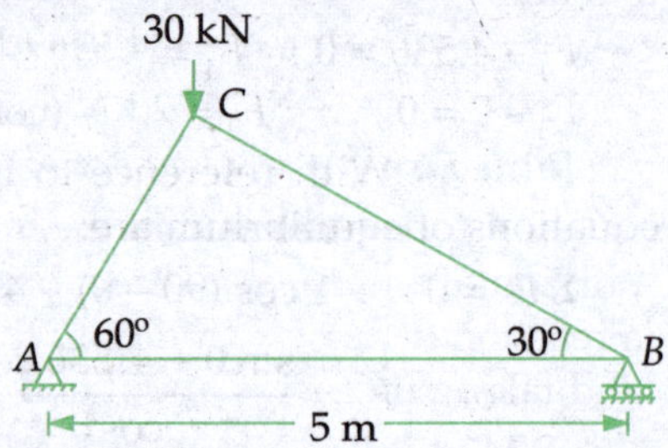

Fig. 5.23

Let the section 1-1 be passed which divides the truss in two parts. The left part of the truss is in equilibrium under the action of :

(*i*) reaction $R_a = 22.5$ kN,

(*ii*) unknown force F_1 in member AC (assumed compressive)

(*iii*) unknown force F_2 in member AB (assumed tensile).

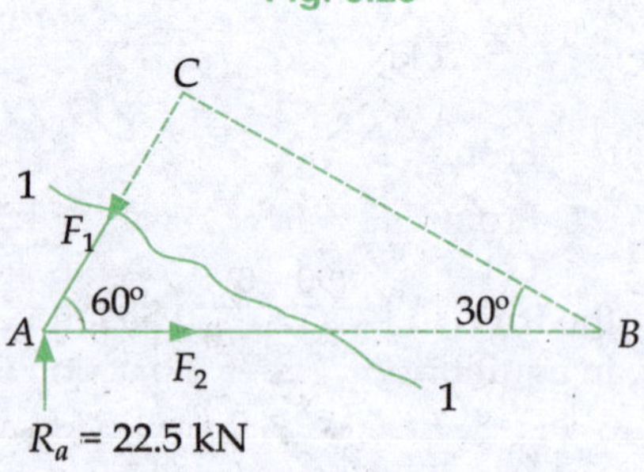

Taking moments of these forces about point B,

$\Sigma M_b = 0$: $\quad R_a \times AB - F_1 \times BC = 0$

$$\therefore \quad F_1 = R_a \times \frac{AB}{BC} = 22.5 \times \frac{5}{4.33}$$

$$= 22.98 \text{ kN (compressive)}$$

Taking moments about point C,

$\Sigma M_c = 0$: $\quad R_a \times AC \cos 60° - F_2 \times AC \sin 60° = 0$

$$\therefore \quad F_2 = R_a \times \frac{AC \cos 60°}{AC \sin 60°} = R_a \times \frac{\cos 60°}{\sin 60°} = 22.5 \times \frac{0.5}{0.866} = 12.99 \text{ kN (tensile)}$$

To determine force in member BC, another section 2-2 is passed which cuts the members AB and BC.

Consider equilibrium of the right part of the truss under the action of force F_2 = 12.99 kN, force F_3 in member BC (assumed compressive) and reaction R_b = 7.5 kN.

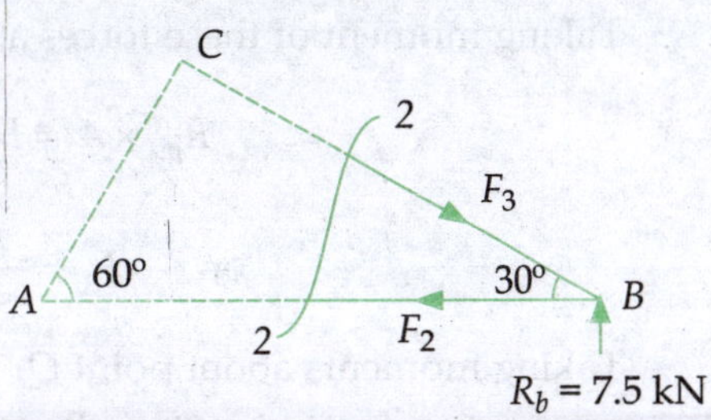

Taking moments about point A,

$$\Sigma M_a = 0: \quad -R_b \times AB + F_3 \times AC = 0$$

$$\therefore \quad F_3 = R_b \times \frac{AB}{AC} = 7.5 \times \frac{5}{2.5} = 15 \text{ kN (compressive)}$$

EXAMPLE 5.16

A frame PQRSTU is hinged to a rigid support at P and is simply supported at T. It is loaded as indicated below in Fig. 5.24. Estimate the magnitude and nature of forces in the members PQ and UT.

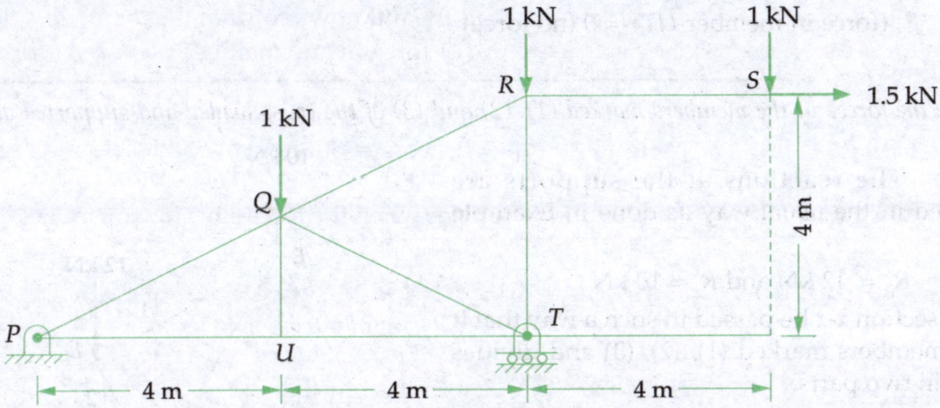

Fig. 5.24

Solution : The reaction at the hinged support P can have two components R_{px} and R_{py} acting in the horizontal and vertical directions respectively. Roller or simple support T is frictionless and provides a reaction R_t at right angles to the roller base. Considering the free body diagram of the entire truss and taking moments about end P,

$$\Sigma M_p = 0: \quad 1 \times 4 + 1 \times 8 + 1 \times 12 + 1.5 \times 4 - R_t \times 8 = 0$$

$$R_t = \frac{4 + 8 + 12 + 6}{8} = 3.75 \text{ kN}$$

From $\Sigma F_y = 0$: $\quad R_{py} = (1 + 1 + 1) - 3.75 = -0.75 \text{ kN} \downarrow$

From $\Sigma F_x = 0$: $\quad R_{px} = 1.5 \text{ kN} \leftarrow$

Let a section 1 – 1 be passed which divides the truss in two parts. The left part of the truss is in equilibrium under the action of

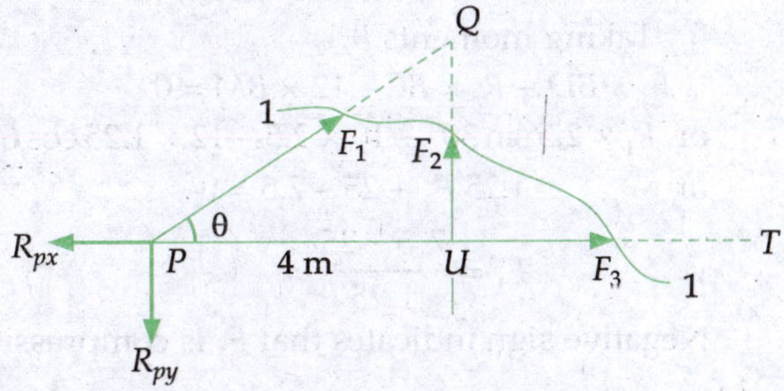

(i) Reactions R_{px} = 1.5 kN and R_{py} = 0.75 kN acting at point in the directions shown

(ii) Unknown force F_1 in member PQ (assumed tensile)

(iii) Unknown force F_2 in member UQ (assumed tensile)

(iv) Unknown force F_3 in member UT (assumed tensile)

Taking moment of these forces about point U,

$$R_{py} \times 4 = F_1 \sin\theta \times 4; \quad \theta = \tan^{-1}\left(\frac{4}{8}\right) = 26.56°$$

$$\text{or} \quad F_1 = \frac{R_{py}}{\sin\theta} = \frac{0.75}{\sin 26.56°} = 1.667 \text{ kN (tension)}$$

Taking moments about point Q,

$$R_{px} \times UQ = R_{py} \times PU + F_3 \times UQ$$

$$\text{or} \quad 1.5 \times (4\tan\theta) = 0.75 \times 4 + F_3 \times 4$$

$$\text{or} \quad F_3 = \frac{1.5 \times 4\tan 26.56° - 0.75 \times 4}{4} = 0$$

Thus, F_1 (force in member PQ) = 1.677 kN (tension)

F_3 (force in member UT) = 0 (no force)

EXAMPLE 5.17

Determine the forces in the members marked (1), (2) and (3) of the truss loaded and supported as shown in Fig. 5.25.

Solution : The reactions at the supports are determined in the usual way as done in Example 5.4

$$R_a = 12 \text{ kN and } R_c = 10 \text{ kN}$$

Let a section x-x be passed in such a way that it cuts the members marked (1), (2), (3) and divides the truss in two parts.

$$CD = BC\cos 30° = 2.5 \times 0.866 = 2.165 \text{ m}$$

$$BD = BC\sin 30° = 2.5 \times 0.5 = 1.25 \text{ m}$$

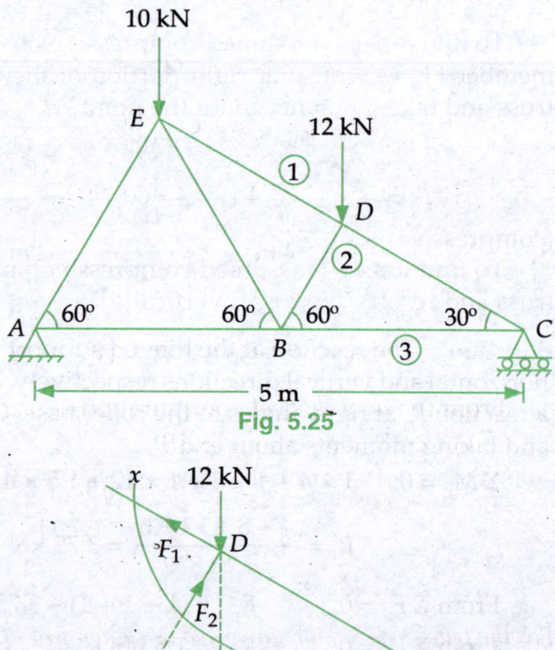

Fig. 5.25

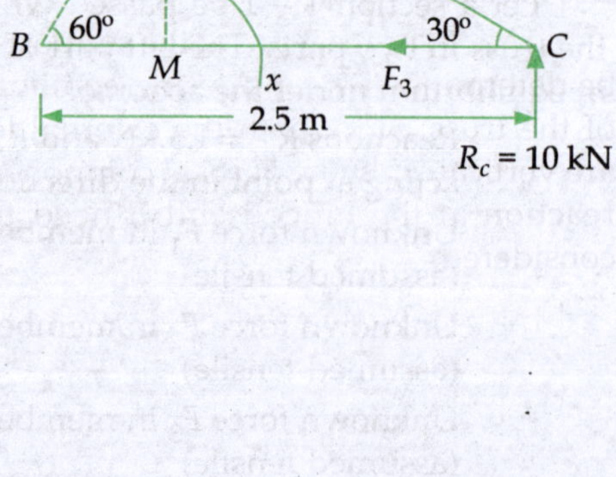

Under equilibrium conditions of right part of the truss,

(*i*) Taking moments about C,

$$F_2 \times CD - 12 \times MC = 0$$

$$\therefore \quad F_2 = 12 \times \frac{MC}{CD} = 12 \times \frac{CD\cos 30°}{CD}$$

$$= 12 \times \cos 30° = 12 \times 0.866$$

$$= \mathbf{10.39 \text{ kN}} \text{ (compression)}$$

(*ii*) Taking moments B,

$$F_1 \times BD + R_c \times BC - 12 \times BM = 0$$

$$\text{or} \quad F_1 \times 2.5\sin 30° + 10 \times 2.5 - 12 \times 1.25\cos 60° = 0$$

$$\text{or} \quad 1.25\,F_1 + 25 - 7.5 = 0$$

$$\therefore \quad F_1 = \frac{7.5 - 25}{1.25} = \mathbf{-14 \text{ kN}}$$

Negative sign indicates that F_1 is compressive.

(*iii*) Taking moments about D,

$$F_3 \times DM - R_c \times MC = 0$$

$$\therefore \quad F_3 = R_c \times \frac{MC}{DM} = R_c \times \frac{CD\cos 30^\circ}{CD\sin 30^\circ}$$

$$= 10 \times \frac{0.866}{0.5} = 17.32 \text{ kN (tensile)}$$

EXAMPLE 5.18

Make calculations for the load W that would produce a force of magnitude 160 kN in the AB member of the cantilever truss shown in Fig. 5.26. For this loading, workout forces in members AE and FE of the truss.

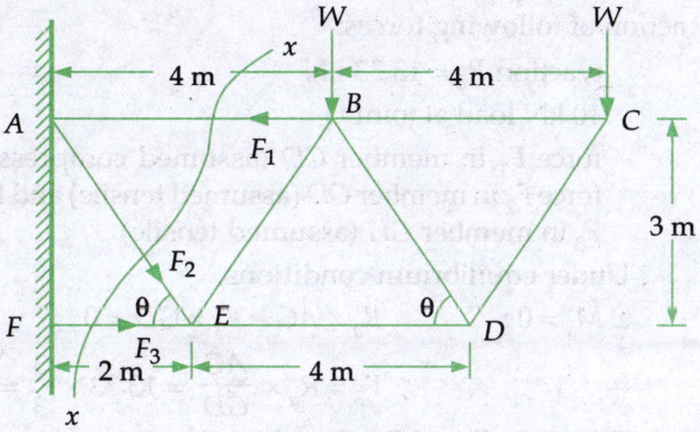

Fig. 5.26

Solution : Let a section x-x be passed in such a way that it cuts the members AB, AE, FE and divides the truss in two parts.

Consider equilibrium of right part of the truss and take moments about the point E,

$$(W \times 2) + (W \times 6) = F_1 \times 3 = 160 \times 3$$

$$\therefore \quad W = \frac{160 \times 3}{8} = 60 \text{ kN}$$

To find force F_3 (assumed compressive) in member FE, we consider right portion of the truss and take moments about the point A,

$$(W \times 8) + (W \times 4) = F_3 \times 3$$

$$\therefore \quad F_3 = \frac{12\,W}{3} = 4\,W = 4 \times 60 = 240 \text{ kN}$$

(compressive).

To find force F_2 (assumed compressive) in member AE, we again consider right portion of the truss and resolve the forces vertically.

$$F_2 \sin\theta = W + W\,; \qquad F_2 = \frac{2\,W}{\sin\theta}$$

From the geometrical configuration,

$$\sin\theta = \frac{AF}{AE} = \frac{3}{\sqrt{3^2 + 2^2}} = 0.832$$

$$\therefore \quad F_2 = \frac{2 \times 60}{0.832} = 144.23 \text{ kN (compressive)}$$

EXAMPLE 5.19

For the truss loaded and supported as shown in Fig. 5.27, determine the forces in members CD, DG and GH. Use the method of sections.

Solution : The reactions at the supports can be determined by considering equilibrium of the truss. Since both the external loads are vertical, only the vertical component of reaction at the hinged end B need to be considered.

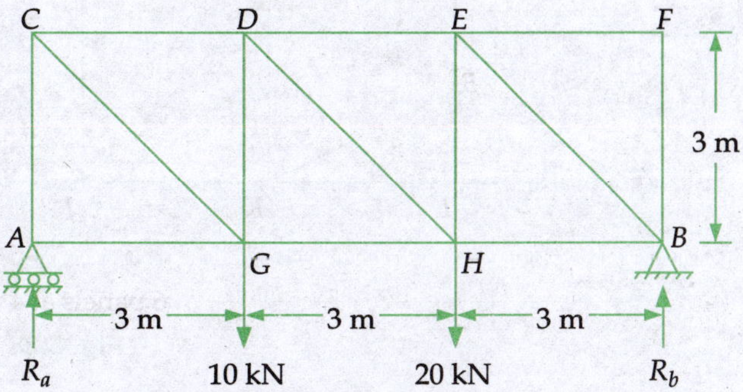

Fig. 5.27

Taking moments about end B,

$\Sigma M_b = 0: \; R_a \times 9 - 10 \times 6 - 20 \times 3 = 0$

$$\therefore \quad R_a = \frac{10 \times 6 + 20 \times 3}{9} = 13.33 \text{ kN}$$

Further, $\Sigma F_y = 0: \quad R_a + R_b - 10 - 20 = 0$

$$\therefore \quad R_b = 30 - R_a = 30 - 13.33 = 16.67 \text{ kN}$$

Let section 1-1 be passed in such a way that it cuts the members *CD*, *DG*, *GH* and divides the truss in two parts.

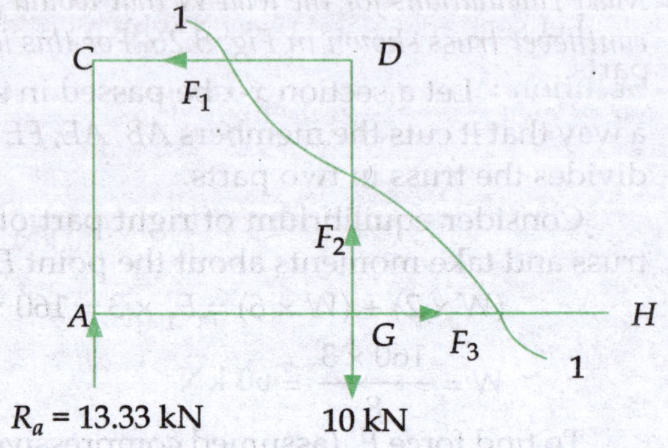

The left part of the truss is in equilibrium under the action of following forces:

(*i*) reaction R_a = 13.33 kN

(*ii*) 10 kN load at joint *G*

(*iii*) force F_1 in member *CD* (assumed compressive), force F_2 in member *GD* (assumed tensile) and force F_3 in member *GH* (assumed tensile)

Under equilibrium conditions,

$\Sigma M_a = 0: \quad R_a \times AG - F_3 \times GD = 0$

$$\therefore \quad F_3 = R_a \times \frac{AG}{GD} = 13.33 \times \frac{3}{3} = 13.33 \text{ kN (tensile)}$$

$\Sigma M_g = 0: R_a \times AG - F_1 \times AC = 0$

$$\therefore \quad F_1 = R_a \times \frac{AG}{AC} = 13.33 \times \frac{3}{3} = 13.33 \text{ kN (compressive)}$$

$\Sigma M_c = 0: 10 \times AG - F_2 \times CD - F_3 \times GD$

$$\therefore \quad F_2 = \frac{10 \times 3 - 13.33 \times 3}{3} = -3.33 \text{ kN}$$

The negative with force F_2 is indicative of the fact that a wrong choice has been made while assuming its direction. Obviously, the force F_2 in member *GD* is 3.33 kN compressive.

EXAMPLE 5.20

Calculate the force in member DJ of the Howe roof truss as shown in Fig. 5.28. Neglect any horizontal component of force at the supports.

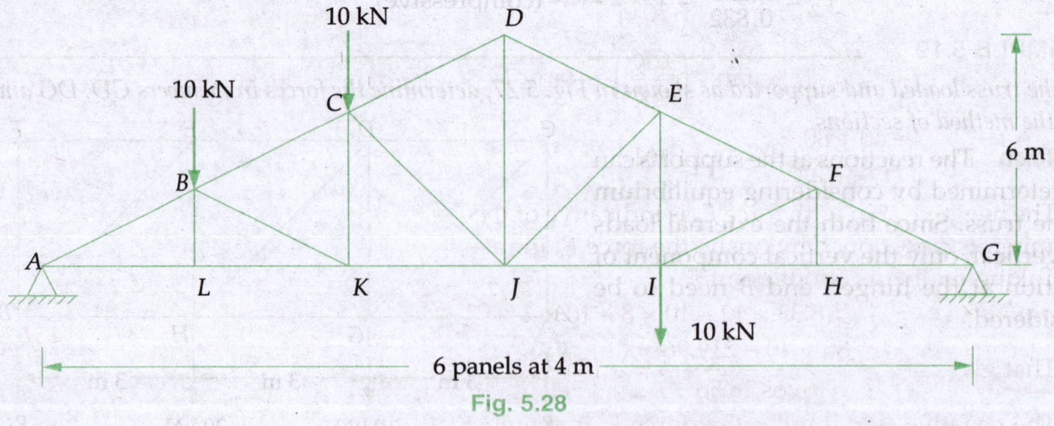

Fig. 5.28

Solution : The reactions at the supports can be determined by considering equilibrium of the entire truss. Since the external loads are vertical and any horizontal component of force at the supports is to be neglected, only the vertical component of the reactions at the support need to be considered.

Taking moments about end A,

$$R_g \times 24 = 10 \times 4 + 10 \times 8 + 10 \times 16 = 280$$

$$\therefore \quad R_g = \frac{280}{24} = 11.67 \text{ kN}$$

and $\quad R_a = (10 + 10 + 10) - 11.67 = 18.33$ kN

Let section 1–1 be passed which cuts the members CD, CJ and KJ and divides the truss in two parts.

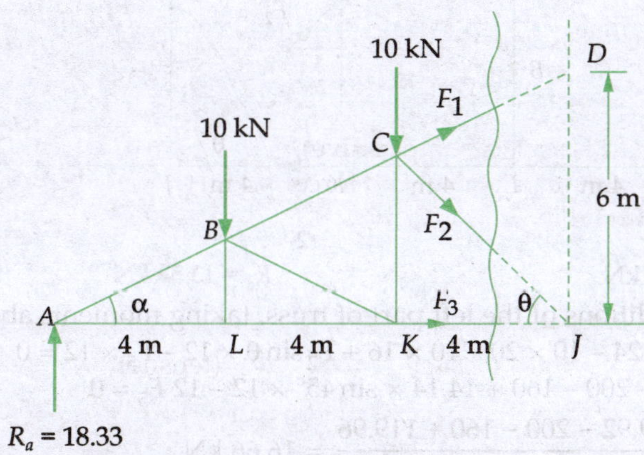

Under equilibrium conditions of left part of the truss

Taking moments about point A,

$$10 \times 4 + 10 \times 8 + F_2 \sin\theta \times 12 = 0$$

or $$F_2 = -\frac{40 + 80}{12 \sin\theta}$$

From the geometrical configuration,

$$\tan\alpha = \frac{CK}{AK} = \frac{DJ}{AJ}; \quad CK = \frac{DJ}{AJ} \times AK = \frac{6}{12} \times 8 = 4 \text{ m}$$

$$\theta = \tan^{-1}\left(\frac{CK}{KJ}\right) = \tan^{-1}\left(\frac{4}{4}\right) = \tan^{-1} 1 = 45°$$

$$\alpha = \tan^{-1}\left(\frac{CK}{AK}\right) = \tan^{-1}\left(\frac{4}{8}\right) = \tan^{-1} 0.5 = 26.56°$$

Then : $F_2 = -\dfrac{40 + 80}{12 \sin 45} = -14.14$ kN

The negative sign with force F_2 is indicative of the fact that a wrong choice has been make while assuming its direction. Obviously, the force F_2 in member CJ is 14.14 kN compressive.

Taking moments about point J,

$$18.33 \times 12 - 10 \times 8 - 10 \times 4 + F_1 \cos 26.56 \times 6 = 0$$

That gives : $F_1 = \dfrac{80 + 40 - 219.96}{6 \cos 26.56} = -18.63$ N

The negative sign implies that force F_1 in member CD is compressive.

Taking moments about point C,

$$18.33 \times 8 - 10 \times 4 - F_3 \times 4 = 0$$

$\therefore$ Force in member KJ,

$$F_3 = \frac{18.33 \times 8 - 10 \times 4}{4} = 26.66 \text{ kN} \quad \text{(tension)}$$

(ii) Now, let a section 2–2 be passed which cuts the member DE, DJ, JC and JK.

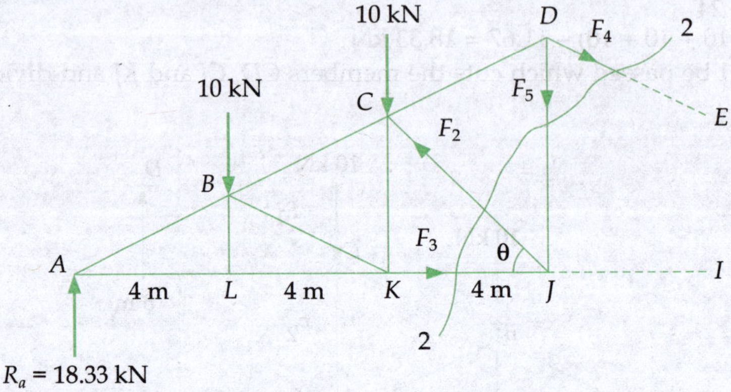

Under equilibrium conditions of the left part of truss, taking moments about point G,

$$18.33 \times 24 - 10 \times 20 - 10 \times 16 + F_2 \sin\theta \times 12 - F_5 \times 12 = 0$$

or $$439.92 - 200 - 160 + 14.14 \times \sin 45° \times 12 - 12\, F_5 = 0$$

$\therefore$ $$F = \frac{439.92 - 200 - 160 + 119.96}{12} = 16.66 \text{ kN}$$

The member DJ is thus subjected to a tensile force of 16.66 kN

EXAMPLE 5.21

Using the method of sections, analyse the struss shown in Fig. 5.29 regarding forces in members ED, DF and FC.

Solution : Let section x-x be passed in such a way that it cuts the member ED, DF, FC and divides the truss in two parts. For considering the equilibrium of the upper part of the truss there is no need to make calculations for the reaction at the supports.

Let F_1, F_2 and F_3 be the forces with their directions assumed as shown.

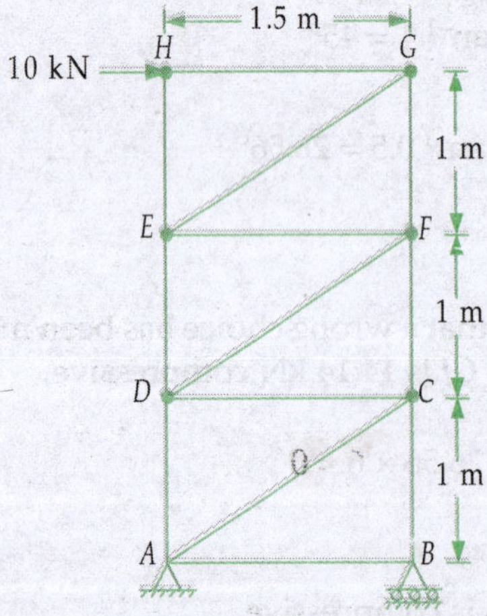

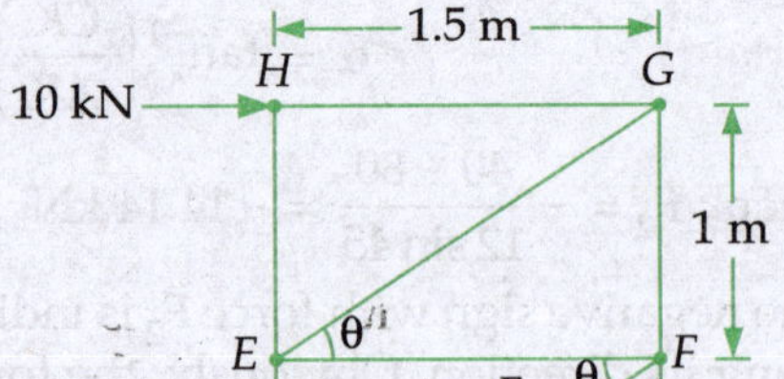

Fig. 5.29

Under equilibrium conditions,

$$\Sigma M_f = 0: \quad 10 \times GF - F_1 \times EF = 0$$

$$\therefore F_1 = 10 \times \frac{GF}{EF} = 10 \times \frac{1}{1.5} = 6.67 \text{ kN (tensile)}$$

$$\Sigma F_d = 0: \quad 10 \times HD - F_3 \times FE = 0$$

$$\therefore F_3 = 10 \times \frac{HD}{FE} = 10 \times \frac{2}{1.5} = 13.33 \text{ kN}$$

(compressive)

Further equating all the forces in horizontal direction we get,

$$10 - F_2 \cos\theta = 0; \qquad F_2 = \frac{10}{\cos\theta}$$

From the geometry $\cos\theta = \frac{1.5}{\sqrt{1.5^2 + 1^2}} = \frac{1.50}{1.80} = 0.833$

$$\therefore \quad F_2 = \frac{10}{0.833} = 12 \text{ kN (tensile)}$$

5.3.3. Graphical methods

The analytical methods, *i.e.*, the method of joints and the method of sections give absolutely correct results regarding forces set up in the members of a perfect truss when it is loaded. However, there arise certain situations where it becomes impossible to get the results from these methods. Recourse is then made to graphical methods which are quite convenient and provide reasonably accurate results.

The steps involved for obtaining graphical solution of a frame are explained below with reference to a triangular truss loaded and supported as shown in Fig. 5.30.

The reactions at the supports are determined in the usual way as done in Example 5.2

$$R_a = 22.5 \text{ kN} \quad \text{and} \quad R_b = 7.5 \text{ kN}$$

- **Making a space diagram:** The given truss or frame is drawn accurately according to some linear scale. The loads and support reactions are also indicated on the frame both in magnitude and direction. The naming of the various members of the frame is then done according to Bow's notation. In this technique, a force is designated by two letters which are written on either side of the line of action of the force.

  In the space diagram drawn as shown in Fig. 5.30, the spaces between the forces have been marked as *P*, *Q*, *R* and *S*.

- **Constructing a vector diagram:**

  (*i*) Take any point *p* and draw a vertical line *pq* to represent force *PQ* = 30 kN. If scale is taken as 1 cm = 5 kN, then *pq* = 6 cm. From point *q*, take *qr* equal to 1.5 cm to represent the reaction R_b = 7.5 kN acting vertically upwards. Then *rp* represents the reaction R_a.

  (*ii*) Choose a joint where not more than two unknown forces are acting (say joint A).

  (*iii*) From point *p*, draw a line parallel to *PS* and from point *q* draw a line parallel to *QS*. These two lines meet at point *s*. Measure *ps* and *sr* in the vector diagram.

  ps = 5.2 cm ∴ Force *PS* (force in member *AC*) = 5.2 × 5 = 26 kN
  sr = 2.6 cm ∴ Force *SR* (force in member *AB*) = 2.6 × 5 = 13 kN

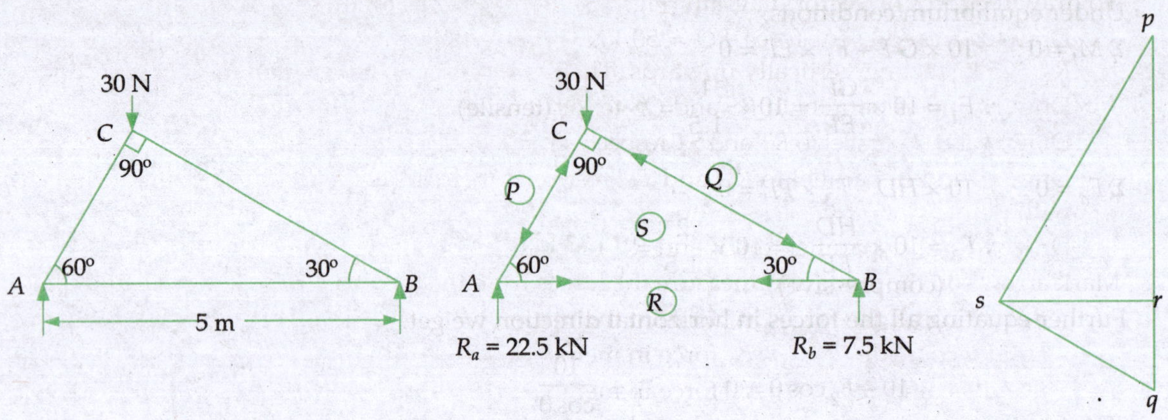

Fig. 5.30

For determining the nature of these forces, we move in clockwise direction round the joint A in space diagram. First PS, then SR and then RP.

In the vector diagram p to s means that the arrow must be marked towards the joint A in space diagram. Obviously then the member AC is subjected to 26 kN compressive force.

Considering SR and the corresponding vector s to r in vector diagram, the joint A is being pulled and as such arrow must be marked directed away from the joint. Obviously then member AC is subjected to 13 kN tensile force.

(iv) Considering joint B: In the vector diagram, the vector sq represents force in member BC designated by SQ in the space diagram

sq = 3 cm $\quad\therefore$ Force SQ (force in member BC) = 3 × 5 = **15 kN**

Again moving round the joint B in clockwise direction, we have first RS, then SQ and then QR. In the vector diagram

rs : arrow is to be marked from B towards A (away from the joint)

sq : arrow is to be marked from C towards B (towards the joint)

Obviously then the member CB is subjected to 15 kN compressive force.

EXAMPLE 5.22

A truss has been loaded and supported as shown in Fig. 5.31. Determine graphically the forces in the various members of the structure shown in the figure.

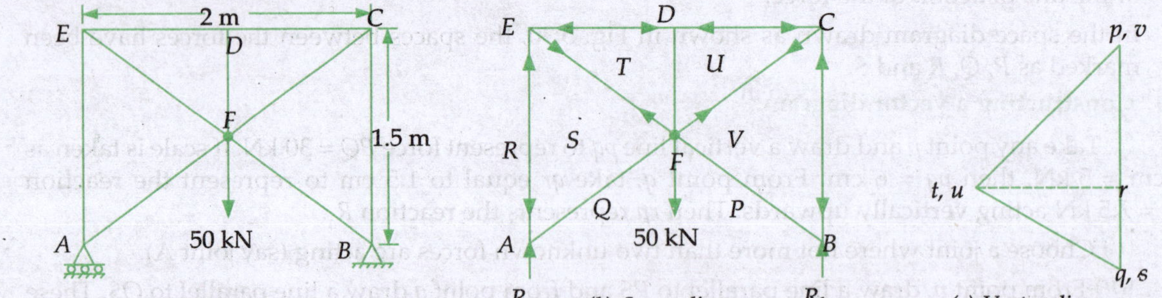

(a) Given truss (b) Space diagram (c) Vector diagram

Fig. 5.31

Solution : Draw the space diagram and name the various forces according to Bow's notation (Fig. 5.31 *b*).

To draw the force or vector diagram (Fig. 5.31 *c*), proceed as follows:

(*i*) Draw *pq* to represent force *PQ* = 50 kN. From point *q*, take *qr* to represent the reaction R_a = 25 kN acting vertically upwards. Then *rp* represents the reaction R_b = 25 kN.

(*ii*) Draw *rs* and *qs* parallel to *RS* and *QS* to locate point *s*. The point *s* coincides with point *q*.

(*iii*) Draw *rt* and *st* parallel to *RT* and *ST* respectively and locate the point *t*.

(*iv*) Draw *ru* and *tu* parallel to *RU* and *TU* respectively to locate point *u*. The point *u* coincides with point *t*.

(*v*) Draw *uv* and *rv* parallel to *UV* and *RV* respectively to locate point *v*.

Mark arrows on the joints and measure the forces from the force diagram. It may be noted that:

q and *s* coincide $\therefore$ force in member *AF* = 0
t and *u* coincide $\therefore$ force in member *DF* = 0
p and *v* coincide $\therefore$ force in member *BF* = 0

The forces in the various members have been tabulated belows:

Member	*AE*	*ED, DC*	*BC*	*BF*	*AF*	*DF*	*EF, FC*
Bow's notation	*RS*	*RT, RU*	*RV*	*VP*	*QS*	*UT*	*ST, UV*
Force (kN)	25	34	30	0	0	0	42
Nature	*C*	*C*	*C*		-	-	*T*

EXAMPLE 5.23

Find graphically or otherwise, the forces in all the members of the truss shown in Fig. 5.32. Present your results in a tabular form, stating whether a member is in tension or compression.

Solution : Since the loading and truss are symmetrical,

$$R_1 = R_2 = \frac{4+2+2}{2} = 4 \text{ kN}$$

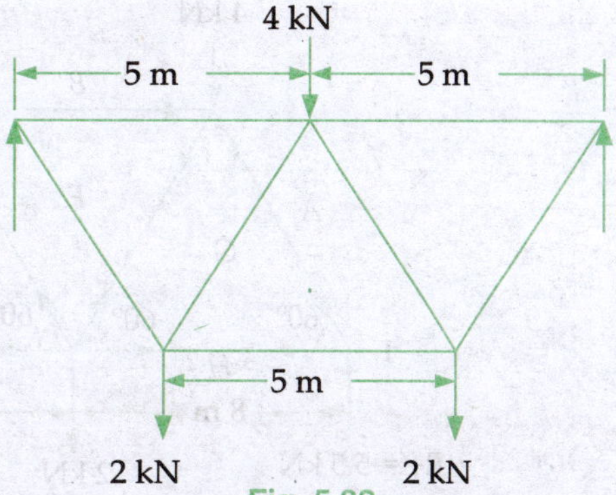

Fig. 5.32

Choosing a convenient scale, draw the space diagram and indicate the forces by using Bow's notation. Proceed to draw the force/vector diagram by considering the joints in succession.

Mark arrows on the joints and measure the magnitude of forces from the vector diagram. The forces in the various members are tabulated below:

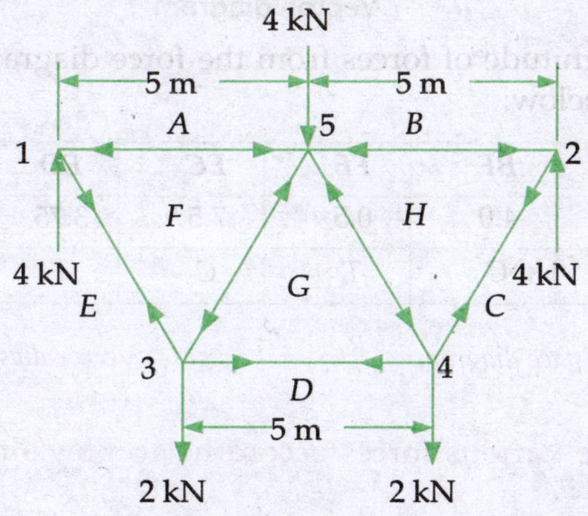

Space diagram

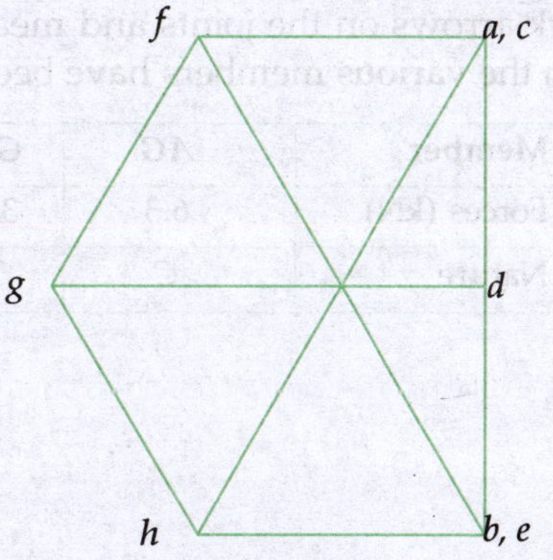

Vector diagram

Member Bow's Notation	*AF*	*BH*	*HC*	*HG*	*GD*	*GF*	*EF*
Force (kN)	2	2	4.5	2.25	3	2.25	4.5
Nature	*C*	*C*	*T*	*C*	*T*	*C*	*T*

EXAMPLE 5.23

Determine the stresses in all members of the truss shown in Fig. 5.33. Use graphical method.

Solution : Let R_1 = reaction of the left-hand support

R_2 = reaction of the right hand support

To find these reactions, we take moments about joint 1. That gives

$$R_2 \times 16 = (6 \times 12) + (2 \times 8) + (4 \times 4)$$

$$= 72 + 16 + 16 = 104$$

$$\therefore \quad R_2 = \frac{104}{16} = 6.5 \text{ kN}$$

$$R_1 = (6 + 2 + 4) - 6.5 = 5.5 \text{ kN}$$

Choosing a convenient scale, draw the space and vector diagrams as shown.

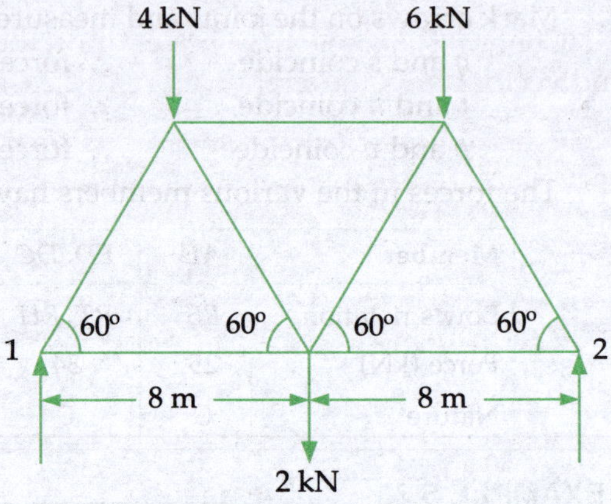

Fig. 5.33

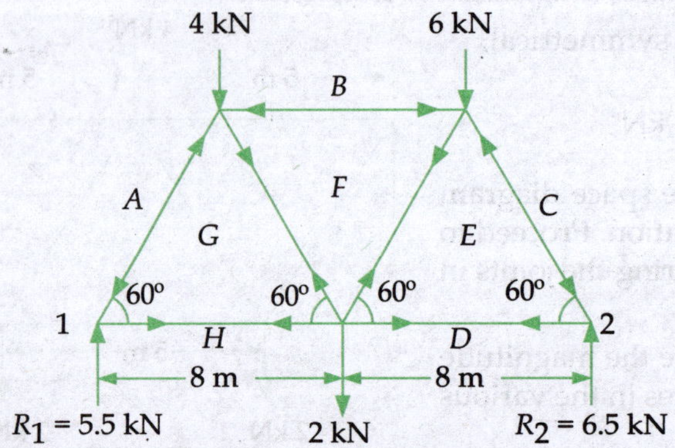

Space diagram

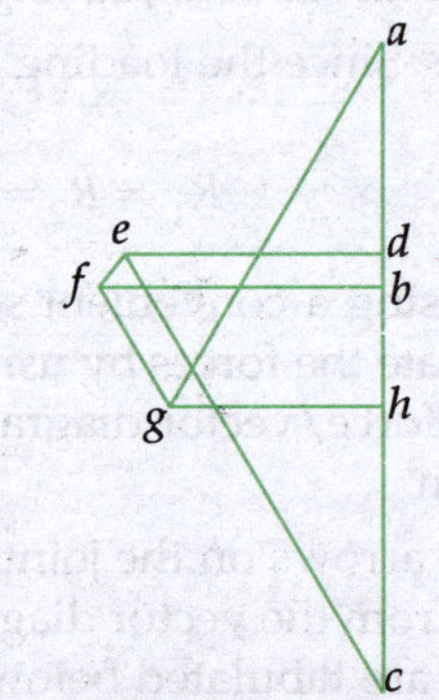

Vector diagram

Mark arrows on the joints and measure the magnitude of forces from the force diagram. The forces in the various members have been tabulated below:

Member	*AG*	*GH*	*GF*	*BF*	*FE*	*EC*	*ED*
Forces (kN)	6.3	3.2	1.7	4.0	0.5	7.5	3.75
Nature	*C*	*T*	*T*	*C*	*T*	*C*	*T*

EXAMPLE 5.25

Determine the stresses in the members of the truss shown in Fig. 5.34. Use graphical method.

Solution : Let,

R_1 = reaction of the left-hand support

R_2 = reaction of the right-hand support

Since the truss and the loads are symmetrical,

$$R_1 = R_2 = \frac{50 + 100 + 200 + 100 + 50}{2} = 250 \text{ kN}$$

Choosing a convenient scale, draw the space and vector diagrams as shown.

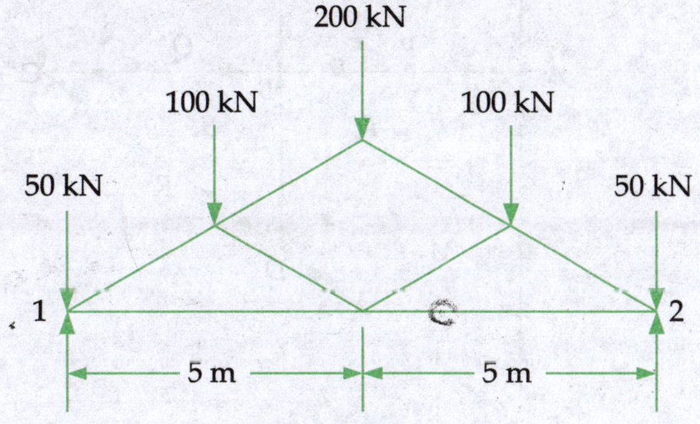

Fig. 5.34

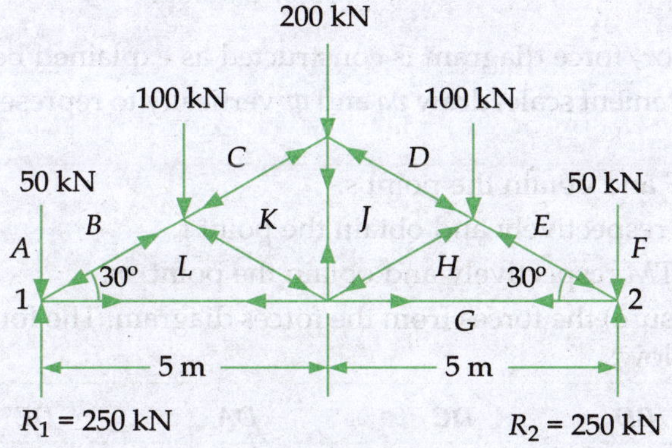

Space diagram

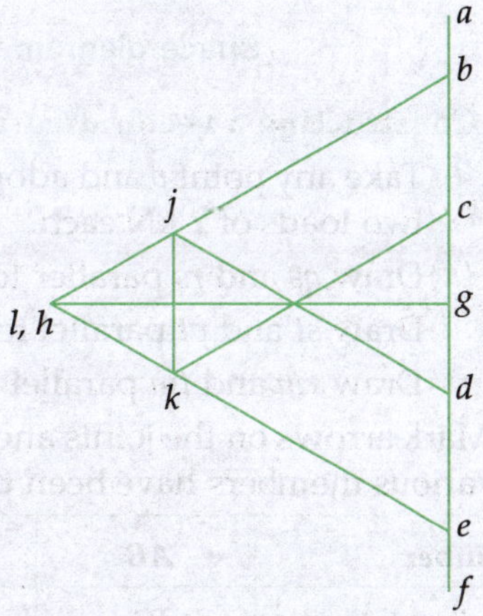

Vector diagram

Mark arrows on the joints and measure the magnitude of forces from the force diagram. The forces in the various members have been tabulated below:

Member	*BL, HE*	*KC, JD*	*LK, JH*	*GL, GH*	*KJ*
Force (kN)	400	300	100	350	100
Nature	*C*	*C*	*C*	*T*	*T*

EXAMPLE 5.26

Using graphical method, determine the magnitude and nature of the forces in all members of the cantilever truss loaded as shown in Fig. 5.35.

Solution : The steps involved for determining graphically the forces in members of the truss are:

- **Making a space diagram:** Draw the space diagram for the truss along with loads of 2 kN at joints *B* and *C*. Name the various members according to Bow's notation.

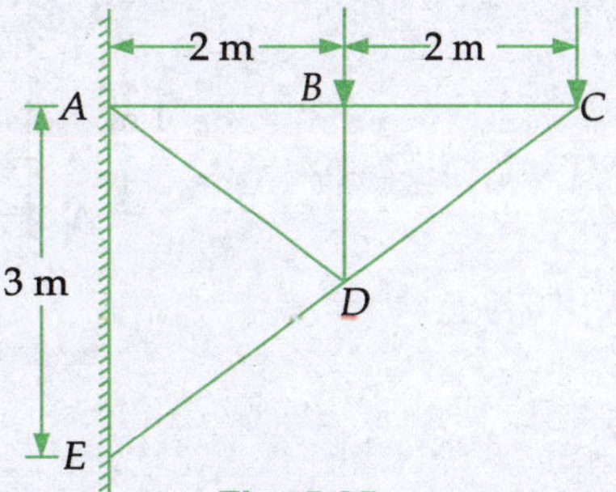

Fig. 5.35

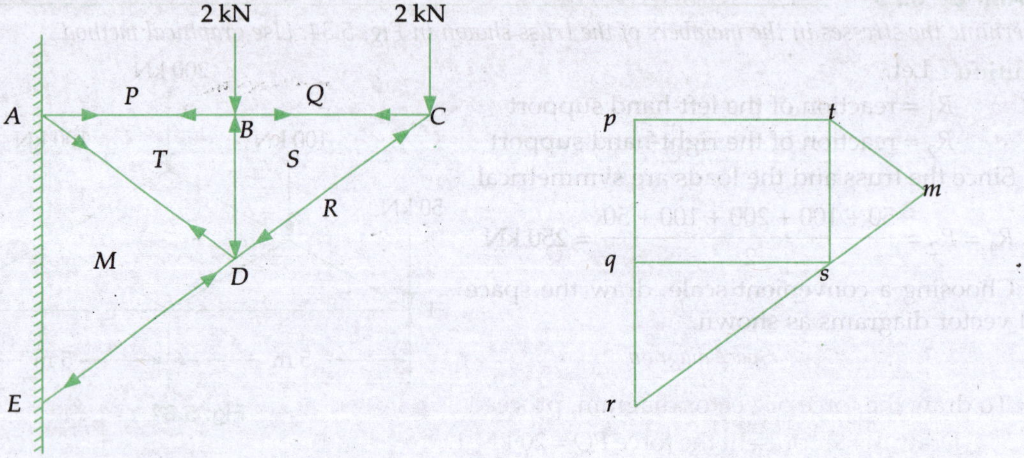

Space diagram Vector diagram

- **Constructing a vector diagram:** The vector/force diagram is constructed as explained below :
  (*i*) Take any point *p* and adopting a convenient scale, draw *pq* and *qr* vertically to represent the two loads of 2 kN each.
  (*ii*) Draw *qs* and *rs* paraller to *QS* and *RS* and obtain the point *s*.
  (*iii*) Draw *st* and *pt* parallel to *ST* and *PT* respectively and obtain the point *t*.
  (*iv*) Draw *rm* and *tm* parallel to *RM* and *TM* respectively and obtain the point *m*.

Mark arrows on the joints and scale (measure) the forces from the forces diagram. The forces in the various members have been tabulated below:

Member	*AB*	*BC*	*BD*	*DC*	*DA*	*DE*
Bow's notation	*PT*	*QS*	*TS*	*SR*	*TM*	*MR*
Force	2.65 kN	2.65 kN	1.65	3.35 kN	2.0 kN	5.0 kN
Nature	tensile	tensile	tensile	compressive	compressive	compressive

EXAMPLE 5.27

Workout graphically the forces in the various members of the cantilever truss shown in Fig. 5.36.

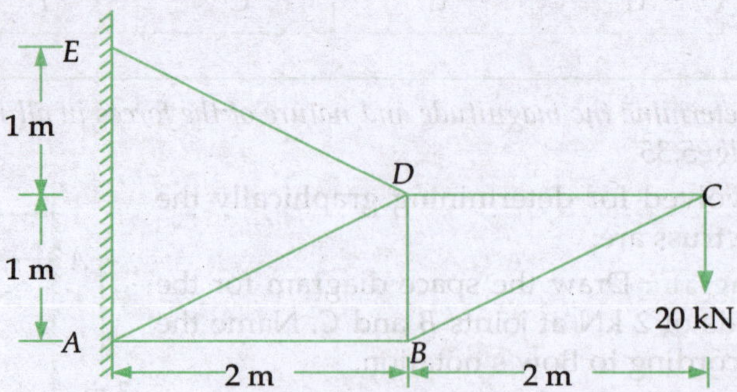

Fig. 5.36. Load truss

Solution : Draw the space diagram and name the various forces following Bow's notation.

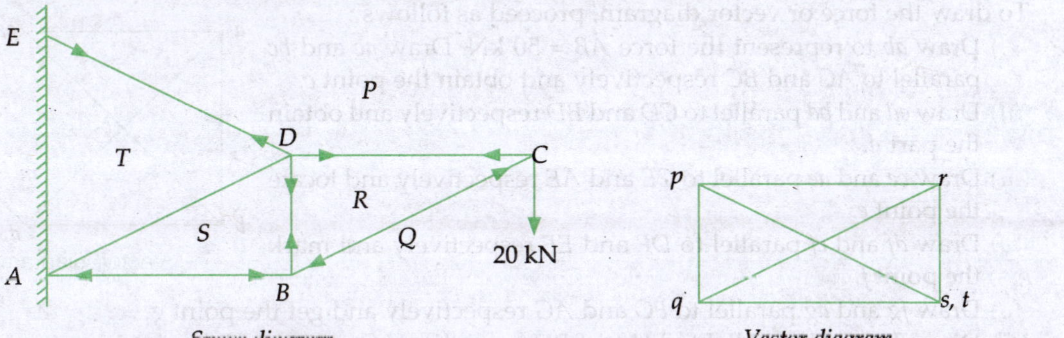

Space diagram *Vector diagram*

To draw the force or vector diagram, proceed as follows:

(*i*) Draw *pq* to represent the force *PQ* = 20 kN. Draw *pr* and *qr* parallel to *PR* and *QR* respectively and obtain the point *r*.

(*ii*) Draw *rs* and *qs* parallel to *RS* and *QS* respectively and get the point *s*.

(*iii*) Draw *pt* and *st* parallel to *PT* and *ST* respectively and get the point *t*. The point *t* coincides with point *s*.

Mark arrows on the joints and measure the forces from the force diagram. The forces in the various members have been tabulated below:

Member	*AB*	*BC*	*CD*	*DA*	*DE*	*DB*
Bow's notation	QS	QR	PR	ST	PT	RS
Force	40	45	40	0	45	20
Nature	C	C	*T*	—	*T*	*T*

EXAMPLE 5.28

The loading and the support conditions for a particular cantilever truss are shown in Fig. 5.37. Determine graphically the force in each member of the truss, stating whether it is in tension or compression.

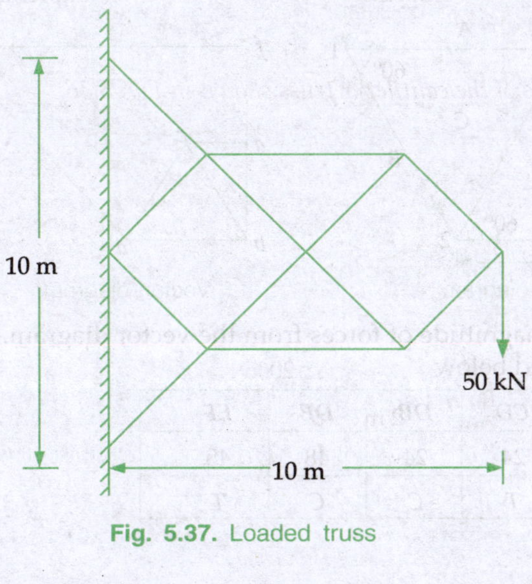

Fig. 5.37. Loaded truss

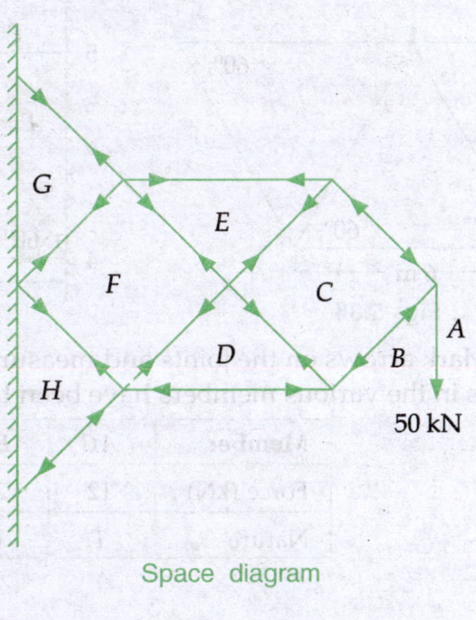

Space diagram

Solution : Draw the space diagram and name the various forces following Bow's notation.

To draw the force or vector diagram, proceed as follows :

(*i*) Draw *ab* to represent the force *AB* = 50 kN. Draw *ac* and *bc* parallel to *AC* and *BC* respectively and obtain the point *c*.

(*ii*) Draw *cd* and *bd* parallel to *CD* and *BD* respectively and obtain the part *d*.

(*iii*) Draw *ce* and *ae* parallel to *CE* and *AE* respectively and locate the point *e*.

(*iv*) Draw *df* and *ef* parallel to *DF* and *EF* respectively and mark the point *f*.

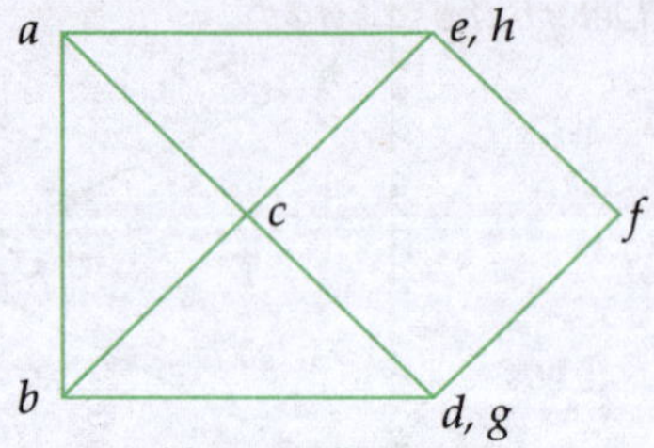

Vector diagram

(*v*) Draw *fg* and *ag* parallel to *FG* and *AG* respectively and get the point *g*.

(*vi*) Draw *fh* and *bh* parallel to *FH* and *BH* respectively and obtain the point *h*.

Mark arrows on the joints and measure the forces from the force diagram. The forces in the various members have been tabulated below:

Member	Force (kN)	Nature	Member	Force (kN)	Nature
AG	72	*T*	*HF*	35	*C*
AE	50	*T*	*FG*	35	*T*
AC	35	*T*	*FE*	35	*T*
BC	35	*C*	*FD*	35	*C*
BD	50	*C*	*EC*	35	*C*
BH	72	*C*	*DC*	35	*T*

EXAMPLE 5.29

Determine graphically the forces in all members of the cantilever truss loaded as shown in Fig. 5.38.

Solution : Choosing a convenient scale, draw the space and vector diagrams. After the figure while drawing the vector diagram, consider the joints in succession.

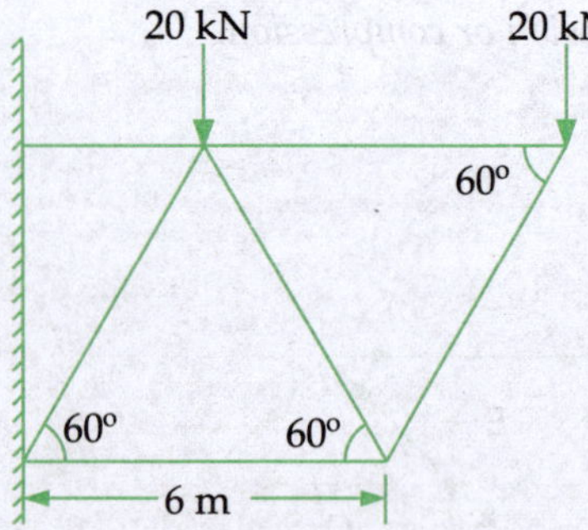

Fig. 5.38

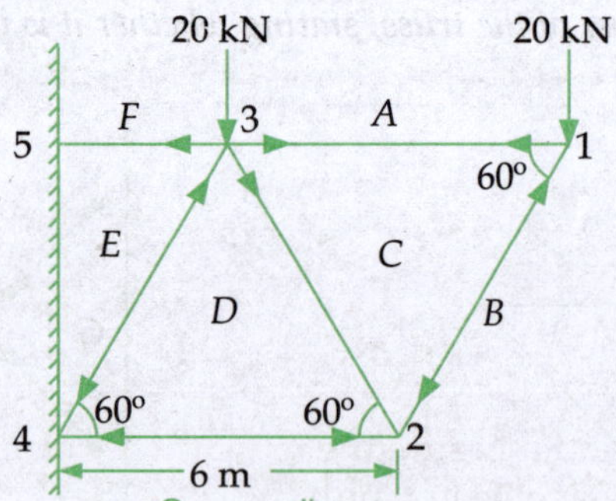

Space diagram

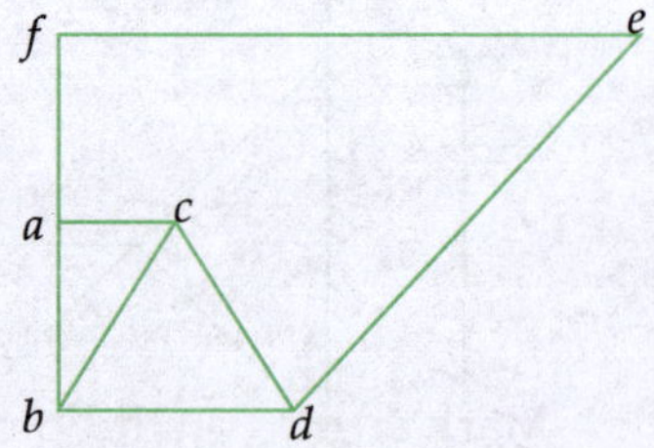

Vector diagram

Mark arrows on the joints and measure the magnitude of forces from the vector diagram. The forces in the various members have been tabulated below :

Member	*AC*	*BC*	*CD*	*DB*	*DE*	*EF*
Force (kN)	12	24	24	24	48	48
Nature	*T*	*C*	*T*	*C*	*C*	*T*

EXAMPLE 5.30

Determine graphically the forces in all members of the cantilever frame loaded as shown in Fig. 5.39.

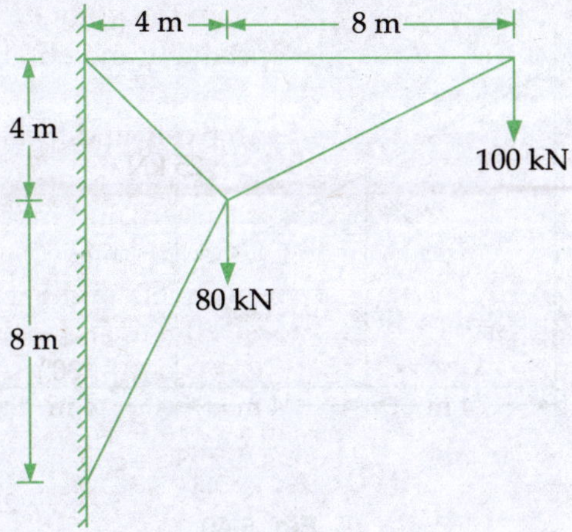

Fig. 5.39

Solution : Choosing a convenient scale, draw the space and vector diagrams as shown. While drawing the vector diagram, the joints are to be considered in succession.

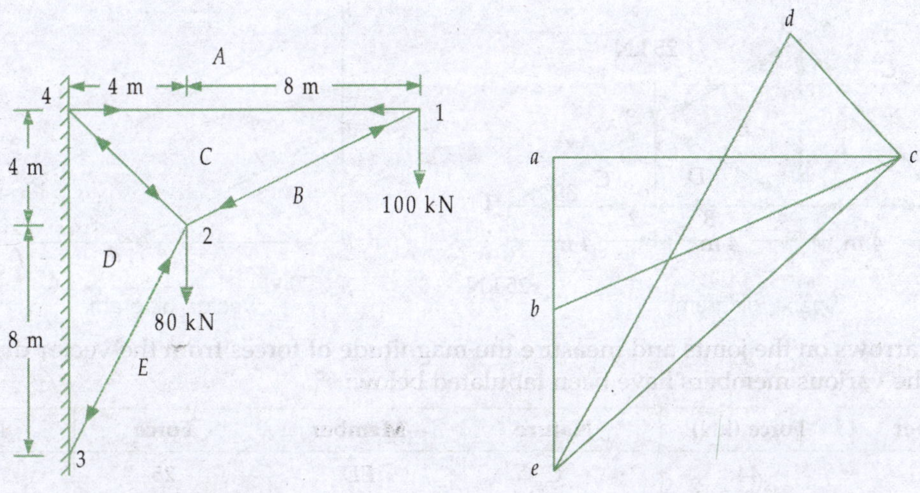

Space diagram Vector diagram

Mark arrows on the joints and measure the magnitudes of forces from the vector diagram. The forces in the various members have been tabulated below:

Member	*AC*	*BC*	*CD*	*DE*
Force (kN)	200	225	100	280
Nature	*T*	*C*	*C*	*C*

EXAMPLE 5.31

A cantilever frame has been loaded as shown in Fig. 5.40. Determine graphically the forces in all the members of the frame.

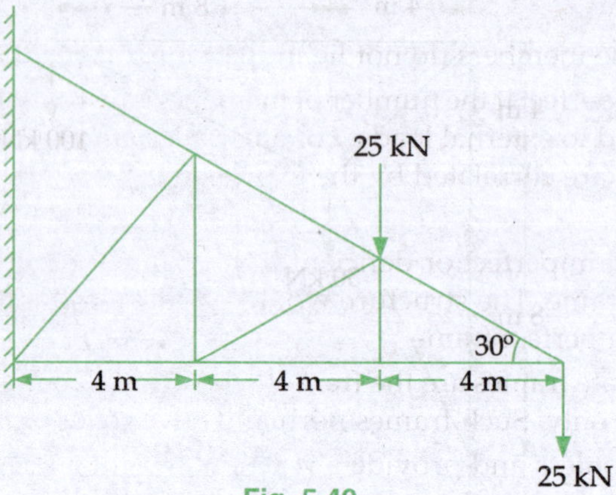

Fig. 5.40

Solution : Choosing a convenient scale, draw the space and vector diagrams as shown. While drawing the vector diagram, consider the joints in succession.

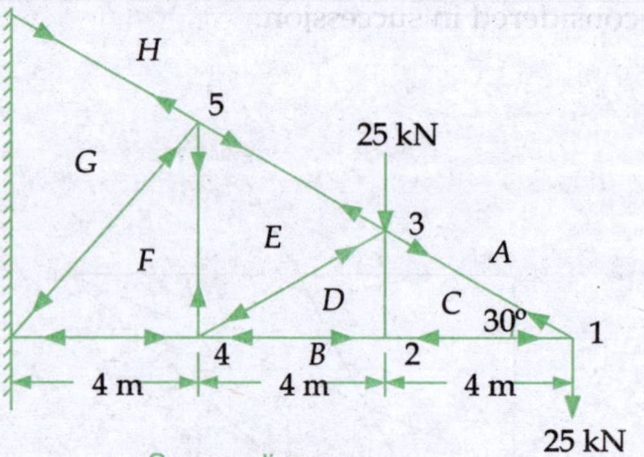

Space diagram

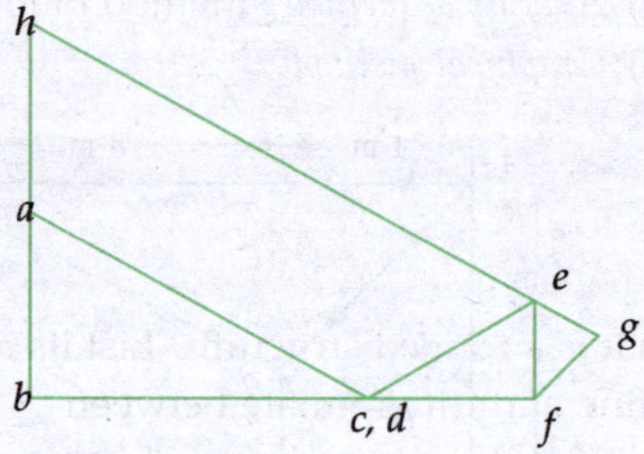

Vector diagram

Mark arrows on the joints and measure the magnitude of forces from the vector diagram. The forces in the various members have been tabulated below:

Member	Force (kN)	Nature	Member	Force	Nature
BC	44	C	*ED*	25	C
BD	44	C	*DC*	0	-
BF	65	C	*CA*	50	T
FG	11.5	C	*HE*	75	T
FE	12.5	T	*HG*	85	T

REVIEW AND SUMMARY

1. A frame is a rigid structure made up of three or more members welded or riveted together. The framed structure finds application in the roofs of the sheds at railway platforms, workshops and industrial buildings, and bridge trusses etc.
2. The bars used in the framed strucutre are slender and are usually of angle section, *T*-section and *I*-section.

3. The bars are called two-force members because they are either in tension or in compression. The member under tension is called a tie and the member under compression is called a strut.
4. The frame is called plane frame if all the members lie in the same plane and the forces act along the plane of the frame.

   A frame in which all the members do not lie in the same plane is called a space frame.
5. A structure is said to be perfect if the number of members is just sufficient to prevent its distorsion of shape when subjected to external loads. For a perfect frame, the number of members (m) and the number of joints (j) are correlated by the expression :

   $$m = 2j - 3$$

   The structure is termed imperfect or deficient if the number of members in it is less than that required for a perfect frame. The structure would be called redundant if it has members more than that required for a perfect frame.
6. The statically determinate frames are the frames whose force analysis can be made by applying the equations of statics only. Such frames normally have roller or hinged supports.

   Roller frames are frictionless and provide a reaction at right angles to the roller base. The line action of the reaction for a hinged support depends upon the load system on the frame.
7. The analysis of a frame involves the determination of reaction at the supports and internal stresses induced in the members due to external loads.
8. The stresses in the members of a framed structure are worked out by the following methods :
   (i) *Analytical method :* method of joints and method of sections
   (ii) *Graphical method*

REVIEW QUESTIONS

A. Conceptual and conventional questions :

1. What is a framed structure? List its applications.
2. Define and differentiate between
   (i) plane frame and space frame,
   (ii) statically determinate and statically indeterminate frame,
   (iii) perfect, deficient and redundant frame.
3. What distinguishes a roller support from a hinged support?
4. State the assumptions made while making an analysis of a framed structure.
5. Name the different methods of finding out the forces in the members of a perfect frame. Which one is used where and why?
6. Which one of the followings is a redundant frame and what is its degree of redundancy?

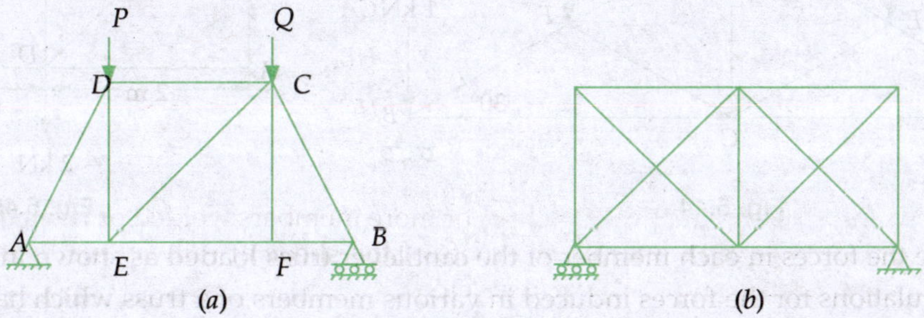

Fig. 5.41

7. Outline the steps involved while making an analysis of a simple truss by the (*i*) method of joints, and (*ii*) method of sections.
8. How will you find the forces in the members of truss joint by graphical method. Illustrate with an example.
9. With reference to the trusses shown in Fig. 5.42, identity the members which will not be subjected to any force.

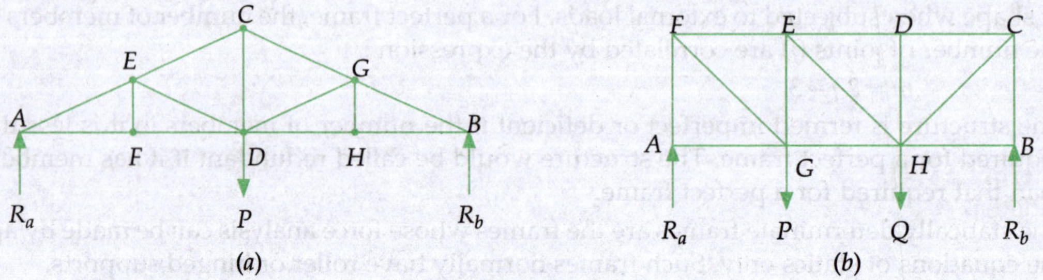

Fig. 5.42

10. Determine the magnitude and nature of forces in members of the truss loaded and supported as shown in Fig. 5.43.

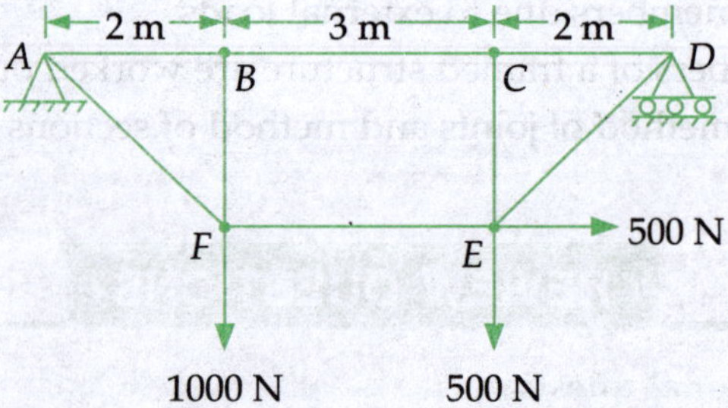

Fig. 5.43

11. Determine the magnitude and nature of forces in the various members of the truss loaded and supported as shown in Fig. 5.44.

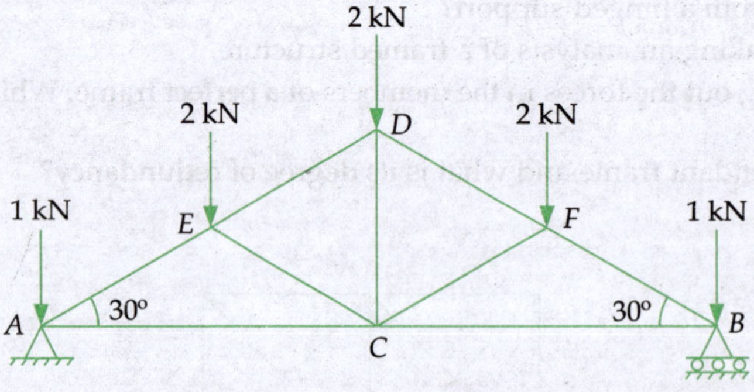

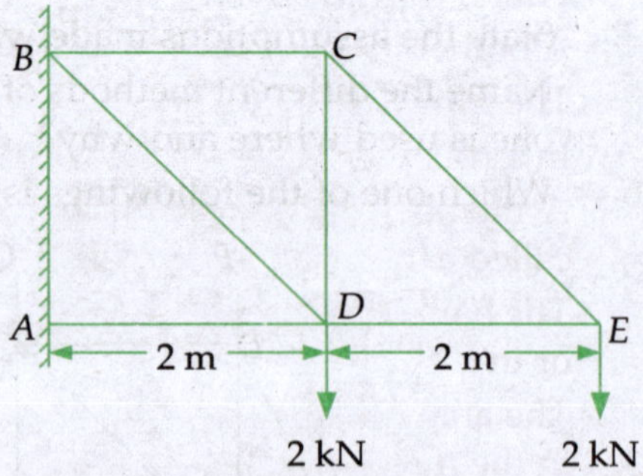

Fig. 5.44

Fig. 5.45

12. Determine the forces in each member of the cantilever truss loaded as shown in Fig. 5.45.
13. Make calculations for the forces induced in various members of a truss which has been loaded and supported as shown in Fig. 5.46.

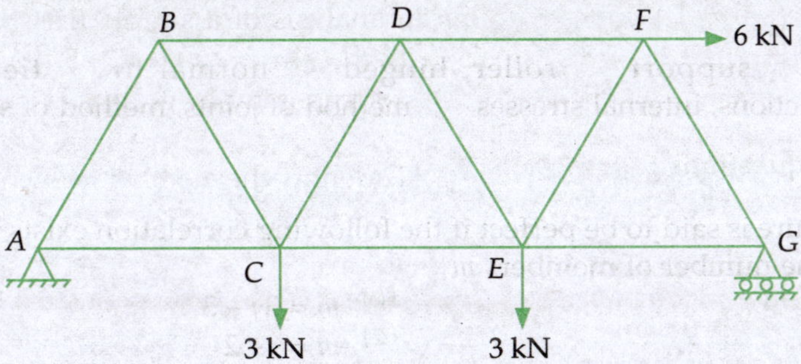

Fig. 5.46

14. Determine the forces in the members of a plane truss which is hinged to the foundation and is loaded as shown in Fig. 5.47.

($F_5 = F_5 = 0$; $F_4 = 2$ kN (*T*) ; $F_3 = 2.83$ kN (*C*) ; $F_2 = 2$ kN (*C*) and $F_1 = 2.83$ (*T*))

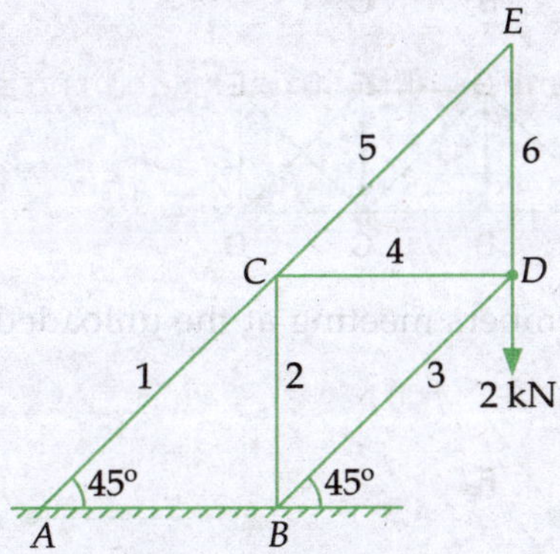

Fig. 5.47

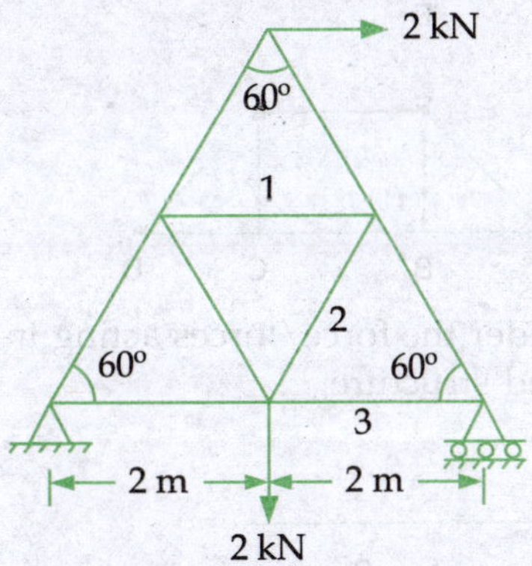

Fig. 5.48

15. A truss is loaded and supported as shown in Fig. 5.48. Make calculations for the axial forces set up in the members 1, 2 and 3.

($F_1 = 1$ kN (*C*); $F_2 = 1$ kN (*T*); and $F_3 = 1.58$ kN (*T*))

B. Complete the following statements with most appropriate word/words :

1. Frame is a .... structure that is used to .... loads
2. A body that supports another body acted upon by a system of forces is called a ....
3. The frames are generally supported on a .... support or a .... support
4. In a roller support, the line of reaction is .... the plane of rollers.
5. A member of a framed structure under tension is called a ...., and under compression is called a ....
6. The bars of a framed structured are called two-force members because they are either in .... or in ....
7. The analysis of a truss involves the determination of .... at the supports and .... induced in the members due to external loads
8. The .... and .... are the analytical methods for finding out the stresses in the members of a framed structure
9. If only a single force is acting at a joint, then for the equilibrium of this joint, the force must be ....

Answers:

1. rigid, support 2. support 3. roller, hinged 4. normal to 5. tie, strut 6. tension, compression 7. reactions, internal stresses 8. method of joints, method of sections 9. zero

C. Multiple choice questions :

1. A framed structure is said to be perfect if the following correlation exists between the number of joints j and the number of members m

(a) $m = 2j - 3$ (b) $m = 3j - 3$
(c) $m = 2j - 1$ (d) $m = j - 2$

2. Which of the structures shown below represents a redundant frame?

(a)
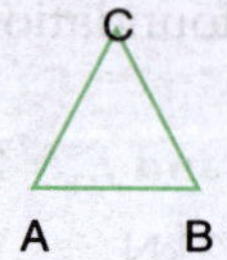

(b)
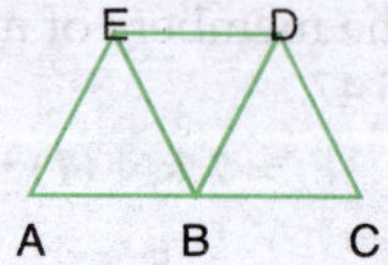

(c)
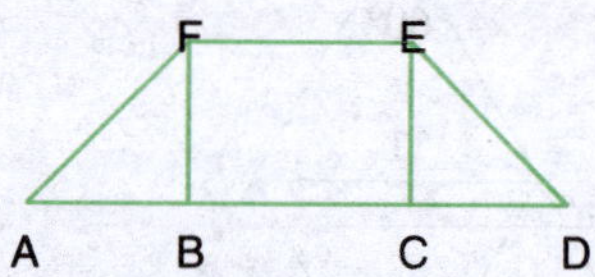

(d)
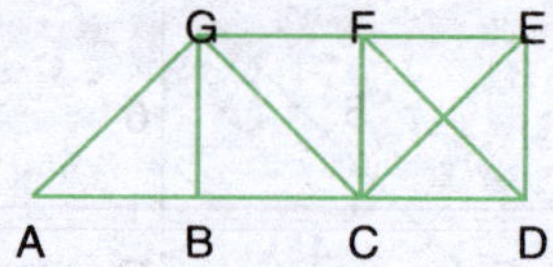

3. Consider the force/forces acting in the member/members meeting at the unloaded joint of a framed structure

1.
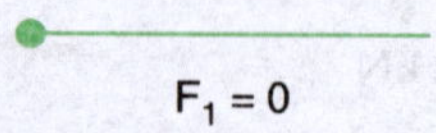

2.
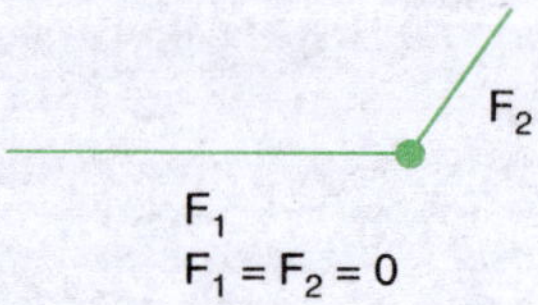

3.
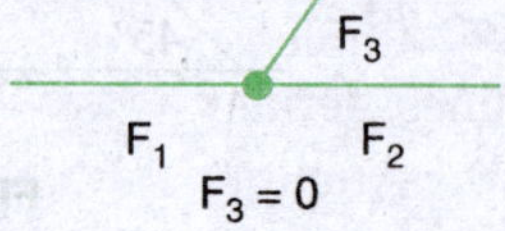

Which of these aspects is correct ?

(a) 1, 2 and 3 (b) 1 and 2 (c) 1 and 3 (d) 2 and 3

4. The figure given below shows a pin-jointed plane truss loaded at the point M by hanging a mass of 100 kg.

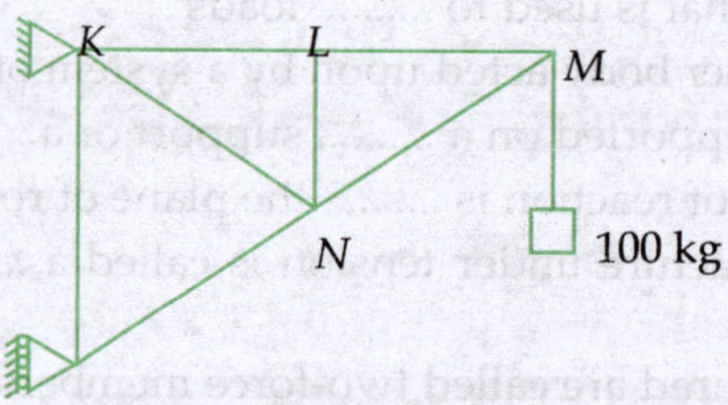

The member LN of the truss is subjected to a load of

(a) zero (b) 490 N is compression
(c) 981 N is compression (d) 981 in tension

Answers :

1. (a) 2. (d) 3. (a) 4. (a)

6 Cables and Strings

Learning objectives : This chapter has been devoted to the study of

- *cables and strings, and their applications*
- *analysis of cables with supports at same or different levels*
- *catenary : shape of the cable under its own weight*
- *analysis of cable with concentrated loads*

Cables and strings are flexible, light and inextensible members capable of withstanding only axial tensile loads. Due to flexibility, resistance to bending and shear force is small and hence neglected. When loaded, the cable takes up a profile that depends on its length and the loads supported by it. Being inextensible, the cable has constant length before and after the application of load.

Cables are used in many engineering applications such as suspension bridges, transmission lines, aerial tramways and guy wires for towers and radio antennas.

Attention has been directed in the following paragraphs to study the behaviour of cables and strings supported at same or different levels and subjected to concentrated or uniformly distributed loads. Necessary relations have been set up between tensions, span, sag, support reactions and the length of the cable.

6.1. CABLES WITH UNIFORMLY DISTRIBUTED LOAD

6.1.1 Cable with supports at same level

Consider a cable or string supported at points A and B at the same level. The cable carries uniformly distributed load of w per unit length over its span and due to loading, the cable hangs in the shape of a curve.

When supported at ends at the same level:

- the horizontal distance l between the supports is referred to as *span* of the cable, and
- the vertical distance y_0 from the supports to the lowest point is called *sag* or *dip*

The coordinates are chosen at the lowest point O of the hanging cable and equilibrium is considered of portion OP of the cable that extends from lowest point O up to point $P(x, y)$. The portion OP of the string is in equilibrium under the action of following set of forces:

(i) load wx acting vertically downwards at midway between O and P

(ii) pull H_0 acting horizontally at the lowest point O

(iii) tension T acting along the tangent at P and making an angle θ with the horizontal

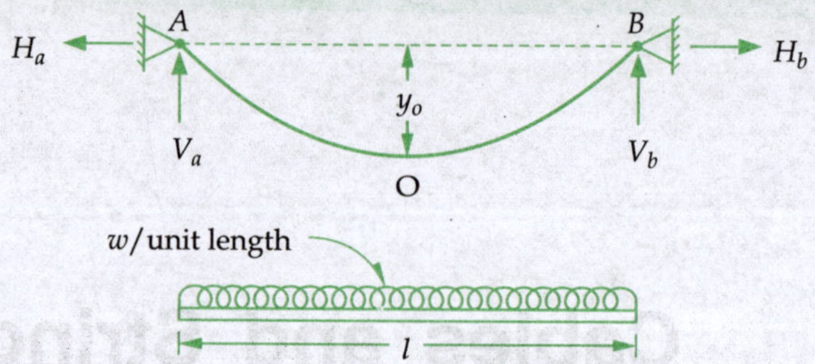

Applying the equilibrium conditions,

$$\Sigma F_x = 0; \qquad T\cos\theta - H_0 = 0$$

$$T\cos\theta = H_0 \qquad ...(i)$$

$$\Sigma F_y = 0; \qquad T\sin\theta - wx = 0$$

$$T\sin\theta = wx \qquad ...(ii)$$

From expression (*i*) and (*ii*) :

$$\tan\theta = \frac{wx}{H_0}$$

Knowing that $\tan\theta = dy/dx$, we may write

$$\frac{dy}{dx} = \frac{wx}{H_0}\ ; \ dy = \frac{wx}{H_0}dx$$

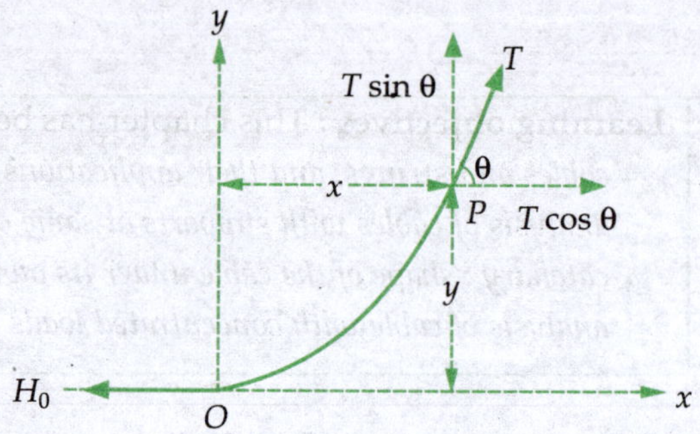

Fig. 6.1

Upon integration: $\qquad y = \dfrac{wx^2}{2H_0} + C$

The constant of integration can be worked out from the condition that $x = 0$ and $y = 0$ at point chosen as origin. That gives $C = 0$ and accordingly

$$y = \frac{wx^2}{2H_0} \qquad ...(6.1)$$

This is the equation of the curve in which the cable hangs: it represents a parabola with vertex at *O*. Hence

"A cable subjected to uniformly distributed load along the horizontal takes the shape of a parabola."

Tension in the cable: From expressions (*i*) and (*ii*)

$$T = \sqrt{(H_0)^2 + (wx)^2} \qquad ...(6.2)$$

The above equation gives the value of tension at any point of the cable. Tension is minimum and equal to H_0; the pull acting horizontally at the lowest point *O* of the hanging cable where $x = 0$. Tension would be maximum where *x* is maximum, i.e., at the support points.

Considering the equilibrium of the entire cable,

$$\Sigma F_x = 0: \qquad H_b - H_a = 0 \quad \text{or} \quad H_a = H_b$$

$$\Sigma F_y = 0: \qquad V_a + V_b - wl = 0 \quad \text{or} \quad V_a + V_b = wl$$

Due to symmetry, the vertical reactions at the point of support are :

$$V_a = V_b = \frac{wl}{2} \qquad ...(6.3)$$

The horizontal pull H_a or H_b can be worked out from the parabolic equation of the hanging cable:

$$y = \frac{wx^2}{2H_0}$$

The coordinates of end support A are $\left(-\frac{l}{2}, y_0\right)$ and therefore

$$y_a = \frac{w\left(-\frac{l}{2}\right)^2}{2H_a} = \frac{wl^2}{8H_a}$$

Since $y_a = y_b = y_0$ (curve being symmetrical), we have

$$H_a = H_b = \frac{wl^2}{8y_0} \qquad \text{...(6.4)}$$

Equation 6.4 gives minimum or horizontal tension in the cable.

Maximum tension occurs at the support and is equal to the resultant reaction

$$T_{\max} = R_a = \sqrt{V_a^2 + H_a^2}$$
$$= R_b = \sqrt{V_b^2 + H_b^2} \qquad \text{...(6.5)}$$

The above relation can be transformed as

$$T_{\max} = \sqrt{\left(\frac{wl}{2}\right)^2 + \left(\frac{wl^2}{8y_0}\right)^2}$$
$$= \frac{wl}{2}\sqrt{1 + \frac{l^2}{16y_0^2}} \qquad \text{...(9.6)}$$

If ϕ is the angle made by the resultant with the horizontal, then

$$\tan\phi = \frac{V_a}{H_a} = \frac{V_b}{H_b} \qquad \text{...(6.7)}$$

Length of cable: The cable hangs in the form of parabola given by

$$y = \frac{wx^2}{2H_0}$$

Upon differentiation with respect to x,

$$\frac{dy}{dx} = \frac{w}{2H_0} \times 2x = \frac{w}{H_0}x$$

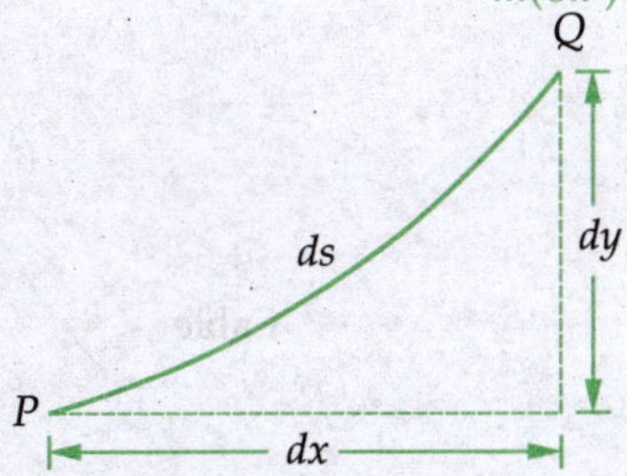

Fig. 6.2

Considering a small portion of length ds (Fig. 6.2) and taking the length of arc PQ equal to the length of the chord PQ, we have

$$ds = \sqrt{(dx)^2 + (dy)^2} = dx\sqrt{1 + (dy/dx)^2}$$
$$= dx\sqrt{1 + \left(\frac{w}{H_0}x\right)^2}$$

When the term inside the square root is expanded by binomial theorem, we get

$$ds = dx\left[1+\frac{1}{2}\frac{w^2x^2}{H_0^2}+\ldots\right] = dx\left[1+\frac{w^2}{2H_0^2}x^2\right] \qquad \text{Neglecting higher order terms}$$

Integrating the above expression between the limits $x = 0$ and $x = l/2$

$$s = \int_0^{l/2}\left[1+\frac{w^2}{2H_0^2}x^2\right]dx = \left[x+\frac{w^2}{2H_0^2}\times\frac{x^3}{3}\right]_0^{l/2} = \frac{l}{2}+\frac{w^2}{2H_0^2}\times\frac{l^3}{24}$$

$$= \frac{l}{2}+\frac{w^2}{48H_0^2}l^3$$

Since the dip point O has been taken as origin and the limits of integration have been from 0 to $l/2$, the above identity gives the length of half of the cable.

$\therefore$ Total length of the cable

$$L = 2\left[\frac{l}{2}+\frac{w^2}{48H_0^2}l^3\right] = l+\frac{w^2}{24H_0^2}l^3$$

Substituting the value of $H_0 = \dfrac{wl^2}{8y_0}$ in the above identity, we get

$$L = l+\frac{w^2}{24}l^3\times\left(\frac{8y_0}{wl^2}\right)^2 = l+\frac{8}{3}\frac{y_0^2}{l} \qquad \ldots(6.8)$$

EXAMPLE 6.1

A baloon exerts an upward force of 400 N on its ground cable at the attachment point as shown in Fig. 6.3. A horizontal wind causes shift to the position shown. If the cable weighs 6 N/m, determine the maximum tension in the cable if its length is 40 m.

Solution : Consider the free body diagram of the cable as shown in Fig. 6.4.

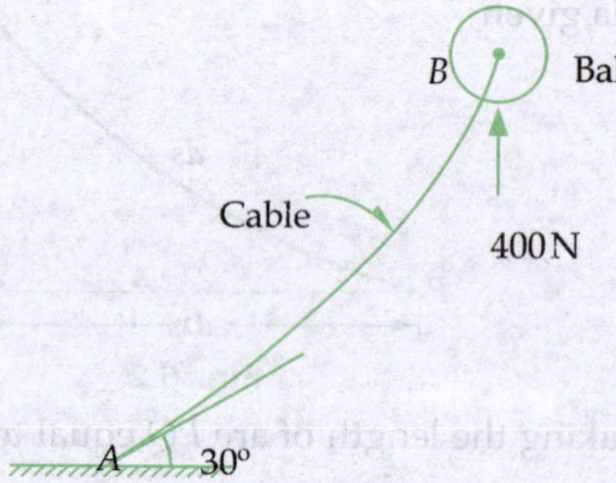

Fig. 6.3

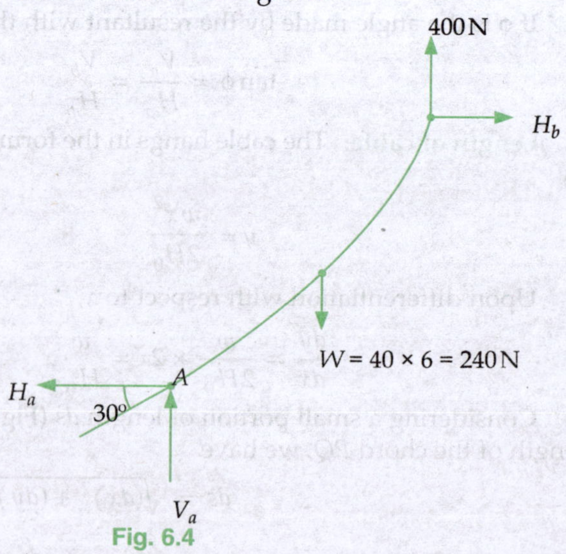

Fig. 6.4

From conditions of equilibrium,

$\Sigma F_x = 0\,;\qquad H_b - H_a = 0$

$\Sigma F_y = 0\,;\qquad 400 + V_a - 240 = 0\,;\qquad V_a = -160$ N

The $-ve$ sign indicates that the reaction V_a is acting in the downward direction.

At point A, the resultant reaction must be tangential to the cable at that point. Accordingly

$$\frac{V_a}{H_a} = \tan 30°\,;\quad H_a = \frac{160}{\tan 30°} = \textbf{277.3 N}$$

$$H_b = H_a = 277.3 \text{ N}$$

Maximum tension in the cable will be at the upper end B, and this is the resultant of buoyant force of 400 N and H_b = 277.3 N

$$T_{max} = \sqrt{400^2 + 277.3^2} = \textbf{486.72 N}$$

Tension at the lower end,

$$T_a = R_a = \sqrt{V_a^2 + H_a^2} = \sqrt{160^2 + 277.3^2} = \textbf{320.15 N}$$

EXAMPLE 6.2

A cable is stretched between two supports which are at the same level having a span of 100 meters. The sag of the cable is 10 metres and the tension at the lowest point of the cable is 600 kN. If the cable carries a uniformly distributed load over its horizontal span, make calculations for

(a) *the reactions at the supports,*

(b) *the minimum and maximum tensions in the cable, and*

(c) *the length of the loaded cable.*

Solution : Fig. 6.5 shows the cable indicating the span, sag, load and the reactions at the supports A and B which are at the same level.

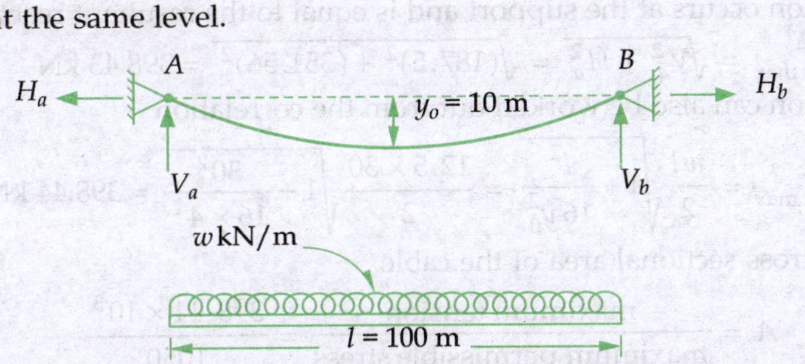

Fig. 6.5

The tension at the lowest point of the cable is given by

$$H_0 = \frac{wl^2}{8y_0} \quad \therefore \quad w = \frac{8y_0 H_0}{l^2}$$

$$= \frac{8 \times 10 \times 600}{100^2} = 4.8 \text{ kN/m}$$

Vertical reactions at the support points are

$$V_a = V_b = \frac{wl}{2} = \frac{4.8 \times 100}{2} = 240 \text{ kN}$$

$$H_a = H_b = H_0 = 600 \text{ kN}$$

Reactions at the support are

$$R_a = R_b = \sqrt{(240)^2 + (600)^2} = \textbf{646.22 kN}$$

Let ϕ be the angle which the resultant makes with the horizontal. Then

$$\tan\phi = \frac{240}{600} = 0.4\,; \qquad \phi = 21.8°$$

(*b*) Maximum tension occurs at the support and is equal to the resultant reaction

$$T_{max} = R_a = R_b = 646.22 \text{ kN}$$

EXAMPLE 6.3

A suspension bridge having 30 m span and 1.25 m wide platform is subjected to an average load of 20 kN/m². The bridge is supported by a pair of cables having central dip of 4 m. If the maximum permissible stress in the cable material is not to exceed 1050 N/mm², make calculations for the necessary cross-section of the cable.

Solution : Total load = average load × area of platform

$$= 20 \times (30 \times 1.25) = 750 \text{ kN}$$

Since the bridge is supported by a pair of cables,

$$\text{load per cable} = \frac{750}{2} = 375 \text{ kN}$$

Load per meter length of the cable $w = \frac{375}{30} = 12.5 \text{ kN/m}$

If A and B are the support points at the same level, then

$$V_a = V_b = \frac{wl}{2} = \frac{12.5 \times 30}{2} = 187.5 \text{ kN}$$

$$H_a = H_b = H_0 = \frac{wl^2}{8y_0} = \frac{12.5 \times 30^2}{8 \times 4} = 351.56 \text{ kN}$$

Maximum tension occurs at the support and is equal to the resultant reaction

$$T_{max} = \sqrt{V_a^2 + H_a^2} = \sqrt{(187.5)^2 + (351.56)^2} = 398.43 \text{ kN}$$

Maximum tension can also be worked out from the correlation

$$T_{max} = \frac{wl}{2}\sqrt{1 + \frac{l^2}{16y_0^2}} = \frac{12.5 \times 30}{2}\sqrt{1 + \frac{30^2}{16 \times 4^2}} = 398.44 \text{ kN}$$

∴ Necessary cross-sectional area of the cable

$$A = \frac{\text{maximum tension}}{\text{maximum permissible stress}} = \frac{398.44 \times 10^3}{1050}$$

$$= 379.47 \text{ mm}^2$$

EXAMPLE 6.4

A steel wire of uniform cross-section and mass density 8000 kg/m³ hangs in the form of a parabola. If the central dip equals 1/10th of span and the stress in the steel wire is limited to 120 MN/m², make calculations for the maximum horizontal span.

Solution : Length of the wire $L = l + \frac{8}{3}\frac{y_0^2}{l} = l + \frac{8}{3}\frac{(l/10)^2}{l} = 1.027\,l$

Total weight of the wire $= (\rho\, g) \times AL$

$$= (8000 \times 9.81) \times A \times 1.027\,l = 80.6 \times 10^3\,Al$$

If A and B are the support points at the same level, then

$$V_a = V_b = \frac{W}{2} = 0.5\,W$$

$$H_a = H_b = H = \frac{wl^2}{8y_0} = \frac{Wl}{8 \times l/10} = 1.25\ W$$

Maximum tension occurs at the support and is equal to the resultant reaction. Therefore

$$T_{max} = R_a = \sqrt{V_a^2 + H_a^2} = \sqrt{(0.5\,W)^2 + (1.25\,W)^2}$$

$$= 1.346\ W = 1.346 \times (80.6 \times 10^3\ Al)$$

The maximum permissible stress f_{max} is given by

$$f_{max} = \frac{T_{max}}{\text{cross} - \text{sectional area}}$$

$$120 \times 10^6 = \frac{1.346 \times (80.6 \times 10^3\ Al)}{A}$$

$\therefore$ Maximum horizontal space $l = \dfrac{120 \times 10^6}{1.346 \times (80.6 \times 10^3)} =$ **1106 m**

EXAMPLE 6.5

A telephone cable is suspended between two poles 120 m apart and on the same level. The cable weighs 75 N/m and can resist a pull of 12 kN. If length of the cable is 150 m, make calculations for the dip of the cable and the total weight which could bc uniformly supported along the cable.

Solution : Length of the cable $L = l + \dfrac{8}{3}\dfrac{y_0^2}{l}$

$$150 = 120 + \frac{8}{3} \times \frac{y_0^2}{120}$$

That gives: Dip of cable $y_0 = \sqrt{\dfrac{(150-120) \times 3 \times 120}{8}} = 36.74$ cm

If A and B are the support points at the same level, then

$$V_a = V_b = \frac{wl}{2} = \frac{w \times 120}{2} = 60\ w$$

$$H_a = H_b = H_0 = \frac{wl^2}{8y_0} = \frac{w \times 120^2}{8 \times 36.74} = 49\ w$$

Maximum tension occurs at the support and is equal to the resultant reaction. Therefore

$$T_{max} = \sqrt{V_a^2 + H_a^2}\ ; \qquad 12 \times 10^3 = \sqrt{(60\,w)^2 + (49\,w)^2}$$

$$= 3600\ w^2 + 2401\ w^2 = \sqrt{6001\,w^2}$$

$\therefore$ Weight per meter length of span, $w = \dfrac{(12 \times 10^3)^2}{6001} = 154.91$ N/m

Self weight of cable = 75 N/m

$\therefore$ External load that can be supported = 154.91 – 75 = 79.91 N/m

Hence external load that can be supported

= 79.91 × 120 = **9589.2 N**

6.1.2 Cable with supports at different levels

Consider a cable supported at different levels and carrying a uniformly distributed load over its horizontal span. O is the lowest point of the cable and is taken as origin.

The analysis is made carrying out the following steps:

(*i*) Portion OB of the cable is extended to OB_1 such that B_1 is at the same level as that at A. Obviously OB_1 is the mirror of OA

For cable AOB_1 :

Span = $2l_1$ and central dip = y_1

$$\text{Horizontal thrust } H_a = \frac{wl^2}{8y_0} = \frac{w(2l_1)^2}{8y_1}$$

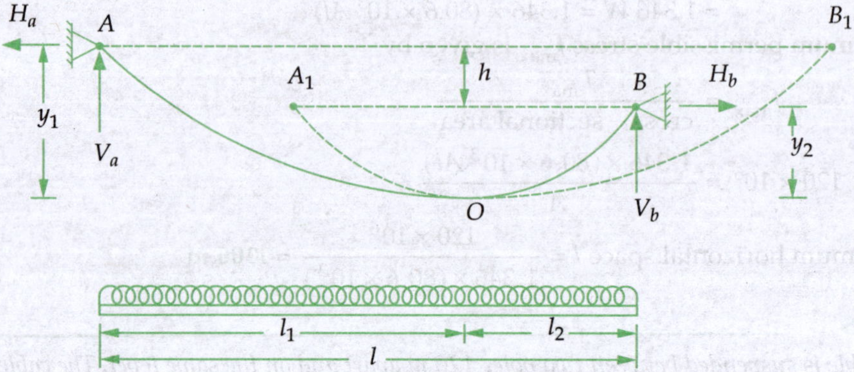

Fig. 6.6

(ii) Portion AO of the cable is cut short to A_1O such that A_1 is at the same level as that at B. Obviously OA_1 is the mirror image of OB.

For cable A_1OB :

Span = $2l_2$ and central dip = y_2

$$\text{Horizontal thrust } H_b = \frac{wl^2}{8y_0} = \frac{w(2l_2)^2}{8y_2}$$

(iii) Since the cable is supporting vertical loads only, the horizontal thrust at A must be equal to horizontal thrust at B.

$$\therefore \quad \frac{w(2l_1)^2}{8y_1} = \frac{2(2l_2)^2}{8y_2}$$

That gives :

$$\frac{l_1}{l_2} = \sqrt{\frac{y_1}{y_2}} \qquad \text{...(6.9)}$$

The above correlation helps to locate the position of the lowest point O which is taken as origin

(iv) Co-ordinates of support $A \equiv -l_1, y_2$

Co-ordinates of support $B \equiv l_2, y_2$

$$\text{Horizontal thrust } H_a = \frac{wx^2}{2y} = \frac{w(-l_1)^2}{2y_1} = \frac{wl_1^2}{2y_1} = \frac{wl_2^2}{2y_2}$$

Vertical reactions at the support ends can be obtained by taking moments about the support end B

$$V_a \times l = wl \times \frac{l}{2} - H_a \times h$$

$$\therefore \quad V_a = \frac{wl}{2} - \frac{H \times h}{l}$$

Similarly $$V_b = \frac{wl}{2} - \frac{H \times h}{l} \qquad (\because \; H_a = H_b = H)$$

Tensions in the string at support points A and B are

$$T_a = \sqrt{V_a^2 + H^2} \quad \text{and} \quad T_b = \sqrt{V_b^2 + H^2}$$

Since the support A is higher than support B, R_a is more than R_b and accordingly the maximum tension in the cable will be at A.

(v) Length of a hanging cable $L = l + \frac{8}{3}\frac{y_0^2}{l}$

$\therefore$ Length of cable AOB_1,

$$L_1 = 2l_1 + \frac{8}{3}\frac{y_1^2}{2l_1} = 2\left[l_1 + \frac{2}{3}\frac{y_1^2}{l_1}\right]$$

Length of cable A_1OB,

$$L_2 = 2l_2 + \frac{8}{3}\frac{y_2^2}{2l_2} = 2\left[l_2 + \frac{2}{3}\frac{y_2^2}{l_2}\right]$$

Actual length of cable AOB,

$$L = \frac{L_1 + L_2}{2} = l_1 + \frac{2}{3}\frac{y_1^2}{l_1} + l_2 + \frac{2}{3}\frac{y_2^2}{l_2}$$

$$= l + \frac{2}{3}\left(\frac{y_1^2}{l_1} + \frac{y_2^2}{l_2}\right) \qquad \text{...(6.10)}$$

EXAMPLE 6.6

A cable of 30 m span is having supports at different levels, the right support is higher than the left by 10 m. The lowest point of the cable is 5 m below the left support. The cable is subjected to a uniformly distributed load of 250 N/m over its horizontal span. Determine the followings:

(a) *the reactions at the supports both in magnitude and direction*

(b) *the maximum and minimum tension occurring in the cable, and*

(c) *the length of the cable.*

Solution : Refer Fig. 6.7 indicating various perameters for the cable. The lowest point O is selected as reference of co-ordinate axes.

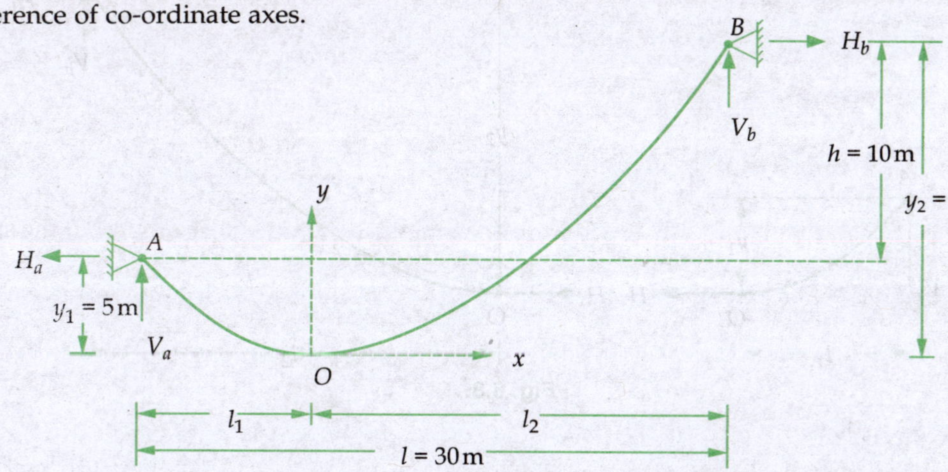

Fig. 6.7

$y_1 = 5$ m and $y_2 = 15$ m are the dip of point O with respect to left and right supports respectively.

$$l_1 + l_2 = 30 \text{ m}$$

and
$$\frac{l_1}{l_2} = \sqrt{\frac{y_1}{y_2}} = \sqrt{\frac{5}{15}} = 0.577\;;\; l_1 = 0.577\, l_2$$

$$\therefore \quad 0.577\, l_2 + l_2 = 30\;; \quad l_2 = 19.02 \text{ m}$$

$$l_1 = 30 - 19.02 = 10.98 \text{ m}$$

From the equilibrium conditions:

$$\Sigma F_x = 0; \quad -H_a + H_b = 0\;; \quad H_a = H_b$$

$$H_a = \frac{w l_1^2}{2y_1} = \frac{250 \times 10.98^2}{2 \times 5} = 3014 \text{ N}$$

$$H_b = \frac{w l_2^2}{2y_2} = \frac{250 \times 19.02^2}{2 \times 15} = 3014 \text{ N}$$

The calculated values satisfy the equilibrium condition $\Sigma F_x = 0$

$$\Sigma F_y = 0: \quad V_a + V_b - 250 \times 30 = 0$$

ΣM about both A and $B = 0$.

$$-V_b \times 30 + H_b \times 10 + (250 \times 30) \times \frac{30}{2} = 0$$

$$V_b \times 30 = 3014 \times 10 + 250 \times 30 \times 15 = 142640$$

$$\therefore \quad V_b = \frac{142640}{30} = 4755 \text{ N}$$

and
$$V_a = 250 \times 30 - V_b = 7500 - 4755 = 2745 \text{ N}$$

Alternatively. The vertical reactions V_a and V_b can be easily worked out by drawing the free body diagrams of portions AO and OB of the cable.

$$V_a = w l_1 \quad \text{and} \quad V_b = w l_2 \text{ for equilibrium in } y\text{-direction}$$

$$V_a = 250 \times 10.98 = 2745 \text{ N}$$

$$V_b = 250 \times 19.02 = 4755 \text{ N}$$

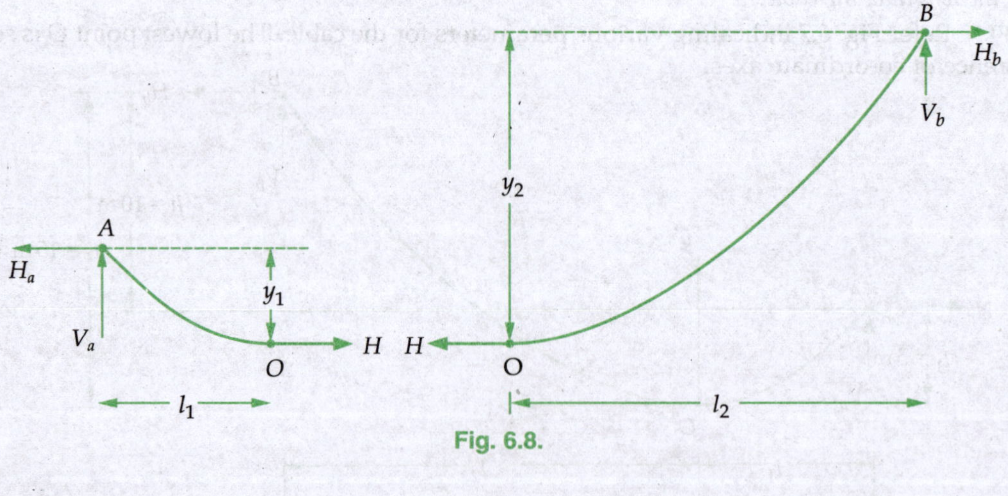

Fig. 6.8.

(*a*) Reaction at support *A*

$$R_a = \sqrt{V_a^2 + H_a^2} = \sqrt{2745^2 + 3014^2} = \mathbf{4077\ N}$$

$$\theta_a = \tan^{-1}\left(\frac{2745}{3014}\right) = \mathbf{42.32°}$$

Reaction at support *B*,

$$R_b = \sqrt{V_b^2 + H_b^2} = \sqrt{4755^2 + 3014^2} = \mathbf{5630\ N}$$

$$0_b - \tan^{-1}\left(\frac{4755}{3014}\right) = \mathbf{57.63°}$$

(*b*) Maximum tension occurs at the higher support and is equal to the resultant reaction at that support.

$$T_{max} = R_b = \mathbf{5630\ N} \quad \text{and} \quad T_{min} = H_a = H_b = \mathbf{3014\ N}$$

(*c*) Length of the loaded cable

$$L = l + \frac{2}{3}\left(\frac{y_1^2}{l_1} + \frac{y_2^2}{l_2}\right) = 30 + \frac{2}{3}\left(\frac{5^2}{10.98} + \frac{15^2}{19.02}\right) = \mathbf{39.404\ m}$$

6.2. CATENARY

When the cable is subjected to a uniformly distributed load per unit length along the cable itself, it hangs freely and it takes up a shape known as Catenary. It is essentially the shape of the cable or string under its own weight.

Consider a cable suspended at two points *A* and *B* which may be at the same or different level. The cable is shown hanging freely under its own weight with C as the lowest point. The cable is flexible but not weightless. Such a situation occurs in cables used in transmission lines.

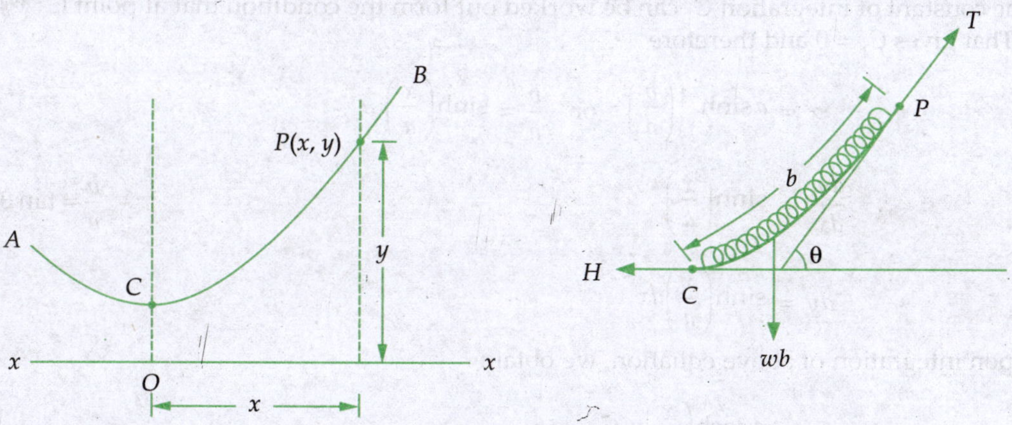

Fig. 6.9

The portion *CP* of the cable is in equilibrium under the action of the following set of forces:

(*i*) self load *wb* acting vertically downwards. Here *b* is the length of portion *CP* of the cable and *w* is the weight per unit length of the cable.

(*ii*) pull *H* acting horizontally at the lowest point C

(*iii*) tension *T* acting along the tangent. The tangent is inclined at an angle θ with the horizontal.

Applying conditions of equilibrium,

$\Sigma F_x = 0$: $\quad T\cos\theta - H = 0$ or $T\cos\theta = H$ ...(*i*)

$\Sigma F_y = 0$: $\quad T\sin\theta - wb$ or $T\sin\theta = wb$ ...(*ii*)

Dividing expression (*ii*) by (*i*), we get

$$\tan\theta = \frac{wb}{H}$$

Let it be presumed that horizontal pull H is equal to weight of length a of the cable such that $H = wa$. Then

$$\tan\theta = \frac{wb}{wa} = \frac{b}{a}\ ; \qquad b = a\tan\theta \qquad ...(6.11)$$

Equation 6.11 is referred to as ***intrinsic equation of the catenary.***

It is considered worthwhile to transform the intrinsic equation into cartesian co-ordinates to understand its various characteristics.

Knowing that :

$$\tan\theta = \frac{dy}{dx} \quad \text{and} \quad db = dx\sqrt{1+(dy/dx)^2} \qquad \text{(see Fig. 6.2)}$$

$$dx = \frac{db}{\sqrt{1+(dy/dx)^2}} = \frac{db}{\sqrt{1+\tan^2\theta}}$$

$$= \frac{db}{\sqrt{1+b^2/a^2}} \qquad (\because \ b = a\tan\theta)$$

Upon integration of above equation, we get

$$x = a\sinh^{-1}\left(\frac{b}{a}\right) + C_1$$

The constant of integration C_1 can be worked out form the condition that at point $C : x = 0$ and $a = 0$. That gives $C_1 = 0$ and therefore

$$x = a\sinh^{-1}\left(\frac{b}{a}\right) \quad \text{or} \quad \frac{b}{a} = \sinh\left(\frac{x}{a}\right) \qquad ...(iv)$$

$$\therefore \quad \frac{dy}{dx} = \sinh\left(\frac{x}{a}\right) \qquad \left(\because \ \frac{b}{a} = \tan\theta = \frac{dy}{dx}\right)$$

$$dy = \sinh\left(\frac{x}{a}\right)dx$$

Upon integration of above equation, we obtain

$$y = a\cosh\left(\frac{x}{a}\right) + C_2$$

The constant of integration C_2 can be worked out from the condition that at point $C : x = 0$ and $a = 0$. That gives $C_2 = 0$ and therefore

$$y = a\cosh\left(\frac{x}{a}\right) \qquad ...(6.12)$$

Equation 6.12 is the equation of catenary in cartesian coordinates. Distance a is called *parameter* of the catenary, point C is *vertex* of the catenary and x - x is the *directrix*

The cartesian equation of the catenary can be re-written as

$$\frac{y}{a} = \cosh\left(\frac{x}{a}\right) \quad \text{or} \quad \frac{y^2}{a^2} = \cosh^2\left(\frac{x}{a}\right) \qquad ...(v)$$

Also from expression (*iv*) : $\frac{b^2}{a^2} = \cosh^2\left(\frac{x}{a}\right)$ ...(*vi*)

Subtracting identity (*vi*) from identity (*v*), we get

$$\frac{y^2}{a^2} - \frac{b^2}{a^2} = \cosh^2\left(\frac{x}{a}\right) - \sinh^2\left(\frac{x}{a}\right)$$

$$= 1 \qquad (\because \cosh^2\theta - \sinh^2\theta = 1)$$

$$\therefore \quad y^2 - b^2 = a^2 \quad \text{or} \quad y = a^2 + b^2$$

$$= a^2 + a^2\tan^2\theta \qquad (\because b = a\tan\theta)$$

$$= a^2(1 + \tan^2\theta) = a^2\sec^2\theta$$

$$\therefore \quad y = a\sec\theta \qquad ...(6.13)$$

From expression (*i*) : $T\cos\theta = H$

$$T = \frac{H}{\cos\theta} = H\sec\theta$$

$$= wa\sec\theta \qquad (\because H = wa)$$

Substituting the value of $a\sec\theta$ as per equation 6.13, we get

$$T = wy \qquad ...(6.14)$$

Thus the tension at any point P is equal to weight of string of length equal to the height of the point above the directrix.

The horizontal co-ordinate of the P is defined by the relation

$$x = 2.3\,a\log_{10}(\sec\theta + \tan\theta) \qquad ...(6.15)$$

Note: When the quantity x/a becomes very small or the cable is very light, the catenary approximates to a parabola.

EXAMPLE 6.7

A 7 m long heavy string ABCDE hangs over two smooth pegs as shown in Fig. 6.10. Locate the position of vertex C along the string with reference to peg B.

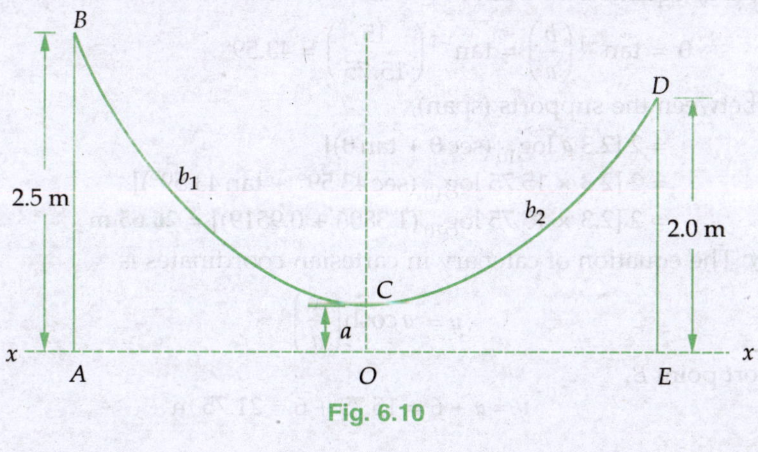

Fig. 6.10

Solution : Length of string $BCD = 7 - 2.5 - 2.0 = 2.5$ m

Let b_1 and b_2 be the lengths of portions BC and CD respectively of the string. Then

$$b_2 = (2.5 - b_1) \text{ m}$$

The equation of catenary is:

$$y^2 = a^2 + b^2$$

For the peg B: $b = b_1$ and $y = 2.5$ m

For the peg A: $b = b_2 = (2.5 - b_1)$ m and $y = 2.0$ m

Substituting these values in the general equation of the catenary

$$(2.5)^2 = a^2 + b_1^2 \text{ or } 6.25 = a^2 + b_1^2 \quad ...(i)$$

and $$(2.0)^2 = a^2 + (2.5 - b_1)^2 \text{ or } 4 = a^2 + 6.25 + b_1^2 - 5b_1 \quad ...(ii)$$

Subtracting expression (*ii*) from expression (*i*), we get

$$2.25 = -6.25 + 5b_1$$

$$\therefore \quad b_1 = \frac{2.25 + 6.25}{5} = \mathbf{1.7 \text{ m}}$$

The vertex of the catenary is thus located at 1.7 m from peg B along the support.

EXAMPLE 6.8

A cable of length 30 m and mass 100 kg hangs between two points at the same level and has a central dip of 6 m. Determine the distance between the two supports and the maximum tension in the cable.

Solution : The situation corresponds to a cable that hangs freely due to its own weight.

The equation of catenary is:

$$y = a^2 + b^2$$

For the support point B:

$$b = \frac{30}{2} = 15 \text{ m} \quad \text{and} \quad y = (a + 6) \text{ m}$$

$$\therefore \quad (a + 6)^2 = a^2 + 15^2$$

or $$a^2 + 36 + 12a = a^2 + 225$$

or $$12a = 225 - 36 = 189; \quad a = \frac{189}{12} = 15.75 \text{ m}$$

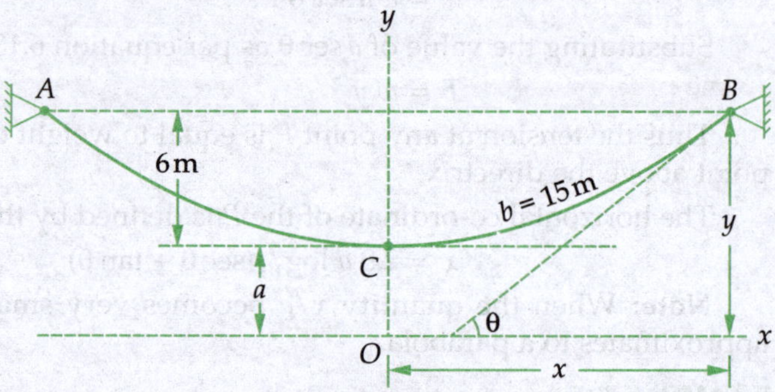

Fig. 6.11

From the catenary equation $b = a \tan\theta$

$$\theta = \tan^{-1}\left(\frac{b}{a}\right) = \tan^{-1}\left(\frac{15}{15.75}\right) = 43.59°$$

(*a*) Distance between the supports (span),

$$= 2\,[2.3\, a \log_{10}(\sec\theta + \tan\theta)]$$

$$= 2\,[2.3 \times 15.75 \log_{10}(\sec 43.59° + \tan 43.59°)]$$

$$= 2\,[2.3 \times 15.75 \log_{10}(1.3806 + 0.9519)] = \mathbf{26.65 \text{ m}}$$

Alternatively The equation of catenary in cartesian coordinates is

$$y = a\cosh\left(\frac{x}{a}\right)$$

For the support point B,

$$y = a + 6 = 15.75 + 6 = 21.75 \text{ m}$$

$$\therefore \quad 21.75 = 15.75 \cosh\left(\frac{x}{15.75}\right)$$

or
$$\frac{x}{15.75} = \cosh^{-1}\left(\frac{21.75}{15.75}\right) = 0.847$$

$$x = 15.75 \times 0.847 = 13.34$$

$\therefore$ Distance between the supports,

$$\text{Span} = 2x = 2 \times 13.34 = \mathbf{26.68\ m}$$

(*b*) Maximum tension in the cable occurs at end supports *A* or *B* (being at the same level)

$$T_{max} = wy$$

where w = self weight per meter length of the cable

$$= \frac{100 \times 9.81}{30} = 32.7 \text{ N/m}$$

$\therefore$ Maximum tension $T_{max} = 32.7 \times (a + 6)$

$$= 32.7 \times (15.75 + 6) = \mathbf{711.22\ N}$$

6.3. CABLE WITH CONCENTRATED LOADS

When the cables are subjected to concentrated loads and supported at same or different levels, calculation need to made in respect of the following parameters:-

(*a*) vertical reactions and horizontal thrusts at the end supports

(*b*) sag at the points of application of concentrated loads, and

(*c*) tensions induced in all he segments of the cables.

The procedure of making such calculations has been illustrated by Example 6.9,

EXAMPLE 6.9

A string has been subjected to the loading system as shown in Fig. 6.12 given below:

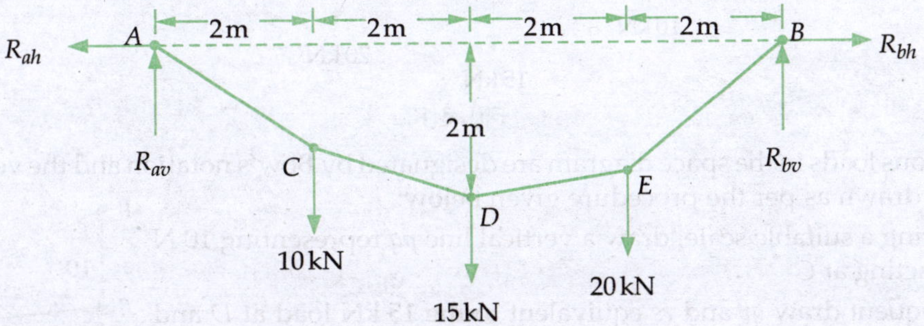

Fig. 6.12

The point D is at a depth of 2 m from the level of supports at A and B. Make calculations for the following parameters:

(*a*) *vertical reactions and horizontal thrusts at the end supports,*

(*b*) *sag at the loading points C and E, and*

(*c*) *tensions induced in all the segments of the string.*

Solution : Let R_{av} and R_{ah} be vertical reaction and horizontal thrust at end support *A*. The corresponding parameters at end support *B* are designated by R_{bv} and R_{bh} respectively.

Taking moments about point *A*, we get

$$R_{bv} \times 8 - 20 \times 6 - 15 \times 4 - 10 \times 2 = 0$$

$$\therefore \quad R_{bv} = \frac{20 \times 6 + 15 \times 4 \times 10 \times 2}{8} = \frac{120 + 60 + 20}{8} = 25 \text{ kN}$$

Also $\quad R_{av} + R_{bv} = 10 + 15 + 20 = 45$

$$\therefore \quad R_{av} = 45 - 25 = 20 \text{ kN}$$

(*b*) Considering equilibrium of string *ACD* and taking moments about point *D* (clockwise positive), we obtain

$$R_{av} \times 4 - R_{ah} \times 2 - 10 \times 2 = 0$$

$$R_{ah} = \frac{4 R_{av} - 20}{2} = \frac{4 \times 20 - 20}{2} = 30 \text{ kN}$$

Also $R_{bh} = 30$ kN

(*c*) Sag at points *C* and *E*

Taking moments of forces acting in the string *AC* about *C*,

$$R_{av} \times 2 - R_{ah} \times y_c = 0$$

$$\therefore \quad y_c = \frac{2 R_{av}}{R_{ah}} = \frac{2 \times 20}{30} = 1.33 \text{ m}$$

Taking moments of forces acting in the string *BE* about *E*,

$$R_{bh} \times y_e - R_{bv} \times 2 = 0$$

$$y_e = \frac{2 R_{bv}}{R_{bh}} = \frac{2 \times 25}{30} = 1.67 \text{ m}$$

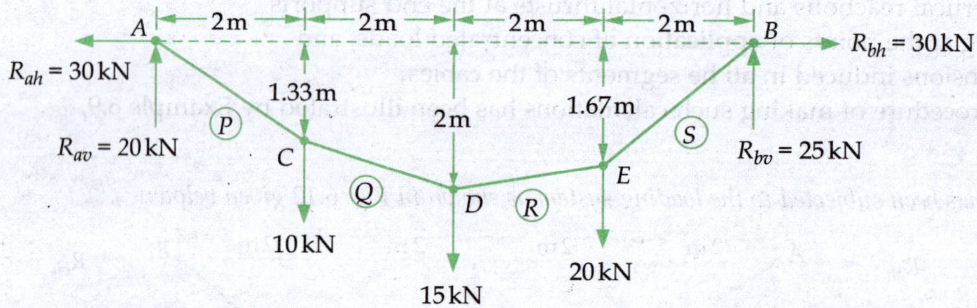

Fig. 6.13

The various loads in the space diagram are designated by Bow's notation and the vector diagram (Fig. 6.14) is drawn as per the procedure given below:

(*i*) Selecting a suitable scale, draw a vertical line *pq* representing 10 N load acting at *C*

(*ii*) Subsequent draw *qr* and *rs* equivalent to the 15 kN load at *D* and 20 kN load at *E*.

(*iii*) From points *p* and *q* draw lines parallel to *AC* and *CD* respectively and let these lines meet at point *O*.

(*iv*) Join *or* and *os*.

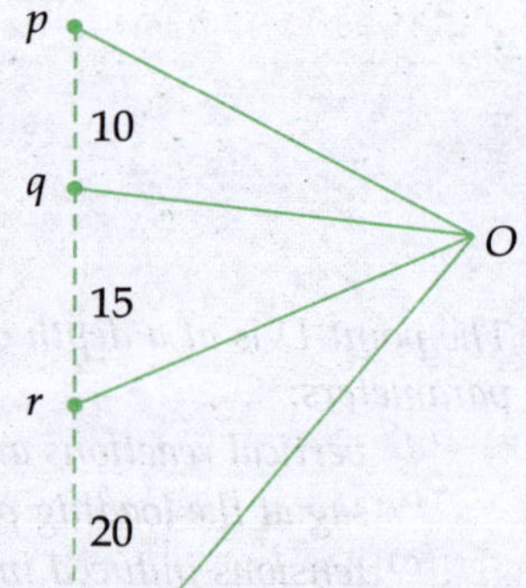

Fig. 6.14

Then by measurements, we have

T_1 (Tension in string *AC*) $\equiv op = 36$ kN

T_2 (Tension in string *CD*) $\equiv oq = 32$ kN

T_3 (Tension in string *DE*) $\equiv or = 30.5$ kn

T_4 (Tension in string *EB*) $\equiv os = 39$ kN

REVIEW AND SUMMARY

1. Cables and strings are flexible, light and inextensible members capable of withstanding only axial tensile loads.
2. When loaded, the cable takes up a profile that depends on its length and the loads supported by it.
3. Cables are used in suspension bridges, transmission lines, aerial tramways and guy wires for towers and radio antennas.
4. When supported at ends at the same level
   (*i*) span represents the horizontal distance between the supports,
   (*ii*) sag or dip represents the vertical distance from the supports to the lowest point of the hanging cable
   (*iii*) tension is maximum at the support point, and is minimum at the lowest point; the pull acts horizontal at the lowest point
   (*iv*) cable takes the shape of parabola when subjected to uniformly distributed load along the span.
5. Catenary is the shape of the cable or string under its own weight.

REVIEW QUESTIONS

A. Conceptual and conventional questions :

1. What is a cable? For what purpose it is used?
2. Show that when a cable carries load which is uniformly distributed along the horizontal, it hangs in the shape of a parabola
3. Where do the maximum and minimum tensions occur in a cable carrying uniformly load along the horizontal?
4. Set up an expression for tension at the two ends of a cable when it is supported at different levels.
5. Define span and sag (dip) in the context of cables.
6. What is a catenary?
7. A cable AB supports three loads as shown in the figure given below:

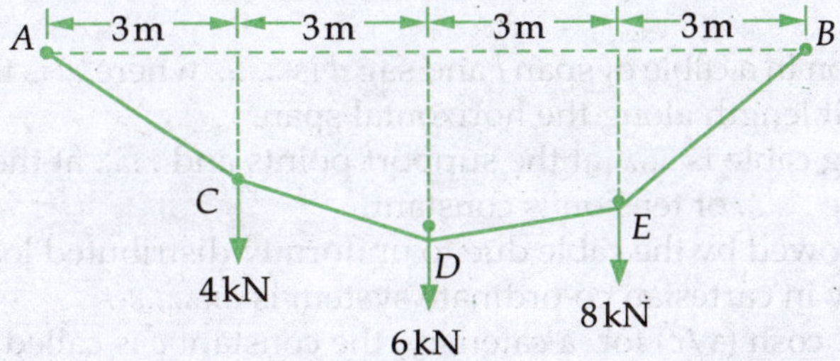

Fig. 6.15

If the dip at the central load is estimated to be 2 m, work out the components of reaction at support *A*, the sag under 4 kN load and the tension in portion *CD* of the cable.

(R_{av} = 8 kN, R_{ah} = 18 kN, y_c = 1.33 m and T_{cd} = 18.5 kN)

8. A wire carrying a load of 5 kN per meter run of horizontal span is stretched between two supports 200 m apart. The supports are at the same level and the wire has sag of 10 m. Determine (*a*) maximum and minimum tension induced in the wire, (*b*) length of the wire.

(2.55 MN, 2.5 MN, 201.33 m)

9. An electric cable is attached to the electric poles 25 m apart. The cable weighs 2.5 N/m and can resist a pull of 300 N. Assuming that the cable takes a parabolic shape, determine the smallest value of sag in the cable. **(0.651 m)**

10. A light cable is attached to a support at A, passes over a small pulley at B and supports a load W as shown in Fig. 6.16.
The cable has a self weight of 8 N/m and this can be assumed as uniformly distributed along the horizontal. The span of the cable is 60 m and the central dip is 1 m. Make calculations for the value of W and the length of the cable.

Fig. 6.16.

(3608 N, 60.044 m)

11. A flexible cable carries a uniformly distributed vertical load w kN/m and is placed between two supports 30 m apart as shown in Fig. 6.17. The right support is 5 m lower than the left support and the lowest point of the cable is 9 m below the left support. If minimum tension in the cable is 360 kN, determine (*a*) the magnitude of uniformly distributed load, (*b*) the vertical reactions at the end supports and (*c*) the maximum tensions in the cables.

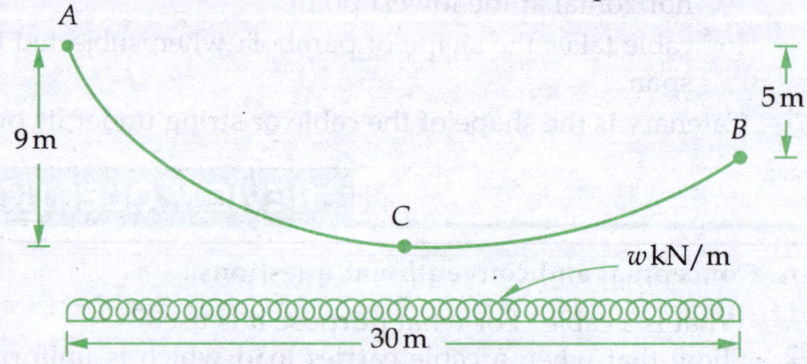

Fig. 6.17.

B. Complete the following statements with most appropriate word/words :

1. Cables and strings are assumed to be .... and ....
2. Cables and strings can withstand only .... loads.
3. A cable subjected to uniformly distributed load along the horizontal takes the shape of a ....
4. For a cable with the supports at the same level, the horizontal distance between the supports is referred to as ....
5. The vertical distance from the support to the lowest point of the hanging cable is called the .... or ....
6. The minimum tension in a cable of span l and sag d is .... where w is the uniformly distributed vertical load per unit length along the horizontal span.
7. Tension in a hanging cable is .... at the support points and .... at the lowest point.
8. At any point of cable .... of tension is constant.
9. .... is a curve followed by the cable due to uniformly distributed load along the cable it self.
10. Equation of catenary in cartesian co-ordinate system is ....
11. In the equation $y = c \cosh (x/c)$ for a catenary, the constant c is called ....
12. Parabolic curve of a cable will be due to uniformly distributed load in the ....

Answers :

1. perfectly flexible, inextensible 2. axial tensile 3. parabola 4. span 5. sag or dip 6. $\dfrac{wl^2}{8d}$
7. maximum, minimum 8. horizontal component 9. catenary 10. $y = c \cosh (x/c)$
11. parameter 12. vertical direction

Centre Points: Centroid and Centre of Gravity 7

Learning objectives : A study of the subject matter presented in this chapter will enable the readers to

- *define centre of gravity, centre of mass and centroidal axis ; differentiate between centre of gravity and centroid*
- *state the equations for locating the centroid/centre of gravity of a body*
- *determine the position of centroid/centre of gravity of different plane sections commonly used in engineering practice*
- *state and appreciate the utility of Pappus and Guldinus theorems*

A body is comprised of a number of particles. Each particle is attracted by force due to gravity of earth towards its centre. Since the body size is negligible in comparison with the earth, the gravitational forces can be assumed to be uniform and parallel to each other. The resultant of all these forces acts at a point which is known as centre of gravity : a point at which the entire weight of the body acts irrespective of the orientation of the body. There are many engineering situations where it is essential to know the position of centre of gravity of bodies. Attention has been focussed in this chapter to discuss the concept of centre of gravity and different aspects related to it.

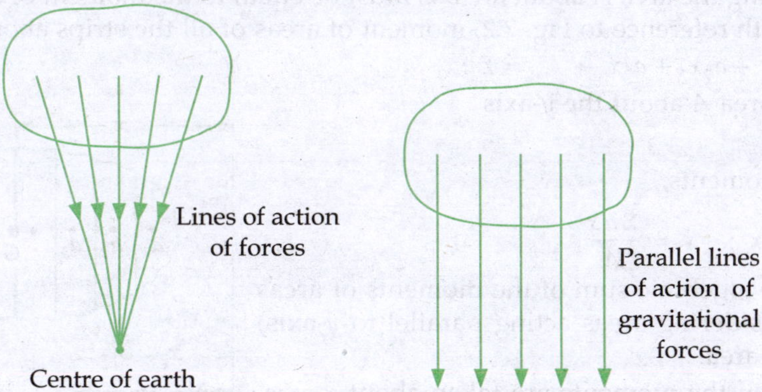

Fig. 7.1

7.1. CENTRE OF GRAVITY, CENTRE OF MASS AND CENTROID

Centre of gravity of a body is defined as the point through which resultant of the gravitational force (weight) acts for any orientation of the body. The position of *CG* depends upon the shape of the body and this may or may not necessarily be within the boundary of the body. Further, (*i*) A body has only one centre of gravity.

(*ii*) Its location does not change even with a change in the orientation of the solid body. Whatever may be the orientation of the body, there is no change in the position of mass particles relative to each other. Obviously then the resultant of the gravitational forces on the body acts through the same point.

(*iii*) It lies in a plane of symmetry, if any, of a body. The plane of symmetry of a body is a plane through the body such that the mass distribution is symmetrical on either side of the plane.

(*iv*) It is an imaginary point which may occur inside or outside the body.

(*v*) *Centre of mass* (CM) is the point where the entire mass of the body is assumed to be concentrated. The weight of body is the product of its mass and gravitational acceleration ($W = mg$) and if the small variations in gravitational acceleration from point to point on earth are neglected, then CM of the body is same as its CG, *i.e.*, CG and CM coincide.

The plane figures such as rectangle/parallelogram, triangle/polygon, circle and line etc. have only the length, area and volume and no mass or weight. The point where the entire length, area or volume is assumed to be concentrated is called the *centroid.*

Centroid of an object is the geometric centre of the object :

(*i*) The centroid of a line is the point at which acts the total length of the line.

(*ii*) The centroid of an area represents the point where the total area of the plane figure is concentrated.

(*iii*) The centroid of volume denotes the point where acts the entire volume of the body.

The term centre of gravity applies to bodies with mass and weight, and the centroid applies to plane figures which have area only but no mass. When thickness, *i.e.*, mass of the body is not considered, the *CG* and centroid are synonymous and pass through the same point. Further if value of gravity is uniform for each part and the solid is of uniform density throughout, then the centroid, centre of gravity and centre of mass are coincident.

7.2. LOCATION OF CENTROID/CENTRE OF GRAVITY

The co-ordinates $(\bar{x}, \bar{y})$ of the centroid are determined from the fact that moments of all small areas dA comprising the area A about an axis must be equal to the moment of area A itself about the same axis. With reference to Fig. 7.2, moment of areas of all the strips about y-axis

$$= a_1x_1 + a_2x_2 + a_3x_3 + \ldots = \Sigma a x$$

moment of total area A about the y-axis

$$= A\bar{x}$$

Equating these moments,

$$A\bar{x} = \Sigma a x \; ; \quad x = \frac{\Sigma a x}{A} \qquad \ldots(7.1)$$

where $\Sigma a x$ is the algebraic sum of the moments of areas about y-axis (considering areas acting parallel to y-axis) and A is the total area.

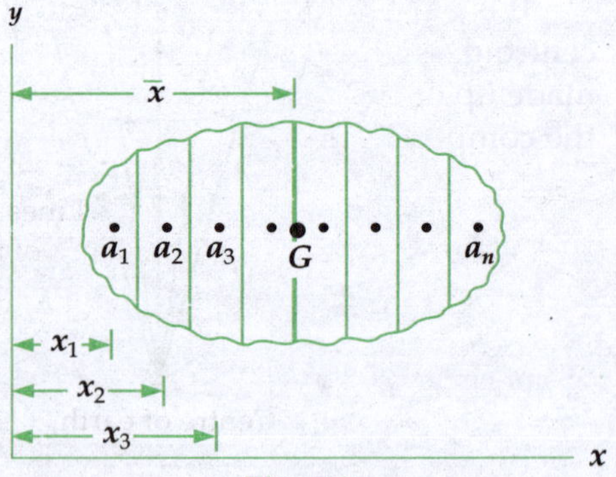

Fig. 7.2

Similarly when the moments are taken about x-axis, we get

$$A\bar{y} = \Sigma a y \; ; \qquad y = \frac{\Sigma a y}{A} \qquad \ldots(7.2)$$

where $\Sigma a y$ is the algebraic sum of the moments of areas about x-axis (considering areas acting parallel to x-axis) and A is the total area.

The location of *CG* of solids is accomplished on the same lines with the exception that moments are taken of masses, rather than of areas, of the individual parts of the solids. The *CG* will then have the co-ordinates

$$\bar{x} = \frac{\Sigma mx}{\Sigma m} \quad \text{and} \quad \bar{y} = \frac{\Sigma my}{\Sigma m} \qquad ...(7.3)$$

If all the sections are of the same material having density ρ, then

$$m_1 = \rho V_1; \quad m_2 = \rho V_2; \quad m_3 = \rho V_3 \quad \text{etc}$$

and $$V = V_1 + V_2 + V_3 + \ldots$$

The identity 7.3 may then be rewritten as

$$\bar{x} = \frac{\Sigma\rho Vx}{\Sigma\rho V} = \frac{\rho\Sigma Vx}{\rho\Sigma V} = \frac{\Sigma Vx}{\Sigma V}$$

$$\bar{y} = \frac{\Sigma\rho Vy}{\Sigma\rho V} = \frac{\rho\Sigma Vy}{\rho\Sigma V} = \frac{\Sigma Vy}{\Sigma V} \qquad ...(7.4)$$

These expressions show that if the sections/bodies are made of the same material and are of the same density through out, their centre of gravity is their centre of volume.

Further, if the segments/bodies are of same cross-section but of different lengths, them

$$V_1 = al_1; \quad V_2 = al_2; \quad V_3 = al_3 \text{ etc}$$

and $$V = al_1 + al_2 + al_3 + \ldots$$

By substitution of these results in expression 7.4, we obtain

$$\bar{x} = \frac{\Sigma alx}{\Sigma al} = \frac{a\Sigma lx}{a\Sigma l} = \frac{\Sigma lx}{\Sigma l}$$

$$\bar{y} = \frac{\Sigma aly}{\Sigma al} = \frac{a\Sigma ly}{a\Sigma l} = \frac{\Sigma ly}{\Sigma l} \qquad ...(7.5)$$

Apparently the centre of gravity of a wire, pipe or rod of constant cross-section corresponds to their centre of lengths.

It is worthwhile to note that the expressions 7.1 to 7.5 are equally valid when calculating the centroid/centre of gravity of composite laminas and solids. A composite figure is the figure made up of basic plane figures like triangle, rectangle, circle, semi-circle and square etc. A few of the composite figures have been illustrated below in Fig. 7.3.

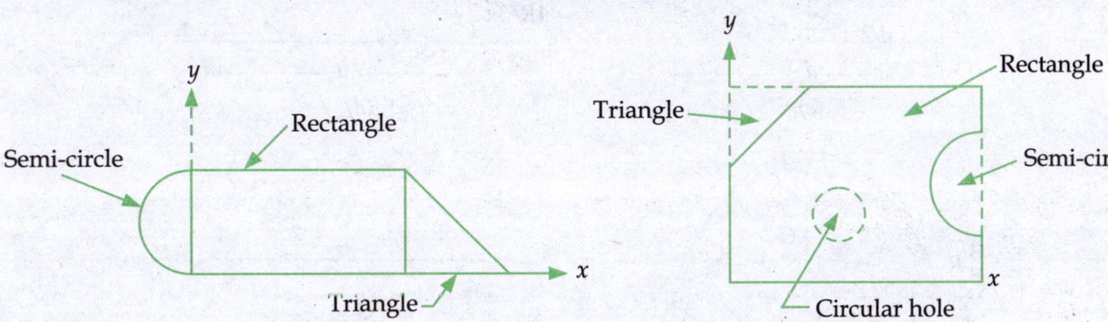

Fig. 7.3. Composite figures

The area of a hole or any cut or punched section existing in the plane figure will be treated as negative. The moment of a negative area will also be negative provided the entire figure lies in the first quadrant.

Information is given below on the position of centroid/centre of gravity of different laminas (plane sections) most commonly used in engineering practice:

(i) For a rectangular section, *CG* lies at the intersection of diagonals, *i.e.*, at the mid point of its breadth as well as the depth. (Fig. 7.4 *a*).

(ii) For the triangular section, *CG* lies at the point of intersection of the three medians. If height of the triangle is h, then its *CG* is at a distance of $h/3$ from the base along vertical axis (Fig. 7.4 *b*).

(iii) For a circular section, *CG* lies at the centre of circle (Fig. 7.4 *c*). The *CG* of a semi-circle of radius R is at a distance of $\frac{4R}{3\pi}$ from its base measured along vertical axis (Fig. 7.4 *d*).

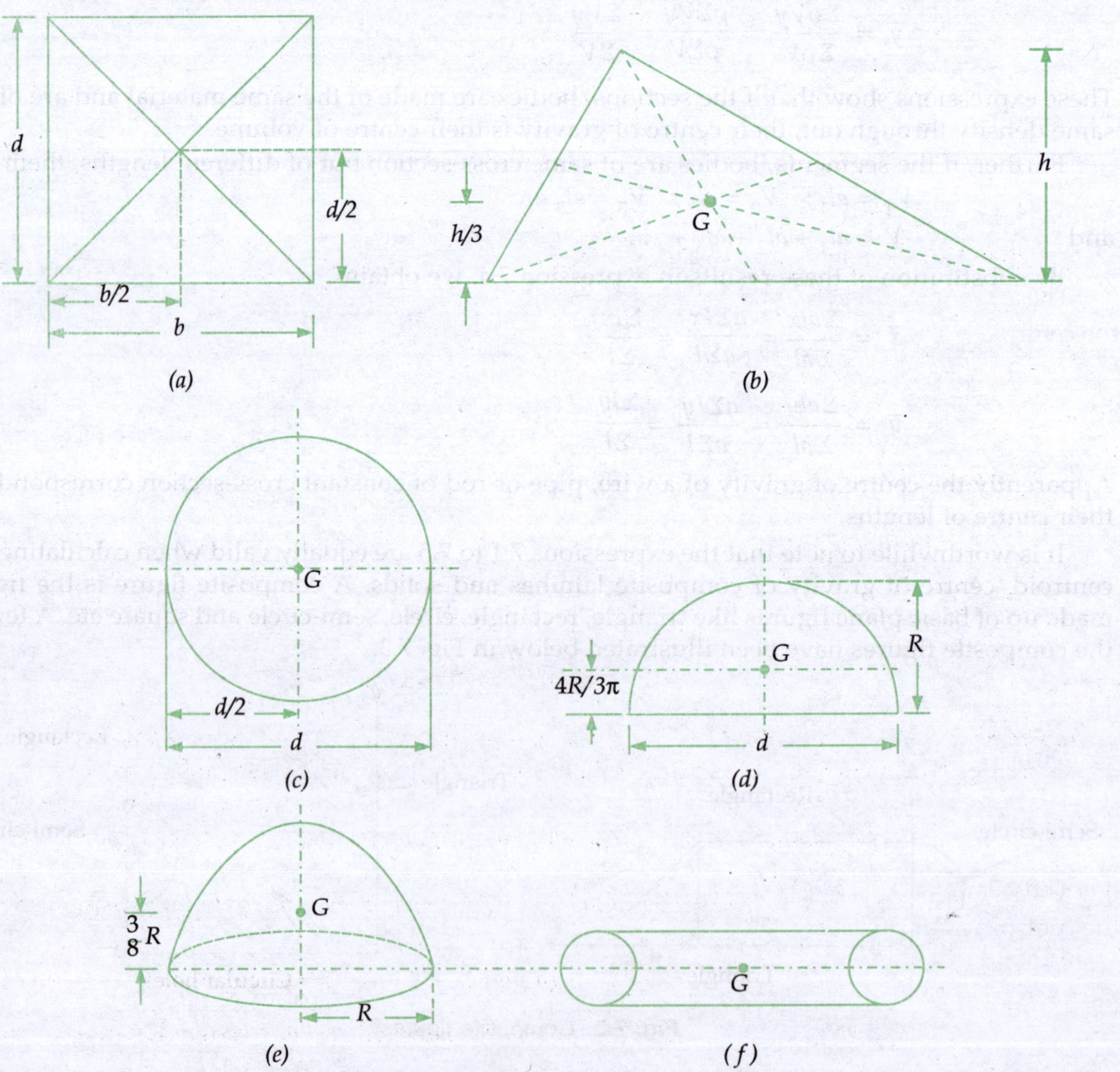

Fig. 7.4

(iv) The CG of a sphere of radius R lies at a distance of R from every point. For the hemi-sphere, CG would be at a distance of $3R/8$ from its base measured along the vertical axis. (Fig. 7.4 *e*).

(v) The CG of a uniform rod is at its mid point (Fig. 7.4 *f*).

EXAMPLE 7.1

Illustrate few examples of composite plane figures which are :

(i) symmetrical about x-axis *(ii) symmetrical about y-axis*

(iii) symmetrical about both axis *(iv) not symmetrical about any axis.*

For each case, comment on the position of centroid.

Solution : *(i)* Symmetrical about x-axis

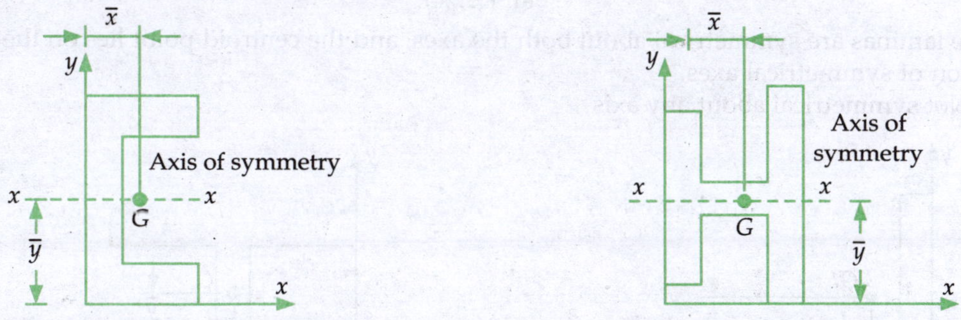

Fig. 7.5. (a)

These laminas are symmetrical about x-axis and the centroid lies on this axis of symmetry. For the centroid point $(\overline{x}, \overline{y})$

(a) the value of $\overline{x}$ is determined by calculations, and

(b) the value of $\overline{y}$ gets determined directly, *i.e.*, it lies in the axis of symmetry x-x.

(ii) Symmetrical about y-axis

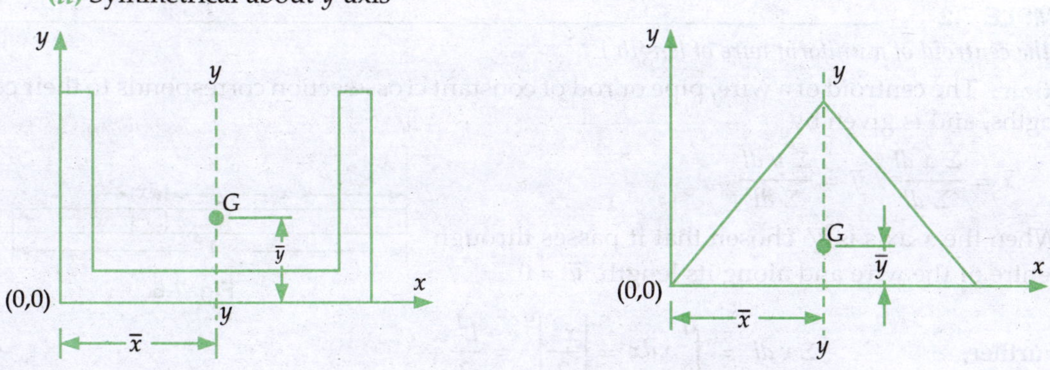

Fig. 7.5. (b)

These laminas are symmetrical about y-axis and the centroid lies on this axis of symmetry. For the centroid point $(\overline{x}, \overline{y})$

(a) the $\overline{x}$ value is determined directly, *i.e.*, it lies on the axis of symmetry y-y, and

(b) the $\overline{y}$ value needs to be determined by calculations.

(*iii*) Symmetrical about both the axes

Fig. 7.5. (c)

These laminas are symmetrical about both the axes, and the centroid point lies on the point of intersection of symmetrical axes.

(*iv*) Not symmetrical about any axis

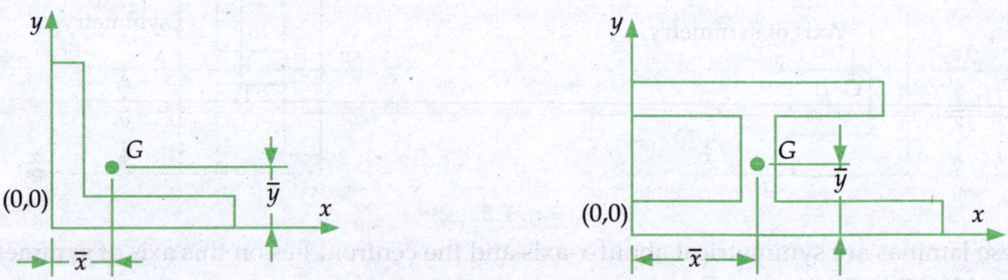

Fig. 7.5. (d)

These laminas have no symmetry about any axes. Hence calculations need to be made for determining both the coordinates $\bar{x}$ and $\bar{y}$ of the centroidal point.

EXAMPLE 7.2

Find the centroid of a uniform wire of length L

Solution : The centroid of a wire, pipe or rod of constant cross-section corresponds to their centre of lengths, and is given by

$$\bar{x} = \frac{\Sigma\, x\, dl}{\Sigma\, dl},\ \bar{y} = \frac{\Sigma\, y\, dl}{\Sigma\, dl}$$

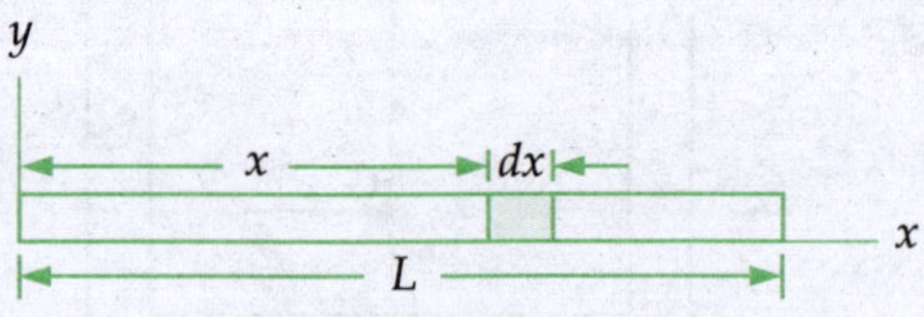

Fig. 7.6

When the x-axis is so chosen that it passes through the centre of the wire and along its length, $\bar{y} = 0$

Further, $$\Sigma\, x\, dl = \int_0^L x\, dx = \left|\frac{x^2}{2}\right|_0^L = \frac{L^2}{2}$$

and $$\Sigma\, dl = \int_0^L dx = L$$

$$\therefore \quad \bar{x} = \frac{L^2/2}{L} = \frac{L}{2}$$

EXAMPLE 7.3

Locate the centroid of the bent wire shown in Fig. 7.7

Solution : The composite wire consists of two parts *AB* and *BC* :

(*i*) Length of portion $AB = l_1 = 100$ mm

Its centroid C_1 is at the centre of the line

$$x_1 = 50 \cos 30^\circ = 43.30 \text{ mm}$$
$$y_1 = 50 \sin 30^\circ = 25 \text{ mm}$$

(*ii*) Length of semi-circular portion *BC*,

$$l_2 = \pi r = \pi \times 50 = 157 \text{ mm}$$

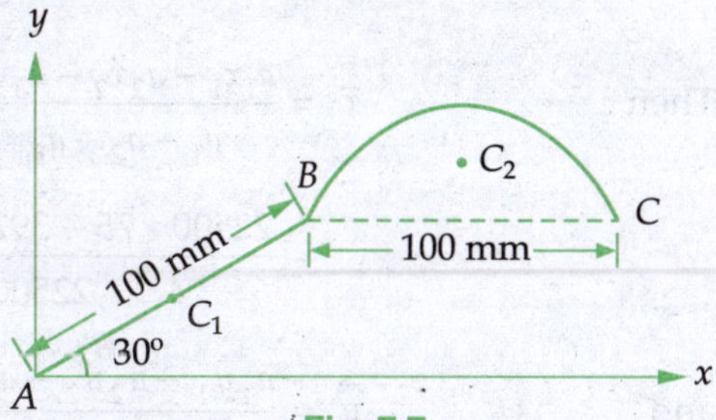

Fig. 7.7

The centroid C_2 of a semi-circular line is at $\frac{2r}{\pi} = \frac{2 \times 50}{\pi} = 31.83$ mm from its base line *BC*. Accordingly the co-ordinates of centroid C_2 of the semi-circular portion are :

$$x_2 = 100 \cos 30^\circ + 50 = 136.6 \text{ mm}$$
$$y_2 = 100 \sin 30^\circ + 31.83 = 81.83 \text{ mm}$$

Let the centroid of the bent wire be located at distance $\bar{x}$ from the *x*-axis and at distance $\bar{y}$ from the *y*-axis. Then

$$\bar{x} = \frac{l_1 x_1 + l_2 x_2}{l_1 + l_2} = \frac{100 \times 43.30 + 157 \times 136.6}{100 + 157} = \mathbf{100.3 \text{ mm}}$$

and

$$\bar{y} = \frac{l_1 y_1 + l_2 y_2}{l_1 + l_2} = \frac{100 \times 25 + 157 \times 81.83}{100 + 157} = \mathbf{59.72 \text{ mm}}$$

EXAMPLE 7.4

Locate the centroid of the area shown in Fig. 7.8. All the dimensions are in mm.

Solution : The given area can be considered as a rectangle of size 150 × 150 mm minus a semicircle of radius 50 mm minus a triangle of base 50 mm and height 100 mm.

The areas and coordinates of centroid of those segments with respect to the given axis are :

Rectangle : $a_1 = 150 \times 150 = 22500 \text{ mm}^2$

$$x_1 = \frac{150}{2} = 75 \text{ mm and}$$
$$y_1 = \frac{150}{2} = 75 \text{ mm}$$

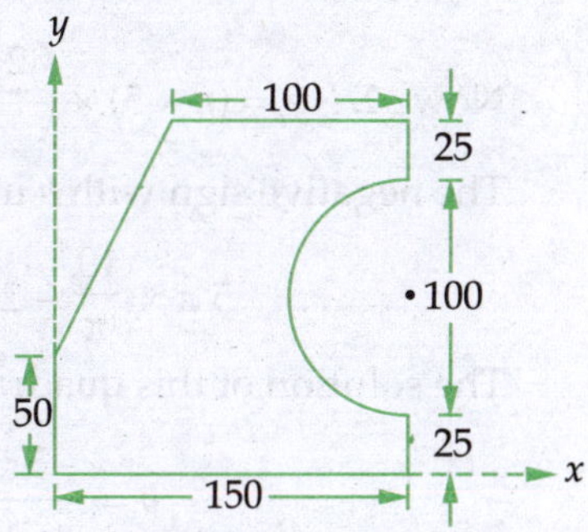

Fig. 7.8

Semicircle :

$$a_2 = \frac{\pi r^2}{2} = \frac{\pi \times 50^2}{2} = 3925 \text{ mm}^2$$
$$x_2 = 150 - \frac{4r}{3\pi} = 150 - \frac{4 \times 50}{3\pi} = 128.77 \text{ m}$$
$$y_2 = 25 + r = 25 + 50 = 75 \text{ mm}$$

Triangle :

$$a_3 = \frac{1}{2} bh = \frac{1}{2} \times 50 \times 100 = 2500 \text{ mm}^2$$
$$x_3 = \frac{b}{3} = \frac{50}{3} = 16.67 \text{ mm}$$

$$y_3 = 150 - \frac{h}{3} = 150 - \frac{100}{3} = 116.67 \text{ mm}$$

Then : $$\bar{x} = \frac{a_1x_1 - a_2x_2 - a_3x_3}{a_1 - a_2 - a_3}$$

$$= \frac{22500 \times 75 - 3925 \times 128.77 - 2500 \times 16.67}{22500 - 3925 - 2500} = \mathbf{70.94 \text{ mm}}$$

and $$\vec{y} = \frac{a_1y_1 - a_2y_2 - a_3y_3}{a_1 - a_2 - a_3}$$

$$= \frac{22500 \times 75 - 3925 \times 75 - 2500 \times 116.67}{22500 - 3925 - 2500} = \mathbf{68.52 \text{ mm}}$$

EXAMPLE 7.5

A slender homogenous wire of uniform cross-section is bent as shown in Fig. 7.9. It is desired that centre of gravity of the wire should coincide with centre C of the semi-circular portion. What should then be the value of dimension a ?

Solution : With C as origin, the configuration of the given wire is symmetrical about y-axis and therefore $\bar{x} = 0$. Further, it is also stated that CG of the wire coincides with the centre C of the semi-circular portion and hence $\bar{y}$ must also be zero, *i.e.*, $\bar{y} = 0$.

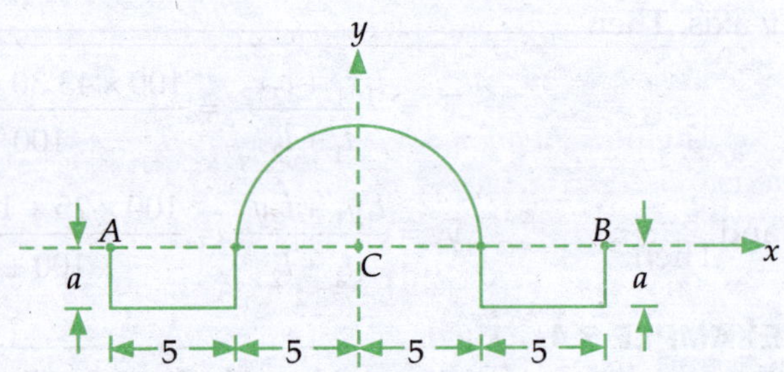

Fig. 7.9

For a composite line segment, $\bar{y}$ is given by

$$\bar{y} = \frac{\Sigma l_i y_i}{\Sigma l_i} = 0 \text{ and so}$$

$$\Sigma l_i y_i = 0$$

$$\text{Now} : \Sigma l_i y_i = (\pi \times 5) \times \left(\frac{2 \times 5}{\pi}\right) + 4b \left(\frac{-b}{2}\right) + 2 \times 5 (-b)$$

The negative sign with b in the above expression is due to these portions lying below x-axis.

$$\therefore \quad 5\pi \times \frac{10}{\pi} - 2b^2 - 10b = 0 \quad \text{or} \quad b^2 + 5b - 25 = 0$$

The solution of this quadratic equation gives :

$$b = \frac{-5 \pm \sqrt{25 - 4 \times 5 \times (-25)}}{2} = \frac{-5 \pm 11.18}{2}$$

Neglecting negative value of b, we get

$$b = \frac{-5 + 11.18}{2} = \mathbf{3.09 \text{ units}}$$

EXAMPLE 7.6

Locate the centroid of the area shown in Fig. 7.10 with respect to the axes indicate in the figure.

Solution : The composite area has been divided into three segments namely a triangle, a rectangle and a semi-circle. The areas and the co-ordinates of centroid of these segments with respect to the given axes are:

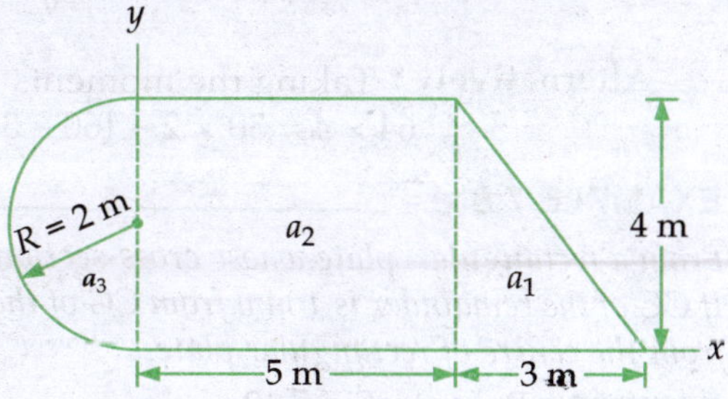

Fig. 7.10

Triangular segment:

$$a_1 = \frac{1}{2} \times 3 \times 4 = 6 \text{ m}^2$$

$$x_1 = 5 + \frac{1}{3} * 3 = 6 \text{ m} \; ; \quad y_1 = \frac{4}{3} = 1.33 \text{ m}$$

Rectangular segment: $a_2 = 5 \times 4 = 20 \text{ m}^2$

$$x_2 = \frac{5}{2} = 2.5 \text{ m} \; ; \quad y_2 = \frac{4}{2} = 2 \text{ m}$$

Semi-circular segment: $a_3 = \frac{1}{2} \times \pi \times 2^2 = 6.28 \text{ m}^2$

$$x_3 = -\frac{4R}{3\pi} = -\frac{4 \times 2}{3 \times \pi} = -0.849 \text{ m}$$

$$y_3 = \frac{4}{2} = 2 \text{ m}$$

The negative with co-ordinate x_3 stems from the fact that this co-ordinate lies on the left of y-axis

Then: $$\bar{x} = \frac{a_1x_1 + a_2x_2 + a_3x_3}{a_1 + a_2 + a_3} = \frac{6 \times 6 + 20 \times 2.5 + 6.28(-0.849)}{6 + 20 + 6.28} = \mathbf{2.5 \text{ m}}$$

and $$\bar{y} = \frac{a_1y_1 + a_2y_2 + a_3y_3}{a_1 + a_2 + a_3} = \frac{6 \times 1.33 + 20 \times 2 + 6.28 \times 2}{6 + 20 + 6.28} = \mathbf{1.875 \text{ m}}$$

EXAMPLE 7.7

A triangular plate in the form of an isosceles triangle ABC has base BC = 10 cm and altitude = 12 cm. From this plate, a portion in the shape of an isosceles triangle OBC is removed. If O is the mid-point of the altitude of triangle ABC, then determine the distance of CG of the remainder section from the base.

Solution : Refer Fig 7.11

For a triangle of height h, the CG lies on the axis at a distance $h/3$ from the base.

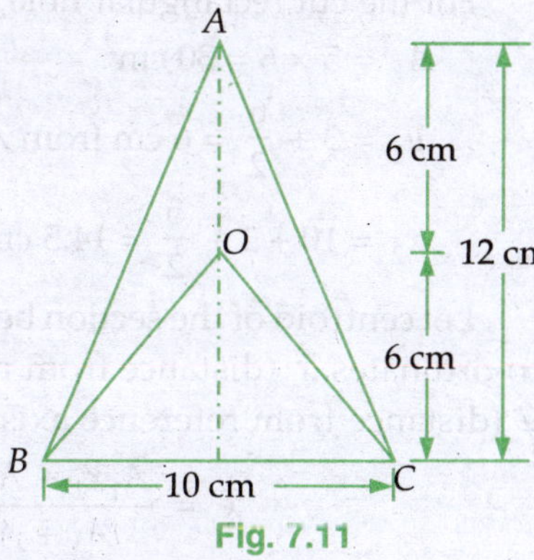

Fig. 7.11

For triangle ABC,

$$\text{Area } A_1 = \frac{1}{2} \times 10 \times 12 = 60 \text{ cm}^2$$

$$y_1 = 12/3 = 4 \text{ cm from base } BC$$

For triangle OBC,

$$A_2 = \frac{1}{2} \times 10 \times 6 = 30 \text{ cm}^2$$

$$y_2 = 6/3 = 2 \text{ cm from base } BC.$$

Let y be the distance of CG of the section $ABOCA$ from the base line BC.

$$y = \frac{A_1y_1 + A_2y_2}{A_1 + A_2} = \frac{60 \times 4 + (-30) \times 2}{60 + (-30)}$$

The negative sign with A_2 stems from the fact that this area has been removed.

$$y = \frac{240 - 60}{60 - 30} = 6 \text{ cm}$$

Alternatively : Taking the moments of different areas about the base line BC, we get

$$64 \times 4 = 30 \times 2 + (60 - 30)\, y \;; \quad y = 6 \text{ cm}$$

EXAMPLE 7.8

From a rectangular plate whose cross-section is 8 cm × 6 cm, a circular disc of 3 cm² in area is removed. If CG of the remainder is 1 mm from CG of the rectangular plate, work out the distance of the centre of disc from the centre of rectangular plate.

Solution : Refer to Fig 7.12.

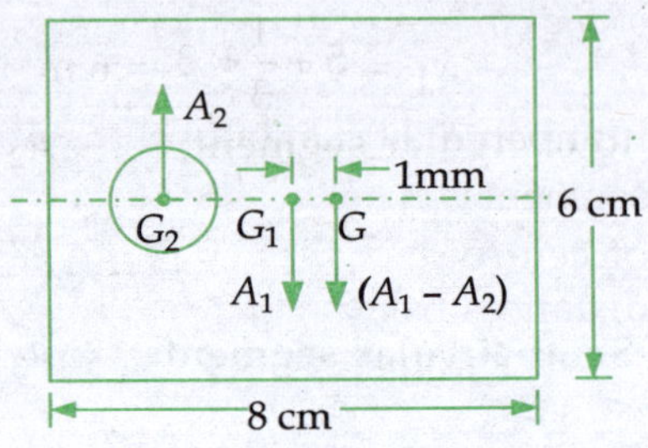

Fig. 7.12

Let G_1 and G_2 be the CG's of the rectangular plate and disc respectively . The centre of gravity G of the remainder lies on that side of G_1 which is opposite to G_2.

Taking moment of areas about G_1,

$$A_1 \times 0 = A_2 \times G_1G_2 + (A_1 - A_2)\, G_1G$$

$$(8 \times 6) \times 0 = 3 \times G_1 G_2 + (48 - 3) \times 0{\cdot}1$$

$$\therefore \quad G_1 G_2 = -\frac{45 \times 0{\cdot}1}{3} = -1{\cdot}5 \text{ cm}$$

The negative sign stems from the fact that G_2 is to the left of G_1.

EXAMPLE 7.9

A rectangular lamina ABCD 20 cm × 25 cm has a rectangular hole of 5 cm × 6 cm as shown in Fig. 7.13. Locate the centroid of the section.

Solution : For the rectangular lamina $ABCD$,

$$A_1 = 25 \times 20 = 500 \text{ cm}^2$$

$$y_1 = \frac{25}{2} = 12.5 \text{ cm from } AB$$

$$x_1 = \frac{20}{2} = 10 \text{ cm from } AD$$

For the cut rectangular hole,

$$A_2 = 5 \times 6 = 30 \text{ cm}^2$$

$$y_2 = 3 + \frac{6}{2} = 6 \text{ cm from } AB$$

$$x_2 = 10 + 2 + \frac{5}{2} = 14.5 \text{ cm from } AD$$

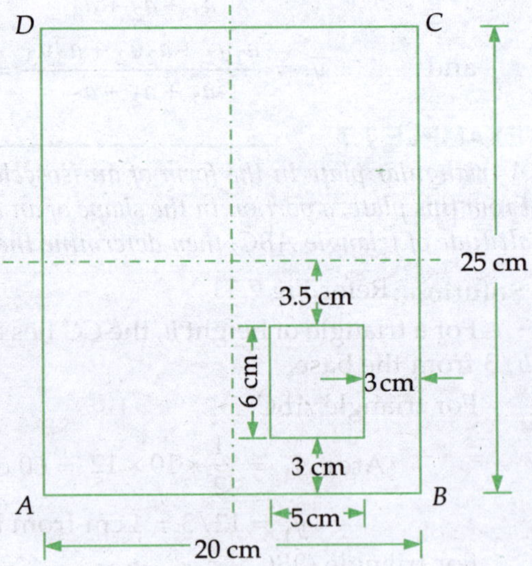

Fig. 7.13

Let centroid of the section be located at a point with co-ordinates $\bar{x}$ (distance from reference axis AD) and $\bar{y}$ (distance from reference axis AB). Then

$$\bar{x} = \frac{A_1x_1 + A_2x_2}{A_1 + A_2}$$

and

$$\bar{y} = \frac{A_1y_1 + A_2y_2}{A_1 + A_2}$$

Considering area A_2 of the cut hole as negative, we get

$$\bar{x} = \frac{500 \times 10 - 30 \times 14 \cdot 5}{500 - 30} = \frac{5000 - 435}{470} = \mathbf{9.71\ cm}$$

$$\bar{y} = \frac{500 \times 12.5 - 30 \times 6}{500 - 30} = \frac{6250 - 180}{470} = \mathbf{12.91\ cm}$$

EXAMPLE 7.10

From a circular lamina of diameter d, a square hole has been punched out. If one diagonal of the square coincides with the radius of the circle, determine the distance of the centre of the remainder from the centre of the circle.

Solution : The section consists of two parts which have been numbered 1 and 2 as shown in Fig. 7.14. Since the remainder segment is symmetrical about x-axis, $\bar{y} = 0$. For finding out location of centroid, the areas of the components and their centroidal distances from line LMN are as given below:

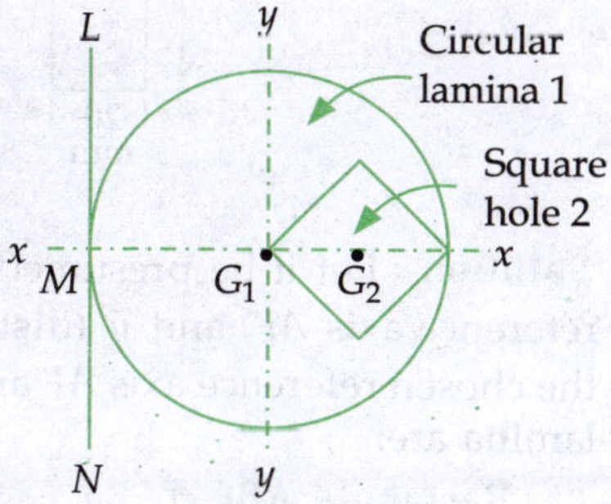

Fig. 7.14

1. Circular lamina

   area $A_1 = \frac{\pi}{4} d^2$

   centroidal distance $x_1 = MG_1 = d/2$

2. Square hole (removed)

   length of diagonal $= d/2$

   $l^2 + l^2 = \frac{d^2}{4}$ where l is the length of each side

   area $A_2 = l^2 = \frac{d^2}{8}$ (it will be negative since it has been removed)

   centroidal distance $x_2 = MG_2 = MG_1 + G_1G_2 = \frac{d}{2} + \frac{d}{4} = \frac{3d}{4}$

If $\bar{x}$ is the location of centroid of the remainder as measured from LMN, then from moment principle

$$(A_1 - A_2)x = A_1x_1 - A_2x_2$$

$$\left(\frac{\pi}{4}d^2 - \frac{d^2}{8}\right)\bar{x} = \frac{\pi}{4}d^2 \times \frac{d}{2} - \frac{d^2}{8} \times \frac{3d}{4}$$

or $$d^2\left(\frac{\pi}{4} - \frac{1}{8}\right)\bar{x} = \frac{\pi}{8}d^3 - \frac{3d^3}{32} = \frac{d^3}{8}\left(\pi - \frac{3}{4}\right)$$

$$\bar{x} = \frac{d}{8}\,\frac{4\pi - 3}{4} \times \frac{8}{2\pi - 1} = \frac{d}{4}\left(\frac{4\pi - 3}{2\pi - 1}\right)$$

When the centroidal distance is measured from the centre of circle, then

$$\bar{x} = \frac{d}{2} - \frac{d}{4}\left(\frac{4\pi - 3}{2\pi - 1}\right) = \frac{d}{4}\left[2 - \frac{4\pi - 3}{2\pi - 1}\right]$$

$$= \frac{d}{4}\left[\frac{4\pi - 2 - 4\pi + 3}{2\pi - 1}\right] = \frac{d}{4(2\pi - 1)}$$

EXAMPLE 7.11

Determine the centroid of the lamina shown in Fig. 7.15 given below:

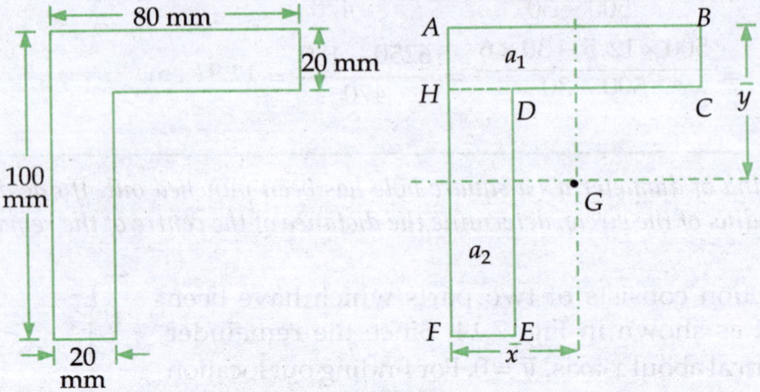

Fig. 7.15

Solution : Let it be presumed that *CG* is located of *G* with the co-ordinates $\bar{x}$ (distance from reference axis *AF*) and $\bar{y}$ (distance from reference axis *AB*). Then areas and co-ordinates (from the chosen reference axis *AF* and *AB*) of the centroids of the two rectangles comprising the given lamina are:

Rectangle ABCH $\quad a_1 = 80 \times 20 = 1600 \text{ mm}^2$

$$x_1 = \frac{80}{2} = 40 \text{ mm}; \quad y_1 = \frac{20}{2} = 10 \text{ mm}$$

Rectangle DEFH $\quad a_2 = 80 \times 20 = 1600 \text{ mm}^2$

$$x_2 = \frac{20}{2} = 10 \text{ mm}; \quad y_2 = 20 + \frac{80}{2} = 60 \text{ mm}$$

Then:

$$\bar{x} = \frac{a_1x_1 + a_1x_2}{a_1 + a_2} = \frac{1600 \times 40 + 1600 \times 10}{1600 + 1600} = 25 \text{ mm}$$

$$\bar{y} = \frac{a_1y_1 + a_2y_2}{a_1 + a_2} = \frac{1600 \times 10 + 1600 \times 60}{1600 + 1600} = 35 \text{ mm}$$

Hence *CG* of the section is at a distance of 25 mm from face *AF* and 35 mm from face *AB*.

EXAMPLE 7.12

Locate the centroid of T-section as shown in Fig 7.16 given below:

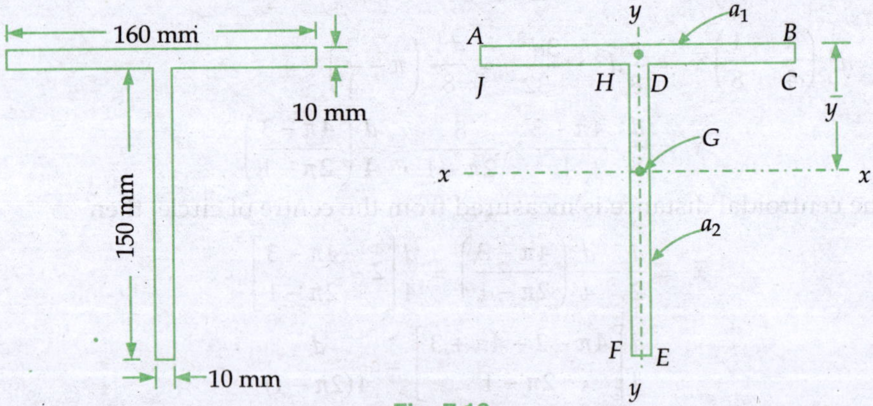

Fig. 7.16

Solution : Let the reference axes be chosen as shown in Fig.7.13. Due to symmetry of the section, its CG lies at point G on the y-axis, *i.e.*, $\bar{x} = 0$.

Further the top fibre AB has been chosen as the reference axis to locate the centroid $\bar{y}$.

The areas and co-ordinates of the centroid of the two rectangles comprising the given section are:

Rectangle ABCJ $\quad a_1 = 160 \times 10 = 1600 \text{ mm}^2$

$$y_1 = \frac{10}{2} = 5 \text{ mm}$$

Rectangle DEFH $\quad a_2 = 150 \times 10 = 1500 \text{ mm}^2$

$$y_2 = 10 + \frac{150}{2} = 85 \text{ mm}$$

Then:
$$\bar{y} = \frac{a_1 y_1 + a_2 y_2}{a_1 + a_2} = \frac{1600 \times 5 + 1500 \times 85}{1600 + 1500} = 43.71 \text{ mm}$$

Hence CG of the given T-section lies on they y-axis ($\bar{x} = 0$) and at 43.71 mm from the top face of its flange.

EXAMPLE 7.13

(*a*) *Find the centroid of the truss Gusset plate shown in Fig. 7.17 given below:*

(*b*) *Find the centroid of a semicircular section having outer and inner radii 200 mm and 160 mm respectively.*

Solution : The given area can be considered as rectangle $ABFE$ minus triangle CFD. The areas and co-ordinates of these segments with respect to given axis are:

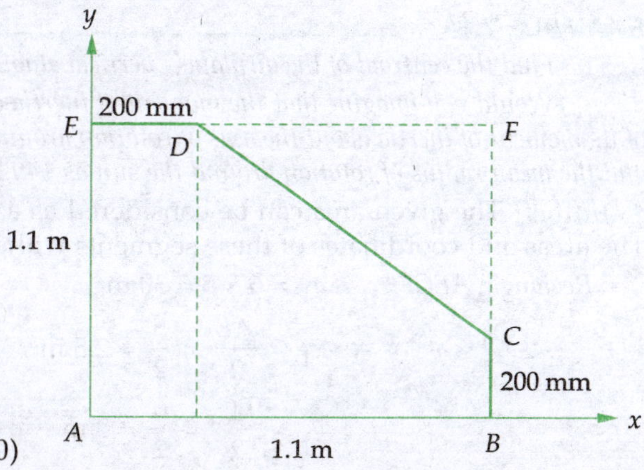

Fig. 7.17

Rectangle: $a_1 = 1100 \times 1100$
$= 121 \times 10^4 \text{ mm}^2$

$$x_1 = \frac{1100}{2} = 550 \text{ mm}$$

$$y_1 = \frac{1100}{2} = 550 \text{ mm}$$

Triangle: $a_2 = \frac{1}{2} \times (1100 - 200) \times (1100 - 200)$
$= 40.5 \times 10^4 \text{ mm}^2$

$$x_2 = ED + \frac{2DF}{3}$$
$$= 200 + \frac{2(900)}{3} = 800 \text{ mm}$$

$$y_2 = BC + \frac{2FC}{3} = 200 + \frac{2(900)}{3} = 800 \text{ mm}$$

Then,
$$\bar{x} = \frac{a_1 x_1 - a_2 x_2}{a_1 - a_2} = \frac{121 \times 10^4 \times 550 - 40.5 \times 10^4 \times 800}{121 \times 10^4 - 40.5 \times 10^4}$$
$$= 424.22 \text{ mm}$$

$$\bar{y} = \frac{a_1 y_1 - a_2 y_2}{a_1 - a_2} = \frac{121 \times 10^4 \times 550 - 40.5 \times 10^4 \times 800}{121 \times 10^4 - 40.5 \times 10^4}$$
$$= 424.22 \text{ mm}$$

(b) With reference to Fig. 7.18.

$$r_1 = 200 \text{ mm}; \quad r_2 = 160 \text{ mm}$$

The areas and co-ordinates of the segments 1 and 2 with respect to axis x and y are:

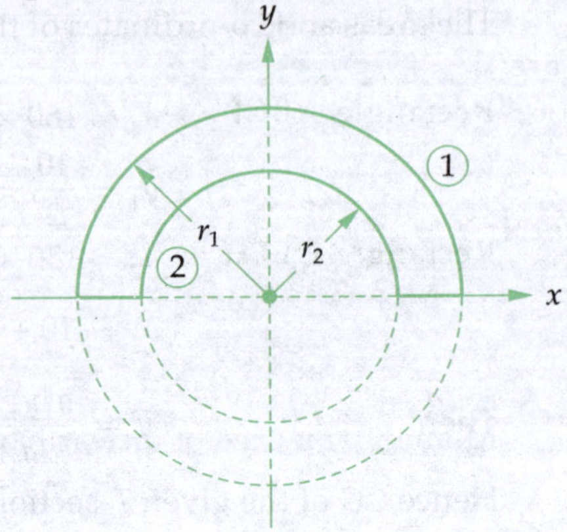

Fig. 7.18

$$a_1 = \frac{1}{2}\left(\pi r_1^2\right) = \frac{\pi}{2} \times 200^2 = 20000\pi \text{ mm}^2$$

$$x_1 = 0 \quad \text{and} \quad y_1 = \frac{4r_1}{3\pi} = \frac{4 \times 200}{3\pi} = 84.92 \text{ m}$$

$$a_2 = \frac{1}{2}\left(\pi r_2^2\right) = \frac{\pi}{2} 160^2 = 12800\pi \text{ mm}^2$$

$$x_2 = 0 \quad \text{and} \quad y_2 = \frac{4r_2}{3\pi} = \frac{4 \times 160}{3\pi} = 67.94 \text{ mm}$$

Then: $$\bar{x} = \frac{a_1 x_1 - a_2 x_2}{a_1 - a_2} = \frac{20000\pi \times 0 - 12800\pi \times 0}{20000\pi - 12800\pi} = 0$$

$$\bar{y} = \frac{a_1 y_1 - a_2 y_2}{a_1 - a_2} = \frac{20000\pi \times 84.92 - 12800\pi \times 67.94}{20000\pi - 12800\pi}$$

$$= \mathbf{115.11 \text{ mm}}$$

EXAMPLE 7.14

(a) Find the centroid of the airplane's vertical stabilizer shown in Fig. 7.19.

(b) Would you imagine that the moment of inertia of the earth around its own axis is negligible fraction of its moment of inertia about the axis of rotation around the sun? Take mean radius of the earth as 6371 km and the mean radius of rotation around the sun as 149.7 × 10^6 km.

Solution : The given area can be considered as a rectangle $AECF$ minus triangles BEC and ADF. The areas and coordinates of these segments with respect to given axis are:

Rectangle AECF: $$a_1 = 6 \times 5 = 30 \text{ m}^2$$

$$x_1 = \frac{AE}{2} = \frac{5}{2} = 2.5 \text{ m}$$

$$y_1 = \frac{EC}{2} = \frac{6}{2} = 3 \text{ m}$$

Triangle BEC: $$a_2 = \frac{1}{2} \times 1 \times 6 = 3 \text{ m}^2$$

$$x_2 = AB + \frac{2BE}{3}$$

$$= 4 + \frac{2 \times 1}{3} = 4.66 \text{ m}$$

$$y_2 = \frac{1}{3} EC = \frac{1}{3} \times 6 = 2 \text{ m}$$

Triangle ADF: $$a_3 = \frac{1}{2} \times 3 \times 6 = 9 \text{ m}^2$$

$$x_3 = \frac{FD}{3} = \frac{3}{3} = 1 \text{ m}$$

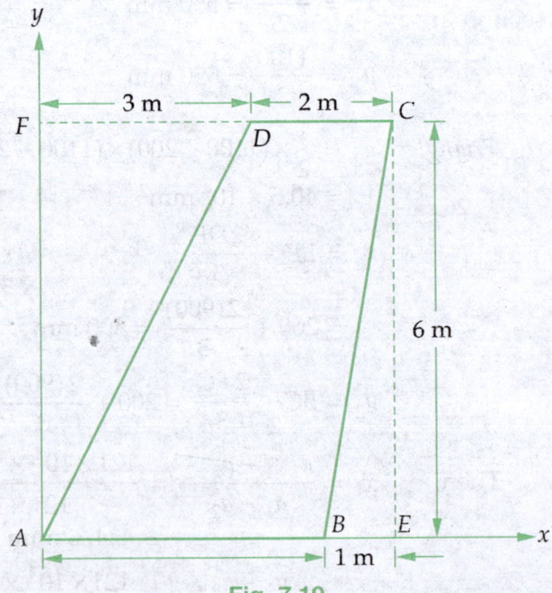

Fig. 7.19

$$y_3 = \frac{2AF}{3} = \frac{2}{3} \times 6 = 4 \text{ m}$$

Then:
$$\bar{x} = \frac{a_1x_1 - a_2x_2 - a_3x_3}{a_1 - a_2 - a_3}$$

$$= \frac{30 \times 2.5 - 3 \times 4.66 - 9 \times 1}{30 - 3 - 9} = \mathbf{2.89\ m}$$

and
$$\bar{y} = \frac{a_1y_1 - a_2y_2 - a_3y_3}{a_1 - a_2 - a_3} = \frac{30 \times 3 - 3 \times 2 - 9 \times 4}{30 - 3 - 9} = \mathbf{2.67\ m}$$

(*b*) Moment of inertia of the earth about its axis,

$$I_1 = \frac{2}{5}MR^2 = \frac{2}{5}M(6371)^2 = 16.23 \times 10^6\, M$$

Moment of inertia of the earth about the axis of rotation around the sun,

$$I_2 = I_1 + Md^2 = 16.23 \times 10^6\, M + M \times (149.7 \times 10^6)^2$$
$$= 16.23 \times 10^6\, M + 22410.09 \times 10^{12}\, M$$
$$= 22410.09002 \times 10^{12}\, M$$

$$\text{Ratio } \frac{I_1}{I_2} = \frac{16.23 \times 10^6}{22410.09002 \times 10^{12}} = 7.24 \times 10^{-10}$$

Since the ratio is negligible, the moment of inertia of the earth around its oun axis can be imagined to be a negligible fraction of its moment of inertia about the axis of rotation around the sun.

EXAMPLE 7.15

Locate the position of the centroid of the plane shaded area depicted in Fig. 7.20 given below:

Solution : Let the reference axes be chosen as shown in Fig. 7.20. Due to symmetry of the section about y-axis, centroid lies on y-axis and hence $\bar{x} = 0$.

The areas and co-ordinates of the centroid of the different sections comprising the given section are:

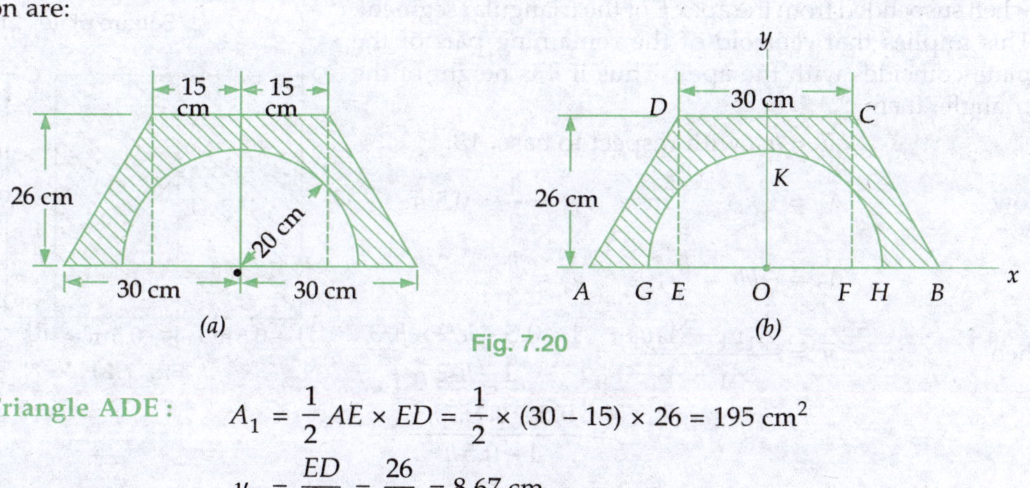

Fig. 7.20

Triangle ADE : $A_1 = \frac{1}{2} AE \times ED = \frac{1}{2} \times (30 - 15) \times 26 = 195 \text{ cm}^2$

$$y_1 = \frac{ED}{3} = \frac{26}{3} = 8.67 \text{ cm}$$

Rectangle EDCF : $A_2 = DC \times ED = 30 \times 26 = 780 \text{ m}^2$

$$y_2 = \frac{ED}{2} = \frac{26}{3} = 13 \text{ cm}$$

Triangle FCB : $A_3 = \frac{1}{2} FB \times FC = \frac{1}{2}(30 - 15) \times 26 = 195 \text{ cm}^2$

$$y_3 = \frac{FC}{3} = \frac{26}{3} = 8.67 \text{ cm}$$

Semi circle GKH : $A_4 = \frac{1}{2}\pi r^2 = \frac{1}{2}\pi \times (20)^2 = 628 \text{ cm}^2$

$$y_4 = \frac{4r}{3\pi} = \frac{4 \times 20}{3\pi} = 8.49 \text{ cm}$$

Let the centroid be located at distance $\bar{y}$ from the x-axis. Then

$$y = \frac{A_1y_1 + A_2y_2 + A_3y_3 + A_4y_4}{A_1 + A_2 + A_3 + A_4}$$

$$= \frac{(195 \times 8{\cdot}67) + (780 \times 13) + (195 \times 8{\cdot}67) + (-628 \times 8{\cdot}49)}{195 + 780 + 195 - 628}$$

The –ve sign with A_4 stems from the fact that this area has been included though it is non-existent

$$\bar{y} = \frac{1690{\cdot}65 + 10140{\cdot}00 + 169.65 - 5331{\cdot}72}{542} = 15.11 \text{ cm}$$

EXAMPLE 7.16

An isosceles triangle is to be cut from one edge of a square plate of side 1 m such that the remaining part of the plate remains in equilibrium in any position when suspended from the apex of the triangle. Calculate the area of the triangle to be removed.

Solution : Refer Fig. 7.21. It is to be noted that

(*i*) the section consists of two parts; square plate (1) and the triangular segment (2) which has been removed.

(*ii*) the remaining part of the plate remains in equilibrium when suspended from the apex *E* of the triangular segment. This implies that centroid of the remaining part of the plate coincides with the apex. Thus if *h* is height of the triangle, then

$$\bar{y} = h \quad \text{with respect to base } AB.$$

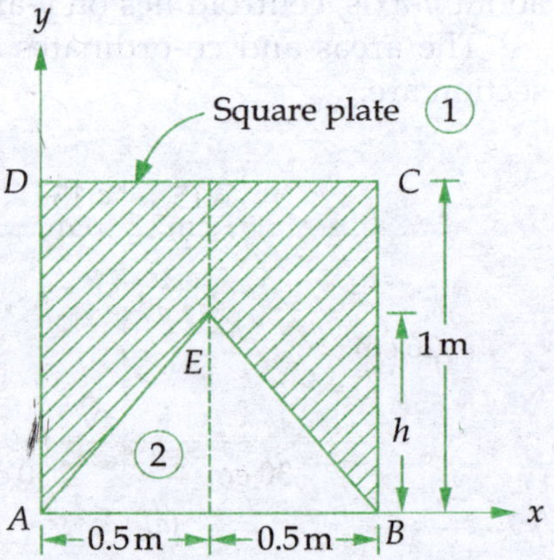

Fig. 7.21

Now, $A_1 = 1 \times 1 = 1 \text{ m}^2; \quad y_1 = \frac{1}{2} = 0.5 \text{ m}$

$$A_2 = \frac{1}{2}bh = \frac{h}{2}; \quad y_1 = \frac{h}{3}$$

Then $\bar{y} = \frac{A_1y_1 - A_2y_2}{A_1 - A_2} = \frac{1 \times 0.5 - h/2 \times h/3}{1 - h/2}$

$$= \frac{0.5 - 0.167h^2}{1 - 0.5h}$$

$$\therefore \quad h = \frac{0.5 - 0.167h^2}{1 - 0.5h} \quad (\because \ \bar{y} = h)$$

Upon simplication: $0.333\,h^2 - h + 0.5 = 0$

Solution of this quadratic equation gives

$$h = \frac{1 \pm \sqrt{1^2 - 4(0.5 \times 0.333)}}{2 \times 0.333} = \frac{1 \pm 0.578}{0.666}$$

$$= 2.37 \text{ m} \quad \text{or} \quad 0.63.4 \text{ m}$$

The value h = 2.37 m is not possible. As such the height of the triangle to be removed is 0.634 m

$$\therefore \quad \text{Area of triangle to be removed} = \frac{1}{2} \times 1 \times 0.634 = \mathbf{0.317\ m^2}$$

EXAMPLE 7.17

Determine the coordinates of the centroid of the lamina as shown in Fig. 7.22(a). The shaded area is the opening in the lamina.

Solution : The given lamina can be split up into the following basic figures as shown in Fig. 7.22 (b)

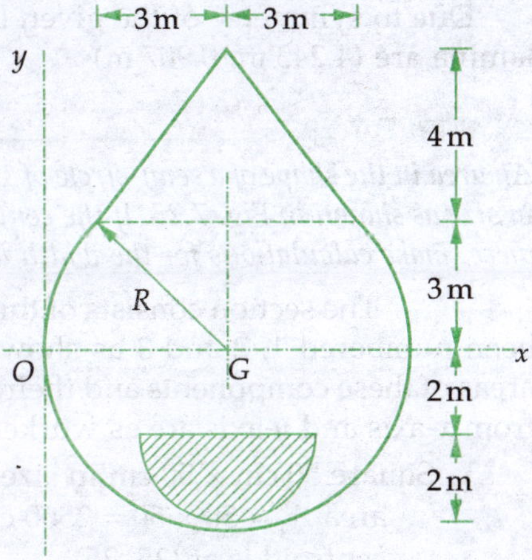

Fig. 7.22 (a)

(*i*) Circle of radius R

$$R = BG = \sqrt{(BC)^2 + (CG)^2} = \sqrt{3^2 + 3^2} = 4.243 \text{ m}$$

$$A_1 = \pi R^2 = \pi \times (4.243)^2 = 56.53 \text{ m}^2$$

$$y_1 = 0$$

(*ii*) Triangle BDG

$$A_2 = \frac{1}{2} bh = \frac{1}{2} \times 6 \times 3 = 9 \text{ m}^2$$

$$y_2 = \frac{2}{3} h = \frac{2}{3} \times 3 = 2 \text{ m}$$

(*iii*) Triangle BDE

$$A_3 = \frac{1}{2} \times 6 \times 4 = 12 \text{ m}^2$$

$$y_3 = 3 + \frac{1}{3} \times 4 = 4.33 \text{ m}$$

(*iv*) Semi circular opening (minus)

$$A_4 = \frac{1}{2} \times (\pi \times 2^2) = 6.28 \text{ m}^2$$

For a semicircular segment, the centroid is at a distance of $4R/3\pi$ from base.

$$\therefore \quad y_4 = -\left[2 + \frac{4R}{3\pi}\right] = -\left[2 + \frac{4 \times 2}{3\pi}\right] = -2.85 \text{ m}$$

It is to be noted that y_4 is negative because it is below x-axis.

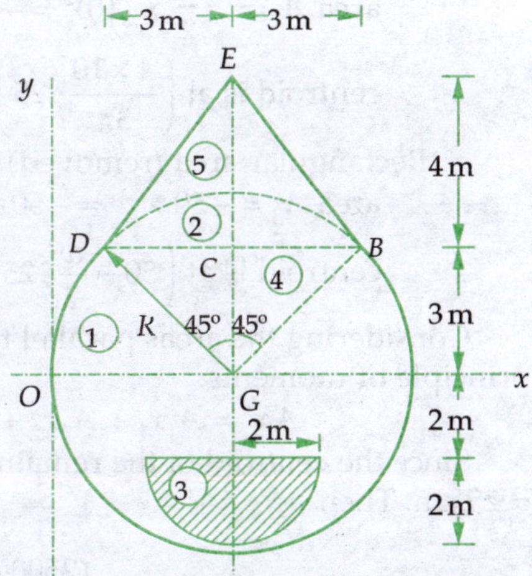

Fig. 7.22 (b)

(*v*) Sector BGD (minus)

$$A_5 = \frac{\pi}{4} R^2 = \frac{\pi}{4} \times (4.243)^2 = 14.14 \text{ m}^2$$

For a sector, the centroid is at a distance of $\frac{2}{3} \frac{R \sin\alpha}{\alpha}$ from the apex point.

α = semi vestex angle in radians $= 45 \times \frac{\pi}{180} = 0.785$

$$\therefore \quad y_5 = \frac{2}{3} \times \frac{4.243 \sin 45°}{0.785} = 2.55 \text{ m}$$

Thus for the given lamina,

$$\bar{y} = \frac{A_1 y_1 + A_2 y_2 + A_3 y_3 - A_4 y_4 - A_5 y_5}{A_1 + A_2 + A_3 - A_4 - A_5}$$

$$= \frac{56.53 \times 0 + 9 \times 2 + 12 + 4.33 - 6.28 \times (-2.85) - 14.14 \times 2.55}{56.53 + 9 + 12 - 6.28 - 14.14}$$

$$= \frac{+18 + 51.96 + 17.9 - 36.06}{57.11} = \frac{51.8}{57.11} = 0.907 \text{ m}$$

Due to symmetry of the given lamina $\bar{x} = R = 4.243$ m. Hence coordinate of centroid of the lamina are (4.243 m, 0.907 m)

EXAMPLE 7.18

An area in the shape of a semi-circle of 10 cm radius has been punched from a square plate of 50 cm × 50 cm in size as shown in Fig. 7.23. If the centroid of the remainder piece is to be kept the same as for the original piece, make calculations for the width w of the area, shown hatched, that needs to be removed.

Solution : The section consists of three parts which have been numbered 1, 2 and 3 as shown in the figure. The areas of these components and their centroidal distances from x-axis and y-axis are as worked out below:

1. Square 50 cm × 50 cm in size
   area $A_1 = 50 \times 50 = 2500 \text{ cm}^2$
   centroid is at (25, 25)
2. Semi-circular area (removed)
   area $A_2 = -\frac{\pi}{2} \times (10)^2 = -157 \text{ cm}^2$
   centroid is at $\left(\frac{4 \times 10}{3\pi}, 25\right)$
3. Rectangular area (removed)
   area $A_3 = -50 \times w = -50w \text{ cm}^2$
   centroid is at $\left(50 - \frac{w}{2}, 25\right)$

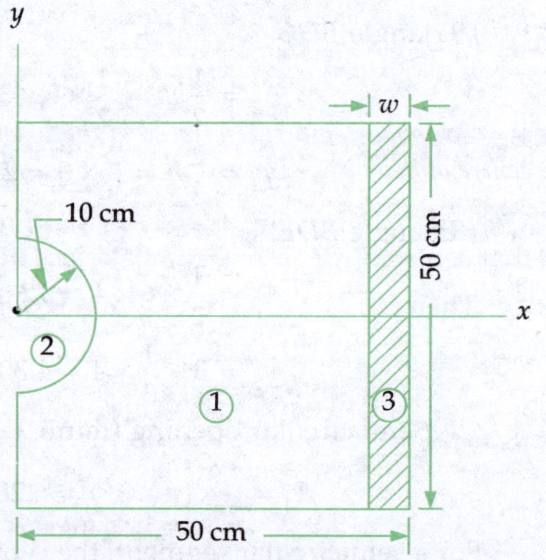

Fig. 7.23

Considering the areas parallel to y-axis, we can write the following identity by applying the principle of moments.

$$A\bar{x} = A_1 x_1 + A_2 x_2 + A_3 x_3$$

Since the centroid of the remainder piece is to be the same as for the original square plate, x = 25 cm. Then

$$(2500 - 157 - 50w) \times 25 = 2500 \times 25 - 157 \times \frac{4 \times 10}{3\pi} - 50w\left(50 - \frac{w}{2}\right)$$

or $\quad 58575 - 1250w = 62500 - 665.67 - 2500w + 25w^2$

or $\quad 25w^2 - 1250w + 3258.33 = 0$

or $\quad w^2 - 50w + 130.33 = 0$

Solution of this quadratic equation gives

$$w = \frac{50 \pm \sqrt{50^2 - 4 \times 130.33}}{2} = \frac{50 \pm 44.48}{2} = 47.24 \text{ cm and } 2.76 \text{ cm}$$

w = 47.24 cm is not possible

$\therefore \quad w$ = **2.76 cm**

EXAMPLE 7.19

A 4 cm diameter hole is to be drilled in a metallic piece of uniform thickness and configuration as depicted in the figure given below. All dimensions are in cm.

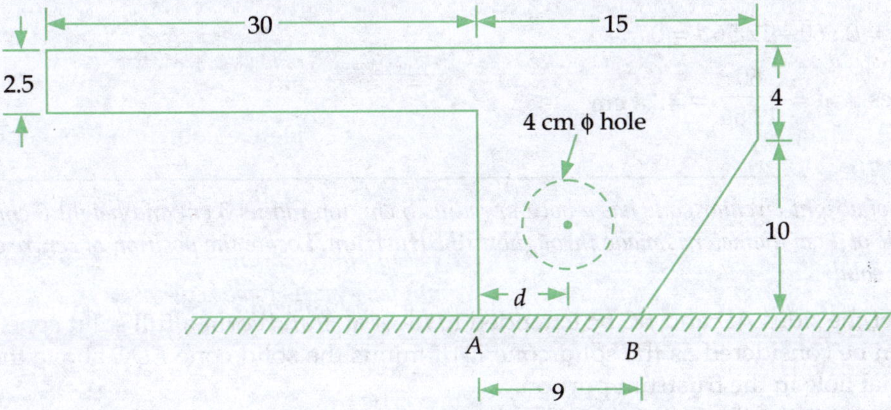

Fig. 7.24 (a)

To avoid tipping of the piece as and when placed on a horizontal floor, what should be the distance d of the centre of hole from the vertical face ?

Solution : For tipping not to occur, the resultant weight should act through the point A. This implies that centroid of the entire metal piece lies on a vertical through it.

Then taking the vertical line through A as the y-axis, $\bar{x}$ is zero.

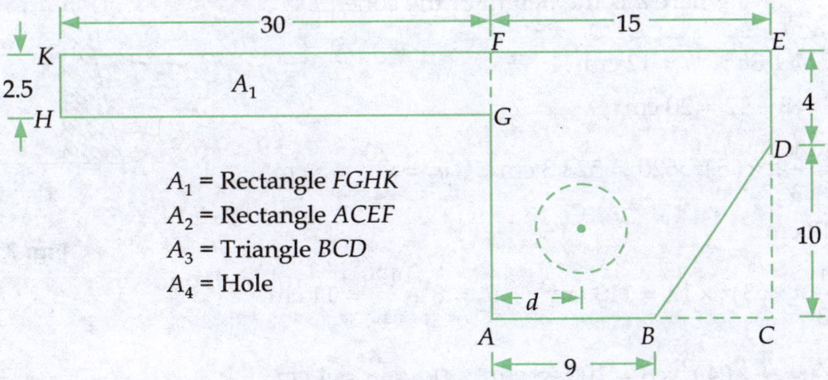

Fig. 7.24 (b)

$$A_1 = 30 \times 2.5 = 75 \text{ cm}^2 \quad \text{and} \quad x_1 = -\frac{30}{2} = -15 \text{ cm}$$

(– *ve* as it lies towards left of y-axis)

$$A_2 = 15 \times 14 = 210 \text{ cm}^2 \quad \text{and} \quad x_2 = \frac{15}{2} = 7.5 \text{ cm}$$

$$A_3 = \frac{1}{2} \times 6 \times 10 = 30 \text{ cm}^2 \quad \text{and} \quad x_3 = 9 + \frac{2}{3} \times 6 = 13$$

$$A_4 = \frac{\pi}{4} \times (4)^2 = 12.56 \text{ cm}^2 \quad \text{and} \quad x_4 = d$$

Now :
$$\bar{x} = \frac{75(-15)+210\times7.5-30\times13-12.56d}{75+210-30-12.56}$$

$$= \frac{-1125+1575-390-12.56d}{214.44} = \frac{60-12.56d}{168.24}$$

Since $\bar{x} = 0 : 60 - 12.56d = 0$

That gives : $d = \dfrac{60}{12.56} = $ **4.78 cm**

EXAMPLE 7.20

The frustrum of a right circular cone has a bottom radius 5 cm, top radius 3 cm and height 8 cm. A co-axial cylindrical hole of 4 cm diameter is made throughout the frustrum. Locate the position of centre of gravity of the remaining solid.

Solution : Let the sides *AD* and *BC* be extended to meet at *E* and form a full solid cone. The given solid can then be considered as the solid cone *EAB* minus the solid cone *EDC* above the frustrum minus the axial hole in the frustrum portion.

Due to symmetry of the section, the *CG* lies at a point on the *y*-axis i.e., $\bar{x} = 0$. Further, let the bottom surface *AB* be chosen as the reference axis to locate the centroid $\bar{y}$.

The volumes and the co-ordinates of the centroid of the different segments comprising the given section are :

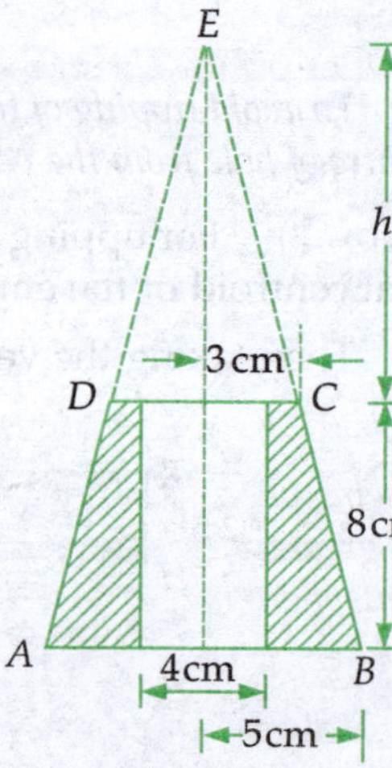

Fig. 7.25

Cone *EAB* : Since the triangles *EAB* and *EDC* are equiangular

$$\frac{5}{3} = \frac{8+h}{h} \quad \text{where } h \text{ is the height of the cone } EDC$$

$$5h = 24 + 3h ; \quad h = 12 \text{ cm}$$

Height of cone *EAB* = 8 + 12 = 20 cm

$$V_1 = \frac{1}{3}\pi \times (5)^2 \times 20 = 523.3 \text{ cm}^3 ; \quad y_1 = \frac{20}{4} = 5 \text{ cm}$$

Cone EDC : $h = 12$ cm

$$V_2 = \frac{1}{3}\pi \times (3)^2 \times 12 = 113 \text{ cm}^3 ; \quad y_2 = 8 + \frac{12}{4} = 11 \text{ cm}$$

Cylindrical hole : $V_3 = \dfrac{\pi}{4} \times (4)^2 \times 8 = 100.48 \text{ cm}^3 ; \quad y_3 = \dfrac{8}{2} = 4 \text{ cm}$

Let the centroid of the given solid be located at distance $\bar{y}$ from the *x*-axis. Then

$$\bar{y} = \frac{V_1y_1 - V_2y_2 - V_3y_3}{V_1 - V_2 - V_3} = \frac{523.3\times5 - 113\times11 - 100.48\times4}{523.3 - 113 - 100.48} = \textbf{3.139 cm}$$

Hence *CG* of the given section lies on the *y*-axis ($\bar{x} = 0$) and at distance 3.139 cm from the bottom face *AB*.

EXAMPLE 7.21

A body consists of a solid hemisphere of radius 4 cm and a right circular solid cone of height 12 cm. The hemisphere and cone have a common base and are made of the same material. Locate the position of CG of the composite body.

Solution : Refer Fig. 7.26 for the composite body. Mass of body equals the product of its density and volume. Since both the sections are made of the same material, calculations can be made on the basis of volume.

For the solid cone *ABC*

$$\text{Volume } V_1 = \frac{1}{3}\pi r^2 h = \frac{1}{3}\pi \times 4^2 \times 12 = 200{\cdot}96 \text{ cm}^3$$

Distance of CG of cone from its apex A,

$$y_1 = 12 - \frac{12}{4} = 9 \text{ cm}$$

The distance y_1 has been so calculated because for a solid cone of height h, the CG lies on the axis at distance $h/4$ from the base.

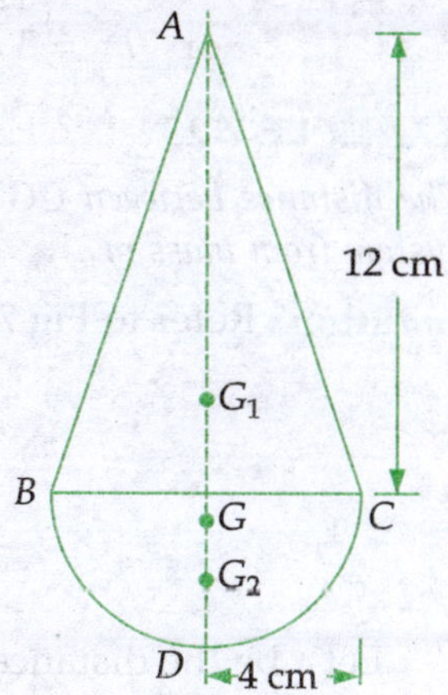

Fig. 7.26

For the hemisphere *BCD*

$$\text{Volume } V_2 = \frac{1}{2}\times\frac{4}{3}\pi r^3 = \frac{1}{2}\times\frac{4}{3}\pi\times 4^3 = 133{\cdot}97 \text{ cm}^3$$

Distance of CG of hemisphere from point A,

$$y_2 = 12 + \frac{3}{8}r = 12 + \frac{3}{8}\times 4 = 13.5 \text{ cm}$$

Let point G represent the CG of the composite body with distance $AG = \bar{y}$

$$\text{Then} \quad \bar{y} = \frac{V_1 y_1 + V_2 y_2}{V_1 + V_2}$$

$$= \frac{(200{\cdot}96\times 9) + (133{\cdot}97\times 13{\cdot}5)}{200{\cdot}96 + 133{\cdot}97} = 10{\cdot}8 \text{ cm for vertex } A$$

EXAMPLE 7.22

A hemisphere and a cone have their bases joined together; the two bases being of same size. What should be the greatest height of the cone in terms of radius of the base so that the combined body of the cone and hemisphere may stand upright?

Solution : With reference to Fig. 7.27

Volume of cone

$$V_1 = \frac{1}{3}\pi r^2 h$$

Position of its centroid from point C,

$$CG_1 = \frac{1}{4}h$$

Volume of hemisphere $V_2 = \dfrac{2}{3}\pi r^3$

Position of its centroid from point C,

$$CG_2 = -\frac{3}{8}r \quad (-\text{ve because it lies below point C})$$

Then from the principle of moments,

$$(V_1 + V_2)\,CC' = V_1 \times CG_1 + V_2 \times CG_2$$

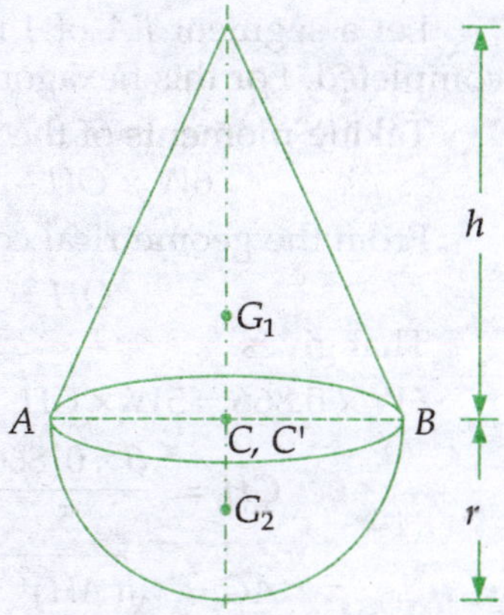

Fig. 7.27

$$= \frac{1}{3}\pi r^2 h \times \frac{1}{4}h + \frac{2}{3}\pi r^3\left(-\frac{3}{8}r\right) = \frac{1}{12}\pi r^2 h^2 - \frac{1}{4}\pi r^4$$

The composite body will stand upright if the position of its centre of gravity coincides with the centre of common base, i.e., $CC' = 0$

That gives

$$0 = \frac{1}{12}\pi r^2 h^2 - \frac{1}{4}\pi r^4$$

or $h^2 = 3r^2$ $\quad\therefore\quad h = \mathbf{1.732\ r}$

EXAMPLE 7.23

The distance between CG of two masses m_1 and m_2 is l. Work out the distance of CG of the composite system from mass m_1.

Solution : Refer to Fig 7.28.

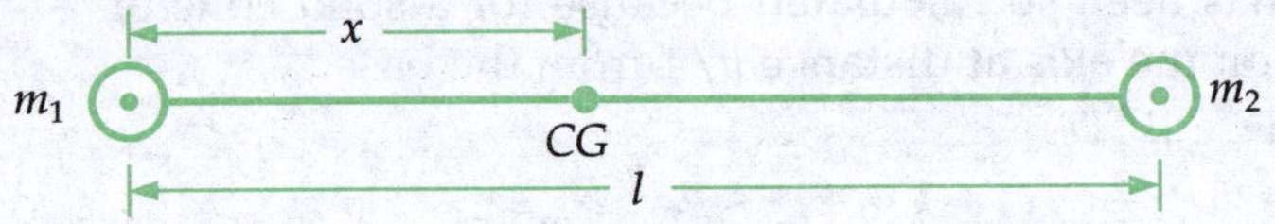

Fig. 7.28

Let x be the distance of CG of the composite system from mass m_1

Taking moments of masses m_1, m_2 and $(m_1 + m_2)$ about the centre of gravity of mass m_1,

$$m_2 l = (m_1 + m_2)\,x$$

$$\therefore\quad x = \frac{m_2}{m_1 + m_2}\,l$$

EXAMPLE 7.24

A rod of length 5 m is bent so as to form five sides of a regular hexagon. Make calculations for the distance of its centre of gravity from either end of the rod.

Solution : Reference Fig 7.29, $ABCDEF$ is the rod in which

$$AB = BC = CD = DE = EF = 1\text{ m}$$

If W is the weight of each portion, then total weight of the rod is $5W$ and this acts at point G which represents the centre of gravity of the rod. The distance AG is to be determined.

Let a segment FA of 1 m length and weight W be introduced and the regular hexagon be completed. For this hexagon, the CG will be at point O.

Taking moments of the weights about line FA, we get

$$6W \times OH - W \times 0 = 5W \times GH$$

From the geometrical configuration,

$$OH = 1\sin 60° = 0.866\text{ m and } AH = 0.5\text{ m}$$

That gives

$$6W \times 0.866 = 5W \times GH$$

$$GH = \frac{6 \times 0{\cdot}866}{5} = 1{\cdot}039\text{ m}$$

$$\therefore\quad AG = \sqrt{(AH)^2 + (GH)^2} = \sqrt{0{\cdot}5^2 + 1{\cdot}039^2} = \mathbf{1.15\ m}$$

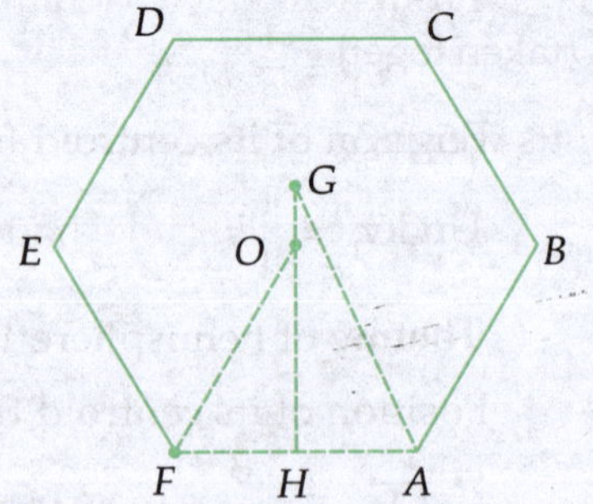

Fig. 7.29

EXAMPLE 7.25

The density at point of a slender rod varies with the first power of the distance of the point from one end of the rod. Locate the mass centre. How does the mass centre compares with the length centroid?

Solution : The mass centroid (centre) of a wire, pipe or rod of constant cross-section is given by

$$\bar{x} = \frac{\Sigma x dm}{\Sigma dm} \text{ and } \bar{y} = \frac{\Sigma y dm}{\Sigma dm}$$

When the x-axis is so chosen that it passes through the centre of the slender rod and along its length, $\bar{y} = 0$

Let the density (taken as mass per unit length) at distance x from the left end of the rod be $\rho = kx$. The mass of an element dx located at distance x from the origin (left end of the rod) is

$$dm = \rho\, dx = kx\, dx$$

Then : $$\Sigma x\, dm = \int_o^l k x^2 dx = k\frac{l^3}{3}$$

$$\Sigma dm = \int_o^l k x dx = \frac{kl^2}{2}$$

$$\therefore \quad \bar{x} = \frac{kl^3/3}{kl^2/2} = \frac{2}{3}l$$

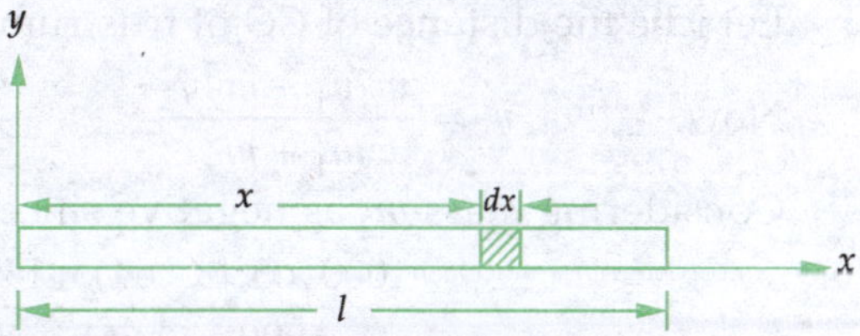

Fig. 7.30

Thus the mass centre of a slender rod is located at two-third of the length from the end chosen as reference.

The centroid of the length of the rod, however, is at its mid point (Refer to Example 7.2).

EXAMPLE 7.26

Three identical boxes, each having length l and weight w, are placed as shown in Fi.g 7.31. Find out the maximum possible distance m through which the top box can extend out from the bottom so that there is no possibility of the toppling of the stack.

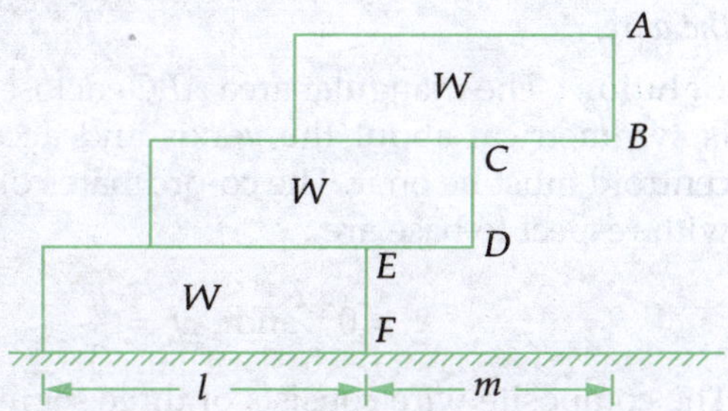

Fig. 7.31

Solution : It is to be noted that to avioid toppling over of the top box, its centre of grovity is in line with the outer edge CD of the middle box. This means that the line of action of the weight W of the top box passes through edge CD of the middle box. Obviously the maximum permissible overhang of the box is $l/2$.

$\therefore$ Distance $CB = l/2$

From free body diagram of middle and bottom box taken together, let x be the overhang of middle slab. Then its weight acts at distance $\left(\frac{l}{2} - x\right)$ from edge EF

Under equilibrium conditions : $\Sigma M_e = 0$

That is : $$Wx - W\left(\frac{l}{2} - x\right) = 0$$

or $$x = \frac{l}{2} - x \; ; x = \frac{l}{4}$$

$$\therefore \quad \text{Distance } m = \frac{l}{2} + \frac{l}{4} = \frac{3l}{4}$$

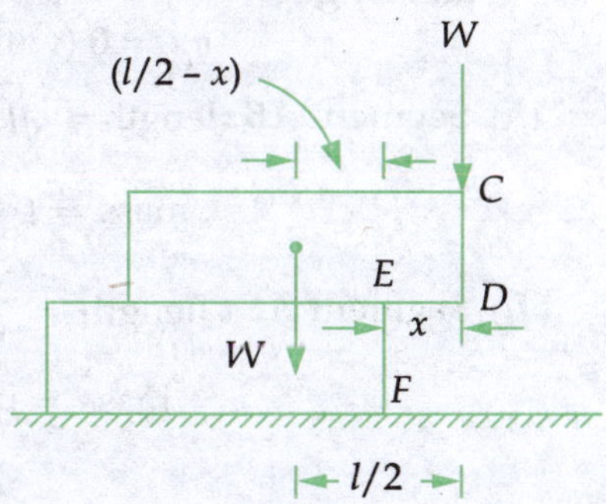

EXAMPLE 7.27

A right circular cone of 20 cm height weighs 1000 N. A cone of 8 cm height and 64 N weight is removed from the top. Determine the distance of CG of the frustrum from the base.

Solution : Refer to Fig 7.32

For a solid cone of height h, the CG lies on the axis at distance $h/4$ above the base.

Distance of CG of full cone from BC

$$y_1 = \frac{20}{4} = 5 \text{ cm}$$

Distance of CG of removed conical section from base line BC

$$y_2 = 12 + \frac{8}{4} = 14 \text{ cm}$$

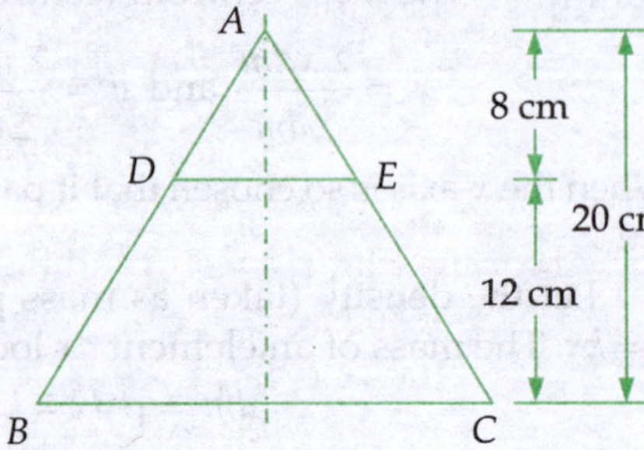

Fig. 7.32

Let y be the distance of CG of frustrum $DECB$ from the base line BC.

Now, $$\bar{y} = \frac{m_1 y_1 + m_2 y_2}{m_1 + m_2}$$

Considering mass m_2 as negative since it has been removed,

$$\bar{y} = \frac{1000 \times 5 + (-64) \times 14}{1000 + (-64)} = \frac{5000 - 896}{936} = \mathbf{4.385 \text{ cm}}$$

Alternatively : Taking moments of the weights of different sections about the base line BC, we get

$$1000 \times 5 = 64 \times 4 + (1000 - 64)\, y\,; \quad y = 4.385 \text{ cm}$$

EXAMPLE 7.28

A thin wire of homogeneous material is bent to form an isosceles triangle ABC as shown in Fig. 7.33. Determine the base angle θ for which the centre of mass of the wire coincides with the centroid of the area enclosed by the wire.

Solution : The triangular area ABC enclosed by the wire is symmetrical about the y-axis and accordingly the centroid must lie on it. The co-ordinates of the centroid with respect to base are :

$$x_c = 0 \quad \text{and} \quad y_c = \frac{h}{3}$$

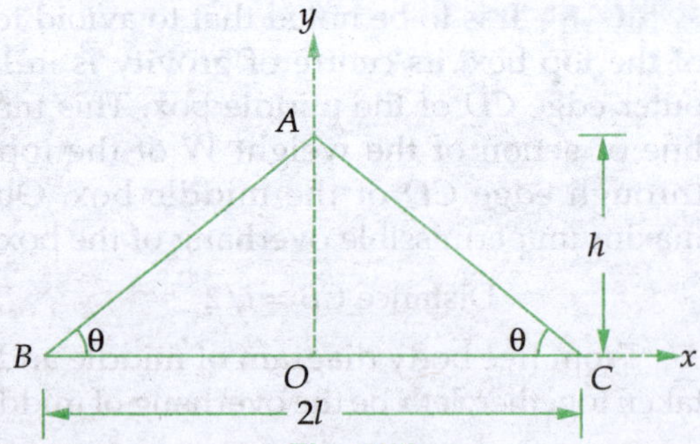

Fig. 7.33

The composite wire consists of three segments

(*i*) **Base BC :** mass $= 2\,\rho l$ where ρ is the density per unit length

$$y_{c_1} = 0 \text{ (with respect to base } BC)$$

(*ii*) **Segment AB :** length $= \sqrt{l^2 + h^2}$

$$\text{mass} = \rho\sqrt{l^2 + h^2} \quad \text{and} \quad y_{c_2} = \frac{h}{2}$$

(*iii*) **Segment AC :** length $= \sqrt{l^2 + h^2}$

$$\text{mass} = \rho\sqrt{l^2 + h^2} \quad \text{and} \quad y_{c_3} = \frac{h}{2}$$

Then y-coordinate of the centre of mass of the wires forming the triangle is

$$y_c = \frac{\Sigma m \times y_c}{\Sigma m} = \frac{2\rho l \times 0 + \rho\sqrt{l^2+h^2} \times \frac{h}{2} + \rho\sqrt{l^2+h^2} \times \frac{h}{2}}{2\rho l + \rho\sqrt{l^2+h^2} + \rho\sqrt{l^2+h^2}}$$

$$= \frac{\rho h\sqrt{l^2+h^2}}{2\rho l + 2\rho\sqrt{l^2+h^2}} = \frac{h\sqrt{l^2+h^2}}{2\left[l + \sqrt{l^2+h^2}\right]}$$

According to the given condition,

$$\frac{h}{3} = \frac{h\sqrt{l^2+h^2}}{2\left[l + \sqrt{l^2+h^2}\right]}$$

or $$2l + 2\sqrt{l^2+h^2} = 3\sqrt{l^2+h^2}$$

or $$2l = \sqrt{l^2+h^2} \quad ; \quad 4l^2 = l^2 + h^2$$

That gives : $h = \sqrt{3}\, l$

$$\therefore \quad \theta = \tan^{-1}\left(\frac{h}{l}\right) = \tan^{-1}(\sqrt{3}) = 60°$$

EXAMPLE 7.29

A wooden plank of size 10 cm width and 1 cm thick is fixed on top of a semi-circular steel plate of radius 5 cm and thickness 1 cm as shown in Fig. 7.34. Determine the maximum height h of the wooden plank so that CG of the composite body lies in the junction of the two. Take density of wood equal to 1/10 that of steel.

Solution : Mass of body equals the product of its density and volume.

$$m_w = \rho_w \times h\,t\,b$$
$$= \rho_w \times h \times 10 \times 100 = 1000\, h\, \rho_w$$

$$m_s = \rho_s \times \frac{\pi}{2} r^2 \times t$$
$$= \rho_s \times \frac{\pi}{2}(50)^2 \times 10 = 39250\, \rho_s$$

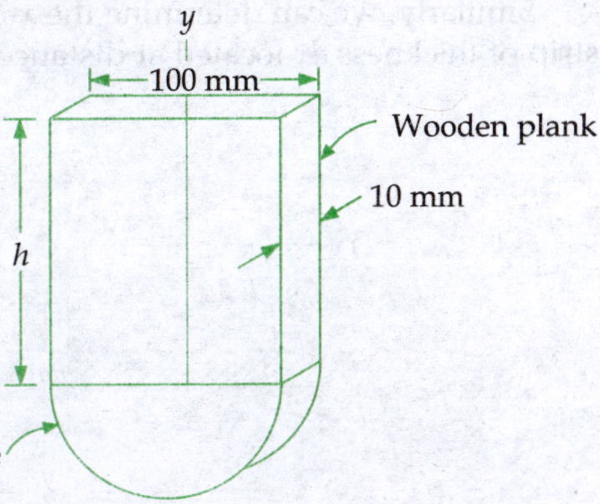

Fig. 7.34

With the co-ordinate axes as shown:

$$y_w = \frac{h}{2} \quad \text{and} \quad y_s = -\frac{4r}{3\pi} = -\frac{4 \times 50}{3\pi} = -21.33 \text{ mm}$$

The negative sign with y_s stems from the fact that centroid for the steel plate lies below the x-axis.

Then $$\bar{y} = \frac{m_w y_w + m_s y_s}{m_w + m_s} = \frac{1000\, h\rho_w \times \frac{h}{2} + 39250\, \rho_s(-21.23)}{1000\, h\rho_w + 39250\, \rho_s}$$

Since CG of the composite body is given to lie on the junction through which passes the x-axis, $\bar{y} = 0$.

That gives: $1000\, h\, \rho_w \times \dfrac{h}{2} + 39250\, \rho_s(-21.23) = 0$

Further $\rho_s = 10\,\rho_w$

$$\therefore \quad 1000\,\frac{h^2}{2}\rho_w + 39250 \times 10\,\rho_w(-21.23) = 0$$

or $$h^2 = \frac{2 \times 39250 \times 10 \times 21.23}{100} = 16665.56$$

$\therefore$ Height of wooden plank, $h = \sqrt{16665.56}$ = **129.09 mm**

EXAMPLE 7.30

Determine the centroid of area of a rectangle of breadth b and height h.

Solution : Let $ABCD$ be a rectangle of breadth b and height h. To determine y-coordinate of the centroid of this rectangle, consider a strip of thickness dy located at distance y from side AB of the rectangle. For this elemental strip,

$$\text{area} = b\,dy$$

$$\text{moment about } x\text{-axis} = b\,dy \times y = by\,dy$$

$$\text{area of rectangle } ABCD = b\,h$$

If y is the distance of the centroid from the base, then from the moment principle

$$bh \times y = \int_o^h b\,y\,dy = b\left|\frac{y^2}{2}\right|_o^h = \frac{bh^2}{2}$$

$$\therefore \quad \bar{y} = \frac{bh^2}{2} \div bh = \frac{h}{2}$$

Thus the centroid of a rectangle of height h is at a distance $h/2$ from the base.

Similarly, we can determine the x-coordinate of the centroid of this rectangle by considering a strip of thickness dx located at distance x from side AD of the rectangle. Then

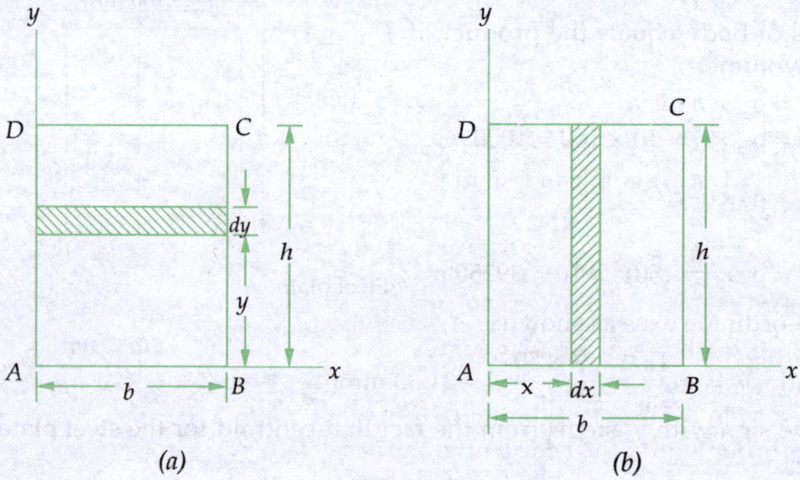

Fi.g. 7.35

$$bh \times \bar{x} = \int_o^b h\,dx \times x = \int_o^b h\,x\,dx = \frac{hb^2}{2}$$

$$\therefore \quad x = \frac{b}{2}$$

EXAMPLE 7.31

Determine the centroid of the area of a triangle with respect to its base.

Solution : Let ABC be the triangle of base width b and height h. Consider an elementary strip of width l, thickness dy and located at distance y from base BC of the triangle. For this elemental strip,

$$\text{area} = l\,dy$$

$$\text{moment about } x\text{-axis} = l\,dy \times y$$

Since the integration is to be done with respect to y within the limits 0 to h, it is necessary to express l in terms of y. For that we obtain the following correlation from the similarity of triangles ADE and ABC,

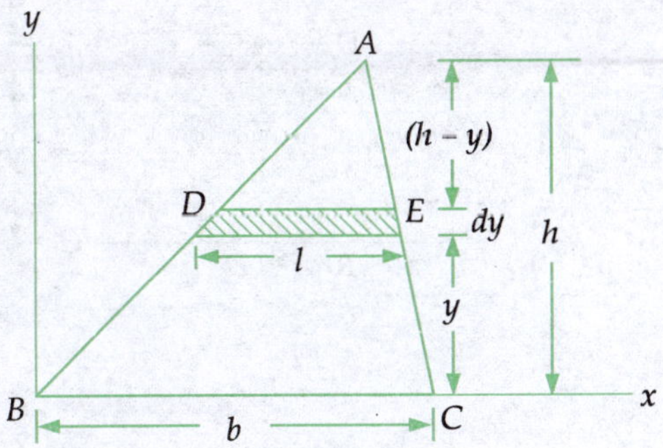

Fig. 7.36

$$\frac{l}{b} = \frac{h-y}{h}\ ; \quad l = b\left(1-\frac{y}{h}\right)$$

$\therefore$ Moment of elemental strip about x-axis

$$= b\left(1-\frac{y}{h}\right)y\,dy$$

$$\text{area of triangle } ABC = \frac{1}{2}bh$$

If $\bar{y}$ is the distance of the centroid from the base, then from the moment principle

$$\frac{1}{2}bh \times \bar{y} = \int_0^h b\left(1-\frac{y}{h}\right)y\,dy = b\int_0^h\left(y-\frac{y^2}{h}\right)dy$$

$$= b\left|\frac{y^2}{2}-\frac{y^3}{3h}\right|_0^h = b\left(\frac{h^2}{2}-\frac{h^2}{3}\right) = \frac{bh^2}{6}$$

$$\therefore \quad \bar{y} = \frac{bh^2}{6}\times\frac{2}{bh} = \frac{h}{3}$$

Thus the centroid of a triangle of height h is at a distance $h/3$ from the base or $2h/3$ from the apex.

EXAMPLE 7.32

Determine the centre of gravity of a solid right circular cone of height h and base circle radius R.

Solution : Let ABC be the right circular cone of height h and base circle radius R.

Consider an elemental ring of radius r between two planes parallel to the base BC and at distance y and $(y + dy)$ from it. Presuming that the material of the cone is homogeneous (*i.e.*, its density is same through out), the centre of gravity will be its centre of volume.

For the elemental ring,

$$\text{volume} = \pi r^2\,dy$$

$$\text{moment about } x\text{-axis} = \pi r^2 dy \times y$$

Since the integration is to be done with reference to y within the limits 0 to h, it is necessary to express r in terms of y. For that we obtain the following correlation from the similarity of triangles ADE and ABC

$$\frac{r}{R} = \frac{h-y}{h}\ ;\quad r = R\left(1-\frac{y}{h}\right)$$

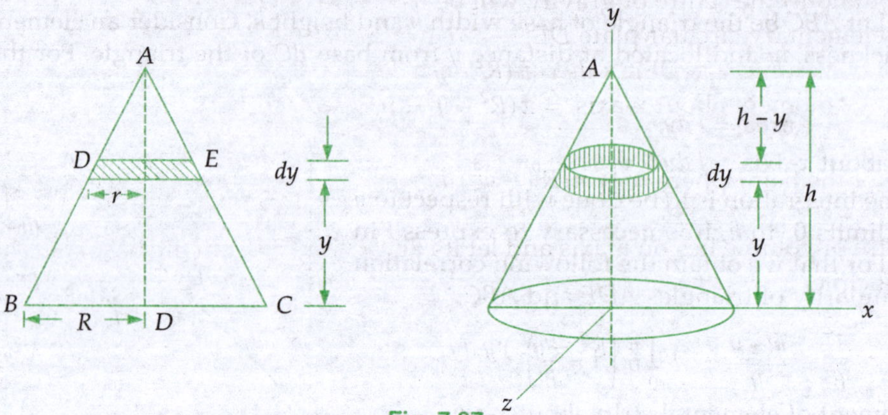

Fig. 7.37

$\therefore$ moment of elemental volume about x-axis $= \pi R^2 \left(1-\frac{y}{h}\right)^2 dy \times y$

$$\text{Volume of cone ABC} = \frac{1}{3}\pi R^2 h$$

If $\bar{y}$ is the distance of the centre of volume from the base, then from the moment principle

$$\frac{1}{3}\pi R^2 h \times \bar{y} = \int_o^h \pi R^2 \left(1-\frac{y}{h}\right)^2 y\,dy = \pi R^2 \int_o^h \left(y + \frac{y^3}{h^2} - 2\frac{y^2}{h}\right) dy$$

$$= \pi R^2 \left| \frac{y^2}{2} + \frac{y^4}{4h^2} - \frac{2}{3}\frac{y^3}{h} \right|_o^h = \pi R^2 \times \frac{h^2}{12}$$

$$\therefore\quad \bar{y} = \left(\pi R^2 \times \frac{h^2}{12}\right) \times \frac{3}{\pi R^2 h} = \frac{h}{4}$$

Thus the centre of gravity of right circular cone of height h is at a distance $h/4$ from the base or $3h/4$ from the apex.

EXAMPLE 7.33

Locate the centre of gravity of a solid hemisphere of radius R. How this value would be affected if the hemisphere is hollow with negligible thickness?

Solution : Let ACB be a solid hemisphere of radius R.

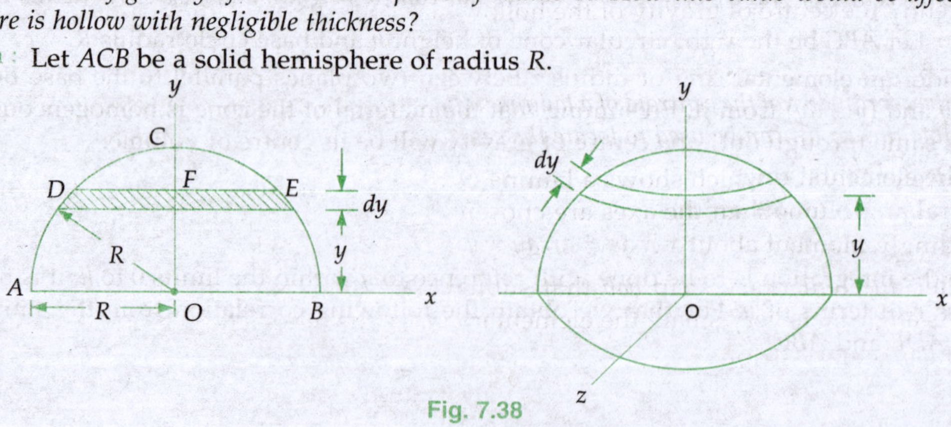

Fig. 7.38

Consider an elementary circular plate DE cut off by the planes parallel to AB and at distances y and $(y + dy)$ from it. Presuming that the material of the hemisphere is homogeneous (*i.e.*, same density throughout) the centre of gravity will be its centre of volume.

For the elementary circular plate DE,

$$\text{volume} = \pi (DF)^2 \times dy = \pi (R^2 - y^2)\, dy$$

$$\text{moment about x-axis} = \pi (R^2 - y^2)\, dy \times y$$

$$\text{Volume of the hemisphere} = \int_0^R \pi(R^2 - y^2)\, dy = \pi \left| R^2 \times y - \frac{y^3}{3} \right|_0^R = \frac{2}{3} \pi R^3$$

The centre of volume lies on y-axis and let its distance be $\bar{y}$ from the base AB. Then from the moment principle,

$$\frac{2}{3} \pi R^3 \times \bar{y} = \int_0^R \pi (R^2 - y^2)\, y\, dy = \pi \int_0^R (R^2 y - y^3)\, dy$$

$$= \pi \left| R^2 \frac{y^2}{2} - \frac{y^4}{4} \right|_0^R = \pi \frac{R^4}{4}$$

$$\therefore \quad \bar{y} = \frac{\pi R^4}{4} \times \frac{3}{2\pi R^3} = \frac{3R}{8}$$

Hence centre of gravity of a solid hemisphere of radius R lies on the central radius at distance $3R/8$ from the plane base.

(*b*) If the hemisphere is hollow of negligible thickness, then DE is ring with

$$\text{elemental area} = 2\pi R\, dy$$

$$\text{moment about } x\text{-axis} = 2\pi R\, dy \times y$$

$$\text{Area of thin hollow hemisphere} = \int_o^R 2\pi R dy = 2\pi R\, |y|_0^R = 2\pi R^2$$

The centre of area lies on y-axis and let its distance from base AB be $\bar{y}$. Then from the moment principle

$$2\pi R^2 \times y = \int_o^R 2\pi R y\, dy = 2\pi R \left| \frac{y^2}{2} \right|_o^R = \pi R^3$$

$$\therefore \quad \bar{y} = \frac{\pi R^3}{2\pi R^2} = \frac{R}{2}$$

Apparently the centre of gravity of the hollow hemisphere bisects the central radius.

EXAMPLE 7.34

Determine the coordinates of the centroid of a lamina in the shape of a circular sector of radius r and central angle 2α. How this result can be used to locate the centroid of a semicircle and quarter of a circle ?

Solution : Refer Fig 7.39 which shows a lamina OAB in the shape of a circular sector of radius r and central angle 2α. When the axes are chosen as shown, the x-axis is an axis of symmetry and accordingly $\bar{y} = 0$

Consider an elementary strip oab that subtends an angle $d\theta$ at the centre and has base $ab = r\, d\theta$. As angle $d\theta$ is very small, the elementary strip can be considered to be a triangle.

area of elementary strip

$$= \frac{1}{2}(r\,d\theta) \times r = \frac{r^2 d\theta}{2}$$

Centroid of this triangular strip lies on a line that joins O to the mid point of ab and at a distance $\frac{2}{3}r$ from O.

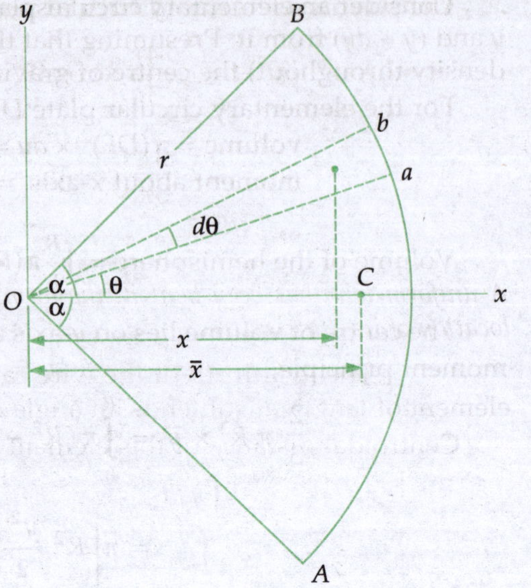

Fig. 7.39

distance x of centroid from y-axis

$$= \frac{2}{3} r \cos\theta$$

moment of area of elementary strip about y-axis

$$= \frac{r^2 d\theta}{2} \times \frac{2}{3} r \cos\theta$$

$$= \frac{1}{3} r^3 \cos\theta\, d\theta$$

If $\bar{x}$ is the coordinate of the centroid of the lamina from y-axis, then from the moment principle,

$$\bar{x} \times \int_{-\alpha}^{+\alpha} \frac{r^2 d\theta}{2} = \int_{-\alpha}^{+\alpha} \frac{1}{3} r^3 \cos\theta\, d\theta$$

It is to be noted that the variable θ varies from $-\alpha$ to $+\alpha$

$$\bar{x} \times \frac{r^2}{2} \left|\theta\right|_{-\alpha}^{+\alpha} = \frac{r^3}{3} \left|\sin\theta\right|_{-\alpha}^{+\alpha}$$

or $$\bar{x} \times \frac{r^2}{2}(2\alpha) = \frac{r^3}{2} \times 2\sin\alpha$$

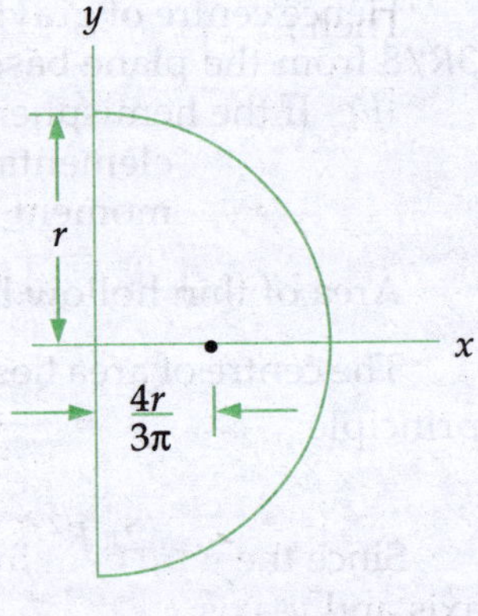

$$\therefore \quad \bar{x} = \left[\frac{r^3}{3} \times 2\sin\alpha\right] \times \left[\frac{2}{r^2(2\alpha)}\right] = \frac{2r\sin\alpha}{3\alpha}$$

This is the location of the centroid of any sector. The angle α has to be taken in radians.

(*b*) **In case of a semicircle**

$$2\alpha = 180^\circ = \pi \text{ radians}$$

$$\bar{x} = \frac{2r\sin\pi/2}{3 \times \pi/2}$$

$$= \frac{4r}{3\pi} = 0.424\, r$$

In case of quadrant of a circle

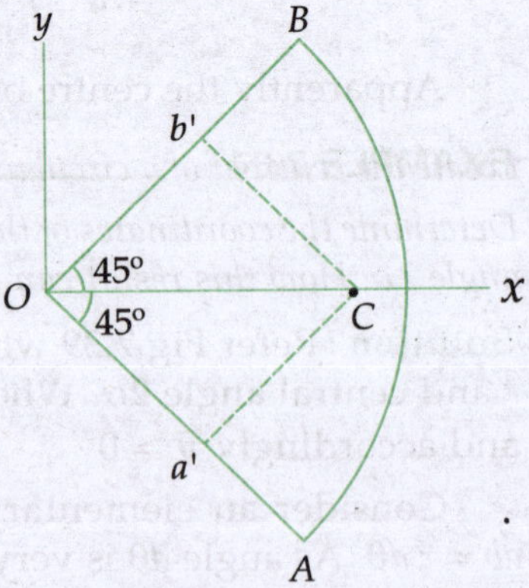

$$2\alpha = 90^\circ = \frac{\pi}{2} \text{ radians}$$

$$\bar{x} = \frac{2r\sin\pi/4}{3 \times \pi/4}$$

$$= \sqrt{2}\,\frac{4r}{3\pi} \qquad \text{(distance } OC\text{)}$$

The position of centroid with respect to the radii OA and OB will be

$$oa' = \sqrt{2}\,\frac{4r}{3\pi} \times \cos 45^{\circ} = \frac{4r}{3\pi}$$

$$ob' = \sqrt{2}\,\frac{4r}{3\pi}\,\cos 45^{\circ} = \frac{4r}{3\pi}$$

EXAMPLE 7.35

A uniform wire has been bent in the form of a quadrant of the arc of a circle of radius r. Workout the location of its centroid.

Solution : The length AB of the wire can be divided into a number of small elements; one such element of length dl subtends an angle $d\theta$ at 0.

Centroidal distance of this element from x-axis is

$$\bar{y} = \frac{\int y\,dl}{\int dl}$$

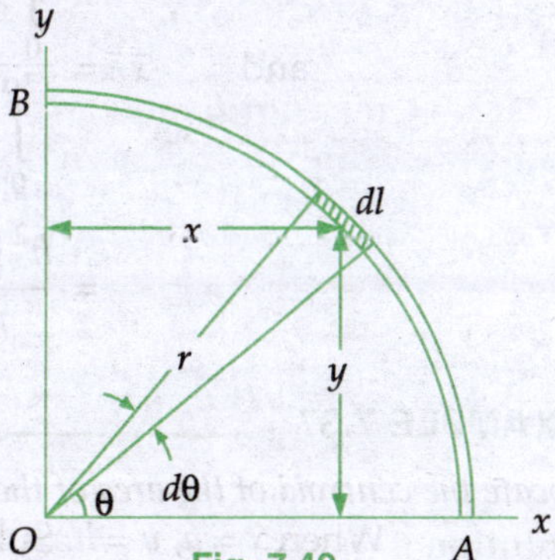

Fig. 7.40

In terms of polar coordinates,

$$y = r\sin\theta \text{ and } dl = r\,d\theta$$

Then : $$\int y\,dl = \int_0^{\pi/2} (r\sin\theta) r\,d\theta = r^2 \int_0^{\pi/2} \sin\theta\,d\theta$$

$$= r^2 \left|\cos\theta\right|_0^{\pi/2} = r^2$$

$$\int dl = \int_0^{\pi/2} r\,d\theta = r\left|\theta\right|_0^{\pi/2} = \frac{r\pi}{2}$$

$$\therefore \quad \bar{y} = \frac{r^2}{\frac{r\pi}{2}} = \frac{2r}{\pi}$$

Since the wire bent in the form of a quadrant of the arc of a circle is symmetrical about the x-axis and y-axis,

$$\bar{x} = \bar{y} = \frac{2r}{\pi}$$

EXAMPLE 7.36

Locate the centroid of a circular arc as shown below in Fig. 7.41.

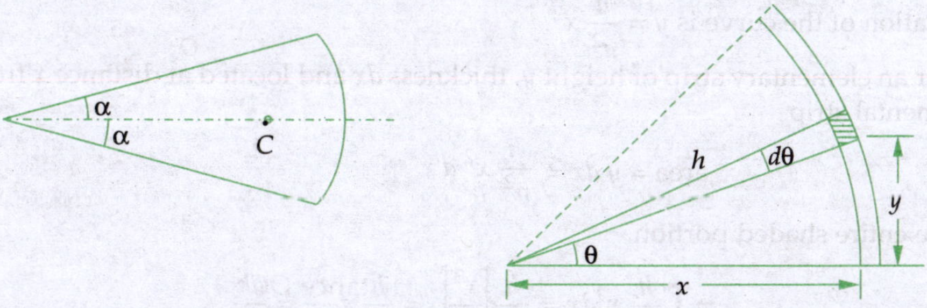

Fig. 7.41

Solution : For a small element with included angle $d\theta$

$$dl = h\, d\theta$$

$$y = h \sin\theta \quad \text{and} \quad x = h\cos\theta$$

$$\bar{y} = \frac{\int_0^{2\alpha} y\, dl}{\int_0^{2\alpha} dl} = \frac{\int_0^{2\alpha} h\sin\theta \times h\, d\theta}{\int_0^{2\alpha} h\, d\theta} = \frac{\int_0^{2\alpha} h^2 \sin\theta\, d\theta}{\int_0^{2\alpha} h\, d\theta}$$

$$= \frac{h^2 |\cos\theta|_0^{2\alpha}}{h|\theta|_0^{2\alpha}} = \frac{h^2 \cos 2\alpha}{h \times 2\alpha} = \frac{h}{2}\frac{\cos 2\alpha}{\alpha}$$

and
$$\bar{x} = \frac{\int_0^{2\alpha} x\, dl}{\int_0^{2\alpha} dl} = \frac{\int_0^{2\alpha} h\cos\theta \times h\, d\theta}{\int_0^{2\alpha} h\, d\theta} = \frac{\int_0^{2\alpha} h^2 \cos\theta\, d\theta}{\int_0^{2\alpha} h\, d\theta}$$

$$= \frac{h^2 |\sin\theta|_0^{2\alpha}}{h|\theta|_0^{2\alpha}} = \frac{h^2 \sin 2\alpha}{h \times 2\alpha} = \frac{h}{2}\frac{\sin 2\alpha}{\alpha}$$

EXAMPLE 7.37

Locate the centroid of the area of the parabolic shaded portion shown in Fig. 7.42.

Solution : When $x = a$, $y = h$. Substituting these values in the equation of the curve, we get,

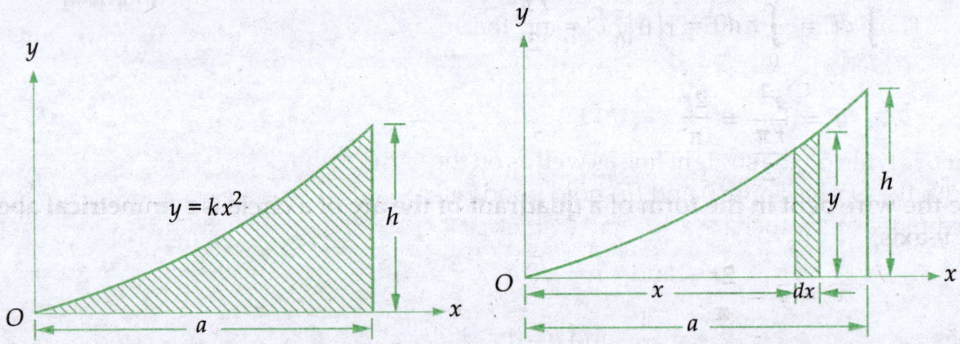

Fig. 7.42

$$h = k a^2 \quad \text{or} \quad k = h/a^2$$

$\therefore$ Equation of the curve is $y = \dfrac{h}{a^2} x^2$

Consider an elementary strip of height y, thickness dx and located at distance x from the y-axis. For this elemental strip

$$\text{area} = y\, dx = \frac{h}{a^2} x^2\, dx$$

$\therefore$ area of the entire shaded portion

$$= \int_o^a \frac{h}{a^2} x^2 dx = \frac{h}{a^2}\left|\frac{x^3}{3}\right|_0^a = \frac{h}{a^2} \times \frac{a^3}{3} = \frac{ah}{3}$$

Position of centroid from y-axis

Moment of the elemental area about y-axis = $y\,dx \times x$

$$= \frac{h}{a^2}x^2\,dx \times x = \frac{h}{a^2}x^3 dx$$

If $\bar{x}$ is the distance of the centroid from y-axis, then from the moment principle,

$$\frac{ah}{3} \times \bar{x} = \int_o^a \frac{h}{a^2}x^3 dx = \frac{h}{a^2}\left|\frac{x^4}{4}\right|_o^a = \frac{h a^2}{4}$$

$$\therefore \quad \bar{x} = \frac{h a^2}{4} \div \frac{a h}{3} = \frac{3}{4}a$$

Position of centroid above x-axis

Moment of the elemental area about x-axis,

$$= y\,dx \times \frac{y}{2} = y^2\frac{dx}{2} = \left(\frac{h}{a^2}x^2\right)^2\frac{dx}{2} = \frac{h^2}{2a^4}x^4 dx$$

If $\bar{y}$ is the distance of the centroid above x-axis, then from the moment principle

$$\frac{a h}{3} \times \bar{y} = \int_o^a \frac{h^2}{2a^4}x^4 dx = \frac{h^2}{2a^4}\left|\frac{x^5}{5}\right|_o^a = \frac{h^2 a}{10}$$

$$\therefore \quad \bar{y} = \frac{h^2 a}{10} \div \frac{a h}{3} = \frac{3}{10}h$$

EXAMPLE 7.38

Locate the centroid of the shaded area bounded by a straight line and a parabola as shown in Fig. 7.43.

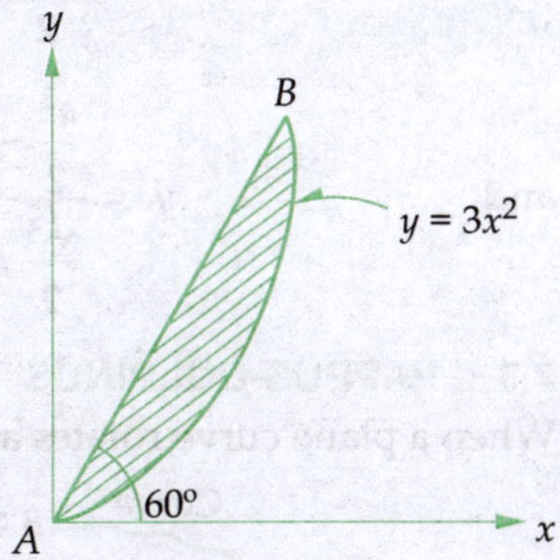

Fig. 7.43

Solution : The equations of the parabola and the straight line are

$$y = 3x^2 \quad \text{...(i)}$$

and $$y = mx = (\tan 60^o) \times x = \sqrt{3}\,x \quad \text{...(ii)}$$

The point B is lying on the straight line as well as on the given parabola. Hence both the expressions (*i*) and (*ii*) hold good for point B. If B has the co-ordinates (a, b), then

$$b = 3a^2 \text{ and } b = \sqrt{3}\,a$$

That gives : $$a = \frac{\sqrt{3}}{3} = \frac{1}{\sqrt{3}} \text{ and } b = 1$$

Hence, point B has the co-ordinates $\left(\frac{1}{\sqrt{3}}, 1\right)$.

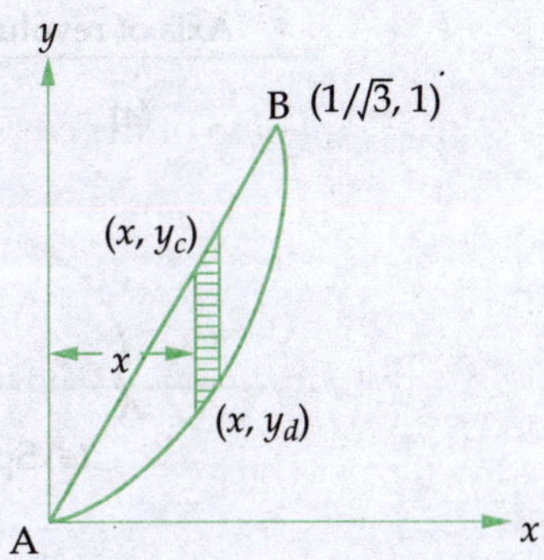

Consider an elemental vertical strip of thickness dx at a distance x from y-axis.

Area of the elemental strip, dA

$$= (y_c - y_d)dx$$

As $$y_c = \sqrt{3}\,x \quad \text{and} \quad y_d = 3x^2$$

$$dA = (\sqrt{3}x - 3x^2)dx$$

The x and y-coordinates of a centroid are prescribed by the relations :

$$\bar{x} = \frac{\int x\,dA}{\int dA} \quad \text{and} \quad \bar{y} = \int \frac{y\,dA}{dA}$$

Now, $$\int dA = \int_o^a (\sqrt{3}x - 3x^2)\,dx = \left|\sqrt{3}\frac{x^2}{2} - 3\frac{x^3}{3}\right|_o^a = \frac{\sqrt{3}}{2}a^2 - a^3$$

$$\int x\,dA = \int_o^a x(\sqrt{3}x - 3x^2)\,dx = \left|\sqrt{3}\frac{x^3}{3} - 3\frac{x^4}{4}\right|_o^a = \frac{a^3}{\sqrt{3}} - \frac{3}{4}a^4$$

$$\int y\,dA = \int \frac{y_c + y_d}{2}\,dA = \frac{1}{2}\int_o^a (\sqrt{3}x + 3x^2)(\sqrt{3}x - 3x^2)\,dx$$

$$= \frac{1}{2}\int_o^a (3x^2 - 9x^4)\,dx = \frac{1}{2}\left|3\frac{x^3}{3} - 9\frac{x^5}{5}\right|_o^a = \frac{a^3}{2} - \frac{9}{10}a^5$$

$$\therefore \quad \bar{x} = \frac{\frac{a^3}{\sqrt{3}} - \frac{3}{4}a^4}{\frac{\sqrt{3}}{2}a^2 - a^3} = \frac{\frac{a}{\sqrt{3}} - \frac{3}{4}a^2}{\frac{\sqrt{3}}{2} - a}$$

Substituting the value $a = \frac{1}{\sqrt{3}}$

$$\bar{x} = \frac{\frac{1}{3} - \frac{3}{4}\times\frac{1}{3}}{\frac{\sqrt{3}}{2} - \frac{1}{\sqrt{3}}} = \frac{\sqrt{3}}{6} = 0.289$$

and $$\bar{y} = \frac{\frac{a^3}{2} - \frac{9}{10}a^5}{\frac{\sqrt{3}}{2}a^2 - a^3} = \frac{\frac{a}{2} - \frac{9}{10}a^3}{\frac{\sqrt{3}}{2} - a} = \frac{\frac{1}{2}\times\frac{1}{\sqrt{3}} - \frac{9}{10}\times\left(\frac{1}{\sqrt{3}}\right)^3}{\frac{\sqrt{3}}{2} - \frac{1}{\sqrt{3}}} = \frac{2}{5} = 0.4$$

7.3. PAPPUS-GULDINUS THEOREMS

When a plane curve rotates about a fixed axis, it generates a surface called the ***surface of revolution***.

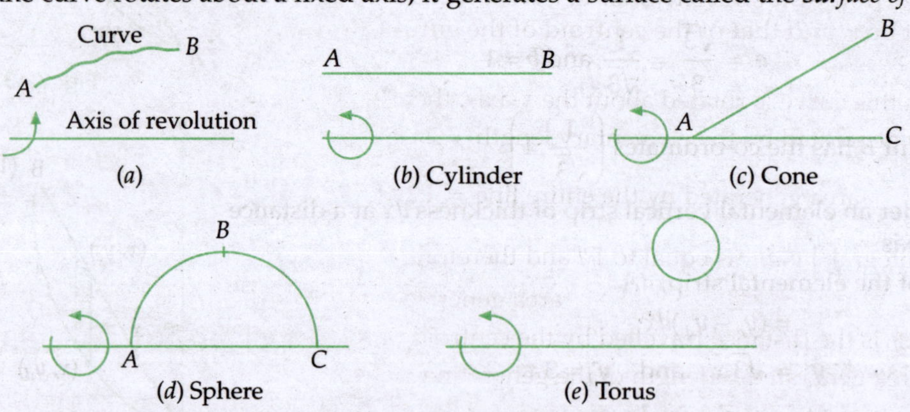

Fig. 7.44 Generation of surface of revolution

Examples (*i*) the surface of a cylinder is generated when straight line *AB* is rotated about an axis.
(*ii*) the surface of a cone gets generated when a straight line *AB* is rotated about an axis *AC*
(*iii*) the surface of a sphere is obtained when a semi-circular arc *ABC* is rotated about the diameter *AC*
(*iv*) the surface of a ring or torus gets formed when the circumference of a circle is rotated about a non-intersecting axis.

The generation/formation of these surfaces of revolution has been indicated in Fig. 7.44.

When a plane area is rotated about a fixes axis, it generates a body called the *body of revolution*.

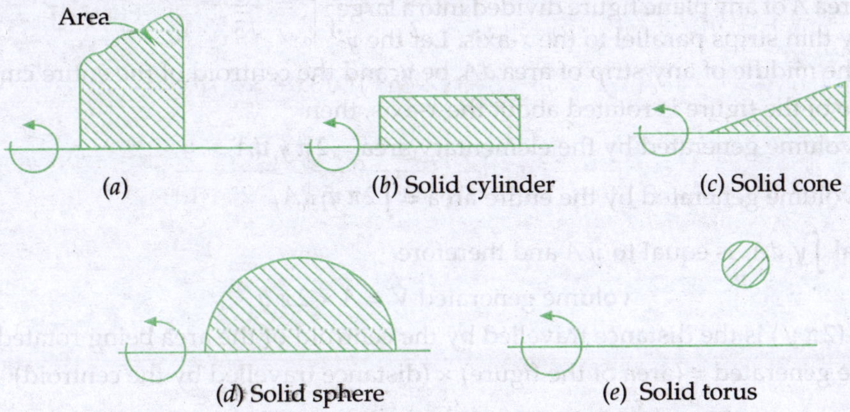

Fig. 7.45 Generation of solid volume of revolution

Examples: A solid cylinder may be generated by rotating a rectangular area, a cone by rotating a triangular area, a solid sphere by rotating a semi-circular area, and a solid torus by rotating a full circular area.

The generation/formation of these bodies of revolution has been indicated in Fig. 7.45.

Theorem 1. The area of the surface generated by rotating any plane curve about a non-intersecting axis in its plane is equal to the product of the length of the curve and the distance travelled by its centroid.

Let *AB* be any plane curve of length l that lies in the x-y plane and does not intersect the x-axis. Consider an elemental length dl_i so short that it may be thought to be a minute portion of a straight line. Let y-coordinate of its mid point be y_i and that of the centroid of the entire curve be $\overline{y}$.

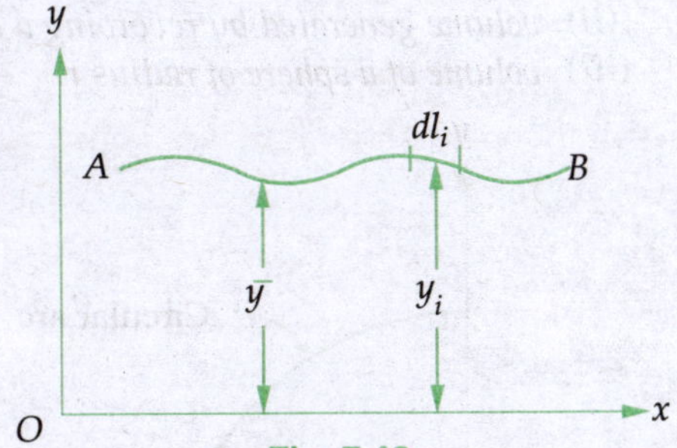

Fig. 7.46

When this curve is rotated about the x-axis, then

area generated by the elementary length $= 2\pi y_i\, dl_i$

area generated by the entire line $= \int 2\pi y_i\, dl_i$

The integral $\int y_i dl_i$ is equal to $\overline{y}\, l$ and therefore

$$\text{area generated} = l \times 2\pi \overline{y}$$

where $2\pi y$ is the distance travelled by the centroid of the curve

$\therefore$ area generated = (length of the generating curve) × (distance travelled by centroid of the curve)

It is to be noted that the generating curve does not cross the axis about which it is rotated. If it does so, then the two sections on either side of the axis would generate areas of opposite signs and the theorem would not apply.

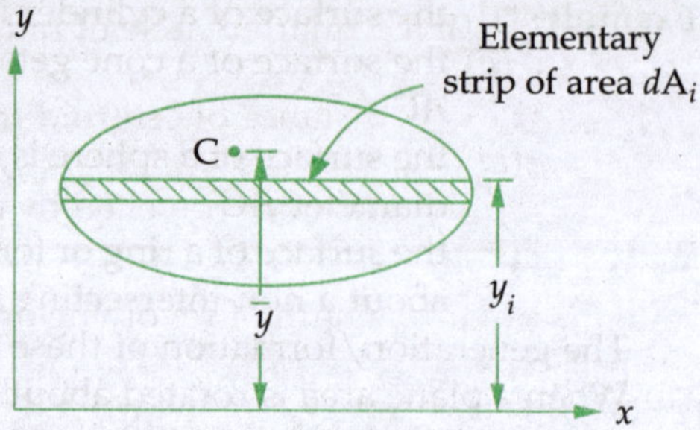

Fig. 7.47

Theorem II. The volume of the solid generated by rotating any plane figure about a non-intersecting axis in its plane is equal to the product of area of the figure and the distance travelled by its centroid.

Consider area A of any plane figure divided into a large number of very thin strips parallel to the x-axis. Let the y-coordinate of the middle of any strip of area dA_i be y_i and the centroid of the entire curve be $\bar{y}$. When the plane of the figure is rotated about the x-axis, then

$$\text{volume generated by the elementary area} = 2\pi y_i \, dA_i$$

$$\text{volume generated by the entire area} = \int 2\pi y_i \, dA_i$$

The integral $\int y_i dA_i$ is equal to $\bar{y} A$ and therefore

$$\text{volume generated } V = A \times 2\pi \bar{y}$$

The parameter $(2\pi \bar{y})$ is the distance travelled by the centroid of the area being rotated.

$\therefore$ volume generated = (area of the figure) × (distance travelled by the centroid)

Utility : The theorems of Pappus- Guldinus offer a simple way to compute the areas of surfaces of revolution and the volumes of bodies of revolution.

Conversely their application lies in determining the centroid of a plane curve when the area of the surface generated by the curve is known or to determine the centroid of a plane area when the volume of the body generated by the area is known.

EXAMPLE 7.39

Using Pappus - Guldinus theorems, calculate the

(i) area of a surface of revolution obtained by revolving a quarter circular arc of radius r about o - y (Fig. 7.48 a)

(ii) volume generated by revolving a quarter ellipse about o - x (Fig. 7.48 b)

(iii) volume of a sphere of radius r.

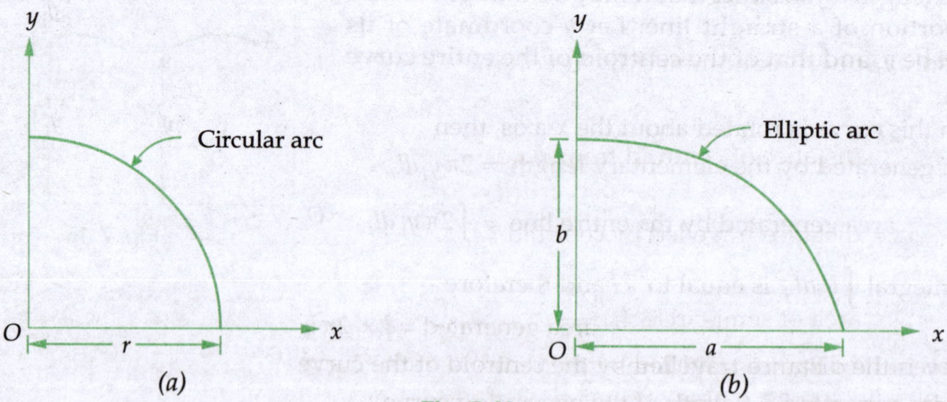

Fig. 7.48

Solution : (*i*) For a quarter circle of radius r,

$$\text{distance of centroid from } y\text{-axis} = \frac{2r}{\pi}$$

$$\text{distance travelled by centroid} = 2\pi \times \frac{2r}{\pi} = 4r$$

$$\text{length of revolving arc} = \frac{\pi r}{2}$$

From Theorem I of Pappus-Guldinus

$$\text{area generated} = \text{length of arc} \times \text{distance travelled by centroid}$$

$$= \frac{\pi r}{2} \times 4r = 2\pi r^2$$

(*ii*) For a quarter ellipse,

$$\text{distance of centroid from } x\text{-axis} = \frac{4b}{3\pi}$$

$$\text{distance travelled by centroid} = 2\pi \times \frac{4b}{3\pi} = \frac{8b}{3}$$

$$\text{area of the surface being revolved} = \frac{\pi ab}{4}$$

From Theorem II of Pappus - Guldinus,

$$\text{volume generated} = \text{area revolved} \times \text{distance travelled by centroid}$$

$$= \frac{\pi ab}{4} \times \frac{8b}{3} = \frac{2\pi ab^2}{3}$$

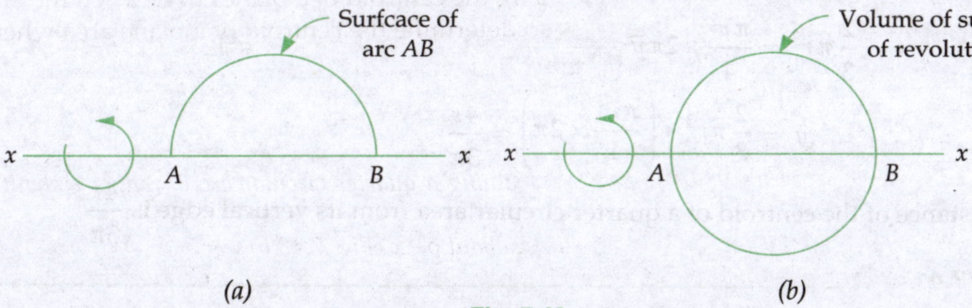

Fig. 7.49

(*iii*) A solid sphere is obtained by revolving a semi-circular area about its diameter AB (Fig. 7.49).

$$\text{distance of centroid from } x\text{-axis} = \frac{4r}{3\pi}$$

$$\text{distance travelled by centroid} = 2\pi \times \frac{4r}{3\pi} = \frac{8r}{3}$$

$$\text{area of semi-circle} = \frac{\pi r^2}{2}$$

From Theorem II of Pappus-Guldinus

$$\text{volume of solid sphere} = \text{area of semi-circle} \times \text{distance travelled by entroid}$$

$$= \frac{\pi r^2}{2} \times \frac{8r}{3} = \frac{4}{3}\pi r^3$$

EXAMPLE 7.40

Apply Pappus-Guldinus theorem to obtain :

(i) the centroid of a semi-circular arc of radius r

(ii) the centroid of a quarter circular area of radius r.

Solution : (*i*) A sphere results when a semi-circular arc is rotated about it axis.

$$\text{area of sphere} = 4\pi r^2$$
$$\text{length of semi-circular arc} = \pi r$$

According to Pappus-Guldinus theorem, the area of the sphere equals the product of the length of the generating semicircle and the distance travelled by its centroid in one revolution. That is

$$4\pi r^2 = \pi r \times 2\pi \bar{y}$$

$$\therefore \quad \bar{y} = \frac{4\pi r^2}{\pi r \times 2\pi} = \frac{2r}{\pi}$$

Thus the centroid of a line in the form of semi-circular arc is $\frac{2r}{\pi}$

(*ii*) A hemisphere results when a quarter circular area is rotated about its axis.

$$\text{volume of hemisphere} = \frac{2}{3}\pi r^3$$

$$\text{area of quarter circle} = \frac{\pi r^2}{4}$$

According to Pappus-Guldinus theorem, the volume generated equals the product of area of the revolving body and the distance travelled by its centroid while it is being revolved. That is

$$\frac{2}{3}\pi r^3 = \frac{\pi r^2}{4} \times 2\pi \bar{y}$$

$$\therefore \quad \bar{y} = \frac{2}{3}\pi r^3 \div \left(\frac{\pi r^2}{4} \times 2\pi\right) = \frac{4r}{3\pi}$$

Thus the distance of the centroid of a quarter circular area from its vertical edge is $\frac{4r}{3\pi}$

EXAMPLE 7.41

A solid circular metal ring has a rectangular cross-section of size 4 cm × 6 cm. If its external and internal radii are 8 cm and 6 cm respectively, calculate the total surface area and volume of the ring.

Solution : According to 1st theorem of Pappus-Guldinus,

Surface area generated = length of revoluing shape × distance travelled by its centroid

= (circumference of rectangular cross-section) × $2\pi \bar{y}$

$$= 2(4+6) \times 2\pi\left(\frac{8+6}{2}\right) = \mathbf{879.2\ cm^2}$$

(*b*) According to 2nd theorem of Pappus-Guldinus,

Volume of ring = area of revolving shape × distance travelled by its centroid

$$= (4 \times 6) \times 2\pi\left(\frac{8+6}{2}\right) = \mathbf{1055.04\ cm^3}$$

EXAMPLE 7.42

(*a*) *A line of length 2l unit is rotated about y-axis as shown in Fig. 7.50. Apply Pappus-Guldinus theorem to determine the surface area of the line.*

(*b*) *The frustrum of a right circular cone has bottom diameter 50 cm, top diameter 20 cm and height 30 cm as shown in Fig. 7.51. Determine the area of its lateral surface.*

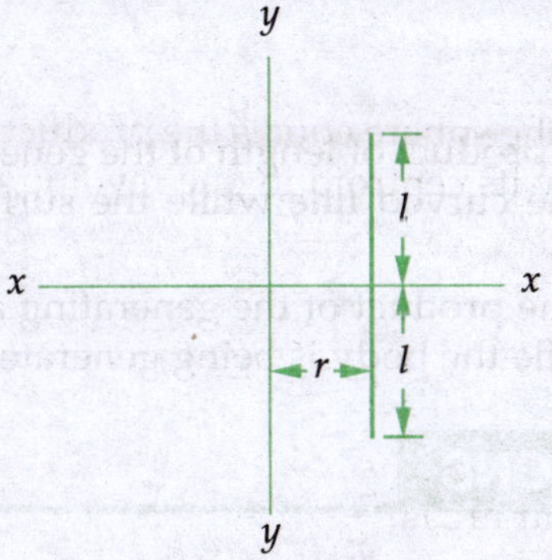

Fig. 7.50

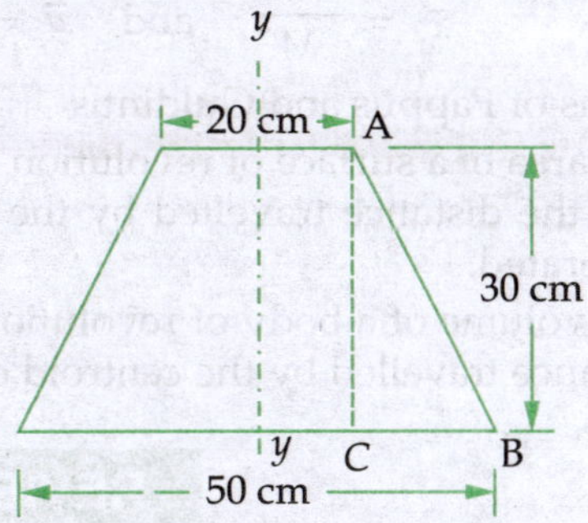

Fig. 7.51

Solution : From 1st theorem of Pappus-Guldinus

Surface area generated = length of revolving line × distance travelled by its centroid

$$= 2l \times (2\pi \times r) = 4\pi rl$$

(*b*) The lateral surface of the frustrum of the cone is generated when the generator line *AB* is revolved about the axis *yy*.

$$\text{length of line } AB = \sqrt{(AC)^2 + (BC)^2} = \sqrt{30^2 + 15^2} = 33.54 \text{ cm}$$

Distance of centroid of the line from the axis of rotation

$$= \frac{10 + 25}{2} = 17.5 \text{ cm}$$

From 1st theorem of Pappus-Guldinus,

area of lateral surface generated = length of the revolving line × distance travelled by its centroid

$$= 33.54 \times (2\pi \times 17.5) = 3686 \text{ cm}^2$$

REVIEW AND SUMMARY

1. Centre of gravity of a body is that point through which the entire weight of the body acts irrespective of its orientation. A body has only one centre of gravity.
2. Centre of mass is the central point for the distribution of mass, *i.e.*, the point at which the entire mass of the body is assumed to be concentrated.
3. The term centre of gravity is applied to solids with mass and weight, while centroid applies to length, area and volume of plane figures which have no mass.
   When thickness, *i.e.*, mass of the body is not considered, the centre of gravity and centroid are synonymous and pass through the same point. Further, centroid and centre of mass coincide if the density of the material is uniform throughout the body.
4. Centroidal axis is that axis which passes through the centre of gravity of the body or through the centroid of an area.

5. The co-ordinates of the centroid of an area are prescribed by the relation.

$$\bar{x} = \frac{\Sigma ax}{A} \quad \text{and} \quad \bar{y} = \frac{\Sigma ay}{A}$$

The centre of gravity of a solid has the co-ordinates

$$\bar{x} = \frac{\Sigma mx}{M} \quad \text{and} \quad \bar{y} = \frac{\Sigma my}{M}$$

6. Theorems of Pappus and Guldinus
   (*i*) The area of a surface of revolution is equal to the product of length of the generating curve and the distance travelled by the centroid of the curved line while the surface is being generated,
   (*ii*) The volume of a body of revolution is equal to the product of the generating area and the distance travelled by the centroid of the area while the body is being generated.

REVIEW QUESTIONS

A. Conceptual and conventional questions :

1. Define the terms : centroid, centre of gravity and centre of mass.
2. Differentiate between centre of gravity and centre of mass. Under what conditions these will coincide?
3. The centroid of an area always lies on the axis of symmetry, if any, of an area. What does the term axis of symmetry imply?
4. The centre of gravity of a body is the point at which the resultant gravitational force on the body acts, irrespective of the orientation of the body. Point out the significance of the phrase 'irrespective of the orientation of the body'.
5. Give the position of centroid of the following standard sections:
   (*i*) rectangle (*ii*) triangle (*iii*) uniform rod and (*iv*) semicircle.
6. Write down the positions of the centre of gravity for the following regular solids: cylinder, sphere, hemisphere and right circular cone.
7. What is meant by an axis/plane of symmetry of a body in the context of centroid/centre of gravity of bodies?
8. Derive from first principles the centroids of
   (*a*) cone and (*b*) hemisphere
9. When will be centroid and centre of mass coincide?
10. Define first moment of an area about an axis
11. Comment on the validity of the statements :
    (*a*) "When a small hole is scooped out from a body and filled with a denser material, its *CG* moves towards the hole."
    (*b*) "If gravitational acceleration is uniform over a planet, the centroid coincides with the centre of gravity of the body."
12. What is the outcome of the first and second theorems of Pappus and Guldinus?
13. Determine the centroid of the area remaining after a circle of diameter *r* is removed from a circle of radius *r* as shown in Fig. 7.52.
    (*r*/6 to the left of *y*-axis)

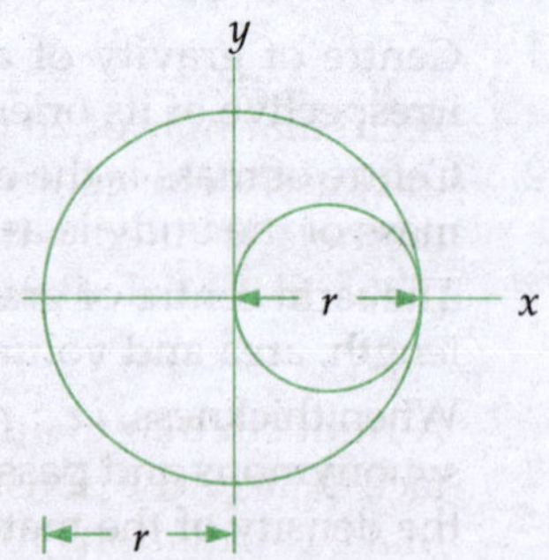

Fig. 7.52

14. A sphere of radius r is cut from a larger sphere of radius R. If the distance between their centres is a, locate the centroid of the remaining volume. $\left(\dfrac{ar^3}{R^3 - r^3}\right)$

15. Locate the centroid of area $ABCDE$ shown in Fig. 7.53. **(7.51 cm, 5.83 cm)**

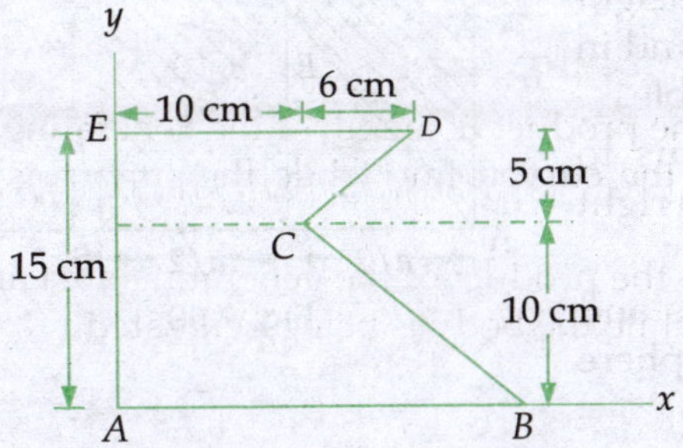

Fig. 7.53

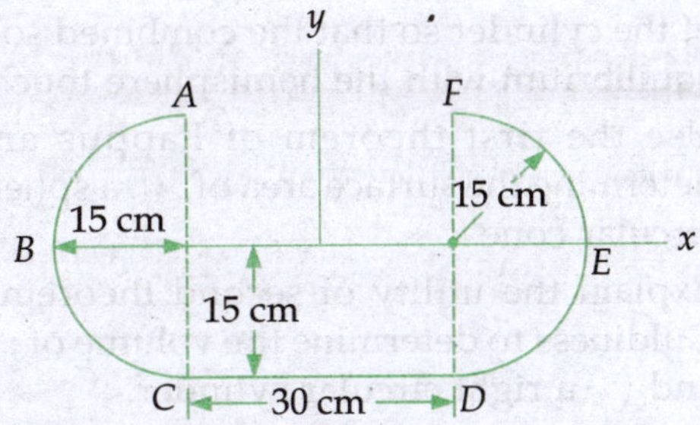

Fig. 7.54

16. A rod has been bent into the shape $ABCDEF$ as shown in Fig. 7.54. Determine the position of its centroid. **(0, – 3.62 cm)**
17. Find the centre of gravity of the channel section whose flanges are 40 mm × 10 mm, and web is 60 mm × 10 mm. The total height of the channel is 80 mm. **(15 mm, 30 mm)**
18. Determine the centroid of z-shaped lamina shown in Fig. 7.55.

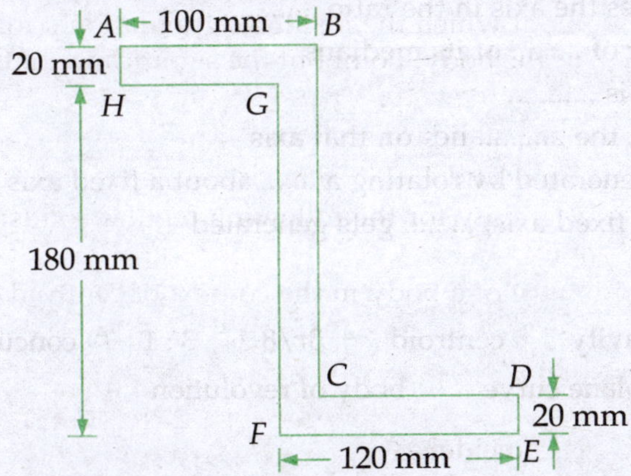

Fig. 7.55

$\left(\begin{array}{l}\bar{x} = 104.74 \text{ mm from DE ;} \\ \bar{y} = 95.26 \text{ mm from FE}\end{array}\right)$

19. A circular hole of 20 mm diameter has been drilled in a 100 diameter circular plate of uniform thickness. The centre of hole lies at 26 mm distance from the centre of plate. Subsequently the hole is filled with a material having density three times that of density of material of plate. Make calculations for the centre of gravity of the composite plate. **(1.93 mm from the centre of plate)**

20. A right circular cylinder of 12 cm diameter is attached to a hemisphere of the same diameter face to face. The density of the material of the hemisphere is twice that of the cylinder material. Make calculations for the greatest height of the cylinder so that the centre of gravity of the composite body lies on the plane joining the hemisphere with the cylinder. **(6 cm)**

21. An isosceles triangle AEB is to be cut from a square ABCD of side a as shown in Fig. 7.56. Compute the height h of this triangle so that its vertex E will be the centroid of the remaining shaded area.

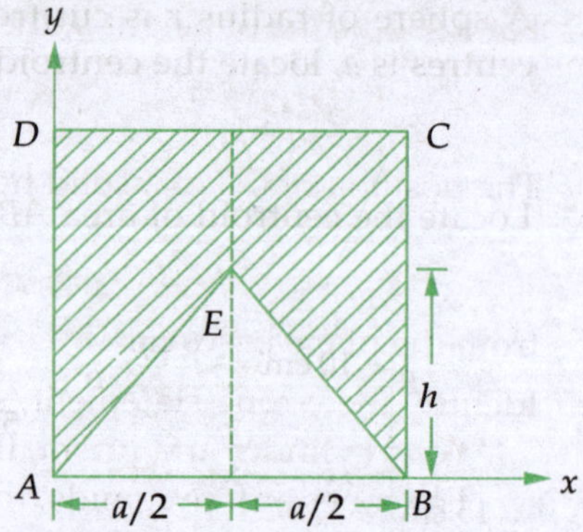

Fig. 7.56

22. A solid consists of a cylinder and a hemisphere of equal radius fixed base to base. Find the ratio of radius to height of the cylinder so that the combined solid may stand in equilibrium with the hemisphere touching the table.
23. Use the first theorem of Pappus and Guldinus to determine the surface area of : (*i*) a sphere and (*ii*) a right circular cone.
24. Explain the utility of second theorem of Pappus and Guldiness to determine the volume of : (*i*) a hemisphere and (*ii*) a right circular cylinder.

B. Complete the following statements with most appropriate word/words :

1. Centroid relates to .... of line or curve, ... of a surface and .... of a body
2. .... is the point where the entire weight of the body is assumed to the concentrated
3. The first moment of area about any axis passing through the .... is zero
4. The distance of the *CG* of a solid hemisphere from the base is .... where *r* is the radius of the sphere
5. The *CG* of a solid circular cone divides the axis in the ratio ....
6. The centroid of a triangle is the point of .... of its medians
7. First moment of an area about *x*-axis is ....
8. If an area is symmetric about an axis, the .... lies on that axis
9. Surface of revolution is the surface generated by rotating a .... about a fixed axis
10. When a plane area is rotated about a fixed axis, ... gets generated

Answers :

1. length, area, volume 2. centre of gravity 3. centroid 4. $3r/8$ 5. 3 : 1 6. concurrence or intersection 7. $\int y dA$ 8. centroid 9. plane curve 10. body of revolution

C. Multiple choice questions :

1. Centre of mass of a body
   (*a*) must lie somewhere inside the body
   (*b*) is located at the geometric centre of the body
   (*c*) lies at the geometric centre of the body provided it is of uniform density
   (*d*) is synonymous with centre of gravity
2. The *CG* of a plane lamina will not be at its geometrical centre if it is a
   (*a*) circle (*b*) square
   (*c*) rectangle (*d*) right angled triangle
   (*e*) equilateral triangle
3. The *CG* of a triangle lies at the point of intersection of
   (*a*) diagonals (*b*) altitudes
   (*c*) bisector of angles (*d*) medians

4. For a solid cone of height h, the CG lies on the axis at a distance above the base equal to

(a) $\frac{h}{4}$ (b) $\frac{h}{3}$ (c) $\frac{2h}{3}$ (d) $\frac{3h}{8}$

5. The position of CG of a solid hemisphere of radius r lies on the central radius at a distance

(a) $\frac{2}{3}r$ (b) $\frac{3}{4}r$ (c) $\frac{3}{8}r$ (d) $\frac{1}{2}r$

from the plane base

6. Identify the wrong statement/statements
   (a) For a cylinder and prism, the CG lies on the axis at half the height from the base
   (b) For cones and pyramids, the CG lies at one-fourth of the height of the axis from the base
   (c) The centroid of a semicircle lies on the radius perpendicular to the base at $4r/3\pi$ where r is the radius of the semicircle
   (d) The CG of a triangular plane figure lies at the point of intersection of altitudes and its distance is $2h/3$ from the base

7. Refer Fig. 7.57 for the geometry of a trapezium. The distance of its CG from base line is

(a) $\frac{h}{4}\left(\frac{2a+b}{a+b}\right)$ (b) $\frac{h}{3}\left(\frac{2a+b}{a+b}\right)$

(c) $\frac{h}{2}\left(\frac{a+2b}{a+b}\right)$ (d) $\frac{h}{3}\left(\frac{a+2b}{a+b}\right)$

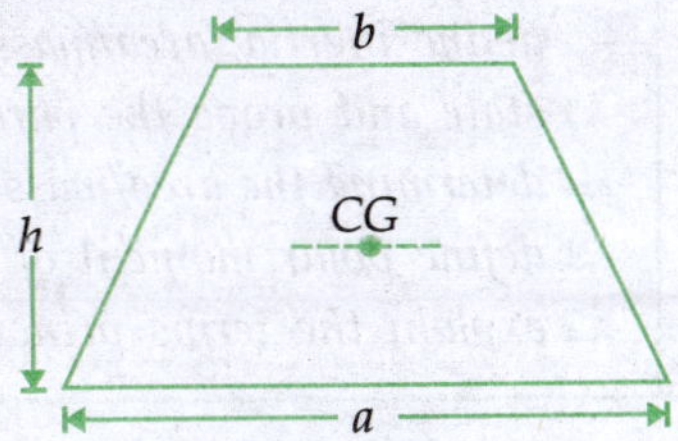

Fig. 7.57

8. If the distance between CG of two masses m_1 and m_2 is l, then the distance of CG of the composite system from mass m_1 will be

(a) $\frac{m_2 l}{m_1 + m_2}$ (b) $\frac{m_1 l}{m_1 + m_2}$

(c) $\frac{(m_1 - m_2) l}{m_1 + m_2}$ (d) $\frac{m_1}{l(m_1 + m_2)}$

9. A right circular cone of 20 cm height weighs 1000 N. A cone of 8 cm height and 64 N weight is removed from the top. The distance of CG of the frustrum from the base is approximately

(a) 2.5 cm (b) 3.25 cm (c) 4.25 cm (d) 6 cm

10. A triangular plate in the form of an isosceles triangle ABC has base BC = 10 cm and altitude = 12 cm. From this plate, a portion in the shape of an isosceles triangle OBC is removed. If O is the mid point of the altitude of triangle ABC, then distance of CG of the remainder section from the base is

(a) 3.5 cm (b) 4.8 cm (c) 6 cm (d) 8 cm

11. For what of the following configurations of plane lamina, the centre of gravity does not lie at its geometrical centre

(a) circle (b) square
(c) rectangle (d) right angled triangle

Answers :

1. (c)	2. (d)	3. (d)	4. (a)	5. (c)	6. (d)
7. (b)	8. (a)	9. (c)	10. (c)	11. (d)	

Moment of Inertia : Area and Mass 8

Learning objectives : After having made a study of the subject matter presented in this chapter, the reader will be able to

- *define inertia, area/mass moment of inertia and radius of gyration of a lamina*
- *state and prove the parallel axis theorem and perpendicular axis theorem*
- *determine the area/mass moment of inertia of lamina of different shapes and composite sections*
- *define polar moment of inertia and state its applications*
- *explain the terms principal moments of inertia and product of inertia*

It may be recalled that the moment of force about a point is the product of force (F) and the perpendicular distance (x) between the point and the line of action of force.

Moment of force = $F x$

If this moment $F x$ is further multiplied by the distance x, then a quantity $F x^2$ is obtained which is referred to as the moment of moment or the second moment of force

Moment of moment = $F x \times x = F x^2$

If the term force F in the above identity is replaced by area or mass of the body, the resulting parameter is called the moment of inertia (*MOI*). Thus

Moment of inertia of a plane area = $A x^2$

Mass moment of inertia of a body = $m x^2$

where A and m respectively denote the area and mass of the body.

Inertia refers to the property of a body by virtue of which the body resists any change in its state of rest or of uniform motion. Area moment of inertia is considered only for plane figures for which the mass is assumed to be negligible. It is essentially a measure of resistance to bending, and is applied while dealing with the deflection or deformation of members in bending.

The mass moment of inertia pertains only to solid bodies having mass. It gives a measure of the resistance that body offers to change in angular velocity and accordingly is used in conjunction with rotation of rigid bodies.

8.1. MOMENT OF INERTIA AND RADIUS OF GYRATION

Moment of inertia (*MOI*) of any lamina is the second moment of all elemental areas dA comprising the lamina. With refence to Fig. 8.1.

$$I_{xx} = \text{moment of inertia about } x\text{-axis} = \Sigma\ (y\, dA)\, y$$

where $y\,dA$ is the first moment of area dA about x-axis and $(y\,dA)\,y$ is the moment of first moment (called second moment) of area dA about x-axis.

$$I_{xx} = \Sigma\, y^2\, dA$$

Likewise:

$$I_{yy} = \text{moment of inertia about } y\text{-axis}$$

$$= \Sigma\, x^2\, dA$$

Obviously moment of inertia of a section about an axis is prescribed by the cummulative product of area and square of the distance from that axis.

The units of moment of MOI are the fourth power of length. When the measurements are in mm, MOI has units of mm^4.

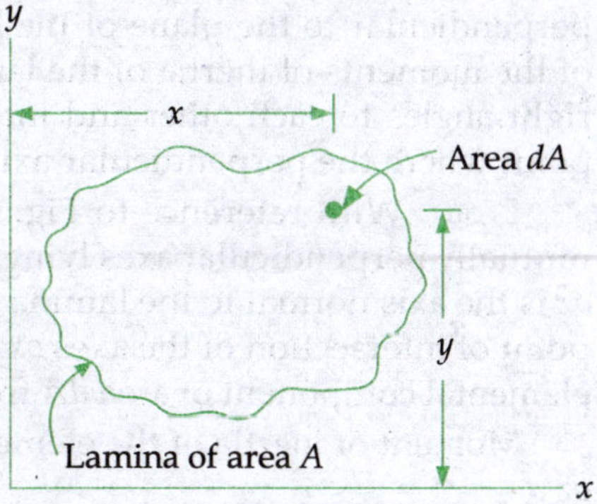

Fig. 8.1

8.1.1. Parallel axis theorem

The moment of inertia of a plane lamina about any axis is equal to the sum of its MOI about a parallel axis through its centre of gravity G and the product of its area (mass) and the square of the distance between the two axes. With reference to Fig. 8.2.

$$I_{AA} = I_{gxx} + A\,h^2 \qquad \text{...(8.3)}$$

where I_{gxx} is MOI of the lamina about an axis x-x passing through its CG and I_{AA} is the MOI about any axis AA which is parallel to x-x and at a distance h from it.

Proof : The lamina consists of an infinite number of small elemental components parallel to the x-axis. Let one such elemental component of area dA be located at distance y from the x-axis. Obviously then its distance from the axis AA will be $(h + y)$.

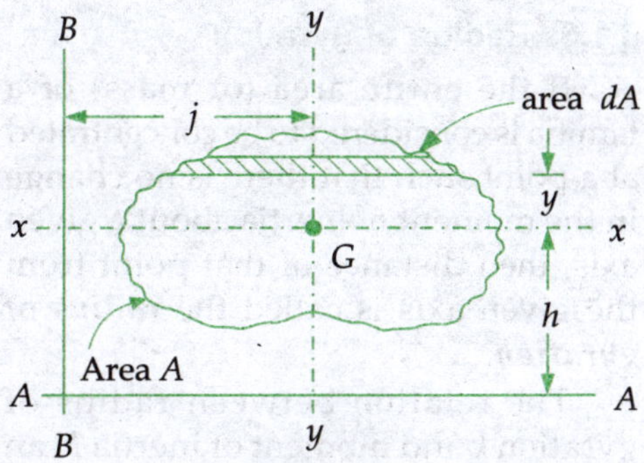

Fig. 8.2

Moment of inertia of the elemental component about axis AA will be

$$= dA\,(h + y)^2$$

Then moment of inertia of the entire lamina about axis AA

$$= \Sigma\, dA(h+y)^2$$

$$= \Sigma\, dA h^2 + \Sigma\, dA y^2 + \Sigma\, dA(2hy)$$

$$= h^2 \Sigma\, dA + \Sigma\, dA y^2 + 2h \Sigma\, dA y$$

Now, $h^2 \Sigma\, dA = Ah^2$ $\qquad (\because\ \Sigma\, dA = A)$

$\Sigma\, dA\, y^2$ = moment of inertia of the lamina about the axis x-x.

$\Sigma\, dA\, y$ = 0 because $x - x$ is centroidal axis.

That gives : $\quad I_{AA} = I_{xx} + Ah^2 \qquad$ ...(8.4)

Also $\quad I_{BB} = I_{yy} + Aj^2 \qquad$ ...(8.5)

8.1.2. Perpendicular axis theorem

The moment of inertia of a plane lamina about an axis perpendicular to the plane of the lamina is equal to the sum of the moments of inertia of the lamina about the two axes at right angles to each other and intersecting each other at the point where the perpendicular axis passes through it.

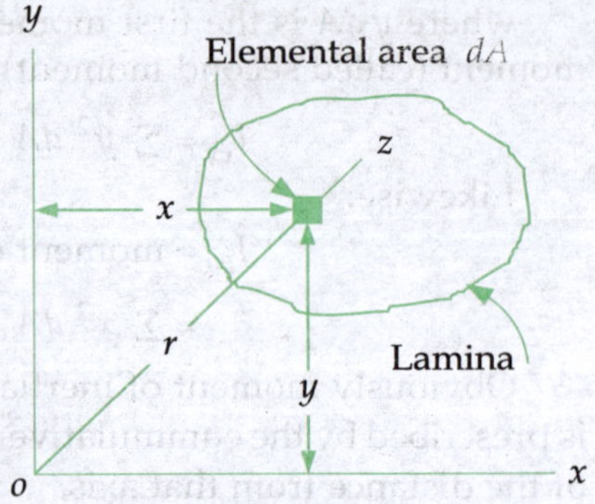

Fig. 8.3

Proof: With reference to Fig. 8.3, *ox* and *oy* are the two mutually perpendicular axes lying in the plane of lamina, and *oz* is the axis normal to the lamina and passing through *o* (the point of intersection of the axes *ox* and *oy*). The distance of an elemental component of area *dA* from *oz*, *i.e.*, from point *o* is *r*.

Moment of inertia of the elemental component about axis *oz*

$$= dA\, r^2 = dA\,(x^2 + y^2)$$

Moment of inertia of the elemental component about axis *oz*

$$I_{zz} = \Sigma\, dA\,(x^2 + y^2) = \Sigma\, dA\, x^2 + \Sigma\, dA\, y^2$$

But, $\Sigma\, dA\, x^2$ = moment of inertia of the lamina about the axis $oy = I_{yy}$

$\Sigma\, dA\, y^2$ = moment of inertia of the lamina about the axis $ox = I_{xx}$

$$\therefore \quad I_{zz} = I_{xx} + I_{yy} \qquad \text{...(8.6)}$$

8.1.3. Radius of gyration

If the entire area (or mass) of a lamina is considered to be concentrated at a point such that there is no change in the moment of inertia about a given axis, then distance of that point from the given axis is called the ***radius of gyration***.

The relation between radius of gyration *k* and moment of inertia *I* can be put in the form

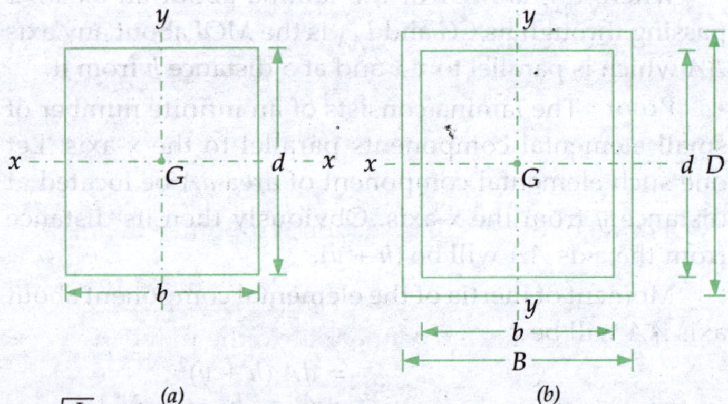

Fig. 8.4

$$I = Ak^2\,; \quad k = \sqrt{\frac{I}{A}} \qquad \text{...(8.7)}$$

Apparently the radius of gyration of a lamina is the square root of the ratio of its moment of inertia to its area.

The moment of inertia of common standard sections are presented below:

(*i*) For a rectangular section with breadth *b* and depth *d* (Fig. 8.4 *a*).

$$I_{xx} = \frac{bd^3}{12}\,; \quad I_{yy} = \frac{db^3}{12}$$

For a hollow rectangular section (Fig. 8.4 *b*)

$$I_{xx} = \frac{BD^3 - bd^3}{12}\,; \quad I_{yy} = \frac{DB^3 - db^3}{12}$$

(*ii*) For a triangular section (Fig. 8.5)

$$I_{xx} = \frac{bh^3}{36}$$

$$I_{AB} = \frac{bh^3}{12} \qquad ...(8.7)$$

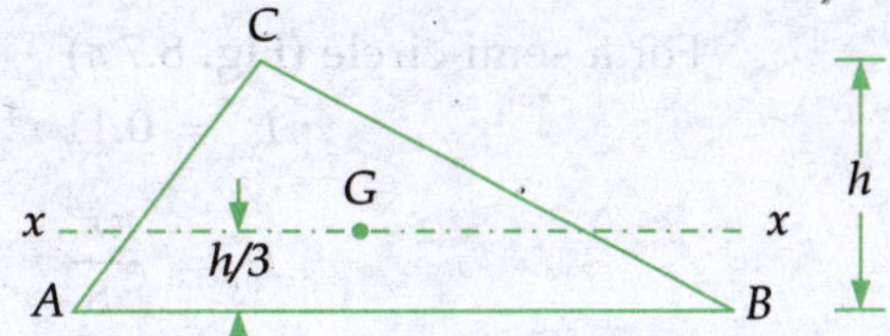

Fig. 8.5

(*iii*) For a circular section of diameter d (Fig. 8.6*a*)

$$I_{xx} = I_{yy} = \frac{\pi}{64} d^4$$

If zz is the axis perpendicular to lamina and passing through CG, then

$$I_{zz} = I_{xx} + I_{yy} = \frac{\pi}{32} d^4$$

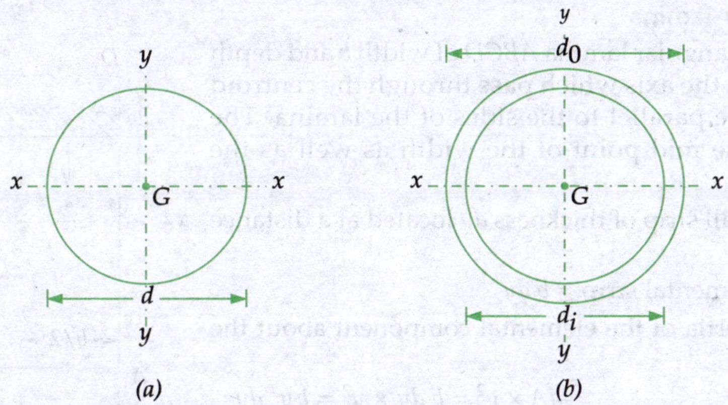

Fig. 8.6

For a hollow circular section with outer diameter d_0 and inner diameter d_i

$$I_{xx} = I_{yy} = \frac{\pi}{64}\left(d_0^4 - d_i^4\right)$$

If zz is the axis perpendicular to the plane of lamina and passing through CG then

$$I_{zz} = I_{xx} + I_{yy} = \frac{\pi}{32}\left(d_0^4 - d_i^4\right)$$

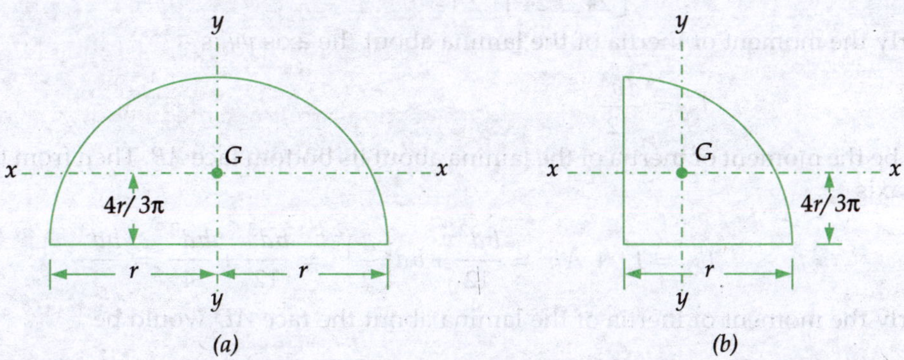

Fig. 8.7

(*iv*) For a semi-circle (Fig. 8.7 *a*)

$$I_{xx} = 0.11\, r^4 \text{ and}$$

$$I_{yy} = \frac{\pi r^4}{8}$$

For a quarter circle (Fig. 8.7*b*)

$$I_{xx} = I_{yy} = 0.055\, r^4$$

(*v*) For an ellipse (Fig 8.8),

$$I_{xx} = \frac{\pi a b^3}{4},\ I_{yy} = \frac{\pi b a^3}{4}$$

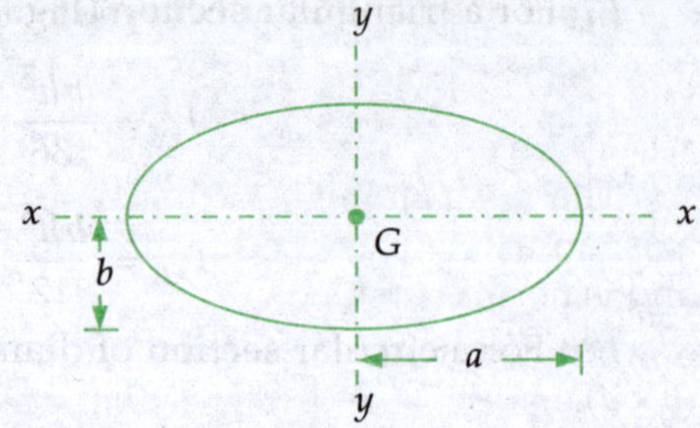

Fig. 8.8

8.2. MOMENT OF INERTIA OF LAMINAE OF DIFFERENT SHAPES

8.2.1. Rectangular lamina

Consider a rectangular lamina *ABCD* of width *b* and depth *d*. Let *xx* and *yy* be the axis which pass through the centroid of the area and are parallel to the sides of the lamina. The centroid lies at the mid point of the width as well as the depth.

Consider a small strip of thickness *dy* located at a distance *y* from the axis *xx*.

Area of the elemental strip = *b dy*

Moment of inertia of the elemental component about the axis *xx*,

$$= dA \times y^2 = b\, dy \times y^2 = by^2\, dy$$

Moment of inertia of the entire lamina about the axis *xx*,

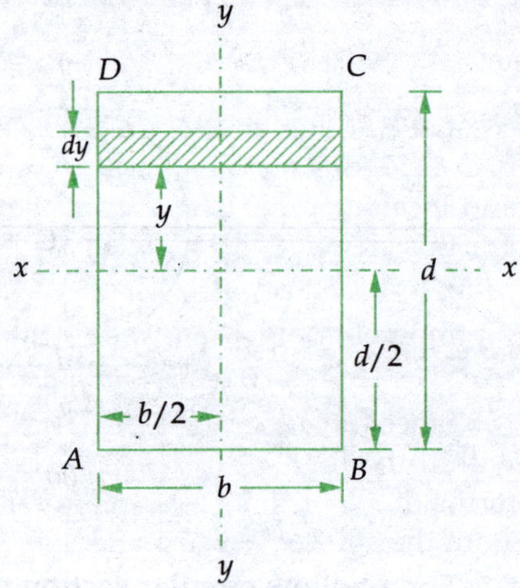

Fig. 8.9

$$I_{xx} = \int_{-\frac{d}{2}}^{\frac{d}{2}} by^2\, dy = b\left|\frac{y^3}{3}\right|_{-\frac{d}{2}}^{\frac{d}{2}}$$

$$= b\left[\frac{d^3}{24} + \frac{d^3}{24}\right] = \frac{bd^3}{12} \qquad \text{...(8.8 a)}$$

Similarly the moment of inertia of the lamina about the axis *yy* is

$$I_{yy} = \frac{db^3}{12} \qquad \text{...(8.8b)}$$

Let I_{AB} be the moment of inertia of the lamina about its bottom face *AB*. Then from the theorem of parallel axis

$$I_{AB} = I_{xx} + Ah^2 = \frac{bd^3}{12} + bd\left(\frac{d}{2}\right)^2 = \frac{bd^3}{12} + \frac{bd^3}{4} = \frac{bd^3}{3} \qquad \text{...(8.9a)}$$

Similarly the moment of inertia of the lamina about the face *AD* would be

$$I_{AD} = \frac{db^3}{3} \qquad \text{...(8.9b)}$$

From the theorem of perpendicular axis, the polar moment of inertia I_p of the lamina is

$$I_p = I_{xx} + I_{yy} = \frac{bd^3}{12} + \frac{db^3}{12} \quad ...(8.10)$$

The polar moment of inertia is the inertia about the polar axis; an axis which passes through the centroid of the lamina and is normal to it.

For a rectangular lamina ($B \times D$) with a rectangular hole ($b \times d$) made centrally (Fig. 8.10) the moment of inertia about any centroidal axis is

= MOI of bigger rectangle

– MOI of smaller rectangle

Thus: $$I_{xx} = \frac{BD^3}{12} - \frac{bd^3}{12} \quad ...(8.11)$$

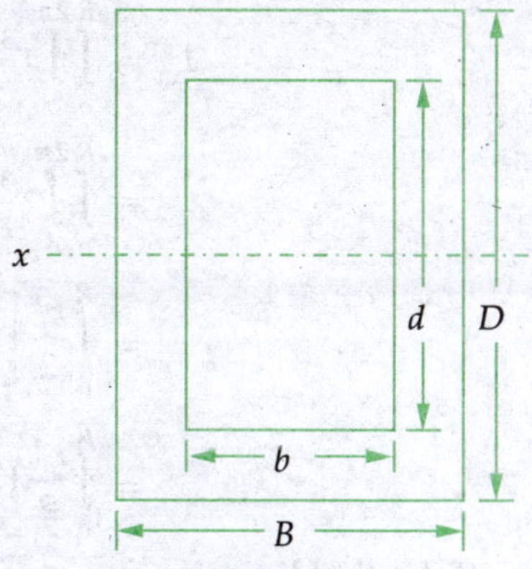

Fig. 8.10

8.2.2. Triangular lamina

Let ABC be the triangle of base width b and height h. Consider an elementary strip of width l, thickness dy and located at distance y from the base of the triangle. For this elemental strip,

area = $l\,dy$

moment of inertia of this strip about base BC

$= y^2\,dA = y^2 l\,dy$

Since the integration is to be done with respect to y within the limits 0 to h, it is necessary to express l in terms of y. For that, we have the following correlation from the similarity of triangles ADE and ABC,

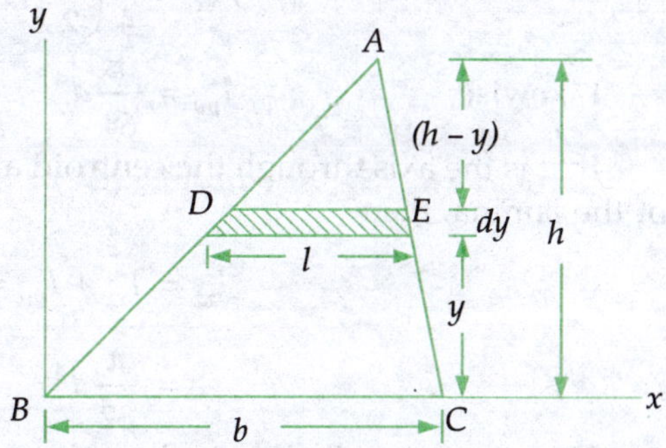

Fig. 8.11

$$\frac{l}{b} = \frac{h-y}{h}; \quad l = b\left(1 - \frac{y}{h}\right)$$

∴ Moment of inertia of the triangle about the base

$$I_{base} = \int_o^h y^2 b\left(1 - \frac{y}{h}\right) dy$$

$$= b\int_o^b \left(y^2 - \frac{y^3}{h}\right) dy = b\left|\frac{y^3}{3} - \frac{y^4}{4h}\right|_o^h = b\left(\frac{h^3}{3} - \frac{h^3}{4}\right) = \frac{bh^3}{12} \quad ...(8.12)$$

For a triangle, the centroidal axis I_{xx} is at a distance of $y_c = h/3$ from the base. Then from the theorem of parallel axis : $I_{base} = I_{xx} + Ay_c^2$, we have

$$\therefore \quad I_{xx} = I_{base} - Ay_c^2 = \frac{bh^3}{12} - \left(\frac{1}{2}bh\right) \times \left(\frac{h}{3}\right)^2 = \frac{bh^3}{12} - \frac{bh^3}{18} = \frac{bh^3}{36} \quad ...(8.13)$$

8.2.3. Circular lamina

Consider an element of sides $r\,d\theta$ and dr within a circular lamina of radius R. Moment of this elemental area about the diametrical axis x-x,

$$= y^2\,dA = (r \sin\theta)^2 \times r d\theta\, dr = r^3 \sin^2\theta\, d\theta\, dr$$

$\therefore$ Moment of inertia of the entire circular lamina

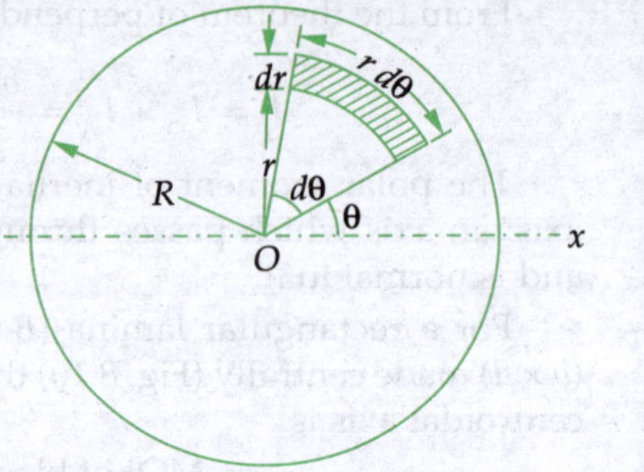

Fig. 8.12

$$I_{xx} = \int_0^R \int_0^{2\pi} r^3 \sin^2\theta \, d\theta \, dr$$

$$= \int_0^R \int_0^{2\pi} r^3 \frac{1-\cos 2\theta}{2} d\theta \, dr$$

$$= \int_0^R \frac{r^3}{2}\left[\theta - \frac{\sin 2\theta}{2}\right]_0^{2\pi} dr$$

$$= \int_0^R \frac{r^3}{2}(2\pi)\, dr = 2\pi \left|\frac{r^4}{8}\right|_0^R = \frac{\pi}{4} R^4$$

If d is the diameter of the circular lamina, then

$$I_{xx} = \frac{\pi}{4}\left(\frac{d}{2}\right)^4 = \frac{\pi}{64} d^4 \qquad ...(8.14)$$

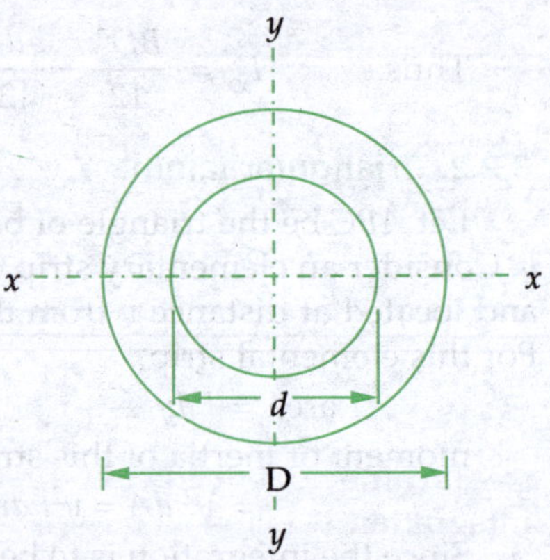

Fig. 8.13

Likewise $\qquad I_{yy} = \frac{\pi}{64} d^4$

If zz is the axis through the centroid and normal to the plane of the lamina, then

$$I_{zz} = I_{xx} + I_{yy} = \frac{\pi}{64} d^4 + \frac{\pi}{64} d^4$$

$$= \frac{\pi}{32} d^4 \qquad ...(8.15)$$

The axis zz is called the polar axis and I_{zz} is referred to as the *polar moment of inertia*.

Polar moment of inertia has application in problems relating to torsion of cylindrical shafts and rotation of slabs.

For a circular lamina of diameter D with a central circular hole of diameter d (Fig. 8.13) the moment of inertia about any centroidal axis is

$$I_{xx} = I_{yy} = \frac{\pi}{64}(D^4 - d^4) \qquad ...(8.16)$$

The corresponding polar moment of inertia is

$$I_{zz} = I_p = \frac{\pi}{32}(D^4 - d^4) \qquad ...(8.17)$$

8.2.4. Semi-circular lamina

The moment of inertia of a circular lamina having diameter d about its diametrical axis AB is

$$= \frac{\pi}{64} d^4$$

For the semi-circular lamina with AB as its base, the moment of inertia about AB would be

$$I_{AB} = \frac{1}{2} \times \left(\frac{\pi}{64} d^4\right) = \frac{\pi}{128} d^4$$

It can be obtained from first principles if the limit of integration is taken as 0 to π instead of 0 to 2π in the derivation of moment of inertia of a circular lamina about its diametrical axis. That is

$$I_{AB} = \int_0^R \int_0^\pi r^3 \sin^2\theta \, d\theta \, dr = \frac{\pi}{8} R^4 = \frac{\pi}{128} d^4 \quad ...(8.18)$$

The distance of centroidal axis xx of the semicircle from its base AB is

$$h = \frac{4R}{3\pi} = \frac{2d}{3\pi}$$

$$\text{Area of semicircle} = \frac{1}{2}\left(\frac{\pi}{4} d^2\right) = \frac{\pi d^2}{8}$$

From parallel axis theorem,

$$\frac{\pi d^4}{128} = I_{xx} + \frac{\pi d^2}{8} \times \left(\frac{2d}{3\pi}\right)^2 = I_{xx} + \frac{d^4}{18\pi}$$

$$\therefore \quad I_{xx} = \frac{\pi d^4}{128} - \frac{d^4}{18\pi} = 0.00686\, d^4 = 0.11 R^4 \quad ...(8.19)$$

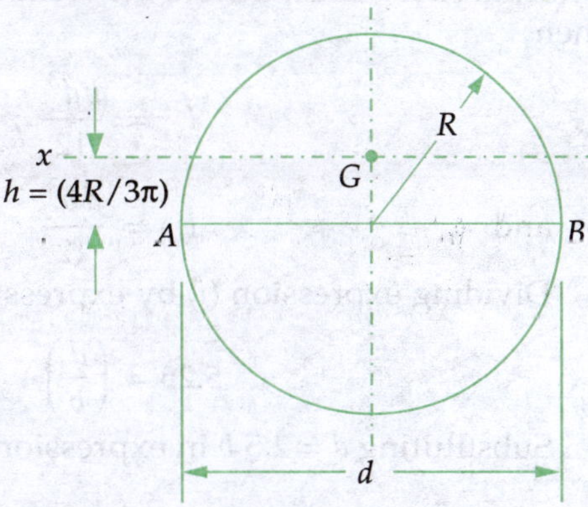

Fig. 8.14

8.2.5. Quarter of a circle

Reference Fig. 8.15, LAB is the quadrant of a circular lamina of diameter d. The moment of inertia of a quadrant equals 1/4th of the moment of inertia of the circular lamina.

$$\therefore \quad I_{AB} = \frac{1}{4} \times \left(\frac{\pi}{64} d^4\right) = \frac{\pi}{256} d^4$$

It can be obtained from first principles if the limit of integration is taken as 0 to $\pi/2$ instead of 0 to 2π in the derivation of moment of inertia of a circular lamina about its diametral axis. That is

$$I_{AB} = \int_0^R \int_0^{\pi/2} r^3 \sin^2\theta \, d\theta \, dr$$

$$= \frac{\pi}{16} R^4 = \frac{\pi}{256} d^4 \quad ...(8.20)$$

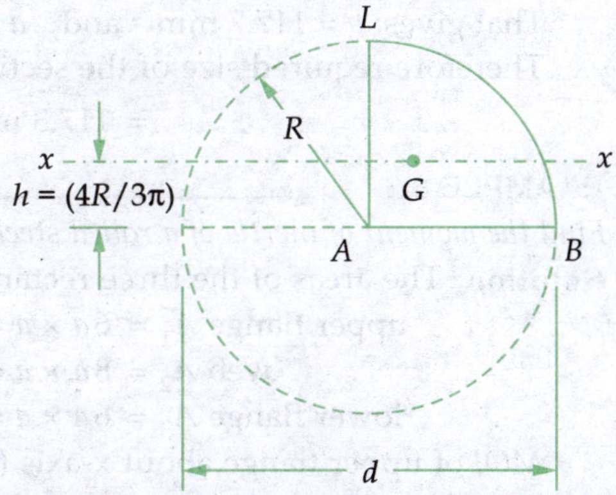

Fig. 8.15

The distance of the centroid of the quadrant LAB from AB is

$$h = \frac{4R}{3\pi} = \frac{2d}{3\pi}$$

$$\text{Area of quadrant} = \frac{1}{4} \times \left(\frac{\pi}{4} d^2\right) = \frac{\pi}{16} d^2$$

From parallel axis theorem,

$$I_{AB} = I_{xx} + Ah^2$$

$$\text{or} \quad \frac{\pi d^4}{256} = I_{xx} + \frac{\pi d^2}{16} \times \left(\frac{2d}{3\pi}\right)^2 = I_{xx} + \frac{d^4}{36\pi}$$

$$\therefore \quad I_{xx} = \frac{\pi d^4}{256} - \frac{d^4}{36\pi} = 0.00343\, d^4 = 0.055 R^4 \quad ...(8.21)$$

EXAMPLE 8.1

The moment of inertia of rectangular section beam about x-x and y-y axes passing though the centroid are 250×10^6 mm^4 and 40×10^6 mm^4 respectively. Calculate the size of the section.

Solution : Let b and d denote the breadth and depth respectively of the rectangular section beam. Then

$$I_{xx} = \frac{bd^3}{12}\ ; \quad 250 \times 10^6 = \frac{bd^3}{12} \qquad ...(i)$$

and

$$I_{yy} = \frac{db^3}{12}\ ; \quad 40 \times 10^6 = \frac{db^3}{12} \qquad ...(ii)$$

Dividing expression (*i*) by expression (*ii*)

$$5.25 = \left(\frac{d}{b}\right)^2 \quad \text{or} \quad \frac{d}{b} = 2.5$$

Substituting $d = 2.5\,b$ in expression (*i*), we get

$$\frac{b}{12}(2.5\,b)^3 = 250 \times 10^6$$

$$\text{or} \quad b^4 = \frac{250 \times 10^6 \times 12}{(2.5)^3} = 1.92 \times 10^8$$

That gives: $b = 117.7$ mm and $d = 2.5 \times 117.7 = 294.25$ mm

Therefore required size of the section is:

$$= 117.3 \text{ mm (breadth)} \times 294.25 \text{ mm (depth)}$$

EXAMPLE 8.2

Find the moment of inertia of a rolled steel joist girder of symmetrical I section shown in Fig. 8.16.

Solution : The areas of the three rectangles comprising the I-section are:

$$\text{upper flange } A_1 = 6a \times a = 6a^2$$

$$\text{web } A_2 = 8a \times a = 8a^2$$

$$\text{lower flange } A_3 = 6a \times a = 6a^2$$

MOI of upper flange about x-axis (using parallel axis theorem)

$$= \frac{6a \times a^3}{12} + 6a^2 \times \left(4a + \frac{a}{2}\right)^2$$

$$= \frac{a^4}{2} + \frac{243a^4}{2} = 122\,a^4$$

MOI of web about x-axis $= \dfrac{a \times (8a)^3}{12} = \dfrac{128\,a^4}{3}$

MOI of lower flange about x-axis (using parallel axis theorem)

$$= \frac{6a \times a^3}{12} + 4a^2\left(4a + \frac{a}{2}\right)^2$$

$$= \frac{a^4}{2} + \frac{243a^4}{2} = 122\,a^4$$

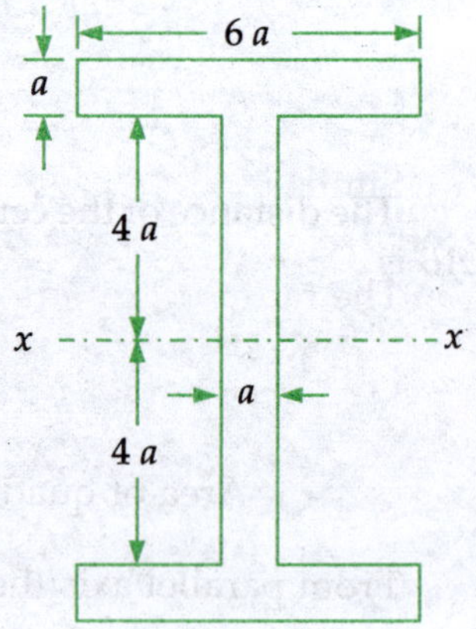

Fig. 8.16

$\therefore$ Total *MOI* of the given I-section about x-axis

$$= 122\,a^4 + \frac{128\,a^4}{3} + 122\,a^4$$

$$= \frac{860}{3}a^4$$

The *MOI* of the given I-section could also be worked out with reference to Fig. 8.17.

$$I_{xx} = I_{x1} - I_{x2}$$

$$= \frac{6a \times (10a)^3}{12} - \frac{5a \times (8a)^3}{12}$$

$$= 500\,a^4 - \frac{640}{3}a^4 = \frac{860}{3}a^4$$

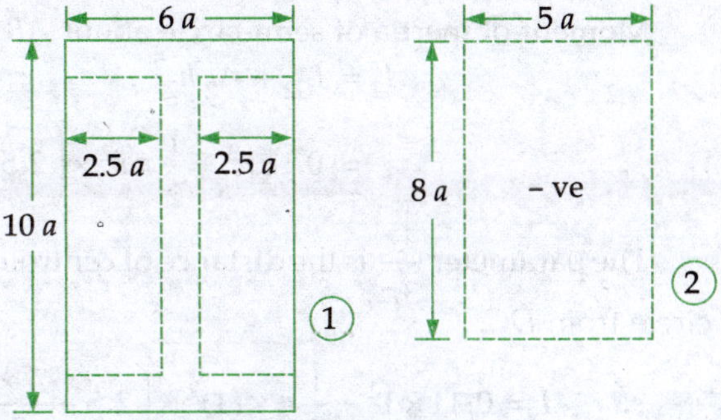

Fig. 8.17

EXAMPLE 8.3

Determine the moment of inertia of the T-section shown in Fig. 7.13 about an axis passing through the centroid and parallel to top most fibre of the section. Proceed to determine the moment of inertia about axis of symmetry and hence find out the radii of gyration.

Solution : From the calculations made in Example 7.10 the CG of the given T-section lies on the y-axis and at distance 43.71 mm from the top face of its flange

$$\bar{x} = 0 \quad \text{and} \quad \bar{y} = 43.71 \text{ mm}$$

Refering to this centroidal axis, the centroid of a_1 is (0.0, 38.71 mm) and that of a_2 is (0.0, 41.29 mm).

Moment of inertia of the section about centroid axis is

$$I_{xx} = MOI \text{ of area } a_1 \text{ about centroidal axis} + MOI \text{ of area } a_2 \text{ about centroidal axis}$$

$$= \left[\frac{160 \times 10^3}{12} + 1600 \times (38.71)^2\right] + \left[\frac{10 \times 150^3}{12} + 1500 \times (41.29)^2\right]$$

$$= 7780672 \text{ mm}^4$$

Similarly $$I_{yy} = \frac{10 \times 160^3}{12} + \frac{150 \times 10^3}{12} = 3425833 \text{ mm}^4$$

The radius of gyration is given by $k = \sqrt{\frac{I}{A}}$

$$\therefore \quad k_{xx} = \sqrt{\frac{7780672}{3100}} = \mathbf{50.1 \text{ mm}}$$

$$k_{yy} = \sqrt{\frac{3425833}{3100}} = \mathbf{34.24 \text{ mm}}$$

EXAMPLE 8.4

Determine the moment of inertia of the area shown shaded in Fig. 8.18 about axis *xx* which coincides with the base edge *AB*.

Solution : The given section comprises the full rectangle *ABCD* minus the semi-circle *DEC*.

Moment of cnertia of rectangle *ABCD* about *AB*

$$I_1 = I_{G1} + A_1 h_1^2$$

$$= \frac{2 \times 2.5^3}{12} + (2 \times 2.5) \times 1.25^2$$
$$= 2.604 + 7.812 = 10.416 \text{ cm}^4$$

Moment of inertia of semi-circle about AB

$$I_2 = I_{G2} + A_2 h_2^2$$
$$= 0.11\, r^2 + \frac{1}{2} \pi r^2 \times \left(2.5 - \frac{4r}{3\pi}\right)^2$$

The parameter $\frac{4r}{3\pi}$ is the distance of centroid of semi-circle from DC.

$$\therefore \quad I_2 = 0.11 \times 1^2 + \frac{1}{2} \pi \times (1)^2 \times \left(2.5 - \frac{4 \times 1}{3\pi}\right)^2$$
$$= 0.11 + 6.76 = 6.87 \text{ cm}^4$$

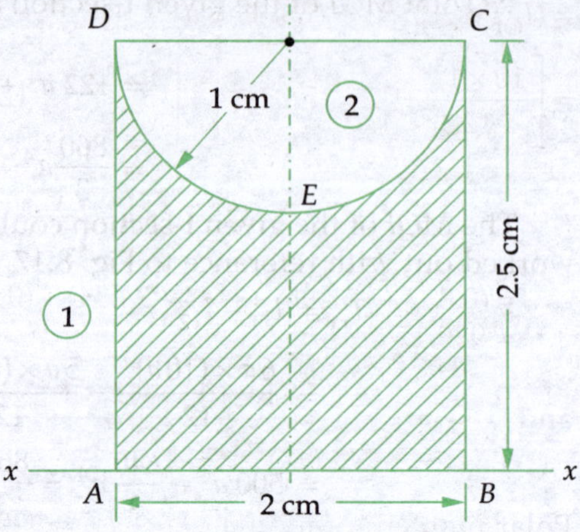

Fig. 8.18

$\therefore$ Moment of inertia of shaded area bout AB = 10.416 – 6.87 = 3.546 cm^4

EXAMPLE 8.5

Determine the polar moment of inertia of the I-section shown in Fig. 8.19. Also make calculations for the radius of gyration with respect to x-axis and y-axis.

Solution : The I-section is symmetrical about y-axis and accordingly its CG lies at point G on the y-axis, i.e., $x = 0$. Further, the bottom fibre of lower flange has been chosen as reference axis to locate the centroid $\bar{y}$.

The areas and co-ordinates of centroids of the three rectangles comprising the given section are:

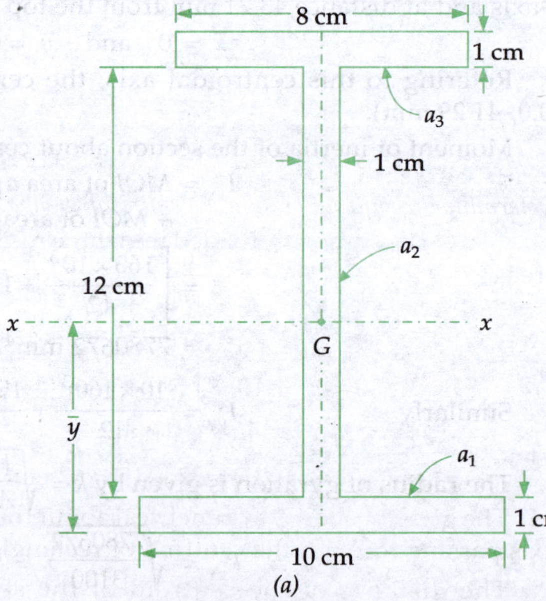

(a)

Fig. 8.19 (a)

Lower flange: $a_1 = 10 \times 1 = 10 \text{ cm}^2$

$$y_1 = \frac{1}{2} = 0.5 \text{ cm}$$

Web: $a_2 = 12 \times 1 = 12 \text{ cm}^2$

$$y_2 = 1 + \frac{12}{2} = 7 \text{ cm}$$

Upper flange: $a_3 = 8 \times 18 \text{ cm}^2$

$$y_3 = 1 + 12 + \frac{1}{2} = 13.5 \text{ cm}$$

Then:
$$\bar{y} = \frac{a_1 y_1 + a_2 y_2 + a_3 y_3}{a_1 + a_2 + a_3}$$
$$= \frac{10 \times 0.5 + 12 \times 7 + 8 \times 13.5}{10 + 12 + 8} = \frac{5 + 84 + 108}{30} = 5.57 \text{ cm}$$

With reference to the centroidal axes, the centroid of the lower flange, web and upper flange are (0, 5.07), (0, 1.43) and (0, 7.93) respectively.

Moment of inertia of the I-section about centroidal axis is

= *MOI* of area a_1 about centroidal axis + *MOI* of area a_2 about centroidal axis + *MOI* of area a_3 about centroidal axis.

$$= (I_{G1} + A_1h_1^2) + (I_{G2} + A_2h_2^2) + (I_{G3} + A_3h_3^2)$$

$$= \left[\frac{10 \times 1^3}{12} + 10 \times (5.07)^2\right] + \left[\frac{1 \times 12^3}{12} + 12 \times (1.43)^2\right] + \left[\frac{8 \times 1^3}{12} + 8 \times (7.93)^2\right]$$

$$= (0.833 + 257.05) + (144 + 24.54) + (0.67 + 503.08)$$
$$= 930.17 \text{ cm}^4$$

and $$I_{yy} = \frac{1 \times 10^3}{12} + \frac{12 \times 1^3}{12} + \frac{1 \times 8^3}{12}$$
$$= 83.33 + 1 + 42.64 = 127 \text{ cm}^4$$

Polar moment of inertia $= I_{xx} + I_{yy} = 930.17 + 127$
$= 1057.17 \text{ cm}^4$

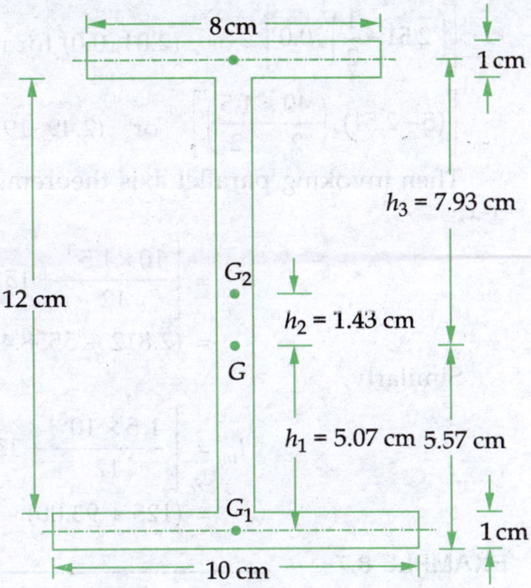

Fig. 8.19 (b)

(*b*) The radius of gyration is given by: $k = \sqrt{\frac{I}{A}}$

$$\therefore \quad k_{xx} = \sqrt{\frac{930.17}{30}} = 5.567 \text{ cm}$$

$$k_{yy} = \sqrt{\frac{127}{30}} = 2.057 \text{ cm}$$

EXAMPLE 8.6

Determine the moment of inertia about centroidal axes x-x and y-y of the channel section shown in Fig. 8.20.

Solution : The section is divided into three rectangles with areas

$$A_1 = 10 \times 1.5 = 15 \text{ cm}^2$$
$$A_2 = (40 - 1.5 - 1.5) \times 1 = 37 \text{ cm}^2$$
$$A_3 = 10 \times 1.5 = 15 \text{ cm}^2$$
$$\Sigma A = A_1 + A_2 + A_3$$
$$= 15 + 37 + 15 = 67 \text{ cm}^2$$

The given section is symmetrical about the horizontal axis passing through the centroid of rectangle A_2.

The distance of the centroid of the section with reference to section 1 – 1 is

$$\frac{\Sigma A x}{\Sigma A} = \frac{(15 \times 5) + \left(37 \times \frac{1}{2}\right) + (15 \times 5)}{67} = 2.51 \text{ cm}$$

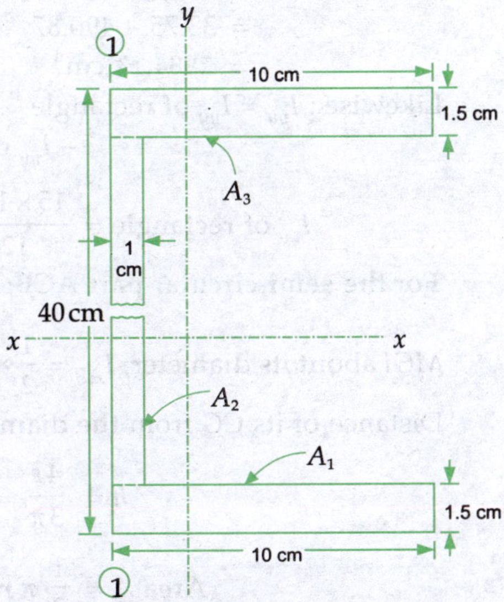

Fig. 8.20

With reference to the centroidal axes *x-x* and *y-y*, the centroids of the rectangles are:

$$\left[(5 - 2.51), \left(\frac{40}{2} - \frac{1.5}{2}\right)\right] \quad \text{or} \quad (2.49, 19.25) \text{ for rectangle } A_1$$

$\left[\left(2.51-\frac{1}{2}\right), 0.0\right]$ or (2.01, 0.0) for rectangle A_2

$\left[(5-2.51), \left(\frac{40}{2}-\frac{1.5}{2}\right)\right]$ or (2.49, 19.25) for rectangle A_3

Then invoking parallel axis theorem, the moment of inertia of areas A_1, A_2, and A_3 about x-x,

$$I_{xx} = \left[\frac{10 \times 1.5^3}{12} + 15 \times 19.25^2\right] + \left[\frac{1 \times 37^3}{12}\right] + \left[\frac{10 \times 1.5^3}{12} + 15 \times 19.25^2\right]$$

$$= (2.812 + 5558.437) + (4221.083) + (2.812 + 558.437) = \mathbf{155343.58\ cm^4}$$

Similarly,

$$I_{yy} = \left[\frac{1.5 \times 10^3}{12} + 15 \times 2.49^2\right] + \left[\frac{37 \times 1^3}{12}\right] + \left[\frac{1.5 \times 10^3}{12} + 15 \times 2.49^2\right]$$

$$= (125 + 93.00) + 3.08 + (125 + 93.00) = \mathbf{439.08\ cm^4}$$

EXAMPLE 8.7

Determine I_{xx} and I_{yy} of the cross-section of a cast iron beam shown in Fig. 8.21.

Solution : The *MOI* of the given sections can be worked out by looking it as a rectangle minus two semi-circles.

$$\therefore \quad I_{xx} = I_{xx} \text{ of rectangle} - I_{xx} \text{ of circular part}$$

$$= \frac{bd^3}{12} - \frac{\pi r^4}{4}$$

$$= \frac{12 \times 15^3}{12} - \frac{\pi \times 5^4}{4}$$

$$= 33.75 - 490.87$$

$$= \mathbf{2884.13\ cm^4}$$

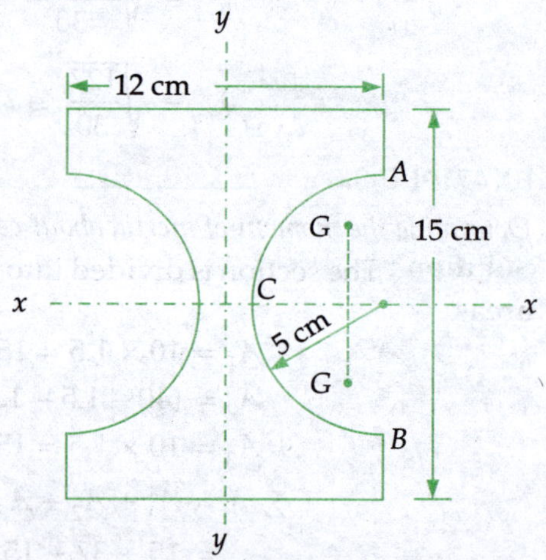

Fig. 8.21

Likewise : $I_{yy} = I_{yy}$ of rectangle $- I_{yy}$ of semi-circular parts

$$I_{yy} \text{ of rectangle} = \frac{15 \times 12^3}{12} = 2160 \text{ cm}^4$$

For the semi-circular part ACB;

$$MOI \text{ about its diameter, } I_{AB} = \frac{1}{2} \times \frac{\pi \times 5^4}{4} = 245.43 \text{ cm}^4$$

Distance of its *CG* from the diameter,

$$h = \frac{4r}{3\pi} = \frac{4 \times 5}{3\pi} = 2.12 \text{ cm}$$

$$\text{Area } A = \frac{1}{2}\pi r^2 = \frac{1}{2}\pi \times 5^2 = 39.27 \text{ cm}^2$$

From the correlation, $I_{AB} = I_{GG} + Ah^2$, the moment of inertia of semi-circular part about its centroidal axis

$$I_{GG} = 245.43 - 39.27 \times (2.12)^2 = 68.94 \text{ cm}^4$$

Again from the parallel axis theorem,

$$I_{yy} = I_{GG} + Ah_1^2$$

where h_1 = distance between axis and G-axis, = 6 – 2.12 = 3.88 cm

$$\therefore \quad I_{yy} = 68.94 + 39.27 \times 3.88^2 = 660.13 \text{ cm}^4$$

Since there are two semi-circular parts,

I_{yy} for two semi-circular parts = 2 × 660.13 = 1320.26 cm^4

∴ I_{yy} for the section = 2160 – 1320.26 = **839.74 cm^4**

EXAMPLE 8.8

Determine the moments of inertia about the x and y centroidal axis of a beam whose cross-sectional area is as shown in Fig. 8.22. All dimensions are in cm.

Solution : The given section has been divided into three segments marked 1, 2 and 3

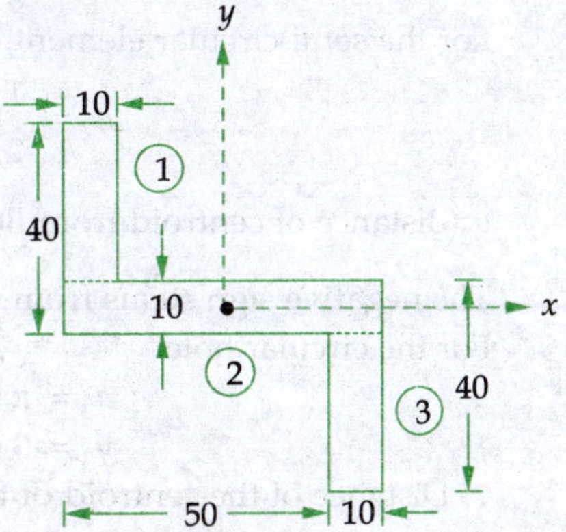

Fig. 8.22

$$(I_{xx})_1 = I_{G_1} + A_1h_1^2 = I_{G_1} + A_1(\bar{y} - y_1)^2$$

$$= \frac{1}{12} \times 10 \times 30^3 + (30 \times 10)(35 - 15)^2$$

$$= 1.425 \times 10^5 \text{ cm}^4$$

$$(I_{xx})_2 = I_{G_2} + A_2h_2^2 = I_{G_2} + A_2(\bar{y} - y_2)^2$$

$$= \frac{1}{12} \times 10^3 \times 60 + (60 \times 10) \times 0$$

$$= 0.05 \times 10^5 \text{ cm}^4$$

$$(I_{xx})_3 = I_{G_3} + A_3h_3^2 = I_{G_3} + A_3(\bar{y} - y_3)^2$$

$$= \frac{1}{12} \times 10 \times 30^3 + (30 \times 10)(35 - 15)^2 = 1.425 \times 10^5 \text{ cm}^4$$

$$\therefore \quad I_{xx} = 1.425 \times 10^5 + 0.05 \times 10^5 + 1.425 \times 10^5 = \mathbf{2.90 \times 10^5 \text{ cm}^4}$$

$$(I_{yy})_1 = I_{G_1} + A_1h_1^2 = I_{G_1} + A_1(\bar{x} - x_1)^2$$

$$= \frac{1}{12} \times 30 \times 10^3 + (30 \times 10) \times (30 - 5)^2 = 1.9 \times 10^5 \text{ cm}^4$$

$$(I_{yy})_2 = I_{G_2} + A_2h_2^2 = I_{G_2} + A_2(\bar{x} - x_2)^2$$

$$= \frac{1}{12} \times 60^3 \times 10 + (60 \times 10) \times 0 = 1.8 \times 10^5 \text{ cm}^4$$

$$(I_{yy})_3 = I_{G_3} + A_3h_3^2 = I_{G_3} + A_3(\bar{x} - x_3)^2$$

$$= \frac{1}{12} \times 30 \times 10^3 + (30 \times 10) \times (30 - 5)^2$$

$$= 1.9 \times 10^5 \text{ cm}^4$$

$$\therefore \quad I_{yy} = 1.9 \times 10^5 + 1.8 \times 10^5 + 1.9 \times 10^5 = \mathbf{5.6 \times 10^5 \text{ cm}^4}$$

EXAMPLE 8.9

Find the moment of inertia about the centroid horizontal axis of the area shown shaded in Fig. 8.23. The section consists of triangle ABC, semi-circle on BC as diameter, and a circular hole of diameter 4 cm with its centre on BC.

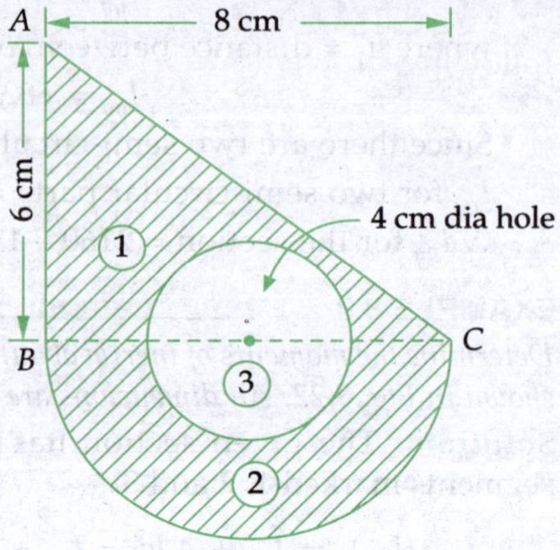

Fig. 8.23

Solution : The shaded area can be considered as a triangle (1), semicircle (2) and a circular hole (3)

Location of Centroid : For the triangular element,

$$a_1 = \frac{1}{2} \times 6 \times 8 = 24 \text{ cm}^2$$

y_1 (distance of centroid from BC)

$$= \frac{6}{3} = 2 \text{ cm}$$

For the semi-circular element,

$$a_2 = \frac{1}{2}\pi r^2 = \frac{1}{2}\pi \times 4^2 = 25.12 \text{ cm}^2$$

$$y_2 \text{(distance of centroid from } BC) = \frac{-4r}{3\pi} = \frac{-4 \times 4}{3\pi} = -1.7 \text{ cm}$$

The negative sign stems from the fact that it lies below BC.

For the circular hole

$$a_3 = \pi r^2 = \pi \times 2^2 = 12.56 \text{ cm}^2 \text{ (this area is removed)}$$

$$y_3 = 0 \text{ (centroid lies on } BC)$$

$\therefore$ Distance of the centroid of the shaded area from BC

$$= \frac{\Sigma ay}{\Sigma a} = \frac{a_1 y_1 + a_2 y_2 - a_3 y_3}{a_1 + a_2 - a_3} = \frac{24 \times 2 + 25.12 \times (-1.7) - 12.56 \times 0}{24 + 25.12 - 12.56} = 0.145 \text{ cm}$$

Moment of Inertia

I_1 = moment of inertia of triangle ABC about base BC

$$= \frac{1}{12} bh^3 = \frac{1}{12} \times 8 \times 6^3 = 144 \text{ cm}^4$$

I_2 = moment of inertia of semi-circle about BC

$$= \frac{1}{128}\pi d^4 = \frac{1}{128} \times \pi \times 8^4 = 100.48 \text{ cm}^4$$

I_3 = moment of inertia of circular hole about BC

$$= \frac{\pi}{64} d^4 = \frac{\pi}{64} \times 4^4 = 12.56 \text{ cm}^2$$

$\therefore$ Moment of inertia of the shaded area about BC

$$= 144 + 100.48 - 12.56 = 231.92 \text{ cm}^4$$

Area of the shaded portion = 24 + 25.12 – 12.56 = 36.56 cm

Invoking parallel axis theorem,

Moment of inertia of shaded area about centroidal axis

$$I_G = I_{BC} - A h^2 = 231.92 - 36.56 \times 0.145^2 = 231.15 \text{ cm}^4$$

EXAMPLE 8.10

Find the horizontal centroidal moment of inertia of the lamina ABCDEFG shown in Fig. 8.24.

Solution : The composite figure is divided into the following simple figures:

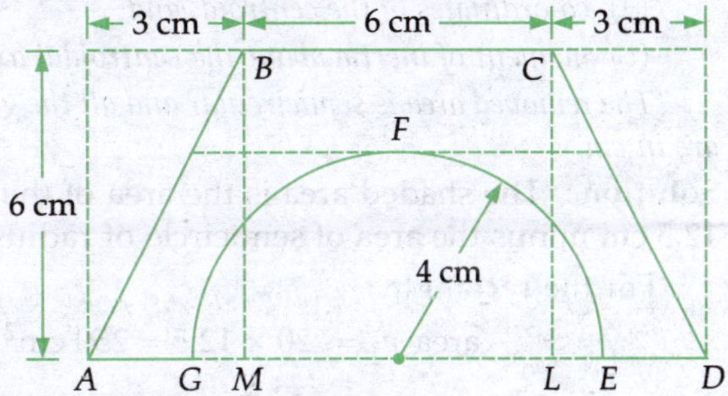

Fig. 8.24

1. A triangle *ABM*: $A_1 = \frac{1}{2} bh$

$$= \frac{1}{2} \times 3 \times 6 = 9 \text{ cm}^2$$

$$I_{G_1} = \frac{bh^3}{36} = \frac{3 \times 6^3}{36} = 18 \text{ cm}^4$$

y_1 (centroidal distance from line *AD*) $= \frac{6}{3} = 2$ cm

2. A square *BCLM*: $A_2 = 6 \times 6 = 36 \text{ cm}^2$

$$I_{G_2} = \frac{6 \times 6^3}{12} = 108 \text{ cm}^4$$

y_2 (centroidal distance from line *AD*) $= \frac{6}{2} = 3$ cm

3. A triangle *CDL*: $A_3 = \frac{1}{2} bh = \frac{1}{2} \times 3 \times 6 = 9 \text{ cm}^2$

$$I_{G_3} = \frac{bh^3}{36} = \frac{3 \times 6^3}{36} = 18 \text{ cm}^4$$

y_3 (centroidal distance from line *AD*) $= \frac{6}{3} = 2$ cm

4. A semi-circle *GFE* to be subtracted: $A_4 = \frac{\pi r^2}{2} = \frac{\pi \times 4^2}{2} = 25.12 \text{ cm}^2$ (– ve)

$$I_{G_4} = 0.11 \text{ r}^4 = 0.11 \times 4^4 = 28.16 \text{ cm}^4$$

y_4 (centroidal distance from line *AD*) $= \frac{4r}{3\pi} = \frac{4 \times 4}{3\pi} = 1.698$ cm

For the composite section

$$\bar{y} = \frac{\Sigma Ay}{\Sigma A} = \frac{(9 \times 2) + (36 \times 3) + (9 \times 2) - (25.12 \times 1.698)}{9 + 36 + 9 - 25.12}$$

$$= \frac{18 + 108 + 18 - 42.65}{28.88} = 3.51 \text{ cm}$$

Then
$$I_{xx} = I_{xx1} + I_{xx2} + I_{xx3} - I_{xx4}$$

$$= [18 + 9 \times (3.51 - 2)^2] + [108 + 36 \times (3.51 - 3)^2] + [18 + 9 \times (3.51 - 2)^2] - [28.16 + 25.12\,(3.51 - 1.698)^2]$$

The above relation has been written by applying the parallel axis theorem:

$$I_{xx} = I_{GG} + Ah^2$$

$\therefore \quad I_{xx} = 38.52 + 117.36 + 38.52 - 110.64 =$ **83.76 cm⁴**

EXAMPLE 8.11

For the shaded area shown in Fig. 8.25, determine :

(*a*) *co-ordinates of the centroid, and*

(*b*) *moment of inertia about the centroidal axes*

The removed area is semicircular and all the given dimensions are in cm.

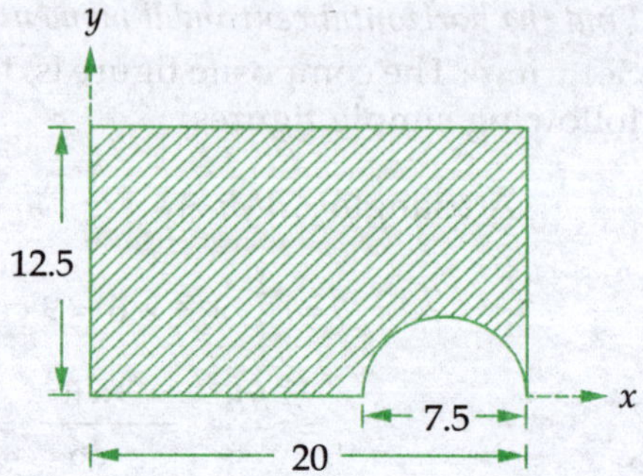

Fig. 8.25

Solution : The shaded area is the area of the rectangle 20 × 12.5 cm minus the area of semicircle of radius 3.75 cm.

For the rectangle :

$$\text{area } a_1 = 20 \times 12.5 = 250 \text{ cm}^2$$

$$y_1 = \frac{12.5}{2} = 6.25 \text{ cm}\,; \quad x_1 = \frac{20}{2} = 10 \text{ cm}$$

For the semicircle :

$$\text{area } a_2 = \frac{1}{2}\pi r^2 = \frac{1}{2} \times \pi \times 3.75^2 = 22.08 \text{ cm}^2$$

$$y_2 = \frac{4r}{3\pi} = \frac{4 \times 3.75}{3\pi} = 1.59 \text{ cm}$$

$$x_2 = (20 - 3.75) = 16.25 \text{ cm}$$

∴ For the shaded section,

$\bar{x}$ (distance of centroid from axis *oy*)

$$= \frac{a_1x_1 - a_2x_2}{a_1 - a_2} = \frac{250 \times 10 - 22.08 \times 16.25}{250 - 22.08} = 9.39 \text{ cm}$$

$\bar{y}$ (distance of centroid from axis *ox*)

$$= \frac{a_1y_1 - a_2y_2}{a_1 - a_2} = \frac{250 \times 6.25 - 22.08 \times 1.59}{250 - 22.08} = 6.70 \text{ cm}$$

Thus the co-ordinates of the centroid are: $\bar{x} = 9.39$ cm and $\bar{y} = 6.70$ cm

Moment of inertia of the shaded section = MOI of rectangle – MOI of semicircle

(*i*) With reference to centroidal *x*-axis

$$\text{MOI of rectangle} = \frac{20 \times 12.5^3}{12} + (20 \times 12.5) \times (6.7 - 6.25)^2$$

$$= 3255.21 + 50.62 = 3305.83 \text{ cm}^4$$

$$\text{MOI of semicircle} = 0.11 r^4 + \frac{1}{2}\pi r^2 (\bar{y} - y_2)^2$$

$$= 0.11 \times 3.75^4 + \frac{1}{2}\pi (3.75)^2 \times (6.70 - 1.59)^2$$

$$= 21.75 + 576.50 = 598.25 \text{ cm}^4$$

∴ MOI of shaded section about centroidal *x*-axis

$$= 3305.83 - 598.25 = 2707.58 \text{ cm}^4$$

(*ii*) With reference to centroidal *y*-axis

$$\text{MOI of rectangle} = \frac{12.5 \times 20^3}{12} + (20 \times 12.5)(10 - 9.39)^2$$
$$= 8333.33 + 93.02 = 8426.35 \text{ cm}^4$$

$$\text{MOI of semicircle} = \frac{\pi r^4}{8} + \frac{1}{2}\pi r^2 (20 - 3.75 - 9.39)^2$$
$$= \frac{\pi \times 3.75^4}{8} + \frac{1}{2}\pi (3.75)^2 \times (6.86)^2$$
$$= 77.62 + 1038.98 = 1116.6 \text{ cm}^4$$

∴ MOI of shaded section about centroidal *y*-axis

$$= 8426.35 - 1116.6 = 7309.75 \text{ cm}^4$$

EXAMPLE 8.12

Determine the moment of inertia of the plane area shown in Fig. 8.26, about its centroidal axis.

Solution : The given section has been divided into three segments marked as 1, 2 and 3.

1. Rectangle

$$a_1 = 8 \times 12 = 96 \text{ cm}^2$$
$$x_1 = 4 + \frac{8}{2} = 8 \text{ cm}$$
$$y_1 = \frac{12}{2} = 6 \text{ cm}$$

2. Triangle

$$a_2 = \frac{1}{2} \times 4 \times 12 = 24 \text{ cm}^2$$
$$x_2 = \frac{2}{3} \times 4 = 2.67 \text{ cm}$$
$$y_2 = \frac{12}{3} = 4 \text{ cm}$$

3. Semicircle

$$a_3 = \frac{1}{2} \times \left[\frac{\pi}{4} \times (4)^2\right] = 6.28 \text{ cm}^2$$
$$x_3 = 4 + 2 + \frac{4}{2} = 8 \text{ cm}$$
$$y_3 = 12 - \left(\frac{4 \times 2}{3\pi}\right) = 11.15 \text{ cm}$$

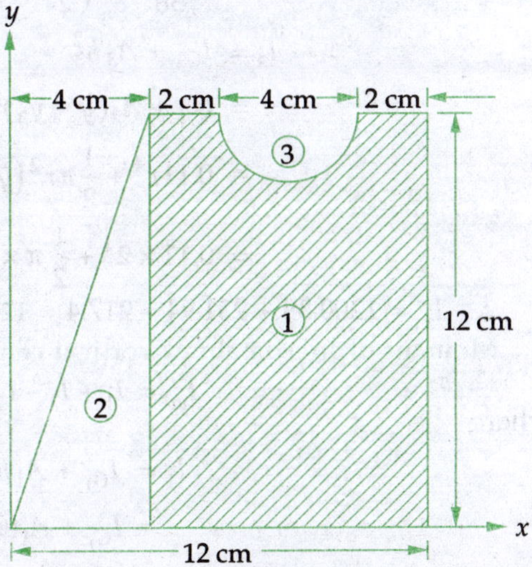

Fig. 8.26

Then the coordinates of the centroid of the given plane area are :

$$\bar{x} = \frac{a_1 x_1 + a_2 x_2 - a_3 x_3}{a_1 + a_2 - a_3} = \frac{96 \times 8 + 24 \times 2.67 - (6.28 \times 8)}{96 + 24 - 6.28} = \frac{768 + 64.08 - 50.24}{113.72}$$
$$= 6.87 \text{ cm}$$

$$\bar{y} = \frac{a_1 y_1 + a_2 y_2 - a_3 y_3}{a_1 + a_2 - a_3} = \frac{96 \times 6 + 24 \times 4 - 6.28 \times 11.15}{96 + 24 - 6.28} = \frac{576 + 96 - 70.02}{113.72}$$
$$= 5.29 \text{ cm}$$

The centroid of the given lamina have been shown in Fig. 8.27.

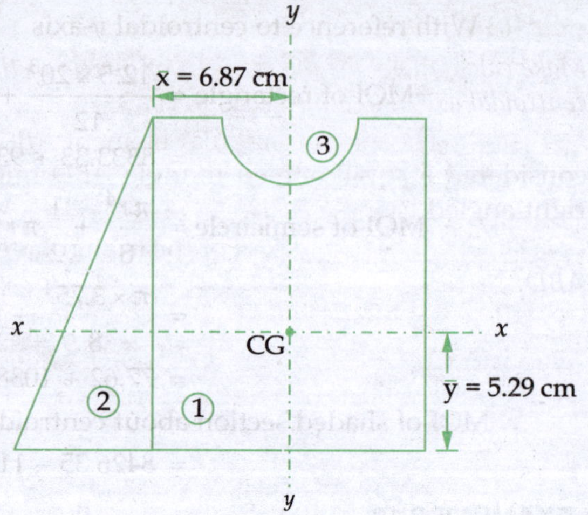

Fig. 8.27

Moment of inertia about horizontal centroidal axis:

$$I_{xx} = I_1 + I_2 - I_3$$

where

$$I_1 = I_{G_1} + A_1 h_1^2$$
$$= I_{G_1} + A_1(\bar{y} - y_1)^2$$
$$= \frac{8\times12^3}{12} + (8\times12)(5.29-6)^2$$
$$= 1200.39 \text{ cm}^4$$
$$I_2 = I_{G_2} + A_2 h_2^2$$
$$= I_{G_2} + A_2(\bar{y} - y_2)^2$$
$$= \frac{4\times12^3}{36} + \left(\frac{1}{2}\times4\times12\right)(5.29-4)^2 = 231.94 \text{ cm}^4$$
$$I_3 = I_{G_3} + A_3 h_3^2$$
$$= I_{G_3} + A_3(\bar{y} - y_3)^2$$
$$= 0.11r^4 + \frac{1}{2}\pi r^2(\bar{y} - y_3)^2$$
$$= 0.11\times2^4 + \frac{1}{2}\pi\times2^2(5.29-11.15)^2 = 217.41 \text{ cm}^4$$

$\therefore\ I_{xx} = 1200.39 + 231.94 - 217.4 =$ **1214.93 cm^4**

Moment of inertia about vertical centroidal axis

$$I_{yy} = I_1 + I_2 - I_3$$

where

$$I_1 = I_{G_1} + A_1 h_1^2$$
$$= I_{G_1} + A_1(\bar{x} - x_1)^2$$
$$= \frac{12\times8^3}{12} + (12\times8)\times(6.87-8)^2 = 634.58 \text{ cm}^4$$
$$I_2 = I_{G_2} + A_2 h_2^2$$
$$= I_{G_2} + A_2(\bar{x} - x_2)^2$$
$$= \frac{12\times4^3}{36} + \left(\frac{1}{2}\times12\times4\right)\times\left(6.87 - \frac{8}{3}\right)^2 = 444.69 \text{ cm}^4$$
$$I_3 = I_{G_3} + A_3 h_3^2$$
$$= \frac{\pi r^4}{8} + \frac{1}{2}\pi r^2(\bar{x} - x_3)^2$$
$$= \frac{\pi\times2^4}{8} + \frac{1}{2}\pi(2)^2(6.87-8)^2 = 14.3 \text{ cm}^4$$

$\therefore\ I_{yy} = 634.58 + 444.69 - 14.3 =$ **1064.97 cm^4**

EXAMPLE 8.13

Make calculations for the moment of inertia of the triangular element ABC shown in Fig. 8.28, about its centroidal axes. All dimensions in cm.

Solution : The given triangular element *ABC* can be considered as right angled triangle *ABD* minus the right angled triangle *ACD*.

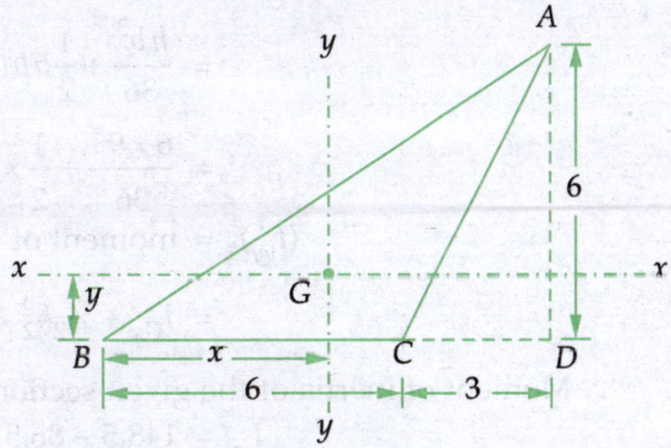

Fig. 8.28

Location of centroid : For the triangular element *ABD*,

$$a_1 = \frac{1}{2} \times 6 \times 9 = 27 \text{ cm}^2$$

$$x_1 \text{ (measured from } B) = \frac{2}{3} \times 9 = 6 \text{ cm}$$

$$y_1 \text{ (measured from base } BC) = \frac{1}{3} \times 6 = 2 \text{ cm}$$

For the triangular element *ACD*,

$$a_2 = \frac{1}{2} \times 3 \times 6 = 9 \text{ cm}^2$$

$$x_2 \text{ (measured from } B) = 6 + \frac{2}{3} \times 3 = 8 \text{ cm}$$

$$y_2 \text{ (measure from base } BC) = \frac{1}{3} \times 6 = 2 \text{ cm}$$

Then for the given triangular section,

$$\bar{x} = \frac{a_1x_1 - a_2x_2}{a_1 - a_2} = \frac{27 \times 6 - 9 \times 8}{27 - 9} = 5 \text{ cm}$$

$$\bar{y} = \frac{a_1y_1 - a_2y_2}{a_1 - a_2} = \frac{27 \times 2 - 9 \times 2}{27 - 9} = 2 \text{ cm}$$

Moment of inertia

$(I_{xx})_1$ = moment of inertia of triangle *ABD* about centroidal axis *xx*

$$= I_{G_1} + A_1h_1^2$$

$$= \frac{bh^3}{36} + \frac{1}{2} bh(\bar{y} - y)^2$$

$$= \frac{9 \times 6^3}{36} + \frac{1}{2} \times 9 \times 6(2 - 2) = 54 \text{ cm}^4$$

$(I_{xx})_2$ = moment of inertia of triangle *ACD* about centroidal axis *xx*

$$= I_{G_2} + A_2h_2^2$$

$$= \frac{3 \times 6^3}{36} + \frac{1}{2} \times 3 \times 6 \times (2 - 2)^2 = 18 \text{ cm}^4$$

∴ Moment of inertia of the given section *ABC* about centroidal axis *xx*

$$I_{xx} = 54 - 18 = 36 \text{ cm}^4$$

$(I_{yy})_1$ = moment of inertia of triangle ABD about centroidal axis

$$= I_{G_1} + A_1h_1^2$$

$$= \frac{hb^3}{36} + \frac{1}{2}bh(\bar{x} - x)^2$$

$$= \frac{6 \times 9^3}{36} + \frac{1}{2} \times 6 \times 9 \times (6-5)^2 = 148.5 \text{ cm}^4$$

$(I_{yy})_2$ = moment of inertia of triangle ACD about centroidal axis yy

$$= I_{G_2} + A_2h_2^2 = \frac{6 \times 3^3}{36} + \frac{1}{2} \times 3 \times 6 \times (8.5)^2 = 85.5 \text{ cm}^4$$

∴ Moment of inertia of the given section ABC about centroidal axis yy

$$I_{yy} = 148.5 - 86.5 = 63 \text{ cm}^4$$

EXAMPLE 8.14

Two equal channels are kept as shown in Fig. 8.29 at a distance d between them to form the cross-section of a column. Calculate the value of distance d if the centroidal moments of inertia I_{xx} and I_{yy} of the cross-section are equal.

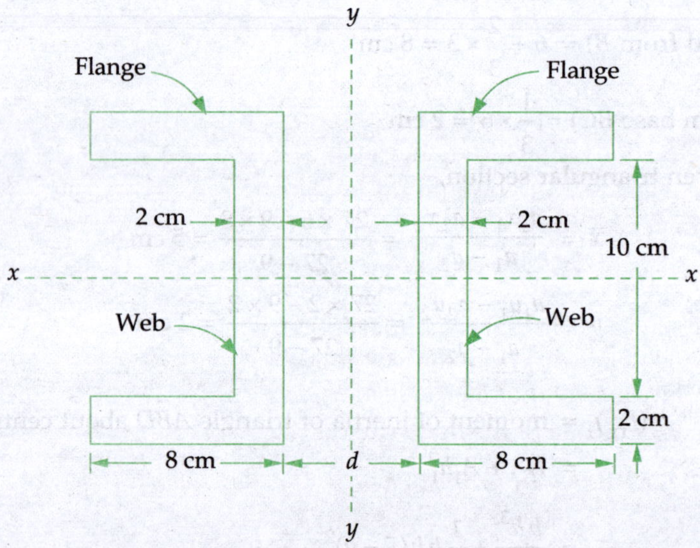

Fig. 8.29

Solution : I_{xx} = moment of inertia of 4 flanges + moment of inertia of 2 webs

$$= 4\left[\frac{8 \times 2^3}{12} + 8 \times 2 \times \left(5 + \frac{2}{2}\right)^2\right] + 2\left[\frac{2 \times 10^3}{12}\right]$$

$$= 4\,(5.33 + 576) + 333.33 = 2658.65 \text{ cm}^4$$

Here (5 + 2/2) is the distance of the centroid of the flange with area (8 × 2) from the x-axis.

I_{yy} = moment of interlia of 4 flanges + moment of inertia of 2 webs

$$= 4\left[\frac{2 \times 8^3}{12} + 8 \times 2\left(4 + \frac{d}{2}\right)^2\right] + 2\left[\frac{10 \times 2^3}{12} + 10 \times 2\left(\frac{2}{2} + \frac{d}{2}\right)^2\right]$$

$$= 4\left[85.33 + 16\left(16 + \frac{d^2}{4} + 4d\right)\right] + 2\left[6.67 + 20\left(1 + \frac{d^2}{4} + d\right)\right]$$

$$= 341.32 + 1024 + 16\,d^2 + 256d + 13.34 + 40 + 5d^2 + 40d$$

$$= 21\,d^2 + 296\,d + 1418.66$$

But it is given that $I_{xx} = I_{yy}$

$$\therefore \qquad 21d^2 + 296d + 1418.66 = 2658.65$$

or $$21d^2 + 296d - 1240 = 0$$

or $$d^2 + 14.09d - 59.05 = 0$$

Solution of the quadratic equation gives

$$d = \frac{-14.09 \pm \sqrt{(14.09)^2 - 4\times 1(-59.05)}}{2\times 1} = \frac{-14.09 \pm 20.85}{2}$$

$$= \mathbf{3.28\ cm}$$

(Taking + *ve* velue)

Hence the distance between the two shannel sections is 3.28 cm.

8.3. MASS MOMENT OF INERTIA

The mass moment of inertia of a body about a particular axis is defined as *"the product of the mass and the square of the distance between the mass centre of the body and the axis"*.

The mass moment of inertia is an important term for the study of the rotational motion of a rigid body. It gives a measure of the resistance that the body offers to change in angular velocity.

The body can be considered to be split up into small masses. Let

$m_1, m_2 \ldots m_n$ be the masses of the various elements of the body

and $r_1, r_2 \ldots r_n$ be the distances of the above mentioned elements from the axis about which mass moment of inertia is to be determined.

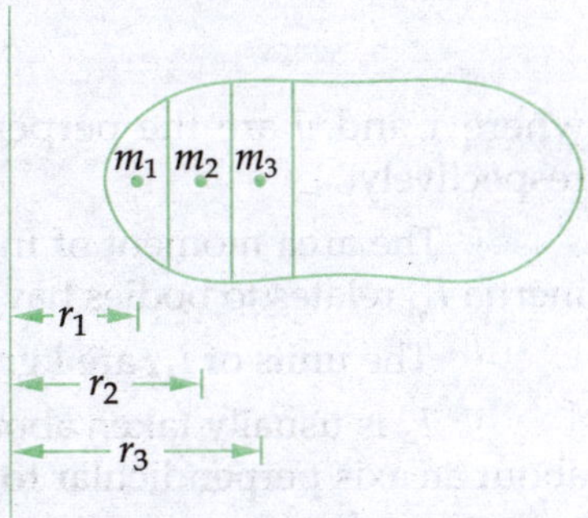

Fig. 8.30

The mass moment of inertia of the body can be written as

$$I = m_1r_1^2 + m_2r_2^2 + \ldots m_nr_n^2$$

$$= \Sigma\, mr^2$$

The summation of a large number of terms in the above expression can be replaced by integration. Consider a small mass *dm* rotating about and at distance *r* from the axis of rotation, then

$$I = \int r^2 dm$$

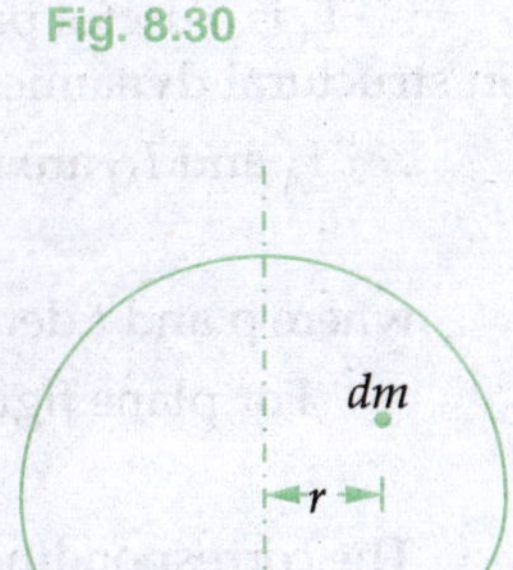

Fig. 8.31

The radius of gyration *k* of the body with respect to the prescribed axis is defined by the relation

$$I = k^2M\,; \quad k = \sqrt{\frac{I}{M}}$$

where *M* is the mass of the body.

The radius of gyration gives a measure of the distance at which the entire mass of the body is assumed to be concentrated.

Unit of mass moment of inertia is kg-m^2 and that of radius of gyration is m.

Parallel axis theorem : The mass moment of inertia of a body about an axis parallel to the centroidal axis is equal to the moment of inertia about the centroidal axis plus the product of mass and the square of the shortest distance between centre of gravity and the axis concerned. With reference to Fig. 8.32.

$$I_{XX} = I_{xx} + mh^2$$

where m is the mass of the body and h is the distance between the two parallel axis.

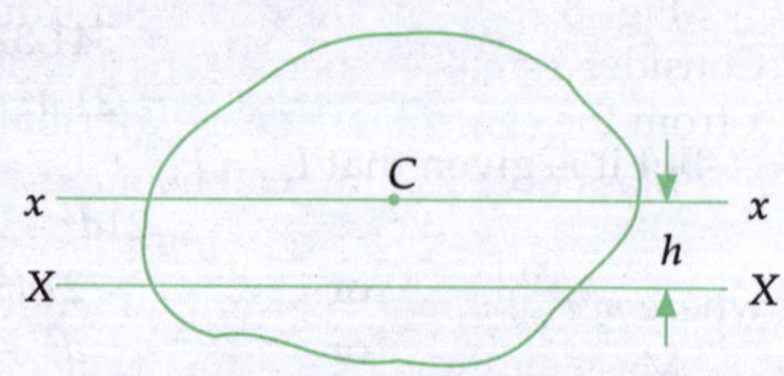

Fig. 8.32

Comparison of mass moment of inertia and area moment of inertia :

(*i*) The mass moment of inertia is mathematically defined as :

$$I_M = \int x^2 \, dm$$

where dm is the mass of a small element of the body at normal distance x from the line or axis about which moment of inertia is desired.

The area moment of inertia I_A is the second moment of area, and is mathematically defined as

$$I_x = \int y^2 \, dA \quad \text{and} \quad I_y = \int x^2 \, dA$$

where x and y are the perpendicular distances of small elemental area dA from y and x-axis respectively.

(*ii*) The area moment of inertia I_A is related to area of plane figures, and the mass moment of inertia I_M relates to bodies having a bulk and hence mass.

(*iii*) The units of I_M are kgm^2. Being the second moment of area, I_A has the units m^4.

(*iv*) I_A is usually taken about an axis in the plane of the figure under consideration. I_M is taken about an axis perpendicular to the plane of the figure.

(*v*) I_A is a useful parameter for the analysis of structures in static conditions. I_M has applications in structural dynamics.

(*vi*) I_A and I_M are related to each other by the expression

$$I_M = \rho \, t \, I_A$$

where ρ and t denote the density and thickness of the material.

(*vii*) For plane figures, the parallel axis theorem is specified as :

$$(I_A)_{AA} = (I_A)_{xx} + A\,h^2$$

The corresponding expression for solid bodies is

$$(I_M)_{AA} = (I_M)_{xx} + M\,h^2$$

(*viii*) The radius of gyration for a plane figure and a solid figure is worked out from the following identities

$$k_A = \sqrt{\frac{I_A}{A}} \quad \text{and} \quad k_M = \sqrt{\frac{I_M}{M}}$$

8.4. MASS MOMENT OF INERTIA FOR SPECIFIED CASES

8.4.1. Thin uniform rod

Fig. 8.33 shows a thin uniform rod of length l. Consider a small elemental length dx located at distance x from the centroidal axis y-y normal to the rod.

Mass of the segmental element

$$dm = m\,dx$$

where m is the mass per unit of length.

Mass moment of inertia of the element about axis y-y

$$= dm\,x^2 = mx^2\,dx$$

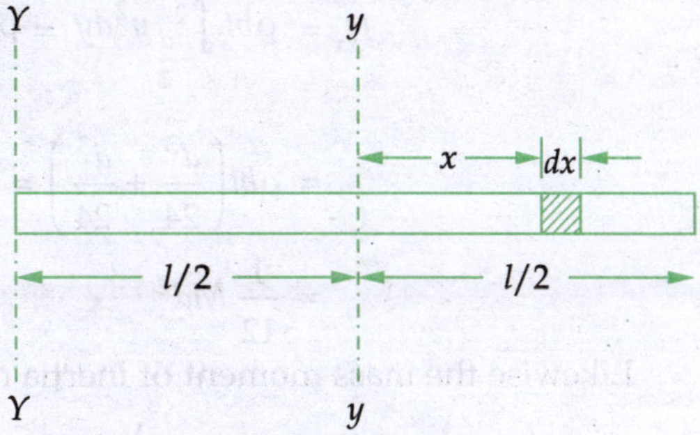

Fig. 8.33

The mass moment of inertia of the entire rod about axis y-y can be worked out by integrating the above expression between the limits $-\frac{l}{2}$ to $\frac{l}{2}$. That is

$$I_{yy} = m\int_{-\frac{l}{2}}^{\frac{l}{2}} x^2\,dx = m\left|\frac{x^3}{3}\right|_{-\frac{l}{2}}^{\frac{l}{2}} = m\left(\frac{l^3}{24}+\frac{l^3}{24}\right)$$

$$= \frac{ml^3}{12} = \frac{Ml^2}{12}$$

where $M = ml$ is the mass of the whole rod.

If it is required to determine the mass moment of inertia of the rod about axis YY at the left end of the rod, we can use the parallel axis theorem

$$I_{YY} = I_{yy} + Mh^2$$

where h is the distance between the axis YY and the centroidal axis yy.

$$I_{YY} = \frac{Ml^2}{12} + M\left(\frac{l}{2}\right)^2 = \frac{Ml^2}{3}$$

8.4.2. Rectangular plate

Figure 8.34, shows a rectangular plate of width b, depth d and uniform thickness t. If ρ is the density of the plate material, then mass of the plate

$$M = \text{density} \times \text{volume}$$
$$= \rho\, b\, t\, d$$

Consider an elemental strip of depth dy located at distance y from the centroidal axis xx

mass of the elementa strip $dm = \rho\, b\, t\, dy$

mass moment of inertial of the strip about axis xx

$$= dm\, y^2 = \rho\, b\, t\, y^2 dy$$

The moment of inertia for the entire mass of plate about axis xx can be worked out by integrating the above expression between the limits $-\frac{d}{2}$ to $\frac{d}{2}$. That is

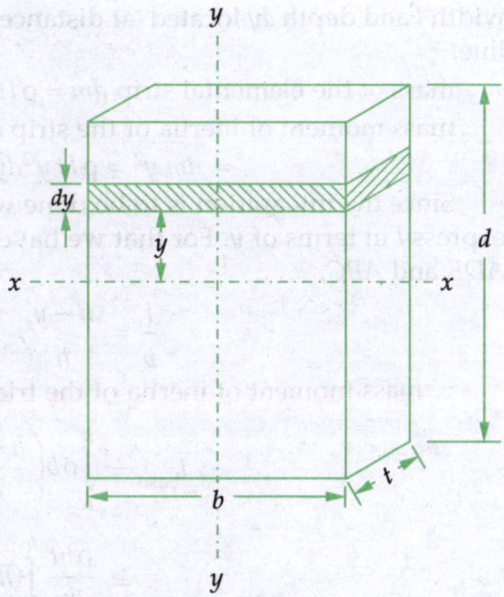

Fig. 8.34

$$I_{xx} = \rho bt \int_{-\frac{d}{2}}^{\frac{d}{2}} y^2 dy = \rho bt \left| \frac{y^3}{3} \right|_{-\frac{d}{2}}^{\frac{d}{2}}$$

$$= \rho bt \left(\frac{d^3}{24} + \frac{d^3}{24} \right) = (\rho bt d) \times \frac{d^2}{12}$$

$$= \frac{1}{12} Md^2 \quad \text{...(8.23 a)}$$

Likewise the mass moment of inertia of the rectangular plate about the centroidal axis yy is

$$I_{yy} = \frac{1}{12} Mb^2 \quad \text{...(8.23 b)}$$

From perpendicular axis theorem, the moment of inertia about axis zz is

$$I_{zz} = I_{xx} + I_{yy}$$

$$= \frac{1}{12} Md^2 + \frac{1}{12} Mb^2 = \frac{1}{12} M(d^2 + b^2) \quad \text{...(8.24)}$$

8.4.3. Triangular plate

Figure 8.35 shows a triangular plate of base width b, height h and uniform thickness t. If ρ is the density of the plate material, then

mass of the plate,

$$M = \text{density} \times \text{volume}$$

$$= \text{density} \times (\text{area} \times \text{thickness})$$

$$= \rho \times \frac{1}{2} bht$$

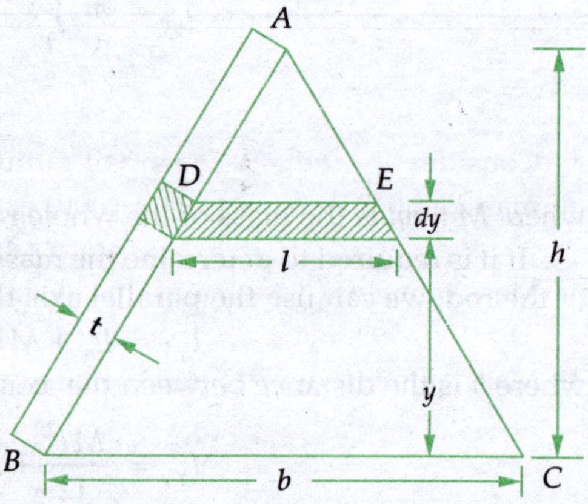

Fig. 8.35

Consider an elemental strip (assumed rectangle) of width l and depth dy located at distance y from the base line.

mass of the elemental strip $dm = \rho l t \, dy$

mass moment of inertia of the strip about base

$$= dm \, y^2 = \rho l t y^2 dy$$

Since the integration is to be done with respect to y within the limits 0 to h, it is necessary to express l in terms of y. For that we have the following correlation from the similarity of triangles ADE and ABC,

$$\frac{l}{b} = \frac{h - y}{h}, \quad l = b\frac{h - y}{h}$$

$\therefore$ mass moment of inertia of the triangular plate about base line

$$I_{\text{base}} = \int_0^h \rho b \left(\frac{h - y}{h} \right) t \, y^2 dy$$

$$= \frac{\rho bt}{h} \int_0^h (h - y) y^2 dy$$

$$= \frac{\rho bt}{h}\left|\frac{hy^3}{3} - \frac{y^4}{4}\right|_o^h = \frac{\rho bt}{h}\left(\frac{h^4}{3} - \frac{h^4}{4}\right) = \frac{\rho bt h^3}{12}$$

$$= \frac{\rho bht}{2} \times \frac{h^2}{6} = \frac{1}{6} M h^2 \qquad \text{...(8.25)}$$

8.4.4. Circular lamina

Figure 8.36 shows a thin circular plate of radius R and uniform thickness t. If ρ is the density of the plate material, then

mass of the plate M = density × volume

$= \text{density} \times (\text{area} \times \text{thickness}) = \rho \pi R^2 t$

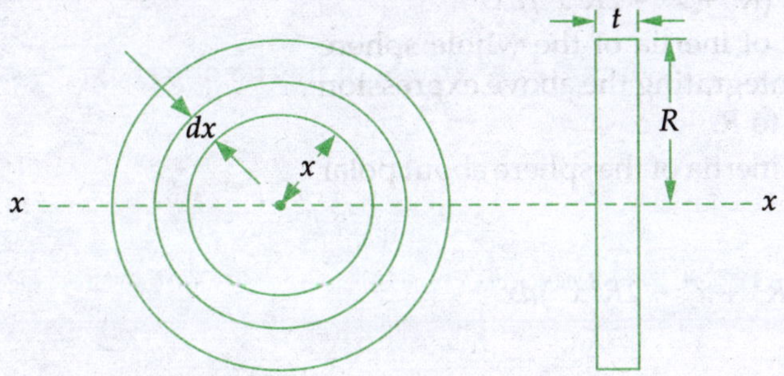

Fig. 8.36

Consider an elementary ring of radius r and width dr.

mass of elemental ring $dm = \rho\,[\pi(r + dr)^2 - \pi r^2]t$

$= \rho\,(2\pi r\,dr)\,t = 2\pi t \rho r\,dr$

Mass moment of inertia of this elementary ring about the polar axis zz

$= dm\,r^2 = 2\pi t \rho r^3\,dr$

Mass moment of inertia of the circular plate about polar axis zz

$$= 2\pi t \rho \int_o^R r^3\,dr = 2\pi t \rho \left|\frac{r^4}{4}\right|_o^R$$

$$= \rho \pi R^2 t \times \frac{R^2}{2} = \frac{1}{2} MR^2 \qquad \text{...(8.26)}$$

where $M = \rho \pi R^2 t$ is the mass of the circular lamina

Invoking the theorem of perpendicular axis, the mass moment of inertia of a circular lamina about xx or yy axis is

$$I_{xx} = I_{yy} = \frac{I_{zz}}{2} = \frac{1}{4} MR^2 \qquad \text{...(8.27)}$$

8.4.5. Solid sphere

Figure 8.37 shows a solid sphere of radius R with O as centre. If ρ is the density of the material of the sphere, then

mass of the sphere = density × volume

$= \rho \times \frac{4}{3} \pi R^3$

Let attention be focussed on a thin disc AB of thickness dx and at radius x from the centre.

radius of the disc $y = \sqrt{R^2 - x^2}$

mass of the disc dm

$$= \rho \times \pi y^2 dx$$
$$= \rho\pi (R^2 - x^2)\, dx$$

Mass moment of inertia of this elementary disc about the polar axis zz

$$dm\, y^2 = \rho\pi (R^2 - x^2)\, dx \times (R^2 - x^2)$$
$$= \rho\pi (R^2 - x^2)^2\, dx$$
$$= \rho\pi (R^4 + x^4 - 2R^2x^2)\, dx$$

The mass moment of inertia of the whole sphere can be worked out by integrating the above expression between the limits $-R$ to R.

$\therefore$ Mass moment of inertia of the sphere about polar axis zz

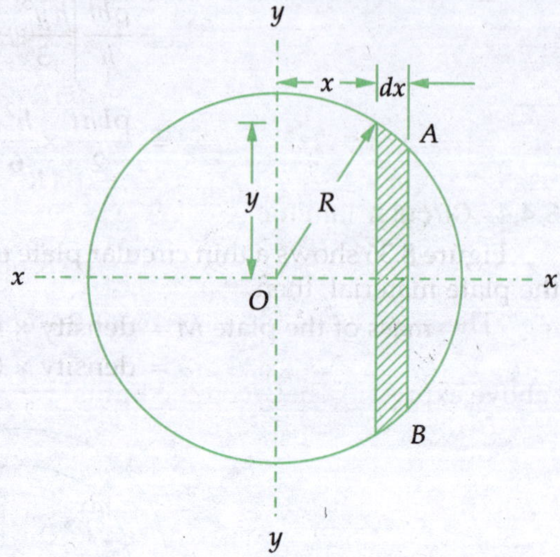

Fig. 8.37

$$I_{zz} = \rho\pi \int_{-R}^{R} (R^4 + x^4 - 2R^2x^2)\, dx$$

$$I_{zz} = \rho\pi \left| R^4x + \frac{x^5}{5} - 2R^2\frac{x^3}{3} \right|_{-R}^{R} = \frac{16\rho\pi R^5}{15} = \frac{4}{5}MR^2 \qquad \text{...(8.28)}$$

where $M = \dfrac{4}{3}\rho\pi R^3$ is the mass of the solid sphere

Invoking the theorem of perpendicular axis, the mass moment of inertia of a solid sphere about xx or yy axis is,

$$I_{xx} = I_{yy} = \frac{I_{zz}}{2} = \frac{2}{5}MR^2$$

8.4.6. Solid cylinder

Figure 8.38 shows a solid cylinder of radius R and height h. If ρ is the density of the material of the cylinder, then

$$\text{mass of the cylinder} = \text{density} \times \text{volume}$$
$$M = \rho \times \pi R^2 h$$

Consider a thin disc of thickness dy located at distance y from the centroidal axis xx.

mass of the elemental disc, $dm = \rho \times \pi R^2 dy$

It may be recalled that mass moment of inertia of a circular lamina about its diametral axis is given by

$$= \frac{1}{4}MR^2$$

$\therefore$ mass moment of inertia of the elemental disc about its diameteral axis is

$$I_{dia} = \frac{1}{4}dm\, R^2$$

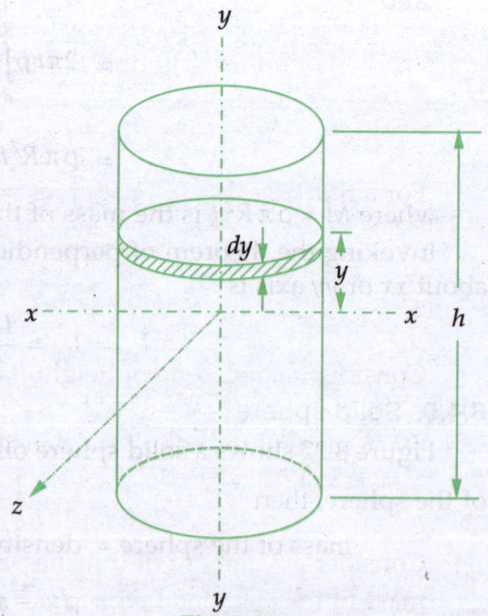

Fig. 8.38

Invoking parallel axis theorem, the mass moment of inertia of elemental disc about axis xx is

$$I_{xx} = I_{dia} + (dm)\,y^2$$

$$= \frac{1}{4} dm R^2 + dm y^2$$

$$= \frac{1}{4} (\rho\pi R^2\, dy) R^2 + (\rho\pi R^2\, dy) y^2$$

$$= \frac{1}{4} \rho\pi R^4\, dy + \rho\pi R^2 y^2\, dy$$

The mass moment of inertia of the entire solid cylinder can be worked out by integrating the above expression between the limits $-\frac{h}{2}$ to $\frac{h}{2}$. Thus,

$$I_{xx} = \frac{1}{4} \rho\pi R^4 \int_{-\frac{h}{2}}^{\frac{h}{2}} dy + \rho\pi R^2 \int_{-\frac{h}{2}}^{\frac{h}{2}} y^2 dy$$

$$= \frac{1}{4} \rho\pi R^4 h + \frac{1}{12} \rho\pi R^2 h^3$$

$$= \rho\pi R^2 h \left(\frac{R^2}{4} + \frac{h^2}{12} \right) = M \left(\frac{R^2}{4} + \frac{h^2}{12} \right) = \frac{1}{12} M (3R^2 + h^2) \quad \text{...(8.29 } a\text{)}$$

where $M = \rho\pi R^2 h$ is the mass of the cylinder.

Similarly

$$I_{yy} = \frac{1}{12} M (3R^2 + h^2) \quad \text{...(8.29 } b\text{)}$$

and $$I_{zz} = I_{xx} + I_{yy} = \frac{1}{6} M (3R^2 + h^2) \quad \text{...(8.30)}$$

Note : For a thin cylinder, $R = 0$. That gives :

$$I_{xx} = I_{yy} = \frac{1}{12} M h^2 = \frac{1}{12} M l^2$$

For a thin disc, $h = 0$. That gives :

$$I_{xx} = I_{yy} = \frac{1}{4} M R^2 \quad \text{and} \quad I_{zz} = \frac{1}{2} M R^2$$

8.4.7. Right circular cone

Consider a solid cone of height h and radius R. If ρ is the density of the material of the cone, then

mass of the cone M = density × volume

$$= \rho \times \frac{1}{3} \pi R^2 h$$

Consider an element of thickness dy and radius r at distance y from the apex A

mass of the elemental strip, $dm = \rho\pi r^2 dy$

mass moment of inertia of the elemental strip about axis yy

$$= \frac{1}{2} \times \text{mass moment of inertia about polar axis}$$

$$= \frac{1}{2} dm\, r^2 = \frac{1}{2} (\rho\pi r^2\, dy) r^2$$

$$= \frac{1}{2} \rho\pi r^4\, dy$$

Since the integration is to be done with respect to y within the limits 0 to h, it is necessary to express r in terms of y. For that we have the following correlation from the similarity of triangles ADE and ABC

$$\frac{r}{R} = \frac{y}{h}, \quad r = R\frac{y}{h}$$

∴ mass moment of inertia of the cone about axis yy

$$I_{yy} = \int_o^h \frac{1}{2} \rho\pi \left(R\frac{y}{h} \right)^4 dy$$

$$= \frac{\rho\pi R^4}{2h^4} \left| \frac{y^5}{5} \right|_o^h$$

$$= \frac{\rho\pi R^4 h}{10} = \frac{\rho\pi R^2 h}{3} \times \frac{3}{10} R^2 = \frac{3}{10} MR^2 \quad \text{...(8.31)}$$

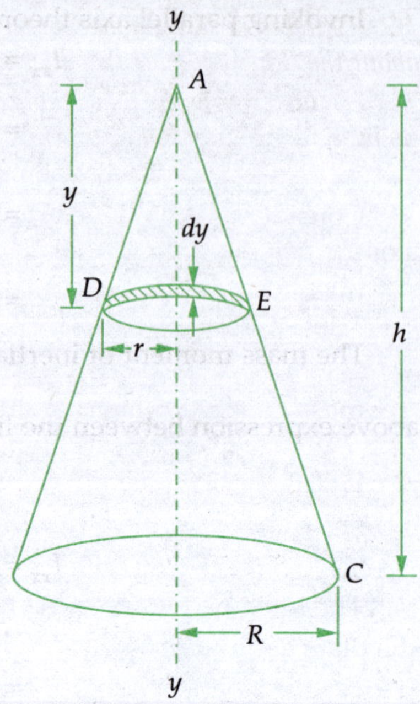

Fig. 8.39

where $M = \frac{1}{3} \pi\rho R^2 h$ is the mass of the right circular cone.

EXAMPLE 8.15

Would you imagine that the moment of inertia of the earth around its own axis is negligible fraction of its moment of inertia about the axis of rotation around the sun? Take mean radius of the earth as 6,371 km and the mean radius of rotation around the sun as 149.7 × 10⁶ km.

Solution : (*b*) Moment of inertia of the earth about its axis,

$$I_1 = \frac{2}{5} MR^2 = \frac{2}{5} M(6371)^2 = 16.23 \times 10^6 M$$

Moment of inertia of the earth about the axis of rotation around the sun,

$$I_2 = I_1 + Md^2 = 16.23 \times 10^6 M + M \times (149.7 \times 10^6)^2$$
$$= 16.23 \times 10^6 M + 22410.09 \times 10^{12} M$$
$$= 22410.09002 \times 10^{12} M$$

$$\text{Ratio } \frac{I_1}{I_2} = \frac{16.23 \times 10^6}{22410.09002 \times 10^{12}} = 7.24 \times 10^{-10}$$

Since the ratio is negligible, the moment of inertia of the earth around its own axis can be imagined to be a negligible fraction of its moment of inertia about the axis of rotation around the sun.

EXAMPLE 8.16

(*a*) *Set up a relation between the mass moment of inertia and the moment of inertia of a plate about an axis lying in the plane of the plate.*

(*b*) *Determine the mass moment of inertia of a steel rectangular plate 8 cm × 16 cm and thickness 1 cm about the centroidal axis parallel to the 8 cm side. Take mass density of steel as 8000 kg/m*^{3}.

(*c*) *Determine the mass moment of inertia of a triangular plate of base 50 cm and height 100 cm about its base. The plate has a uniform thickness of 1 cm and density of the plate material is 7500 kg/m*^{3}.

Solution : Figure 8.40 shows a thin plate of uniform thickness t and homogeneous mass density ρ.

Consider an elemental area dA located at distance r from the axis xx

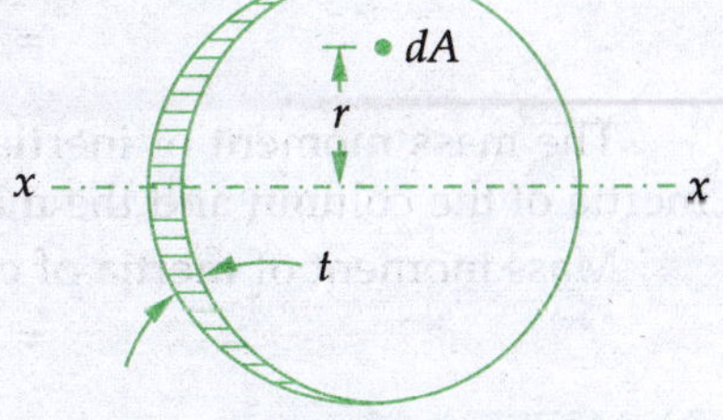

Fig. 8.40

mass of the elemental area, dm = density × volume

$$= \rho \times (dA \times t)$$
$$= \rho t\, dA$$

mass moment of inertia of this elemental area

$$= dm \times r^2 = \rho t r^2\, dA$$

For the entire plate,

$$(I_{xx})_{\text{mass}} = \int \rho t r^2\, dA = \rho t \int r^2\, dA$$

The integral $\int r^2 dA$ is the moment of inertia of the area of the plate about the axis xx

$$\therefore \quad (I_{xx})_{\text{mass}} = \rho t\, (I_{xx})_{\text{area}}$$

(*b*) Refer Fig. 8.41 for the configuration of the rectangular plate

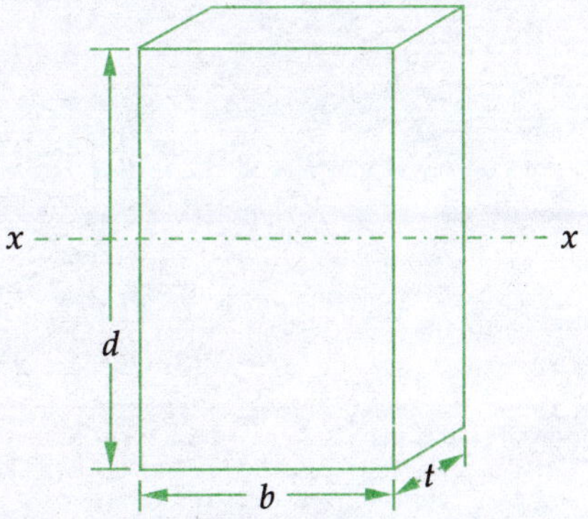

Fig. 8.41

b = 8 cm ; d = 16 cm and t = 1 cm

mass of the plate M= density × volume

$$= 8000 \times \left(\frac{8 \times 16 \times 1}{10^6}\right)$$
$$= 1.024 \text{ kg}$$

Mass moment of inertia about axis xx is

$$I_{xx} = \frac{1}{12} M d^2 = \frac{1}{12} \times 1.024 \times \left(\frac{16}{100}\right)^2 = 2.184 \times 10^{-3} \text{ kg m}^2$$

(*c*) For the triangular plate,

base width b = 50 cm; height h = 100 cm and thickness t = 1 cm

mass of the triangular plate, M = density × volume

$$= \text{density} \times (\text{area} \times \text{thickness})$$
$$= 7500 \times \left(\frac{1}{2} \times 0.5 \times 1\right) \times 0.01 = 18.75 \text{ kg}$$

Mass moment of inertia of a triangular plate about its base is

$$= \frac{1}{6} M h^2 = \frac{1}{6} \times 18.75 \times 1^2 = 3.125 \text{ kg m}^2$$

EXAMPLE 8.17

A rectangular column of section 30 cm × 50 cm and height 4 m is centrally cast over a concrete bed which measures 3 m × 5 m and thickness 40 cm. If the mass density of concrete is 2500 kg/m^3*, make calculations for the mass moment of inertia of the column and bed combination about a pole.*

Solution : Mass of column M_c = density × volume

$$= 2500 \times (0.3 \times 0.5 \times 4) = 1500 \text{ kg}$$

mass moment of inertia of the column about the pole

$$= \frac{1}{2} M(d^2 + b^2) = \frac{1}{12} \times 1500 \times (0.3^2 + 0.5^2) = 42.5 \text{ kg m}^2$$

Mass of the concrete bed = 2500 × (3 × 5 × 0.4) = 15000 kg

mass moment of inertia of concrete bed about the pole

$$= \frac{1}{12} \times 15000 \times (3^2 + 5^2) = 42500 \text{ kg m}^2$$

The mass moment of inertia of the composite body is given by the sum of mass moment of inertia of the column and the mass moment of inertia of the bed.

Mass moment of inertia of composite section

$$= 42.5 + 42500 = 42542.5 \text{ kg m}^2$$

EXAMPLE 8.18

Compute I_{xx}, I_{yy} and I_{zz} for the homogeneous rectangular parallelopiped shown below in Fig. 8.42.

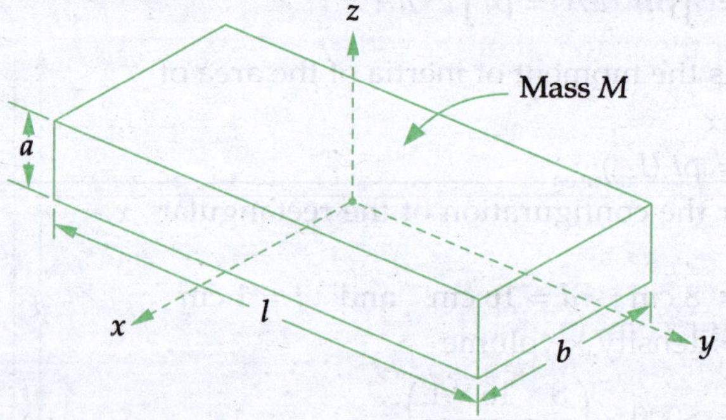

Fig. 8.42

Solution :

$$I_{xx} = \rho \iiint_V (y^2 + z^2)\, dv$$

$$= \rho \int (y^2 + z^2)\, dx\, dy\, dz \qquad \text{where } dv = dx\, dy\, dz$$

$$= \rho \int_{x=-b/2}^{b/2} \int_{y=-l/2}^{l/2} \int_{z=-a/2}^{a/2} (y^2 + z^2)\, dx\, dy\, dz$$

$$= \rho \int_{y=-l/2}^{l/2} \int_{z=-a/2}^{a/2} (y^2 + z^2)\, dy\, dz \left| \frac{x}{1} \right|_{-b/2}^{b/2}$$

$$= \rho b \int_{y=-l/2}^{l/2} \int_{z=-a/2}^{a/2} (y^2 + z^2)\, dy\, dz$$

$$= \rho b \int_{z=-a/2}^{a/2} \left| \frac{y^3}{3} + z^2 y \right|_{-l/2}^{l/2} dz = \rho b \int_{z=-a/2}^{a/2} \left(\frac{l^3}{12} + z^2 l \right) dz$$

$$= \rho b \left| \frac{l^3}{12} z + \frac{z^3}{3} l \right|_{-a/2}^{a/2} = \frac{\rho abl}{12} (l^3 + a^2) = \frac{M}{12} (l^2 + a^2)$$

where $M = \rho\,(abl) = \rho V$ is the mass of the parallelopiped.

Following the same steps, I_{yy} and I_{zz} can be worked out as

$$I_{yy} = \rho \iiint_V (x^2 + z^2)\,dV = \frac{M}{12}(b^2 + a^2)$$

and $$I_{zz} = \rho \iiint_V (x^2 + y^2)\,dV = \frac{M}{12}(b^2 + l^2)$$

EXAMPLE 8.19

The flat surface of hemisphere of radius R is connected to one flat surface of a cylinder of radius R and length l and made of same material. If the total mass be M, show that the moment of inertia of combination about the axis of cylinder is

$$\frac{MR^2(l/2 + 4R/15)}{l + 2R/3}$$

Solution : Refer Fig. 8.43, for the arrangement of the hemisphere and the cylinder.

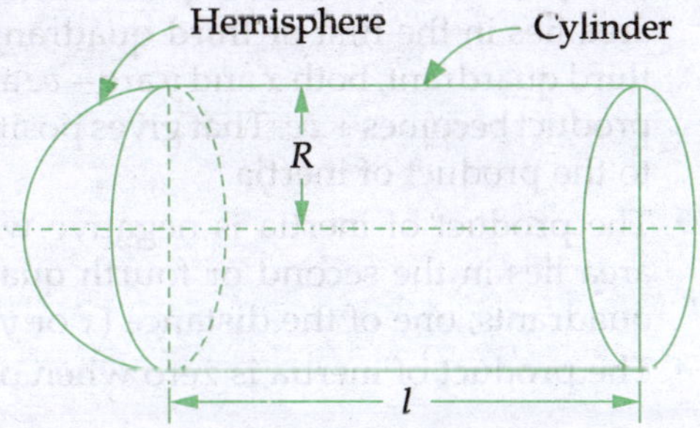

Fig. 8.43

Let ρ be the density of the material of the given solids. Then mass of the system

$$M = \frac{2}{3}\pi R^2 \rho + \pi R^2 l \rho$$

or $$\rho = \frac{M}{\pi R^2 (l + 2R/3)}$$

Moment of inertia of the combination,

$$I_{xx} = I_{xx} \text{ of hemisphere} + I_{xx} \text{ of solid cylinder}$$

For a cylinder of mass m and radius $r : I = \frac{1}{2}mr^2$ and

For a sphere of mass m and radius $r : I = \frac{2}{5}mr^2$

$$\therefore \quad I_{xx} = \frac{2}{5}\left(\frac{2}{3}\pi R^3 \rho\right)R^2 + \frac{1}{2}\left(\pi R^2 \rho l\right)R^2 = \left(\frac{4}{15}\pi R^5 + \frac{1}{2}\pi R^4 l\right)\rho$$

$$= \left(\frac{4}{15}\pi R^5 + \frac{1}{2}\pi R^4 l\right) \times \frac{M}{\pi R^2 (l + 2R/3)} = \left[\frac{4}{15}MR^3 + \frac{1}{2}MR^2 l\right] \times \frac{1}{l + 2R/3}$$

$$= \frac{MR^2(4R/15 + l/2)}{(l + 2R/3)}$$ which is the required expression.

8.5. PRODUCT OF INERTIA

Consider an elemental area dA which represents one of the small areas that comprise the plane area A. Let this elemental area be at a distance x and y from y-axis and x-axis respectively.

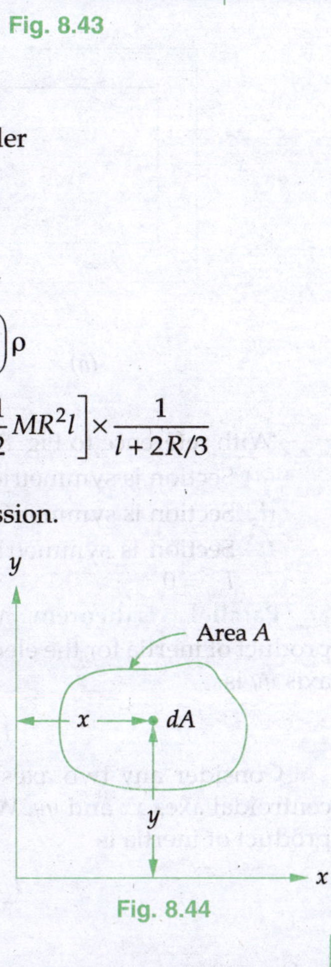

Fig. 8.44

The moment of this area about x-axis is $y\,dA$. Further, the moment of ydA about y-axis is $xy\,dA$. Then the term $xy\,dA$ is called the *product of inertia* of area dA with respect to x-axis and y-axis. The integral $\int xy\,dA$ is called the product inertia of the entire area A about the x and y axis. The product of inertia is denoted by I_{xy}. Thus,

$$I_{xy} = \int xy\, dA \qquad ...(8.32)$$

The unit for product of inertia is m^4.

Salient features :

- The moment of inertia I_{xx}, I_{yy} is always positive.
- The product of inertia I_{xy} may be either positive or negative or it may be zero depending on the location of the area relative to the axis.
- The product of inertia is positive when the area lies in the first or third quadrant. In the third quardrant, both x and y are – *ve* and their product becomes + *ve*. That gives positive sign to the product of inertia.

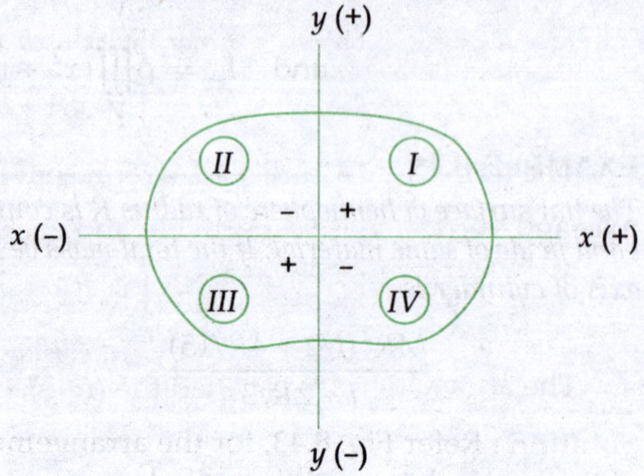

Fig. 8.45

- The product of inertia is negative when the area lies in the second or fourth quadrant. This aspect stems form the fact that in these quadrants, one of the distance (x or y) is –*ve* and that makes the product xy negative.
- The product of inertia is zero when one or both of the x and y are axis of symmetry.

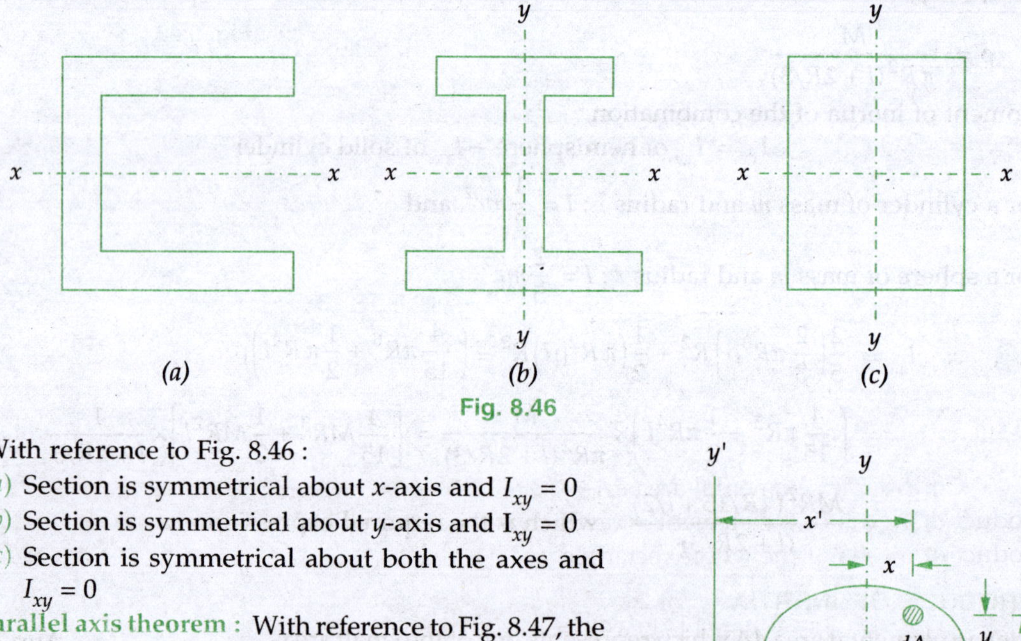

Fig. 8.46

With reference to Fig. 8.46 :

(*a*) Section is symmetrical about x-axis and $I_{xy} = 0$

(*b*) Section is symmetrical about y-axis and $I_{xy} = 0$

(*c*) Section is symmetrical about both the axes and $I_{xy} = 0$

Parallel axis theorem : With reference to Fig. 8.47, the product of inertia for the elemental area about axis xx and axis yy is

$$I_{xy} = \int xy\, dA$$

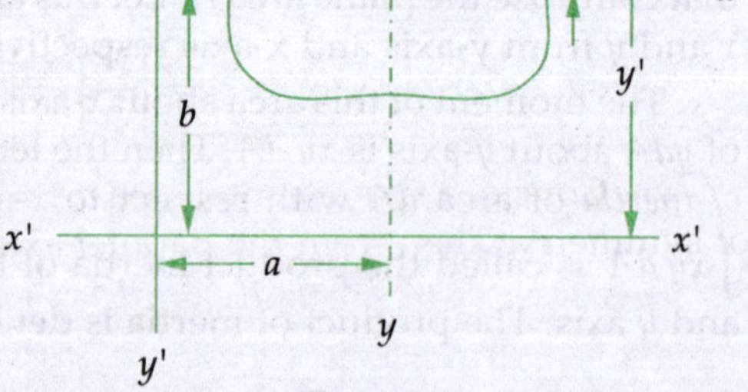

Fig. 8.47

Consider any two axes $x'x'$ and $y'y'$ parallel to the centroidal axes xx and yy. With respect to these axes, the product of inertia is

$$I_{x'y'} = \int x'\, y'\, dA$$

It may be noted from Fig. 8.47, $x' = x + a$ and $y' = y + b$

$$\therefore \quad I_{x'y'} = \int (x+a)(y+b)\,dA$$
$$= \int (xy + xb + ay + ab)\,dA$$
$$= \int xy\,dA + b\int x\,dA + a\int y\,dA + ab\int dA$$

The terms $\int x\,dA$ and $\int y\,dA$ represent the first moment of area about the centroidal horizontal axis and vertical axis respectively and each equals zero. That gives

$$I_{x'y'} = \int xy\,dA + ab\int dA$$
$$= I_{xy} + abA \qquad \ldots(8.34)$$

The above identity is called the *transfer formula* for the product of inertia.

EXAMPLE 8.20

(a) Set up a relation for the product of inertia of a rectangle of width b and height h with respect to axes shown in Fig. 8.48.

(b) A right angled triangle has width b and height h. Show that its product of inertia with respect to x and y axes shown in Fig. 8.49 is given by $b^2h^2/2y$.

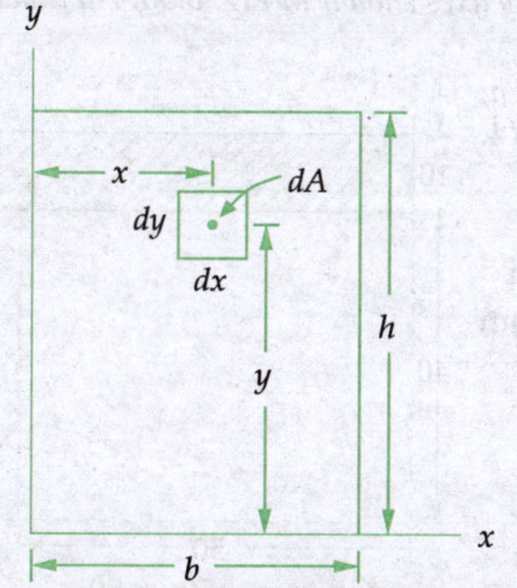

Fig. 8.48

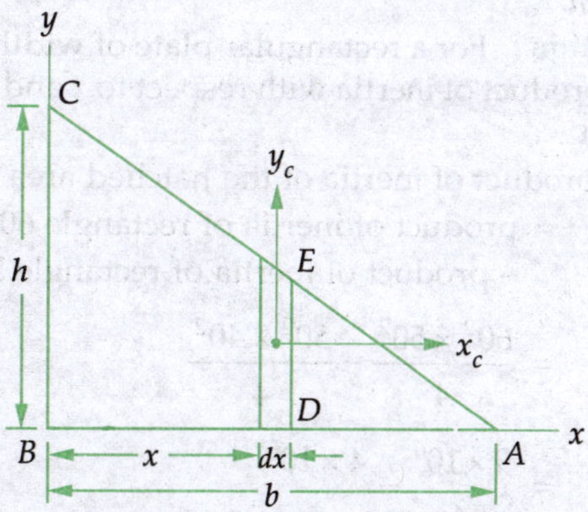

Fig. 8.49

Solution : Consider an elemental area $dA = dx\,dy$ of the rectangle with coordinates x and y.

Product of inertia of the element $= xy\,dA = xy\,dx\,dy$

Product of inertia of the whole rectangle

$$= \int_0^b\int_0^h xy\,dx\,dy = \int_0^b x\,dx \times \int_0^h y\,dy$$

$$= \left|\frac{x^2}{2}\right|_0^b \times \left|\frac{y^2}{2}\right|_0^h = \frac{b^2}{2} \times \frac{h^2}{2} = \frac{1}{4}b^2h^2$$

(b) Consider an elemental area $dA = y\,dx$ of the triangle. For this element x_c and y_c are the axes of symmetry. Then from the parallel axis theorem

$$dI_{xy} = dI_{x_c y_c} + x_c y_c\,dA$$
$$= 0 + \left(x \times \frac{y}{2}\right)(y\,dx) = \frac{xy^2}{2}\,dx$$

Since the integration is to be done with respect to x within the limit 0 to b, it is neces... express y in terms of x. For that we have the following correlation from the similarity of triangle ADE and ABC.

$$\frac{y}{h} = \frac{b-x}{b}; \quad y = h\left(1 - \frac{x}{b}\right)$$

Then $$dI_{xy} = \frac{x}{2} h^2 \left(1 - \frac{x}{b}\right)^2 dx = \frac{h^2}{2}\left(x + \frac{x^3}{b^2} - \frac{2x^2}{b}\right) dx$$

$\therefore$ $$I_{xy} = \frac{h^2}{2}\int_0^b \left(x + \frac{x^3}{b^2} - \frac{2x^2}{b}\right) dx$$

$$= \frac{h^2}{2}\left|\frac{x^2}{2} + \frac{x^4}{4b^2} - \frac{2x^3}{3b}\right|_0^b$$

$$= \frac{h^2}{2}\left(\frac{b^2}{2} + \frac{b^2}{4} - \frac{2b^2}{3}\right) = \frac{1}{24} b^2 h^2$$

EXAMPLE 8.21

Determine the product of inertia of the sectioned area about xy axes shown in Fig. 8.50. All dimensions are in mm.

Solution : For a rectangular plate of width b and height h, the product of inertia with respect to x and y axes is $b^2h^2/4$. Then

product of inertia of the hatched area
= product of inertia of rectangle 60 mm × 50 mm
– product of inertia of rectangle 50 mm × 40 mm

$$= \frac{60^2 \times 50^2}{4} - \frac{50^2 \times 40^2}{4}$$

$$= \frac{9 \times 10^6}{4} - \frac{4 \times 10^6}{4}$$

$= 1.25 \times 10^6 \text{ mm}^4$

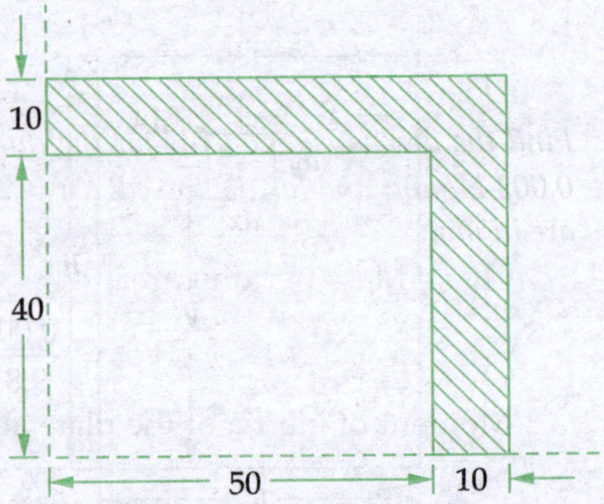

Fig. 8.50

EXAMPLE 8.22

Determine the product of inertia of the channel section (Fig. 8.51) with respect to its centroidal axis. All dimensions are in mm. Comment on the result.

Solution : The section consists of three rectangular segments which have been numbered as 1, 2 and 3

Rectangle 1 :

$$a_1 = 40 \times 10 = 400 \text{ mm}^2; \quad x_1 = \frac{40}{2} = 20 \text{ mm}$$

Rectangle 2 :

$$a_2 = 100 \times 10 = 1000 \text{ mm}^2; \quad x_2 = \frac{10}{2} = 5 \text{ mm}$$

Rectangle 3 :

$$a_3 = 40 \times 10 = 400 \text{ mm}^2; \quad x_3 = \frac{40}{2} = 5 \text{ mm}$$

Then $\bar{x} = \dfrac{a_1x_1 + a_2x_2 + a_3x_3}{a_1 + a_2 + a_3}$

$$= \frac{400 \times 20 + 1000 \times 5 + 400 \times 20}{400 + 1000 + 400}$$

$$= \frac{21000}{1800} = 11.67 \text{ mm}$$

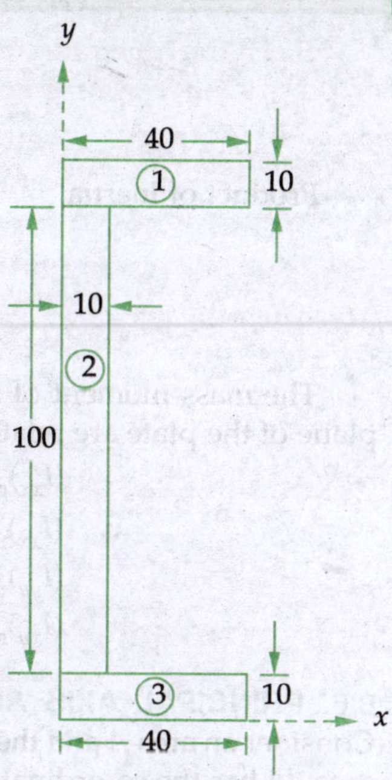

Fig. 8.51

Further the given channel section is symmetrical about x-axis and therefore

$$\bar{y} = \frac{100 + 10 + 10}{2} = 60 \text{ mm}$$

Product of inertia about centroidal axis

$= a_1(x_1 - \bar{x})(y_1 - \bar{y}) + a_2(x_2 - \bar{x})(y_2 - \bar{y}) + a_3(x_3 - \bar{x})(y_3 - \bar{y})$

$= 400(20 - 11.67)\ (115 - 60) + 1000\ (5 - 11.67)(60 - 60)$
$+ 400(20 - 11.67)(5 - 60)$

$= 183260 + 0 - 183260 = 0$

Comment : The section is symmetrical about x-axis. Hence the product of inertia has to be zero.

EXAMPLE 8.23

Find the inertia components for the thin plate shown in Fig. 8.52 given below. The weight of the plate is 0.002 N/mm² and for the top edge $y = 2\sqrt{x}$ where dimensions x and y are in mm.

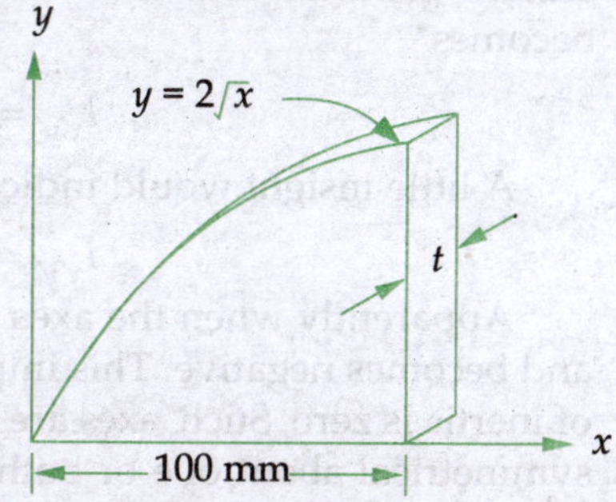

Fig. 8.52

Solution : Mass per unit area,

$$\rho t = \frac{0.002}{9.81} = 0.000204 \text{ kg/mm}^2$$

Moment of inertia of the plate about x-axis,

$$(I_{xx})_{\text{area}} = \int_0^{100} \int_{y=0}^{y=2\sqrt{x}} y^2 dy\, dx$$

$$= \int_0^{100} \left| \frac{y^3}{3} \right|_0^{2\sqrt{x}} dx = \int_0^{100} \frac{8}{3} x^{3/2}\, dx$$

$$= \frac{8}{3} \left| \frac{x^{5/2}}{5/2} \right|_0^{100} = \frac{8}{3} \times \frac{2}{5} (100)^{5/2} = 1.067 \times 10^5 \text{ mm}^4$$

Moment of inertia of the plate about y-axis,

$$(I_{yy})_{\text{area}} = \int_0^{100} \int_{y=0}^{y=2\sqrt{x}} x^2 dy\, dx = \int_0^{100} x^2 \left| \frac{y}{1} \right|_0^{2\sqrt{x}} dx = \int_0^{100} x^2\, 2\sqrt{x}\, dx$$

$$= \int_0^{100} 2\, x^{5/2}\, dx = 2 \left| \frac{x^{7/2}}{7/2} \right|_0^{100}$$

$$= 2 \times \frac{2}{7} \times (100)^{7/2} = 5.71 \times 10^6 \text{ mm}^4$$

Product of inertia, $$I_{xy} = \int_0^{100} \int_{y=0}^{y=2\sqrt{x}} xy \, dx \, dy = \int_0^{100} x \times \left| \frac{y^2}{2} \right|_0^{2\sqrt{x}} dx$$

$$= \int_0^{100} 2x^2 \, dx = 2 \left| \frac{x^3}{3} \right|_0^{100} = 6.67 \times 10^5 \text{ mm}^4$$

The mass moment of inertia and the moment of inertia of a plate about an axis lying in the plane of the plate are related as

$$(I_{xx})_{mass} = \rho t \, (I_{xx})_{area}; \quad (I_{yy})_{mass} = \rho t \, (I_{yy})_{area} \quad \text{and} \quad (I_{xy})_{mass} = \rho t \, (I_{xy})_{area}$$

$$\therefore \quad (I_{xx})_{mass} = 0.000204 \times (1.067 \times 10^5) = \mathbf{21.77 \text{ kg mm}^2}$$

$$(I_{yy})_{mass} = 0.000204 \times (5.71 \times 10^6) = \mathbf{1164.84 \text{ kg mm}^2}$$

$$(I_{xy})_{mass} = 0.000204 \times (6.67 \times 10^5) = \mathbf{136.1 \text{ kg mm}^2}$$

8.6. PRINCIPAL AXIS AND PRINCIPAL MOMENT OF INERTIA

Consider an area A and the co-ordinate axis xx and yy passing through the centroid C. An elemental area dA has the co-ordinates x and y with respect to these axes. The product of inertia of the plane area is then given by

$$I_{xy} = \int xy \, dA$$

When the axes are rotated through 90 degree anti-clockwise, then elemental area shifts to the second quadrant and its ordinates take up the positions x' and y'. The product of inertia then becomes

$$I_{x'y'} = \int x'y' \, dA$$

A little insight would indicate that $x' = y$ and $y' = -x$. That gives

$$I_{x'y'} = -\int xy \, dA = -I_{xy}$$

Apparently when the axes are rotated, the product of inertia for the plane area changes sign and becomes negative. This implies that there are certain directions of axes for which the product of inertia is zero. Such axes are referred to as the *principal axes* of the area. When the plane figure is symmetrical about one or both the axes, the product of inertia about its centroidal axes is zero. Obviously the centroidal axes are the principal axes.

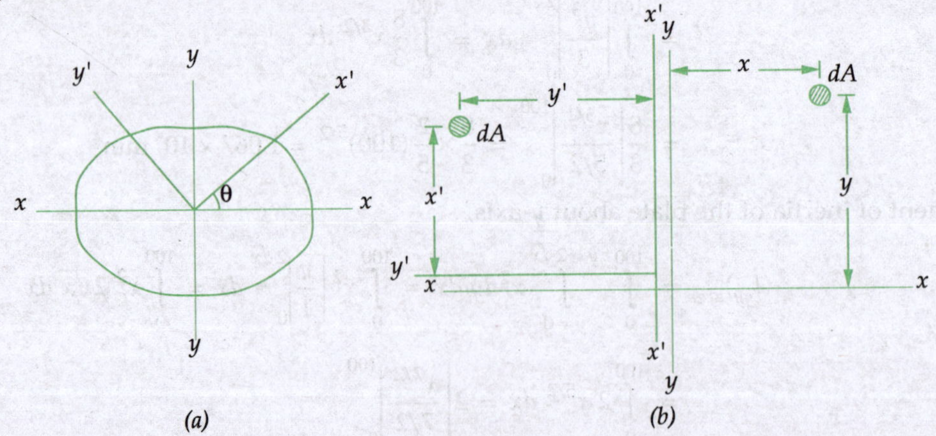

Fig. 8.53

There will always be two principal axes at a given point in the area and they will be mutually perpendicular to each other. The moment of inertia about the principal axes is called the *principal moment of inertia*. The moment of inertia about one of the axis will be maximum and the other will be minimum. The maximum moment of inertia is called the major principal moment of inertia, and the minimum moment of inertia is called the minor principal moment of inertia.

8.6.1. Principal moment of inertias and the location of principal axes

With reference to Fig. 8.54,

- *xx* and *yy* are the axes through the centroid *C* of the plane area *A*
- *x′x′* and *y′y′* are another coordinate axes inclined at angle θ to the axes *xx* and *yy*
- *dA* is an elemental area with coordinates (*x*, *y*) with respect to axes *xx* and *yy*, and coordinates (*x′*, *y′*) with respect to axes *x′x′* and *y′y′*.

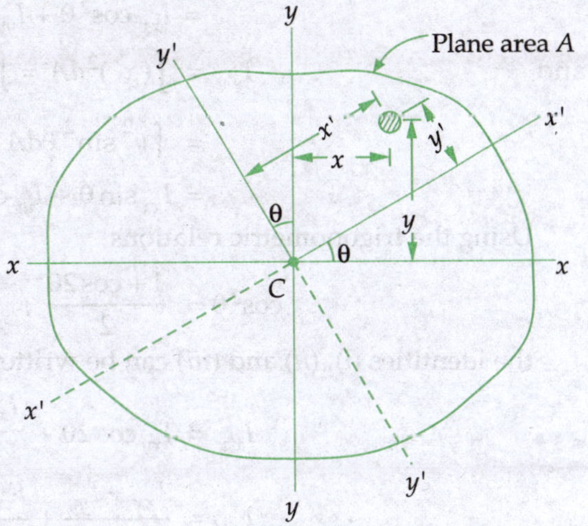

Fig. 8.54.

The following correlations apply to *x*, *y*, *x′*, *y′* and angle of rotation θ (Fig. 8.55)

$x' = EB + BP = CF + BP = y\sin\theta + x\cos\theta$

$y' = CE = BF = AF - AB = y\cos\theta - x\sin\theta$

We know:

$$I_{x'y'} = \int x'y'\,dA = \int (x\cos\theta + y\sin\theta)(y\cos\theta - x\sin\theta)\,dA$$

$$= \int xy\cos^2\theta\,dA - \int x^2\sin\theta\cos\theta\,dA + \int y^2\sin\theta\cos\theta\,dA - \int xy\sin^2\theta\,dA$$

Since $\int y^2\,dA = I_{xx}$; $\int x^2\,dA = I_{yy}$ and $\int xy\,dA = I_{xy}$

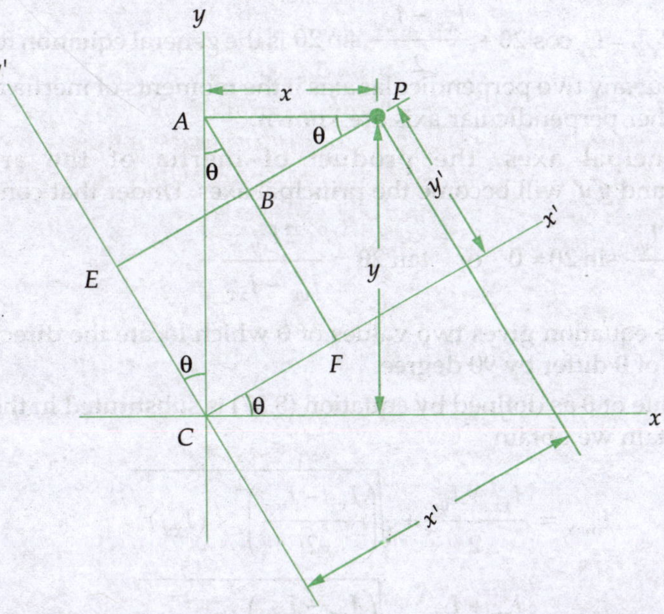

Fig. 8.55

we may write

$$I_{x'y'} = I_{xy}\cos^2\theta - I_{yy}\sin\theta\cos\theta + I_{xx}\sin\theta\cos\theta - I_{xy}\sin^2\theta \quad ...(i)$$

and
$$I_{x'x'} = \int (y')^2 dA = \int (y\cos\theta - x\sin\theta)^2 dA$$
$$= \int y^2\cos^2\theta\, dA + \int x^2\sin^2\theta\, dA - \int 2\,xy\sin\theta\cos\theta\, dA$$
$$= I_{xx}\cos^2\theta + I_{yy}\sin^2\theta - 2I_{xy}\sin\theta\cos\theta \quad ...(ii)$$

and
$$I_{y'y'} = \int (x')^2 dA = \int (y\sin\theta - x\cos\theta)^2 dA$$
$$= \int y^2\sin^2\theta\, dA + \int x^2\cos^2\theta\, dA - \int 2xy\sin\theta\cos\theta\, dA$$
$$= I_{xx}\sin\theta + I_{yy}\cos^2\theta + 2I_{xy}\sin\theta\cos\theta \quad ...(iii)$$

Using the trigonometric relations

$$\cos^2\theta = \frac{1+\cos 2\theta}{2}\ ; \quad \sin^2\theta = \frac{1-\cos 2\theta}{2} \quad \text{and} \quad 2\sin\theta\cos\theta = \sin 2\theta,$$

the identities (*i*), (*ii*) and (*iii*) can be written as

$$I_{x'y'} = I_{xy}\cos 2\theta + \frac{I_{xx} - I_{yy}}{2}\sin 2\theta \quad ...(8.34)$$

$$I_{x'x'} = \frac{I_{xx} + I_{yy}}{2} + \frac{I_{xx} - I_{yy}}{2}\cos 2\theta - I_{xy}\sin 2\theta \quad ...(8.35)$$

and
$$I_{y'y'} = \frac{I_{xx} + I_{yy}}{2} - \frac{I_{xx} - I_{yy}}{2}\cos 2\theta + I_{xy}\sin 2\theta \quad ...(8.36)$$

- It is observed that $I_{x'x'} + I_{y'y'} = I_{xx} + I_{yy}$. Apparently the sum of moments of inertia about any two perpendicular axes remains constant. This characteristic is called the *invariant property*.
- The identity $I_{x'y'} = I_{xy}\cos 2\theta + \frac{I_{xx} - I_{yy}}{2}\sin 2\theta$ is the general equation for finding the product of inertia about any two perpendicular axis if the moments of inertia and product of inertia about two other perpendicular axis are known.

About the principal axes, the product of inertia of the area is zero. When $I_{x'y'} = 0$, the axes $x'x'$ and $y'y'$ will become the principal axes. Under that condition

$$I_{xy}\cos 2\theta + \frac{I_{xx} - I_{yy}}{2}\sin 2\theta = 0 \quad \text{or} \quad \tan 2\theta = \frac{2I_{xy}}{I_{yy} - I_{xx}} \quad ...(8.37)$$

Solution to above equation gives two values of θ which locate the directions of the principal axes. The two values of θ differ by 90 degree.

- When the value of θ as defined by equation (8.37) is substituted in the identities (8.35) and (8.36), we obtain we obtain

$$I_{max} = \frac{I_{xx} + I_{yy}}{2} + \sqrt{\left(\frac{I_{xx} - I_{yy}}{2}\right)^2 + (I_{xy})^2}$$

$$I_{min} = \frac{I_{xx} + I_{yy}}{2} - \sqrt{\left(\frac{I_{xx} - I_{yy}}{2}\right)^2 + (I_{xy})^2} \quad ...(8.38)$$

where I_{max} and I_{min} are the major and minor principal moment of inertias.

8.6.2. Mohr's circle for principle moments of inertia

With reference to Fig. 8.56, the following procedure is adopted for the construction of Mohr's circle for principal moments of inertia :

(*i*) Taking O as origin, measure $OL = I_{yy}$ and $OM = I_{xx}$ (according to some scale) along the horizontal I-axis.

(*ii*) At point L, draw upward perpendicular if I_{xy} is positive and measure $LP = I_{xy}$. If I_{xy} is negative, draw the perpendicular downward.

(*iii*) Bisect LM at N.

(*iv*) With N as centre and NP as radius, draw a circle meeting OL and M extended at points A and B respectively.

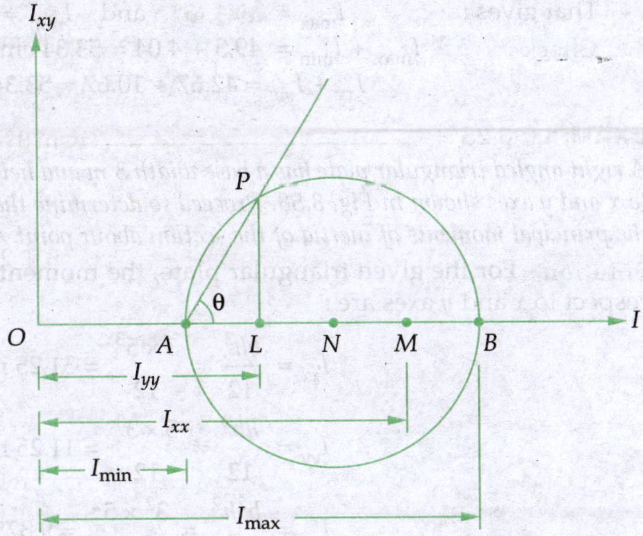

Fig. 8.56

Then OA gives the minor principal moment of inertia and OB gives major principal moment of inertia. The direction of minor principal axis lies along AP and the direction of major principal axis is at right angle to it.

EXAMPLE 8.24

Find the product of inertia of the rectangular section shown in Fig. 8.57, with respect to x and y axes. Proceed to calculate the location of the principal axes and the values of the principal moments of inertia of the section about point A.

Solution : For the given rectangular plate, the moments of inertia and the product of inertia with respect to x and y axes are :

$$I_{xx} = \frac{bh^3}{3} = \frac{2 \times 4^3}{3} = 42.67 \text{ m}^4$$

$$I_{yy} = \frac{hb^3}{3} = \frac{4 \times 2^3}{3} = 10.67 \text{ m}^4$$

$$I_{xy} = \frac{b^2h^2}{4} = \frac{2^2 \times 4^2}{4} = 16 \text{ m}^4$$

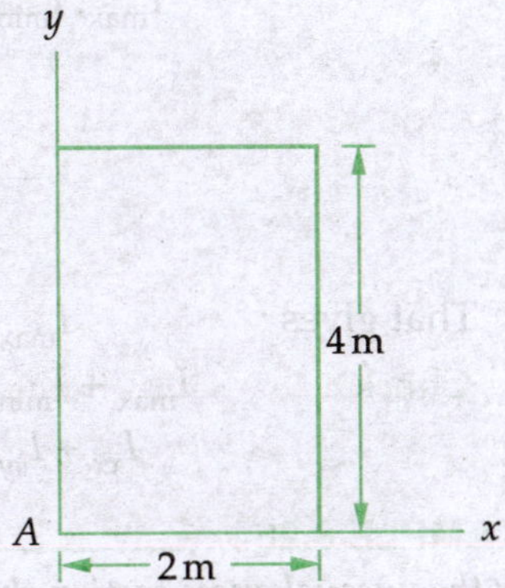

Fig. 8.57

Location of principal axes :

$$\tan 2\theta = \frac{2I_{xy}}{I_{yy} - I_{xx}} = \frac{2 \times 16}{10.67 - 42.67} = -1.0$$

$$2\theta = 315^\circ ; \quad \theta_1 = 157.5^\circ$$

$$\theta_2 = 157.5^\circ - 90^\circ = 67.5^\circ$$

Principal moments of inertia :

$$I_{max}, I_{min} = \frac{I_{xx} + I_{yy}}{2} \pm \sqrt{\left(\frac{I_{xx} - I_{yy}}{2}\right)^2 + \left(I_{xy}\right)^2}$$

$$= \frac{42.67 + 10.67}{2} \pm \sqrt{\left(\frac{42.67 - 10.67}{2}\right)^2 + 16^2}$$

$$= 26.67 \pm 22.63$$

That gives : $I_{max} = 49.3 \text{ m}^4$ and $I_{min} = 4.04 \text{ m}^4$

Check : $I_{max} + I_{min} = 49.3 + 4.04 = 53.34 \text{ cm}^4$

$I_{xx} + I_{yy} = 42.67 + 10.67 = 53.34 \text{ cm}^4$

EXAMPLE 8.25

A right angled triangular plate has a base width 3 m and height 5 m. Find the product of inertia with respect to x and y axes shown in Fig. 8.58. Proceed to determine the location of the principal axes and the values of the principal moments of inertia of the section about point A.

Solution : For the given triangular plate, the moments of inertia and the product of inertia with respect to x and y axes are :

$$I_{xx} = \frac{bh^3}{12} = \frac{3 \times 5^3}{12} = 31.25 \text{ m}^4$$

$$I_{yy} = \frac{hb^3}{12} = \frac{5 \times 3^3}{12} = 11.25 \text{ m}^4$$

$$I_{xy} = \frac{b^2h^2}{24} = \frac{3^2 \times 5^2}{24} = 9.375 \text{ m}^4$$

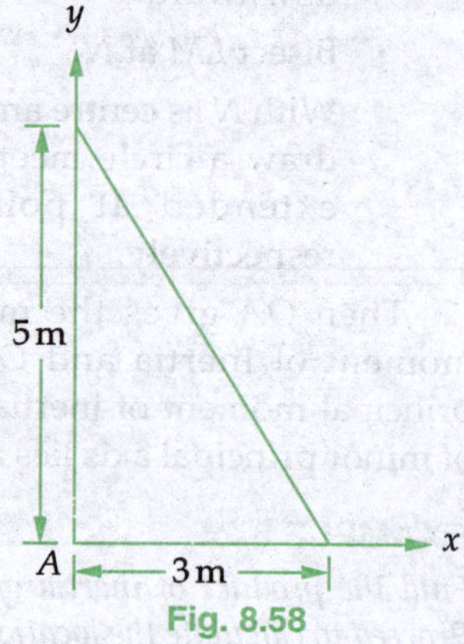

Fig. 8.58

Location of principal axes :

$$\tan 2\theta = \frac{2 I_{xy}}{I_{yy} - I_{xx}} = \frac{2 \times 9.375}{11.25 - 31.25} = -0.9375$$

$$2\theta = 316.85^\circ ; \quad \theta_1 = 158.42^\circ$$

$$\theta_2 = 158.42^\circ - 90^\circ = 68.42^\circ$$

Principal moments of inertia :

$$I_{max}, I_{min} = \frac{I_{xx} + I_{yy}}{2} \pm \sqrt{\left(\frac{I_{xx} - I_{yy}}{2}\right)^2 + (I_{xy})^2}$$

$$= \frac{31.25 + 11.25}{2} \pm \sqrt{\left(\frac{31.25 - 11.25}{2}\right)^2 + (9.375)^2}$$

$$= 21.25 \pm 13.71$$

That gives : $I_{max} = 34.96 \text{ m}^4$ and $I_{min} = 7.54 \text{ m}^4$

Check : $I_{max} + I_{min} = 34.96 + 7.54 = 52.50 \text{ cm}^4$

$I_{xx} + I_{yy} = 31.25 + 11.25 = 52.50 \text{ cm}^4$

EXAMPLE 8.26

For the unequal angle section shown in Fig. 8.59, determine the location of principal axes and the principal moments of inertia.

Solution : The given section is split into two parts marked 1 and 2.

Location of centroidal axis (with respect to faces AB and BC)

$$\bar{x} = \frac{a_1x_1 + a_2x_2}{a_1 + a_2} = \frac{10 \times 5 + 14 \times 0.5}{10 + 14} = 2.375 \text{ cm}$$

$$\bar{y} = \frac{a_1 y_1 + a_2 y_2}{a_1 + a_2} = \frac{10 \times 0.5 + 14 \times 8}{10 + 14} = 4.875 \text{ cm}$$

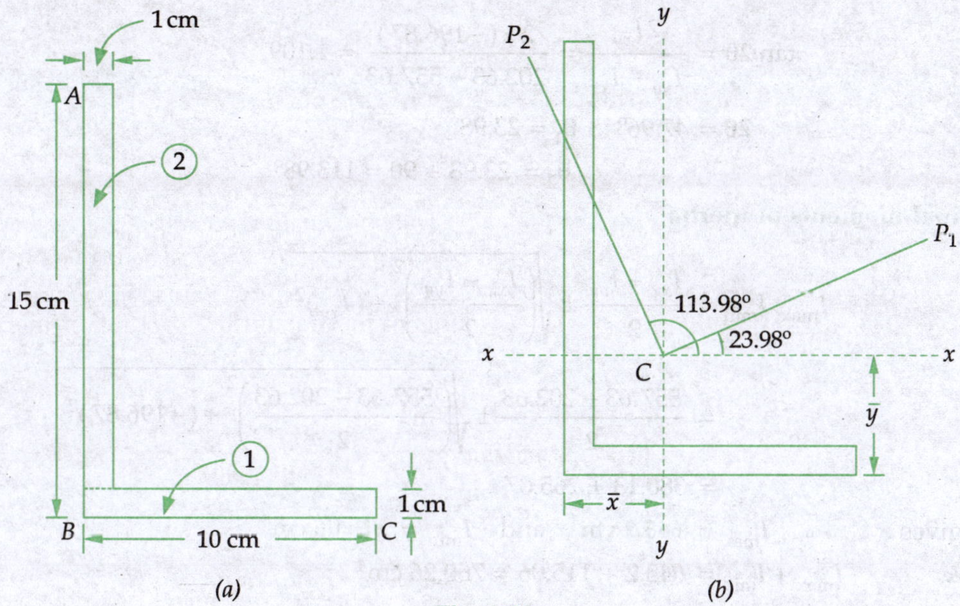

Fig. 8.59

Calculations for moments of inertia and product of inertia

$$(I_{xx})_1 = \frac{b h^3}{12} + A(\bar{y} - y_1)^2$$

$$= \frac{1}{12} \times 10 \times 1^3 + (10 \times 1) \times (4.875 - 0.5)^2 = 192.24 \text{ cm}^4$$

$$(I_{xx})_2 = \frac{1}{12} \times 1 \times 14^3 + (14 \times 1) \times (4.875 - 8)^2 = 365.39 \text{ cm}^4$$

That gives : $I_{xx} = 192.24 + 365.39 = 557.63 \text{ cm}^4$

$$(I_{yy})_1 = \frac{b h^3}{12} + A(\bar{x} - x)^2$$

$$= \frac{1}{12} \times 1 \times 10^3 + (10 \times 1) \times (2.375 - 5)^2 = 152.24 \text{ cm}^4$$

$$(I_{yy})_2 = \frac{1}{12} \times 1^3 \times 14 + (14 \times 1) \times (2.375 - 0.5)^2 = 50.39 \text{ cm}^4$$

That gives : $I_{yy} = 152.24 + 50.39 = 202.63 \text{ cm}^4$

$$(I_{xy})_1 = 0 + (10 \times 1) \times (5 - 2.375)(0.5 - 4.875) = -114.84 \text{ cm}^4$$

$$(I_{xy})_2 = 0 + (14 \times 1) \times (8 - 4.875) \times (0.5 - 2.375) = -82.03 \text{ cm}^4$$

The above results for product of inertia have been written by using parallel axis theorem and noting that product of inertia vanishes if any one of the axes is the axis of symmetry.

That gives : $I_{xy} = -114.84 - 82.03 = -196.87 \text{ cm}^4$

Location of principal axes

$$\tan 2\theta = \frac{2I_{xy}}{I_{yy} - I_{xx}} = \frac{2 \times (-196.87)}{202.63 - 557.63} = 1.109$$

$$2\theta = 47.96^\circ\,; \quad \theta_1 = 23.98^\circ$$

$$\theta_2 = 23.98 + 90 = 113.98^\circ$$

Principal moments of inertia

$$I_{max}, I_{min} = \frac{I_{xx} + I_{yy}}{2} \pm \sqrt{\left(\frac{I_{xx} - I_{yy}}{2}\right)^2 + (I_{xy})^2}$$

$$= \frac{557.63 + 202.63}{2} \pm \sqrt{\left(\frac{557.63 - 202.63}{2}\right)^2 + (-196.87)^2}$$

$$= 380.13 \pm 265.07$$

That gives : $I_{max} = $ **645.2 cm^4** and $I_{min} = $ **115.06 cm^4**

Check : $I_{max} + I_{min} = 645.2 + 115.06 = 760.26 \text{ cm}^4$

$I_{xx} + I_{yy} = 557.63 + 202.63 = 760.26 \text{ cm}^4$

EXAMPLE 8.27

For the centrally symmetrical z section shown in Fig. 8.60, make calculations for the followings :

(*a*) *product of inertia of the section with respect to the x and y axes,*

(*b*) *principal axes and the principal moments of inertia of the section about point P.*

All dimensions are in cm

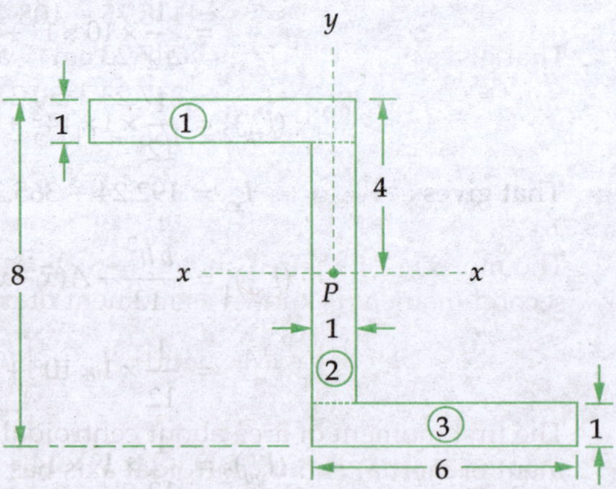

Fig. 8.60

Solution : The section can be split up into three rectangles marked 1, 2 and 3.

Moments of inertia

$$(I_{xx})_1 = \frac{1}{12} \times 60 \times 1^3 + (6 \times 1) \times (4 - 0.5)^2$$

$$= 74 \text{ cm}^4$$

$$(I_{xx})_2 = \frac{1}{12} \times 1 \times 6^3 = 18 \text{ cm}^4$$

$$(I_{xx})_3 = \frac{1}{12} \times 6 \times 1^3 + (6 \times 1) \times (4 - 0.5)^2$$

$$= 74 \text{ cm}^4$$

$$\therefore \quad I_{xx} = (I_{xx})_1 + (I_{xx})_2 + (I_{xx})_3 = 74 + 18 + 74$$

$$= 166 \text{ cm}^4$$

$$(I_{yy})_1 = \frac{1}{12} \times 1 \times 6^3 + (6 \times 1) \times (3 - 0.5)^2 = 55.5 \text{ cm}^4$$

$$(I_{yy})_2 = \frac{1}{12}\times 6\times 1^3 = 0.5 \text{ cm}^4$$

$$(I_{yy})_3 = \frac{1}{12}\times 1\times 6^3 + (6+1)\times(3-0.5)^2 = 55.5 \text{ cm}^4$$

$$\therefore\ I_{yy} = (I_{yy})_1 + (I_{yy})_2 + (I_{yy})_3$$

$$= 55.5 + 0.5 + 55.5 = 111.5 \text{ cm}^2$$

$$(I_{xy})_1 = 0 + (6\times 1)\times(3-5.5)(4-0.5) = -52.5 \text{ cm}^4$$

$$(I_{xy})_2 = 0$$

$$(I_{xy})_3 = (6\times 1)\times(5.5-3)\times(0.5-4) = -52.5 \text{ cm}^4$$

$$\therefore\ I_{xy} = (I_{xy})_1 + (I_{xy})_2 + (I_{xy})_3$$

$$= -52.5 + 0 - 52.5 = -105 \text{ cm}^4$$

(*b*) Position of principal axis

$$\tan 2\theta = \frac{2I_{xy}}{I_{yy}-I_{xx}} = \frac{2\times(-105)}{111.5-166} = 3.853$$

$$2\theta = 75.45^\circ;\quad \theta_1 = 37.72^\circ$$

$$\theta_2 = 37.72 + 90^\circ = 127.72^\circ$$

Principal moments of inertia

$$I_{max}, I_{min} = \frac{I_{xx}+I_{yy}}{2} \pm \sqrt{\left(\frac{I_{xx}-I_{yy}}{2}\right)^2 + (I_{xy})^2}$$

$$= \frac{166+111.5}{2} \pm \sqrt{\left(\frac{166-111.5}{2}\right)^2 + (-105)^2}$$

$$= 138.75 \pm 108.48$$

That gives : $I_{max} = 247.23 \text{ cm}^4$ and $I_{min} = 30.27 \text{ cm}^4$

Check : $I_{max} + I_{min} = 247.23 + 30.27 = 277.5 \text{ cm}^4$

$$I_{xx} + I_{yy} = 166 + 111.5 = 277.5 \text{ cm}^4$$

REVIEW AND SUMMARY

1. The moment of an area about any axis in its plane is called the first moment of area and the second moment is known as moment of inertia of the area. Mathematically

$$I_{xx} = \int_A x^2\,dA \quad \text{and} \quad I_{yy} = \int_A y^2\,dA$$

2. The first moment of area about centroidal axis is zero, while the second moment of area (moment of inertia) about centroidal axis has a minimum non-zero value.
3. The mass moment of inertia about an axis is the product of elemental mass and the square of the distance between the mass centre of the elemental mass and the axis.

$$\text{Mass moment of inertia} = \int x^2\,dm$$

4. Radius of gyration (*k*) of a body or a given body (or lamina) is defined as the distance from the axis of reference to where the whole mass (or area) of a body is assumed to be concentrated, and that is not to make any change in the moment of inertia about the given axis.

$$k_x = \sqrt{I_{xx}/A}\ ;\quad k_y = \sqrt{I_{yy}/A}$$

5. The modulus of section or section modulus of a figure represents the quantity that results when the moment of inertia of the figure about its centre of gravity is divided by the distance of the extremes fibre from the centroidal axis.
6. Parallel axis theorem: The moment of inertia of a plane lamina about any axis is equal to the sum of its MOI about a parallel axis through its centre of gravity and product of area (or mass) and the square of the distance between the two axis.

$$I_{AA} = I_{gxx} + Ah^2$$

7. Perpendicular axis theorem: The moment of inertia of a plane lamina about the axis perpendicular to the plane of the lamina is equal to the sum of moments of inertia of the lamina about the two axes at right angles and intersecting each other at the point where the perpendicular axis passes through it

$$I_{zz} = I_{xx} + I_{yy}$$

8. Polar moment of inertia is defined as the moment of inertia of the lamina or plane about the axis perpendicular to the plane of section. Invoking perpendicular axis theorem,

$$\text{Polar moment of inertia } I_{zz} \text{ (or } J) = I_{xx} + I_{yy}$$

Polar moment of inertia has application in problems relating to torsion of cylindrical shafts and rotation of slabs.
9. The product of inertia (I_{xy}) of an area with respect to x and y-axes is given by the integral of the product of elemental area dA and its central coordinates. That is

$$I_{xy} = \int xy \, dA$$

The salient properties of product of inertia are:
   (*i*) I_{xy} may be either positive or negative
   (*ii*) I_{xy} of the given area with respect to its principal axes is zero.
10. The principal axes of an area about a given point are the two mutually perpendicular axes about which the moments of inertia have the maximum and minimum values.

$$I_{max}, I_{min} = \sqrt{\frac{I_{xy} + I_{yy}}{2} \pm \left[\frac{I_x - I_y}{2}\right]^2 + I_{xy}^2}$$

I_{max} and I_{min} are respectively called the maximum and minimum principal moments of inertia of the area about the given point.

REVIEW QUESTIONS

A. Conceptual and conventional questions :

1. Define moment of inertia of an area. Why is it called the second moment of area?
2. Define radius of gyration. How is it related to moment of inertia?
3. State and explain
   (*a*) parallel axes theorem
   (*b*) perpendicular axes theorem concerning moment of inertia for a plane area.
4. What is meant by polar moment of inertia? State its applications.
5. Define moment of inertia of a mass. How radius of gyration of a body is expressed in terms of its mass moment of inertia
6. State parallel axis theorem as applied to mass moment of inertia.
7. (*a*) Define product of inertia. Mention its salient properties.
   (*b*) State and explain the parallel axis theorem for product of inertia.
   (*c*) State and explain the concepts of principal axis and principal moment of inertia.

8. Set up an expression for the mass moment of inertia of a uniform rod of length l about its
   (*i*) centroidal axis normal to the rod
   (*ii*) axis at the end of the rod and normal to it
9. Consider a semi-circular area of diameter 4 cm. Determine its moment of inertia about a tangent parallel to the diameter (Fig. 8.61).
   ($10.15\ cm^4$)

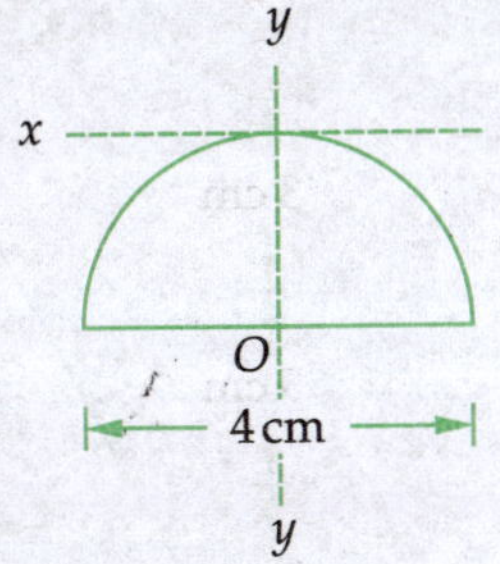

Fig. 8.61

10. Consider a square section *ABCD* having each side equal to 10 cm. What will be the ratio of the moment of inertia of this section about centroidal axis parallel to the side to that about the diagonal *AC*.
    (1)
11. Determine the moment of inertia of the section shown in Fig. 8.62, about the centroidal axis parallel to *yy*-axis. All dimensions are in cm.

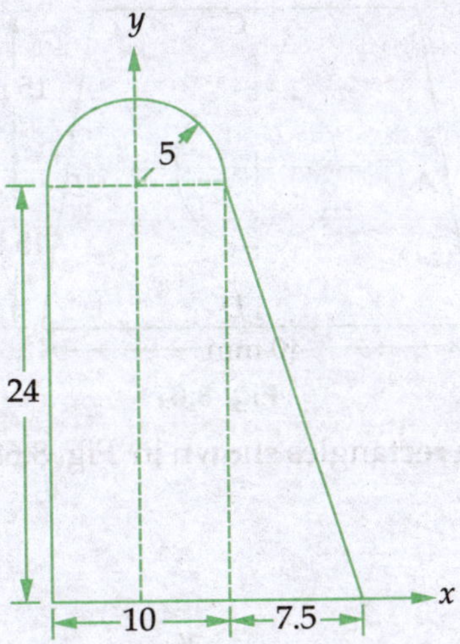

Fig. 8.62

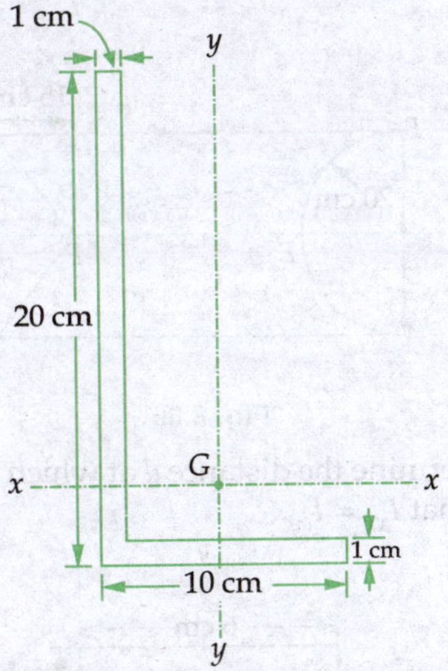

Fig. 8.63

12. Determine the moment of inertia about the centroidal axis for the unequal angle section shown in Fig. 8.63.
    ($1227.6\ cm^4$, $218.6\ cm^4$)
13. Determine the moment of inertia of I-section about the centroidal axis parallel to the flange.

    Top flange = 100 mm × 10 mm
    Bottom flange = 200 mm × 10 mm
    Web = 100 mm × 10 mm
    ($\bar{y}$ = 45.25 mm, I = $843 \times 10^4\ mm^4$)
14. A rectangular hole is made in a triangular section as shown in Fig. 8.64. Determine the moment of inertia of the section :
    (*a*) about base *AB*
    (*b*) about the centroidal axis parallel to base *AB*.
15. Determine the moment of inertia of the z-section (Fig. 8.65) about the centroidal axis *x* - *x* and *y* - *y*. All dimensions are in cm.
16. For the area shown in Fig. 8.66 calculate the moment of inertia about x-axis. What will the corresponding radius of gyration?
17. Find the moment of inertia of the lamina with a circular hole of 15 mm diameter about the axes *AB* and *CD* as shown in Fig. 8.67. All the dimensions are in mm.

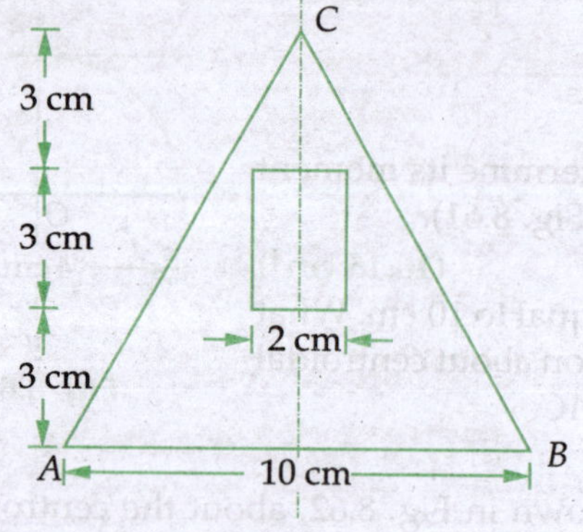

Fig. 8.64

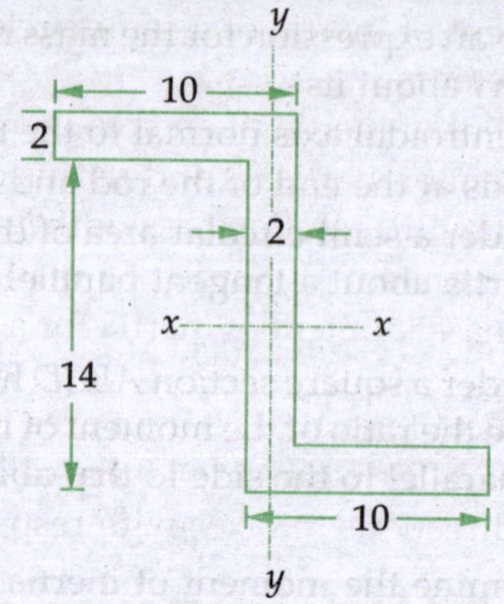

Fig. 8.65

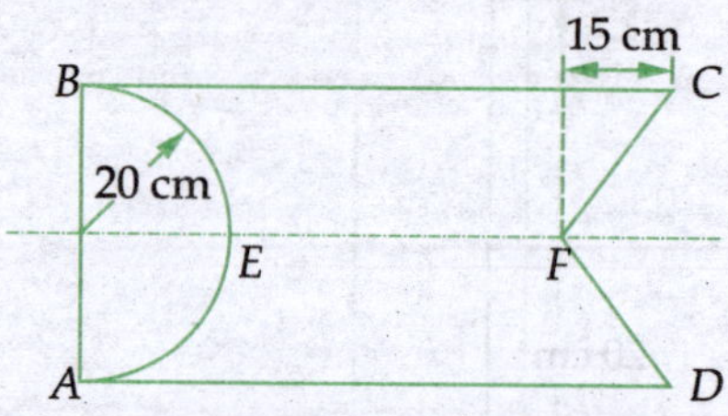

Fig. 8.66

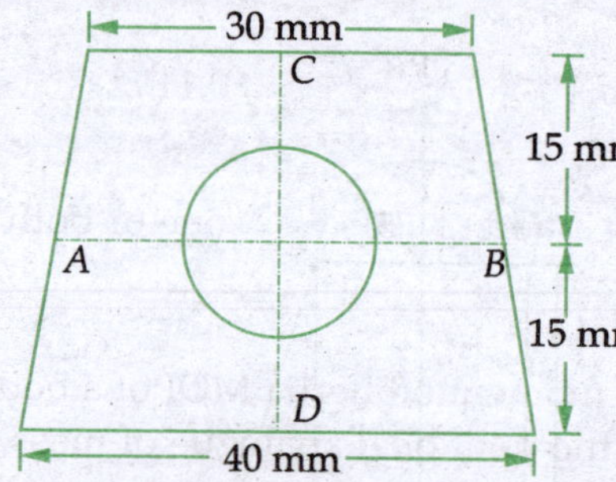

Fig. 8.67

18. Determine the distance d at which the two 2 cm × 6 cm rectangles shown in Fig. 8.68 be spaced so that $I_{xx} = I_{yy}$. **(5.3 cm)**

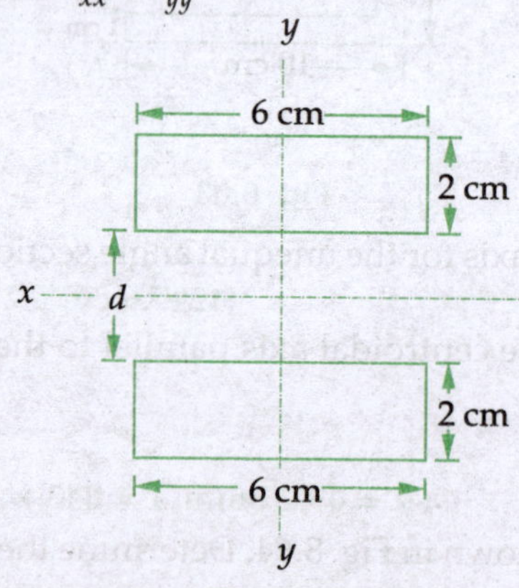

Fig. 8.68

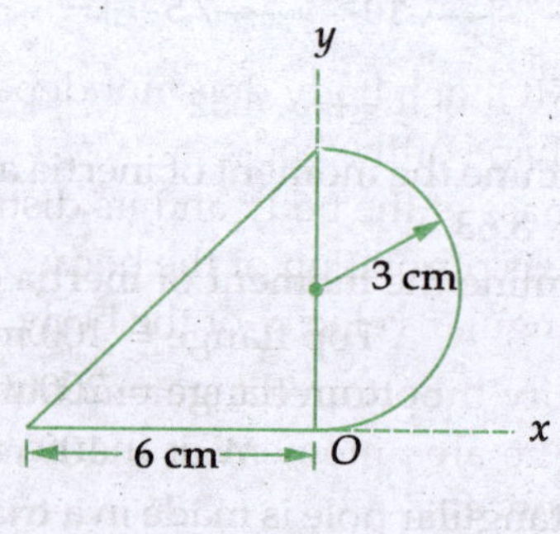

Fig. 8.69

19. For the area shown in Fig. 8.69, determine moments of inertia and product of inertia for axes with origin at O. Proceed to calculate the maximum and minimum moments of inertia.
(I_{xx} = 267 cm^4, I_{yy} = 135 cm^4, I_{xy} = 0, I_{max} = 267 cm^4, I_{min} = 135 cm^4)

B. Complete the following statements with most appropriate word/words :

1. Moment of inertia of an area dA at a distance r from an axis is ....
2. Moment of inertia of a triangle of base (b) and height (h) about an axis through the apex and parallel to the base is ....

3. Moment of inertia of a semi-circle of radius (r) with respect to centroidal axis parallel to the diameter is ....
4. Moment of inertia of an area is minimum about its .... axis
5. Moment of inertia (I) and radius of gyration (k) are related by the expression ....
6. A rectangular section has width b and depth d. Its polar moment of inertia would be ....
7. Mass moment of inertia of a mass dm at a distance r from the axis is ....
8. The mass moment of inertia for a thin spherical shell is .... and that for a solid sphere is ....
9. Product of inertia I_{xy} for an area is zero if it has ....
10. The axes for which $I_{xy} = 0$ are called ....
11. The directions for axes with respected to which the product of inertia is zero are given by tan 2α =

Answers:

1. $\int r^2\,dA$ 2. $\dfrac{bh^3}{4}$ 3. $0.11\,r^4$ 4. centroidal 5. $I = Ak^2$ 6. $\dfrac{bd^3}{12} + \dfrac{db^3}{12}$ 7. $\int r^2\,dm$

8. $\dfrac{2}{3}mr^2$ and $\dfrac{2}{5}mr^2$ 9. one or both axes of symmetry 10. principal axes 11. $\dfrac{2I_{xy}}{I_{xx} - I_{yy}}$

C. Multiple choice questions :

1. The moment of inertia MOI of a body is
   (*a*) the moment of its inertia
   (*b*) the rotational analogue of mass
   (*c*) the rotational moment acting on the body
   (*d*) the inertial moment acting on the body
2. A body of mass M consists of elemental portions of masses m_1, m_2 .... and these masses lie at distances x_1, x_2 .... from a fixed straight line called axis. The MOI of the body about that axis is
   (*a*) $\sum m\,x$ (*b*) $\sum m\,x^2$ (*c*) $\dfrac{\sum m\,x^2}{\sum m\,x}$ (*d*) $\dfrac{\sum m^2\,x}{\sum m}$
3. The MOI of a body does not depend upon
   (*a*) shape of the body
   (*b*) mass of the body and its distribution within the body
   (*c*) axis of rotation of the body
   (*d*) angular velocity of the body
4. Identify the incorrect statement/statements:
   (*a*) The area moment of inertia and the mass moment of inertia are two entirely unrelated concepts
   (*b*) An axis passing through the centroid and lying in the plane is called centroidal axis
   (*c*) Moment of inertia of a body equals twice the kinetic energy of rotation when angular velocity of the body is unity
   (*d*) Radius of gyration is the effective distance where the entire mass may be considered to be located with respect to the axis of rotation
   (*e*) None of the above
5. The moment of inertia of an area is always least with respect to
   (*a*) centroidal axis
   (*b*) vertical axis
   (*c*) radius of gyration
   (*d*) depends upon configuration of the area

6. The moment of inertia of rectangular element ($b \times h$) about an axis coincident with side b would be

(a) $bh^3/12$ (b) $bh^3/6$ (c) $bh^3/3$ (d) $2bh^3/3$

7. For a rectangular element with sides b and h, the moment of inertia about its centroidal axis parallel to side b would be

(a) $bh^3/3$ (b) $bh^3/12$ (c) $bh^3/24$ (d) $bh^3/36$

8. The moment of inertia of a triangle of base width b and height h with respect to its base would be

(a) $bh^3/8$ (b) $bh^3/12$ (c) $bh^3/24$ (d) $bh^3/36$

9. The ratio of moment of inertia of a rectangle and that of a triangle, having same base and height, with respect to their bases would be

(a) 2 : 1 (b) 3 : 1 (c) 4 : 1 (d) 6 : 1

10. The ratio of the moment of inertia of a triangle of base width b and height h with respect to an axis coinciding with its base and a centroidal axis parallel to the base would be

(a) 2 : 1 (b) 3 : 1 (c) 1 : 4 (d) 6 : 1

11. The radius of gyration of a circular area of radius r with respect to centroidal axis is

(a) 0.1 r (b) 0.2 r
(c) 0.5 r (d) 0.7 r

12. The ratio of moment of inertia of a circular body about its x-axis to that about y-axis is

(a) 0.25 (b) 0.5 (c) 1.0 (d) 2.0

13. The second moment of a plane area about any axis compared to its second moment about the neutral axis is always

(a) more (b) less (c) equal
(d) relationship depends upon the dimensions of the plane area

14. The axis X-X and Y-Y for a plane area A are separated by distance r. Then according to parallel axis theorem:

(a) $I_x I_y = Ar^2$ (b) $\dfrac{I_y}{I_x} = Ar^2$ (c) $I_y = I_x + Ar^2$ (d) $I_x + I_y = Ar^2$

15. If I_x and I_y denote the moment of inertia of a lamina about x-axis and y-axis respectively, then the moment of inertia of the lamina about z-axis would be

(a) $I_x + I_y$ (b) $I_x \times I_y$ (c) $\sqrt{I_x^2 + I_y^2}$ (d) $I_x - I_y$

16. Polar moment of inertia

(a) applies to masses whereas moment of inertia is applicable to areas only
(b) is the moment of inertia for an area with respect to a line or axis parallel to centroidal axis
(c) is the moment of inertia for an area with respect to a line or axis perpendicular to the plane of the area
(d) is the moment of inertia for an area relative to a line or axis lying outside the plane of area.

Answers :

1. (c)	2. (b)	3. (d)	4. (e)	5. (a)	6. (c)
7. (b)	8. (b)	9. (c)	10. (b)	11. (c)	12. (c)
13. (a)	14. (c)	15. (a)	16. (c)		

9 Friction

Learning objectives : Attention has been focussed in this chapter to make the readers conversant with:

- *friction and its characteristics; dry and fluid friction; static and dynamic friction*
- *laws of solid friction*
- *terms related to friction such as coefficient of friction, angle of friction, angle of respose and cone of friction*
- *ladder friction, rope and belt friction and wedge friction*

When a body slides over another body, a force is exerted at the surface of contact by the stationary body on the moving body. This resisting force is called the force of friction and acts in a direction opposite to the direction of motion. The frictional forces are present throughout in nature and exist to a considerable extent in all machines; no matter how accurately they are made. Friction is quite undesirable and needs to be mitigated in some machines and processes such as

- power screws
- bearings and gears
- flow of fluids in pipes

Its presence would cause loss of power, wearing out of parts and huge economic losses. However, the working of many devices such as

— friction brakes and clutches
— belt and rope drives
— holding and fastening devices

depends on friction and there the presence of friction forces is advantageous. Obviously friction is both a liability and a necessity, and is often referred to as necessary evil.

9.1. FRICTION AND ITS CHARACTERISTICS

Consider a block resting on a horizontal plane. Experience and observations indicate that the mating surfaces are generally rough and have irregularities. The irregularities are somewhat like hills and valleys. The valleys of one surface mesh with the hills of the other surface. Obviously, there is some degree of interlocking and wedging action between the contact surfaces.

When a force P is applied to the block in a direction parallel to the surface of contact, a force F automatically gets induced, acts in a direction opposite to that in which motion is sought and offers resistance to the movement of block. This resisting force acts tangentially to the contact surface and is known as friction force. Such forces always develop whenever there exists a tendency for the sliding of contact surfaces. The friction is attributed to the interlocking of minute projections on the surface which oppose the relative motion.

Refer Fig. 9.2 for the idealised plot of friction force F and the tractive (applied) force P. Its salient features are:

(i) When static conditions prevail i.e., when there is no motion of the block, balance between applied force $\boldsymbol{P}$ and the friction force $\boldsymbol{F}$ is maintained. The frictional force increases in direct proportion to the applied force and this is depicted by line 0-1 which slopes at 45° with the axis of applied force. Line 0-1 represents the *zone of static friction* and point 1 corresponds to the maximum static friction.

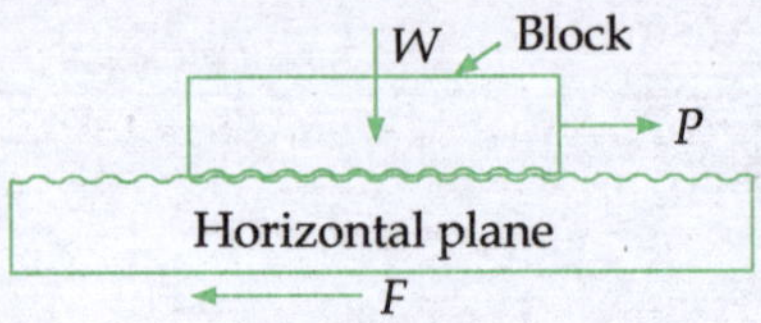

Fig. 9.1

(ii) There comes a limit beyond which frictional force cannot increase and the block is set into motion. The friction force at this instant is maximum and is called the limiting friction.

The *limiting friction* is the maximum friction force at the time of impending motion, i.e., when the motion is about to begin.

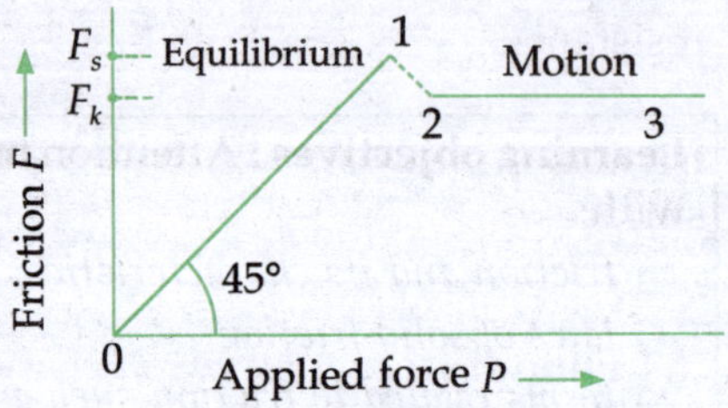

Fig. 9.2

(iii) When the motion of block commences, the friction force drops to somewhat low value called the kinetic friction. The decrease may be atribtuted to the fact that the chance of meshing of hills and valleys of the two surfaces becomes less. Such a stage is represented by straight line 2-3 of the graph and it is known as zone of *kinetic friction*:

Between points 1 and 2, the variation is uncertain and is shown by a dotted line.

The main characteristics of friction force are:

- Friction always opposes the relative motion of the body and is tangential to the surface of contact.
- Friction is a passive force; it exists as long as the tractive force acts.
- Friction is a self-adjusting force; only that much comes into play as is just sufficient to prevent motion.
- Both the limiting static friction and kinetic friction are proportional to the normal force.

9.2. NATURE OF FRICTION

There are essentially two types of friction, namely, dry friction and fluid friction.

9.2.1. Dry friction

Dry friction (also called *coulomb friction*) manifests when the contact sufaces are dry and there is tendency for relative motion.

Dry friction is further subdivided into:

- Sliding friction: friction between two surfaces when one surface slides over another.
- Rolling friction: friction between two surfaces which are separated by balls or rollers.

Rolling friction is always less than sliding friction.

It may also be pointed out that rolling friction also occurs because of deformation to a surface under a rolling load. Refer Fig. 9.3 which depicts a small surface depression caused by a wheel, and the wheel being pulled out of this depression. A horizontal pull P is applied to the wheel and it tends to move up at point A.

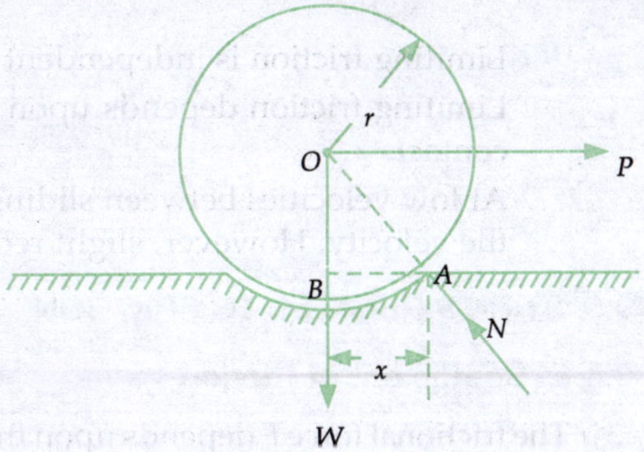

Fig. 9.3. Rolling friction

Under equilibrium conditions :

$$\Sigma M_A = 0 \quad \text{and that gives}$$

$$W \times x - P \times OB = 0$$

where W is the weight of wheel, and the distance OB for a small depression equals the radius of wheel ($OB \simeq r$). Then

$$Pr = Wx$$

Note :

(i) The horizontal component of reaction N equals P and is called the *rolling resistance*. The distance x represents the term called the *coefficient of rolling resistance*.

(ii) The magnitude of rolling friction due to surface depression is greater than the sliding friction. It implies that a greater force is required to roll a wheel than slide it on a rough horizontal surface.

9.2.2. Fluid friction

Fluid friction manifests when a lubricating fluid is introduced between the contact surfaces of two bodies.

If the thickness of the lubricant or oil between the mating surfaces is small, then the friction between the surfaces is called *greasy* or *non-viscous friction*. The oil gets absorbd by the surfaces and the contact between them is no more a metal to metal contact. Instead the contact is through thin layer of oil and that ultimately results is less friction.

When the two surfaces are separated by a thick film of lubricant, metallic contact is entirely non-existent. The friction is due to viscosity of the oil, or the shear resistance between the layers of the oil rubbing against each other. Obviously then there occurs a great reduction in friction. This frictional force is knwon as *viscous* or *fluid friction*.

9.3. STATIC AND DYNAMIC FRICTION

The *static friction* is the frictional force that develops between mating surfaces when subjected to external forces but there is no relative motion between them.

The *dynamic friction* is the frictional force that develops between mating surfaces when subjected to external forces and there is relative motion between them. The dynamic friction is also known as *kinetic friction*.

The kinetic friction is always somewhat less than the maximum static friction force.

9.4. LAWS OF SOLID FRICTION (STATIC OR DYNAMIC)

Based upon experimental evidence, the following laws of friction have been established for dry contact surfaces.

(i) Friction acts tangential to the surfaces in contact and is in a direction opposite to that in which motion is to impend, i.e., take place.

(ii) Friction force is maximum at the instant of impending motion. Its variation from zero to maximum value (limiting friction) depends upon the resultant force tending to cause motion.

(iii) The magnitude of limiting friction bears a constant ratio to the normal reaction between the mating surfaces.

This ratio drops to a slightly lower value when the motion starts.

(iv) Limiting friction is independent of the area and shape of contact surfaces.

(v) Limiting friction depends upon the nature (roughness or smoothness) of the surfaces in contact.

(vi) At low velocities between sliding surfaces, the friction force is practically independent of the velocity. However, slight reduction in friction occurs when the speeds are high.

9.5. TERMS RELATED TO FRICTION

9.5.1. Coefficient of friction

The frictional force F depends upon the wedging action between irregularities of contact surfaces and becomes more pronounced with increase in normal reaction R between them. Mathematically $F \propto R$. The ratio F/R is called *coefficient of friction* and is denoted by μ.

The coefficient of friction is defined as the ratio of force of friction to the normal reaction between the contact surfaces.

The coefficient of friction would be high if the contact surfaces are rough. Obviously, the coefficient of friction measures the roughness between a pair of contact surfaces and incorporates the geometric property of the mating contours.

(i) When the system is in a state of impending motion, the frictional force has the limiting (maximum) value F_s and the ration F_s/R is called the coefficient of static friction μ_s. Thus $\mu_s = F_s/R$.

(ii) When motion starts, the maximum friction falls to a lower value F_k called kinetic friction. Experimental results indicate that the friction force developed during motion remains fairly constant. The ratio F_k/R is called the coefficient of kinetic friction μ_k. Thus $\mu_k = F_k/R$.

The value of μ_k is found to be less than μ_s for the same pair of contact surfaces.

(iii) The approximate values of static friction coefficients for pair of contacting surfaces are:

- wood on wood = 0.3 to 0.5
- metal on metals = 0.15 to 0.25
- rubber tyre on concrete = 0.7 to 0.9
- wood on metals = 0.2 to 0.6
- masonry on clay = 0.3 to 0.5

9.5.2. Total reaction

The block A resting on a rough surface is in equilibrium under the action of following forces:

(i) weight W of the block, acting downwards through its centre of gravity

(ii) normal reaction R between the block and surface

(iii) frictional force $F = \mu R$ where μ is the coefficient of friction

(iv) tractive force P

The forces R and F act at right angles to each other. Their resultant equals $\sqrt{R^2 + F^2}$ and is called the total reaction or resultant reaction.

9.5.3. Angle of friction

The angle of friction ϕ is a measure of the limiting position of total reaction between the two contacting surfaces. It is defined as the angle which the resultant of normal reaction and limiting force of friction makes with the normal reaction.

With reference to Fig. 9.4

R = normal reaction;

F = limiting force of friction

$$S = \sqrt{R^2 + F^2}$$

= total or resultant reaction

$$\tan\phi = \frac{F}{R} \text{ where } \phi \text{ is the angle of friction}$$

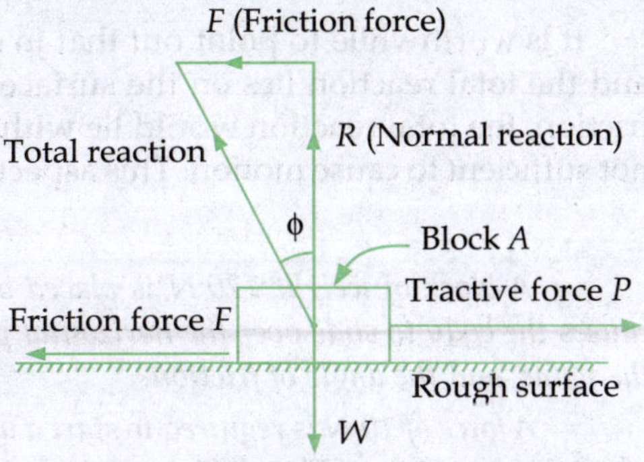

Fig. 9.4

The ratio F/R is also called the coefficient of friction, μ.

$$\therefore \quad \mu = \tan\phi \qquad \text{...(9.1)}$$

The above correlation does suggest that angle of friction in radians is equal to the coefficient of friction provided the angle has a small value.

$$\mu = \tan\phi \simeq \phi \text{ in radians}$$

9.5.4. Angle of repose

Consider a block of weight W resting on an inclined plane OA making an angle α with the horizontal. Let the angle α be increased gradually till the block is just at the point of sliding. The block is then in equilibrium state under the influence of following set of forces:

(i) weight W of the block acting vertically downwards,

(ii) normal reaction R acting at right angles to the inclined plane, and

(iii) limiting forces of friction, $F = \mu R$; acting up the plane as the block is to slide downwards.

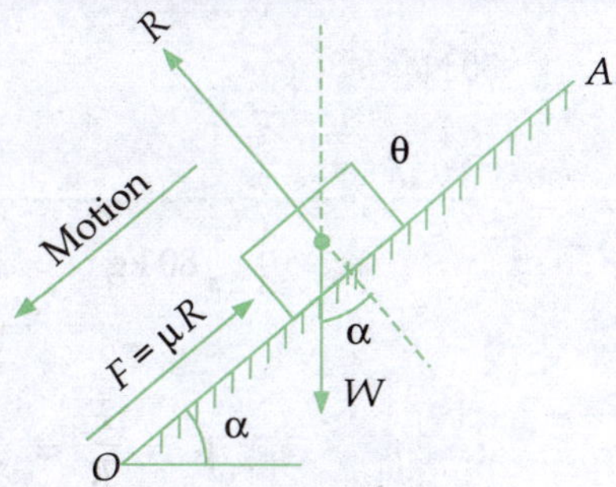

Fig. 9.5

Resolving these forces along and perpendicular to the plane

$$\mu R = W \sin\alpha$$

$$R = W \cos\alpha$$

These two expressions give : $\mu = \tan\alpha$

In terms of angle of friction ϕ, the coefficient of friction is given as: $\mu = \tan\phi$

$$\therefore \quad \tan\phi = \tan\alpha \quad \text{or} \quad \phi = \alpha \qquad \text{...(9.2)}$$

The angle α of the inclined plane at which a block resting on it is about to slide down the plane is called the angle of repose, and it is equal to the angle of friction between the block and the inclined plane.

9.5.5. Cone of friction

With reference to Fig. 9.6, the lines of action of normal reaction R and the frictional force F meet at point O. The magnitude of total reaction is $\sqrt{F^2 + R^2}$ and it makes an angle ϕ with the normal reaction. Obviously O is the vertex of a cone whose axis is R and semi-vertex angle is ϕ. Such a right circular cone is called the cone of friction.

Apparently the cone of friction is the imaginary cone AOB generated in the case of non-coplanar forces by revolving the static resultant reaction S about the normal OR.

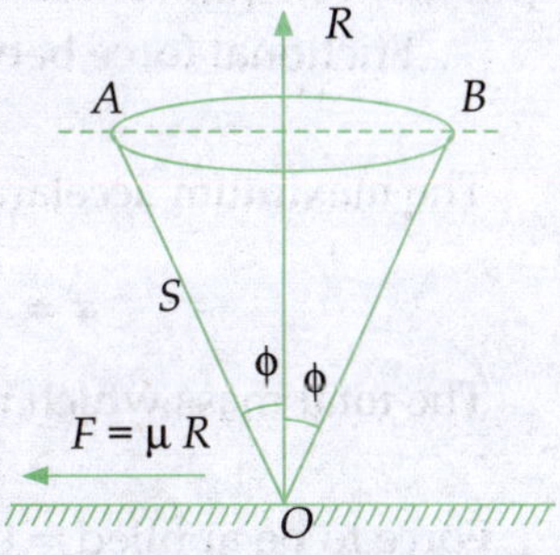

Fig. 9.6

It is worthwhile to point out that in case of impending motion, the friction force is maximum and the total reaction lies on the surface of cone. When the friction force is less than the limiting friction, the total reaction would lie within the cone. In that case, the forces acting on the body are not sufficient to cause motion. This aspect forms the working principle for self-locking mechanisms.

EXAMPLE 9.1

(a) *A block of weight 120 N is placed on a rough horizontal plane. If a horizontal force of 50 N just causes the body to slide over the horizontal plane, calculate the coefficient of friction between the block and the plane, and the angle of friction.*

(b) *A force of 10 N is required to start a uniform plank of wood slide over a bench for which the coefficient of friction is constant. The plank is cut into three equal pieces and these pieces are stacked one above the other. What force will be now needed to start movement of the pile?*

(c) *A block of mass 80 kg rests on a floor and the static coefficient of friction for the contact surface is 0.15. It is acted upon by a horizontal force of 50 N as shown in Fig. 9.7. What frictional force will be developed along the contact surface?*

(d) *Refer Fig. 9.8 which shows a mass m_1 = 0.5 kg resting on a frictionless surface and another mass m_2 = 0.25 kg placed above m_1; the coefficient of friction between m_1 and m_2 being 0.3. If m_2 is to slide on m_1, what amount of force is needed to pull mass m_1?*

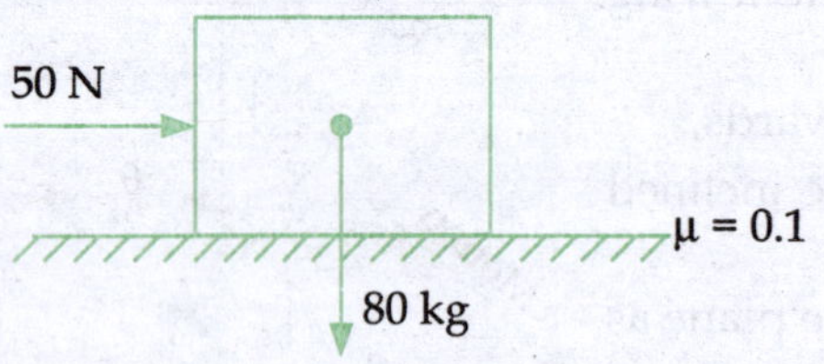

Fig. 9.7

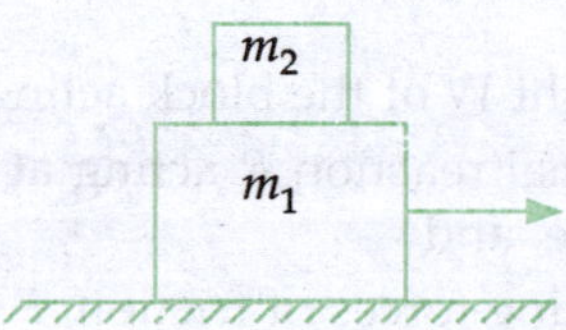

Fig. 9.8

Solution : (a) $\mu = \dfrac{F}{R} = \dfrac{50}{120} =$ **0.4167**

Also $\mu = \tan\phi$ where ϕ is the angle of friction

$\phi = \tan^{-1}(\mu) = \tan^{-1}(0.4167) =$ **22.62°**

(b) $F = \mu R$. The coefficient of friction μ and the normal reaction $R = mg$ remain constant. As such the same force of 10 N would be needed to start the movement of pile system.

(c) Maximum value of static friction $= \mu R = \mu mg$

$= 0.15 \times (80 \times 9.81) = 117.72$ N

When the applied force is less than this maximum value, the friction force is obtained by the equations of equilibrium. $\Sigma F_y = 0$ and that gives $F = 50$ N.

(d) Frictional force between mass m_1 and $m_2 = \mu R = \mu\ (m_2 g)$

$= 0.3 \times (0.25 \times 9.81) = 0.736$ N

The maximum accelaration with which it can move without sliding

$$a = \frac{0.736}{0.25} = 2.944 \text{ N} \qquad (\because \quad F = ma)$$

The total mass which the applied force has to move

$= 0.25 + 0.5 = 0.75$ kg

Force to be applied $= 0.75 \times 2.944 =$ **2.208 N**

EXAMPLE 9.2

A block of mass 25 kg is at rest on a horizontal floor. To slide the block, a force P is applied along the horizontal direction. The coefficient of static friction and dynamic friction between the block and floor are stated to be 0.36 and 0.3 respectively. For each value of P given below, work-out the corresponding value of frictional force F and complete the table.

Applied force P (N)	20	40	60	80	88.3	90
Frictional Force F (N)						

Solution : Weight of the block $W = 25 \times 9.81 = 245.25$ N

Normal reaction $R = W = 245.25$ N

Limiting value of force of friction $F = \mu_s R = 0.36 \times 245.25 =$ **88.3 N**

The block will remain at rest till the applied force is less than 88.3 N

When the applied force exceeds μR, the block starts sliding. The dynamic conditions then prevail and the frictional force reduces to

$$F = \mu_d R = 0.3 \times 245.25 = 73.57 \text{ N}$$

Applied force (N)	20	40	60	80	88.3	90
Frictional Force (N)	20	40	60	80	88.3	73.57

EXAMPLE 9.3

A body weighing 300 N is resting on a rough horizontal table. A pull of 100 N applied at an angle of 15 degree with the horizontal just causes the body to slide over the table. Make calculations for the normal reaction and the coefficient of friction.

Solution : In the limiting equilibrium, the forces are balanced. That is

$$\Sigma F_x = 0; \quad F - 100\cos 15^\circ = 0$$

$$\therefore \quad F = 100\cos 15^\circ = 96.59 \text{ N}$$

$$\Sigma F_y = 0; \quad R + 100\sin 15^\circ - 300 = 0$$

$$\therefore \quad R = 300 - 100\sin 15^\circ = 300 - 25.88 = 274.12 \text{ N}$$

$$\mu = \frac{F}{R} = \frac{96.59}{274.12} = 0.352$$

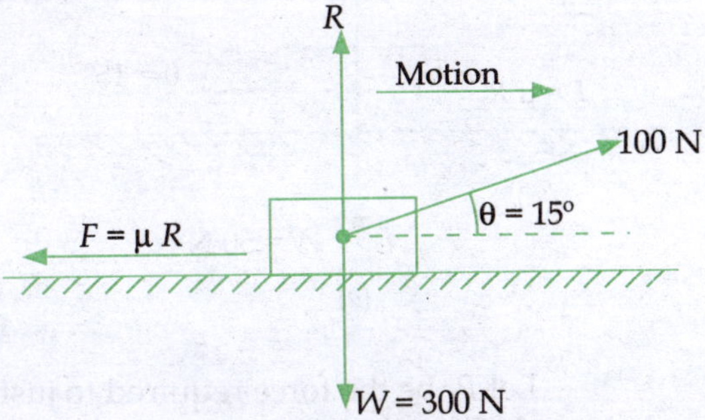

Fig. 9.9

EXAMPLE 9.4

A body of weight 100 N rests on a rough horizontal surface ($\mu = 0.3$) and is acted upon by a force applied at an angle of 30 degree to the horizontal. What force is required to just cause the body to slide over the surface?

(b) Proceed to determine the inclination and magnitude of minimum force required to set the block into impending motion.

Solution : In the limiting equilibrium, the forces are balanced. That is

$$\Sigma F_x = 0; \quad F = P\cos\theta$$

$$\Sigma F_y - 0; \quad R - W - P\sin\theta$$

$$\text{Also} \quad F = \mu R$$

$$\therefore \quad \mu(W - P\sin\theta) = P\cos\theta$$

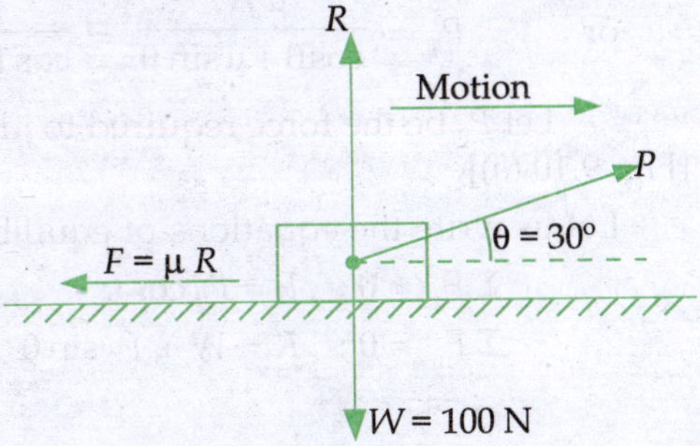

Fig. 9.10

$$\text{or } P = \frac{\mu W}{\cos\theta + \mu \sin\theta} = \frac{0.3 \times 100}{\cos 30^\circ + 0.3 \sin 30^\circ} = \frac{0.3 \times 100}{0.866 + 0.15} = 29.53 \text{ N}$$

(b) Let α be the inclination of the minimum force required to just cause the body to slide over the surface. Then

$$P = \frac{\mu W}{\cos\alpha + \mu \sin\alpha} \quad \text{...}(i)$$

For force P to be minimum, the denominator must be maximum. For that

$$\frac{d}{dx}(\cos\alpha + \mu \sin\alpha) = 0$$

$$\text{or} \quad -\sin\alpha + \mu\cos\alpha = 0; \quad \tan\alpha = \mu = 0.3$$

$$\therefore \quad \alpha \tan^{-1}(0.3) = 16.7^\circ$$

Substituting $\alpha = 16.7^\circ$ in identity (i) above, we get

$$P_{min} = \frac{0.3 \times 100}{\cos 16.7^\circ + 0.3 \sin 16.7^\circ} = \frac{30}{0.958 + 0.086} = 28.73 \text{ N}$$

EXAMPLE 9.5

A wooden block of weight 50 N rests on a horizontal plane. Determine the force required to just (a) pull it, and (b) push it. Take coefficient friction $\mu = 0.4$ between the mating surfaces. Comment on the result.

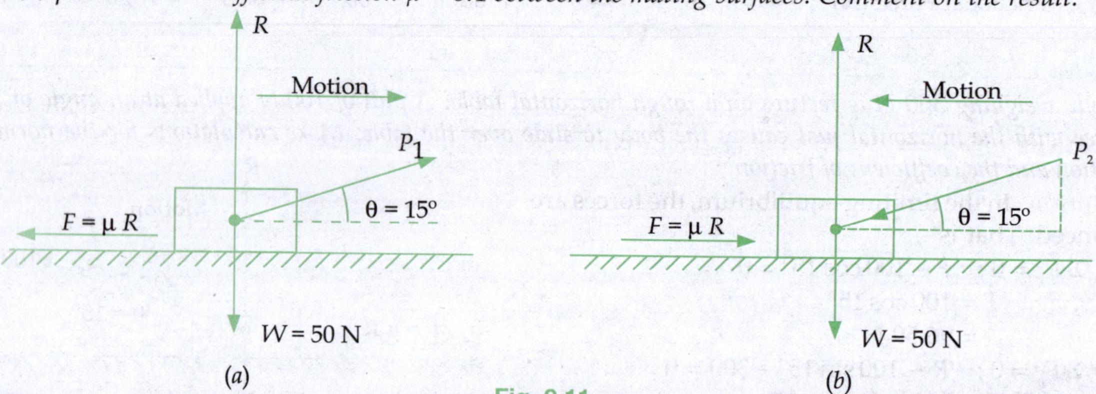

Fig. 9.11

Solution : Let P_1 be the force required to just pull the block. In the limiting equilibrium, the forces are balanced. That gives

$$\Sigma F_x = 0; \quad F = P_1 \cos\theta$$

$$\Sigma F_y = 0; \quad R = W - P_1 \sin\theta$$

$$\text{Also} \quad F = \mu R \quad \therefore \quad \mu(W - P_1 \sin\theta) = P_1 \cos\theta$$

$$\text{or} \quad P_1 = \frac{\mu W}{\cos\theta + \mu\sin\theta} = \frac{0.4 \times 50}{\cos 15^\circ + 0.4 \sin 15^\circ} = \frac{0.4 \times 50}{0.966 + 0.103} = 18.70 \text{ N}$$

(b) Let P_2 be the force required to just push the block. With reference to the free body diagram [Fig. 9.10 (b)],

Let us write the equations of equilibrium,

$$\Sigma F_x = 0; \quad F = P_2 \cos\theta$$

$$\Sigma F_y = 0; \quad R = W + P_2 \sin\theta$$

Also $F = \mu R$

$\therefore \quad \mu (W + P_2 \sin\theta) = P_2 \cos\theta$

or $$P_2 = \frac{\mu W}{\cos\theta - \mu\sin\theta} = \frac{0.4 \times 50}{\cos 15^\circ - 0.4 \sin 15^\circ} = \frac{0.4 \times 50}{0.966 - 0.103} = \mathbf{23.17\ N}$$

Comments. It is easier to pull the block than push it.

EXAMPLE 9.6

A body resting on a rough horizontal plane required a pull of 24 N inclined at 30° to the plane just to move it. It was also found that a push of 30 N at 30° to the plane was just enough to cause motion to impend. Make calculations for the weight of body and the coefficient of friction.

Solution : Refer to Fig. 9.12 for the system of forces when pull and push are applied to the body. Force of friction is always opposite to the direction of motion.

In the limiting equilibrium, the forces are balanced. With reference to the free body diagram (Fig. 9.12 (*a*) when pull is applied)

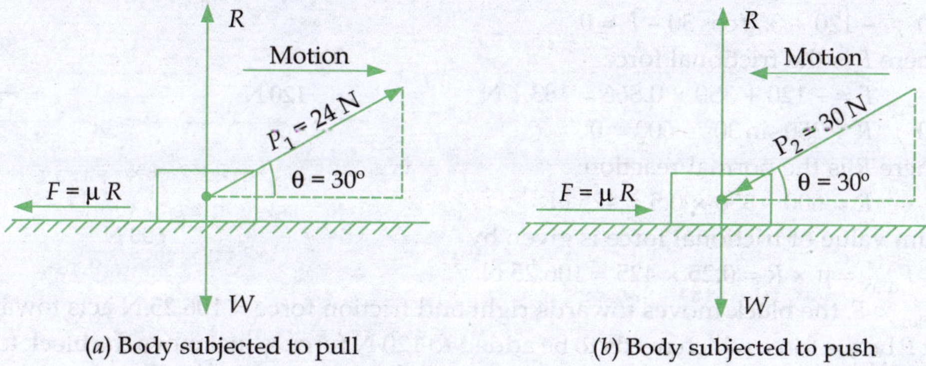

Fig. 9.12

$$\Sigma F_x = 0 ; \quad F = P_1 \cos\theta$$
$$\Sigma F_y = 0; \quad R = W - P_1 \sin\theta$$

Also $F = \mu R$

$\therefore \quad \mu (W - P \sin\theta) = P_1 \cos\theta$

or $$P_1 = \frac{\mu W}{\cos\theta + \mu \sin\theta} \quad \text{...}(i)$$

With reference to the free body diagram (Fig. 9.12 (*b*) when push is applied)

$$\Sigma F_x = 0 ; \quad F = P_2 \cos\theta$$
$$\Sigma F_y = 0 ; \quad R = W + P_2 \sin\theta$$

Also $F = \mu R \quad \therefore \quad \mu (W + P_2 \sin\theta) = P_2 \cos\theta$

or $$P_2 = \frac{\mu W}{\cos\theta - \mu \sin\theta} \quad \text{...}(ii)$$

From expression (*i*) and (*ii*),

$$\frac{P_1}{P_2} = \frac{\cos\theta - \mu\sin\theta}{\cos\theta + \mu\sin\theta}$$

or $$\frac{24}{30} = \frac{\cos 30^\circ - \mu \sin 30^\circ}{\cos 30^\circ + \mu \sin 30^\circ} = \frac{0.866 - 0.5\mu}{0.866 + 0.5\mu}$$

or $$0.6928 + 0.4\,\mu = 0.866 - 0.5\,\mu$$

$$\therefore \quad \mu = \frac{0.866 - 0.6928}{0.9} = 0.192$$

For expression (*i*)

$$W = \frac{P_1(\cos\theta + \mu\sin\theta)}{\mu} = \frac{P_1(\cos 30^\circ + \mu\sin 30^\circ)}{\mu}$$

$$= \frac{24\,(0.866 + 0.192 \times 0.5)}{0.192} = 120.25 \text{ N}$$

EXAMPLE 9.7

A block weighing 600 N has been subjected to the load system as shown in Fig. 9.13. Work-out the frictional force for the block and state whether it is in equilibrium or in motion. What additional force needs to be added to 120 N force so that the block just moves to the left?

Solution : In the limiting condition, the forces are balanced. That is

$\Sigma F_x = 0 \;;\; -120 + 350\cos 30 - F = 0$

where F is the frictional force

$F = -120 + 350 \times 0.866 = 183.1$ N

$\Sigma F_y = 0 \;;\; R + 350\sin 30^\circ - 600 = 0$

where R is the normal reaction

$R = 600 - 350 \times 0.5 = 425$ N

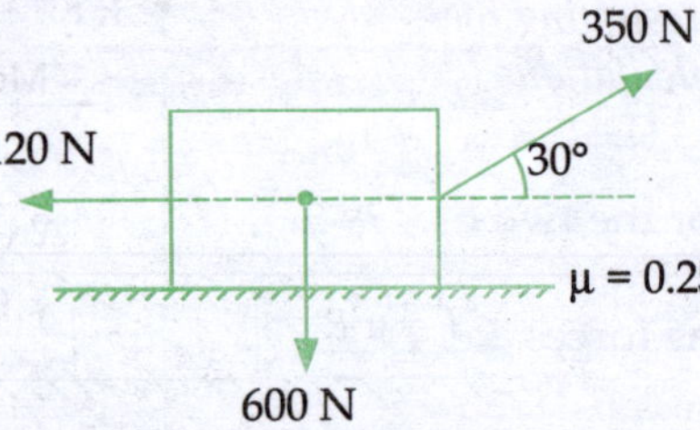

Fig. 9.13

Maximum value of frictional force is given by

$$F_{max} = \mu \times R = 0.25 \times 425 = 106.25 \text{ N}$$

Since $F_{max} < F$, the block moves towards right and friction force = 106.25 N acts towards left.

(*b*) Let P be the force which needs to be added to 120 N force to just move the block to the left. When the block moves towards left, the friction force will act towards right. Then from the relation $\Sigma F_x = 0$,

$$-(120 + P) + 350\cos 30^\circ + F_{max} = 0$$

$$-(120 + P) + 350 \times 0.866 + 106.25 = 0$$

$$\therefore \quad P = -120 + 350 \times 0.866 + 106.25 = 289.35 \text{ N}$$

EXAMPLE 9.8

Consider a system of three blocks resting upon one another as shown in Fig. 9.14. The block A weighs 150 N, B weighs 50 N and C weighs 100 N. The coefficients of friction are 0.3 between A and B, 0.2 between B and C, and 0.1 between C and the ground. Determine the least horizontal force P necessary to start motion of any part of the system.

Solution : Limiting force of friction between contacting surfaces,

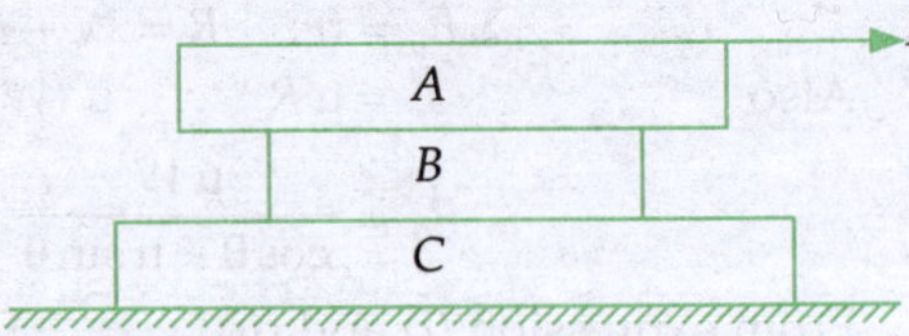

Fig. 9.14

$$F = \mu R = \mu W$$

$\therefore$ Limiting force of friction between A and B,

$$F_{ab} = \mu_{ab} \times W_a = 0.3 \times 150 = 45 \text{ N}$$

Limiting force of friction between B and C,

$$F_{bc} = \mu_{bc}\,(W_a + W_b) = 0.2\,(150 + 50) = 40 \text{ N}$$

Limiting force of friction between C and ground

$$F_{cg} = \mu_{cg}\,(W_a + W_b + W_c) = 0.1\,(150 + 50 + 100) = 30 \text{ N}$$

With gradual increase in applied force P, the frictional force increases till it attains the maximum (limiting) values. Any further increase in P sets the body in motion. Now

(*i*) When P = 30 N, the block A exerts a force of 30 N on block B. However this force is less than the limiting friction between A and B, and obviously there will be no relative motion betwen A and B.

(*ii*) The block B also exerts a force of 30 N on block C. This force is again less than limiting friction between B and C. Obviously no relative motion between B and C.

(*iii*) The block C too is subjected to a pull of 30 N. This equals the limiting friction between block C and the ground. As such the block C will be on the point of motion.

The above remarks tell us that when a horizontal P equal to 30 N is applied, all the three blocks would be in a state of impending motion as a single body.

EXAMPLE 9.9

Two blocks A and B of weight 4 kN and 2 kN respectively are in equilibrium position as shown in Fig. 9.15. Presuming that coefficient of friction between the two blocks as well as between block and floor is 0.25, make calculations for the force P required to move the block A.

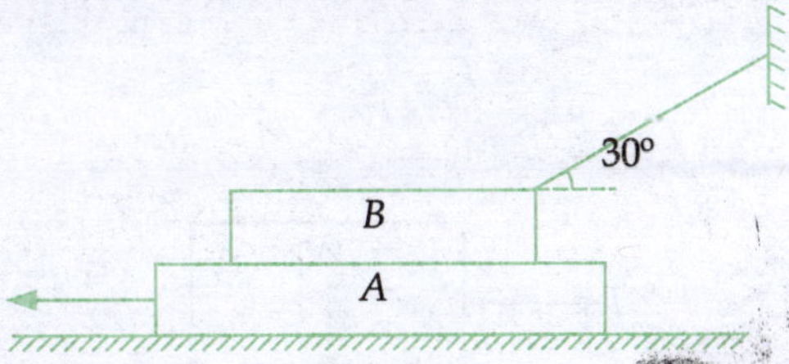

Fig. 9.15

Solution : Refer Fig. 9.16 for the free body diagrams for the two blocks.

Considering equilibrium of block B. Resolving the forces along the horizontal and vertical directions,

$$T\cos 30^\circ - \mu R_b = 0\,; \qquad T\cos 30^\circ = \mu R_b$$

$$R_b + T\sin 30^\circ - W_b = 0; \quad T\sin 30^\circ = W_b - R_b$$

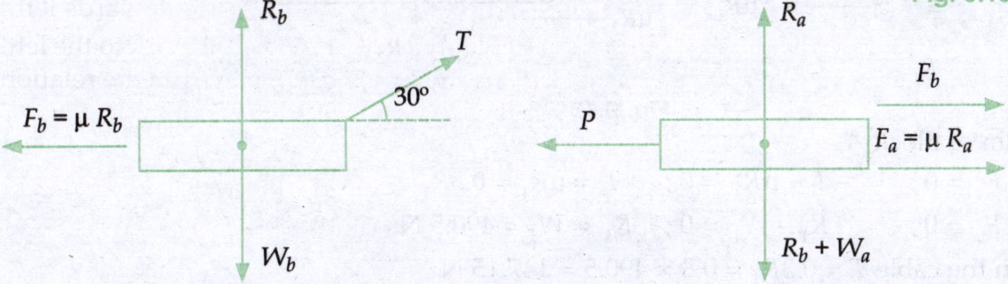

Fig. 9.16

$$\therefore \qquad \tan 30^\circ = \frac{W_b - R_b}{\mu R_b} = \frac{2 - R_b}{0.25\, R_b}$$

or

$$0.5773 = \frac{2 - R_b}{0.25\, R_b}\,; \quad 0.1443\, R_b = 2 - R_b$$

$$\therefore \qquad R_b = \frac{2}{1.1443} = 1.748 \text{ N}$$

$$F_b = \mu R_b = 0.25 \times 1.748 = 0.437 \text{ N}$$

Considering equilibrium of block A. Resolving the forces along the horizontal and vertical directions,

$$F_b + \mu R_a - P = 0\,; \quad P = F_b + \mu R_a$$

$$R_a - R_b - W_a = 0\,; \quad R_a = R_b + W_a = 1.748 + 4 = 5.748$$

$$\therefore \qquad P = 0.437 + 0.25 \times 5.748 = \mathbf{1.874\ N}$$

EXAMPLE 9.10

Two blocks A and B weighing 50 kg and 80 kg respectively are in equilibrium in the position shown in Fig. 9.17. Calculate the force P required to move the lower block B and tension in the cable. Take coefficient of friction at all contact surfaces to be 0.3.

Solution : Weight of block $A = 50 \times 9.81$

$= 490.5$ N

Weight of block $B = 80 \times 9.81$

$= 784.8$ N

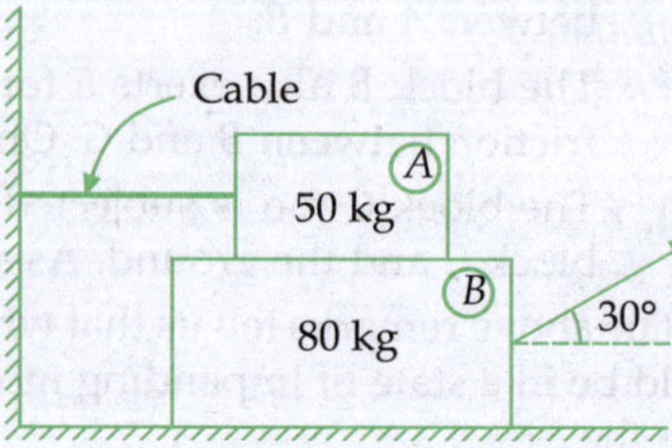

Fig. 9.17

As the block B tends to move towards right, there will be rubbing action between the surfaces of blocks A and B. The rubbing action will cause the force of friction to act between the two surfaces. The block A is tied to a cable, the other end of which is fixed to a vertical plane. When the block B tends to move towards right, the block A moves towards left with respect to block B. Hence the force of friction $F = \mu R_1$ on the lower surface of block A will act towards right as shown in free body diagram of block A (Fig. 9.18)

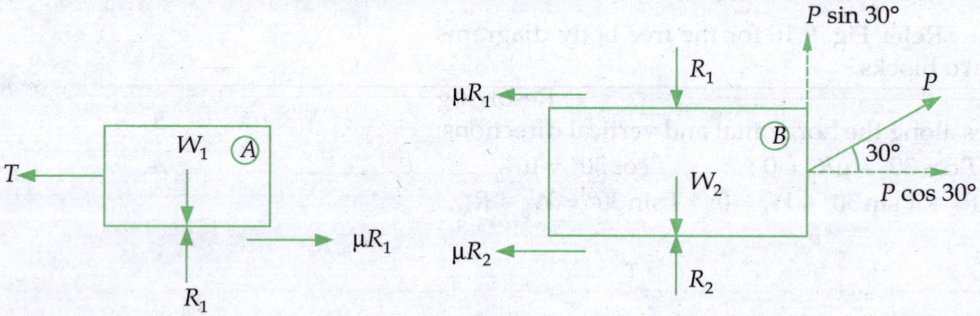

Fig. 9.18

For equilibrium of block A,

$$\Sigma F_x = 0; \quad -T + \mu R_1 = 0; \quad T = \mu R_1 = 0.3R_1$$

$$\Sigma F_y = 0; \quad R_1 - W_1 = 0; \quad R_1 = W_1 = 490.5 \text{ N}$$

∴ Tension in the cable $T = 0.3R_1 = 0.3 \times 490.5 = 147.15$ N

Considering free body diagram of block B (Fig. 9.17),

$$\Sigma F_x = 0; \quad -\mu R_1 - \mu R_2 + P\cos 30° = 0$$

or $$-0.3 \times 490.5 - 0.3R_2 + 0.866\,P = 0 \qquad \ldots(i)$$

$$\Sigma F_y = 0; \quad R_2 - W_2 - R_1 + P\sin 30° = 0$$

or $$R_2 - 784.8 - 490.5 + 0.5P = 0$$

or $$R_2 = 1275.3 - 0.5P$$

Then from identity (*i*)

$$-0.3 \times 490.5 - 0.3(1275.3 - 0.5P) + 0.866P = 0$$

or $$-147.15 - 382.59 + 0.15P + 0.866P = 0$$

∴ Force required to move the lower block,

$$P = \frac{382.59 + 147.15}{0.15 + 0.866} = \mathbf{521.4\ N}$$

EXAMPLE 9.11

Two blocks A and B weighing 50 N and 80 N respectively are positioned as shown in Fig. 9.19. The coefficient of friction between ground and block B is 0.1 and that between block B and block A is 0.28. State whether B is stationary with respect to ground and A moves or B is stationary with respect to A. Proceed to determine the minimum value of weight W in the pan so that motion starts.

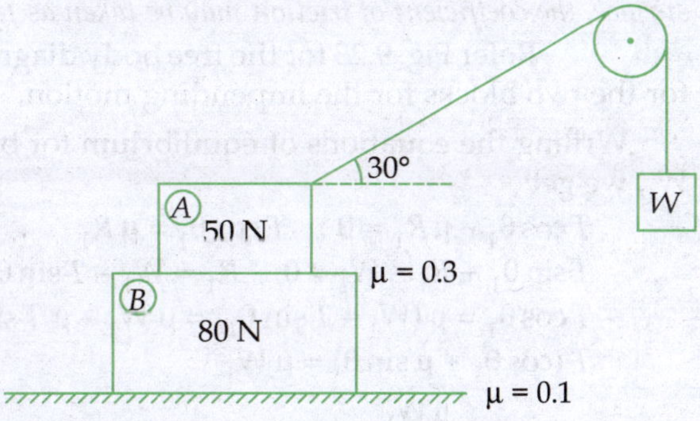

Fig. 9.19

Solution : The coefficient of friction between the block *B* and ground (μ = 0.1) is much less than that between the blocks *A* and *B* (μ = 0.28). Obviously when weight *W* is put in the pan, motion will start of both the blocks together. As such the block *B* will be stationary with respect to block *A*.

With reference to free body diagram of the whole system (Fig. 9.20), the conditions for limiting equilibrium are :

$$\Sigma F_x = 0 : \quad -0.1\,R + W\cos 30^\circ = 0$$

$$\therefore \quad R = \frac{W\cos 30^\circ}{0.1} = 8.66\ \text{W}$$

$$\Sigma F_y = 0 : \quad R - (50 + 80) + W\sin 30^\circ = 0$$

$$\text{or} \quad 8.66\,W - 130 + 0.5\,W = 0$$

$$\therefore \quad W = \frac{130}{8.66 + 0.5} = \mathbf{14.21\ N}$$

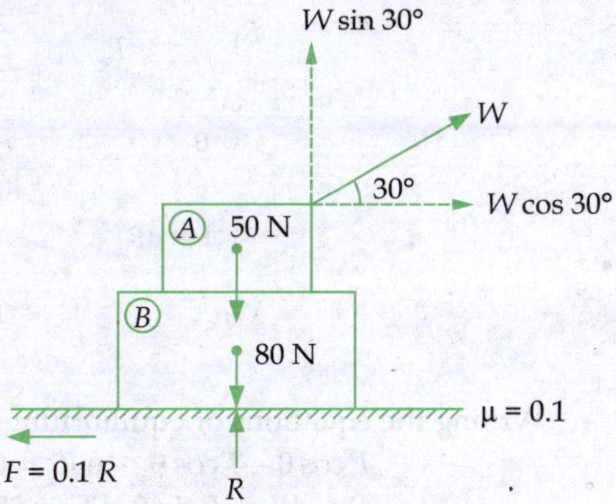

Fig. 9.20

This is the minimum weight in the pan required to start the motion.

(b) When the weight in the pan exceeds beyond a certain limit, the block *A* may move with respect to *B*. The block *B* will then remain stationary with respect to ground.

Considering free body diagram of block *A*, the conditions of limiting equilibrium are :

$$\Sigma F_x = 0 : \quad W\cos 30^\circ - 0.3\,R = 0$$

$$R = \frac{W\cos 30^\circ}{0.3} = 2.887\ W$$

$$\Sigma F_y = 0 : \quad R - 50 + W\sin 30^\circ = 0$$

$$\text{or} \quad = 2.887\ W - 50 + 0.5\ W = 0$$

$$\therefore \quad W = \frac{50}{2.887 + 0.5} = \mathbf{14.76\ N}$$

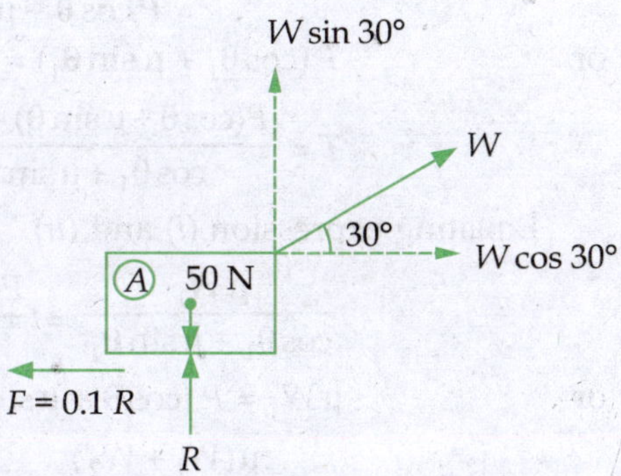

Fig. 9.21

EXAMPLE 9.12

Two blocks of weight W_1 and W_2 are connected by a string and rest on a rough horizontal plane as shown in Fig. 9.22. Set up an expression for the magnitude and direction of the least force P that needs to be applied to the upper block to induce sliding. For each block and the surface, the coefficient of friction may be taken as μ.

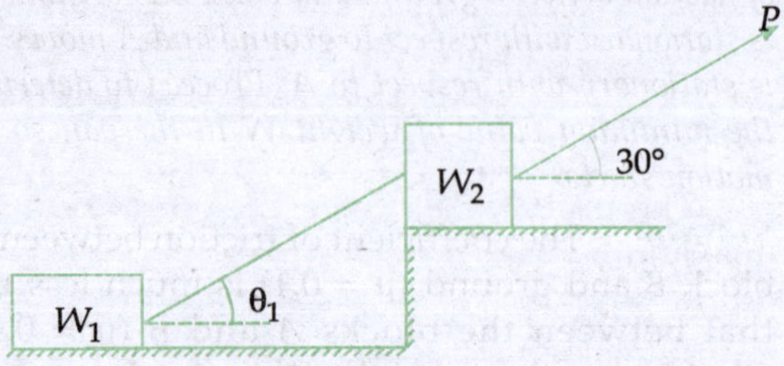

Fig. 9.22

Solution : Refer Fig. 9.23 for the free body diagrams for the two blocks for the impending motion.

Writing the equations of equilibrium for block W_1, we get

$$T\cos\theta_1 - \mu R_1 = 0\,; \quad T\cos\theta_1 = \mu R_1$$

$$T\sin\theta_1 + R_1 - W_1 = 0; \quad R_1 = W_1 - T\sin\theta_1$$

$$\therefore \quad T\cos\theta_1 = \mu(W_1 - T\sin\theta_1) = \mu W_1 - \mu T\sin\theta_1$$

$$T(\cos\theta_1 + \mu\sin\theta) = \mu W_1$$

$$T = \frac{\mu W_1}{\cos\theta_1 + \mu\sin\theta_1} \qquad \ldots(i)$$

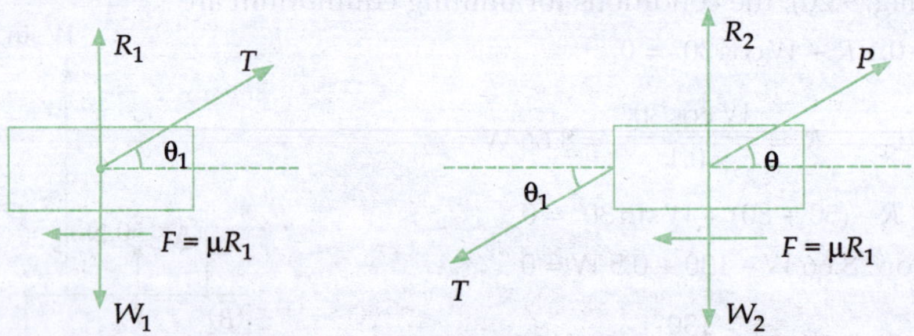

Fig. 9.23

Writing the equations of equilibrium for block W_2, we get

$$P\cos\theta - T\cos\theta_1 - \mu R_2 = 0\,; \quad T\cos\theta_1 = P\cos\theta - \mu R_2$$

$$R_2 - W_2 + P\sin\theta - T\sin\theta_1 = 0\,; \quad R_2 = W_2 - P\sin\theta + T\sin\theta_1$$

$$\therefore \quad T\cos\theta_1 = P\cos\theta - \mu(W_2 - P\sin\theta + T\sin\theta_1)$$

$$= P\cos\theta - \mu W_2 + \mu P\sin\theta - \mu T\sin\theta_1$$

or

$$T(\cos\theta_1 + \mu\sin\theta_1) = P(\cos\theta + \mu\sin\theta) - \mu W_2$$

$$T = \frac{P(\cos\theta + \mu\sin\theta) - \mu W_2}{\cos\theta_1 + \mu\sin\theta_1} \qquad \ldots(ii)$$

Equating expression (*i*) and (*ii*),

$$\frac{\mu W_1}{\cos\theta_1 + \mu\sin\theta_1} = \frac{P(\cos\theta + \mu\sin\theta) - \mu W_2}{\cos\theta_1 + \mu\sin\theta_1}$$

or

$$\mu W_1 = P(\cos\theta + \mu\sin\theta) - \mu W_2$$

$$\therefore \quad P = \frac{\mu(W_1 + W_2)}{\cos\theta + \mu\sin\theta}$$

Substituting $\mu = \tan\phi = \dfrac{\sin\phi}{\cos\phi}$ where ϕ is the angle of friction,

$$P = \frac{\frac{\sin\phi}{\cos\phi}(W_1 + W_2)}{\cos\theta + \frac{\sin\phi}{\cos\phi}\sin\theta}$$

$$= \frac{\sin\phi(W_1 + W_2)}{\cos\phi\,\cos\phi + \sin\phi\,\sin\theta} = \frac{(W_1 + W_2)\sin\phi}{\cos(\theta - \phi)} \qquad ...(9.3)$$

The pull P will be least if the demoninator is maximum. The parameter $\cos(\theta - \phi)$ has a maximum value equal to 1.

$$\cos(\theta - \phi) = 1 = \cos 0°$$

$$\theta - \phi = 0\,; \quad \theta = \phi \qquad ...(9.4)$$

Thus the tractive force P will be minimum if its inclination with the horizontal is equal to the angle of friction. Further

$$P_{min} = (W_1 + W_2)\sin\phi \qquad ...(9.5)$$

EXAMPLE 9.13

A homogeneous block of weight W and height b rests on a horizontal plane on its square base of side a. A force P is applied horizontally to the block at height h. If the coefficient of friction is μ, determine the maximum value of h so that the block slides without toppling.

Solution : Refer to Fig. 9.24 for the force system and the free body diagram.

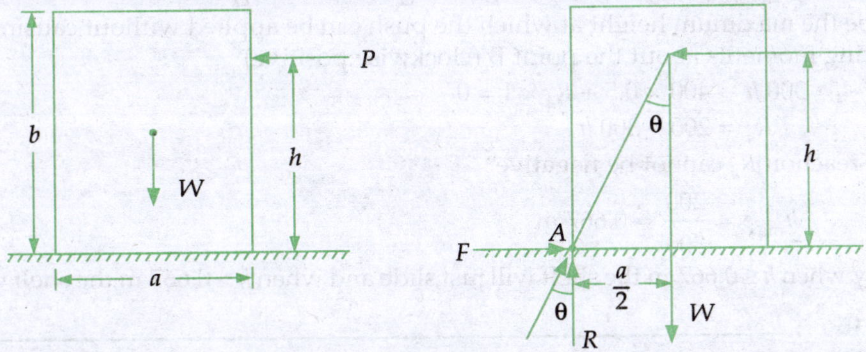

Fig. 9.24

At the instant of toppling, the entire reaction between the plane and the block will be at point A. The body is acted upon by the following set of forces:

(*i*) Weight W of the block

(*ii*) Force applied P

(*iii*) Resultant force due to the frictional force F and the normal reaction

Since the system is in equilibrium, these forces would be concurrent and pass through a point. Then

$$\tan\theta = \frac{\frac{a}{2}}{h} = \frac{a}{2h}\,; \quad h = \frac{a}{2\tan\theta}$$

Also $\quad \mu = \tan\theta \qquad \therefore\ h = \dfrac{a}{2\mu}$

This is the maximum height at which a horizontal force can be applied to slide the block without toppling.

EXAMPLE 9.14

A steel shelf 1.5 m high × 1.0 m wide and weighing 400 N is mounted on bushes A and B (Fig. 9.25). These bushes donot rotate when the shelf is moved along the floor. Assuming that coefficient of friction between the bushes and floor is 0.75, work-out the force required just to cause the shelf to move. If the shelf is not to tip over, determine the maximum height at which the force can be applied.

Solution : Due to symmetry of the system, the reactions R_a and R_b at the rollers are equal

$$R_a = R_b = \frac{400}{2} = 200 \text{ N}$$

Let P be the push that needs to be applied to the shelf to just set it into motion towards right. Force of friction is always opposite to the direction of motion. Accordingly frictional force at the rollers would act towards left and it equals

$$F_a = F_b = \mu R = 0.75 \times 200 = 150 \text{ N}$$

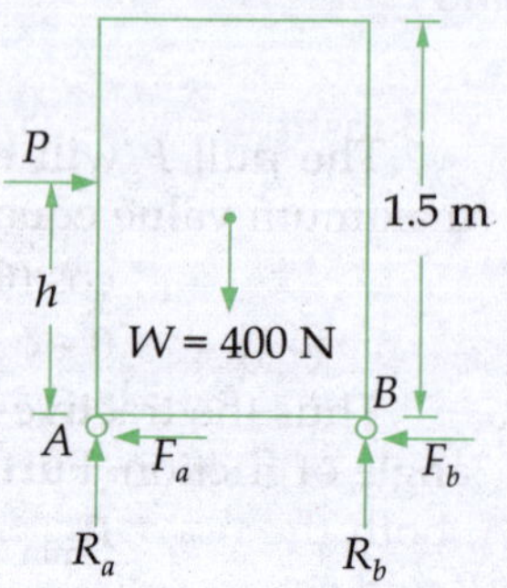

Fig. 9.25

In the limiting equilibrium (impending motion), the forces are balanced. That is

$$\Sigma F_x = 0\,; \quad P - F_a - F_b = 0$$
$$P = F_a + F_b = 150 + 150 = 300 \text{ N}$$

Thus a force of 300 N is required for impending motion to right.

(*b*) Let h be the maximum height at which the push can be applied without causing tip over of the shelf. Taking moments about the point B (clockwise positive)

$$300\,h - 400 \times 0.5 + R_a \times 1 = 0$$
$$R_a = 200 - 300\,h$$

Since the reaction R_a cannot be negative

$$h_{\max} = \frac{200}{300} = 0.667 \text{ m}$$

Obviously when $h \le 0.667$ m the shelf will just slide and when $h > 0.667$ m the shelf will tip over.

EXAMPLE 9.15

A rectangular block, 1.25 m high × 1 m wide, weighing 200 N rests on a rough horizontal plane ($\mu = 0.25$). A pull P is applied to the block at its left hand top corner at an angle of 25° with the horizontal. Make calculations for the value of P if the block is to

(i) slide over the horizontal surface, and

(ii) tip about the left hand bottom edge.

Comment on the results.

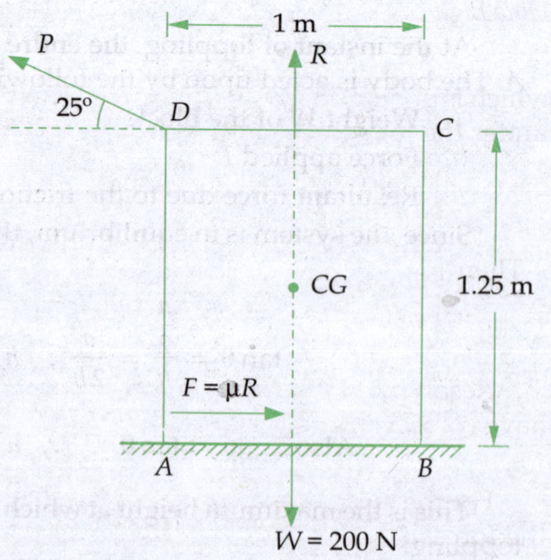

Fig. 9.26

Solution : When the block slides, it is in equilibrium under the following set of forces:

(*i*) Weight W of block acting vertically downwards through its centre of gravity CG.

(*ii*) the normal reaction R acting upward along the line of action of weight.

(*iii*) Friction force $F = \mu R$ acting along the bottom side towards right (opposite to the direction in which the block slides).

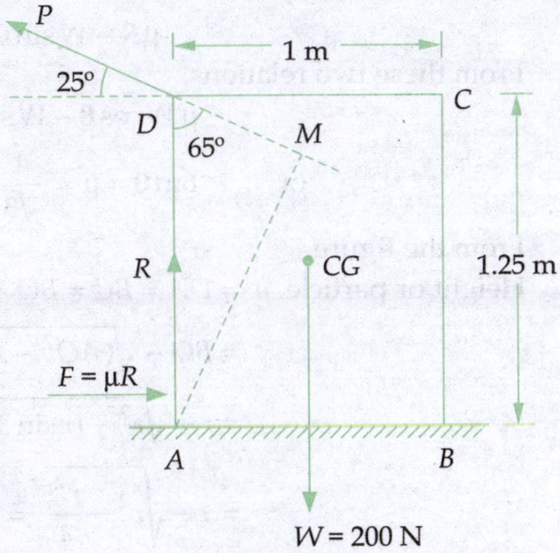

Fig. 9.27

Under equilibrium, the components of all forces along horizontal and vertical directions are zero. That is

$$F - P\cos 25° = 0; \quad \mu R = P\cos 25° = 0.906\,P \quad ...(i)$$

and $\quad R - W + P\sin 25° = 0; R = W - P\sin 25°$

$$= 200 - 0.423\,P$$

Substituting this value of R in identity (i), we get

$$0.25(200 - 0.423P) = 0.906P$$

or $\quad 50 - 0.106P = 0.906P$

$\therefore$ Magnitude of pull, $P = \dfrac{50}{0.906 + 0.106} = $ **49.41 N**

(ii) When the block is to tip about the corner point A, both the normal reaction R and frictional force $F = \mu R$ will act at this point.

At equilibrium condition of the block, the moment of all forces about any point is zero. Then taking moments of forces about point A at which the block is to tip, we get

$$P \times AM - W \times \frac{AB}{2} = 0 \qquad \text{(moments of } R \text{ and } F \text{ are each equal to zero)}$$

where AM is the perpendicular drawn on the line of action of the required force.

$$P \times 1.25\sin 65° - 200 \times \frac{1}{2} = 0$$

$\therefore$ Magnitude of pull, $P = \dfrac{100}{1.25\sin 65°} = $ **88.3 N**

Comments: When the pull P is applied gradually, the block will slide before tipping. This stems from the fact that when P reaches a value of 49.41 N, sliding starts and it continues till P reaches 88.3 N. At $P = 88.3$ N, the block tips about the corner point A.

EXAMPLE 9.16

How high can a particle rest inside a hollow sphere of radius r if the coefficient of friction be $\dfrac{1}{\sqrt{3}}$?

Solution : With reference to Fig. 9.28, let the particle be at rest in limiting equilibrium at point A which is at height $h = AD$ from the lowest point of the sphere. Further, the particle is in equilibrium under the following set of forces:

(i) Weight W of particle acting vertically downwards,

(ii) Normal reaction R which is at right angle to the tangent AE and passes through the centre O of the sphere,

(iii) Friction force $F = \mu R$ which acts along the tangent AE.

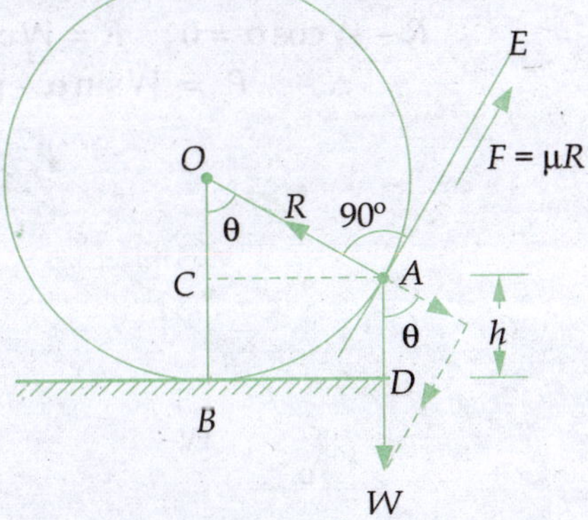

Fig. 9.28

Resolving the forces along and perpendicular to AE, we have

$$\mu R - W\sin\theta \quad \text{and} \quad R - W\cos\theta$$

For the system to be in equilibrium, the resolved parts of the forces along any two mutually perpendicular directions are separately equal to zero. Accordingly

$$\mu R - W\sin\theta = 0 \quad \text{and} \quad R - W\cos\theta = 0$$

From these two relations,

$$\mu W\cos\theta - W\sin\theta = 0$$

or $$\tan\theta = \mu = \frac{1}{\sqrt{3}}; \quad \theta = 30°$$

From the figure,

Height or particle, $h = DA = BC = BO - CO$

$$= BO - \sqrt{(AO^2 - AC^2)}$$

$$= r - \sqrt{r^2 - (r\sin 30°)^2}$$

$$= r - \sqrt{r^2 - \frac{r^2}{4}} = r - r\sqrt{1 - \frac{1}{4}}$$

$$= r\left(1 - \frac{\sqrt{3}}{2}\right) = 0.135\,r$$

9.6. INCLINED PLANES

An inclined plane is a plane that makes a certain angle to the horizontal. Such planes are quite common, e.g., steep hill road and ramp in a building.

EXAMPLE 9.17

A body of weight W is placed on a rough inclined plane of inclination α. Find the force which should be applied parallel to the plane so that the body is on the point of motion: (i) down the plane (ii) up the plane.

Solution : Refer to Fig. 9.29 which shows a body of weight W resting on a rough inclined plane of inclination α with the horizontal. The body is subjected to a force P acting along the plane:

(i) When tendency of the body is to move downwards (Fig. 9.30):

The friction force on the body acts up the plane, since the body has impending motion down the plane. The various forces which keep the body in equilibrium and as shown in the free body diagram are:

(i) weight W of the body, (ii) reaction R of the plane,
(iii) tractive force P, and (iv) frictional force $F = \mu R$

Resolving these forces along and perpendicular to the plane, we get

$$P + \mu R - W\sin\alpha = 0; \quad P = W\sin\alpha - \mu R$$

$$R - W\cos\alpha = 0; \quad R = W\cos\alpha$$

$$\therefore \quad P = W\sin\alpha - \mu W\cos\alpha$$

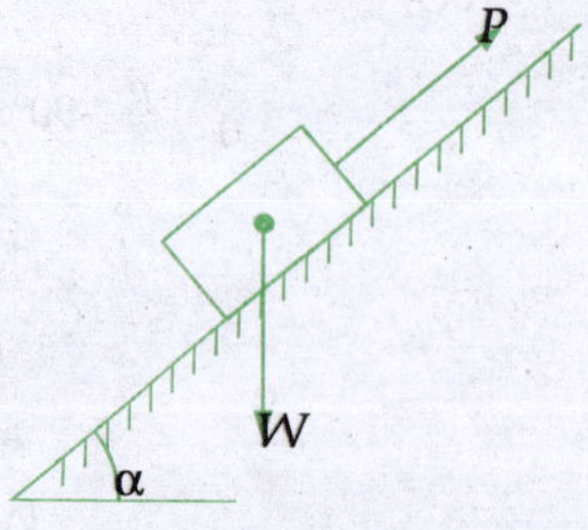

Fig. 9.29

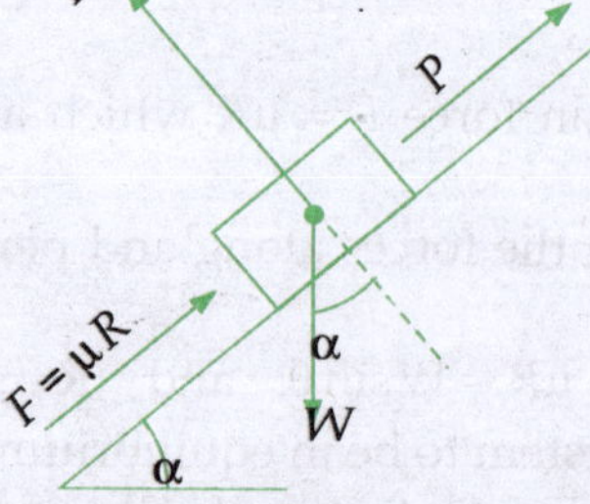

Fig. 9.30

Substituting $\mu = \tan\phi$ where ϕ is the angle of friction

$$P = W\sin\alpha - \tan\phi\, W\cos\alpha = W\sin\alpha - \frac{\sin\phi}{\cos\phi}\, W\cos\alpha$$

$$= \frac{W}{\cos\phi}(\sin\alpha\cos\phi - \sin\phi\cos\alpha) = \frac{W}{\cos\phi}\sin(\alpha - \phi) \quad \text{...(9.6)}$$

(*ii*) **When tendency of the body is to move upwards (Fig. 9.31):**

The friction force on the body acts down the plane, since the body has impending motion up the plane. The various forces which keep the body in equilibrium and as shown in the free body diagram are:

(*i*) weight W of the body,
(*ii*) reaction R of the plane,
(*iii*) tractive force P, and
(*iv*) frictional force $F = \mu R$

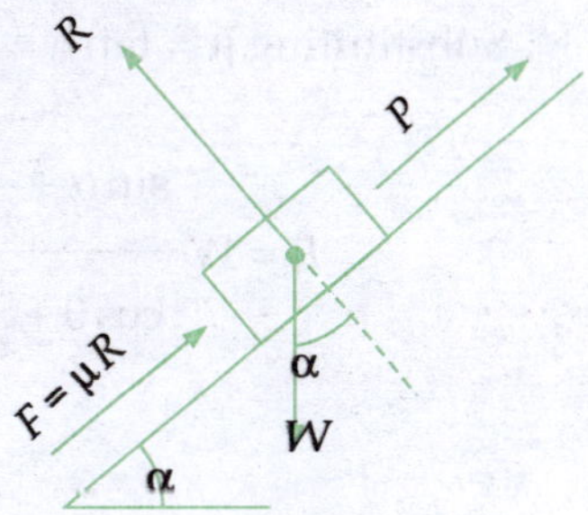

Fig. 9.31

Resolving these forces along and perpendicular to the plane, we get

$$P - \mu R - W\sin\alpha = 0$$
$$P = W\sin\alpha + \mu R$$
$$R - W\cos\alpha = 0;\quad R = W\cos\alpha$$
$$\therefore \quad P = W\sin\alpha + \mu W\cos\alpha$$

Substituting $\mu = \tan\phi$ where ϕ is the angle of friction

$$P = W\sin\alpha + \tan\phi\, W\cos\alpha = W\sin\alpha + \frac{\sin\phi}{\cos\phi}\, W\cos\alpha$$

$$= \frac{W}{\cos\phi}(\sin\alpha\cos\phi + \sin\phi\cos\alpha) = \frac{W}{\cos\phi}\sin(\alpha + \phi) \quad \text{...(9.7)}$$

EXAMPLE 9.18

Find the least force required to drag a block of weight W placed on a rough inclined plane having inclination α with the horizontal. The tractive force applied to the block makes an angle θ to the inclined plane. Consider the following cases: (i) the block is to move up the plane, (ii) the block is to move down the plane.

Solution : Refer Fig. 9.32 which shows a block of weight W resting on a rough inclined plane of inclination α with the horizontal. The block is subjected to a force P acting at angle θ with the inclined plane.

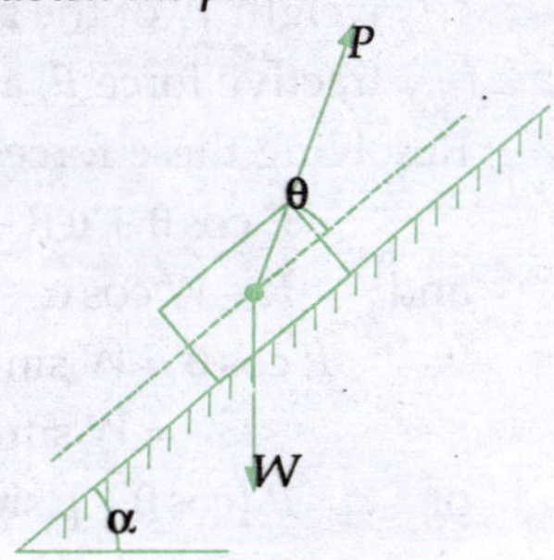

Fig. 9.32

(*i*) **When tendency of the block is to move up the plane (Fig. 9.33)**

Since the block has impending motion up the plane, the friction force would act down the plane. The various forces which keep the block in equilibrium and as shown in the free body diagram are:

(*i*) weight W of the block,
(*ii*) reaction R of the plane,
(*iii*) tractive force P, and
(*iv*) friction force $F = \mu R$

Resolving these forces along and perpendicular to the plane, we get

$$P\cos\theta - \mu R - W\sin\alpha = 0$$

or $\quad P\cos\theta = W\sin\alpha + \mu R$

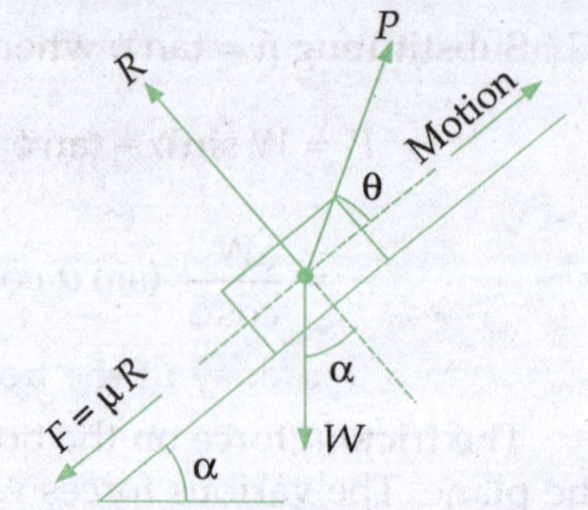

Fig. 9.33

$$\text{and} \quad R - W\cos\alpha + P\sin\theta = 0$$

$$\text{or} \quad R = W\cos\alpha - P\sin\theta$$

$$\therefore \quad P\cos\theta = W\sin\alpha + \mu\,(W\cos\alpha - P\sin\theta)$$

$$= W\sin\alpha + \mu W\cos\alpha - \mu P\sin\theta$$

$$\text{or} \quad P\cos\theta + \mu P\sin\theta = W\sin\alpha + \mu W\cos\alpha$$

$$\text{or} \quad P\,(\cos\theta + \mu\sin\theta) = W\,(\sin\alpha + \mu\cos\alpha)$$

$$\therefore \quad P = W\,\frac{\sin\alpha + \mu\cos\alpha}{\cos\theta + \mu\sin\theta}$$

Substituting $\mu = \tan\phi = \dfrac{\sin\phi}{\cos\phi}$, where ϕ is the angle of friction.

$$P = W\,\frac{\sin\alpha + \dfrac{\sin\phi}{\cos\phi}\cos\alpha}{\cos\theta + \dfrac{\sin\phi}{\cos\phi}\sin\theta} = W\,\frac{\sin\alpha\cos\phi + \sin\phi\cos\alpha}{\cos\theta\cos\phi + \sin\theta\sin\phi}.$$

$$= W\,\frac{\sin(\alpha + \phi)}{\cos(\theta - \phi)} \qquad \text{...(9.8)}$$

The pull P will be least if the denominator $\cos(\theta - \phi)$ is maximum, and the maximum value of $\cos(\theta - \phi)$ is equal to 1

$$\cos(\theta - \phi) = \phi\,; \quad \cos(\theta - \phi) = \cos 0$$

$$\theta - \phi = 0\,; \quad \theta = \phi \qquad \text{...(9.9)}$$

Thus the tractive force P will be minimum if its inclination with the inclined plane is equal to the angle of friction. Further

$$P_{\min} = W\sin(\alpha + \phi)$$

(*ii*) When tendency of the block is to move down the plane (Fig. 9.34)

Since the block has impending motion down the plane, the friction force would act up the plane. The various forces which keep the block in equilibrioum and as shown in the free body diagram are:

(*i*) weight W of the block, (*ii*) reaction R of the plane,
(*iii*) tractive force P, and (*iv*) frictional force $F = \mu R$

Resolving these forces along and perpendicular to the plane, we get

$$P\cos\theta + \mu R - W\sin\alpha = 0 \quad \text{or} \quad P\cos\theta = W\sin\alpha - \mu R$$

$$\text{and} \quad R - W\cos\alpha + P\sin\theta = 0 \quad \text{or} \quad R = W\cos\alpha - P\sin\theta$$

$$\therefore \quad P\cos\theta = W\sin\alpha - \mu\,(W\cos\alpha - P\sin\theta)$$

$$= W\sin\alpha - \mu W\cos\alpha + \mu P\sin\theta$$

$$\text{or} \quad P\,(\cos\theta - \mu\sin\theta) = W\,(\sin\alpha - \mu\cos\alpha)$$

$$\therefore \quad P = W\,\frac{\sin\alpha - \mu\cos\alpha}{\cos\theta - \mu\sin\theta}$$

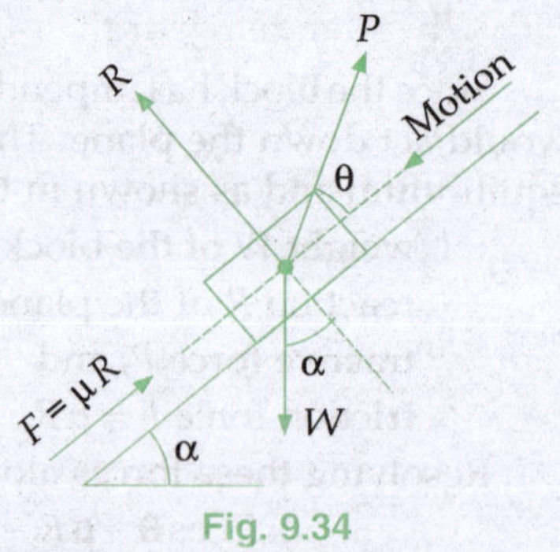

Fig. 9.34

Substituting $\mu = \tan\phi = \dfrac{\sin\phi}{\cos\phi}$ where ϕ is the angle of friction

$$P = W \frac{\sin\alpha - \dfrac{\sin\phi}{\cos\phi}\cos\alpha}{\cos\theta - \dfrac{\sin\phi}{\cos\phi}\sin\theta}$$

$$= W \frac{\sin\alpha\cos\phi - \sin\phi\cos\alpha}{\cos\theta\cos\phi - \sin\theta\sin\phi} = W \frac{\sin(\alpha - \phi)}{\cos(\theta + \phi)} \qquad ...(9.10)$$

EXAMPLE 9.19

A block of weight W rests on a rough inclined plane of inclination α with the horizontal. The block is subjected to a force P acting horizontally. Set up the following relations for the equilibrium of the block:

$P = W\tan(\alpha - \phi)$ *when the block is at the point of sliding down*

$= W\tan(\alpha + \phi)$ *when the block is at the point of sliding up*

Here φ is the angle of friction.

Solution : Refer to Fig. 9.35 which shows a block of weight W resting on a rough inclined plane of inclination α with the horizontal. The block is subjected to a force P that acts horizontally.

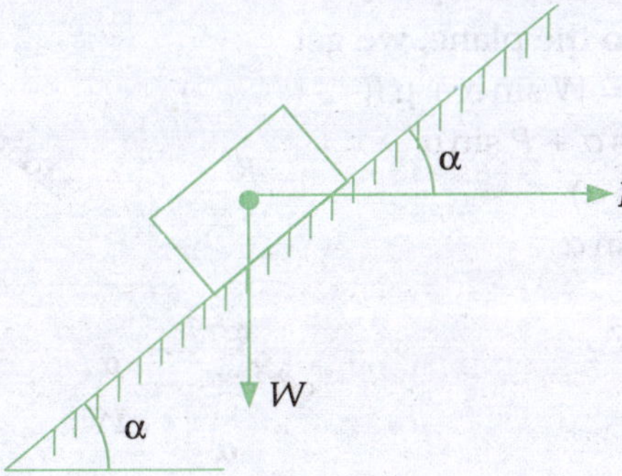

Fig. 9.35

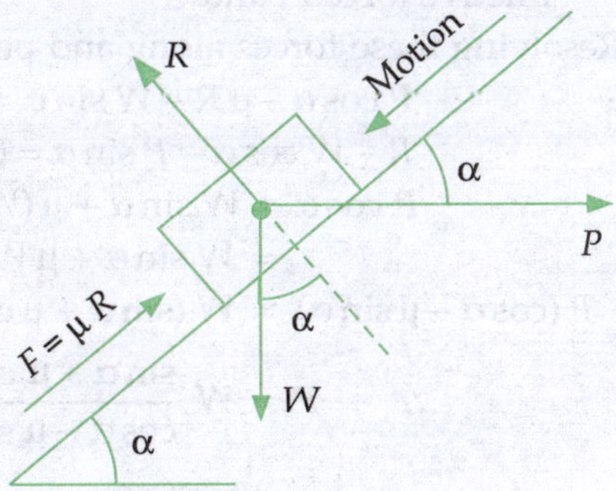

Fig. 9.36

(i) When tendency of the block is to move down the plane (Fig. 9.36)

Since the block has impending motion down the plane, the friction force would act up the plane. The various forces which keep the block in equilibrium and as shown in the free body diagram are:

(i) weight W of the block, (ii) reaction R of the plane,
(iii) tractive force P, and (iv) frictional force $F = \mu R$

Resolving these forces along and perpendicular to the plane, we get

$$P\cos\alpha + \mu R - W\sin\alpha = 0$$

or $$P\cos\alpha = W\sin\alpha - \mu R$$

and $$R - W\cos\alpha - P\sin\alpha = 0\,;\quad R = W\cos\alpha + P\sin\alpha$$

∴ $$P\cos\alpha = W\sin\alpha - \mu(W\cos\alpha + P\sin\alpha)$$
$$= W\sin\alpha - \mu W\cos\alpha - \mu P\sin\alpha$$

or $$P(\cos\alpha + \mu\sin\alpha) = W(\sin\alpha - \mu\cos\alpha)$$

∴ $$P = W\frac{\sin\alpha - \mu\cos\alpha}{\cos\alpha + \mu\sin\alpha}$$

Substituting $\mu = \tan\phi = \dfrac{\sin\phi}{\cos\phi}$ where φ is the angle of friction

$$P = W \frac{\sin\alpha - \dfrac{\sin\phi}{\cos\phi}\cos\alpha}{\cos\alpha + \dfrac{\sin\phi}{\cos\phi}\sin\alpha} = W \frac{\sin\alpha\cos\phi - \sin\phi\cos\alpha}{\cos\alpha\cos\phi + \sin\alpha\sin\phi}$$

$$= W \frac{\sin(\alpha - \phi)}{\cos(\alpha - \phi)} = W \tan(\alpha - \phi) \qquad \text{...(9.11)}$$

Note : The above relation is essentially valid when $\alpha > \phi$. In case $\phi > \alpha$, this relation takes the form

$$P = W \tan(\phi - \alpha)$$

(ii) **When tendency of the block is to move up the plane (Fig. 9.37)**

Since the block has impending motion up the plane, the friction force would act down the plane. The various forces which keep the block in equilibrium and as shown in the free body diagram are:

(i) weight W of the block, *(ii)* reaction R of the plane
(iii) tractive force P, and *(iv)* frictional force $F = \mu R$

Resolving these forces along and perpendicular to the plane, we get

$$P\cos\alpha - \mu R - W\sin\alpha = 0\,; \quad P\cos\alpha = W\sin\alpha + \mu R$$

$$R - W\cos\alpha - P\sin\alpha = 0\,; \quad R = W\cos\alpha + P\sin\alpha$$

$$\therefore \quad P\cos\alpha = W\sin\alpha + \mu(W\cos\alpha + P\sin\alpha)$$

$$= W\sin\alpha + \mu W\cos\alpha + \mu P\sin\alpha$$

or $$P(\cos\alpha - \mu\sin\alpha) = W(\sin\alpha + \mu\cos\alpha)$$

$$\therefore \quad P = W \frac{\sin\alpha + \mu\cos\alpha}{\cos\alpha - \mu\sin\alpha}$$

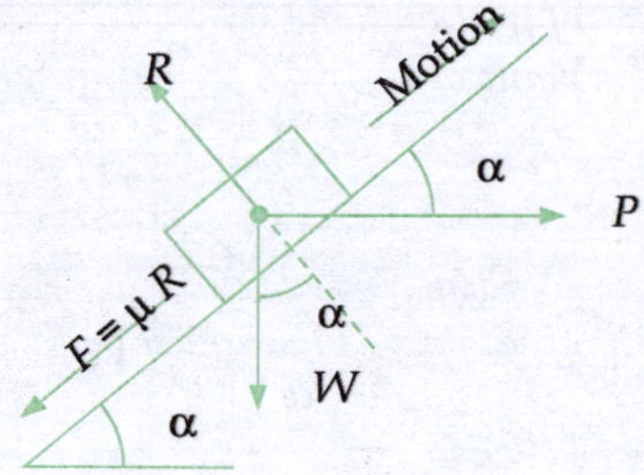

Fig. 9.37

Substituting $\mu = \tan\phi = \dfrac{\sin\phi}{\cos\phi}$ where ϕ is the angle of friction

$$P = W \frac{\sin\alpha + \dfrac{\sin\phi}{\cos\phi}\cos\alpha}{\cos\alpha - \dfrac{\sin\phi}{\cos\phi}\sin\alpha} = W \frac{\sin\alpha\cos\phi + \sin\phi\cos\alpha}{\cos\alpha\cos\phi - \sin\phi\sin\alpha}$$

$$= W \frac{\sin(\alpha + \phi)}{\cos(\alpha + \phi)} = W \tan(\alpha + \phi) \qquad \text{...(9.12)}$$

EXAMPLE 9.20

A block, in the shape of a rectangular prism rests on a rough inclined plane, as shown in Fig. 9.38. The block is tied up by a horizontal string which has a tension of 10 N. If the block weighs 35 N, determine:

(i) the frictional force on the block,
(ii) the normal reaction of the inclined plane, and
(iii) the coefficient of friction between the contacting surfaces.

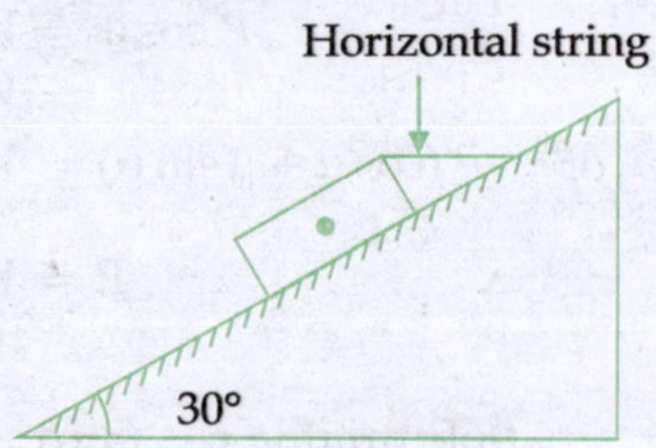

Fig. 9.38

Solution : The various forces acting on the block are:

(i) weight $W = 35$ N of the block,
(ii) tension $T = 10$ N in the string,
(iii) normal reaction R at the contact surface, and
(iv) friction force $F = \mu R$

Refer to Fig. 9.39 for the free body diagram of the force system under equilibrium conditions,

$$\Sigma F_x = 0 \text{ (along the plane)}$$

$$\mu R - W \sin 30^\circ + T \cos 30^\circ = 0$$

$$\mu R = W \sin 30^\circ - T \cos 30^\circ$$

$$= 35 \sin 30^\circ - 10 \cos 30^\circ$$

$$= 35 \times 0.5 - 10 \times 0.866$$

$$= 17.5 - 8.66 = 8.84 \text{ N} \quad ...(i)$$

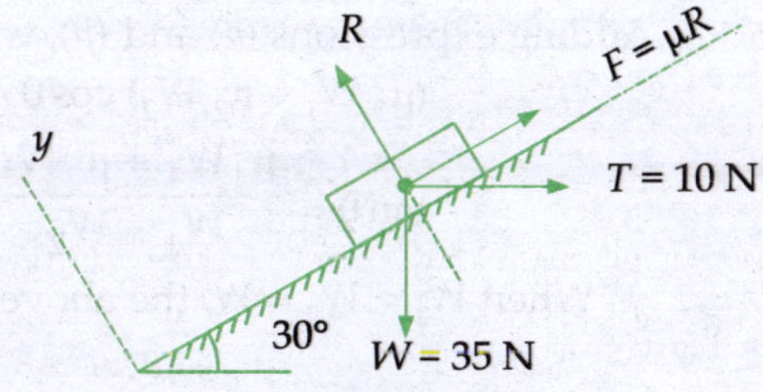

Fig. 9.39

$$\Sigma F_y = 0 \text{ (perpendicular to the plane)}$$

$$R - W \cos 30^\circ - T \sin 30^\circ = 0$$

$$R = W \cos 30^\circ + T \sin 30^\circ$$

$$= 35 \times 0.866 + 10 \times 0.5$$

$$= 30.31 + 5 = 35.31 \text{ N} \quad ...(ii)$$

From identities (*i*) and (*ii*)

$$\mu = \frac{8.84}{35.31} = 0.25$$

EXAMPLE 9.21

Two blocks of weight W_1 and W_2 rest on a rough inclined plane and are connected by a string. The coefficients of friction between these blocks and the plane are μ_1 and μ_2 respectively. Show that the blocks will be on the point of motion when the inclination of the plane with the horizontal is given by

$$\tan \theta = \frac{\mu_1 W_1 + \mu_2 W_2}{W_1 + W_2}$$

What will be this inclination if the blocks are of equal weight and the friction coefficients are $\mu_1 = 1/2$ and $\mu_2 = 1/3$?

Solution : Refer Fig. 9.40 for the arrangement and the free body diagrams for the two blocks.

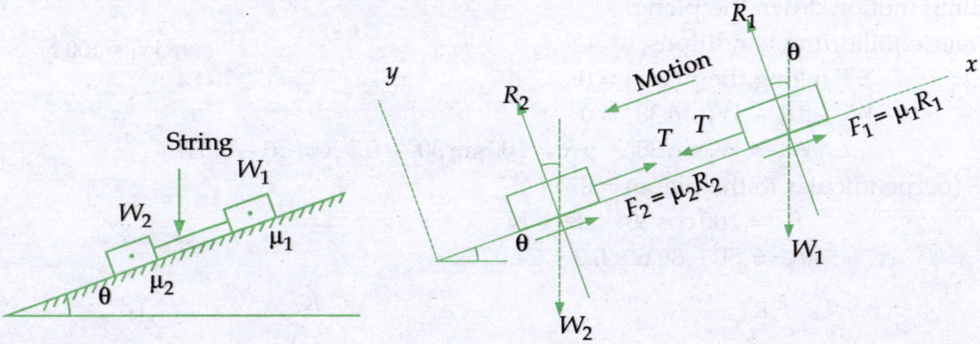

Fig. 9.40

Let T be the tension in the string and θ be the inclination of the plane with the horizontal.

Considering equilibrium for block of weight W_1

$$\Sigma F_x = 0 \text{ (along the plane)}$$

$$\mu_1 R_1 - T - W_1 \sin \theta = 0 \quad ...(i)$$

$$\Sigma F_y = 0 \text{ (perpendicular to the plane)}$$

$$R_1 - W_1 \cos \theta = 0 ; \quad R_1 = W_1 \cos \theta \quad ...(ii)$$

From identities (*i*) and (*ii*)

$$\mu_1 W_1 \cos \theta - W_1 \sin \theta - T = 0 \quad ...(a)$$

Similarly, considering equilibrium for block of weight W_2, we may write

$$\mu_2 R_2 - W_2 \sin\theta + T = 0$$

$$R_2 - W_2 \cos\theta = 0; \quad R_2 = W_2 \cos\theta$$

or $$\mu_2 W_2 \cos\theta - W_2 \sin\theta + T = 0 \qquad ...(b)$$

Adding expressions (*a*) and (*b*), we get

$$(\mu_1 W_1 + \mu_2 W_2)\cos\theta - (W_1 + W_2)\sin\theta = 0$$

$$\therefore \quad \tan\theta = \frac{\mu_1 W_1 + \mu_2 W_2}{W_1 + W_2}$$ which is the required expression.

(*b*) When $W_1 = W_2 = W$, the above expression reduces to

$$\tan\theta = \frac{\mu_1 + \mu_2}{2}$$

Substituting $\mu_1 = 1/2$ and $\mu_2 = 1/3$,

$$\tan\theta = \frac{1/2 + 1/3}{2} = \frac{5}{12}$$

$\therefore$ Inclination of plane, $\theta = \tan^{-1}(5/12) = 22.62°$

EXAMPLE 9.22

A block of weight W_1 = 100 N rests on an inclined plane and another weight W_2 is attached to the first weight through a string as shown in Fig. 9.41. If the coefficient of friction between the block and plane is 0.3, determine the maximum and minimum values of W_2 so that equilibrium can exist.

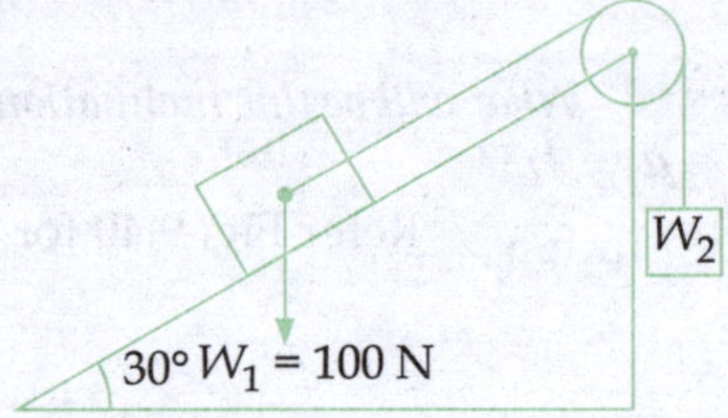

Fig. 9.41

Solution : There will be two values for W_2

(*i*) When weight W_1 = 100 N is about to slide down the plane

(*ii*) When weight W_1 = 100 N is about to slide up the plane

Case (*i*): Refer Fig. 9.42 (*a*) for the free body diagram for impending motion down the plane.

Under equilibrium conditions,

$$\Sigma F \text{ (along the plane)} = 0$$

$$W_2 + \mu R - W_1 \sin 30° = 0$$

$$W_2 = W_1 \sin 30° - \mu R = 100 \sin 30° - 0.3R = 50 - 0.3R$$

ΣF (perpendicular to the plane) = 0

$$R = 100 \cos 30° = 86.6 \text{ N}$$

$$\therefore \quad W_2 = 50 - 86.6 \times 0.3 = 24.02 \text{ N} \simeq 24 \text{ N}$$

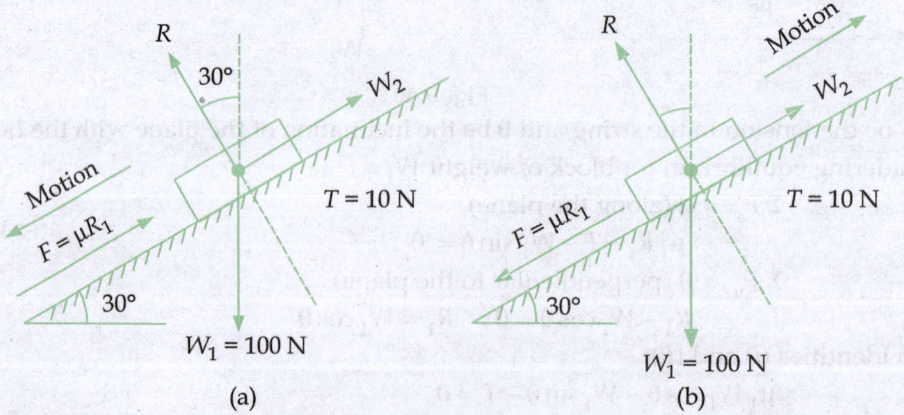

Fig. 9.42

Case (*ii*): Refer Fig. 9.42 (*b*) for the free body diagram for impending motion up the plane. Under equilibrium conditons,

$$\Sigma F \text{ (along the plane)} = 0$$

$$W_2 - W_1 \sin 30^\circ - \mu R = 0$$

$$W_2 = \mu R + W_1 \sin 30^\circ = 0.3\,R + 100 \sin 30^\circ = 0.3\,R + 50$$

$$\Sigma F \text{ (perpendicular to plane)} = 0$$

$$R = 100 \cos 30^\circ = 86.6$$

$$\therefore \quad W_2 = 0.3 \times 86.6 + 50 = 75.98 \text{ N} \simeq 76 \text{ N}$$

Therefore the limiting values of weight *W* are 24 N and 76 N.

EXAMPLE 9.23

Two blocks of weight 200 N and 300 N and connected by a string passing over a frictionless pulley rest on rough surfaces; block of weight 200 N on horizontal surface and the other on an inclined surface as shown in Fig. 9.43.

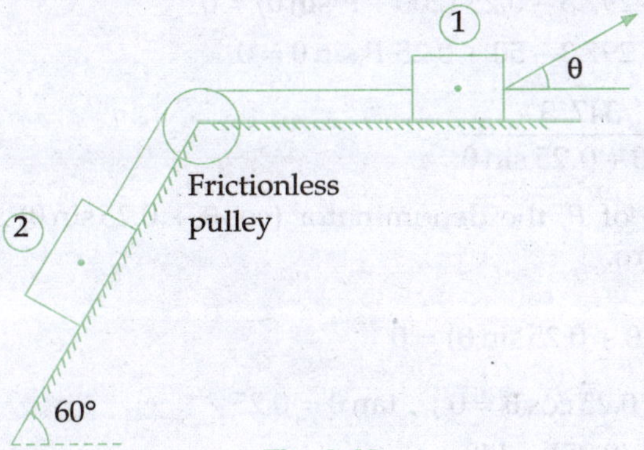

Fig. 9.43

For both the surfaces the coefficients of friction $\mu = 0.25$. Find out the minimum value of force, both in magnitude and direction, for the motion to impend.

Solution : Refer to Fig. 9.44 for the various forces acting on the two blocks:

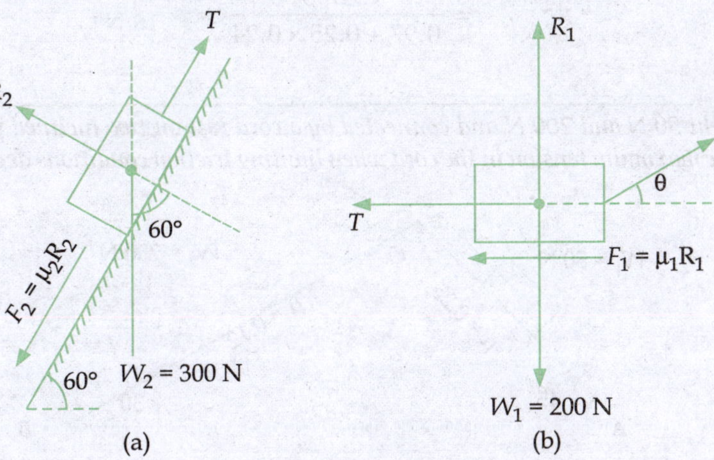

Fig. 9.44

Considering equilibrium of block 2

ΣF (along the plane) = 0 ; $T - W_2 \sin 60^\circ - \mu_2 R_2 = 0$

or $T = W_2 \sin 60^\circ + \mu_2 R_2 = 300 \times 0.866 + 0.25\, R_2$

ΣF (perpendicular to the plane) = 0 ; $R_2 - W_2 \cos 60^\circ = 0$

or $R_2 = W_2 \cos 60^\circ = 300 \times 0.5 = 150$ N

$\therefore$ $T = 300 \times 0.866 + 0.25 \times 150 = 259.8 + 37.5 =$ **297.3 N**

Considering equilibrium of block 1

ΣF (along the plane) = 0 ; $P \cos\theta - T - \mu_1 R_1 = 0$

or $P\cos\theta - 297.3 - 0.25\, R_1 = 0$ ...(i)

ΣF (perpendicular to the plane) = 0 ; $W_1 - R_1 - P\sin\theta = 0$

or $R_1 = W_1 - P\sin\theta = 200 - P\sin\theta$

Substituting this value of R_1 in expression (i), we get

$P\cos\theta - 297.3 - 0.25\,(200 - P\sin\theta) = 0$

or $P\cos\theta - 297.3 - 50 + 0.25\,P\sin\theta = 0$

or $$P = \frac{347.3}{\cos\theta + 0.25\sin\theta}$$

For minimum value of P, the denominator $(\cos\theta + 0.25\sin\theta)$ should be maximum, i.e., its differential should be zero.

$$\frac{d}{d\theta}(\cos\theta + 0.25\sin\theta) = 0$$

or $-\sin\theta + 0.25\cos\theta = 0$; $\tan\theta = 0.25$

$\therefore$ $\theta = \tan^{-1}(0.25) = 14^\circ$

$\therefore$ $$\text{Minimum value of } P = \frac{347.3}{\cos 14^\circ + 0.25\sin 14^\circ}$$

$$= \frac{347.3}{0.97 + 0.25 \times 0.24} = \mathbf{337.18\ N}$$

EXAMPLE 9.24

Two blocks of weight 50 N and 200 N and connected by a cord rest on two inclined planes as shown in Fig. 9.45. Determine the maximum tension in the cord when limiting friction conditions develop for both the blocks.

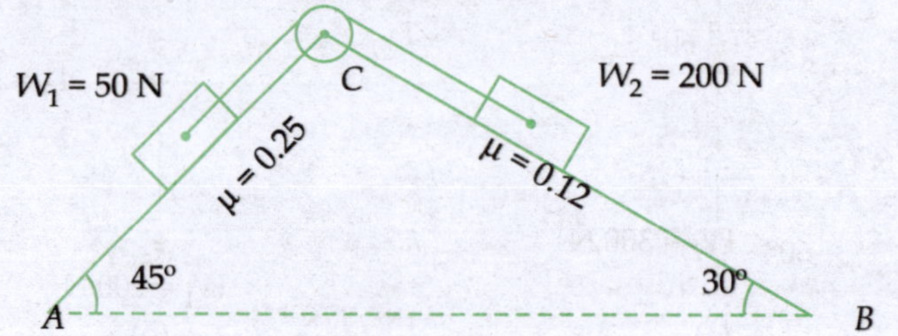

Fig. 9.45

Solution : Refer Fig. 9.46 for the free body diagram of the two blocks

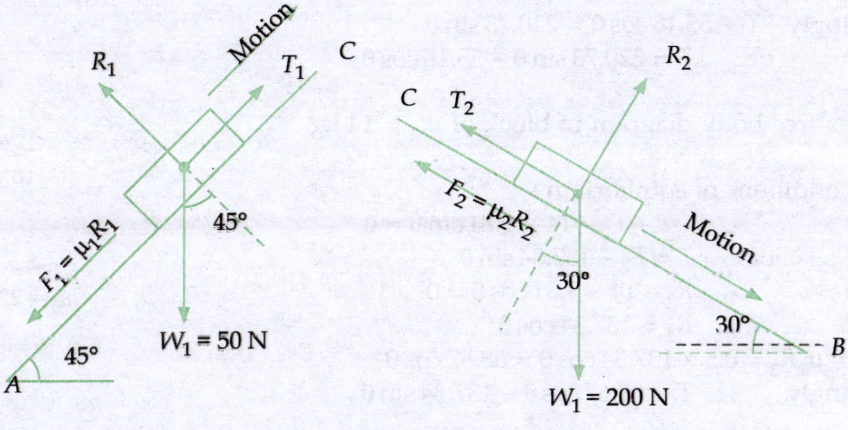

Fig. 9.46

Considering equilibrium of block resting on the plane of 45° inclination:

Resolving forces along and perpendicular to the plane AC

$$\mu_1 R_1 + W_1 \sin 45^\circ - T_1 = 0$$

$$R_1 - W_1 \cos 45^\circ = 0\,; \quad R_1 = 50 \cos 45^\circ = 35.35 \text{ N}$$

$$\therefore \quad T_1 = \mu_1 R_1 + W_1 \sin 45^\circ$$

$$= 0.25 \times 35.35 + 50 \sin 45^\circ$$

$$= 8.84 + 35.35 = 44.19 \text{ N}$$

Considering equilibrium of block resting on the plane of 30° inclination:

Resolving forces along and perpendicular to plane BC

$$T_2 + \mu_2 R_2 - W_2 \sin 30^\circ = 0$$

$$R_2 - W_2 \cos 30^\circ = 0\,; \quad R_2 = 200 \cos 30^\circ = 173.2 \text{ N}$$

$$\therefore \quad T_2 = W_2 \sin 30^\circ - \mu_2 R_2$$

$$= 200 \sin 30^\circ - 0.12 \times 173.2$$

$$= 100 - 20.78 = 79.22 \text{ N}$$

Hence cord will be subjected to a maximum tension of 79.22 N

EXAMPLE 9.25

Two masses $m_1 = 22.5$ kg and $m_2 = 14$ kg are tied together by a rope parallel to the inclined plane surface as shown in Fig. 9.47. The coefficient of friction between m_1 and the plane is 0.25 while that of mass m_2 and the plane is 0.5. Determine :

(a) *the value of the inclination of the plane surface θ for which the masses will just start sliding downwards, and*

(b) *the tension is the rope.*

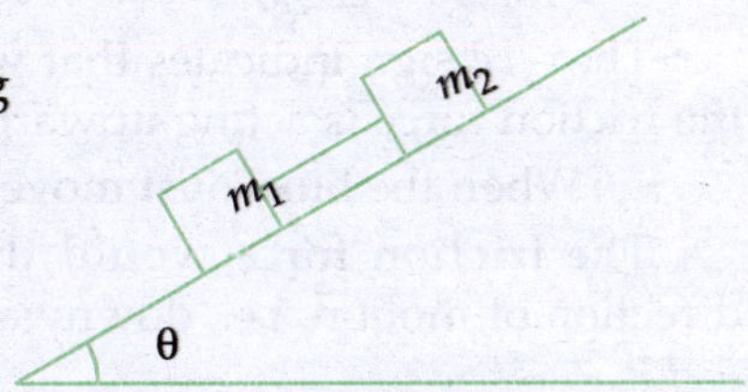

Fig. 9.47

Solution : Consider free body diagram of block of $m_1 = 22.5$ kg mass.

Applying conditions of equilibrium :

$\Sigma F_x = 0\,;$ $\quad T + F_1 - 22.5 \times 9.81 \sin\theta = 0$

or $\quad T + F_1 = 220.73 \sin\theta$

$\Sigma F_y = 0\,;$ $\quad R_1 - 22.5 \times 9.81 \cos\theta$

or $\quad R_1 = 220.73 \cos\theta$

But $F_1 = \mu_1 R_1 = 0.25 \times 220.73 \cos\theta = 55.18 \cos\theta$

Accordingly $T + 55.18 \cos\theta = 220.73 \sin\theta$

or $T = 220.73 \sin\theta - 55.18 \cos\theta$ ...(a)

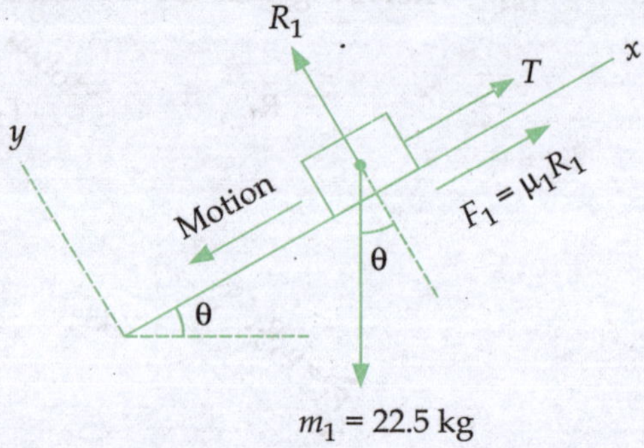

Fig. 9.48

Consider free body diagram of block of $m_2 = 14$ kg mass.

Apply conditions of equilibrium :

$\Sigma F_x = 0$; $-T + F_2 - 14 \times 9.81 \sin\theta = 0$

or $T = F_2 - 137.34 \sin\theta$

$\Sigma F_y = 0$; $R_2 - 14 \times 9.81 \cos\theta = 0$

or $R_2 = 137.34 \cos\theta$

But $F_2 = \mu_2 R_2 = 0.5 \times 137.34 \cos\theta = 68.67 \cos\theta$

Accordingly $T = 68.67 \cos\theta - 137.34 \sin\theta$ ...(b)

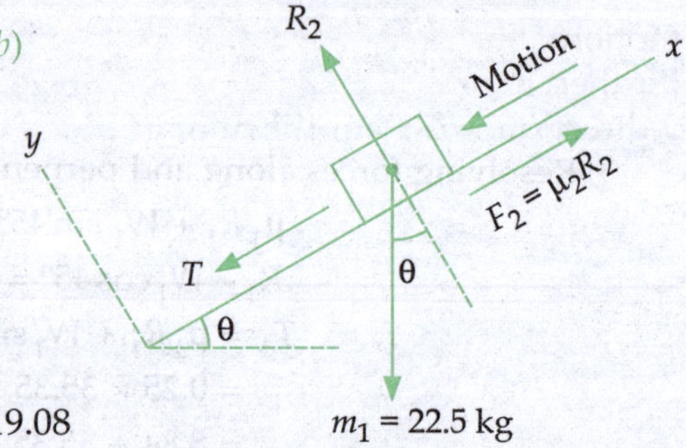

Fig. 9.49

From expressions (a) and (b)

$220.73 \sin\theta - 55.18 \cos\theta = 68.67 \cos\theta - 137.34 \sin\theta$

or $358.07 \sin\theta = 123.85 \cos\theta$

or $\tan\theta = \dfrac{123.85}{358.07} = 0.3459$

$\therefore$ $\theta = \tan^{-1}(0.3459) = 19.08°$

(b) Substituting $\theta = 19.08°$ in expression (a),

Tension in the rope $T = 220.73 \sin 19.08 - 55.18 \cos 19.08$

$= 72.15 - 52.15 =$ **20 N**

EXAMPLE 9.26

The coefficient of static and kinetic friction between the 100 kg block and inclined plane (Fig. 9.50) are 0.3 and 0.2 respectively. Determine :

(a) *the friction force F acting on the block when P is applied with a magnitude of 200 N to the block at rest,*

(b) *the force P required to initiate motion up the incline from rest, and*

(c) *the friction force F acting on the block if P = 600 N.*

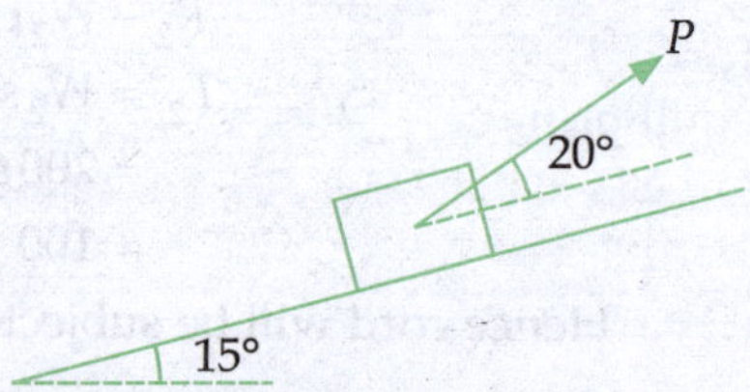

Fig. 9.50

Solution : Consider free body diagram of the block when $P = 200$ N is applied to the block at rest.

From condition of equilibrium,

$\Sigma F_x = 0$:

$200 \cos 20° - F - 100 \times 9.81 \sin 15° = 0$

$F = 200 \times 0.9397 - 100 \times 9.81 \times 0.2588$

$= 187.94 - 253.88 =$ **– 65.94 N**

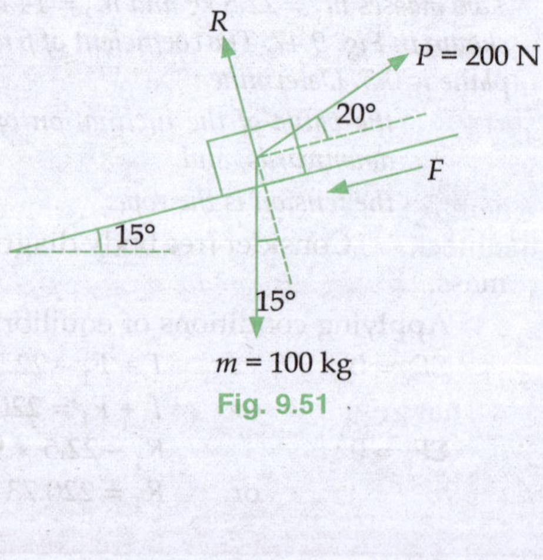

Fig. 9.51

The – *ve* sign indicates that when the block is at rest, the friction force is acting upwards along the incline.

(b) When the block just moves up the plane

The friction force would then act opposite to the direction of motion, i.e., downwards along the plane.

$F_x = 0$: $\quad P \cos 20° - F - 100 \times 9.81 \sin 15° = 0$

$$P \cos 20° = F + 100 \times 9.81 \times 0.259$$

$$0.9397\, P = 0.3\, R + 253.88 \quad \text{...}(i)$$

$F_y = 0$: $\quad R + P \sin 20° - 100 \times 9.81 \times \cos 15° = 0$

$$R = 947.57 - 0.342P \quad \text{...}(ii)$$

From expressions (*i*) and (*ii*)

$$0.9397P = 0.3\,(947.57 - 0.342P) + 253.88$$

$$= 284.27 - 0.1026P + 253.88 = 538.15 - 0.1026P$$

$$\therefore \quad P = \frac{538.15}{0.9397 + 0.1026} = 516.31 \text{ N}$$

(*c*) Since the applied force P = 600 N is more than 516.31 N (a force which just initiates the motion), the block will move up along the plane and the friction force would act down the slope. Further, the kinetic friction would prevail with $\mu = 0.2$. Then from condition of equilibrium in the *y*-direction,

$\Sigma F_y = 0$: $\quad R + 600 \sin 20° - 100 \times 9.81 \cos 15° = 0$

$$R = 100 \times 9.81 \cos 15° - 600 \sin 20°$$

$$= 947.57 - 205.21 = 742.36$$

Friction force $F = \mu R = 0.2 \times 742.36 =$ **148.47 N**

EXAMPLE 9.27

A block A weighing W newtons is placed on a rough inclined plane having $\mu = 0.2$ and is held in position by means of a horizontal rod hinged to the block B which presses against a rough vertical wall having $\mu = 0.4$, as shown in Fig. 9.52. If the block B weighs 500 N, find the minimum value of W consistent with equilibrium.

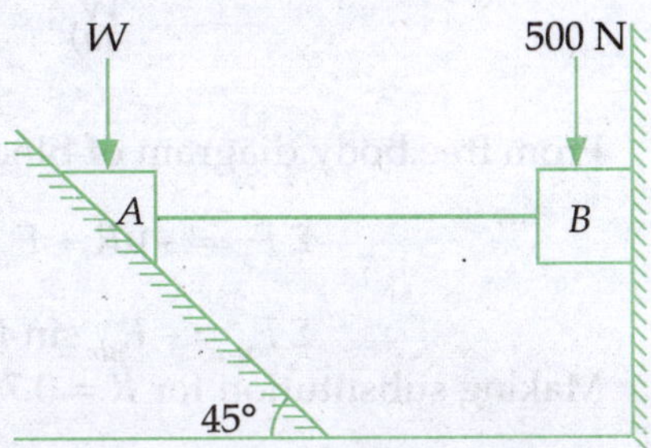

Fig. 9.52

Solution : When the block B just slides down, the frictional force $F = \mu R_b$ at the wall surface acts upwards. Under equilibrium conditions,

$\Sigma F_x = 0$: $\quad P - R_b = 0\,; \quad P = R_b$

$\Sigma F_y = 0$: $\quad \mu R_b - 500 = 0$

$$R_b = \frac{500}{0.4} = 1250 \text{ N}$$

Hence horizontal force P in the rod is 1250 N.

With block B sliding downwards, the block A has a tendency to move up the plane. The following relationship then holds good between W and P,

$$P = W \tan(\alpha + \phi)$$

where α is the inclination of the plane and ϕ is the angle of friction for the block

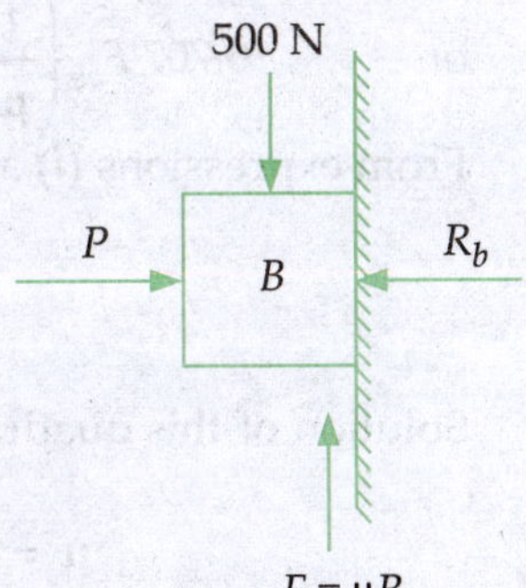

Fig. 9.53

$$\tan \phi = \mu = 0.2\,; \quad \phi = 11.30°$$

$$\therefore \quad 1250 = W \tan(45 + 11.30) = 1.5\, W$$

That gives : $W = \dfrac{1250}{1.5} = 833.3$ N

Thus the minimum value of W consistent with equilibrium is 833.3 N.

EXAMPLE 9.28

Two identical blocks A and B are connected by a rod and rest respectively against vertical wall and horizontal floor as shown in Fig. 9.54. The sliding motion of the blocks impends when rod makes an angle of 45° with the horizontal. Estimate the coefficient of friction assuming it to be same both at the floor and wall.

Solution : Let F_{ab} be the force in the rod. With reference to free body diagram of block A (Fig. 9.55 *a*) and for its limiting equilibrium

$$\Sigma F_x = 0 \quad \text{and} \quad \Sigma F_y = 0$$

$$\therefore \quad \Sigma F_x = R - F_{ab} \cos 45^\circ = 0$$

$$\text{or} \quad R = 0.707\, F_{ab}$$

$$\Sigma F_y = F - W - F_{ab} \sin 45^\circ = 0$$

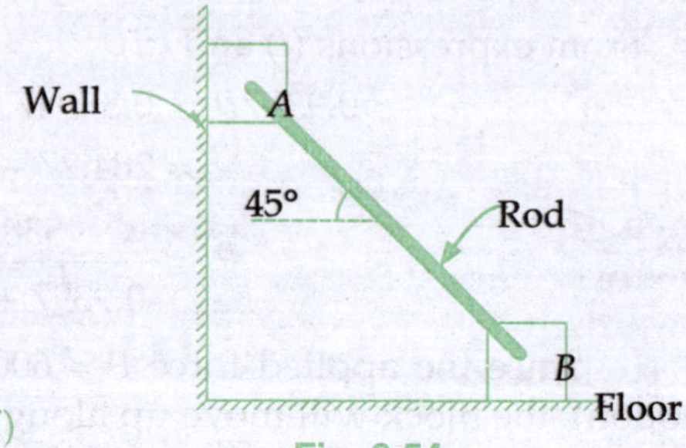

Fig. 9.54

Making substitution for $F = \mu R = \mu \times 0.707\, F_{ab}$, we have

$$\mu \times 0.707\, F_{ab} - W - 0.707\, F_{ab} = 0$$

$$F_{ab} = \frac{W}{0.707\,(1+\mu)} \quad \ldots(i)$$

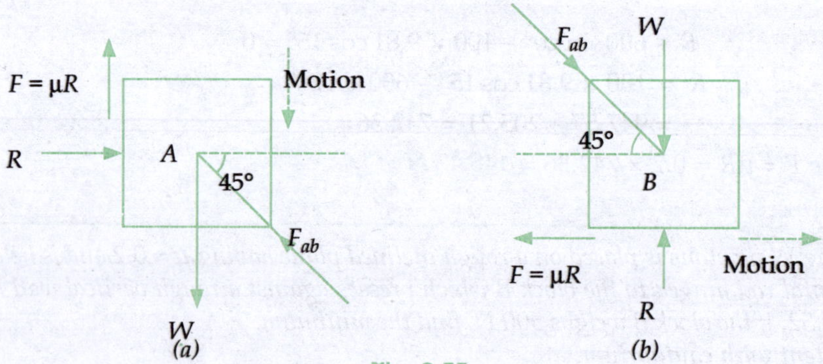

Fig. 9.55

From free body diagram of block B (Fig. 9.55 *b*), the equations for limiting equilibrium are

$$\Sigma F_x = -\mu R + F_{ab} \cos 45^\circ = 0; \quad R = \frac{0.707\, F_{ab}}{\mu}$$

$$\Sigma F_y = -F_{ab} \sin 45^\circ - W + R = 0$$

Making substitution for $R = 0.707\, F_{ab}/\mu$, we get

$$-0.707\, F_{ab} - W + \frac{0.707\, F_{ab}}{\mu} = 0$$

$$\text{or} \quad 0.707\, F_{ab} \left(\frac{1}{\mu} - 1 \right) = W\,; \quad F_{ab} = \frac{W}{0.707} \frac{\mu}{1-\mu} \quad \ldots(ii)$$

From expressions (*i*) and (*ii*)

$$\frac{W}{0.707\,(1+\mu)} = \frac{W}{0.707} \frac{\mu}{1-\mu}$$

$$\text{or} \quad (1+\mu)\,\mu = 1 - \mu\,; \quad \mu^2 + 2\mu - 1 = 0$$

Solution of this quadratic equation gives :

$$\mu = \frac{-2 \pm \sqrt{2^2 - 4 \times (-1) \times 1}}{2 \times 1} = \frac{-2 \pm 2.828}{2}$$

Neglecting – *ve* value which is impossible, we obtain

$$\mu = 0.414$$

9.7. LADDER FRICTION: NON-CONCURRENT FORCE SYSTEM

A ladder is an arrangement used for climbing on the walls or roofs. It essentially consists of two long uprights of wood or iron and connected by a number of cross bars. These cross bars are called rungs and provide steps for climbing.

Figure 9.56 shows a ladder *AB* with its end *A* resting on the ground and end *B* leaning against a wall. The ladder is acted upon by the following set of forces:

(*i*) weight *W* acting down-wards at its mid point.

(*ii*) normal reaction R_b and friction force $F_b = \mu R_b$ at the end *B* leaning against the wall. Since the ladder has a tendency to slip downwards, the friction force will be acting upwards. If the wall is smooth ($\mu = 0$), the friction force will be zero.

(*iii*) normal reaction R_a and friction force $F_a = \mu R_a$ at the end *A* resting on the floor. Since the ladder, upon slipping, tends to move away from the wall, the direction of friction force will be towards the wall.

Applying equilibrium conditions, the algebraic sum of the horizontal and vertical component of forces would be zero.

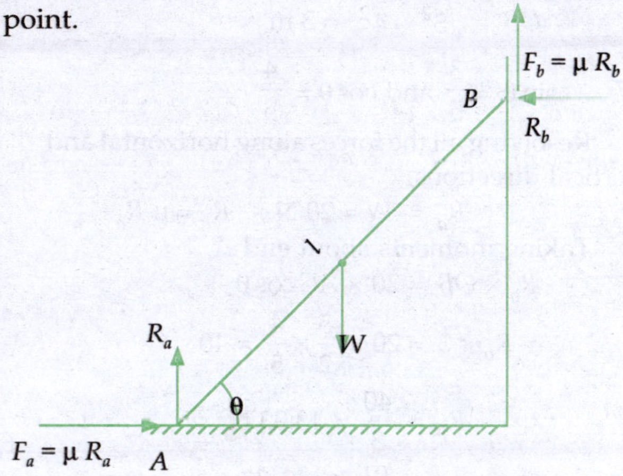

Fig. 9.56

EXAMPLE 9.29

Determine the minimum angle θ at which a uniform ladder can be placed against a wall without slipping under its own weight. The coefficient of friction for all surfaces is 0.2.

Solution : Refer Fig. 9.56 for the ladeder *AB* leaning against a wall and various forces acting on it.

Resolving all the forces along horizontal and vertical directions,

$$R_b = \mu R_a \text{ and } R_a + \mu R_b = W$$

$$\therefore \quad R_a + \mu (\mu R_a) = W \quad \text{or} \quad R_a = \frac{W}{(1+\mu^2)}$$

Taking moments about end *B*,

$$R_a l \cos\theta = W \times \frac{l}{2}\cos\theta + \mu R_a l \sin\theta$$

where *l* is the length of the uniform ladder

Substituting $R_a = \dfrac{W}{1+\mu^2}$ in the momentum equation, we get

$$\frac{Wl}{1+\mu^2}\cos\theta = \frac{Wl}{2}\cos\theta + \mu\frac{W}{1+\mu^2} l \sin\theta$$

$$\text{or} \quad \frac{\mu}{1+\mu^2}\sin\theta = \left(\frac{1}{1+\mu^2} - \frac{1}{2}\right)\cos\theta = \frac{1-\mu^2}{2(1+\mu^2)}\cos\theta$$

$$\text{or} \quad \tan\theta = \frac{1-\mu^2}{2\mu}\ ; \quad \theta = \tan^{-1}\left(\frac{1-\mu^2}{2\mu}\right)$$

$$= \tan^{-1}\left(\frac{1-0.2^2}{2\times 0.2}\right) = \mathbf{67.38^\circ}$$

EXAMPLE 9.30

A uniform ladder of length 5 m and weighing 20 N is placed against a smooth vertical wall with its lower end 4 m away from the wall. If the ladder is just to slip, determine: (a) the co-efficient of friction between the ladder and floor: (b) the frictional force acting on the ladder at the point of contact between ladder and floor.

Solution : Refer Fig. 9.57 for the ladder and various forces acting on it.

$$OA = 4 \text{ m}; \quad AB = 5 \text{ m}$$

$$OB = \sqrt{5^2 - 4^2} = 3 \text{ m}$$

$$\sin\theta = \frac{3}{5} \text{ and } \cos\theta = \frac{4}{5}$$

Resolving all the forces along horizontal and vertical directions,

$$R_a = W = 20 \text{ N}; \quad R_b = \mu R_a$$

Taking moments about end A,

$$R_b \times OB = 20 \times AC\cos\theta$$

$$R_b \times 3 = 20 \times \frac{5}{2} \times \frac{4}{5} = 40$$

$$\therefore \quad R_b = \frac{40}{3} = 13.33 \text{ N}$$

$$\mu = \frac{R_b}{R_a} = \frac{13.33}{20} = 0.667$$

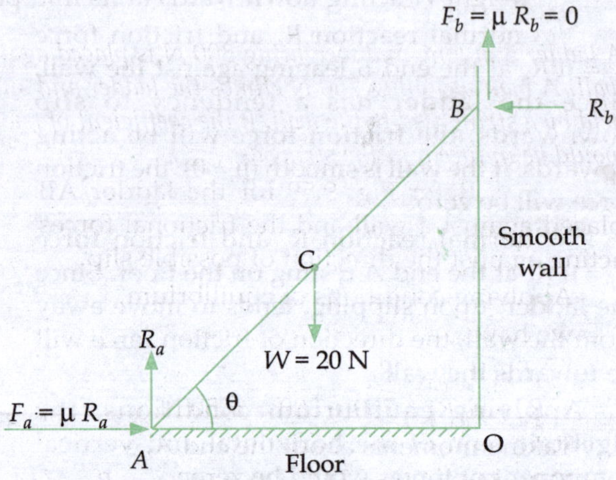

Fig. 9.57

Friction force between the ladder and the floor,

$$F_a = \mu R_a = 0.667 \times 20 \simeq 13.4 \text{ N}$$

EXAMPLE 9.31

A uniform ladder of weight 250 N and length 5 m is placed against a vertical wall in a position where its inclination to the vertical is 30°. A man weighing 800 N climbs the ladder. At what position will he induce slipping? Take coeficient of friction $\mu = 0.2$ at both the contact surfaces of the ladder.

Solution : Refer Fig. 9.58 for the ladder and the forces acting on it. Let x be the distance from end A along the ladder where the man will induce slipping.

Resolving all the forces along horizontal and vertical directions

$$R_b = \mu R_a; \quad R_a = \frac{R_b}{\mu} = \frac{R_b}{0.2} = 5R_b$$

and $\quad R_a + \mu R_b = 250 + 800 = 1050$

$\therefore \quad 5R_b + 0.2 R_b = 1050$

$$R_b = \frac{1050}{5.2} = 201.92 \text{ N}$$

Taking moments about end A,

$$R_b \times OB + \mu R_b \times OA = 250 \times EA + 800 \times FA$$

From the geometrical configuration,

$$OB = 5\cos 30^\circ = 4.33 \text{ m}; \quad OA = 5\sin 30^\circ = 2.5 \text{ m}$$

$$EA = CA\sin 30^\circ = 2.5\sin 30^\circ = 1.25 \text{ m}$$

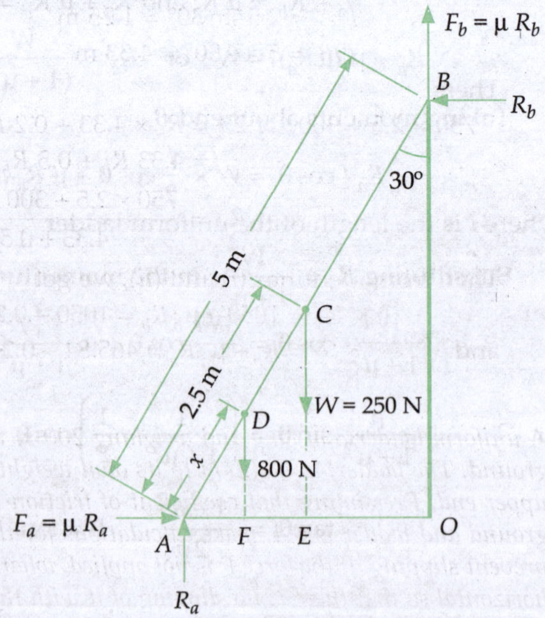

Fig. 9.58

$$FA = x \sin 30° = 0.5x$$

Substituting the above values in moment equation, we get

$$(201.92 \times 4.33) + (0.2 \times 201.92 \times 2.5) = 250 \times 1.25 + (800 \times 0.5x)$$

or $$874.31 + 100.96 = 312.5 + 400x$$

$$\therefore \quad x = \frac{874.31 + 100.96 - 312.5}{400} = \mathbf{1.66\ m}$$

EXAMPLE 9.32

A ladder 5 m long and weighting 300 N is placed against a vertical wall with its lower end 2.5 m from the wall. A man weighing 750 N climbs the ladder and sits at its top. To avoid slipping of ladder, its bottom is held by a string tied to the wall. If the coefficient of friction for both the contact surfaces is 0.2, what tension would be induced in the string?

Solution : Refer Fig. 9.59 for the ladder AB placed against a wall and the frictional forces acting against the direction of possible slip.

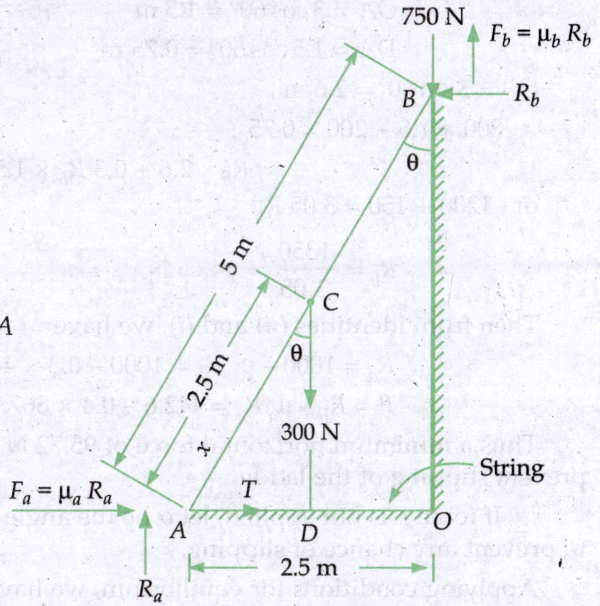

Fig. 9.59

Applying conditions of equilibrium, we have,

$$T + \mu_a R_a = R_b \quad \ldots(i)$$

$$R_a + \mu_b R_b = 750 + 300 = 1050 \text{ N} \quad \ldots(ii)$$

Taking moments about the end A,

$$750 \times OA + 300 \times DA = R_b \times OB + \mu_b R_b + OA$$

From the geometry of given arrangement,

$$\theta = \sin^{-1}\left(\frac{AO}{AB}\right)$$

$$= \sin^{-1}\left(\frac{2.5}{5}\right) = \sin^{-1}(0.5) = 30°$$

$$DA = 2.5 \sin 30° = 1.25 \text{ m}$$

$$OB = 5 \cos 30° = 4.33 \text{ m}$$

Then:

$$750 \times 2.5 + 300 \times 1.25 = R_b \times 4.33 + 0.2\, R_b \times 2.5$$

$$= 4.33\, R_b + 0.5\, R_b$$

$$\therefore \quad R_b = \frac{750 \times 2.5 + 300 \times 1.25}{4.33 + 0.5} = 465.84 \text{ N}$$

Then from identities (*i*) and (*ii*), we get

$$R_a = 1050 - \mu_b R_b = 1050 - 0.2 \times 465.84 = \mathbf{956.83\ N}$$

and $$T = R_b - \mu_a R_a = 465.84 - 0.2 \times 956.83 = \mathbf{274.47\ N}$$

EXAMPLE 9.33

A uniform ladder 3 m long and weighing 200 N is placed against a wall making an angle of 60° with the ground. The ladder, in addition to its own weight has to support a 800 N weight of a man standing at its upper end. Presuming that coefficient of friction between the wall and ladder is 0.3 and that between the ground and ladder is 0.4, make calculations for the horizontal force P to be applied at the ground level to prevent slipping. If the force P is not applied, what should be the minimum inclination of the ladder with the horizontal so that there is no slipping of it with the man standing its upper end.

Soluton : Refer Fig. 9.60 for the ladder AB placed against a wall and various forces ac is the horizontal force which has been applied at the ground level to prevent slipping.

Resolving all the forces along horizontal and vertical directions,

$$P + \mu_a R_a = R_b \quad ...(i)$$

$$R_a + \mu_b R_b = W_1 + W_2 = 1000 \quad ...(ii)$$

Taking moments about the end A,

$W_2 \times OA + W_1 \times DA = R_b \times OB + \mu_b R_b \times OA$

From the geometrical configuration,

$$OA = 3 \cos 60^\circ = 1.5 \text{ m}$$

$$DA = 1.5 \cos 60^\circ = 0.75 \text{ m}$$

$OB = 3 \sin 60^\circ = 2.6$ m

$\therefore$ $800 \times 1.5 + 200 \times 0.75$

$$= R_b \times 2.6 + 0.3 R_b \times 1.5$$

or $1200 + 150 = 3.05 R_b$

$$R_b = \frac{1350}{3.05} = \mathbf{442.6\ N}$$

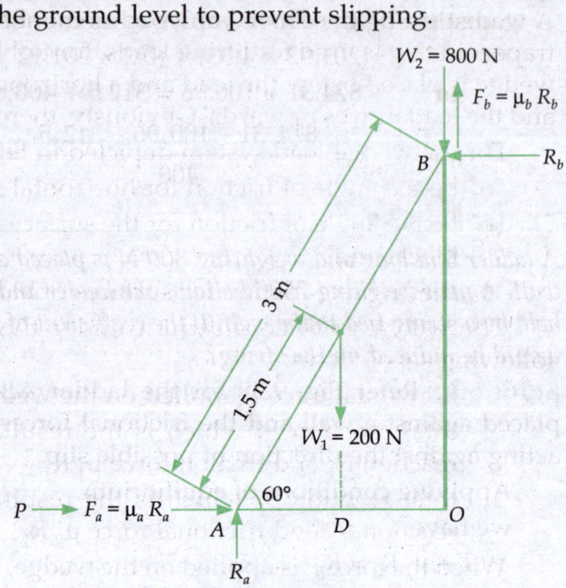

Fig. 9.60

Then from identities (*ii*) and (*i*), we have

$$R_a = 1000 - \mu_b R_b = 1000 - 0.3 \times 442.6 = 867.2 \text{ N}$$

$$P = R_b - \mu_a R_a = 442.6 - 0.4 \times 867.2 = \mathbf{95.72\ N}$$

Thus a minimum horizontal force of 95.72 N needs to be applied at the ground level to prevent slipping of the ladder.

(*b*) If force P is not applied, let α be the angle that the ladder should make with the horizontal to prevent any chance of slipping.

Applying conditions for equilibrium, we have

$$R_b = \mu_a R_a = 0.4 R_a$$

$$R_a + \mu_b R_b = W_1 + W_2 = 1000$$

These identities give

$$R_a + 0.3 \times (0.4 R_a) = 1000 ; \quad R_a = \frac{1000}{1 + 0.3 \times 0.4} = 892.8 \text{ N}$$

and $$R_b = 0.4 R_a = 0.4 \times 892.8 = 357.1 \text{ N}$$

Taking moments about the end A,

$$W_2 \times OA + W_1 \times DA = R_b \times OB + \mu_b R_b \times OA$$

or $(800 \times 3 \cos\alpha) + (200 \times 1.5 \cos\alpha) = 357.1 \times 3 \sin\alpha + (0.3 \times 357.1 \times 3 \cos\alpha)$

or $2400 \cos\alpha + 300 \cos\alpha = 1071.3 \sin\alpha + 321.4 \cos\alpha$

or $2378.6 \cos\alpha = 1071.3 \sin\alpha ; \quad \tan\alpha = \dfrac{2378.6}{1071.3} = 2.22$

$\therefore$ $\alpha = \tan^{-1}(2.22) = \mathbf{65.75^\circ}$

9.8. WEDGE FRICTION

A wedge is a piece of metal or wood in the shape of a prism whose section is usually triangular or trapezoidal. It is used for lifting loads, for tightening fits or keys for shafts. When lifting loads, the wedge is placed below the load and a horizontal force P is applied. The wedge moves towards left and the load moves upwards. Obviously, there is sliding at the surfaces AB, CD and GH. (Fig. 9.61)

For the wedge-load system depicted in Fig. 9.62, let

ϕ_1 be the angle of friction for horizontal surface and bottom surface of the wedge,

ϕ_2 be the angle of friction for the surfaces of wedge and block, and

ϕ_3 the angle of friction for surfaces of block and wall.

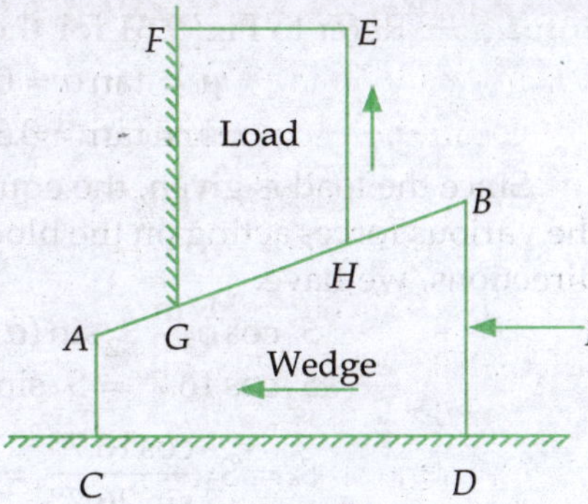

Fig. 9.61

The wedge is in equilibrium under the action of the following set of forces:

(i) horizontal force P applied on the wedge

(ii) total reaction S_1 which is the resultant of normal reaction R_1 and friction force $\mu_1 R_1$

(iii) total reaction S_2 which is the resultant of normal reaction R_2 and frictional force $\mu_2 R_2$

When the force P is applied on the wedge, the surface CD moves towards left and hence frictional force acts towards right. Likewise, the frictional force on surface AB would act upwards along AB.

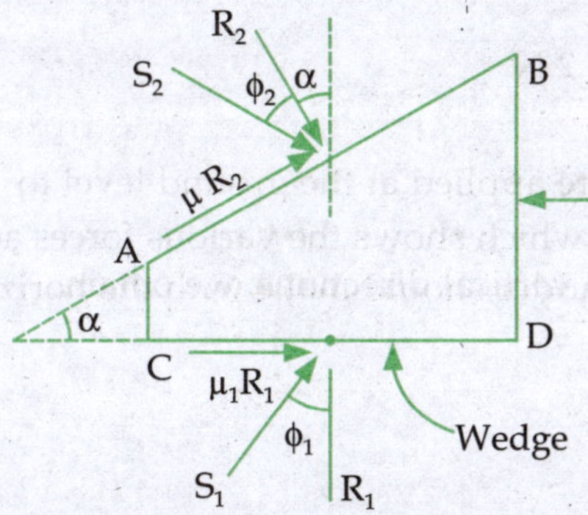

Fig. 9.62

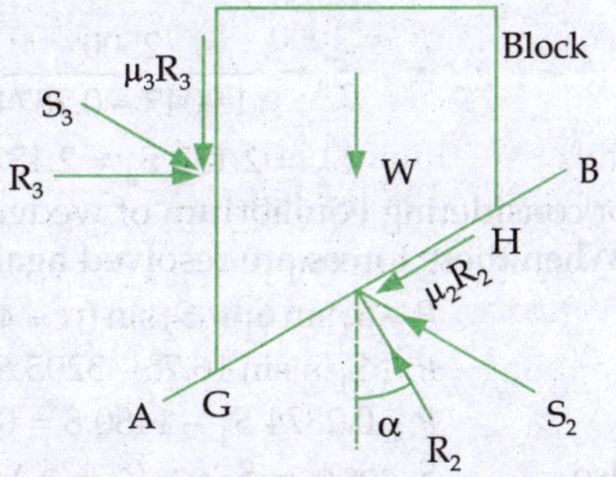

Fig. 9.63

Resolving all the forces horizontally and vertically, we get

$$P - S_1 \sin \phi_1 - S_2 \sin (\alpha + \phi_2) = 0 \qquad ...(i)$$

$$S_1 \cos \phi_1 - S_2 \cos (\alpha + \phi_2) = 0 \qquad ...(ii)$$

The **block** (Fig. 9.63) is in equilibrium under the action of following set of forces:

(i) W (the weight of block)

(ii) S_2 (a total reaction of normal reaction R_2 and frictional force $\mu_2 R_2$)

(iii) S_3 (a total reaction of normal reaction R_3 and frictional force $\mu_3 R_3$)

When wedge is pushed, the block is raised. The surface of the block is in contact with surface of wedge and therefore the total reaction S_2 will be equal and opposite to that of S_2 which was acting on wedge. Surface moves up and so the frictional force $\mu_3 R_3$ acts downward.

Resolving all the forces along horizontal and vertical directions, we get:

$$S_3 \cos \phi_3 - S_2 \sin (\alpha + \phi_2) = 0 \qquad ...(iii)$$

$$S_3 \sin \phi_3 + W - S_2 \cos (\alpha + \phi_2) = 0 \qquad ...(iv)$$

The solution of the identities (*i*) to (*iv*) helps to determine the values of unknown para...

As wedges are subjected to concurrent force system, solution can be obtained by applying Lami's theorem.

EXAMPLE 9.34

A block weighing 2500 N, overlying a 10-degree wedge on a horizontal floor and leaning against a vertical wall, is to be raised by applying a horizontal force to the wedge. The coefficient of friction between all surfaces in contact is 0.3. Make calculations for the minimum horizontal force to be applied to raise the block.

Solution : Refer to Fig. 9.61 for the wedge-load system

$$\mu = \tan\phi = 0.3 \quad \text{(given)}$$

$$\therefore \quad \phi = \tan^{-1} 0.3 = 16.7^\circ$$

Since the load is given, the equilibrium of the block would be considered first. Fig. 9.63 shows the various forces acting on the block. When these forces are resolved along horizontal and vertical directions, we have:

$$S_3 \cos\phi_3 - S_2 \sin(\alpha + \phi_2)\,; \quad \phi_2 = \phi_3 = 16.7^\circ$$

$$S_3 \cos 16.7^\circ = S_2 \sin(10 + 16.7)$$

$$\therefore \quad S_2 = S_3 \frac{\cos 16.7^\circ}{\sin 26.7^\circ} = S_3 \times \frac{0.9578}{0.4493} = 2.132\, S_3$$

Also:

$$S_3 \sin\phi_3 + W - S_2 \cos(\alpha + \phi_2) = 0$$

$$S_3 \times \sin 16.7^\circ + 2500 - 2.132\, S_3 \times \cos(10 + 16.7) = 0$$

$$0.2874\, S_3 + 2500 - 1.9047\, S_3 = 0$$

$$\therefore \quad S_3 = \frac{2500}{1.9047 - 0.2874} = 1545.8 \text{ N}$$

$$S_2 = 2.132\, S_3 = 2.132 \times 1545.8 = 3295.6 \text{ N}$$

For considering equilibrium of wedge, refer to Fig. 9.62 which shows the various forces acting on it. When these forces are resolved against horizontal and vertical directions, we obtain

$$P - S_1 \sin\phi_1 - S_2 \sin(\alpha + \phi_2) = 0$$

$$P - S_1 \times \sin 16.7^\circ - 3295.6 \sin(10 + 16.7) = 0$$

$$P - 0.2874\, S_1 - 1480.8 = 0$$

Also

$$S_1 \cos\phi_1 - S_2 \cos(\alpha + \phi_2) = 0$$

$$S_1 = S_2 \frac{\cos(\alpha + \phi_2)}{\cos\phi_1} = S_2 = \frac{\cos(10^\circ + 16.7^\circ)}{\cos 16.7^\circ}$$

$$= 3295.6 \times \frac{\cos 26.7^\circ}{\cos 16.7^\circ} = 3074 \text{ N}$$

$$\therefore \quad P = 0.2874\, S_1 + 1480.8 = 0.2874 \times 3074 + 1480.8 = 2364.3 \text{ N}$$

Alternatively : As wedges are subjected to concurrent force system, solution can also be obtained by applying Lami's theorem.

The block is in equilibrium under the action of forces: weight W and total reactions S_2 and S_3 (Fig. 9.64)

$$\therefore \quad \frac{S_2}{\sin(90 - \phi)} = \frac{S_3}{\sin\{180 - (\alpha + \phi)\}} = \frac{W}{\sin\{\phi + 90 + (\alpha + \phi)\}}$$

or

$$\frac{S_2}{\cos\phi} = \frac{S_3}{\sin(\alpha + \phi)} = \frac{W}{\sin(90 + \alpha + 2\phi)}$$

$$\text{or} \quad \frac{S_2}{\cos 16.7^\circ} = \frac{S_3}{\sin 26.7^\circ} = \frac{W}{\sin 133.4^\circ}$$

$$\therefore \quad S_2 = W \frac{\cos 16.7^\circ}{\sin 133.4^\circ} = 2500 \times \frac{0.9578}{0.7265} = 3295.9 \text{ N}$$

$$S_3 = W \frac{\sin 26.7^\circ}{\sin 133.4^\circ} = 2500 \times \frac{0.4493}{0.7265} = 1546.1 \text{ N}$$

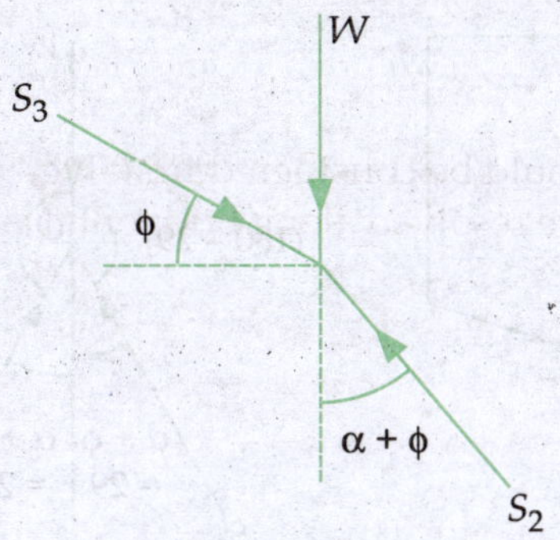

Fig. 9.64

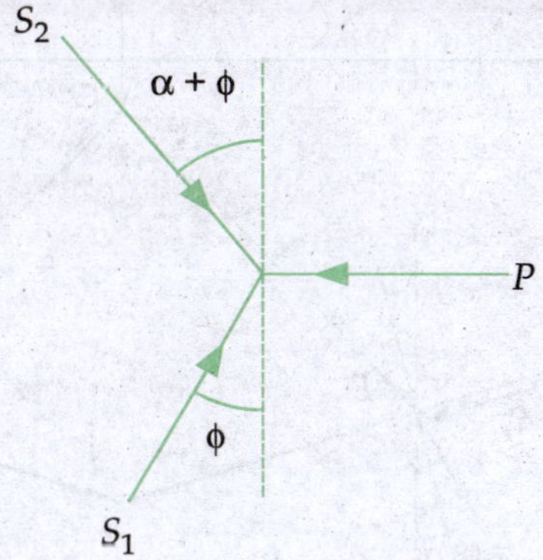

Fig. 9.65

The wedge is in equilibrium under the action of forces: P, reactions S_1 and S_2 (Fig. 9.65).

$$\therefore \quad \frac{P}{\sin\{180 - \phi - (\alpha + \phi)\}} = \frac{S_2}{\sin(90 + \phi)} = \frac{S_1}{\sin\{90 + (\alpha + \phi)\}}$$

$$\text{or} \quad \frac{P}{\sin[180 - 16.7 - (10 + 16.7)]} = \frac{S_2}{\cos\phi} = \frac{S_2}{\cos 16.7}$$

$$\therefore \quad P = S_2 \times \frac{\sin 136^\circ}{\cos 16.7^\circ}$$

$$= 3295.9 \times \frac{0.687}{0.9578} = 2364 \text{ N}$$

EXAMPLE 9.35

A block C of 60 kN is to be raised by means of two wedges A and B as shown in Fig. 9.65. The limiting coefficient of friction $\mu = 0.25$ for all surfaces. Neglecting the weight of the wedge blocks, determine the value of force P for upward impending motion of block C.

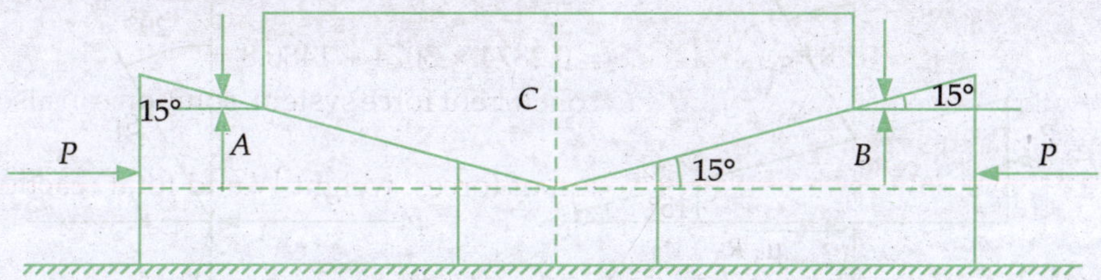

Fig. 9.66

Solution : $\mu = \tan\phi = 0.25$ where ϕ is the angle of friction

$\phi = \tan^{-1}(0.25) = 14^\circ$

The **block** C is in equilibrium under the action of

(*i*) weight $W = 60$ kN of block C

(*ii*) total reaction S_1 which is the resultant of normal reaction R_1 and friction force $\mu_1 R_1$ at point E

(*iii*) total reaction S_2 which is the resultant of normal reaction R_2 and friction force $\mu_2 R_2$ at point F

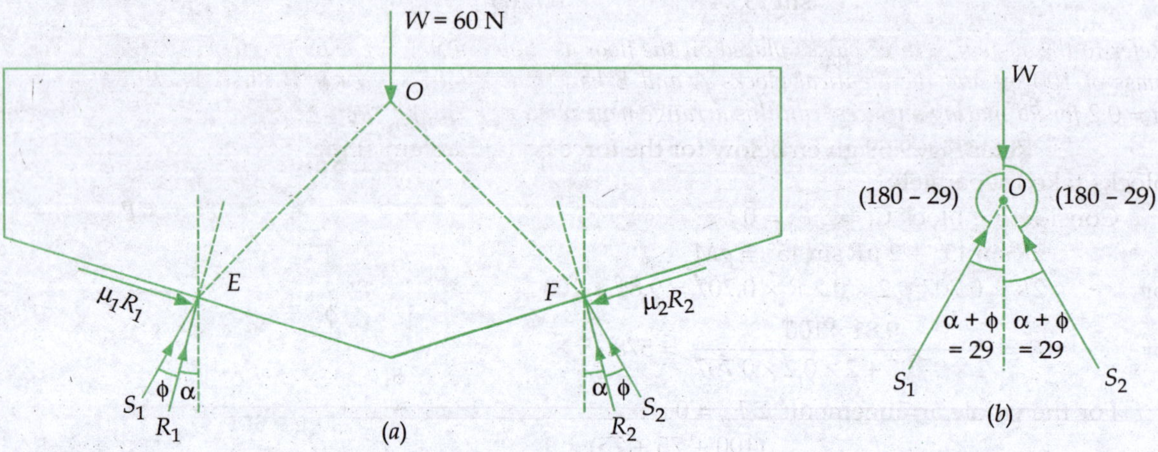

Fig. 9.67

These forces meet at point O ∴

Applying Lami's theorem

$$\frac{W}{\sin 58^\circ} = \frac{S_1}{\sin(180-29)} = \frac{S_2}{\sin(180-29)}$$

$$\therefore \quad S_1 = S_2 = W \times \frac{\sin 151^\circ}{\sin 58^\circ} = 60 \times \frac{0.4849}{0.848} = \mathbf{34.3\ kN}$$

The block A is in equilibrium under the action of following forces (Fig. 9.68)

(*i*) horizontal force P,

(*ii*) reaction S_1 from block C, and

(*iii*) total reaction S_3 which is the resultant of normal reaction R_3 and friction force $\mu_3 R_3$ at the horizontal surface.

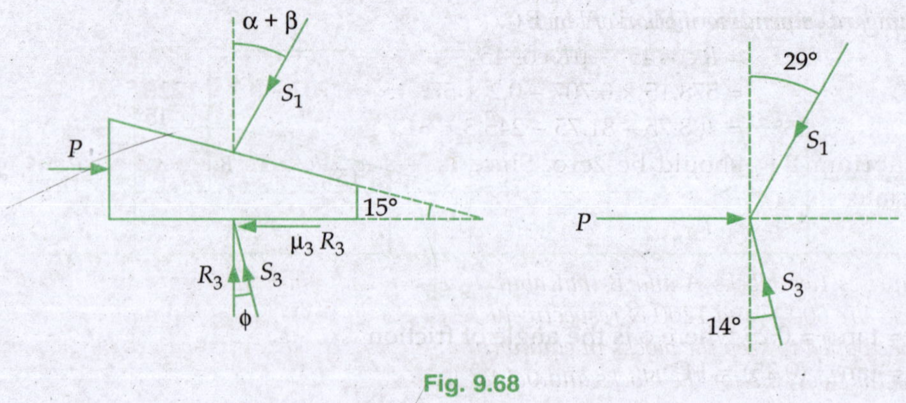

Fig. 9.68

Applying Lami's theorem,

$$\frac{P}{\sin(180-29-14)} = \frac{S_1}{\sin(90+14)} = \frac{S_3}{\sin(29+180+14)}$$

$$\therefore \quad P = S_1 \times \frac{\sin 137^\circ}{\sin 104^\circ} = 34.3 \times \frac{0.682}{0.97} = \textbf{24.12 kN}$$

EXAMPLE 9.36

Refer to the system of three blocks placed on the floor as shown in the Fig. 9.69 given below. Block C has a mass of 100 kg, and the identical blocks A and B have masses of 75 kg each. If the coefficient of friction $\mu = 0.2$ for all mating surfaces, can this arrangement remain in equilibrium?

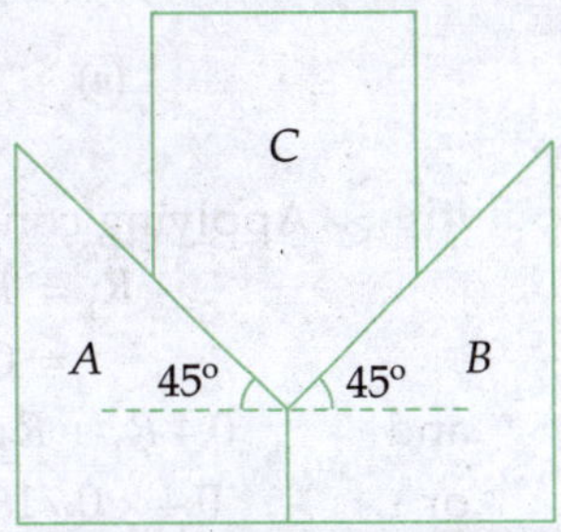

Fig. 9.69

Solution : Refer Fig. 9.69 given below for the force body diagram of the blocks taken separately.

Considering block C: $= \Sigma F_y = 0$

$$2R\sin 45^\circ + 2\,\mu R \sin 45^\circ = gM_c$$

or $\quad 2R \times 0.707 + 2 \times 0.2\,R \times 0.707 = 9.81 \times 100$

or $\quad R = \dfrac{9.81 \times 100}{2 \times 0.707 + 2 \times 0.2 \times 0.707} = 578.15 \text{ N}$

For the whole arrangement: $\Sigma F_y = 0$

$$(100 + 75 + 75) \times 9.81 = 2\,R_1$$

or $\quad R_1 = \dfrac{250 \times 9.81}{2} = 1226.25 \text{ N}$

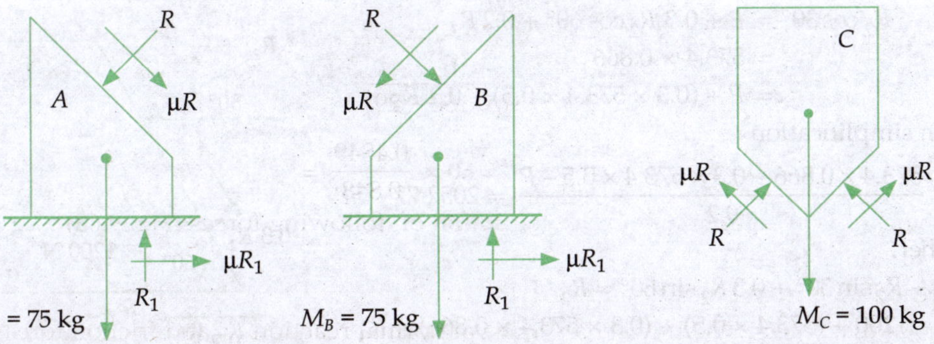

Fig. 9.70

Considering equilibrium of block A or B

$$\Sigma F_x = R\cos 45^\circ - \mu R \cos 45^\circ - \mu R_1$$
$$= 578.15 \times 0.707 - 0.2 \times 578.15 \times 0.707 - 0.2 \times 1226.5$$
$$= 408.75 - 81.75 - 245.3 = 81.7 \text{ N}$$

For equilibrium ΣF_x should be zero. Since it is not, the arrangement will not remain in equilibrium state.

EXAMPLE 9.37

A system comprises two blocks A and B that abut against a wall and a floor as shown in Fig. 9.71. The weights of blocks are 600 N and 1200 N respectively. Make calculations for the minimum horizontal force P that needs to be applied to keep the blocks in equilibrium. Assume the following values for the coefficients of friction: 0.2 for the floor, 0.3 for the blocks and 0.4 for the wall.

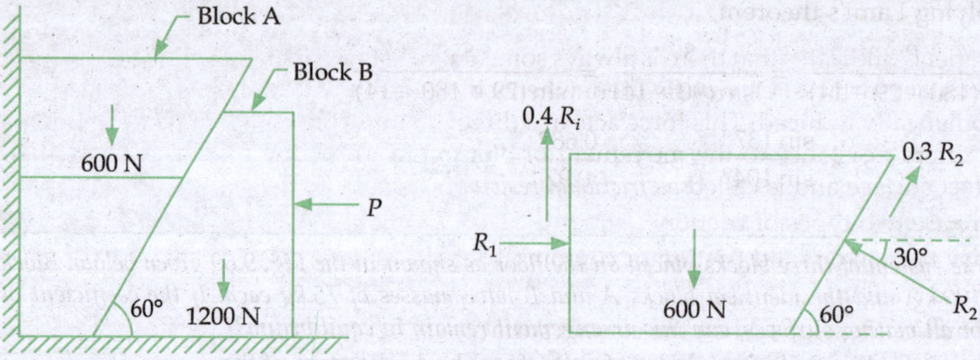

(a) (b)

Fig. 9.71.

Solution : Applying conditions of equilibrium ($\Sigma F_x = 0$ and $\Sigma F_y = 0$) for block A, (Fig. 9.71 b)

$$R_1 = R_2 \cos 30° - 0.3\, R_2 \cos 60°$$
$$= 0.866\, R_2 - 0.15\, R_2 = 0.716\, R_2$$

and $$0.4\, R_1 + R_2 \sin 30° + 0.3\, R_2 \sin 60° = 600$$

or $$0.4 \times 0.716\, R_2 + 0.5\, R_2 + 0.26\, R_2 = 600$$

That gives: $$R_2 = \frac{600}{(0.4 \times 0.716) + 0.5 + 0.26} = 573.4 \text{ N}$$

Applying conditions of equilibrium ($\Sigma F_x = 0$ and $\Sigma F_y = 0$) for block B,

$$R_2 \cos 30° = P + 0.3\, R_2 \cos 60° + 0.2\, R_3$$

or $$= 573.4 \times 0.866$$
$$= P + (0.3 \times 573.4 \times 0.5) + 0.2\, R_3$$

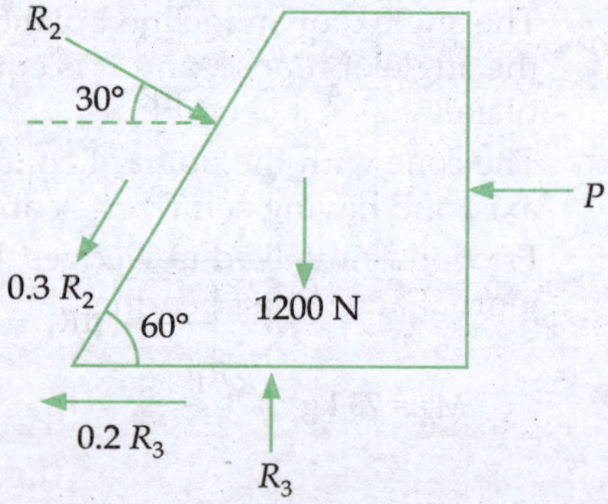

Fig. 9.71. (c)

Upon simplification

$$R_3 = \frac{573.4 \times 0.866 - 0.3 \times 573.4 \times 0.5 - P}{0.2} = 2052.77 - 5P$$

Further,

$$1200 + R_2 \sin 30° + 0.3 R_2 \sin 60° = R_3$$

or $$1200 + (573.4 \times 0.5) + (0.3 \times 573.4 \times 0.866) = R_3$$

or $$R_3 = 1200 + 286.7 + 148.97 = 1635.67$$

$\therefore$ Required minimum horizontal force,

$$P = \frac{2052.77 - R_3}{5}$$
$$= \frac{2052.77 - 1635.67}{5} = \mathbf{83.42 \text{ N}}$$

9.9. SCREW JACK

Refer Chapter on Lifting Machines

9.10. ROPE AND BELT FRICTION

Refer Chapter on Power Transmission: Methods and Devices

REVIEW AND SUMMARY

1. Experience indicates that there is always some degree of interlocking and wedge action between the contact surfaces. Obviously then, when a block is made to slide over a surface, a force gets automatically induced. This force acts in a direction opposite to that in which motion is sought and offers resistance to the movement of blocks. This resisting force acts tangentially to the contact surface and is called as *frictional resistance* or *force of friction.*
2. There are two types of friction
   (*a*) dry friction or solid friction or coulomb friction
   (*b*) fluid friction
   Dry friction develops when the contact surfaces are dry (no lubrication) and there is tendency for relative motion. Fluid friction manifests when a lubricating film is introduced between the contact surfaces and is due to the adjacent layers of fluid moving at different velocities.
3. The friction experienced by the block when at rest is called static friction. The limiting friction is the maximum friction force at the time of impending motion, *i.e.*, when the motion is about to commence. When the motion of block commences, the friction force drops to some what low value called the dynamic or kinetic friction.
4. There are two types of dynamic frictions namely, the sliding friction and the rolling friction. Sliding friction is the friction experienced by a body when it slides over another body. Rolling friction manifests when a body rolls over another body.
5. The ratio of the force of friction F to the normal reaction R between the contact surfaces is called the coefficient of friction μ. That is $\mu = F/R$.
   The angle of friction ϕ is the angle which the resulatant of normal reaction R and limiting force of friction F makes with the normal reaction. That is $\tan \phi = F/R = \mu$.
   The angle α of an inclined plane at which a block resting on it is about to slide down is called the angle of repose, and it is equal to the angle of friction between the block and the inclined plane.
   The cone with the point of contact as the vertex and the normal at the point of contact as the axis, and having semi-vertex angle ϕ is known as cone of friction.
6. Friction is involved in wedges, ladders, belts and ropes, square threaded screws (screw jack), journal and thrust bearings.

REVIEW QUESTIONS

A. Conceptual and conventional questions :

1. Define the term friction. How does it come into play?
2. Print out the characteristics of frictional force.
3. Cite some engineering applications where (*a*) friction is undesirable, (*b*) friction is quite necessary.
4. Friction is a necessary evil. Comment on the validity of this statement.
5. Explain, with the help of neat diagram, the concept of limiting friction.
6. Differentiate between (*a*) static and kinetic friction (*b*) sliding and rolling friction.
7. How will you distinguish between viscous friction and non-viscous friction?
8. State the main laws of dry (solid) friction.
9. Define the following terms:
   (*a*) coefficient of friction, (*b*) angle of friction,
   (*c*) angle of repose, and (*d*) cone of friction.

10. (*a*) Can we talk of friction coefficient for a single surface?
    (*b*) What is wrong with the following statement:
    "coefficient of static friction for concrete is 0.6."
11. What is cone of friction? State its significance.
12. A weight of 200 N is resting on a rough horizontal plane, and can be just moved by a force of 80 N applied horizontally. Determine the coefficient of friction, and the direction and magnitude of the resultant reaction of the plane.
    (0.4, 215.4 N at an angle of 68.2° with horizontal)
13. A body of weight 200 N is placed on a rough horizontal plane. Determine the coefficient of friction if a horizontal force of 130 N is just sufficient to cause the body slide over the plane. Also calculate the resultant reaction. **(0.65, 238 N)**
14. Find the least friction required to pull a body of weight W placed on a rough horizontal plane, when the force is applied at an angle θ with the horizontal. **($W \sin\theta$)**
15. A body of weight 450 N is pulled up an inclined plane, by a force of 300 N. The inclination of the plane is 30° to the horizontal and the force is applied parallel to the plane. Find the coefficient of friction. **(0.192)**
16. The force required to pull a body of weight 100 N on a rough horizontal plane is 30 N. Determine the coefficient of friction if the force applied is at an angle of 15° with the horizontal.
    (0.313)
17. A wooden block of 5 kg mass rests on a rough horizontal plane. What pull inclined at 15 degree to the horizontal is required to make the block just move? How this result will be affected if push is applied along the same line of action? Comment on the change. Take coefficient of friction between the block and the plane, μ = 0.4.
18. A body resting on a rough horizontal plane can be just moved by a force inclined to the horizontal. If friction coefficient is 0.5, compare the force with the weight of the body.
    (0.45)
19. A body of 400 N is pulled up along an inclined plane having inclination 30° to the horizontal at a steady speed. If the coefficient of friction between the body and the plane is 0.3 and force is applied parallel to the inclined plane, find the force required. Find also the work done on the body if the distance travelled by the body is 120 m along the plane.
20. A cord connects two bodies of weight 500 N and 1000 N. The two bodies are placed on an inclined plane and cord is parallel to the inclined plane. The coefficient of friction for the weight of 500 N is 0.2 and that of 1000 N is 0.4. Determine the inclination of the plane to the horizontal and tension in the cord when the motion is about to take place, down the inclined plane. The body weighing 500 N is below the body weighing 1000 N. **(18.417°, 63.086 N)**
21. A block of weight 500 N rests on a surface for which the coefficient of friction is 0.2. Work out the range of angle α such that there is no movement of the block when a force of 100 N is applied as shown in Fig. 9.72. **(135°)**
22. A uniform ladder rests in limiting equilibrium with one end on a rough horizontal plane and the other against a smooth vertical wall. Show that a man ascending the ladder cannot go more than halfway up.

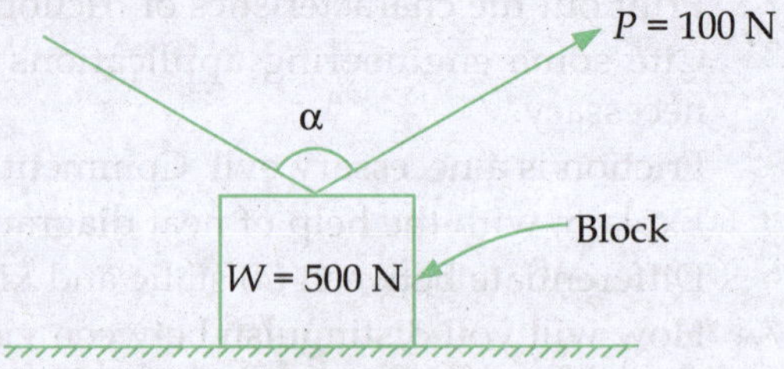

Fig. 9.72

23. A uniform ladder of length 13 m and weighing 30 N is placed against a smooth vertical wall with its lower end 10 m from the wall. In this position the ladder is just to slip. Determine the
   (*a*) coefficient of friction between ladder and the floor,
   (*b*) frictional force acting on the ladder at the point of contact between ladder and floor.

((*a*) 0.602, (*b*) 18.06 N)

24. A uniform ladder of weight 800 N and of length 7 m rests on a horizontal ground and leans against a smooth vertical wall. The angle made by the ladder with the horizontal is 60°. When a man of weight 600 N stands on the ladder at a distance of 4 m from the top of ladder, the ladder is at the point of sliding. Determine the coefficient of friction between ladder and the floor. **(0.237)**

25. A uniform ladder of length 15 m rests against a vertical wall making an angle of 60° with the horizontal Coefficient of friction between wall and the ladder, and the ground and ladder are 0.3 and 0.25 respectively. A man weighing 500 N ascends the ladder. How high will he be able to go before the ladder slips? Find the weight necessary to be put at the bottom of the ladder so as to just sufficient to permit the man to go the top. Assume weight of ladder is 850 N.

(6.39 m, 607.6 m)

26. A uniform ladder 3 m long weighs 18 N. It is placed against a wall 60° with floor as shown in Fig. 9.73 The coefficient of friction between the wall and the ladder is 0.25 and that between the floor and the ladder is 0.35. The ladder, in addition to its own weight has to support a man weighing 90 N at its top at A. Calculate

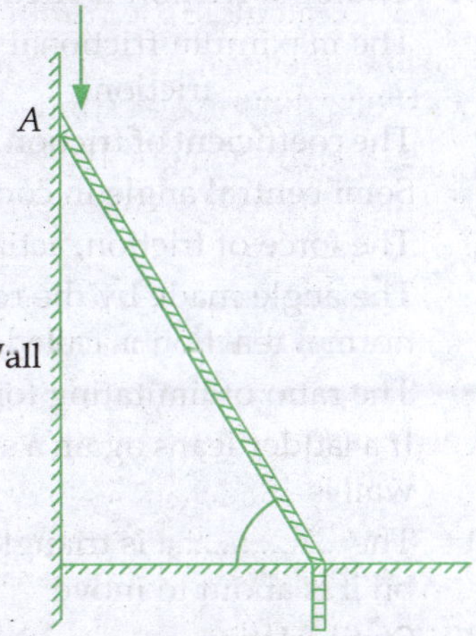

Fig. 9.73

   (*a*) The horizontal force F to be applied to the ladder at the floor level to prevent slipping.
   (*b*) If the force F is not applied what would be the minimum inclination of the ladder with the horizontal so that there is no slipping of it with the man at the top?

((*a*) 16.6 N, (*b*) 69°)

27. A wedge A of 50 N weight is to be driven between an inclined support and a block B of 2000 N weight (Fig. 9.74). If coefficient of friction between all surfaces of contact is 0.3, make calculations for the magnitude of the force P required to start moving the wedge.

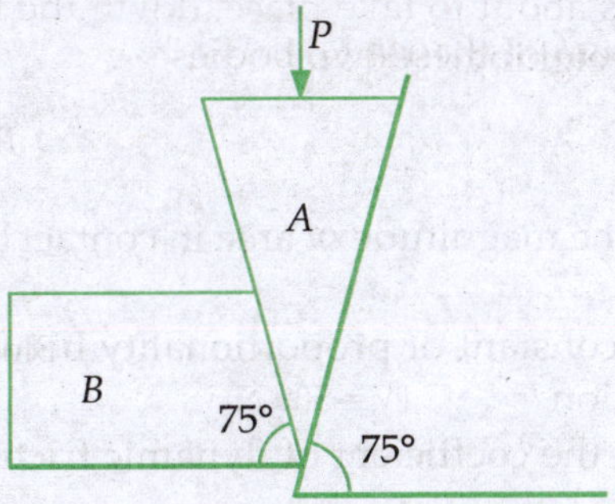

Fig. 9.74

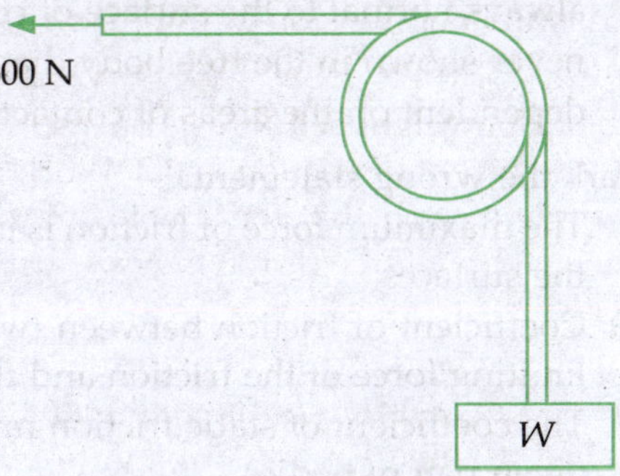

Fig. 9.75

28. A pull of 600 N is exerted on a rope which is wrapped 1.25 turns round a stationary horizontal drum as shown in Fig. 9.73 Presuming that coefficient of friction between the brake and drum is 0.3, make calculations for the weight that can be supported when
   (*a*) impending motion is upward,
   (*b*) impending motion is downward. **(56.85 N, 6330 N)**

B. Complete the following statements with most appropriate word/words :

1. Frictional force is .... of area of contact and acts in .... direction to the surfaces in contact.
2. .... friction is another name for dynamic friction.
3. Coulomb friction is friction between .... surfaces.
4. The maximum frictional force that acts when a body is about to slide over a surface is called .... friction.
5. The coefficient of friction μ and the angle of friction ϕ are related by the expression ....
6. Semi-central angle in cone of friction is equal to ....
7. The force of friction, acting on a moving body is called the ....
8. The angle made by the resultant of normal reaction and the limiting force of friction with the normal reaction is called the ....
9. The ratio of limitating force of friction to the normal reaction is known as ....
10. If a ladder leans again a smooth vertical wall, then the force of friction between the ladder and wall is ....
11. The .... is triangle made by an inclined surface with the horizontal, when body placed on it is about to move.
12. Static friction .... with applied force.

Answers:

1. independent, tangential 2. kinetic 3. dry 4. limiting 5. μ = tan θ 6. angle of friction
7. dynamic friction 8. angle of friction 9. coefficient of friction 10. zero
11. angle of respose 12. increases

C. Multiple choice questions :

1. The force of friction between two bodies in contact is
   (*a*) a function of the relative velocity between them
   (*b*) always normal to the surface of contact
   (*c*) never shown in the free body diagram of the system of these two bodies
   (*d*) dependent on the areas of contact
2. Mark the wrong statement:
   (*a*) The maximum force of friction is independent of the magnitude of area in contact between the surfaces
   (*b*) Coefficient of friction between two surfaces is a constant of proportionality between the limiting force of the friction and the normal reaction
   (*c*) The coefficient of static friction must be less than the coefficient of dynamic friction for a given pair of bodies
   (*d*) If a block slides over an inclined plane, the reaction by the plane on the block may not be in line with the weigth force

3. Which of the following statements about friction is true ?
   (a) friction can be reduced to zero
   (b) frictional force can accelerate a body
   (c) frictional force is proportional to the area of contact between the two surfaces
   (d) kinetic friction is always greater than rolling friction

4. The coefficient of friction between two surfaces is a constant of proportionality between the applied tangential force and the normal reaction
   (a) at the instant of the application of force
   (b) at any instant when the body is at rest
   (c) at the instant of impending motion
   (d) at an instant after motion takes place

5. The magnitude of coefficient of static friction between two bodies in contact
   (a) always lies between zero and unity
   (b) may exceed unity
   (c) is less than the coefficient of dynamic friction between the same pair of bodies
   (d) depends upon areas of contact

6. Coefficient of kinetic friction between a moving body and the surface on which it moves
   (a) is dependent on the velocity of body
   (b) depends only on the nature of the surfaces
   (c) depends on the nature of surfaces and the relative velocity between the surface and body
   (d) depends on the nature of the surfaces, weight of the body and its velocity

7. Identify the wrong statement:
   (a) The maximum inclination of the plane at which a body remains in equilibrium over the inclined plane by the frictional force alone is called the angle of repose
   (b) Angle of friction is the angle between the normal reaction and the resultant of limiting friction and normal reaction
   (c) Angle of friction equals the angle of repose
   (d) The coefficient of friction μ and the angle of friction ϕ are related by the expression $\mu = \sin\phi$

8. The coefficient of friction between two surfaces is 0.2. Then the angle of friction as
   (a) $\sin^{-1}(0.2)$ (b) $\cos^{-1}(0.2)$ (c) $\tan^{-1}(0.2)$ (d) $\cot^{-1}(5)$

9. A zero angle of friction implies that
   (a) frictional force is infinite (b) frictional force is zero
   (c) frictional force acts normal to the plane
   (d) frictional force acts along the direction of motion.

10. Brakes are applied to a car travelling at 90 km/hr and it comes to stop 5 seconds there-after. The coefficient of friction between the wheels and road surface is
   (a) 0.18 (b) 0.25 (c) 0.36 (d) 0.5
   Take $g = 10\ m/s^2$.

11. Consider the following statements
   1. Friction acts in a direction opposite to that in which motion is intended
   2. Friction comes only that much into play as is just sufficient to prevent motion
   3. Friction exists as long as the tractive force acts

Which of these statements is/are correct?

(a) 1 only (b) 1 and 2 (c) 1 and 3 (d) 1, 2 and 3

12. All of the following statements are correct, except
   (a) Coulomb friction manifests when the contact surfaces are dry and there is tendency for relative motion
   (b) Friction force acts normal to the surfaces in contact and is maximum at the instant of impending motion
   (c) A great reduction in friction occurs when lubricating film is introduced between the contact surfaces
   (d) The kinetic friction is always somewhat less than the limiting static friction

13. Which of the following statements is not correct about the limiting friction?
   (a) it is the maximum friction force at the time of impending motion
   (b) it depends upon the roughness or smoothness of the surfaces in contact
   (c) it is affected by the area and shape of the contact surfaces
   (d) its magnitude bears a constant ratio to the normal reaction between the mating surfaces

14. A block resting on an inclined plane begins to slide down the plane when the angle of inclination is gradually increased to 30°. Then the coefficient of friction between the block and plane is closest to

   (a) 0.50 (b) 0.58 (c) 0.72 (d) 0.87

15. Identify the correct statement
   (a) angle of repose = angle of limiting friction
   (b) limiting friction = static friction
   (c) coefficient of friction = tangent of angle of friction
   (d) tangent of angle of repose = coefficient of kinetic friction

16. A 1 kg block is resting on a surface with coefficient of friction $\mu = 0.1$. A force of 0.8 N is applied to the block as shown in the figure given below :

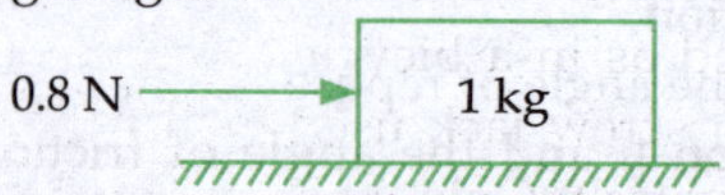

The friction force is

   (a) zero (b) 0.8 N (c) 0.98 N (d) 1.2 N

Answers :

1. (c)	2. (c)	3. (d)	4. (a)	5. (c)	6. (c)
7. (d)	8. (d)	9. (b)	10. (b)	11. (d)	12. (b)
13. (c)	14. (b)	15. (c)	16. (b)		

Lifting Machines 10

Learning objectives : A study of this chapter will enable the readers to know about the

- *simple machines and purpose served by them*
- *terms commonly used while dealing with machines; velocity ratio, mechanical advantage and efficiency*
- *reversible and irreversible machine; law of machine*
- *construction and operational details of lifting machines*

A ***machine*** may be defined as a device or a contrivance that receives energy in some available form and uses it for doing a particular useful work. For example, an IC engine converts the energy provided by gasoline (petrol or diesel) into motion of translation. Essentially a machine enables us

(*i*) to multiply force as in crowbar,

(*ii*) to apply force at a convenient point as in fire tongs for lifting a burning piece of coal,

(*iii*) to apply force in a convenient direction as in a single fixed pulley and levers,

(*iv*) to obtain a gain in speed as in a bicycle,

(*v*) to lift heavy loads as in screw jack used to lift the weight of a car or a lorry.

A machine does not work by itself. Only when energy is supplied to it, it does some useful work. The six basic simple machines are :

(*i*) lever	(*ii*) pulley	(*iii*) inclined plane
(*iv*) screw	(*v*) wedge	(*vi*) wheel and axle

The vast variety of machines that have been created or developed since the time of our primitive ancestors are made up of one or more of these six simple machines.

The machines which are used to lift heavy loads are called ***lifting machines.*** These machines can lift loads by applying a small effort in a convenient direction. Pulleys used to lift water from a well and screw jacks used to lift motor car (when the tyre is being changed) are the common examples of lifting machines.

10.1. BASIC CONCEPTS AND DEFINITIONS

Before studying the constructional and operational details of lifting machines, it is appropriate to define the following terms commonly used while dealing with them :

Let an effort P be applied through a distance y to lift a load W through distance x. Then

- **Load (W) :** any weight which is lifted by the machine.

- **Effort (P)** : a force required to lift or displace the load.
- **Mechanical advantage (MA)** : the ratio of weight lifted to effort applied

$$MA = \frac{\text{weight lifted}}{\text{effort applied}} = \frac{W}{P} \qquad ...(10.1)$$

The force or effort applied is generally smaller than the load lifted and that gives mechanical advantage greater than one.

- **Velocity ratio (VR)** : the ratio of distance moved by the effort to the distance moved by the load

$$VR = \frac{\text{distance moved by effort}}{\text{distance moved by load}} = \frac{y}{x} \qquad ...(10.2)$$

- **Input** : Work expended on the machine and it is measured by the product of the effort and the distance through which it moves.

$$\text{Input} = \text{effort} \times \text{distance moved by the effort} = P\,y \qquad ...(10.3)$$

- **Output** : Useful work done by the machine and it is measured by the product of the load lifted and the distance through which it is lifted.

$$\text{Output} = \text{load lifted} \times \text{distance moved by the load} = W\,x \qquad ...(10.4)$$

- **Efficiency of machine** : ratio between useful work done by the machine and the work expended on the machine

$$\eta = \frac{\text{useful work done by the machine}}{\text{work expended on the machine}}$$

$$= \frac{\text{output of the machine}}{\text{input of the machine}} = \frac{W\,x}{P\,y} = \frac{W}{P} \times \frac{1}{y/x}$$

$$= \frac{\text{mechanical advantage}}{\text{velocity ratio}} = \frac{MA}{VR} \qquad ...(10.5)$$

Machine efficiency is a fraction less than one. No device is available in the universe whose efficiency is 100%. That implies that work input is always greater than the useful work obtained. That is

$$P\,y > W\,x \;;\; \frac{y}{x} > \frac{W}{P}$$

or $$VR > MA$$

Since mechanical advantage is usually greater than one, the velocity ratio would also be greater than one. This fact leads to the statement :

"In a simple machine, a small force when applied through a large distance overcomes a large force through a small distance".

- **Ideal machine** : The friction losses of energy are negligibly small in an ideal machine. If the friction losses are neglected, the work put in by the effort equals the work done on the load. That is

$$P\,y = W\,x \;;\; \frac{W}{P} = \frac{y}{x} \quad \text{or} \quad MA = VR$$

and $$\eta = \frac{W\,x}{P\,y} = 1$$

Such a machine where η = 100% and mechanical advantage = velocity ratio is termed as an **ideal machine**.

- **Ideal effort and ideal load :** Consider an ideal machine (no friction at the mating parts) for which efficiency is 100% and

$$MA = VR \; ; \; \frac{W}{P} = VR$$

$$\therefore \quad \text{Ideal effort } P_{ideal} = \frac{W}{VR}$$

For an actual machine in which friction is present

$$\eta = \frac{MA}{VR} = \frac{W}{P} \times \frac{1}{VR}$$

$$\therefore \text{ Actual effort } P_{actual} = \frac{W}{\eta} \times \frac{1}{VR}$$

Thus the extra effort required or the effort wasted in friction while lifting a given load is

$$P_{friction} = P_{actual} - P_{ideal} = \frac{W}{\eta} \times \frac{1}{VR} - \frac{W}{VR} = \frac{W}{VR}\left(\frac{1}{\eta} - 1\right) \qquad ...(10.6)$$

Equation 10.6. represents friction in terms of effort for a prescribed load.

Similarly, for a given effort P,

$$W_{ideal} = \text{ideal load lifted or the ideal resistance overcome}$$
$$= P \times VR$$

and, W_{actual} = actual load lifted or the actual resistance overcome

$$= P\,\eta \times VR$$

Then the decrease in load lifted due to friction is

$$W_{friction} = W_{ideal} - W_{actual} = P \times VR - P\,\eta \times VR$$
$$= P \times VR\,(1 - \eta) \qquad ...(10.7)$$

Equation 10.7 represents the frictional force of the machine in terms of load for a prescribed effort.

EXAMPLE 10.1

An effort of 50 N is required by a machine to lift a load of 500 N. The distance moved by the effort is 63 cm and the corresponding load movement is 6 cm. Make calculations for the mechanical advantage, velocity ratio and efficiency of the machine.

Solution : Mechanical advantage $(MA) = \frac{\text{load lifted}}{\text{effort applied}} = \frac{500}{50} = 10$

$$\text{Velocity ratio } (VR) = \frac{\text{distance moved by effort}}{\text{distance moved by load}} = \frac{63}{6} = 10.5$$

$$\text{Efficiency of machine } (\eta) = \frac{MA}{VR} = \frac{10}{10.5} = 0.952 \text{ or } 95.2\%$$

EXAMPLE 10.2

The velocity ratio of a machine is 15 and its efficiency is 65%. Determine the load which can be raised on application of an effort of 50 N.

Solution : From the relation $\eta = \frac{MA}{VR}$, we have

$$MA = \eta \times VR = 0.65 \times 15 = 9.75$$

The mechanical advantage MA is also given by

$$MA = \frac{W}{P}$$

$\therefore$ Load lifted $W = P \times MA = 50 \times 9.75 =$ **487.5 N**

EXAMPLE 10.3

In a lifting machine, 25 per cent of the effort applied to raise a load is lost in friction. What load can be lifted through a height of 0.8 m if an effort of 200 N is moved through 10 m. Also determine the mechanical advantage and efficiency of the machine.

Solution : Velocity ratio $(VR) = \frac{\text{distance moved by effort}}{\text{distance moved by load}} = \frac{10}{0.8} = 12.5$

Effort lost in friction $= 0.25 \times 200 = 50$ N

The extra effort required or the effort wasted in friction while lifting a given load is

$$P_{friction} = P_{actual} - P_{ideal} = P_{actual} - \frac{W}{VR}$$

$$\therefore \quad 50 = 200 - \frac{W}{12.5}$$

That gives : $W = (200 - 50) \times 12.5 = 1875$ N

Mechanical advantage $(MA) = \frac{W}{P} = \frac{1875}{200} = 9.375$

Machine efficiency $(\eta) = \frac{MA}{VR} = \frac{9.375}{12.5} = 0.75$ or 75%

EXAMPLE 10.4

An effort of 60N is applied to a machine to lift a load of 900 N. If the velocity ratio of the machine is 20, determine :

(*a*) *efficiency of the machine,*

(*b*) *frictional force in terms of effort, and*

(*c*) *frictional force in terms of load.*

Solution : (*a*) Mechanical advantage

$$MA = \frac{W}{P} = \frac{90}{60} = 15$$

Machine efficiency $\eta = \frac{MA}{VR} = \frac{15}{20} = 0.75$ or 75%

(*b*) The frictional force in terms of effort is

$$P_{friction} = P_{actual} - P_{ideal} = P_{actual} - \frac{W}{VR} = 60 - \frac{900}{20} = 15 \text{ N}$$

(*c*) The frictional force in terms of load is

$$W_{friction} = W_{ideal} - W_{actual} = P \times VR - W_{actual} = 60 \times 20 - 900 = 300 \text{ N}$$

10.2. REVERSIBLE AND IRREVERSIBLE MACHINE

Let an effort P be applied through a distance y to lift a load W through a distance x. On removal of effort P, the following two conditions are likely to occur :

(*i*) The work done by the machine is in reverse direction and the load falls. The machine is then called ***reversible machine***. A pulley used to draw water from a well with the help of bucket is a reversible machine because the bucket falls back when the effort to pull it up is removed.

(*ii*) The load does not fall, *i.e.*, the work is not done by the machine in reverse direction. The machine is then said to be ***irreversible*** or ***self-locking***. A screw jack used to lift the motor car is a self-locking type lifting machine because it holds the car at the same position even when the application of effort is stopped.

In an irreversible machine, some work done is lost due to friction and is given by

$$\text{Friction work} = \text{input} - \text{output} = P\,y - W\,x$$

On removal of effort, the load will not fall if the friction work is more than the output of machine. That is

$$(P\,y - W\,x) > W\,x$$

or $$P\,y > 2\,W\,x \quad \text{or} \quad \frac{W\,x}{P\,y} < \frac{1}{2}$$

But $$\frac{W\,x}{P\,y} = \eta$$

$$\therefore \qquad \eta < \frac{1}{2} \quad \text{or} \quad 50\%$$

Thus the condition for irreversibility or self-locking of a machine is that the efficiency of machine should be less than 50 per cent. If the efficiency exceeds 50%, the machine would be reversible.

EXAMPLE 10.5

A machine with velocity ratio 25 can lift a load of 200 N on application of an effort of 20 N. Comment on the reversibility of machine.

Also make calculations for the friction loss of machine.

Solution : Mechanical advantage $MA = \frac{W}{P} = \frac{200}{20} = 10$

Machine efficiency $\eta = \frac{MA}{VR} = \frac{10}{25} = 0.4$ or 40%

Since the efficiency of machine is less than 40%, the machine is irreversible or self-locking.

(*b*) The frictional loss in terms of load is

$$W_{friction} = W_{ideal} - W_{actual} = P \times VR - W_{actual} = 20 \times 25 - 200 = 300 \text{ N}$$

10.3. LAW OF MACHINE

The law of machine is an equation that prescribes the relationship between the effort applied and the load lifted by a machine.

Let P be the effort required to lift W. Then for a machine with constant velocity ratio, the law of the machine is given as

$$P = mW + C \quad \text{(straight line)} \qquad \text{...(10.8)}$$

where, m = slope of straight line = $\frac{\Delta P}{\Delta W}$

C = intercept of the line on P-axis

This relationship has been depicted graphically in Fig. 10.1 both for the ideal and actual machine.

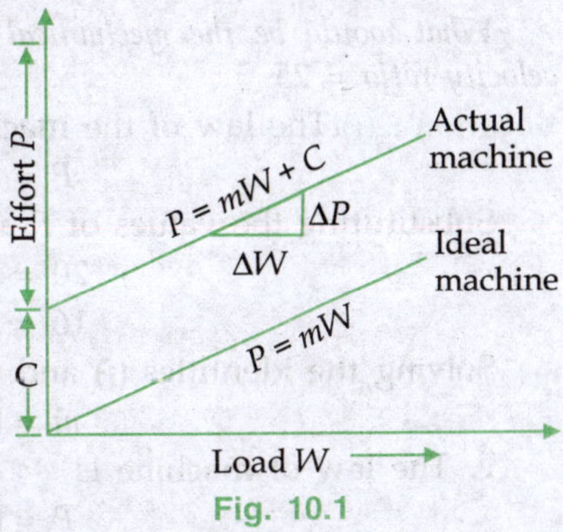

Fig. 10.1

(*i*) For an ideal machine, the straight line passes through the origin and the intercept $C = 0$.

(*ii*) For an actual machine, the straight line has an intercept C on the P-axis. This in represents the effort required to overcome friction. If the effort applied is less than C, then the load will not be lifted.

The slope of the line is constant and equal to the coefficient of friction. It may be recalled that

$$\text{Mechanical advantage } MA = \frac{W}{P} = \frac{W}{mW + C} = \frac{1}{m + C/W} \quad ...(10.9)$$

By increasing W, the value of factor C/W decreases, and that in turn increases the mechanical advantage. The maximum or limiting value of the mechanical advantage will be $1/m$ when C/W approaches zero.

$$\text{Further, } \eta = \frac{MA}{VR} = \frac{W}{P \times VR} = \frac{W}{(mW + C) \times VR} = \frac{1}{(m + C/W) \times VR} \quad ...(10.10)$$

Obviously the relationship between η and W is hyperbolic, and this has been depicted in Fig. 10.2. By increasing W, the value of factor C/W decreases and that in turn increases the efficiency. The maximum or limiting value of the efficiency of machine will be

$$\eta_{max} = \frac{1}{m \times VR} \quad ...(10.11)$$

and it occurs when C/W approaches zero.

The relation 10.11 is a straight line parallel to W-axis, and the hyperbolic curve prescribed by the relation 10.11 continuously approaches it. In other words the maximum efficiency line is asymptotic to the curve defined by equation 10.11.

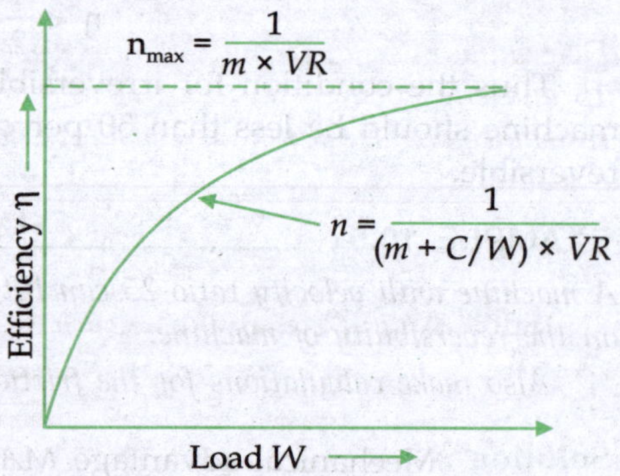

Fig. 10.2

EXAMPLE 10.6

In an experimental test conducted on a hoisting machine, it was observed that an effort of 20 kN was applied to lift a load 90 kN, where as an effort of 16 kN was required to lift a load of 70 kN. Determine the following :

(*i*) *law of the machine,*

(*ii*) *the limiting mechanical advantage,*

(*iii*) *the limiting efficiency, and*

(*iv*) *the effort required to lift a load of 15 kN.*

What would be the mechanical advantage and efficiency of the machine at this moment ? Take velocity ratio = 25

Solution : (*i*) The law of the machine is

$$P = mW + C$$

Substituting the values of P and W as obtained from the test run,

$$20 = m \times 90 + C \quad ...(i)$$

$$16 = m \times 70 + C \quad ...(ii)$$

Solving the identities (*i*) and (*ii*), we get

$$m = 0.2 \quad \text{and} \quad C = 2$$

$\therefore$ The law of machine is

$$P = 0.2\,W + 2$$

(*ii*) Limiting value of mechanical advantage (*MA*)

$$(MA)_{max} = \frac{1}{m} = \frac{1}{0.2} = 5$$

(*iii*) Limiting value of efficiency

$$\eta_{max} = \frac{1}{m \times VR} = \frac{1}{0.2 \times 25} = 0.2 \text{ or } 20\%$$

(*iv*) Effort required to lift a load of 15 kN

$$P = 0.2 \times 15 + 2 = 5 \text{ kN}$$

$$MA = \frac{W}{P} = \frac{15}{5} = 3$$

$$\eta = \frac{MA}{VR} = \frac{3}{25} = 0.12 \text{ or } 12\%$$

EXAMPLE 10.7

Determine the effort required to lift a load of 150 N if the lifting machine is stated to have velocity ratio 15 and efficiency 60 per cent. Also set up an equation prescribing the law of machine if the frictional resistance of the machine is constant. Proceed to find the effort to run this machine at : (i) no load, (ii) a load of 100 N.

Solution : Machine efficiency $\eta = \frac{MA}{VR} = \frac{W/P}{VR}$

Substituting the given values for η, *VR* and *W*, we obtain

$$0.6 = \frac{150/P}{15} \quad \text{or} \quad 0.6 \times 15 = \frac{150}{P}$$

$$\therefore \quad \text{Effort } P = \frac{150}{0.6 \times 15} = 16.67 \text{ N}$$

(*ii*) The law of the machine is

$$P = mW + C$$

The intercept *C* represents the effort required to overcome friction and equals the effort wasted in friction while lifting a given load. That is

$$C = P_{friction} = P_{actual} - P_{ideal} = P_{actual} - \frac{W}{VR} = 16.67 - \frac{150}{15} = 6.67$$

Further, when effort *P* = 16.67 N, the load lifted *W* = 150 N.

Substituting these values

$$16.67 = m \times 150 + 6.67$$

$$\therefore \quad \text{Slope of line } m = \frac{16.67 - 6.67}{150} = 0.0667$$

Hence the law of machine is

$$P = 0.0667\,W + 6.67$$

Effort required at no load,

$$P = 0.0667 \times 0 + 6.67 = 6.67 \text{ N}$$

Effort required at a load of 100 N,

$$P = 0.0667 \times 100 + 6.67 = 13.34 \text{ N}$$

EXAMPLE 10.8

In the case of a lifting machine, efforts required to lift loads of 50 N and 80 N were 12 N and 18 N respectively. If velocity ratio for the machine was 6, determine :

(a) *relation between the load and effort (law of the machine),*

(b) *the efficiency of the machine and the effort lost in friction at 50 N load,*

(c) *the maximum efficiency which can be expected from this machine.*

Solution : The relationship between the effort applied and the load lifted is given by

$$P = mW + C$$

Substituting the given values of P and W,

$$12 = m \times 50 + C \quad \ldots(i)$$

$$18 = m \times 80 + C \quad \ldots(ii)$$

Solution of the identities (*i*) and (*ii*) gives

$$m = 0.2 \quad \text{and} \quad C = 2$$

Hence the law of machine is

$$P = 0.2\,W + 2$$

(b) When a load of 50 N is to be lifted

$$MA = \frac{W}{P} = \frac{50}{12} = 4.167$$

$$VR = 6 \text{ (given)}$$

$$\eta = \frac{MA}{VR} = \frac{4.167}{6} = 0.694 \quad \text{or} \quad 69.4\%$$

Effort lost in friction,

$$P_{friction} = P_{actual} - P_{ideal} = P_{actual} - \frac{W}{VR} = 12 - \frac{50}{6} = 3.67 \text{ N}$$

(c) Maximum possible efficiency of any machine

$$\eta_{max} = \frac{1}{m \times VR} = \frac{1}{0.2 \times 6} = 0.833 \quad \text{or} \quad 83.3\%$$

EXAMPLE 10.9

The law of a machine is prescribed by the relation

$$P = 0.04\,W + 7.5$$

where P is the effort required to lift load W, both expressed in newtons. Work out the mechanical advantage and efficiency of the machine when the load is 1.5 kN and velocity ratio is 35. What is the maximum efficiency of the machine ?

If $P_{friction}$ is lost in friction, proceed to establish the relation between $P_{friction}$ and W. What will be the value of $P_{friction}$ when a load of 1.5 kN is to be lifted ?

Solution : Substituting the value W = 1.5 kN = 1500 N in the law of machine, we get

$$P = 0.04 \times 1500 + 7.5 = 67.5 \text{ N}$$

$$\text{Mechanical advantage } MA = \frac{W}{P} = \frac{1500}{67.5} = 22.22$$

$$\text{Machine efficiency } \eta = \frac{MA}{VR} = \frac{22.22}{35} = 0.635 \text{ or } 63.5\%$$

The maximum efficiency of the machine is given by

$$\eta_{max} = \frac{1}{m \times VR}$$

Comparing $P = 0.04\ W + 7.5$ with the law of machine given by the relation $P = mW + C$, we have $m = 0.04$ and $C = 7.5$

$$\therefore \quad \eta_{max} = \frac{1}{0.04 \times 35} = 0.714 \text{ or } 71.4\%$$

(b) Effort lost in friction

$$P_{friction} = P_{actual} - P_{ideal} = P_{actual} - \frac{W}{VR} = (0.04\ W + 7.5) - \frac{W}{VR}$$

$$= W\left(0.04 - \frac{1}{VR}\right) + 7.5 = W\left(0.04 - \frac{1}{35}\right) + 7.5 = 0.0114\ W + 7.5$$

Substituting $W = 1500$ N in the above relation, we get

$$P_{friction} = 0.0114 \times 1500 + 7.5 = 34.6 \text{ N}$$

10.4. LEVERS

A lever is essentially a rigid straight bar which rests on and can turn about a point, called fulcrum. It enables a small effort to overcome a large load. It may be used independently or as an integral part of some mechanism or machine *either to gain advantage in the effort applied or in the distance to be moved by the effort.*

With reference to adjoining Fig. 10.3., the fulcrum point O lies between the points of application of forces at ends A and B. An effort P is applied at point B of the lever to overcome a load W at the point A.

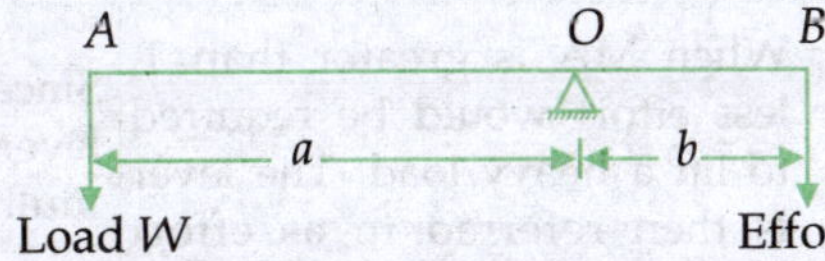

Fig. 10.3

The perpendicular distance of point A, at which load is applied, from the fulcrum (O) is called *load arm* ($OA = a$).

The perpendicular distance of point B, at which effort is applied, from the fulcrum (O) is called *effort arm* ($OB = b$).

When the lever is in equilibrium, $\Sigma M = 0$. Taking moments about the fulcrum point O, we get

$$W \times a = P \times b \quad ; \quad \frac{W}{P} = \frac{b}{a} = \frac{\text{length of effort arm}}{\text{length of load arm}}$$

The above relation is known as the *law of lever*. This relation has been setup with the assumptions :

(i) the lever is weightless, (ii) the friction is neglected.

Since the ratio of load lifted to effort applied, *i.e.*, W/P represents mechanical advantage, we may state.

"The mechanical advantage of a lever is equal to the ratio of the length of effort arm to the length of load arm".

For greater mechanical advantage, that is, to lift a greater load with less effort, the effort arm should be as large as possible.

The ratio of length of effort arm to length of load arm (b/a) is called leverage.

10.4.1. Classification of levers

A lever is classified as a lever of first, second or third kind depending on the relative position of its fulcrum, load and effort points.

A brief summary of the different kinds of levers is given below in Table 10.1.

Table 10.1. Salient features of different kinds of Levers

Lever of first kind	Lever of Second Kind	Lever of third kind
1. Fulcrum is between the load and the effort. 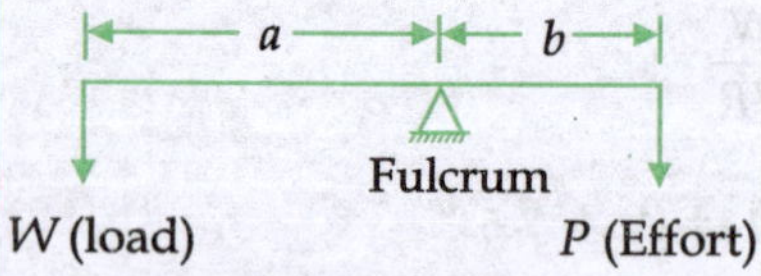	Load is between the effort and fulcrum. 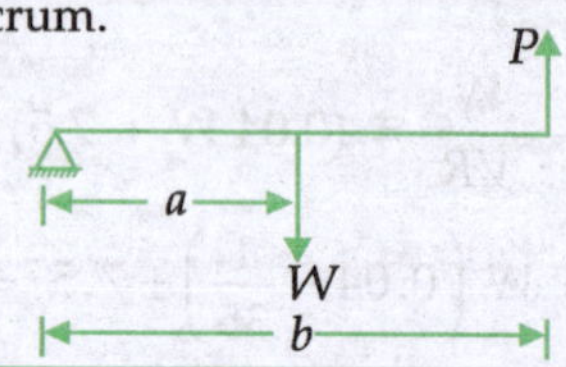	Effort is between the fulcrum and the load. 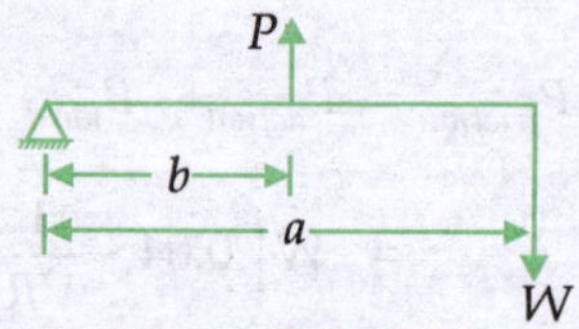
2. Mechanical advantage $= \dfrac{\text{effort arm}}{\text{load arm}} = \dfrac{b}{a}$ *MA* can be more than 1, equal to 1 or less than 1.	$MA = \dfrac{b}{a}$ *MA* is always greater than 1.	$MA = \dfrac{b}{a}$ *MA* is always less than 1.
3. *MA* increases with movement of fulcrum towards the load.	*MA* increases by moving the load towards the fulcrum.	*MA* cannot be made greater than 1 by any movement of load point.
4. When MA is greater than 1, less effort would be required to lift a heavy load. The lever is then referred to as effort multiplier lever.	Since *MA* is always greater than 1, lever of the second kind is an effort multiplexer lever.	Since *MA* is always less than 1, lever of the third kind is only a speed multiplier lever. With these levers, there is no gain in froce but there is gain in speed. This implies that such levers cannot lift heavy loads but provide an increase in the speed of lifting.
5. **Examples :** Sea saw, scissor, nail extractor, plier, handle of a water pump.	**Examples :** Wheel barrow, nut cracker, lemon crusher, oar of a boat	**Examples :** Fire tongs, knife, human arm, spade used for turning soil.

Apart from gaining advantage in force, a lever can also be used to gain advantage in displacement. With reference to Fig. 10.4, the lever *AOB* is capable of rotating about its fulcrum point *O*. Corresponding to rotation α of the rod,

displacement of end $A = a\alpha$

displacement of end $B = b\alpha$

$$\therefore \quad \frac{\text{displacement of end } B}{\text{displacment of end } A} = \frac{b\alpha}{a\alpha} = \frac{b}{a}$$

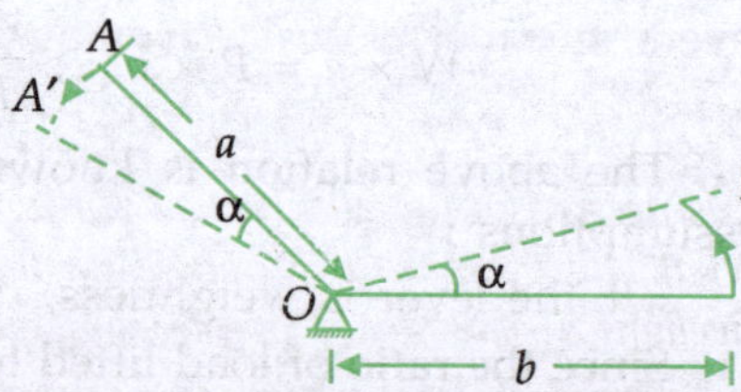

Fig. 10.4

With an increase in the ratio b/a, a small displacement to end *A*, results into a large displacement of end *B*. This aspect has been put to an extensive use in the construction of measuring instruments.

10.4.2. Compound levers

A compound lever is combination of simple levers linked with one another. Such levers are used to obtain higher mechanical advantage; an advantage which is rather difficult to get by the use of simple lever.

With reference of Fig. 10.5., *AB* is a simple lever connected to another simple lever *CD* with the help of a link *BC*. *P* is the effort applied at end *A* to lift a load *W* acting at point *F*.

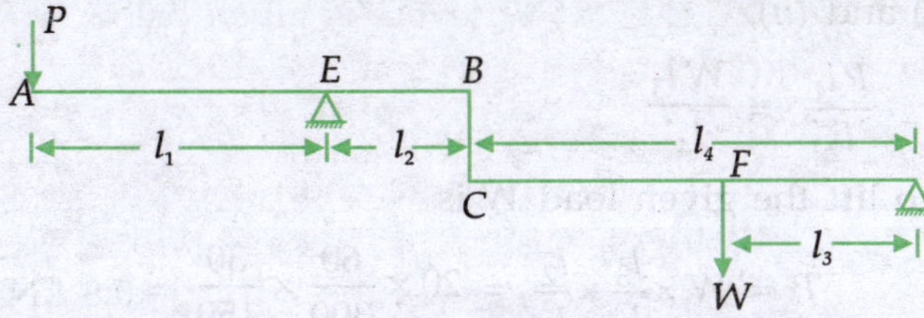

Fig. 10.5

From free body diagram (Fig. 10.6*a*) of lever *AB*, we have

$Pl_1 = Ql_2$ (Taking moments about *E*)

$$Q = \frac{Pl_1}{l_2} \quad ...(i)$$

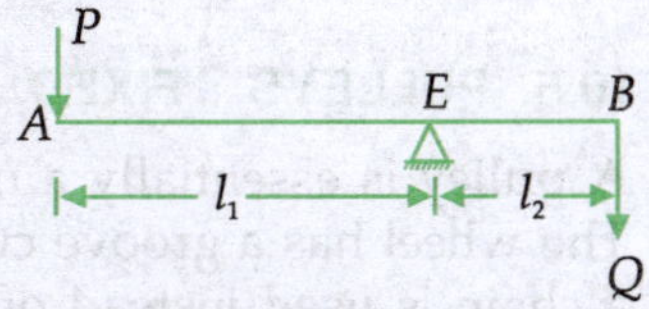

Fig. 10.6. (a)

where *Q* is the force in link *BC*. From free body diagram (fig. 10.6*b*) of lever *CD*, we have

$Ql_4 = Wl_3$ (Taking moments about *D*)

$$Q = \frac{Wl_3}{l_4} \quad ...(ii)$$

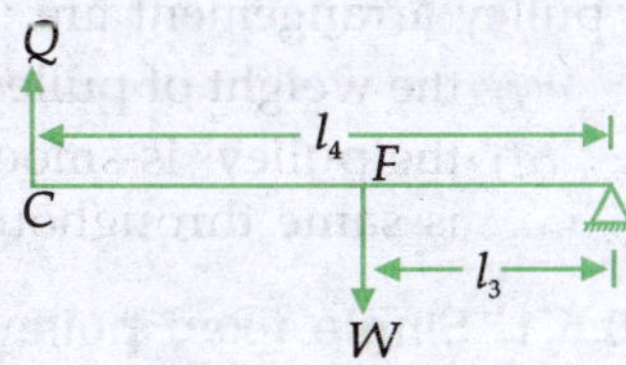

Fig. 10.6. (b)

From expressions (*i*) and (*ii*)

$$\frac{Pl_1}{l_2} = \frac{Wl_3}{l_4}$$

∴ Mechanical advantage of the compound lever,

$$MA = \frac{W}{P} = \frac{l_1}{l_2} \times \frac{l_4}{l_3}$$

Let $\frac{l_1}{l_2} = \frac{l_4}{l_3} = 10$

If only the lever *AB* is used, the mechanical advantage would be 10. By combining two levers, the mechanical advantage gets increased to 10 × 10 = 100.

EXAMPLE 10.10

It is desired to lift 20 kN load acting at point F with the help of a system of levers as shown in Fig. 10.5. What effort should be applied at end A of the lever so that the load just gets lifted. Also determine the mechanical advantage of the composite lever.

Take l_1 = 150 mm; l_2 = 30 mm; l_3 = 60 mm, and l_4 = 300 mm.

Solution : Refer Fig. 10.6. for the free body diagrams of the levers *AB* and *CD*.

For lever *AB*, taking moments about point *E* and equating it to zero,

$P \times l_1 - Q \times l_2 = 0$; *Q* is the force in link *B*

$$Q = \frac{Pl_1}{l_2} \quad ...(i)$$

For lever *CD*, taking moments about point *D* and equating it to zero,

$$Q \times l_4 - W \times l_3 = 0$$

$$Q = \frac{Wl_3}{l_4} \quad ...(iii)$$

From expressions (*i*) and (*ii*),

$$\frac{P l_1}{l_2} = \frac{W l_3}{l_4}$$

∴ Effort P needed to lift the given load W is

$$P = W \times \frac{l_3}{l_4} \times \frac{l_2}{l_1} = 20 \times \frac{60}{300} \times \frac{30}{150} = 0.8 \text{ kN}$$

$$\text{Mechanical advantage } (MA) = \frac{W}{P} = \frac{20}{0.8} = 25$$

10.5. PULLEYS : FIXED AND MOVABLE

A pulley is essentially a metallic or wooden wheel which is capable of rotation about an axis. The wheel has a groove cut along its periphery and a rope is made to rest in the groove. When a chain is used instead of rope, sprocket teeth are cut on the periphery of the wheel.

Pulleys are used to lift the weights. Every one has seen a pulley being used to take out water from a well.

Pulleys are of two types : fixed pulley and movable pulley. Assumption made for analysing a pulley arrangement are :

(*i*) the weight of pulley is small compared to the weight to be lifted and hence is neglected.

(*ii*) the pulley is smooth, *i.e.*, the tension of the string or rope passing around the pulley is same throughout.

10.5.1. Single fixed pulley

The block or axle supporting the pulley is fixed; its position does not change when the chain or rope passing around its periphery is moved. The weight W is attached to one end of the of the rope and the effort P is applied at the other end. The effort can be applied in any convenient direction; vertically downward or inclined as it does not change the value of tension in the rope.

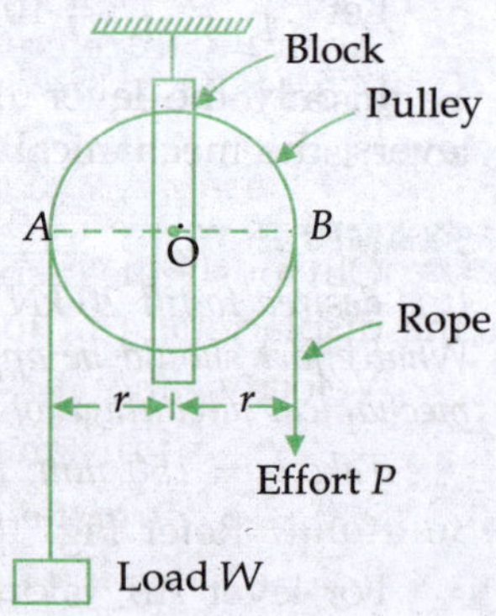

Fig. 10.7. Fixed Pulley

From the equilibrium condition $\Sigma M = 0$, we have

$$P \times r - W \times r = 0 ; \qquad P = W$$

$$\therefore \text{ Mechanical advantage } MA = \frac{W}{P} = 1$$

In the absence of friction,

$$\text{Velocity ratio, } VR = MA = 1$$

$$\text{and machine efficiency } \eta = \frac{MA}{VR} = 1 \text{ or } 100\%.$$

From the above results, it is apparent that with a fixed pulley, there is no gain in the effort applied. The system only helps in affecting a change in the direction of the application of the effort. Accordingly a fixed pulley is used

(*i*) to change the direction of applied force which is always easier to apply in the downward direction.

(*ii*) to raise a load in upward direction by applying effort in downward direction.

10.5.2. Single movable pulley

A movable pulley changes its position when the work is being done. Load to be raised is attached to the pulley itself and the axle of the pulley rises and descends with the load.

With reference to Fig. 10.8, the rope is fixed to a support at its one end and passes round the movable pulley from below. The effort P is applied at the free end of the rope and the load W to be lifted is hung form the hook attached to the movable pulley.

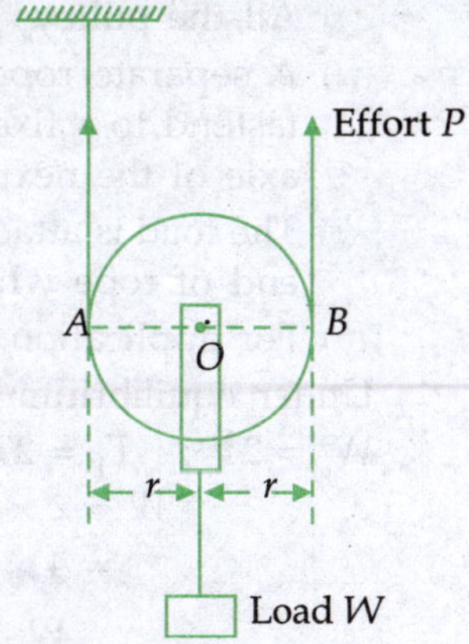

Fig. 10.8. Movable Pulley

Under equilibrium conditions $\Sigma M = 0$

Taking moments of all force about the axle,

$$Q \times r - P \times r = 0 ; \quad P = Q \quad ...(i)$$

Taking moments about point A,

$$W \times r - P \times 2r - 0 ; \quad P - \frac{W}{2} \quad ...(ii)$$

From expressions (*i*) and (*ii*),

$$P = Q = \frac{W}{2}$$

That is, the weight W is equally shared by the two portions of the rope.

$$\text{Mechanical advantage } MA = \frac{W}{P} = \frac{W}{\frac{W}{2}} = 2$$

It is customary in practical applications to use a combination of a fixed pulley and movable pulley as shown in Fig. 10.9. In such an arrangement

(*i*) The movable pulley doubles the effort applied

$$T = \frac{W}{2}$$

(*ii*) The fixed pulley changes the direction of effort applied (downward).

Again, with reference to Fig. 10.9, let the load W be raised to point O' through distance y. The string or rope is then pulled by the effort through distance

$$AA' + BB' = 2y$$

Obviously the displacement of the effort P is twice the displacement of load. Then

Velocity ratio (VR)

$$= \frac{\text{distance moved by the effort}}{\text{distance moved by the load}}$$

$$= \frac{2y}{2} = 2$$

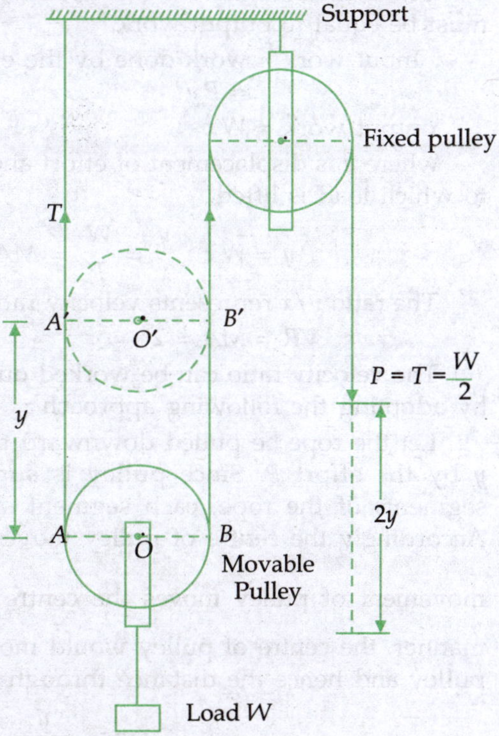

Fig. 10.9. A single fixed and movable pulley

10.6. SYSTEM OF PULLEYS

Quite often, a number of pulleys are so arranged that the composite system results into a gain in mechanical advantage. There are essentially three systems of pulleys, *i.e.*, the first, second and third system.

10.6.1. First system of pulleys

Fig. 10.10 shows first system of pulleys using three pulleys. The salient features of the arrangement are :

(*i*) All the pulleys 1, 2 and 3 are movable pulleys.

(*ii*) A separate rope passes around the periphery of each pulley. One end of the rope is fastend to a fixed support and the other end (called the effort end) is connected to the axle of the next upper pulley.

(*iii*) The load is attached to the bottom most pulley where as the effort is applied to the effort end of rope which passes round the upper most pulley.

(*iv*) For application of effort in a convenient direction, a fixed pulley may also be used.

Under equilibrium conditions,

$W = 2T_i$; $T_1 = 2T_2$; $T_2 = 2T_3$ and $P = T_3$

$$\therefore \quad W = 2 \times 2T_2 = 4T_2 = 4 \times 2T_3 = 8T_3$$

$$MA = \frac{W}{P} = \frac{8T_3}{T_3} = 8 = 2^3$$

Under ideal conditions (no friction) the input work must be equal to output work.

Input work = work done by the effort P $= Py$

output work $= Wx$

where y is displacement of effort and x is the height to which load is lifted.

$$\therefore \quad Py = Wx \; ; \; \frac{y}{x} = \frac{W}{P} = MA$$

The ratio y/x represents velocity ratio of the system.

$$\therefore \quad VR = MA = 2^3$$

The velocity ratio can be worked out independently by adopting the following approach :

Let the rope be pulled downward through distance y by the effort P. Since pulley is supported by two segments of the rope, each segment shortens by $y/2$. Accordingly the centre of pulley moves up by a vertical distance equal to y/2. This upward movement of pulley moves the centre of pulley by $\frac{1}{2}$ of $\frac{y}{2} = \frac{y}{2^2}$. Proceeding in the same manner, the centre of pulley would move up by a distance $y/2^3$. The weight W is attached to pulley and hence the distance through which load is lifted is $y/2^3$.

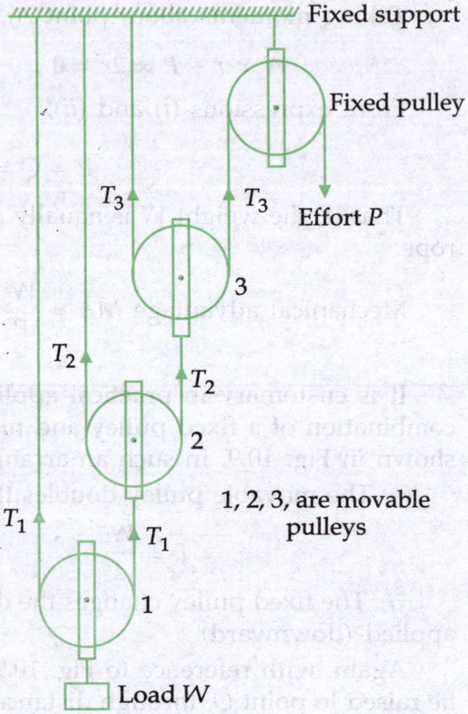

Fig. 10.10. First system of pulleys

$$\therefore \quad VR = \frac{y}{y/2^3} = 2^3$$

In general, if there are n number of pulleys

$$VR = MA = 2^n$$

10.6.2. Second system of pulleys

Figure 10.11 shows an arrangement for the second system of pulleys.

The salient features of the arrangement are :

(*i*) The system has three pulleys in the upper block and two pulleys in the lower block.

(*ii*) The upper block is fixed to a support and the lower block is movable.

(iii) One end of the rope is fixed to the upper pulley in the lower block and the same rope goes round the various pulleys.

(iv) The weight W is attached to the lower block and the effort is applied at the free end of the rope.

For determining velocity ratio of the system, refer to free body of the lower block. The load W is being supported by 5 segments of string and each segment has the same tension T induced in it. Therefore for equilibrium of the lower block,

$$W = 5T = 5P \qquad (\because T = P)$$

$$\therefore \text{ Mechanical advantage } MA = \frac{W}{P} = 5$$

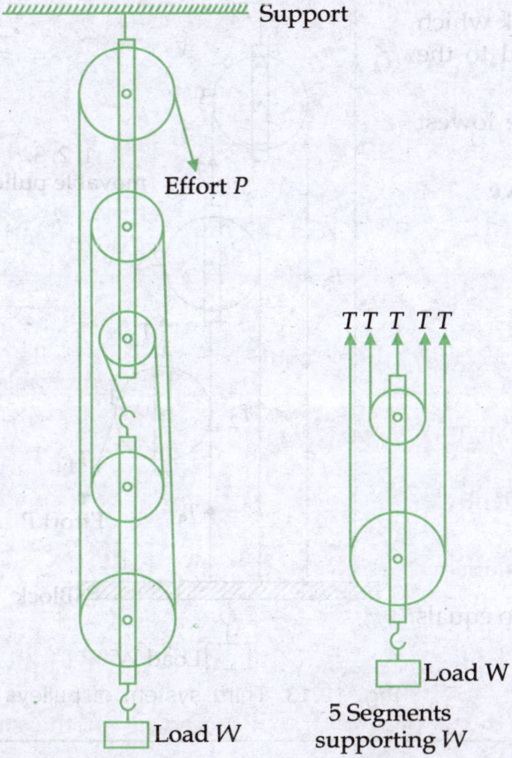

Fig. 10.11. Second system of pulleys

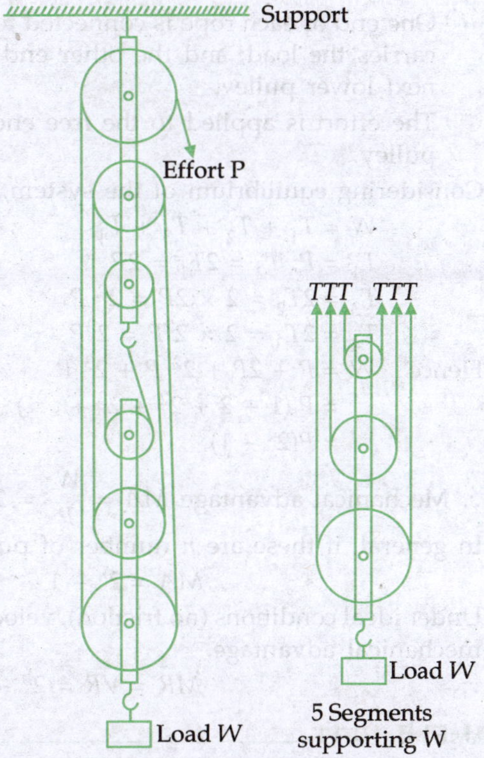

Fig. 10.12. Second system of pulleys

Under ideal conditions (no friction), the velocity ratio equals the mechanical advantage.

$$\therefore \qquad VR = MA = 5$$

If both the upper and lower block have the same number of pulleys, then start is made from one end of the rope fixed to the lower most pulley in the upper block (Fig. 10.12). Equilibrium of the lower block then gives,

$$W = 6T \qquad (\because \text{ there are 6 segments of string})$$

$$= 6P \qquad (\because T = P)$$

$\therefore$ Mechanical advantage

$$MA = \frac{W}{P} = 6$$

For the ideal system (no friction) the velocity ratio equals the mechanical advantage

$\therefore \quad VR = MA = 6$

From the results obtained above, it may be concluded that in second system of pulleys

$$VR = \text{number of pulleys}$$

10.6.3. Third system of pulleys

Figure 10.13 shows the arrangement of third system of pulleys.

The salient features of this arrangement are :

(i) Several movable pulleys are used and the top most pulley is kept fixed.

(ii) There is same number of ropes as is the number of pulleys.

(iii) One end of each rope is connected to a block which carries the load; and the other end is fixed to the next lower pulley.

(iv) The effort is applied to the free end of the lowest pulley.

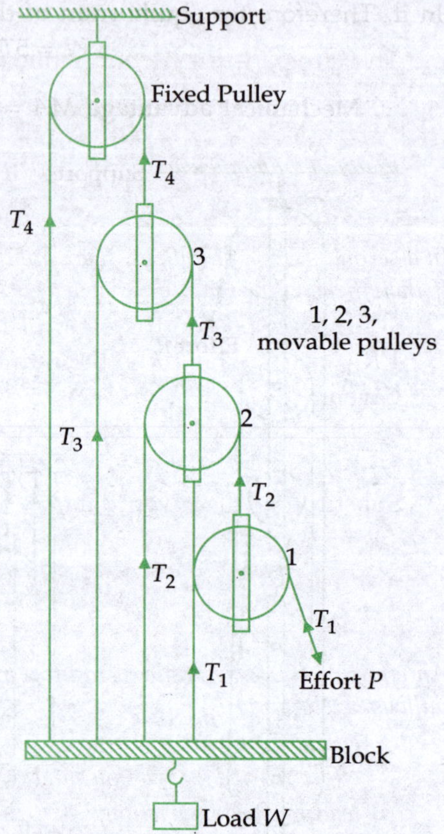

Fig. 10.13. Third system of pulleys

Considering equilibrium of the system, we have

$$W = T_1 + T_2 + T_3 + T_4$$
$$T_1 = P;\ T_2 = 2T_1 = 2P$$
$$T_3 = 2T_2 = 2 \times 2P = 2^2 P$$
$$T_4 = 2T_3 = 2 \times 2^2 P = 2^3 P$$

Hence
$$W = P + 2P + 2^2 P + 2^3 P$$
$$= P(1 + 2 + 2^2 + 2^3 + \ldots)$$
$$= P(2^4 - 1)$$

$\therefore$ Mechanical advantage $MA = \dfrac{W}{P} = 2^4 - 1$

In general, if these are n number of pulleys

$$MA = 2^n - 1$$

Under ideal conditions (no friction), velocity ratio equals the mechanical advantage.

$$\therefore \quad MR = VR = 2^n - 1$$

EXAMPLE 10.11

In a system of pulleys with one string there are five segments of the string at the lower block. What is the velocity ratio of the pulley arrangement? If a force of 200 N is required to lift a load of 600 N, calculate the efficiency of the system.

Solution : Since there is only one string, the arrangement corresponds to second system of pulleys for which

VR = number of segments of the string supporting the lower block = 5

$$MA = \frac{\text{load}}{\text{effort}} = \frac{600}{200} = 3$$

$$\eta = \frac{MA}{VR} = \frac{3}{5} = 0.6 \quad \text{or} \quad 60\%$$

EXAMPLE 10.12

For a third order pulley system having three pulleys, an effort of 200 N is required to lift a load of 1000 N. Calculate the efficiency of the system and the effort lost in friction.

Solution : For a third order pulley system,

$VR = 2^n - 1$ where n is the number of pulleys in the system

$$= 2^3 - 1 = 7$$

$$MA = \frac{\text{load}}{\text{effort}} = \frac{1000}{200} = 5$$

$$\eta = \frac{MA}{VR} = \frac{5}{7} = 0.714 \text{ or } 71.4\%$$

The effort lost in friction while lifting a given load is

$$P_{friction} = P_{actual} - P_{ideal}$$

$$= P_{actual} - \frac{W}{VR} = 200 - \frac{1000}{7} = 57.14 \text{ N}$$

EXAMPLE 10.13

In a second system of pulleys, there are three pulleys in the upper block and two pulleys in the lower block. If the efficiency of the pulley system is 70 per cent, calculate the effort required to lift a load of 300 N.

Solution : For a second, system of pulleys

$VR = n$ where n is the number of pulleys in the system

Given : $n = 3 + 2 = 5$; $VR = 5$

$$\eta = \frac{MA}{VR} = \frac{W/P}{VR}$$

Substituting the given data,

$$0.7 = \frac{3000/P}{5} \; ; \qquad P = \frac{3000}{0.7 \times 5} = 857 \text{ N}$$

EXAMPLE 10.14

An effort of 100 N is required to lift a load of 2500 N by the first system of pulleys which has 5 movable pulleys. Determine :

(*a*) *efficiency of the machines,*

(*b*) *effort wasted in friction, and*

(*c*) *load wasted in friction.*

Solution : For a first system of pulleys,

$VR = 2^n$ where n is the number of movable pulleys

$$= 2^5 = 32$$

$$MA = \frac{W}{P} = \frac{2500}{100} = 25$$

$$\eta = \frac{MA}{VR} = \frac{25}{32} = 0.781 \text{ or } \mathbf{78.1\%}$$

(*b*) The effort lost in friction is

$$P_{friction} = P_{actual} - P_{ideal}$$

$$= P_{actual} - \frac{W}{VR} = 100 - \frac{2500}{32} = 21.875 \text{ N}$$

10.6.4. Block and tackle

In the first, second and third system of pulleys, the pulleys are arranged in a casing one upon another vertically. Quite often, the pulleys are arranged on the same horizontal axis and then encased. Such an arrangement is called a ***block***. A block may have one, two or three pulleys as shown in Fig. 10.14.

A ***tackle*** refers to blocks of pulleys (fixed and movable) which are grouped together by winding a rope or chain around them. The tackle may be used either for lifting purpose or for tightening the guy ropes of a supporting structure.

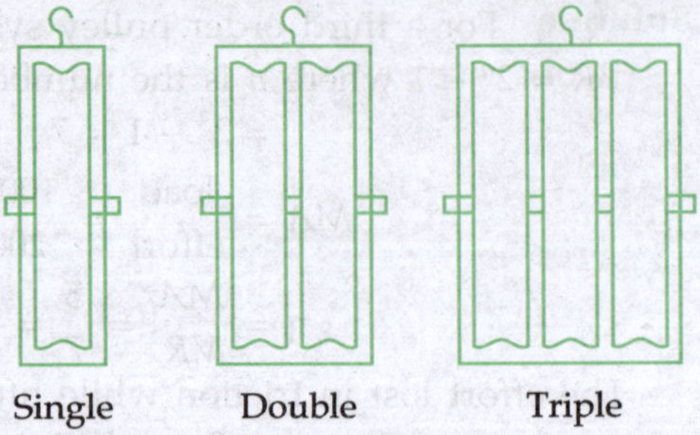

Fig. 10.14. Pulley block

10.7. SIMPLE WHEEL AND AXLE

A simple wheel and axle unit consists of a wheel A of larger diameter and an axle B of small diameter; both are keyed to the same spindle C. The entire assembly is mounted on ball bearing so that the wheel and axle can be rotated. The load W to be lifted is attached to a string which is wound round the axle. Another string is wound round the wheel and the effort P is applied to it. These two strings are wound in opposite directions. This aspect makes the load move upward when the effort is applied downward.

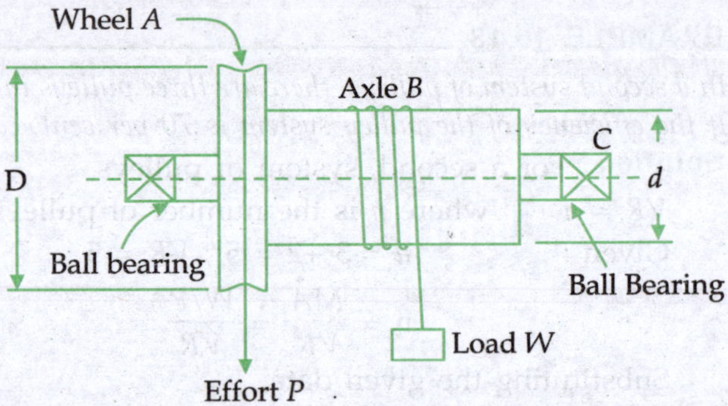

Fig. 10.15. Simple wheel and axle

The wheel and axle are keyed to the same spindle and therefore when the wheel makes one revolution, the axle would also turn one revolution. Let

D = diameter of the wheel and d = diameter of the axle.

Then in one revolution of wheel, the effort moves through πD and the load is lifted through distance πd.

$$\text{Velocity ratio } (VR) = \frac{\text{distance moved by effort}}{\text{distance moved by load}} = \frac{\pi D}{\pi d} = \frac{D}{d}$$

If t_1 and t_2 represent the thickness of string on the wheel and axle, then

$$VR = \frac{D + t_1}{D + t_2}$$

Neglecting friction at the contact surfaces, the forces acting on the system are shown in Fig. 10.16. For equilibrium $\Sigma M = 0$ and therefore taking moments of all forces about O, we get

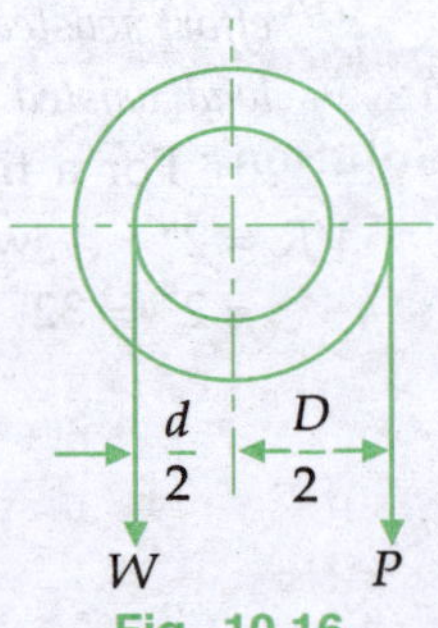

Fig. 10.16

$$P \times \frac{D}{2} - W \times \frac{d}{2} \simeq 0\,; \qquad PD = Wd$$

$$\therefore \text{ Mechanical advantage } (MA) = \frac{W}{P} = \frac{D}{d}$$

The mechanical advantage is equal to velocity ratio. This has to be so because the friction has been neglected. Obviously then the efficiency of the system will be 100 per cent.

Effect of friction,

Let P' be the effort required to lift the load W when friction is taken into account. Then for one revolution of the wheel and axle, work input (work done by the effort) = $P' \times \pi D$

work output (work done by the load) = $W \times \pi d$

$$\text{Efficiency of the machine } \eta = \frac{\text{work output}}{\text{work input}} = \frac{W}{P'} \times \frac{\pi d}{\pi D} = \frac{W}{P'} \times \frac{d}{D}$$

From the above expression,

$$P' = W\frac{d}{D} \times \frac{1}{\eta}$$

$$MA \text{ (actual)} = \frac{W}{P'} = \frac{D}{d} \times \eta$$

EXAMPLE 10.15

In a simple wheel and axle machine, a load of 25 N is raised by an effort of 5 N at an efficiency of 80%. If the diameter of wheel is 375 mm, determine the diameter of the axle.

Solution : $$MA = \frac{W}{P} = \frac{25}{5} = 5$$

$$\eta = \frac{MA}{VR}; \; VR = \frac{MA}{\eta} = \frac{5}{0.8} = 6.25$$

For a simple wheel and axle arrangement, the velocity ratio is also given by

$VR = \dfrac{D}{d}$ where D and d are the diameters of wheel and axle respectively.

$$\therefore \quad 6.25 = \frac{375}{d} \quad \text{and} \quad d = \frac{375}{6.25} = \textbf{60 mm}$$

EXAMPLE 10.16

In a wheel and axle arrangement, the diameter of the wheel is 600 mm and that of the axle drum is 200 mm. The thickness of cord on the wheel is 5 mm and that on the axle is 10 mm. What is the velocity ratio of the machine?

(b) This machine is required to lift 1500 N load with a velocity of 20 m/min. If the machine has an efficiency of 75 per cent, estimate the work that needs to be supplied to the machine to perform the task.

Solution : For the wheel and axle arrangement, the velocity ratio is

$$VR = \frac{D + t_1}{d + t_2} = \frac{600 + 5}{200 + 10} = 2.88$$

(b) From the relation :

$$\eta = \frac{MA}{VR} = \frac{W/P}{VR}, \text{ we have}$$

$$\text{Effort, } P = \frac{W}{\eta \times VR} = \frac{1500}{0.75 \times 2.88} = \textbf{694.5 N}$$

Again, from the definition of velocity ratio

$$VR = \frac{\text{distance moved by effort}}{\text{distance moved by load}}$$

Distance moved by effort = 2.88 × 20 = 57.6 m/min

Work supplied = effort × distance moved by effort

= 694.5 × 57.6 = 40003 Nm/min = 6666.7 Nm/s = **6666.7 W**

10.8. DIFFERENTIAL WHEEL AND AXLE

The unit consists of a wheel *A* and two axles *B* and *C*. The wheel and the two axles are keyed to the same shaft (spindle) which is supported in ball bearings. The effort is applied to the string which is wound round the wheel. Another string is wound on the two axles and it carries the load through a pulley. The strings on the wheel and smaller axle are wound in the same direction where as winding of string on the bigger axle is in the opposite direction.

When the effort *P* is applied in the downward direction, there is unwinding of the strings on the wheel and smaller axle. The string winds on the bigger axle at the same time and the load *W* is lifted upward. Let

D = diameter of wheel

d_1 and d_2 = diameter of bigger and smaller axle

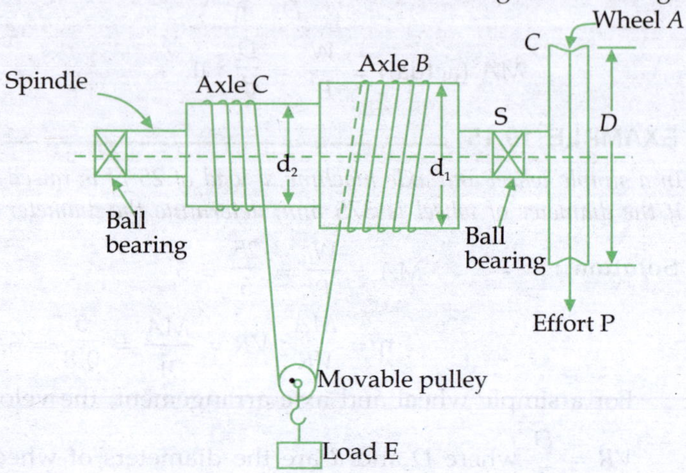

Fig. 10.17. Differential wheel and axle

Then in one revolution of wheel and axle (all are keyed to same shaft),

distance moved by effort = πD (downward)

length of string that winds on bigger axle = πd_1

length of string that unwinds on smaller axle = πd_2

∴ Net length of string which will get wound on bigger axle = $\pi d_1 - \pi d_2$

Since the string carries the load through pulley, the load will get lifted by distance

$$= \frac{\pi d_1 - \pi d_2}{2} = \frac{1}{2}\pi(d_1 - d_2)$$

$$\therefore \quad \text{Velocity ratio } (VR) = \frac{\text{distance moved by effect}}{\text{distance moved by load}}$$

$$= \frac{\pi D}{\frac{1}{2}\pi(d_1 - d_2)} = \frac{2D}{d_1 - d_2}$$

For a larger velocity ratio, d_1 and d_2 are made nearly equal.

Neglecting friction at the contact surfaces, the forces acting on the system are shown in Fig. 10.18. Let it be assumed that segments *ab* and *cd* of the string supporting the load to be vertical. Then from equilibrium condition $\Sigma V = 0$, the tension in each of the segments equals $W/2$. Further, from moment equilibrium $\Sigma M = 0$.

Taking moments of all forces about 0, we get

$$P \times \frac{D}{2} - \frac{W}{2} \times \frac{d_1}{2} + \frac{W}{2} \times \frac{d_2}{2} = 0$$

$$\text{or} \quad PD = \frac{W}{2}(d_1 - d_2)$$

$$\therefore \text{ Mechanical advantage } (MA) = \frac{W}{P} = \frac{2D}{d_1 - d_2}$$

The mechanical advantage is equal to velocity ratio. This has to be so because the friction has been neglected. Obviously then efficiency of the system will be 100 per cent.

Effect of friction

Let P' be the effort required to lift the load W when friction is taken into account. Then for are revolution of the wheel and axles,

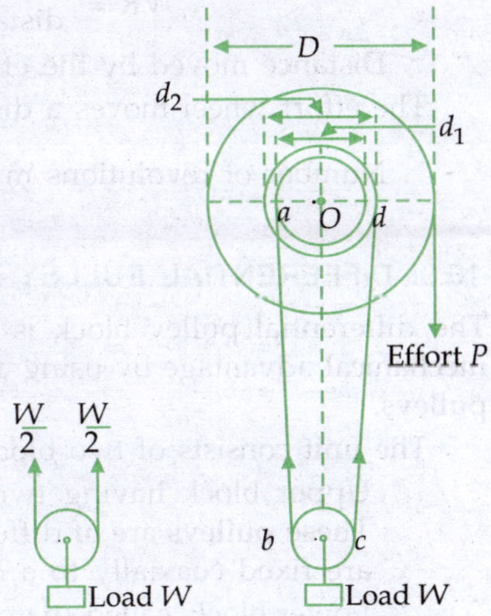

Fig. 10.18

Work input (work done by the effort) = $P' \times \pi D$

Work output (work done by the load) = $W \times \frac{\pi}{2}(d_1 - d_2)$

$$\text{Efficiency of machine } \eta = \frac{\text{work output}}{\text{work input}} = \frac{W \times \frac{\pi}{2}(d_1 - d_2)}{P' \times \pi D}$$

$$= \frac{W}{P'} \times \frac{d_1 - d_2}{2D}$$

From the above expression,

$$P' = W\,\frac{d_1 - d_2}{2D} \times \frac{1}{\eta}$$

and $$MA\ (\text{actual}) = \frac{W}{P'} = \frac{2D}{d_1 - d_2} \times \eta$$

EXAMPLE 10.17

A weight of 800 N is to be lifted by means of a differential wheel and axle arrangement. The diameter of wheel is 250 mm, and the diameters of the larger and smaller axles are 100 mm and 75 mm respectively. If the effort applied to the wheel is 50 N, make calculations for the velocity ratio, mechanical advantage and efficiency of the machine.

Solution : For a differential wheel and axle arrangement,

$$VR = \frac{2D}{d_1 - d_2} = \frac{2 \times 250}{100 - 75} = 20$$

$$MA = \frac{W}{P} = \frac{800}{50} = 16$$

$$\eta = \frac{MA}{VR} = \frac{16}{20} = 0.8 \text{ or } 80\%.$$

EXAMPLE 10.18

In a differential wheel and axle arrangement, the diameter of the wheel is 15 cm and that of the axles are 8 cm and 5 cm. If the efficiency of the machine is 80%. calculate the effort required to lift a load of 300 N. Also calculate the number of revolutions the machine would make in lifting the load through a distance of 25 cm.

Solution : For a differential wheel and axle arrangement,

$$VR = \frac{2D}{d_1 - d_2} = \frac{2 \times 15}{8 - 5} = 10$$

From the relation: $\eta = \frac{MA}{VR} = \frac{W/P}{VR}$, we have

$$\text{Effort } P = \frac{W}{\eta \times VR} = \frac{300}{0.8 \times 10} = 37.5 \text{ N}$$

(*b*) From the definition of velocity ratio,

$$VR = \frac{\text{distance moved by effort}}{\text{distance moved by load}}$$

∴ Distance moved by the effort = 10 × 25 = 250 cm

The effort wheel moves a distance of $\pi D = (\pi \times 15)$ cm in one revolution.

∴ Number of revolutions made by the effort wheel $= \frac{250}{\pi \times 15} =$ **5.31 revolutions**

10.9. DIFFERENTIAL PULLEY BLOCK

The differential pulley block is used to lift very heavy loads and provides a good gain in mechanical advantage by using a far less number of pulleys than that employed in systems of pulleys.

The unit consists of two blocks :

(*i*) Upper block having two pulleys *A* and *B*. These pulleys are of different diameters and are fixed coaxially to a common spindle.

(*ii*) Lower block, called snatch block, that carries a movable pulley to which the load is attached.

An endless chain passes over pulley *A*, then under the snatch block and finally over pulley *B*. The effort *P* is applied to the part of the chain which passes over pulley *A*. The slipping of the chain is prevented by providing teeth on the pulleys.

In one revolution of the pulley, a length πD of chain passes over *A* towards right (unwinding of chain from pulley *A*). This pull represents displacement of the effort. Simultaneously a length πd passes over towards the left (winding of chain on pulley *B*). The lengthening of chain on the effort side and its shortening on the load side results in the lifting of weight :

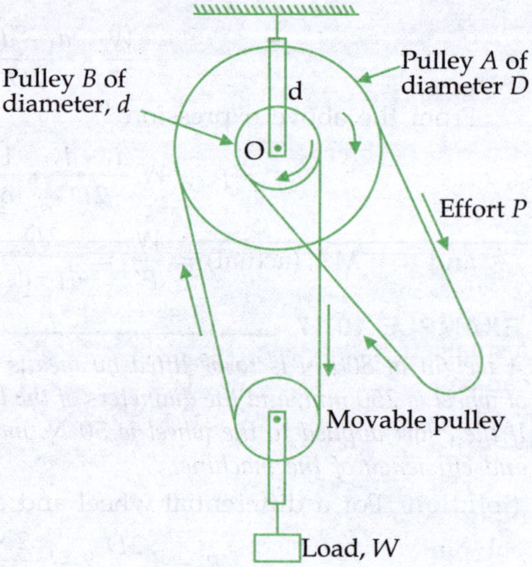

Fig. 10.19. Differential pulley block

Net shortening of the chain $= \pi D - \pi d = \pi(D - d)$

The shortening of the chain is divided equally between two segments of the chain supporting the pulley in the lower block. Hence displacement of the load is $\frac{\pi(D-d)}{2}$.

$$\text{Velocity ratio } (VR) = \frac{\text{distance moved by effort}}{\text{distance moved by load}} = \frac{\pi D}{\frac{1}{2}\pi(D-d)}$$

$$= \frac{2D}{D-d}$$

Neglecting friction at the contact surfaces, the forces acting on the system are shown in Fig. 10.20. Let it be assumed that segments of the chain supporting the load to be vertical. Then from equilibrium condition $\Sigma V = 0$; the tension in each of the segments equals $W/2$. Further from moment equilibrium $\Sigma M = 0$. Taking moments of all force about the axis of upper block,

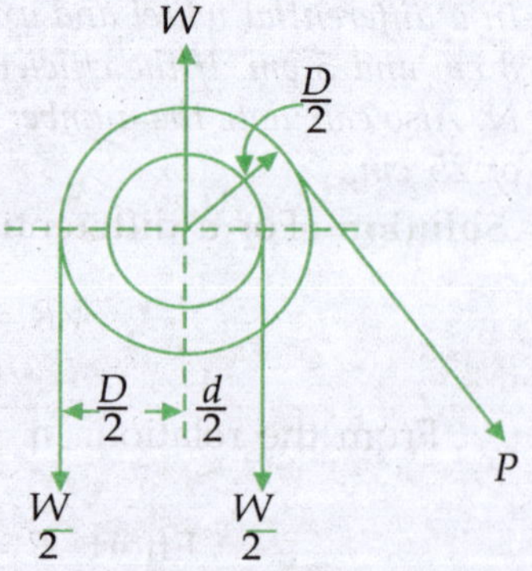

Fig. 10.20

$$P \times \frac{D}{2} + \frac{W}{2} \times \frac{d}{2} - \frac{W}{2} \times \frac{D}{2} = 0$$

or $$\frac{PD}{2} = \frac{W}{4}(D-d)$$

$\therefore$ Mechanical advantage $(MA) = \dfrac{W}{P} = \dfrac{2D}{D-d}$

The mechanical advantage is equal to velocity ratio. This has to be so because friction has been neglected. Obviously the efficiency of the system will be 100 per cent.

EXAMPLE 10.19

The upper block of the differential pulley block has two pulleys of diameter 250 mm and 200 mm. What load will be lifted by this machine by the application of an effort equal to 20 N? Take efficiency of the system as 60 per cent.

Solution : For a differential pulley block,

$$VR = \frac{2D}{D-d} = \frac{2 \times 250}{250-200} = 10$$

From the relation :

$$\eta = \frac{MA}{VR} = \frac{W/P}{VR}, \text{ we have}$$

Load lifted, $W = \eta \times VR \times P = 0.6 \times 10 \times 20 =$ **120 N**

10.10. WINCH CRABS

Winch crabs are the machines used for lifting rather heavy loads from a hole by the application of a small effort. These machines employ a gear system for increasing the velocity ratio. A winch crab can be a single purchase (one set of gears is used) or a double purchase (two sets of gears are used).

10.10.1. Single purchase winch crab

The unit consists of two axles called the effort axle and the load axle. Both the axles are mounted on rigid supports called strands. The effort axle carries a toothed wheel called pinion. The pinion meshes with a spur wheel which is keyed to the drum of load axle. Pinion is always smaller in size and it contains less number of teeth as compared to that on the spur wheel. A rope is wound on the drum and the load is attached to the free end of rope. The effort is applied at the end of a lever which communicates with the effort axle.

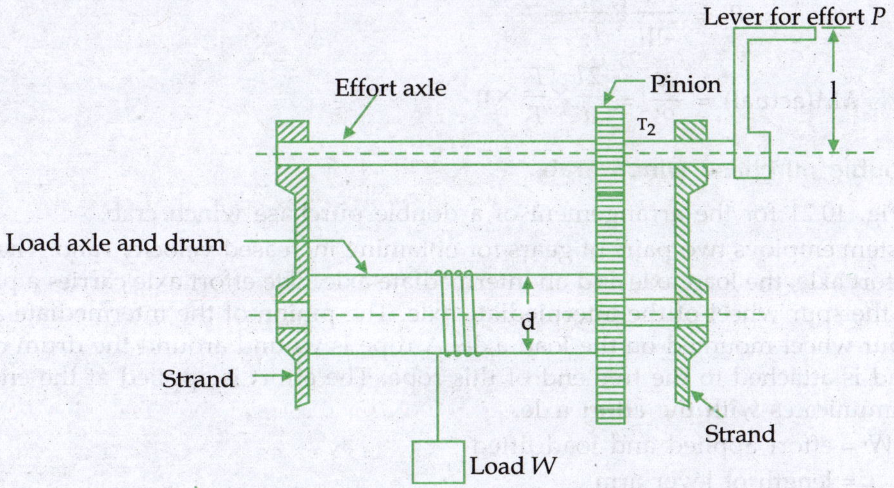

Fig. 10.21. Single purchase winch crab

Let P, W = effort applied and load lifted

T_1, T_2 = number of teeth on pinion and spur gear

l = length of the lever arm, and

d = diameter of drum on load axle

When the lever completes one revolution, the distance moved by the effort is $2\pi l$. For one revolution of pinion, the number of revolutions undergone by the spur wheel is T_1/T_2. As the spur wheel and the drum are fixed on the same axle, the number of revolutions undergone by the drum will be equal to that undergone by the wheel, *i.e.*, T_1/T_2

$$\therefore \text{ Displacement of load } = \pi d \times \frac{T_1}{T_2}$$

$$\text{Velocity ratio } (VR) = \frac{\text{displacement of effort}}{\text{displacement of load}} = \frac{2\pi l}{\pi d \times T_1/T_2} = \frac{2l}{d} \times \frac{T_2}{T_1}$$

Neglecting friction, the work done by effort equals the work done by the load. Therefore

Effort × displacement of effort = Load × displacement of load

$$P \times 2\pi l = W \times (\pi d \times T_1/T_2)$$

$$\therefore \text{ Mechanical advantage } (MA) = \frac{W}{P} = \frac{2l}{d} \times \frac{T_2}{T_1}$$

The mechanical advantage is equal to velocity ratio. This has to be so because friction has been neglected. Obviously the efficiency of the system will be 100 per cent.

Effect of friction :

Let P' be the effort required to lift the load W when friction is taken into account.

Work input (work done by the effort) = $P' \times 2\pi l$

Work output (work done by the load) = $W \times (\pi d \times T_1/T_2)$

$$\text{Efficiency of machine } \eta = \frac{\text{work output}}{\text{work input}} = \frac{W \times (\pi d \times T_1/T_2)}{P' \times 2\pi l}$$

From the above expression

$$P' = \frac{Wd}{2\eta l} \times \frac{T_1}{T_2}$$

and

$$MA(\text{actual}) = \frac{W}{P'} = \frac{2l}{d} \times \frac{T_2}{T_1} \times \eta$$

10.10.2. Double purchase winch crab

Refer Fig. 10.21 for the arrangement of a double purchase winch crab.

The system employs two pairs of gears for obtaining increased velocity ratio. There are three axles: an effort axle, the load axle and an intermediate axle. The effort axle carries a pinion which gears with the spur wheel of the intermediate axle. The pinion of the intermediate axle meshes with the spur wheel mounted on the load axle. A rope is wound around the drum on load axle and the load is attached to the free end of this rope. The effort is applied at the end of a lever which communicates with the effort axle.

Let P, W = effort applied and load lifted

l = length of lever arm

d = diameter of drum and load axle.

T_1 , T_3 = number of teeth on pinion of effort axle and intermediate axle

T_2 , T_4 = number of teeth on spur wheel of intermediate axle and load axle.

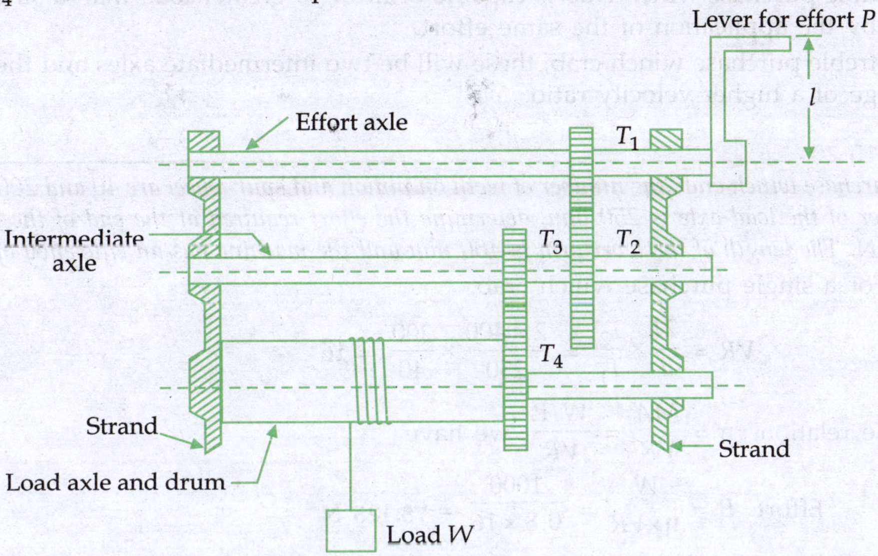

Fig. 10.22. Double purchase winch crab

When the lever is given one revolution, the distance moved by the effort P is

$$= 2\pi l$$

The effort axle and the pinion (teeth T_1) mounted on it also turn one revolution. This pinion engages with spur wheel (teeth T_2) fixed onto the intermediate axle. Then the revolutions made by the intermediate axle are

$$= 1 \times \frac{T_1}{T_2}$$

The pinion (teeth T_3) mounted on the intermediate axle will also turn through the same number of revolutions. This pinion meshes with spur wheel (teeth T_4) of the load axle. As such, the revolutions made by the load axle and the drum mounted on it are

$$= 1 \times \frac{T_1}{T_2} \times \frac{T_3}{T_4}$$

$\therefore$ Distance moved by the load

$$= \pi d \times \left(1 \times \frac{T_1}{T_2} \times \frac{T_3}{T_4}\right) = \pi d \times \frac{T_1}{T_2} \times \frac{T_3}{T_4}$$

$$\text{Velocity ratio } (VR) = \frac{\text{displacement of effort}}{\text{displacement of load}} = \frac{2\pi l}{\pi d \times T_1/T_2 \times T_3/T_4} = \frac{2l}{d} \times \frac{T_2}{T_1} \times \frac{T_4}{T_3}$$

Neglecting loss of energy due to friction, the mechanical advantage equals the velocity ratio. That is

$$MA = \frac{2l}{d} \times \frac{T_2}{T_1} \times \frac{T_4}{T_3}$$

If loss of energy due to friction is taken into account, then

$$MA = \frac{2l}{d} \times \frac{T_2}{T_1} \times \frac{T_4}{T_3} \times \eta$$

where η is the efficiency of the system.

Note:

(*i*) A double purchase winch crab is capable of lifting a greater load than a single purchase winch crab by the application of the same effort.

(*ii*) In a treble purchase winch crab, these will be two intermediate axles and the system has the advantage of a higher velocity ratio.

EXAMPLE 10.20

In a single purchase winch crab, the number of teeth on pinion and spur wheel are 40 and 200 respectively. If the diameter of the load axle is 250 mm, determine the effort required at the end of the lever to lift a load of 1000 N. The length of the lever arm is 400 mm and the machine has an efficiency of 80 per cent.

Solution : For a single purchase winch crab,

$$VR = \frac{2l}{d} \times \frac{T_2}{T_1} = \frac{2 \times 400}{250} \times \frac{200}{40} = 16$$

From the relation : $\eta = \frac{MA}{VR} = \frac{W/P}{VR}$, we have

$$\text{Effort } P = \frac{W}{\eta \times VR} = \frac{1000}{0.8 \times 16} = \mathbf{78.125\ N}$$

EXAMPLE 10.21

The following data refers to a double purchase winch crab :

length of the lever (crank) arm = 400 mm

diameter of the drum = 128 mm

number of teeth for the pinion on effort axle = 15

number of teeth for the spur wheel on intermediate axle = 80

number of teeth for the pinion on intermediate axle = 20

number of teeth for the spur wheel on load axle = 75

This machine can lift loads of 100 N and 150 N by app`lying efforts of 1.75 N and 2.5 N respectively. Determine:

(*a*) *law of the machine,*

(*b*) *effort required to lift a load of 200 N and efficiency at this load, and*

(*c*) *maximum efficiency of the machine.*

Solution : The law of the machine is

$$P = mW + C$$

Substituting the given values of P and W,

$$1.75 = m \times 100 + C \quad \ldots(i)$$

$$2.5 = m \times 150 + C \quad \ldots(ii)$$

Solving the identities (*i*) and (*ii*), we get

$$m = 0.015 \quad \text{and} \quad C = 0.25$$

$\therefore$ The law of machine is

$$P = 0.015\ W + 0.25$$

(*b*) Effort required to lift a load of 200 N,

$$P = 0.015 \times 200 + 0.25 = \mathbf{3.25\ N}$$

In the case of a double winch crab,

$$VR = \frac{2l}{d} \times \frac{T_2}{T_1} \times \frac{T_4}{T_3} = \frac{2 \times 400}{128} \times \frac{80}{15} \times \frac{75}{20} = 125$$

$$MA = \frac{W}{P} = \frac{200}{3.25} = 61.54$$

$$\eta = \frac{MA}{VR} = \frac{61.54}{125} = 0.492 \text{ or } \mathbf{49.2\%}$$

(*c*) Limiting value of efficiency,

$$\eta_{max} = \frac{1}{m \times VR} = \frac{1}{0.015 \times 125} = 0.533 = \mathbf{53.3\%}$$

10.11. INCLINED PLANE

Fig. 10.23 shows an inclined plane *AB* making an angle α with the horizontal. *A* body of weight *W* is placed at point *C* and it is connected to an inextensible string which passes over a pulley positioned at *B*. An effort is applied at the free end of the string.

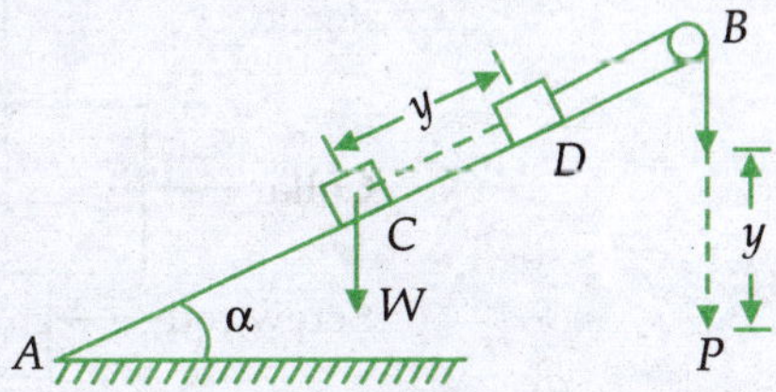

Fig. 10.23. Inclined plane

$$MA = \frac{W}{P}$$

Let the effort *P* move downward through a vertical distance *y*. Since the string is inextensible, the weight *W* would also move upward along the plane through the same distance *y*. The corresponding vertical movement of the load is $y \sin\alpha$

$$\therefore \quad VR = \frac{\text{distance moved by the effort}}{\text{distance moved by the load}} = \frac{y}{y \sin\alpha} = \text{cosec}\,\alpha$$

EXAMPLE 10.22

An effort of 60 N is required to lift a load of 100 N lying on an inclined plane having an inclination of 30° with the horizontal. Determine the velocity ratio, mechanical advantage and efficiency. How this efficiency would be affected if coefficient of friction between the load and inclined surface is 0.3? Also calculate the effort lost in friction.

Solution : $VR = \text{cosec}\,\alpha = \text{cosec}\,30° = 2$

$$MA = \frac{W}{P} = \frac{100}{60} = 1.67$$

$$\eta = \frac{MA}{VR} = \frac{1.67}{2} = 0.835 \text{ or } 83.5\%$$

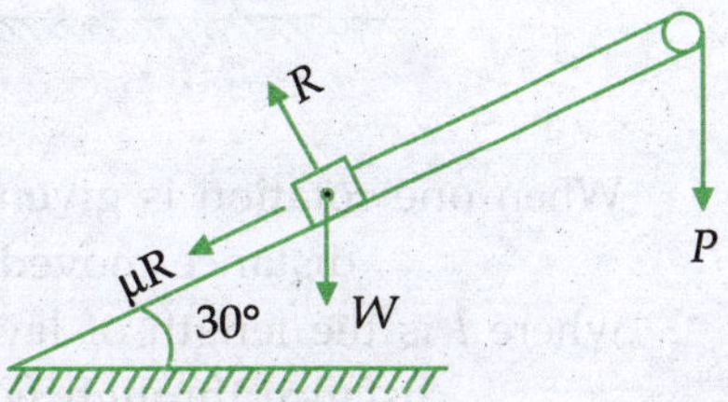

Fig. 10.24

(*b*) Refer Fig. 10.24. Resolving the forces perpendicular and parallel to the inclined plane,

$$R = 100 \cos 30° = 86.6 \text{ N}$$

$$\mu R + 100 \sin 30° = P' \quad (P' = \text{actual effort})$$

$$\therefore \quad P' = 0.30 \times 86.6 + 100 \times 0.5 = 75.9 \text{ N}$$

$$\eta = \frac{MA}{VR} = \frac{W/P'}{VR} = \frac{100}{75.9 \times 2} = 0.658 \text{ or } \mathbf{65.8\%}$$

Effort lost in friction = 75.9 − 60 = **15.9 N**

10.12. SCREW JACK

Screw jack is a simple machine used for lifting heavy loads, through short distances, with the help of small effort applied at its handle. The most common application is the raising of the front or rear portion of a vehicle for the purpose of changing the wheel or tyre.

Construction: The unit consists of a base which is in the shape of the frustrum of a hollow cone. The frustrum has a hole which is threaded to receive a strong square threaded screw. A collar is fitted onto the top of the screw in such a manner that the screw can rotate freely in it. The load to be raised or lowered is placed on the top of the collar. A lever or handle, often called tommy bar, can be inserted into a hole provided in the collar. When an effort is applied at the end of the lever, the top of the screw goes either up or down depending on the direction in which the lever is turned.

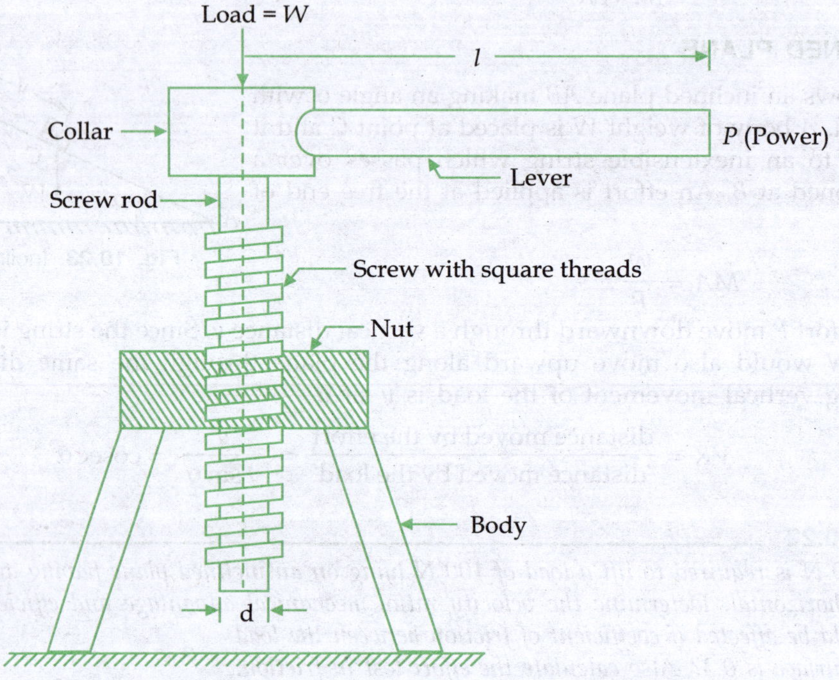

Fig. 10.25. Screw jack

When one rotation is given to the handle,

distance moved by the effort = $2\pi l$

where l is the length of lever arm from the centre of screw.

distance through which load is lifted = pitch of the thread = p

$$\therefore \text{ Velocity ratio } VR = \frac{\text{distance moved by effort}}{\text{distance moved by the load}} = \frac{2\pi l}{p}$$

It may be pointed out that for a multiple threaded screw, the pitch equals np where

n = number of starts, and

p = distance between two consective threads (apparent pitch)

Analysis : The working principle of a screw jack is the same as that of the inclined plane.

Refer Fig. 10.26 which shows development of one complete turn of a screw thread. Distance AB will be equal to the circumference and distance BC will be equal to pitch of the screw. The inclination of the thread is given by

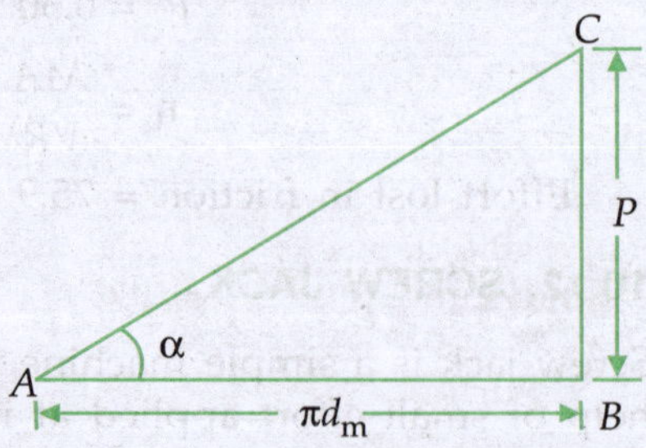

Fig. 10.26

$$\tan \alpha = \frac{p}{\pi d_m} \quad \text{...(10.12)}$$

where d_m is the mean diameter of the thread. The angle α is referred to as the helix angle.

(a) Effort required to lift the load: The situation corresponds to that of lifting a load W up an inclined plane by a horizontal force P as shown in Fig. 10.27.

Since the block has impending motion up the plane, the friction force would act down the plane. The various forces which keep the block in equilibrium and as shown in the free body diagram are:

(i) weight W of the block,
(ii) reaction R of the plane
(iii) tractive force P, and
(iv) frictional force $F = \mu R$

Resolving these forces along and perpendicular to the plane, we get

$$P \cos\alpha - \mu R - W \sin\alpha = 0 \quad ; \quad P\cos\alpha = W\sin\alpha + \mu R$$

$$R - W\cos\alpha - P\sin\alpha = 0 \quad ; \quad R = W\cos\alpha + P\sin\alpha$$

$$\therefore \quad P\cos\alpha = W\sin\alpha + \mu(W\cos\alpha + P\sin\alpha)$$

$$= W\sin\alpha + \mu W\cos\alpha + \mu P\sin\alpha$$

or $\quad P(\cos\alpha - \mu\sin\alpha) = W(\sin\alpha + \mu\cos\alpha)$

$$\therefore \quad P = W\frac{\sin\alpha + \mu\cos\alpha}{\cos\alpha - \mu\sin\alpha}$$

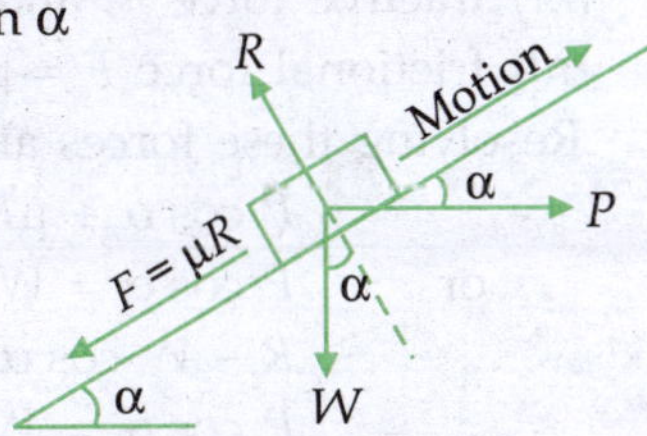

Fig. 10.27

Substituting $\mu = \tan\phi = \dfrac{\sin\phi}{\cos\phi}$ where ϕ is the angle of friction

$$P = W\frac{\sin\alpha + \dfrac{\sin\phi}{\cos\phi}\cos\alpha}{\cos\alpha - \dfrac{\sin\phi}{\cos\phi}\sin\alpha} = W\frac{\sin\alpha\cos\phi + \sin\phi\cos\alpha}{\cos\alpha\cos\phi - \sin\phi\sin\alpha}$$

$$= W\frac{\sin(\alpha+\phi)}{\cos(\alpha+\phi)} = W\tan(\alpha+\phi) \quad \text{...(10.12 a)}$$

Here P is the effort applied at the mean radius of screw jack. But in practice, the effort is applied at the end of the handle of the jack.

Let P_h = horizontal force applied at the end of the handle

l = length of the handle

Taking moments of P_h and P about the axis of screw jack, we get

$$P_h \times l = P \times \frac{d_m}{2} = W\tan(\alpha+\phi)\frac{d_m}{2}$$

$$\therefore \quad P_h = \frac{W d_m}{2l}\tan(\alpha+\phi) \quad \text{...(10.13)}$$

The above identity establishes the relation between the effort required at the end of the handle and the load lifted.

Further, $\quad P_h = \dfrac{W d_m}{2l}\tan(\alpha+\phi) = \dfrac{W d_m}{2l}\dfrac{\tan\alpha + \tan\phi}{1 - \tan\alpha\tan\phi}$

Substituting $\tan\alpha = \dfrac{p}{\pi d_m}$ and $\tan\phi = \mu$, we get

$$P_h = \frac{W d_m}{2l}\left[\frac{\dfrac{p}{\pi d_m}+\mu}{1-\dfrac{p}{\pi d_m}\mu}\right] = \frac{W d_m}{2l}\left[\frac{p+\mu\pi d_m}{\pi d_m - p\mu}\right] \quad \text{...(10.14)}$$

The above relation gives the value of P_h in terms of coefficient of friction and pitch of the screw.

(*b*) **Effort required to lower the load:** Since the block has impending motion down the plane, the friction force would act up the plane. The various forces which keep the block in equilibrium and as shown in the free body diagram are:

(*i*) weight W of the block,
(*ii*) reaction R of the plane,
(*iii*) tractive force P, and
(*iv*) frictional force $F = \mu R$

Resolving these forces along and perpendicular to the plane, we get

$$P\cos\alpha + \mu R - W\sin\alpha = 0$$

or $$P\cos\alpha = W\sin\alpha - \mu R$$

$$R - W\cos\alpha - P\sin\alpha = 0 \quad \text{or} \quad R = W\cos\alpha + P\sin\alpha$$

$\therefore$ $$P\cos\alpha = W\sin\alpha - \mu(W\cos\alpha + P\sin\alpha)$$
$$= W\sin\alpha - \mu W\cos\alpha - \mu P\sin\alpha$$

or $$P(\cos\alpha + \mu\sin\alpha) = W(\sin\alpha - \mu\cos\alpha)$$

$\therefore$ $$P = W\frac{\sin\alpha - \mu\cos\alpha}{\cos\alpha + \mu\sin\alpha}$$

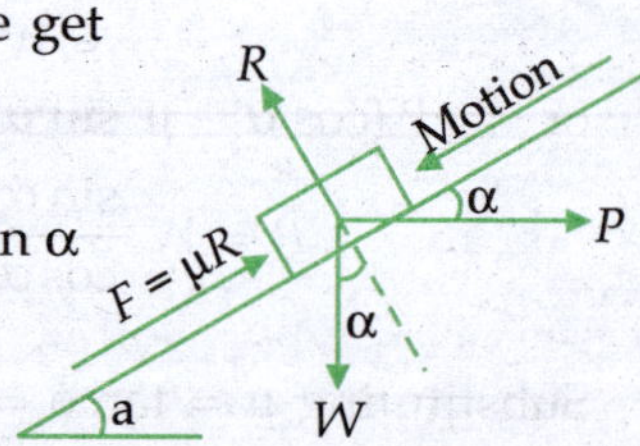

Fig. 10.28

Substituting $\mu = \tan\phi = \dfrac{\sin\phi}{\cos\phi}$ where ϕ is the angle of friction

$$P = W\frac{\sin\alpha - \dfrac{\sin\phi}{\cos\phi}\cos\alpha}{\cos\alpha + \dfrac{\sin\phi}{\cos\phi}\sin\alpha} = W\frac{\sin\alpha\cos\phi - \sin\phi\cos\alpha}{\cos\alpha\cos\phi + \sin\alpha\sin\phi} = W\frac{\sin(\alpha-\phi)}{\cos(\alpha-\phi)}$$

$$= W\tan(\alpha - \phi) \quad \text{...(10.15)}$$

Here P is the effort applied at the mean radius of the jack. But in actual practice, an effort P_h is applied at the end of the handle of the jack.

Equating the moments of P and P_h about the axis of the jack,

$$P_h \times l = P \times \frac{d_m}{2} = W\tan(\alpha - \phi) \times \frac{d_m}{2}$$

$\therefore$ $$P_h = W\frac{d_m}{2l}\tan(\alpha - \phi) \quad \text{...(10.16)}$$

The above identity establishes the relation between the effort required at the end of the handle to lower the load.

Further, $$P_h = W\frac{d_m}{2l}\,\frac{\tan\alpha - \tan\phi}{1 + \tan\alpha\tan\phi}$$

Substituting $\tan\alpha = \dfrac{p}{\pi d_m}$ and $\tan\phi = \mu$, we have

$$P_h = \frac{W d_m}{2l}\left[\frac{\dfrac{p}{\pi d_m} - \mu}{1 + \dfrac{p}{\pi d_m}\mu}\right] = \frac{W d_m}{2l}\left[\frac{p - \mu\pi d_m}{\pi d_m + \mu p}\right] \quad \text{...(10.17)}$$

The above relation gives the value of P_h in terms of coefficient of friction and pitch of the screw.

Efficiency of screw jack: The effort required at the mean radius of screw jack to lift the load is given by

$$P = W\tan(\alpha + \phi)$$

In the absence of friction, ϕ will be zero. In that case the ideal effort would equal

$$P_0 = W\tan\alpha$$

$$\therefore \quad \text{Efficiency } \eta = \frac{\text{ideal effort}}{\text{actual effort}} = \frac{W\tan\alpha}{W\tan(\alpha+\phi)} = \frac{\tan\alpha}{\tan(\alpha+\phi)} \quad \text{...(10.18)}$$

The condition for maximum efficiency can be worked out by differentiating the above expression with respect to α and setting the derivative equal to zero. That is

$$\frac{d\eta}{d\alpha} = \frac{d}{d\alpha}\left[\frac{\tan\alpha}{\tan(\alpha+\phi)}\right] = 0$$

or $$\frac{\sec^2\alpha\tan(\alpha+\phi) - \sec^2(\alpha+\phi)\tan\alpha}{\tan^2(\alpha+\phi)} = 0$$

or $$\sec^2\alpha\tan(\alpha+\phi) = \sec^2(\alpha+\phi)\tan\alpha$$

or $$\frac{1}{\cos^2\alpha}\times\frac{\sin(\alpha+\phi)}{\cos(\alpha+\phi)} = \frac{1}{\cos^2(\alpha+\phi)}\times\frac{\sin\alpha}{\cos\alpha}$$

or $$\sin(\alpha+\phi)\cos(\alpha+\phi) = \sin\alpha\cos\alpha$$

or $$\sin 2(\alpha+\phi) = \sin 2\alpha$$

$$\therefore \quad 2(\alpha+\phi) = \pi - 2\alpha\ ;\ \alpha = \frac{\pi}{4} - \frac{\phi}{2} = 45^\circ - \frac{\phi}{2}$$

Maximum efficiency can be worked out by substituting $\alpha = 45^\circ - \dfrac{\phi}{2}$ in the expression for efficiency

$$\eta_{max} = \frac{\tan(45^\circ - \phi/2)}{\tan(45^\circ + \phi/2)} = \frac{\dfrac{1-\tan\phi/2}{1+\tan\phi/2}}{\dfrac{1+\tan\phi/2}{1-\tan\phi/2}} = \frac{(1-\tan\phi/2)^2}{(1+\tan\phi/2)^2}$$

$$= \frac{\left[\dfrac{\cos\phi/2 - \sin\phi/2}{\cos\phi/2}\right]^2}{\left[\dfrac{\cos\phi/2 + \sin\phi/2}{\cos\phi/2}\right]^2} = \frac{(\cos\phi/2 - \sin\phi/2)^2}{(\cos\phi/2 + \sin\phi/2)^2}$$

$$= \frac{\cos^2\phi/2 + \sin^2\phi/2 - 2\cos\phi/2\sin\phi/2}{\cos^2\phi/2 + \sin^2\phi/2 + 2\cos\phi/2\sin\phi/2}$$

Since $\sin^2 \phi/2 + \cos^2 \phi/2 = 1$ and $2 \sin \phi/2 \cos \phi/2 = \sin \phi$, we can write

$$\eta_{max} = \frac{1 - \sin \phi}{1 + \sin \phi} \quad \text{...(10.19)}$$

A jack is self locking if its efficiency is less than 50 percent or $\alpha < \phi$. If the jack is not self-locking, the weight resting on the collar starts coming down. An effort equal to $P = W \tan (\alpha - \phi)$ has then to be applied at the mean radius of the jack to balance the load which is descending.

The efficiency of a screw jack is independent of load and depends only on the helix angle of screw and the angle of friction ϕ.

$$\eta = \frac{\tan \alpha}{\tan (\alpha + \phi)}$$

Further $\mu = \tan \phi$ and $\tan \alpha = \dfrac{P}{\pi d_m}$

The following observations may thus be made:

(*i*) efficiency increases with increase in helix angle, *i.e.*, with increase in pitch of screw thread.

(*ii*) higher the angle of friction ϕ, lower will be the efficiency. The efficiency drops with increase in coefficient of friction between the nut and screw.

(*iii*) in case of V-thread with included angle 2β, the coefficient of friction μ, as compared to that for square thread is higher and is given by

$$\mu_1 = \frac{\mu}{\cos \beta}$$

Apparently the efficiency of a V-threaded screw jack would be less.

(*iv*) the maximum efficiency of screw jack is

$$\eta_{max} = \frac{1 - \sin \phi}{1 + \sin \phi}$$

The maximum efficiency is independent of load, helix angle and depends only on the angle of friction. Higher the angle of friction, lower will the efficiency. Since $\mu = \tan \phi$, the efficiency drops with increase in coefficient of friction between the fixed nut and screw.

EXAMPLE 10.23

A power screw moves a nut against a force of 10 kN at a speed of 6 mm/s. The screw has a single square thread of 6 mm pitch on a major diameter of 45 mm. Calculate the power of the driving motor. The coefficient of friction at the screw thread is 0.15.

Solution: Thickness of threads $t = \dfrac{p}{2} = \dfrac{6}{2} = 3$ mm

Outer diameter of screw $d_o = 45$ mm (given)

Diameter at the base of screw $d_i = d_0 - 2t$

$$= 45 - 2 \times 3 = 39 \text{ mm}$$

Mean diameter of screw $= \dfrac{d_0 + d_i}{2} = \dfrac{45 + 39}{2} = 42$ mm

Now $\tan \alpha = \dfrac{p}{\pi d_m} = \dfrac{6}{\pi \times 42} = 0.0455; \quad \alpha = 2.605°$

$\tan \phi = \mu = 0.15; \quad \phi = 8.53°$

The force needed to raise or lower the nut,

$$P = W \tan(\alpha + \phi) = 100 \tan(2.605 + 8.53) = 19.68 \text{ kN}$$

Power of driving motor = $P \times V$

$$= 19.68 \times (6 \times 10^{-3}) = 0.1181 \text{ kN m/s} = \mathbf{0.118\ kW}$$

EXAMPLE 10.24

A load of 500 N is required to be lifted by a screw jack whose spindle has the square threads with the following specifications:

outer diameter = 8 cm; threads in 1.6 cm = 2

thread thickness = half the pitch

coefficient of friction between nut and screw = 0.1

What force needs to be applied at the end of a handle of 50 cm length?

Solution: For 2 square threads in 1.6 cm length,

$$\text{pitch of screw } p = \frac{1.6}{2} = 0.8 \text{ cm}$$

$$\text{thickness of threads } t = \frac{1}{2} \times 0.8 = 0.4 \text{ cm}$$

$$\text{outer diameter of screw } d_1 = 8 \text{ cm}$$

$$\text{diameter at the base of screw } d_2 = d_1 - 2t = 8 - 2 \times 0.4 = 7.2 \text{ cm}$$

$$\text{mean diameter of screw } d_m = \frac{d_1 + d_2}{2} = \frac{8 + 7.2}{2} = 7.6 \text{ cm}$$

Now

$$\tan \alpha = \frac{p}{\pi d_m} = \frac{0.8}{3.14 \times 7.6} = 0.0335$$

$$\alpha = \tan^{-1}(0.0335) = 1.92^\circ$$

$$\tan \phi = \mu = 0.1 \quad \text{and} \quad \phi = \tan^{-1}(0.1) = 5.72^\circ$$

The effort P applied at the mean radius of the jack in lifting a given load is given by

$$P = W \tan(\alpha + \phi) = 500 \tan(1.92 + 5.72) = 67.07 \text{ N}$$

If P_h is the effort applied at the end of a handle of length l, then equating the moments of P and P_h about the axis of jack, we get

$$P_h \times l = P \times \frac{d_m}{2} \quad ; \quad P_h = P \times \frac{d_m}{2l} = 67.07 \times \frac{7.6}{2 \times 30} = \mathbf{5.097\ N}$$

EXAMPLE 10.25

A square threaded screw of 7.5 cm mean diameter and 1.5 cm pitch is utilized to lift a weight of 600 N. Determine

(*a*) *the tangential force which needs to be applied to it at the end of a handle of length 40 cm.*

(*b*) *the efficiency of jack and state whether the jack is self locking or not.*

If the jack is self locking, calculate the torque necessary to lower the load. If not, what amount of torque needs to be applied to keep the load from descending?

Assume coefficient of friction at the screw thread = 0.05

Solution: The tangential force applied at the mean radius of the screw is given by

$$P = W \tan(\alpha + \phi) = W \left[\frac{\tan \alpha + \tan \phi}{1 - \tan \alpha \tan \phi} \right]$$

Now, $\tan\alpha = \frac{p}{\pi d_m} = \frac{1.5}{\pi \times 7.5} = 0.0637 \quad ; \quad \alpha = 3.639^\circ$

$\tan\phi = \mu = 0.05 \quad ; \quad \phi = 2.862^\circ$

$\therefore \quad P = 600\left[\frac{0.0637 + 0.05}{1 - 0.0637 \times 0.05}\right] = 68.44 \text{ N}$

The effort P_h to be applied at the end of a handle of length can be worked out from the correlation

$$P_h \times l = P \times \frac{d_m}{2} \quad ; \quad P_h = \frac{P d_m}{2l} = \frac{68.44 \times 7.5}{2 \times 40} = \mathbf{6.416\ N}$$

(*b*) Efficiency of the screw jack is given by

$$\eta = \frac{\tan\alpha}{\tan(\alpha + \phi)} = \frac{\tan 3.639}{\tan(3.639 + 2.862)} = \frac{0.0637}{0.11395} = 0.558 \text{ or } 55.8\%$$

Since the efficiency of the screw jack is more than 50%, it is not self locking. The weight W = 600 N will than start coming down. The tangential force required at the mean radius to prevent the load from descending is

$$P_1 = W \tan(\alpha - \phi) = 600 \tan(3.639 - 2.862) = 8.137 \text{ N}$$

$\therefore$ Torque to be applied, $T = P_1 \times \frac{d_m}{2} = 8.137 \times \frac{7.5}{2}$

$= 30.51 \text{ N cm} = \mathbf{0.3051\ Nm}$

EXAMPLE 10.26

The spindle of a screw jack has single start square threads of 60 mm root diameter and 8 mm pitch. The jack supports a load of 500 N on its head and that revolves with the screw. If the coefficient of friction at the screw threads is 0.1, determine:

(*a*) *the effort to be applied at the end of a handle of 50 cm length to raise the load,*
(*b*) *work done in lifting the load through 10 mm,*
(*c*) *efficiency of the screw, and*
(*d*) *ratio between efforts required to raise and lower the load.*

Solution: The effort P applied at the mean radius of the jack in lifting a given load is given by

$$P = W \tan(\alpha + \phi)$$

Now, $\tan\alpha = \frac{P}{\pi d_m} = \frac{8}{\pi \times 60} = 0.04246 \quad ; \quad \alpha = \tan^{-1}(0.04246) = 2.43^\circ$

$\tan\phi = 0.1 \quad ; \quad \phi = \tan^{-1}(0.1) = 5.72^\circ$

$\therefore \quad P = 500 \tan(2.43 + 5.72) = 71.60 \text{ N}$

If P_h is the effort applied at the end of a handle of length l, then equating the moments of P and P_h about the axis of the jack, we get

$$P_h \times l = P \times \frac{d_m}{2}; \; P_h = \frac{P d_m}{2l} = \frac{71.60 \times 60}{2 \times 50} = \mathbf{4.296\ N}$$

(*b*) Torque required at the mean radius of the jack,

$$T = P \times \frac{d_m}{2} = 71.60 \times \frac{60}{2} = 2148 \text{ N mm} = 2.148 \text{ Nm}$$

Number of turns required to lift the load through 100 mm,

$$= \frac{\text{distance moved}}{\text{pitch of thread}} = \frac{100}{8} = 12.5$$

Since one turn ≡ 2π radians, we have

total angular displacement = 12.5 × 2π = 78.5 radians

Work done = torque required × angular displacement

= 2.148 × 78.5 = **168.62 Nm**

(c) Efficiency of screw jack $= \dfrac{\text{work done by screw}}{\text{work done on screw}}$

$$= \frac{\text{load lifed} \times \text{distance moved}}{\text{work done on screw}} = \frac{500 \times 0.1}{168.62}$$

= 0.2965 = **29.65%**

The efficiency of a screw jack can also be worked out from the correlation

$$\eta = \frac{\tan \alpha}{\tan(\alpha + \phi)} = \frac{\tan 2.43}{\tan(2.43 + 5.72)} = \frac{0.04246}{0.1432} = 0.2965 = 29.65\%$$

(d) The effort required at the mean radius to lower the load is given by

$$P = W \tan(\alpha - \phi) \quad \text{if } \alpha > \phi$$

$$= W \tan(\phi - \alpha) \quad \text{if } \alpha < \phi$$

For the given case : α = 2.43° and ϕ = 5.72°

$\therefore$ P = 500 tan (5.72 – 2.43) = 28.74 N

and Required ratio $= \dfrac{71.60}{28.74} =$ **2.49**

EXAMPLE 10.27

The threads used in a turnbuckle are specified to be single start square threads having a mean diameter of 50 mm and pitch 12 mm. If the coefficient of friction between the screw and the nut is 0.25, make calculations for the turning moment necessary to (a) tighten and (b) slacken the wire when the pull is 1200 N. It may be assumed that no twising occurs in the rope.

Solution: $\tan \alpha = \dfrac{P}{\pi d_m} = \dfrac{12}{\pi \times 50} = 0.0764 \; ; \; \alpha = \tan^{-1}(0.0764) = 4.37^\circ$

$$\tan \phi = \mu = 0.15 \; ; \; \phi = \tan^{-1}(0.15) = 8.55^\circ$$

(a) Tightening of the rope is equivalent to moving the load up the inclined plane. For that, the effort applied at the mean radius of the jack is

$$P = W \tan(\alpha + \phi) = 1200 \tan(4.37 + 8.55) = 275.28 \text{ N}$$

Torque required at the mean radius of jack,

$$T = P \times \frac{d_m}{2} = 275.28 \times \frac{0.05}{2} = \mathbf{6.882 \ Nm}$$

(b) Slackening of the rope is equivalent to moving the load down the inclined plane. For that, the effort applied at the mean radius of the jack is

$$P = W \tan(\phi - \alpha) \qquad (\because \alpha < \phi)$$

$$= 1200 \tan(8.55 - 4.37) = 87.70 \text{ N}$$

Torque required at the mean radius of the jack,

$$T = P \times \frac{d_m}{2} = 87.70 \times \frac{0.05}{2} = \mathbf{2.192 \ Nm.}$$

10.13. DIFFERENTIAL SCREW JACK

Fig. 10.29 shows a differential screw jack which provides an increased velocity ratio compare. to that of a simple screw jack.

The unit consists of two threaded elements A and B of different diameters and pitches, but having threads in the same direction. The element A is essentially a cylinder with threads both on its outer and inner surface. The threads on its outer surface mesh with the thread nut C; this nut also functions as body of the jack. The segment B has threads only on its outer surface and these engage with the internal threads of the element A. Obviously the element A behaves as a screw for the nut and also as a nut for the element B.

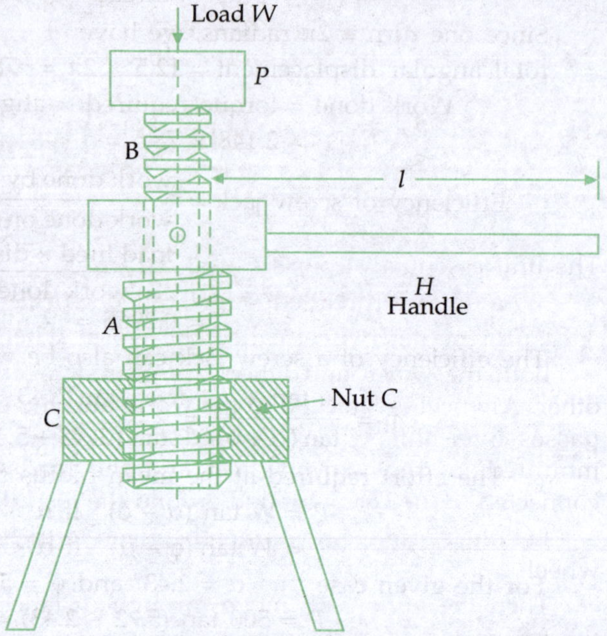

Fig. 10.29. Differential screw jack

The weight W to be raised or lowered is placed on the top of a platform (collar). The platform does not rotate but moves in the vertical direction only. The effort P is applied at the end of a lever or handle which is inserted into a hole provided in the collar.

Let p_a and p_b be the pitches of screws A and B respectively. When effort is applied and the lever is made to turn one complete rotation, the screw A lifts up by pitch length p_a and in the same time the screw B comes done by pitch length p_b. Then

Net distance through which load is lifted $= p_a - p_b$

If l is the length of lever, then

distance moved by the effort $= 2\pi l$

$$\therefore \quad \text{Velocity ratio } (VR) = \frac{\text{distance moved by the effort}}{\text{distance moved by the load}} = \frac{2\pi l}{p_a - p_b}$$

$$\text{Mechanical advantage } (MA) = \frac{W}{P}, \quad \text{and}$$

$$\text{Efficiency, } \eta = \frac{MA}{VR}$$

With the same effort a differential screw jack is capable of lifting larger loads than a simple screw jack.

EXAMPLE 10.28

A differential screw jack has the following specifications :

pitch of larger screw = 12 mm

pitch of smaller screw = 8 mm

length of lever arm from centre of screw = 400 mm.

Find its velocity ratio. If an effort of 75 N applied at the end of the lever lifts a load of 10 KN, determine the efficiency of the jack.

Solution : For a differential screw jack,

$$VR = \frac{2\pi l}{p_a - p_b} = \frac{2\pi \times 400}{12-8} = 628$$

$$MA = \frac{W}{P} = \frac{10 \times 10^3}{75} = 133.3$$

$$\eta = \frac{MA}{VR} = \frac{133.3}{628} = 0.212 \text{ or } 21.2\%$$

10.14. WORM AND WORM WHEEL

The unit essentially consists of :

(*i*) a square threaded screw called worm, and

(*ii*) a toothed wheel called the worm wheel.

Both the worm and wheel are geared together and their axes are at right angles to each other. A wheel is attached to the worm and the effort P is applied to one end of the rope which passes over this wheel. A load drum is securely mounted on the worm wheel and the load W is connected to the string wound around the load drum.

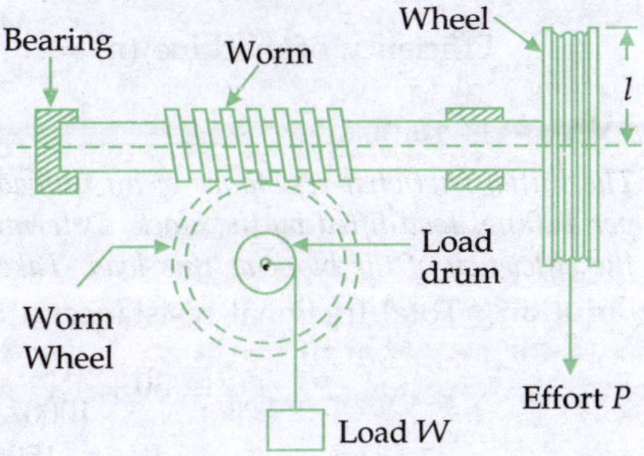

Fig. 10.30. Worm and worm wheel

Let one complete revolution be given to the effort wheel.

Then the effort will move through distance $2\pi l$ where l is radius of the wheel.

If the worm is *single threaded*, then for one revolution of the effort wheel, the worm would push the worm wheel through one teeth. If the number of teeth on the worm wheel is T, then a push of one teeth implies that the load drum will move through $\frac{1}{T}$ revolution. Therefore,

$$\text{Distance moved by load} = 2\pi r \times \frac{1}{T} = \frac{2\pi r}{T} \text{ where } r \text{ is the radius of load drum.}$$

$$\text{Velocity ratio } (VR) = \frac{\text{distance moved by effort}}{\text{distance moved by load}} = \frac{2\pi l}{2\pi r/T} = \frac{lT}{r}$$

In case the loss of energy due to friction is neglected, then

$$MA = VR = \frac{lT}{r}$$

If the loss of energy due to friction is taken into consideration, then

$$MA = \frac{W}{P} \quad \text{and} \quad \eta = \frac{MA}{VR}$$

When worm is *double threaded*, then for one revolution of effort wheel,

movement of wheel = 2 teeth

$$\text{movement of load drum} = \frac{2}{T} \text{ revolution}$$

$$\text{distance moved by load} = 2\pi r \times \frac{2}{T} = 2 \times \frac{2\pi r}{T}$$

$$\text{velocity ratio } (VR) = \frac{2\pi l}{2 \times 2\pi r/T} = \frac{lT}{2r}$$

In general, when the worm is n-threaded

$$\text{Velocity ratio } (VR) = \frac{lT}{nr}$$

EXAMPLE 10.29

The number of teeth on the worm wheel of a single threaded worm and worm wheel arrangement is 60 and this machine needs an effort of 20 N while lifting 500 N load. If the diameter of effort wheel and load drum are 250 mm and 125 mm respectively, make calculations for the velocity ratio, mechanical advantage and efficiency of the machine.

Solution :

$$\text{Velocity ratio } (VR) = \frac{lT}{nr} = \frac{(250/2 \times 60)}{1 \times (125/2)} = 120$$

$$\text{Mechanical advantage } (MA) = \frac{W}{P} = \frac{500}{20} = 25$$

$$\text{Efficiency of machine } (\eta) = \frac{MA}{VR} = \frac{25}{120} = 0.208 \text{ or } 20.8\%$$

EXAMPLE 10.30

The initial frictional resistance of an unloaded pulley block is 30 N and it increases at the rate of 12 N per 1000 N load lifted by the block. Determine the effort required to lift a load of 15000 N and calculate the efficiency of the block at this load. Take velocity ratio as 20.

Solution : Total frictional resistance at a load of 15000 N

$$P_{friction} = 30 + \frac{12}{1000} \times 15000 = 210 \text{ N}$$

$$P_{ideal} = \frac{W}{VR} = \frac{15000}{20} = 750 \text{ N}$$

The effort wasted in friction while lifting a given load is

$$P_{friction} = P_{actual} - P_{ideal}$$

$$210 = P_{actual} - 750$$

$$\therefore \text{Actual effort } P_{actual} = 960 \text{ N}$$

$$\eta = \frac{MA}{VR} = \frac{W/P}{VR} = \frac{W}{P \times VR} = \frac{15000}{960 \times 20}$$

$$= 0.781 \text{ or } 78.1\%$$

REVIEW AND SUMMARY

1. Machine is a device or contrivance that receives energy in some available form, and does some useful work

   A machine may enable to
   - multify force,
   - apply force at a convenient point and convenient direction,
   - obtain a gain in speed, and
   - life heavy loads

2. The basic simple machines are :
   (*i*) lever (*ii*) pulley (*iii*) inclined plane
   (*iv*) screw (*v*) wedge (*vi*) wheel and axle
3. Machines which are used to lift heavy loads are called lifting machines, *e.g.*, pulley and screw jacks
4. Some of the terms associated with machines are :
   (*i*) weight to be lifted is called load (W) and the force applied to lift or displace the load is called effort (P)
   (*ii*) mechanical advantage (MA) represents the ratio of load lifted to effort applied :
   $$MA = W/P$$
   (*iii*) the ratio of distance (y) moved by the effort to the distance moved by the load (x) is the velocity ratio : $VR = y/x$
   (*iv*) the ratio between the useful work done by the machine and useful work expended on machine
   $$\eta = \frac{\text{output}}{\text{input}} = \frac{W\,x}{P\,y} = \frac{W}{P} \times \frac{1}{y/x} = \frac{MA}{VR}$$
   Neglecting friction losses, the work input to the machine equals the work done in displacing the load. In that case $\eta = 1$.
   Such a machine where $\eta = 100\%$, and mechanical advantage equals velocity ratio is called an ideal machine.
5. In a reversible machine, work is done by the machine in reverse direction and the load falls, *e.g.*, pulley used to draw water from the well.
   When the work is not done by the machine in reverse direction, the load does not then fall, *e.g.*, screw jack used to lift the motor car.
6. The law of machine prescribes the relation between the effort applied and load lifted. For a machine with constant velocity ratio : $P = mW + C$
   (*i*) for an ideal machine, the straight line passes through the origin and intercept $C = 0$
   (*ii*) for an actual machine, the straight line has intercept on the P-axis. This intercept represents the effort required to overcome friction. If the effort applied is less than C, then the load will not be lifted.

REVIEW QUESTIONS

A. Conceptual and conventional questions :

1. What is a machine ? How does an ideal machine differ from an actual machine ?
2. How does a machine help to achieve gain in force ?
3. State the reasons for the output work to be always less than the input work in an actual machine.
4. Define velocity ratio, mechanical advantage and efficiency of a machine.
5. State the condition for the reversibility and self locking of a machine.
6. Which of the parameters, *i.e.*, velocity ratio and mechanical advantage is affected by friction between the contact surfaces?
7. What is an irreversible machine? State the efficiency condition for a machine to be irreversible.
8. Set up expressions for the frictional loss in terms of load and effort.
9. How efficiency of a machine is related to its mechanical advantage and velocity ratio?

10. Differentiate between a block and a tackle.
11. Derive expressions for the velocity ratio for the following lifting machines :
    (*i*) Wheel and axle
    (*ii*) First system of pulleys
    (*iii*) Single purchase winch crab
12. Describe the first, second and third system of pulleys.
13. What are winch crabs?
    Sketch a single purchase winch crab and set up expression for its velocity ratio and mechanical advantage. How the mechanical advantages would be affected when friction is taken into account?
14. In case of a lifting machine, efforts of 16 N and 20 N were required to lift loads of 75 N and 90 N respectively. If the relation between load W and effort P conforms to the straight line relationship $P = mW + C$, find the values of constants m and C.
    If the velocity ratio of the machine is 15, determine its efficiency at a load of 120 N. **(0.2, 2, 27.76%)**
15. A single fixed pulley is used to lift a load of 400 N by the application of an effort of 480 N through a vertical height of 5 m in 10 seconds. Determine the velocity ratio, mechanical advantage and efficiency of the system. Why is the efficiency not 100 percent?
    (*b*) What is the energy gained by the load in 10 seconds and how much is the power developed by the effort?
    (*c*) The effort applied is greater than the load. Then what is the justification for using such a machine? **(1, 0.833, 83.3%, 2000 *J*, 240 *W*)**
16. In a differential wheel and axle, the diameter of the wheel is 750 mm, and that of the axles are 250 mm and 300 mm. Determine the effort that needs to be applied to raise a load of 25 kN. Assume the efficiency of machine to be 60%. **(1388.89 N)**
17. A differential wheel and axle is used to lift a box of mass 4000 kg. The larger and smaller radii of the beam are 40 cm and 30 cm respectively. The length of the crank arm is 50 cm. If the efficiency of the machine is 80 per cent, determine the motive force required to lift the load, the velocity ratio and the mechanical advantage gained. **(2452.5 N, 20, 16)**
18. Sketch the arrangement of a differential pulley block and explain its working.
    For this set up, there are 10 recesses on the larger pulley and 9 recesses on the smaller pulley. Determine the velocity ratio and efficiency at a load of 15 kN if the effort required to raise this load is 1.25 kN. **(20, 60%)**
19. A machine of mass 1500 kg is to be lifted by using a differential pulley block. The radii of the bigger and smaller pulleys of the fixed block are 400 mm and 200 mm respectively. If the efficiency of the pulley block is 50 per cent, determine the effort required to raise the load. **(3678.75 N)**
20. Describe, with neat sketch the construction and working of a single purchase winch crab. The following are the specifications for a single purchase winch crab:

    length of lever arm = 500 mm
    diameter of drum = 200 mm
    diameter of rope = 15 mm
    number of teeth on pinion = 30
    number of teeth on wheel = 150

    If the efficiency of the winch crab is 40 per cent, what load will be lifted by an effort of 250 N applied at the end of the lever ? **(2325 N)**

21. For a double purchase crab, the following details are available :

number of teeth on pinions = 20 and 25
number of teeth on spur wheels = 50 and 45
length of effort handle = 500 mm
effective diameter of drum = 160 mm

What load will be lifted by a effort of 300 N applied at the end of the handle? Assume that the winch has an efficiency of 50%. **(4218.75 N)**

B. Complete the following statements with most appropriate word/words :

1. In a simple machine, a small force through a large distance overcomes a .... force through a .... distance
2. A pulley used to draw water from a well with the help of a bucket is a .... machine, whereas the screw jack used to lift the motor car is said to be .... machine
3. The law of machine prescribes the relationship between the .... and the ....
4. In a lever of second kind, the mechanical advantage is always .... than one and it increases by moving the load towards the ....
5. A single fixed pulley is used to move load in the .... direction by applying errort in the .... direction
6. If there are n number of pulleys in the first system of pulleys, then under ideal conditions (no friction)

   velocity ratio = mechanical advantage = ...
7. A .... refers to the blocks of pulleys (fixed and movable) which are grouped together by winding a rope or chain around them
8. The working principle of screw jack is same as that of ....
9. A screw jack is said to be .... if its efficiency is less than 50 percent.
10. The winch crabs employ a .... for increasing the velocity ratio
11. The upper block in a differential pulley block has .... pulleys which are fixed .... to a common spindle
12. The effort in a differential wheel and axle arrangement is applied to the string which is wound round the ....

Answers :

1. large, small 2. reversible, irreversible 3. effort applied, load lifted 4. greater, fulcrum
5. upward, downward 6. 2^n 7. tackle 8. inclined plane 9. self-locking 10. gear system
11. two, coaxially 12. wheel

C. Multiple choice questions :

1. The velocity ratio in a simple machine is the ratio of
   (*a*) value of load lifted to that of effort applied
   (*b*) magnitude of load lifted to that of effort applied
   (*c*) distance moved by effort to the distance moved by the load
   (*d*) efficiency of machine to its mechanical advantage
2. In a lifting machine
   (*a*) effort applied is always greater than the load lifted
   (*b*) mechanical advantage represents the ratio of effort applied to load lifted

(c) velocity ratio is always less than unity
(d) large loads can be lifted by applying a smaller effort when velocity ratio is greater than unity

3. For an ideal machine
(a) the velocity ratio is practically zero
(b) the mechanical advantage tends to infinity
(c) the mechanical advantage equals the velocity ratio
(d) the load falls to its original position as soon as the effort is removed

4. The law of machines is an equation that prescribes the relation between
(a) efficiency and velocity ratio
(b) effort applied and the load lifted
(c) mechanical advantage and velocity ratio
(d) effort and velocity ratio

5. Which of the following conditions is necessary for an irreversible or self-locking machine?
(a) the friction coefficient exceeds 0.5
(b) the efficiency tends to infinity
(c) the efficiency is less than 50%
(d) the load falls to its original position on the removed of effort applied

6. Which of the following is not a lever of first kind?
(a) sea saw (b) handle of a water pump
(c) plier (d) bar of a boat

7. If there are n number of moveable pulleys in the first system of pulleys, then the velocity ratio is
(a) 2^n (b) $2^n - 1$ (c) $2^n + 1$ (d) n^2

8. In a single purchase winch crab, the number of teeth in the pinion and spur wheel are T_1 and T_2 respectively. The velocity ratio of the system is proportional to
(a) $\frac{T_1}{T_2}$ (b) $\frac{T_2}{T_1}$ (c) $\sqrt{T_1 T_2}$ (d) $\frac{T_2 - T_1}{T_2 + T_1}$

9. The pitch difference of the two screws in a differential screw jack is designed to be made smaller. With this the velocity ratio of the jack
(a) increases (b) decreases
(c) remains unaffected
(d) may increase or decrease depending upon effort applied and load lifted

10. In a worm and worm wheel arrangement, the velocity ratio is
(a) directly proportional to radius of load axle (drum)
(b) inversely proportional to radius of load axle
(c) inversely proportional to number of teeth on the worm wheel
(d) inversely proportional to radius of effort wheel

Answers :

1. (c)	2. (d)	3. (c)	4. (b)	5. (c)	6. (d)
7. (a)	8. (b)	9. (a)	10. (b)		

11 Shear Force and Bending Moment

Learning objectives : The main objective of the subject matter included in this chapter is to familiarize the readers with the

- *types of beams and loads*
- *concept of shear force and bending moment*
- *relation between load intensity, shear force and bending moment*
- *shear force and bending moment diagrams for standard cases*
- *point of contraflexure in a loaded beam*

A beam is a structural member whose longitudinal dimension is large compared to the transverse dimension. The beam is supported along its length and is acted upon by a system of loads at right angles to its axis. Due to external loads and couples, shear force and bending moment develop at any section of the beams. For the design of beams, information about the shear force and bending moment is desired. Accordingly, it is appropriate to learn about the variation of shear force and bending moment along the length of the beam.

11.1. SHEAR FORCE AND BENDING MOMENT

Consider a simply supported beam acted upon by different point loads as shown in Fig. 11.1.

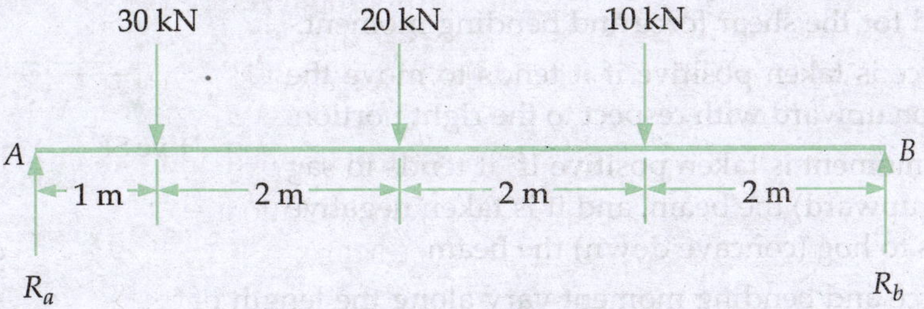

Fig. 11.1

The loads are transferred to supports and for equilibrium conditions

$$R_a + R_b = W_1 + W_2 + W_3 = 30 + 20 + 20 = 60 \text{ kN}$$

Also ΣM_a (moments about support A) = 0. That gives

$$R_b \times 7 - 10 \times 5 + 20 \times 3 + 30 \times 1 = 140$$

$$\therefore \quad R_b = \frac{140}{7} = 20 \text{ kN and } R_a = 60 - 20 = 40 \text{ kN}$$

Imagine the beam to be cut at an arbitrary section xx at distance $x = 4$ m from the end A, and draw separately the free body diagrams of the two portions (Fig. 11.2).

- Considering equilibrium of forces on each portion of the beam, the net resultant vertical forces are:

$$F_{left} = 30 + 20 - 40 = 10 \text{ kN} \quad \text{(downward)}$$

$$F_{right} = 10 - 20 = -10 \text{ kN} \quad \text{(upward)}$$

It is to be noted that forces on the left and right sides of the section xx are equal in magnitude but opposite in direction.

Obviously, section xx is subjected to a force of 10 kN which is trying to shear the beam. The force of 10 kN is called as **shear force** at section xx.

"Shear force at a section in a beam is the force that is trying to shear off the section. It is obtained as algebraic sum of all the forces acting normal to the axis of beam ; either to the left or to the right of the section."

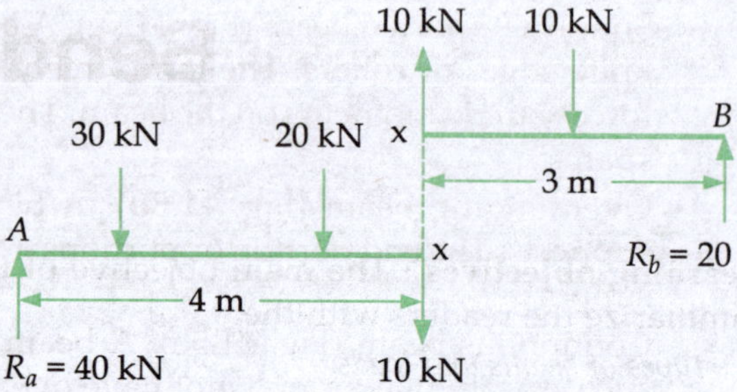

Fig. 11.2

- Considering equilibrium of moments on each portion of the beam,

$$M_{left} = 40 \times 4 - 30 \times 3 - 20 \times 1 = 50 \text{ kNm} \quad \text{(clockwise)}$$

$$M_{right} = 20 \times 3 - 10 \times 1 = 50 \text{ kNm} \quad \text{(anti-clockwise)}$$

It is to be noted that moments on the left and right sides of the section xx are equal in magnitude but opposite in direction. Obviously section xx is acted upon by a moment of 50 kNm. Since this moment is trying to bend the beam, it is called **bending moment**.

"Bending moment at a section in a beam is the moment that tends to bend the beam and is obtained as algebraic sum of moment of all the forces about the section, acting either to the left or to the right of the section."

Sign Conversion: The following sign conventions are normally adopted for the shear force and bending moment.

(*i*) Shear force is taken positive if it tends to move the left portion upward with respect to the right portion.

(*ii*) Bending moment is taken positive if it tends to sag (concave upward) the beam, and it is taken negative if it tends to hog (concave down) the beam.

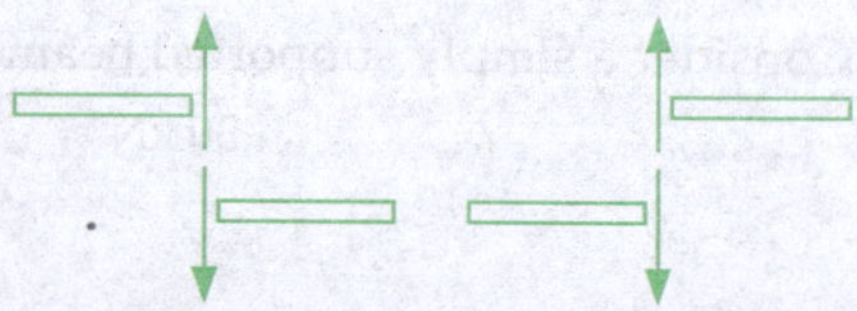

Fig. 11.3

The shear force and bending moment vary along the length of the beam and this variation is represented graphically. The plots are known as shear force and bending moment diagrams. In these diagrams, the abscissa indicates the position of section along the beam, and the ordinate represents the value of SF and BM respectively. These plots help to determine the maximum value of each of these quantities.

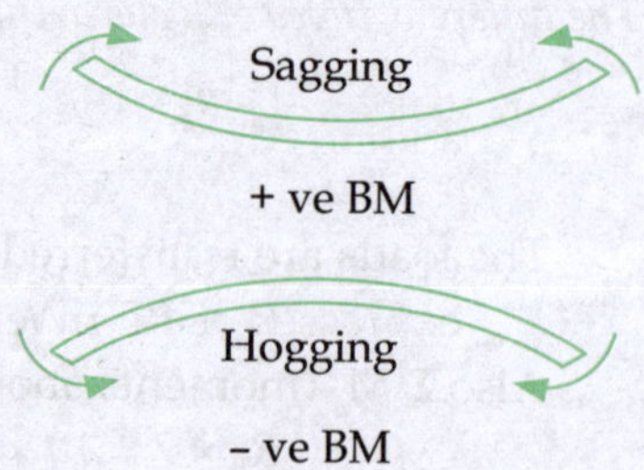

Fig. 11.4

11.2. TYPES OF BEAMS AND LOADS

The shear force and the bending moment that develop at any cross-section of the beam depend upon the types of beams and the types of loads acting on them.

Beam are generally classified as:

- Cantilever beam (Fig. 11.5a): A beam having its one end fixed or built-in and the other end free to deflect. There is no deflection or rotation at the fixed end.
- Fixed beam (Fig. 11.5b): A beam having both of its ends fixed or built-in.
- Simply supported beam (Fig. 11.5c): A beam made to freely rest on supports which may be knife edges or rollers. The term 'freely supported' implies that these supports exert only forces but no moments on the beam. The horizontal distance between the supports is called span.
- Overhanging beam (Fig. 11.5d): A beam having one or both ends extended over the supports. The end portion or portions extend in the form of cantilever beyond the support/supports.
- Continuous beam (Fig. 11.5e): A beam provided with more than two supports. Further such a beam may or may not have overhang.

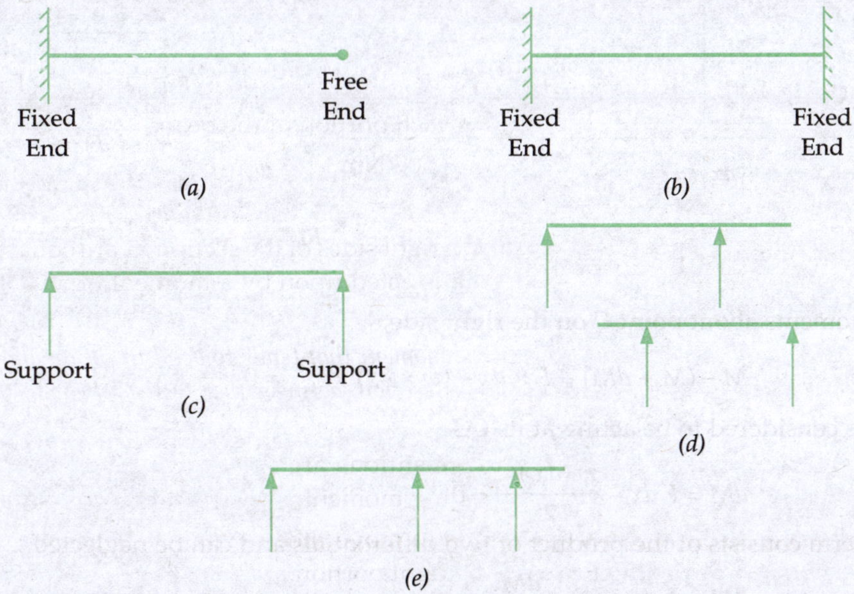

Fig. 11.5

The different types of loads acting on a beam are:

- **Concentrated load:** The load acts at a point on the beam. This point load is applied through a knife edge.
- **Uniformly distributed load:** The load is evenly distributed over a part or the entire length of the beam. The total udl is assumed to act at the centre of gravity of the load. The udl is expressed as N/m length of beam.
- **Uniformly varying load:** The load whose intensity varies linearly along the length of beam over which it is applied.
- A beam may be loaded by a couple whose magnitude is expressed as Nm.

A beam may carry any one of the above load systems or combination of two or more loads at a time.

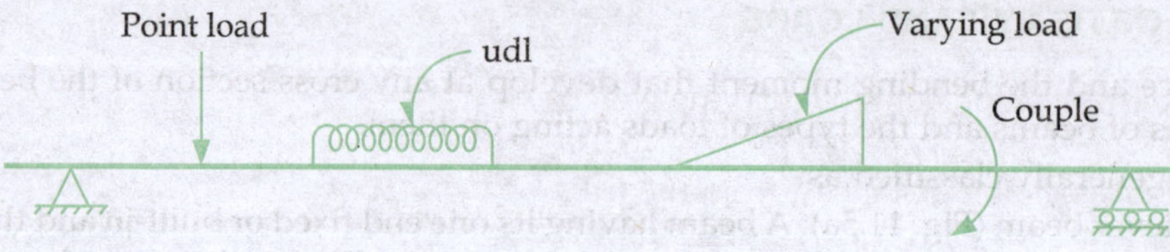

Fig. 11.6

11.3. RELATION BETWEEN LOAD INTENSITY, SF AND BM

Consider a beam subjected to any type of transverse load of the general form shown in Fig. 11.7. Isolate from the beam an element of length dx at a distance x from left end and draw its free body diagram as shown in Fig. 11.7. Since the element is of extremely small length, the loading over the beam can be considered to be uniform and equal to w kN/m. The element is subject to shear force F on its left hand side and shear force $(F + dF)$ on its right hand side. Further, the bending moment M acts on the left side of the element and it changes to $(M + dM)$ on the right side.

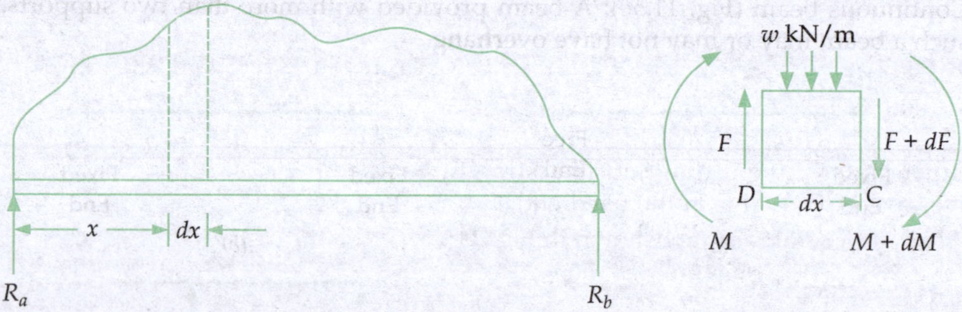

Fig. 11.7

Taking moments about point C on the right side,

$$\Sigma M_c = 0: \quad M - (M + dM) + F \times dx - (w \times dx) \times \frac{dx}{2} = 0$$

The udl is considered to be acting at its CG

$$dM = F\,dx - \frac{w\,(dx)^2}{2} = 0$$

The last term consists of the product of two differentials and can be neglected.

$$\therefore \quad dM = F\,dx \text{ or } F = \frac{dM}{dx}$$

Thus the shear force is equal to the rate of change of bending moment with respect to x.

Applying the condition $\Sigma F_y = 0$ for equilibrium, we obtain

$$F - w\,dx - (F + dF) = 0$$

$$\text{or} \quad w = \frac{dF}{dx}$$

That is the intensity of loading is equal to rate of change of shear force with respect to x.

11.4. SHEAR FORCE AND BENDING MOMENT DIAGRAMS FOR A FEW STANDARD CASES

The method of drawing shear force and bending moment diagrams for a few standard cases is explained below.

11.4.1. Cantilever subjected to concentrated load at free end

Consider a cantilever of length l carrying weight W at its free end.

At a section x-x at any distance x from the free end.

Shear force $\Sigma F = -W$

Apparently the shear force

(*i*) is negative as the left segment of the beam moves downward.

(*ii*) does not vary with x and has a constant value (equal to load W) through-out the beam length

Bending moment $BM = -Wx$

Apparently the bending moment

(*i*) is negative as it tends to hog the beam (concave downward)

(*ii*) has a linear variation with distance x from the free end.

At $x = 0$ $BM = 0$

At $x = l$ $BM = -Wl$

Thus bending moment diagram is a straight line which gives maximum value at the fixed end.

The SF and BM diagrams are shown in Fig. 11.8.

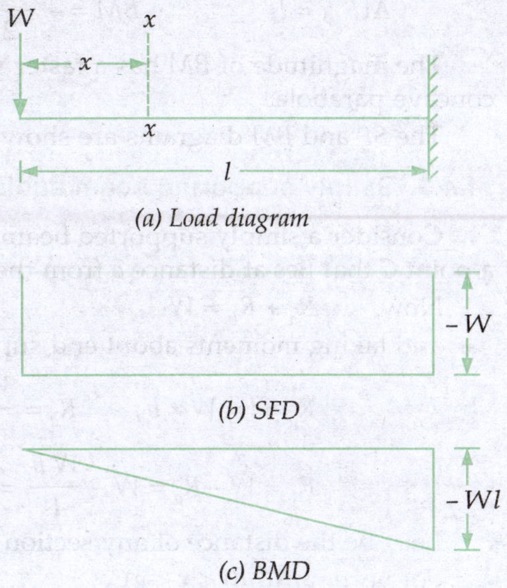

Fig. 11.8

11.4.2. Cantilever subjected to *udl* over its entire span

Consider a cantilever of length l carrying uniformly distributed load of intensity w/unit length over the entire span.

At a section x-x at any distance x from the free end

Shear force $SF = -wx$

Apparently the shear force

(*i*) is negative as the left segment of the beam moves downward

(*ii*) has a linear variation with distance x from the free end

At $x = 0$ $SF = 0$

At $x = l$ $SF = -wl$

Thus the SF diagram is a straight line with maximum value at the fixed end.

Bending moment $BM = -wx \times \frac{x}{2} = -w\frac{x^2}{2}$

udl is considered to be acting at its centre.

Apparently the bending moment

(*i*) is negative as its tends to hog the beam (concave downward)

(*ii*) has a parabolic variation with distance x from the free end.

At $x = 0$ $BM = 0$

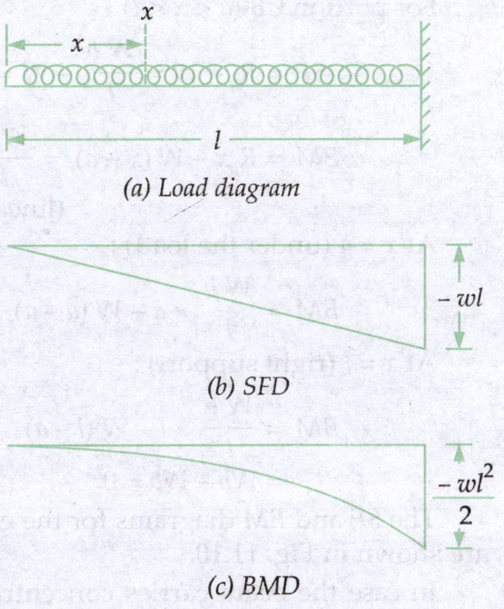

Fig. 11.9

At $x = l$ $\quad BM = -\dfrac{wl^2}{2}$

The magnitude of BM has a faster variation with x and accordingly its variation conforms to concave parabola.

The SF and BM diagrams are shown in Fig. 11.9.

11.4.3. Simply supported beam subjected to concentrated load

Consider a simply supported beam AB with span l and subjected to a concentrated load W at a point C that lies at distance a from the left and support A.

Now, $\quad R_a + R_b = W$

and taking moments about end support B,

$$R_a \times l = W \times b; \qquad R_a = \frac{W\,b}{l}$$

$$R_b = W - R_a = W - \frac{W\,b}{l} = \frac{W\,(l-b)}{l} = \frac{W\,a}{l}$$

Let x be the distance of any section considered from the end support A. Then

For portion AC $(0 < x < a)$

$$SF = R_a = \frac{W\,b}{l} \qquad \text{(constant)}$$

$$BM = R_a x = \frac{W\,b}{l}\,x \qquad \text{(linear variation)}$$

At $x = 0$ (left support) : $BM = 0$

At $x = a$ (under the load) : $BM = \dfrac{Wab}{l}$

For portion CB $(a < x < l)$

$$SF = R_a - W = \frac{W\,b}{l} - W = W\left(\frac{b}{l} - 1\right) = \frac{-W\,a}{l} \qquad \text{(constant)}$$

$$BM = R_a x - W\,(x-a) = \frac{W\,b}{l}x - W\,(x-a)$$

(linear variation)

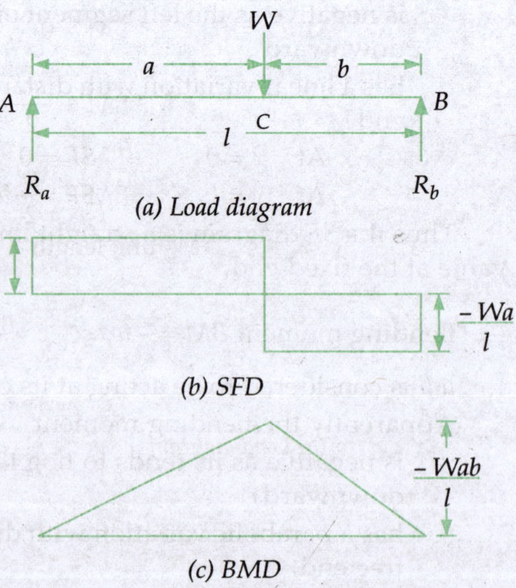

(a) Load diagram

(b) SFD

(c) BMD

Fig. 11.10

At $x = a$ (under the load) :

$$BM = \frac{W\,b}{l} \times a - W\,(a-a) = \frac{Wab}{l}$$

At $x = l$ (right support) :

$$BM = \frac{W\,b}{l} \times l - W\,(l-a)$$

$$= Wb - Wb = 0$$

The SF and BM diagrams for the entire beam are shown in Fig. 11.10.

In case the beam carries concentrated load at mid span, then $a = b = \dfrac{l}{2}$ and accordingly SF and BM under the load are:

$$SF = \frac{Wb}{l} = \frac{W \times \frac{l}{2}}{l} = \frac{W}{2}$$

$$BM = \frac{Wab}{l} = \frac{W \times \frac{l}{2} \times \frac{l}{2}}{l} = \frac{Wl}{4}$$

11.4.4. Simply supported beam subjected to udl over its entire span

Consider a simply supported beam of span l carrying uniformly distributed load of intensity w/unit length over the entire span.

Because of symmetrical loading each vertical reaction equals half the total load on the span. That is

$$R_a = R_b = \frac{wl}{2}$$

At a section at distance x from the left and support A,

$$\text{Shear force } SF = R_a - wx = \frac{wl}{2} - wx \quad \text{(linear variation)}$$

$$\text{At } x = 0; \quad SF = \frac{wl}{2}$$

$$\text{At } x = l; \quad SF = \frac{wl}{2} - wl = \frac{wl}{2}$$

Apparently the shear force changes uniformly from $\frac{wl}{2}$ at end A to $\frac{-wl}{2}$ at end B. Further, the shear force will be zero at $x = \frac{l}{2}$ (mid span).

$$\text{Bending moment } BM = R_a x - wx \times \frac{x}{2} = \frac{wl}{2}x - \frac{wx^2}{2} \quad \text{(parabolic variation)}$$

The *udl* is considered to be acting at its *CG*.

$$\text{At } x = 0; \quad BM = 0$$

$$\text{At } x = \frac{l}{2}; \quad BM = \frac{wl}{2} \times \frac{l}{2} - \frac{wl^2}{8} = \frac{wl^2}{8}$$

$$\text{At } x = l; \quad BM = \frac{wl}{2} \times l - \frac{wl^2}{2} = 0$$

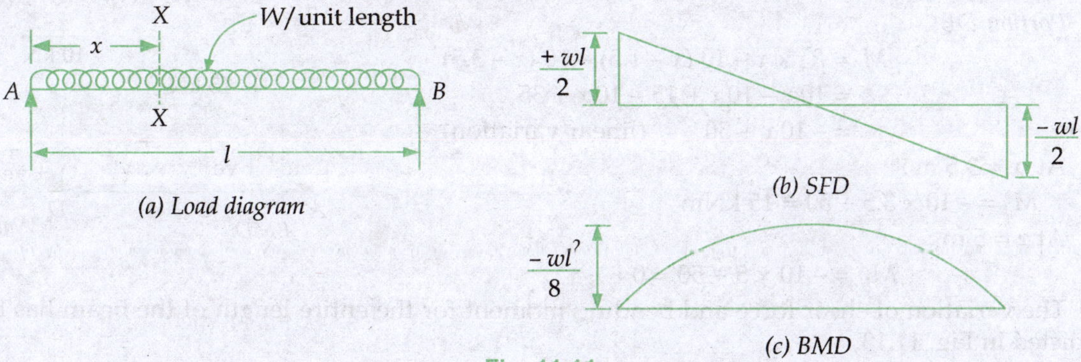

Fig. 11.11

A reduction in the value of *BM* becomes faster as *x* increases, and accordingly the bending moment variation conforms to convex parabola. Further, it is to be noted that maximum bending moments occurs at the mid span where shear force is zero.

EXAMPLE 11.1

Draw the shear force and bending moment diagrams for the beam loaded and supported as shown in Fig. 11.12.

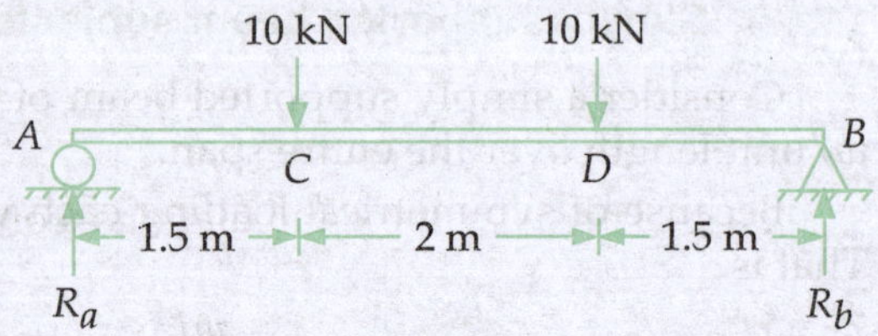

Fig. 11.12

Solution: The line of action of the reaction will be at right angles to the roller base at end *A*. The reaction at a hinge can have two components acting in the horizontal and the vertical directions. Since there is no horizontal external force acting on the beam, the reaction at the hinged end *B* will be only in the vertical direction.

Due to symmetry of loading,

$$R_a = R_b = \frac{10 + 10}{2} = 10 \text{ kN}$$

Shear force

SF at *A* = 10 kN
SF just on left of *C* = 10 kN
SF just on right of *C* = 10 – 10 = 0
SF just on left of *D* = 0
SF just on right of *D* = 0 – 10 = – 10 kN
SF just on left of *B* = – 10 kN
SF just on right of *B* = – 10 + 10 = 0

Bending moment. Taking a section at distance *x* from end *A* and considering forces on left side.

Portion AC:

$$M = R_a \times x = 10x \qquad \text{(linear variation)}$$

At $x = 0$: $M_a = 0$

At $x = 1.5$ m : $M_c = 10 \times 1.5 = 15$ kNm

Portion CD:

$$M = R_a \times x - 10(x - 1.5) = 10x - 10x + 15 = 15 \text{ kNm}$$

The bending moment remains constant at 15 kNm within the portion *CD*.

Portion DB:

$$M = R_a \times x - 10(x - 1.5) - 10(x - 3.5) = 10x - 10x + 15 - 10x + 35 = -10x + 50 \qquad \text{(linear variation)}$$

At $x = 3.5$ m :

$$M_a = -10 \times 3.5 + 50 = 15 \text{ kNm}$$

At $x = 5$ m :

$$M_b = -10 \times 5 + 50 = 0$$

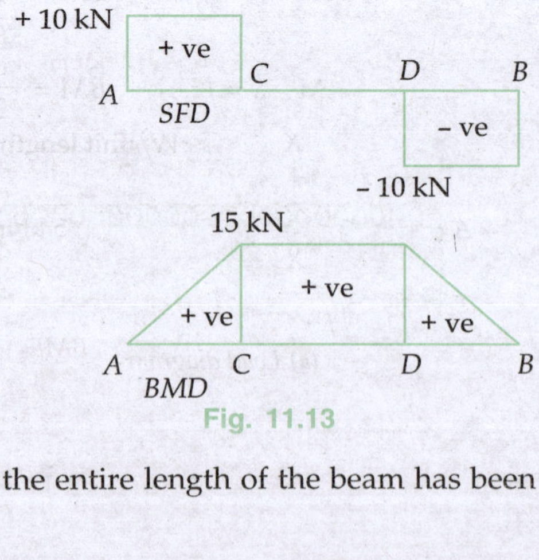

Fig. 11.13

The variation of shear force and bending moment for the entire length of the beam has been depicted in Fig. 11.13.

EXAMPLE 11.2

Construct the shear force and bending moment diagram for the cantilever beam loaded as shown in Fig. 11.14.

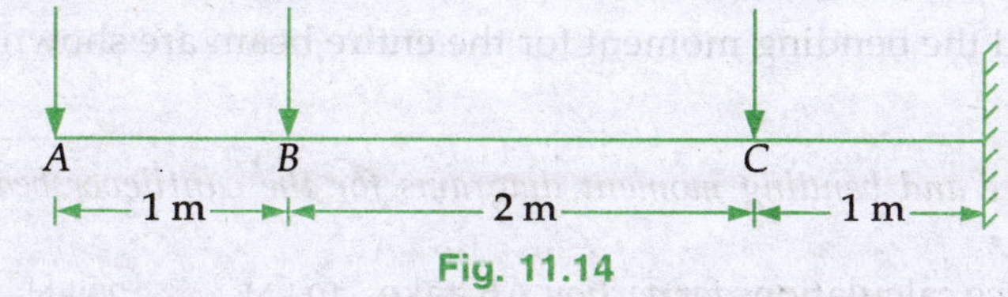

Fig. 11.14

Solution: For shear force calculations, consider any section at distance x from the free end A

At $x = 0$: $SF = -5$ kN

The shear force is being taken – ve because it tends to move the left portion downward with respect to the right portion.

At $x = 1$ m

just left of B : $SF = -5$ kN just right of B : $SF = -5 - 4 = -9$ kN

At $x = 3$ m

just left of C : $SF = -9$ kN just right of C : $SF = -9 - 3 = -12$ kN

Bending moment

Portion AB: Imagine a section between A and B, and at distance x from end A. Then

$M_x = -5x$ (linear variation)

At $x = 0$: $M_a = 0$

At $x = 1$ m : $M_b = -5 \times 1 = -5$ kNm

Portion BC: Consider the section to be between B and C, and at distance x from end A. Then

$M_x = -5x - 4(x - 1)$ (linear variation)

At $x = 1$ m : $M_b = -5 \times 1 - (1 - 1) = -5$ kNm as calculated above

At $x = 3$ m : $M_c = -5 \times 3 - 4(3 - 1) = -23$ kNm

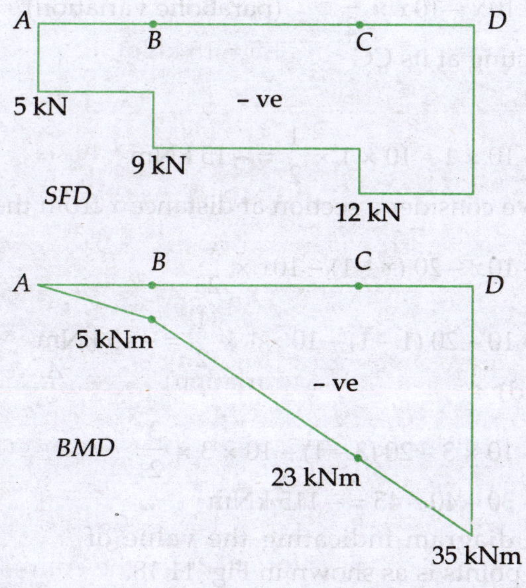

Fig. 11.15

Portion CD: Consider the section to be between C and D, and at distance x from end A.

$$M_x = -5x - 4(x-1) - 3(x-3) \quad \text{(linear variation)}$$

At $x = 3$ m : $M_c = -5 \times 3 - 4(3-1) - 3(3-3) = -23$ kNm

At $x = 4$ m : $M_d = -5 \times 4 - 4(4-1) - 3(4-3) = -35$ kNm

The shear force and the bending moment for the entire beam are shown in Fig. 11.15.

EXAMPLE 11.3

Construct the shear force and bending moment diagrams for the cantilever beam loaded as shown in Fig. 11.16.

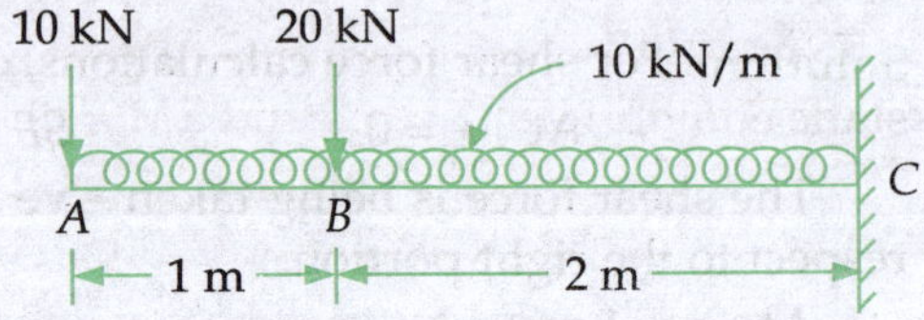

Fig. 11.16

Solution: For shear force calculations *for portion AB*, take section at distance x from end A.

$SF = -10 - 10x$ (linear variation)

At $x = 0$; $SF = -10$ kN

At $x = 1$ m (just to left of point B);

$SF = -10 - 10 = -20$ kN

For *portion BC*, again we consider a section at distance x from the end A,

$$SF = -10 - 20 - 10x \quad \text{(linear variation)}$$

At $x = 1$ m (just to left of point B);

$SF = -10 - 20 - 10 = -40$ kN

At $x = 3$ m (fixed end) ;

$SF = -10 - 20 - 10 \times 3 = -60$ kN

The shear force diagram indicating the values of shear force at salient points is as shown in Fig. 11.17.

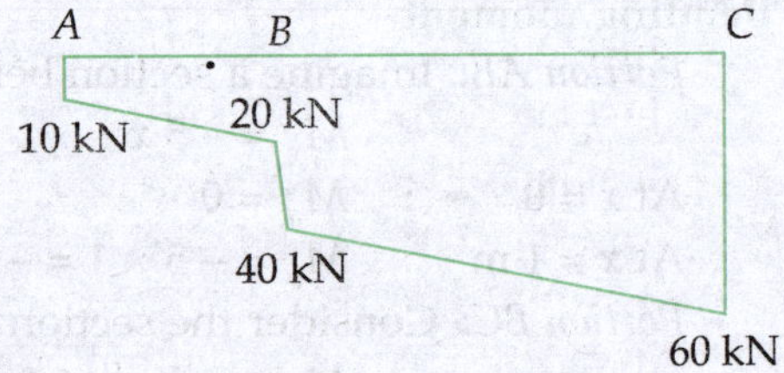

Fig. 11.17

(b) For bending moment for *portion AB*, take section at distance x from the free end A.

$$BM = -10x - 10x \times \frac{x}{2} \quad \text{(parabolic variation)}$$

The *udl* is taken to be acting at its CG

At $x = 0$; $BM = 0$

At $x = 1$ m ; $BM = -10 \times 1 - 10 \times 1 \times \frac{1}{2} = -15$ kNm

For portion BC, again we consider a section at distance x from the end A

$$BM = -10x - 20(x-1) - 10x \times \frac{x}{2}$$

At $x = 1$ m ; $BM = -10 - 20(1-1) - 10 \times 1 \times \frac{1}{2} = -15$ kNm

At $x = 3$ m (fixed end) :

$$BM = -10 \times 3 - 20(3-1) - 10 \times 3 \times \frac{3}{2}$$

$$= -30 - 40 - 45 = -115 \text{ kNm}$$

The bending moment diagram indicating the value of bending moment at salient points is as shown in Fig. 11.18.

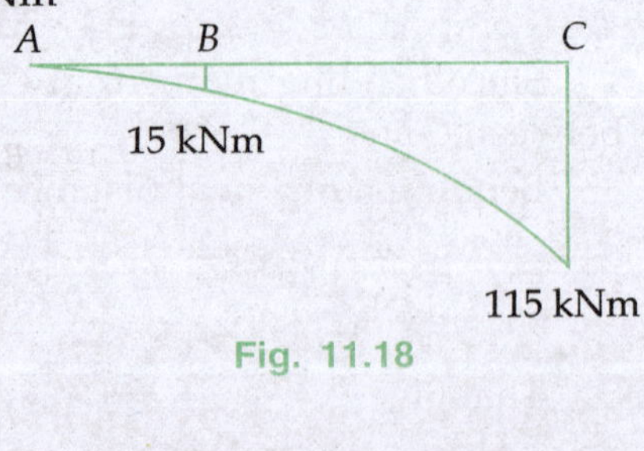

Fig. 11.18

EXAMPLE 11.4

Determine the reactions and construct the shear force and bending moment diagrams for the beam loaded as shown in Fig. 11.19. Also find the point of contraflexure, if any.

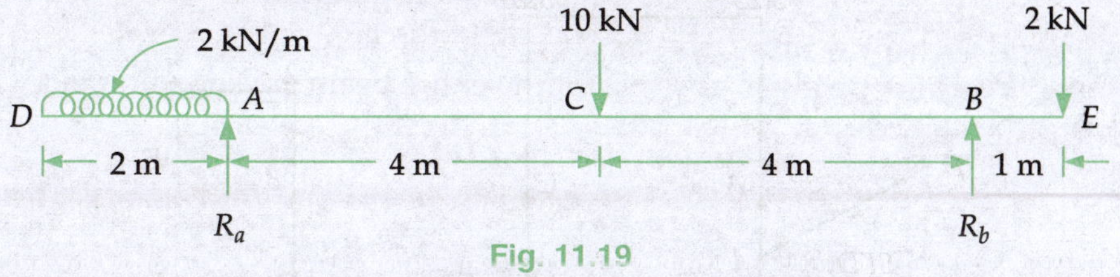

Fig. 11.19

Solution: A point of contraflexure is a point where bending moment is zero. From conditions of static equilibrium ($\Sigma V = 0$ and $\Sigma M = 0$), we have

$$R_a + R_b = 2 \times 2 + 10 + 2 = 16 \quad ...(i)$$

$$-2 \times 2 \times 10 + R_a \times 9 - 10 \times 5 + R_b \times 1 = 0\ ;\ 9R_a + R_b = 90 \quad ...(ii)$$

The *udl* is considered to be concentrated at its *CG*.

From expression (*i*) and (*ii*) : $R_a = 9.25$ kN and $R_b = 6.75$ kN

Shear Force:

At $D = 0$

Just left of $A = -2 \times 2 = -4$ kN ; Just right of $A = -4 + 9.25 = 5.25$ kN

Just left of $C = 5.25$ kN ; Just right of $C = 5.25 - 10 = -4.75$ kN

Just left of $B = -4.75$ kN ; Just right of $B = -4.75 + 6.75 = 2$ kN

Just left of $E = 2$ kN ; Just right of $E = 2 - 2 = 0$ kN

Bending moment

$$M_D = 0$$

At distance x from D (within portion DA)

$$M_x = -2x \times \frac{x}{2} = -x^2$$

$\therefore$ M (at x = 1m) = 1 and M (at x = 2m) = – 4

$$M_A = -4 \text{ kNm}$$

$$M_C = -2 \times 2 \times 5 + 9.25 \times 4 = -20 + 37 = 17 \text{ kNm}$$

Apparently there is a point of contraflexure between A and C as bending moment changes sign between A and C.

Bending moment at x between A and C with x measured from D

$$M_x = -4(x-1) + 9.25(x-2) = 5.25x - 14.5$$

$\therefore$ $5.25x - 14.5 = 0$ for point of contraflexure

That gives $x = \dfrac{14.5}{5.25} = 2.76$ m

$M_B = -2 \times 1 = -2$ kNm (considering the segment EB from right hand side)

Since bending moment at C is + ve and at B is – ve, there is also a point of contraflexure between C and B.

Bending moment at distance x measured from end E towards left,

$$M_x = -2x + 6.75(x-1) = 4.75x - 6.75$$

$\therefore$ $4.75x - 6.75 = 0$ for point of contraflexure.

That gives $x = \dfrac{6.75}{4.75} = 1.42$ m

The shear force and bending moment diagrams for the entire beam are shown in Fig along with position of points of contraflexture.

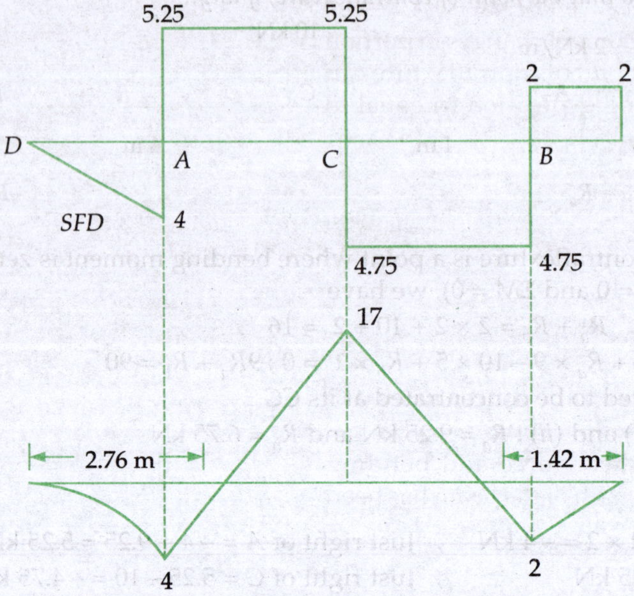

Fig. 11.20

EXAMPLE 11.5

A cantilever beam has been loaded as shown in Fig. 11.21. Draw the shear force and bending moment diagrams, and locate the position of the point of contraflexture.

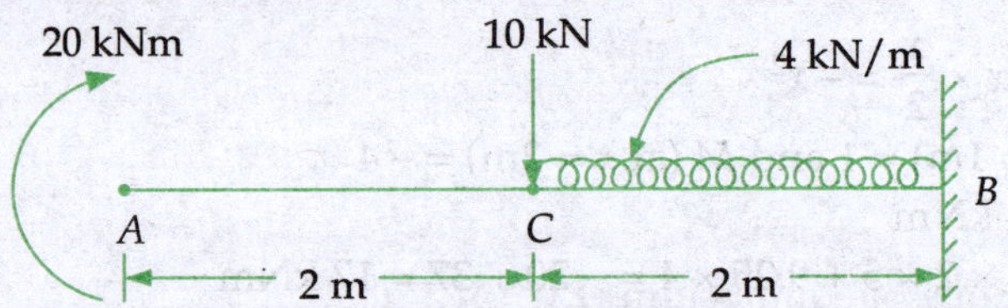

Fig. 11.21

Solution: Consider any section at distance x from the free end

Shear force:

Portion AC : The shear force is zero in this section.

Portion CB : $SF = 10 + 4\,(x - 2)$ (linear variation)

At $x = 2$ m : $SF = 10$ kN

$x = 4$ m : $SF = 10 + 4\,(4 - 2) = 18$ kN

Bending moment:

Portion AC: The bending moment has a constant value of 20 kNm

Portion CB: $$M_x = 20 - 10\,(x - 2) - 4\,(x - 2) \times \frac{(x-2)}{2} \quad \text{...(i)}$$

At $x = 2$ m: $$M_c = 20 - 10\,(2 - 2) - 4 \times \frac{(2-2)^2}{2} = 20 \text{ kNm}$$

$x = 4\text{ m}: \quad M_b = 20 - 10\,(4-2) - 4 \times \dfrac{(4-2)^2}{2}$

$= 20 - 20 - 4 \times 2 = -8\text{ kNm}$

Since the bending moment changes in portion *CB*, the location of the point of contraflexture can be determined by setting expression (*i*) equal to zero. That is

$20 - 10\,(x-2) - 2\,(x-2)^2 = 0$

Upon simplification, we get

$x^2 + x - 16 = 0$

$\therefore \quad x = \dfrac{-1 \pm \sqrt{1+64}}{2}$

$= 3.53\text{ m}$

Thus the point of con-traflexture is located at 3.53 m distance from the free end.

The variation of shear force and bending moment for the entire length of the beam has been depicted in Fig. 11.22.

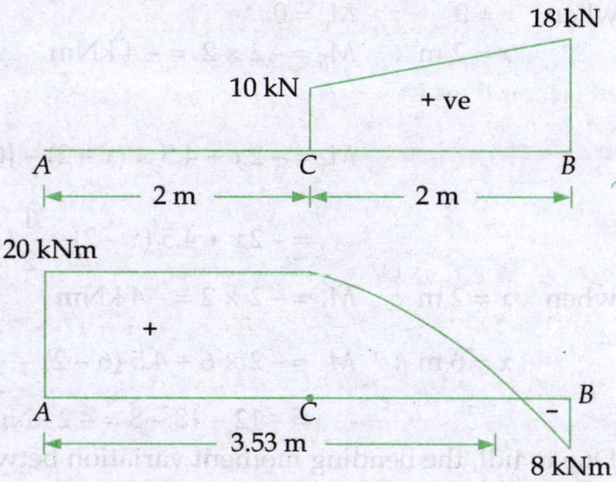

Fig. 11.22

EXAMPLE 11.6

Draw the shear force and bending moment diagram for the beam loaded and supported as shown in Fig. 11.23. Also locate the point where SF changes sign in the section BC.

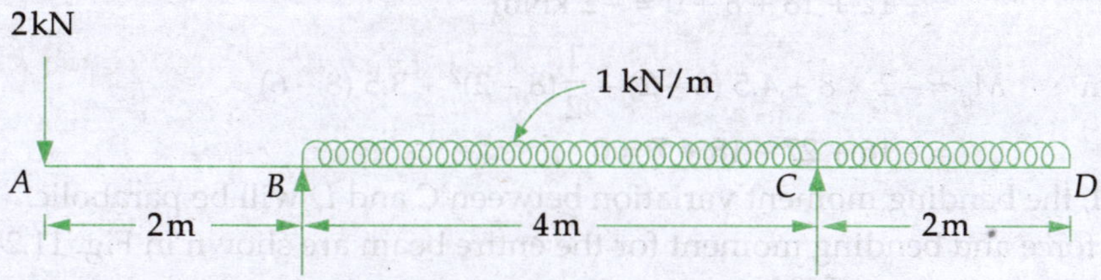

Fig. 11.23

Solution : From conditions of equilibrium ($\Sigma V = 0$ and $\Sigma M = 0$), we have

$$R_b + R_c = 2 + (4 + 2) \times 1 = 8 \qquad ...(i)$$

Taking moments about end point *A* (clockwise moments positive),

$$-R_b \times 2 + 6 \times (2 + 3) - R_c \times 6 = 0$$

The udl is considered to be concentrated at its CG

$$R_b + 3R_c = 15 \qquad ...(ii)$$

Solving expression (*i*) and (*ii*),

$$R_b = 4.5\text{ kN and } R_c = 3.5\text{ kN}$$

Shear force

At point *A*	:	$SF = -2$ kN
Just left of *B*	:	$SF = -2$ kN
Just right of *B*	:	$SF = -2 + 4.5 = 2.5$ kN
Just left of *C*	:	$SF = 2.5 - 4 \times 1 = -1.5$ kN
Just right of *C*	:	$SF = -1.5 + 3.5 = 2$ kN
At point *D*	:	$SF = 2 - 2 \times 1 = 0$

Bending moment : Considering any section at distance x from end A

Portion AB

$$M_x = -2 \times x = -2x \quad \text{(linear variation with } x\text{)}$$

when $x = 0$: $M_a = 0$

$x = 2$ m : $M_b = -2 \times 2 = -4$ kNm

Between B and C

$$M_x = -2x + 4.5 \times (x-2) - \{(x-2) \times 1\} \times \frac{x-2}{2}$$

$$= -2x + 4.5\,(x-2) - \frac{1}{2}(x-2)^2$$

when $x = 2$ m : $M_b = -2 \times 2 = -4$ kNm

$x = 6$ m : $M_c = -2 \times 6 + 4.5\,(6-2) - \frac{1}{2}(6-2)^2$

$= -12 - 18 - 8 = -2$ kNm

Due to udl, the bending moment variation between B and C will be parabolic.

Between C and D

$$M_x = -2x + 4.5\,(x-2) - \frac{1}{2}(x-2)^2 + 3.5\,(x-6)$$

when $x = 6$ m : $M_c = -2 \times 6 + 4.5\,(6-2) - \frac{1}{2}(6-2)^2 + 3.5\,(6-6)$

$= -12 + 18 + 8 + 0 = -2$ kNm

$x = 8$ m : $M_d = -2 \times 8 + 4.5\,(8-2) - \frac{1}{2}(8-2)^2 + 3.5\,(8-6)$

$= -16 + 27 - 18 + 7 = 0$

Due to udl, the bending moment variation between C and D will be parabolic.

The shear force and bending moment for the entire beam are shown in Fig. 11.24.

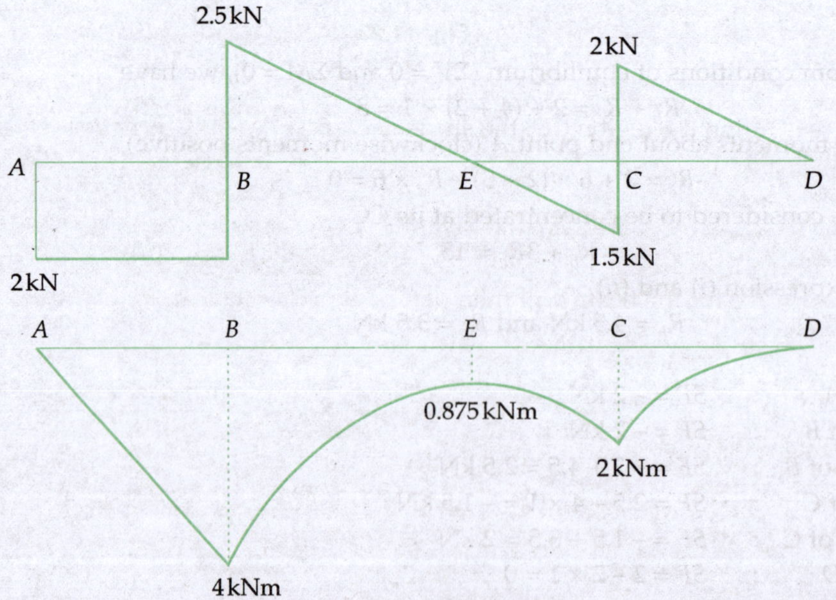

Fig. 11.24

The shear force charges sign at point E and its location with respect to point B is

$$\frac{2.5}{BE} = \frac{1.5}{EC}; \qquad \frac{2.5}{7} = \frac{1.5}{4-y}$$

$$10 - 2.5y = 1.5y; \quad y = 2.5 \text{ m}$$

The point E is thus located at distance $x = 2 + 2.5 = 4.5$ m from point A

Bending moment at point E

$$M_e = -2 \times 4.5 + 4.5(4.5-2) - \frac{1}{2}(4.5-2)^2$$

$$= -9 + 11.25 - 3.125 = -0.875 \text{ kNm}$$

EXAMPLE 11.7

Draw shear force and bending moment diagram for the cantilever beam loaded as shown in Fig. 11.25.

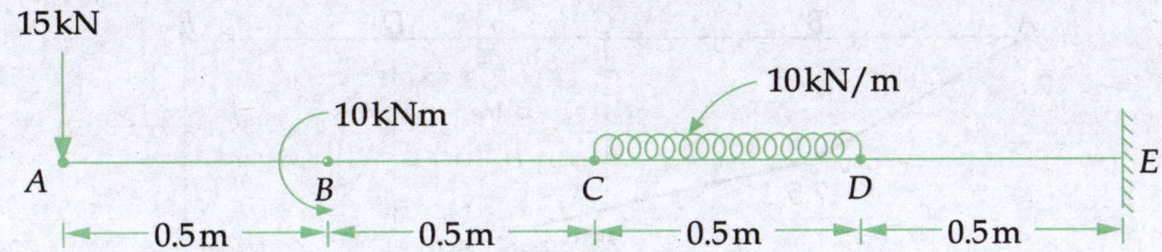

Fig. 11.25

Solution : Consider any section of distance x from end A.

Shear Force

(*i*) **Portion** ***AC***

At A: $SF = -15$ KN

This shear force is taken – ve because it tends to move the left point downward with respect to right portion.

Since there is no load on the beam in the portion AC, the shear force is constant at – 15 kN from A to B.

(*ii*) **Portion** ***CD***

$$SF = -15 - 10 \times (x-1)$$

At point C ($x = 1$) : $SF = -15 - 10(1-1) = -15$ kN

At point D ($x = 1.5$) : $SF = -15 - 10(1.5-1) = -20$ kN

Since there is udl in the region CD, the shear force changes linearly from – 15 kN to – 20 kN in span CD of the beam.

(*iii*) **Portion** ***DE***

Since there is no load on the beam in the portion DE, the shear force is constant at – 20 kN from D to E.

There shear force for the beam will then plot as shown in Fig. 11.26.

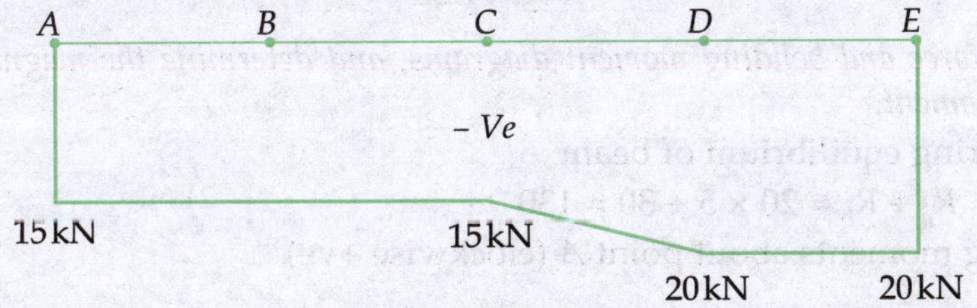

Fig. 11.26

Bending Moment :

$$M_a = 0$$
$$M_b = -15 \times 0.5 - 10 = -17.5 \text{ kNm}$$
$$M_c = -15 \times 1 - 10 = -25 \text{ kNm}$$

At any cross-section xx in portion CD and at distance x from end A

$$M_x = -15x - 10 - 10(x-1) \times \frac{x-1}{2}$$
$$= -15x - 10 - 5(x-1)^2$$

At C ($x = 1$) : $M_c = -15 \times 1 - 10 - 5(1-1)^2 = -25$ kNm

At D ($x = 1.5$) : $M_d = -15 \times 1.5 - 10 - 5(1.5 - 1)^2$
$$= -22.5 - 10 - 1.25 = -33.75 \text{ kN}$$

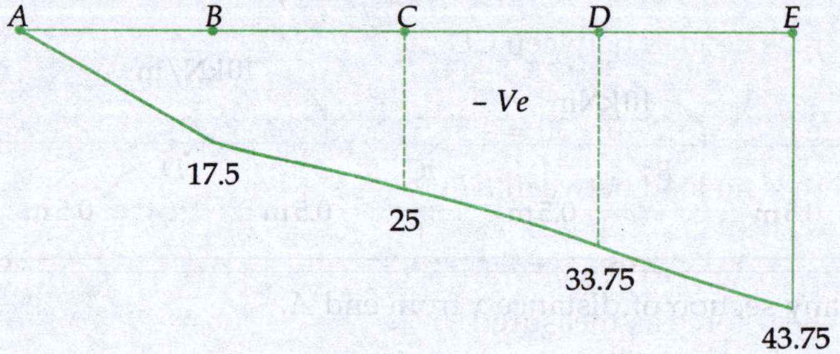

Fig. 11.27

Due to udl, the variation of bending moment in portion CD of the beam will be parabolic

$$M_e = -15 \times 2 - 10 - 10 \times 0.5 \times 0.75$$
$$= -30 - 10 - 3.75 = -43.75 \text{ kN}$$

Since there is no load on the beam in portion DE, the bending moment will vary linearly from – 33.75 kN at point D to – 43.75 kN at point E.

The bending moment for the entire beam will then plot as shown in Fig. 11.27.

EXAMPLE 11.8

A simply supported beam is loaded and supported as shown in the figure given below :

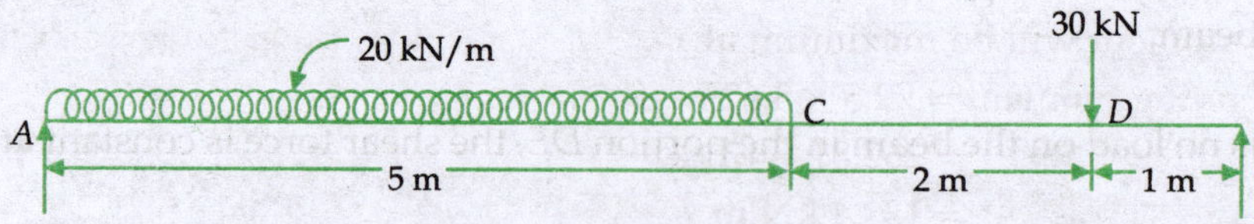

Fig. 11.28

Draw the shear force and bending moment diagrams, and determine the magnitude and location of maximum bending moment.

Solution : Considering equilibrium of beam

$\Sigma F_y = 0 :$ $R_a + R_b = 20 \times 5 + 30 = 130$

$\Sigma M = 0 :$ Taking moments about point A (clockwise +ve)

$$-R_b \times 8 + 30 \times 7 + 20 \times 5 \times \frac{5}{2} = 0$$

or $$R_b = \frac{210 + 250}{8} = 57.5 \text{ kN}$$

and $$R_a = 130 - 57.5 = 72.5 \text{ kN}$$

Consider any section at distance x from end A.

Shear Force :

***Portion AC* :** $SF = R_u - wx = 72.5 - 20x$ (linear variation)

At point A, $x = 0$ and $SF = 72.5$ kN

At point B, $x = 5$ and $SF = 72.5 - 20 \times 5$
$= -27.5$ kN

***Portion CD* :** Since there is no load between CD, the shear force remains constant at – 27.5 kN just upto left of point D.

SF just on right side of $D = -27.5 - 30 = -57.5$ kN

***Portion DB* :** Since there is no load between DB, the shear force remains constant at – 57.5 kN just upto left of point B.

SF just on right side of support $B = -57.5 + 57.5 = 0$

The point of zero shear force as measured from end A and lying between AB can be worked out form the relation

$$72.5 - 20x = 0\ ; \quad x = \frac{72.5}{20} = 3.625 \text{ m}$$

Bending Moment :

***Portion AC* :** $BM = R_a x - \dfrac{wx^2}{2}$

$= 72.5x - 10x^2$ (parabolic variation)

At point A, $x = 0$ and $BM = 0$

At point B, $x = 5$ and $BM = 72.5 \times 5 - 10 \times 5^2$
$= 112.5$ kNm

The bending moment will be maximum at $x = 3.625$ m where shear force is zero.

Maximum bending moment $= 72.5 \times 3.625 - 10 \times 3.625^2$
$= 262.81 - 131.41$
$= 131.4$ kN m

***Portion CD* :** As there is no loading in this section, the bending moment will have a linear variation and at point D ($x = 7$), it will have the value.

$$= 72.5 \times 7 - 20 \times 5 \times \left(\frac{5}{2} + 2\right) = 57.5 \text{ kNm}$$

***Portion DB* :** As there is no loading in this section, the bending moment will have a linear variation and it will drop from 57.5 kNm at point D to zero at support point B.

The variations in shear force and bending moment for the entire beam are shown in Fig. 11.29.

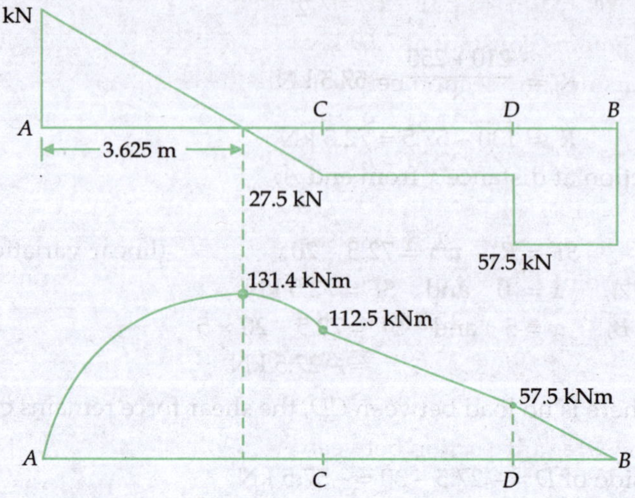

Fig. 11.29

EXAMPLE 11.9

A simply supported beam with 8 m span is loaded as shown in the figure given below:

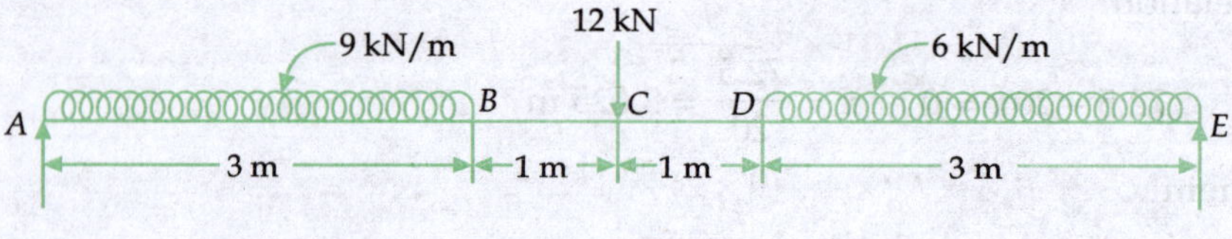

Fig. 11.30

Draw the shear force and bending moment diagrams. Also determine the magnitude and position of maximum bending moment on the beam.

Solution : Considering equilibrium of beam

$\Sigma F_y = 0:$ $\quad R_a + R_e = (9 \times 3) + 12 + (6 \times 3) = 57$ kN

$\Sigma M = 0:$ Taking moments about end point A (clockwise moments +ve)

$$27 \times 1.5 + 12 \times 4 + 18 \times 6.5 - R_e \times 8 = 0$$

The *udl* is considered to be concentrated at *CG*.

$$R_e = \frac{27\times1.5+12\times4+18\times6.5}{8}$$

$$= \frac{40.5+48+117}{8} = 25.69 \text{ kN}$$

and $\quad R_a = 57 - 25.69 = 31.31$ kN

Shear Force:

Portion AB **:** Consider any section at distance x from end support A

$$SF = 31.31 - 9x \qquad \text{(linear variation)}$$

At point A, $\quad x = 0$ and $SF = 31.31$ kN

At point B, $\quad x = 3$ m and $SF = 31.31 - 9 \times 3 = 4.31$ kN

Portion BCD **:** The shear force remains constant at 4.32 kN between B and just left of C.

Just right of CD; $SF = 4.31 - 12 = 7.69$ kN

The shear force remains constant at 7.69 kN between and just left of D.

Portion DE **:** Consider any section between DE and at distance x from end support A.

$$SF = 31.31 - 9 \times 3 - 12 - (x - 5) \times 6$$
$$= 22.31 - 6x$$

At point D, $x = 5$ m

and $SF = 22.31 - 6 \times 5 = -7.69$ kN

At just left of point E, $x = 8$ m

and $SF = 22.31 - 6 \times 8 = -25.69$ kN

At point E, $SF = 25.69 - 25.69 = 0$

Bending Moment :

Portion AB **:** Consider any section between AB at distance x from the end support A.

$$BM = 31.31x - 9\,\frac{x^2}{2} \qquad \text{(parabolic variation)}$$

At point A, $x = 0$ and $BM = 0$

At point B, $x = 3$ m

and $BM = 31.31 \times 3 - \dfrac{9\times 3^2}{2} = 53.43$ kN m

Portion BCD **:**

At point C, $BM = R_a \times 4 - (9 \times 2) \times (1.5 + 1)$

$= 31.31 \times 4 - 67.5$

$= 57.74$ kN m

At point D, $BM = R_a \times 5 - (9 \times 3) \times (1.5 + 2)$

$= 31.31 \times 5 - 94.5 - 12$

$= 50.05$ kN m

Portion DE **:** Consider any section within DE at distance x from the end support A.

$$BM = 31.31x - (9 \times 3) \times (x - 1.5) - 12 \times (x - 4) - 6x\,(x - 5) \times \frac{x-5}{2}$$
$$= 31.31x - 27 \times (x - 1.5) - 12\,(x - 4) - 3\,(x - 5)^2$$

BM at D (at $x = 5$)

$$= 31.31 \times 5 - 27\,(5 - 1.5) - 12\,(5 - 4) - 3\,(5 - 5)^2$$
$$= 50.05 \text{ kN m}$$

BM at E (at $x = 8$)

$$= 31.31 \times 8 - 27\,(8 - 1.5) - 12\,(8 - 4) - 3\,(8 - 5)^2 = 0$$

Since there is *udl* in the segment DE, the variation in bending moment is parabolic.

The variation in shear force and bending moment for the entire beam are as shown in Fig 11.31.

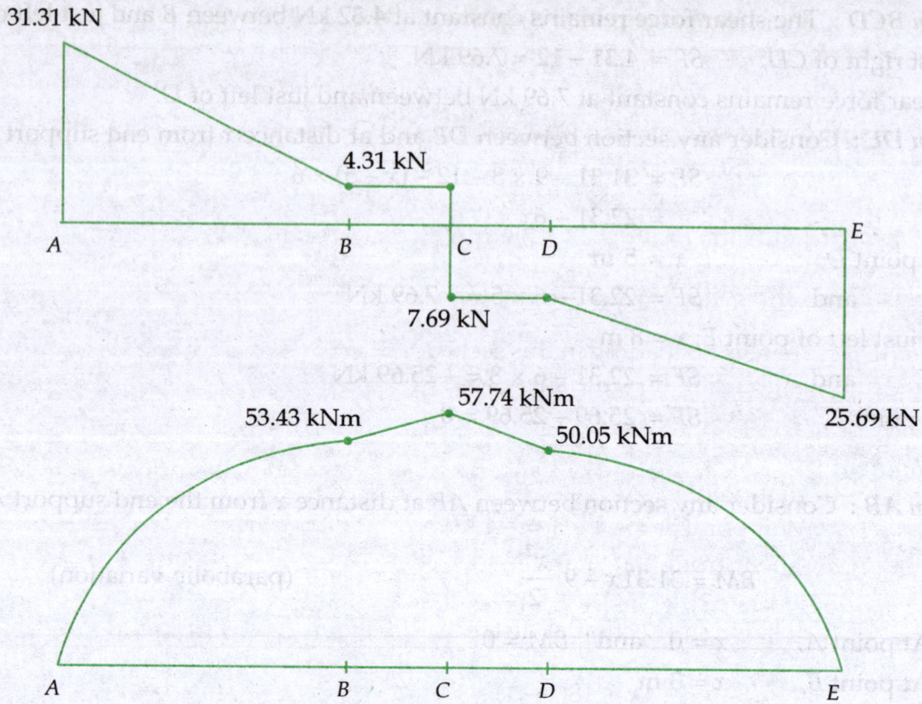

Fig. 11.31

EXAMPLE 11.10

Explain clearly the terms shearing force and bending moment for a beam.

Draw the shear force and bending moment diagrams for the simply supported overhanging beam shown in the figure given below :

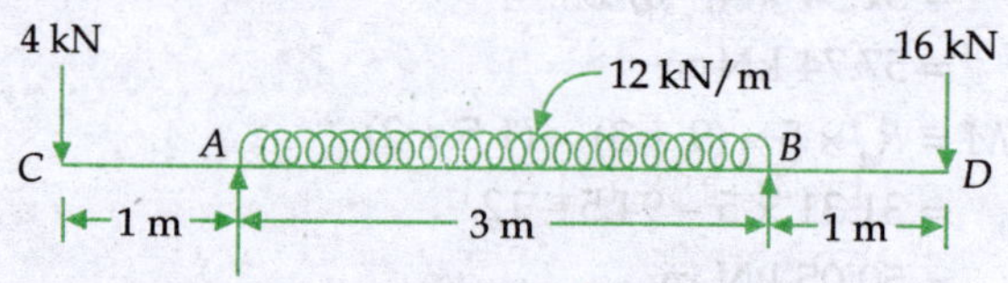

Fig. 11.32

Also locate the position of the points of contraflexure, if any

Solution : Considering equilibrium of beam

$\Sigma F_y = 0$: $R_a + R_b = 4 + 12 \times 3 + 16 = 56$ kN

$\Sigma M = 0$: Taking moments about point A (clockwise +ve)

$$16 \times 4 - R_b \times 3 + 12 \times 3 \times \frac{3}{2} - 4 \times 1 = 0$$

or $$R_b = \frac{64 + 54 - 4}{3} = 38 \text{ kN}$$

and $$R_a = 56 - 38 = 18 \text{ kN}$$

Shear Force :

Portion CA : At point C,

$SF = -4$ kN and it remains constant upto just left of point A because there is no loading between this segment.

SF just on right side of $A = -4 + 18 = 14$ kN

Portion AB : Consider any section at distance x from point C.

$$SF = -4 + 18 - (x - 1) \times 12$$
$$= 26 - 12x$$

SF just on left side of B, *i.e.*, at $x = 4$ m

$$SF = 26 - 12 \times 4$$
$$= -22 \text{ kN}$$

SF just right side of $B = -22 + 38 = 16$ kN

The shear force will be zero at $x = \frac{26}{12} = 2.1666$ m

Portion BD: As there is no loading in this section, the shear force remains constant at 16 kN just upto left of D and just to right of D, the shear foce will become

$$16 - 16 = 0$$

Bending Moment :

Portion CA : BM at point C = 0

BM at point $A = -4 \times 1 = -4$ kNm

The bending moment from 0 at A and – 4 kNm at B will have linear variation as there is no loading in this segment.

Portion AB : Consider any section at distance x from the end point C.

$$BM = -4x + 18(x - 1) - 12(x - 1) \times \frac{x-1}{2}$$
$$= -4x + 18x - 18 - 6(x - 1)^2$$
$$= 14x - 18 - 6(x - 1)^2 \quad ...(i)$$

At point A, $x = 1$ m

and $BM = 14 \times 1 - 18 - 6(1 - 1)^2 = -4$ kNm

The bending moment at $x = 2.166$ m (point E)

$$= 14 \times 2.166 - 18 - 6\,(2.166 - 1)$$

At point B, $x = 4$ m

and $BM = 14 \times 4 - 18 - 6(4 - 1)^2$

$$= -16 \text{ kNm}$$

The variation of bending moment from – 4 kN m at point A to – 16 kN m at point B will be parabolic as there is *udl* in this segment.

Portion BD : As there is no loading in this section, the bending moment will have a linear variation and it will rise from – 16 kNm at point B to zero at end point D.

The variations is shear force and bending moment for the entire beam are shown in Fig. 11.33.

Points of Contraflexure : The bending moment is zero at the points of contraflexure.

Since the bending moment changes sign in portion AB, the location of the points of contraflexure can be determined by setting expression (*i*) equal to zero. That is

$$14x - 18 - 6(x - 1)^2 = 0$$

Upon simplification : $3x^2 - 13x + 12 = 0$

The solution of this quadratic equation gives

$$x = \frac{13 \pm \sqrt{(-13)^2 - 4 \times 3 \times 12}}{2 \times 3}$$
$$= \frac{13 \pm 5}{5} = 3 \text{ m} \quad \text{and} \quad 1.33 \text{ m}$$

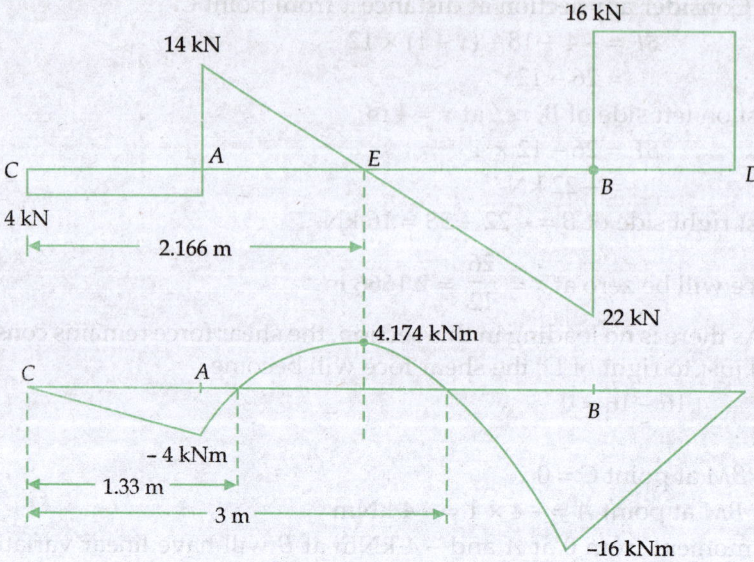

Fig. 11.33

Thus the points of contraflexure are located at 1.33 m and 3 m from the end point C.

EXAMPLE 11.11

A horizontal beam 10 m long carries a uniformly distributed load of 8 kN/m together with concentrated loads of 40 kN at the left end and 60 kN at the right end. The beam is supported at two points 6 m, so chosen that reaction is the same at the each support. Determine the position of props and show the variation of shear force and bending moment over the entire length of the beam.

Solution: Refer Fig. 11.34 for the beam loaded and supported as per the statement. Let the prop C be at distance a from end A.

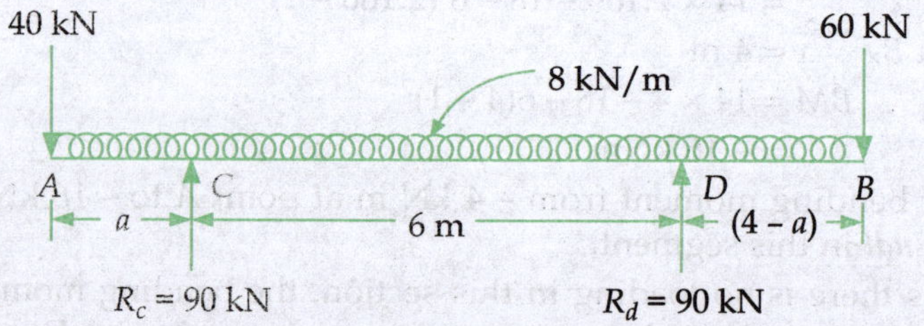

Fig. 11.34

Then the prop D is at distance $(4 - a)$ from end B.

Total load on the beam = 40 + 60 + (10 × 8) = 180 kN. Since reaction is the same at each support,

$$R_c = R_d = \frac{180}{2} = 90 \text{ kN}$$

Taking moments about end A,

$$60 \times 10 + (8 \times 10) \times \frac{10}{2} = 90 \times a + 90\,(6 + a)$$

or $$600 + 400 = 90a + 540 + 90a$$

$$\therefore \quad a = \frac{(600 + 400) - 540}{180} = 2.55 \text{ m}$$

Thus the left support is 2.55 m from A and the right support is $(4 - 2.55) = 1.45$ m from B.

Shear force:

SF at $A = -40$ kN

SF just on left side of $C = -40 - 8 \times 2.55 = -60.40$ kN

SF just on right side of $C = -60.40 + 90 = 29.60$ kN

SF just on left side of $D = 29.60 - 8 \times 6 = -18.40$ kN

SF just on right side of $D = -18.40 + 90 = 71.60$ kN

SF just on left side of $B = 71.60 - 8 \times 1.45 = 60$ kN

SF just on right side of $B = 60 - 60 = 0$

The point of zero shear stress as measured from end A and lying between CD can be worked out from the equation.

$$-40 + 90 - 8x = 0\ ; \quad x = \frac{50}{8} = 6.25 \text{ m}$$

Bending moment :

BM at $A = 0$

$$BM \text{ at } C = -40 \times 2.55 - (8 \times 2.55) \times \frac{2.55}{2} = -128 \text{ kNm}$$

$$BM \text{ at } D = -40 \times 8.55 - (8 \times 8.55) \times \frac{8.55}{2} + 90 \times 6$$

$$= -342 - 292.4 + 540 = -94.4 \text{ kNm}$$

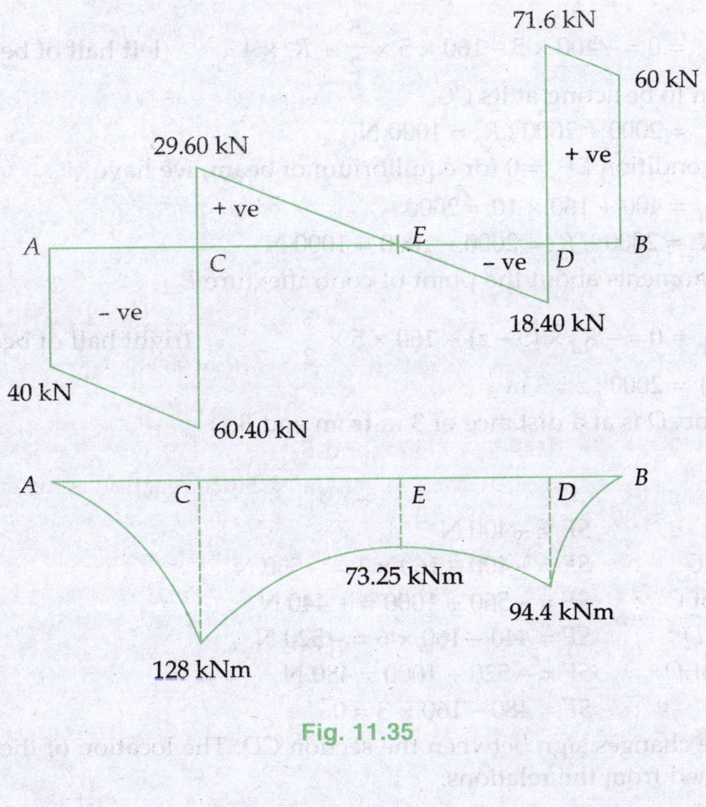

Fig. 11.35

BM at a distance of 6.25 m from *A*,

$$= -40 \times 6.25 - (8 \times 6.25) \times \frac{6.25}{2} + 90 \times (6.25 - 2.55)$$

$$= -250 - 156.25 + 333 = -73.25 \text{ kNm}$$

The variation of shear force and bending moment length of the beam has been depicted in Fig 11.35.

EXAMPLE 11.12

A horizontal beam AB of span 10 m carries a uniformly distributed load of intensity 160 N/m and a point load of 400 N at the left end A. The beam is supported at a point C which is 1m from A and at D which is on the right half of the beam. If the point of contraflexture is at the mid point of the beam, determine the distance of support at D from the end B of the beam. Proceed to draw the shear force and bending moment diagrams for the arrangement.

Solution:

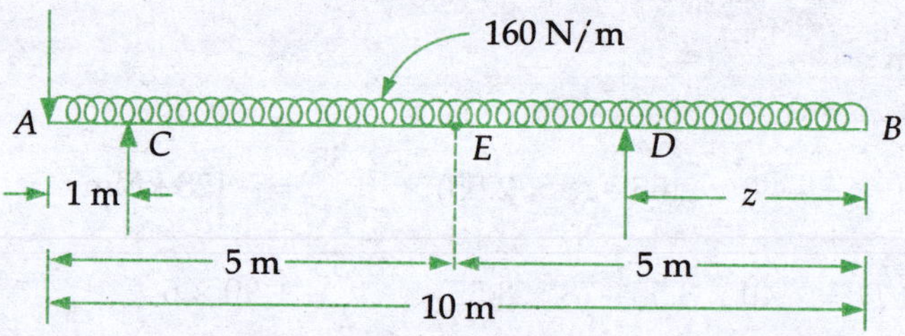

Fig. 11.36

The bending moment is zero at the point of contraflexture. Therefore

$$M_e = 0 = -400 \times 5 - 160 \times 5 \times \frac{5}{2} + R_c \times 4 \qquad \text{(left half of beam)}$$

The *udl* is taken to be acting at its *CG*.

or $\quad 4R_c = 2000 + 2000 \,;\, R_c = 1000 \text{ N}$

Applying the condition $\Sigma F_y = 0$ for equilibrium of beam, we have

$$R_c + R_d = 400 + 160 \times 10 = 2000$$

$$\therefore \quad R_d = 2000 - R_c = 2000 - 1000 = 1000 \text{ N}$$

Again taking moments about the point of contraflexture *E*,

$$M_e = 0 = -R_d \times (5 - z) + 160 \times 5 \times \frac{5}{2} \qquad \text{(right half of beam)}$$

$1000 \times (5 - z) = 2000 \,;\, z =$ **3 m**

Thus the support *D* is at a distance of 3 m from end *B*.

Shear Force

Portion AC:

At *A*	:	$SF = -400$ N
Just left of *C*	:	$SF = -400 - 160 \times 1 = -560$ N
Just right of *C*	:	$SF = -560 + 1000 = +440$ N
Just left of *D*	:	$SF = 440 - 160 \times 6 = -520$ N
Just right of *D*	:	$SF = -520 + 1000 = 480$ N
At point *B*	:	$SF = 480 - 160 \times 3 = 0.$

The shear force changes sign between the section *CD*. The location of the point of zero shear stress can be obtained from the relations:

$-400 - 160x + 1000 = 0\,; \quad x = 3.75 \text{ m}$

Bending moment: Considering any section at distance x from end A,

Portion AC:

$$M_x = -400x - 160x \times \frac{x}{2} = -400 \times -80x^2$$

when $x = 0 : M_a = 0$

$x = 1 \text{ m} : M_c = -400 - 80 = -480 \text{ Nm}$

Between C and D

$$M_x = -400x - 160x \times \frac{x}{2} + R_b\,(x-1)$$

$$= -400x - 80x^2 + 1000\,(x-1).$$

when $x = 1 \text{ m} : M_b = -400 \times 1 - 80 \times 1^2 + 1000\,(1-1) = -480 \text{ Nm}$

$x = 3.75 \text{ m} : M = -400 \times 3.75 - 80 \times 3.752 + 1000\,(3.75 - 1) = 125 \text{ Nm}$

$x = 5 \text{ m}: M_e = -400 \times 5 - 80 \times 52 + 1000\,(5-1) = 0$

$x = 7 \text{ m} : M_d = -400 \times 7 - 80 \times 7^2 + 1000\,(7-1) = -720 \text{ Nm}$

Between D and B:

$$M_x = -400x - 80x^2 + 1000 \times (x-1) + 1000\,(x-7)$$

At $x = 7 \text{ m} : M_d = -400 \times 7 - 80 \times 7^2 + 1000 \times (7-1) + 1000\,(7-7) = -720 \text{ Nm}$

$x = 10 \text{ m} : M_b = -400 \times 10 - 80 \times 10^2 + 1000\,(10-1) + 1000\,(10-7)$

$= -4000 - 8000 + 9000 + 3000 = 0.$

The shear force and bending moment for the entire beam are shown in Fig. 11.37.

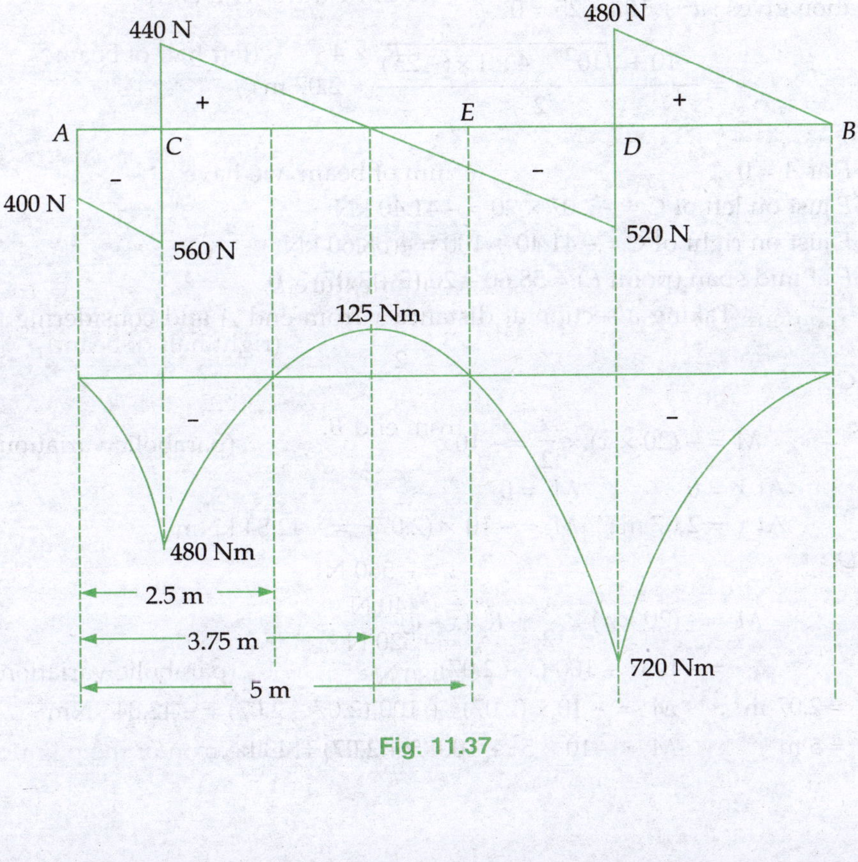

Fig. 11.37

EXAMPLE 11.13

A girder 10 m long rests on two supports with equal overhangs on either side and carries a uniformly distributed load of 20 kN/m over the entire length. Calculate the overhangs if the maximum bending moment, positive or negative, is to be as small as possible. Proceed to draw the shear force and bending moment diagrams for the arrangement.

Solution: Refer to Fig. 11.38 for the space diagram of the loaded girder. The overhang on each side has been indicated as *a*.

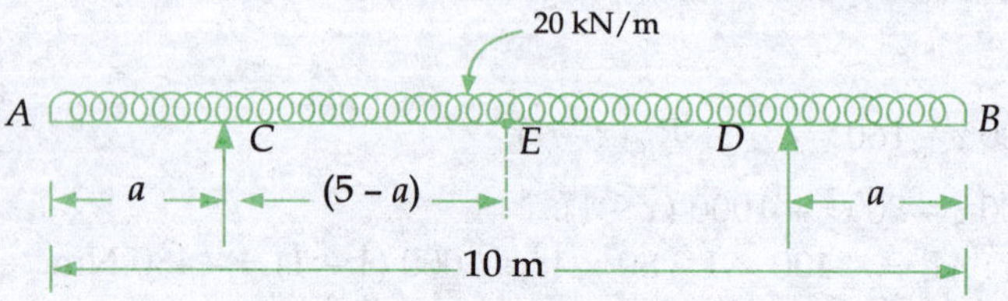

Fig. 11.38

Due to symmetrical arrangement, the total load on the beam will be shared equally between the two supports.

$$\therefore \quad R_c = R_d = \frac{20 \times 10}{2} = 100 \text{ kN}$$

The maximum positive moment would occur at the mid span (point *E*) and the maximum negative would occur at the supports. Since these moments are stated to be equal in magnitude, we have:

$$(20 \times a) \times \frac{a}{2} = 100\,(5 - a) - (20 \times 5) \times \frac{5}{2}$$

Simplification gives : $a^2 + 10a - 25 = 0$

$$\therefore \quad a = \frac{-10 + \sqrt{10^2 - 4 \times 1 \times (-25)}}{2} = 2.07 \text{ m}$$

Shear force:

SF at $A = 0$

SF just on left of $C = -2.07 \times 20 = -41.40$ kN

SF just on right of $C = -41.40 + 100 = +58.60$ kN

SF at mid span (point *E*) $= 58.60 - 20\,(5 - 2.07) = 0$

Bending moment: Taking a section at distance *x* from end *A* and considering forces on left hand side.

Portion AC:

$$M = -(20 \times x) \times \frac{x}{2} = -10x^2 \qquad \text{(parabolic variation)}$$

At $x = 0$: $M_a = 0$

At $x = 2.07$ m : $M_c = -10 \times (2.07)^2 = -42.84$ kNm

Portion CD:

$$M = -(20 \times x) \times \frac{x}{2} + R_c\,(x - a)$$

$$= -10x^2 + 100\,(x - 2.07) \qquad \text{(parabolic variation)}$$

At $x = 2.07$ m : $M_c = -10 \times (2.07)^2 + 100\,(2.07 - 2.07) = -42.84$ kNm

At $x = 5$ m : $M_e = -10 \times 5^2 + 100\,(5 - 2.07) = 43$ kNm

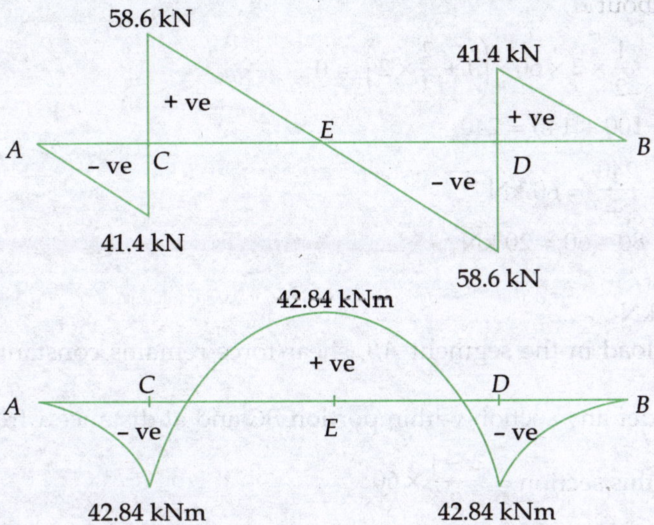

Fig. 11.39

The slight variation in the magnitude of bending moment at the support (point B) and at the centre (point E) is due to rounding off.

For locating the position of the point of contraflexure, we have

$$-10x^2 + 100(x - 2.07) = 0$$

or $$x^2 - 10x + 20.7 = 0$$

$$\therefore \quad x = \frac{10 \pm \sqrt{10^2 - 4 \times 20.7}}{2} = 2.97 \text{ m and } 7.07 \text{ m}$$

The shear force and the bending moment diagrams for the entire span of the girder are shown in Fig. 11.39.

Note: The *SF* and *BM* for the right half has been drawn making use of symmetry.

EXAMPLE 11.14

Draw the shear force and bending moment diagrams for the overhanging beam loaded as shown in the figure given below.

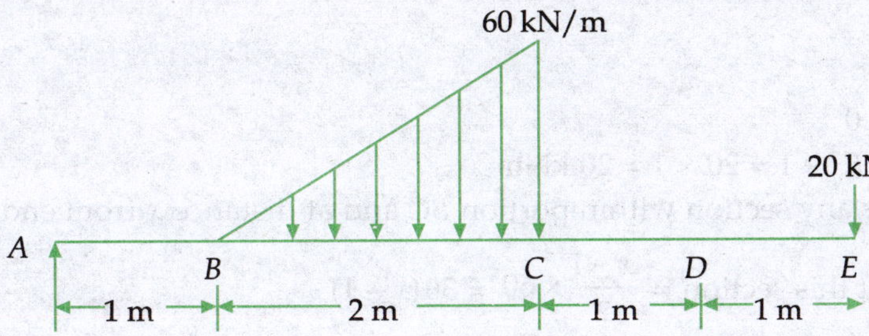

Fig. 11.40

Indicate values at salient points of the beam. Locate the position of the point of contraflexure if any.

Solution: From conditions of static equilbrium ($\Sigma Fy = 0$ and $\Sigma M = 0$), we have

$$R_a + R_d = \left(\frac{1}{2} \times 2 \times 60\right) + 20 = 80$$

Taking moment about A,

$$\text{or } R_d \times 4 - 20 \times 5 - \frac{1}{2} \times 2 \times 60 \times \left(1 + \frac{2}{3} \times 2\right) = 0$$

$$\text{or} \quad 4R_d = 100 + 140 = 240$$

$$R_d = \frac{240}{4} = 60 \text{ kN}$$

$$\text{and} \quad R_a = 80 - 60 = 20 \text{ kN}$$

Shear Force:

SF at $A = R_a = 20$ kN

Since there is no load in the segment AB, shear force remains constant at 20 kN within this portion of the beam.

Portion BC: Consider any section within portion BC and at distance x from end A,

$$\text{Load intensity at this section} = \frac{x-1}{2} \times 60$$

$$SF = 20 - \frac{1}{2}(x-1) \times \left\{\frac{x-1}{2} \times 60\right\}$$

$$= 20 - 15\,(x-1)^2 \qquad \text{(parabolic variation)}$$

At point B : $x = 1$ m and $SF = 20 - 15(1-1)^2 = 20$ kN

At point C : $x = 3$ m and $SF = 20 - 15(3-1)^2 = -40$ kN

The location of zero shear force can be worked out from the relation.

$$20 - 15(x-1)^2 = 0$$

$$\text{or} \quad x - 1 = \sqrt{\frac{20}{15}} = 1.54 \qquad \therefore x = 2.154 \text{ m}$$

Since there is no load in portion CD of the beam, the shear force from point C to just left of point D will remain constant at – 40 kN (the shear force at point C).

SF just an right side of $D = -40 + 60 = 20$ kN

This value of shear force remains constant within portion DE (because of no loading) and at point E, it takes the values

$$20 - 20 = 0 \text{ kN}$$

Bending Moment

Portion AB

BM at point $A = 0$

BM at point $B = R_a \times 1 = 20 \times 1 = 20$ kNm

Portion BC: Consider any section within portion BC and at distance x from end A.

$$\text{Load intensity at this section} = \frac{x-1}{2} \times 60 = 30(x-1)$$

$$BM = 20x - \left[\frac{1}{2}(x-1) \times 30(x-1)\right] \times \frac{x-1}{3}$$

Here $\frac{x-1}{3}$ is the distance of CG of triangular load from the section.

$$BM = 20x - 5(x-1)^3 \qquad \text{(cubic variation)}$$

At point B : $x = 1$ m and $BM = 20 \times 1 - 5(1-1)^3 = 20$ kNm

At point C : $x = 3$ m and $BM = 20 \times 3 - 5(3-1)^3 = 20$ kNm

The bending moment will be maximum at the point where shear force is zero, *i.e.*, at $x = 2.154$ m

Maximum bending moment

$$= 20 \times 2.154 - 5(2.154 - 1)^3 = 35.40 \text{ kNm}$$

Portion CD:

$$BM \text{ at point } C = 20 \text{ kNm (calculated above)}$$

$$BM \text{ at point } D = 20 \times 4 - \left(\frac{1}{2} \times 2 \times 60\right) \times \left(1 + \frac{1}{3} \times 2\right)$$

$$= 80 - 100 = -20 \text{ kNm}$$

Since shear forces remains constant due to no load in this section, the bending moment will have linear variation from 20 kNm (at point C) to – 20 kNm (at point D).

Since the bending moment changes sign in the portion CD, there is a point of contraflexure in this portion and its location with respect to point A can be worked out from the relation

$$20 \times x - \left(\frac{1}{2} \times 2 \times 60\right)\left(x - 3 + \frac{2}{3}\right) = 0$$

or $$20x - 60x + 180 - 40 = 0$$

That gives $$x = \frac{-180 + 40}{-40} = 3.5 \text{ m}$$

Thus the point of contraflexure is located at 3.5 m from the end A.

Portion DE

BM at point D = – 20 kNm (calculated above)

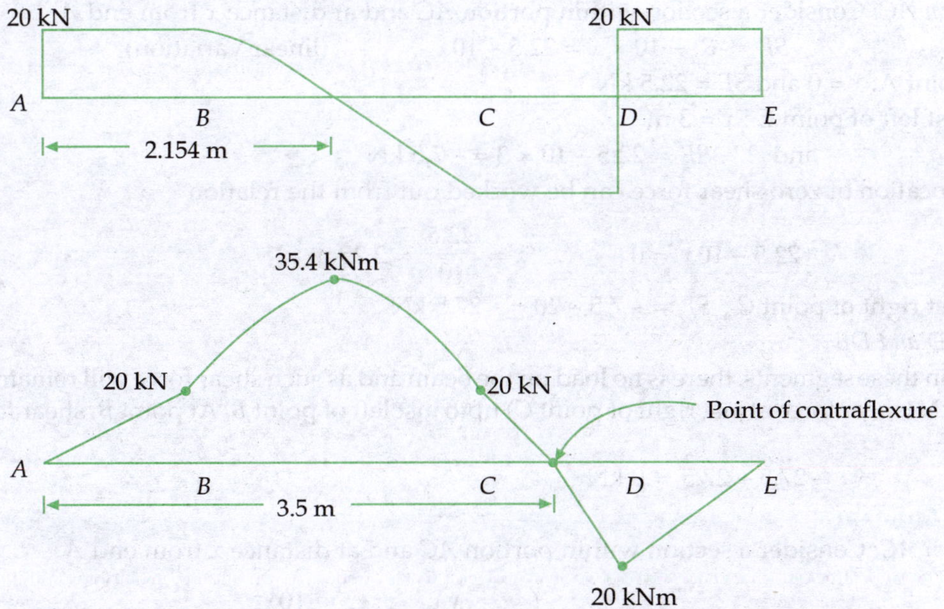

Fig. 11.41

The bending moment at the free end E will be zero. Since shear force remains constant due to no load in this section, the bending moment will have linear variation from – 20 kNm (at point D) to zero (at point E).

The variation of shear force and bending moment for the entire length of the beam has been shown in Fig. 11.41.

EXAMPLE 11.15

A simply supported beam with supports at A and B, 6 metres apart is loaded as shown in the figure given below.

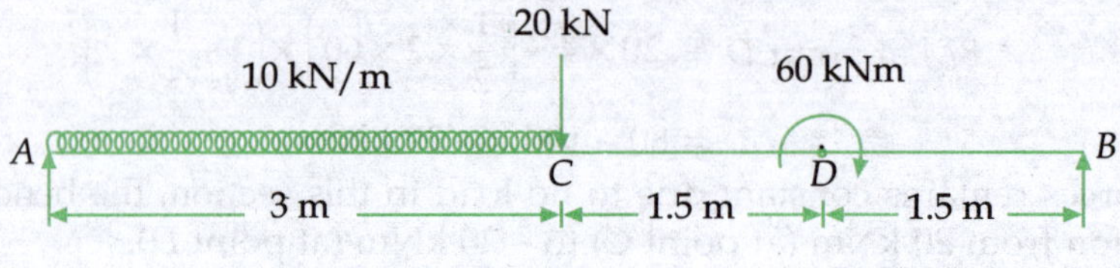

Fig. 11.42

Draw the shear force and bending moment diagram l and indicate there upon the corresponding values at salient points of the beam. Locate the position of the point of contraflexure, if any.

Solution : From conditions of static equilibrium ($\Sigma F_y = 0$ and $\Sigma M = 0$), we have

$$R_a + R_b = (10 \times 3) + 20 = 50$$

Taking moments about support point B,

$$R_a \times 6 - (10 \times 3) \times (1.5 + 3) - 20 \times 3 + 60 = 0$$

That gives: $$R_a = \frac{135 - 60 + 60}{6} = 22.5 \text{ kN}$$

and $$R_b = 50 - 22.5 = 27.5 \text{ kN}$$

Shear Force

Portion AC: Consider a section within portion AC and at distance x from end A,

$$SF = R_a - 10 \times x = 22.5 - 10x \qquad \text{(linear variation)}$$

At point A: $x = 0$ and $SF = 22.5$ kN

At just left of point C : $x = 3$ m

and $SF = 22.5 - 10 \times 3 = -7.5$ kN

The location of zero shear force can be worked out from the relation

$$22.5 - 10x = 0 \quad \therefore \quad x = \frac{22.5}{10} = 2.25 \text{ m}$$

At just right of point C : $SF = -7.5 - 20 = -27.5$ kN

Portion CD and DB

Within these segments, there is no load on the beam and as such shear force will remain constant at – 27.5 kN (shear force at just right of point C) upto just left of point B. At point B, shear force takes the value

$$-27.5 + 27.5 = 0 \text{ kN}$$

Bending Moment

Portion AC: Consider a section within portion AC and at distance x from end A,

$$BM = R_a \times x - \left(wx \times \frac{x}{2}\right) = 22.5x - \frac{10x^2}{2} \qquad \text{(parabolic variation)}$$

At point *A*: $x = 0$ and $BM = 0$

At point C : $x = 3$ m and

$$BM = 22.5 \times 3 - \frac{10 \times 3^2}{2} = 22.5 \text{ kNm}$$

Bending moment will be maximum at the point where shear force is zero, *i.e.*, at $x = 2.25$ m.

$$\text{Maximum bending moment} = 22.5 \times 2.25 - \frac{10 \times 2.25^2}{2} = 25.31 \text{ kN}$$

Portion CD: Consider a section within portion *CD* and at distance *x* from end *A*,

$$\begin{aligned} BM &= 22.5x - 10 \times 3\,(x - 1.5) - 20\,(x - 3) \\ &= 22.5x - 30x + 45 - 20x + 60 \\ &= -27.5x + 105 \quad \text{(linear variation)} \end{aligned}$$

At point *C*: $x = 3$ m and $BM = -27.5 \times 3 + 105 = 22.5$ kNm

At just left of point *D* ($x = 4.5$ m)

$$BM = -27.5 \times 4.5 + 105 = -18.75 \text{ kNm}$$

At just right of point *D*: $BM = -18.75 + 60 = 41.25$ kNm

Since the bending moment changes sign in the portion *CD*, there is point of contaflexure in this portion and its location with respect to point A can be worked out from the relation

$$-27.5x + 105 = 0 \quad \therefore \quad x = \frac{105}{27.5} = 3.818 \text{ m}$$

Thus the point of contraflexure is located at 3.818 m from the end A.

Portion DB

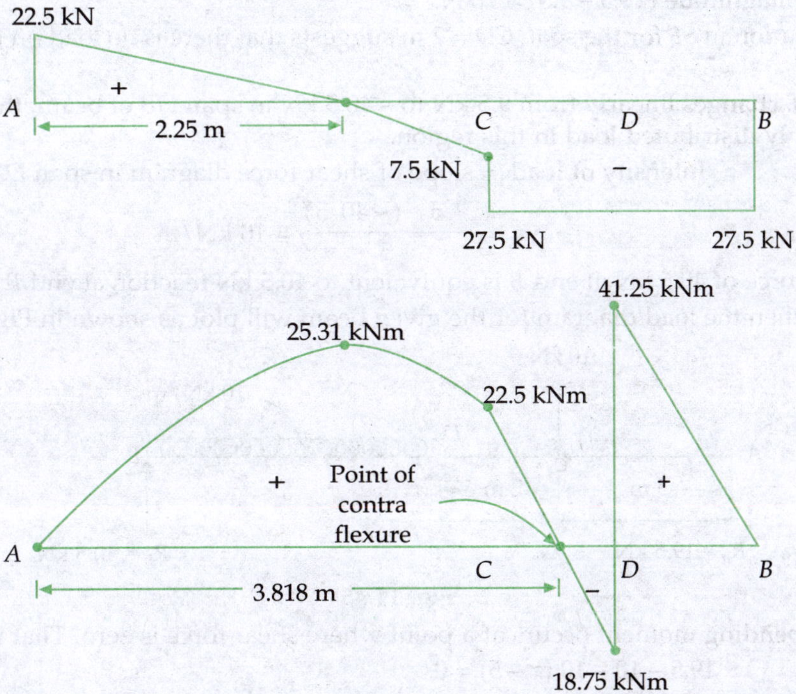

Fig. 11.43

BM at point $D = 41.25$ kNm (calculated above). The bending moment at the support point B has to zero. Since shear force remains constant in this region due to no load, the bending moment will have linear variation from 41.25 kNm (at point D) to zero (at point E).

The variation of shear force and bending moment for the entire length of the beam has been shown in Fig. 11.43.

EXAMPLE 11.16

The shear force diagram for a simply supported beam of 10 m span is shown in Fig. 11.44. Draw the corresponding load diagram. Proceed to calculate the maximum bending moment and its location.

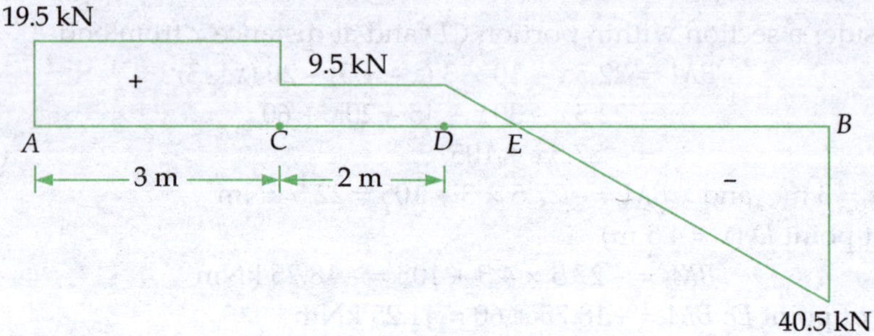

Fig. 11.44

Solution :

(*i*) Shear force of 19.5 kN at end A is equivalent to 19.5 kN reaction at end A.

(*ii*) No variation in SF for the span $AC = 3$ m suggests that there is no load on the beam in this section.

(*iii*) At C, there is an abrupt change in SF. Hence at C, there must be a concentrated vertical load of magnitude $(19.5 - 9.5) = 10$ N.

(*iv*) No variation in SF for the span $CD = 2$ m suggests that there is no load on the beam in this region.

(*v*) Since SF changes linearly from 9.5 kN to – 40.5 kN in span DB of beam, there should be a uniformly distributed load in this region.

$$\text{Intensity of load} = \text{slope of shear force diagram in span } DB$$

$$= \frac{9.5 - (-40.5)}{5} = 10 \text{ kN/m}$$

(*vi*) Shear force of 40.5 kN at end B is equivalent to 40.5 kN reaction at end B.

Obviously then the load diagram for the given beam will plot as shown in Fig. 11.35.

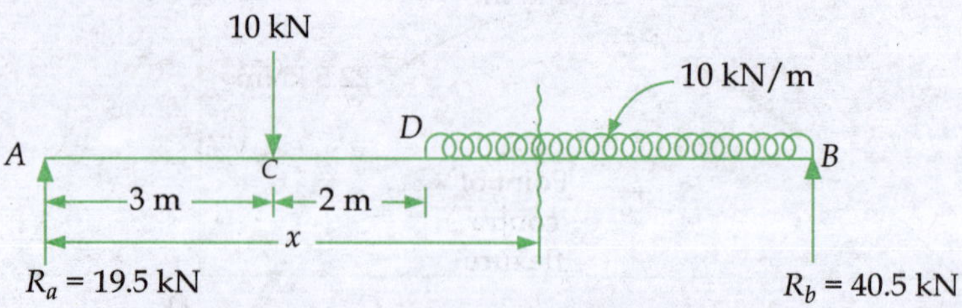

Fig. 11.45

Maximum bending moment occurs at a point where shear force is zero. That is

$$19.5 - 10 - 10\,(x - 5) = 0$$

$$\therefore \quad x = 5.95 \text{ m from end } A$$

Bending moment at the chosen section,

$$M = 19.5x - 10\,(x - 3) - 10 \times (x - 5) \times \frac{x - 5}{2}$$

$$M_{max} = 19.5 \times 5.95 - 10\,(5.95 - 3) - 10\,(5.95 - 5) \times \frac{(5.95 - 5)}{2}$$

$$= 116.02 - 29.50 - 4.51 = 82.01 \text{ kNm}$$

EXAMPLE 11.17

Fig. 11.46 shows the shear force diagram of a beam ABCD. Find the position of the supports and the magnitude of reactions.

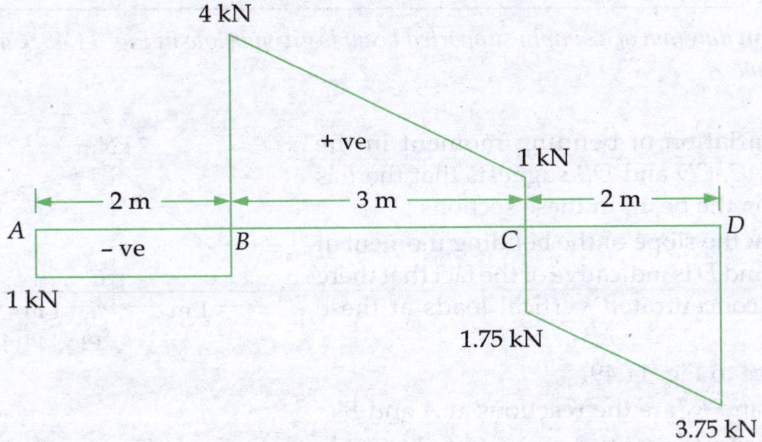

Fig. 11.46

Solution:

(*i*) Corresponding to shear force of 1 kN at *A*, there must be point load of 1 kN at *A*.

(*ii*) *SF* is constant from *A* to *B* and obviously there is no load on the beam in this region.

(*iii*) SF changes sign at *B* and accordingly there will be a support at *B* with

Reaction $R_b = 1 + 5 = 5$ kN

(*iv*) Shear force consists of inclined line for the portion *BC* and that indicates the presence of *udl* in this portion.

Intensity of load = slope of shear force diagram in span *BC*

$$= \frac{4 - 1}{3} = 1 \text{ kN/m}$$

(*v*) *A* point C, the *SF* changes from + 1 kN to – 1.75 kN, and accordingly there is a point load at C.

Magnitude of point load at C = 1 + 1.75 = 2.75 kN

(*vi*) Shear force changes linearly form 1.75 kN to 3.75 kN in span *CD* of beam; there should be a uniformly distributed load in this region.

Intensity of load = slope of shear force diagram in span *CD*

$$= \frac{3.75 - 1.75}{2} = 1 \text{ kN/m}$$

(*vii*) Shear force of – 3.75 kN at end *D* is equivalent to 3.75 kN reaction at end *D*.

The load diagram for the given beam will plot as shown in Fig. 11.47.

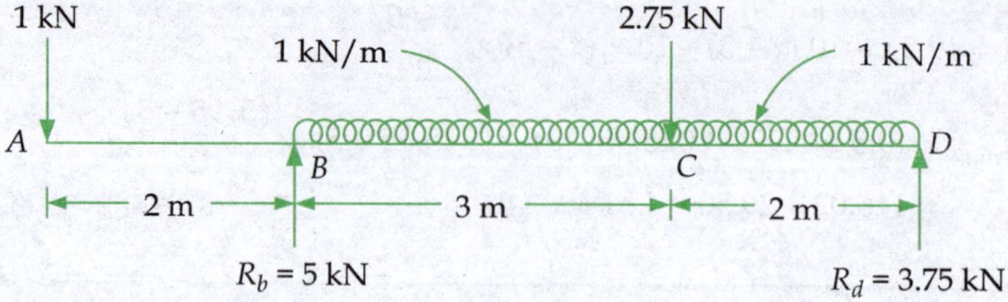

Fig. 11.47

EXAMPLE 11.18

The bending moment diagram of a simply supported beam is given below in Fig. 11.48. Calculate the support reactions of the beam.

Solution:

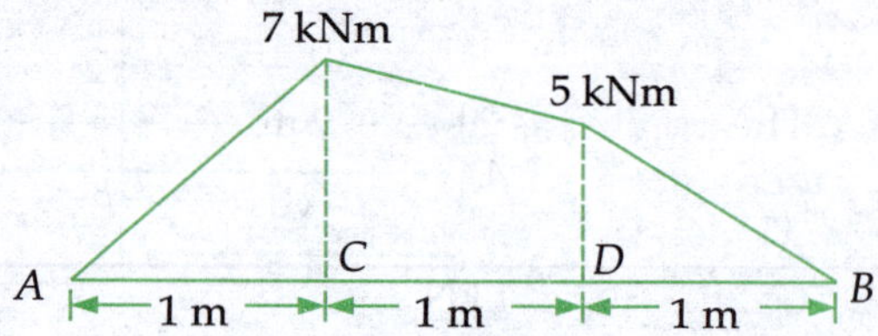

Fig. 11.48

(*i*) Linear variation of bending moment in the sections *AC*, *CD* and *DB* suggests that there is no load on the beam in these sections

(*ii*) Change in the slope of the bending moment at points *C* and *D* is indicative of the fact that there must be concentrated vertical loads at these points.

With reference to Fig. 11.49,

R_a and R_b are the reactions at *A* and *B*

W_c and W_d are the concentrated point loads at *C* and *D*.

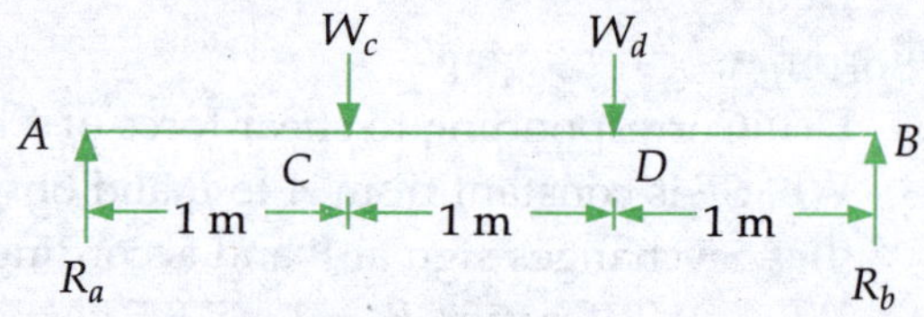

Fig. 11.49

Moment at $C : R_a \times 1 = 7 \quad \therefore \quad R_a = 7$ kN

Moment at $D : R_a \times 2 - W_c \times 1 = 5$

or $\quad 7 \times 2 - W_c \times 1 = 5\,;\; W_c = 9$ kN

Moment at $B : R_a \times 3 - W_c \times 2 - W_d \times 1 = 0$

or $\quad 7 \times 3 - 9 \times 2 - W_d \times 1 = 0\,;\; W_d = 3$

Considering equilibrium of forces in the vertical directions,

$$R_a + R_b = W_c + W_d$$

$$\therefore \quad R_b = W_c + W_d - R_a = 9 + 3 - 7 = 5 \text{ kN}$$

Thus the reactions at the end supports are:

$$R_a = 7 \text{ kN and } R_b = 5 \text{ kN}$$

REVIEW AND SUMMARY

1. A ***beam*** is a supported structural member sufficiently long as compared to its lateral dimensions and subjected to loads transverse to longitudinal axis.
2. The ends of a beam can be simply supported (ends where slope changes but the vertical displacement of deflection remains zero), fixed (ends where there is no vertical displacement and deflection too remains zero) or free (ends which are free to take any slope and any position).
3. With two simply supported ends, the beam is said to be simply supported and the distance between the supports is called span of the beam. A beam with one end free and the other end fixed is called cantilever A beam which is supported on more than two supports is called continuous beam.

4. ***Shear force*** at any cross-section of a beam is the unbalanced vertical force to the left or right of the section. Resultant moment of forces on the right or on the left of a section is called **bending moment**.
5. For a beam carrying transverse point loads or distributed loads, the rate of change of shear force is equal to the rate of loading $\left(\frac{dF}{dx} = w\right)$ and the rate of change of bending moment is equal to the shear force $\left(\frac{dM}{dx} = F\right)$ at a particular section.
6. The shear force
   (*i*) changes suddenly at a section where there is a vertical point load.
   (*ii*) remains constant between any two vertical point loads.
   (*iii*) has a straight line variation for a uniformly distributed load.
7. The bending moment
   (*i*) has parabolic variation for a uniformly distributed load
   (*ii*) is maximum at a section where shear force is zero after changing its sign.
8. The point of contraflexure or the point of inflexion is the point where the bending moment is zero after changing its sign.
9. At a section of the beam where couple is applied, the shear force remains unaltered but there occurs a sudden change in the bending moment.

REVIEW QUESTIONS

A. Conceptual and conventional questions :

1. Define a beam. What is a cantilever, a fixed beam and an overhang beam?
2. List the various types of loads to which a beam can be subjected.
3. Define shear force and bending moment.
4. What sign conventions are normally adopted while ploting the shear force and bending moment diagrams?
5. What are sagging and hogging moments.
6. Describe how you would proceed to draw shear force and bending moment in case of a:
   (*i*) simply supported beam loaded with a uniformly distributed load over its entire span
   (*ii*) concentrated load being applied at the mid span of a simply supported beam.
7. What is the point of contraflexture?
8. Draw the shear force and bending moment diagrams for the simply supported overhanging beam loaded as shown in Fig. 11.50. Also determine the location of the points of contraflexture, if any.

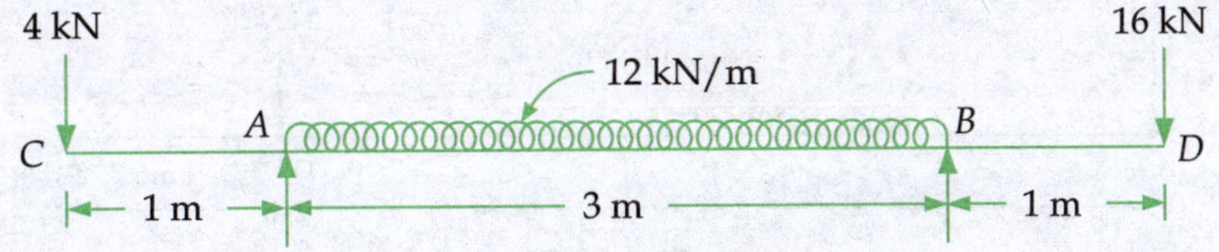

Fig. 11.50

(R_a = 18 kN, R_b = 38 kN, x = 1.33 m and 3.0 m from C)

9. Determine the reactions and construct the shear force and bending moment diagrams for the simply supported beam loaded as shown in Fig. 11.51. Also determine the position and magnitude of maximum bending moment.

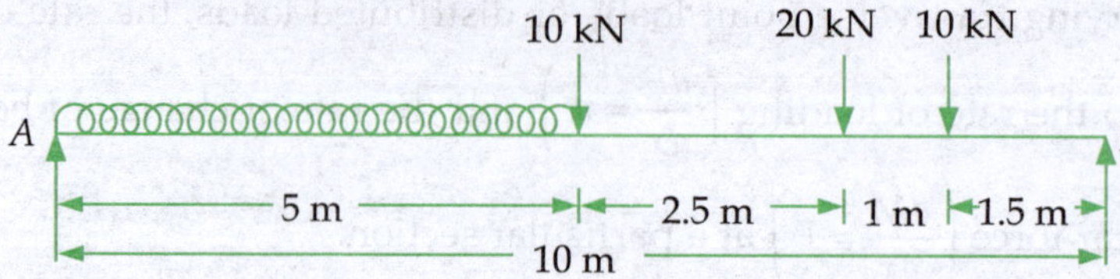

Fig. 11.51

(M_{max} = 120.25 kNm at 4.9 m from *A*)

10. For a symmetrically loaded overhang beam shown in Fig. 11.52, make calculations for the value of load *W* such that the bending moment becomes zero at the mid span of the beam.

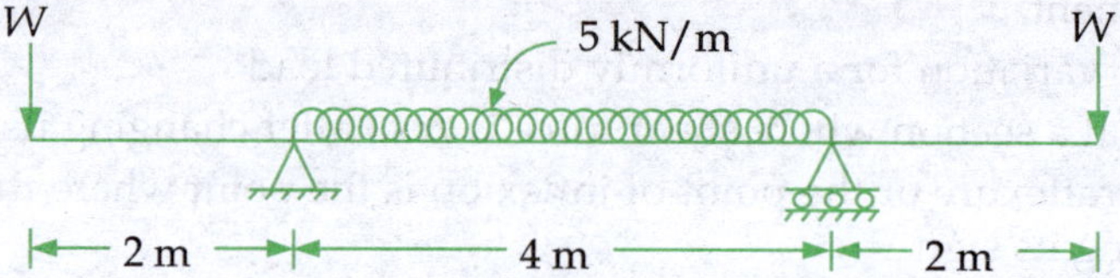

Fig. 11.52

(*W* = 5 kN)

11. Draw the shear force and bending moment diagrams for the beam loaded and supported as shown in Fig. 11.53.

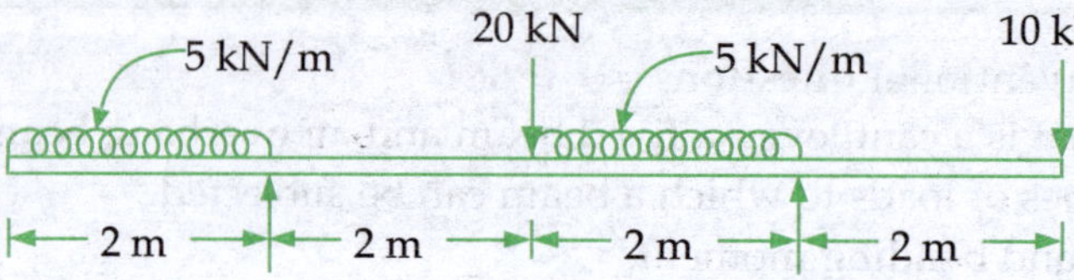

Fig. 11.53

12. Draw the shear force and bending moment diagram for the beam loaded as shown in Fig. 11.54.

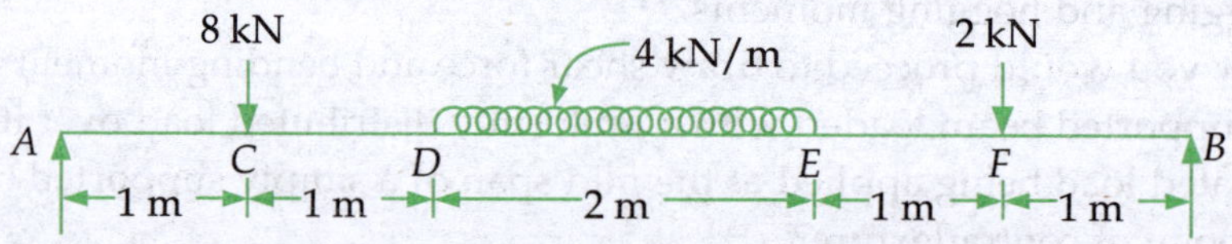

Fig. 11.54

13. Draw the shear force and bending moment diagrams for a cantilever beam loaded as shown in Fig. 11.55 given below

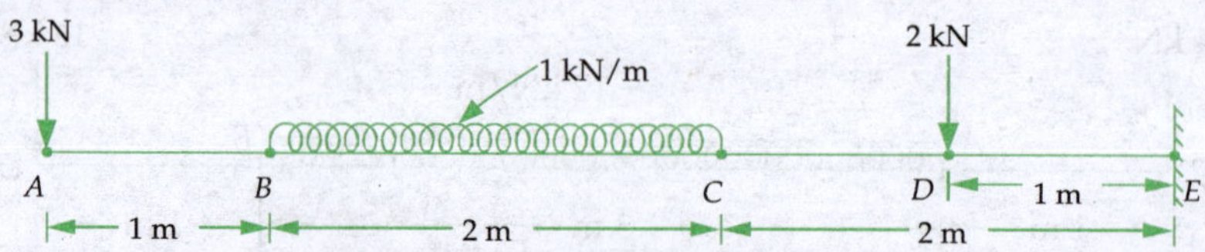

Fig. 11.55

Locate the position for maximum bending moment and determine its value.

B. Complete the following statements with most appropriate word/words :

1. Shear force at any point on a beam is the algebraic sum of .... on either side of the point.
2. The .... changes suddenly at a point where there is vertical point load.
3. In a beam where the shear force changes sign, the bending moment will be ....
4. The rate of change of bending moment dM/dx at any section is equal to .... at that section.
5. The shear force at certain section of a beam is stated to be zero. The bending momnent at that section will be ....
6. The bending moment for a certain section of the beam is constant. For that section, shear force will be ....
7. The shear force and bending moment diagrams for a cantilever beam carrying a concentrated load at free end will respectively be .... and ....
8. For a centilever beam carrying several concentrated loads, the shear force is maximum at the ....
9. A beam is said to be .... if it has more than two supports.
10. The point of centraflexure in a loaded beam refers to the section where the .... changes sign.
11. The point of contra-flexure occurs only in .... beams.
12. A simply supported beam having equal overhang lengths and carries equal concentrated loads at ends. Bending moment over length between the supports is ....

Answers :

1. loads 2. shear force 3. maximum 4. shear force 5. either minimum or maximum
6. zero 7. rectangle and triangle 8. support 9. continuous 10. bending moment
11. over hanging 12. non-zero constant

C. Multiple choice questions :

1. A simply supported beam of span l and carrying a load W concentrated at the mid span will have a maximum bending moment of

   (*a*) $\frac{Wl}{8}$ (*b*) $\frac{Wl}{4}$ (*c*) $\frac{Wl}{2}$ (*d*) Wl

2. For a beam of length l simply supported at its ends and carrying uniformly distributed load w per unit lenth, maximum bending moment occurs at the centre of beam and is given by

   (*a*) $\frac{wl^2}{2}$ (*b*) $\frac{wl^2}{4}$ (*c*) $\frac{wl^2}{8}$ (*d*) $\frac{wl^2}{16}$

3. A simply supported beam carries two equal concentrated loads W at a distance $l/3$ from either support. The value of maximum bending moment any where in the section will be

   (*a*) $\frac{Wl}{2}$ (*b*) $\frac{Wl}{3}$ (*c*) $\frac{Wl}{4}$ (*d*) $\frac{Wl}{6}$

4. If a simply supported beam carries a uniformly distributed load on the entire span, the shear force

   (*a*) has a maximum value at the mid span (*b*) has a linear variation along the span
   (*c*) is constant at all the sections (*d*) is not zero at any section of the beam

5. A simply supported beam carries several concentrated loads. The shear force would be maximum
(a) at mid span
(b) at either of the supports
(c) under the smallest load
(d) under the largest load

6. A concentrated load P acts on a simply supported beam of span l at a distance $\frac{l}{3}$ from the left support. The bending moment at the point of application of load is given by
(a) $\frac{Pl}{3}$
(b) $\frac{2Pl}{3}$
(c) $\frac{Pl}{9}$
(d) $\frac{2Pl}{9}$

7. If a beam is subjected to a constant bending moment along its length, then the shear force will
(a) also have a constant value every where along its length
(b) be zero at all sections along the beam
(c) be maximum at the centre and zero at the ends
(d) be zero at the centre and maximum at the ends

8. The shear force at certain section of a beam is stated as zero. The bending moment at the section will be
(a) minimum
(b) maximum
(c) either minimum or maximum
(d) zero
(e) no correlation exists between the two

9. The bending moment for a certain portion of the beam is constant. For that section, shear force would be
(a) zero
(b) increasing
(c) decreasing
(d) constant

10. A simply supported beam has equal overhanging lengths and carries equal concentrated loads at ends. Bending moment over the length between the supports
(a) is zero
(b) is a non-zero constant
(c) is maximum at mid span
(d) varies uniformly from one support to the other.

11. The maximum bending moment in a simply supported beam of length l loaded by a concentrated load W at the mid point is given by
(a) Wl
(b) $\frac{1}{2}Wl$
(c) $\frac{1}{4}Wl$
(d) $\frac{1}{8}Wl$

12. The point of contraflexture in a loaded beam refers to the section where
(a) the bending moment is maximum
(b) the shear force is maximum
(c) the shear force is zero
(d) the bending moment changes sign

13. Which of the following beams is likely to have the point of contraflexture ?
(a) cantilever beam
(b) simply supported beam
(c) beam with over hangs
(d) beam fixed at both ends

14. The shear force and bending moment diagrams for a cantilever beam carrying a concentrated load at free end will respectively be
(a) triangle and rectangle
(b) triangle and parabola
(c) rectangle and triangle
(d) rectangle and parabola

15. For the cantilever beam loaded as shown in figure given below :

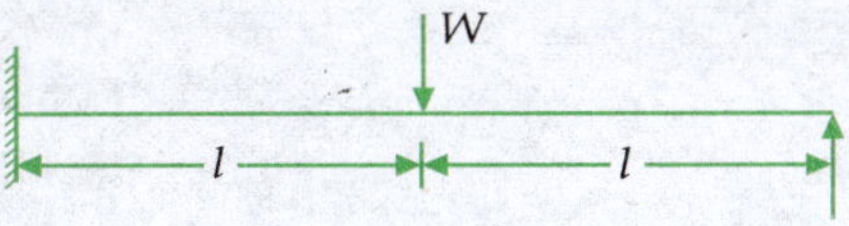

Fig. 11.56

The maximum bending moment will occur at

(a) mid span of beam *i.e.*, under the load
(b) the free end supported with a prop
(c) between the fixed end and the mid span
(d) the fixed end
(e) between mid span and the propped free end.

16. The figure given below shows the shear force diagram for the beam *ABCD*.
The bending moment in the portion *BC* of the beam

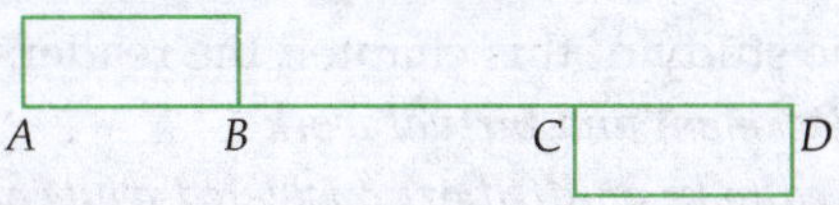

Fig. 11.57

(a) is non-zero constant
(b) is zero
(c) varies linearly from *B* to *C*
(d) varies parabolically from *B* to *C*

17. A 2 m long beam *BC* carries a single concentrated load at its mid span and is simply supported at its ends by two cantilevers *AB* = 1 m long and *CD* = 2 m long as shown in the figure given below

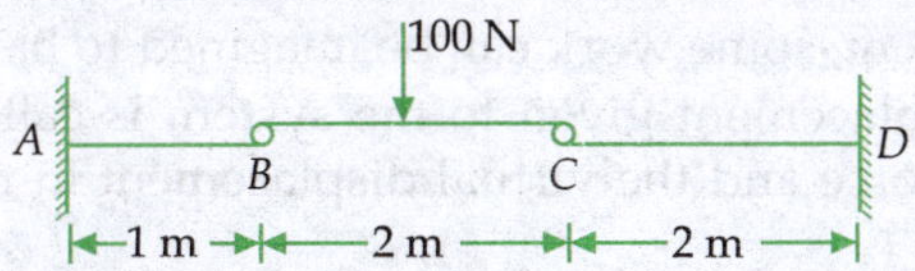

Fig. 11.58

The shear force at end *A* of the cantilever *AB* will be

(a) zero (b) 40 N (c) 50 N (d) 60 N

18. A simply supported beam is loaded as shown in the Fig. 11.59.

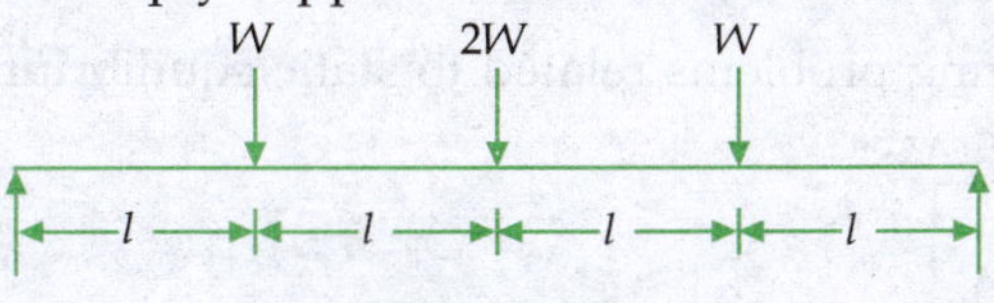

Fig. 11.59

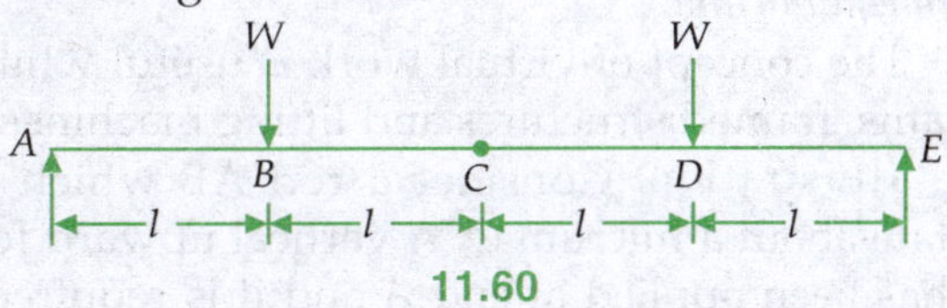

11.60

The maximum shear force in the beam will be

(a) zero (b) *W* (c) 2*W* (d) 4*W*

19. Consider a simply supported beam loaded as shown in the Fig. 11.60.
Which portion of this beam will be subjected to pure bending ?

(a) *AB* (b) *DE* (c) *AE* (d) *BD*

Answers :

1. (b)	2. (c)	3. (b)	4. (b)	5. (b)	6. (d)
7. (b)	8. (c)	9. (a)	10. (b)	11. (a)	12. (d)
13. (c)	14. (c)	15. (a)	16. (a)	17. (c)	18. (c)
19. (d)					

Virtual Work 12

Learning objectives : After the study of this chapter, the readers will be able to :

- *define the terms virtual displacement and virtual work*
- *understand the forces that need to be omitted or considered while evaluating the virtual work*
- *express the various virtual displacements in terms of a single virtual displacement*
- *write the virtual-work equation and solve it for unknown force, couple etc.*

When a system or a body is in equilibrium under the action of a set of forces, there is no displacement and accordingly no work in done. However if the system is imagined to undergo any arbitrary but small displacement, some work can be imagined to have been done.

The imaginary small displacement given to the system is called virtual displacement. The product of the net resultant force and the virtual displacement in the direction of force is called virtual work.

The principle of virtual work may be stated as: "*The work done on a rigid body or a system of rigid bodies in equilibrium, when acted upon a set of forces, is zero for any virtual displacement compatible with the constraints on the system.*"

Conversely: "*If the algebraic sum of the virtual work for every virtual displacement is zero, the system is in equilibrium.*"

The concept of virtual work is useful while solving problems related to static equilibrium of beams, framed structures and lifting machines.

Illustration: Consider a rod AB which can rotate about a fulcrum O. A vertical upward force P has been applied at end A and it is required to find the vertical downward force that need to be applied at end B to keep the rod horizontal.

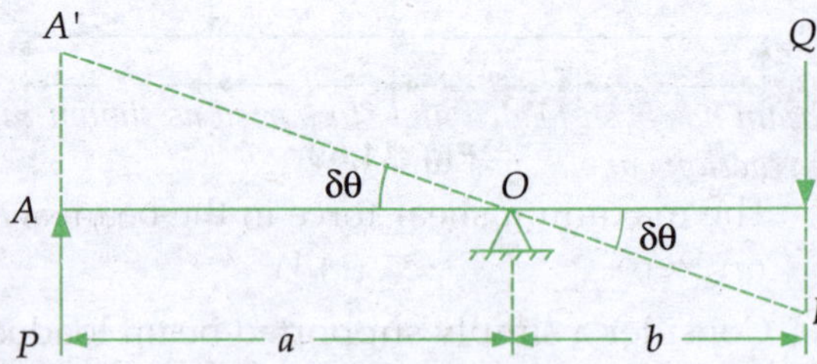

Fig. 12.1

Let it be imagined that the rod rotates through an angle $\delta\theta$ and takes the position $A'B'$. The displacement of A is AA' and that of B is BB'. For small angular rotation of the rod, these displacements can be taken perpendicular to AB and accordingly

$$AA' = a\,\delta\theta \quad \text{and} \quad BB' = b\,\delta\theta$$

Work done by force $P = P \times a\,\delta\theta$

Work done by force $Q = -Q \times b\,\delta\theta$

Total work $= Pa\,\delta\theta - Qb\,\delta\theta$

According to the principle of virtual work, the total work done has to be zero.

$$\therefore \quad Pa\,\delta\theta - Qb\,\delta\theta = 0$$

$$\text{That gives: } Q = P \times \frac{a}{b}$$

The following forces do no work and are omitted while applying the principle of virtual work:

(*i*) tension in a light inextensible string

(*ii*) forces normal to the direction of displacement

(*iii*) reaction between smooth bodies; this is because such a reaction is perpendicular to the direction of displacement

(*iv*) reaction at smooth pins and hinges which do not move

(*v*) action and reaction between two bodies whose equilibrium is being considered together

(*vi*) reaction at a point or an axis fixed in space around which the body is constrained to turn.

Further, the following sign conventions are adopted while writing the expressions for virtual work:

1. Upward force, forces acting towards right and the forces in the clockwise direction are considered positive.
2. Downward forces, forces acting towards left and the forces in the anticlockwise direction are taken negative.

The virtual work becomes positive when the displacement is in the same direction as the force and negative when displacement is opposite to the direction of force.

Furthermore, it is generally assumed that there is no friction in hinges, bearings or sliding surfaces.

Virtual work versus real work

(*i*) The displacements in virtual work (VW) are small and imaginary.

The displacements in real work (RW) are neither imaginary nor kept deliberately small.

(*ii*) The calculations for certain unknown forces are made by equating the sum of VW by all forces to zero. However, the sum of real work is generally not equated to zero.

(*iii*) The concept of VW is applied for making an analysis of the structures in equilibrium.

The application of RW pertains to the situations when there is movement of a body or structure.

EXAMPLE 12.1

A beam has been loaded and supported as shown in Fig. 12.2. Use the method of virtual work to determine the reactions at the end supports.

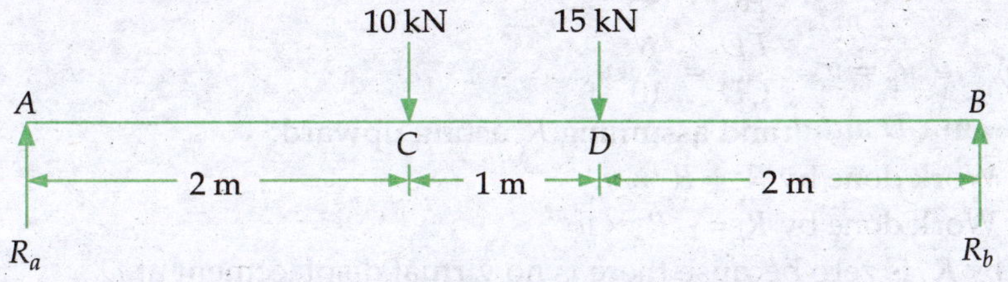

Fig. 12.2

Solution: Let the beam be given a vertical virtual displacement δ_b at end B holding the end A in position.

From the similarity of triangles

$$\frac{\delta_c}{2} = \frac{\delta_d}{3} = \frac{\delta_b}{5}$$

$$\therefore \quad \delta_c = 0.4\,\delta_b \quad \text{and} \quad \delta_d = 0.6\,\delta_b.$$

$$\text{Total work} = R_b \times \delta_b - 15 \times \delta_d - 10 \times \delta_c + R_a \times 0$$

While writing the above expression, the upward forces have been considered positive, and the down forces as negative.

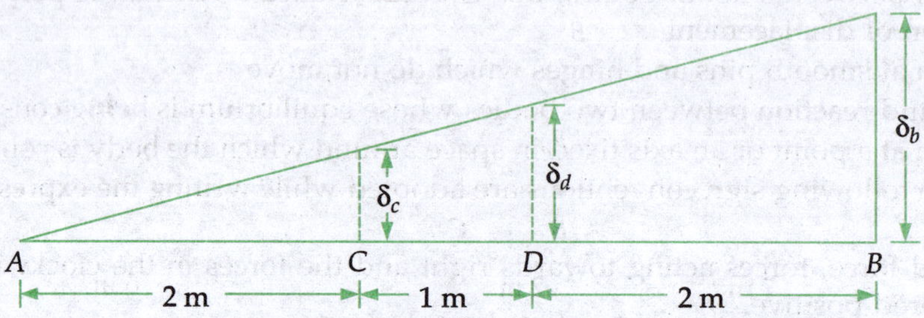

Fig. 12.3

According to the principle of virtual work, the total work done by all the forces must be zero. That gives

$$R_b \times \delta_b - 15 \times \delta_d - 10 \times \delta_c + R_a \times 0 = 0$$

or $\quad R_b \times \delta_b - 15 \times 0.6\,\delta_b - 10 \times 0.4\,\delta_b = 0$

or $\quad R_b - 9 - 4 = 0\,;\ R_b =$ **13 kN**

$$R_a = (10 + 15) - 13 = \textbf{12 kN}$$

EXAMPLE 12.2

Two beams AC and CD of lengths 9 m and 10 m respectively are hinged at C. These are supported on rollers on the ends A and D, and a hinged support at B which lies at distance 7 m from A. The beam is acted upon by a concentrated load of 700 N at point E located 6 m from end A. Determine the reactions at the hinge point and at the support B. Use the principle of virtual work.

Solution: Refer Fig. 12.4 for the given system of beams.

Let the system be given a vertical upward virtual displacement y_c at the point C holding the ends *A* and *D* in position.

From the geometrical configuration of the system after virtual displacement,

$$y_b = y_c \times \frac{AB}{AC} = \frac{7}{9} y_c$$

$$y_e = y_c \times \frac{ED}{CD} = \frac{6}{10} y_c$$

Consider beam *CD* alone and assuming R_c acting upward.

Work done by $R_c = R_c \times y_c$

Work done by $R_e = -R_e \times y_e$

Work done by R_d is zero because there is no virtual displacement at *D*

$\therefore \quad$ Total work $= R_c y_c - R_e y_e$

Using the principle of virtual work,

$$R_c y_c - R_e y_e = 0 \quad \text{or} \quad R_c y_c - 700 \times \frac{6}{10} y_c = 0\,; \quad R_c = 420 \text{ kN}$$

The positive sign with R_c suggests that the assumption of R_c acting upward on beam CD is correct.

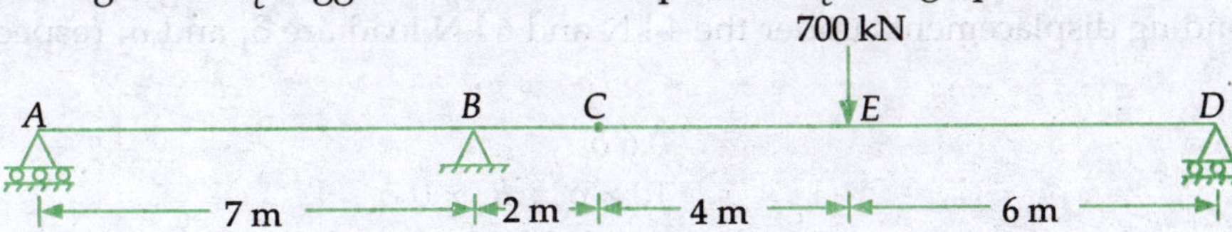

Fig. 12.4

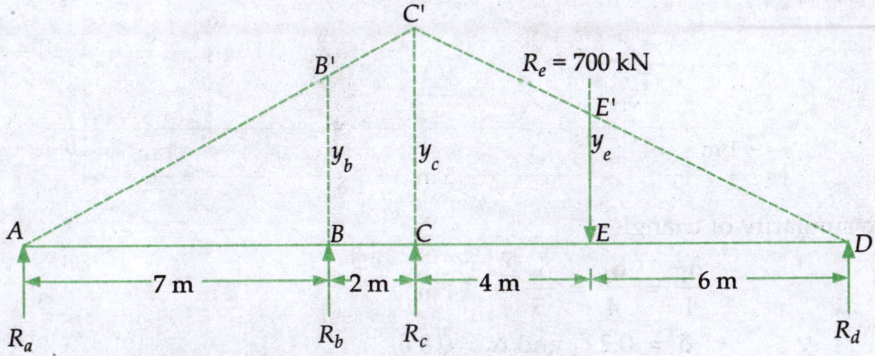

Fig. 12.5

Consider beam AC alone:

Since R_c is upward on CD, it must be downward on AC

Work done by $R_b = R_b \times y_b$

Work done by $R_c = - R_c \times y_c$

Work done by R_a is zero since there is no virtual displacement at A

$\therefore$ Total work $= R_b y_b - R_c y_c$

Using the principle of work,

$$R_b y_b - R_c y_c = 0 \quad \text{or} \quad R_b \times \frac{7}{9} y_c - 420 \times y_c = 0\,; \quad R_b = 540 \text{ kN}$$

EXAMPLE 12.3

A beam of span 5 m is supported at A and B and is subjected to a load system shown in the Fig. 12.6. Using the principle of virtual work, find the reactions at A and B.

Solution: The given load of 2 kN/m is uniformly distributed for 2 m length between AC. Its equivalent load is $2 \times 2 = 4$ kN and it acts at its centre of gravity at a distance of 1 m form end A. Likewise the equivalent load corresponding to *u dl* of 3 kN/m between DB is 6 kN and it acts at 1 m distance from end B.

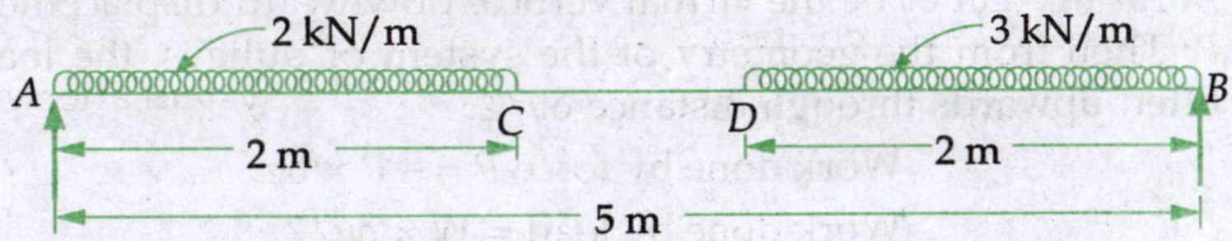

Fig. 12.6

The equivalent load diagram for the given beam is as then shown below:

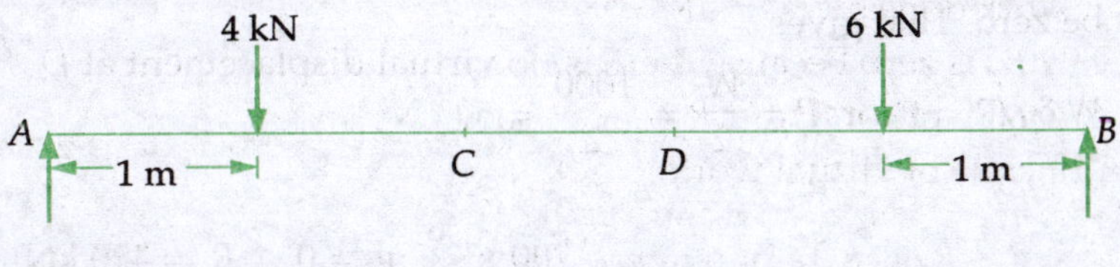

Let the beam be given a vertical virtual displacement δ_b at end B holding the end A in posi. The corresponding displacements under the 4 kN and 6 kN load are δ_1 and δ_2 respectively.

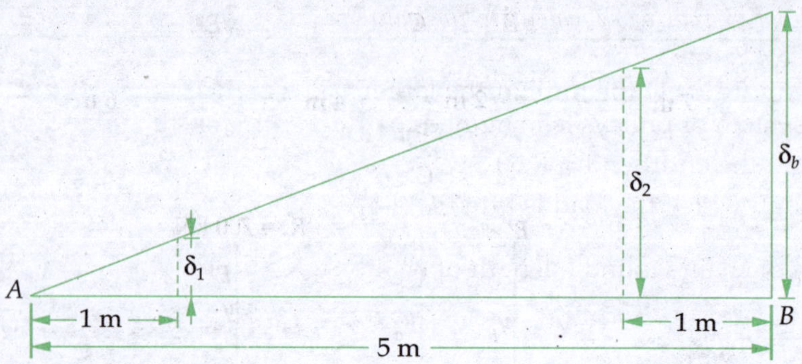

From the similarity of triangles

$$\frac{\delta_1}{1} = \frac{\delta_2}{4} = \frac{\delta_b}{5}$$

$$\therefore \quad \delta_1 = 0.2\,\delta_b \text{ and } \delta_2 = 0.8\,\delta_b$$

$$\text{Total work} = R_b \times \delta_b - 6\,\delta_2 - 4\,\delta_1 + R_a \times 0$$

While writing the above expression, the upward forces have been taken positive, and downward forces as negative.

According to the principle of virtual work, the total work done by all the forces is zero. That gives

$$R_b \times \delta_b - 6\,\delta_2 - 4\,\delta_1 + R_a \times 0 = 0$$

$$\text{or} \quad R_b \times \delta_b - 6 \times 0.8\,\delta_b - 4 \times 0.2\,\delta_b = 0$$

$$\text{or} \quad R_b - 4.8 - 0.8 = 0; \quad R_b = \mathbf{5.6\ kN}$$

$$\text{and} \quad R_a = (4 + 6) - 5.6 = \mathbf{4.4\ kN}$$

EXAMPLE 12.4

A weight W of 1000 N is to be raised by a system of pulleys as shown in Fig. 12.7. Using the principle of virtual work, find the value of force P which can hold the system in equilibrium.

Solution: Let δy be the virtual vertical downward displacement of force P. Then from the geometry of the system of pulleys, the load will get lifted upwards through distance $\delta y/2$.

$$\text{Work done by force } P = -P \times \delta y$$

$$\text{Work done by load} = W \times \delta y/2$$

While writing the above expressions, the upward displacements have been taken as positive and the down ward displacements as negative.

$$\text{Total work} = -P\,\delta y + W\,\delta y/2$$

According to the principle of virtual work, the total work done by all the forces must be zero. That gives

$$-P\,\delta y + W\,\delta y/2 = 0 \text{ or } P = \frac{W}{2} = \frac{1000}{2} = \mathbf{500\ N}$$

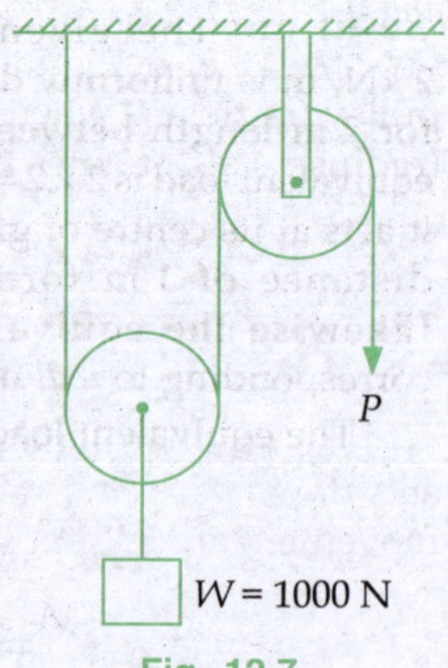

Fig. 12.7

EXAMPLE 12.5

The diameters of the two pulleys of a differential pulley block (Fig. 12.8) are 30 cm and 25 cm respectively. Using the principle of virtual work, calculate the value of effort P to lift a load of 3000 N. Neglect any frictional effects.

Solution: Let R and r denote the radii of the larger and smaller pulleys of the differential pulley block. When the system of pulleys is imagined to undergo a small angular displacement $\delta\theta$, then

displacement at the effort end = $R\,\delta\theta$

distance through which the load is lifted

$$= \frac{1}{2}\text{ (length of chain wound – length of chain unwound)}$$

$$= \frac{1}{2}(R\,\delta\theta - r\,\delta\theta)$$

$$= \frac{1}{2}(R - r)\,\delta\theta$$

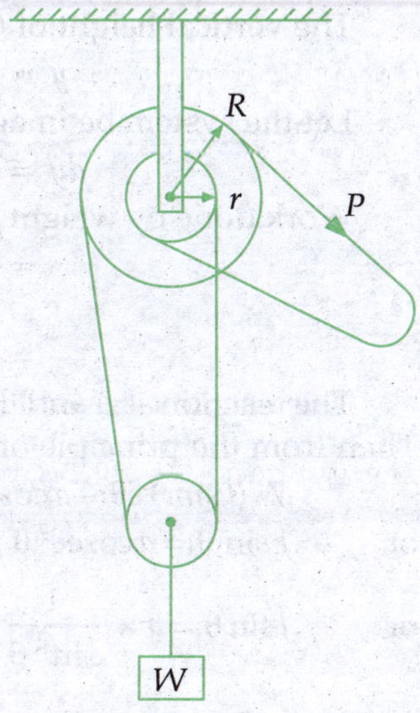

Fig. 12.8

That gives:

work done during downward movement of effort = $-P \times R\,\delta\theta$

work done by upward movement of load = $W \times \frac{1}{2}(R - r)\,\delta\theta$

Algebraic sum of the work done = $W \times \frac{1}{2}(R - r)\,\delta\theta - P \times R\,\delta\theta$

According to the principle of virtual work,

$$W \times \frac{1}{2}(R - r)\,\delta\theta - P \times R\,\delta\theta = 0$$

$$\therefore \quad P = \frac{1}{2}W\frac{R - r}{R} = \frac{1}{2} \times 3000 \times \frac{15 - 12.5}{15} = 250 \text{ N}$$

EXAMPLE 12.6

The pulley arrangement shown in Fig. 12.9 is used for hoisting a load W. Find the applied force P in terms of load W for equilibrium of the system. The radii of the two steps of the pulley are r_1 and r_2. Neglect friction and use the virtual work principle.

Solution: Let $\delta\theta$ be the small virtual angular displacement of the larger pulley (radius r_2) when pull P is applied. By this action, the smaller pulley (radius r_1) also rotates by angle $\delta\theta$. Then

$$\text{displacement of } W = r_1\,\delta\theta$$

$$\text{displacement of } P = r_2\,\delta\theta - r_1\,\delta\theta = (r_2 - r_1)\,\delta\theta$$

By using the principle of virtual work: $\Sigma W = 0$

$$P(r_2 - r_1)\,\delta\theta - W r_1\,\delta\theta = 0$$

The negative sign with the second term is due to the fact that weight acts downward and its lift is upwards (weight force and direction of its movement are in opposite directions)

$$P = \frac{W r_1\,\delta\theta}{(r_2 - r_1)\delta\theta} = W\frac{r_1}{r_2 - r_1}$$

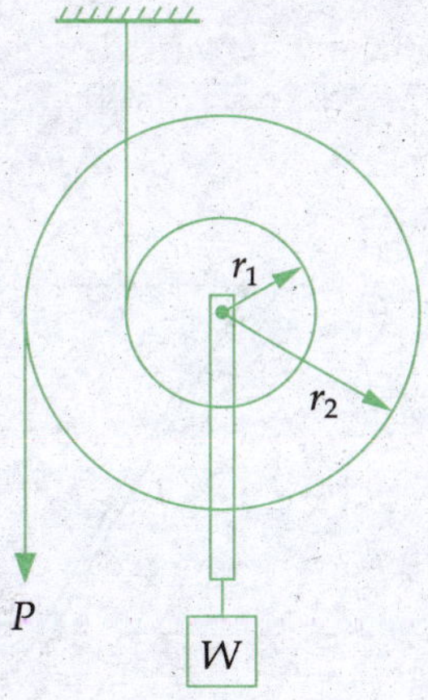

Fig. 12.9

EXAMPLE 12.7

A uniform rod of length 2 l rests in equilibrium against a smooth vertical wall and upon a smooth peg P at distance a from the wall (Fig. 12.10). Show that, in the position of equilibrium, the inclination of rod with the wall should be $\theta = \sin^{-1}\left(\frac{a}{l}\right)^{1/3}$.

Solution: Let P be the peg and G be the centre of gravity CG of the rod. The weight W of the rod acts vertically downwards at its CG.

The vertical height of G above the peg is given by

$$y = AG\cos\theta - a\cot\theta = l\cos\theta - a\cot\theta$$

Let the system be imagined to rotate through small angle $\delta\theta$ so that θ becomes $(\theta + \delta\theta)$.

$$dy = -l\sin\theta\,\delta\theta + a\,\text{cosec}^2\theta\,\delta\theta$$

Work done by weight W of the rod,

$$= -W\,dy \qquad \text{(– ve sign due to load acting downwards)}$$

$$= W(l\sin\theta\,\delta\theta - a\,\text{cosec}^2\theta\,\delta\theta)$$

The reaction at A and that at P do not contribute towards work.

Then from the principle of virtual work

$$W(l\sin\theta\,\delta\theta - a\,\text{cosec}^2\theta\,\delta\theta) = 0$$

or $\quad l\sin\theta = a\,\text{cosec}^2\theta$;

or $\quad l\sin\theta = a \times \dfrac{1}{\sin^2\theta}$; $\quad \sin^3\theta = \dfrac{a}{l}\dfrac{1}{\sin\theta}$

$$\therefore \quad \theta = \sin^{-1}\left(\frac{a}{l}\right)^{1/3}$$

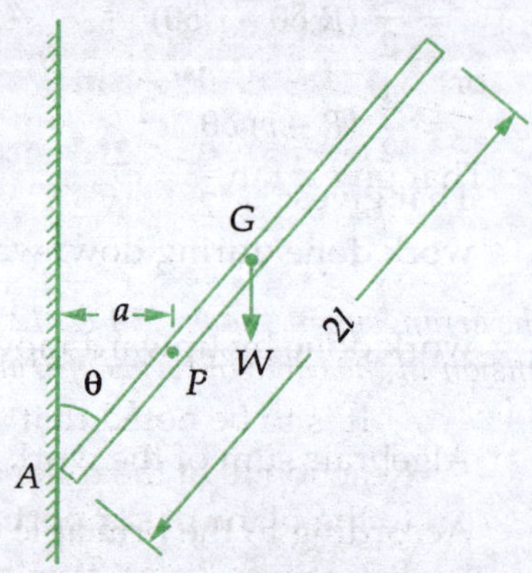

Fig. 12.10

EXAMPLE 12.8

Each of the two uniform hinged bars has a weight W and length l, and is supported and loaded as shown in Fig. 12.11. Using the principle of virtual work, determine the angle θ for a given force P.

Solution: Taking A as origin (Fig. 12.12), the position and displacement at points D and E are

$$y_d = -\frac{l}{2}\cos\frac{\theta}{2} : \delta y_d = \frac{l}{4}\sin\frac{\theta}{2}\,\delta\theta$$

$$y_e = -\frac{l}{2}\cos\frac{\theta}{2} : \delta y_e = \frac{l}{4}\sin\frac{\theta}{2}\,\delta\theta$$

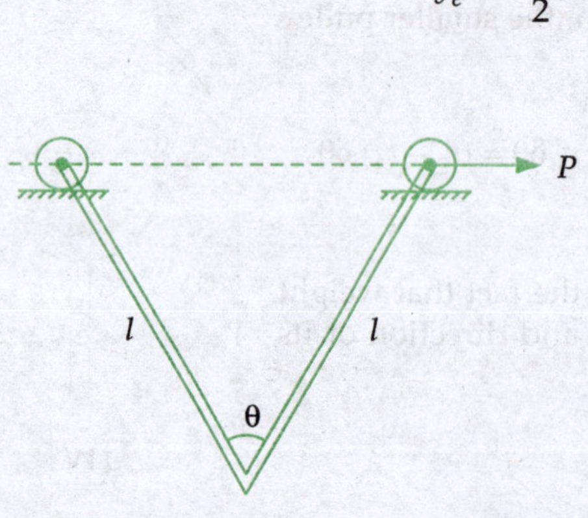

Fig. 12.11

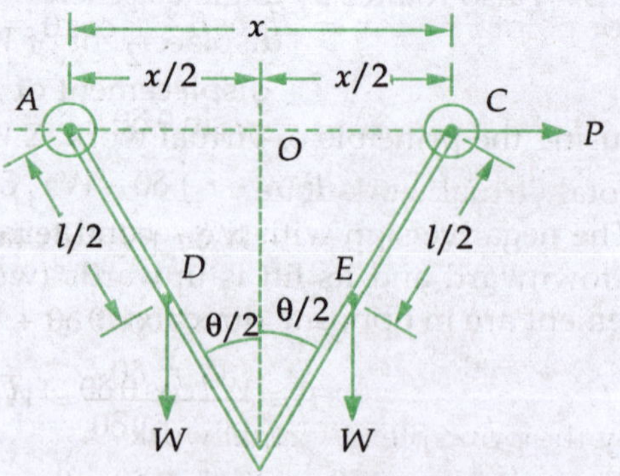

Fig. 12.12

D and E are the points of centre of gravity of the bars.

$$y_c = AC = 2l \sin\frac{\theta}{2}$$

$$\delta y_c = l\cos\frac{\theta}{2}\,\delta\theta$$

The virtual displacement δy_c of point C is in the direction of force and accordingly the virtual work done by force P will be positive. Further, when the point C moves towards right, the points D and E move will upwards *i.e.*, opposite to the direction of loads and as such their work will be negative. Then by the principle of virtual work.

$$-W\,\delta y_d - W\,\delta y_e + P\,\delta y_c = 0$$

or
$$-W\times\frac{l}{4}\sin\frac{\theta}{2}\,\delta\theta - W\times\frac{l}{4}\sin\frac{\theta}{2}\delta\theta + p\times l\frac{\cos\theta}{2}\delta\theta$$

or
$$-W\frac{l}{2}\sin\frac{\theta}{2} = Pl\cos\frac{\theta}{2}$$

That gives: $\tan\frac{\theta}{2} = \frac{2l}{W}$

EXAMPLE 12.9

The arrangement shown in Fig. 12.13 is required to remain in state of equilibrium. Set up an expression for tension in the cable in terms of θ and load W. Use the method of virtual work.

Solution: It is to be noted that

(*i*) Reaction R_b at the roller support is perpendicular to the plane of support.

(*ii*) Reaction R_a at the hinged support has a component R_{ah} in the horizontal direction and a component R_{av} in the vertical direction.

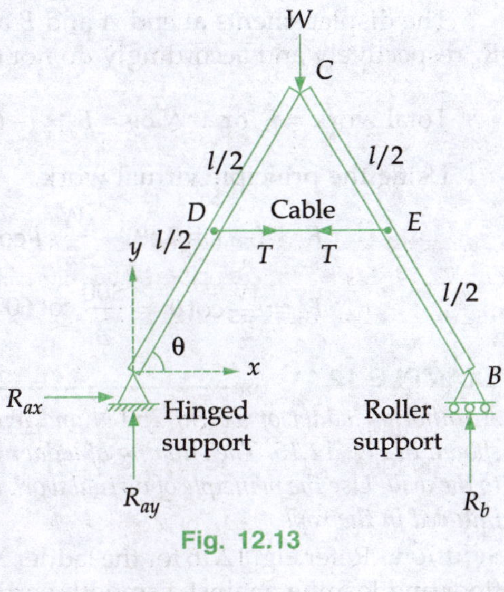

Fig. 12.13

Since the hinged support has no translatory motion and there is no virtual displacement in the direction of reaction R_b, the reactions at the supports will have no contribution towards work. As such the work will be done only by the forces acting at points C, D and E.

For point C : $y = l\sin\theta$; $\delta y = l\cos\theta\,\delta\theta$

For point D : $x = \frac{l}{2}\cos\theta$; $\delta x = -\frac{l}{2}\sin\theta\,\delta\theta$

For point E : $x = 2l\cos\theta - \frac{l}{2}\cos\theta = \frac{3}{2}l\cos\theta$

$$\delta x = -\frac{3}{2}l\sin\theta\,\delta\theta$$

Total virtual work done

$$= -W\,\delta y + T\times\delta x \text{ (at D)} - T\times\delta x \text{ (at E)}$$

$$= -W\times l\cos\theta\,\delta\theta + T\left(-\frac{l}{2}\sin\theta\right)\delta\theta - T\left(-\frac{3l}{2}\sin\theta\right)\delta\theta$$

$$= -W\,l\cos\theta\,\delta\theta + T\,l\sin\theta\,\delta\theta$$

By the principle of virtual work

$$-W\,l\cos\theta\,\delta\theta + T\,l\sin\theta\,\delta\theta = 0$$

$$W\cos\theta = T\sin\theta\,;\ T = W\cot\theta$$

EXAMPLE 12.10

A uniform ladder of 300 N weight rests against a smooth vertical wall and a rough horizontal floor making an angle of 60° with the horizontal. Use the method of virtual work to find the frictional force between the foot of ladder and the rough horizontal floor.

Solution: Refer Fig. 12.14 for the ladder *AB* leaning against a wall and various forces acting on it.

If l is the length of the ladder, then height of *G* above base line *AC*,

$$y = AG \sin\theta = \frac{l}{2}\sin\theta$$

Further, let $x = AC = l\cos\theta$

When the end A moves leftwards through very small distance δx, the weight of ladder goes downwards through a small distance δy.

$$\delta x = -\,l\sin\theta\,\delta\theta \quad \text{and} \quad \delta y = \frac{l}{2}\cos\theta\,\delta\theta$$

With end A moving left wards, the friction force F_a would act towards right as shown.

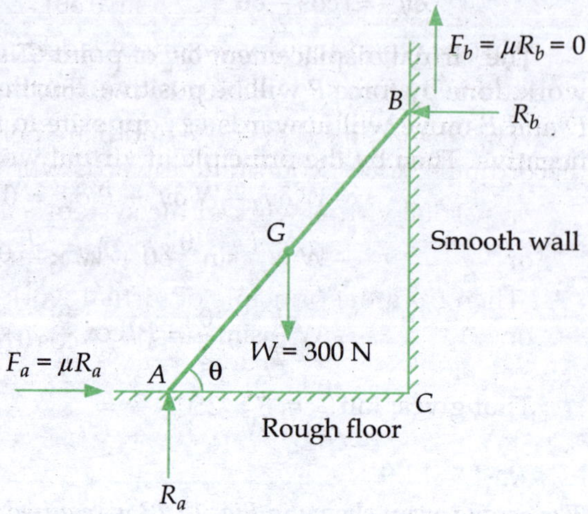

Fig. 12.14

Work done by frictional force at $A = F_a \times \delta x$

Work done by weight of ladder $= W \times \delta y$

The displacements at end *A* and *B* are perpen-dicular to the line of action of reactions R_a and R_b respectively and accordingly do not contribute towards work.

$$\text{Total work} = F_a\,\delta x + W\,\delta y = F_a \times (-\,l\sin\theta\,\delta\theta) + W \times \frac{l}{2}\cos\theta\,\delta\theta$$

Using the principle virtual work,

$$F_a \times (-\,l\sin\theta\,\delta\theta) + \frac{W}{2}\,l\cos\theta\,d\theta = 0$$

$$\therefore \quad F_a = \frac{W}{2}\cot\theta = \frac{300}{2}\cot 60^\circ = \mathbf{86.6\ N}$$

EXAMPLE 12.11

A uniform ladder of weight 250 N and length 6 m rests on a smooth floor and against a smooth wall as shown in Fig. 12.15. The slipping of ladder is prevented by connecting a horizontal rope from its mid point to the wall. Use the principle of virtual work to calculate the tension induced in the rope.

Solution: Refer Fig. 12.15 for the ladder *AB* resting on smooth floor and leaning against a smooth vertical wall.

Length of the ladder

$$AB = \sqrt{(AC)^2 + (BC)^2} = \sqrt{3^2 + 4^2} = 5\text{ m}$$

The horizontal rope *DE* joins the mid points of ladder (*D*) and wall (*E*). The weight of the ladder acts at point *D* which is its centre of gravity.

Let the co-ordinates of point *D* be (x, y) and the ladder be given a small displacement such that its lower end *A* moves towards left.

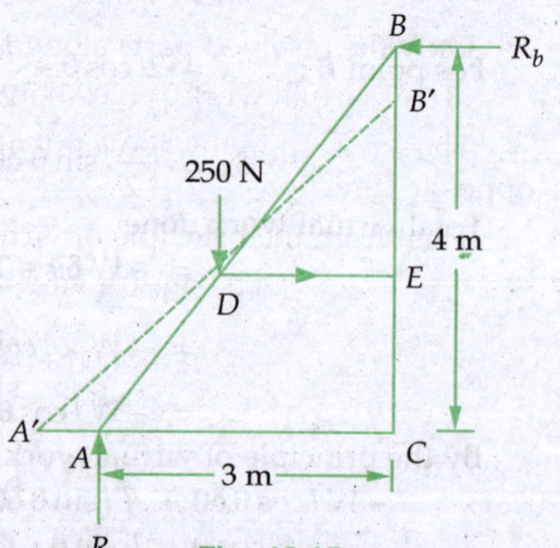

Fig. 12.15

$$x = AD\cos\theta = 2.5\cos\theta\,; \quad \delta x = -2.5\sin\theta\,\delta\theta$$

$$y = AD\sin\theta = 2.5\sin\theta\,; \quad \delta y = 2.5\cos\theta\,\delta\theta$$

Virtual work done by tension T in the rope

$$= T(-2.5\sin\theta\,\delta\theta)$$

Virtual work done by weight of ladder (250 N)

$$= 250\,(2.5\cos\theta\,\delta\theta)$$

The displacement at end A and B are perpendicular to line of action of reactions R_a and R_b and accordingly do not contribute towards work.

$\therefore$ Total virtual work of the system

$$= T(-2.5\sin\theta\,\delta\theta) + 250(2.5\cos\theta\,\delta\theta)$$

Then from the principle of virtual work

$$T(-2.5\sin\theta\,\delta\theta) + 250\,(2.5\cos\theta\,\delta\theta) = 0$$

$$T = 250\cot\theta = 250 \times \frac{3}{4} = \mathbf{187.5\ N}$$

EXAMPLE 12.12

A block of weight 2000 N rests on a smooth inclined plane that makes an angle of 30° with the horizontal. This block is supported by load P lying on another smooth plane of inclination 60° as shown in Fig. 12.16. The block and the load have been connected by an inelastic string. Determine the value of load P by using the method of virtual work.

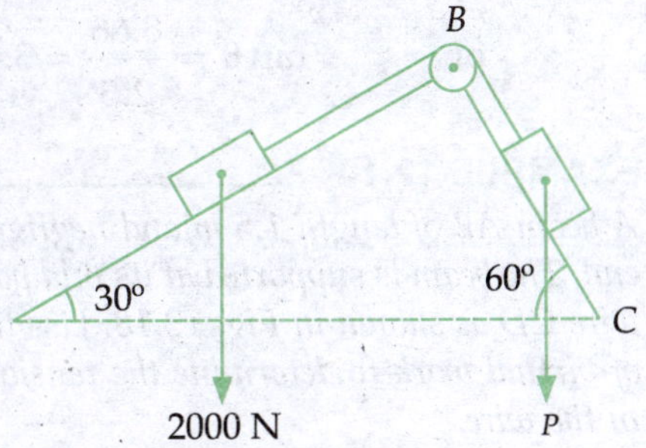

Fig. 12.16

Solution: It is to be noted that:

(*i*) Small displacement δx of 2000 N parallel to plane AB equals the displacement δy of load P along the plane BC.

(*ii*) When the load 2000 N moves down the plane, the load P moves up the plane.

(*iii*) The resolved parts of 2000 N parallel and perpendicular to plane AB are:

$$2000\sin 30^\circ \quad \text{and} \quad 2000\cos 30^\circ$$

The resolved parts of P parallel and perpendicular to plane BC are:

$$P\sin 60^\circ \quad \text{and} \quad P\sin 60^\circ$$

Only the resolved parts along the planes will contribute towards work.

Total virtual work $= (2000\sin 30^\circ) \times \delta x - (P\sin 60^\circ) \times \delta y$

The – *ve* sign in the above identity stems from the fact that movement of the two loads are in opposite directions.

Then from the principle of virtual work,

$$(2000\sin 30^\circ) \times \delta x - (P\sin 60^\circ) \times \delta y = 0$$

$$\therefore \quad P = 2000 \times \frac{\sin 30^\circ}{\sin 60^\circ}$$

$$= 2000 \times \frac{0.5}{0.866} = \mathbf{1154.73\ N}$$

EXAMPLE 12.13

Two spheres P and Q of weights 5 N and 10 N respectively can slide freely along the bars AC and BC which are arranged as shown in Fig. 12.17. An extensible string DE connects these spheres. Calculate the value of θ which defines the position of equilibrium. Use the principle of virtual work.

Solution: If l is the length of string ; then

$$DE = l\cos\theta\,; \qquad CE = l\sin\theta$$

Accordingly when a virtual displacement $d\theta$ is given to the string DE, we have

$$\delta_{DE} = -\,l\sin\theta\,\delta\theta \quad \text{and} \quad \delta_{CE} = l\cos\theta\,\delta\theta$$

The resolved parts of the weights along the planes are

$$5\sin 30^\circ \quad \text{and} \quad 10\sin 60^\circ$$

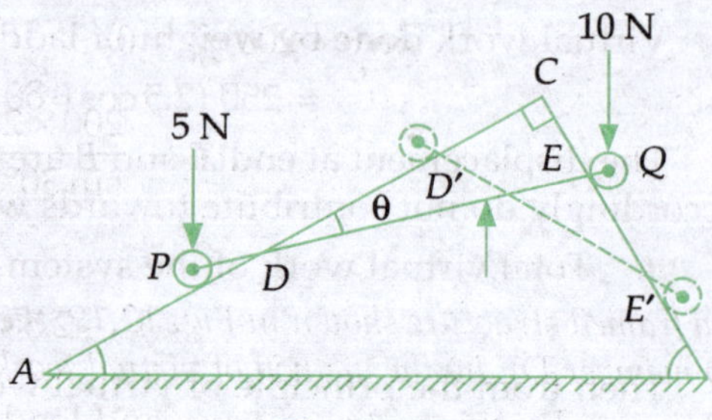

Fig. 12.17

Only these resolved parts contribute towards work.

Work done at point $D = (5\sin 30^\circ) \times \delta_{DE}$

$= 5\sin 30^\circ \times (-\,l\sin\theta\,\delta\theta) = -\,2.5\,l\sin\theta\,\delta\theta$

Work done at point $E = 10\sin 60^\circ \times l\cos\theta\,\delta\theta = 8.66\,l\cos\theta\,\delta\theta$

Total virtual work $= -\,2.5\,l\sin\theta\,\delta\theta + 8.66l\cos\theta\,\delta\theta$

Then from the principle of virtual work,

$$-\,2.5\,l\sin\theta\,\delta\theta + 8.66\,l\cos\theta\,\delta\theta = 0$$

or $\quad \tan\theta = \dfrac{8.66}{2.5} = 3.464 \qquad \therefore\ \theta = \tan^{-1}(3.464) = 73.9^\circ$

EXAMPLE 12.14

A beam AB of length 1.5 m and negligible weight is hinged at end A and carries a 200 N weight at its free end. The beam is supported at its mid point C by a wire CD as shown in Fig. 12.18. Use the method of virtual work to determine the tension induced in the wire.

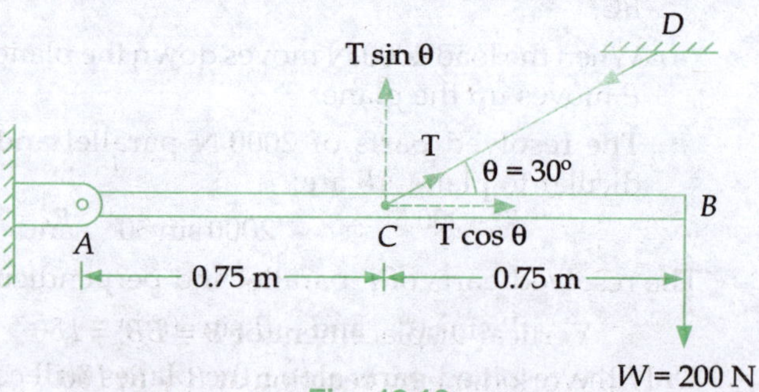

Fig. 12.18

Solution: Let the beam be given a vertical virtual displacement y_b at end B holding the hinged end A in position. The corresponding displacement at C would be

$$y_c = y_b \times \frac{AC}{AB} = y_b \times \frac{0.75}{1.5} = \frac{y_b}{2}$$

If T is the tension in the string, then its vertical and horizontal components are

$$T_v = T\sin\theta \quad \text{and} \quad T_h = T\cos\theta$$

Work done by load W at end $B = -\,W \times y_b = -\,200\,y_b$

(– ve sign due to load acting downwards)

Work done by component T_v of tension in the string

$$= T\sin\theta \times y_c = T\sin\theta \times \frac{y_b}{2} = \frac{T\,y_b}{2}\sin\theta$$

The horizontal component $T\cos\theta$ will have no contribution towards work done because its line of action is normal to the direction of displacement.

Algebraic sum of work done = $\frac{T\,y_b}{2}\sin\theta - 200\,y_b$

According to the principle of virtual work, the total work done by all the forces must be zero. That gives

$$\frac{T\,y_b}{2}\sin\theta - 200\,y_b = 0$$

$$T = \frac{200\times 2}{\sin 30^\circ} = \frac{200\times 2}{0.5} = 800\text{ N}$$

EXAMPLE 12.15

In a framed structure shown in Fig. 12.19, the length of all the members is equal. Determine the force in the top member DE by the method of virtual work.

Solution: By symmetry of truss and loading

$R_a = R_b = 200$ N

The member DE is cut and replaced by an axial compressive force P in it. The system is then made movable by imagining the portion BCE to rotate through an angle $\delta\theta$. Then B gets displaced to B' and E to E'.

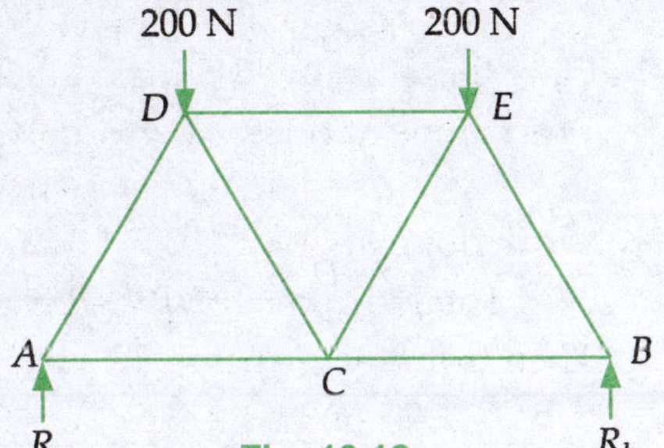

Fig. 12.19

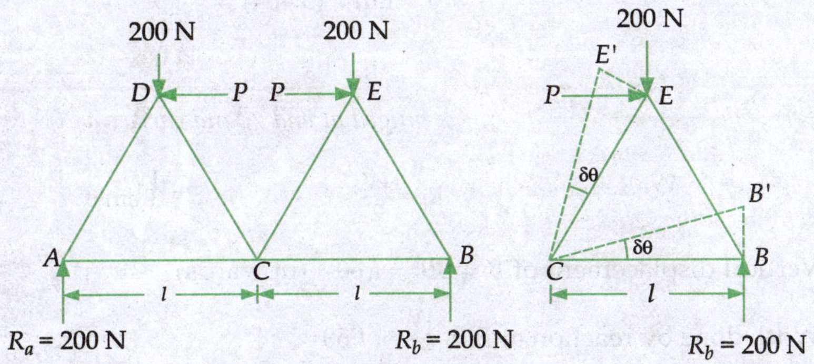

Fig. 12.20

Vertical displacement of $B = BB' = l\,\delta\theta$

Work done by reaction at $B = R_b\,l\,\delta\theta = 200\,l\,\delta\theta$

The horizontal and vertical displacements at E are:

$EE'\sin 60^\circ = l\,\delta\theta \times 0.866$ (towards left)

and $EE'\cos 60^\circ = l\,\delta\theta \times 0.5$ (upwards)

Work done by axial force $P = -P\,l\,\delta\theta \times 0.866$

Work done by load of 200 N at $E = -200\,l\,\delta\theta \times 0.5$

Total work done $= 200\,l\,\delta\theta - 0.866\,P\,l\,\delta\theta - 100\,l\,\delta\theta$

According to the principle of virtual work,

$$200\,l\,\delta\theta - 0.866\,P\,l\,\delta\theta - 100\,l\,\delta\theta = 0$$

or $0.866\,P = 100$;

$$P = \frac{100}{0.866} = 115.47\text{ N} \quad \text{(Compression)}$$

EXAMPLE 12.16

Using the method of virtual work, determine the force in the top member of the simple truss consisting of equilateral triangles as shown in Fig. 12.21.

Solution: By symmetry of truss and loading

$$R_a = R_b = \frac{W}{2}$$

The member DE is cut and replaced by an axial compressive force P in it. The system is then made movable by giving a virtual angular displacement of $\delta\theta$ about C. Then B gets displaced to B' and E to E'.

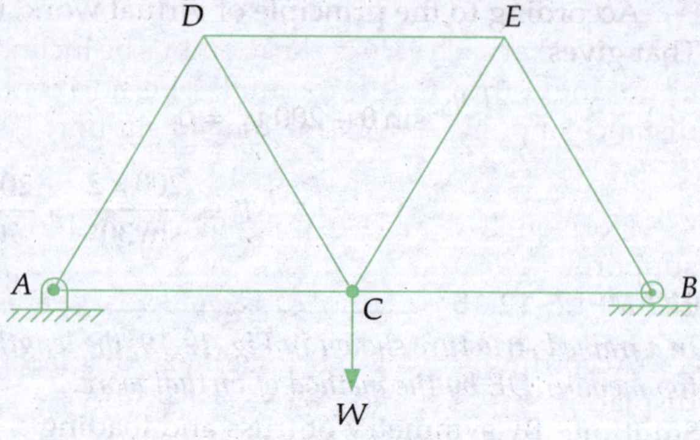

Fig. 12.21

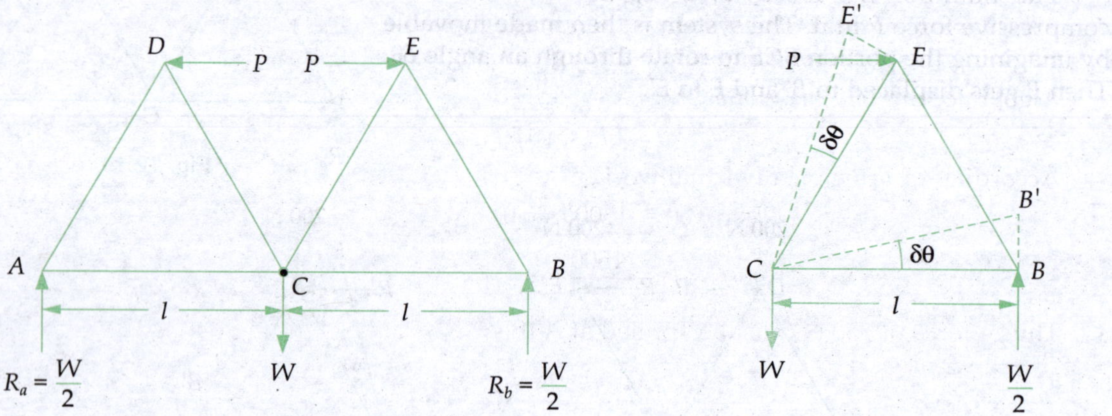

Vertical displacement of $B = BB' = l\,\delta\theta$ (upwards)

Work done by reaction at $B = \frac{W}{2} \times l\,\delta\theta$

Horizontal displacement of point $E = EE' \times \sin 60°$

$= l\,\delta\theta \times 0.866$ (towards left)

Work done by axial force $P = -P \times l\,\delta\theta \times 0.866$

The negative sign stems from the fact that force P and displacement of point E are opposite to each other.

According to the principle of virtual work, the work done by all the forces must by zero. That gives

$$\frac{W}{2} \times l\,\delta\theta - P\,l\,\delta\theta \times 0.866 = 0$$

or $$P = \frac{W}{2 \times 0.866} = \mathbf{0.577\ W}$$

EXAMPLE 12.17

A number of bars are hinged together to form three identical rhombuses as shown in Fig. 12.22. If a horizontal force of 250 N is applied at B, determine the force P which must be applied at A to preserve equilibrium.

Solution: Let each side of the rhombus be inclined at an angle θ to the horizontal. The horizontal distance of point A from the hinged support

$$x = 2\,l\cos\theta$$

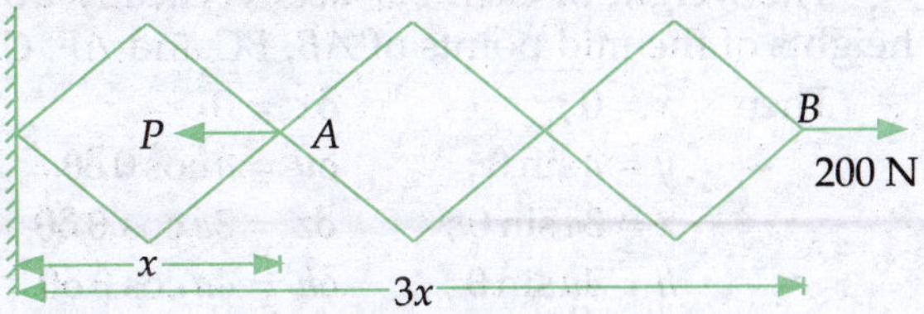

Fig. 12.22

Horizontal distance of point B from the hinged support

$$3x = 6\,l\cos\theta$$

Accordingly when an imaginary displacement $\delta\theta$ is given to the bars, then corresponding displacements at A and B are

$$\delta x = -2\,l\sin\theta\,\delta\theta \quad \text{and} \quad 3\,\delta x = -6\,l\sin\theta\,\delta\theta$$

That gives:

$$\text{Work done at } A = -P \times (-2l\sin\theta\,\delta\theta)$$
$$\text{Work done at } B = 250 \times (-6l\sin\theta\,\delta\theta)$$

Algebraic sum of work done

$$= 2\,Pl\sin\theta\,\delta\theta - 1500\,l\sin\theta\,\delta\theta$$

According to principle of virtual work,

$$2\,Pl\sin\theta\,\delta\theta - 1500\,l\sin\theta\,\delta\theta = 0$$

$$\therefore \quad P = \frac{1500}{2} = 750 \text{ N}$$

Thus a force of 750 N must be applied at A if equilibrium is to be maintained.

EXAMPLE 12.18

A hexagonal frame ABCDE is made up of six bars of equal length and weight. One of the bars is fixed on a horizontal plane and the system lies in the vertical plane as shown in Fig. 12.23. The mid points of the two upper non-horizontal bars are connected by a string PQ. Using the principle of virtual work, show that the tension in the string is 6 W cot θ where W is the weight of each bar and θ is the inclination of the inclined bars with the horizontal.

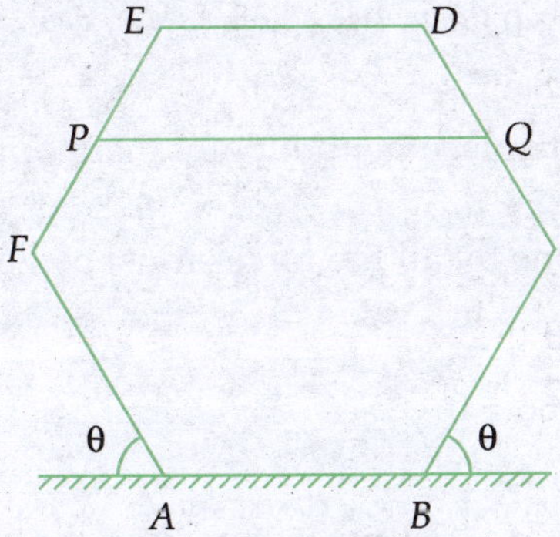

Fig. 12.23

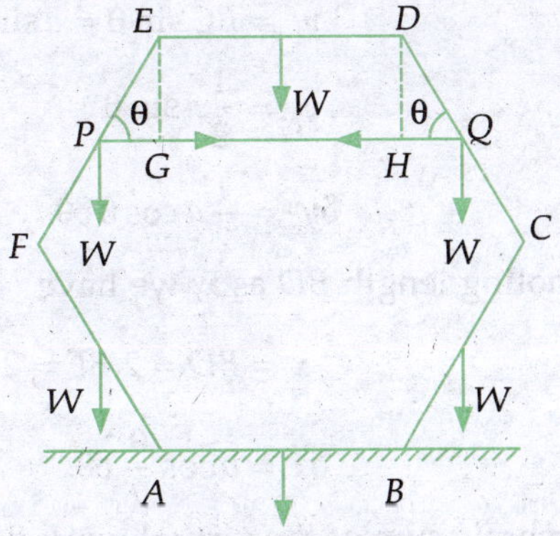

Fig. 12.24

Solution: Let $2a$ be the length of each side of the hexagonal frame.

Length of string $PQ = PG + GH + HQ = a\cos\theta + 2a + a\cos\theta = 2a + 2a\cos\theta$

$$\therefore \quad \delta_{PQ} = -2a\sin\theta\,\delta\theta$$

The weight of each bar acts vertically downwards at its mid point. Let x, y, z and h be the heights of the mid points of AB, BC and AF, CD and FE, and ED as measured above AB.

Then
$$x = 0; \qquad \delta x = 0$$
$$y = a\sin\theta; \qquad \delta y = a\cos\theta\,\delta\theta$$
$$z = 3a\sin\theta; \qquad \delta z = 3a\cos\theta\,\delta\theta$$
$$h = 4a\sin\theta; \qquad \delta h = 4a\cos\theta\,\delta\theta$$

By the principle of virtual work,

$$W \times \delta x + 2W \times \delta y + 2W\,\delta z + W\,\delta h + T \times \delta_{PQ} = 0$$

or $$W \times 0 + 2W \times a\cos\theta\,\delta\theta + 2W \times 3a\cos\theta\,\delta\theta + W \times 4a\cos\theta\,\delta\theta + T \times (-2a\sin\theta\,\delta\theta) = 0$$

or $$12W\,a\cos\theta\,\delta\theta = 2T\,a\sin\theta\,\delta\theta$$

$\therefore$ Tension in the string PQ

$$T = \frac{12\,W\,a\cos\theta\,\delta\theta}{2a\sin\theta\,\delta\theta} = 6W\frac{\cos\theta}{\sin\theta} = 6\,W\cot\theta$$

EXAMPLE 12.19

Four uniform rods, each of length a and weight W are hinged together at the extremities and form a rhombus. A string of length b joins the shorter diagonal of the rhombus so formed. If one of the rods is supported in the horizontal position, show that the tension T of the string is

$$T = \frac{2W(2a^2 - b^2)}{a\sqrt{4a^2 - b^2}}$$

Solution: Refer Fig. 12.25 which shows the rhombus formed by four uniform rods. The side AB is supported in horizontal position and the shorter diagonal BD is joined by a string. The centre of gravity of the four rods will be at point E which is the point of intersection of the diagonals AC and BD of the rhombus.

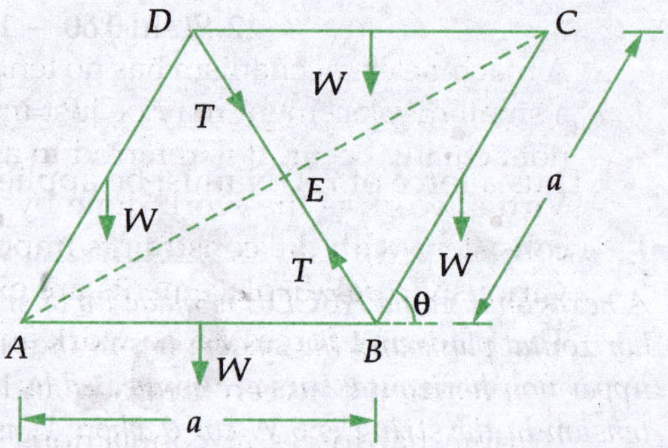

Fig. 12.25

Let angle $BCD = \theta$. Then position of points C and E with respect to rod AB are;

$$y_c = BC\sin\theta = a\sin\theta$$

$$y_e = \frac{1}{2}a\sin\theta$$

$$\delta y_e = \frac{1}{2}a\cos\theta\,\delta\theta$$

Denoting length BD as x, we have

$$x = BD = 2\,BE = 2\,BC\sin\frac{\theta}{2} = 2b\sin\frac{\theta}{2}$$

$$\delta x = b\cos\frac{\theta}{2}\,\delta\theta$$

Algebraic sum of the virtual work done by tension T in the rod BD, and by the weights of the four rods is

$$= -T\,\delta x + 4W\,\delta y_e$$

By the principle of virtual work,

$$-T\,\delta x + 4W\,\delta y_e = 0$$

or

$$T = 4W\frac{\delta y_e}{\delta x} = 4W \times \frac{\frac{1}{2}a\cos\theta\,\delta\theta}{b\cos\frac{\theta}{2}\delta\theta} = \frac{2W\cos\theta}{\cos\frac{\theta}{2}}$$

From the geometry of the rhombus,

$$\sin\frac{\theta}{2} = \frac{BE}{BC} = \frac{b/2}{a} = \frac{b}{2a}$$

From trigonometric relations,

$$\cos\frac{\theta}{2} = \sqrt{1-\sin^2\frac{\theta}{2}} = \sqrt{1-\frac{b^2}{4a^2}} = \frac{\sqrt{4a^2-b^2}}{2a}$$

$$\cos\theta = 1-2\sin^2\frac{\theta}{2} = 1-2\frac{b^2}{4a^2} = \frac{2a^2-b^2}{2a^2}$$

$\therefore$ Tension T in the string $= 2W \times \dfrac{(2a^2-b^2)/2a^2}{\sqrt{4a^2-b^2}\;/2a} = \dfrac{2W\,(2a^2-b^2)}{a\sqrt{4a^2-b^2}}$

which is the required expression.

REVIEW AND SUMMARY

1. A machine or mechanism has no tendency to move when it is in a state of equilibrium. However, a small displacement may be just imagined to have been caused. Since this displacement does not actually occur, it is referred to as virtual displacement.
2. Virtual work is the work done by a force or couple during a virtual displacement which is consistent with the constraints imposed on the system. While computing the virtual work, the various virtual displacements are expressed in terms of a single virtual displacement.
3. The following forces do no work and are omitted in forming the equation for virtual work :
   (*i*) reaction between the body and any smooth surface with which it is in contact
   (*ii*) internal forces at connections (mutual action and reaction between two bodies whose equilibrium is being considered)
   (*iii*) forces exerted by inextensible cords and cables.
4. The principle of virtual work states that : The total virtual work done by all the forces and couples acting on a rigid body for any virtual displacement is zero.

REVIEW QUESTIONS

A. Conceptual and conventional questions :

1. State the principle of virtual work. Explain it giving a suitable example.
2. Mention the forces which are generally omitted while applying the principle of virtual work.
3. A load W is being lifted by a pulley system as shown in Fig. 12.26, by applying a pull at the end of a cord supporting the pulleys having the same diameter. Using the principle of virtual, set up the relationship between P and W.

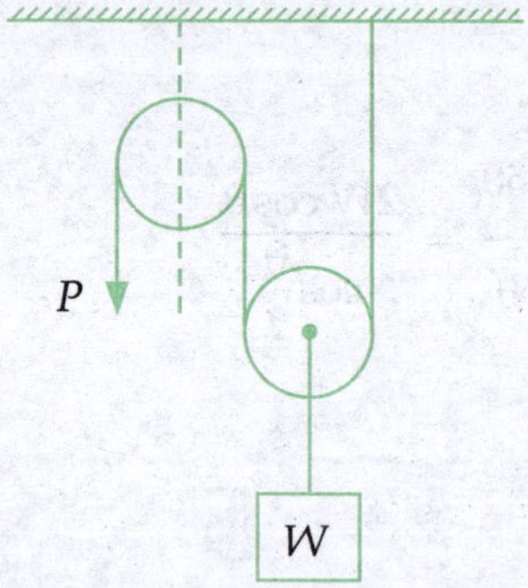

Fig. 12.26

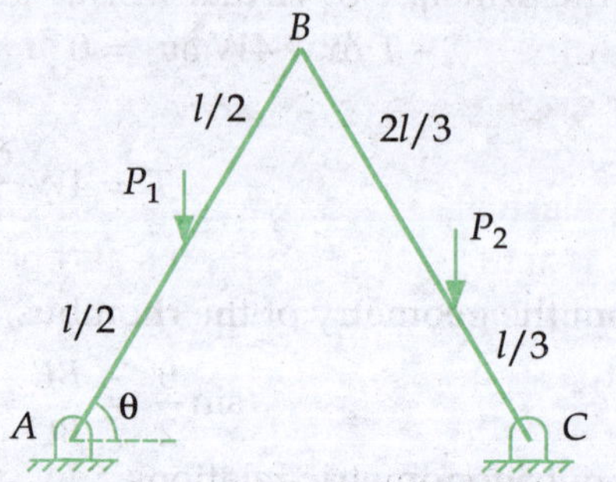

Fig. 12.27

4. Using the method of virtual work, setup the following expression for horizontal component of reaction at C in respect of the frame supported and loaded as shown in Fig. 12.27.

$$R_{ch} = \frac{P_1 P_2}{24} \cot\theta$$

How this expression would be affected if force P_2 acts horizontally to the right instead of vertically?

$$\left(\frac{P_1}{4}\cot\theta + \frac{5}{6}P_2\right)$$

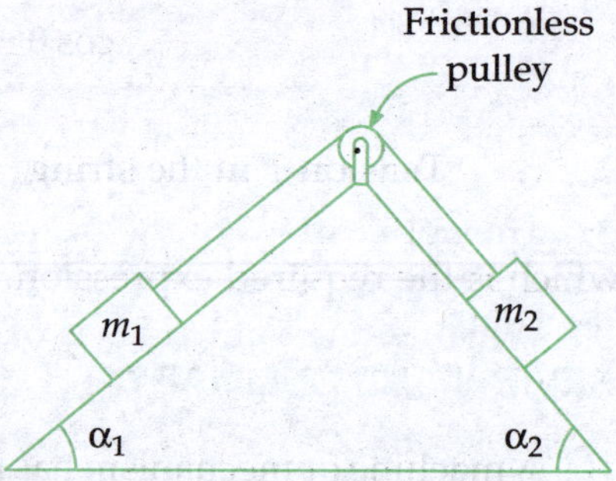

Fig. 12.28

5. A friction less double-incline with angle α_1 and α_2 as shown in Fig. 12.28, carries a set of sliding masses m_1 and m_2 connected with an inextensible string and passing over a frictionless pulley. By using the principle of virtual work, obtain the relationship between α_1 and α_2 in terms of m_1 and m_2 and hence determine α_2 if $\alpha_1 = 30°$ and $m_1 = 2m_2$

6. A regular hexagonal frame $ABCDEF$ is made up of six uniform identical bars each weighing W, joined together (Fig. 12.29). The frame rests in a vertical plane and the bar AB is fixed in horizontal position. The mid points of AB and DE are connected by a light string. Using the method of virtual work, find tension in the string.

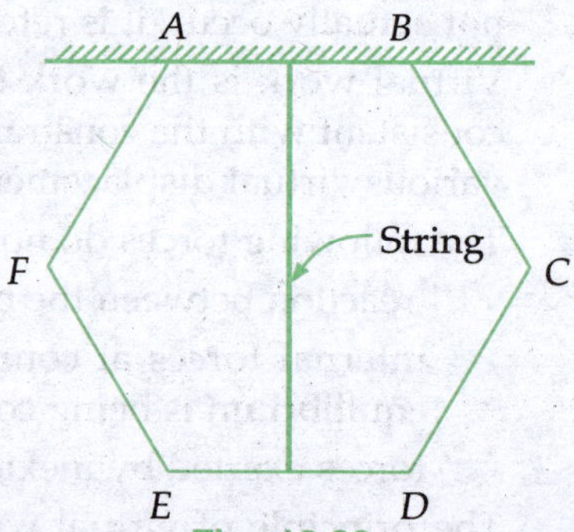

Fig. 12.29

Hint: Let $2a$ be the length of each side of the hexagonal frame. The weight of each bar acts vertically downwards at its mid point. Let x, y, z and h be the heights of the mid points of AB, BC and AF, CD and FE, and ED as measured below AB. Then

$$x = 0\,; \qquad \delta x = 0$$
$$y = a\sin\theta\,; \qquad \delta y = a\cos\theta\,\delta\theta$$
$$z = 3a\sin\theta\,; \qquad \delta z = 3a\cos\theta\,\delta\theta$$
$$h = 4a\sin\theta\,; \qquad \delta h = 4a\cos\theta\,\delta\theta$$

Length of the string $l = 4a\sin\theta\,;\quad \delta l = 4a\cos\theta\,\delta\theta$

By the principle of virtual work,

$$W \times 0 + 2W \times a\cos\theta\,\delta\theta + 2W \times 3a\cos\theta\, d\theta$$
$$+ W \times 4a\cos\theta\,\delta\theta - T \times 4a\cos\theta\,\delta\theta = 0$$

$$\therefore \quad \text{Tension in the string, } T = \frac{12W\,a\cos\theta\,\delta\theta}{4a\cos\theta\,\delta\theta} = 3\,W$$

B. Multiple choice questions :

1. Identify the forces which are omitted while applying the principle of virtual work
   (*a*) gravity forces
   (*b*) conservative forces
   (*c*) elastic forces
   (*d*) reactions at smooth pins and hinges
2. Which amongst the following is the most appropriate statement pertaining to the concept of virtual work?
   (*a*) no work is done by any force
   (*b*) work is done by the forces and the structure is displaced
   (*c*) each force performs some work and the sum total of all such works is equated to zero.
   (*d*) the structure remains in equilibrium state
3. If a system is in equilibrium and the position of the system depends upon many independent variables, the principle of virtual work states that the partial derivatives of its total potential energy with respect to each of the independent variable must by
   (*a*) – 1.0 (*b*) zero
   (*c*) 1.0 (*d*) infinite
4. The applications of the principle of virtual work include solution to problems in :
   A. beams B. framed structures C. lifting machines
   Which of the followings is true ?
   (*a*) A and B (*b*) B and C
   (*c*) A and C (*d*) A, B and C

Answers :

1. (*d*) 2. (*c*) 3. (*b*) 4. (*d*)

❑❑❑

Dynamics: Kinematics and Kinetics

When several forces of different magnitudes and directions act upon a body, they can be reduced to a resultant force-couple system. The body remains in static equilibrium (*i.e.*, absolute or relative rest) when the resultant force and couple are zero. When the resultant force-couple system is not zero, a change occurs in the state of rest or motion of the body. The analysis of the bodies in motion is dealt through dynamics which constitutes a part of the subject of mechanics. Dynamics is further divided into Kinematics and Kinetics.

Kinematics is concerned with the description of motion of objects independent of causes of motion. Here study is made of the relationships between displacement, velocity, acceleration and time of a given motion without considering the forces that cause motion.

Kinetics relates to action of forces and the resulting motion. Here study is made of the relationship between the mass and motion of the body and the forces acting on it. Such a study helps to predict the motion caused by a given force system or to determine the forces needed to cause a prescribed motion.

Attention has been directed in the following chapters to explain the basic terms and general principles related to dynamics of bodies.

13 Rectilinear and Curvilinear Motion

Learning objectives : The main objective of the subject matter presented in this chapter is to make the reader familiar with following aspects of kinematics

- *rectilinear and curvilinear motion*
- *displacement, velocity, acceleration and graphical representation of their interrelationship*
- *equation of rectilinear motion with uniform acceleration and motion under gravity*
- *motion along a curved path, normal and tangential components of acceleration*
- *relative velocity*

When a body changes its position with respect to other bodies, then the body is said to be in motion. Thus, a relative change in the position of a body is its motion. Motion involves both space and time.

Trajectory is a path followed by a body during its motion. It may be a straight line or a curve. Straight line path is observed in *rectilinear motion*, whereas curved path is observed in *curvilinear motion*. If the curved path lies in a plane, its is called plane curvilinear motion. *Rotary* or *circular* motion is a special case of curvilinear motion where all particles of a body move in a concentric circle.

13.1. DISPLACEMENT, VELOCITY AND ACCELERATION

Displacement: *"The change of position of a particle or a body with respect to a certain fixed reference point is termed as displacement."*

Consider a body that moves along a curved path and takes time t to move from position A to B. The total distance covered by the body along the path followed is called *distance* and the linear (shortest) distance between the two positions is called *displacement*.

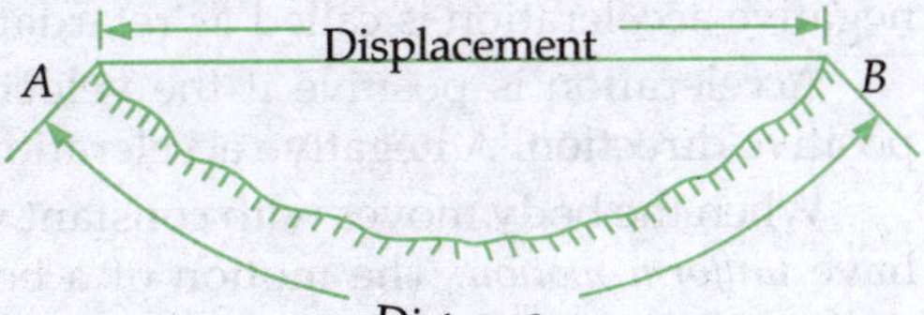

Fig. 13.1

Again, with reference to Fig. 13.2, let the body move from reference point O to position A and then from A to position B.

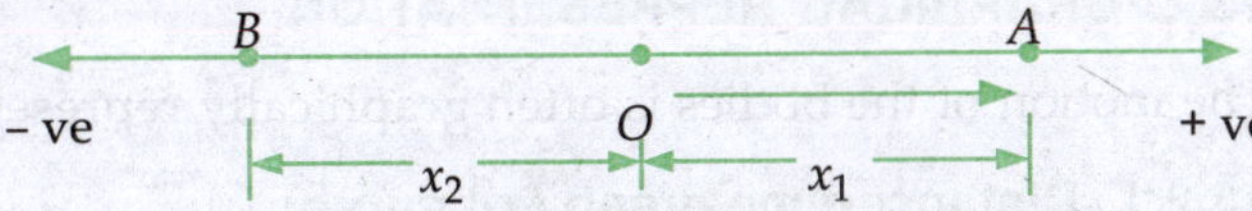

Fig. 13.2

Then:

Displacement of body $= -x_2$

Distance travelled by the particle

$$= x_1 + x_1 - x_2 = 2x_1 - x_2$$

It is to be noted that:

(i) Displacement is the shortest distance covered from the particle from the reference point.

(ii) Displacement is taken positive to the right of reference point and negative to its left.

(iii) Displacement has both magnitude and direction and so is a vector quantity. Distance is scalar quantity since it has only magnitude.

Velocity: *The rate of change of position of a body with respect to time is called velocity.*

With reference to Fig. 13.3, the position A of a body at time t is at distance x from the reference point (origin O). At time $(t + \Delta t)$, the body shifts to position B which is at a distance $(x + \Delta x)$ from O. Then average velocity of the body over time interval Δt is

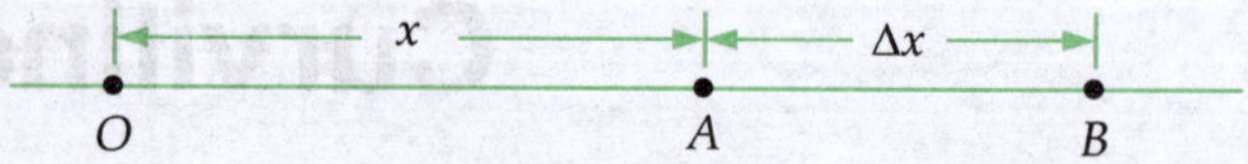

Fig. 13.3

$$v_{av} = \frac{\Delta x}{\Delta t} \quad ...(13.1)$$

The velocity at a particular instant of time is called instantaneous velocity. Choosing the time interval Δt and the displacement Δx to be very very small,

$$\text{Instantaneous velocity } v = \lim_{\Delta x \to 0} \frac{\Delta x}{\Delta t} = \frac{dx}{dt} \quad ...(13.2)$$

It is to be noted that velocity of a body is a measure of the rate of change of its position in a particular direction, and hence is a vector quantity. Speed is defined as rate of change of position of the body (with respect to its surroundings) irrespective of direction; speed is a scalar quantity.

The units of velocity or speed for defining their magnitude are taken as m/s or km/hr.

Acceleration: *The rate of change of velocity of a body with respect to time is called acceleration.*

Let v be the velocity of the body at any time t, and let it become $(v + \Delta t)$ at a later time $(t + \Delta t)$. Then average acceleration of the body over time interval Δt is

$$\text{Average acceleration } a = \frac{\Delta v}{\Delta t} \quad ...(13.3)$$

If the time interval Δt and the change in velocity Δv are chosen to be very very small, then

$$\text{acceleration } a = \lim_{\Delta t \to 0} \frac{\Delta v}{\Delta t} = \frac{dv}{dt} \quad ...(13.4)$$

The units for defining the magnitude of acceleration are m/s^2 or km/hr^2. The acceleration may be positive or negative. The positive acceleration is simply referred to as acceleration and the negative acceleration is called as retardation or deceleration.

Acceleration is positive if the velocity is increasing and the body is moving further in the positive direction. A negative acceleration implies a slow down in the motion of the body.

When the body moves with constant velocity (*i.e.* when acceleration is zero) the body is said to have *uniform motion*. The motion of a body moving with constant acceleration is referred to as *uniformly accelerated motion.*

13.2. GRAPHICAL REPRESENTATION

The motion of the bodies is often graphically represented by the following plots:

13.2.1. Distance-time graph (x-t curve)

The displacement of the body is plotted with respect to time; displacement x lies along ordinate and time t is taken along abscissa;

Since $v = dx/dt$, *the slope of x - t curve at any instant gives the velocity of the body at that instant.* Then, with reference to Fig. 13.4.

(*i*) At time t_1, the curve has a positive slope and accordingly the velocity is positive. A positive velocity implies that the displacement is increasing and that the body is moving to the right of origin. If the slope is positive and constant, the displacement would increase by equal amount in equal intervals of time, *i.e.*, the body would be moving with a uniform velocity.

(*ii*) At time 2, the slope is zero and so is the velocity. The displacement is then not changing with time and the body is at rest.

(*iii*) At time 3, the slope is negative and so is the velocity.

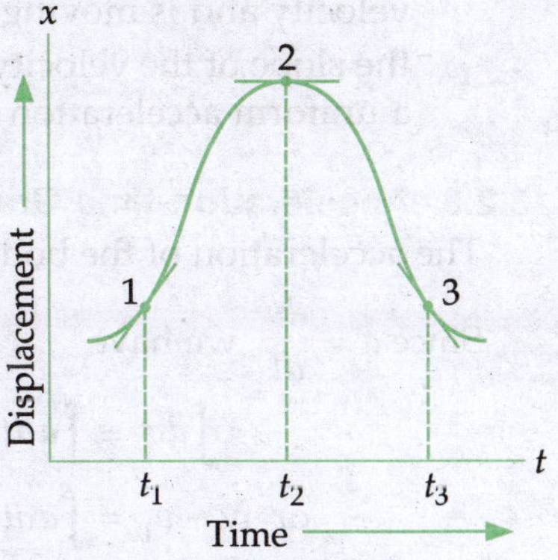

Fig. 13.4

13.2.2. Velocity-time graph (v – t curve)

The velocity of the body is plotted as a function of time. Since $a = dv/dt$, the slope of the $v-t$ curve at any instant gives the acceleration of the body at that instant.

Further, $$v = \frac{dx}{dt}$$

$$\int dx = \int v\,dt$$

$$(x_2 - x_1) = \int v\,dt \qquad \text{...(13.5)}$$

The above identity implies that *area under the v – t curve corresponding to a given time interval gives the change in displacement or the distance travelled by the body during the same time interval.*

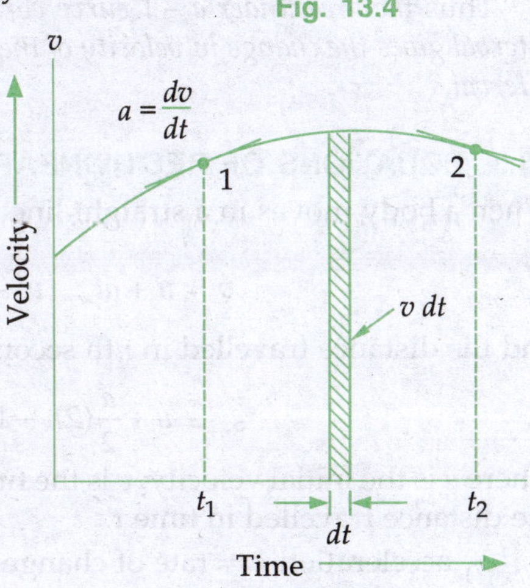

Fig. 13.5

With reference to Fig. 13.6:

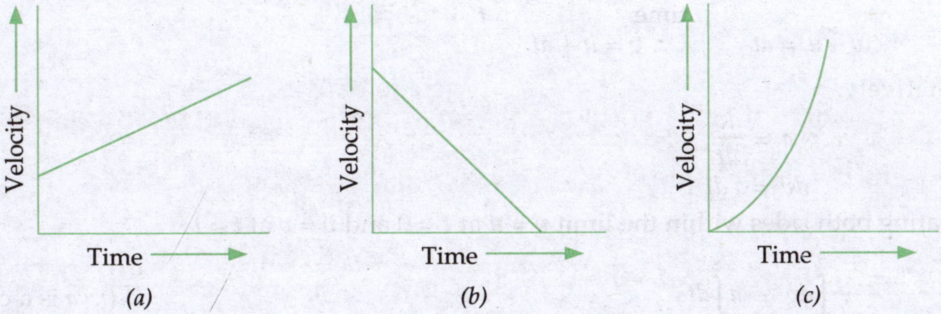

Fig. 13.6

(*a*) the velocity increases linearly with time. Since the $v-t$ curve has a constant slope, the velocity increases by equal amounts at equal intervals of time. The body is then moving with a constant acceleration. At the start (when $t = 0$), the body has a finite velocity. The $v-t$ curve depicted in Fig. 13.6 (*a*) thus corresponds to a body which has finite initial velocity and is moving with a constant acceleration.

(*b*) the body has a finite initial velocity. With passage of time, the velocity decreases linearly with time and finally becomes zero, *i.e.*, the body comes to rest.

The $v - t$ curve depicted in Fig. 13.6 (*b*) thus corresponds to a body which has finite initial velocity and is moving with a constant retardation (negative slope of graph).

(*c*) the slope of the velocity-time graph is different at different times. The body does not have a uniform acceleration as its velocity is not changing at constant time.

13.2.3. Acceleration-time Graph (a - t curve)

The acceleration of the body is plotted as function of time.

Since $a = \dfrac{dv}{dt}$, we have

$$\int dv = \int a\,dt$$

$$\text{or } v_2 - v_1 = \int a\,dt \qquad ...(13.6)$$

Thus *the area under a – t curve corresponding to a given time interval gives the change in velocity of the body during the some time interval.*

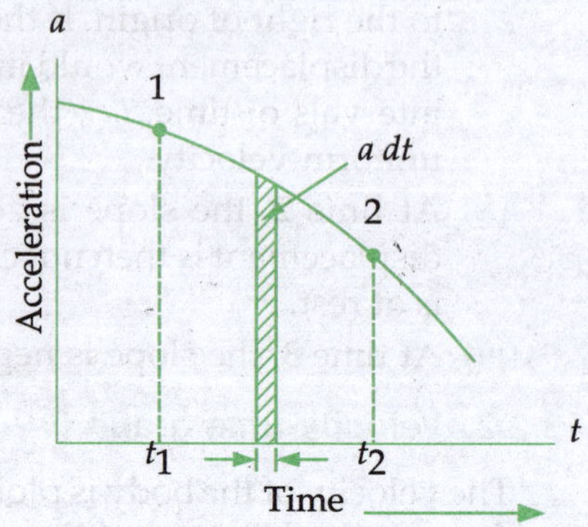

Fig. 13.7

13.3. EQUATIONS OF RECTILINEAR MOTION

When a body moves in a straight line with uniform acceleration, the equations of motion are:

$$v = u + at\ ; \quad v^2 - u^2 = 2as\ ; \quad s = ut + \frac{1}{2}at^2$$

and the distance travelled in nth second is

$$s_n = u + \frac{a}{2}(2n - 1)$$

where u is the initial velocity, v is the final velocity, a is the acceleration of the body and s represents the distance travelled in time t.

(*i*) acceleration a = rate of change of velocity

$$= \frac{\text{change of velocity}}{\text{time}} = \frac{v - u}{t}$$

$$v - u = at \qquad \therefore v = u + at \qquad ...(13.7)$$

Alternatively:

$$a = \frac{dv}{dt}$$

$$dv = a\,dt$$

Integrating both sides within the limit $v = u$ at $t = 0$ and $v = v$ at $t = t$

$$\int_u^v dv = a\int_0^t dt \qquad (\because a \text{ is a constant})$$

$$v - u = at \qquad \therefore v = u + at$$

(*ii*) Distance travelled = average velocity × time

$$s = \frac{u + v}{2} \times t = \frac{u + (u + at)}{2} \times t \qquad (\because v = u + at)$$

$$= \left(u + \frac{at}{2}\right) \times t = ut + \frac{1}{2}at^2 \qquad ...(13.8)$$

Alternatively:

$$v = \frac{dx}{dt}$$

$$dx = v\,dt = (u + at)\,dt$$

Upon integration

$$\int_0^s dx = \int_0^t (u + at)\,dt \qquad \text{assuming } x = 0 \text{ at } t = 0 \text{ and } x = s \text{ at } t = t$$

$$\text{or} \quad s = ut + \frac{1}{2}at^2$$

(*iii*) Distance = average velocity × time

$$s = \frac{u+v}{2} \times t = \frac{u+v}{2} \times \frac{v-u}{a} \qquad \left(\because \; v = u + at, \; t = \frac{v-u}{a}\right)$$

$$= \frac{v^2 - u^2}{2a} \qquad \therefore \quad v^2 - u^2 = 2as \qquad \text{...(13.9)}$$

Alternatively:

$$a = \frac{dv}{dt} = \frac{dv}{dx} \times \frac{dx}{dt} = \frac{dv}{dx} \times v$$

$$\text{or} \quad a\,dx = v\,dv$$

Upon integration

$$a\int_0^s dx = \int_u^v v\,dv \qquad \text{assuming that } v = u \text{ at } x = 0 \text{ and } v = v \text{ at } x = s$$

$$as = \left|\frac{v^2}{2}\right|_u^v ; \quad as = \frac{v^2 - u^2}{2}$$

$$\therefore \quad v^2 - u^2 = 2as$$

(*iv*) Distance covered in n seconds,

$$s_n = un + \frac{1}{2}an^2$$

Distance covered in $(n - 1)$ seconds

$$s_{n-1} = u(n-1) + \frac{1}{2}a(n-1)^2 = u(n-1) + \frac{1}{2}a(n^2 + 1 - 2n)$$

$\therefore$ Distance covered in nth second,

$$s_{nth} = s_n - s_{n-1}$$

$$= un + \frac{1}{2}an^2 - u(n-1) - \frac{1}{2}a(n^2 + 1 - 2n)$$

$$= un + \frac{1}{2}an^2 - un + u - \frac{1}{2}an^2 - \frac{1}{2}a + an$$

$$= u - \frac{a}{2} + an = u + \frac{a}{2}(2n - 1) \qquad \text{...(13.10)}$$

EXAMPLE. 13.1

A car travels from one station to another along a straight road. First half of the distance is covered with velocity 60 km/hr and the second half is covered with velocity 90 km/hr. Determine the average speed of the motor.

Solution: Let t be the total time taken to cover the entire distance s. Then

$$t = t_1 + t_2 = \frac{s/2}{v_1} + \frac{s/2}{v_2} = \frac{s}{2}\left(\frac{v_1 + v_2}{v_1 v_2}\right)$$

$$v_{av} = \frac{s}{t} = s \div \frac{s}{2}\left(\frac{v_1 + v_2}{v_1 v_2}\right) = \frac{2v_1 v_2}{v_1 + v_2} = \frac{2 \times 60 \times 90}{60 + 90} = \mathbf{72\ km/hr}$$

EXAMPLE 13.2

A train crosses a tunnel in 30 seconds time. The speeds of the train at entry to and at exit from the tunnel are 36 km/hr and 54 km/hr respectively. If the acceleration remains constant, then make calculations for the length of tunnel.

Solution: Average speed $= \dfrac{36 + 54}{2} = 45 \text{ km/hr}$

$$= \frac{45 \times 1000}{3600} = 12.5 \text{ m/s}$$

Length of tunnel = average speed x time = 12.5 × 30 = **375 m**

EXAMPLE 13.3

A toy car accelerates from rest at a constant rate of 2 m/s² for some time. Then it retards at a constant rate of 4 m/s² and comes to rest. If the car remains in motion for 3 seconds, determine the maximum speed attained and the total distance travelled by the car.

Solution: Let V be the maximum speed attained by the car. Then

$$V = a_1 t_1 = a_2 t_2$$

where t_1 and t_2 are time periods for acceleration and retardation respectively.

$$\therefore \quad t_1 = \frac{a_2}{a_1} t_2 = \frac{4}{2} t_2 = 2\, t_2$$

Also

$$t = t_1 + t_2$$

$$3 = 2t_2 + t_2; \quad t_2 = 1\,s \quad \text{and} \quad t_1 = 2\,s$$

$$\therefore \quad V = a_1 t_1 = 2 \times 2 = 4 \text{ m/s}$$

Now applying the correlation $s = ut + \frac{1}{2}at^2$, we get

$$s_1 = 0 \times 2 + \frac{1}{2} \times 2 \times 2^2 = 4 \text{ m}$$

$$s_2 = 4 \times 1 + \frac{1}{2} \times (-4) \times 1^2 = 2 \text{ m}$$

$\therefore$ Total distance covered = 4 + 2 = 6 m

EXAMPLE. 13.4

Two cars A and B move at 54 km/hr in the same direction and the car B is 300 m ahead of car A. If the car A is accelerated at 6 m/s² while the car B continues to move with the same velocity, calculate the time taken by car A to overtake car B.

Solution: Speed of cars A and B $= 54$ km/hr

$$= \frac{54 \times 1000}{3600} = 15 \text{ m/s}$$

Let the cars A and B meet at location C after time t covering distances s_a and s_b respectively.

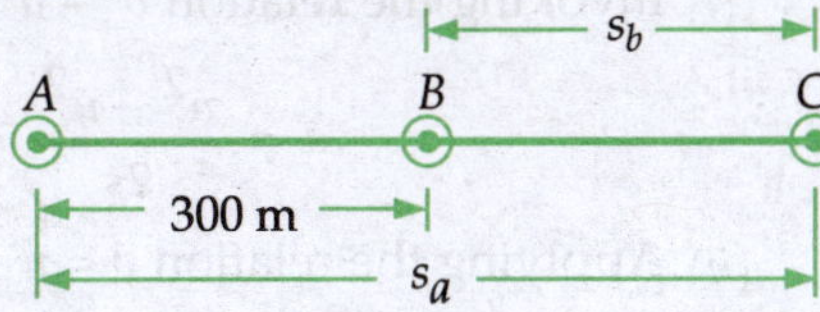

Fig. 13.8

Then

$$s_a = s_b + 300$$

Applying the correlation $s = ut + \frac{1}{2}at^2$ for accelerated motion of car A and $s = vt$ for car B moving with the same velocity, we get

$$15t + \frac{1}{2}(6)t^2 = 15t + 300$$

$$3t^2 = 300 \; ; \; t = \sqrt{100} = \mathbf{10\ s}$$

EXAMPLE 13.5

A particle moves with uniform acceleration along a straight line ABC. The speeds of the particle at positions A and C are 5 cm/s and 25 cm/s respectively. If the point B lies midway between A and C, what will be the ratio of times taken by the particle to travel distances AB and BC ?

Solution: Refer Fig. 13.9

Invoking the relation $v^2 - u^2 = 2as$ where a is the uniform acceleration

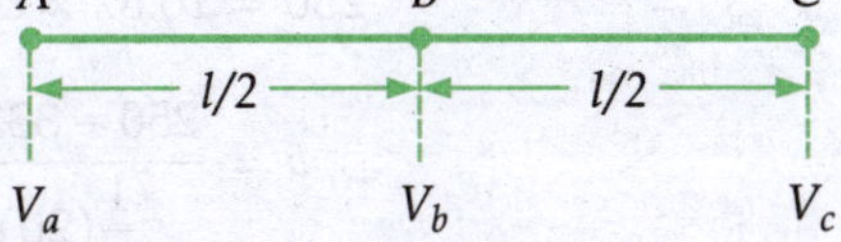

Fig. 13.9

$$V_c^2 - V_a^2 = 2al \; ;$$

$$25^2 - 5^2 = 2al \text{ or } a = \frac{300}{l}$$

$$V_b^2 - V_a^2 = 2a \times \frac{l}{2} \; ; \; V_b^2 - 5^2 = 2 \times \frac{300}{l} \times \frac{l}{2} = 300$$

$$V_b = \sqrt{325} \simeq 18 \text{ m/s}$$

The times t_1 and t_2 to cover distances AB and BC are given by

$$t_1 = \frac{V_b - V_a}{a} = \frac{18-5}{a} = \frac{13}{a}$$

$$t_2 = \frac{V_c - V_b}{a} = \frac{25-13}{a} = \frac{12}{a}$$

$$\therefore \text{ Required ratio } \frac{t_1}{t_2} = \frac{13}{12} \simeq 1:1$$

EXAMPLE 13.6

On turning a corner, a motorist driving at 54 km/hr finds a child on the road 50 m ahead. He instantaneously stops the engine, applies the brakes and the vehicle is brought to halt within 5 m of the child. Make calculations for the uniform rate of deceleration and the time for which the brakes were applied.

Solution: Speed of vehicle $u = 54$ km/hr $= \frac{54 \times 1000}{3600} = 15$ m/s

$$s = 50 - 5 = 45 \text{ m} \; ; \quad v = 0$$

(*i*) Invoking the relation $v^2 - u^2 = 2as$

$$a = \frac{v^2 - u^2}{2s} = \frac{0^2 - 15^2}{2 \times 45} = -2.5 \text{ m/s}^2 \text{ (retardation)}$$

(*ii*) Applying the relation $v = u + at$

$$t = \frac{v - u}{a} = \frac{0 - 15}{(-2.5)} = 6 \text{s}$$

EXAMPLE 13.7

A motorist is driving his car at 60 km/hr when he observes that a traffic light 250 m ahead turns red. The traffic light is timed to remain red for 20 seconds before it turns green. The motorist wishes to pass the traffic lights without stopping to wait for it to turn green. Make calculations for : (a) the required uniform acceleration of the car, and (b) the speed of the car as it passes the traffic light.

Solution: Speed of car $u = 60 \text{ km/hr} = \dfrac{60 \times 1000}{3600} = 16.67 \text{ m/s}$

$$s = 250 \text{ m and } t = 20 \text{ s}$$

(*a*) Applying the relation $s = ut + \frac{1}{2}at^2$

$$250 = 16.67 \times 20 + \frac{1}{2}a(20)^2$$

$$a = \frac{250 - 333.34}{\frac{1}{2}(20)^2} = -0.417 \text{ m/s}^2 \text{ (retardation)}$$

(*b*) Using the relation $v = u + at$

$$v = 16.67 + (-0.417) \times 20 = 8.33 \text{ m/s}$$

$$= \frac{8.33 \times 3600}{1000} \simeq 30 \text{ km/hr}$$

EXAMPLE 13.8

A miscreant snatches a purse from a lady and runs away on a motor cycle at an acceleration of 1.5 m/s². A police party standing near by starts the chase 15 seconds later in the police van driven at 60 m/s. What time would be taken by the police party to catch up with the miscreant?

Solution: Let t seconds be the time taken by the police party to catch up with the miscreant. Then

Distance covered by the snatcher in $(t + 15)$ seconds

$$s = ut + \frac{1}{2}at^2$$

$$= 0 + \frac{1}{2} \times 1.5\,(t + 15)^2$$

$$= 0.75\,(t^2 + 225 + 36\,t) \qquad (\because u = 0)$$

The police van has to cover this distance in t seconds when driven at uniform speed of 60 m/s : $s = 60\,t$

$$\therefore \quad 60\,t = 0.75\,(t^2 + 225 + 30\,t)$$

Upon simplification: $t^2 - 50\,t + 225 = 0$

Solution of this quadratic equation gives,

$$t = \frac{-(-50) \pm \sqrt{(-50)^2 - 4 \times 1 \times 225}}{2}$$

$$= \frac{50 \pm 40}{2} = 45 \text{ or } 5 \text{ seconds}$$

The police van will catch up with miscreant at time smaller of the two calculated values. That is: $t = 5$ seconds.

EXAMPLE 13.9

A truck carrying unauthorised goods and moving at a speed of 10 m/s crosses the police check post without stopping. After 5 seconds, the police party starts the jeep from rest and chases the truck with an acceleration of 1.2 m/s². Calculate the time taken by the police jeep to catch the truck.

Solution: Let t seconds be the time taken by the police party to catch the truck after the police jeep starts from rest.

Distance travelled by the truck in $(t + 5)$ seconds

$$S = \text{uniform speed} \times \text{time} = 10\,(t + 5)$$

To catch up the truck, the police jeep has to cover the same distance in t seconds

$$S = ut + \frac{1}{2}at^2$$

$$= 0 + \frac{1}{2} \times 1.2 \times t^2 = 0.6\,t^2 \qquad (\because\ u = 0)$$

$$\therefore \qquad 10\,(t + 5) = 0.6\,t^2$$

$$\text{or} \qquad 0.6\,t^2 - 10\,t - 50 = 0$$

The solution of this quadratic equation gives

$$t = \frac{-(-10) \pm \sqrt{(-10)^2 - 4 \times 0.6 \times (-50)}}{2 \times 0.6}$$

$$= \frac{10 \pm \sqrt{100 + 120}}{1.2} = \frac{10 \pm 14.83}{1.2}$$

Taking positive sign, $t = \dfrac{24.83}{1.2} = 20.69$ seconds.

The police jeep will catch up the truck 20.69 sec after start from rest.

EXAMPLE 13.10

Three marks A, B and C spaced at a distance of 100 m are made along a straight road. A car starting from rest and accelerating uniformly passes the mark A and takes 10 seconds to reach the mark B, and further 8 seconds to reach the mark C. Make calculations for

(a) *the magnitude of the acceleration of the car,*

(b) *the velocity of the car at A,*

(c) *the velocity of the car at B, and*

(d) *the distance of the mark A from the starting point.*

Solution: Refer Fig. 13.10 for the position of the starting point O and the marks A, B and C.

Let u_a and u_b be the velocities at A and B respectively.

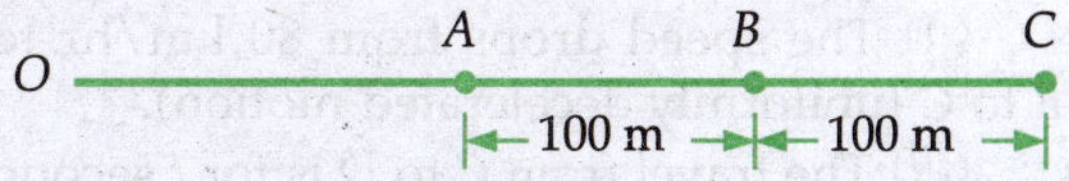

Fig. 13.10

Considering motion from A to B,

$$s = u_a t + \frac{1}{2}at^2$$

$$100 = u_a \times 10 + \frac{1}{2} a \times (10)^2$$

$$100 = 10u_a + 50a \quad ...(i)$$

Considering motion from A to C

$$200 = u_a \times 18 + \frac{1}{2} a \times (18)^2$$

$$200 = 18u_a + 162a \quad ...(ii)$$

Solving identities (*i*) and (*ii*), we get

$$u_a = 8.61 \text{ m/s} \quad \text{and} \quad a = 0.278 \text{ m/s}^2$$

(*c*) Considering motion from B to C and using the relation $s = ut + \frac{1}{2}at^2$

$$100 = (u_b \times 8) + \left(\frac{1}{2} \times 0.278 \times 8^2\right) = 8u_b + 8.896$$

∴ Velocity of car at mark B, $u_b = \frac{100 - 8.896}{8} = $ **11.388 m/s**

(*d*) Considering motion from O to A and using the relation $v^2 - u^2 = 2as$

$$u_a^2 - u_0^2 = 2as$$

$$8.61^2 - 0^2 = 2 \times 0.278 \times s$$

∴ Distance of mark A from the starting point 0,

$$s = \frac{8.61^2}{2 \times 0.278} = \textbf{133.3 m}$$

EXAMPLE 13.11

A car travels along a straight road between two stops which are 10 km apart. The car travels first for t seconds time at 80 km/hr and subsequently brakes are applied for 5 seconds to reduce the speed from 80 km/hr to 60 km/hr. There after, the car travels with uniform speed of 60 km/hr for t seconds and reaches the final stop. Make calculations for the time interval.

Solution: Refer Fig. 13.11 which depicts the travel of the car between the stops A and D.

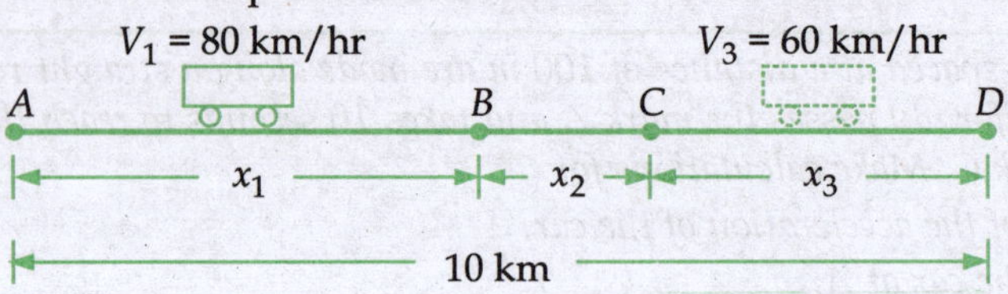

Fig. 13.11.

It may be noted that :

(*i*) The travel from A to B is with 80 km/hr speed for t seconds (uniform motion).

(*ii*) The speed drops from 80 km/hr to 60 km/hr in 5 second period during travel from B to C (uniformly decelerated motion).

(*iii*) The travel from C to D is for t seconds with uniform speed of 60 km/hr.

For uniform motion from A to B and from C to D,

$$x_1 = V_1 t_1 \quad \text{and} \quad x_3 = V_3 t_3$$

$$\therefore \quad \frac{x_1}{x_3} = \frac{V_1 t_1}{V_3 t_3} = \frac{V_1}{V_3} = \frac{80}{60} \quad (\because\ t_1 = t_3)$$

or $$x_3 = 0.75\, x_1 \qquad \ldots(i)$$

For uniformly decelerated motion from B to C

$$x_2 = \text{average speed} \times \text{time}$$

$$= \left(\frac{80+60}{2} \times \frac{1000}{3600}\right) \times 5 = 97.22 \text{ m} \qquad \ldots(ii)$$

Now : $$x_1 + x_2 + x_3 = 10 \times 1000 \text{ metre}$$

$$x_1 + 97.22 + 0.75\, x_1 = 10000$$

That gives : $$x_1 = \frac{10000}{1.75} = 5714.3 \text{ m} \simeq 5.714 \text{ km}$$

and time of travel from A to B,

$$t_1 = \frac{\text{distance}}{\text{speed}} = \frac{5.714}{80} = 0.0714 \text{ hr}$$

$$= \mathbf{4.284 \text{ minute}}$$

EXAMPLE 13.12

Two railway stations A and B are 60 km apart: The 150 m long Himalayan non-stop express train passes station A at some instant of time with uniform speed of 80 km per hour toward station B. After 12 minutes, the 180 m long Shatabdi train passes station B with uniform speed of 100 km/hr and heads towards station A. Both the trains run on parallel tracks. Make calculation for :

(a) *the time taken for the engines to cross each other after the Shatabdi train passes the station B*

(b) *the time taken to cross each other's last coach*

Solution: Let t minutes be the time taken for the engines to cross each other. Then

$$H_{12} + H_t + S_t = \text{distance between the stations } A \text{ and } B$$

Here : H_{10} = distance covered by Himalayan train in 12 minutes

$$= \frac{80}{60} \times 12 = 16 \text{ km}$$

H_t = distanced covered by Himalayan train in t minutes.

$$= \frac{80}{60} \times t = 1.33\,t \text{ km}$$

S = distance covered by Shatabdi train in t minutes

$$= \frac{100}{60} \times t = 1.67\,t \text{ km}$$

Then : $$16 + 1.33\,t + 1.67t = 60$$

$$\therefore \quad t = \frac{60-16}{3} = 14.67 \text{ minutes}$$

The two engines would thus be crossing each other 14.67 minutes after the Shatabdi train leaves station B.

(b) Relative velocity of Shatabdi train with respect to Himalayan train

$$V = 100 + 80 = 180 \text{ km/hr}$$

Total length of the two trains

$$= 150 + 180 = 330 \text{ m} = 0.33 \text{ km}$$

The time taken to cross each other's last coach is then given as

$$t = \frac{\text{distance to be covered}}{\text{relative velocity}} = \frac{0.33}{180} \text{ hr}$$

$$= \frac{0.33}{180} \times 3600 = \textbf{6.6 sec}$$

EXAMPLE 13.13

A vehicle having rectilinear motion is moving with a velocity of 36 km/hr and accelerates uniformly to 72 km/hr over a distance of 200 m. Calculate the acceleration and the time taken. How much distance will be covered by the vehicle in the 5th second?

Solution: $u = 36 \text{ km/hr} = \dfrac{36 \times 1000}{3600} = 10 \text{ m/s}$

$$v = 72 \text{ km/hr} = \frac{72 \times 1000}{3600} = 20 \text{ m/s}$$

If a is the acceleration and t is the time taken, then using the relations $v^2 - u^2 = 2as$ and $v = u + at$, we get

$$20^2 - 10^2 = 2a \times 200\,; \quad a = \frac{400 - 100}{2 \times 200} = \textbf{0.75 m/s}^2$$

$$20 = 10 + 0.75t\,; \quad t = \frac{20 - 10}{0.75} = \textbf{13.33 s}$$

(*b*) Distance moved in the 5th second,

$$s_n = u + \frac{a}{2}(2n - 1) = 10 + \frac{0.75}{2}(2 \times 5 - 1) = \textbf{13.375 m}$$

EXAMPLE 13.14

A car moving in a straight line starts from rest, accelerates uniformly at 10 m/s² for the first 3s, then travels at uniform speed for the next 2s and finally decelerates uniformly at 5 m/s² for the next 4s. Draw the acceleration-time, velocity-time graphs and determine the distance travelled by the car during the given time period.

Solution: Refer to Fig. 13.12 for the acceleration time plot

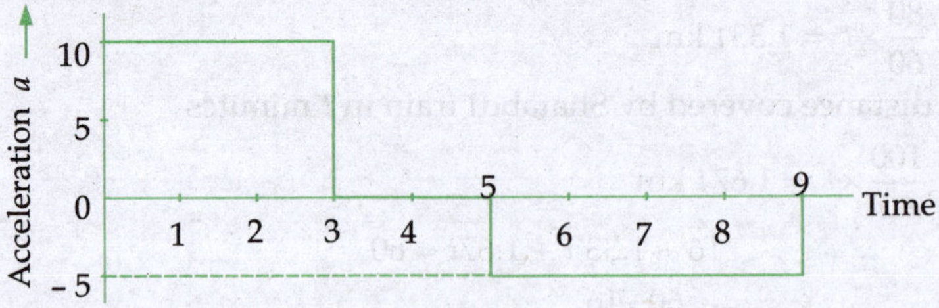

Fig. 13.12

Using the relation $v = u + at$

velocity after $3s = 0 + 10 \times 3 = 30$ m/s

velocity after $9s = 30 + (-5) \times 4 = 10$ m/s

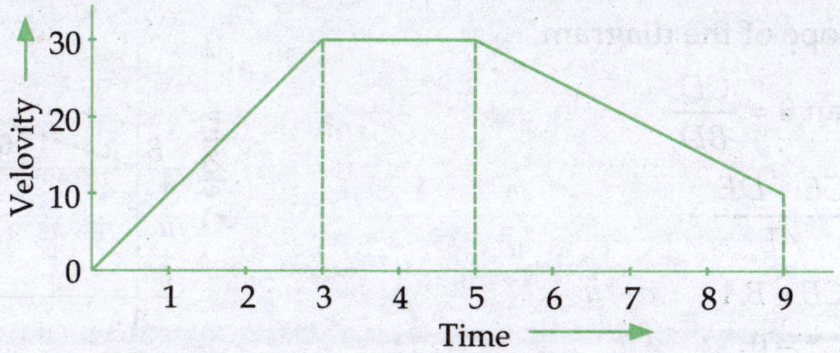

Fig. 13.13

Refer to Fig. 13.12 for the velocity-time graph.

Area under $v - t$ graph gives the distance travelled by the moving car.

$$\therefore \quad \text{Distance travelled} = \left(\frac{1}{2} \times 3 \times 30\right) + (30 \times 2) + \frac{30 + 10}{2} \times 4$$

$$= 45 + 60 + 80 = \mathbf{185\ m}$$

EXAMPLE 13.15

Two stations A and B lie a distance 3.6 km apart. An automobile starts from rest from station P and speeds upto 12 m/s with a constant acceleration of 1.2 m/s². The speed is maintained constant for a certain time and then the brakes are applied. The automobile finally comes to rest at station B with a uniform deceleration of 1.5 m/s². Determine the total time required to perform this journey between the given stations.

Solution: Maximum velocity attained by the automobile during the course of motion is 12 m/s.

Refer Fig. 13.14 which represents velocity time graph for the automobile. The area under the velocity-time graph represents the distance travelled.

$$S = S_1 + S_2 + S_3$$

$$3600 = \left(\frac{1}{2} \times 12 \times t_1\right) + (12 \times t_2) + \left(\frac{1}{2} \times 12 \times t_3\right) = 6t_1 + 12t_2 + 6t_3 \quad ...(i)$$

Considering the acceleration period (*A* to *C*)

$$v = u + at\,; \quad 12 = 0 + 1.2t, \quad \therefore\ t_1 = 10\,s$$

Considering the deceleration period (*D* to *B*)

$$v = u + at\,; \quad 0 = 12 + (-1.5) \times t_3 \quad \therefore\ t_3 = 8\,s$$

Substituting $t_1 = 10\,s$ and $t_3 = 8\,s$ in expression (*i*), we get

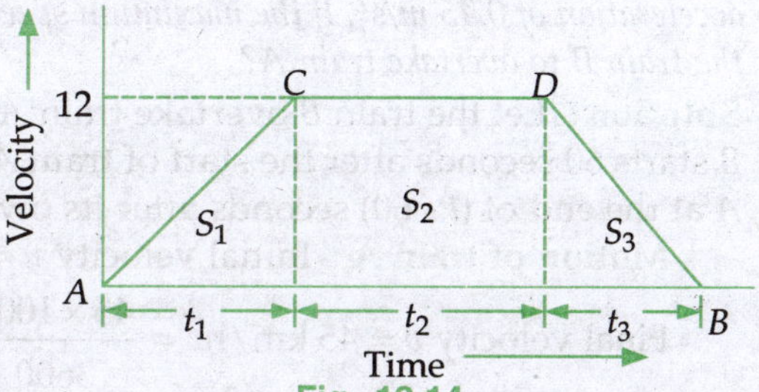

Fig. 13.14

$$3600 = 6 \times 10 + 12t_2 + 6 \times 8$$

$$t_2 = \frac{3600 - 60 - 48}{12} = 291\ \text{s}$$

$\therefore$ Total time taken to cover the distance = 10 + 291 + 8 = **309 s**

EXAMPLE 13.16

Establish the equations of linear motion by refering to velocity-time diagram. Presume that the motion is with uniform acceleration.

Solution: Refer Fig. 13.15 for the velocity-time diagram. At $t = 0$, the velocity has a finite value u and it becomes v at time t. Since the acceleration is uniform, the velocity has a linear variation *i.e.*, the slope of the curve is constant.

acceleration a = slope of the diagram

$$= \tan\theta = \frac{CD}{BD}$$

$$= \frac{CE - DE}{AE}$$

$$= \frac{CE - BA}{t - 0} = \frac{v - u}{t}$$

$$\therefore \quad v = u + at$$

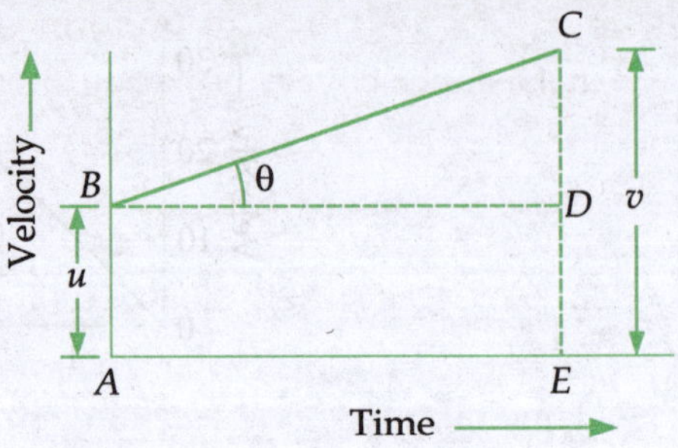

Fig. 13.15

(*ii*) The area under the velocity time graph represents the distance travelled

$$s = \text{area ABCE}$$

$$= \text{area of rectangle ABDE + area of triangle BDC}$$

$$= (AE \times AB) + \frac{1}{2} BD \times CD = (AE \times AB) + \frac{1}{2} BD \times BD \tan\theta$$

$$= (t \times u) + \frac{1}{2} t \times t \times a = ut + \frac{1}{2} at^2$$

(*iii*) $$s = \text{area } ABCE = \frac{AB + EC}{2} \times AE = \frac{u + v}{2} \times t = \frac{u + v}{2} \times \frac{v - u}{a} \qquad \left(\because \; v = u + at, \; t = \frac{v - u}{a}\right)$$

$$= \frac{v^2 - u^2}{2a}$$

$$\therefore \quad v^2 - u^2 = 2as$$

EXAMPLE 13.17

Train A starts with uniform acceleration of 0.25 m/s², attains a speed of 45 kmph which subsequently remains constant. One minute after the start of train A, another train B steams off on a parallel track with uniform acceleration of 0.45 m/s². If the maximum speed gained by train is 60 km/hr, how much time does it take for the train B to overtake train A?

Solution : Let the train B overtake train A after t seconds from the start of train A. Since the train B starts 60 seconds after the start of train A, it also implies that the train B will catch up with train A at the end of (t - 60) seconds after its own start.

Motion of train A : Initial velocity $u = 0$; acceleration $a = 0.25 \text{ m/s}^2$

Final velocity $v = 45 \text{ km/hr} = \dfrac{45 \times 1000}{3600} = 12.5 \text{ m/s}$

Time taken to attain speed of 12.5 m/s,

$$t_1 = \frac{v - u}{a} = \frac{12.5 - 0}{0.25} = 50 \text{ sec}$$

For 50 seconds of its travel, the train is having accelerated motion. Distance travelled during this period is

$$s_1 = ut_1 + \frac{1}{2} at_1^2 = \frac{1}{2} at_1^2 \qquad (\because u = 0)$$

$$= \frac{1}{2} \times 0.25 \times 50^2 = 312.5 \text{ m}$$

Subsequently the train moves with uniform velocity of 12.5 m/s for (t – 50) seconds when it is overtaken. Distance travelled during this period is

s_2 = uniform velocity × time = 12.5 (t – 50)

∴ Distance covered by train A before it is overtaken by train B

$$= 312.5 + 12.5(t - 50) \quad ...(a)$$

Motion of train B :

$$u = 0, \quad a = 0.45 \text{ m/s}^2$$

$$v = \frac{60 \times 1000}{3600} = 16.67 \text{ m/s}$$

Time taken to attain speed of 16.67 m/s,

$$t_2 = \frac{16.67 - 0}{0.45} = 37.04 \text{ sec}$$

Distance travelled during this period of accelerated motion

$$= \frac{1}{2} \times 0.45 \times (37.04)^2 = 308.69 \text{ m}$$

Subsequently the train moves with uniform velocity of 16.67 m/s for [t – 60 – 37.04] sec when it overtakes train A. Distance travelled during this period is

$$= 16.67 \times (t - 97.04)$$

Distance covered by train B when it overtakes train A

$$= 308.69 + 16.67\,(t - 97.04) \quad ...(b)$$

When the train B overtakes train A, the distance covered by both the trains will be equal. That is

$$312.5 + 12.5(t - 50) = 308.69 + 16.67(t - 97.04)$$

or

$$312.5 + 12.5t - 625 = 308.69 + 16.67t - 1617.6$$

That gives $t = \dfrac{312.5 - 625 - 308.69 + 1617.6}{16.67 - 12.5} = $ **238.95 sec**

The solution can be obtained easily from velocity-time graphs for trains A and B shown in Fig. 13.16

V_a = maximum speed attained by train A = 12.5 m/s

t_1 = time taken by train A to attain speed of 12.5 m/s = 50 sec

V_b = maximum speed of train B = 60 km/hr = 16.67 m/s

t_2 = time taken by train B to attain speed of 16.67 m/s = 37.04 sec

At the time of overtake, the distance travelled by both train A and B is same. The distance covered is given by the area under velocity time graph. Accordingly

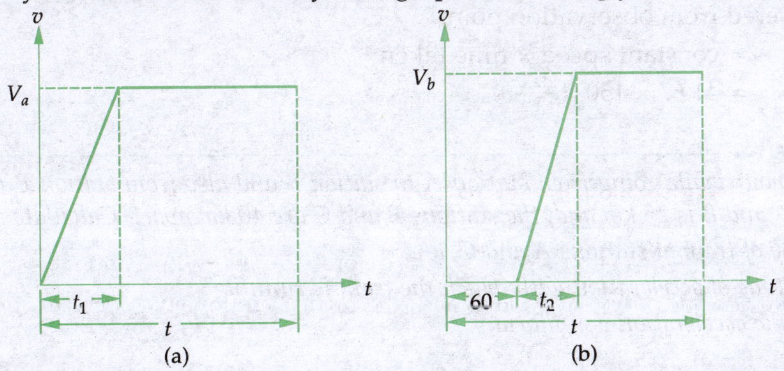

Fig. 13.16

area for car A (Fig. 13.16 a) = area for car B (Fig. 13.16 b)

$$\frac{1}{2} \times t_1 \times V_a + (t - t_1) V_a = \frac{1}{2} t_2 \times V_b + (t - t_2 - 60) V_b$$

or $$\frac{1}{2} \times 50 \times 12.5 + (t - 50) \times 12.5 = \frac{1}{2} \times 37.04 \times 16.67 + (t - 37.04 - 60) \times 16.67$$

or $$312.5 + 12.5t - 625 = 308.73 + 16.67t - 1617.66$$

$$\therefore \quad t = \frac{312.5 - 625 - 308.73 + 1617.66}{16.67 - 12.5} = \textbf{238.95 sec}$$

Hence car B will overtake car A at $t = 238.95$ sec

EXAMPLE 13.18

A truck carrying smuggled goods and speeding at 100 km/hr is observed by a radar equipped police van. The police party starts the pursuit 40 seconds after the observation and the police van accelerates to 150 km/hr in 20 seconds time. How far from the observation point the chase would end if the speeds of the truck and police van are maintained constant on straight road? Obtain your solution by velocity – time graph.

Solution: Speed of truck which is constant throughout

$$V_t = 100 \text{ km/hr} = \frac{100 \times 10^3}{3600} = 27.78 \text{ m/s}$$

For the first 20 seconds, the car accelerates and then maintains constant speed equal to

$$V_c = 150 \text{ km/hr} = \frac{150 \times 10^3}{3600} = 41.67 \text{ m/s}$$

If the pursuit is over at $t = t_1$, then the $v - t$ diagram will be as shown in Fig. 13.17. The velocity chase would end when the truck and the car travel equal distance, and the distance travelled by a moving vehicle is given by area under $v - t$ curve.

$\therefore$ area for the $v - t$ curve of truck

= area for the $v - t$ curve of car

or $$27.78\, t_1 = \frac{1}{2}(60 - 40) \times 41.67 + (t_1 - 60) \times 41.67$$

$$= 416.7 + 41.67 t_1 - 2500.2$$

That gives : $$t_1 = \frac{2500.2 - 416.7}{41.67 - 27.78} = \textbf{150 sec}$$

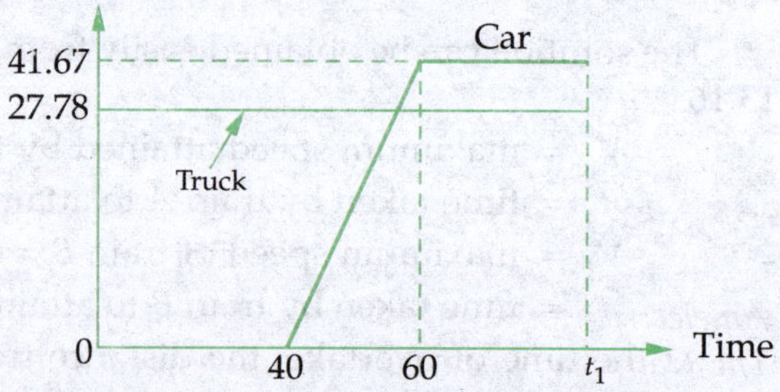

Fig. 13.17

Distance covered from observation point

= constant speed × time taken

= 41.67 × 150 = **6250.5 metre**

EXAMPLE 13.19

A train takes 2/3 hour while going from station A to station B and also from station B to C. The distance between stations A and B is 25 km and, the stations B and C are 40 km apart. Calculate :

(a) *the velocity of train at stations A and C, and*

(b) *the magnitude of acceleration with which the train is running*

Assume that the acceleration is uniform.

Solution : Refer to Fig 13.18 for the velocity-time graph

The area of the velocity-time graph gives the distance travelled during any time interval. Thus,

$$s_1 = \text{area of trapezium } OABE$$

$$= \frac{1}{2} \times \text{sum of parallel sides} \times \text{altitude}$$

$$= \frac{1}{2}(V_a + V_b) \times t_1$$

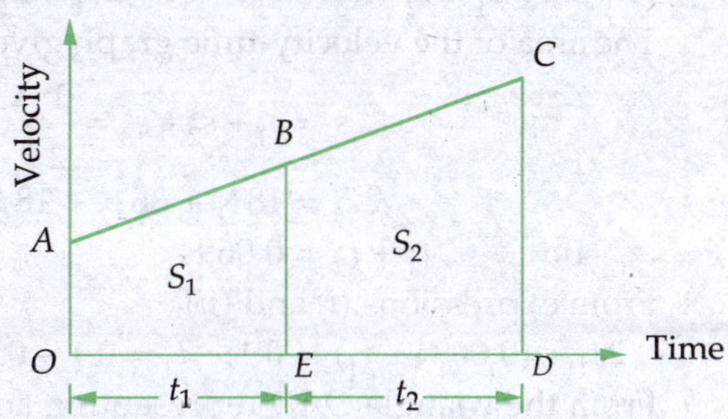

Fig. 13.18

where V_a and V_b are the velocities of the running train as it passes stations A and B respectively.

$$\therefore \quad 25 = \frac{1}{2}(V_a + V_b) \times \frac{2}{3} \quad ; \quad V_a + V_b = 75 \qquad ...(i)$$

Likewise :

$$40 = \frac{1}{2}(V_b + V_c) \times \frac{2}{3} \quad ; \quad V_b + V_c = 120 \qquad ...(ii)$$

Subtracting expression (*i*) from expression (*ii*),

$$V_c - V_a = 45 \qquad ...(iii)$$

Further, $$s_1 + s_2 = \frac{1}{2}(V_a + V_c) \times (t_1 + t_2)$$

$$25 + 40 = \frac{1}{2}(V_a + V_c) \times \frac{4}{3} \quad ; \quad V_a + V_c = 97.5 \qquad ...(iv)$$

Adding expressions (*iii*) and (*iv*), we get

$2V_c = 142.5$; V_c = **71.25 km/hr**

$V_a = 71.25 - 45 =$ **26.25 km/hr**

(*b*) Acceleration a = rate of change of velocity

$$= \frac{V_c - V_a}{t_1 + t_2} = \frac{71.25 - 26.25}{2/3 + 2/3} = \mathbf{33.75\ km/hr^2}$$

EXAMPLE 13.20

An electric train runs between two stations which are 2.5 km apart and takes 6 minutes from start to stop. The train has a constant running speed of 36 km/hr between the end of acceleration and beginning of retardation. If the acceleration and retardation are both uniform and numerically equal to each other, make calculations for their values.

Solution : Refer to Fig 13.19 for the velocity-time graph.

The slope of velocity-time graph represents acceleration.

Since acceleration and retardation are numerically equal to each other, we have :

$$\tan\theta_1 = \tan\theta_2 \, ; \quad \theta_1 = \theta_2$$

Obviously the triangles OAD and BCE are similar in all respects. That gives

$$OD = EC, \quad \text{i.e.,} \quad t_1 = t_3$$

The total travel time from start to stop is given to be 6 minutes, i.e., 0.1 hour

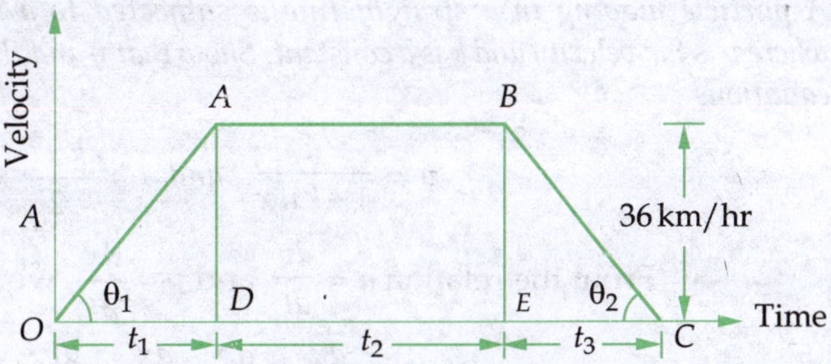

Fig. 13.19

$$\therefore \quad t_1 + t_2 + t_3 = 0.1 \quad \text{or} \quad 2t_1 + t_2 = 0.1 \qquad \text{...}(i)$$

The area of the velocity-time graph gives the distance travelled during any time interval. Thus

$$s = s_1 + s_2 + s_3 = \frac{1}{2} \times 36 \times t_1 + 36 \times t_2 + \frac{1}{2} \times 36 \times t_3$$

$$\therefore \quad 2.5 = 18t_1 + 36t_2 + 18t_3 = 36t_1 + 36t_2 \qquad (\because t_3 = t_1)$$

$$\text{or} \quad t_1 + t_2 = 0.0694 \qquad \text{...}(ii)$$

From expressions (*i*) and (*ii*),

$$2t_1 + (0.0694 - t_1) = 0.1; \quad t_1 = 0.1 - 0.0694 = 0.0306 \text{ hr}$$

From the triangle *OAD* representing accelerating motion,

$$\text{acceleration } a = \text{rate of change of velocity}$$

$$= \frac{V_a - V_o}{t_1} = \frac{36 - 0}{0.0306} = \mathbf{1176.47 \text{ km/hr}^2}$$

Since retardation is stated to be numerically equal to acceleration, the magnitude of retardation is also 1176.47 km/hr^2.

EXMAPLE 13.21

A particle starts with velocity v_o and the acceleration-velocity relationship is prescribed as $a = -kv$ where k is a constant. Set up an expression that prescribes the displacement-time relation for the particle.

Solution : $a = -kv; \quad \frac{dv}{dt} = -kv \quad \text{or} \quad \int_{v_o}^{v} \frac{dv}{v} = -\int_{0}^{t} k\,dt$

$$\text{or} \quad \log_e \frac{v}{v_o} = -kt \quad \therefore \quad v = v_o e^{-kt}$$

Further $v = \frac{dx}{dt}$ and therefore

$$\frac{dx}{dt} = v_o e^{-kt} \quad \text{or} \quad \int_0^x dx = v_o \int_0^t e^{-kt} dt$$

$$\text{or} \quad x = \frac{v_o}{-k} \left| e^{-kt} \right|_0^t = \frac{v_o}{-k}(e^{-kt} - 1)$$

That gives the displacement-time relationship as

$$x = \frac{v_0}{k}(1 - e^{-kt})$$

EXAMPLE 13.22

A particle moving in a straight line is subjected to a resistance which produces a retardation of kv^3 where v is the velocity and k is a constant. Show that v and the time t are given in terms of x (the distance) by the equations

$$v = \frac{u}{1 + kux} \quad \text{and} \quad t = \frac{kx^2}{2} + \frac{x}{u}.$$

Solution : From the relation $a = \frac{dv}{dt}$ and $v = \frac{dx}{dt}$, we have

$$a = \frac{dv}{dt} = \frac{dv}{dx} \times \frac{dx}{dt} = \frac{dv}{dx} \times v$$

$\therefore \quad v\dfrac{dv}{dx} = -kv^3$; – ve sign because of retardation

or $\quad \dfrac{dv}{v^2} = -k\,dx$

upon integration : $\quad \left|-\dfrac{1}{v}\right|_u^v = -kx$

or $\quad -\dfrac{1}{v}+\dfrac{1}{u} = -kx \;;\quad \dfrac{1}{v} = \dfrac{1}{u}+kx = \dfrac{1+kux}{u}$

$\therefore \quad v = \dfrac{u}{1+kux}$

Further $v = \dfrac{dx}{dt}$ and therefore

$$(1 + kux)\,dx = u\,dt$$

Upon integration : $\quad x + \dfrac{kux^2}{2} = ut$

Dividing throughout by u, we get

$$t = \frac{kx^2}{2} + \frac{x}{u}$$

EXAMPLE 13.23

A projectile enters a resisting medium at x = 0 with an initial velocity V_0 = 0.4 m/s and travels 0.1 m before coming to rest. The velocity of the projectile varies with distance according to the relation $V = V_0 - k\,x$ where k is a constant. Calculate the time taken by the projectile to penetrate 0.08 m into the resisting medium.

Solution : $V_0 = 0.4$ m/s and $V = 0$ at $x = 0.1$. Accordingly from the given relation $V = V_0 - k\,x$, we have

$$0 = 0.4 - k \times 0.1 \quad ; \quad k = 4$$

$\therefore \quad V = 0.4 - 4x$

At x = 0.08 m : $V = 0.4 - 4 \times 0.08 = 0.02$ m/s

Now, $\quad a = \dfrac{dV}{dt} = -4\dfrac{dx}{dt} = -4V$. That gives

$$\frac{dV}{V} = -4\,dt$$

Integrating both sides within the prescribed limits

$$\int_{0.4}^{V} \frac{dV}{V} = -4\int_0^t dt$$

or $\quad \left|\log_e V\right|_{0.4}^{V} = -4t \quad ; \quad \log_e V - \log_e 0.4 = -4t$

$\therefore \quad \log_e V = -4t + \log_e 0.4 = -4t - 0.916$

$\therefore \quad \log_e(0.02) = -4t - 0.916$

or $\quad -3.912 = -4t - 0.916 \;;\quad t =$ **0.749 sec**

EXAMPLE 13.24

The rectilinear motion of a motor car starting from rest is prescribed by the relation $a = \dfrac{6}{1.5\,V + 2}$ *where a is the acceleration in* m/s^2 *and V is the velocity in m/s. Calculate the time taken and distance covered by the motor car to attain the velocity of 6 m/s.*

Solution : Using the relation $a = dV/dt$, the given equation can be written as

$$\frac{dV}{dt} = \frac{6}{1.5V + 2}; \quad (1.5\,V + 2)\,dV = 6\,dt$$

Integrating both sides within the limits

At $t = 0$, $V = 0$ and at $t = t$, $V = 6$ m/s, we get

$$\int_o^6 (1.5\,V + 2)dV = 6\int_o^t dt \quad \text{or} \quad \left|1.5\frac{V^2}{2} + 2V\right|_0^6 = 6t$$

or $$1.5 \times \frac{6^2}{2} + 2 \times 6 = 6t; \quad t = \textbf{6.5 sec}$$

(*b*) Using the relation $a = V\,dV/dx$, the given equation can be written as

$$V\frac{dV}{dx} = \frac{6}{1.5\,V + 2} \quad \text{or} \quad (1.5V^2 + 2V)\,dV = 6\,dx$$

Integrating both sides within the limits

At $x = 0$, $V = 0$ and at $x = x$, $V = 6$ m/s

$$\int_o^6 (1.5\,V^2 + 2V)\,dV = 6\int_o^x dx \quad \text{or} \quad \left|1.5\frac{V^3}{3} + 2\frac{V^2}{2}\right|_o^6 = 6x$$

or $$1.5 \times \frac{6^3}{3} + 2 \times \frac{6^2}{2} = 6x \;; \quad x = \textbf{24 m}$$

Thus the distance travelled by the motor car is 24 m

EXAMPLE 13.25

A train travels along a straight track such that the distance travelled is directly proportional to the cube of the time interval. If it travels 480 m during the first minute, determine the displacement, velocity and acceleration 15 seconds after the start.

Solution : Since the distance travelled is directly proportional to cube of time interval, we may write

$$x \propto t^3 \;; \quad x = ct^3 \quad \text{where } c \text{ is constant}$$

Substituting the given data : $x = 480$ m at $t = 1$ min $= 60$ sec

$$480 = c \times 60^3 \;; \quad c = 2.22 \times 10^{-3}$$

Thus the relationship between displacement and time is

$$x = 2.22 \times 10^{-3}\,t^3$$

Differentiating with respect to time t,

$$\frac{dx}{dt} = v = 2.22 \times 10^{-3} \times (3t^2) = 6.66 \times 10^{-3}\,t^2$$

and differentiating once again,

$$\frac{d^2x}{dt^2} = \frac{dv}{dt} = a = 6.66 \times 10^{-3} \times (2t) = 13.32 \times 10^{-3} t$$

The expressions for the displacement, velocity and acceleration are, therefore,

Displacement $x = 2.22 \times 10^{-3} t^3$; velocity $v = 6.66 \times 10^{-3} t^2$

and acceleration $a = 13.32 \times 10^{-3} t$

Substituting $t = 15$ seconds after the start

$$x = 2.22 \times 10^{-3} \times (15)^3 = \mathbf{7.492\ m}$$

$$v = 6.66 \times 10^{-3} \times (15)^2 = \mathbf{1.4985\ m/s}$$

$$a = 13.32 \times 10^{-3} \times (15) = \mathbf{0.1998\ m/s^2}$$

EXAMPLE 13.26

A particle moves along a straight line and its motion is represented by the equation

$$s = 16t + 4t^2 - 3t^3$$

where s is in metres and t is in seconds. Determine :

(a) *displacement, velocity and acceleration 2 seconds after start*

(b) *displacement and acceleration when velocity is zero*

(c) *displacement and velocity when acceleration is zero*

Solution : $s = 16t + 4t^2 - 3t^3$

Differentiating with respect to t,

$$\frac{ds}{dt} = 16 + 8t - 9t^2 ; \quad v = 16 + 8t - 9t^2$$

Again differentiating with respect to t,

$$\frac{d^2s}{dt^2} = 8 - 18t ; \quad a = 8 - 18t$$

(a) when $t = 2$, i.e., 2 seconds after start

$$s = 16 \times 2 + 4 \times (2)^2 - 3 \times (2)^3 = 32 + 16 - 24 = \mathbf{24\ m}$$

$$v = 16 + 8 \times 2 - 9 \times (2)^2 = 16 + 16 - 36 = \mathbf{-4\ m/s}$$

$$a = 8 - 18 \times 2 = 8 - 36 = \mathbf{-28\ m/s^2}$$

(b) At condition of zero velocity,

$$0 = 16 + 8t - 9t^2 ; \quad 9t^2 - 8t - 16 = 0$$

$$t = \frac{8 \pm \sqrt{64 - 4 \times 9 \times (-16)}}{2 \times 9} = \frac{8 \pm 25.3}{18}$$

The particle will have zero velocity at $t = \frac{8 + 25.3}{18} = 1.85\,\text{s}$

Displacement at 1.85 second after start

$$= 16 \times 1.85 + 4 \times (1.85)^2 - 3 \times (1.85)^3$$

$$= 29.6 + 13.69 - 18.99 = \mathbf{24.3\ m}$$

Acceleration at 1.85 second after start,

$$= 8 - 18 \times 1.85 = 8 - 33.3 = \mathbf{-25.3\ m/s^2}$$

(c) Corresponding to zero acceleration

$$8 - 18t = 0 ; \quad t = 0.444\,\text{s}$$

Displacement at 0.444 second after start

$$= 16 \times 0.444 + 4 \times (0.444)^2 - 3 \times (0.444)^3$$
$$= 7.104 + 0.788 - 0.262 = \mathbf{7.63\ m}$$

Velocity at 0.444 s after start

$$= 16 + 8 \times 0.444 - 9 \times (0.444)^2 \quad = 16 + 3.552 - 1.774 = \mathbf{17.775\ m/s}$$

It is worthwhile to remark that the particle will have maximum velocity when acceleration is zero. Thus

$$V_{max} = 17.775 \text{ m/s}$$

EXAMPLE 13.27

A particle moves along a straight line with an acceleration prescribed by the relation :

$$a = (4t^2 - 3t + 2) \text{ where } a \text{ is in m/s}^2 \text{ and } t \text{ is in seconds}$$

The particle has a velocity of 10 m/s at t = 3 seconds, and it is located 12 m to the right of origin at t = 2 seconds. Determine the position and velocity of the particle after 5 seconds.

Solution : From the relation $a = dv/dt$, we have

$$dv = a\,dt = (4t^2 - 3t + 2)\,dt$$

$$v = \int (4t^2 - 3t + 2)dt = 4\frac{t^3}{3} - 3\frac{t^2}{2} + 2t + C_1$$

From the given condition : $v = 10$ m/s at $t = 3$ seconds, we have

$$10 = 4 \times \frac{3^3}{3} - 3 \times \frac{3^2}{2} + 2 \times 3 + C_1 = 36 - 13.5 + 6 + C_1$$

or

$$C_1 = 10 - 36 + 13.5 - 6 = -18.5$$

$$\therefore \quad v = \frac{4}{3}t^3 - \frac{3}{2}t^2 + 2t - 18.5 \qquad ...(i)$$

(*a*) When $t = 5$ seconds, the velocity of the particle becomes

$$v = \frac{4}{3} \times (5)^3 - \frac{3}{2} \times (5)^2 + 2 \times 5 - 18.5$$
$$= 166.67 - 37.5 + 10 - 18.5 = \mathbf{120.67\ m/s}$$

(*b*) Using the relation $v = dx/dt$, equation (*i*) can be written as

$$dx = \left(\frac{4}{3}t^3 - \frac{3}{2}t^2 + 2t - 18.5\right)dt$$

Upon integration: $x = \int \left(\frac{4}{3}t^3 - \frac{3}{2}t^2 + 2t - 18.5\right)dt = \frac{t^4}{3} - \frac{t^3}{2} + t^2 - 18.5t + C_2$

From the given condition $x = 12$ m at $t = 2$ seconds, we have

$$12 = \frac{2^4}{3} - \frac{2^3}{2} + 2^2 - 18.5 \times 2 + C_2 = 5.33 - 4 + 4 - 37 + C_2$$

or

$$C_2 = 12 - 5.33 + 4 - 4 + 37 = 43.67$$

$$\therefore \quad x = \frac{t^4}{3} - \frac{t^3}{2} + t^2 - 18.5t + 43.67$$

Hence when $t = 5$ seconds, the displacement of the particle from origin becomes

$$x = \frac{5^4}{3} - \frac{5^3}{2} + 5^2 - 18.5 \times 5 + 43.67$$
$$= 208.33 - 62.5 + 25 - 92.5 + 43.67 = \mathbf{122\ m}$$

EXAMPLE 13.28

The acceleration of a particle is expressed as a = 10 – x. The particle starts with no initial velocity at the position x = 0. Determine (i) the velocity of the particle when x = 8 m, (ii) the position of the particle where the velocity is again zero, and (iii) the velocity of the particle when acceleration becomes zero.

Solution : From the relation $a = \frac{dV}{dt}$ and $V = \frac{dx}{dt}$, we have

$$a = \frac{dV}{dt} = \frac{dV}{dx} \times \frac{dx}{dt} = \frac{dV}{dx} \times V$$

$$\therefore \quad V\,dV = a\,dx = (10 - x)\,dx$$

upon integration : $\frac{V^2}{2} = \left[10x - \frac{x^2}{2}\right] + C$

The constant of integration C can be worked out from the condition that the particle starts from rest. That is

At t = 0, $V = 0$ and $x = 0$

That gives C = 0

$$\therefore \quad V^2 = 20x - x^2 \quad \text{and} \quad a = 10 - x$$

(*i*) Velocity of the particle when $x = 8$ m

$$V^2 = 20 \times 8 - 8^2 = 96 \,; \quad V = 9.798 \text{ m/s}$$

(*ii*) Position of the particle when the velocity becomes zero

$$V^2 = 20x - x^2 = x(20 - x) = 0$$

$$\therefore \quad x = 0 \quad \text{and} \quad x = 20 \text{ m}$$

(*iii*) Velocity of the particle when acceleration is zero

$$a = 10 - x = 0 \quad ; \quad x = 10 \text{ m}$$

$$V^2 = 20 \times 10 - 10^2 = 100 \;; \quad V = 10 \text{ m/s}$$

EXAMPLE 13.29

A particle moves along horizontal direction and its position at any instant is prescribed by the relation

$$x = 3t^3 - 5t^2$$

where x is in metres and t is in seconds.

Make calculations for the following parameters :

(*a*) *displacement during t = 2 sec to 5 sec*

(*b*) *average velocity during t = 2 sec to 5 sec and instantaneous velocity at t = 2 sec and 5 sec.*

(*c*) *average acceleration during t = 2 sec to 5 sec and instantaneous acceleration at t = 2 sec and 5 sec.*

(*d*) *distance travelled in the first four seconds (t = 0 to t = 4 sec).*

Solution : $x = 3t^3 - 5t^2$

(*a*) Displacement during $t = 2$ sec to 5 sec

$$x_5 - x_2 = (3 \times 5^3 - 5 \times 5^2) - (3 \times 2^3 - 5 \times 2^2)$$

$$= (375 - 125) - (24 - 20) = 246 \text{ m}$$

(*b*) Average velocity and instantaneous velocity

Let subscripts *f* and *i* denote the final and initial values

$$V_{av} = \frac{x_f - x_i}{t_f - t_i} = \frac{246}{5 - 2} = 82 \text{ m/s}$$

Upon differentiation of displacement x with respect to time t, one obtains the expression for instantaneous velocity.

$$V = \frac{dx}{dt} = \frac{d}{dt}(3t^3 - 5t^2) = 9t^2 - 10t$$

Instantaneous velocity at $t = 2$ sec and $t = 5$ sec then works out as

$$V_2 = 9 \times 2^2 - 10 \times 2 = \mathbf{16\ m/s}$$

$$V_5 = 9 \times 5^2 - 10 \times 5 = \mathbf{175\ m/s}$$

(c) Average acceleration and instantaneous acceleration

$$a_{av} = \frac{V_f - V_i}{t_f - t_i} = \frac{175 - 16}{5 - 2} = \mathbf{53\ m/s^2}$$

Upon differentiation of velocity V with respect to time t, one obtains the expression for instantaneous acceleration :

$$a = \frac{dV}{dt} = \frac{d}{dt}(9t^2 - 10t) = 18t - 10$$

Instantaneous acceleration at $t = 2$ sec and $t = 5$ sec then works out as

$$a_2 = 18 \times 2 - 10 = \mathbf{26\ m/s^2}$$

$$a_5 = 18 \times 5 - 10 = \mathbf{80\ m/s^2}$$

(d) Distance travelled in the first 4 seconds :

From the given expression $x = 3t^3 - 5t^2$, it is evident that upto certain value of t, the displacement is negative. At the instant when change in sign occurs, dx/dt is zero. That gives

$$9t^2 - 10t = 0; \quad t = 0, 1.11$$

$$x \text{ at } 1.11 = 3 \times (1.11)^3 - 5 \times (1.11)^2 = -2.057 \text{ m}$$

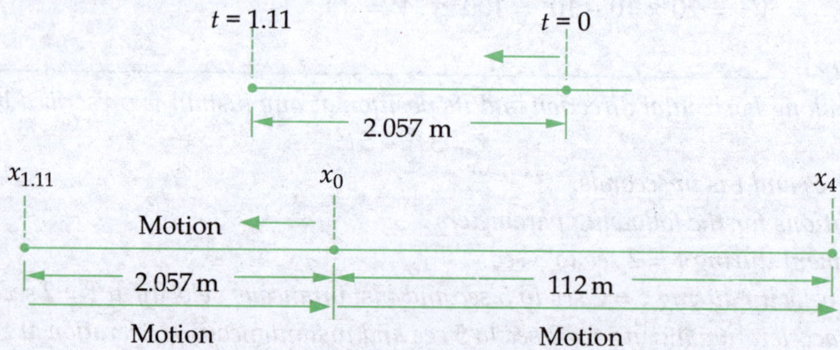

Fig. 13.20

From $t = 0$ to $t = 1.11$, (dx/dt) is negative, hence the motion is towards the left.

When $t > 1.11$, (dx/dt) is positive; hence the motion is towards right.

At $t = 4 : x = 3 \times 4^3 - 5 \times 4^2 = 112$ m

$\therefore$ Total distance travelled = 2.057 + 2.057 + 112 = **116.114 m**

EXAMPLE 13.30

The velocity-time graph for a motor car on a straight road is as shown in Fig. 13.21. What will be the acceleration of the car at s = 25 m and s = 75 m?

Solution : Acceleration $a = \frac{dv}{dt} = \frac{dv}{ds} \times \frac{ds}{dt}$

Since $ds/dt = v$, we have

$$a = v\frac{dv}{ds}$$

Further dv/ds represents slope of v-s curve

(*i*) At $s = 25$ m : $v = 5$ m/s and

$$\frac{dv}{ds} = \frac{5-0}{25-0} = 0.2$$

$\therefore$ acceleration $a = 5 \times 0.2 =$ **1 m/s²**

(*ii*) At $s = 75$ m : $v = 5$ m/s

and $$\frac{dv}{ds} = \frac{0-5}{100-75} = -0.2$$

$\therefore$ acceleration $a = 5 \times (-0.2) = -1$ m/s²

The negative sign implies that at $s = 75$ m, the motor car will be undergoing deceleration.

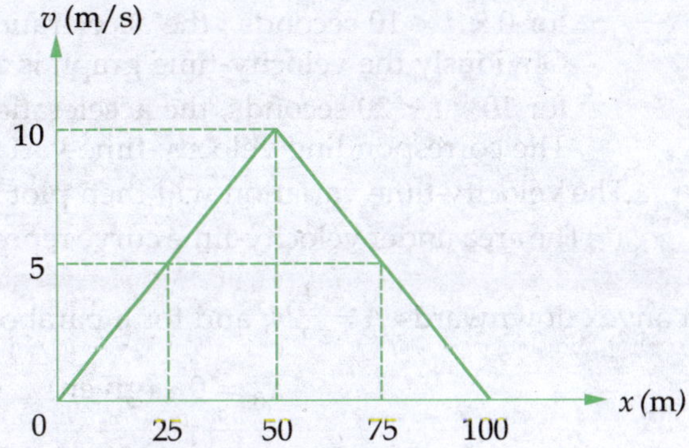

Fig. 13.21

EXMAPLE 13.31

The variation of acceleration with time for a particle undergoing motion is as depicted in Fig. 13.22. The particle starts from rest and moves along a straight horizontal path.

Plot the velocity-time and the displacement-time graphs. Proceed to calculate the maximum speed attained and the maximum distance traversed by the particle.

Solution : The area under acceleration-time curve represents change in velocity.

$$V_0 = 0 \quad \text{(given)}$$

$$V_{10} - V_0 = \frac{1}{2} \times 10 \times 6 = 30$$

$$\therefore \quad V_{10} = 30 \text{ m/s}$$

$$V_{20} - V_{10} = \frac{1}{2} \times (20-10) \times 5 = 30$$

$$\therefore \quad V_{20} = 60 \text{ m/s}$$

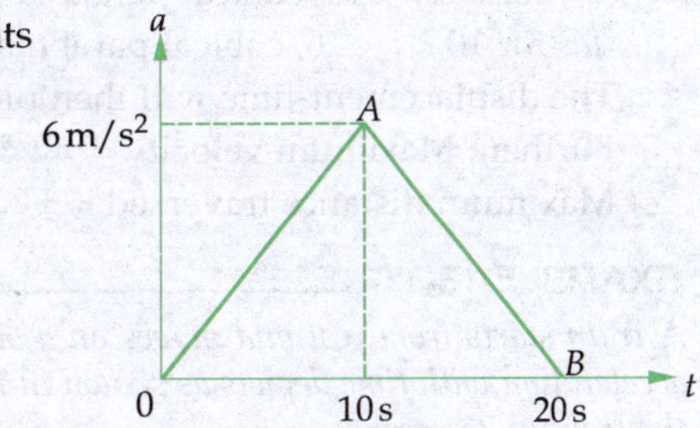

Fig. 13.22

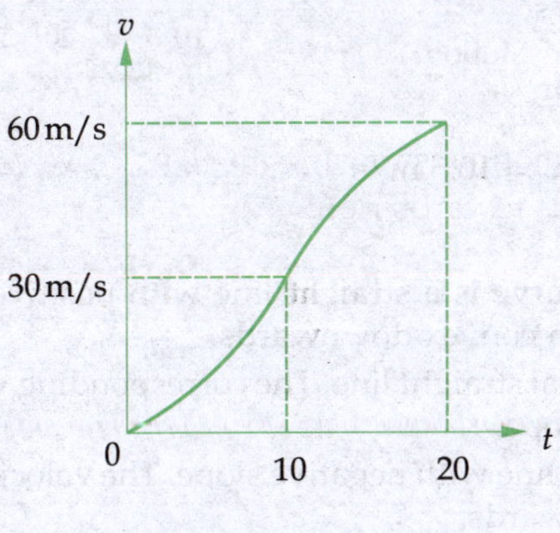

Fig. 13.23

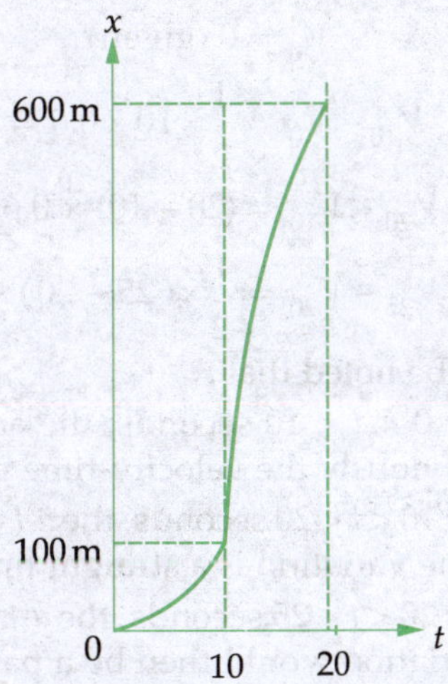

Fig. 13.24

It is to be noted that :

(*i*) for $0 < t < 10$ seconds ; the acceleration-time curve is a straight line with positive slope. Obviously the velocity-time graph is a parabola convex downwards.

(*ii*) for $10 < t < 20$ seconds, the acceleration-time curve is a straight line with negative slope. The corresponding velocity-time variation is a parabola convex upwards.

The velocity-time variation will then plot as shown in Fig. 13.23.

(*b*) The area under velocity-time curve represents change in displacement. Further for a parabola convex downwards $A = \frac{1}{3}bh$ and for a parabola convex upwards $A = \frac{2}{3}bh$

$$x_0 = 0 \quad \text{(given)}$$

$$x_{10} - x_0 = \frac{1}{3} \times 10 \times 30 = 100$$

$$\therefore \quad x_{10} = 100 \text{ m}$$

$$x_{20} - x_{10} = \frac{2}{3} \times 10 \times 30 + 10 \times 30 = 500$$

$$\therefore \quad x_{20} = 600 \text{ m}$$

It is to be noted that :

(*i*) for $0 < t < 10$ seconds and for $10 < t < 20$, the slope of *v-t* curve is positive

(*ii*) for $0 < t < 10$, cubical parabola is convex downwards

(*iii*) for $10 < t < 20$, cubical parabola is convex upwards

The displacement-time will then plot as shown in Fig. 13.24.

Further : Maximum velocity = **60 m/s**

Maximum distance traversed = **600 m**

EXAMPLE 13.32

A train starts from rest and moves on a straight road along the horizontal direction. The variation of its acceleration with time depicts as shown in Fig. 13.25. Calculate the speed attained and distance covered by the train in 25 seconds.

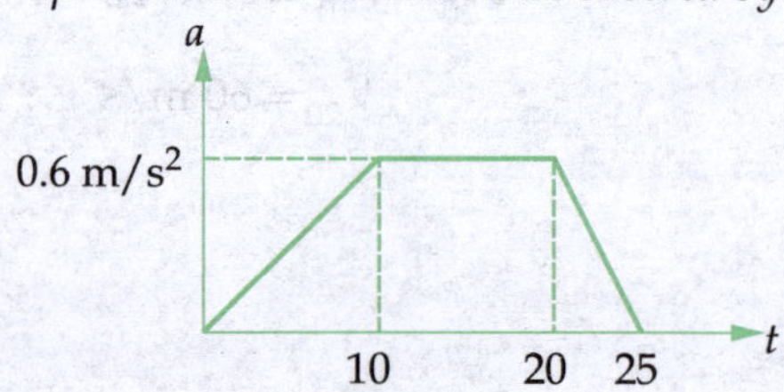

Fig. 13.25

Solution : The area under acceleration-time curve represents change in velocity,

$$V_0 = 0 \text{ (given)}$$

$$V_{10} - V_0 = \frac{1}{2} \times 10 \times 0.6 = 3 \quad \therefore \quad V_{10} = 3 \text{ m/s}$$

$$V_{20} - V_{10} = (20 - 10) \times 0.6 = 6 \quad \therefore \quad V_{20} = 9 \text{ m/s}$$

$$V_{25} - V_{20} = \frac{1}{2} \times (25 - 20) \times 0.6 = 1.5 \quad \therefore \quad V_{25} = 10.5 \text{ m/s}$$

It is to be noted that :

(*i*) for $0 < t < 10$ seconds, the acceleration-time curve is a straight line with positive slope. Obviously the velocity-time graph is a parabola convex downwards.

(*ii*) for $10 < t < 20$ seconds, the *a-t* curve is a horizontal straight line. The corresponding velocity time variation is a straight line with positive slope.

(*iii*) for $20 < t < 25$ seconds, the *a-t* curve is a straight line with negative slope. The velocity-time variation would then be a parabola convex upwards.

The velocity time variation would then plot as shown in Fig. 13.26.

Distance travelled = area of velocity-time graph

$$S = s_1 + s_2 + s_3$$

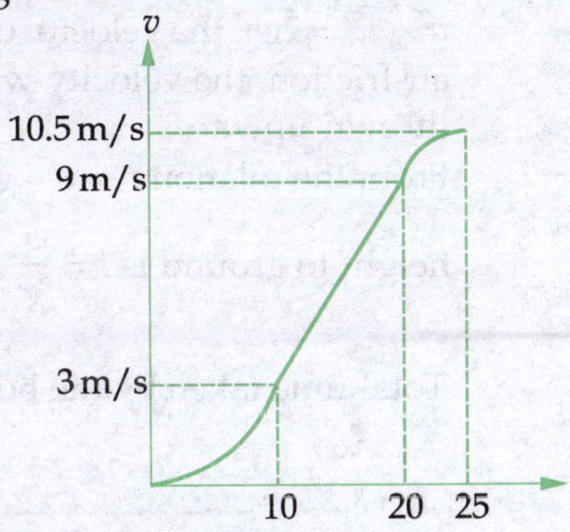

Fig. 13.26

Recalling that for a parabola convex downwards $A = \frac{1}{3} bh$

and for a parabola convex upwards $A = \frac{2}{3} bh$, we have

$$s_1 = \frac{1}{3} \times 10 \times 3 = 10 \text{ m}$$

$$s_2 = \frac{1}{2} \times (20 - 10) \times (9 - 3) + (20 - 10) \times 3 = 60 \text{ m}$$

$$s_3 = \frac{2}{3} \times (25 - 20) \times (10.5 - 9) + (25 - 20) \times (10.5 - 9) = 12.5$$

∴ Total distance travelled by the train 25 seconds = 10 + 60 + 12.5 = 82.5 m

13.4. MOTION UNDER GRAVITY

When a body falls freely, then its velocity increases as it approaches the earth. The increase in the velocity of falling body is due to gravitational acceleration g whose value is normally taken to be 9.81m/s^2. When motion is considered under gravity, the acceleration, a in the equations of motion for rectilinear motion is replaced by g and we obtain

$$v = u + gt \,;\, h = ut + \frac{1}{2} gt^2 \,; \quad v^2 - u^2 = 2gh$$

and distance travelled in the n th second is

$$s_n = u + \frac{g}{2}(2n - 1)$$

(*i*) When the body falls from height h, then $u = 0$ and hence

$$v = gt \,; \quad h = \frac{1}{2} gt^2 \,; \quad v^2 = 2gh$$

and $$s_n = \frac{g}{2}(2n - 1)$$

(*ii*) When the bodies are thrown vertically upwards, then g becomes a retarding parameter. The velocity of the body then decreases gradually till it becomes zero. For a body thrown in vertically upward direction, the equations of motion are

$$v = u - gt \,; \quad h = ut - \frac{1}{2} gt^2 \,; \quad v^2 - u^2 = -2gh$$

and $$s_n = u - \frac{g}{2}(2n - 1)$$

(*iii*) At the point of maximum height attained by a body thrown vertically upwards, the velocity v of body becomes zero. Then from expression $v = u - gt$ and $v^2 - u^2 = -2gh$, we have

Time taken in attaining maximum height = $\dfrac{u}{g}$

Maximum height attained = $\dfrac{u^2}{2g}$

(*iv*) After attaining the maximum height, the body starts falling down. Then from the relation $v^2 - u^2 = 2gh$, the velocity of the body on reaching the ground will be $v = \sqrt{2gh}$. Neglecting air friction, the velocity with which the body reaches the ground is same with which it is thrown upwards.

(*v*) From the relation $v = u - gt$, the time taken by the body to fall from position of maximum height to ground is $t = \frac{v}{g} = \frac{u}{g}$

$\therefore$ Total time taken by the body to fall back to ground is $\frac{2u}{g}$

EXAMPLE 13.33

A body is allowed to fall from the top of a tower. It falls through half the height in 2 seconds. What will be the total time to reach the ground?

Take $g = 10$ m/s^2.

Solution: $s = ut + \frac{1}{2}gt^2$ and $u = 0$

$$\frac{s}{2} = 0 \times t + \frac{1}{2} \times 10 \times (2)^2 = 20$$

$\therefore$ Height of tower $s = 40$ m

Again, $$40 = \frac{1}{2} \times 10 \times t^2 \quad \therefore \quad t \simeq 2.8 \text{ s}$$

EXAMPLE 13.34

A stone takes 6 seconds to reach the ground after it is dropped from the tower. If the stone is stopped after 2 seconds of its fall and then released again, how much more time it will take to reach the ground. Take $g = 10$ m/s^2.

Solution: $s = ut + \frac{1}{2}gt^2$ and $u = 0$

$\therefore$ Height of tower $s = 0 \times 6 + \frac{1}{2} \times 10 \times 6^2 = 180$ m

Distance covered in 2 seconds $= \frac{1}{2} \times 10 \times 2^2 = 20$ m

Let t be the time taken by the stone to cover the remaining 160 m, starting from rest.

$$160 = \frac{1}{2} \times 10 \times t^2 \quad \therefore \quad t = 5.6 \text{ s}$$

EXAMPLE 13.35

A body is dropped from rest at height h. It covers a distance of 9h/25 in the last second. Determine the height h. Take $g = 10$ m/s^2.

Solution: Let t be the time taken to reach the ground level. Then

$$h = \frac{1}{2}gt^2 \qquad (\because u = 0)$$

Also $$\left(1 - \frac{9}{25}\right)h = \frac{1}{2}g(t-1)^2$$

Dividing the above identities and solving for t, we get

$$\frac{16}{25} = \frac{(t-1)^2}{t^2} \; ; \quad \frac{4}{5} = \frac{t-1}{t} \quad \therefore \; t = 5 \text{ s}$$

$$\therefore \quad \text{Height } h = \frac{1}{2} \times 10 \times 5^2 = 125 \text{ m}$$

EXAMPLE 13.36

Water drips from a tap at uniform rate onto the floor which is 1.5 m below. The first drop strikes the floor at the instant the fourth drop is on the verge of falling from the tap. What would be the location of the different drops when the first drop hits the floor?

Solution : The drops fall at regular intervals of time say t. When the first drop D_1 strikes the floor, the fourth drop D_4 begins to fall from the tap. This implies that the time interval between the first and 4th drop is $3t$, between the first and third drop is $2t$ and between the first and second drop is t.

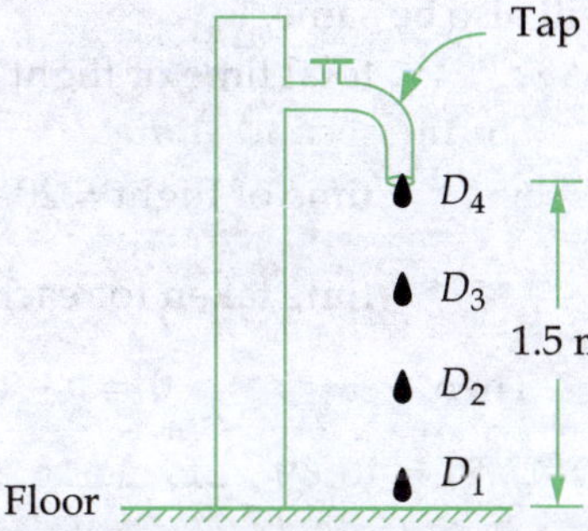

Fig. 13.27

Applying the relation $s = ut + \frac{1}{2}gt^2$ and with $u = 0$,

we have

$$\text{Vertical distance } D_1D_4 = 1.5 = \frac{1}{2} \times 9.81 \times (3t)^2$$

$$t = \sqrt{\frac{2 \times 1.5}{9 \times 9.81}} = 0.1843 \text{ s}$$

Then : $$D_3D_4 = \frac{1}{2} \times 9.81 \times (0.1843)^2 = 0.167 \text{ m}$$

$$D_2D_4 = \frac{1}{2} \times 9.81 \times (2 \times 0.1843)^2 = 0.667 \text{ m}$$

Accordingly : height of drop D_3 above the floor = (1.5 – 0.167) = 1.333 m

height of drop D_2 above the floor = (1.5 – 0.667) = 0.833 m

EXAMPLE 13.37

A ball of mass 2 kg was thrown vertically upwards and at a height of 5 m above the ground, its kinetic energy was equal to its potential energy. How much total time does the ball take to get back to the point from where it was thrown? Take $g = 10$ m/s^2.

Solution: At a height of 5 m above the ground;

$$mgh = \frac{1}{2}mv^2 \; ; \quad v = \sqrt{2gh} = \sqrt{2 \times 10 \times 5} = 10 \text{ m/s}$$

Also $$v^2 - u^2 = 2gh \; ; \quad 10^2 - u^2 = 2 \times (-10) \times 5$$

That gives initial velocity of ball $u = 10\sqrt{2}$ m/s

At maximum height, the ball will have zero velocity. Accordingly

$$0 = u - gt \; ; \quad t = \frac{u}{g} = \frac{10\sqrt{2}}{10} = \sqrt{2} \text{ second}$$

Hence time taken in ascending the maximum height will be $\sqrt{2}$ second. The time taken for return path will also be same. Hence total time taken will be $2\sqrt{2}$ seconds.

EXAMPLE 13.38

A stone was thrown vertically upwards from the ground with a velocity of 100 m/s. After 2 seconds, another stone was thrown vertically upwards from the same place. If both the stones strike the ground at the same time, calculate the velocity with which the second stone was thrown. Take $g = 10$ m/s².

Solution: At maximum height, the stone will have zero velocity. Accordingly considering upward motion of first stone.

$$0 = u - gt; \quad t = \frac{u}{g} = \frac{100}{10} = 10 \text{ s}$$

Thus the stone will take 10 s in ascending to maximum height. The time taken for return path will also be same.

$\therefore$ total time of flight = 10 + 10 = 20 s

For the second stone:

time of flight = 20 − 2 = 18 s

time taken to reach maximum height $= \frac{18}{2} = 9$ s

Then $0 = u - 10 \times 9; \quad u =$ **90 m/s**

EXAMPLE 13.39

A stone is dropped into a well and the sound of splash is heard after 4 seconds. Assuming velocity of sound to be 350 m/s, make calculations for the depth of well.

Solution: Let h = the depth of well

t_1 = time taken by stone to strike water

t_2 = time taken by sound to reach from surface of water to top of well

Then $t_1 + t_2 = 4$ ...(*i*)

Considering downward motion of stone and using the relation $s = ut + \frac{1}{2}gt^2$, we have

$$h = 0 \times t_1 + \frac{1}{2} \times 9.81 \times t_1^2; \quad h = 4.905\, t_1^2$$

Considering the motion of sound, the time taken by the sound to reach from surface of water to top of well is given by

$$t_2 = \frac{\text{depth of well}}{\text{speed of sound}} = \frac{h}{350} = \frac{4.905 t_1^2}{350} \qquad (\because h = 4.905 t_1^2)$$

Then from identity (*i*),

$$t_1 + \frac{4.905 t_1^2}{350} = 4; \quad 350 t_1 + 4.905 t_1^2 = 1400$$

$$\therefore \quad 4.905\, t_1^2 + 350 t_1 - 1400 = 0$$

Solution of this quadratic equation gives

$$t_1 = \frac{-350 \pm \sqrt{350^2 + 4 \times 4.905 \times 1400}}{2 \times 4.905} = \frac{-350 \pm 387.26}{9.81}$$

Taking the + ve root; $t_1 = 3.798$ s

$\therefore$ Depth of well $h = 4.905\, t_1^2 = 4.905 \times 3.798^2 =$ **70.75 m**

EXAMPLE 13.40

A stone is dropped from the top of a tower 40 m height. At the same instant, another stone is thrown upward from the foot of tower with an initial velocity of 20 m/s. At what distance from the top and after how much time the two stones cross each other ? Further proceed to calculate the relative velocity with which the stones cross.

Solution: Let the two stones cross each other at distances h from the top of tower after t seconds.

Applying the relation $s = ut + \frac{1}{2}at^2$

For the downward motion of first stone,

$$u = 0; \quad s = h \text{ and } a = g = 9.81 \text{ m/s}^2$$

$$\therefore \quad h = 0 \times t + \frac{1}{2} \times 9.81 \times t^2$$

$$= 4.905\,t^2 \qquad ...(i)$$

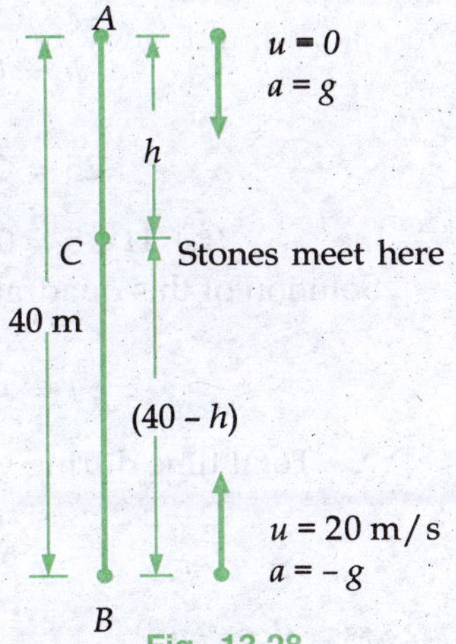

Fig. 13.28

For the upward motion of second stone

$$u = 20 \text{ m/s};$$

$$s = (40 - h)$$

and $\quad a = -g = -9.81 \text{ m/s}^2$

$$\therefore \quad 40 - h = 20\,t + \frac{1}{2} \times (-9.81)t^2$$

$$= 20\,t - 4.905\,t^2 \qquad ...(ii)$$

Adding expressions (*i*) and (*ii*), we get

$$40 = 20\,t; \quad \therefore \; t = \textbf{2 second}$$

$$h = 4.905\,t^2 = 4.905 \times 2^2 = \textbf{19.62 m}$$

The stones thus cross each other at a distance of 19.62 m from the top of the tower.

(*b*) velocity of first stone at the crossing point

$$v = u + gt = 0 + 9.81 \times 2 = 19.62 \text{ m/s} \downarrow$$

velocity of second stone at the crossing point

$$v = u - gt = 20 - 9.81 \times 2 = 0.38 \text{ m/s} \uparrow$$

Relative velocity $= 19.62 - (-0.38) =$ **20 m/s**

EXAMPLE 13.41

A stone is thrown vertically upwards with a velocity 20 m/s from the top of a tower 25 m high.. Make calculations for the following parameters:

(*i*) *the maximum height to which the stone will rise in its flight*

(*ii*) *velocity of the stone during its downward travel at a point in the same level as the point of projection,*

(*iii*) *time required for the stone to reach the ground.*

Take $g = 10$ m/s²

Solution:

(*i*) **Motion B to C:** Let t be the time taken by the stone to reach the maximum height (level C).

Using the relation: $v = u + at$

$$u = 20 \text{ m/s}, \quad v = 0 \quad \text{and} \quad a = -g = -10 \text{ m/s}^2$$

$$\therefore \quad 0 = 20 - 10\,t; \quad t = 2 \text{ s}$$

Also $\quad v^2 - u^2 = 2\,ah$

$$0^2 - 20^2 = 2 \times (-10) \times h; \quad h = 20 \text{ m}$$

Total height from ground = 20 + 25 = **45 m**

(ii) **Motion C to B**

The ball took 2 s while ascending from level B to maximum height (level C). It will take same time during return journey from level C to B.

Velocity at B, $v_b = u + gt$

$= 0 + 10 \times 2 =$ **20 m/s**

Apparently the velocity with which the stone reaches level B (top of tower) is same with which it was thrown upwards.

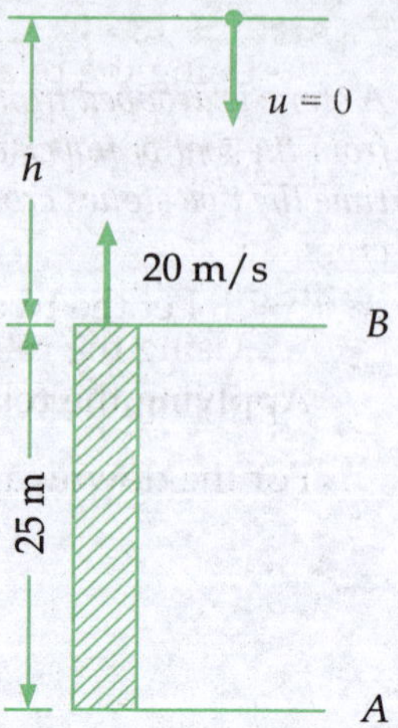

Fig. 13.29

(iii) **Motion B to A**

$$h = ut + \frac{1}{2}gt^2$$

$$25 = 20\,t + \frac{1}{2} \times 10 \times t^2 \;;\quad 25 = 20\,t + 5\,t^2$$

$$\therefore \quad t^2 + 4t - 5 = 0$$

Solution of this quadratic equation gives

$$t = \frac{-4 \pm \sqrt{4^2 - 4 \times 1 \times (-5)}}{2} = \frac{-4 \pm 6}{2} = 1\,\text{s}$$

$\therefore$ Total time during which the stone is in flight

= time from B to C + time from C to B + time from B to A

= 2 + 2 + 1 = **5 s**

EXAMPLE 13.42

A particle projected vertically upwards is seen at P after t_1 seconds. If it takes t_2 second more from this point P to the point of projection, prove that

(i) the height of P above the point of projection $= \frac{1}{2} g\, t_1 t_2$

(ii) the maximum height attained $= \frac{1}{2} g \left[\frac{t_1 + t_2}{2} \right]^2$

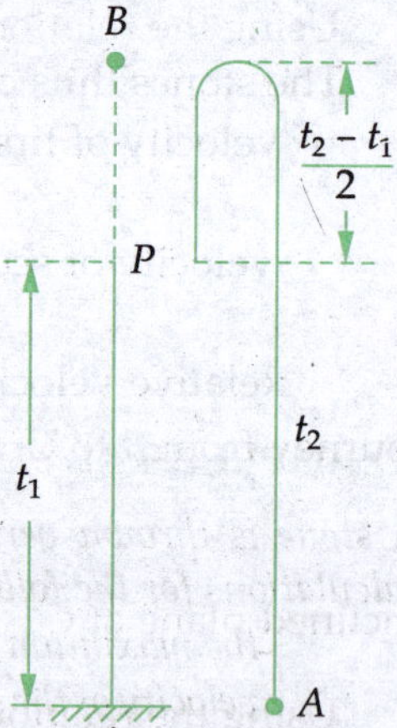

Fig. 13.30

Solution: Time of travel from A to $P = t_1$

Time of travel from P to B to $A = t_2$

Since time of ascent equals the time of descent,

Time of travel from P to $B = \frac{t_2 - t_1}{2}$

$\therefore$ Time taken by the particle in ascending to maximum height

$$= t_1 + \frac{t_2 - t_1}{2}$$

$$= \frac{t_1 + t_2}{2}$$

At position B corresponding to maximum height, the particle will have zero velocity. Accordingly considering upward motion of particle from A to B,

$$0 = u - g\left(\frac{t_1 + t_2}{2}\right); \quad u = g\left(\frac{t_1 + t_2}{2}\right)$$

(*i*) Using the relation $s = ut - \frac{1}{2}gt^2$,

Height of P above the point of projection A

$$s = g\left(\frac{t_2 + t_2}{2}\right)t_1 - \frac{1}{2}g\,t_1^2 = \frac{g\,t_1^2}{2} + \frac{g\,t_1 t_2}{2} - \frac{g\,t_1^2}{2} = \frac{1}{2}g\,t_1 t_2$$

(*ii*) Using the relation $v_2 - u^2 = -2gs$,

Maximum height attained by the particle,

$$s = \frac{u^2}{2g} \qquad (\because \quad v = 0 \text{ at the position of maximum height})$$

$$= g^2\left(\frac{t_1 + t_2}{2}\right)^2 \times \frac{1}{2g} = \frac{1}{2}g\left(\frac{t_1 + t_2}{2}\right)^2$$

EXAMPLE 13.43

Two stones are dropped simultaneously from rest from two points both 200 m above the ground. One of the stone during its downward motion hits a fixed inclined plane located at 80 m distance from the ground. The plane is inclined at an angle to the horizontal as shown in Fig. 13.31. As a result of this impact, the direction of velocity of the stone becomes horizontal. Compare the times of fall of the two stones to reach the ground.

Solution :

Motion of stone A :

$u = 0; \quad s = 200$ m

t = time taken to reach the ground

Using the relation : $s = ut + \frac{1}{2}gt^2$, we have

$$200 = 0 \times t + \frac{1}{2} \times 9.81 \times t^2;$$

$$t = \sqrt{\frac{200 \times 2}{9.81}} = 6.385 \text{ sec}$$

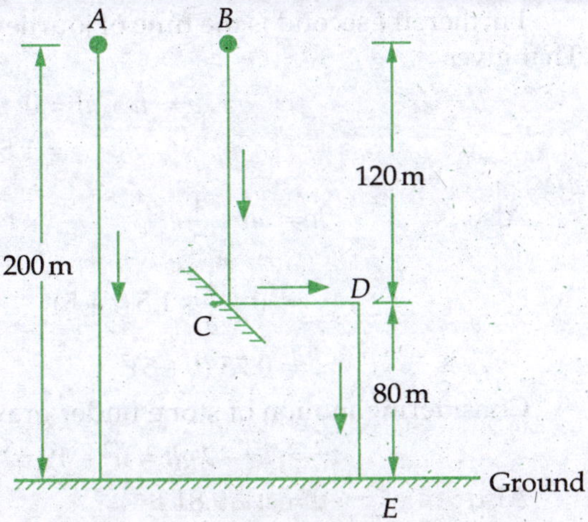

Fig. 13.31

Motion of stone B : From the first part of the journey from B to C,

$u = 0; \quad s = 120$ m

t_1 = time taken to reach the inclined plane at C

Using the relation : $s = ut + \frac{1}{2}gt^2$, we have

$$120 = 0 \times t_1 + \frac{1}{2} \times 9.81 \times t_1^2; \quad t_1 = \sqrt{\frac{120 \times 2}{9.81}} = 4.946 \text{ sec}$$

After impact with the inclined plane at point C, the direction of motion of stone B becomes horizontal. This implies that

(*i*) the velocity of stone B becomes zero in the vertical direction.

(*ii*) the journey for the remaining distance of (200 – 120) = 80 m is performed with zero initial velocity.

$$80 = 0 \times t_2 + \frac{1}{2} \times 9.81 \times t_2^2; \quad t_2 = \sqrt{\frac{80 \times 2}{9.81}} = 4.038 \text{ sec}$$

$\therefore$ Time taken by the stone B to reach the ground

$$= 4.946 + 4.038 = 8.984 \text{ sec}$$

Hence ratio of times taken by stones B and A to reach the ground

$$= \frac{8.984}{6.385} = 1.407$$

EXAMPLE 13.44

A balloon rises from the ground with constant acceleration of 1.5 m/s. Five seconds later, a stone is thrown upwards from the same point on the ground. Determine:

(a) *the velocity that needs to be imparted to the stone so that it just touches the ascending balloon.*

(b) *the height of the stone from the ground when it touches the balloon.*

Solution: Both the balloon and stone will have the same velocity when the stone just touches the ascending balloon.

Considering uniform accelerating motion of the balloon,

$$v^2 = u^2 + 2ah \quad \text{or} \quad v^2 = 0 + 2 \times (1.5)h$$
$$= 3h \qquad \ldots(i)$$

Further, if t second is the time of journey of stone, travel time for balloon will be $(t + 5)$ second. That gives

$$v = u + at = 0 + 1.5 \times (t + 5)$$
$$= 1.5\,(t + 5) \qquad \ldots(ii)$$

Also : $\quad h = ut + \frac{1}{2}at^2 \qquad t = (t + 5)$ for balloon

$$= 0 + \frac{1}{2} \times 1.5\,(t + 5)^2$$
$$= 0.75\,(t + 5)^2 \qquad \ldots(iii)$$

Considering motion of stone under gravity

$$v^2 = u^2 - 2gh = u^2 - 19.62h \qquad \ldots(iv)$$

Also : $\quad v = u - 9.81\,t \qquad \ldots(v)$

Equating the expressions for v^2 as per identities (*i*) and (*iv*),

$$3h = u^2 - 19.62h \quad \therefore \quad u^2 = 22.62h$$

Since, $h = 0.75\,(t + 5)^2$ as per identity (*iii*), we get

$$u^2 = 22.62 \times 0.75\,(t + 5)^2 \quad \text{or} \quad u = 4.12\,(t + 5)$$

Substituting this value of u in identify (v), we obtain

$$v = 4.12\,(t + 5) - 9.81\,t$$

Equating the above expression for v and that as per identify (*ii*), we get

$$1.5\,(t + 5) = 4.12\,(t + 5) - 9.81\,t$$

or $\quad 1.5\,t + 7.5 = 4.12t + 20.6 - 9.81\,t$

Solution gives: $t = \dfrac{20.6 - 7.5}{1.5 - 4.12 + 9.81} = $ **1.82 sec**

and $u = 4.12\,(t + 5) = 4.12\,(1.82 + 5) = 28.1 \text{ m/s}$

$h = 0.75\,(t + 5)^2 = 0.75\,(1.82 + 5)^2 = $ **34.88 m**

EXAMPLE 13.45

A ball is thrown vertically upward with velocity 15 m/s from the 10 m level in an elevator shaft. At the same instant, the elevator is stated to pass the 5 m level moving upward with a constant velocity of 2 m/s. Determine:

(a) when and where the ball will hit the elevator,

(b) relative velocity of the ball with respect to the elevator when the ball hits it (the elevator).

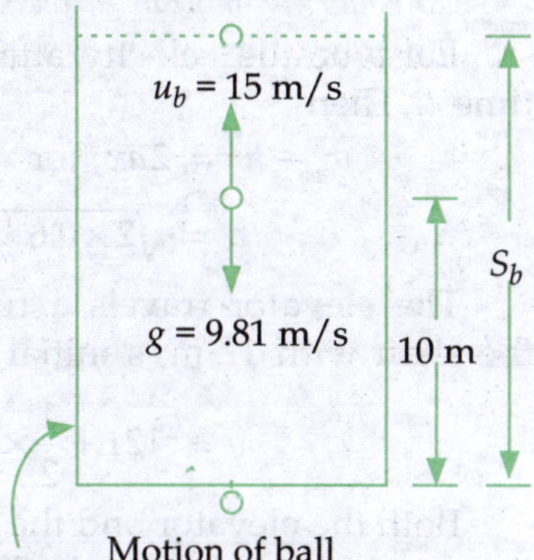

Fig. 13.32

Solution: Let the ball hit the elevator after time t and when it is at distance S_b from the origin O at ground level.

With uniform accelerated motion of the ball,

$$S_b = 10 + u_b t + \frac{1}{2} a t^2$$

$$= 10 + 15t + \frac{1}{2} \times (-9.81)\, t^2$$

$$= 10 + 15t - 4.905t^2 \quad \text{...(i)}$$

With uniform motion of the elevator,

$$S_e = 5 + u_e t = 5 + 2t \quad \text{....(ii)}$$

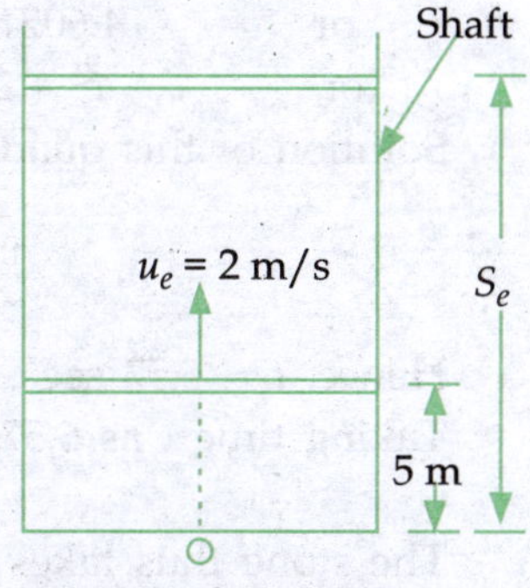

Fig. 13.33

Both the elevator and the ball would be at the same distance from the origin point at ground level when the ball hits the elevator. Thus

$$10 + 15t - 4.905t^2 = 5 + 2t$$

or $\quad 4.905t^2 - 13t - 5 = 0$

or $\quad t^2 - 2.65t - 1.02 = 0$

Solution of this quadratic equation gives

$$t = \frac{-(-2.65) \pm \sqrt{(-2.65)^2 - 4 \times 1 \times (-1.02)}}{2}$$

$$= \frac{2.65 \pm 3.33}{2}$$

Hence $t = 2.99$ sec and -0.34 sec

Taking time t as 2.99 sec and its substitution in identity (*ii*) gives :

$S_e = 5 + 2 \times 2.99 = $ **10.98 m**

The ball thus takes 2.99 sec and is at a distance of 10.98 m from the origin at ground.

(*b*) The relative velocity of ball with respect to the elevator is

$$V_{b-e} = u_b - u_e = (15 - 9.81\,t) - 2$$

$= (15 - 9.81 \times 2.99) - 2 = $ **– 16.33 m/s**

EXAMPLE 13.46

An open plat-form type elevator moves down a mini shaft with an acceleration of 0.6 m/s². After the elevator has travelled 120 m from the top of shaft, a stone is dropped from the top of shaft. Determine:

(a) The time taken by the stone to hit the elevator,

(b) The distance travelled by the stone

Solution: Let t be the time taken by the stone to hit the elevator.

Considering motion of stone:

$$u = 0 \quad \text{and} \quad g = 9.81 \text{ m/s}^2$$

$$S_s = ut + \frac{1}{2}gt^2$$

$$= \frac{1}{2} \times 9.81 \times t^2 = 4.905t^2 \quad \ldots(i)$$

Considering motion of elevator:

Let v be the velocity attained by the elevator in time t. Then

$$v^2 - u^2 = 2ax \quad \text{or} \quad v^2 = 2ax \quad (\because u = 0)$$

$$\therefore \quad v = \sqrt{2 \times 0.6 \times 120} = 12 \text{ m/s}$$

The elevator travels a further distance y down the shaft with 12 m/s initial velocity. Then

$$y = 12t + \frac{1}{2} \times 0.6t^2 = 12t + 0.3t^2$$

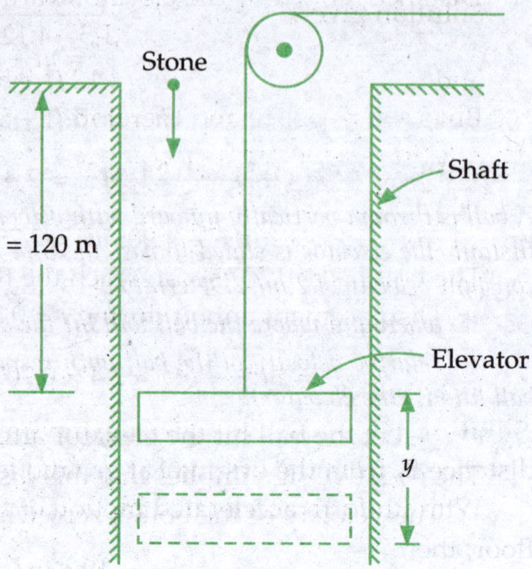

Fig. 13.34

Both the elevator and the stone would be at the same distance from the top of the shaft when the stone hits the elevator. Thus

$$4.905t^2 = 120 + (12t + 0.3t^2)$$

or $$4.605t^2 - 12t - 120 = 0$$

or $$t^2 - 2.606t - 26.06 = 0$$

Solution of this quadratic equation gives :

$$t = \frac{-(-2.606) \pm \sqrt{(2.606)^2 - 4 \times 1 \times (-2.606)}}{2} = \frac{2.606 \pm 10.537}{2}$$

Hence: $t = 6.57$ sec and -3.96 sec

Taking time t as 6.57 sec and its substitution in identity (*i*) gives

$$S_t = 4.905 \times 6.57^2 = \mathbf{211.72 \text{ m}}$$

The stone thus takes 6.57 sec and travels a distance of 211.72 m to hit the elevator.

EXAMPLE 13.47

An elevator whose floor to ceiling height is 2.75 m starts ascending with constant upward acceleration of 1.2 m/s². Two seconds after the start, a loose bolt drops from the ceiling towards the floor of the elevator. Calculate :

(*a*) *time taken by the bolt to reach the elevator floor,*

(*b*) *displacement of the bolt during that time, and*

(*c*) *the distance travelled by the bolt during the same time.*

Solution : Using kinematic relation $V = u + at$, the velocity of the elevator at two seconds after the start is

$$V = 0 + 2 \times 1.2 = 2.4 \text{ m/s} \quad \text{(upwards)}$$

The bolt travels up with a velocity of 2.4 m/s and then drops under gravity. If t denotes the time to reach the floor, then the distance moved by the bolt is

$$s_1 = ut - \frac{1}{2}gt^2 = 2.4t - \frac{1}{2} \times 9.81t^2$$

Distance moved by the elevator during the same time t is

$$s_2 = ut + \frac{1}{2}gt^2 = 2.4t + \frac{1}{2} \times 1.2t^2$$

But $s_2 = (s_1 + 2.75)$ and therefore

$$2.4t + \frac{1}{2} \times 1.2t^2 = 2.4t - \frac{1}{2} \times 9.81t^2 + 2.75$$

or $$(0.6 + 4.905)t^2 = 2.75 \quad ; \quad t = 0.707 \text{ sec}$$

The bolt takes 0.707 sec to reach the floor.

(*b*) Displacement of bolt during $t = 0.707$ sec

$$= ut - \frac{1}{2}gt^2 = 2.4 \times 0.707 - \frac{1}{2} \times 9.81 \times 0.707^2$$

$$= 1.697 - 2.452 = -0.755 \text{ m}$$

The negative sign implies that this displacement is downwards with respect to initial position.

(*c*) The velocity of the bolt is zero when it reaches the floor. If t is the time taken to reach the floor, then

$$v = u - gt \quad ; \quad 0 = 2.4 - 9.81t \quad ; \quad t = 0.2446$$

Distance moved by bolt during this time is

$$s = ut - \frac{1}{2}gt^2 = 2.4 \times 0.2446 - \frac{1}{2} \times 9.81 \times (0.2446)^2$$

$$= 0.2935 \text{ m}$$

Time during which the bolt moves up = (0.707 – 0.2446) = 0.4624 sec

Distance moved by the bolt during this period is

$$s = ut - \frac{1}{2}gt^2 = 0 \times 0.4624 + \frac{1}{2}(-9.81) \times (0.4624)^2$$

$$= -1.0487 \text{ m} = 1.049 \text{ m in magnitude}$$

∴ Total distance moved by the bolt

$$= 0.2935 + 1.0487 = 1.3422 \text{ m}$$

13.5. CURVILINEAR MOTION

The motion of a particle along a curved path other than a straight line is known as curvilinear motion.

Examples of curvilinear motion are :

(*i*) an automobile vehicle negotiating a turn on the road

(*ii*) projectile motion of a bullet fired from a gun

(*iii*) motion of bob of pendulum oscillating in vertical plane

(*iv*) motion of satellite around the earth.

The curvilinear motion may be two-dimensional (plane curvilinear) or three-dimensional (space curvilinear). The study of the plane curvilinear motion is made with reference to the following coordinate systems :

13.5.1. Rectangular-cartesian x, y coordinates

The position of the particle on the curved path at any instant is defined by the position vector $\vec{r}$.

$$\vec{r} = xi + yj \text{ where } i \text{ and } j \text{ are the unit vectors}$$

Magnitude $r = |\vec{r}| = \sqrt{x^2 + y^2}$

Velocity vector $\vec{V} = \frac{d}{dt}(\vec{r}) = \frac{d}{dt}(x\boldsymbol{i} + y\,\boldsymbol{j})$

Since $\boldsymbol{i}$ and $\boldsymbol{j}$ are constant terms, we get

$$\vec{V} = \boldsymbol{i}\frac{dx}{dt} + \boldsymbol{j}\frac{dy}{dt} = V_x\boldsymbol{i} + V_y\boldsymbol{j}$$

Resultant velocity $V = \sqrt{V_x^2 + V_y^2}$

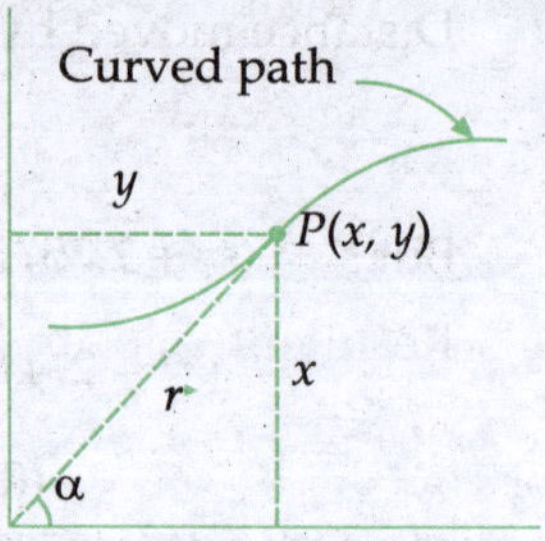

Fig. 13.35

The direction of velocity is tangential to the path of motion of the particle. If α is the angle made by the resultant with x-axis, then

$$\tan\alpha = \frac{V_y}{V_x} \quad ; \quad \alpha = \tan^{-1}\left(\frac{V_y}{V_x}\right)$$

Further, acceleration $\vec{a} = \frac{d\vec{v}}{dt} = \frac{d}{dt}(V_x\boldsymbol{i} + V_y\,\boldsymbol{j}) = \frac{d}{dt}\left(\frac{dx}{dt}\boldsymbol{i} + \frac{dy}{dt}\boldsymbol{j}\right)$

$$= \frac{d^2x}{dt^2}\boldsymbol{i} + \frac{d^2y}{dt^2}\boldsymbol{j} = a_x\boldsymbol{i} + a_y\boldsymbol{j}$$

Resultant acceleration $a = \sqrt{a_x^2 + a_y^2}$

If β is the angle made by resultant acceleration with x-axis, then

$$\tan\beta = \frac{a_y}{a_x} \quad ; \quad \beta = \tan^{-1}\left(\frac{a_y}{a_x}\right)$$

For a curvilinear motion in space, we may write

$$\vec{r} = x\boldsymbol{i} + y\boldsymbol{j} + z\boldsymbol{k}$$

$$\vec{V} = V_x\boldsymbol{i} + V_y\boldsymbol{j} + V_z\boldsymbol{k}$$

$$\vec{a} = a_x\boldsymbol{i} + a_y\boldsymbol{j} + a_z\boldsymbol{k}$$

EXAMPLE 13.48

The motion of a particle is defined by the relations : $x = t^2 + 3t$ and $y = t^3 - 8t^2 + 3$ where x and y are in meters and y is in seconds. (a) Write the equations defining the motion of the particle in vectorial form (b) Calculate the velocity and acceleration of the particle at time $t = 2$ seconds.

Solution : $x = t^2 + 3t \qquad y = t^3 - 8t^2 + 3$

$$V_x = \frac{dx}{dt} = 2t + 3 \qquad V_y = \frac{dy}{dt} = 3t^2 - 16t$$

$$a_x = \frac{dV_x}{dt} = 2 \qquad a_y = \frac{dV_y}{dt} = 6t - 16$$

At any time t, the motion of the particle in vectorial form can be written as

Position vector $\vec{r} = x\boldsymbol{i} + y\boldsymbol{j} = (t^2 + 3t)\boldsymbol{i} + (t^3 - 8t^2 + 3)\boldsymbol{j}$

Velocity vector $\vec{V} = V_x\boldsymbol{i} + V_y\boldsymbol{j} = (2t + 3)\boldsymbol{i} + (3t^2 - 16t)\boldsymbol{j}$

Acceleration $\vec{a} = a_x\boldsymbol{i} + a_y\boldsymbol{j} = 2\boldsymbol{i} + (6t - 16)\boldsymbol{j}$

(b) At time $t = 2$ seconds

$$\vec{V} = (2 \times 2 + 3)\,i + (3 \times 2^2 - 16 \times 2)\,j = 7i - 20j = V_x i + V_y j$$

$\therefore \quad V_x = 7$ m/s and $V_y = -20$ m/s.

Since V_x is positive and V_y is negative, the resultant lies in the fourth quadrant.

Resultant velocity $V = \sqrt{V_x^2 + V_y^2} = \sqrt{7^2 + (-20)^2} =$ **21.19 m/s**

$$\tan\alpha = \frac{V_y}{V_x} = \frac{20}{7} = 2.857 \quad ; \quad \alpha = 70.71^\circ$$

Further, $\vec{a} = 2i + (6 \times 2 - 16)\,j = 2i - 4j = a_x i + a_y j$

$\therefore \quad a_x = 2\text{ m/s}^2$ and $a_y = -4\text{ m/s}^2$

Since a_x is positive and a_y is negative, the resultant acceleration lies in the fourth quadrant.

Resultant acceleration $a = \sqrt{a_x^2 + a_y^2} = \sqrt{2^2 + (-4)^2} =$ **4.47 m/s²**

$$\tan\beta = \frac{a_y}{a_x} = \frac{4}{2} = 2 \quad ; \quad \beta = 63.43^\circ$$

The velocity and acceleration vectors have been shown below in Fig. 13.36

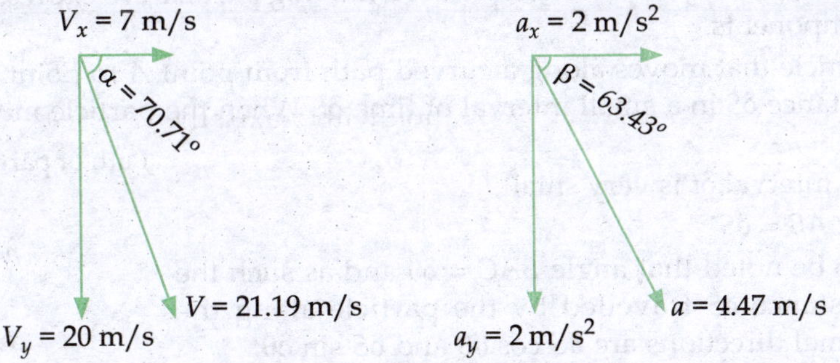

Fig. 13.36

EXAMPLE 13.49

A particle moves in x-y plane and its coordinates are defined by the relations

$$x = -3\sin t \quad \text{and} \quad y = 2 - 3\cos t$$

where x and y are in metres and time t is in seconds:

(a) *State the equation of path along which the particle moves, and*

(b) *Write expressions for position vector, velocity vector and acceleration vector for the particle at the instant when it starts motion.*

Solution : $x = -3\sin t$ and $y = 2 - 3\cos t$

or $\quad \sin t = -\dfrac{x}{3}$ and $\cos t = \dfrac{2-y}{3}$

$$\sin^2 t + \cos^2 t = \left(\frac{-x}{3}\right)^2 + \left(\frac{2-y}{3}\right)^2$$

$$\therefore \quad 1 = \frac{x^2}{3^2} + \frac{(2-y)^2}{3^2} \quad \text{or} \quad x^2 + (2-y)^2 = 3^2$$

Apparently the curved path is a circle of radius 3 m and origin (0, 2) (*b*) Position vector $\vec{r} = xi + yj = (-3\sin t)i + (2 - 3\cos t)j$

Velocity vector $\vec{V} = \dfrac{d\vec{V}}{dt} = \dfrac{d}{dt}\left[(-3\sin t)\,i + (2 - 3\cos t)j\right]$

$= (-3\cos t)\,i + (3\sin t)\,j$

Acceleration vector $\vec{a} = \dfrac{d\vec{V}}{dt} = \dfrac{d}{dt}\left[(-3\cos t)i + (3\sin t)\,j\right]$

$= (3\sin t)\,i + (3\cos t)\,j$

At the instant of start, *i.e.* at $t = 0$

$\vec{r} = (-3\sin 0)\,i + (2 - 3\cos 0)\,j = -1j$

$\vec{V} = (-3\cos 0)\,i + (3\sin 0)\,j = -3i$

$\vec{a} = (3\sin 0)\,i + (3\cos 0)\,j = 3j$

Obviously at start, the particle has a velocity of -3 m/s^2 along $-ve$ x-axis and an acceleration of 3 m/s^2 along the $+ve$ y-axis.

13.5.2. Tangential and normal coordinates

In this system, the velocity and acceleration of a moving particle are expressed in tangential (t) and normal (n) components.

Consider a particle that moves along a curved path from point A to point B and traverses an infinitely small distance δS in a small interval of time δt. When the particle moves, the distance S changes with time.

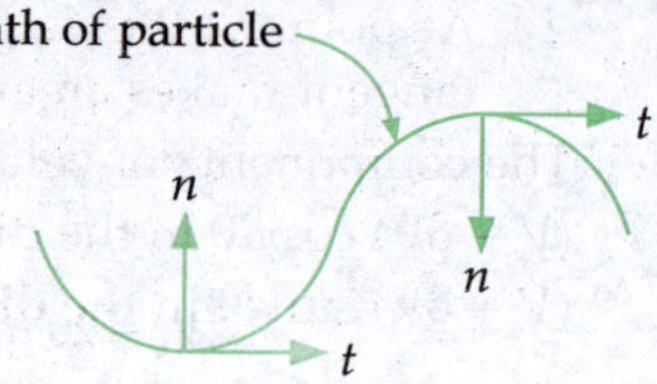

When the time interval δt is very small,

chord AB = arc $AB = \delta S$

Further, it is to be noted that angle $BAC = \delta\theta$ and as such the components of distance δS travelled by the particle along the tangential and normal directions are $\delta S \cos d\theta$ and $\delta S \sin \delta\theta$.

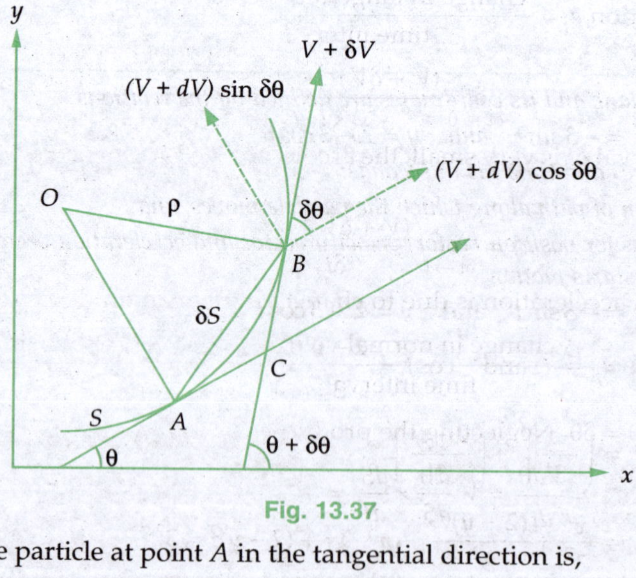

Fig. 13.37

The velocity of the particle at point A in the tangential direction is,

$$V_t = \frac{\text{distance moved along the tangential direction}}{\text{time interval}}$$

$$= \underset{\delta t \to 0}{\delta t} \frac{\delta S \cos \delta\theta}{\delta t} = \underset{\delta t \to 0}{\delta t} \frac{\delta S}{\delta t} \cos \delta\theta$$

As $\delta\theta$ is small, $\cos \delta\theta \to 1$ and therefore

$$V_t = \underset{\delta t \to 0}{\delta t} \frac{\delta S}{\delta t} = \frac{dS}{dt}$$

The velocity of the particle in the normal direction is,

$$V_n = \frac{\text{distance moved along the tangential direction}}{\text{time interval}}$$

$$= \underset{\delta t \to 0}{\delta t} \frac{\delta S \sin \delta\theta}{\delta t}$$

As $\delta\theta$ is small, $\sin \delta\theta \to \delta\theta$. Neglecting the product $(\delta S \times \delta\theta)$, we get

$$V_n = 0$$

Thus the normal component of velocity of the moving particle is zero. It implies that the velocity of a particle moving along a curved path is equal to the tangential velocity and that the direction of velocity is tangential to the path at any point.

Further, it may be noted that

(*i*) At point A, the velocity V is directed along the tangent that is inclined at an angle θ with the horizontal. The velocity in a direction normal to the tangent is zero.

(*ii*) At point B, the velocity $(V + \delta V)$ is again directed along the tangent at that point. This tangent makes an angle $\delta\theta$ with the tangent at point B.

The components of velocity vector $(V + \delta V)$ are;

$(V + \delta V) \cos d\theta$ in the direction of tangent at B, and

$(V + \delta V) \sin d\theta$ in the direction normal to tangent at B

$$\text{Tangential acceleration } a_t = \frac{\text{change in tangential velocity}}{\text{time interval}}$$

$$= \underset{\delta t \to 0}{\delta t} \frac{(V + \delta V)\cos \delta\theta - V}{\delta t}$$

When the time interval δt is very small, the arc AB will be very small and so will be $\delta\theta$ and then $\cos \delta\theta = 1$

$$\therefore \quad a_t = \underset{\delta t \to 0}{\delta t} \frac{(V + \delta V) - V}{\delta t} = \frac{dV}{dt}$$

Thus the tangential acceleration is due to change in the magnitude of velocity.

$$\text{Normal acceleration } a_n = \frac{\text{change in normal velocity}}{\text{time interval}} = \underset{\delta t \to 0}{\delta t} \frac{(V + \delta V)\sin \delta\theta - 0}{\delta t}$$

For small $d\theta$, $\sin \delta\theta = \delta\theta$. Neglecting the product $(\delta V \times \delta\theta)$ we may write

$$a_n = \frac{V d\theta}{dt} = V \frac{d\theta}{dS} \times \frac{dS}{dt} = \frac{V\, dS/dt}{dS/d\theta} = \frac{V \times V}{\rho} = \frac{V^2}{\rho}$$

where $\rho = \dfrac{dS}{d\theta}$ is the radius of curvature of the path. When the body moves in a circular orbit of radius r, then

$$a_n = \frac{V^2}{r}$$

The normal acceleration is due to change in the direction of velocity. It is directed towards the centre of curvature and is known as centripetal acceleration.

Resultant acceleration $a = \sqrt{a_t^2 + a_n^2}$

If the resultant acceleration makes an angle ϕ with the normal, then

$$\tan\phi = \frac{a_t}{a_n}$$

It is worthwhile to note that

(*i*) If the magnitude of velocity is constant between two points, then the tangential acceleration $a_t = 0$.

(*ii*) If the direction of motion of particle does not change, *i.e.*, if the motion of the particle is along a straight line, then $\rho \to \infty$. In that case the normal acceleration $a_n = 0$.

(*iii*) When the particle is moving along a circle, the radius of curvature ρ is equal to radius of the circle r. Hence the normal acceleration is

$$a_n = \frac{v^2}{r}$$

The normal acceleration is directed towards the centre of curvature and hence is known as **centripetal acceleration**.

EXAMPLE 13.50

A car tarvels with constant speed of 60 km/hr along a depression in a road. If the depression is defined by the equation $x^2 = 150\,y$, make calculations for:

(*a*) *the acceleration when the car is at the deepest point in the depression,*

(*b*) *the radius of curvature at the deepest point in the depression.*

Solution : The tangential component of acceleration a_t is the rate of change of speed. Since the speed remains constant throughout travel of car along the curve, $a_t = 0$. Hence the total acceleration at the deepest point O of the depression is the normal component of acceleration a_n at point O.

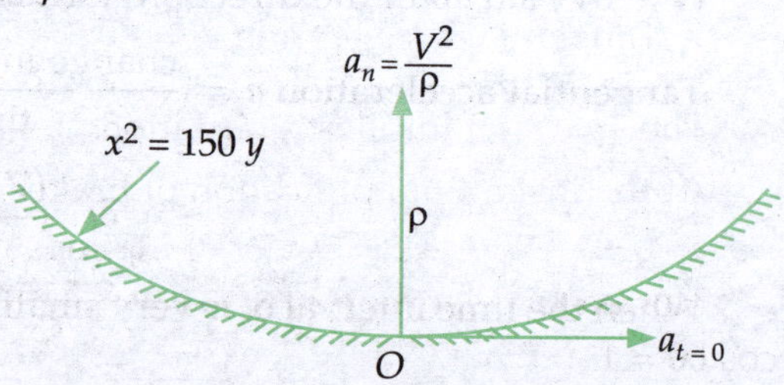

Fig. 13.38

The normal component of acceleration is worked out from the relation

$$a_n = \frac{V^2}{\rho} \qquad \ldots(i)$$

where $V = 60 \text{ km/hr} = \dfrac{60 \times 10^3}{3600} = 16.67 \text{ m/s}$

and ρ is the radius of curvature defined by the relation,

$$\rho = \frac{\left[1 + \left(\dfrac{dy}{dx}\right)^2\right]^{1/2}}{\dfrac{d^2y}{dx^2}}$$

From the equation $x^2 = 150\,y$ that defines the curve for depression

$$\frac{dy}{dx} = \frac{2x}{150} = \frac{x}{75} \quad \text{and} \quad \frac{d^2y}{dx^2} = \frac{1}{75}$$

At the origin O : $\frac{dy}{dx} = 0$ and $\frac{d^2y}{dx^2} = \frac{1}{75}$

$$\therefore \quad \rho = \frac{(1+0)^{1/2}}{\frac{1}{75}} = 75 \text{ m}$$

Then from expression (*i*)

$$a_n = \frac{V^2}{\rho} = \frac{16.67^2}{75} = 3.705 \text{ m/s}^2$$

$\therefore$ Resultant acceleration

$$a = a_n = \mathbf{3.705\ m/s^2}$$

EXAMPLE 13.51

A motorist is driving at 80 km/hr on the curved position of a highway of 400 m radius. He suddenly applies the brakes and that causes the speed to decrease to 45 km/hr at a constant rate in 8 seconds. Determine the tangential and normal components of acceleration immediately after the application of brakes and 4 seconds later.

Solution : Initial velocity of vehicle $u = 80 \text{ km/hr} = \frac{80 \times 1000}{3600} = 22.22 \text{ m/s}$

Final velocity of vehicle $v = \frac{45 \times 1000}{3600} = 12.5 \text{ m/s}$

Tangential acceleration is the rate of change of velocity

$\therefore$ Tangential acceleration $a_t = \frac{12.5 - 22.22}{8} = -1.215 \text{ m/s}^2$

The – ve sign implies retardation

At the instant of application of brakes, $V = 22.22$ m/s

Normal acceleration $a_n = \frac{V^2}{r} = \frac{22.22^2}{r} = 1.234 \text{ m/s}^2$

$\therefore$ Total acceleration $a = \sqrt{a_n^2 + a_t^2} = \sqrt{(1.234)^2 + (-1.215)^2} = \mathbf{1.732\ m/s^2}$

(*b*) Velocity of vehicle 4 seconds after the application of brakes,

Since the tangential component of acceleration is uniform, we can use the equation of motion. That is

$$V = u + at = 22.22 + (-1.215 \times 4) = 17.36 \text{ m/s}$$

$$a_n = \frac{(17.36)^2}{400} = 0.753 \text{ m/s}^2 \quad ; \quad a_t = -1.215 \text{ m/s}^2$$

$$\therefore \quad a = \sqrt{(0.753)^2 + (-1.215)^2} = \mathbf{1.429\ m/s^2}$$

EXAMPLE 13.52

An automobile enters a curved road at 30 km/hr and then leaves at 48 km/hr. The curved road is in the form of quarter of a circle and has a length of 400 m. If the car travels at constant acceleration along the curve, calculate the resultant acceleration at both ends of the curve.

Solution : With reference to Fig. 13.39, let the automobile travel from A to B.

$$V_a = 30\ \text{km/hr} = \frac{30 \times 1000}{3600} = 8.33\ \text{m/s}$$

$$V_b = 48\ \text{km/hr} = \frac{48 \times 1000}{3600} = 13.33\ \text{m/s}$$

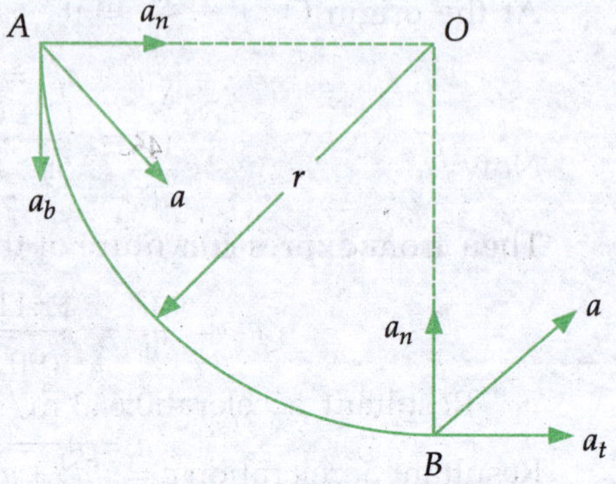

Fig. 13.39

Using the kinematic equation $v^2 - u^2 = 2as$,

$$\text{Tangential acceleration } a_t = \frac{13.33^2 - 8.33^2}{2 \times 400} = 0.135\ \text{m/s}^2$$

If r is the radius of the circular track, then

$$\frac{1}{4} \times (2\pi r) = 400 \quad ; \quad r = \frac{400 \times 2}{\pi} = 254.77\ \text{m}$$

At end A :

$$a_n = \frac{V_a^2}{r} = \frac{8.33^2}{254.77} = 0.272\ \text{m/s}^2$$

$$a_t = 0.135\ \text{m/s}^2$$

$$\therefore \text{ Resultant acceleration } a = \sqrt{0.272^2 + 0.135^2} = \mathbf{0.304\ m/s^2}$$

If α is the direction of resultant with tangential acceleration, then

$$\tan \alpha = \frac{a_n}{a_t} = \frac{0.272}{0.135} = 2.0148 \quad ; \quad \alpha = \mathbf{63.60}$$

At end B :

$$a_n = \frac{V_b^2}{r} = \frac{13.33^2}{254.77} = 0.697\ \text{m/s}^2$$

$$a_t = 0.135\ \text{m/s}^2$$

$$\text{Resultant acceleration } a = \sqrt{0.697^2 + 0.135^2} = \mathbf{0.71\ m/s^2}$$

If α is the direction of resultant with tangential acceleration, then

$$\tan \alpha = \frac{a_n}{a_t} = \frac{0.697}{0.135} = 5.163 \quad ; \quad \alpha = \mathbf{79.04^\circ}$$

EXAMPLE 13.53

On a bumpy road, the motorist applies brakes and the speed of the vehicle reduces uniformly from 60 km/hr at point A to 40 km/hr at point B; the points A and B are 120 m apart. If the vehicle has a resultant acceleration of 2.5 m/s² at point A, estimate the radius of curvature of the road at this point. Proceed to calculate the acceleration of the vehicle at point B where the road has a radius of curvature equal to 90 m.

Solution : $V_a = 60\ \text{km/hr} = \frac{60 \times 1000}{3600} = 16.67\ \text{m/s}$

$$V_b = \frac{40 \times 1000}{3600} = 11.11\ \text{m/s}$$

Using the kinematic relation $v^2 - u^2 = 2as$,

Tangential acceleration $a_t = \dfrac{11.11^2 - 16.67^2}{2 \times 120} = -0.643 \text{ m/s}^2$

The – ve sign implies retardation

At point A : $a = 2.5 \text{ m/s}^2$; $a_t = -0.643 \text{ m/s}^2$

$$\therefore \quad a_n = \sqrt{a^2 - a_t^2} = \sqrt{(2.5)^2 - (-0.643)^2} = 2.416 \text{ m/s}^2$$

Now $a_n = \dfrac{V^2}{r}$ or $r = \dfrac{V^2}{a_n} = \dfrac{16.67^2}{2.416} = 115$ **m**

Thus the radius of curvature of the road at point A is 115 m

At point B : $a_n = \dfrac{V^2}{r} = \dfrac{11.11^2}{90} = 1.371 \text{ m/s}^2$

$a_t = -0.643 \text{ m/s}^2$

Resultant acceleration $a = \sqrt{a_n^2 + a_t^2} = \sqrt{(1.371)^2 + (-0.643)^2} = 1.514 \text{ m/s}^2$

Hence the acceleration of the vehicle at point B is 1.514 m/s^2.

EXAMPLE 13.54

A motorist is driving at a constant speed of 36 km/hr on the curved portion of a road of 200 m radius. Determine the velocity and acceleration component of the vehicle along (a) x and y directions (b) normal and tangential directions 10 seconds after it has entered the curved portion of the road.

Solution: Refer to Fig. 13.40.

$$\text{Speed } V = 36 \text{ km/hr} = \frac{36 \times 1000}{3600} = 10 \text{ m/s}$$

$$\text{Angular velocity } \omega = \frac{V}{r} = \frac{10}{200} = 0.05 \text{ rad/s}$$

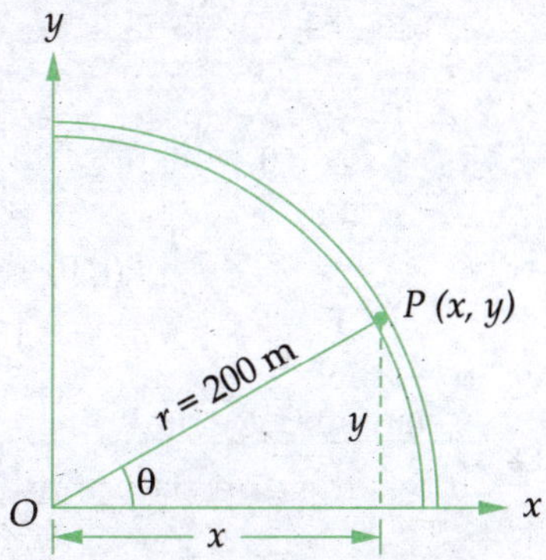

Fig. 13.40

At any time instant t, the vehicle has the coordinates

$$x = r\cos\theta = r\cos\omega t$$
$$y = r\sin\theta = r\sin\omega t$$

The velocity and acceleration components of the vehicle in x and y-directions at $t = 10$ second then work out as

$$V_x = \frac{dx}{dt} = \frac{d}{dt}(r\cos\omega t) = -\omega r\sin\omega t$$

$$V_y = \frac{dy}{dt} = \frac{d}{dt}(r\sin\omega t) = \omega r\cos\omega t$$

$$a_x = \frac{dV_x}{dt} = \frac{d}{dt}(-\omega r\sin\omega t) = -\omega^2 r\cos\omega t$$

$$a_y = \frac{dV_y}{dt} = \frac{d}{dt}(\omega r\cos\omega t) = -\omega^2 r\sin\omega t$$

Substituting $t = 10$ s, $r = 200$ m and $\omega = 0.05$ rad/s

$$= 0.05 \times \frac{180}{\pi} = 2.866 \text{ degrees}$$

We get: $V_x = -0.05 \times 200\ [\sin(2.866 \times 10)] = -4.796$ m/s (←)

$V_y = 0.05 \times 200\ [\cos(2.866 \times 10)] = 8.775$ m/s (↑)

$a_x = -(0.05)^2 \times 200\ [\cos(2.866 \times 10)] = -0.439$ m/s^2 (←)

$a_y = -(0.05)^2 \times 200\ [\sin(2.866 \times 10)] = -0.24$ m/s^2 (↓)

(*ii*) The velocity and acceleration components in tangential and normal directions are:

$$V_n = 0 \quad \text{and} \quad V_t = V = 10 \text{ m/s}$$

$$a_n = \frac{V^2}{r} = \frac{10 \times 10}{200} = 0.5 \text{ m/s}^2$$

$$a_t = \frac{dV}{dt} = 0 \qquad (\because V \text{ is constant})$$

Angular position θ of the vehicle after 10 seconds

$$= \omega t = 2.866 \times 10 = 28.66°$$

13.5.3. Radial and transverse co-ordinates

In this system, the position of the particle is defined by the polar coordinates r and θ. Further, the velocity and acceleration of the particle are resolved into components along and perpendicular to the position vector $\vec{r}$. These components are called :

(*i*) radial components denoted by $\vec{r}$

(*ii*) transverse components denoted by $\vec{\theta}$

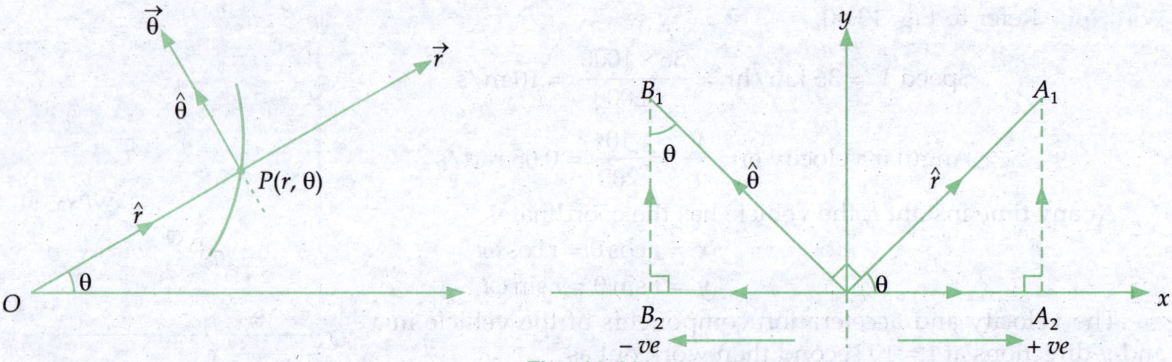

Fig. 13.41

The unit vectors of $\vec{r}$ and $\vec{\theta}$ are represented by $\hat{r}$ and $\hat{\theta}$ respectively. The meanings of these vectors can be understood and appreciated by referring to Fig. 13.41. Here A_1A_2 and B_1B_2 are perpendiculars onto the x-axis, and are drawn from the end points A_1 and B_1 of the unit vectors $\hat{r}$ and $\hat{\theta}$ respectively.

Further : $PA_1 = |\hat{r}| = 1$ (unity) and $PB_1 = |\hat{\theta}| = 1$ (unity)

That gives: $A_1A_2 = \sin\theta$ and $B_1B_2 = \cos\theta$

$PA_2 = \cos\theta$ and $PB_2 = \sin\theta$

Applying the triangle law of addition of vectors

$$\vec{r} = \overrightarrow{PA_1} = \overrightarrow{PA_2} + \overrightarrow{A_2A_1} = (\cos\theta)i + (\sin\theta)j$$

$$\vec{\theta} = \overrightarrow{PB_1} = \overrightarrow{PB_2} + \overrightarrow{B_2B_1} = -(\sin\theta)i + (\cos\theta)j$$

Velocity components: When the above identities are differentiated with respect to time t, we get

$$\frac{d}{dt}(\hat{r}) = [(-\sin\theta)i + (\cos\theta)j]\frac{d\theta}{dt} = \dot{\theta}\,\hat{\theta} \quad \ldots(i)$$

$$\frac{d}{dt}(\hat{\theta}) = [(-\cos\theta)i + -(\cos\theta)j]\frac{d\theta}{dt} = -\dot{\theta}\,\hat{r} \quad \ldots(ii)$$

where $\dot{\theta} = \dfrac{d\theta}{dt}$

Denoting $|\vec{r}|$ as r, we can write $\vec{r} = r\,\hat{r}$

In terms of position vector $\vec{r}$, the velocity vector is defined as

$$\vec{V} = \frac{d}{dt}(\vec{r}) = \frac{d}{dt}(r\,\hat{r}) = \frac{dr}{dt}\hat{r} + r\frac{d\hat{r}}{dt}$$

Replacing $\dfrac{d\hat{r}}{dt}$ by $\dot{\theta}\,\hat{\theta}$ as worked out above, we get

$$\vec{V} = \frac{dr}{dt}\hat{r} + r(\dot{\theta}\,\hat{\theta})$$

$$= \dot{r}\,\hat{r} + (r\,\dot{\theta})\,\hat{\theta} \qquad \text{where } \dot{r} = \frac{dr}{dt}$$

Thus the radial and transverse components of velocity are :

Radial component $V_r = \dot{r} = \dfrac{dr}{dt}$

Transverse component $V_\theta = r\,\dot{\theta} = r\dfrac{d\theta}{dt}$

Acceleration components: The acceleration of the particle is given by

$$\vec{a} = \frac{d}{dt}(\vec{V}) = \frac{d}{dt}\left[\dot{r}\,\hat{r} + (r\,\dot{\theta})\,\hat{\theta}\right]$$

$$= \frac{d\dot{r}}{dt}\hat{r} + \dot{r}\frac{d\hat{r}}{dt} + \frac{d}{dt}(r\dot{\theta})\,\hat{\theta} + r\dot{\theta}\frac{d\hat{\theta}}{dt}$$

$$= \frac{d\dot{r}}{dt}\hat{r} + \dot{r}\frac{d\hat{r}}{dt} + \left(r\frac{d\dot{\theta}}{dt} + \dot{\theta}\frac{dr}{dt}\right)\hat{\theta} + r\dot{\theta}\frac{d\hat{\theta}}{dt}$$

Substituting the values for $\dfrac{d}{dt}(\hat{r})$ and $\dfrac{d}{dt}(\hat{\theta})$ as given above in expressions (*i*) and (*ii*), we may write

$$\vec{a} = \ddot{r}\,\hat{r} + \dot{r}(\dot{\theta}\,\hat{\theta}) + \left(r\frac{d\dot{\theta}}{dt} + \dot{\theta}\frac{dr}{dt}\right)\hat{\theta} + r\dot{\theta}(-\dot{\theta}\,\hat{r})$$

$$= (\ddot{r} - r\,\dot{\theta}^2)\,\hat{r} + (r\,\ddot{\theta} + 2\,\dot{r}\,\dot{\theta})\,\hat{\theta}$$

Apparently then, the radial an transverse components of accelerations are :
Radial component of acceleration

$$\alpha_r = \ddot{r} - r\,\dot{\theta}^2 = \frac{d^2 r}{dt^2} - r\left(\frac{d\theta}{dt}\right)^2$$

Transverse component of acceleration

$$\alpha_\theta = r\,\ddot{\theta} + 2\,\dot{r}\,\dot{\theta} = r\frac{d^2\theta}{dt^2} + 2\frac{dr}{dt}\frac{d\theta}{dt}$$

$$= \frac{1}{r}\frac{d}{dt}\left(r\frac{d^2\theta}{dt^2}\right)$$

It may be noted that:

(i) When a particle moves along a circle, r = constant and then $\dot{r} = \ddot{r} = 0$. The acceleration components then take the values

$$\alpha_r = -r\,\dot{\theta}^2 = -r\left(\frac{d\theta}{dt}\right)^2$$

$$\alpha_\theta = r\,\ddot{\theta} = r\frac{d^2\theta}{dt^2}$$

(ii) When only r varies and θ is constant, then the motion of the particle corresponds to a rectilinear motion along a fixed θ – direction. Accordingly θ is constant and

$$\dot{\theta} = \ddot{\theta} = 0$$

$$V_r = \dot{r} = \frac{dr}{dt} \quad \text{and} \quad V_\theta = 0$$

$$\alpha_r = \ddot{r} = \frac{d^2 r}{dt^2} \quad \text{and} \quad \alpha_\theta = 0$$

(iii) In three dimensional cylinderical polar coordinates, the position of a particle is described by the coordinates (r, θ, z). The velocity and acceleration vectors can be written as,

$$\vec{V} = \frac{d\vec{r}}{dt} = \dot{r}\,\hat{r} + (r\,\dot{\theta})\,\hat{\theta} + \dot{z}\,\hat{z}$$

$$\vec{a} = \frac{d^2\vec{r}}{dt} = (\ddot{r} - r\,\dot{\theta}^2)\,\hat{r} + (2\,\dot{r}\,\dot{\theta} + r\,\ddot{\theta})\,\hat{\theta} + \ddot{z}\,\hat{z}$$

EXAMPLE 13.55

A particle moves along a curved path defined by

$$r = 5\theta \quad \text{and} \quad \theta = \frac{t^2}{3}$$

where r is in metres and t is in seconds.

Make calculations for the velocity and acceleration of the particle when $\theta = \pi/2$.

Solution :

$$\theta = \frac{t^2}{3} \qquad r = 5\theta = 5 \times \frac{t^2}{3} = \frac{5}{3}t^2$$

$$\frac{d\theta}{dt} = \frac{2}{3}t \qquad \frac{dr}{dt} = \frac{5}{3} \times 2t = \frac{10}{3}t$$

$$\frac{d^2\theta}{dt^2} = \frac{2}{3} \qquad \frac{d^2r}{dt^2} = \frac{10}{3}$$

When $\theta = \pi/2$; $\frac{\pi}{2} = \frac{t^2}{3}$ and $t = \sqrt{\frac{3\pi}{2}} = 2.17$ sec

Velocity of particle at $t = 2.17$ sec

$$\text{radial velocity } V_r = \frac{dr}{dt} = \frac{10}{3}t = \frac{10}{3} \times 2.17 = 7.23 \text{ m/s}$$

$$\text{transverse velocity } V_\theta = r\frac{d\theta}{dt} = \frac{5}{3}t^2 \times \frac{2}{3}t = \frac{10}{9}t^3 = \frac{10}{9} \times 2.17^3 = 11.35 \text{ m/s}$$

$$\text{resultant velocity } V = \sqrt{V_r^2 + V_\theta^2} = \sqrt{7.23^2 + 11.35^2} = \mathbf{13.46 \text{ m/s}}$$

The angle made by the resultant velocity with the radial direction is

$$\tan\alpha = \frac{11.35}{7.23} = 1.57 ; \quad \alpha = 57.5^\circ$$

Acceleration of particle at $t = 2.17$ sec

$$\text{radial acceleration } a_r = \frac{d^2r}{dt^2} - r\left(\frac{d\theta}{dt}\right)^2 = \frac{10}{3} - \frac{5}{3}t^2 \times \left(\frac{2}{3}t\right)^2 = \frac{10}{3} - \frac{20}{27}t^4$$

$$= \frac{10}{3} - \frac{20}{27}(2.17)^4 = -13.09 \text{ m/s}^2$$

$$\text{transverse acceleration } a_\theta = 2\frac{dr}{dt}\frac{d\theta}{dt} + r\frac{d^2\theta}{dt^2} = 2 \times \frac{10}{3}t \times \frac{2}{3}t + \frac{5}{3}t^2 \times \frac{2}{3} = \frac{50}{9}t^2$$

$$= \frac{50}{9} \times (2.17)^2 = 26.19 \text{ m/s}^2$$

$$\text{resultant acceleration } a = \sqrt{a_r^2 + a_\theta^2} = \sqrt{-13.09^2 + 26.16^2} = 29.25 \text{ m/s}^2$$

The angle made by the resultant acceleration with the radial direction is

$$\tan\beta = \frac{29.25}{13.09} = 2.234 \quad ; \quad \beta = 65.88^\circ$$

EXAMPLE 13.56

The position of a particle has been described by the co-ordinates (r, θ). Write expressions for the velocity and acceleration. Extend these expressions to the three-dimensional cylindrical polar co-ordinates where the position is described by the co-ordinates (r, θ, z)

(b) A straight bar hinged at one end rotates in a horizontal plane. Its position at any time t is defined by the angle $\theta = 2\pi(t - 1.5t^2)$ radians with the x-axis. A collar sliding on this bar has a mass of 5 kg and its position on the bar at time t is defined by the distance $r = (2t^2 - 0.9\,t^3)$ metres from the hinge. Find the radial and transverse components of force acting on the collar at t = 2.5 sec.

Solution: The velocity components along the radial (r) and transverse (θ) directions are given by

$$\text{radial velocity } v_r = \frac{dr}{dt} = \dot{r}$$

transverse velocity $v_\theta = r\dfrac{d\theta}{dt} = r\,\dot{\theta}$

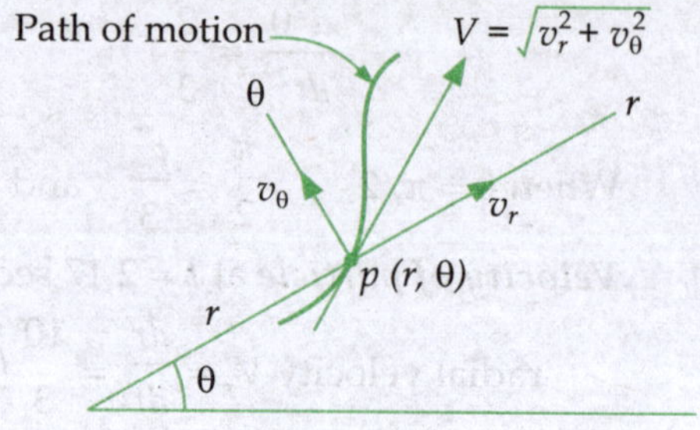

Fig. 13.42

The acceleration components along the radial and transverse directions are:

$$\text{radial acceleration } a_r = \frac{d^2r}{dt^2} - r\left(\frac{d\theta}{dt}\right)^2 = \ddot{r} - r(\dot{\theta})^2$$

$$\text{transverse acceleration } a_\theta = 2\frac{dr}{dt}\frac{d\theta}{dt} + r\frac{d^2\theta}{dt^2}$$

$$= 2\,\dot{r}\,\dot{\theta} - r\,\ddot{\theta}$$

In vector form, the velocity and acceleration of the particle can be written as

$$\bar{v} = \frac{d\,\bar{r}}{dt} = \dot{r}\,\hat{r} + (r\,\dot{\theta})\hat{\theta}$$

$$\bar{a} = \frac{d^2 r}{dt^2} = (\ddot{r} - r\,\dot{\theta}^2)\hat{r} + (2\,\dot{r}\,\dot{\theta} - r\ddot{\theta})\hat{\theta}$$

where $\hat{r}$ and $\hat{\theta}$ are the unit vectors along the r and θ directions respectively.

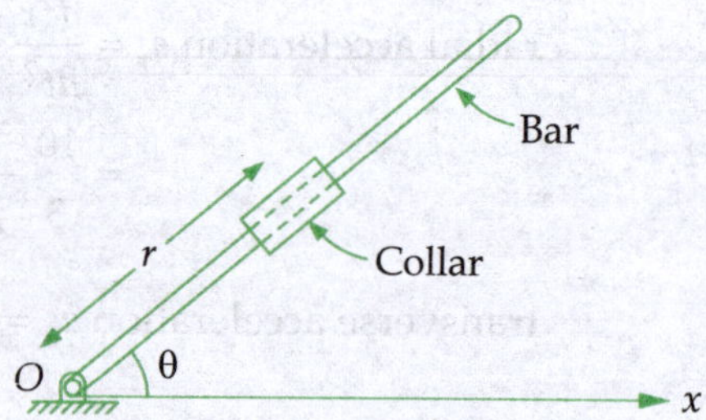

Fig. 13.43

In cylindrical polar coordinates where the position is described by the coordinates (r, θ, z), the velocity and acceleration can be written as

$$\bar{v} = \frac{d\,\bar{r}}{dt} = \dot{r}\,\hat{r} + (r\,\dot{\theta})\,\hat{\theta} + \dot{z}\,\hat{z}$$

and

$$\bar{a} = \frac{d^2\,\bar{r}}{dt^2} = (\ddot{r} - r\,\dot{\theta}^2)\hat{r} + (2\,\dot{r}\,\dot{\theta} + r\,\ddot{\theta})\hat{\theta} + \ddot{z}\,\hat{z}$$

(*b*)

	At time instant $t = 2.5$ sec
$r = 2t^2 - 0.9\,t^3$	$r = 2 \times 2.5^2 - 0.9 \times 2.5^3 = -1.56$
$\dot{r} = 4\,t - 2.7\,t^2$	$\dot{r} = 4 \times 2.5 - 2.7 \times 2.5^2 = -6.875$
$\ddot{r} = 4 - 5.4\,t$	$\ddot{r} = 4 - 5.4 \times 2.5 = -9.5$
$\theta = 2\pi\,(t - 1.5\,t^2)$	$\theta = 2\,\pi(2.5 - 1.5 \times 2.5^2) = -43.2$
$\dot{\theta} = 2\,\pi - 6\,\pi t$	$\dot{\theta} = 2\,\pi - 6\pi \times 2.5 = -40.84$
$\ddot{\theta} = -6\,\pi$	$\ddot{\theta} = -6\,\pi = -18.84$

Now, radial component of acceleration

$$a_r = \ddot{r} - r(\dot{\theta})^2 = -9.5 - (-1.56) \times (-40.84)^2$$

$$= -9.5 + 2601.93 = 2592.43 \text{ m/s}^2$$

and transverse component of acceleration

$$a_\theta = r\,\ddot{\theta} + 2\dot{r}\,\dot{\theta} = -1.56 \times (-18.85) + 2(-6.87)\,(-40.84)$$

$$= 29.41 + 561.14 = 590.55 \text{ m/s}^2$$

Then, radial force $F_r = m\,a_r = 5 \times 2592.43 =$ **12962 N**

transverse force $F_\theta = m\,a_\theta = 5 \times 590.55 =$ **2953 N**

EXAMPLE 13.57

A wheel of radius 0.5 m is turned to advance upon a right handed screw of pitch 1 cm. At an instant, wheel is turned at a rotational speed of 2 rad/sec, determine the velocity and acceleration of the hand held at wheel. If the wheel was accelerated rotationally at 0.6 rad/s^2, what would be the velocity and acceleration of wheel ?

Solution: In the cylindrical polar coordinate system (r, θ, z), the velocity and acceleration are prescribed by the relations:

$$\bar{v} = \frac{d\bar{r}}{dt} = \dot{r}\,\hat{r} + (r\,\dot{\theta})\hat{\theta} + \dot{z}\,\hat{z}$$

$$\bar{a} = \frac{d^2\bar{r}}{dt^2} = (\ddot{r} - r\,\dot{\theta}^2)\hat{r} + (2\dot{r}\,\dot{\theta} + r\,\ddot{\theta})\hat{\theta} + \ddot{z}\,\hat{z}$$

where $\hat{r}$, $\hat{\theta}$ and $\hat{z}$ are the unit vectors in respective directions.

Now, radial velocity $v_r = \dot{r} = \frac{dr}{dt} = 0$ as there is no change in radius of the wheel

transverse velocity $v_\theta = r\,\dot{\theta} = r\,\omega = 0.5 \times 2 = 1$ m/s

axial velocity $v_z = \dot{z}$ = pitch × rotational speed in rps

$$= p \times \frac{\omega}{2\pi} = 0.01 \times \frac{2}{2\pi} = 0.00318 \text{ m/s}$$

Resultant velocity $V = \sqrt{v_r^2 + v_\theta^2 + v_z^2} = \sqrt{0^2 + 1^2 + 0.00318^2}$

$\simeq$ **1 m/s**

Further, radial acceleration $a_r = \ddot{r} - r\,\dot{\theta}^2$

$$= \frac{d^2r}{dt^2} - r\left(\frac{d\theta}{dt}\right)^2 = 0 - r\,\omega^2$$

$$= -0.5 \times 2^2 = -2 \text{ m/s}^2$$

transverse acceleration $a_\theta = 2\dot{r}\,\dot{\theta} + r\,\ddot{\theta}$

$$= 2\frac{dr}{dt}\frac{d\theta}{dt} + r\frac{d^2\theta}{dt^2}$$

$$= 2\frac{dr}{dt}\omega + r\frac{d\omega}{dt} = 0 + 0 = 0$$

This is because $\frac{dr}{dt} = 0$ (as there is no change in radius)

and $\frac{d\omega}{dt} = 0$ (as speed of rotation of wheel is constant)

axial acceleration $a_z = \ddot{z} = \frac{d\dot{z}}{dt} = \frac{d}{dt}\left(p \times \frac{\omega}{2\pi}\right)$

$$= \frac{p}{2\pi}\frac{d\omega}{dt} = 0 \qquad (\because \omega = \text{constant})$$

Resultant acceleration $a = \sqrt{a_r^2 + a_\theta^2 + a_z^2}$

$= \sqrt{2^2 + 0^2 + 0^2} = $ **2 m/s²**

(ii) When the wheel is accelerated rotationally, the velocity components will remain same and so will be the resultant velocity.

$$a_r = \frac{d^2r}{dt^2} - r\omega^2 = 2 \text{ m/s}^2 \quad \text{(no change)}$$

$$a_\theta = 2\frac{dr}{dt}\omega + r\frac{d\omega}{dt} = 2\frac{dr}{dt}\omega + r\alpha$$

$$= 0 + 0.5 \times 0.6 = 0.3 \text{ m/s}^2$$

$$a_z = \frac{d}{dt}\left(p \times \frac{\omega}{2\pi}\right) = \frac{p}{2\pi}\frac{d\omega}{dt} = \frac{p}{2\pi}\alpha$$

$$= \frac{0.01}{2\pi} \times 0.6 = 0.000954 \text{ m/s}^2$$

Resultant acceleration $= \sqrt{2^2 + 0.3^2 + 0.000954^2}$

$=$ **2.022 m/s²**

13.6. RELATIVE VELOCITY

Consider two trains A and B moving in parallel tracks and in the same direction with velocities 60 km/hr and 45 km/hr respectively. These are the absolute velocities with respect to a fixed frame of reference. To a passenger sitting in train B, the train A will appear to be moving with a speed of (60 – 45) = 15 km/hr.

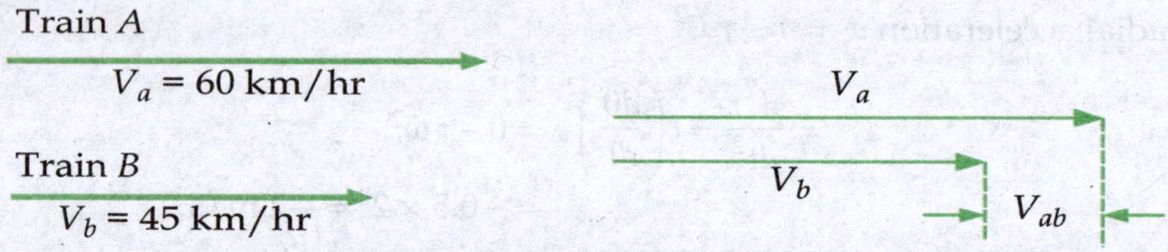

Fig. 13.44

This implies that relative velocity of A with respect to B is

$$V_{ab} = V_a - V_b = 60 - 45 = 15 \text{ km/hr}$$

When the motions of the trains are along parallel tracks but in opposite directions, then relative velocity of A with respect to B is

$$V_{ab} = V_a - (-V_b) = 60 - (-45) = 105 \text{ km/hr}$$

Quite often the motion of two bodies is along different directions rather than being along parallel lines (Fig. 13.46 *a*). However, the relative velocity of A with respect to B will always be equal to the difference between the velocities of A of B. Since velocity is a vector quantity, the relative velocity V_{ab} will be the vector difference between V_a and V_b. This vectorial difference can be determined graphically as shown in Fig. 13.46 *b* and 13.46 *c*. It is to be noted that V_{ab} (relative velocity of A with respect to B) and V_{ba} (relative velocity of B with respect to A) have the same magnitude but opposite directions.

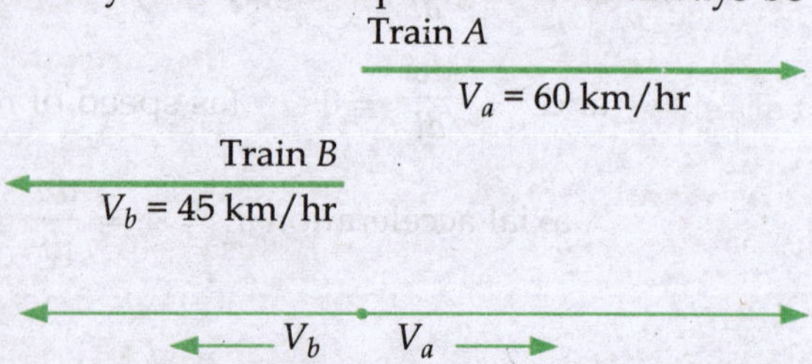

Fig. 13.45

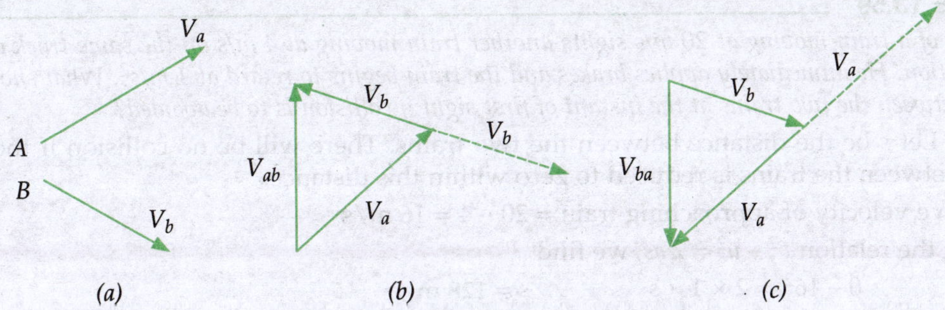

Fig. 13.46

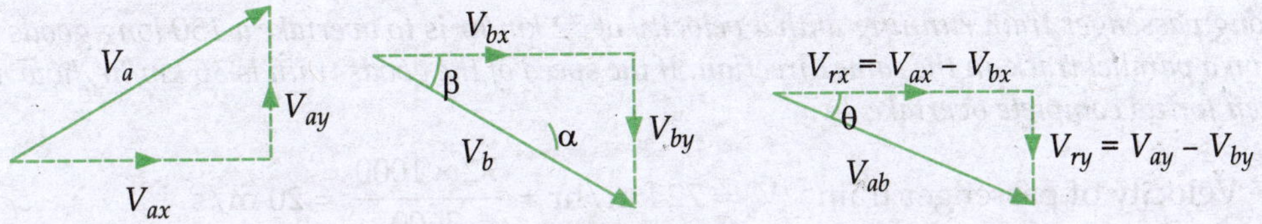

Fig. 13.47

The method of resolution of vectors can also be used to determine the magnitude and direction of relative velocity. The steps involved in this analytical method are:

(*i*) Workout the horizontal and vertical components of the given velocity vectors.

$$V_{ax} = V_a \cos\alpha\,; \quad V_{ay} = V_a \sin\alpha$$
$$V_{bx} = V_b \cos\beta\,; \quad V_{by} = V_b \sin\beta$$

(*ii*) Obtain the components of relative velocity by taking the difference in the component velocities in each direction

$$V_{rx} = V_{ax} - V_{bx} \quad \text{and} \quad V_{ry} = V_{ay} - V_{by}$$

The relative velocity V_{ab} is then determined by combining the components of relative velocity. That is

$$V_{ab} = \sqrt{V_{rx}^2 + V_{ry}^2}$$

and its inclination with horizontal axis is

$$\theta = \tan^{-}\left(\frac{V_{ry}}{V_{rx}}\right)$$

EXAMPLE 13.58

Two cars are moving in the same direction with a speed of 45 km/hr and a distance of 10 km separates them. A car coming from the opposite direction meets these two cars at an interval of 6 minutes. Determine the speed of this car.

Solution: Let V be the speed of the car coming from the opposite direction. Then velocity of this car relative to either of the two cars is

$$V - (-45) = (V + 45) \text{ km/hr}$$

$$\therefore \quad 10 = (V + 45) \times \frac{6}{60} \qquad (\because\ s = vt)$$

$$V = \frac{10 \times 60}{6} - 45 = \textbf{55 km/hr}$$

EXAMPLE 13.59

The driver of a train moving at 20 m/s sights another train moving at 4 m/s on the same track and in the same direction. He immediately applies brakes and the train begins to retard at 1 m/s^2. What should be the distance between the two trains at the instant of first sight if collision is to be avoided?

Solution: Let s be the distance between the two trains. There will be no collision if the relative velocity between the trains is reduced to zero within this distance.

Relative velocity of approaching train = 20 – 4 = 16 m/s

Using the relation $v^2 - u^2 = 2as$, we find

$$0 - 16^2 = 2 \times 1 \times s \qquad \therefore \; s = \mathbf{128\ m}$$

EXAMPLE 13.60

A 200 m long passenger train running with a velocity of 72 km/hr is to overtake a 150 long goods train that is moving on a parallel track in the same direction. If the speed of the goods train is 36 km/hr, how much time will be taken for its complete overtake ?

Solution: Velocity of passenger train $V_a = 72 \text{ km/hr} = \dfrac{72 \times 1000}{3600} = 20 \text{ m/s}$

Velocity of goods train $V_b = 36 \text{ km/hr} = \dfrac{36 \times 1000}{3600} = 10 \text{ m/s}$

Relative velocity of passenger train with respect to goods train,

$$V_{ab} = V_a - V_b = 20 - 10 = 10 \text{ m/s}$$

Total distance to be covered = 200 + 150 = 350 m

Then $\quad 350 = 10 \times t \qquad (\because \; s = vt)$

$$\therefore \quad t = \mathbf{35\ s}$$

Hence the passenger train will take 35 seconds to pass the goods train.

EXAMPLE 13.61

Two train A (length 125 m) and B (length 100 m) move in opposite directions along parallel tracks. At the instant of complete pass over, the train A is moving at 10 m/s with a constant acceleration of 0.1 m/s^2, and train B has a uniform speed. If the trains take 12 seconds to pass one another, determine the uniform speed of train B.

Solution: Total distance to be passed = 125 + 100 = 225 m

From the relation, $s = ut + \dfrac{1}{2}at^2$

$$225 = V \times 12 + \frac{1}{2} \times 0.1 \times 12^2 = 12V + 7.2$$

where V is the relative velocity of train A with respect to train B

$$V = \frac{225 - 7.2}{12} = 18.15 \text{ m/s}$$

If V_b is the velocity of train B, them

$$V = V_a - (-V_b)\,; \quad 18.5 = 10 + V_b$$

$$\therefore \quad V_b = 18.5 - 10 = \mathbf{8.5\ m/s}$$

EXAMPLE 13.62

Two roads cross at 90° angle at the intersection point O as shown in Fig. 13.48. At an instant of time, a car B passes the intersection travelling east at a constant speed of 36 km/hr. At the same instant, a car A at a distance of 35 m north of the intersection starts from the rest and moves south with a constant acceleration of 1.2 m/s². Determine the position, velocity and acceleration of car B relative to car A six seconds after car A crosses the intersection of loads.

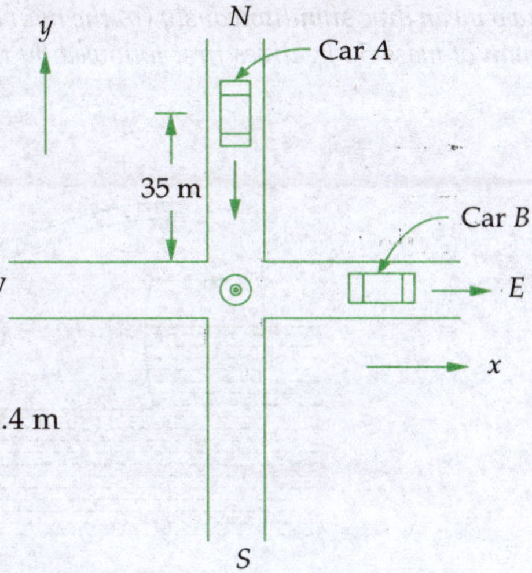

Fig. 13.48

Solution:

Car A: $u = 0, a = 1.2 \text{ m/s}$ and $t = 6 \text{ s}$

$$v = u + at = 0 + 1.2 \times 6 = 7.2 \text{ m/s}$$

$$s = ut + \frac{1}{2}at^2 = 0 + \frac{1}{2} \times 1.2 \times 6^2 = 21.6 \text{ m}$$

The car A will thus be at a distance of $(35 - 21.6) = 13.4$ m from the origin along y-axis.

Car B: speed $= 36 \text{ km/hr} = \dfrac{36 \times 1000}{3600}$

$$= 10 \text{ m/s} \quad \text{and} \quad u = 0$$

$$s = ut + \frac{1}{2}at^2 = 10 \times 6 = 60 \text{ m}$$

The car B will thus be at a distance of 60 m from the origin along x-axis.

(*i*) Relative position of car A with respect to car B

$$\bar{r}_{ab} = \bar{y}_a - \bar{x}_b$$

and this is prescribed by vector $\overline{BA}$

$$\overline{BA} = \sqrt{OA^2 + OB^2}$$

$$= \sqrt{13.4^2 + 60^2} = 61.478$$

$\therefore \quad \bar{r}_{AB} =$ **61.478 m** and

$$\alpha = \tan^{-1}\frac{13.4}{60} = \mathbf{12.587°}$$

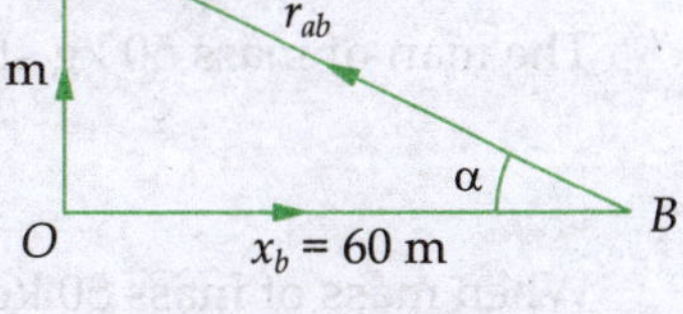

(*ii*) Relative velocity of car A with respect to car B

$$\bar{V}_{ab} = \bar{v}_a - \bar{v}_b$$

and this is represented by vector $\overline{BA}$

$$\overline{BA} = \sqrt{OA^2 + OB^2} = \sqrt{7.2^2 + 10^2} = 12.32$$

$\therefore \quad \bar{V}_{ab} = 12.32 \text{ m/s}$ and

$$\beta = \tan^{-1}\frac{7.2}{10} = 35.75°$$

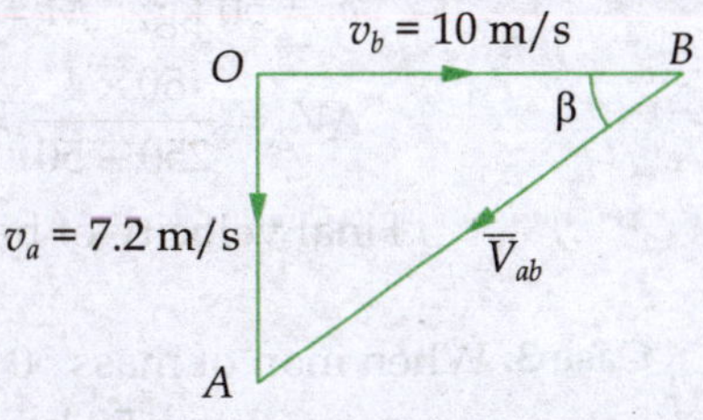

(*iii*) Relative acceleration of car A with respect to car B

$$\bar{a}_{ab} = \bar{a}_a - \bar{a}_b = 1.2 - 0 = 1.2 \text{ m/s}^2$$

EXAMPLE 13.63

Two men, M_1 of mass 50 kg and M_2 of mass 75 kg, dive off the ends of a boat of mass $M = 250$ kg so that their relative velocity with respect to the boat is 4 m/sec. If the boat is initially at rest, find its final velocity if (a) two mean dive simultaneously (b) the man of mass 75 kg dives first followed by the man of mass 50 kg (c) the man of mass 50 kg dives first followed by the man of mass 75 kg.

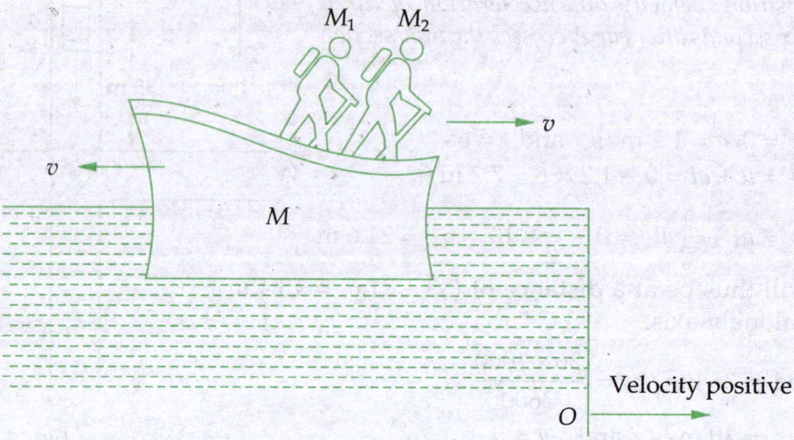

Fig. 13.49

Solution: When a person of mass m with velocity v jumps from a trolley or boat of mass M, there occurs an increase in velocity of boat given by

$$\Delta V = \frac{mv}{M+m}$$

The increase ΔV in the velocity of boat is independent of the initial velocity V of the boat.

Case 1: Both men dive simultaneously

$$m = (50 + 75) = 125 \text{ kg}; \quad M = 250 \text{ kg} \quad \text{and} \quad v = 4 \text{ m/s}$$

$$\therefore \quad \Delta V = \frac{125 \times 4}{250 + 125} = 1.33 \text{ m/s}$$

Initially the boat is at rest, *i.e.*, $V = 0$

$\therefore$ Final velocity of boat $= V + \Delta V = 0 + 1.33 =$ **1.33 m/s**

Case 2: When man of mass 75 kg dives first

$$m = 75 \text{ kg}; \quad v = 4 \text{ m/s}$$

The man of mass 50 kg still remains in the boat and so $M = 250 + 50 = 300$ kg

$$\Delta V_1 = \frac{mv}{M+m} = \frac{75 \times 4}{300 + 75} = 0.8 \text{ m/s}$$

When mass of mass 50 kg dives next

$$m = 50 \text{ kg}; \quad M = 250 \text{ kg (only boat remains) and } v = 4 \text{ m/s}$$

$$\Delta V_2 = \frac{50 \times 4}{250 + 50} = 0.667 \text{ m/s}$$

$\therefore$ Final velocity of boat $= V + \Delta V_1 + \Delta V_2$
$= 0 + 0.8 + 0.667 =$ **1.467 m/s**

Case 3: When man of mass 50 kg dives first

$$m = 50 \text{ kg}; \quad M = (250 + 75) = 325 \text{ kg} \quad \text{and} \quad v = 4 \text{ m/s}$$

$$\Delta V_1 = \frac{mv}{M+m} = \frac{50 \times 4}{325+50} = 0.533 \text{ m/s}$$

When man of mass 75 kg dives next

$$m = 75 \text{ kg}; \quad M = 250 \text{ kg} \quad \text{and} \quad v = 4 \text{ m/s}$$

$$\Delta V_2 = \frac{75 \times 4}{250+75} = 0.923 \text{ m/s}$$

$$\therefore \quad \text{Final velocity of boat } = V + \Delta V_1 + \Delta V_2$$

$$= 0 + 0.533 + 0.923 = \mathbf{1.456 \text{ m/s}}$$

EXAMPLE 13.64

A car moving along a South-East road at 70 km/hr crosses a railway crossing. At the same time a train also passes the same crossing at 50 km/hr along the railway track which is due East. Determine the magnitude and direction of the apparent velocity of car as observed by a passenger in the train.

Solution: Refer Fig. 13.50

A velocity equal and opposite to that of passenger in the train is superimposed both on the car as well as the train. By doing so, the velocity of train reduces to zero. The car will then have a resultant velocity V_{ct} due to its own velocity V_c and the superimposed velocity V_t. This resultant is then the apparent velocity of car with respect to passenger in the train.

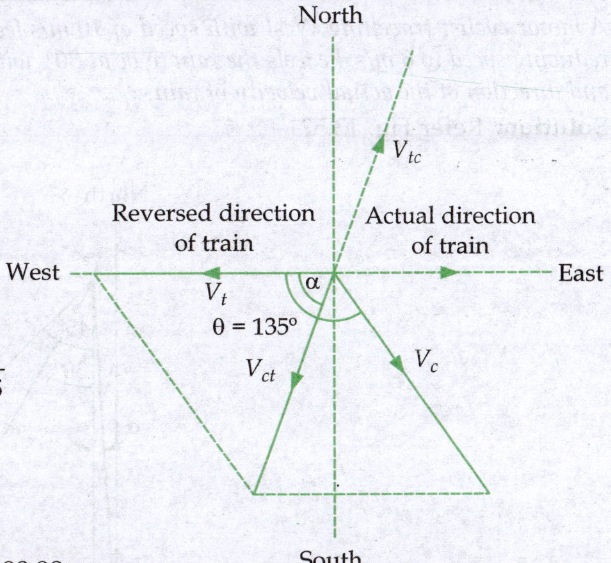

Fig. 13.50

From the parallelogram law of vectors,

$$V_{ct} = \sqrt{V_c^2 + V_t^2 + 2V_cV_t \cos\theta}$$

$$= \sqrt{70^2 + 50^2 + 2 \times 70 \times 50 \times \cos 135}$$

$$= \mathbf{49.50 \text{ m/s}}$$

$$\tan \alpha = \frac{V_c \sin\theta}{V_t + V_c \cos\theta}$$

$$= \frac{70 \sin 135^\circ}{50 + 70 \cos 135} = \frac{70 \times 0.707}{50.0 - 49.50} = 98.98$$

or $\alpha = 89.42°$ with West, *i.e.*, 0.58° West of South

Note : The apparent velocity of train as observed by the driver of the car will have the same magnitude but opposite in direction (shown dotted)

Magnitude = 49.50 m/s

Direction = 0.58° East of North.

EXAMPLE 13.65

A passenger sitting in a train moving at 54 km/hr is hit by a stone thrown at right angles to it with a velocity of 18 km/hr. Calculate the velocity and the direction with which the stone appears to hit the passenger.

Solution: Refer Fig 13.51

Velocity of train $V_t = 54$ km/hr

$$= \frac{54 \times 1000}{3600} = 15 \text{ m/s}$$

Velocity of stone $V_s = 18$ km/hr

$$= \frac{18 \times 1000}{3600} = 5 \text{ m/s}$$

A velocity equal and opposite to that of train is superimposed both on the stone and the train. By doing so, the velocity of train reduces to zero. The stone will then have a resultant velocity V_{st} due to its own velocity V_s and the superimposed velocity V_t. This resultant is then the apparent velocity with which the stone hits the passenger.

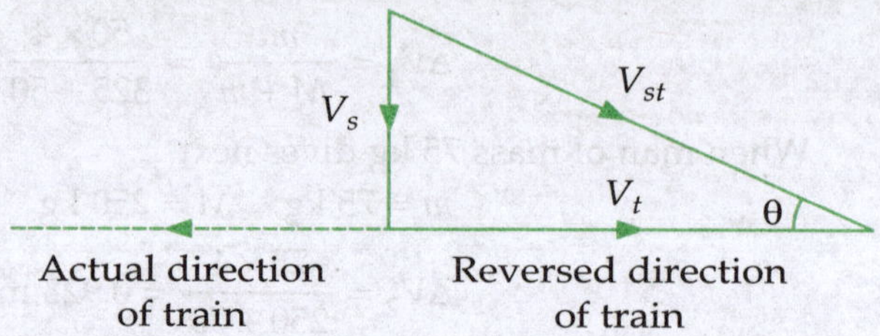

Fig. 13.51

$$V_{st} = \sqrt{V_s^2 + V_t^2} = \sqrt{5^2 + 15^2} = \sqrt{250}$$

$$\theta = \tan^{-1}\frac{V_s}{V_t} = \tan^{-1}\frac{5}{15} = \mathbf{18.43^\circ}$$

EXAMPLE 13.66

A motor cyclist travelling West with speed of 10 m/s feels the rain coming down at 45° to the vertical. On reducing speed to 6 m/s, he feels the rain to be at 30° with the vertical. Make calculations for the magnitude and direction of the actual velocity of rain.

Solution: Refer Fig. 13.52

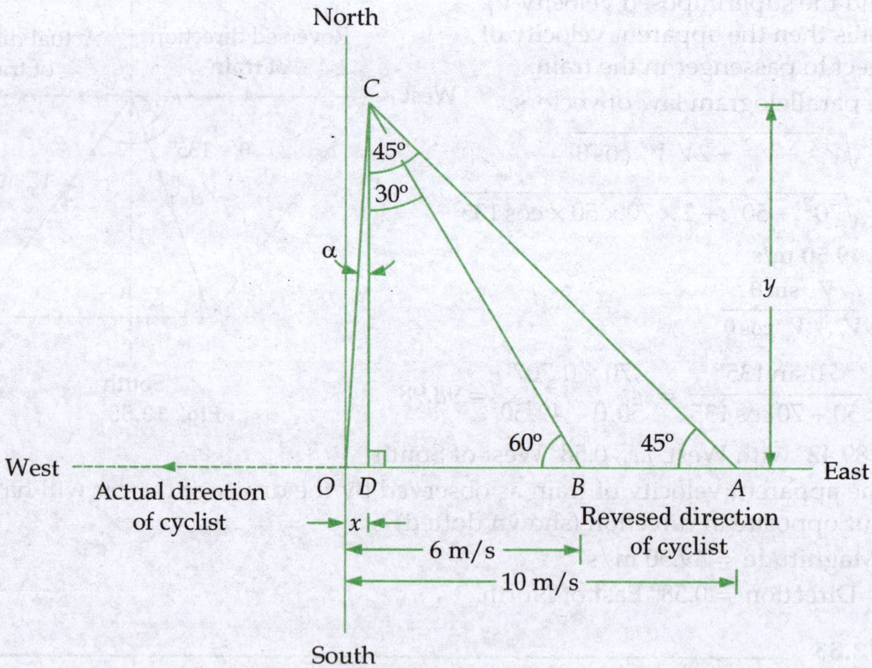

Fig. 13.52

Let CO represent the magnitude and direction of the actual velocity of rain.

From ΔADC :

$$\tan 45^\circ = \frac{DC}{AD}$$

$$1 = \frac{y}{10 - x} \;;\quad y = 10 - x \qquad \text{...}(i)$$

From ΔBDC:

$$\tan 60^\circ = \frac{DC}{BD}$$

$$1.732 = \frac{y}{6 - x} \;;\quad y = 10.392 - 1.732\,x \qquad \text{...}(ii)$$

From expressions (*i*) and (*ii*)

$$10 - x = 10.392 - 1.732\,x$$

$$x = \frac{10.392 - 10}{0.732} = 0.535 \text{ m/s}$$

Substituting $x = 0.535$ m/s in expression (*i*), one obtains

$$y = 10 - 0.535 = 9.465 \text{ m/s}$$

$$\tan\alpha = \frac{x}{y} = \frac{0.535}{9.465} = 0.0565 \;;\quad \alpha = \mathbf{3.23^\circ}$$

Magnitude of velocity of rain,

$$OC = \sqrt{OD^2 + CD^2}$$

$$= \sqrt{0.535^2 + 9.465^2} = \sqrt{0.286 + 89.586} = \mathbf{9.48 \text{ m/s}}$$

EXAMPLE 13.67

Two ships, S_1 and S_2, at a given instant are 3 km apart; S_2 being south-east of S_1. The ship S_1 is travelling 12 km/hr due east and ship S_2 is travelling 20 km/hr due north. Make calculations for (a) the velocity of ship S_2 relative to ship S_1, (b) the shortest distance between the two ships, and (c) the time when the two ships are nearest.

Solution: Refer Fig. 13.53

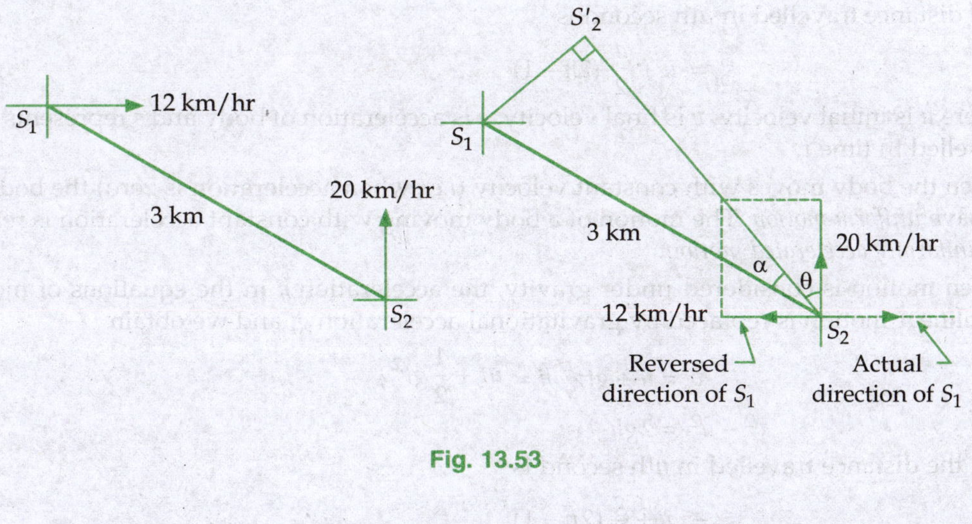

Fig. 13.53

Velocity of ship S_2 relative to ship S_1

$$V_r = \sqrt{20^2 + 12^2} = \sqrt{544} = 23.32 \text{ km/hr}$$

$$\tan\theta = \frac{12}{20} = 0.6; \quad \theta = 30.97^\circ$$

The ship S_2 will appear to travel at 23.32 km/hr along a direction 30.97° west of north to a passenger on ship S_1.

(*b*) $\alpha = 45 - 30.97 = 14.03^\circ$

The shortest distance between the two ships is the perpendicular distance from the apparently stationary ship S_1 to the apparent direction of ship S_2

$$\text{Shortest distance} = S_1 S_2' = S_1 S_2 \sin\alpha$$

$$= 3\sin 14.03^\circ = \mathbf{0.7275 \text{ km}}$$

(*c*) Apparent (relative) velocity of ship S_2 = 23.32 km/hr

The apparent distance travelled by ship S_2 when it reaches the shortest distance point S_2' is

$$S_2 S_2' = S_1 S_2 \cos\alpha = 3\cos 14.03^\circ = 2.911 \text{ km}$$

∴ Interval of time when the two ships are at the shortest distance,

$$t = \frac{s}{v} = \frac{2.911}{23.32} = 0.1248 \text{ hr} = \mathbf{7.488 \text{ min}}$$

REVIEW AND SUMMARY

1. When a body changes its position with respect to its surroundings, then the body is said to be in motion. Thus relative change in position is motion.
2. Trajectory is a path followed by the body during its motion. It may be a straight line or a curve. Straight line path is observed in rectilinear motion, whereas curved path is observed in curvilinear motion.
3. When a body moves in a straight line with uniform velocity, the equations of motion are:

$$v = u + at; \quad v^2 - u^2 = 2as; \quad s = ut + \frac{1}{2}at^2$$

and distance travelled in *n*th second is

$$s_n = u + \frac{a}{2}(2n - 1)$$

where *u* is initial velocity, *v* is final velocity, *a* is acceleration of body and *s* represents distance travelled in time *t*.

4. When the body moves with constant velocity (*i.e.*, when acceleration is zero) the body is said to have *uniform motion*. The motion of a body moving with constant acceleration is referred to as *uniformly accelerated motion*.
5. When motion is considered under gravity, the acceleration *a* in the equations of motion for rectilinear motion is replaced by gravitational acceleration *g*, and we obtain

$$v = u + gt; \quad h = ut + \frac{1}{2}gt^2;$$

$$v^2 - u^2 = 2gh$$

and the distance travelled in *n*th second is

$$s_n = u + \frac{g}{2}(2n - 1)$$

6. When a body moves in a circular orbit of radius r, then

Normal acceleration $a_n = \dfrac{V^2}{r}$

The normal acceleration is due to change in the direction of velocity. It is directed towards the centre of curvature and is known as centripetal acceleration.

The tangential acceleration is due to change in magnitude of velocity and it equals

$$a_t = \frac{\text{change in tangential velocity}}{\text{time interval}} = \frac{dV}{dt}$$

Resultant acceleration $a = \sqrt{a_t^2 + a_n^2}$

It the resultant acceleration makes an angle ϕ with the normal, then

$$\tan\phi = \frac{a_t}{a_n}$$

It is worthwhile to note that:

(*i*) If the magnitude of velocity is contant between two points, then tangential acceleration $a_t = 0$

(*ii*) If the direction of motion of the particle does not change, *i.e.*, the motion of the particle is along a straight line, then $r \to \infty$. In that case normal acceleration $a_n = 0$.

7. Relative velocity is the velocity of an object as observed from another moving object, and it equals as vector difference of the velocities of the two objects.
8. Resultant velocity is the combined effect obtained as the vector addition of the velocities caused by different forces acting on an object.

REVIEW QUESTIONS

A. Conceptual and conventional questions :

1. Differentiate between kinematics and kinetics.
2. When is the motion of a particle called rectilinear, and when curvilinear?
3. Differentiate between (*a*) displacement and distance, (*b*) average velocity and instantaneous velocity, (*c*) average acceleration and instantaneous acceleration.
4. How a uniform motion differs from a uniformly accelerated motion?
5. What parameters define rectilinear motion? State the relation between those parameters?
6. Draw and interpret the following graphs pertaining to rectilinear motion
   (*a*) Displacement-time curve, (*b*) Velocity-time curve and
   (*c*) Acceleration-time curve
7. What is meant by relative motion?
8. Define normal acceleration and tangential acceleration in connection with curvilinear motion.
9. A car travels along a straight track between two points at a mean speed of 80 m/s. It returns at a mean speed of 60 km/hr. What is the mean speed for the round trip?

(66.55 km/hr)

10. A train accelerates from rest with constant acceleration α m/s^2 and after some time decelerates at a uniform rate of β m/s^2 till it comes to rest. If the total time taken between the two stations is t, work out an expression for the maximum velocity acquired by the train during the course of its motion. $\left(V_{max} = \dfrac{\alpha\beta}{\alpha + \beta} t\right)$

11. The velocity-time graph of a particle moving along a straight line is as shown in Fig. 13.54. What would be the ratio of the distances covered during the time intervals *BD* and *DC*? **(3 : 1)**

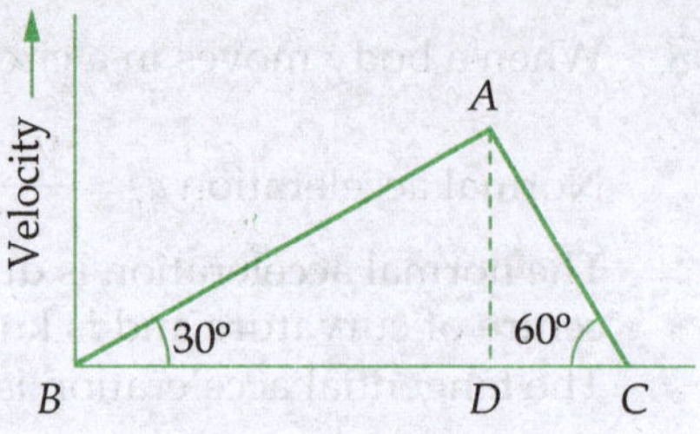

Fig. 13.54

12. The displacement time graph for two particles A and B are straight lines inclined at angles of 30° and 45° with the time axis. What will be the ratio of their velocities? **($V_a/V_b = 0.577$)**
13. A vehicle having rectilinear motion is moving with a velocity of 54 km/hr. The vehicle is brought to halt in a distance of 100 m by the application of brakes. Determine the uniform rate of deceleration and the time for which the brakes were applied? **(1.125 m/s², 13.33s)**
14. A body in reclinear motion is found to travel 15 m in 5th second and 25 m in the 10th second. What distance will it travel in 15 seconds from the staring point if its acceleration is constant throughout? **(315 m)**
15. Two trains *A* and *B* leave the same station on parallel tracks. The train *A* starts with a uniform acceleration of 2.0 m/s² and attains a speed of 45 km/hr which remains constant. The train *B* leaves 1 minute later, with a uniform acceleration of 0.4 m/s² to attain a maximum speed of 72 km/hr. When will the train *B* overtatke train *A*? **(114.6 s)**
16. A ball is projected upwards with a velocity of 24 m/s. Two seconds later, a second ball is projected vertically upwards with a velocity of 18 m/s. At what point above the surface of earth will they meet? **(15.95 m)**
17. To a man walking with a speed of 2 m/s, the rain appears to fall vertically. If the velocity of rain is 4 m/s, calculate its direction with respect to the direction of man's velocity.
18. A car is travelling east at a constant speed 7 m/s. At the same time, another car starts from rest towards north with a constant acceleration of 2 m/s². Determine the position, velocity and acceleration of second car relative to the first car after 5 seconds. **(s_{21} = 43 m at 35.5°; V_{21} = 12.2 m/s at 55°; a_{21} = 2 m/s²)**
19. A particle moves in the x-y plane and its *x*-coordinate is defined as
$$x = 5t^3 - 105t$$
where *x* is in cm and *t* is in seconds. When *t* = 2 sec, the total acceleration equals 75 cm/s². Presuming that *y*-component of acceleration is constant and the particle starts from rest at the origin when *t* = 0, make calculation for the velocity when *t* = 4 second.
20. A particle is projected to move along a parabola $y^2 = 4x$. At a certain instant, when passing through a point *P* (4, 4), its speed is 5 m/s and the rate of increase of its speed is 3 m/s² along the path. Express the velocity and acceleration in terms of rectangular coordinates.

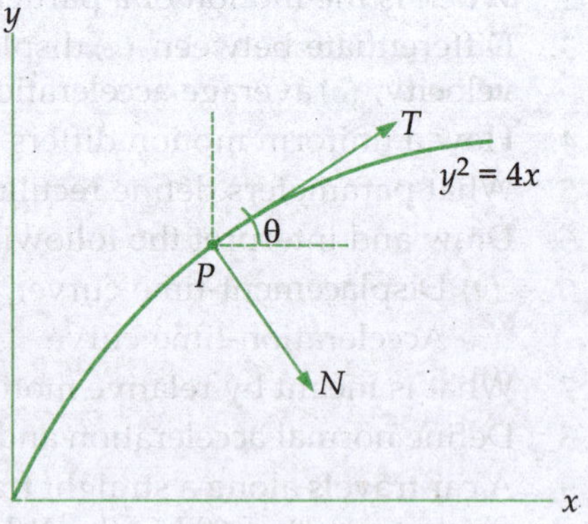

Fig. 13.55

Hint: $\frac{dy}{dx} = \frac{2}{y} = \tan\theta$

$\therefore\quad \theta \text{ at } P\,(4, 4) = \tan^{-1}\left(\frac{2}{4}\right) = 26.56°$

Components of tangential velocity (V_t)

$$V_x = 5\cos\theta = 5\cos 26.56° = 4.47 \text{ m/s}$$
$$V_y = 5\sin\theta = 5\sin 26.56° = 2.34 \text{ m/s}$$

Components of tangential acceleration (a_t)

$$a_x = 3\cos 26.56 = 2.68 \text{ m/s}^2$$
$$a_y = 3\sin 26.56 = 1.34 \text{ m/s}^2$$

Normal acceleration $\quad a_n = \dfrac{V^2}{R}$

$$R = \frac{\left[1 + (dy/dx)^2\right]^{3/2}}{d^2y/dx^2} = 22.36 \text{ m}$$

Then $\quad (a_n)_x = \dfrac{V^2}{R}\cos(90 - \theta) = 0.5 \text{ m/s}^2$

$$(a_n)_y = -\frac{V^2}{R}\cos(90 - \theta) = -1.0 \text{ m/s}^2$$

B: Complete the following statements with most appropriate word/words :

1. Motion of particle on a straight line is termed as .... motion and on curved path as .... motion.
2. Plane motion means motion with .... acceleration.
3. The change in the position of a particle with respect to a certain fixed reference is termed as ....
4. The motion of a body moving with constant acceleration is called .... motion.
5. The slope of distance-time graph at any instant gives the .... of the body at that instant.
6. The area of velocity-time graph with time axis represents the .... by the body.
7. The area under the acceleration-time graph corresponding to a given time interval gives the change in .... of the body during the same time interval.
8. For a body thrown vertically upwards with velocity u, the maximum height attained is ....
9. For a body thrown vertically upwards with velocity u, the total time taken by the body to fall back to ground is ....
10. A stone is made to fall from rest from the top of a tower. The distance travelled in the first, second and third second would be in the ratio ....
11. The normal acceleration of a body moving in a circular orbit is directed .... the centre of curvature and is called .... acceleration.
12. For a body moving in a circular orbit, the magnitude of velocity remains constant. Then its .... will be zero.
13. Relative velocity is the velocity of an object as observed from another moving object and it equals as ....of the velocities of the two objects.
14. Starting from rest when a body moves with uniform acceleration, then the distances travelled by the body in successive seconds are in the ratio ....

Answers:

1. linear, curvilinear 2. uniform 3. displacement 4. uniformly accelerated 5. velocity 6. distance covered 7. velocity 8. $\dfrac{u^2}{g}$ 9. $\dfrac{2u}{g}$ 10. 1 : 3 : 5 11. inwards towards, centripetal 12. tangential acceleration 13. vector difference 14. 1 : 4 : 9

C. Multiple choice questions :

1. For first half of time a car travels with velocity v_1 and for the second half of time it travels with velocity v_2. The average velocity of the car is

(a) $\frac{v_1 + v_2}{2}$ (b) $\frac{2 v_1 v_2}{v_1 + v_2}$ (c) $\sqrt{v_1 v_2}$ (d) $\frac{v_1 v_2}{2(v_1 + v_2)}$

2. Sandeep goes to office at a speed of 6 km/hr and returns to his home at a speed of 4 km/hr. If the total time taken is 2 hr, the distance between his office and home is closest to

(a) 5 km (b) 8 km (c) 10 km (d) 12 km

3. Mandeep is late for his office by 16 minutes when he walks at 3/4 of his usual speed. What is then the usual time taken by him to cover the stipulated distance ?

(a) 30 min (b) 36 min (c) 48 min (d) 60 min

4. A passenger train takes 2 hours less for a journey of 300 km if its speed is increased by 5 km/hr over its usual speed. What is then the usual speed?

(a) 15 km/hr (b) 29 km/hr (c) 25 km/hr (d) 30 km/hr

5. A cyclist starts from rest with uniform acceleration. If he covers 2.6 m in the 7th second, the uniform acceleration is

(a) 0.3 m/s^2 (b) 0.4 m/s^2 (c) 0.5 m/s^2 (d) 0.6 m/s^2

6. A particle starts from rest and moving with a constant acceleration covers a distance x_1 in the 3rd second and distance x_2 in the 5th second. The ratio x_1/x_2 is

(a) 3/5 (b) 5/9 (c) 9/25 (d) 25/81

7. A car moving with uniform acceleration covers 450 m in a 5 second interval, and covers 700 m in the next 5 second interval. The acceleration of the car is

(a) 7 m/s^2 (b) 10 m/s^2 (c) 25 m/s^2 (d) 50 m/s^2

8. A car moving with speed u can be stopped in minimum distance x when brakes are applied. If the speed becomes n times, the minimum distance over which the car can be stopped would take the value

(a) x/n (b) nx (c) x/n^2 (d) n^2x

9. When two bodies move uniformly towards each other, the distance between them decreases by 6 m/s. If both bodies move in the same direction with the same speeds, the distance between them increases by 4 m/s. Then the speeds of the two bodies are:

(a) 3 m/s and 3 m/s (b) 4 m/s and 2 m/s
(c) 5 m/s and 1 m/s (d) 8 m/s and 5 m/s

10. A particle starts from rest with a constant acceleration α m/s^2 and after some time it decelerates at a uniform rate of β m/s^2 till it comes to rest. If the total time taken between two rest positions is t, the maximum velocity acquired by the particle would be

(a) $\frac{\alpha + \beta}{2} t$ (b) $\frac{\alpha - \beta}{2} t$ (c) $\left(\frac{\alpha\beta}{\alpha + \beta}\right) t$ (d) $\left(\frac{\alpha + \beta}{\alpha - \beta}\right) t$

11. A train accelerates from rest at a constant rate α for some time and then it retards to rest at constant rate β. If total distance covered by the train is x, then maximum velocity attained by the train during the course of its motion would be

(a) $\left[\frac{\alpha + \beta}{2\alpha\beta} x\right]^{1/2}$ (b) $\left[\frac{2\alpha\beta}{\alpha + \beta} x\right]^{1/2}$ (c) $\left[\frac{\alpha + \beta}{\alpha - \beta} x\right]^{1/2}$ (d) $\left[\frac{\alpha\beta}{\alpha - \beta} x\right]^{1/2}$

12. A body is allowed to fall from the top of a tower. It falls through half the height in 2 seconds. The total time taken to reach the ground is nearly

(a) 4.5 sec (b) 4 sec (c) 3.2 sec (d) 2.8 sec

13. A body is dropped from rest at height h. It covers a distance $9/25\ h$ in the last second. The height h is about

(a) 100 m (b) 115 m (c) 125 m (d) 150 m

14. A stone takes 6 seconds to reach the ground after it is dropped from a tower. If the stone is stopped after 2 seconds of its fall and then released again, it will take .... more seconds to reach the ground.

(a) 3.4 (b) 4.0 (c) 4.8 (d) 5.6

Take $g = 10\ m/s^2$

15. Two metallic blocks having masses in the ratio 2 : 3 are made to slide down a frictionless inclined plane starting initially from rest position. When these blocks reach the bottom of the inclined plane, they will have their kinetic energies in the ratio

(a) 2 : 3 (b) 4 : 9 (c) 3 : 2 (d) 9 : 4

16. The time-acceleration graph of a body moving with uniform acceleration is a straight line

(a) passing through the origin (b) inclined to the time axis
(c) parallel to time axis (d) parallel to acceleration axis

17. The displacement-time graph for two particles A and B are straight lines inclined at angles of 30° and 45° with the time axis. The ratio of velocities $V_a : V_b$ will be about

(a) 0.33 (b) 0.5 (c) 0.58 (d) 0.87

18. The time variation of the position of a particle in rectilinear motion is given by

$$x = 2t^3 + t^2 + 2t$$

If v is the velocity and a the acceleration of particle in consistent units, the motion started with

(a) $v = 0 ; a = 0$ (b) $v = 0 ; a = 2$ (c) $v = 2 ; a = 0$ (d) $v = 2, a = 2$

19. A boatman can row at 6 km/hr in still water. If the river flows steadily at 3 km/hr, along which direction with the bank of river the boatman should row in order to reach a point on the other bank directly opposite to the point from where he started ?

(a) 30° (b) 60° (c) 120° (d) 150°

The angle θ is measured from the river bank in counter clockwise direction.

20. A body moves with a speed of 10 m/s in the curved path of 25 m radius of curvature. If the tangential acceleration is $3\ m/s^2$, then the total acceleration for the body will be

(a) $3.3\ m/s^2$ (b) $4\ m/s^2$ (c) $5\ m/s^2$ (d) $5.8\ m/s^2$

21. The distance covered by a particle travelling on a circular path and starting from rest is given by the equation: $s = kt^2$ where k is a constant. The tangental acceleration of the particle will be

(a) $k/2$ (b) k (c) $2k$ (d) $4k$

Answers :

1. (a)	2. (a)	3. (c)	4. (c)	5. (b)	6. (b)
7. (b)	8. (d)	9. (b)	10. (c)	11. (b)	12. (d)
13. (c)	14. (d)	15. (a)	16. (c)	17. (c)	18. (d)
19. (c)	20. (c)	21. (c)			

Projectiles 14

Learning objectives : Attention has been focussed in this chapter to familiarize the reader with

- *projectile and its motion*
- *trajectory or the path of a projectile*
- *projectile terms and their relationship with the velocity and angle of projection*
- *projectile motion up and down on an inclined plane*

A projectile refers to any object which is thrown into the space and whose subsequent motion is subjected to only gravitational forces. A projectile travels both in the horizontal and vertical directions and traces a curvilinear path which is called trajectory.

The necessary conditions for a body to be called a projectile are:

(*i*) it should be given some initial velocity

(*ii*) it should move under the action of gravity alone; there should not be any other driving force.

Examples of projectile motion are:

(*i*) a bullet fired from a gun,

(*ii*) throwing of a cricket ball from one player to another,

(*iii*) a stone thrown horizontally from the top of a tower, and

(*iv*) a free liquid jet issuing from the nozzle into atmosphere.

14.1. TERMS RELATED TO PROJECTILE MOTION

(1) *Velocity of projection* is the velocity V with which the body is projected into space. The angle α which this velocity makes with the horizontal is called the *angle of projection*.

(2) *Range* is the horizontal distance between the point of projection and the point where the projectile hits the ground. Apparently range is the horizontal distance which the projectile travels during its flight.

(3) *Time of flight* is the time interval during which the projectile remains in motion.

(4) *Maximum height* is the maximum vertical distance of projectile from the point of projection.

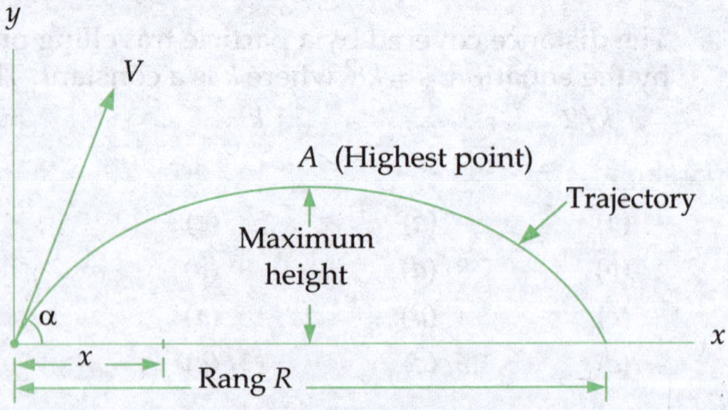

Fig. 14.1

14.2. EQUATION OF PROJECTILE PATH

Consider the motion of an object which is projected from point O with velocity V inclined at an angle α with the horizontal ground surface. After attaining the maximum height H, it descends and finally hits the ground at point B.

The following observations may be made:

(*i*) the velocity at any point on the trajectory consists of two components; velocity u in the x-direction and velocity v in the y-direction.

$$u = V\cos\alpha \quad \text{and} \quad v = V\sin\alpha$$

(*ii*) the object does not experience any acceleration in the x-direction, the horizontal component does not change during the entire flight. Apparently horizontal component of velocity any point on the trajectory equals the horizontal component $V\cos\alpha$ of the initial velocity.

(*iii*) the vertical motion of the body is influenced by gravitation effects. The vertical component of velocity starts diminishing from the initial value $V\sin\alpha$, attains zero value at the top most point of trajectory, thereafter changes direction and gains velocity.

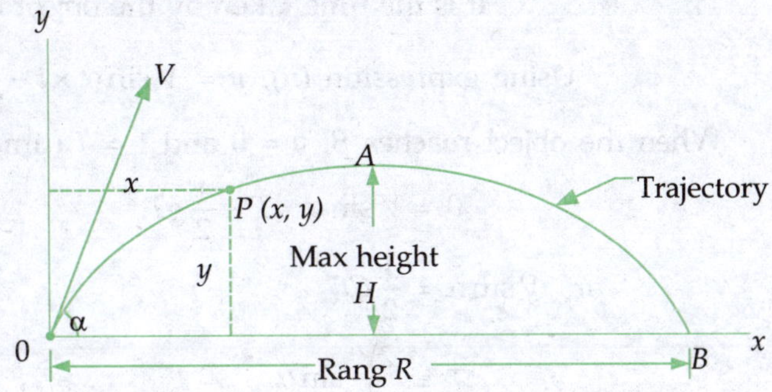

Fig. 14.2

If $P(x, y)$ is the position of the object after a time t, then the horizontal and vertical distances travelled by the object are:

$$x = V\cos\alpha \times t \qquad ...(i)$$

$$y = V\sin\alpha \times t - \frac{1}{2}gt^2 \qquad ...(ii)$$

From expression (*i*), $t = \dfrac{x}{V\cos\alpha}$. Substituting this value in expression (*ii*), we get

$$y = V\sin\alpha \times \frac{x}{V\cos\alpha} - \frac{1}{2}g\left(\frac{x}{V\cos\alpha}\right)^2 = x\tan\alpha - \frac{1}{2}\frac{gx^2}{V^2\cos^2\alpha}$$

But $\dfrac{1}{\cos^2\alpha} = \sec^2\alpha = 1 + \tan^2\alpha$

$$\therefore \quad y = x\tan\alpha - \frac{1}{2}\frac{gx^2}{V^2}(1+\tan^2\alpha) \qquad ...(14.1)$$

Equation 14.1 gives the variation of y with the square of x, and essentially represents a parabola. Thus the path of a projectory is a parabola.

The analysis made above is based on the following assumptions:

(*i*) air resistance is negligible

(*ii*) constant gravitational acceleration

(*iii*) particle representation of the body is adequate

(*iv*) smooth surface of the body

The following information can be obtained by inspection of the equation 14.1.

- **Maximum height attained:** Using the relation $v^2 - u^2 = 2as$,

  u = initial velocity = $V \sin \alpha$

  $v = 0$ at the highest point

  $a = -g$; object is moving upward and gravity is acting in the downward direction.

  $0 - (V \sin \alpha)^2 = 2 \times (-g) \times H$

  where H is the maximum vertical height attained by the object

$$\therefore \quad H = \frac{V^2}{2g} \sin^2 \alpha \qquad \text{...(14.2)}$$

- **Time of flight:** It is the time taken by the object in reaching from position O to B (Fig. 14.2).

  Using expression (*ii*), $y = V \sin \alpha \times t - \frac{1}{2} g t^2$

  When the object reaches B, $y = 0$ and $t = T$ (time of flight)

$$0 = V \sin \alpha \times T - \frac{1}{2} g T^2$$

$$\text{or} \quad V \sin \alpha = \frac{1}{2} g T$$

$$\therefore \quad T = \frac{2V}{g} \sin \alpha \qquad \text{...(14.3)}$$

- **Time to reach maximum height:** Using the relation $v = u + at$ at point A where the object reaches maximum height,

$$0 = V \sin \alpha - gt$$

$$t = \frac{V}{g} \sin \alpha \qquad \text{...(14.4)}$$

  The time to reach maximum height is thus half the time of flight.

- **Horizontal range:** During the time of flight, the body moves in horizontal direction with uniform velocity $V \cos \alpha$

  Range = horizontal velocity × time of flight

$$R = V \cos \alpha \times \frac{2V}{g} \sin \alpha = \frac{V^2}{g}(2 \sin \alpha \cos \alpha) = \frac{V^2}{g} \sin 2\alpha \qquad \text{...(14.5)}$$

  Range will be maximum for a given velocity of projection (V) when $\sin 2\alpha$ is maximum or when

$$\sin 2\alpha = 1 \quad \text{or} \quad \sin 2\alpha = \sin 90^\circ$$

$$2\alpha = 90^\circ \quad \text{or} \quad \alpha = 45^\circ$$

$$\text{Maximum range } R_{max} = \frac{V^2}{g} \sin 2\alpha = \frac{V^2}{g} \sin 90^\circ = \frac{V^2}{g} \qquad \text{...(14.6)}$$

  Thus the projectile will cover a maximum range when it is directed at an angle of 45°.

- **Angle of projection for a given range**

  From the expression for the range of projectile,

$$R = \frac{V^2}{g} \sin 2\alpha = \frac{V^2}{g} \sin(\pi - 2\alpha) \quad [\because \sin 2\theta = \sin(\pi - 2\theta)]$$

Obviously, for a given value of the initial velocity V, there are two angles of projection which provide the same range. These are α and $\frac{\pi - 2\alpha}{2}$ or $\left(\frac{\pi}{2} - \alpha\right)$. These two angles of projection are equally inclined to the direction of projection for maximum range.

EXAMPLE 14.1

A projectile is fired with a velocity of 300 m/s at an angle 30° with horizontal. Workout the velocity and direction of the projectile after 10 seconds of its firing.

Solution: **At the instant of firing,** the horizontal and vertical components of velocity are

$$V_x = V\cos\alpha = 300\cos 30^\circ = 259.8 \text{ m/s}$$

$$V_y = V\sin\alpha = 300\sin 30^\circ = 150 \text{ m/s}$$

After 10 seconds of firing,

The horizontal velocity V_x of the projectile remains constant as no acceleration acts in the horizontal direction.

The initial velocity in the vertical direction V_y goes on decreasing because of the constant deceleration due to gravity ($a = -g$).

$$\text{Vertical velocity after 10 seconds} = 150 - 9.81 \times 10 = 51.9 \text{ m/s}$$

$$\text{Resultant velocity} = \sqrt{259.8^2 + 51.9^2} = 264.9 \text{ m/s}$$

If the resultant makes an angle θ with horizontal direction, then

$$\tan\theta = \frac{51.9}{259.8} = 0.199 \quad \therefore \quad \theta = 12.3^\circ$$

EXAMPLE 14.2

An object is projected at an angle of 60° with the horizontal. If the horizontal range of the object is 2.5 km, make calculations for the velocity of projection and the maximum height attained by the object.

Solution: Using the relation for horizontal range,

$$R = \frac{V^2}{g}\sin 2\alpha$$

$$2.5 \times 1000 = \frac{V^2}{9.81}\sin(2 \times 60)$$

$$\therefore \quad \text{Velocity of projection } V = \sqrt{\frac{2.5 \times 1000 \times 9.81}{\sin 120^\circ}} = 168.28 \text{ m/s}$$

Invoking the relation for maximum height attained by the object;

$$H = \frac{V^2}{2g}\sin^2\alpha = \frac{168.28^2}{2 \times 9.81} \times \sin^2 60^\circ = 1082.5 \text{ m}$$

EXAMPLE 14.3

Two balls are projected from the same point in directions inclined at 60° and 30° with the horizontal. Calculate the ratio of their velocities of projection if (a) the balls attain the same maximum height, (b) the ranges are equal for the balls.

Solution: Let suffix 1 and 2 refer to the two balls.

(*a*) Relation for maximum height is: $H = \frac{V^2}{2g}\sin^2\alpha$

Equating the two maximum heights,

$$\frac{V_1^2}{2g}\sin^2\alpha_1 = \frac{V_2^2}{2g}\sin^2\alpha_2$$

$$\frac{V_1}{V_2} = \sqrt{\frac{\sin^2\alpha_2}{\sin^2\alpha_1}} = \frac{\sin\alpha_2}{\sin\alpha_1} = \frac{\sin 30^\circ}{\sin 60^\circ} = \frac{0.5}{0.866} = \mathbf{0.577}$$

(*b*) Relation for the range is: $R = \frac{V^2}{g}\sin 2\alpha$

Equating the two ranges,

$$\frac{V_1^2}{g}\sin 2\alpha_1 = \frac{V_2^2}{g}\sin 2\alpha_2$$

$$\left(\frac{V_1}{V_2}\right)^2 = \frac{\sin 2\alpha_2}{\sin 2\alpha_1} = \frac{\sin 60}{\sin 120} = \frac{0.886}{0.866} = 1$$

$$\therefore \quad \frac{V_1}{V_2} = \mathbf{1}$$

EXAMPLE 14.4

A projectile has a horizontal range S. If y_1 and y_2 are the greatest heights in the two paths possible, show that $S = 4\sqrt{y_1 y_2}$

Solution: The horizontal range and the maximum height attained by a projectile are given by

$$R = \frac{V^2}{g}\sin 2\alpha \quad \text{and} \quad H = \frac{V^2}{2g}\sin^2\alpha$$

For a prescribed range, there can be two possible angles of projection, *i.e.*, α and (90 – α). Accordingly

$$y_1 = \frac{V^2}{2g}\sin^2\alpha$$

$$y_2 = \frac{V^2}{2g}\sin^2(90-\alpha) = \frac{V^2}{2g}\cos^2\alpha$$

$$y_1 y_2 = \frac{V^2}{2g}\sin^2\alpha \times \frac{V^2}{2g}\cos^2\alpha = \left(\frac{V^2}{2g}\right)^2 (\sin\alpha\cos\alpha)^2$$

$$\therefore \quad \sqrt{y_1 y_2} = \frac{V^2}{2g}\sin\alpha\cos\alpha \qquad \ldots(i)$$

Further

$$S = \frac{V^2}{g}\sin 2\alpha = \frac{V^2}{g}\times 2\sin\alpha\cos\alpha = 4\frac{V^2}{2g}\sin\alpha\cos\alpha \qquad \ldots(ii)$$

From expression (*i*) and (*ii*)

$S = 4\sqrt{y_1 y_2}$ which was required to be proved.

EXAMPLE 14.5

A projectile is projected from the base of a hill whose shape is that of a right circular cone with axis vertical. The projectile grazes the vertex and strikes the hill again at a point on the base. Set up the following relations:

$$\tan\alpha = 2\cot\theta \text{ and } V^2 = gh\left(2+\frac{1}{2}\tan^2\theta\right)$$

where h is the height of cone and θ is its semivertical angle; V is the velocity of projection inclined at angle α with the horizontal, and g is the acceleration due to gravity.

Solution: Refer Fig 14.3, for the projected profile and the configuration of the hill.

The relations for the maximum height h and horizontal range R are

$$h = \frac{V^2}{2g}\sin^2\alpha$$

$$R = \frac{V^2}{g}\sin 2\alpha$$

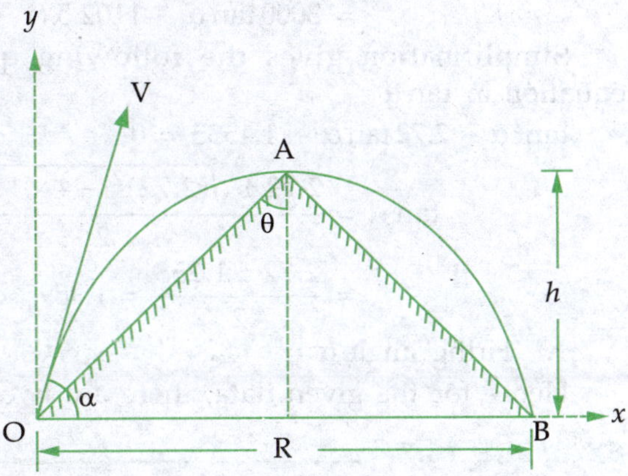

Fig. 14.3

From the configuration of hill,

$$\cot\theta = \frac{h}{R/2}$$

$$= \frac{V^2}{2g}\sin^2\alpha \times \frac{2g}{V^2\sin 2\alpha}$$

$$= \frac{\sin^2\alpha}{\sin 2\alpha} = \frac{\sin^2\alpha}{2\sin\alpha\cos\alpha} = \frac{\sin\alpha}{2\cos\alpha}$$

That gives: $\cot\theta = \frac{\tan\alpha}{2}$ or $\tan\alpha = 2\cot\theta$

(*b*) From the relation for maximum height

$$V^2 = \frac{2gh}{\sin^2\alpha} = 2gh\,\mathrm{cosec}^2\alpha = 2gh\,(1+\cot^2\alpha) \quad \text{...(i)}$$

From the identity $\tan\alpha = 2\cot\theta$; we have

$$\frac{1}{\tan\alpha} = \frac{1}{2\cot\theta} \quad \text{or} \quad \cot\alpha = \frac{1}{2}\tan\theta$$

Substituting $\cot^2\alpha = \frac{1}{4}\tan^2\theta$ in identity (*i*), we get

$$V^2 = 2gh\left(1+\frac{1}{4}\tan^2\theta\right) = gh\left(2+\frac{1}{2}\tan^2\theta\right)$$

EXAMPLE 14.6

A projectile is fired with an initial velocity of 200 m/s to hit a target located 500 m above the level of gun point and at a horizontal distance of 3000 m. Neglecting the air resistance, determine the firing angle.

Solution: Refer Fig. 14.4. O is the gun point and $P(x, y)$ is the position of target after time t. Then

$$x = V\cos\alpha \times t; \quad t = \frac{x}{V\cos\alpha}$$

$$y = V\sin\alpha \times t - \frac{1}{2}gt^2 = V\sin\alpha \times \frac{x}{V\cos\alpha} - \frac{1}{2}g \times \left(\frac{x}{V\cos\alpha}\right)^2$$

$$= x\tan\alpha - \frac{1}{2}\frac{gx^2}{V^2}\sec^2\alpha \qquad \left(\because \frac{1}{\cos^2\alpha} = \sec^2\alpha\right)$$

Substituting the given data, we get

$$500 = 3000\tan\alpha - \frac{1}{2} \times \frac{9.81 \times 3000^2}{200^2}\sec^2\alpha$$

$$= 3000\tan\alpha - 1102.5\sec^2\alpha$$

$$= 3000\tan\alpha - 1102.5\,(1 + \tan^2\alpha)$$

Simplification gives the following quadratic equation in $\tan\alpha$

$$\tan^2\alpha - 2.72\tan\alpha + 1.4535 = 0$$

$$\tan\alpha = \frac{2.72 \pm \sqrt{(2.72)^2 - 4 \times 1 \times 1.4535}}{2}$$

$$= \frac{2.72 \pm 1.258}{2} = 1.989 \quad \text{or} \quad 0.731$$

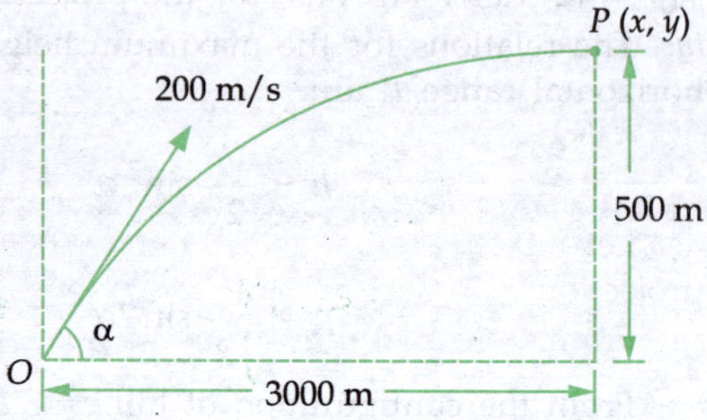

Fig. 14.4

$\therefore$ Firing angle α = **63.33° or 36.17°**

Hence for the given data, there are two angles of projection.

EXAMPLE 14.7

Two adjacent guns having the muzzle velocity of 400 m/s fire simultaneously at angles of α_1 and α_2 for the same target at a range 4800 m. Calculate the time difference between the hits. Assume gravitational acceleration g = 9.80 m/s^2.

Solution: Relation for the range is $R = \dfrac{V^2}{g}\sin 2\alpha$

Equating the two ranges for the same velocity of projection,

$$\frac{V^2}{g}\sin 2\alpha_1 = \frac{V^2}{g}\sin 2\alpha_2 = 4800 \qquad ...(i)$$

$$\text{or} \quad \sin 2\alpha_1 = \sin 2\alpha_2 = \sin(\pi - 2\alpha_2)$$

$$\therefore \quad 2\alpha_1 = \pi - 2\alpha_2 \quad \text{or} \quad \alpha_1 = \frac{\pi}{2} - \alpha_2 \qquad ...(ii)$$

Further from expression (*i*)

$$\frac{400^2}{9.80} \times \sin 2\alpha_1 = 4800$$

$$\sin 2\alpha_1 = \frac{4800 \times 9.80}{400^2} = 0.294$$

$$2\alpha_1 = 17.1^\circ \quad \text{or} \quad \alpha_1 = 8.55^\circ$$

and from expression (*ii*),

$$\alpha_2 = 90 - \alpha_1 = 90 - 8.55 = 81.45^\circ$$

Time of flight for the two projectiles are,

$$t_1 = \frac{2V}{g}\sin\alpha_1 \quad \text{and} \quad t_2 = \frac{2V}{g}\sin\alpha_2$$

$$\text{Time difference} = t_2 - t_1 = \frac{2V}{g}(\sin\alpha_2 - \sin\alpha_1) = \frac{2\times 400}{9.80}(\sin 81.45° - \sin 8.55°)$$

$$= \frac{2\times 400}{9.80}(0.989 - 0.1487) = 68.53 \text{ s}$$

EXAMPLE 14.8

A jet of water discharging from a nozzle hits a vertical screen placed at a distance of 6 m from the nozzle at a height of 4 m. When the screen is shifted by 4 m away from the nozzle from its initial position, the jet hits the screen again at the same point. Determine the angle and velocity with which the jet issues from the nozzle.

Solution: With reference to Fig. 14.5,

$OA = 6$ m; $AC = 4$ m

$AB = CD = 6$ m

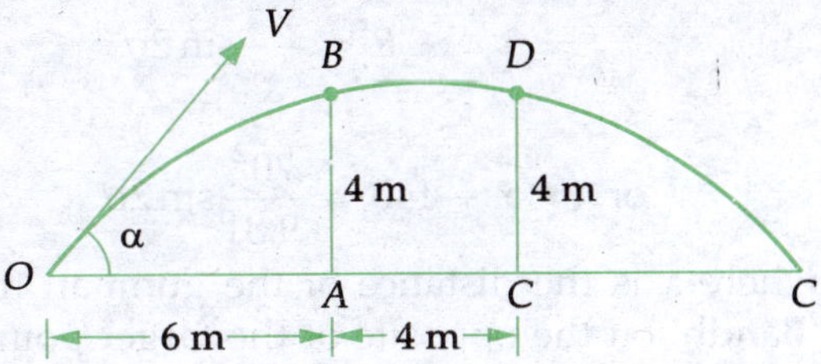

Fig. 14.5

Let V be the initial velocity and α be the angle of projection.

From the equation of trajectory y

$$y = x\tan\alpha \quad \frac{gx^2}{2V^2\cos^2\alpha}$$

we have

$$4 = 6\tan\alpha - \frac{g\times 6^2}{2(V\cos\alpha)^2} \quad ...(i)$$

and

$$4 = 10\tan\alpha - \frac{g\times 10^2}{2(V\cos\alpha)^2} \quad ...(ii)$$

From identities (*i*) and (*ii*),

$$10\tan\alpha - \frac{100g}{2(V\cos\alpha)^2} = 6\tan\alpha - \frac{36g}{2(V\cos\alpha)^2}$$

or

$$4\tan\alpha = \frac{64\times g}{2(V\cos\alpha)^2}; \quad (V\cos\alpha)^2 = \frac{78.48}{\tan\alpha} \quad ...(iii)$$

Substitution of this value of $(V\cos\alpha)^2$ in identity (*i*) gives:

$$4 = 6\tan\alpha - \frac{g\times 6^2}{2}\times\frac{\tan\alpha}{78.48}$$

$$= 6\tan\alpha - 2.25\tan\alpha = 3.75\tan\alpha$$

$\therefore$ Angle of projection $\alpha = \tan^{-1}\left(\frac{4}{3.75}\right) = 46.85°$

Substituting in identify (*iii*),

$$\text{Velocity of projection} \quad V = \sqrt{\frac{78.48}{\sin\alpha\cos\alpha}}$$

$$= \sqrt{\frac{78.48}{\sin 46.85\cos 46.85}} = 12.54 \text{ m/s}$$

EXAMPLE 14.9

It is observed that the enemy has concentrated its armour behind a 6 m height 'Bandh' at a horizontal distance of 140 m. At what distance and at what inclination the guns should be fixed so that the shot may pass just clear of the top of 'Bandh' and exactly hit the enemy position? The muzzle velocity of the gun shot be taken as 70 m/s.

Solution: Let V be the initial velocity and α be the angle of projection. The top of the 'Bandh' AC must lie on the path of projectile.

Range of the bullet,

$$R = \frac{V^2}{g}\sin 2\alpha$$

or $$x + 140 = \frac{70^2}{9.81}\sin 2\alpha$$

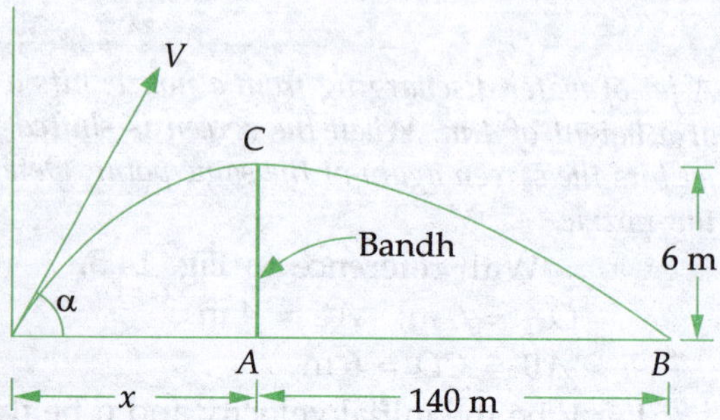

Fig. 14.6

where x is the distance of the gunman from the 'Bandh' on the opposite of the target point B and on the same horizontal plane as the target. Further, since $\sin 2\alpha = 2 \sin \alpha \cos \alpha$, we can write

$$x + 140 = \frac{70^2}{9.81}\times 2\sin\alpha\cos\alpha = 998.98\sin\alpha\cos\alpha \qquad ...(a)$$

or $$\frac{x+140}{\cos^2\alpha} = 998.98\frac{\sin\alpha\cos\alpha}{\cos^2\alpha}$$

or $$\frac{x+140}{\cos^2\alpha} = 998.98\tan\alpha$$

That gives: $$\tan\alpha = \frac{x+140}{998.98\cos^2\alpha}$$

From the equation of trajectory,

$$y = x\tan\alpha - \frac{gx^2}{2V^2\cos^2\alpha}$$

Substituting the appropriate values:

$$6 = x\left(\frac{x+140}{998.98\cos^2\alpha}\right) - \frac{9.81\times x^2}{2\times 70^2\cos^2\alpha}$$

$$= x\left(\frac{x+140}{998.98\cos^2\alpha}\right) - \frac{x^2}{998.98\cos^2\alpha}$$

$$= \frac{x^2+140x-x^2}{998.98\cos^2\alpha} = \frac{140x}{998.98\cos^2\alpha}$$

That gives: $$x = \frac{6\times 998.98\cos^2\alpha}{140} = 42.81\cos^2\alpha \qquad ...(b)$$

Substituting this value of x in identity (a), we obtain

$$42.81\cos^2\alpha + 140 = 998.98\sin\alpha\cos\alpha$$

$$\text{or} \qquad 42.81 + 140\sec^2\alpha = 998.98\tan\alpha$$

$$\text{or} \qquad 42.81 + 140(1 + \tan^2\alpha) = 998.98\tan\alpha$$

Upon simplification, we get the following quadratic equation

$$\tan^2\alpha - 7.14\tan\alpha + 1.31 = 0$$

and its solution gives

$$\tan\alpha = 6.95 \text{ and } 0.188$$

$$\therefore \qquad \alpha = 81.82° \text{ and } 10.65°$$

Then from the relation $x = 42.18\cos^2\alpha$ as per identify (*b*), we get

$$x = 42.81\cos 281.82° = \mathbf{0.87\ m}$$

$$\text{or} \qquad x = 42.81\cos^2 10.65° = \mathbf{41.36\ m}$$

EXAMPLE 14.10

A projectile is aimed at a target which lies in the horizontal plane through the point of projection. It falls a metres short when the angle of projection is α while it overshoots the target by b metres when the angle of projection is β. Show that the proper angle of projection to hit the target is

$$\theta = \frac{1}{2}\sin^{-1}\left[\frac{a\sin 2\beta + b\sin 2\alpha}{a+b}\right]$$

It may be presumed that the velocity of projection is same in all the three cases and there is no resistance

Solution: Let V be the velocity of projection in each case, θ be the correct angle of projection and the target T be located at a distance R for the point of projection.

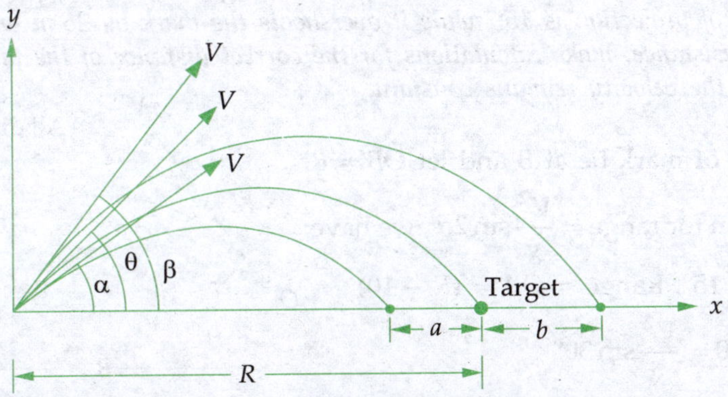

Fig. 14.7

$$\text{Now,} \qquad R = \frac{V^2}{g}\sin 2\theta \qquad ...(i)$$

$$R - a = \frac{V^2}{g}\sin 2\alpha \qquad ...(ii)$$

$$R + b = \frac{V^2}{g}\sin 2\beta \qquad ...(iii)$$

Subtracting expression (*i*) from expression (*iii*)

$$b = \frac{V^2}{g}(\sin 2\beta - \sin 2\alpha)$$

$$\text{or} \quad \frac{V^2}{g} = \frac{b}{\sin 2\beta - \sin 2\theta} \quad ...(iv)$$

Subtracting expression (*ii*) from expression (*iii*),

$$a + b = \frac{V^2}{g}(\sin 2\beta - \sin 2\alpha)$$

$$\text{or} \quad \frac{V^2}{g} = \frac{a+b}{\sin 2\beta - \sin 2\alpha} \quad ...(v)$$

From expressions (*iv*) and (*v*)

$$\frac{\sin 2\beta - \sin 2\theta}{b} = \frac{\sin 2\beta - \sin 2\alpha}{a+b}$$

or $\quad a\sin 2\beta + b\sin 2\beta - a\sin 2\theta - b\sin 2\theta = b\sin 2\beta - b\sin 2\alpha$

or $\quad (a + b)\sin 2\theta = a\sin 2\beta + b\sin 2\alpha$

$$\text{or} \quad \sin\theta = \frac{a\sin 2\beta + b\sin 2\alpha}{a+b}$$

$$\text{That gives:} \quad \theta = \frac{1}{2}\sin^{-1}\left[\frac{a\sin 2\beta + b\sin 2\alpha}{a+b}\right]$$

which is the required expression.

EXAMPLE 14.11

A projectile is aimed at a mark on the horizontal plane through the point of projection and falls 10 m short when the angle of projection is 15° while it overshoots the mark by 25 m when the inclination is 40°. Neglecting air resistance, make calculations for the correct distance of the target and the required angle of projection if the velocity remains constant.

Solution:

Let the position of mark be at B and let $OB = R$.

Using the relation for range $= \dfrac{V^2}{g}\sin 2\alpha$, we have:

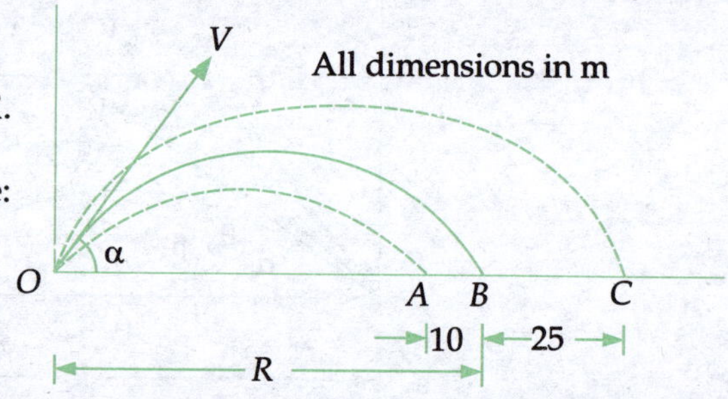

Fig. 14.8

First case : $\alpha = 15°$, Range $= OA = (R - 10)$

$$\therefore \quad R - 10 = \frac{V^2}{g}\sin 30°$$

$$R - 10 = \frac{V^2 \times 0.5}{g} = \frac{0.5V^2}{g} \quad ..(i)$$

Second case : $\alpha = 40°$, Range $OC = (R + 25)$

$$\therefore \quad R + 25 = \frac{V^2}{g}\sin 80° = \frac{0.985\,V^2}{g} \quad ...(ii)$$

Dividing expression (*i*) by expression (*ii*), we get

$$\frac{R-10}{R+25} = \frac{0.5\,V^2}{g} \times \frac{g}{0.985V^2} = 0.5076$$

or $\quad R - 10 = 0.5076R + 12.69$

$$\therefore \quad R = \frac{12.69 + 10}{1 - 0.5076} = \textbf{46.08 m}$$

From expression (*i*)

$$V^2 = \frac{(R-10)\times g}{0.5} = \frac{(46.08-10)\times 9.80}{0.5} = 707.89$$

$$\therefore \quad V = \sqrt{707.89} = \textbf{26.61 m/s}$$

Further, $R = \dfrac{V^2}{g}\sin 2\alpha$

$$\sin 2\alpha = \frac{46.08}{26.61^2}\times 9.80 = 0.6377$$

$$\therefore \quad 2\alpha = 39.62^\circ \; ; \quad \alpha = 19.81^\circ \text{ and } (90 - 19.81) = 70.19^\circ.$$

EXAMPLE 14.12

There is a 3 m high wall 3 m in front of enemy target. If the gunman is 5 m away from the wall on the opposite side of the target and on the same horizontal plane as the target, find the angle of projection and the least velocity of projection so that the shell strikes the target.

Solution: Let V be the initial velocity required and α be the angle of projection.

Range = 5 + 3 = 8 m

The top of wall AC must lie on the path of projectile. The co-ordinates of the top of the wall are : x = 5 m and y = 3 m.

Range of projectile,

$$R = \frac{V^2}{g}\sin 2\alpha \; ; \quad 8 = \frac{V^2}{g}\sin 2\alpha$$

$$\therefore \quad V^2 = \frac{8g}{\sin 2\alpha} \qquad \text{...}(i)$$

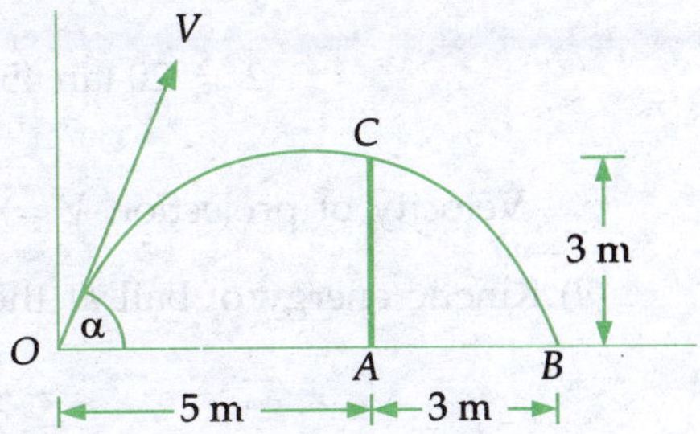

Fig. 14.9

From the equation of the trajectory

$$y = x\tan\alpha - \frac{gx^2}{2V^2\cos^2\alpha}$$

$$3 = 5\tan\alpha - \frac{g\times 5^2}{2V^2\cos^2\alpha} \qquad \text{...}(ii)$$

Substituting for $V^2 = \dfrac{8g}{\sin 2\alpha}$ from expression (*i*),

$$3 = 5\tan\alpha - \frac{5^2}{2\cos^2\alpha}\times\frac{\sin 2\alpha}{8} = 5\tan\alpha - \frac{25}{16}\times\frac{2\sin\alpha\cos\alpha}{\cos^2\alpha}$$

$$= 5\tan\alpha - \frac{25}{8}\tan\alpha = 1.875\tan\alpha$$

$$\tan\alpha = \frac{3}{1.875} = 1.6 \qquad \therefore \quad \alpha = 58^\circ$$

From expression (*i*), $V^2 = \dfrac{8g}{\sin 2\alpha} = \dfrac{8\times 9.81}{\sin 116^\circ} = 87.39$

$$\therefore \quad V = \sqrt{87.39} = \textbf{9.348 m/s}$$

EXAMPLE 14.13

The following data pertains to a sporting event for shot-put

mass of the ball = 6 kg

height and inclination of throw = 2 m and 45° with

horizontal distance covered on ground = 20 m

Determine: (a) kinetic energy of the ball at the instant of projection

(b) total time elapsed during the event

(c) maximum height attained

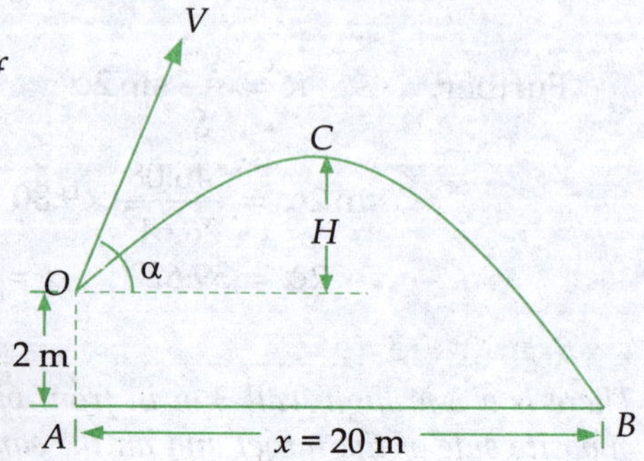

Fig. 14.10

Solution : With reference to Fig. 14.10.

O is the point of projection

C is the highest point reached

B is the point of land

The equation of trajectory of an object projected at angle α with an initial velocity V from point O is given by

$$y = x \tan\alpha - \frac{1}{2}\frac{gx^2}{V^2\cos^2\alpha}$$

$$-2 = 20\tan 45° - \frac{1}{2}\times\frac{9.81\times 20^2}{V^2\cos^2 45°} = 20 - \frac{3925.2}{V^2}$$

$\therefore$ Velocity of projection $V = \sqrt{\dfrac{3925.2}{22}} = 13.36$ m/s

(*a*) Kinetic energy of ball at the instant of projection

$$= \frac{1}{2}mV^2 = \frac{1}{2}\times 6\times 13.36^2 = 535.45 \text{ J}$$

(*b*) Height $H = \dfrac{V^2}{2g}\sin^2\alpha = \dfrac{13.36^2}{2\times 9.81}\times\sin^2 45° = 4.55$ m

$\therefore$ Maximum height reached above the ground

$$= 2 + 4.55 = 6.55 \text{ m}$$

(*c*) The total time elapsed during the event can be calculated by using the relation

$$x = V\cos\alpha \times t \quad\text{or}\quad 20 = 13.36\cos 45° \times t$$

That gives : $t = \dfrac{20}{13.36}\times\cos 45° = 2.12$ **sec**

EXAMPLE 14.14

The records for the shot put and discus throw for a particular sports meet are stated to be 18 m and 65 m respectively. Assuming their respective masses to be 6 kg and 2 kg, compare the work done by the champions in making their record throws. Each trajectory starts at an elevation of 1.8 m above the ground and has an inclination of 45° with the horizontal. Neglect air resistance and the spin of discus throw.

Solution : The equation of the trajectory of an object projected at angle α with an initial velocity V from the point O is given by

$$y = x\tan\alpha - \frac{gx^2}{2V^2\cos^2\alpha}$$

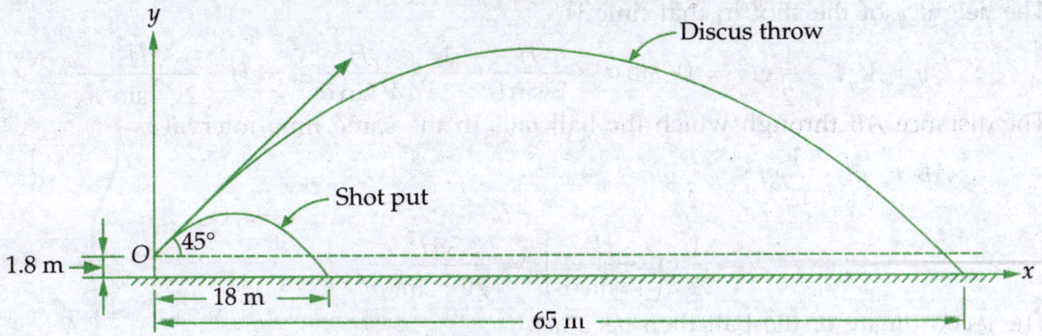

Fig. 14.11

For the shot put:

$$-1.8 = 18\tan 45^\circ - \frac{9.81\times 18^2}{2V^2\cos^2 45^\circ} = 18\times 1 - \frac{3179.4}{V^2}$$

$\therefore$ Velocity of projection $V = \sqrt{\dfrac{3179.4}{19.8}} = 12.67$ m/s

The kinetic energy at the time of projection of shot put

$$= \frac{1}{2}mV^2 = \frac{1}{2}\times 6\times 12.67^2 = \mathbf{481.64\ J}$$

For the discus throw:

$$-1.8 = 65\tan 45^\circ - \frac{9.81\times 65^2}{2\times V^2\cos^2 45} = 65\times 1 - \frac{41459.77}{V^2}$$

$\therefore$ Velocity of projection $V = \sqrt{\dfrac{41459.77}{66.8}} = 24.91$ m/s

The kinetic energy at the time of projection of discus throw

$$= \frac{1}{2}mV^2 = \frac{1}{2}\times 2\times (24.91)^2 = \mathbf{620.58\ J}$$

Remarks : Work done equals the kinetic energy imparted to the object at the time of its projection. Obviously, the champion of discus throw does more work than the champion of shot put.

EXAMPLE 14.15

A gun is fired along the line of sight OA to hit a ball lying at position A as shown in Fig. 14.12. Simultaneously, the ball is made to fall freely under gravity. Show that the shot will hit the ball regardless of the initial velocity of the shot and the distances.

Proceed to find the height from ground at which the ball will be intercepted.

Solution : The horizontal distance x travelled by the shot in time t is

$$x = V_x t = (V\cos\alpha)\,t$$

or

$$t = \frac{x}{V\cos\alpha} = \frac{H/\tan\alpha}{V\cos\alpha} = \frac{H}{V\sin\alpha}$$

The height y of the shot in that time is

$$y = V_y t - \frac{1}{2}gt^2 = (V \sin\alpha) \times \frac{H}{V \sin\alpha} - \frac{1}{2}g\left(\frac{H}{V \sin\alpha}\right)^2 = H - \frac{gH^2}{2V^2 \sin^2\alpha} \quad ...(i)$$

The distance AB through which the ball falls in the same time interval is

$$AB = ut + \frac{1}{2}gt^2 = \frac{1}{2}gt^2 \qquad (\because u = 0)$$

$$= \frac{1}{2}g\left(\frac{H}{V \sin\alpha}\right)^2 = \frac{gH^2}{2V^2 \sin^2\alpha}$$

The y-coordinate of the ball then becomes

$$y = AC - AB = H - \frac{gH^2}{2V^2 \sin^2\alpha} \quad ...(ii)$$

The expressions (*i*) and (*ii*) show that the positions of the bullet and ball are identical at time t. Obviously the shot would hit the ball irrespective of initial coordinates of the ball with respect to the gun and initial velocity of the shot.

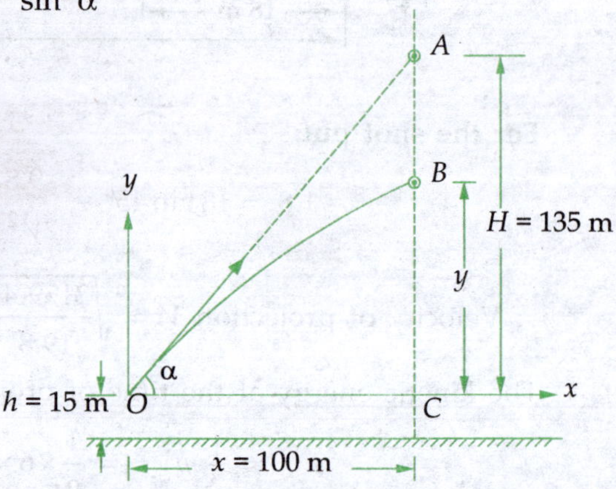

Fig. 14.12

(*b*) From the data shown in the given figure,

$$\tan\alpha = \frac{135}{100} = 1.35\,; \quad \alpha = 53.47^\circ$$

$$t = \frac{x}{V \cos\alpha} = \frac{100}{45 \cos 53.47} = 3.73 \text{ sec}$$

$$AB = \frac{1}{2}gt^2 = \frac{1}{2} \times 9.81 \times (3.73)^2 = 68.24 \text{ m}$$

$\therefore$ Height of ball above x-axis = 135 − 68.24 = 66.76 m

Height of ball above the ground = 66.76 + 15 = **81.76 m**

The ball would be thus intercepted at height of 81.76 m above the ground.

EXAMPLE 14.16

A particle P is projected from a point A with an initial velocity of 50 m/s at an angle 30° with the horizontal. At the same instant, a second particle Q is projected from the opposite direction with initial velocity of 40 m/s from point B which is level with and at distance 80 m from point A. Make calculations for the angle of projection of particle Q if the particles collide and the time at which the collision occurs.

Solution : Refer Fig. 14.13. Let the particles be projected from points A and B which are at level with and 80 m apart.

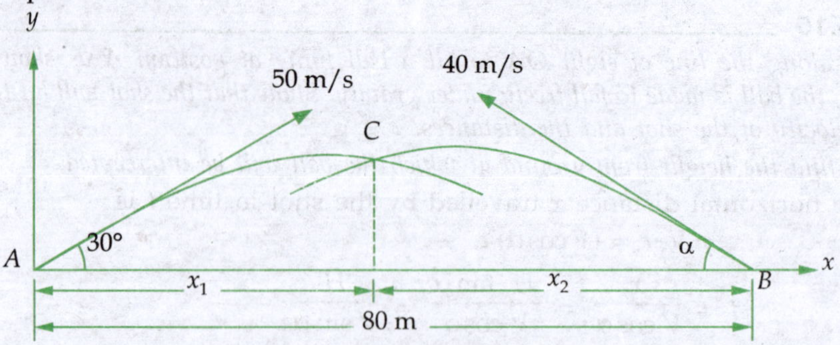

Fig. 14.13

Let the particles collide t seconds after they were projected.

Motion of particle P: At the instant of projection, the horizontal and vertical components of velocity are

$$V_x = V\cos\alpha = 50 \times \cos 30^\circ = 43.3 \text{ m/s}$$
$$V_y = V\sin\alpha = 50 \times \sin 30^\circ = 25 \text{ m/s}$$

At time t seconds after projection,

$$x_1 = 43.3t \quad \text{and} \quad y_1 = 25t - \frac{1}{2}\times 9.81t^2 \qquad \text{...(i)}$$

Motion of particle Q:

$$V_x = 40\cos\alpha \quad \text{and} \quad V_y = 40\sin\alpha$$

Then: $$x_2 = (40\cos\alpha)\,t \quad \text{and} \quad y_2 = (40\sin\alpha)\,t - \frac{1}{2}\times 9.81t^2 \qquad \text{...(ii)}$$

The particles will collide if at a particular instant, they are at the same point. For that $y_1 = y_2$ and therefore

$$25t - \frac{1}{2}\times 9.81t^2 = (40\sin\alpha)\,t - \frac{1}{2}\times 9.81t^2$$

or $$25 = 40\sin\alpha \quad \therefore \quad \alpha = \sin^{-1}\left(\frac{25}{40}\right) = \mathbf{38.68^\circ}$$

Thus for collision to occur, the particle Q has to be projected with 38.68° inclination to the horizontal.

Further: $$x_1 + x_2 = 80$$

or $$43.3t + (40\cos 38.68^\circ)t = 80$$

or $$43.3t + 31.22t = 80 \qquad \therefore \quad t = \frac{80}{43.3 + 31.22} = \mathbf{1.073 \text{ sec}}$$

Thus the particles would collide 1.073 seconds after their projection.

EXAMPLE 14.17

A projectile P is fired with a velocity of 200 m/s at an angle of 60° with the horizontal. After some time a missile M is shot from the same point to destroy the projectile. The angle of projection and the velocity for the missile are 45° and 2000 m/s respectively. Calculate the height, horizontal distance and time with respect to firing of P at which the destruction takes place.

Solution : Refer Fig. 14.14. At the instant when the destruction takes place, the x and y coordinates of the projectile P and the missile M are the same.

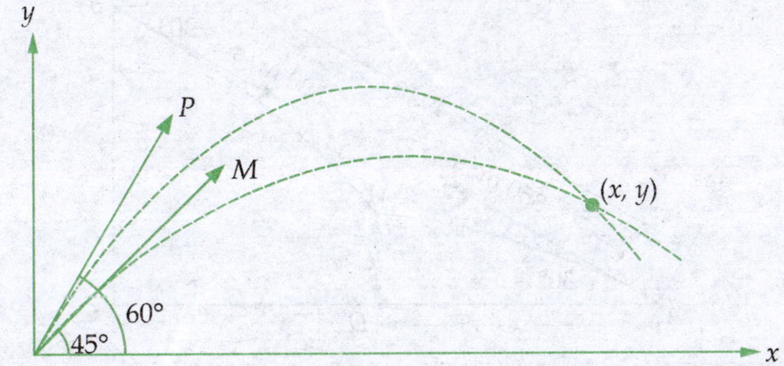

Fig. 14.14

From the equation of trajectory,

$$y = x\tan\alpha - \frac{1}{2}\frac{gx^2}{V^2\cos^2\alpha}, \text{ we have :}$$

$$y = x\tan 60^\circ - \frac{1}{2}\frac{gx^2}{200^2\times\cos^2 60^\circ} \quad \text{for the projectile} \quad ...(i)$$

and $$y = x\tan 45^\circ - \frac{1}{2}\frac{gx^2}{2000^2\cos^2 45^\circ} \quad \text{for the missile} \quad ...(ii)$$

Therefore at the instant of destruction

$$x\tan 60 - \frac{1}{2}\frac{gx^2}{200^2\cos^2 60^\circ} = x\tan 45^\circ - \frac{1}{2}\frac{gx^2}{2000^2\cos^2 45^\circ}$$

or $$1.732\,x - 4.905\times 10^{-4}x^2 = x - 0.0245\times 10^{-4}x^2$$

$$\therefore \quad x = \frac{1.732-1}{(4.905-0.0245)\times 10^{-4}} = 1499.85 \simeq \textbf{1500 m}$$

This is the horizontal distance form the firing point.

Then from expression (*i*),

$$y = 1500\tan 60^\circ - \frac{1}{2}\times\frac{9.81\times 1500^2}{200^2\cos^2 60^\circ}$$

$$= 2598.08 - 1103.62 = 1494.46 \text{ m} \simeq \textbf{1495 m}$$

This is the height from the base line.

The horizontal distance is also given by : $x = (V\cos\alpha)\times t$

$$\therefore \quad 1500 = (200\cos 60^\circ)\times t = 100t; \quad t = 15 \text{ sec}$$

This is the time with respect to firing of projectile P at which destruction occurs.

EXAMPLE 14.18

Two guns are pointed at each other, one upward at an angle of 30° and the other at the same angle of depression. The muzzles of the guns are 40 m apart. If the guns are shot with velocities of 350 m/s upwards and 300 m/s downwards respectively, determine when and where the shots will meet.

Solution: Refer Fig. 14.15. The shots have been fired upward from point A and downwards from point B, and they are shown meeting at point M after time t seconds

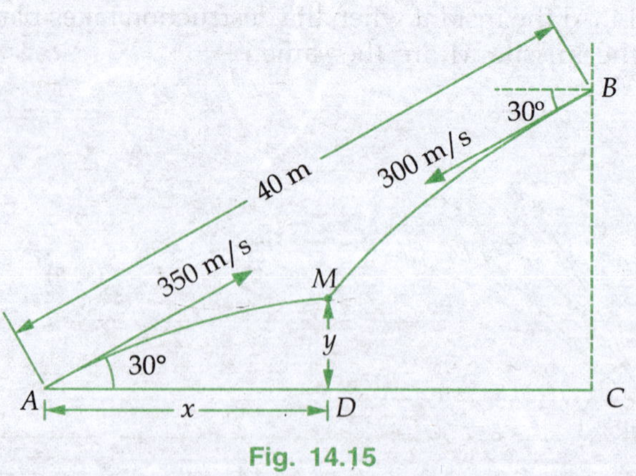

Fig. 14.15

Horizontal distance between A and B,

$$AC = 40 \cos 30° = 34.64 \text{ m}$$

Horizontal distance covered by shot fired from point A in time t second

$$x = AD = 350 \cos 30° \times t = 303.1\, t \quad ...(i)$$

Horizontal distance covered by shot fired from point B in time t second

$$DC = (34.64 - x) = 300 \cos 30° \times t = 259.8\, t \quad ...(ii)$$

Adding expressions (i) and (ii)

$$34.64 = (303.1 + 259.8)t = 562.7\, t$$

$$\therefore \quad t = \frac{34.64}{562.9} = \mathbf{0.0615\ s}$$

The co-ordinates of the meeting point M are then worked out as

$$x = \text{horizontal component of velocity at } A \times \text{time}$$
$$= V \cos\alpha \times t = 350 \cos 30° \times 0.0615 = \mathbf{18.64\ m}$$

$$y = \text{initial velocity in vertical direction} \times t - \frac{1}{2} g t^2$$
$$= V \sin\alpha \times t - \frac{1}{2} g t^2$$
$$= 350 \sin 30° \times 0.0615 - \frac{1}{2} \times 9.81 \times (0.0615)^2$$
$$= 10.762 - 0.0185 = \mathbf{10.68\ m}$$

EXAMPLE 14.19

A projectile is sent from point A with a velocity V. Find the least value of V so that it passes through points B and C. The coordinates of point A, B and C are (0, 0), (4.8, 3.6) and (8.4, 0) metres.

Solution: The relation for the range R of projectile is

$$R = \frac{V^2}{g} \sin 2\alpha = \frac{V^2}{g} 2 \sin\alpha \cos\alpha$$

where V is the velocity of projection at angle α to the horizontal as shown in Fig. 14.16.

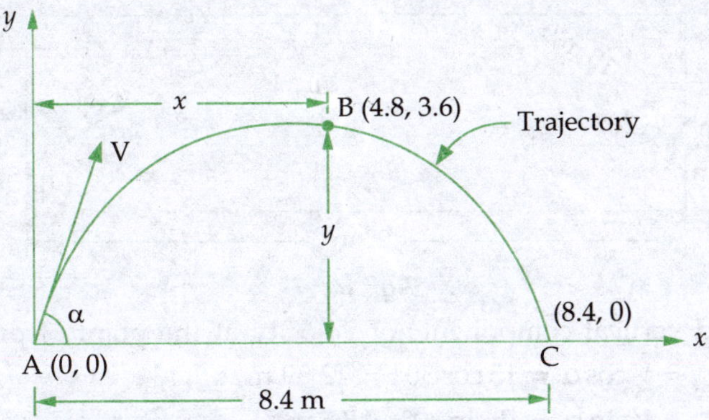

Fig. 14.16

Substituting the value of range $R = 8.4$ m, we have

$$V^2 = \frac{gR}{2 \sin\alpha \cos\alpha} = \frac{9.81 \times 8.4}{2 \sin\alpha \cos\alpha} = \frac{41.20}{\sin\alpha \cos\alpha} \quad ...(i)$$

The equation of the path of projectile is given by

$$y = x\tan\alpha - \frac{gx^2}{2V^2}(1+\tan^2\alpha)$$

Substituting the co-ordinates of point B (4.8, 3.6) in the above equation, we get

$$3.6 = 4.8\tan\alpha - \frac{9.81\times 4.8^2}{2V^2}(1+\tan^2\alpha)$$

Substituting $V^2 = \frac{41.2}{\sin\cos\alpha}$ as obtained above in expression (*i*) and replacing

$1 + \tan^2\alpha = \sec^2\alpha = \frac{1}{\cos^2\alpha}$, we get

$$3.6 = 4.8\tan\alpha - \frac{9.81\times 4.8^2}{2\times 41.20}\times\frac{\sin\alpha\cos\alpha}{\cos^2\alpha}$$

$$= 4.8\tan\alpha - 2.743\tan\alpha = 2.057\tan\alpha$$

$$\therefore \quad \tan\alpha = \frac{3.6}{2.057} = 1.75 \quad \text{and} \quad \alpha = 60.2°$$

Now, using the expression (*i*)

$$V^2 = \frac{41.20}{\sin 60.2°\cos 60.2°} = \frac{41.20}{0.8677\times 0.4977} = 95.54$$

$\therefore$ Velocity of projection $V = \sqrt{95.54} =$ **9.77 m/s**

EXAMPLE 14.20

A ball is thrown from one end of a room of 6 m height. The ball is thrown at 2 m height above the floor with a speed of 15 m/s and 30 degree angle of elevation. Determine:

(*a*) *whether the ball will hit the ceiling, and*

(*b*) *the height at which the ball would hit the opposite wall which is at a horizontal distance of 20 m from the place where it is thrown.*

Solution: Refer Fig. 14.17.

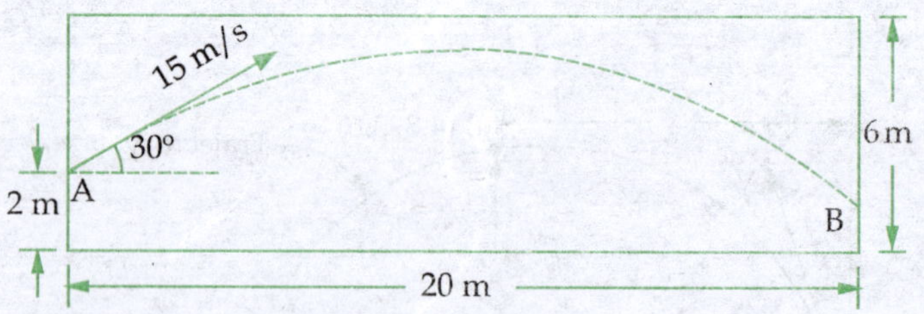

Fig. 14.17

The horizontal and vertical components of velocity at the point of projection A are:

$$V_{x_0} = V\cos\alpha = 15\cos 30° = 12.99 \text{ m/s}$$

$$V_{y_0} = V\sin\alpha = 15\sin 30° = 7.5 \text{ m/s}$$

Let t be time after projection when the ball reaches the maximum height. At the point of maximum height the vertical component of velocity would become zero. That is

$$V_y = V_{y_0} - gt = 7.5 - 9.81\,t = 0$$

$$\therefore \quad t = \frac{7.5}{9.81} = 0.764 \text{ s}$$

Maximum height above the point of projection A,

$$y_{max} = V_{y_0} t - \frac{1}{2} g t^2 = 7.5 \times 0.764 - \frac{1}{2} \times 9.81 \times (0.764)^2$$
$$= 5.73 - 2.86 = 2.87 \text{ m}$$

Maximum height above floor = 2 + 2.87 = 4.87 m

Since this height is less than the height of the room, the ball will not hit the ceiling.

(b) Let t be the time when the ball hits the opposite wall at point B. Then

$$20 = V_{x_0} \times t = 12.99 \times t \qquad \therefore \qquad t = 1.54 \text{ s}$$

$$y = V_{y_0} t - \frac{1}{2} g t^2 = 7.5 \times 1.54 - \frac{1}{2} \times 9.81 \times (1.54)^2$$
$$= 11.55 - 11.63 = -0.08 \text{ m}$$

The –ve sign suggests that the ball strikes 0.08 m below the projection point A.

Height of strike point above the floor level = 2 – 0.08 = **1.92 m**

EXAMPLE 14.21

An object of mass 5 kg is projected with a velocity of 20 m/s at an angle of 60° to the horizontal. At the highest point of its path, the projectile explodes and breaks up into two fragments of masses 1 kg and 4 kg and these fragments separate horizontally. The explosion releases internal energy such that the kinetic energy of the system at the highest point is doubled. Calculate the separation between two fragments when they reach the ground.

Solution: At the instant of being projected, the horizontal component of velocity is

$$v = 20 \cos 60° = 10 \text{ m/s}$$

Let v_1 and v_2 be the velocities of the two fragments of masses 1 kg and 4 kg respectively. These will be the velocities in the horizontal direction as the fragments separate horizontally after the explosion.

As per law of conservation of momentum

$$mv = m_1 v_1 + m_2 v_2$$
$$5 \times 10 = 1 \times v_1 + 4 \times v_2 \qquad \therefore \qquad v_1 + 4 v_2 = 50 \qquad ...(i)$$

Kinetic energy of the object before explosion,

$$= \frac{1}{2} m v^2 = \frac{1}{2} \times 5 \times 10^2 = 250 \text{ J}$$

Kinetic energy after explosion,

$$= \frac{1}{2} m_1 v_1^2 + \frac{1}{2} m_2 v_2^2 = \frac{1}{2} \times 1 \times v_1^2 + \frac{1}{2} \times 4 \times v_2^2 = \frac{1}{2} v_1^2 + 2 v_2^2$$

As per the given condition,

$$\frac{1}{2} v_1^2 + 2 v_2^2 = 2 \times 250 \qquad \therefore \qquad v_1^2 + 4 v_2^2 = 1000 \qquad ...(ii)$$

From expressions (i) and (ii),

$$(50 - 4 v_2)^2 + 4 v_2^2 = 1000$$

or $\quad 2500 + 16 v_2^2 - 400 v_2 + 4 v_2^2 = 1000$

or $\quad v_2^2 - 20 v_2 + 75 = 0$

Solution of this quadratic equation gives :

$$v_2 = \frac{-(-20) \pm \sqrt{(-20)^2 - 4 \times 75 \times 1}}{2} = \frac{20 \pm 10}{2} = 15 \text{ m/s or } 5 \text{ m/s}$$

$$v_1 = 50 - 4 v_2 = -10 \text{ m/s} \quad \text{or} \quad 30 \text{ m/s}$$

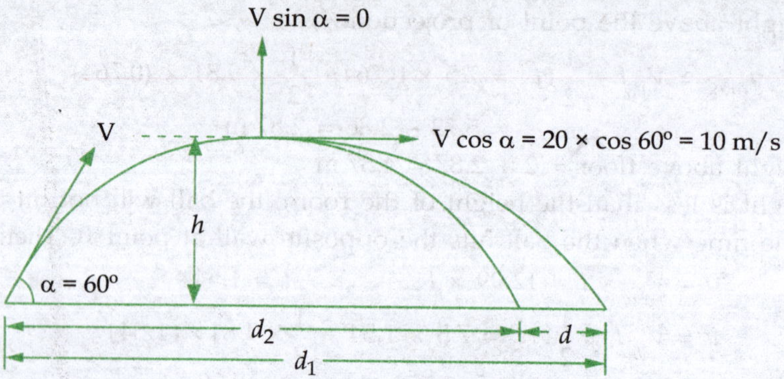

Fig. 14.18

Maximum height reached by the object,

$$h = \frac{V^2}{2g}\sin^2\alpha = \frac{20^2}{2 \times 9.81} \times \sin^2 60° = 15.29 \text{ m}$$

Time taken to reach the ground can be obtained by using the relation:

$$s = ut + \frac{1}{2}gt^2$$

$$15.29 = 0 \times t + \frac{1}{2} \times 9.81t^2 \quad \therefore \quad t = 1.77 \text{ sec}$$

The reparation between the two fragments refers to the difference between the horizontal ranges.

Separation between two fragments for first set

$$d = d_1 - d_2 = 15 \times 1.77 - (-10) \times 1.77 = \mathbf{44.25 \text{ m}}$$

EXAMPLE 14.22

An aeroplane is flying at height h with horizontal velocity U. A shell is fired from the ground exactly at the instant when the aeroplane is vertically above the gun. Setup expressions for the minimum initial velocity V of shell and its inclination α with horizontal so that it hits the plane.

Solution: In order to hit the plane, the initial velocity V of shell and its inclination α with horizontal should be such that

(*i*) horizontal component of initial velocity equals the speed of aeroplane that is

$$V \cos\alpha = U$$

$$\therefore \quad \alpha = \cos^{-1}\frac{U}{V}$$

(*ii*) Vertical component of initial velocity is such that under the effect of gravitational retardation, it can go upto height h. Then considering vertical motion and using the relation $v^2 - u^2 = 2as$, we have

$$0^2 - (V\sin\alpha)^2 = -2gh$$

$$\text{or} \quad V^2\sin^2\alpha = 2gh;$$

$$V^2(1 - \cos^2\alpha) = 2gh$$

$$\text{or} \quad V^2\left(1 - \frac{U^2}{V^2}\right) = 2gh; \quad V^2 - U^2 = 2gh$$

$$\therefore \quad V = \sqrt{U^2 + 2gh}$$

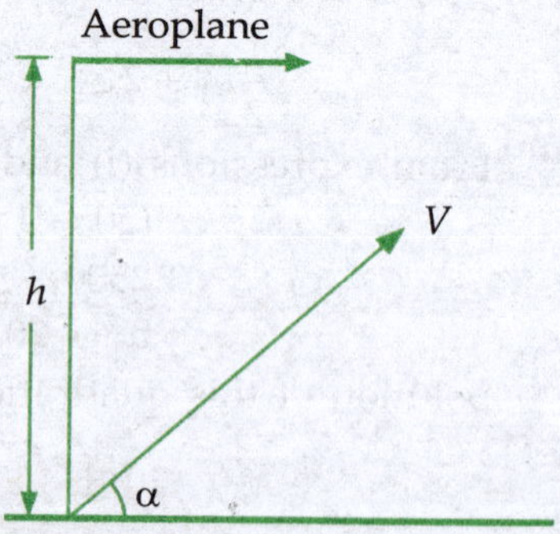

Fig. 14.19

EXAMPLE 14.23

A pilot flying his bomber at a height of 600 m with uniform horizontal velocity of 480 km/hr wants to strike a target which lies on the ground in the direction of flight. At what angle below the horizontal should the pilot see the target at the time of bombing in order to hit the target. Neglect air resistance.

Solution: At any instant of time, the bomb is subjected to

(*i*) horizontal motion with constant horizontal velocity

$$u = 480 \text{ km/hr} = \frac{480 \times 1000}{3600} = 133.33 \text{ m/s}$$

(*ii*) vertical motion with initial velocity zero and acceleration due to gravity

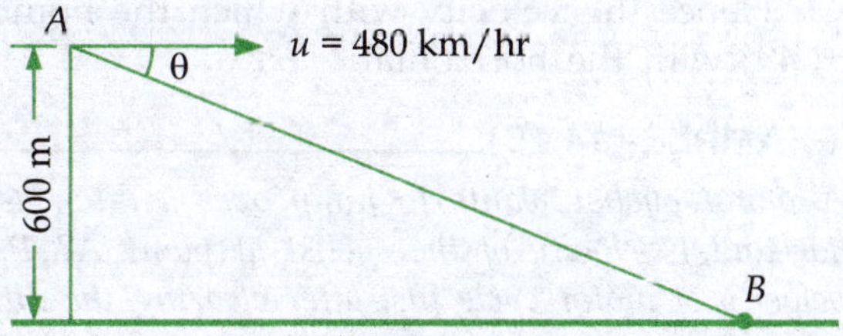

Fig. 14.20

Considering vertical motion: $h = \frac{1}{2}gt^2$

$$\therefore \quad \text{time of flight} = \sqrt{\frac{2h}{g}}$$

$$= \sqrt{\frac{2 \times 600}{9.81}} = 11.06 \text{ seconds}$$

During this period, horizontal distance travelled by the bomb,

$$= \text{horizontal velocity} \times \text{time}$$

$$= 133.33 \times 11.06 = 1471 \text{ m}$$

$\therefore$ The angle θ below the horizontal at which the pilot must see the target while releasing the bomb, is given by

$$\tan\theta = \frac{600}{1471} = 0.4079 \qquad \therefore \quad \theta = 22.19°$$

EXAMPLE 14.24

An aircraft flying at a height of 800 m with uniform horizontal velocity of 360 km/hr releases a bomb which hits a target on the ground. Find (a) the time required for the bomb to reach the target, and (b) the horizontal distance of the aircraft from the target when the bomb was released (c) direction and the velocity with which the bomb hits the target.

Solution: At any instant of time, the bomb is subjected to

(*i*) horizontal motion with constant horizontal velocity

$$u = 360 \text{ km/hr}$$

$$= \frac{360 \times 1000}{3600} = 100 \text{ m/s}$$

(*ii*) Vertical motion with initial velocity zero and acceleration due to gravity.

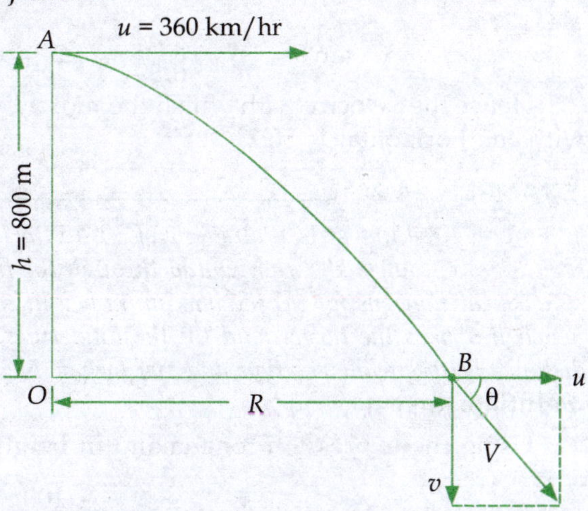

Fig. 14.21

Considering vertical motion: $h = \frac{1}{2}gt^2$

$$\therefore \quad \text{time of flight } t = \sqrt{\frac{2h}{g}}$$

$$= \sqrt{\frac{2 \times 800}{9.81}} = 12.77 \text{ s}$$

During this period, the horizontal distance travelled by the bomb,

$$= \text{horizontal velocity} \times \text{time} = 100 \times 12.77 = \textbf{1277 m}$$

Let v be the vertical velocity of the bomb when it hits the target

$$v = gt = 9.81 \times 12.77 = 125.27 \text{ m/s}$$

$$\text{Resultant velocity } V_r = \sqrt{u^2 + v^2} = \sqrt{100^2 + 125.27^2} = \textbf{160.29 m/s}$$

$$\tan\theta = \frac{125.27}{100} = 1.2527; \quad \theta = 51.49^\circ$$

Hence the velocity with which the bomb hits the target is 160.29 m/s and its direction is 51.49° with the horizontal.

EXAMPLE. 14.25

A motor cyclist wants to jump over a ditch as shown in Fig. 14.22. What should be the minimum horizontal velocity of the cyclist at point A? Proceed to workout the magnitude and direction of the velocity of motor cycle just after clearing the ditch.

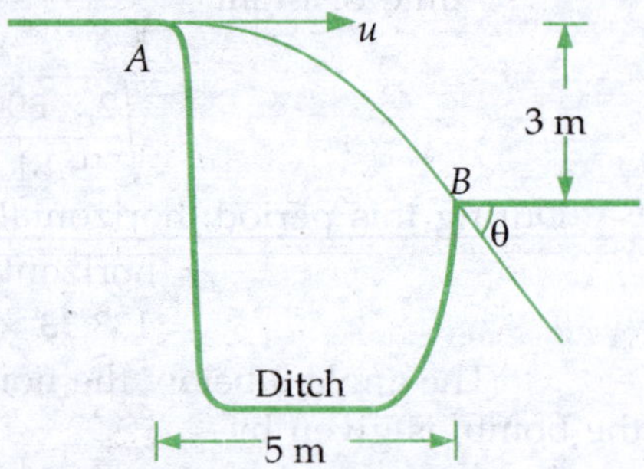

Fig. 14.22

Solution: At any instant of time, the cyclist is subjected to

(i) horizontal motion with constant horizontal velocity u

(ii) vertical motion with initial velocity zero and acceleration due to gravity.

From these two motion, we have

$$5 = ut \quad \text{and} \quad 3 = \frac{1}{2}gt^2$$

That gives:
$$3 = \frac{1}{2}g\left(\frac{5}{u}\right)^2 = \frac{1}{2} \times 9.81 \times \frac{25}{u^2}$$

$$\therefore \quad u = \sqrt{\frac{9.81 \times 25}{2 \times 3}} = 6.39 \text{ m/s} = \frac{6.39 \times 3600}{1000} = \textbf{23 km/hr}$$

At point B: $u = 6.39$ m/s

$$v = \sqrt{2gh} = \sqrt{2 \times 9.81 \times 3} = 7.67 \text{ m/s}$$

$$\text{Resultant velocity} = \sqrt{u^2 + v^2} = \sqrt{6.39^2 + 7.67^2} = \textbf{9.98 m/s}$$

$$\tan\theta = \frac{v}{u} = \frac{7.67}{6.39} = 1.2; \quad \theta = 50.2^\circ$$

Hence the velocity with which the motor cyclist clears the ditch is 9.98 m/s and its inclination with the horizontal is 50.2°.

EXAMPLE 14.26

A bullet is fired upwards at an angle of 30° to the horizontal from a point on a hill and it strikes the target which is 80 m lower than horizontal through the firing point. If the initial velocity of the bullet is 100 m/s, make calculations for (a) the maximum height to which the bullet will rise, (b) the actual velocity with which it strikes the target, and (c) the total time required for the flight of the bullet. (d) the horizontal distance between hill position and the target. Neglect the air resistance.

Solution: Refer Fig. 14.23.

Using the expression for maximum height

$$H = \frac{V^2}{2g}\sin^2\alpha = \frac{100^2}{2 \times 9.81}\sin^2 30 = 127.42 \text{ m}$$

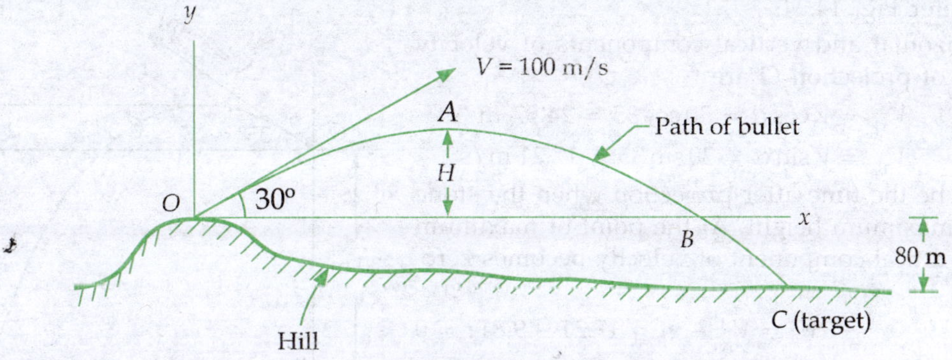

Fig. 14.23.

(*b*) At the point of firing *O*, the horizontal and vertical components of velocity are

$$u = V \cos \alpha \;=\; 100 \cos 30^\circ \;=\; 86.6 \text{ m/s}$$

$$v = V \sin \alpha = 100 \times \sin 30^\circ = 50 \text{ m/s}$$

The horizontal component remains constant during the entire flight whereas the vertical component is affected by gravity. Further at the point of maximum height, the vertical component is zero.

Considering motion from to *A* to *C* and using the correlation $v^2 - u^2 = 2as$ for the vertical motion,

$$v^2 - 0 = 2 \times 9.81 \times (80 + 127.42) \; ; \quad v = 63.79 \text{ m/s}$$

This is the vertical component of velocity with which the bullet strikes the targent

Horizontal component = 86.6 m/s

$\therefore$ Resultant velocity $= \sqrt{63.79^2 + 86.6^2} =$ **107.59 m/s**

(*c*) Let t_1 = time of flight from the point of projection *O* to the highest point on the trajectory. Using the equation $v = u + at$, we have

$$0 = 86.6 - 9.81\, t_1 \qquad (a = -g \text{ as the motion is upward})$$

$$t_1 = \frac{86.6}{9.81} = 5.096 \text{ s}$$

Let t_2 = time of flight from point *A* to the target point C. Using the equation $v = u + at$, we have

$$63.79 = 0 + 9.81\, t_2 \qquad (a = g \text{ as the motion is downward})$$

$$t_2 = \frac{63.79}{9.81} = 6.50 \text{ s}$$

$\therefore$ Total time of flight $t = t_1 + t_2 = 5.096 + 6.50 = 11.596$ s

(*c*) The horizontal distance between hill position and the target,

$$s = u \times t = 86.6 \times 11.596 = \mathbf{1004.2 \text{ m}}$$

EXAMPLE 14.27

A stone is projected upwards from the top of a 30 m high building with a velocity of 30 m/s at angle of 35° with horizontal. Make calculations for :

(*a*) *the greatest height reached by the stone above the ground,*

(*b*) *the horizontal distance from the point of projection to the point where the stone strikes the ground.*

(*c*) *the velocity with which it strikes the ground, and*

(*d*) *the time of flight*

Solution: Refer Fig. 14.24.

The horizontal and vertical components of velocity at the point of projection O are:

$$V_{x_0} = V\cos\alpha = 30\cos 35 = 24.57 \text{ m/s}$$
$$V_{y_0} = V\sin\alpha = 30\sin 35 = 17.21 \text{ m/s}$$

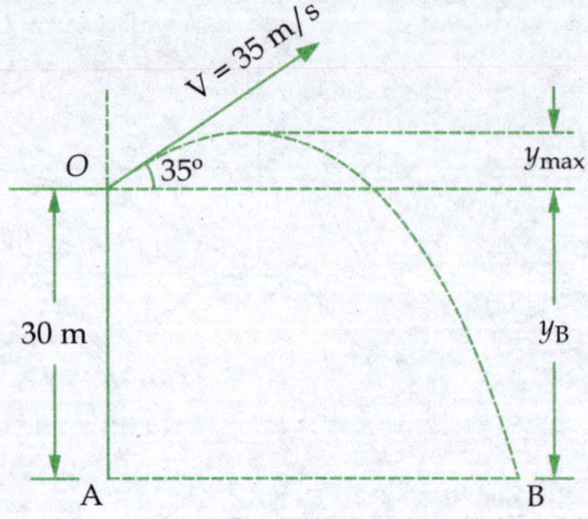

Fig. 14.24

(*a*) Let t be the time after projection when the stone reaches the maximum height. At the point of maximum height, the vertical component of velocity becomes zero. That is

$$V_y = V_{y_0} - gt = 17.21 - 9.81t = 0$$
$$\therefore \quad t = \frac{17.21}{9.81} = 1.75 \text{ s}$$

Maximum height attained by the stone above the point of projection O,

$$y_{max} = V_{y_0}t - \frac{1}{2}gt^2 = 17.21 \times 1.75 - \frac{1}{2} \times 9.81 \times (1.75)^2$$
$$= 30.12 - 15.02 = 15.1 \text{ m}$$

Maximum height attained by the stone above the ground

$$= 30 + 15.1 = \mathbf{45.1 \text{ m}}$$

(*b*) Let B be the point at which the stone strikes the ground after time t. Then

$$y_B = -30 \text{ m} \quad \text{and} \quad y_B = V_{yo}t - \frac{1}{2}gt^2$$
$$-30 = 17.21t - \frac{1}{2} \times 9.81 \times t^2 = 17.21t - 4.905t^2$$

or
$$4.905t^2 - 17.21t - 30 = 0$$

$$\therefore \quad t = \frac{17.21 \pm \sqrt{(17.21)^2 - 4 \times 4.905 \times (-30)}}{2 \times 4.905} = \frac{17.21 \pm 29.74}{9.81} = 4.786 \text{ s}$$

The horizontal distance AB is then given by

$$AB = V_{x_0} \times t = 24.57 \times 4.786 = \mathbf{117.59 \text{ m}}$$

(*c*) At point B where the stone strikes the ground, the horizontal and vertical components of velocity are

$$V_x = V_{x_0} = 24.57 \text{ m/s}$$
$$V_y = V_{y_0} - gt = 17.21 - 9.81 \times 4.786$$
$$= -29.74 \text{ m/s}$$
$$V_B = \sqrt{V_x^2 + V_y^2}$$
$$= \sqrt{(24.57)^2 + (29.74)^2} = \mathbf{38.57 \text{ m/s}}$$
$$\theta = \tan^{-1}\left(\frac{V_y}{V_x}\right) = \tan^{-1}\left(\frac{29.74}{24.57}\right) = 50.42°$$

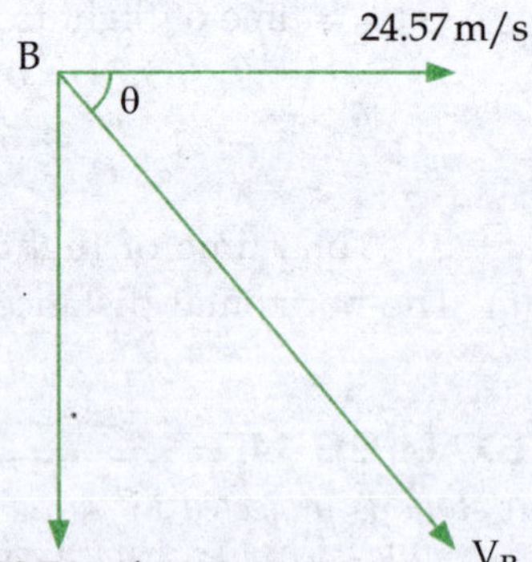

Fig. 14.25

(*d*) Time of flight = time of trajectory from the point of projection (O) to the point where it strikes the ground (B)

$$= \mathbf{4.786 \text{ s}}$$

14.3. PROJECTION ON AN INCLINED PLANE

Let a body be projected with velocity V making an angle α with the horizontal plane $o - x$, OA is an inclined plane making angle β with the horizontal plane. The body after following the parabolic path strikes the inclined plane at point A; the distance OA represents the range R of the projectile on the inclined plane. The co-ordinates of point A with reference to $o - x$ and $o - y$ are:

$$x = OA\cos\beta = R\cos\beta$$

$$y = OA\sin\beta = R\sin\beta$$

The equation of the parabolic path (trajectory) followed by the projectile during its flight is

$$y = x\tan\alpha - \frac{gx^2}{2V^2\cos^2\alpha}$$

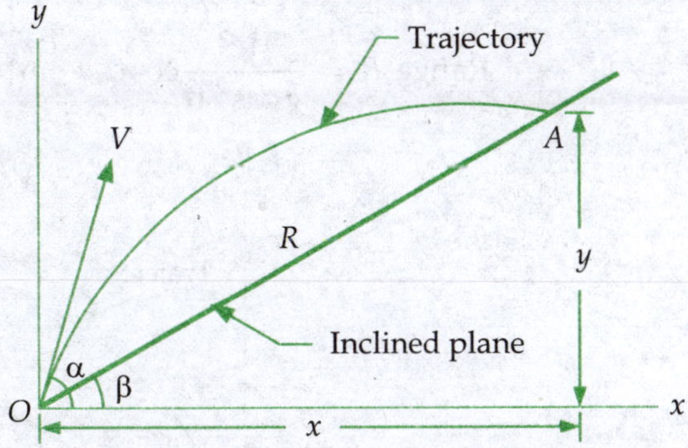

Fig. 14.26

Substituting the values of x and y in the above equation, we get

$$R\sin\beta = R\cos\beta\tan\alpha - \frac{gR^2\cos^2\beta}{2V^2\cos^2\alpha}$$

or $$\frac{gR^2\cos^2\beta}{2V^2\cos^2\alpha} = R(\sin\beta - \cos\beta\tan\alpha)$$

$$\therefore \quad \text{Range } R = \frac{2V^2\cos^2\alpha}{g\cos^2\beta}\left[\frac{\cos\beta\sin\alpha - \sin\beta\cos\alpha}{\cos\alpha}\right]$$

$$= \frac{2V^2\cos\alpha}{g\cos^2\beta}\sin(\alpha - \beta) = \frac{V^2}{g\cos^2\beta}[2\cos\alpha\sin(\alpha - \beta)]$$

Since $\sin A - \sin B = 2\cos\frac{A+B}{2}\sin\frac{A-B}{2}$, we have

$$2\cos\alpha\sin(\alpha - \beta) = \sin(2\alpha - \beta) - \sin\beta$$

$$\therefore \quad R = \frac{V^2}{g\cos^2\beta}[\sin(2\alpha - \beta) - \sin\beta] \qquad ...(14.7)$$

Range R will be maximum when

$$\sin(2\alpha - \beta) = 1 = \sin 90^\circ$$

$$2\alpha - \beta = 90^\circ \qquad ...(14.8)$$

and $$R_{max} = \frac{V^2}{g\cos^2\beta}(1 - \sin\beta) = \frac{V^2}{g(1 - \sin^2\beta)} \times (1 - \sin\beta) = \frac{V^2}{g(1 + \sin\beta)} \qquad ...(14.9)$$

(ii) Time of flight

Resolving initial velocity V along and perpendicular to the inclined plane,

velocity along the plane = $V\cos(\alpha - \beta)$

velocity perpendicular to the plane= $V\sin(\alpha - \beta)$

Acceleration due to gravity along and perpendicular to the plane are $g\sin\beta$ and $g\cos\beta$ respectively.

Time of flight T is the time taken by the object in reaching from position O to A. For motion normal to the inclined plane, the net distance moved is zero. Then using the equation

$$s = ut + \frac{1}{2}at^2, \text{ we have}$$

$$0 = V\sin(\alpha - \beta) \times T - \frac{1}{2} g\cos\beta T^2$$

$\therefore$ Time of flight $T = \dfrac{2V}{g}\dfrac{\sin(\alpha-\beta)}{\cos\beta}$ ...(14.10)

(iii) When the body is projected down the plane (Fig. 14.27), β = – β and accordingly

- Range $R = \dfrac{2V^2}{g\cos^2\beta}\cos\alpha \times \sin[\alpha - (-\beta)] = \dfrac{2V^2}{g}\cos^2\beta\cos\alpha\sin(\alpha+\beta)$ ...(14.11)

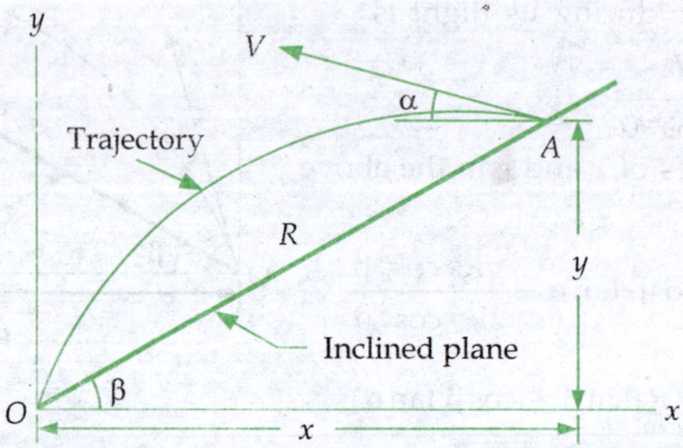

Fig. 14.27

- Maximum range $R_{max} = \dfrac{V^2}{g[1+\sin(-\beta)]} = \dfrac{V^2}{g}(1-\sin\beta)$ ...(14.12)
- Condition for maximum range

$$2\alpha - (-\beta) = 90^\circ ; 2\alpha + \beta = 90^\circ$$ ...(14.13)

- Time of flight $T = \dfrac{2V\sin[\alpha-(-\beta)]}{g\cos\beta} = \dfrac{2V}{g}\dfrac{\sin(\alpha+\beta)}{\cos\beta}$ ...(14.14)

EXAMPLE. 14.28

A man standing on an incline with slope 1 in 4 fires two bullets; one up the incline and the other down the incline. Both the bullets are fired with velocity 200 m/s and at angles of 45 degree with horizontal. Make calculations for the total range of the bullets along the plane.

Solution: Refer Fig. 14.28 showing the projectory followed by the two bullets when fired from position *P* on the incline.

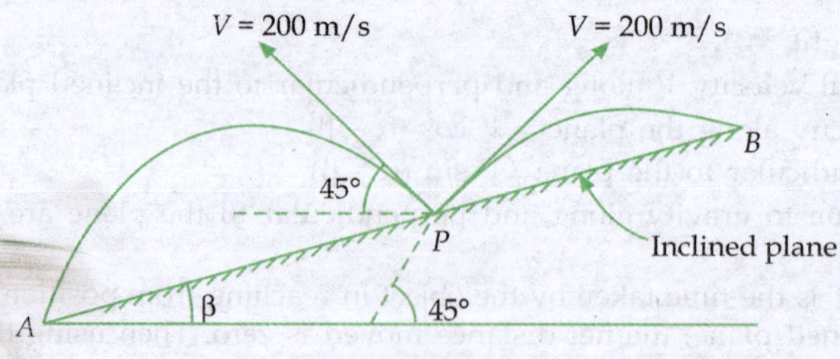

Fig. 14.28.

Here angle of projection $\alpha = 45°$, velocity of projection $V = 200$ m/s and angle of inclined plane $B = \tan^{-1}\left(\frac{1}{4}\right) = 14°$. Then

$$\text{Range down the plane } PA = \frac{2V^2}{g}\frac{\cos\alpha\sin(\alpha+\beta)}{\cos^2\beta}$$

$$= \frac{2\times 200^2}{9.81}\frac{\cos 45°\sin(45+14)°}{\cos^2 14°} = 5250.6 \text{ m}$$

$$\text{Range up the plane } PB = \frac{2V^2}{g}\frac{\cos\alpha\sin(\alpha-\beta)}{\cos^2\beta}$$

$$= \frac{2\times 200^2}{9.81}\frac{\cos 45°\sin(45-14)°}{\cos^2 14°} = 3154.3 \text{ m}$$

Total range of the bullets is along the incline

$$AB = PA + PB = 5250.6 + 3154.3 \simeq \mathbf{8405 \text{ m}}$$

EXAMPLE. 14.29

A projectile is shot from a point on an inclined plane with a velocity of 30 m/s. The angle of plane and that of projection are 20° and 55° to the horizontal respectively. Is the condition satisfied for maximum range? If so, find the maximum range and the time of flight. How the time of flight would be affected when the projectile is shot downwards?

Solution: Range is maximum when

$$\sin(2\alpha-\beta) = 1 = \sin\frac{\pi}{2}$$

$$\text{or} \quad 2\alpha - \beta = \frac{\pi}{2}; \quad \alpha = \frac{\pi}{4}+\frac{\beta}{2} \quad ...(i)$$

where α and β are the angle of projection and angle of plane to the horizontal respectively.

Substituting $\alpha = 55°$ and $\beta = 20°$ in expression, (*i*) we find

$$55° = \frac{180}{4}+\frac{20}{2} = 55° \quad \text{which is true.}$$

Obviously the condition for maximum range is satisfied

$$R_{max} = \frac{V^2}{g(1+\sin\beta)} = \frac{30^2}{9.81(1+\sin 20°)} = \mathbf{68.36 \text{ m}}$$

$$\text{Time of flight } T = \frac{2V}{g}\frac{\sin(\alpha-\beta)}{\cos\beta} = \frac{2\times 30}{9.81}\times\frac{\sin(55-20)}{\cos 20} = \mathbf{3.733 \text{ s}}$$

When the projectile is shot downwards,

$$\text{Time of flight} = \frac{2V}{g}\frac{\sin(\alpha+\beta)}{\cos\beta} = \frac{2\times 30}{9.81}\times\frac{\sin(55+20)}{\cos 20°} = \mathbf{6.287 \text{ s}}$$

EXAMPLE 14.30

A projectile is fired from a point on an inclined plane with a velocity of 50 m/s. The angle of projection and the angle of plane are 60° and 25° to the horizontal respectively. If the motion of the projectile is up the plane, determine:

(*a*) *time of flight*

(*b*) *range up the plane*

(*c*) *condition for maximum range, and*

(*d*) *maximum range up the plane.*

Solution: Refer Fig. 14.27 for the inclined plane and the trajectory of the projectile; O represents the point of projection.

$$\alpha = 60° \quad \text{and} \quad \beta = 25°$$

(*a*) Time of flight $T = \frac{2V}{g}\frac{\sin(\alpha-\beta)}{\cos\beta} = \frac{2\times50}{9.81}\frac{\sin(60-25)}{\cos 25} = \frac{2\times50}{9.81}\times\frac{0.5736}{0.9063} = \mathbf{6.45\ s}$

(*b*) Range up the plane $R = \frac{2V^2}{g\cos^2\beta}\cos\alpha\sin(\alpha-\beta) = \frac{2\times50^2}{9.81\cos^2 25}\cos 60\sin(60-25)$

$$= \frac{2\times50^2}{9.81\times(0.9063)^2}0.5\times0.5736 = \mathbf{177.96\ m}$$

(*c*) Condition for maximum range

$$2\alpha - \beta = 90°$$

$$\alpha = \frac{90+\beta}{2} = \frac{90+25}{2} = \mathbf{57.5°}$$

(*d*) Maximum range up the plane

$$= \frac{V^2}{g}(1+\sin\beta) = \frac{50^2}{9.18}(1+\sin 25°) = \frac{50^2}{9.81(1+0.4226)} = \mathbf{179.14\ m}$$

REVIEW AND SUMMARY

1. A projectile refers to any object which is thrown into space and whose subsequent motion is subjected to gravitational forces.
2. The projectile travels both in the horizontal and vertical directions and traces a curvilinear path which is called trajectory.
3. The necessary conditions for a body to be called a projectile are :
   (*i*) it should be given some initial velocity, and subsequently
   (*ii*) it should be under the action of gravity alone; there should not be any other driving force.
4. Examples of projecticle motion are :
   (*i*) a bullet from a gun
   (*ii*) throwing of cricket ball from one player to another
   (*iii*) a stone thrown horizontally from the top of a tower, and
   (*iv*) a free liquid jet issuing from the nozzle into atmosphere.
5. Consider the motion of an object which is projected from point O with velocity V inclined at an angle α with the horizontal ground surface. After attaining the maximum height H, it descends and finally hits the ground at point B.
   The terms related to projectile motion are:
   1. ***Velocity of projection*** is the velocity V with which the body is projected into space. The angle α which this velocity makes with the horizontal is called the ***angle of projection.***
   2. ***Trajectory*** is the curvilinear path traced by the projectile during the period of its motion. The path of trajectory is a parabola prescribed by the relation:

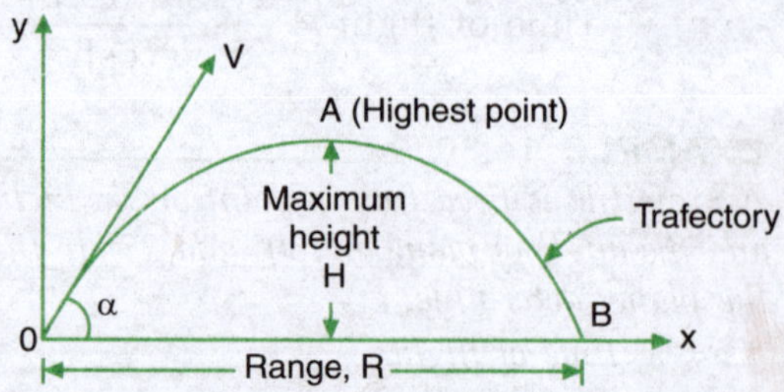

Fig. 14.29.

$$y = x\tan\alpha - \frac{1}{2}\frac{gx^2}{V^2}(1+\tan^2\alpha)$$

where (x, y) is the position of the object on the trajectory.

3. ***Maximum height*** is the maximum vertical distance of the object from the point of projection

$$H = \frac{V^2}{2g}\sin^2\alpha$$

4. ***Time of flight*** is the time interval during which the object remains in motion (time taken by the object in reaching from position O to B).

$$T = \frac{2V}{g}\sin\alpha$$

The time taken by the object to reach maximum height (point A) is half the time of flight

$$t = \frac{V}{g}\sin\alpha$$

5. ***Range*** is the horizontal distance between the point of projection and the point where the projectile hits the ground (distance OB). Apparently range is the horizontal distance which the projectile travels during its flight

$$R = \frac{V^2}{g}\sin 2\alpha$$

Range will be maximum for a given velocity of projection V when $\sin 2\alpha$ is maximum or when $\alpha = 45°$

$$R_{max} = \frac{V^2}{g}$$

For a given value of initial velocity, there are two angles of projection which provide the same range. These are α and $\left(\frac{\pi}{2} - \alpha\right)$.

6. For a projectile up on an inclined plane

$$T = \frac{2V}{g}\frac{\sin(\alpha-\beta)}{\cos\beta}\ ; \quad R = \frac{2V^2}{g}\frac{\cos\alpha\sin(\alpha-\beta)}{\cos^2\beta}$$

For maximum rnage : $2\alpha - \beta = 90°$

and $$R_{\max} = \frac{V^2}{g(1+\sin\beta)}$$

where u is the velocity of approach, α is angle projection with horizontal and β is angle of inclined plane with horizontal.

7. For a projectile down an inclined plane

$$T = \frac{2V}{g}\frac{\sin(\alpha+\beta)}{\cos\beta}\ ; \quad R = \frac{2V^2}{g}\frac{\cos\alpha\sin(\alpha+\beta)}{\cos^2\beta}$$

For maximum range : $2\alpha + \beta = 90°$

and $$R_{\max} = \frac{V^2}{g(1-\sin\beta)}$$

REVIEW QUESTIONS

A. Conceptual and conventional questions :

1. Define projectile motion.
2. List the necessary conditions for a body to be called a projectile.
3. List a few situations which constitute projectile. Does a rocket fired into space conform to a projectile?
4. Set up an expression showing that path of a projectile is a parabola.
5. State the assumptions made in the study of motion of a projectile.
6. Define the following terms as related to projectile motion :
   (*a*) velocity and angle of projection
   (*b*) range and (*c*) time of flight.
7. A gun has a maximum horizontal large *R*. What should be the firing angle so that it hits a target at a distance *R*/2 on the same level?
8. A launcher projects a missile with an initial speed of 60 m/s. If the missile lands 3230 m away at the same level, what must have been the angle of elevation of the launcher? Calculate the corresponding time of flight and the maximum height attained.

   (33.35°; 6.18 s; 46.84 m) or (59.65; 10.55 s; 136.64 m)
9. An aircraft moving horizontally at 400 km/hr accidentally loses a rivet when it is at a height of 1800 m from a point *A* on the ground level. Determine the location of point *B* with respect to *A* where the rivet will land.

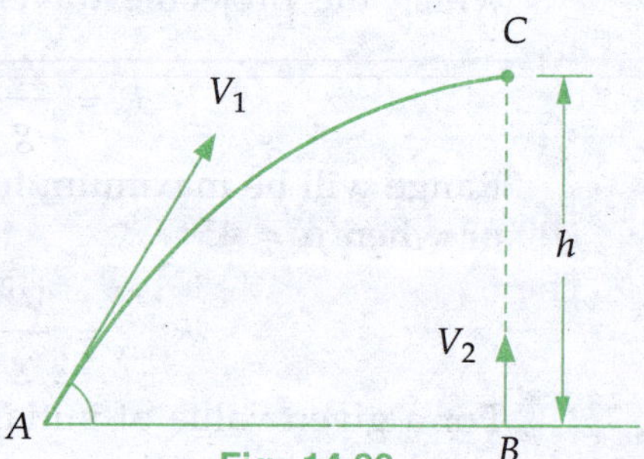

Fig. 14.30

   (21.28.5 m)
10. A projectile has a range *R* on a horizontal plane and it rises to maximum height *h* during the course of its flight. Show that the maximum horizontal range with the same velocity of projection is given by

$$h_{max} = 23h + \frac{R^2}{8h}$$

11. A shell is fired horizontally from cliff 45 m above the sea level with an initial velocity of 300 m/s to hit a target at sea level. Determine : (*a*) the time taken by the shell to hit the target, (*b*) the horizontal distance of cliff from the target, and (*c*) the velocity with which the shell strikes the target. **(3s, 900m, 301.5m/s)**
12. Two tall buildings are 100 m apart. A ball thrown horizontally from a window 150 m above the ground in one building enters a window 25 m from the ground in the other building. What is the initial velocity of the ball?

Hint: Vertical distance travelled by the ball = 150 – 25 = 125 m

$$s = \frac{1}{2}gt^2 \quad (\because \text{ initial velocity in the vertical direction is zero}).$$

$$t = \sqrt{\frac{2\times 125}{9.81}} = 5.048 \text{ s}$$

Horizontal distance travelled by the ball = 100 m

$\therefore$ Initial horizontal velocity $= \dfrac{100}{5.048} = 19.80$ m

13. A particle is projected with velocity V_1 from a point A making an angle of 60° with the horizontal. Simultaneously another particle is thrown vertically upwards with velocity V_2 from point B. Both the particles reach the highest point on the parabolic part of the first particle at the same time. Find the ratio of velocities V_2/V_1.

Hint: The time taken by the first particle to reach the highest point C on its parabolic path

$$t_1 = \frac{V_1}{g}\sin\alpha = \frac{V_1}{g}\sin 60° = \frac{\sqrt{3}}{2}\frac{V_1}{g}$$

The time taken by the second particle to reach the highest point C

$$t_2 = \frac{V_2}{g} \quad (v = u + at;\ u = v_2 \text{ and } v = 0)$$

$$\therefore \quad \frac{\sqrt{3}}{2}\frac{V_1}{g} = \frac{V_2}{g};\ \frac{V_2}{V_1} = \sqrt{3} : 2$$

14. A plane has a maximum rise of 5 in 12. A shot is projected with a velocity of 200 m/s at an elevation of 30°. Find the range of shot, if (a) it is fired up the plane, (b) it is fired down the plane. **(1065 m, 6590 m)**

B. Complete the following statements with most appropriate word/words :

1. The curved path followed by a projectile is known as ....
2. The equation of trajectory of a projectile is

$$y = x\tan\alpha - \text{..................}$$

3. The directions of velocity and acceleration are .... to each other at the top position of trajectory of a projectile.
4. A projectile covers a maximum range when it is directed at an angle of ....
5. The horizontal range remains the same when the angle of projection α is replaced by ....
6. At the point of maximum height attained, the projectile has .... velocity in the vertical directions.
7. For a projectile, time of flight is .... the time to reach the greatest height.
8. The maximum horizontal range is .... the maximum height attained by the projectile.

Answers:

1. trajectory 2. $y = x\tan\alpha - \dfrac{gx^2}{2V^2}(1+\tan^2\alpha)$ 3. perpendicular 4. 45° 5. $\left(\dfrac{\pi}{2} - \alpha\right)$

6. zero 7. twice 8. half

C. Multiple choice questions :

1. Which one of the following situations constitutes a projectile?
   (a) lifting of a balloon
   (b) taking off an aircraft
   (c) throwing of a cricket ball from one player to another
   (d) launching of a satellite

2. The equation of a projectile is

$$y = \sqrt{3}\,x - \frac{1}{2}gx^2$$

The angle of projection is given by

(a) $\tan\theta = \frac{1}{\sqrt{3}}$ (b) $\tan\theta = \sqrt{3}$ (c) 60° (d) zero

3. Two stones are projected from the same point with the same speed but at angles $\pi/3$ and $\pi/6$ respectively. If their ranges are R_1 and R_2, then

(a) $R_1 = 2R_2$ (b) $R_1 = R_2$ (c) $R_1 = 0.5R_2$ (d) $R_1 = 0.25R_2$

4. A projectile can have the same range for two angles of projection. If t_1 and t_2 are the times of flight in the two cases, then the product of the two times of flight is proportional to

(a) R^2 (b) $1/R^2$ (c) R (d) $1/R$

5. Two stones are projected with the same velocity (in magnitudes) but making different angles with the horizontal. If their ranges are equal and the angle of projection of one is 60°, then the ratio of the maximum heights attained (y_2/y_1) will be

(a) 3 : 1 (b) 2 : 1 (c) 1 : 2 (d) 1 : 3

6. Two balls are projected from the same point making angles of 60° and 30° with the vertical axis. If both the balls are to attain the same height, the ratio of the speeds of projection V_1/V_2 should be

(a) 1 : 2 (b) 1 : 1 (c) $1 : \sqrt{3}$ (d) $\sqrt{2} : 1$

7. A projectile has a range R in the gravitational field of earth. Under identical conditions of velocity and acceleration, its range on moon would be:

(a) $6\,R$ (b) $\sqrt{6}\,R$ (c) $R/6$ (d) $R/\sqrt{6}$

Take g on moon equal to one-sixth of that on earth.

8. If the time of flight of a bullet over a horizontal range R is T seconds, the inclination of the direction of projection with the horizontal is

(a) $\tan^{-1}\left(gT^2/2R\right)$ (b) $\tan^{-1}\left(gT/2R\right)$

(c) $\tan^{-1}\left(gT^2/4R\right)$ (d) $\tan^{-1}\left(2\,gT/R\right)$

9. For a projectile, the maximum horizontal range R_m and the maximum height H attained during the course of flight conform to the identity :

(a) $R_m = H$ (b) $R_m = 2H$ (c) $R_m = 3H$ (d) $R_m = 4H$

10. Which of the following remains constant during flight of a projectile ?

(a) vertical component of velocity

(b) horizontal component of velocity

(c) angle of projection

(d) sum of its kinetic energy and potential energy

11. A particle is projected at an angle θ to the horizontal and it attains a maximum height H. The time taken by the projectile to reach the highest point of its path is

(a) $\sqrt{H/g}$ (b) $\sqrt{2H/g}$ (c) $\sqrt{2\,H\sin\theta/g}$ (d) $\sqrt{2\,H/g\sin\theta}$

12. A particle is projected with velocity V_1 from point A making an angle of 60° with the horizontal. Simultaneously another particle is thrown vertically upwards with velocity V_2 from point B. Both the particles reach the highest point on the parabolic path of the first particle at the same time.

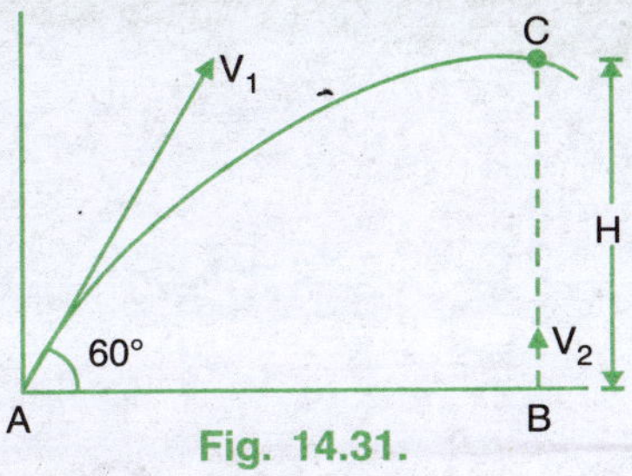

Fig. 14.31.

The ratio V_2/V_1 is

(a) $1 : \sqrt{3}$ (b) $\sqrt{2} : 1$

(c) $\sqrt{3} : 2$ (d) $2 : 1$

13. A projectile projected on a horizontal plans has $\mu = 1$ m/s and $\alpha = 30°$. The maximum height attained is

(a) $\frac{1}{g}$ (b) $\frac{1}{2g}$ (c) $\frac{1}{4g}$ (d) $\frac{1}{8g}$

14. The ratio of the ranges on an inclined plane with motion upward and motion downward for a given velocity is

(a) $\frac{\cos(\alpha-\beta)}{\cos(\alpha+\beta)}$ (b) $\frac{\tan(\alpha-\beta)}{\tan(\alpha+\beta)}$ (c) $\frac{\sin(\alpha-\beta)}{\sin(\alpha+\beta)}$ (d) $\frac{\sin(\alpha+\beta)}{\cos(\alpha-\beta)}$

15. Examine the following statements and identify the statement which is not true :

(a) A particle projected vertically upward can be called a projectile
(b) A projectile always has a parabolic trajectory
(c) The altitude of the flight of an aircraft pilot equals the horizontal distance of the target on ground at an angle of 45-degree with the horizontal
(d) In case of a projectile shot on a horizontal plane, there are two angles of projection for the same range

Answers :

1. (c)	2. (b)	3. (b)	4. (c)	5. (d)	6. (c)
7. (a)	8. (a)	9. (d)	10. (b)	11. (b)	12. (c)
13. (d)	14. (c)	15. (b)			

Rotation and General Plane Motion 15

Learning objectives : The key concepts included in this chapter relate to the following aspects of rotation and general plane motion

- *angular displacement, velocity and acceleration; relationship between circular motion and linear motion*
- *equations of motion is a circular path with uniform angular acceleration*
- *general plane motion and Chasle's theorem*
- *instantaneous centre and its applications*
- *Coriolis' law*

The motion of a particle in curvilinear motion is on a curved path. The particle does not repeat the journey once it has passed over a point on the curved path. Its centre of rotation also goes on changing.

In the rotary (circular) motion, the movement of the particle is along a circular path. The particle repeats its journey along the same circular path about the same centre of rotation which remains fixed. The shafts, pulleys and flywheels etc., undergo circular motion when they rotate about their geometric axis.

15.1 TERMS RELATED TO CIRCULAR MOTION

The terms associated with circular motion are:

(i) *Angular displacement:* The displacement of a body in rotation is called angular displacement, and it is measured in terms of the angle through which the body moves from the initial state. With reference to Fig. 15.1, a body starts from initial position A and while moving in circular path reaches the position B after a certain time interval. Then $\angle AOB = \theta$ is a measure of the angular displacement of the body. This displacement has magnitude as well as direction, and apparently displacement is a vector quantity. The direction is a rotation-either clockwise or anti-clockwise.

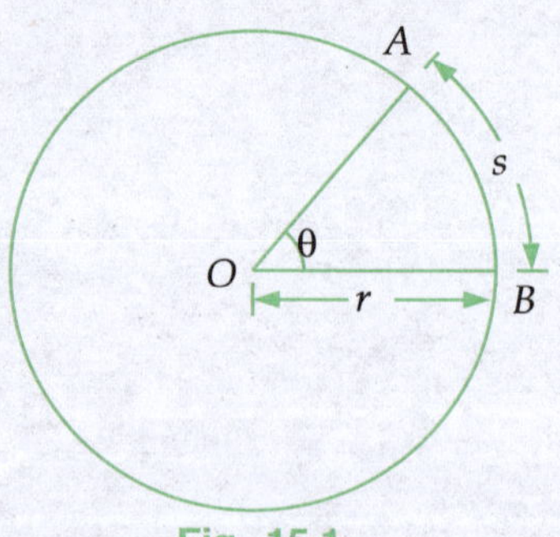

Fig. 15.1

Angular displacement θ can be measured in radians, degrees or revolutions. Various units of angular displacement are connected by the relation:

$$1 \text{ revolution} = 2\pi \text{ radians} = 360 \text{ degrees}$$

(ii) *Angular velocity:* The rate of change of angular displacement of a body with respect to time is called angular velocity.

If the body traverses angular distance $d\theta$ over a time interval dt, then the average angular velocity ω is

$$\omega = \frac{d\theta}{dt}$$

The unit of angular velocity for defining its magnitude is usually taken as radians per second (rad/s). Many a times angular velocity is expressed in terms of revolutions per minute (rpm). If the body turns N rev/min, then

$$\text{number of revolutions per sec} = \frac{N}{60}$$

In one revolution, the angular displacement of the body is 360 degrees or 2π radians

Angle traversed by the body is one second

= (angle covered in one revolution) × (number of revolutions per second)

$$= 2\pi \times \frac{N}{60} = \frac{2\pi N}{60}$$

But the angle traversed by the body in one second is a measure of the angular velocity w. Then

$$\omega = \frac{2\pi N}{60} \text{ rad/sec}$$

(*iii*) *Angular acceleration:* The rate of change of angular velocity of a body with respect to time is called angular acceleration.

angular acceleration = rate of change of angular velocity

$$\alpha = \frac{d\omega}{dt} = \frac{d}{dt}\left(\frac{\theta}{t}\right) = \frac{d^2\theta}{dt^2}$$

The angular acceleration is sometimes expressed as

$$\alpha = \frac{d\omega}{dt} = \frac{d\omega}{d\theta} \times \frac{d\theta}{dt} = \omega\frac{d\omega}{d\theta} \qquad \left(\because \frac{d\theta}{dt} = \omega\right)$$

The unit of angular acceleration for defining its magnitude is usually taken as radian per second per second (rad/sec^2)

Relationship between circular motion and linear motion: When a body moves from A to B (Fig. 15.1), the distance travelled by it is s. If r is the distance of the body from the centre of rotation, then

$$s = r\theta$$

and $$\frac{ds}{dt} = \frac{d}{dt}(r\theta) = r\frac{d\theta}{dt} \qquad (\because r \text{ is constant})$$

Noting that ds/dt = linear velocity v and $d\theta/dt$ = angular velocity ω, we may write

$$v = \omega r$$

That is, the linear velocity is r-times the angular velocity

When the expression $v = \omega r$ is differentiated with respect to time, we get

$$\frac{dv}{dt} = \frac{d}{dt}(\omega r) = r\frac{d\omega}{dt}$$

Since dv/dt = linear acceleration a and $d\omega/dt$ = angular acceleration α, we may write

$$a = \alpha r$$

That is, the linear acceleration is r-times the angular acceleration.

It is worthwhile to point out that when a body moves along a curved path (curvilinear motion), the linear velocity is referred to as tangential velocity. Further, the body undergoes tangential and normal components of acceleration given by

$$\text{tangential acceleration } a_t = \alpha r$$

$$\text{normal acceleration } a_n = \frac{v^2}{r} = \omega^2 r$$

The tangential acceleration is due to change in magnitude of the velocity and the normal acceleration is due to change in direction of the body.

Fig. 15.2 illustrates these components for any point P. The normal component is directed towards the centre of the circle. The total acceleration a is the vector sum of the two components.

$$a = \sqrt{\alpha_n^2 + \alpha_t^2}$$

and the angle between the total acceleration and the radius is

$$\phi = \tan^{-1}\left(\frac{\alpha_t}{\alpha_n}\right)$$

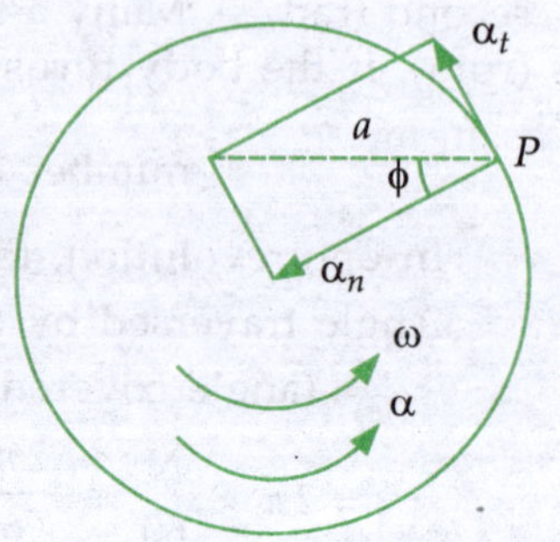

Fig. 15.2

15.2. EQUATIONS OF CIRCULAR MOTION

When a body moves in a circular path with uniform angular acceleration, the equations of motion are:

$$\omega = \omega_0 + \alpha t; \quad \omega^2 - \omega_0^2 = 2\,\alpha\theta; \quad \theta = \omega_0 t + \frac{1}{2}\alpha t^2$$

where ω_0 is the initial angular velocity, ω is the final angular velocity, α is angular acceleration of the body and θ represents the angle moved in time t.

(*i*) angular acceleration α = rate of change of angular velocity

$$= \frac{\text{change of angular velocity}}{\text{time}} = \frac{\omega - \omega_0}{t}$$

$$\omega - \omega_0 = \alpha t$$

$$\therefore \quad \omega = \omega_0 + \alpha t$$

Alternatively:

$$\alpha = \frac{d\omega}{dt} \quad ; \quad d\omega = \alpha\, dt$$

Integrating both sides within the limit $\omega = \omega_0$ at $t = 0$ and $\omega = \omega$ at $t = t$

$$\int_{\omega_0}^{\omega} d\omega = \alpha \int_0^t dt \qquad (\because \alpha \text{ is constant})$$

$$\omega - \omega_0 = \alpha t$$

$$\therefore \quad \omega = \omega_0 + \alpha t \qquad \text{...(15.1)}$$

(*ii*) Angle travelled = average angular velocity × time

$$\theta = \frac{\omega_0 + \omega}{2} \times t = \frac{\omega_0 + \omega_0 + \alpha t}{2} \times t \qquad (\because \omega = \omega_0 + \alpha t)$$

$$= \left(\omega_0 + \frac{\alpha t}{2}\right) t = \omega_0 t + \frac{1}{2}\alpha t^2$$

Alternatively:

$$\omega = \frac{d\theta}{dt} \quad ; \quad d\theta = \omega\, dt = (\omega_0 + \alpha t)\, dt$$

Upon integration

$$\int_0^\theta d\theta = \int_0^t (\omega_0 + \alpha t) dt$$

assuming $\theta = 0$ at $t = 0$ and $\theta = \theta$ at $t = t$

$$\theta = \left| \omega_0 t + \frac{\alpha t^2}{2} \right|_0^t = \omega_0 t + \frac{1}{2}\alpha t^2 \qquad \text{...(15.2)}$$

(iii) Angle moved = average angular velocity × time

$$\theta = \frac{\omega_0 + \omega}{2} \times \frac{\omega - \omega_0}{\alpha} = \frac{\omega^2 - \omega_0^2}{2\alpha} \qquad \left(\because \omega = \omega_0 + \alpha t \quad t = \frac{\omega - \omega_0}{\alpha} \right)$$

$$\therefore \quad \omega^2 - \omega_0^2 = 2\alpha\theta$$

Alternatively:

$$\alpha = \frac{d\omega}{dt} = \frac{d\omega}{d\theta} \times \frac{d\theta}{dt} = \frac{d\omega}{d\theta} \times \omega$$

or $\quad \alpha\, d\theta = \omega\, d\omega$

upon integration and assuming $\omega = \omega_0$ at $\theta = 0$ and $\omega = \omega$ at $\theta = \theta$

$$\alpha \int_0^\theta d\theta = \int_{\omega_0}^\omega \omega\, d\omega$$

$$\alpha\theta = \left| \frac{\omega^2}{2} \right|_{\omega_0}^{\omega} \quad ; \quad \alpha\theta = \frac{\omega^2 - \omega_0^2}{2}$$

$$\therefore \quad \omega^2 - \omega_0^2 = 2\alpha\theta \qquad \text{...(15.3)}$$

EXAMPLE 15.1.

A flywheel rotates freely on friction less bearings at 240 rev/min. How many revolutions will it make in 10 seconds after the start? Proceed to determine the angular speed if this wheel turns 500 revolutions in 15 seconds.

Solution: There will be no resistance to motion in frictionless bearings. Obviously the wheel will then rotate with uniform motion for which $\theta = \omega t$

(i) $\omega = 240$ rev/min; $t = 10$ seconds

$$\therefore \quad \theta = \omega t = 240 \times \frac{10}{60} = 40 \text{ revolutions} = 40 \times 2\pi = 80\pi \text{ radians}$$

(ii) $\theta = 500$ revolutions ; $t = 15$ seconds

$$\therefore \quad \omega = \frac{\theta}{t} = \frac{500}{(15/60)} = 2000 \text{ rev/min} = 2000 \times \frac{2\pi}{60} = \textbf{209.33 rad/s}$$

EXAMPLE 15.2.

The speed of a flywheel changes from 10 rad/s to 30 rad/s in 5 seconds time. Determine the angular acceleration of the wheel.

How many revolutions the wheel would turn to attain a speed of 600 rev/min?

Solution: Using the kinematic relation $\omega = \omega_0 + \alpha t$, we have

$$\text{angular acceleration } \alpha = \frac{\omega - \omega_0}{t} = \frac{30 - 10}{5} = 4 \text{ rad/s}^2$$

(*b*) $$\omega = \frac{2\pi N}{60} = \frac{2\pi \times 600}{60} = 62.8 \text{ rad/s}$$

Then from the relation, $\omega^2 - \omega_0^2 = 2\alpha\theta$,

$$\text{total angular displacement } \theta = \frac{\omega^2 - \omega_0^2}{2\alpha} = \frac{62.8^2 - 10^2}{2 \times 4} = 480.48 \text{ radians.}$$

Now, 1 revolution ≡ 2π radians

$$\therefore \quad \text{Total revolutions made by the wheel} = \frac{480.48}{2\pi} = 76$$

EXAMPLE 15.3.

A roller of a motorised mortar mixer rotates for 5 seconds with a uniform angular acceleration and describes 120 radians during this time. It then rotates with constant angular velocity and covers 100 radians during the next 5 seconds. Make calculations for the initial angular velocity and the angular acceleration.

Solution: Let ω be the constant angular velocity during the later 5 seconds period of motion. Then

$$\omega = \frac{\text{angular displacement}}{\text{time}} = \frac{100}{5} = 20 \text{ rad/s}$$

If ω_0 is the initial angular velocity and α is the constant angular acceleration during the first 5 seconds of motion, then from the kinematic relations

$$\omega = \omega_0 + \alpha t \text{ and } \theta = \omega_0 t + \frac{1}{2}\alpha t^2$$

we have $$20 = \omega_0 + 5\alpha \quad \text{and} \quad 120 = 5\omega_0 + \frac{1}{2} \times \alpha \times 25$$

From these two expressions, we get

$$120 = 5(20 - 5\alpha) + \frac{1}{2}\alpha \times 25 = 100 - 25\alpha + \frac{25\alpha}{2}$$

or $$\frac{-25\alpha}{2} = 20$$

$$\therefore \quad \alpha = -\frac{2 \times 20}{25} = -1.6 \text{ rad/s}^2$$

and $$\omega_0 = 20 - 5\alpha = 20 - 5(-1.6) = 28 \text{ rad/s}$$

EXAMPLE 15.4.

A grinding wheel is attached to the shaft of an electric motor of rated speed 1800 rev/min. When the power is switched on, the unit attains the rated speed in 5 seconds. Assuming uniformly accelerated motion, determine the number of revolutions the wheel turns to attain the rated speed.

Solution: $\omega_0 = 0$; $\omega = \dfrac{2\pi N}{60} = \dfrac{2\pi \times 1800}{60} = 188.4$ rad/s

From the kinematic relation: $\omega = \omega_0 + \alpha t$, we have

$$\text{angular acceleration } \alpha = \frac{\omega - \omega_0}{t} = \frac{188.4 - 0}{5} = 37.68 \text{ rad/s}^2$$

Now from the relation: $\omega_2 - \omega_0^2 = 2\alpha\theta$

$$\text{total angular displacement } \theta = \frac{\omega^2 - \omega_0^2}{2\alpha} = \frac{188.4^2 - 0}{2 \times 37.68} = 471 \text{ radians}$$

Since, 1 revolution $\equiv 2\pi$ radians

$$\text{total revolutions turned by the wheel} = \frac{471}{2\pi} = 75$$

EXAMPLE 15.5.

A 5 m long slender bar is made to rotate through its one end in the horizontal plane about a vertical axis. The bar accelerates uniformly from 1000 rev/min to 1500 rev/min in 5 second period. Determine:

(a) linear velocity of the mid-point of the bar at the beginning and at the end of the time interval.

(b) normal and tangential accelerations of the mid point of the bar 3 seconds after the start of acceleration.

Solution: The mid-point M of the slender bar describes a circular path and its linear velocity is given by: $v = wr$. Let the subscripts b and e denote the beginning and end of the time interval.

At the start of 5*s* interval,

$$v_b = \frac{2\pi \times 1000}{60} \times 3 = 314 \text{ m/s} \qquad \left(\because \omega = \frac{2\pi N}{60}\right)$$

At the end of 5*s* interval,

$$v_e = \frac{2\pi \times 1500}{60} \times 3 = 417 \text{ m/s}$$

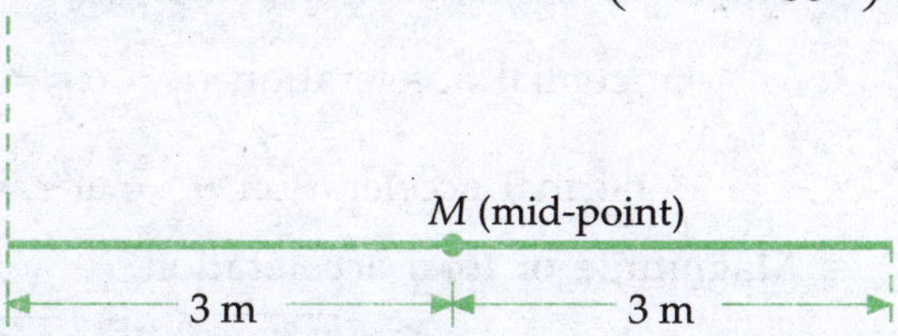

Fig. 15.3

$$\text{Angular acceleration } \alpha = \frac{\omega - \omega_o}{t}$$

$$= \frac{1}{5}\left[\frac{2\pi \times 1500}{60} - \frac{2\pi \times 1000}{60}\right] = 10.47 \text{ rad/s}^2$$

$$\text{After 3 seconds,} \quad \omega = \omega_o + \alpha t = \frac{2\pi \times 1000}{60} + 10.47 \times 3$$

$$= 136.08 \text{ rad/s}$$

The tangential and normal components of acceleration are:

$$\alpha_t = \alpha r = 10.47 \times 3 = 31.41 \text{ m/s}^2$$

$$\alpha_n = \omega^2 r = (136.08)^2 \times 3 = 55553.3 \text{ m/s}^2$$

EXAMPLE 15.6.

The piston of an engine is stated to make 360 strokes per minute. Determine the angular and linear speed of a point on the periphery of 1.8 m diameter flywheel fitted to the crank shaft of the engine.

Solution: Corresponding to two strokes of the piston, there is one revolution of the crank shaft.

$$\therefore \quad \text{Number of revolutions made by the crank shaft} = \frac{360}{2} = 180 \text{ rev/min}$$

Since the flywheel is rigidly fixed to the crankshaft,

Number of revolutions made by the flywheel = 180 rev/min

$$\therefore \quad \text{Angular speed of the flywheel } \omega = \frac{2\pi N}{60} = \frac{2\pi \times 180}{60} = \mathbf{18.84\ rad/s}$$

The linear velocity and the angular velocity are connected by the relation:

$$v = \omega r$$

$$\therefore \quad \text{Linear speed } v = 18.84 \times \left(\frac{1.8}{2}\right) = \mathbf{16.96\ m/s}$$

EXAMPLE 15.7.

A flywheel 0.5 m in diameter accelerates uniformly from rest to 360 rpm in 12 seconds. Determine the velocity and acceleration of a point on the rim of the flywheel 0.1 second after it has started from rest.

Solution: $\omega_0 = 0$; $\omega = \frac{2\pi N}{60} = \frac{2\pi\times 360}{60} = 37.68$ rad/s

From the kinematic relation: $\omega = \omega_0 + \alpha t$, we have

$$\text{angular acceleration } \alpha = \frac{\omega - \omega_0}{t} = \frac{37.68 - 0}{12} = 3.14 \text{ rad/s}$$

Angular velocity of the wheel 1 second after starting

$$\omega = \omega_0 + \alpha t = 0 + 3.14 \times 1 = 3.14 \text{ rad/s}$$

The linear velocity and the angular velocity are connected by the relation:

$$v = wr$$

$$\therefore \quad \text{linear speed } v = 3.14 \times \frac{0.5}{2} = 0.785 \text{ m/s}$$

The linear velocity will be tangential to the rim.

The components of acceleration are:

$$\text{tangential acceleration } \alpha_t = \alpha r = 3.14 \times \frac{0.5}{2} = 0.785 \text{ m/s}^2$$

$$\text{normal acceleration } \alpha_n = \omega^2 r = (3.14)^2 \times \frac{0.5}{2} = 1.232 \text{ m/s}^2$$

Magnitude of total acceleration,

$$\alpha = \sqrt{\alpha_t^2 + \alpha_n^2} = \sqrt{0.785^2 + 1.232^2} = \mathbf{1.461\ m/s^2}$$

Let ϕ be the angle between the total acceleration vector and the radius

$$\phi = \tan^{-1}\frac{\alpha_t}{\alpha_n} = \tan^{-1}\left(\frac{0.785}{1.232}\right) = \mathbf{32.5^\circ}$$

EXAMPLE 15.8

A shaft supported between two bearings carries a wheel of 1.5 m diameter. A constant moment is applied at the rim of the wheel and it attains a speed of 120 revolutions per minute in 10 minutes from rest. Determine:

(a) the peripheral velocity of the wheel,

(b) the angular acceleration of the wheel, and

(c) the number of revolutions made by the wheel

Solution: Angular velocity, $\omega = \frac{2\pi N}{60} = \frac{2\pi \times 120}{60} = 12.56$ rad/s

$$\text{Radius of the wheel, } r = \frac{1.5}{2} = 0.75 \text{ m}$$

The angular and peripheral velocities are connected by the identity: $v = \omega r$

$\therefore$ peripheral speed of the wheel $v = 12.56 \times 0.75 =$ **9.42 m/s**

(*b*) From the kinematic relation $\omega = \omega_0 + \alpha t$, we have

$$\text{angular acceleration } \alpha = \frac{\omega - \omega_0}{t} = \frac{12.56 - 0}{10 \times 60} = \mathbf{0.0209\ rad/s^2}$$

(*c*) From the kinematic relation $\omega^2 - \omega_0^2 = 2\alpha\theta$

$$\text{angular displacement } \theta = \frac{\omega^2 - \omega_0^2}{2\alpha} = \frac{12.56^2 - 0^2}{2 \times 0.0209} = 3774 \text{ radians}$$

Since 2π radians $\equiv$ 1 revolution

$$\text{Total number of revolutions make by the wheel} = \frac{3774}{2\pi} = \mathbf{601}$$

EXAMPLE 15.9.

The rotor of an electric motor accelerates uniformly to its rated speed of 3600 rev/min from rest in 15 seconds. Subsequently the power is turned off and the rotor decelerates to stop. If a total time of 75 seconds elapses from start to stop, determine the revolutions turned during acceleration and deceleration. Comment on the greater time taken during stopping.

Solution: During acceleration phase

$$\omega_0 = 0 \quad \text{and} \quad \omega = \frac{2\pi N}{60}\ \frac{2\pi \times 3600}{60} = 376.8 \text{ rad/s}$$

From the relation $\omega = \omega_0 + \alpha t$, we have

$$\alpha = \frac{\omega - \omega_0}{t} = \frac{376.80}{15} = 25.12 \text{ rad/s}$$

Angle θ covered in 15 seconds,

$$\theta = \omega_0 + \frac{1}{2}\alpha t^2$$

$$= 0 + \frac{1}{2} \times 25.12 \times 15^2 = 2826 \text{ rad}$$

$\therefore$ Number of revolutions taken to attain the rated speed,

$$n_{\text{accn}} = \frac{\theta}{2\pi} = \frac{2826}{2\pi} = 450 \text{ revolutions}$$

(*b*) During deceleration phase,

$$\omega_0 = 376.8 \text{ rad/s}; \quad \omega = 0 \quad \text{and} \quad t = 75 - 15 = 60 s$$

Again from the relation $\omega = \omega_0 + \alpha t$, we have

$$\alpha = \frac{\omega - \omega_0}{t} = \frac{0 - 376.8}{60} = -6.28 \text{ rad/s} \quad \text{(retardation)}$$

Angle turned during deceleration can be worked out from the relation

$$\omega^2 - \omega_0^2 = 2\alpha\theta$$

$$\text{That gives: } \theta = \frac{\omega^2 - \omega_0^2}{2\alpha} = \frac{0 - 376.8^2}{2 \times (-6.28)} = 11304 \text{ rad}$$

The corresponding revolutions taken by the rotor to come to stop

$$n_{\text{retd}} = \frac{11304}{2\pi} = 1800 \text{ revolutions}$$

The taking of greater time during stopping maybe attributed to the following facts:

- the starting velocity is more in the second case than in the first case
- the rotation of the rotor continues during stopping phase due to inertia of the rotor.

EXAMPLE 15.10.

A wheel accelerates uniformly from rest to a speed of 180 rpm in 0.5 seconds. It then rotates at that speed for 2 seconds before decelerating to rest in 0.3 seconds. Workout the revolutions made by the wheel during the entire time interval.

Solution: (*i*) During uniform acceleration (from t = 0 to t = 0.5 sec)

$$\omega_1 = \frac{2\pi \times 180}{60} = 18.84 \text{ rad/s}$$

$$\alpha_1 = \frac{\omega_1 - \omega}{t} = \frac{18.84 - 0}{0.5} = 37.68 \text{ rad/s}^2$$

From the kinematic relation $\omega^2 - \omega_0^2 = 2\alpha\theta$, we have

$$\theta_1 = \frac{\omega_1^2 - \omega_0^2}{2\alpha_1} = \frac{18.84^2 - 0^2}{2 \times 37.68} = 4.7 \text{ radians}$$

(*ii*) During uniform speed (from t = 0.5 sec to 2.5 sec)

$$\omega_2 = 18.84 \text{ rad/s}$$

$$\theta_2 = \omega_2 t = 18.84 \times 2 = 37.68 \text{ radians.}$$

(*iii*) During uniform deceleration (from t = 2.5 sec to 2.8 sec)

$$\omega_2 = 18.84 \text{ rad/s} \quad ; \quad \omega_3 = 0$$

$$\alpha_3 = \frac{\omega_3 - \omega_2}{t} = \frac{0 - 18.84}{0.3} = -62.8 \text{ rad/s}$$

From the kinematic relation $\omega^2 - \omega_0^2 = 2\alpha\theta$, we have

$$\theta_3 = \frac{\omega_3^2 - \omega_2^2}{2\alpha} = \frac{0^2 - 18.84^2}{2(-62.8)} = 2.83 \text{ radians}$$

$\therefore$ Total angular displacement $= \theta_1 + \theta_2 + \theta_3$

$$= 4.7 + 37.68 + 2.83 = 45.22 \text{ radians}$$

Since 1 revolution $\equiv 2\pi$ radians

$$\text{Total revolutions turned by the wheel} = \frac{45.22}{2\pi} = 7.2 \text{ revolutions}$$

Alternatively: Refer Fig. 15.4 which represents the velocity-time graph for the wheel. The area under the velocity-time graph gives the angle turned during the entire time-interval.

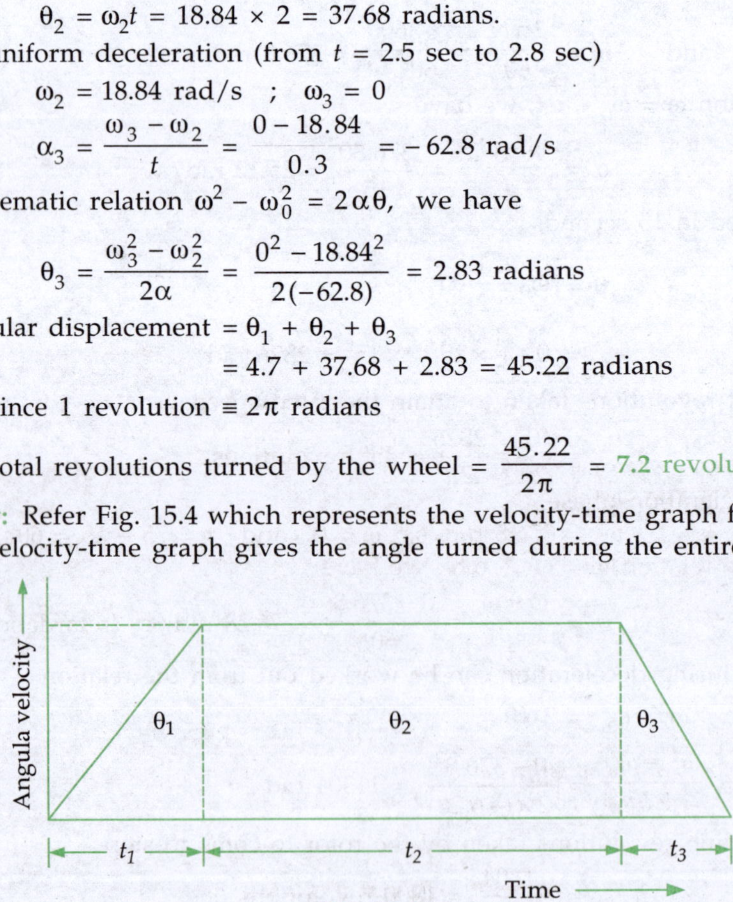

Fig. 15.4

$$\theta = \theta_1 + \theta_2 + \theta_3 = \left(\frac{1}{2} \times 0.5 \times 18.84\right) + (2 \times 18.84) + \left(\frac{1}{2} \times 0.3 \times 18.84\right)$$

$$= 4.71 + 37.68 + 2.83 = 45.22 \text{ radians}$$

$\therefore$ Total revolutions turned by the wheel $= \dfrac{45.22}{2\pi} =$ **7.2** revolutions

EXAMPLE 15.11.

A flywheel had an initial angular speed of 3000 rev/min in clockwise direction. When a constant turning moment was applied to the wheel, it got subjected to a uniform anticlockwise angular acceleration of 3 rev/sec^2. Determine the angular velocity of the wheel after 20 seconds, and the total number of revolutions made during this period.

Solution: Initial angular velocity, $\omega_0 = \dfrac{2\pi N}{60} = \dfrac{2\pi \times 3000}{60} = 314$ rad/s (clockwise)

Angular acceleration $\alpha = 3$ rev/s^2 $= 3 \times 2\pi = 18.84$ rad/s^2 (anticlockwise)

From the kinematic relation, $\omega = \omega_0 + \alpha t$, the angular velocity of the wheel after 20 seconds is

$$\omega = 314 + (-18.84) \times 20 = -62.8 \text{ rad/s}^2$$

The negative sign indicates that the wheel is turning anticlockwise.

Let the wheel change its direction of rotation after t_1 seconds. At the instant of change of direction, the velocity becomes zero. Accordingly

$$0 = 314 - 18.84 t_1 \; ; \quad t_1 = \frac{314}{18.84} = 16.67 \text{ s}$$

For $t_1 = 16.67$ seconds, the wheel would turn clockwise and for

$t_2 = (20 - t_1) = (20 - 16.67)$

$= 3.33$ seconds, the wheel would turn anticlockwise.

Clockwise rotation $\theta_1 = \omega_0 t_1 + \dfrac{1}{2}\alpha t_1^2 = 314 \times 16.67 + \dfrac{1}{2}(-18.84) \times 16.67^2$

$= 2617.67$ radians

$= \dfrac{2617.67}{2\pi} = 416.83$ revolutions

Anticlockwise rotation $\theta_2 = 0 - \dfrac{1}{2}(-18.84) \times 3.33^2$

$= 104.46$ radians $= \dfrac{104.46}{2\pi} = 16.63$ revolutions

$\therefore$ Gross number of revolutions $= 416.83 + 16.63 =$ **433.46 rev**

Net revolutions $= 416.83 - 16.63 =$ **400.20 rev (clockwise)**

EXAMPLE 15.12.

A flywheel rotating freely at 1800 rev/min in the clockwise direction is subjected to a variable counter clockwise torque which is intially applied at t = 0. The torque produces a counter-clockwise acceleration prescribed by the relation $\alpha - 3t$ rad/s^2 where t is the time in second. Make calculations for:

(a) the time required for the flywheel to reduce its clockwise angular speed to 100 rev/min;

(b) the time required for the flywheel to reverse its direction of motion.

Solution: Let it be presumed that the counter clockwise direction is positive.

$$\omega_1 = -\frac{2\pi \times 1800}{60} = -188.4 \text{ rad/s} \qquad \left(\because \omega = \frac{2\pi N}{60}\right)$$

$$\omega_2 = -\frac{2\pi \times 1000}{60} = -104.67 \text{ rad/s}$$

Angular acceleration $\alpha = \frac{d\omega}{dt} = 3t$ or $d\omega = 3t\,dt$. Integrating both sides within the prescribed limits.

$$\int_{-188.4}^{-104.67} d\omega = 3\int_0^t t\,dt$$

or $$-104.67 + 188.4 = \frac{3t^2}{2} \quad \therefore \quad t = \sqrt{\frac{2}{3} \times 83.73} = \mathbf{7.47\ sec}$$

(*b*) At reversal of direction, the angular speed will be zero. Therefore,

$$\int_{-188.4}^{0} d\omega = 3\int t\,dt$$

or $$188.4 = \frac{3t^2}{2} \quad \therefore \quad t = \sqrt{\frac{2}{3} \times 188.4} = \mathbf{11.21\ sec}$$

EXAMPLE 15.13.

A shaft is accelerated uniformly from 600 rev/min to 900 rev/min in 2 seconds. It continues accelerating at this rate for a further period of 4 seconds and then continues to rotate at the maximum speed attained. Make calculations for the time taken to complete the first 180 revolutions.

Solution: (*i*) During uniform acceleration from 600 rpm to 900 rpm (t = 2 sec)]

$$\omega_0 = \frac{2\pi \times 600}{60} = 62.8 \text{ rad/s} \qquad \left(\because \omega = \frac{2\pi N}{60}\right)$$

$$\omega_1 = \frac{2\pi \times 900}{60} = 94.2 \text{ rad/s}$$

$$\alpha_1 = \frac{\omega_1 - \omega_0}{t} = \frac{94.2 - 62.8}{2} = 15.7 \text{ rad/s}^2$$

The angular displacement θ_1 for the entire accelerating period of 6 seconds can be worked out by using the kinematic relation $\theta = \omega t + \frac{1}{2}\alpha t^2$. That is

$$\theta_1 = 62.8 \times 6 + \frac{1}{2} \times 15.7 \times 6^2 = 659.4 \text{ radians}$$

Since 1 revolution ≡ 2π radians,
Revolutions turned by the wheel during the accelerating period

$$= \frac{659.4}{2\pi} = 105 \text{ revolutions.}$$

(*ii*) Revolutions to be made by the shaft at uniform speed attained at the end of accelerating period = 180 – 105 = 75 revolutions

Corresponding angular displacement = 75 × 2π = 471 radians

Angular velocity at the end of acceleration period,

$$= \omega_0 + \alpha t = 62.8 + 15.7 \times 6 = 157 \text{ rad/s}$$

Time taken to turn 471 radians at uniform speed of 157 rad/s

$$t = \frac{\theta}{\omega} = \frac{471}{157} = 3 \text{ seconds}$$

$\therefore$ Required time = 2 + 4 + 3 = 9 seconds

EXAMPLE 15.14.

The motion of a gear is defined by the relation $\theta = 2t^3 - 5t^2 + 8t + 20$ where θ is in radians and t is in seconds.

Determine the angular displacement, the angular velocity and the angular acceleration at time t = 3 seconds.

Solution: Given that $\theta = 2t^3 - 5t^2 + 8t + 20$ ...(*i*)

Differentiating the above expression with respect to t, we get

$$\frac{d\theta}{dt} = 6t^2 - 10t + 8 \quad \text{or} \quad \omega = 6t^2 - 10t + 8 \qquad ...(ii)$$

where $\omega = \frac{d\theta}{dt}$ is the angular velocity of the rotating gear.

Differentiation of identity (*ii*) with respect to t gives:

$$\frac{d\omega}{dt} = 12t - 10 \quad \text{or} \quad \alpha = 12t - 10 \qquad ...(iii)$$

where α = angular acceleration of the rotating gear.

Then at t = 5 seconds we have:

Angular displacement $\theta = 2 \times (5)^3 - 5 \times (5)^2 + 8 \times 5 + 20$

$= 250 - 125 + 40 + 20 =$ **185 radians**

Angular velocity $\omega = 6 + (5)^2 - 10 \times 5 + 8 = 150 - 50 + 8 =$ **108 rad/s**

Angular acceleration $\alpha = 12 \times 5 - 10 =$ **50 rad/s^2**

15.3 GENERAL PLANE MOTION

Generally a body undergoes the following three types of plane motions :

• **Translation :** The particles constituting the rigid body move in parallel planes and travel the same distance. During translation, the particles have the same velocity and acceleration, and a straight line drawn on the moving body remains parallel to its original position at any time.

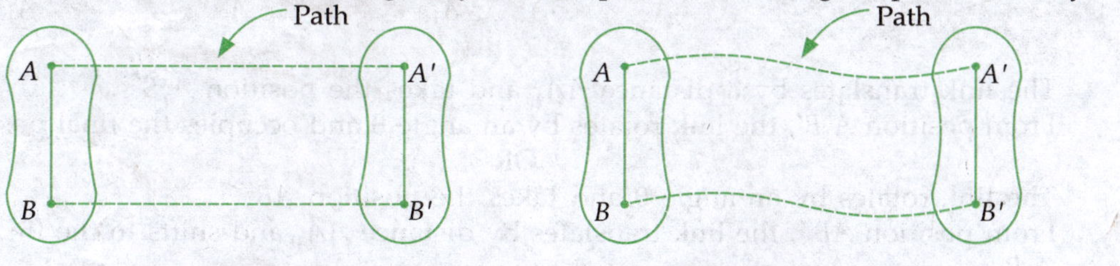

(a) Rectilinear translation *(b) Curvilinear translation*

Fig. 15.5

If the path traced by the particle during motion is a straight line, then the motion is said to be rectilinear translation. If the particle traces a curved path, the motion is called curvilinear translation.

• **Rotation :** The body rotates about a fixed point and all the particles constituting the body move in a circular path. The fixed point about which the body rotates is called the point of rotation and the axis passing through the point of rotation is called the axis of rotation. A point lying on the axis of rotation has a zero velocity and zero acceleration.

The point of rotation may lie inside the body (Fig. 15.6 *a*) or outside the body (Fig. 15.6 *b*)

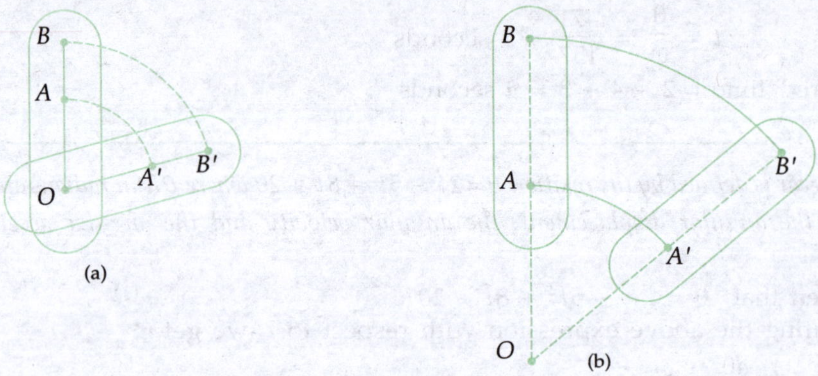

Fig. 15.6

• **General plane motion** (combined motion of translation and rotation) : There exist certain situations where a body possesses both motions of translation and rotation simultaneously at a particular instant. Examples of such a combined motion are :

— motion of roller without slipping
— motion of the wheel of a locomotive train, truck and car etc.
— a rod sliding against a wall at one end and floor at the other end.

The combination of rotation and translation motion is usually referred to as general plane motion.

With reference to Fig. 15.7, a link AB shifts to a new position at any instant. A link is a rigid piece of material joining two parts of a machine. This shift of the link can be considered in two parts:

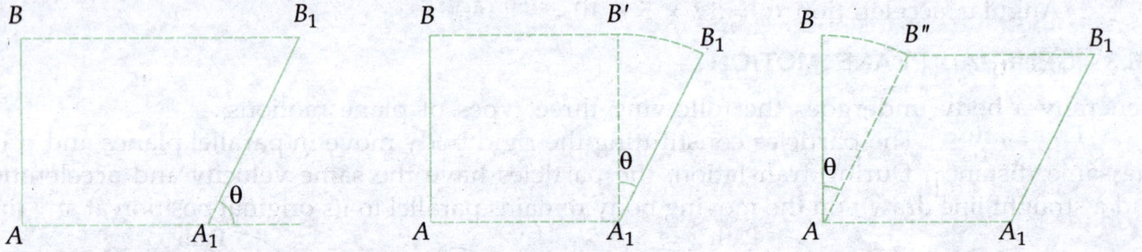

Fig. 15.7.

(*i*) The link translates by a distance AA_1 and takes the position A_1B'
(*ii*) From position A_1B', the link rotates by an angle θ and occupies the final position A_1B_1

OR

(*a*) The link rotates by an angle θ and takes the position AB''
(*b*) From position AB'', the link translates by distance AA_1 and shifts to the final position A_1B_1.

It is to be noted that the shift of the link from the initial to final position will remain same irrespective of the fact whether the motion of translation or that of rotation takes place first.

Chasle's theorem states *that any general displacement of a rigid body can be represented by a combination of translatory motion and rotational motion*.

EXAMPLE 15.15.

Consider a wheel rolling on a horizontal surface, and a bar sliding against wall at one end and the floor at the other end. Illustrate how these acts can be split into motions of translation and rotation.

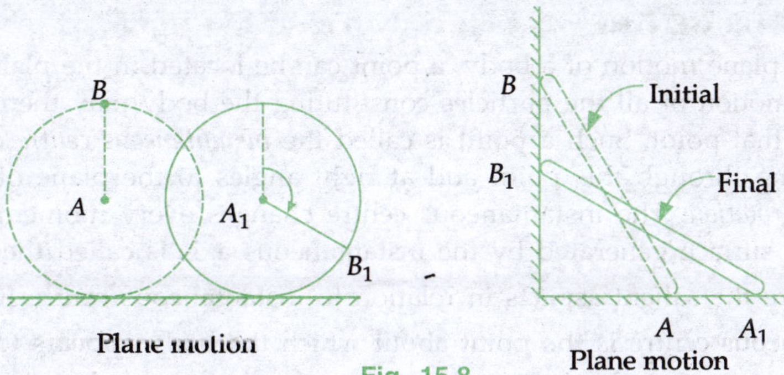

Fig. 15.8

Solution : Case (*a*) : Wheel rolling on a horizontal surface

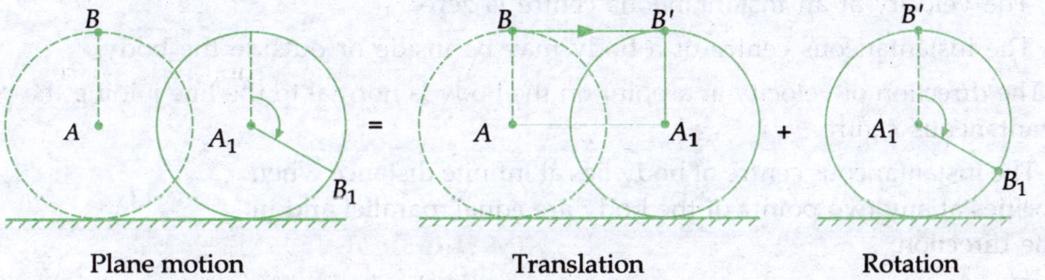

Fig. 15.9

With reference to Fig. 15.9, the general plane motion of a wheel rolling on a horizontal surface from initial position AB to final position A_1B_1 can be replaced by the following motions taken in order

(*i*) motion of translation from AB to A_1B'

(*ii*) rotation about A_1 from position A_1B' to A_1B_1

Case (*b*) : Bar sliding against wall at one end and floor at the other end

With reference to Fig. 15.10, the general plane motion of a bar sliding against wall at one end and floor at the other end from initial position AB to A_1B_1 can be replaced by the following motions taken in order

(*i*) motion of translation of the bar from position AB to A_1B'

(*ii*) rotation about A_1 from position A_1B' to A_1B_1

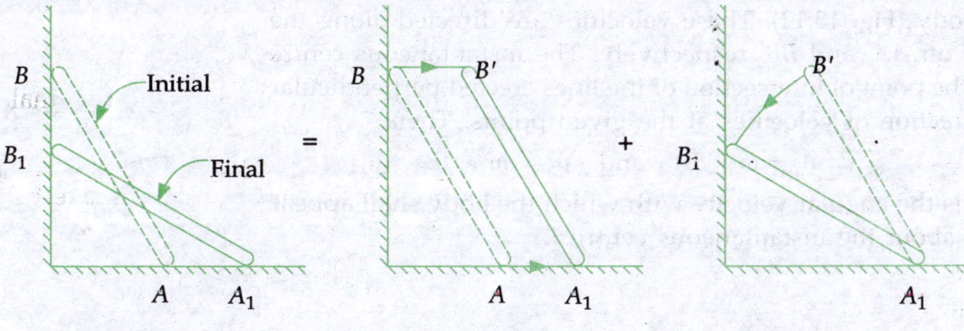

Fig. 15.10

15.4 INSTANTANEOUS CENTRE

While analysing plane motion of a body, a point can be located in the plane which has zero velocity. The plane motion of all the particles constituting the body may then be considered as pure rotation about that point. Such a point is called the *instantaneous centre* or *virtual centre of body*. The axis passing through this point and at right angles to the plane of motion is called *instantaneous axis of rotation*. The instantaneous centre changes every moment, and its locus is called centrode. The surface generated by the instantaneous axis is called the *axode*.

The following are the salient aspects in relation to instantaneous centre IC.

(*i*) The instantaneous centre is the point about which the body appears to rotate.

(*ii*) The instantaneous centre is not a fixed point but changes from one instant to another as the body rotates.

(*iii*) The velocity at an instantaneous centre is zero.

(*iv*) The instantaneous centre of a body may be inside or outside the body.

(*v*) The direction of velocity at a point on the body is normal to the line joining the point and the instantaneous centre.

(*vi*) The instantaneous centre of body lies at infinite distance when the velocities at any two points of the body are equal, parallel and in the same direction.

(*vii*) For a centre moving over a plane, instantaneous centre is the point of contact of the cylinder with the plane.

Depending on the given situation, following are some of the methods used to determine the instantaneous centre:

(*i*) Let v_a and ω be the linear and angular velocities of a point *A* on a rigid body (Fig. 15.11). Since ω and v_a are connected by the expression $v = \omega r$, the instantaneous centre *I* then lies at a distance $\frac{v_a}{\omega}$ along the perpendicular to the direction of velocity v_a at point *A*.

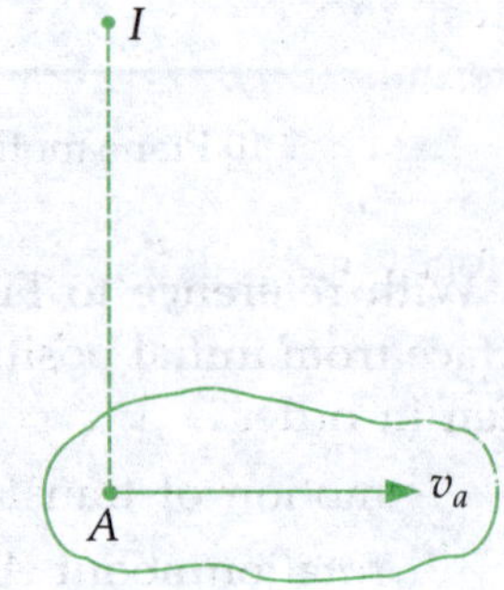

Fig. 15.11

$$IA = \frac{v_a}{\omega}$$

(*ii*) Let v_a and v_b be the linear velocities at points *A* and *B* on a rigid body (Fig. 15.12). These velocities are directed along the direction on *AA*′ and *BB*′ respectively. The instantaneous centre *I* is then the point of intersection of the lines erected perpendicular to the direction of velocities at the given points. Then

$$v_a = \omega \times IA \quad \text{and} \quad v_b = \omega \times IB$$

where ω is the angular velocity with which the body shall appear to rotate about the instantaneous centre *I*.

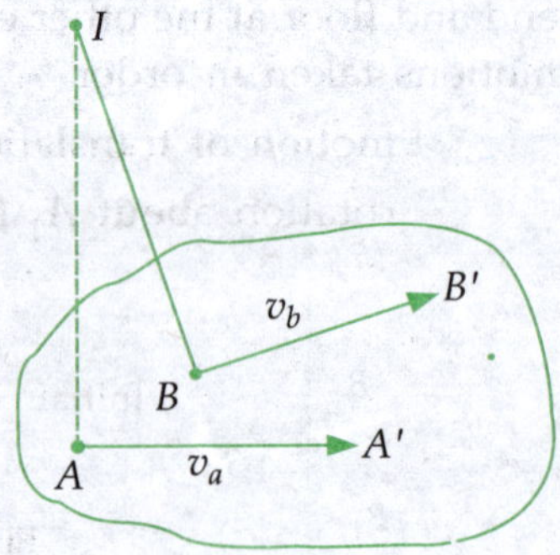

Fig. 15.12

(*iii*) Consider a solid circular cylinder in contact with two horizontal conveyer belts running with different velocities in the same or opposite directions as shown in Fig. 15.13.

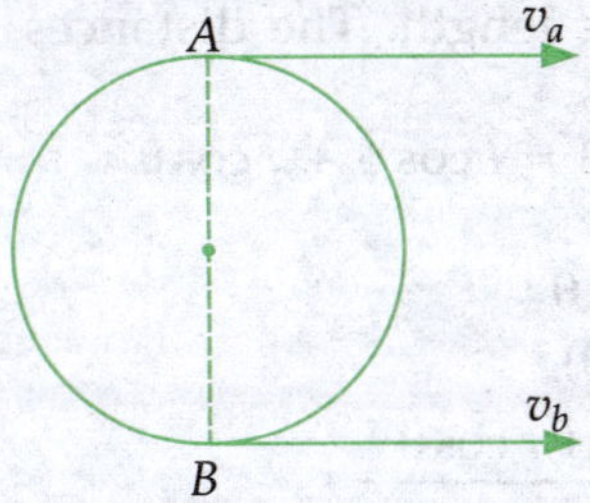

Fig. 15.13

The instantaneous centre is found by determining the point of intersection of line *AB* with the line joining the extremities of the velocities vectors (Fig. 15.14). It is to be noted that when the velocities v_a and v_b are in the same direction, *I* lies outside *AB* and divides it externally in the ratio v_a to v_b.

When velocities v_a and v_b are in opposite directions, *I* lies between *A* and *B*, and divides *AB* internally in the ratio v_a to v_b.

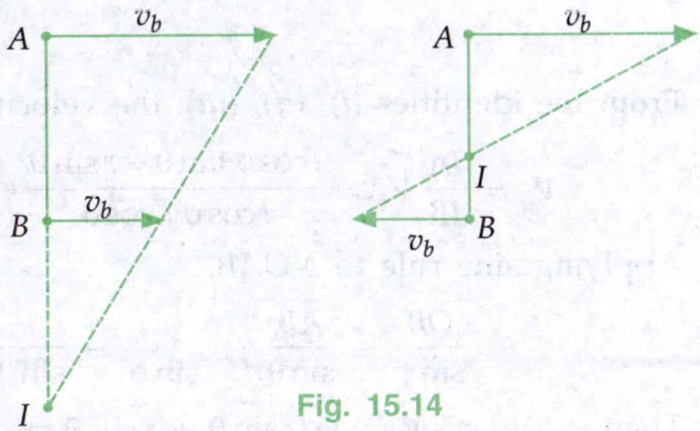

Fig. 15.14

Every point on the cylinder shall appear to rotate about the instantaneous centre *I* with an angular velocity ω, and therefore

$$\omega = \frac{v_a}{IA} = \frac{v_b}{IB}$$

(*iv*) For the piston-connecting rod assembly (Fig. 15.15)

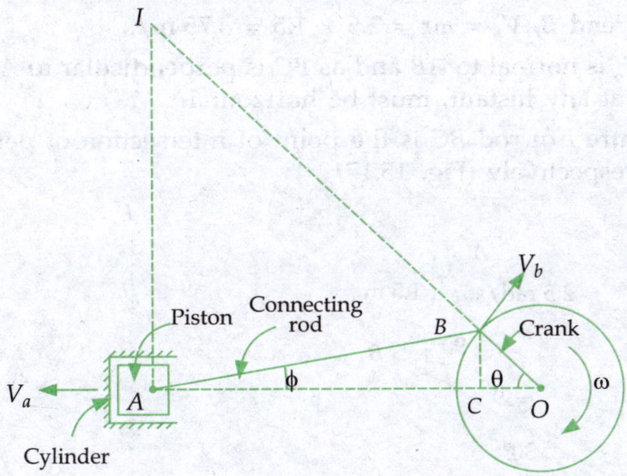

Fig. 15.15

The point *B* moves in a circle whose centre is *O*, and the direction of V_b is perpendicular to *OB*. The velocity V_a of the piston, at any instant, must be horizontal. The instantaneous centre *I* of the connecting rod *AB* is the point of intersection of perpendiculars to V_a and V_b through *A* and *B* respectively. Then angular velocity ω_{ab} of the connecting rod about *I* is

$$\omega_{ab} = \frac{V_a}{IA} = \frac{V_b}{IB} = \frac{\omega r}{IB} \quad \text{...}(i)$$

where ω is angular velocity of the crank and r is its length. The distances IA and IB can be located as under :

$$AO = AC + CO = AB\cos\phi + BO\cos\theta = l\cos\phi + r\cos\theta$$

where l is the length of connecting rod.

$$IA = AO\tan\theta = (l\cos\phi + r\cos\theta)\tan\theta$$
$$= (l\cos\phi\tan\theta + r\sin\theta) \quad \text{...}(ii)$$

$$IB = IO - BO = \frac{AO}{\cos\theta} - BO = \frac{l\cos\phi + r\cos\theta}{\cos\theta} - r$$

$$= \frac{l\cos\phi}{\cos\theta} + r - r = \frac{l\cos\phi}{\cos\theta} \quad \text{...}(iii)$$

From the identities (*i*), (*ii*), (*iii*), the velocity V_a of the piston can be written as

$$V_a = \frac{IA}{IB}V_b = \frac{l\cos\phi\tan\theta + r\sin\theta}{l\cos\phi/\cos\theta} \times \omega r = \left(\cos\theta\tan\theta + \frac{r}{l}\frac{\sin\theta\cos\theta}{\cos\phi}\right)\omega r$$

Applying sine rule to $\Delta\, OAB$,

$$\frac{OB}{\sin\phi} = \frac{AB}{\sin\theta}; \quad \frac{r}{\sin\phi} = \frac{l}{\sin\theta}; \quad \frac{r}{l} = \frac{\sin\phi}{\sin\theta}$$

Then $\quad V_a = \omega\,(r\sin\theta + r\cos\theta\tan\phi) = \omega\,(l\sin\phi + r\cos\theta\tan\phi) \quad \text{...(15.4)}$

EXAMPLE 15.16

At the instant shown in Fig. 15.16, the rod AB is rotating clockwise at 2.5 rad/sec. If end C of the rod BC is free to move on a horizontal surface, make calculations for the angular velocity of rod BC and the velocity of its end point C.

Solution : Velocity of end B, $V_b = \omega r = 2.5 \times 1.5 = 3.75$ m/s

The direction of V_b is normal to AB and as BC is perpendicular to AB, it lies along BC. The velocity V_c of end C, at any instant, must be horizontal.

The instantaneous centre I of rod BC is the point of intersection of perpendiculars to V_b and V_c through B and C respectively (Fig. 15.17)

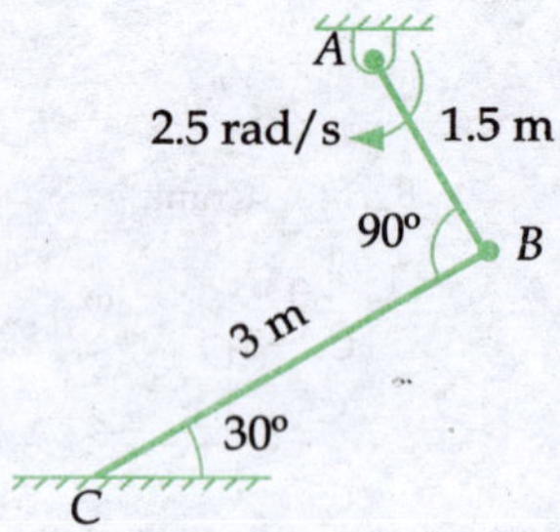

Fig. 15.16

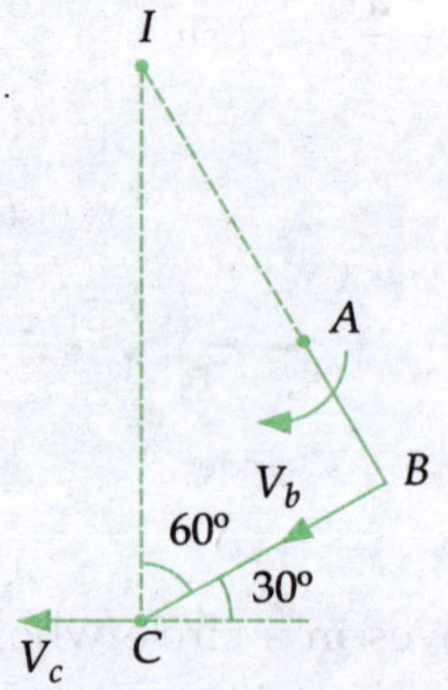

Fig. 15.17

$$IC = \frac{BC}{\cos 60^\circ} = \frac{3}{0.5} = 6 \text{ m}$$

$$IB = IC\sin 60^\circ = 6 \times 0.866 = 5.196 \text{ m}$$

Angular velocity of rod *BC*,

$$\omega = \frac{V_b}{IB} = \frac{V_c}{IC}$$

$$\omega = \frac{3.75}{5.196} = 0.722 \text{ rad/s (clockwise)}$$

and $\quad V_c = \omega \times IC = 0.722 \times 6 = 4.33 \text{ m/s}$

EXAMPLE 15.17.

In a crank and connecting rod mechanism, the crank is 300 mm long and the connecting rod 1500 mm long. If the crank rotates uniformly at 300 rpm, find the velocity of the cross-head when the crank is inclined at 30° with the inner dead centre.

Solution: Refer Fig. 15.18 for the crank and connecting rod mechanism.

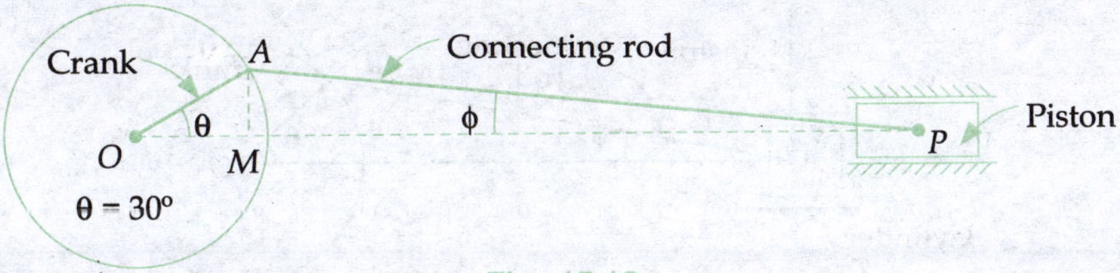

Fig. 15.18.

Crank radius $r = 300$ mm

Length of connecting rod $l = 1500$ mm

From the law of triangles: $\dfrac{l}{\sin\theta} = \dfrac{r}{\sin\phi}$

$$\sin\phi = \frac{r\sin\theta}{l} = \frac{300 \times \sin 30^\circ}{1500} = 0.1$$

That gives: $\phi = 5.74^\circ$

Let x denote the distance between the crank centre O and the piston point P. Then

$$x = OM + OP = r\cos\theta + l\cos\phi$$

$$\frac{dx}{dt} = r\sin\theta \times \frac{d\theta}{dt} + l\sin\phi \times \frac{d\phi}{dt}$$

$$= r\sin\theta\frac{d\theta}{dt} + l\sin\phi\,\frac{d\phi}{d\theta} \times \frac{d\theta}{dt}$$

$$= r\sin\theta \times \omega + l\sin\phi \times \left(\frac{r}{l}\frac{\cos\theta}{\cos\phi}\right) \times \omega \qquad \text{...}(a)$$

Angular velocity $\omega = \dfrac{2\pi N}{60} = \dfrac{2\pi \times 300}{60} = 31.4$ rad/s

Substituting the appropriate values in expression (*a*), we get velocity of cross-head,

$$\frac{dx}{dt} = 300\sin 30^\circ \times 31.4 + 1500 \times \sin 5.74^\circ \times \left(\frac{300}{1500} \times \frac{\cos 30^\circ}{\cos 5.74^\circ}\right) \times 31.4$$

$$= 4710 + 820 = 5530 \text{ mm/s} = 5.53 \text{ m/s}$$

EXAMPLE 15.18.

In a reciprocating engine mechanism (Fig. 15.21), the lengths of the crank OB and connecting rod AB are 30 cm and 1 metre respectively. If the crank is rotating at a constant angular velocity of 200 rev/min, determine (a) the angular velocity of connecting rod and (b) the velocity of the piston when the crank makes an angle of θ = 45° with the horizontal.

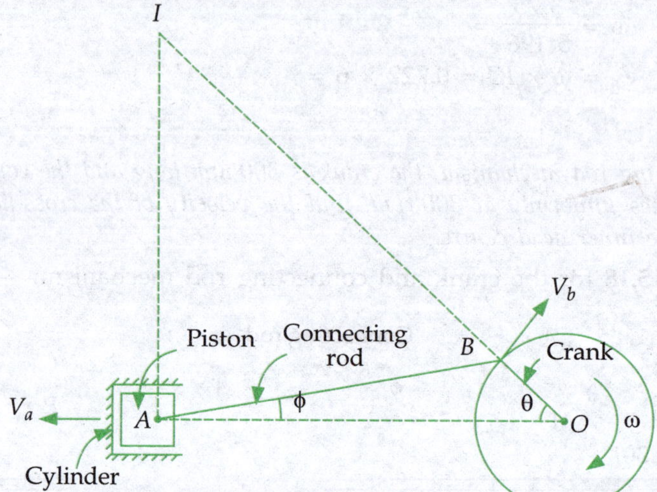

Fig. 15.19

Solution : Angular velocity of crank $\omega = \dfrac{2\pi N}{60} = \dfrac{2\pi \times 200}{60} = 20.93$ rad/s

$$V_b = \omega r = 20.93 \times 0.3 = 6.279 \text{ m/s}$$

The point *B* moves in a circle whose centre is *O*, and the direction of V_a is perpendicular to *OB*. The velocity V_a of the piston, at any instant must be horizontal. The instantaneous centre *I* of the connecting rod *AB* is the point of intersection of perpendiculars to V_a and V_b through *A* and *B* respectively. Then angular velocity ω_{ab} of the connecting rod about *I* is

$$\omega_{ab} = \frac{V_a}{IA} = \frac{V_b}{IB} = \frac{\omega r}{IB} \qquad \text{...}(i)$$

Let ϕ be the angle between the connecting *AB* and the line *AO*. Then applying sine rule to Δ *OAB*

$$\frac{OB}{\sin\phi} = \frac{AB}{\sin\theta}; \quad \sin\phi = \frac{OB}{AB}\sin\theta = \frac{0.3}{1} \times \sin 45^\circ = 0.2121$$

$$\therefore \quad \phi = 12.25^\circ$$

The distance *IA* and *IB* are worked out from the relations

$$IA = l\cos\phi\,\tan\theta + r\sin\theta = (1 \times \cos 12.25^\circ \times \tan 45^\circ) + 0.3\sin 45^\circ$$

$$= 0.9772 + 0.2121 = 1.1843 \text{ m}$$

$$\text{and} \quad IB = \frac{l\cos\phi}{\cos\theta} = 1 \times \frac{\cos 12.25^\circ}{\cos 45^\circ} = 1.382 \text{ m}$$

Then from identity (*i*)

angular velocity of connecting rod, $\omega_{ab} = \dfrac{6.279}{1.382} = $ **4.543 rad/s**

velocity of piston $V_a = \omega_{ab} \times IA = 4.543 \times 1.1843 = $ **5.38 m/s**

EXAMPLE 15.19.

A cylinder of diameter 3 m rolls without slipping along a horizontal surface PQ as shown in Fig. 15.20. If its centre has a uniform velocity of 30 m/s, determine the velocities of the points B and D lying on the rim of the cylinder.

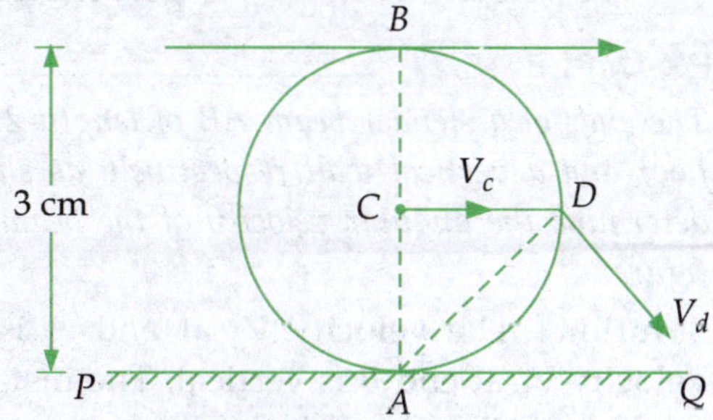

Fig. 15.20

Solution : When the cylinder rolls without slipping, its point of contact with the horizontal surface at any instant has zero velocity. Thus the point A is instantaneously at rest and every other point on the cylinder has angular velocity about A. Hence A is instantaneous centre of the cylinder. Then

$$V_c = \omega \times AC; \quad V_B = \omega \times AB; \quad V_d = \omega \times AD$$

$$\text{Angular velocity of the cylinder } \omega = \frac{V_c}{AC} = \frac{30}{3/2} = 20 \text{ rad/s}$$

$$\text{Distance } AD = \sqrt{(DC)^2 + (AC)^2} = \sqrt{1.5^2 + 1.5^2} = 2.121 \text{ m}$$

∴ Velocity of point B, $V_b = \omega \times AB = 20 \times 3 =$ **60 m/s**

Velocity of point D, $V_d = \omega \times AD = 20 \times 2.121 =$ **42.42 m/s**

EXAMPLE 15.20.

A 30 cm diameter cylinder rolls without slipping along a horizontal floor. The cylinder has its contact with the floor at point D, and its centre point C has a uniform velocity of 4.5 m/s. Determine the velocities of points A and B on the circumference of the cylinder (Fig. 15.21).

Solution : Choosing centre of cylinder C as the pole, angular velocity ω of the wheel is

$$\omega = \frac{V_c}{r} = \frac{4.5}{0.15} = 30 \text{ rad/s}$$

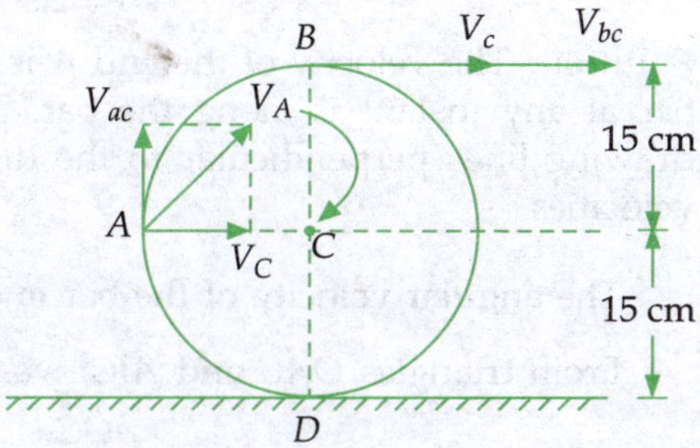

Fig. 15.21

Velocity $\vec{V_A}$ at point A

= velocity of C + velocity of A with respect to C

$$= \vec{V_C} + \overrightarrow{(\omega \times CA)} = \overrightarrow{4.5} + \overrightarrow{(30 \times 0.15)}$$

$$= \overrightarrow{4.5} + \overrightarrow{4.5} = \sqrt{4.5^2 + 4.5^2} = 6.36 \text{ m/s}$$

The inclination α of this velocity with horizontal is

$$\alpha = \tan^{-1}\left(\frac{4.5}{4.5}\right) = 45°$$

$$\text{Likewise: } \vec{V_B} = \vec{V_c} + \vec{V_{bc}} = \vec{V_c} + \overrightarrow{(\omega + CB)}$$

$$= \overrightarrow{4.5} + \overrightarrow{(30 \times 0.15)} = \overrightarrow{4.5} + \overrightarrow{4.5} = 9.0 \text{ m/s}$$

Here addition has been done mathematically becuase both the velocity components lie along the same direction. The velocity vector V_B is along the horizontal direction.

Alternatively: The cylinder rolls wihtout slipping and accordingly at any instant, the contact point D has zero velocity and represents the instantaneous centre. Therefore,

$$V_A = \omega \times (DA) = 30 \times \sqrt{0.15^2 + 0.15^2} = 30 \times 0.212$$
$$= 6.36 \text{ m/s}$$

The velocity vector V_A is perpendicular to DA and at 45° to the horizontal.

and $$V_B = \omega \times (DB) = 30 \times 0.3 = 9.0 \text{ m/s}.$$

The velocity vector V_B is perpendicular to DB, *i.e.*, along the horizontal.

EXAMPLE 15.21.

The ends of a slender beam AB of length 2.5 m are constrained to remain in contact with a horizontal floor and a vertical wall respectively as shown in Fig. 15.22. If the end A has a velocity of 1.5 m/s, determine the angular velocity of the beam and the velocity of its end B at the position shown in the figure.

Solution : The velocity V_a at end A is horizontal and the velocity V_b at end B is vertical. The instantaneous centre I of the beam AB is the point of intersection of perpendicular to V_a and V_b through A and B respectively.

$$IA = AB \sin 60^\circ = 2.5 \times 0.866 = 2.165 \text{ m}$$

$$IB = AB \cos 60^\circ = 2.5 \times 0.5 = 1.25 \text{ m}$$

Then angular velocity ω_{ab} of the beam is given by

$$\omega_{ab} = \frac{V_a}{IA} = \frac{1.5}{2.165} = 0.693 \text{ rad/s}$$

$\therefore$ Velocity of end B is

$$V_b = \omega_{ab} \times IB = 0.693 \times 1.25$$

$$= \mathbf{0.866 \text{ m/s}}$$

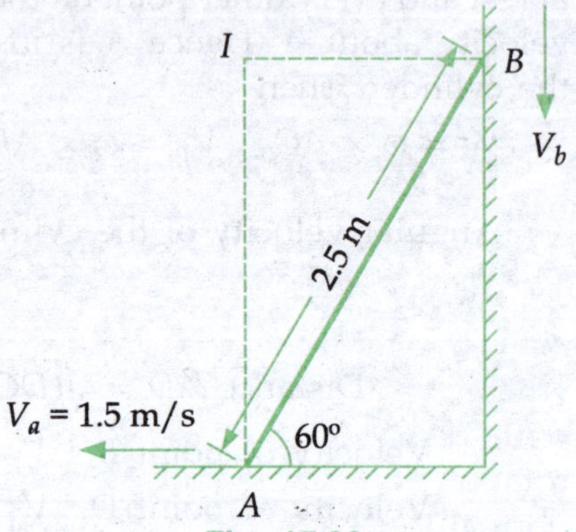

Fig. 15.22

EXAMPLE 15.22.

What is kinetics? Define the terms impulse and angular momentum.

A bar AB rests at the edge of a wall at some point C with its end A resting on a horizontal floor as shown in Fig. 15.23. If the end A moves with a constant velocity V_a, set up an expression for the angular velocity of the bar in terms of h and θ. Proceed to calculate the angular velocity if

$$V_a = 2.5 \text{ m/s}; \quad h = 4 \text{ m} \quad \text{and} \quad \theta = 30^\circ$$

Solution: The velocity of the end A is along the horizontal line OA and the velocity of C on the bar at any instant is along the bar. The position of instantaneous centre I is determined by drawing lines perpendicular to the direction of these velocities.

The angular velocity of the bar $\omega = \dfrac{V_a}{IA}$ ...(*i*)

From triangles OAC and AIC, we have

$$AC = \frac{OC}{\sin\theta};$$

$$AC = IA \cos(90 - \theta)$$

$$= IA \sin\theta$$

$$\therefore \quad \frac{OC}{\sin\theta} = IA \sin\theta$$

or $$IA = \frac{OC}{\sin^2\theta} = \frac{h}{\sin^2\theta}$$

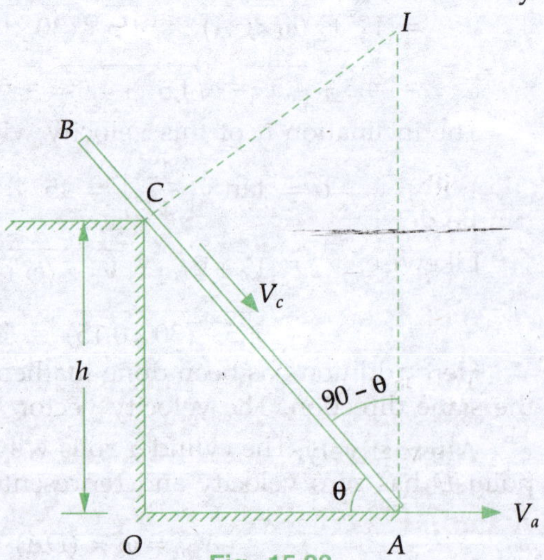

Fig. 15.23

Substituting $IA = \dfrac{h}{\sin^2 \theta}$ in expression (*i*), we get

$$\omega = \frac{V_a \sin^2 \theta}{h} = \frac{2.5 \sin^2 30}{4} = 0.1562 \text{ rad/s}$$

EXAMPLE 15.23.

A bar of length 2 m has its ends A and B constrained to move horizontally and vertically as shown in Fig. 15.22. The end A moves with a constant velocity of 6 m/s horizontally. Make calculations for:

(*a*) *the angular velocity of the bar AB,*

(*b*) *the velocity of the end B, and*

(*c*) *the velocity of the midpoint of the bar at the instant when the bar makes an angle of 30° with the horizontal.*

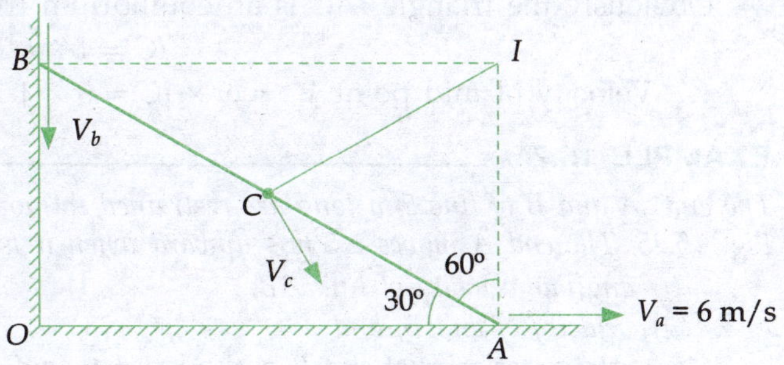

Fig. 15.24

Solution: Velocity of end A, $V_a = 6$ m/s

Velocity of end B, $V_b = V_a + V_{ba}$

where V_{ba} is the relative velocity of end B with respect to end A and it is due to rotation of bar AB about A. This relative is normal to bar AB at point B and it equals ωl.

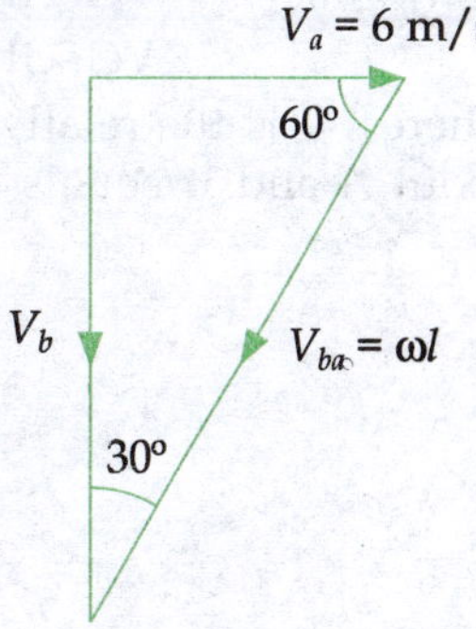

From the vector diagram,

$$\frac{V_a}{\sin 30°} = \frac{V_{ba}}{\sin 90°} = \frac{V_b}{\sin 60°}$$

(*i*) $V_b = \dfrac{V_a}{\sin 30°} \times \sin 60° = \dfrac{6}{0.5} \times 0.866$

$= 10.39$ m/s

(*ii*) $V_{ba} = \dfrac{V_a}{\sin 30°} \times \sin 90° = \dfrac{6}{0.5} \times 1 = 12$ m/s

$\therefore$ Angular velocity of bar, $\omega = \dfrac{V_{ba}}{l} = \dfrac{12}{2} = 6$ rad/s

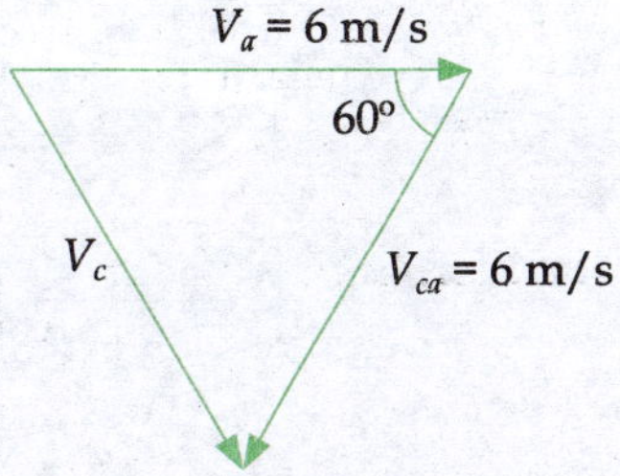

(*iii*) Velocity of midpoint C of the bar

$$\vec{V}_c = \vec{V}_a + \vec{V}_{ca}$$

$$V_a = 6 \text{ m/s and } V_{ca} = \omega \frac{l}{2} = 6 \times \frac{2}{2} = 6 \text{ m/s}$$

With reference to vector diagram, the magnitude of velocity V can be determined by using the cosine relation

$$V_c = \sqrt{V_a^2 + V_{ca}^2 - 2V_a V_{ca} \cos 60°}$$

$$= \sqrt{6^2 + 6^2 - 2 \times 6 \times 6 \times 0.5} = \sqrt{36} = 6 \text{ m/s}$$

This is also apparent from the vector diagram which is an equilateral triangle.

Alternatively the problem can be worked out by the instantaneous centre method.

The velocity V_a at end A is horizontal and the velocity V_b at end B is vertical. The instantaneous centre I of the bar AB is the point of intersection of perpendicular to V_a and V_b through A and B respectively

$$IA = BO = AB\sin 30° = 2\sin 30° = 1 \text{ m}$$
$$IB = AO = AB\cos 30° = 2\cos 30° = 1.732 \text{ m/s}$$

Then the angular velocity of the bar is given by

$$\omega_{ab} = \frac{V_a}{IA} = \frac{6}{1} = 6 \text{ rad/s}$$

(*i*) Velocity of end $B = \omega \times IB = 6 \times 1.732$

$= $ **10.392 m/s**

(*ii*) In triangle $IAC; AC = 1$ m; $IA = 1$ m (calculated above) and $\angle IAC = 60°$

Obviously the triangle IAC is an equilibrium triangle. That gives

$$IC = IA = 1 \text{ m}$$

∴ Velocity of mid point $V_c = \omega \times IC = 6 \times 1 =$ **6 m/s**

EXAMPLE 15.24.

The ends A and B of line 5 m long are restrained to move in vertical and horizontal guides as shown in Fig. 15.25. The end A moves 2.5 m/s upward when it is 3 m above C. Determine:

(*i*) *angular velocity of link AB,*
(*ii*) *velocity of end B, and*
(*iii*) *relative velocity of end B with respect to end A.*

Solution: Velocity of end A: $V_a = 2.5$ m/s

Velocity of end B: $V_b = V_a + V_{ba}$

where V_{ba} is the relative velocity of end B with respect to end A and it is due to rotation of link about A and it equals ωl.

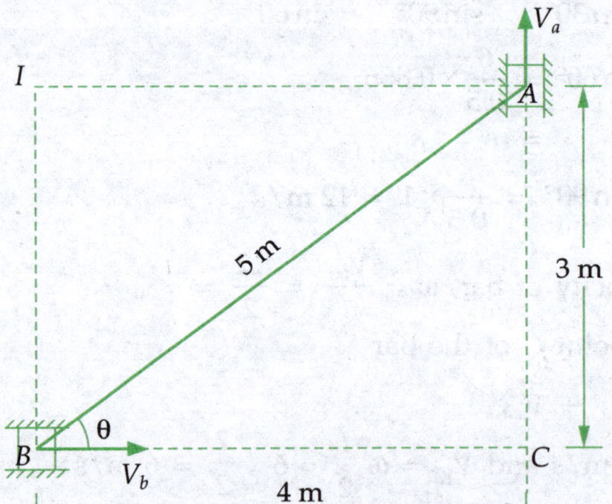

Fig. 15.25

From the vector diagram,

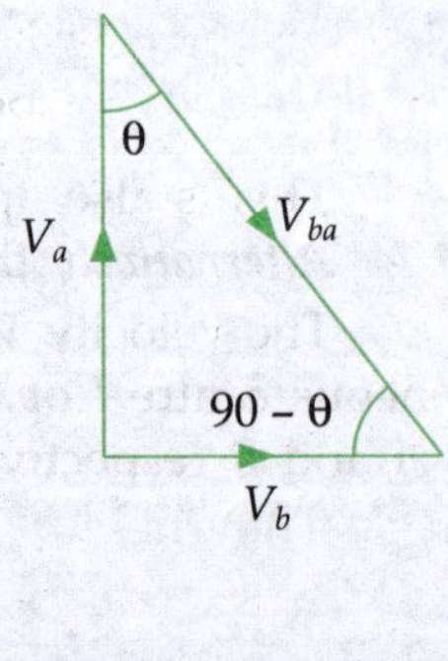

$$\frac{V_a}{\sin(90-\theta)} = \frac{V_{ba}}{\sin 90°} = \frac{V_b}{\sin\theta}$$

or
$$\frac{V_a}{\cos\theta} = \frac{V_{ba}}{\sin 90°} = \frac{V_b}{\sin\theta}$$

From the given geometry: $\sin\theta = \frac{3}{5} = 0.6$ and $\cos\theta = \frac{4}{5} = 0.8$

Then: (i) $V_b = \dfrac{V_a}{\cos\theta} \times \sin\theta = \dfrac{2.5}{0.8} \times 0.6 = $ **1.875 m/s**

(ii) $V_{ba} = \dfrac{V_a}{\cos\theta} \times \sin 90° = \dfrac{2.5}{0.8} \times 1 = $ **3.125 m/s**

(iii) Angular velocity of bar,

$$\omega_{ab} = \frac{V_{ba}}{l} = \frac{3.125}{5} = 0.625 \text{ rad/s}$$

Alternatively, the problem can be worked out by instantaneous centre method

The velocity at end A is horizontal and that at end B is vertical. The point of intersection of the normals to these velocities V_a and V_b locates the instantaneous centre I.

$$IA = 4 \text{ m} \quad \text{and} \quad IB = 3 \text{ m}$$

Then the angular velocity of the bar is given by

$$\omega_{ab} = \frac{V_a}{IA} = \frac{2.5}{4} = 0.625 \text{ rad/s}$$

Velocity of end $B = \omega \times IB = 0.625 \times 3 = 1.875$ m/s

Velocity of end B relative to end A,

$$V_{ba} = \omega \times l = 0.625 \times 5 = 3.125 \text{ m/s}$$

EXAMPLE 15.25.

A compound wheel of configuration illustrated in Fig. 15.26 rolls without slipping on a horizontal surface PQ. If the velocity at the centre of wheel is 1 m/s, determine the velocities at the points A, B and D indicated in the figure.

Solution : When the wheel rolls without slipping, its point of contact with the horizontal surface at any instant has zero velocity. Thus the point E is instantaneously at rest and every other point on the wheel has angular velocity about E. Hence E is the instantaneous centre of the wheel. Then

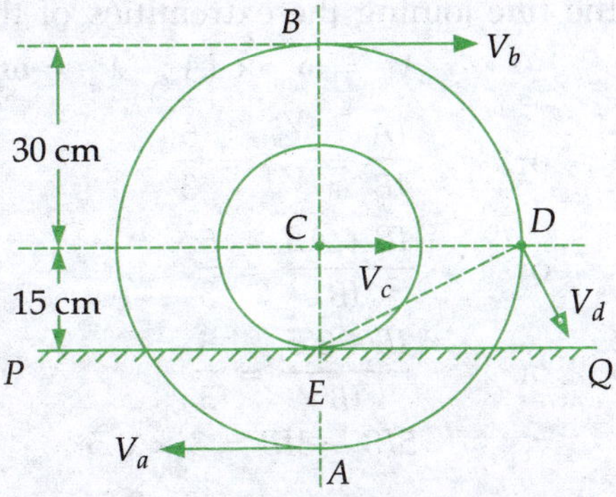

Fig. 15.26

$$V_c = \omega \times CE\,; \quad V_a = \omega \times AE$$
$$V_b = \omega \times BE \quad \text{and} \quad V_d = \omega \times DE$$

Angular velocity of the wheel $\omega = \dfrac{V_c}{CE}$

$$= \frac{1}{0.15} = 6.67 \text{ rad/s}$$

Velocity of point A: $V_a = \omega \times AE = 6.67 \times 0.15 = 1$ m/s

Velocity of point B: $V_b = \omega \times BE = 6.67 \times 0.45 = 3$ m/s

$$\text{Distance } DE = \sqrt{(CE)^2 + (CD)^2}$$

$$= \sqrt{(0.15)^2 + (0.3)^2} = 0.364 \text{ m}$$

$\therefore$ Velocity of point $D: V_d = \omega \times DE = 6.67 \times 0.364 = $ **2.43 m/s**

The directions of these velocities at the given points would be as indicated in the figure.

EXAMPLE 15.26.

A cylindrical roller, 50 cm in diameter, is in contact with two horizontal conveyor belts running at uniform speeds of 5 m/s and 3 m/s as shown in Fig 15.27.

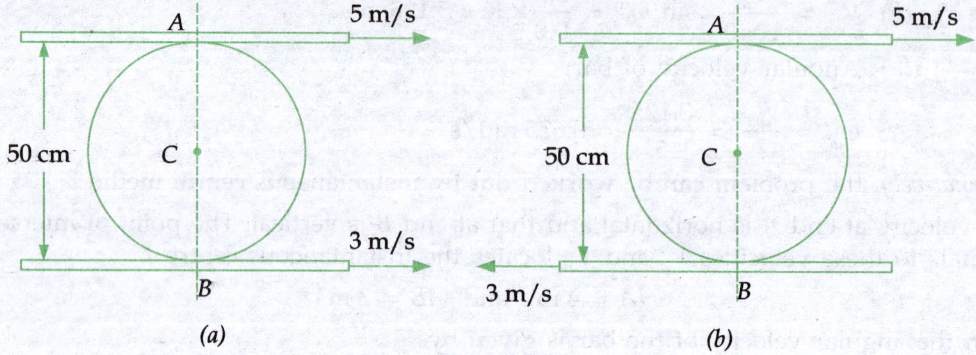

Fig. 15.27

Assuming that there is no slip at the points of contact, determine

(i) *the position of the instantaneous centre of the roller,*

(ii) *the linear velocity of the centre C, and*

(iii) *the angular velocity of the roller.*

(b) *How these parameters would be affected if the velocities of the belts are in opposite direction (Fig. 15.27 b)*

Solution : Case (*a*) : Velocities V_a and V_b are in the same direction.

The instantaneous centre is found by determining the point of intersection of line *AB* with the line joining the extremities of the velocity vectors (Fig. 15.28), then

$$V_a = \omega_{ab} \times IA\,; \quad V_b = \omega_{ab} \times IB$$

or $$\frac{IA}{IB} = \frac{V_a}{V_b} = \frac{5}{3}$$

or $$\frac{IB + AB}{IB} = \frac{5}{3}$$

or $$\frac{IB + 0.5}{IB} = \frac{5}{3}$$

∴ $$5IB - 3IB = 3 \times 0.5$$

or $$IB = \frac{3 \times 0.5}{2} = 0.75 \text{ m} = 75 \text{ cm}$$

$$IA = IB + AB = 75 + 50 = 125 \text{ cm}$$

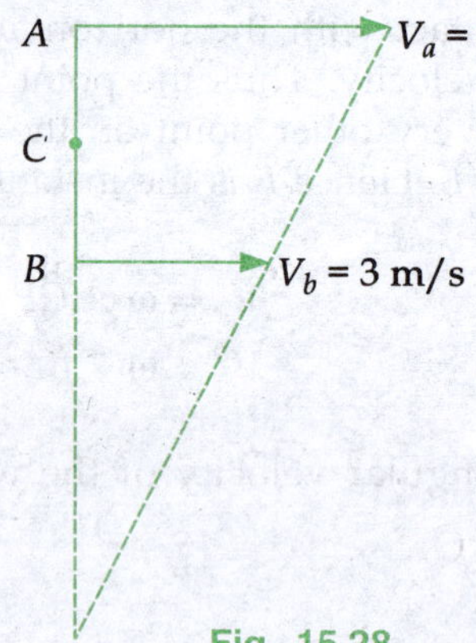

Fig. 15.28

Angular velocity of the roller = angular velocity of *A* about *I*

$$\omega = \frac{V_a}{IA} = \frac{5}{1.25} = \textbf{4 rad/s}$$

Linear velocity of the centre of roller,

$$V_c = \omega \times IC = 4 \times (1.25 - 0.25) = \textbf{4 m/s}$$

(b) When velocities V_a and V_b are in opposite directions, *I* lies between *A* and *B* (Fig. 15.29), and divides *AB* internally in the ratio V_a to V_b. That is,

$$\frac{IA}{IB} = \frac{V_a}{V_b} = \frac{5}{3}$$

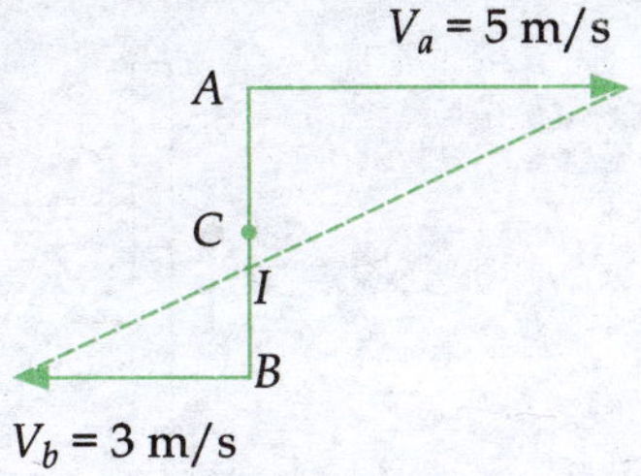

Fig. 15.29

or $$\frac{0.5 - IB}{IB} = \frac{5}{3};$$

$$IB = 0.1875 \text{ m} = 18.75 \text{ cm}$$

$$IA = 50 - 18.75 = 31.25 \text{ cm}$$

Angular velocity of the roller

$$\omega = \frac{V_a}{IA} = \frac{5}{0.3125} = 16 \text{ rad/s}$$

Linear velocity of the centre of roller,

$$V_c = \omega \times IC = \omega \times (IA - AC) = 16 \times (0.3125 - 0.25) = 1 \text{ m/s}$$

15.5 CORIOLIS' LAW

If a point moves along a path that has a rotational motion, the absolute acceleration of the point is given by the vector sum of three accelerations:

(*i*) acceleration of the coincident point relative to which the point under consideration is moving,
(*ii*) acceleration of the point relative to the coincident point, and
(*iii*) coriolis component of acceleration

We know that any plane motion can be replaced by a translation defined by the motion of an arbitrary reference point O and simultaneous rotation about O. Then the absolute acceleration of any point A can be written as

$$\vec{a}_A = \vec{a}_O + \vec{a}_{AO}$$

That is, the absolute acceleration of point A is the sum of its acceleration $\bar{a}_{AO}$ relative to the frame of translation and the acceleration $\bar{a}_O$ of a point of that frame. Now, if the frame is made to rotate with an angular velocity ω, then

$$\vec{a}_A = \vec{a}_B + \vec{a}_{A\omega} + \vec{a}_C$$

where $\vec{a}_A$ = absolute acceleration of particle at point A

$\vec{a}_B$ = acceleration of point B of moving frame ω coinciding with A

$\vec{a}_{A\omega}$ = acceleration of A relative to moving frame ω, and

$\vec{a}_C$ = Coriolis acceleration also called the complementary acceleration

Consider a collar (link 2) which slides at constant velocity along a rod OM (link 1) towards point Q. Simultaneously the rod rotates at a constant angular velocity ω about O as shown in Fig. 15.30. At an instant, the point A on the rod is coincident with point B on the collar. When the collar moves on the rod to point Q in the dt, the rod rotates through an angle $d\theta$ during the same time interval. Then the final position of point A is A_1 and that of point B is B_2.

The motion of the collar from B to B_2 can be considered to be accomplished in the following three stages:

(*i*) B to A_1 due to rotation of rod
(*ii*) A_1 to B_1 due to outward velocity of collar on the rod, and
(*iii*) B_1 to B_2 due to coriolis component of acceleration

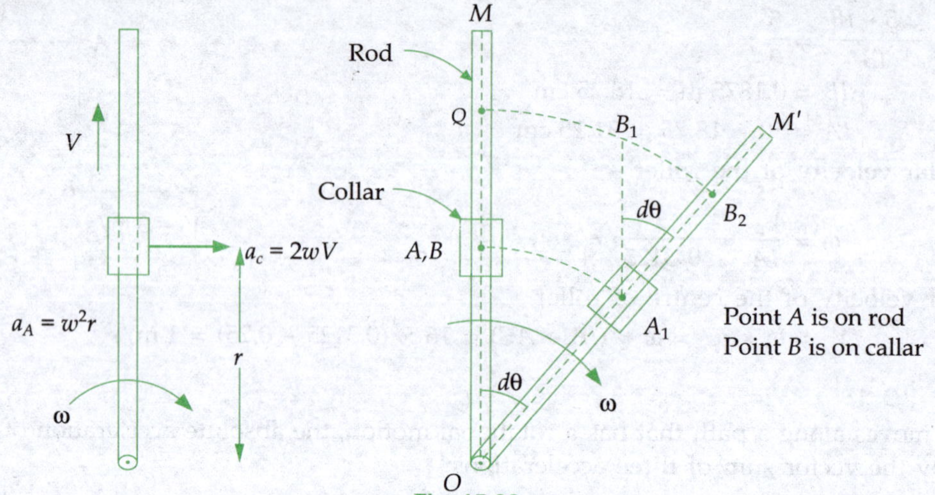

Fig 15.30

The coriolis component acts in the direction of relative velocity vector for the two coincident points (*A* and *B*) rotated by 90° in the direction of angular velocity of the rotation of rod, and it is given by

$$\vec{a}_C = 2\omega V_{ba}$$

where V_{ba} is the velocity of *B* relative to *A*

$$\text{Coriolis force} = m \times \vec{a}_C$$

where *m* is the mass of the collar

It is worthwhile to mention that:

(*i*) The angular velocity of the rod is constant and hence $\vec{a}_A$ reduces to its normal component of magnitude $\omega^2 r$

(*ii*) Since the collar slides along the rod with constant relative speed, the relative acceleration is zero.

(*iii*) Coriolis acceleration is a vector perpendicular to rod *OM* and is of magnitude $2\omega V_{ba}$ and is directed as shown.

Obviously the acceleration of the collar consists of two vectors $\vec{a}_C$ and $\vec{a}_A$

The concept of coriolis acceleration is very useful in the study of long range projectiles and of the bodies whose motions are appreciably affected by the rotation of earth.

REVIEW AND SUMMARY

1. In a circular (rotary) motion, the movement of the particle is along a circular path. The particle repeats its journey along the same circular path about the centre of rotation which remains fixed.

   The shafts, pulleys and flywheels etc. undergo circular motion when they rotate about their geometric axis.

2. The rate of change of angular displacement of a body is called angular velocity.

$$\omega = \frac{d\theta}{dt}$$

If the body turns *N* revolutions per minute, then

$$\omega = \frac{2\pi N}{60} \text{ rad/s}$$

The linear velocity V and the angular velocity ω are related by the expression: $V = \omega r$

3. The rate of change of angular velocity of a body is called angular acceleration.

$$\alpha = \frac{d\omega}{dt} = \frac{d}{dt}\left(\frac{d\theta}{dt}\right) = \frac{d^2\theta}{dt^2}$$

The linear acceleration a and the angular acceleration α are related by the expression: $a = \alpha r$.

4. When a body moves along a curved path, the linear velocity is referred to as tangential velocity. Further, the body undergoes tangential and normal components of acceleration given by

$$a_t = \alpha r \quad \text{and} \quad a_n = \frac{v^2}{r} = \omega^2 r$$

The tangential acceleration is due to change in magnitude of velocity and the normal acceleration is due to change in direction of the body.

5. When a body moves in a circular path with uniform acceleration, the equations of motion are :

$$\omega = \omega_0 + \alpha t \,;\; \omega^2 - \omega_0^2 = 2\,\alpha\theta$$

and $$\theta = \omega_0 t + \frac{1}{2}\alpha t^2$$

where ω_0 is the initial angular velocity, ω is the final angular velocity, α is angular acceleration of the body and θ represents the angle moved in time t.

6. A body is said to have motion of translation when the particles constituting the body move in parallel planes and travel the same distance.
7. A body is said to have motion of rotation when the body moves about a fixed point and all the particles constituting the body move in a circular path. The fixed point about which the body rotates is called the point of rotation and the axis passing through the point of rotation is called the axis of rotation. A point lying on the axis of rotation has a zero velocity and zero acceleration.
8. The combination of rotation and translation motion is usually referred to as general plane motion. The body can then be assumed to be rotating about a certain point which is known as instantaneous centre of rotation.
9. *Chasle's theorem* states that any general displacement of a rigid body can be represented by a combination of translatory motion and rotational motion.

REVIEW QUESTIONS

A. Conceptual and conventional questions :

1. State the difference between curvilinear motion and circular (rotary motion)
2. Define the following terms associated with circular motion :
   angular displacement, angular velocity and angular acceleration
3. Set up the $a = \alpha r$ relation between linear acceleration (a) and angular acceleration (α) where r is the distance of the body from the centre of rotation.
4. Explain the concept of general plane motion. List some examples of such a motion.

5. State and explain Chasle's theorem.
6. Define the term instantaneous centre or virtual centre in plane motion of rigid bodies.
7. Set up a relation for the velocity of position of a reciprocating heat engine in terms of angular speed of crank, crank angle and ratio of lengths of connecting rod and that of crank
8. A flywheel starting from rest rotates at uniform angular acceleration, and after having rotated 400 revolutions its angular velocity is 600 rpm. Find the angular acceleration and the interval of time to complete 400 revolutions. **(0.785 rad/s^2, 80 sec)**
9. The speed of a flywheel increase from 300 to 600 rev/min in 10 seconds. If the diameter of the wheel is 2 m, determine the angular acceleration and number of revolutions made during this period of 10 seconds. Also workout the normal and tangential acceleration at the rim of the wheel at the end of 10 second. **(0.5 rev/s^2, 75 rev, 394 m/s^2, 3.14 m/s^2)**
10. In a crank and connecting rod mechanism, the lengths of crank and connecting rod are 15 cm and 65 cm respectively. The crank turns 310 rev/min in the clockwise direction. What will be the velocity of the piston when the crank angle is 60° with respect to the inner dead centre position. **(4.7 m/s)**
11. A roller of radius 50 mm rides between two horizontal bars moving in opposite directions as shown in Fig. 15.29 Assuming no slip at point of contacts A and B, workout the distance a that defines the position of the path of the instantaneous centre of rotation of the roller. **(42.86 mm)**

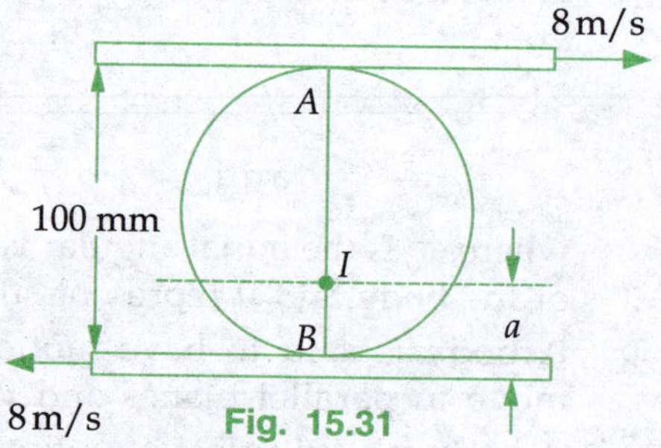

Fig. 15.31

12. A horizontal rod AB is rotating about the vertical axis through the end A at a uniform angular velocity. The linear velocity at the end B is 10 m/s and the linear speed of a point C, 2 metres from A, is 5 m/s. Find the length of the rod and its angular velocity.

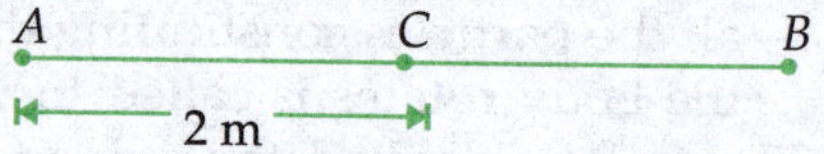

Hint: $V_b = \omega \times AB = 10$

$V_c = \omega \times AC = 5$

$\therefore \quad \omega = \frac{5}{AC} = \frac{5}{2} =$ **2.5 rad/s** and $AB = \frac{10}{\omega} = \frac{10}{2.5} =$ **4 m**

B. Complete the following statements with most appropriate word/words :

1. The shafts, pulleys and flywheels etc., undergo .... motion when they rotate about their .... axis.
2. The linear acceleration a, the angular acceleration α and the distance r of the body from the centre of rotation are correlated by the identity ....
3. The tangential accelertion is due to change in .... of velocity and the normal acceleration is due to change in .... of the body.
4. A body is said to undergo .... when the particles constituting the rigid body move in parallel planes and travel the same distance.
5. The combination of rotation and translation motion of a body is referred to as .... motion.
6. A rod sliding against a wall at one end and floor at the other end represents a combination of .... motion.

7. .... of the body represents the point about which the plane motion of all the particles constituting the body can be considered as pure rotation.
8. In the absense of external torque, the angular momentum is ....
9. The direction of velocity vector at any point on the circular orbit is .... to the circle at that point.
10. When a bar rotates in horizontal plane, the .... velocity at any point on the bar depends upon its distance from the centre of rotation.

Answers :

1. circular, geometric 2. $a = \alpha r$ 3. magnitude, direction 4. motion of translation 5. general plane motion 6. rotation and translation motion 7. instantaneous centre 8. conserved 9. tangential 10. linear

C. Multiple choice questions :

1. A body moving around a fixed axis constitutes :
   (*a*) curvilinear motion (*b*) circular motion
   (*c*) plane motion (*d*) simultaneous translation and rotation
2. Which of the following units do not conform to the concept of rigid body rotation ?
   (*a*) blades of a fan (*b*) pulley of a belt
   (*c*) phonograph records (*d*) door revolving about a hinge
   (*e*) wheels of an automobile
3. With reference to the adjoining figure, the person A is standing at the centre of a rotating platform facing person B who is riding a bicycle; heading east. The relevant speeds and distances are as given in the figure

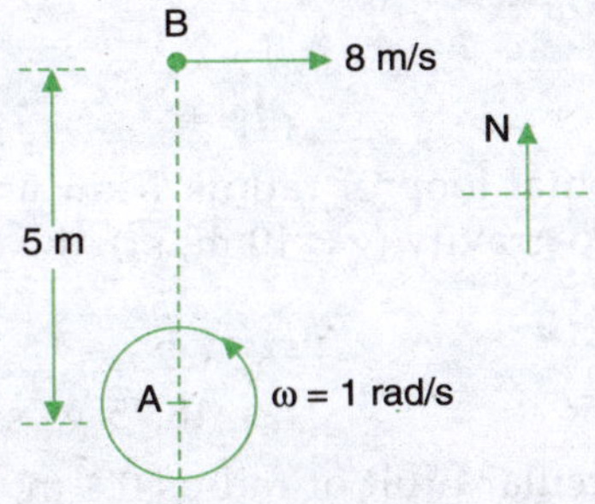

Fig. 15.32.

   At the instant under consideration, the apparent velocity of B as seen by A would be
   (*a*) 3 m/s heading east (*b*) 3 m/s heading west
   (*c*) 8 m/s heading east (*d*) 13 m/s heading east
4. The angular speed of seconds hand of a clock is
   (*a*) π rad/s (*b*) $\pi/6$ rad/s
   (*c*) $\pi/15$ rad/s (*d*) $\pi/30$ rad/s
5. Upon rotation from initially rest position, the radius vector of a body executes angular rotation θ prescribed by the relation $\theta = 4t^3 - 3t^2 + 2t + 6$
   The angular acceleration of the body at time $t - 2$ sec is
   (*a*) 38 rad/s^2 (*b*) 42 rad/s^2
   (*c*) 52 rad/s^2 (*d*) 30 rad/s^2

6. A stone tied to the end of a string 100 cm long is whirled in a horizontal circle with a constant speed. If the stone makes 15 revolutions in 30 seconds, the acceleration of the stone is closest to
   (a) 315 cm/s^2 (b) 630 cm/s^2
   (c) 985 cm/s^2 (d) 3950 cm/s^2
7. The shaft of a motor starts from rest and attains full speed of 1800 rpm in 10 seconds. The shaft has an angular acceleration of
   (a) 3π rad/s^2 (b) 6π rad/s^2
   (c) 12π rad/s^2 (d) 24π rad/s^2
8. Two cars are going with constant speeds, round concentric circles of radii r_1 and r_2, and take the same time to complete their circular paths. Their angular speeds will correspond to the ratio
   (a) $\sqrt{\frac{r_1}{r_2}}$ (b) $\frac{r_1}{r_2}$
   (c) $\left(\frac{r_1}{r_2}\right)^2$ (d) 1
9. The speed of a particle moving in a circle of radius 10 cm changes from 1.5π rad/s to 5π rad/s in 2.2 second time. The corresponding linear acceleration of the particle is
   (a) 25 cm/s^2 (b) 32.5 cm/s^2
   (c) 50 cm/s^2 (d) 65 cm/s^2
10. Starting from rest, a particle travels on a circular path and the distance covered is prescribed by the relation $s = kt^2$; where k is constant and t is the time. The particle then has a tangential acceleration of
   (a) $k/2$ (b) k
   (c) $2k$ (d) $4k$
11. An aircraft executes a horizontal loop of radius 1 km with a steady speed of 900 km/hr. In terms of acceleration due to gravity (g = 10 m/s^2), the centripetal acceleration of aircraft would be
   (a) 3.12 g (b) 6.25 g
   (c) 12.5 g (d) 18.75 g
12. For a particle moving in a circular orbit of radius 0.4 m, the angular velocity and angular acceleration at a particular instant are 2 rad/s and 5 rad/s^2. The particle then has a total linear acceleration of
   (a) 1.9 m/s^2 (b) 2.69 m/s^2
   (c) 3.8 m/s^2 (d) 7.24 m/s^2

Answers :

1. (b)	2. (d) and (e)	3. (d)	4. (d)	5. (b)	6. (c)
7. (b)	8. (d)	9. (c)	10. (c)	11. (b)	12. (b)

Simple Harmonic Motion and Vibrations 16

Learning objectives : This chapter has been devoted to the study of

- *simple harmonic motion and related terms*
- *type of vibrations; longitudinal, transverse and torsional vibrations*
- *causes, harmful effects and remedies for eliminating/reducing vibrations*
- *springs in series and parallel; equivalent spring constant*

Vibration is oscillating motion of a particle or body about a fixed reference. Such motions may be simple harmonic or complex.

16.1 SIMPLE HARMONIC MOTION

The two main characteristics of simple harmonic motion are:

(i) It is a periodic or cyclic motion, i.e., it repeats all its characteristics after a certain fixed interval of time. The body acquires the same position, and moves with the same velocity in the same direction.

(ii) The acceleration of the body is always proportional to its displacement from a fixed point and is always directed towards that point.

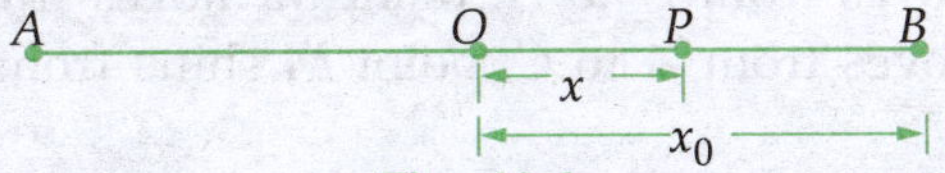

Fig. 16.1

With referece to Fig. 16.1, *O* is the fixed or reference point. The particle first moves from the reference point *O* to *B* (extreme position on the right), and then from *B* to *O*, *O* to *A* (extreme position on left) and *A* to *O*. Thereafter the motion is repeated.

The distances *OA* and *OB* represent the maximum displacement from the reference point. The maximum displacement of the particle from its mean (reference) position of motion is called the **amplitude**.

A simple harmonic motion is prescribed by the relation

$$\frac{d^2x}{dt^2} = -K^2x \qquad ...(16.1)$$

where K^2 is a constant of proportionality. This equation satisfies both conditions of simple harmonic motion:

(*i*) acceleration (d^2x/dt^2) is proportional to x (displacement) at any instant.

(*ii*) the – *ve* sign indicates that the acceleration is directed opposite to displacement x.

Examples of simple harmonic motion are:

- reciprocating motion of piston in the gas cylinder
- oscillation of simple pendulum
- vibration of simple mass and spring system
- light and sound waves due to vibratory disturbances of the medium particles.

16.2 VELOCITY AND ACCELERATION OF A PARTICLE MOVING WITH SIMPLE HARMONIC MOTION (SHM)

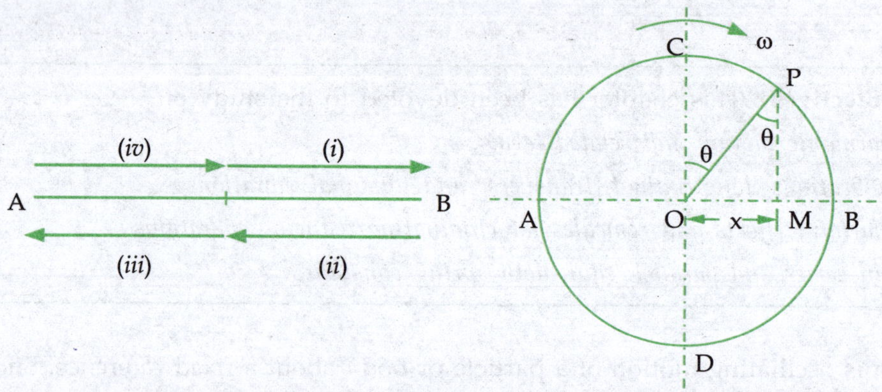

Fig. 16.2

Consider a particle moving clockwise with a constant angular velocity ω along the circumference of a circle of radius r. The particle is initially at position C and at any time t shifts to position P making angle θ with the vertical. M is the foot of perpendicular drawn from P on the horizontal diameter AB of the circular orbit. It is to be noted that

(*i*) When the particle moves from C to B, the projection point M on diameter AB moves from O to B,

(*ii*) When the particle moves from B to D, point M shifts from B to O,

(*iii*) When the particle moves from D to A, point M moves from O to A, and

(*iv*) When the particle moves from A to C, point M shifts from A to O and that completes the cycle.

The motion of M along AB is repetitive and thus simple harmonic.

The displacement at any instant is given by

$$x = OM = r\sin\theta = r\sin\omega t \qquad ...(16.2)$$

When $\omega t = 0 \quad : \quad x = 0$

$\omega t = 90° \quad : \quad x = r$

$\omega t = 180° : \quad x = 0$

The variation of displacement (x) with angle of rotation ($\theta = \omega t$) has been depicted in Fig. 16.3 *a*

Maximum displacement equals the radius of circular orbit and this occurs at $\omega t = 90° = \pi/2$.

Velocity :

When expression 16.2 is differentiated with respect to time t, one obtains,

$$\frac{dx}{dt} = v = \omega r\cos\omega t = \omega r\cos\theta \qquad ...(16.3)$$

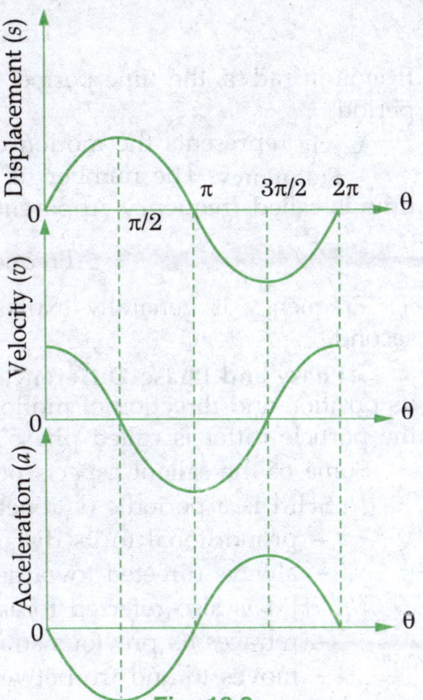

Fig. 16.3

The above relation expresses the velocity at any instant.

When $\omega t = 0$: $v = \omega r$

$\omega t = 90°$: $v = 0$

$\omega t = 180°$: $v = \omega r$

The variation of velocity v with angle of rotation $\theta = \omega t$ has been depicted in Fig. 16.3 *b*

Maximum velocity $v_{max} = \omega r$ and this occurs when

$$\omega t = 0 \text{ or } 180°$$

From the right angled triangle *OPM* (Fig. 16.2)

$$\cos\theta = \frac{PM}{OP} = \frac{\sqrt{OP^2 - OM^2}}{OP} = \frac{\sqrt{r^2 - x^2}}{r}$$

Then for equation 16.3

$$v = \omega\sqrt{r^2 - x^2} \quad ...(16.4)$$

Apparently the velocity will be maximum when $x = 0$, *i.e.*, when the projection point lies at the mean position. Equation 16.4 also indicates that $v_{max} = \omega r$

Acceleration :

Differentiating equation 16.3 with respect to time t,

$$\frac{dv}{dt} = a = -\omega^2 r \sin\omega t = -\omega^2 r \sin\theta = -\omega^2 x$$

The acceleration is thus directly proportional to distance x from the mean position. Further, the negative sign indicates that the direction of acceleration is opposite to the direction in which x increases, i.e., the direction is always directed towards the mean position.

When $\omega t = 0$: $a = 0$

$\omega t = 90°$: $a = -\omega^2 r$

$\omega t = 180°$: $a = 0$

The variation of acceleration a with angle of rotation $\theta = \omega t$ is as shown in Fig. 16.3 c. Maximum acceleration $a_{max} = -\omega^2 r$ and it occurs at $\omega t = 90°$, i.e., when the projection point is farthest from the mean position.

16.3 SHM RELATED TERMS

Some of the terms related to SHM are:

- **Oscillation:** The to and fro motion of a body from its mean position to one extreme position, then to the other extreme position and back to mean position is called oscillation. With reference to Fig. 16.2, the motion from $O-B-A-O$ represents oscillation. One half of the oscillation is called beat
- **Amplitude:** The maximum displacement of the particle from its mean position of motion is called amplitude. With reference to Fig. 16.2,

$$\text{amplitude} = OA = OB = r$$

Obviously, amplitude equals radius of the circular orbit.

- **Time period:** Time taken by a particle to undergo one complete oscillation is called periodic time or time period. Evidently it is equal to the time in which a particle completes one round, i.e., travels 2π radian

$$\text{Time period } T = \frac{2\pi}{\omega}$$

If ω is in rad/s, the time period T will be in seconds. The motion repeats itself after the time period.

Cycle represents the motion completed during one time period.

• **Frequency:** The number of cycles/oscillations of motion completed in a unit interval of time is called frequency Apparently frequency is reciprocal of time period.

$$\text{Frequency } f = \frac{1}{T} = \frac{\omega}{2\pi}$$

Frequency is generally expressed in Hertz (Hz) which equals the number of cycles per second.

• **Phase and Phase Difference:** The state or condition of the vibrating particle in relation to its position and direction of motion is referred to as phase. The amount by which the phases of the particle differ is called phase difference.

Some of the salient aspects of simple harmonic motion (SHM) are :

(*i*) SHM is a periodic or repetitive type of linear motion in which the acceleration is
  - proportional to its distance from a fixed point
  - always directed towards the fixed point

(*ii*) SHM is also referred to as oscillatory motion as the particle
  - retraces its previous straight line path after reaching an extreme position on one side
  - moves to and fro between two bounds (the bounds refer to the extreme positions)

(*iii*) A particle moving with SHM is said to complete one vibration or one oscillation.

(*iv*) One cycle, one oscillation or one vibration are the synonymous terms

(*v*) The time period for one oscillation = $\frac{2\pi}{\omega}$. Hertz is the scientific name for one cycle/sec.

(*vi*) The common examples of bodies that move with SHM are :
  - movement of the piston of a reciprocating engine
  - motion of a pendulum
  - mechanical vibrations of a spring

EXAMPLE 16.1.

For a particle undergoing SHM, the displacement is prescribed by the cosine function given by

$$x = 4\cos 3t$$

Determine (a) position of the particle at t = 1.25 s

(b) amplitude and frequency of the particle.

Solution : $x = 4\cos 3t$

$$\frac{dx}{dt} = v = -12\sin 3t$$

$$\frac{d^2x}{dt^2} = -36\cos 3t = -\omega^2 x$$

(*a*) At $t = 1.25$: $x = 4\cos \times (3 \times 1.25) = 4\cos\left(\frac{180}{\pi} \times 3 \times 1.25\right)^0 = \mathbf{-3.278\ m}$

(*b*) Comparing the given function with $x = r\cos\omega t$

amplitude $r = 4$ m

rotational speed $\omega = 3$ radian/sec

$\therefore$ frequency $f = \frac{\omega}{2\pi} = \frac{3}{2 \times 3.14} = \mathbf{0.478\ Hz}$

EXAMPLE 16.2.

A particle is moving with its acceleration directed to and proportional to its distance from a fixed point. When the distance of the particle from equilibrium position has values of 1.2 m and 1.5 m, the corresponding values of velocity are 5 m/s and 3 m/s. Determine:

(a) *amplitude and time period of oscillation*

(b) *maximum velocity and maximum acceleration*

Set up the relation of the acceleration to its distance from the fixed (mean) point.

Solution : Since the acceleration of the particle is directed and proportional to its distance from the fixed point, it is simple harmonic motion.

For SHM : $V_x = \omega\sqrt{r^2 - x^2}$

Substituting the given values

$$5 = \omega\sqrt{r^2 - 1.2^2} \quad ...(i)$$

and

$$3 = \omega\sqrt{r^2 - 1.5^2} \quad ...(ii)$$

Dividing (*i*) and (*ii*) and upon squaring

$$\left(\frac{5}{3}\right)^2 = \frac{r^2 - 1.2^2}{r^2 - 1.5^2}$$

or

$$\frac{25}{9} = \frac{r^2 - 1.44}{r^2 - 2.25}$$

or

$$9r^2 - 12.96 = 25r^2 - 56.25$$

That gives : $r = \sqrt{\dfrac{56.25 - 12.96}{25 - 9}} = 1.645$ m

$\therefore$ amplitude r = **1.645 m**

From expression (*i*),

$$5 = \omega\sqrt{1.645^2 - 1.2^2} = 1.125\,\omega \quad \therefore \quad \omega = 4.444 \text{ rad/s}$$

Time period of oscillation $T = \dfrac{2\pi}{\omega} = \dfrac{2\pi}{4.444} = \mathbf{1.413\ s}$

(b) Velocity will be maximum when $x = 0$, *i.e.*, at mean position

$$V_{max} = \omega r = 4.444 \times 1.645 = \mathbf{7.31\ m/s}$$

Acceleration will be maximum when x is maximum, *i.e.*, at dead ends

$$a_{max} = \omega^2 x = \omega^2 r = (4.444)^2 \times 1.645 = \mathbf{32.49\ m/s^2}$$

(c) Relation of acceleration and distance x

$$a_x = -\omega^2 x = -(7.31)^2 x = \mathbf{-53.44\ x}$$

EXAMPLE 16.3.

A reciprocating pump plunger is driven by a crank of radius 10 cm and delivers water at the rate of 13 cycles/sec. Determine :

(a) *maximum velocity and acceleration of the piston and rotational speed of the motor driving the piston through crank and connecting rod mechanism.*

(b) *position, velocity and acceleration of the piston 0.025 second after the start from the mean position.*

Solution : Since the piston is driven by crank and connecting rod mechanism, the situation corresponds to simple harmonic motion.

A cycle is repeated after 1 revolution of motor and therefore

Rotational speed of motor $N = 3 \times 60 =$ **180 rev/min**

Angular velocity $\omega = 2\pi \times 3 = 18.84$ rad/s

For SHM :

$$v_x = \omega\sqrt{(r^2 - x^2)} \quad \text{and} \quad a_x = \omega^2 x$$

Velocity will be maximum when $x = 0$, *i.e.*, at mean position

$$v_{max} = \omega r = 18.84 \times 0.1 = \mathbf{1.884\ m/s}$$

Acceleration will be maximum when $x = r$, *i.e.*, at dead ends

$$a_{max} = \omega^2 r = (18.84)^2 \times 0.1 = \mathbf{35.49\ m/s^2}$$

(*b*) At 0.025 sec after start from mean position

For a body performing simple harmonic motion, the displacement time relationship is

$$x = r \sin \omega t$$

$$\frac{dx}{dt} = v = \omega r \cos \omega t$$

$$\frac{d^2x}{dt^2} = a = -\omega^2 r \sin \omega t = -\omega^2 x$$

$$\omega t = 18.84 \times 0.025 = 0.471 \text{ rad} = 27°$$

$\therefore \quad x = 0.1 \sin 27° = \mathbf{0.045\ m}$ $\qquad (r = 10 \text{ cm} = 0.1 \text{ cm})$

$$v = 18.84 \times 0.1 \cos 27° = \mathbf{1.679\ m/s}$$

$$a = -(18.84)^2 \times 0.0454 = \mathbf{-16.11\ m/s}$$

EXAMPLE 16.4.

In a connecting rod and crank mechanism of a steam engine, the stroke length is 0.5 m and the crank turns 150 revolutions per minute. If the piston is assumed to have simple harmonic motion, determine

(*a*) *velocity and acceleration of the piston when it is 0.15 m from one of the dead centres,*

(*b*) *maximum velocity and acceleration of the piston and position of their occurrence, and*

(*c*) *maximum accelerating force the crank is required to exert if mass of the piston in 10 kg.*

Solution : Angular speed of the crank $\omega = \dfrac{2\pi N}{60} = \dfrac{2\pi \times 150}{6} = 15.7$ rad/s

$$\text{radius of the crank} = \frac{1}{2} \times \text{stroke length}$$

$$r = \frac{1}{2} \times 0.5 = 0.25 \text{ m}$$

For simple harmonic motion,

$$\text{Velocity of piston } v = \omega\sqrt{r^2 - x^2}$$

(*a*) When the piston is 0.15 m from dead ends, *i.e.*, from either end of stroke

$$x = 0.25 - 0.15 = 0.1 \text{ m from the mean position}$$

$$v = 15.7\sqrt{0.25^2 - 0.1^2} = 15.7\sqrt{0.0625 - 0.01} = \mathbf{3.6\ m/s}$$

$$a = -\omega^2 x = -15.7^2 \times 0.1 = \mathbf{-24.65\ m/s^2}$$

(*b*) Maximum velocity occurs when $x = 0$, *i.e.*, at mean position

$$v_{max} = \omega r = 15.7 \times 0.25 = 3.925 \text{ m/s}$$

Acceleration will be maximum at dead ends, *i.e.*, when x is maximum ($x = r$)

$$a_{max} = -\omega^2 x = -\omega^2 r = -15.7^2 \times 0.25 = \mathbf{-61.62 \text{ m/s}^2}$$

(*c*) Maximum accelerating force is given by

$$P_{max} = m \times a_{max} = 10 \times 61.62 = \mathbf{616.2 \text{ N}}$$

EXAMPLE 16.5.

A body moving with simple harmonic motion has an amplitude of 1 m and a period of oscillation of 2 seconds. What will be its velocity and acceleration 0.4 seconds after passing an extreme position?

Solution : Rotational speed $\omega = \frac{2\pi}{T} = \frac{2\pi}{2} = 3.14 \text{ rad/s}$

With O as the mean position, B represents an extremity of motion

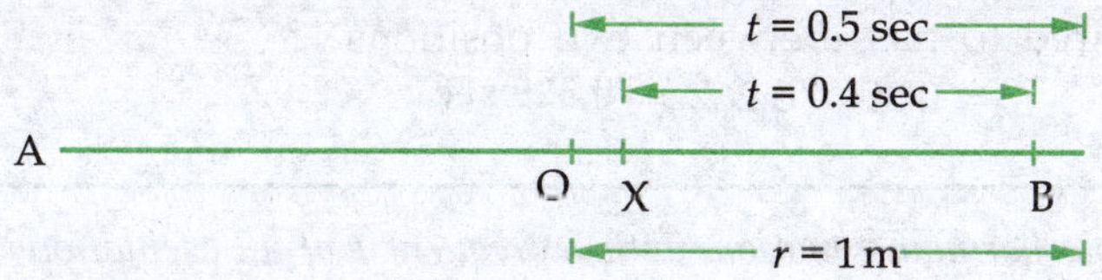

Fig. 16.4

Time required by the body to travel from O to B

$$= \frac{1}{4} \times \text{period of oscillation} = \frac{1}{4} \times 2 = 0.5 \text{ sec}$$

Let x be the position of the body 0.4 sec from extremity B. Then time required by the body to travel from O to x

$$t = 0.5 - 0.4 = 0.1 \text{ sec}$$

For the body executing simple harmonic motion, let the displacement from the mean position be prescribed by the relation

$$x = r \sin \omega t$$

Then
$$\frac{dx}{dt} = v = \omega r \cos \omega t = 3.14 \times 1 \times \cos(3.14 \times 0.1)$$

$$= 3.14 \times 1 \times \cos\left(\frac{180}{\pi} \times 3.14 \times 0.1\right)^0 = \mathbf{2.986 \text{ m/s}}$$

$$\frac{d^2x}{dt^2} = a = -\omega^2 r \sin \omega t = -(3.14)^2 \times 1 \times \sin(3.14 \times 0.1)$$

$$= -(3.14)^2 \times 1 \times \sin\left(\frac{180}{\pi} \times 3.14 \times 0.1\right)^0 = \mathbf{-3.047 \text{ m/s}^2}$$

The –ve acceleration implies that the body is moving towards the mean position O.

EXAMPLE 16.6.

A body performing simple harmonic motion has amplitude of 5 m and periodic time 4 seconds. The body is to pass between two points which are at distance of 4 m and 2 m from the centre of force and are on the same side of it. Determine the time the body will take to accomplish this task.

Solution: Angular velocity $\omega = \frac{2\pi}{T} = \frac{2\pi}{4} = 1.57 \text{ rad/s}$

For a body performing simple harmonic motion, the displacement from the mean position O is given by

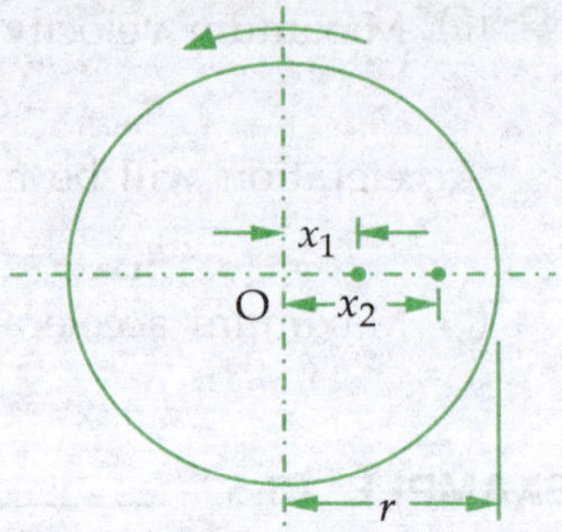

Fig. 16.5

$$x = r \sin \omega t$$

(i) $x_1 = r \sin \omega t_1 ; \quad 4 = 5 \sin \omega t_1$

$$\omega t_1 = \sin^{-1}\left(\frac{4}{5}\right) = 53.13^\circ = 53.13 \times \frac{\pi}{180} = 0.927 \text{ radian}$$

$$\therefore \quad t_1 = \frac{0.927}{1.57} = 0.59 \text{ sec}$$

(ii) $x_2 = r \sin \omega t_2 ; \quad 2 = 5 \sin \omega t_2$

$$\omega t_2 = \sin^{-1}\left(\frac{2}{5}\right) = 23.58^\circ = 23.58 \times \frac{\pi}{180} = 0.411 \text{ radian}$$

$$\therefore \quad t_2 = \frac{0.411}{1.57} = 0.262 \text{ sec}$$

$\therefore$ Time required to pass between two positions

$$= 0.59 - 0.262 = \mathbf{0.328 \text{ sec}}$$

EXAMPLE 16.7.

A particle executes simple harmonic motion with a frequency of 12 oscillations per minute. When the particle lies at a distance of 10 cm from the mean position, its velocity equals 65 percent of maximum velocity. Determine

(a) *the distance between two extreme positions of the particle*

(b) *the maximum acceleration of the particle*

(c) *velocity of the particle and the time that elapses when it is at a distance of 8 cm from the centre of oscillation*

Solution : Frequency $f = \frac{12}{60} = 0.2$ oscillations per second

$$\omega = 2\pi f = 2 \times 3.14 \times 0.2 = 1.256 \text{ rad/s}$$

For simple harmonic motion:

$$v = \omega\sqrt{r^2 - x^2} \quad \text{and} \quad a = \omega^2 x$$

Maximum velocity occurs when $x = 0$, *i.e.*, at mean position

$$v_{max} = \omega r$$

Given: when $x = 10$ cm ; $v = 0.65\, v_{max}$

$$\therefore \quad 0.65\, \omega r = \omega\sqrt{r^2 - 0.1^2}$$

$$\text{or} \quad (0.65\, r)^2 = r^2 - 0.1^2$$

$$\text{or} \quad 0.4225\, r^2 = r^2 - 0.01 ; \quad r = \left(\frac{0.01}{1 - 0.4225}\right)^{1/2} = 0.1318 \text{ m}$$

Thus the motion of the particle has an amplitude of 0.1318 m

$\therefore$ Distance between two extreme positions

$$= 2r = 2 \times 0.1318 \text{ m} = \mathbf{0.2636 \text{ m}}$$

(b) Acceleration will be maximum when the distance from the mean position is maximum, i.e., when $x = r$

$$\therefore \quad a_{max} = \omega^2 x = \omega^2 r = 1.256^2 \times 0.1318 = \mathbf{0.2079 \text{ m/s}^2}$$

(*c*) Velocity of the particle when $x = 8$ cm

$$v = \omega\sqrt{r^2 - x^2} = 1.256 \times \sqrt{0.1318^2 - 0.08^2}$$

$$= 1.256 \times \sqrt{0.1737 - 0.0064} = \mathbf{0.1315\ m/s}$$

Let t be the time that elapses when the particle is 8 cm from the centre of oscillation. Then from the correlation $x = r \sin \omega t$; we have

$$0.08 = 0.1318 \sin \omega t \qquad \therefore \quad \sin \omega t = 0.6070$$

$$\omega t = \sin^{-1} 0.607 = 37.37° = 37.37 \times \frac{\pi}{180} = 0.652 \text{ radian}$$

$$\therefore \quad t = \frac{0.652}{1.256} = \mathbf{0.519\ s}$$

16.4 TYPES OF VIBRATIONS

Vibrations are generally classified on the basis of (*a*) actuating force on the body and (*b*) stresses induced in the supporting medium.

(*a*) The actuating forces give rise to free (or natural) vibrations and forced vibrations. These free and forced vibrations may be further damped or undamped.

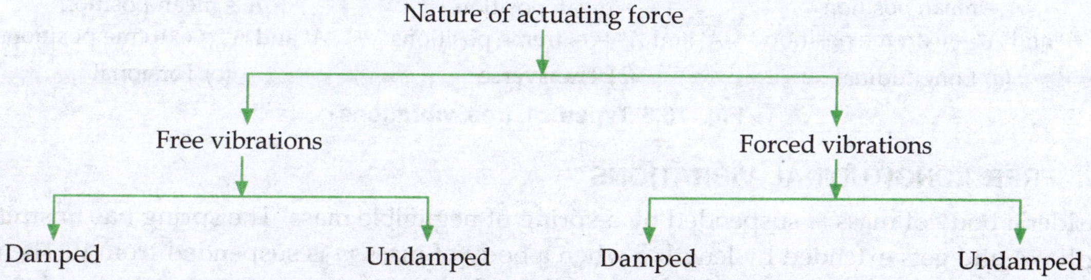

Due to air resistance, all the free and forced vibrations are essentially forced damped vibrations. Free vibrations stop after a long time where as the forced vibration continue since the resistance offered by air is not sufficient to stop the high amplitude and high frequency forced vibrations.

(*b*) Consider a heavy disc suspended from a spring (or spindle) of negligible mass. Depending upon the nature of stresses induced, the system may execute longitudinal, transverse or torsional stresses.

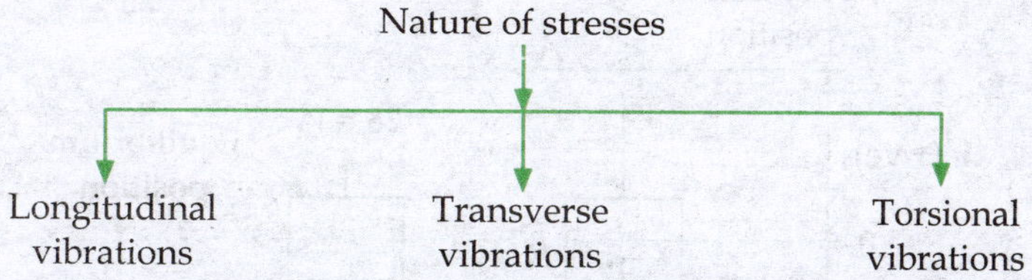

The system is said to execute **longitudinal** vibrations when the disc moves up and down about a mean position. (Fig. 16.6 *a*) The spindle gets elongated and shortened alternately. This deformation induces direct tensile and compressive stress alternately in the shaft. It is to be noted that the particles of the spindle and disc move parallel to the axis of the system.

Transverse vibrations manifest when the disc moves approximately perpendicular to the axis of spindle (Fig. 16.6 *b*). The spindle is straight, bends alternately and bending stresses are induced.

When the particles of the disc mounted at the end of a spindle circle about the spindle axis (Fig. 16.6 *c*) the system is said to excute **torsional** vibrations. In such vibrations, the spindle is twisted and untwisted alternately, and torsional shear stresses are induced in the spindle.

Within elastic limits, vibrations are essentially simple harmonic in character and the restoring force is directly proportional to displacement from the mean position.

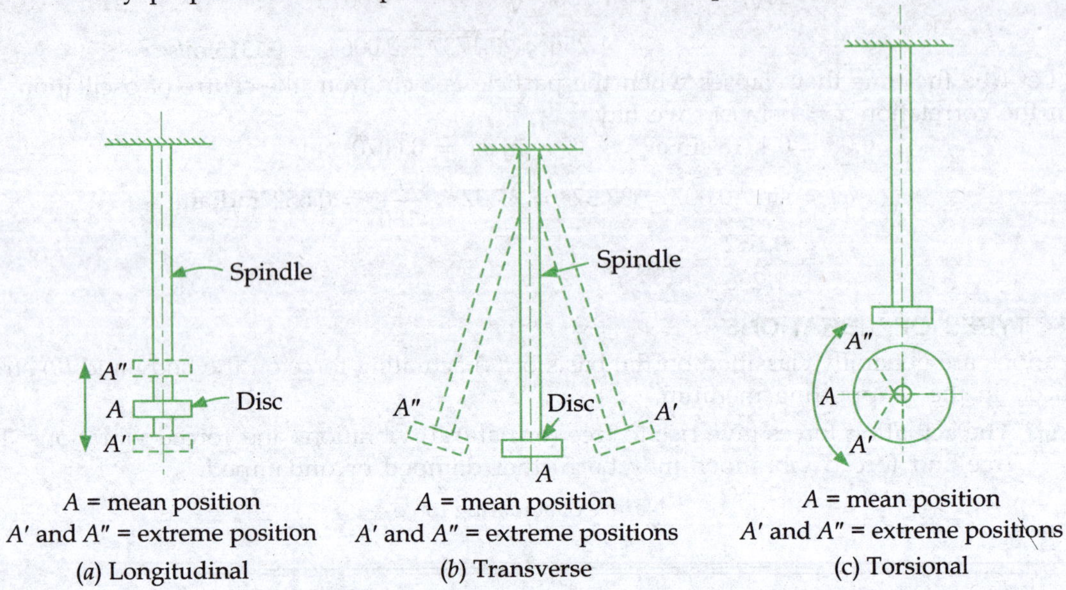

Fig. 16.6 Types of free vibrations

16.5 FREE LONGITUDINAL VIBRATIONS

Consider a body of mass m suspended by a spring of negligible mass. The spring has unstretched length OA and gets extended by length δ when a body of mass m is suspended from it. The point B is called the position of rest or equilibrium. All displacements of mass are measued with respect to this position and the downward displacements are taken positive.

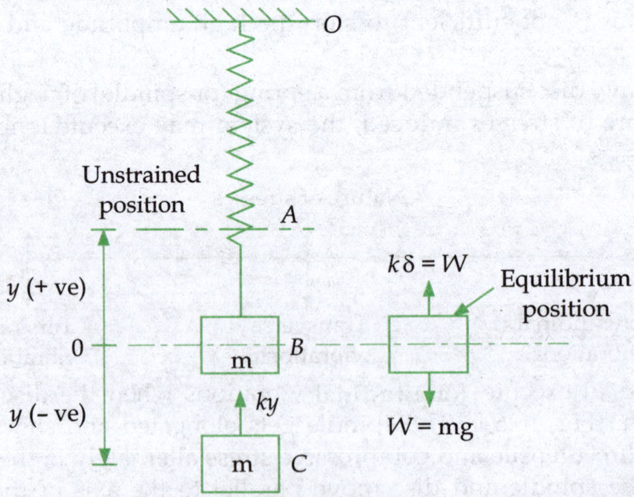

Fig. 16.7

When occupying the mean position, the weight is balanced by the force of spring. That is

$$W = mg = k\delta$$

where k is the stiffness of the spring. The spring stiffness is a measure of the weight required to produce unit deflection; it is expressed in N/m.

When the body is displaced downwards from its mean position by an external force and subsequently released, it moves up and down about its mean position and thus it executes simple harmonic motion. Let C be the position of the body at any time at distance y from the mean postion. Then

$$\text{Restoring force} = -ky$$

The negative sign accounts for the fact that the restoring force acts in the upward direction.

For translational systems, Newtons 2nd law states that

$$\Sigma F = \text{mass} \times \text{acceleration}$$

$$-ky = m\frac{d^2y}{dt^2}$$

or

$$m\frac{d^2y}{dt^2} + ky = 0 \qquad \text{...}(i)$$

For a body executing simple harmonic motion, the fundamental equation of motion is

$$\frac{d^2y}{dt^2} + \omega^2 y = 0 \qquad \text{...}(ii)$$

Comparing equations (*i*) and (*ii*), we get

$$\omega^2 = \frac{k}{m} \quad \text{or} \quad \omega = \sqrt{\frac{k}{m}}$$

the parameter ω is called the angular velocity.

If T is the time period of the oscillation, then a cycle of motion is completed when $\omega T = 2\pi$. That gives.

Natural time period of oscillation,

$$T = \frac{2\pi}{\omega} = 2\pi\sqrt{\frac{m}{k}}$$

Further, natural frequency f_n is the reciprocal of time period.

$$\therefore \text{ Natural frequency } f_n = \frac{1}{T} = \frac{1}{2\pi}\sqrt{\frac{k}{m}}$$

Using the relation $mg = k\delta$ and $\frac{k}{m} = \frac{g}{\delta}$, we may write

$$f_n = \frac{1}{2\pi}\sqrt{\frac{g}{\delta}}$$

Taking the value of g as 9.81 m/s^2 and measuring static displacement δ in metres,

$$f_n = \frac{1}{2\pi}\sqrt{\frac{9.81}{\delta}} = \frac{0.4985}{\sqrt{\delta}} \text{ Hz}$$

It may be noted that while setting upon the above relations, the mass m_s of the spring was not considered. The effect of mass inertia of spindle can be accounted for by adding one-third fraction of its mass to the mass attached at the free end. That would give

$$T = 2\pi\sqrt{\frac{m + \frac{m_s}{3}}{k}} \quad \text{and} \quad f_n = \frac{1}{2\pi}\sqrt{\frac{k}{m + \frac{m_s}{3}}}$$

EXAMPLE 16.8.

A helical spring fixed at its upper end carries a mass of 5 kg attached to it at its lower end. The mass is set into vibration and is measured to make two oscillations per second. Calculate the stiffness of spring.

Solution: Period time $T = \dfrac{1}{\text{Number of oscillations per second}} = \dfrac{1}{2} = 0.5\,\text{s}$

$$\therefore \quad 0.5 = 2\pi\sqrt{\frac{m}{k}} = 2\pi\sqrt{\frac{5}{k}}$$

That gives

$$(2\pi)^2 \times \frac{5}{k} = (0.5)^2$$

$$\text{Spring stiffness, } k = \frac{(2\pi)^2 \times 5}{(0.5)^2} = \mathbf{788.8\ N/m}$$

EXAMPLE 16.9.

A railway wagon weighs 0.25 MN when empty and 0.6 MN when loaded. The springs of the wagon get compressed by 8 cm when it is full. Determine the natural period of vibration when the wagon is empty and when loaded.

(*b*) *When the wagon is empty, it is set in motion with 10 cm amplitude and its displacement is 3 cm from the position of static equilibrium. Calculate the corresponding velocity.*

Solution:

$$\text{Stiffness of spring } k = \frac{\text{load}}{\text{deformation}} = \frac{0.6}{8} = 0.075\ \text{MN/cm}$$

Deformation of springs due to empty wagon

$$\delta_e = \frac{0.25}{0.075} = 3.33\ \text{cm}$$

$\therefore$ Period of vibration when empty

$$T_e = 2\pi\sqrt{\frac{\delta_e}{g}} = 2\pi \times \sqrt{\frac{0.0333}{9.81}} = 0.366\ \text{s}$$

Period of vibration when loaded

$$T = 2\pi\sqrt{\frac{\delta}{g}} = 2\pi \times \sqrt{\frac{0.08}{9.81}} = 0.567\ \text{s}$$

(*b*) When the wagon is empty

$$\text{Angular velocity } \omega_e = \frac{2\pi}{T_e} = \frac{2\pi}{0.366} = 17.16\ \text{rad/s}$$

$$\text{Velocity } v = -\omega_e\sqrt{r^2 - x^2}$$

$$= -17.16\sqrt{0.1^2 - 0.03^2} = \mathbf{1.637\ m/s}$$

EXAMPLE 16.10.

A helical spring of negligible mass is found to extend 0.15 mm under a mass of 0.5 kg. This spring is made to carry a mass of 40 kg at its lower end. Subsequently the spring and the mass system is displaced vertically through 100 mm and released. If the system executes simple harmonic motion, make calculations for the time period of oscillation and its natural frequency.

Solution : Spring stiffness (constant) k = load per unit deflection

$$= \frac{0.5 \times 9.81}{0.15} = 32.7 \text{ N/mm} = 32.7 \times 10^3 \text{ N/m}$$

Mass carried by the spring = 40 kg

$$\text{Time period } T = 2\pi\sqrt{\frac{m}{k}} = 2\pi\sqrt{\frac{40}{32.7 \times 10^3}} = 0.2196 \text{ s}$$

The natural frequency is the reciprocal of time period. Therefore

$$f_n = \frac{1}{T} = \frac{1}{0.2196} = \mathbf{4.55\ Hz}$$

EXAMPLE 16.11.

A vertical spindle of 3 cm diameter and 4 m length is fixed to a ceiling at one end carries a 250 N weight at the other end. The spindle is set into longitudinal vibrations by pulling the block down and then releasing it. Determine the frequency of vibrations by assuming that the modulus elasticity of the material of spindle is 20 × 10⁶ N/m².

Solution : Noting that

(*i*) Stress σ is force per unit area of cross-section (A)

(*ii*) Modulus of elasticity (E) is the ratio of stress (σ) to strain (e)

(*iii*) Strain (e) is the ratio of deformation (δ) to original length (l)

Accordingly

$$\sigma = \frac{F}{A} = \frac{250}{\frac{\pi}{4}(0.03)^2} = 353857 \text{ N/m}^2$$

$$e = \frac{\sigma}{E} = \frac{353857}{20 \times 10^6} = 0.177$$

$$\delta = l \times e = 4 \times 0.0177 = 0.0708 \text{ m}$$

In terms of static deflection δ, the time period is given by

$$T = 2\pi\sqrt{\frac{\delta}{g}} = 2\pi \times \sqrt{\frac{0.0708}{9.81}} = 0.533 \text{ sec}$$

and frequency of vibration $f = \frac{1}{T} = \frac{1}{0.533} = \mathbf{1.876\ Hz}$

EXAMPLE 16.12.

A vertically mounted spring is found to deflect by 15 mm when a body of 30 N weight is suspended from it. How many oscillations are made by the body? Subsequently the body is displaced through a distance of 30 mm from the mean position and then released. Calculate the maximum force to which the spring would be subjected.

Solution : Time period $T = 2\pi\sqrt{\frac{m}{k}}$

Spring stiffness k = weight per unit deflection = $\frac{mg}{\delta}$

$$\therefore \quad T = 2\pi\sqrt{\frac{\delta}{g}} = 2\pi\sqrt{\frac{0.015}{9.81}} = 0.2456 \text{ sec}$$

$$k = \frac{mg}{\delta} = \frac{W}{\delta} = \frac{30}{0.015} = 2000 \text{ N/m}$$

$$\text{Number of oscillations, } n = \frac{1}{T} = \frac{1}{0.2456} = \mathbf{4.072 \text{ vib/sec}}$$

$$\text{Angular velocity } \omega = \frac{2\pi}{T} = \frac{2 \times 3.14}{0.2456} = 25.57 \text{ rad/s}$$

$$\text{Maximum acceleration} = \omega^2 x = (25.57)^2 \times 0.03 = 19.61 \text{ m/s}^2$$

$$\text{Inertia force} = \text{mass} \times \text{acceleration} = \frac{30}{9.81} \times 19.61 = 60 \text{ N}$$

∴ Maximum force on the spring = weight of body + inertia force

= 10 + 60 = **90 N**

Alternatively: The maximum force (tension) in the spring occurs when the displacement of the weight is maximum.

Total extension of the spring

$$= \begin{bmatrix}\text{static extension} \\ \text{due to weight}\end{bmatrix} + \begin{bmatrix}\text{maximum displacement of weight from} \\ \text{the mean position}\end{bmatrix}$$

= 15 + 30 = 45 mm = 0.045 m

∴ Maximum force on the spring

= spring stiffness × total extension of spring

= 2000 × 0.045 = **90 N**

16.6 SPRINGS IN SERIES AND PARALLEL: EQUIVALENT SPRING CONSTANT

If a combination of springs is replaced by a single spring having an equivalent effect, that single spring is called as equivalent spring and its spring constant (or stiffness) as equivalent spring constant (or stiffness).

Consider two springs connected in **series** and carrying a load W at its lower end. (Fig. 16.8). For such a composite system:

(*i*) each spring is subjected to the same load

$$W_1 = W_2 = W$$

(*ii*) the total extension is the sum of the extension produced in each spring

$$\delta = \delta_1 + \delta_2$$

or
$$\frac{W}{k} = \frac{W_1}{k_1} + \frac{W_2}{k_2}$$

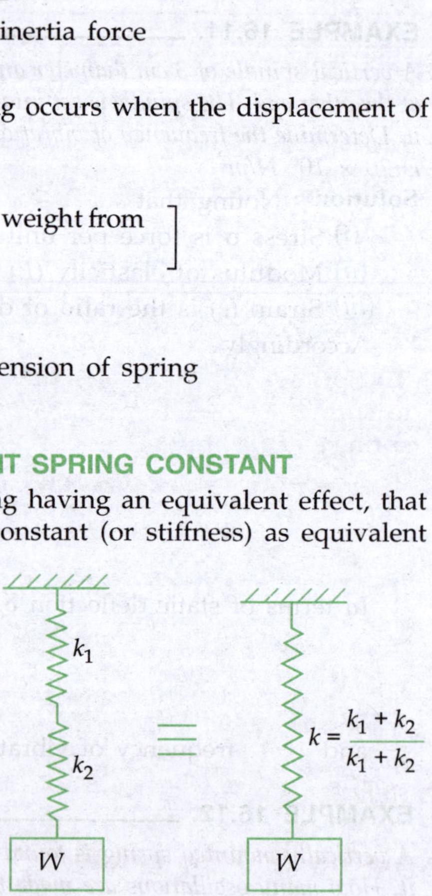

Fig. 16.8.

where k is the stiffness of an equivalent spring. Since $W_1 = W_2 = W$,

$$\frac{1}{k} = \frac{1}{k_1} + \frac{1}{k_2} = \frac{k_1 + k_2}{k_1 k_2}$$

$$\therefore \quad k = \frac{k_1 k_2}{k_1 + k_2}$$

Thus when two springs of stiffnesses k_1 and k_2 are connected in series, the combination is equivalent to a single spring of stiffness $k_1k_2/(k_1 + k_2)$.

Refer to Fig. 16.9 which shows two springs of stiffnesses k_1 and k_2 connected in **parallel** and a mass attached to them. For such a composite system,

(*i*) the axial load is divided between the springs

$$W = W_1 + W_2$$

(*ii*) the axial deflection is same for each spring.

$$\delta_1 = \delta_2 = \delta$$

$$\therefore \quad \frac{W}{\delta} = \frac{W_1}{\delta_1} + \frac{W_2}{\delta_2}$$

$$k = k_1 + k_2$$

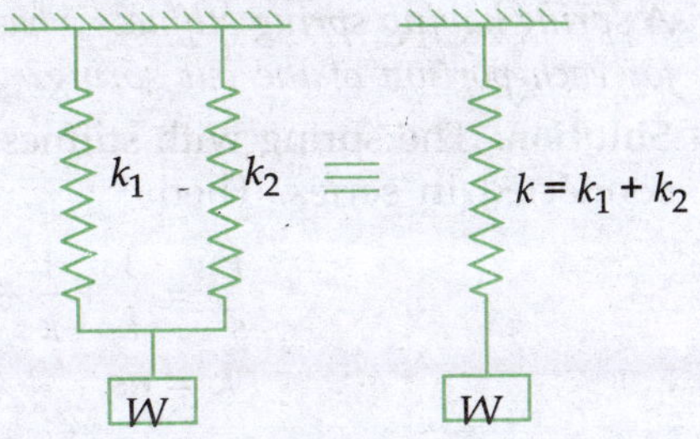

Fig. 16.9.

Thus when two springs of stiffnesses k_1 and k_2 are connected in parallel, the combination is equivalent to a single spring of stiffness k.

EXAMPLE 16.13.

Two identical springs connected in parallel are in series arrangement with another identical spring (Fig. 16.10). If the system is loaded by a weight W, show that the deflection is given by:

$$\delta = \frac{3}{2}\frac{W}{k}$$

Solution: As all the three springs are identical, they will have the same value of spring stiffness. That is

$$k_a = k_b = k_c = k \text{ (say)}$$

Springs A and B are in parallel and their combination is equivalent to spring stiffness $(k_a + k_b) = 2k$. This combination is in series with spring C.

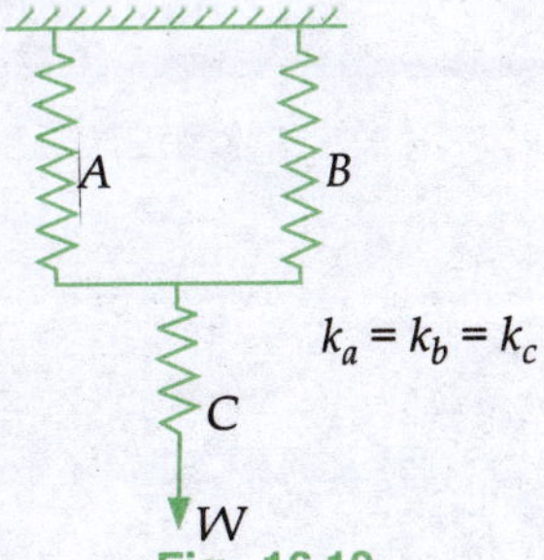

Fig. 16.10.

$\therefore$ Equivalent spring constant (stiffness) of the system is

$$= \frac{2k \times k_c}{2k + k_c} = \frac{2k \times k}{2k + k} = \frac{2}{3}k$$

$$\therefore \quad \text{Deflection } \delta = \frac{\text{weight}}{\text{equivalent spring stiffness}} = \frac{W}{\frac{2}{3}k} = \frac{3}{2}\frac{W}{k}$$

EXAMPLE 16.14.

A series arrangement of two identical springs is conneced in series with parallel combination of two similar springs. Show that overall equivalent spring stiffness is 0.4 k where k is the stiffness of each spring of the stiffness.

Solution: Refer to Fig. 16.11 for the arrangement. For springs A and B arranged in series, the equivalent spring stiffness is

$$= \frac{k_a k_b}{k_a + k_b} = \frac{k \times k}{k + k} = 0.5k$$

For springs C and D arranged in parallel, the equivalent spring stiffness is

$$= k_c + k_d = k + k = 2k$$

Now, the springs with stiffness $0.5k$ and $2k$ are in series.

$\therefore$ Equivalent spring constant of the system is

$$= \frac{0.5k \times 2k}{0.5k + 2k} = \mathbf{0.4\ k}$$

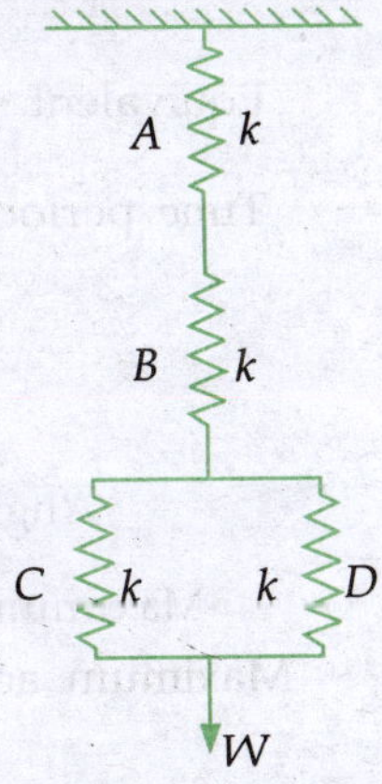

Fig. 16.11.

EXAMPLE 16.15.

A spring having spring stiffness s has been cut into n equal parts. What will be the value of spring stiffness for each portion of the cut spring?

Solution: The spring with stiffness s may be thought to be made of n springs (each of stiffness k) connected in series. Then

$$\frac{1}{s} = \frac{1}{k}+\frac{1}{k}+\frac{1}{k} = \frac{n}{k}$$

$$\therefore \quad k = ns$$

EXAMPLE 16.16.

A block of 50 kg mass is carried by a spring system comprising two springs connected in series. Make calculations for the period of vibration, and maximum velocity and maximum acceleration of the block when the block is pulled by a distance of 2.5 cm from its equilibrium position and then released. The spring constants for the springs are stated to be 4 kN/m and 6 kN/m².

How these results would be affected when the springs are connected in parallel?

Solution: Refer to Fig. 16.12 for the arrangement corresponding to connection of springs first in series and then in parallel.

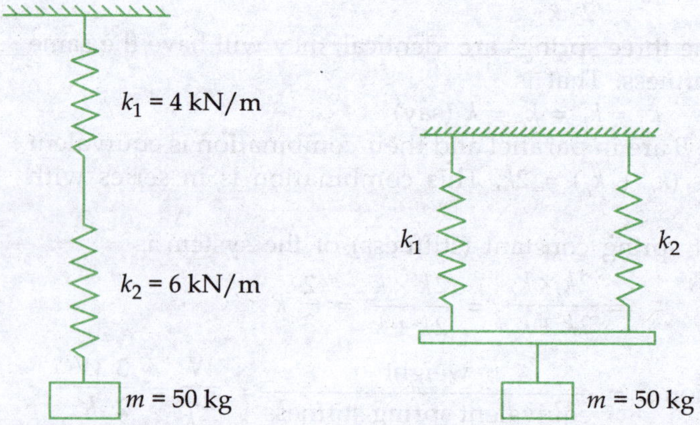

(*a*) Series connection (b) Parallel connection

Fig. 16.12

Case 1: When the springs are connected in series,

$$\text{Equivalent spring stiffness } k_e = \frac{k_1 k_2}{k_1 + k_2} = \frac{4\times 6}{4+6} = \mathbf{2.4\ kN/m}$$

Time period of oscillation (vibration),

$$T = 2\pi\sqrt{\frac{m}{k_e}} = 2\pi\times\sqrt{\frac{50}{2.4\times 10^3}} = \mathbf{0.906\ sec}$$

$$\text{Angular velocity } \omega = \frac{2\pi}{T} = \frac{2\pi}{0.906} = 6.93 \text{ rad/s}$$

$$\text{Maximum velocity } v_{max} = \omega r = 6.93 \times 0.025 = \mathbf{0.173\ m/s}$$

$$\text{Maximum acceleration } a_{max} = \omega^2 r = 6.93^2 \times 0.025 = \mathbf{1.2\ m/s^2}$$

Case 2: When the springs are connected in parallel,

$$k_e = k_1 + k_2 = 4 + 6 = 10 \text{ kN/m}$$

$$T = 2\pi\sqrt{\frac{m}{k_e}} = 2\pi\sqrt{\frac{50}{10 \times 10^3}} = \mathbf{0.444 \text{ sec}}$$

$$\omega = \frac{2\pi}{T} = \frac{2\pi}{0.444} = 14.14 \text{ rad/s}$$

$$v_{max} = \omega r = 14.14 \times 0.025 = \mathbf{0.35 \text{ m/s}}$$

$$a_{max} = \omega^2 r - 14.14^2 \times 0.025 = \mathbf{4.95 \text{ m/s}^2}$$

EXAMPLE 16.17.

A tray of mass m has been placed on three identical springs arranged in parallel, and its period of vibration was noted as 0.6 seconds. Subsequently a mass of 2.5 kg was placed in the tray and the time period was observed to increase to 0.75 seconds. Make calculations for the mass of tray and stiffness of spring.

Solution: Refer to Fig. 16.13 for the arrangement. Let k be the stiffness of individual spring.

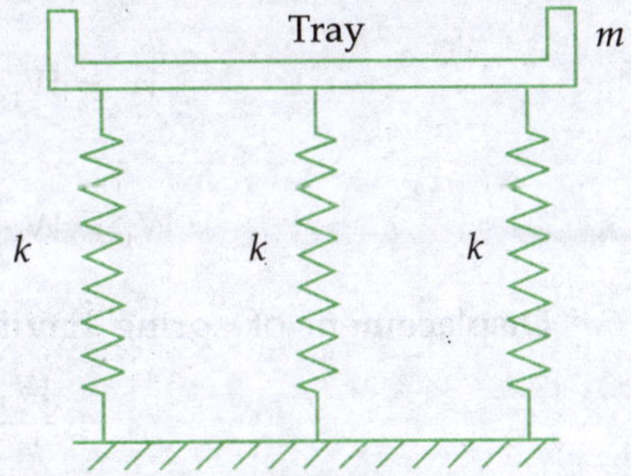

Fig. 16.13

∴ Equivalent spring stiffness $k_e = 3k$

Invoking the relation for time period of oscillation,

$$T = 2\pi\sqrt{\frac{m}{k_e}}$$

$$0.6 = 2\pi\sqrt{\frac{m}{3k}} \quad \text{and} \quad 0.75 = 2\pi\sqrt{\frac{m+2.5}{3k}}$$

Upon dividing and squaring, we get

$$\left(\frac{0.6}{0.75}\right)^2 = \frac{m}{m+2.5}; \quad \frac{0.36}{0.5625} = \frac{m}{m+2.5}$$

or $\quad 0.36\,m + (0.36 \times 2.5) = 0.5625\,m$

$$\therefore \quad \text{mass of tray } m = \frac{0.36 \times 2.5}{0.5625 - 0.36} = \mathbf{4.45 \text{ kg}}$$

Substitute $m = 4.45$ kg in any one of the expressions for time period. Thus

$$0.6 = 2\pi\sqrt{\frac{m}{3k}}; \quad 0.6 = 2\pi\sqrt{\frac{4.45}{3k}}$$

$$0.6^2 = (2\pi)^2 \times \frac{4.45}{3k}$$

$$\therefore \quad \text{Spring stiffness } k = \frac{(2\pi)^2 \times 4.45}{3 \times 0.6^2} = 175.96 \text{ N/m}$$

EXAMPLE 16.18.

Determine the equivalent spring stiffness and natural frequency of the system in which

(a) a 25 kg mass is fixed in between two springs of stiffnesses $k_1 = 12$ N/mm and $k_2 = 20$ N/mm [Fig. 16.14 (a)]

(b) a 25 kg mass is fixed to the mid point of a spring of stiffness $k = 12$ N/mm [Fig. 16.14 (b)].

Solution: (*a*) Let W_1 and W_2 be the loads caried by the upper and lower springs respectively. Then

$$W = W_1 + W_2 \qquad ...(i)$$

Further,

$$\text{Extension of upper spring} = \frac{W_1}{k_1}$$

$$\text{Compression of lower spring} = \frac{W_2}{k_2}$$

Since there is no change in total length of spring,

$$\frac{W_1}{k_1} = \frac{W_2}{k_2}; \quad W_2 = W_1 \frac{k_2}{k_1} \qquad ...(ii)$$

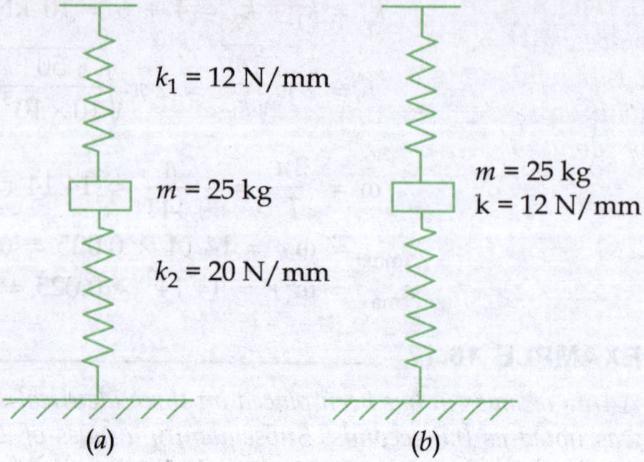

Fig. 16.14

From expressions (*i*) and (*ii*),

$$W = W_1 + W_1 \frac{k_2}{k_1} = W_1\left(1 + \frac{k_2}{k_1}\right) = W_1 \frac{k_1 + k_2}{k_1}$$

$$W_1 = W \frac{k_1}{k_1 + k_2}$$

Displacement of spring having stifness k_1

$$= \frac{W_1}{k_1} = \frac{W k_1}{k_1 + k_2} \div k_1 = \frac{W}{k_1 + k_2}$$

∴ Equivalent spring stiffnes $k_e = k_1 + k_2$

Obviously, the given system corresponds to parallel arrangement of two springs.

$$k_e = 12 + 20 = 32 \text{ N/mm}$$

$$\text{Period of oscillation } T = 2\pi\sqrt{\frac{\delta}{g}} = 2\pi\sqrt{\frac{W}{g\, k_e}} = 2\pi\sqrt{\frac{m}{k_e}}$$

$$\text{Natural frequency } f_n = \frac{1}{T} = \frac{1}{2\pi}\sqrt{\frac{k_e}{m}}$$

$$= \frac{1}{2\pi}\sqrt{\frac{32 \times 10^3}{25}} = \mathbf{5.697 \text{ Hz}}$$

(*b*) The stiffness of a spring is inversely proportional to the number of coils in it. Since the weight is fixed at the mid point, each of the upper and lower segments of the spring will have half of the total number of coils.

$$\therefore \quad \text{Stiffness of each segment, } k_1 \text{ and } k_2 = \frac{k}{\frac{1}{2}} = 2k$$

The system then corresponds to two springs arranged in parallel with equivalent spring stiffness

$$k_e = 2k + 2k = 4k = 4 \times 12 = \mathbf{48 \text{ N/mm}}$$

$$\text{Natural frequency } f_n = \frac{1}{2\pi}\sqrt{\frac{k_e}{m}} = \frac{1}{2\pi}\sqrt{\frac{48}{25}} = \mathbf{6.98 \text{ Hz}}$$

EXAMPLE 16.19.

A rigid weightless beam of length l_1 is hinged to a wall by frictionless connection at its left end. A block of weight W is suspended at its right end by means of a spring of stiffness k_1. The beam is kept in horizontal position by connecting it to the ceiling through another spring of stiffness k_2. The connection of this spring with the beam is at a distance l_2 from the hinged end. Set up the following expression for the frequency of vibration of the system.

$$f_n = \frac{1}{2\pi}\sqrt{\frac{g\,k_1 k_2\,(l_2/l_1)^2}{W\{k_1 + k_2\,(l_2/l_1)^2\}}}$$

Solution: Refer to Fig. 16.17 for the arrangement of the system.

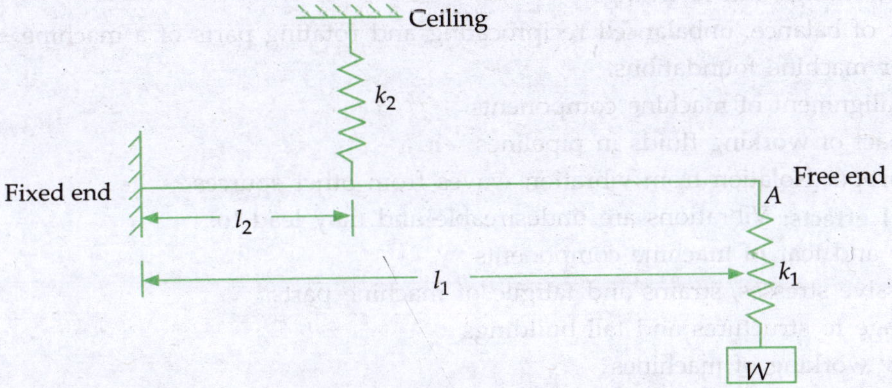

Fig. 16.15

Let P be the force in the spring attached to the ceiling. Then taking moments about the hinged end, we get

$$P \times l_2 = W \times l_1\,; \quad P = W\frac{l_1}{l_2}$$

$$\text{Extension of spring } k_1 = \frac{W}{k_1}$$

$$\text{Extension of spring } k_2 = \frac{P}{k_2} = \frac{W}{k_2}\frac{l_1}{l_2} \text{ at point B.}$$

Corresponding to this extension at point B, the extension at the free end A of the beam would be

$$= \left(\frac{W}{k_2}\frac{l_1}{l_2}\right) \times \frac{l_1}{l_2} = \frac{W}{k_2}\left(\frac{l_1}{l_2}\right)^2$$

$\therefore$ Total downward displacement of weight W,

$$= \frac{W}{k_1} + \frac{W}{k_2}\left(\frac{l_1}{l_2}\right)^2 = \frac{W}{k_1 k_2}\left[k_2 + k_1\left(\frac{l_1}{l_2}\right)^2\right]$$

$$= \frac{W}{k_1 k_2}\left(\frac{l_1}{l_2}\right)^2\left[k_1 + k_2\left(\frac{l_2}{l_1}\right)^2\right]$$

For a body executing simple harmonic motion, the frequency of vibration is given by

$$f_n = \frac{1}{2\pi}\sqrt{\frac{\text{acceleration}}{\text{displacement}}} = \frac{1}{2\pi}\sqrt{\frac{g}{\delta}} = \frac{1}{2\pi}\sqrt{\frac{g\,k_1 k_2 \left(\frac{l_2}{l_1}\right)^2}{W\left\{k_1 + k_2\left(\frac{l_2}{l_1}\right)^2\right\}}}$$

16.7 CAUSES, HARMFUL EFFECTS AND REMEDIES

Causes: Vibrations in machines and structures are caused by:

— loose fittings and fastenings
— lack of balance; unbalanced reciprocating and rotating parts of a machine
— poor machine foundations
— misalignment of machine components
— impact of working fluids in pipelines
— improper isolation from vibration waves from other sources

Harmful effects: Vibrations are undesireable and may lead to:

- wear and tear of machine components
- excessive stresses, strains and fatigue of machine parts
- damage to structures and tall buildings
- faulty working of machines
- loss of energy
- poor finish of work piece if there are vibrations in machine tools

Vibrations have adverse effects on human beings too. The primary effects are discomfort in the form of physical and mental strains, motion sickness, breathing and speech disturbance and interference in task performance.

Remedies: Vibrations are harmful and need to be eliminated or reduced. Possible remedies are:

— proper design
— balancing of reciprocating and rotating parts of a machine
— timely replacement of worn out moving parts
— machine foundations to be on compact ground
— providing vibration insulation pads in machine foundations
— isolation from vibrations of other sources lying nearby
— proper tightening and locking of fastenings.

REVIEW AND SUMMARY

1. A particle is said to execute a simple harmonic motion (SHM) if it has the following characteristics :

   (*i*) **Periodic :** motion repeats itself regularly after equal intervals of time

   (*ii*) **Oscillatory :** repeated to and fro motion about a fixed point

2. A restoring force always acts on the particle so as to bring it back to its mean position. This restoring force on the particle at any point is always directed towards a fixed point (mean position) and is proportional to the distance between the moving particle and the mean position.

3. For a particle moving around a circle of radius r with uniform angular velocity ω, y (displacement from mean position at any time t)

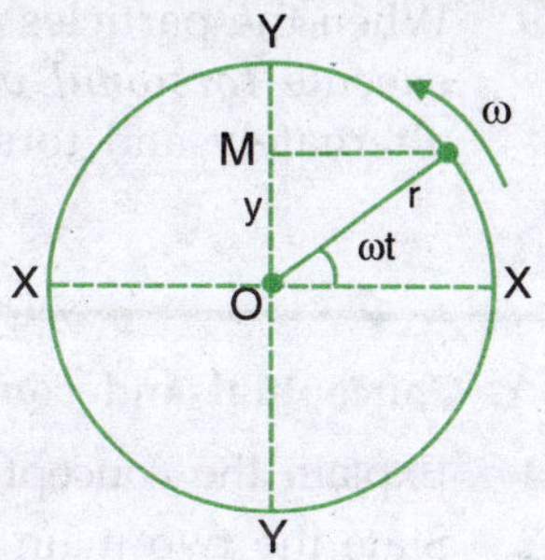

Fig. 16.16

$$y = r \sin \omega t$$

$$\text{velocity, } v = \frac{dy}{dt} = \omega r \cos \omega t$$

$$\text{acceleration, } a = \frac{dv}{dt} = -\omega^2 r \sin \omega t = -\omega^2 y$$

where ω is the angular velocity, r is the maximum displacement or amplitude. For such a motion,

$$\text{periodic time, } T = \frac{2\pi}{\omega}$$

frequency = number of cycles per second.

4. Springs are elastic bodies which absorb energy due to resilience. A spring distorts when loaded and recovers to original shape when the load is removed

   The load required to produce unit deflection is called stiffness of spring.

5. The springs may be connected in series or in parallel.

   In series arrangement,

   (*i*) each spring is subjected to the same load

   (*ii*) total deflection equals the sum of the deflections of the springs

   (*iii*) equivalent spring stiffness $s = \frac{s_1 s_2}{s_1 + s_2}$

   In parallel arrangement,

   (*i*) the axial load is divided between the springs

   (*ii*) axial deflection is same for each spring

   (*iii*) equivalent spring stiffness $s = s_1 + s_2$

6. ***Vibration*** is oscillating (to and pro) motion of a particle or body about a fixed reference. Vibrations are generally classified on the basis of (*a*) actuating force on the body and (*b*) stress induced in the supporting medium.

   The actuating forces give rise to free (or natural) vibrations and forced vibrations. These free and forced vibrations may be further damped or undamped.

   (*i*) Elastic vibrations in which there are no friction and external forces, after giving it an initial displacement, are known as free or natural vibrations, and their frequency is called free or natural frequency.

   (*ii*) When a repeated force continuously acts on the system, the vibrations are said to be *forced*. The frequency of forced vibrations is same as that of applied force and is independent of their own natural frequency.

7. A system is said to execute ***longitudinal vibration*** when a shaft or disc moves up and down parallel to its axis. Such a movement results when the shaft is elongated and shortened alternately. This deformation induces direct tensile and compressive stress in the shaft.

8. ***Transverse vibrations*** manifest when the shaft moves approximately perpendicular to its axis. The shaft is straight, bends alternately and bending stresses are induced in the shaft.

9. When the particles of the shaft move in a circle along the shaft axis, the system is said to execute ***torsional vibrations***. In torsional vibrations, the shaft is twisted and untwisted alternately and torsional shear stresses are induced in the shaft.

REVIEW QUESTIONS

A. Conceptual and conventional questions :

1. Explain the concept of simple harmonic motion.
2. State the two main characteristics of simple harmonic motion.
3. Define the terms amplitude, periodic time and frequency when applied to simple harmonic motion.
4. What type of motion will be caused when a spring is compressed and released?
5. When is a particle subjected to maximum acceleration when it moves in simple harmonic motion.
6. Draw sketches showing variation in displacement, velocity and acceleration of a particle executing simple harmonic motion.
7. How are vibrations classified?
8. What are free, forced and damped vibrations?
9. Classify the vibratory motion according to (*a*) the actuating force (*b*) the stresses induced in the supporting medium.
10. When is the phenomenon of resonance observed?
11. Show that when a particle moves with simple harmonic motion, its time for complete oscillation is independent of the amplitude of its motion.
12. A body moving with simple harmonic motion has an amplitude of 1m and the period of one oscillation is 2 second. Determine the angular speed and acceleration of the body at time instant of 0.4 second after passing through the mid-point **(3.14 rad/s ; 9.4 m/s²)**
13. A body executing simple harmonic motion has velocity of 12 m/s and 3 m/s at distances of 50 mm and 100 mm from the mid-point of its motion. Make calculations for : (a) frequency and amplitude of the motion, (b) acceleration of the body when its displacement from the mid position is 75 mm.

    If the body has a mass of 2 kg, what maximum force will act on it in the direction of motion. **(21.35 cps, 102.5 mm, 1348 m/s², 369.38 N)**
14. Set up the following expression for the frequency of free longitudinal vibrations

$$f_n = \frac{1}{2\pi}\sqrt{\frac{g}{\delta}} = \frac{4.987}{\sqrt{\delta}} \text{ cycles/sec}$$

    where δ is the static deflection of the shaft under the load and expressed in cm.
15. Two springs of stiffness k_1 and k_2 are arranged (*a*) in series, and (*b*) in parallel. Write the expressions for the stiffness of equivalent spring.
16. A weight of 50 N is attached to a vertical spring and is set into vibratory motion. If the weight is found to make 4 oscillations per second, determine the stiffness of the spring. **(32.26 N/cm)**
17. A weight of 30 N suspended from a spring vibrates vertically with an amplitude of 5 cm and a frequency of 1 oscillation per second. Make calculations for (a) the stiffness of spring, (b) maximum velocity of the weight, and (c) the maximum tension induced in the spring. **(120.6 N/m; 0.314 m/s, 36 N)**

18. The deflection of a spring with 20 active turns under a load of 100 N is 10 mm. The spring is cut into two pieces each of 10 active coils and placed in parallel. What would be deflection under the same load?

19. Determine the equivalent spring stifness and natural frequency of the vibrating systems depicted in Fig. 16.17. Take k_1 = 5 N/mm ; k_2 = 8 N/mm and m = 10 kg

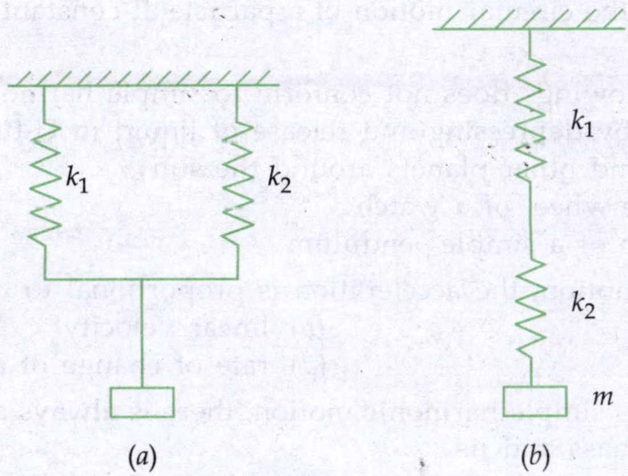

Fig. 16.17

20. A block of mass 50 kg is suspended from two springs arranged in parallel. The spring and the mass system is displaced vertically through 40 mm and then released. If the spring constants are 4 N/mm and 6 N/mm, make calculations for the period of vibrations, maximum velocity and acceleration of the block. **(0.44 sec, 0.57 m/s, 8.16 m/s^2)**

21. List the causes of vibrations in machines and structures, point out their harmful effects and suggest suitable remedies for reducing/removing them.

B. Complete the following statements with most appropriate word/words :

1. A .... motion repeats all its characteristics after a fixed certain interval of time.
2. A .... would result when a spring is compressed and released.
3. In SHM, the maximum displacement of the particle from its mean position is called ....
4. The term .... in SHM denotes the number of cycles/oscillations of motion completed in a unit interval of time.
5. The acceleration in SHM is .... proportional to displacement from the mean position, and is always directed .... the mean position.
6. With in elastic limits, vibrations are essentially .... in character.
7. The actuating forces give rise to free (or natural) vibrations and .... vibrations.
8. Torsional vibrations are said to occur when the particles of a disc mounted at the end of a spindle circle about the ....
9. The load required to produce a unit deflection is called .... of the spring.
10. For two springs connected in ...., the axial load is divided between the springs but the axial deflection is same for each spring.

Answers :

1. periodic 2. simple harmonic motion 3. amplitude 4. frequency 5. directly, towards 6. simple harmonic 7. forced 8. spindle axis 9. stiffness 10. parallel

C. Multiple choice questions :

1. Simple harmonic motion
   (*a*) is another name for periodic motion
   (*b*) is the motion of a point in a circle
   (*c*) is a projection of the circular motion of a particle
   (*d*) is a projection of the circular motion of a particle at constant speed on a diameter of the circle.

2. Which one of the followings does not conform to simple harmonic motion ?
   (*a*) motion followed by depressing and release of liquid in U-tube
   (*b*) motion of earth and other planets around the sun
   (*c*) motion of balance wheel of a watch
   (*d*) motion of the bob of a simple pendulum

3. In simple harmonic motion, the acceleration is proportional to
   (*a*) displacement (*b*) linear velocity
   (*c*) angular velocity (*d*) rate of change of angular velocity

4. When a body executes simple harmonic motion, there is always a constant ratio between the displacement of mass and its
   (*a*) frequency (*b*) velocity
   (*c*) time period (*d*) acceleration

5. The time period of a simple harmonic motion is given by :
$$T = 2\pi\left(\frac{\text{displacement}}{\text{acceleration}}\right)^{1/2}$$
   If displacement increases, the time period would
   (*a*) increase
   (*b*) decrease
   (*c*) remain unchanged
   (*d*) increase or decrease depending upon speed of the particle

6. A particle executes simple harmonic motion with an amplitude of 3 cm and time period of 6 seconds. Its maximum velocity will be
   (*a*) $\pi/2$ (*b*) π
   (*c*) 2π (*d*) 3π cm/s

7. The particle executing simple harmonic motion of amplitude 1 cm has a velocity 10 cm/s when passing through the mean position. The frequency of motion is
   (*a*) $5/\pi$ (*b*) $5/2\pi$
   (*c*) $\sqrt{5}/\pi$ (*d*) $\sqrt{10}/\pi$

8. The maximum velocity and maximum acceleration of a particle executing simple harmonic motion are 2 m/s and 20 m/s^2 respectively. The time period of motion equals
   (*a*) π (*b*) $\pi/5$
   (*c*) $\pi/10$ (*d*) $\pi/3$

9. The speed of particle executing simple harmonic motion with amplitude *a* is half of the maximum speed. At that instant, displacement of the particle is
   (*a*) $a/2$ (*b*) $\sqrt{3}/2\,a$
   (*c*) $2a/\sqrt{3}$ (*d*) $3\sqrt{2}\,a$

10. Which of the following statements is/are wrong about a body executing simple harmonic motion ?
(*a*) periodic time is inversely proportional to angular velocity
(*b*) frequency is inversely proportional to periodic time
(*c*) acceleration is proportional to displacement
(*d*) velocity will be maximum when the amplitude is zero
(*e*) momentum is conserved

11. A closed coil helical spring of stiffness 30 N/mm is arranged in series with another such spring of stiffness 60 N/mm. The stiffness of composite unit would be
(*a*) 20 N/mm (*b*) 30 N/mm
(*c*) 45 N/mm (*d*) 90 N/mm

12. Two closed coil helical springs of stiffness s_1 and s_2 are connected in parallel. The combination is equivalent to a single spring of stiffness
(*a*) $\sqrt{s_1 s_2}$ (*b*) $\frac{s_1 + s_2}{2}$
(*c*) $s_1 + s_2$ (*d*) $\frac{s_1 s_2}{s_1 + s_2}$

13. The particles of a disc mounted at the end of a shaft circle about the shaft axis. Such an effect is caused by
(*a*) longitudinal vibrations (*b*) transverse vibrations
(*c*) torsional vibrations (*d*) periodically forced vibrations

14. Choose the wrong statement in the context of transverse vibrations
(*a*) particles of the shaft move approximately perpendicular to the axis of the shaft
(*b*) shaft is straight and bent alternately
(*c*) bending stresses are induced in the shaft
(*d*) the restoring force need not be directly proportional to the displacement of the particles from the mean position

15. A mass on the end of a spring is displaced from its position and then released. The period of the resulting vibrations is certain to increase if
(*a*) spring constant is increased (*b*) mass is decreased
(*c*) mass is increased (*d*) initial displacement is increased

16. Identify the wrong statement
(*a*) vibrations in machines may be due to loose fittings and lack of balance
(*b*) free vibrations have external force applied at the ends only
(*c*) forced vibrations are independent of natural frequency of vibration
(*d*) a vibrating system is said to be stable if the amplitude of vibration decreases with time

Answers :

1. (*d*)	2. (*b*)	3. (*a*)	4. (*d*)	5. (*c*)	6. (*b*)
7. (*a*)	8. (*b*)	9. (*b*)	10. (*e*)	11. (*a*)	12. (*c*)
13. (*c*)	14. (*d*)	15. (*c*)	16. (*b*)		

Impact: Collision of Elastic Bodies 17

Learning objectives : A study of the subject matter included in this chapter will make the reader conversant with the

- *phenomenon of collision of bodies*
- *type of impacts : direct and indirect or oblique impacts*
- *terms like period of collision (period of collision + period of restitution) and coefficient of restitution*
- *principles used for the analysis of motion of two elastic bodies*
- *loss of energy during impact*

Collision means contact between two bodies for a short duration. During collision, the bodies produce impulsive forces on each other, and this impulsive force is much larger than any other finite force that may be acting. The phenomenon of collision between two bodies which occurs in a very short interval of time and during which the bodies exert relatively large forces on each other is called an **impact**.

17.1 TYPES OF IMPACT

The line joining the centres of colliding bodies and passing through the point of contact is called the *line of impact*. This line passes through the point at which collision takes place between the bodies. Essentially it represents the common normal to the surfaces of the bodies in contact (Fig. 17.1).

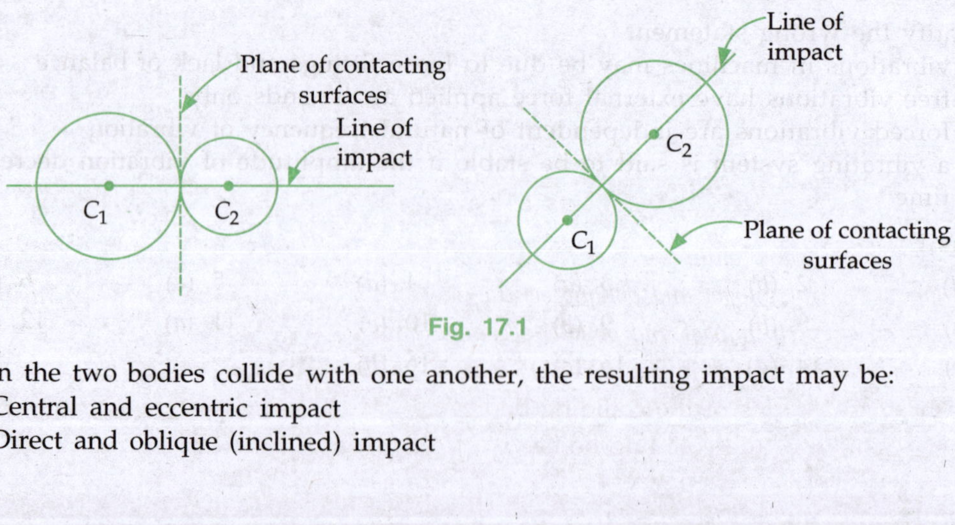

Fig. 17.1

When the two bodies collide with one another, the resulting impact may be:

(*i*) Central and eccentric impact

(*ii*) Direct and oblique (inclined) impact

The impact is called **central** when the mass centres of the colliding bodies are located on the line of impact. Fig. 17.1 represents central impact; the mass centres C_1 and C_2 of the colliding balls lie on the line of impact.

When the mass centres of the colliding bodies are not located on the line of impact, the impact is referred to as *eccentric impact*. Fig. 17.2 depicts eccentric impacts; the mass centres C_1 and C_2 of the colliding bodies donot lie on the line of impact.

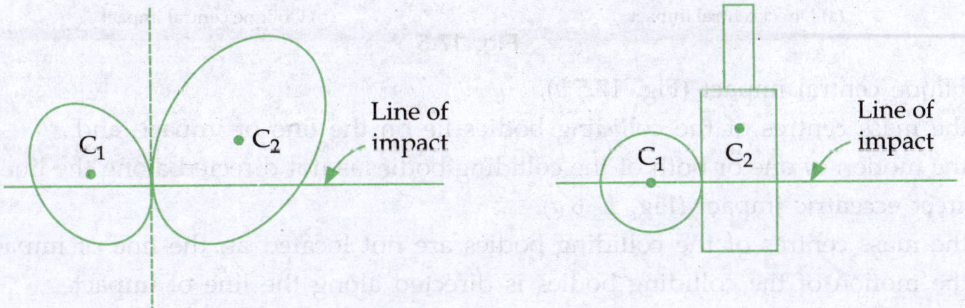

Fig. 17.2. Eccentric impact

The impact is said to be *direct* if before impact the bodies are moving along the line of impact, *i.e.*, the motion of the colliding bodies is directed along the line of impact (Fig. 17.3).

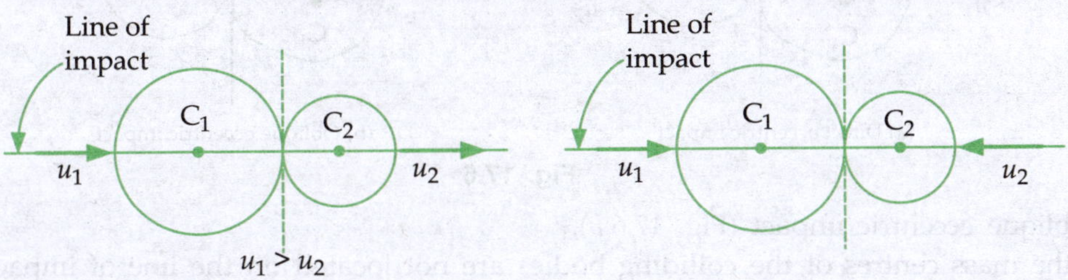

Fig. 17.3. Direct impact

The impact is *indirect* or *oblique* if the motion of one or both of colliding bodies, before impact, is not directed along the line of impact (Fig. 17.4).

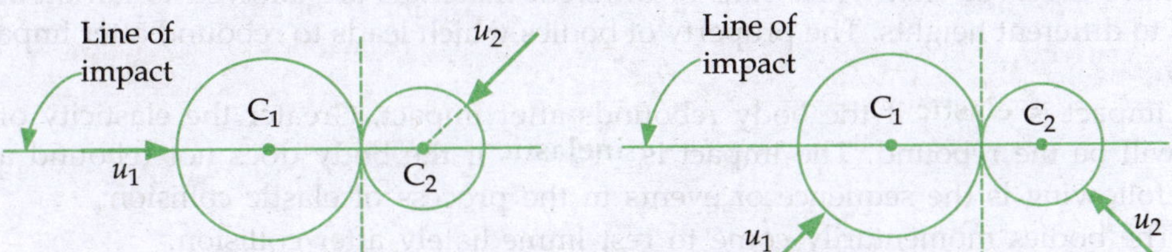

Fig. 17.4. Oblique impact

EXAMPLE 17.1.

Sketch a system of bodies with:

(i) Direct central impact and oblique central impact

(ii) Direct eccentric impact and oblique eccentric impact

Solution : In direct central impact (Fig. 17.5 *a*)

(*i*) the mass centres of the colliding bodies are on the line of impact, and

(*ii*) the motion of the colliding bodies is directed along the line of impact.

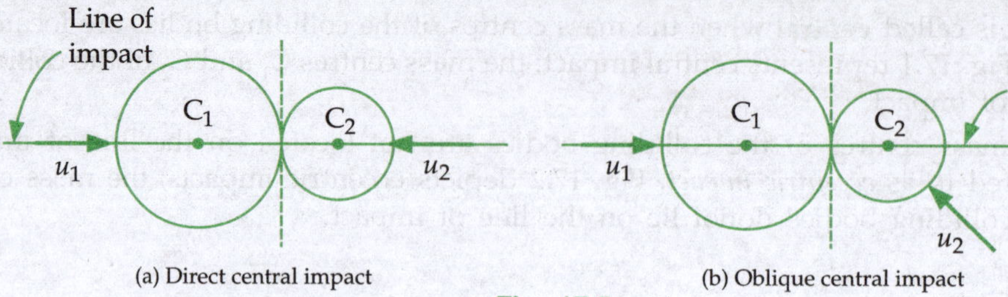

Fig. 17.5

In oblique central impact (Fig. 17.5 *b*),

(*i*) the mass centres of the colliding bodies lie on the line of impact, and

(*ii*) the motion of one or both of the colliding bodies is not directed along the line of impact.

In direct eccentric impact (Fig. 17.6 *a*),

(*i*) the mass centres of the colliding bodies are not located an the line of impact, and

(*ii*) the motion of the colliding bodies is directed along the line of impact.

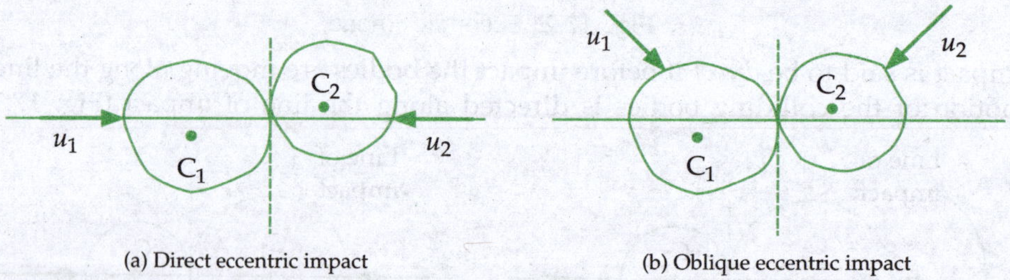

Fig. 17.6

In oblique eccentric impact (Fig. 17.6 *b*),

(*i*) the mass centres of the colliding bodies are not located on the line of impact, and

(*ii*) the motion of one or both of the colliding bodies is not directed along the line of impact.

17.2 ELASTIC AND INELASTIC IMPACT

We have observed that when balls of different materials are allowed to fall on a floor, they rebound to different heights. The property of bodies which leads to rebound after impact is called elasticity.

The impact is **elastic** if the body rebounds after impact. Greater the elasticity of the body, greater will be the rebound. The impact is **inelastic** if the body does not rebound at all.

The following is the sequence of events in the process of elastic collision:

(*i*) The bodies momentarily come to rest immediately after collision.

(*ii*) The bodies tend to compress each other and this compression continues till maximum compression occurs. The compression depends on the magnitude of impact forces and also upon the materials of the bodies. The time that elapses from the instant of initial contact to the instant of maximum deformation is called the *period of deformation.*

(*iii*) After having undergone maximum compression, the bodies tend to regain their original shapes due to their elasticity. The process of regaining the original shape is called *restitution*. The time that elapses from the instant of maximum deformation to the instant the bodies region original shape is called the *period of restitution*.

The sum total of the period of deformation and the period of restitution is called the time of collision, period of contact or period of impact.

17.3 CONSERVATION OF MOMENTUM

Consider two bodies A and B of mass m_1 and m_2 respectively. Let these bodies be moving with respective velocities of u_1 and u_2 before impact, and v_1 and v_2 after impact.

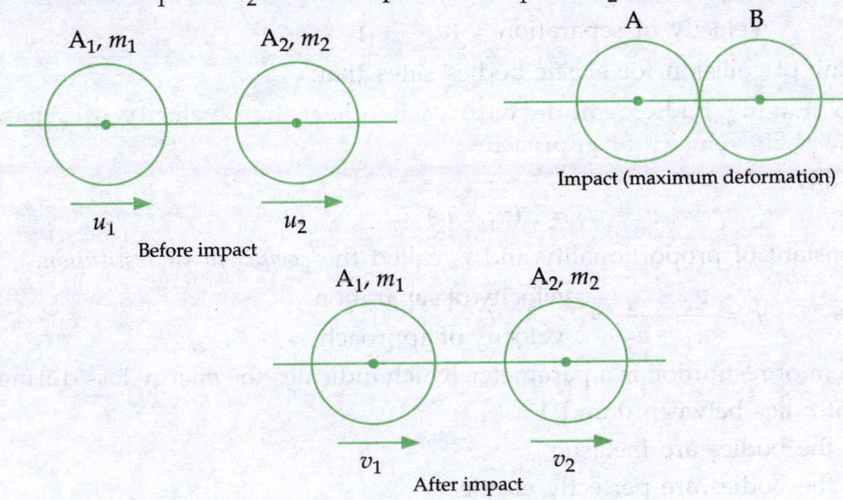

Fig. 17.7.

During collision, there is an impulse $F \times t$ exerted by body A on body B. This impulse on body B is measured by the change in its momentum. That is

Impulse on body B = change in momentum of body B

$$F \times t = m_2 v_2 - m_2 u_2 \quad ...(i)$$

According to Newton's third law of motion, action and reaction between the colliding bodies is equal in magnitude and opposite in direction, and it acts for the same time. Then impulse on body A will be $(-F \times t)$. Accordingly

$$-F \times t = m_1 v_1 - m_1 u_1$$

or

$$F \times t = m_1 u_1 - m_1 v_1 \quad ...(ii)$$

From expressions (i) and (ii)

$$m_1 u_1 - m_1 v_1 = m_2 v_2 - m_2 u_2$$

$$m_1 u_1 + m_2 u_2 = m_1 v_1 + m_2 v_2 \quad ...(17.1)$$

Thus the momentum of the system before collision is equal to the momentum of the system after collision.

If the two bodies moving with velocity u_1 and u_2 before impact get coupled after collision and move together in the same direction with velocity v, then

$$m_1 u_1 + m_2 u_2 = (m_1 + m_2) v \quad ...(17.2)$$

Thus it may be concluded that whenever collision of two bodies takes place, the total momentum of the system taken together is unchanged. However, each body taken separately undergoes a change of momentum of equal magnitude but opposite in sense; one body gains momentum at the cost of other body.

17.4. NEWTON'S LAW OF COLLISION: COEFFICIENT OF RESTITUTION

Consider two bodies A and B of mass m_1 and m_2 respectively. Let these bodies be moving with respective velocities u_1 and u_2 before impact (Fig. 17.7). The impact will take place only if $u_1 > u_2$

$\therefore$ Velocity of approach $= (u_1 - u_2)$

After a short period of contact, the bodies will separate and will start moving with velocities v_1 and v_2 respectively. The separation will occur only if $v_2 > v_1$

$\therefore$ Velocity of separation $= (v_2 - v_1)$

Newton's law of collision for elastic bodies sates that:

"When two moving bodies collide with each other, their velocity of separation bears constant ratio to their velocity of approach".

Mathematically:

$$v_2 - v_1 = e\,(u_1 - u_2)$$

where e is a constant of proportionality and is called the *coefficient* of *restitution*.

$$e = \frac{v_2 - v_1}{u_1 - u_2} = \frac{\text{velocity of separation}}{\text{velocity of approach}} \quad \text{...(17.3)}$$

The coefficient of restitution is a parameter which indicates the energy loss during an impact. The value of e lies between 0 and 1

- If $e = 0$, the bodies are inelastic
- If $e = 1$, the bodies are perfectly elastic

The value of coefficient of restitution depends not only on the material but it also depends on the shape and size of the body.

Note: While using the momentum equation and the relation for the coefficient of restitution, care should be taken to assign proper sign to the velocities.

17.5 LOSS OF KINETIC ENERGY DURING IMPACT

Consider two bodies A and B which experience direct impact. Let

m_1, u_1, v_1 = mass, initial velocity and final velocity for body A

m_1, u_2, v_2 = mass, initial velocity and final velocity for body B

Then: Kinetic energy of the two masses before impact $= \frac{1}{2}m_1u_1^2 + \frac{1}{2}m_2u_2^2$

Kinetic energy of the two masses after impact $= \frac{1}{2}m_1v_1^2 + \frac{1}{2}m_2v_2^2$

$\therefore$ Loss of kinetic energy during impact

$$= \frac{1}{2}\left[(m_1u_1^2 + m_2u_2^2) - (m_1v_1^2 + m_2v_2^2)\right]$$

$$= \frac{1}{2(m_1+m_2)}\left[(m_1+m_2)(m_1u_1^2 + m_2u_2^2) - (m_1+m_2)(m_1v_1^2 + m_2v_2^2)\right]$$

$$= \frac{1}{2(m_1+m_2)}\left[\{m_1^2u_1^2 + m_2^2u_2^2 + m_1m_2(u_1^2+u_2^2)\} - \{m_1^2v_1^2 + m_2^2v_2^2 + m_1m_2(v_1^2+v_2^2)\}\right]$$

It is to be noted that:

$$m_1^2u_1^2 + m_2^2u_2^2 = (m_1u_1 + m_2u_2)^2 - 2m_1m_2u_1u_2$$

and

$$m_1m_2(u_1^2 + u_2^2) = m_1m_2(u_1 - u_2)^2 + 2m_1m_2u_1u_2$$

$$\therefore m_1^2u_1^2 + m_2^2u_2^2 + m_1m_2(u_1^2+u_2^2) = (m_1u_1+m_2u_2)^2 - 2\,m_1m_2u_1u_2 + m_1m_2(u_1-u_2)^2 + 2\,m_1m_2u_1u_2$$

$$= (m_1u_1 + m_2u_2)^2 + m_1m_2(u_1 - u_2)^2$$

Similarly

$$m_1^2 v_1^2 + m_2^2 v_2^2 + m_1 m_2 (v_1^2 + v_2^2) = (m_1 v_1 + m_2 v_2)^2 + m_1 m_2 (v_1 - v_2)^2$$

$\therefore$ Loss in kinetic energy

$$\Delta E = \frac{1}{2(m_1 + m_2)}[(m_1 u_1 + m_2 u_2)^2 + m_1 m_2 (u_1 - u_1)^2 - (m_1 v_1 + m_2 v_2)^2 - m_1 m_2 (v_1 - v_2)^2]$$

From the law of conservation of momentum,

$$m_1 u_1 + m_2 u_2 = m_1 v_1 + m_2 v_2$$

or $$(m_1 u_1 + m_2 u_2)^2 = (m_1 v_1 + m_2 v_2)^2$$

$$\therefore \quad \Delta E = \frac{1}{2(m_1 + m_2)} [m_1 m_2 (u_1 - u_2)^2 - m_1 m_2 (v_1 - v_2)^2]$$

The relation for coefficient of restitution gives

$$e = \frac{v_2 - v_1}{u_1 - u_2}; \qquad v_2 - v_1 = e(u_1 - u_2)$$

$$\therefore \quad \Delta E = \frac{1}{2(m_1 + m_2)} [m_1 m_2 (u_1 - u_2)^2 - m_1 m_2 e^2 (u_1 - u_2)^2]$$

$$= \frac{m_1 m_2}{2(m_1 + m_2)} (u_1 - u_2)^2 (1 - e^2) \quad ...(17.4)$$

Case (*i*) : When the colliding bodies are inelastic, *e.g.*, impact between two putty balls ($e = 0$)

$$\Delta E = \frac{m_1 m_2}{2(m_1 + m_2)} (u_1 - u_2)^2 \quad ...(17.5)$$

The term $\frac{m_1 m_2}{2(m_1 + m_2)}$ represents harmonic mean of m_1 and m_2, and $(u_1 - u_2)$ represents the relative velocity of the colliding bodies before impact. Thus the loss of kinetic energy is equal to kinetic energy of body whose mass is harmonic mean between masses of the colliding bodies and whose velocity is equal to their relative velocity before impact.

Case (*ii*) : When the colliding bodies are perfectly elastic, *e.g.*, impact between two hardened and polished steel balls ($e = 1$)

$$\Delta E = \frac{m_1 + m_2}{2(m_1 + m_2)} (u_1 - u_2)^2 \times (1 - 1) = 0 \quad ...(17.6)$$

which shows that

kinetic energy before impact = kinetic energy after impact.

Thus in case of perfectly elastic impact both the momentum and kinetic energy are conserved.

The silent features/characteristics of different types of impacts are summarised below:

A. Perfectly elastic impact

(*i*) momentum is conserved, *i.e.*,

$$m_1 u_1 + m_2 u_2 = m_1 v_1 + m_2 v_2$$

(*ii*) kinetic energy is conserved, *i.e.*,

$$\frac{1}{2} m_1 u_1^2 + \frac{1}{2} m_2 u_2^2 = \frac{1}{2} m_1 v_1^2 + \frac{1}{2} m_2 v_2^2$$

(*iii*) the two bodies seperate after the impact

(*iv*) the coefficient of restitution is unity

(*v*) there is no permanent deformation in the bodies during the impact

B. Partially elastic impact

(*i*) momentum is conserved, *i.e.*,

$$m_1 u_1 + m_2 u_2 = m_1 v_1 + m_2 v_2$$

(*ii*) there is loss of kinetic energy during the impact.

(*iii*) the two bodies seperate after the impact

(*iv*) the coefficient of restitution is less than unity, i.e., $0 < e < 1$

(*v*) some permanent deformation remains in the body

C. Plastic impact

(*i*) the two bodies move together with common velocity after the impact

(*ii*) momentum is conserved, *i.e.*,

$$m_1 u_1 + m_2 u_2 = (m_1 + m_2)v$$

where v is the common velocity after impact.

(*iii*) there is loss in kinetic energy after the impact

(*iv*) the coefficient of restitution is zero

D. Impact with a fixed surface

The situation occurs when a body of certain mass and moving with a certain velocity strikes a wall.

(*i*) there is change in momentum, i.e., the momentum is not conserved.

(*ii*) there is loss in kinetic energy

(*iii*) the coefficient of restitution is

$$e = -\frac{v}{u}$$

where u and v are velocities of the body before and after the impact.

EXAMPLE 17.2.

Show that :

(*a*) *during elastic impact of two equal masses, there occurs exchange of velocities.*

(*b*) *when a body strikes an immovable body of very large mass, it rebounds with the same velocity with which it strikes*

(*c*) *when a body strikes another body of equal mass at rest, it imparts its entire velocity to the other mass and itself stops.*

Solution :

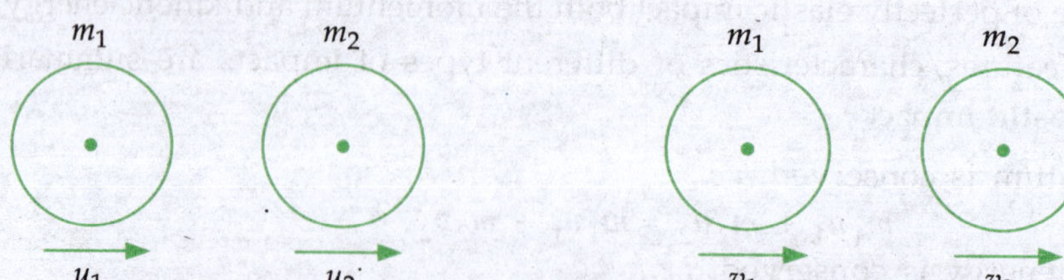

Fig. 17.8

Let u_1, u_2 = velocities of colliding bodies before impact $(u_1 > u_2)$

v_1, v_2 = velocities of the colliding bodies after impact

From the law of conservation of momentum

$$m_1 u_1 + m_2 u_2 = m_1 v_1 + m_2 v_2$$

Since the impact is between equal masses; $m_1 = m_2 = m$. Then

$$u_1 + u_2 = v_1 + v_2 \qquad ...(i)$$

The relation for coefficient of restitution gives

$$e = \frac{v_2 - v_1}{u_1 - u_2}$$

Since the impact is perfectly elastic $e = 1$ and accordingly

$$u_1 - u_2 = v_1 - v_2 \qquad ...(ii)$$

Solution of expressions (*i*) and (*ii*) gives: $v_2 = u_1$ and $v_1 = u_2$ which shows that the masses exchange velocities after elastic impact.

(*b*) Consider a body having direct impact on a fixed plane say a ball dropped on a floor

From the law of conservation of momentum

$$m_1 u_1 + m_2 u_2 = m_1 v_1 + m_2 v_2$$

or $$\frac{m_1}{m_2} u_1 + u_2 = \frac{m_1}{m_2} v_1 + v_2$$

when m_2 is very large and $u_2 = 0$, we get : $v_2 = 0$

The relation for coefficient of restitution gives

$$e = \frac{v_2 - v_1}{u_1 - u_2}$$

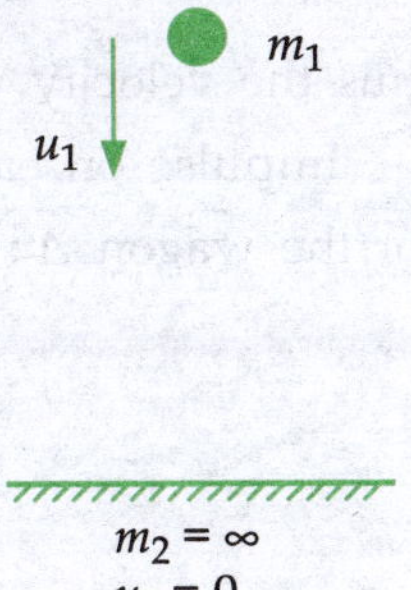

Fig. 17.9

Since the impact is perfectly elastic $e = 1$ and accordingly

$$u_1 - u_2 = v_2 - v_1$$

$$u_1 - 0 = 0 - v_1 \quad \therefore \quad u_1 = -v_1$$

Hence the ball rebounds with the same velocity with which it strikes the floor.

(*c*) Let u_1, u_2 = velocities of the colliding bodies before impact

v_1, v_2 = velocities of the colliding bodies after impact

From the law of conservation of momentum

$$m_1 u_1 + m_2 u_2 = m_1 v_1 + m_2 v_2$$

Since the impact is between equal masses and one of the mass is stationary

$$m_1 = m_2 = m \quad \text{and} \quad u_2 = 0$$

$\therefore$ $$mu_1 + m \times 0 = mv_1 + mv_2$$

or $$v_1 + v_2 = u_1 \qquad ...(i)$$

The relation of coefficient of restitution gives

$$e = \frac{v_2 - v_1}{u_1 - u_2}$$

Since the impact is perfectly elastic $e = 1$ and accordingly

$$v_2 - v_1 = u_1 - u_2 \qquad ...(ii)$$

Solution of expression (*i*) and (*ii*) gives : $v_2 = u_1$ and $v_1 = 0$ which shows that when a body strikes another body of equal mass, it stops after imparting its entire velocity to the other mass.

EXAMPLE 17.3.

Wagon A of mass 100 tonnes moving at 5 km/hr collides with the back of another wagon B of mass 40 tonnes and moving in the same direction at 1.5 km/hr.

After impact, the wagon B sets moving with a velocity of 7.5 m/s. Determine the velocity of wagon A after the impact and the impulse between the two wagons.

Solution: Let suffix 1 and 2 refer to wagon A and wagon B respectively.

From the law of conservation of momentum,

momentum before impact = momentum after impact

$$m_1 u_1 + m_2 u_2 = m_1 v_1 + m_2 v_2$$

or $$100 \times 5 + 40 \times 1.5 = 100 v_1 + 40 \times 7.5$$

or $$500 + 60 = 100 v_1 + 300$$

$$\therefore \quad v_1 = \frac{500 + 60 - 300}{100} = 2.6 \text{ km/hr}$$

Thus the velocity wagon A after the impact is 2.6 km/hr.

(*b*) Impulse on each wagon equals the change in its momentum

For the wagon A: $m = 100 \text{ tonnes} = 100 \times 10^3 \text{ kg}$

$$u_1 = \frac{5 \times 1000}{3600} = 1.39 \text{ m/s}$$

$$v_1 = \frac{2.6 \times 1000}{3600} = 0.72 \text{ m/s}$$

impulse = mass × change in velocity

$$= 100 \times 10^3 \times (1.39 - 0.72) = 67000 \text{ kgm/s} = \mathbf{67000 \text{ Ns}}$$

EXAMPLE 17.4.

A sphere of 1.5 kg mass moves to the right with velocity 12 m/s and strikes the vertical face of a block which is at rest and is of 3 kg mass. The block is supported on rollers and is attached to a spring of 4500 N/m spring constant as shown in Fig. 17.10. Determine the maximum shortening in the length of spring due to impact. Neglect friction and take coefficient of restitution for the block and sphere equal to 0.75.

Solution : Let suffix 1 and 2 refer to the sphere and block respectively. Then

$$m_1 = 1.5 \text{ kg} \quad \text{and} \quad u_1 = 12 \text{ m/s}$$

$$m_2 = 3 \text{ kg} \quad \text{and} \quad u_2 = 0$$

From the law of conservation of momentum,

momentum before impact = momentum after impact

$$m_1 u_1 + m_2 u_2 = m_1 v_1 + m_2 v_2$$

or $$1.5 \times 12 + 3 \times 0 = 1.5 v_1 + 3 v_2$$

$$\therefore \quad v_1 + 2 v_2 = 12 \quad \ldots(i)$$

The relation for coefficient of restitution gives

$$e = \frac{v_2 - v_1}{u_1 - u_2} \quad \text{or} \quad 0.75 = \frac{v_2 - v_1}{12}$$

$$\therefore \quad v_2 - v_1 = 9 \quad \ldots(ii)$$

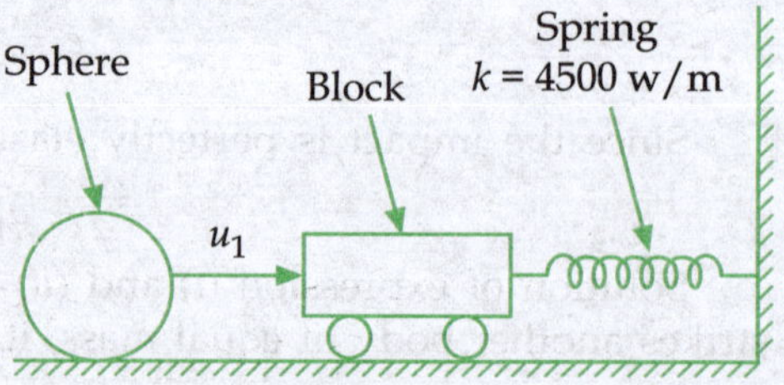

Fig. 17.10

Adding identities (*i*) and (*ii*), we get

$$3v_2 = 21 \qquad \therefore \qquad v_2 = 7 \text{ m/s}$$

The compression of spring can be calculated by applying the work-energy principle. That gives

$$\frac{1}{2}kx^2 = \frac{1}{2}m_2 v_2^2$$

$$\therefore \quad \text{Compression of spring}, \; x = \left[\frac{m_2 v_2^2}{k}\right]^{\frac{1}{2}} = \left[\frac{3 \times 7^2}{4500}\right]^{\frac{1}{2}}$$

$$= 0.1807 \text{ m} = \textbf{180.7 mm}$$

EXAMPLE 17.5.

A ball of mass 2 kg moving with a velocity of 3 m/s impinges directly on a ball of mass 4 kg at rest. After impact, the 2 kg mass ball comes to rest. Determine the velocity of 4 kg ball after striking and the coefficient of restitution between the two balls.

Solution : Let suffix 1 and 2 denote the 2 kg and 4 kg mass ball respectively.

$$m_1 = 2 \text{ kg}; \quad u_1 = 3 \text{ m/s and } v_1 = 0$$

$$m_2 = 4 \text{ kg and } u_2 = 0$$

From the law of conservation of momentum,

momentum before impact = momentum after impact

$$m_1 u_1 + m_2 u_2 = m_1 v_1 + m_2 v_2$$

or $$(2 \times 3) + (4 \times 0) = (2 \times 0) + (4 \times v_2)$$

$$\therefore \quad v_2 = \frac{6 + 0 - 0}{4} = \textbf{1.5 m/s}$$

Thus the velocity of 4 kg ball after the impact is 1.5 m/s.

(*b*) From the relation for coefficient of restitution,

$$e = \frac{v_2 - v_1}{u_1 - u_2} = \frac{1.5 - 0}{3 - 0} = \textbf{0.5}$$

Thus the coefficient of restitution between the balls is 0.5

EXAMPLE 17.6.

A sphere collides directly with a similar sphere which is at rest. Show that their velocities after the impact conform to the ratio (1 – e)/(1 + e) where e is the coefficient of restitution.

Solution : Let suffix 1 and 2 denote the first and second sphere respectively.

From the law of conservation of momentum,

$$m_1 u_1 + m_2 u_2 = m_1 v_1 + m_2 v_2$$

Since the spheres are similar : $m_1 = m_2 = m$ and accordingly

$$mu_1 + (m \times 0) = m(v_1 + v_2)$$

$$\therefore \quad v_1 + v_2 = u_1 \qquad \text{...}(i)$$

From the relation for coefficient of restitution,

$$e = \frac{v_2 - v_1}{u_1 - u_2} = \frac{v_2 - v_1}{u_1 - 0} = \frac{v_2 - v_1}{u_1}$$

$$\therefore \quad v_2 - v_1 = eu_1 \qquad ...(ii)$$

Adding expressions (*i*) and (*ii*)

$$2v_2 = u_1(1 + e) \quad \therefore \quad v_2 = \frac{u_1(1+e)}{2}$$

Then for expression (*i*),

$$v_1 = u_1 - v_2 = u_1 - \frac{u_1(1+e)}{2} = u_1\left[1 - \frac{1+e}{2}\right] = \frac{u_1(1-e)}{2}$$

$$\text{Ratio } \frac{v_1}{v_2} = \frac{u_1(1-e)/2}{u_1(1+e)/2} = \frac{1-e}{1+e}$$

EXAMPLE 17.7.

A ball impinges directly on a similar ball at rest. Due to impact, the first ball comes to rest and half of the initial kinetic energy gets lost. Make calculations for the coefficient for restitution.

Solution : Let suffix 1 and 2 refer to the first and second ball respectively

$$\text{Kinetic energy before impact } E_1 = \frac{1}{2}m_1u_1^2 + \frac{1}{2}m_2u_2^2 = \frac{1}{2}m_1u_1^2 \qquad (\because \; u_2 = 0)$$

$$\text{Kinetic energy after impact } E_2 = \frac{1}{2}m_1v_1^2 + \frac{1}{2}m_2v_2^2 = \frac{1}{2}m_2v_2^2 \qquad (\because \; v_1 = 0)$$

$$\text{Loss of kinetic energy } \Delta E = \frac{1}{2}m_1u_1^2 - \frac{1}{2}m_2v_2^2$$

Since the balls are similar $m_1 = m_2 = m$ and therefore

$$\Delta E = \frac{1}{2}m(u_1^2 - v_2^2)$$

$$\text{By the given condition } \Delta E = \frac{1}{2} \times \left(\frac{1}{2}m_1u_1^2\right)$$

$$\therefore \qquad \frac{1}{2}m(u_1^2 - v_2^2) = \frac{1}{2} \times \left(\frac{1}{2}mu_1^2\right)$$

$$\text{or} \qquad 2u_1^2 - 2v_2^2 = u_1^2 \quad \therefore \quad u_1^2 = 2v_2^2$$

$$\text{That gives : } \frac{v_2}{u_1} = \sqrt{0.5} = 0.707$$

From the relation for coefficient of restitution

$$e = \frac{v_2 - v_1}{u_1 - u_2} = \frac{v_2 - 0}{u_1 - 0} = \frac{v_2}{u_1} = 0.707$$

EXAMPLE 17.8.

A stone of mass 2 kg is thrown upward from the foot of a tower with a velocity of 20 m/s. One second later, another stone also of mass 2 kg is dropped from the top of the tower. If the collision between the stones takes place elastically, make calculations for the velocities of the stones just after collision. Take the height of tower as 20 m.

Solution : Let the two stones collide at distance *h* from the foot of tower after time *t* seconds.

$$\text{Apply the relation} \quad s = ut + \frac{1}{2}at^2$$

For the upward motion of first stone,

$$u = 20 \text{ m/s}, \quad s = h \quad \text{and} \quad a = -g = 9.81 \text{ m/s}^2$$

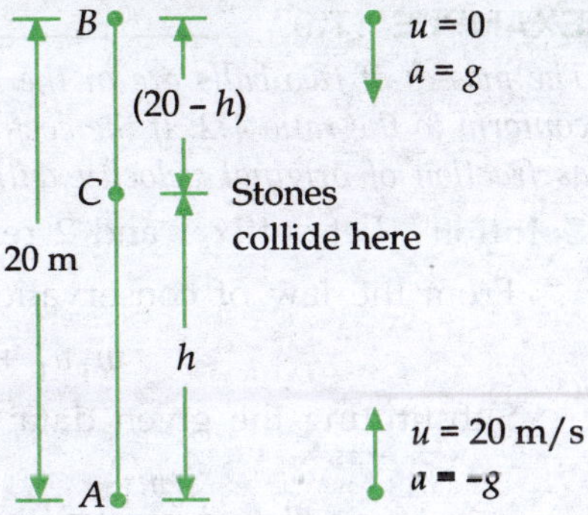

Fig. 17.11

$$\therefore \quad h = 20t - \frac{1}{2}\times 9.81 t^2$$

$$= 20t - 4.905t^2 \qquad ...(i)$$

For the downward motion of second stone

$$u = 0, \quad s = (20 - h)$$

and $\quad a = g = 9.81 \text{ m/s}^2$

$$\therefore \quad 20 - h = 0\times t + \frac{1}{2}\times 9.81\times(t-1)^2$$

$$= 4.905\,(t^2 - 2t - 1) \qquad ...(ii)$$

Adding expressions (*i*) and (*ii*), we get

$$20 = (20t - 4.905t^2) + (4.905t^2 - 9.81t + 4.905)$$

$$= 20t - 9.81t + 4.905$$

That gives : $\quad t = \dfrac{20 - 4.905}{20 - 9.81} = 1.48 \text{ sec}$

Velocity of first stone at the collision point,

$$v = u - gt = 20 - 9.81 \times 1.48 = 5.48 \text{ m/s upward}$$

Velocity of second stone at the collision point

$$v = u + g(t - 1) = 0 + 9.81\,(1.48 - 1)$$

$$= 4.71 \text{ m/s downward}$$

These are the velocities just before collision and hence we write

$$u_1 = 5.48 \text{ m/s} \uparrow \quad \text{and} \quad u_2 = 4.71 \text{ m/s} \downarrow$$

Designating the corresponding velocities after collision as v_1 and v_2 , the coefficient of restitution is

$$e = \frac{v_2 - v_1}{u_1 - u_2} \quad \text{or} \quad 1 = \frac{v_2 - v_1}{5.48 - (-4.71)}$$

The coefficient of restitution has been taken as unity as the collision is elastic. The above expression gives:

$$v_2 - v_1 = 5.48 - (-4.71) = 10.19 \qquad ...(a)$$

From the principle of conservation of momentum,

$$m_1 u_1 + m_2 u_2 = m_1 v_1 + m_2 v_2$$

or $\quad u_1 + u_2 = v_1 + v_2 \qquad (\because \; m_1 = m_2 = 2 \text{ kg})$

$$\therefore \quad v_1 + v_2 = 5.48 + (-4.71) = 0.67 \qquad ...(b)$$

Adding expressions (*a*) and (*b*), we get

$$2v_2 = 10.19 + 0.67$$

or $\quad v_2 = \dfrac{10.19 + 0.67}{2} = 5.43$

and $\quad v_1 = v_2 - 10.19 = 5.43 - 10.19 = -4.76$

Thus the velocities after collision are:

$$v_1 = 4.76 \text{ m/s (downward)}$$

and $\quad v_2 = 5.43 \text{ m/s (upward)}$

EXAMPLE 17.9.

The masses of two balls are in the ratio 2 : 1 and before impact, their velocities in the opposite direction conform to the ratio 1:2. If the coefficient of restitution for the direct impact is 5/6, determine the velocity as fraction of original velocity with which each ball would move after the impact.

Solution : Let suffix 1 and 2 refer to the first and second ball.

From the law of conservation of momentum,

$$m_1 u_1 + m_2 u_2 = m_1 v_1 + m_2 v_2$$

Substituting the given data: $m_2 = m_1/2$ and $u_2 = -2u_1$

$$m_1 u_1 + \frac{m_1}{2}(-2u_1) = m_1 v_1 + \frac{m_1}{2} v_2$$

or $$2\,m_1 u_1 - 2\,m_1 u_1 = 2\,m_1 v_1 + m_1 v_2$$

or $$0 = 2v_1 + v_2\,; \quad v_2 = -2v_1 \qquad ...(i)$$

From the relation for coefficient of restitution,

$$e = \frac{v_2 - v_1}{u_1 - u_2}$$

$$\frac{5}{6} = -\frac{v_2 - v_1}{u_1 - (-2u_1)}\,; \quad v_2 - v_1 = \frac{5}{2}u_1 \qquad ...(ii)$$

From expressions (*i*) and (*ii*)

$$-2v_1 - v_1 = \frac{5}{2}u_1\,; \quad v_1 = \frac{-5}{6}u_1$$

The minus sign indicates that the direction of v_1 is opposite to that of u_1. This implies that the first ball will move in opposite direction with 5/6th of its original velocity.

Substituting the values of v_1 in expression (*i*), we get

$$v_2 = -2v_1 = -2 \times \left(-\frac{5}{6}u_1\right) = -\frac{5}{6}(-2u_1) = -\frac{5}{6}u_2$$

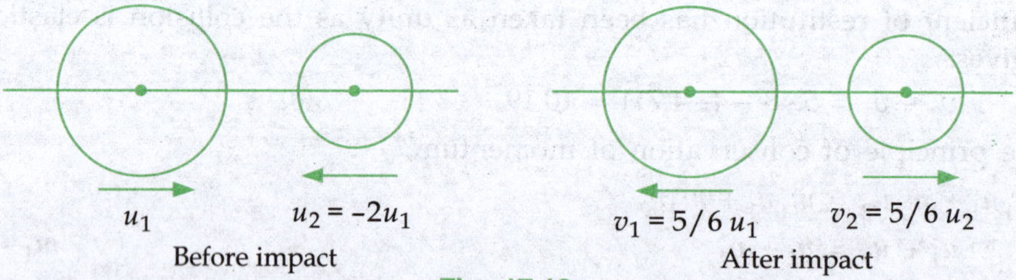

Fig. 17.12

The above identity shows that the second ball will also move in opposite with 5/6th of its original velocity.

Obviously the direction of both the balls gets reversed after the impact.

EXAMPLE 17.10.

Three perfectly elastic balls A, B and C of masses 2 kg, 4 kg and 8 kg respectively move along the same straight line and in the same direction with velocities of 4 m/s, 1 m/s and 0.75 m/s. The ball A collides with ball B and subsequently the ball B impinges with ball C. Make calculations to show that the balls B and C will be brought to rest after the impacts.

Solution : Consider collision between balls A and B.

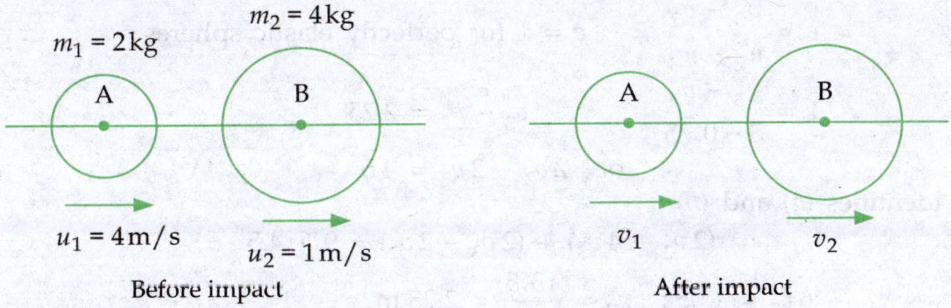

Fig. 17.13

From the law of conservation of momentum

$$m_1 u_1 + m_2 u_2 = m_1 v_1 + m_2 v_2$$

$$2 \times 4 + 4 \times 1 = 2v_1 + 4v_2 ; \quad v_1 + 2v_2 = 6 \qquad ...(i)$$

From the relation for coefficient of restitution,

$$e = \frac{v_2 - v_1}{u_1 - u_2} ;$$

Since the balls are perfectly elastic, $e = 1$ and that gives

$$1 = \frac{v_2 - v_1}{4 - 1} ; \quad v_2 - v_1 = 3 \qquad ...(ii)$$

Adding expressions (*i*) and (*ii*),

$$(v_1 + 2v_2) + (v_2 - v_1) = 6 + 3$$

$$3v_2 = 9; \quad v_2 = 3 \text{ m/s}$$

and $$v_1 = v_2 - 3 = 3 - 3 = 0$$

Thus the ball A will come to rest after the impact

(*ii*) Consider collision between balls B and C

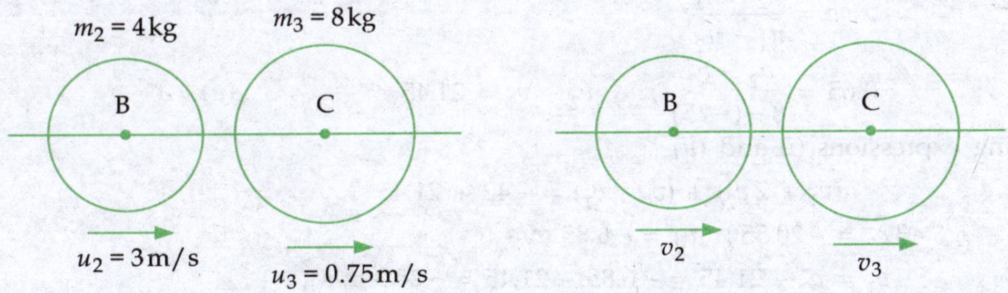

Fig. 17.14

After collision with ball A, the ball B starts moving with initial velocity $u_2 = 3$ m/s.

From the law of conservation of momentum,

$$m_2 u_2 + m_3 u_3 = m_2 v_2 + m_3 v_3$$

or $$4 \times 3 + 8 \times 0.75 = 4v_2 + 8v_3$$

or $$2v_2 + 4v_3 = 9 \qquad ...(a)$$

From the relation for coefficient of restitution

$$e = \frac{v_3 - v_2}{u_2 - u_3}\ ; \qquad e = 1 \text{ for perfectly elastic spheres}$$

$$1 = \frac{v_3 - v_2}{3 - 0.75} \qquad \therefore \quad v_3 - v_2 = 2.25$$

$$\text{or} \quad 2v_3 - 2v_2 = 4.5 \qquad ...(b)$$

Adding identities (*a*) and (*b*)

$$(2v_2 + 4v_3) + (2v_3 - 2v_2) = 9 + 4.5$$

$$6v_3 = 13.5\ ; \quad v_3 = \frac{13.5}{6} = 2.25 \text{ m/s}$$

and $$v_2 = \frac{9 - 4v_3}{2} = \frac{9 - 4 \times 2.25}{2} = 0$$

Obviously the ball *B* comes to rest after the impact.

EXAMPLE 17.11

A body of 10 kg mass moving towards right with a speed of 8 m/s strikes with another body of 20 kg mass moving towards left with 25 m/s. Determine:

(*a*) *final velocity of the two bodies,*

(*b*) *loss in kinetic energy due to impact, and*

(*c*) *impulse acting an either body during impact*

Take coefficient of restitution between the bodies as 0.65.

Solution : Let suffix 1 and 2 refer to bodies of 10 kg mass and 20 kg mass respectively

From the law of conservation of momentum,

$$m_1 u_1 + m_2 u_2 = m_1 v_1 + m_2 v_2$$

or $$10 \times 8 + 20 \times (-25) = 10v_1 + 20v_2$$

or $$80 - 500 = 10v_1 + 20v_2$$

or $$v_1 + 2v_2 = -42 \qquad ...(i)$$

From the relation for coefficient of restitution,

$$e = \frac{v_2 - v_1}{u_1 - u_2}$$

$$0.65 = \frac{v_2 - v_1}{8 - (-25)}\ ; \qquad v_2 - v_1 = 21.45 \qquad ...(ii)$$

Adding expressions (*i*) and (*ii*)

$$(v_1 + 2v_2) + (v_2 - v_1) = -42 + 21.45$$

$$3v_2 = -20.55\ ; \quad v_2 = \mathbf{-6.85 \text{ m/s } (\leftarrow)}$$

and $$v_1 = v_2 - 21.45 = -6.85 - 21.45 = \mathbf{-28.3 \text{ m/s } (\leftarrow)}$$

Both the bodies will move towards left after the impact.

(*b*) Kinetic energy before impact $E_1 = \frac{1}{2}m_1 u_1^2 + \frac{1}{2}m_2 u_2^2 = \frac{1}{2} \times 10 \times 8^2 + \frac{1}{2} \times 20 \times (-25)^2$

$$= 320 + 6250 = 6570 \text{ Nm}$$

Kinetic energy after impact $E_2 = \frac{1}{2}m_1 v_1^2 + \frac{1}{2}m_2 v_2^2 = \frac{1}{2} \times 10 \times (-28.3)^2 + \frac{1}{2} \times 20 \times (-6.85)^2$

$$= 4004.45 + 469.22 = 4473.67 \text{ Nm}$$

$\therefore$ Loss in kinetic energy due to impact

$$\Delta E = E_1 - E_2 = 6570 - 4473.67 = \mathbf{2096.33\ Nm}$$

Alternatively

$$\Delta E = \frac{m_1 m_2}{2(m_1 + m_2)} (u_1 - u_2)^2 (1 - e^2)$$

$$= \frac{10 \times 20}{2(10+20)} \{8-(-25)\}^2 \{1 - 0.65^2\} = \mathbf{2096.33\ Nm}$$

(*c*) Impulse during impact equals change in momentum. Considering body of 10 kg mass,

$$\text{Impulse} = m_1 v_1 - m_1 u_1 = 10 \times (-28.3) - 10 \times 8 = \mathbf{-363\ Ns}$$

Considering body of 20 kg mass,

$$\text{Impulse} = m_2 v_2 - m_2 u_2 = 20 \times (-6.85) - 20 \times (-25) = \mathbf{363\ Ns}$$

Impulse experienced by each body is equal in magnitude but opposite in direction.

EXAMPLE 17.12.

In a pile driving equipment, a ram of 800 kg mass is released from rest 2.5 m above the top of a pile of 2000 kg mass. The ram rebounds to a height of 0.15 m after hitting the pile. Calculate:

(*a*) *the velocity of pile after the impact,*
(*b*) *the coefficient of restitution, and*
(*c*) *the percentage of energy loss due to impact.*

Solution : Let the suffix 1 and 2 refer to ram and pile respectively.

The velocities of ram before and after the impact are:

$$u_1 = \sqrt{2 \times 9.81 \times 2.5} = 7\ \text{m/s}\ (\downarrow)$$

$$v_1 = \sqrt{2 \times 9.81 \times 0.15} = 1.716\ \text{m/s}(\uparrow).$$

Initially the pile is stationary and its velocity before impact is zero.

That is $u_2 = 0$

From the law of conservation of momentum,

$$m_1 u_1 + m_2 u_2 = m_1 v_1 + m_2 v_2$$

or

$$800 \times 7 + 2000 \times 0 = 800 \times (-1.716) + 2000\ v_2$$

$\therefore$ velocity of pile after the impact $v_2 = \dfrac{800 \times 7 + 800 \times 1.716}{2000} = 3.486\ \text{m/s}$

(*b*) Coefficient of restitution $e = \dfrac{v_2 - v_1}{u_1 - u_2} = \dfrac{3.486 - (-1.716)}{7 - 0} = 0.743$

(*c*) Kinetic energy before impact $E_1 = \dfrac{1}{2} m_1 u_1^2 + \dfrac{1}{2} m_2 u_2^2 = \dfrac{1}{2} \times 800 \times 7^2$

$$= 19600\ \text{Nm} \qquad (\because\ u_2 = 0)$$

Kinetic energy after impact $E_2 = \frac{1}{2}m_1 v_1^2 + \frac{1}{2}m_2 v_2^2 = \frac{1}{2} \times 800 \times (1.716)^2 + \frac{1}{2} \times 2000 \times (3.486)^2$

$$= 1177.8 + 12152.2 = 13330 \text{ Nm}$$

Percentage loss in kinetic energy $= \frac{19600 - 13330}{19600} =$ **31.99 %**

EXAMPLE 17.13.

A ball is dropped from a height of 10 m on a smooth floor and it rebounds to a height of 7 m. Determine the coefficient of restitution between the ball and floor and expected height of second rebounce.

Solution : Let u and v be the velocity of ball at first impact (strike and bounce) with the floor. Then

$$u = \sqrt{2gh}\ (\downarrow) \quad \text{and} \quad v = \sqrt{2gh_1}$$

The velocity of floor is zero before and after every impact, i.e., $u_f = 0$ and $v_f = 0$

From the relation for coefficient of restitution

$$e = \frac{v_f - v}{u - u_f} = \frac{-v}{u} \qquad (\because\ u_f = v_f = 0)$$

Since u and v are in opposite directions,

$$e = \frac{v}{u} = \frac{\sqrt{2gh_1}}{\sqrt{2gh}} = \left(\frac{h_1}{h}\right)^{1/2} \quad ...(i)$$

$$= \left(\frac{7}{10}\right)^{1/2} = \mathbf{0.837}$$

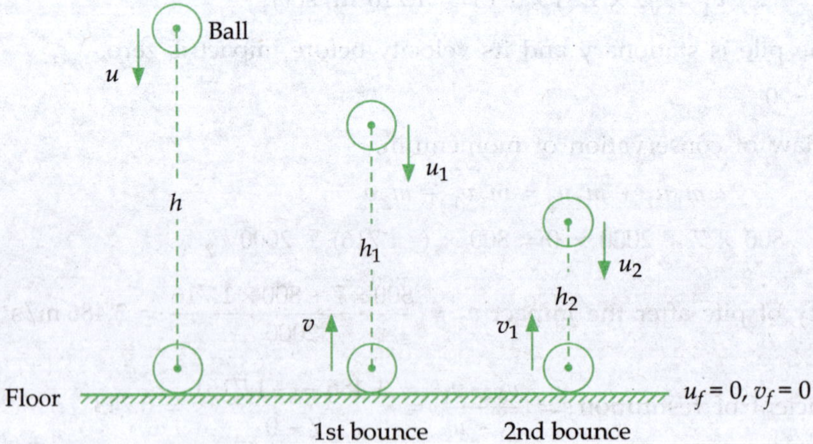

Fig. 17.15

(*b*) From expression (*i*),

$$h_1 = e^2 h$$

Like wise : $h_2 = e^2 h_1 = e^2 \times e^2 h = e^4 h$

$\therefore$ Height at second rebounce $h_2 = 0.337^4 \times 10 =$ **4.9 m**

EXAMPLE 17.14.

From what height must a heavy elastic ball be dropped on a floor so that after rebounding thrice, it will reach a height of 10 m?

Take coefficient of restitution $e = (0.5)^{1/3}$

Solution : Let u and v be the velocity of ball at first impact (strike and bounce) with the floor. Then

$$u = \sqrt{2gh} \ (\downarrow) \quad \text{and} \quad v = \sqrt{2gh_1} \ (\uparrow)$$

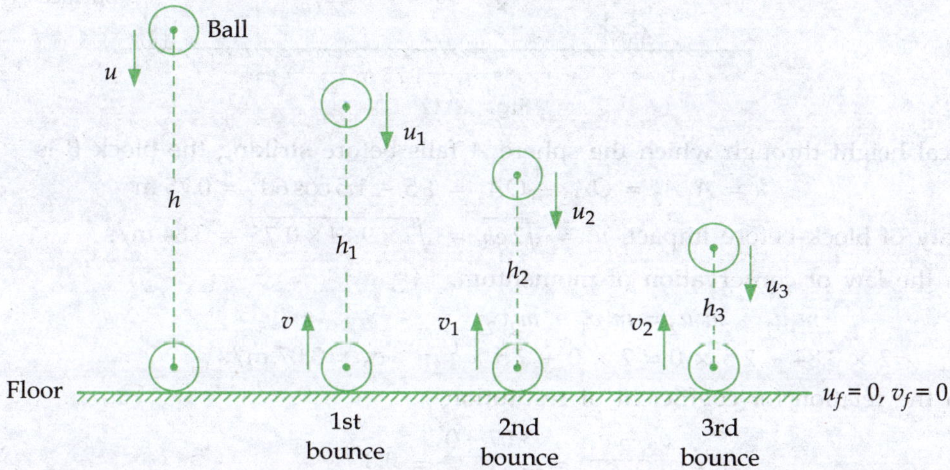

Fig. 17.16

The velocity of floor is zero before and after every impact, i.e., $u_f = 0$ and $v_f = 0$

From the relation for coefficient of restitution,

$$e = \frac{v_f - v}{u - u_f} = \frac{-v}{u} \qquad (\because \ u_f = v_f = 0)$$

Since u and v are in opposite direction

$$e = \frac{v}{u} = \frac{\sqrt{2gh_1}}{\sqrt{2gh}} = \left(\frac{h_1}{h}\right)^{1/2}$$

or $\quad h_1 = e^2 h$

Like wise: $\quad h_2 = e^2 h_1 = e^2 \times e^2 h = e^4 h$

$$h_3 = e^2 h_2 = e^2 \times e^4 h = e^6 h$$

Substituting the given data : $h_3 = 10$ m and $e = (0.5)^{1/3}$

$$h = \frac{h_3}{e^6} = \frac{10}{(0.5)^2} = \mathbf{40 \ m}$$

Thus the ball must be dropped from a height of 40 m.

EXAMPLE 17.15.

A sphere of 2 kg mass is released from rest when $\theta = 60°$ (Fig. 17.17) and it strikes a block B of 2.5 kg mass. The block B is initially resting on a rough horizontal surface and when struck moves through a distance of 1.75 m before again coming to rest. Make calculations for the coefficient of restitution and co-efficient of friction. Take velocity of sphere as zero after the impact.

Solution: Let suffix 1 and 2 denote the sphere and block respectively.

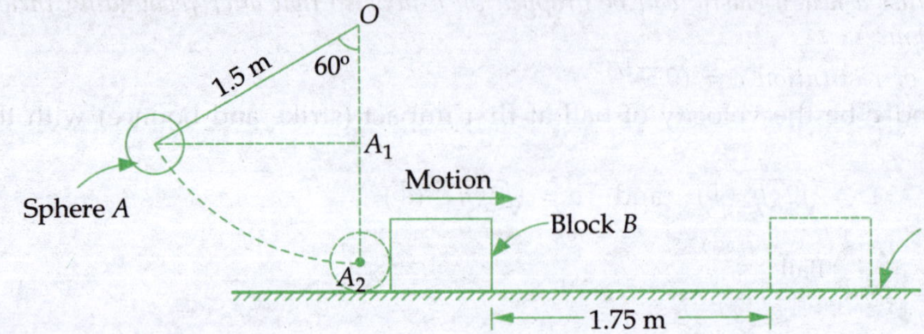

Fig. 17.17.

Vertical height through which the sphere A falls before striking the block B is

$$h = A_1A_2 = OA_2 - OA_1 = 1.5 - 1.5\cos 60^\circ = 0.75 \text{ m}$$

velocity of block before impact, $u_1 = \sqrt{2gh} = \sqrt{2 \times 9.81 \times 0.75} = 3.84 \text{ m/s}$

From the law of conservation of momentum,

$$m_1u_1 + m_2u_2 = m_1v_1 + m_2v_2$$

$$2 \times 3.84 + 2.5 \times 0 = 2 \times 0 + 2.5 \times v_2; \quad v_2 = 3.07 \text{ m/s}$$

From the relation for coefficient of restitution,

$$e = \frac{v_2 - v_1}{u_1 - u_2} = \frac{3.07 - 0}{3.84 - 0} = 0.8$$

(b) Considering motion of block

$$(KE)_1 = \frac{1}{2} \times 2.5 \times (3.07)^2 = 11.78 \text{ Nm}$$

$(KE)_2 = 0$; it is because the block finally comes to rest.

Change in kinetic energy $= (KE)_1 - (KE)_2 = 11.78$ Nm

Work done by frictional force $= \mu R \times$ distance moved

$$W_f = (\mu \times mg) \times d$$

Here $R = mg$ is the normal reaction

$$W_f = (\mu \times 2.5 \times 9.81) \times 1.75 = 42.92\,\mu$$

Then from the work – energy principle

$$42.92\,\mu = 11.78$$

$\therefore$ coefficient of friction, $\mu = \dfrac{11.78}{42.92} = 0.274$

EXAMPLE 17.16.

Two blocks A and B, each of 1 kg mass and resting on a rough inclined plane are released from the position as shown in Fig. 17.18. The static friction between block A and surface of incline is 0.12, and that between block B and surface of incline is 0.3. Make calculations, for the time at which impact would occur, and the velocity of blocks immediately after impact. Take coefficient of restitution equal to 0.75.

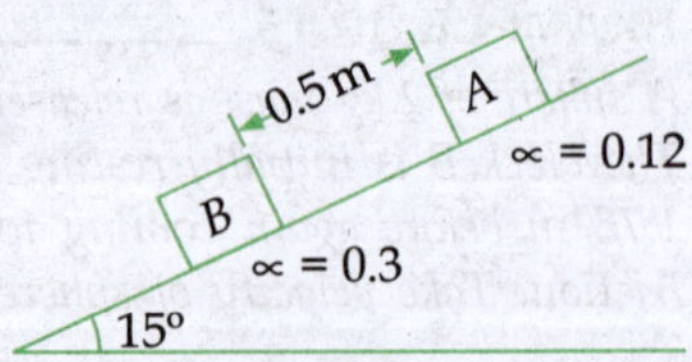

Fig. 17.18

Solution: For block B, $\mu = 0.3$

Angle of repose $\phi = \tan^{-1} 0.3 = 16.7°$

Inclination of the plane (15°) is less than the angle of repose and as such the block B does not move and remains in the same position at rest.

Considering free body diagram of block A,

$$\text{Normal reaction } R = W \cos\theta$$
$$= (1 \times 9.81) \times \cos 15° = 9.476 \text{ N}$$
$$\text{Force of friction } F = \mu R = 0.12 \times 9.476 = 1.137 \text{ N}$$

$$\text{Unbalanced force causing motion} = W \sin 15° - F_f$$
$$= (1 \times 9.81) \times \sin 15° - 1.137$$
$$= 1.402 \text{ N}$$

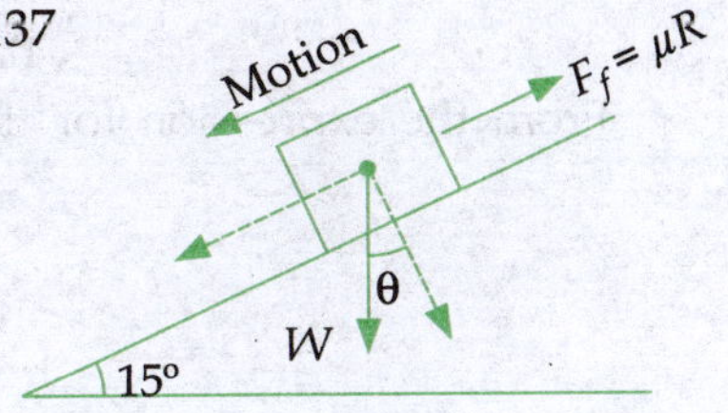

Then from Newton's second law of motion, $F = ma$

$$a = \frac{F}{M} = \frac{1.402}{1} = 1.402 \text{ m/s}^2$$

From kinematic equation $v^2 - u^2 = 2as$,

$$v = \sqrt{2as} = \sqrt{2 \times 1.402 \times 0.5} \quad (\because u = 0)$$
$$= 1.184 \text{ m/s}$$

This is the velocity of block A just before impact. Again from the kinematic equation $v = u + at$

$$t = \frac{v-u}{a} = \frac{1.184 - 0}{1.402} = \mathbf{0.844\ sec}$$

(*b*) From the law of conservation of momentum

$$m_a u_a + m_b u_b = m_a v_a + m_b v_b$$
$$1 \times 1.84 + 1 \times 0 = 1 \times v_a + 1 \times v_b$$
$$\therefore \quad v_a + v_b = 1.184$$

From the relation for coefficient of restitution,

$$e = \frac{v_b - v_a}{u_a - u_b} \quad \text{or} \quad 0.75 = \frac{v_b - v_a}{1.184 - 0}$$
$$\therefore \quad v_b - v_a = 0.888 \quad ...(ii)$$

Solving expressions (*i*) and (*ii*), we get

$$v_b = \frac{1.184 + 0.888}{2} = \mathbf{1.036\ m/s}$$

and $v_a = v_b - 0.888 = \mathbf{0.148\ m/s}$

EXAMPLE 17.17.

A spherical ball A of 100 gm mass, when released from rest, slides down the surface of a smooth container of height 30 cm. It collides with a spherical ball B of mass 25 gm resting at the bottom of the container as shown in Fig. 17.19. What should be the height h from which the ball A be released so that after the impact the ball B just leaves the container.

Assume coefficient of restitution = 0.8.

Solution: Let u_1 and v_1 = velocity of ball A before and after the impact

u_2 and v_2 = velocity of ball B before and after the impact.

From the given data : $u_1 = \sqrt{2gh} = 4.43\sqrt{h}$; $u_2 = 0$

The ball B, after the impact, should attain velocity v_2 just sufficient to rise to a height of 30 cm and leave the container.

$$v_2 = \sqrt{2g \times 0.3} = 2.43 \text{ m/s}$$

From the principle of conservation of momentum

$$m_1 u_1 + m_2 u_2 = m_1 v_1 + m_2 v_2$$

or
$$0.1 \times 4.43\sqrt{h} + 0.025 \times 0 = 0.1 v_1 + 0.025 \times 2.43$$

or
$$0.443\sqrt{h} = 0.1 v_1 + 0.0607 \qquad ...(i)$$

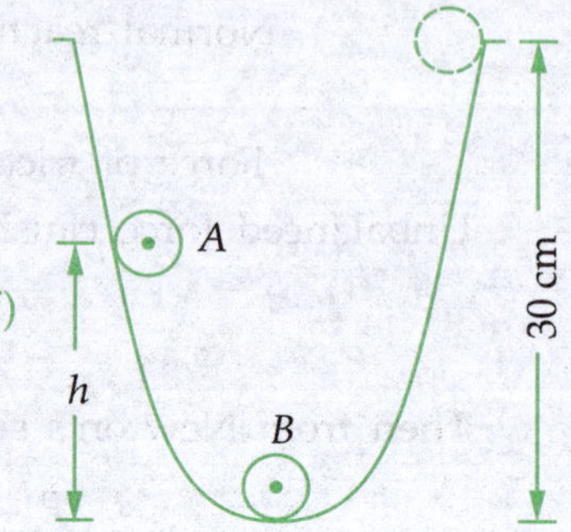

Fig. 17.19

From the expression for the coefficient for restitution,

$$e = \frac{v_2 - v_1}{u_1 - u_2}$$

$$0.8 = \frac{2.43 - v_1}{4.43\sqrt{h}}$$

$$3.544\sqrt{h} = 2.43 - v_1$$

Multiplying both sides by 0.1, we get

$$0.3544\sqrt{h} = 0.243 - 0.1 v_1 \qquad ...(ii)$$

Adding expressions (*i*) and (*ii*)

$$(0.443 + 0.3544)\sqrt{h} = 0.243 + 0.0607$$

or
$$0.7974\sqrt{h} = 0.3037 ; \quad \sqrt{h} = \frac{0.3037}{0.7974} = 0.3809$$

$$\therefore \quad h = (0.3809)^2 = 0.1451 \text{ m}$$

$$= \mathbf{14.51 \text{ cm}}$$

EXAMPLE 17.18.

Two steel balls A and B of mass m and 3 m respectively are suspended by fine threads so that they are just touching each other and are in a state of equilibrium. The ball A is pulled to angle θ as shown in Fig. 17.20 and released. To what angle ball A will swing after its second impact with the ball B. Assume impact to be perfectly elastic.

Solution: Let u_1 = velocity the ball A acquires in swinging downward to the point of impact

u_2 = velocity of ball B just before impact

v_1, v_2 = velocities of ball A and B after the impact

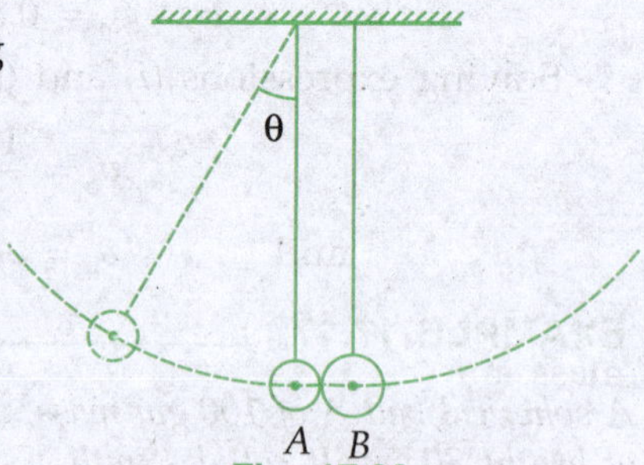

Fig. 17.20

From the principle of conservation of momentum,

$$m_1 u_1 + m_2 u_2 = m_1 v_1 + m_2 v_2$$

or
$$m u_1 + 3m \times 0 = m v_1 + 3m v_2$$

or
$$v_1 + 3v_2 = u_1 \qquad ...(i)$$

From the expression for the coefficient of restitution,

$$e = \frac{v_2 - v_1}{u_1 - u_2}$$

Since the impact is perfectly elastic ; $e = 1$ and therefore

$$1 = \frac{v_2 - v_1}{u_1 - 0} \; ; \quad v_2 - v_1 = u_1 \qquad ...(ii)$$

Adding expressions (*i*) and (*ii*)

$$(v_1 + 3v_2) + (v_2 - v_1) = u_1 + u_1$$

$$\therefore \; 4v_2 = 2u_1 \; ; \qquad v_2 = \frac{u_1}{2}$$

$$\text{and } v_1 = v_2 - u_1 = \frac{u_1}{2} - u_1 = -\frac{u_1}{2}$$

Apparently after the first impact, ball *A* rebounds with half of its striking velocity and an equal velocity in opposite direction is obtained by ball *B*.

The masses behave as pendulum and would return to impact point with same velocity. Accordingly at the instant of second impact,

$$u_1' = \frac{u_1}{2} \quad \text{and} \quad u_2' = -\frac{u_1}{2}$$

From the principle of conservation of momentum,

$$m \times \frac{u_1}{2} + 3m \times \left(-\frac{u_1}{2}\right) = mv_1' + 3mv_2'$$

or

$$v_1' + 3v_2' = -u_1 \qquad ...(a)$$

From the expression for the coefficient of restitution and for perfectly elastic impact,

$$e = 1 = \frac{v_2' - v_1'}{u_1' - u_2'} = \frac{v_2' - v_1'}{\frac{u_1}{2} + \frac{u_1}{2}}$$

or

$$v_2' - v_1' = u_1 \qquad ...(b)$$

From identities (*a*) and (*b*)

$$v_2' = 0 \quad \text{and} \quad v_1' = -u_1$$

Apparently after the second impact, ball *B* comes to rest, ball *A* rebounds with velocity u_1 and accordingly will swing back to the same angular position from which it was released.

EXAMPLE 17.19.

A sphere of 2 kg mass is released from rest and strikes another sphere of 4 kg mass which is at rest (Fig. 17.21). Presuming that the first sphere comes to rest after the impact, determine:

(*a*) *the coefficient of restitution*

(*b*) *the inclination of second sphere corresponding to the highest position to which it will rise.*

Take gravitational acceleration $g = 10\ m/s^2$.

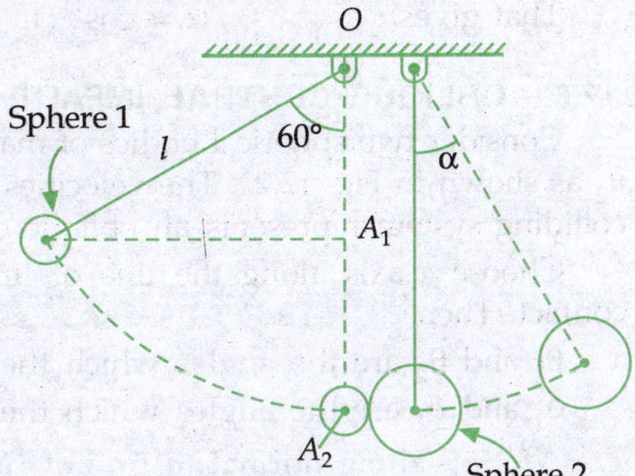

Fig. 17.21

Solution: Let suffix 1 and 2 denote the sphere of 2 kg mass and sphere of 4 kg mass respectively.

Vertical height through which the sphere 1 falls before striking the sphere 2 is

$$h = A_1A_2 = OA_2 - OA_1 = l - l\cos 60^\circ = 0.5\,l$$

$$\text{Work done} = m_1gh = 2 \times 10 \times 0.5\,l = 10\,l$$

Further: $(KE)_i = 0$ (the sphere is initially at rest)

$$(KE)_f = \frac{1}{2} m_1 V^2 = \frac{1}{2} \times 2 \times V^2 = V^2$$

Change in kinetic enegry $= V^2 - 0 = V^2$

Employing the work-energy principle,

Work done = change in kinetic energy

$$10l = V^2 \qquad \therefore \qquad V = \sqrt{10l}$$

At impact, $V = \sqrt{10l}$ is the velocity of sphere 1 before impact, and initial velocity of sphere 2 is zero.

Thus, we may write

$$u_1 = \sqrt{10l} \quad \text{and} \quad u_2 = 0$$

Designating the corresponding velocities after impact as v_1 and v_2, the conservation of momentum may be written as

$$m_1 u_1 + m_2 u_2 = m_1 v_1 + m_2 v_2$$

or

$$2 \times \sqrt{10l} + 0 = 0 + 4 v_2$$

That gives : $v_2 = 0.5\sqrt{10l}$

The coefficient of restitutions is

$$e = \frac{v_2 - v_1}{u_1 - u_2} = \frac{0.5\sqrt{10l} - 0}{\sqrt{10l} - 0} = 0.5$$

(*b*) For the application of work-energy principle to sphere 2, we have

$$v_2 = 0.5\sqrt{10l} \quad \text{and} \quad v_2' \text{ (velocity at highest position)} = 0$$

$$\text{work done} = -\, m_2 g(l - l\cos\alpha)$$

$$\therefore \qquad -\, m_2 g l(1 - \cos\alpha) = 0 - \frac{1}{2} m_2 (0.5\sqrt{10l})^2$$

or

$$4 \times 10l(1 - \cos\alpha) = \frac{1}{2} \times 4 \times 0.5^2 \times 10l$$

or

$$1 - \cos\alpha = 0.125$$

That gives :

$$\alpha = \cos^{-1}(1 - 0.125) = 28.96°$$

17.6. OBLIQUE CENTRAL IMPACT

Consider two spherical bodies of masses m_1 and m_2 moving respectively with velocities u_1 and u_2 as shown in Fig. 17.22. The velocities are not directed along the line of impact and as such the colliding system represents an oblique central impact.

Choose x-axis along the line of impact and y-axis along the tangent line at the point of contact. Then

θ_1 and θ_2 are the angles which the initial velocities u_1 and u_2 make with the x-axis, and

ϕ_1 and ϕ_2 are the angles which the final velocities v_1 and v_2 make with the x-axis.

Motion along normal to line of impact (y – axis) : If the contact surfaces are presumed to be smooth, then there is no frictional force and hence no impulse in the y-direction.

The momentum is accordingly conserved in y-direction for each body taken individually. That is

$$m_1 u_1 \sin\theta_1 = m_1 v_1 \sin\phi_1 \quad \text{or} \quad u_1 \sin\theta_1 = v_1 \sin\phi_1 \qquad ...(i)$$

and
$$m_2 u_2 \sin\theta_2 = m_2 v_2 \sin\phi_2 \quad \text{or} \quad u_2 \sin\theta_2 = v_2 \sin\phi_2 \qquad ...(ii)$$

From the above relations it is evident that component of velocity in the y-direction for each body remains unchanged.

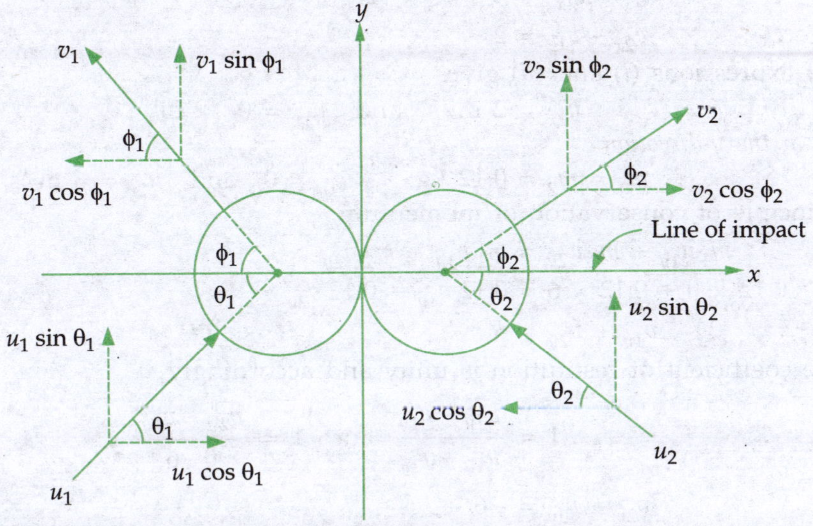

Fig. 17.22

Motion along line of impact (x-axis) : From the principle of conservation of momentum for the entire system (both bodies taken together) we have

$$m_1 u_1 \cos\theta_1 + m_2 u_2 \cos\theta_2 = m_1 v_1 \cos\phi_1 + m_2 v_2 \cos\phi_2 \qquad ...(iii)$$

Newton's law of collision of elastic bodies is equally valid for oblique impacts. That gives

$$e = \frac{v_2 \cos\phi_2 - v_1 \cos\phi_1}{u_1 \cos\theta_1 - u_2 \cos\theta_2} \qquad ...(iv)$$

The identities (*i*) to (*iv*) as written above can be solved to get the four unknowns v_1, v_2, ϕ_1 and ϕ_2. Care needs to be taken to give proper algebraic sign for the velocity with respect to direction of the x-axis chosen.

EXAMPLE 17.20.

A smooth spherical ball of 120 grams is moving from left to right with a velocity of 2 m/s in a horizontal plane. Another identical ball travelling in a perpendicular direction with a velocity of 6 m/s collides with ball A as shown in Fig. 17.23. Determine the velocity of both the balls after the impact.

Solution : For determining the velocities of balls after the impact, we first need to calculate the x and y-components of the velocities.

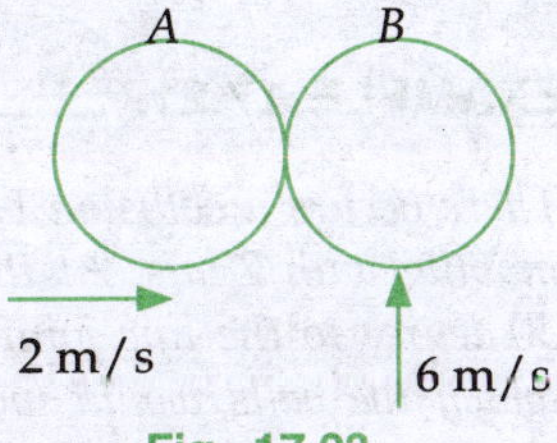

Fig. 17.23

For motion in the x-direction

$$m_1 = m_2 = 0.12 \text{ kg}; \quad u_{1x} = 2 \text{ m/s} \quad \text{and} \quad u_{2x} = 0$$

Then from the principle of conservation of momentum,

$$m_1 u_{1x} + m_2 u_{2x} = m_1 v_{1x} + m_2 v_{2x}$$

or $\quad 0.12 \times 2 + 0.12 \times 0 = 0.12\, v_{1x} + 0.12\, v_{2x}$

$\therefore \quad v_{1x} + v_{2x} = 2 \quad \ldots(i)$

For the smooth balls the impact is perfectly elastic and for that the coefficient of restitution is unity

Therefore, $\dfrac{v_{2x} - v_{1x}}{u_{1x} - u_{2x}} = 1$ or $\dfrac{v_{2x} - v_{1x}}{2 - 0} = 1$

$\therefore \quad v_{2x} - v_{1x} = 2 \quad \ldots(ii)$

Solution of expressions (*i*) and (*ii*) gives

$$v_{2x} = 2 \text{ m/s} \quad \text{and} \quad v_{1x} = 0$$

For motion in the y-direction

$$m_1 = m_2 = 0.12 \text{ kg}; \quad u_{1y} = 0 \quad \text{and} \quad u_{2y} = 6 \text{ m/s}$$

For the principle of conservation of momentum,

$$m_1 u_{1y} + m_2 u_{2y} = m_1 v_{1y} + m_2 v_{2y}$$

or $\quad 0.12 \times 0 + 0.12 \times 6 = 0.12\, v_{1y} + 0.12\, v_{2y}$

$\therefore \quad v_{1y} + v_{2y} = 6 \quad \ldots(iii)$

Again , the coefficient of restitution is unity and accordingly

$$1 = \frac{v_{2y} - v_{1y}}{u_{1y} - u_{2y}} \quad \text{or} \quad 1 = \frac{v_{2y} - v_{1y}}{0 - 6}$$

$\therefore \quad v_{2y} - v_{1y} = -6 \quad \ldots(iv)$

Solving expressions (*iii*) and (*iv*), we get

$$v_{1y} = 6 \text{ m/s} \quad \text{and} \quad v_{2y} = 0$$

For ball *A*, the *x*- and *y*-components of velocity after the impact are:

$$v_{1x} = 0 \quad \text{and} \quad v_{1y} = 6$$

$$\therefore \quad V_A = \sqrt{v_{1x}^2 + v_{1y}^2} = \sqrt{0^2 + 6^2} = 6 \text{ m/s}$$

with inclination $\alpha_A = \tan^{-1}\left(\dfrac{v_{1y}}{v_{1x}}\right) = \tan^{-1}(\infty) = \dfrac{\pi}{2}$

For ball *B*, the *x*- and *y*-component of velocity after the impact are

$$v_{2x} = 2 \text{ m/s} \quad \text{and} \quad v_{2y} = 0$$

$$\therefore \quad V_B = \sqrt{v_{2x}^2 + v_{2y}^2} = \sqrt{2^2 + 0^2} = 2 \text{ m/s}$$

with inclination $\alpha_B = \tan^{-1}\left(\dfrac{v_{2y}}{v_{2x}}\right) = \tan^{-1}\left(\dfrac{0}{2}\right) = 0$

EXAMPLE 17.21.

There occurs a collision between a 2 kg mass body moving with a velocity of 8 m/s and a 4 kg mass body moving with 2 m/s. At the instant of impact, the velocities of the two bodies are parallel and inclined at 30 degree to the line joining their centres. Determine the velocity, both in magnitude and direction, with which the balls would move after the impact.

Take coefficient of restitution e = 0.6

Solution : Refer Fig. 17.24 for the two bodies undergoing oblique central impact.

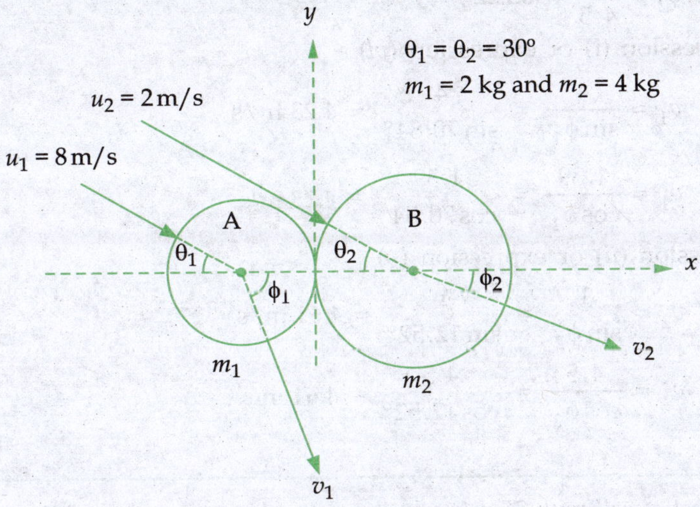

Fig. 17.24

Motion along y-axis : The component of velocity along the y-direction for each body remains unchanged. That is

$$8 \sin 30° = v_1 \sin \phi_1 ; \quad v_1 \sin \phi_1 = 4 \quad ...(i)$$

and
$$2 \sin 30° = v_2 \sin \phi_2 ; \quad v_2 \sin \phi_2 = 1 \quad ...(ii)$$

Motion along x-axis : From the principle of conservation of momentum for the entire system (both bodies taken together), we have

$$m_1 u_1 \cos \theta_1 + m_2 u_2 \cos \theta_2 = m_1 v_1 \cos \phi_1 + m_2 v_2 \cos \phi_2$$

or
$$2 \times 8 \cos 30° + 4 \times 2 \cos 30° = 2\, v_1 \cos \phi_1 + 4\, v_2 \cos \phi_2$$

or
$$13.86 + 6.93 = 2\, v_1 \cos \phi_1 + 4\, v_2 \cos \phi_2$$

or
$$v_1 \cos \phi_1 + 2\, v_2 \cos \phi_2 = 10.39 \quad ...(iii)$$

From the relation for the coefficient of restitution

$$e = \frac{v_2 \cos \phi_2 - v_1 \cos \phi_1}{u_1 \cos \theta_1 - u_2 \cos \theta_2}$$

$$0.6 = \frac{v_2 \cos \phi_2 - v_1 \cos \phi_1}{8 \cos 30° - 2 \cos 30°} = \frac{v_2 \cos \phi_2 - v_1 \cos \phi_1}{6.928 - 1.732}$$

or
$$v_2 \cos \phi_2 - v_1 \cos \phi_1 = 3.12 \quad ...(iv)$$

Adding identities (*iii*) and (*iv*),

$$(v_1 \cos \phi_1 + 2 v_2 \cos \phi_2) + (v_2 \cos \phi_2 - v_1 \cos \phi_1) = 10.39 + 3.12$$

$$3 v_2 \cos \phi_2 = 13.51; \quad v_2 \cos \phi_2 = 4.5 \quad ...(v)$$

Substituting the above value in identity (*iii*),

$$v_1 \cos \phi_1 + 2 \times 4.5 = 10.39 ; \quad v_1 \cos \phi_1 = 1.39 \quad ...(vi)$$

Dividing (*i*) by (*vi*) and dividing (*ii*) by (*v*)

$$\tan \phi_1 = \frac{4}{1.39} = 2.878 ; \quad \phi_1 = 70.84°$$

$$\tan \phi_2 = \frac{1}{4.5} = 0.222\,; \qquad \phi_2 = \mathbf{12.52°}$$

Then from expression (i) or expression (*vi*)

$$v_1 = \frac{4}{\sin \phi_1} = \frac{4}{\sin 70.84°} = 4.23 \text{ m/s}$$

or $$v_1 = \frac{1.39}{\cos \phi_1} = \frac{1.39}{\cos 70.84°} = \mathbf{4.23 \text{ m/s}}$$

and from expression (*ii*) or expression (*v*),

$$v_2 = \frac{1}{\sin \phi_2} = \frac{1}{\sin 12.52°} = 4.61 \text{ m/s}$$

or $$v_2 = \frac{4.5}{\cos \phi_2} = \frac{4.5}{\cos 12.52°} = \mathbf{4.61 \text{ m/s}}$$

EXAMPLE 17.22.

A ball is thrown against a wall with 25 m/s velocity and making an angle of 30° with the wall as shown in Fig. 17.25. Determine the velocity both in magnitude and direction, of the ball after the impact.

Assume the wall to be smooth and take coefficient of restitution between the ball and wall as 0.6

Solution: The wall is fixed and can be considered as infinite mass. Thus

$$u_2 = 0 \quad \text{and} \quad v_2 = 0$$

Consider motion of ball parallel to the wall (*i.e.*, at right angles to the line of impact). Since wall is smooth, the component of velocity parallel to the wall is conserved and we have

$$u_1 \sin \theta_1 = v_1 \sin \phi_1\,; \quad \theta_1 = 90 - 30 = 60°$$

$$25 \times 0.866 = v_1 \sin \phi_1\,; \;\; v_1 \sin \phi_1 = 21.65 \qquad ...(i)$$

Consider motion of ball perpendicular to the wall (*i.e.*, along the line of impact). The relation for the coefficient of restitution can be written as

$$e = \frac{(v_2)_x - (v_1)_x}{(u_1)_x - (u_2)_x} = \frac{0 - v_1 \cos \phi_1}{u_1 \cos \theta_1 - 0}$$

Since $v_1 \cos \phi_1$ and $u_1 \cos \theta_1$ are in opposite directions

$$e = \frac{v_1 \cos \phi_1}{u_1 \cos \theta_1}$$

or $$v_1 \cos \phi_1 = e \times u_1 \cos \theta_1 = 0.6 \times 25 \cos 60° = 7.5 \qquad ...(ii)$$

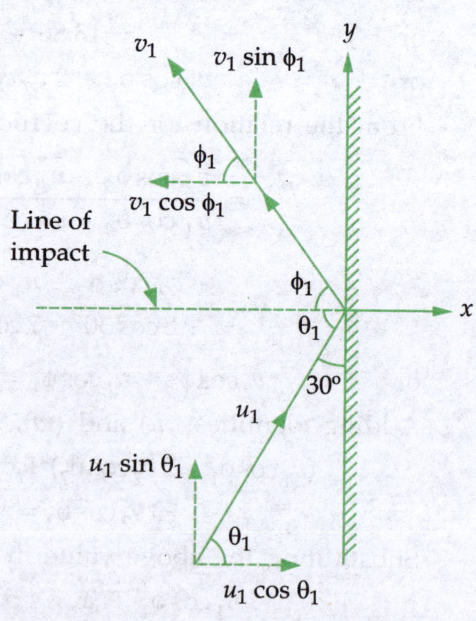

Fig. 17.25

Dividing expression (*i*) by expression (*ii*)

$$\tan \phi_1 = \frac{21.65}{7.5} = 2.887\,; \qquad \phi_1 = \mathbf{70.89°}$$

From expression (*i*) or (*ii*)

$$v_1 = \frac{21.65}{\sin 70.89} = \frac{21.65}{0.945} = \mathbf{22.91 \text{ m/s}}$$

or $$v_1 = \frac{7.5}{\cos 70.89} = \frac{7.5}{0.327} = 22.93 \text{ m/s}$$

EXAMPLE 17.23.

A vertically suspended sphere of 160 kg mass is hit by a 8 kg block moving with a velocity of 10 m/s on a rough horizontal floor as shown in Fig. 17.26. Make calculations for :

(*a*) *angular displacement of sphere from place of impact*
(*b*) *impulsive force*
(*c*) *final position of block with reference to place of impact*
(*d*) *maximum and minimum tensions in the string*

The following data may be taken into consideration while making the calculations:

Coefficient of friction between block and floor = 0.18
Coefficient of restitution for impact between block and sphere = 0.8
length of suspension string of sphere = 2.5 m and
duration of impact = 0.01 sec

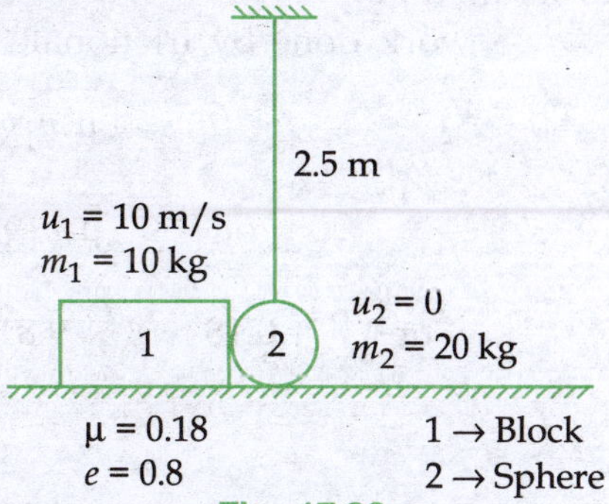

Fig. 17.26

Solution : Let suffix 1 and 2 represent the block and sphere respectively.

Considering impact between the block and the sphere, the principle of conservation of momentum can be written as,

$$m_1 u_1 + m_2 u_2 = m_1 v_1 + m_2 v_2$$
$$8 \times 10 + 16 \times 0 = 8v_1 + 16v_2$$
$$\therefore \quad v_1 + 2v_2 = 10 \qquad \ldots(i)$$

From the expression for the coefficient of restitution, we have

$$e = \frac{v_2 - v_1}{u_1 - u_2} \quad \text{or} \quad 0.8 = \frac{v_2 - v_1}{10 - 0}$$
$$\therefore \quad v_2 + v_1 = 8 \qquad \ldots(ii)$$

Solution of expression (*i*) and (*ii*) gives:

$$v_2 = 6 \text{ m/s } (\rightarrow) \quad \text{and} \quad v_1 = -2 \text{ m/s } (\leftarrow)$$

(*a*) Let α be the angle of swing when the sphere rises through height *h* after impact

Initial and final kinetic engergies of the sphere are :

$$(KE)_i = \frac{1}{2} m_2 v_2^2$$
$$= \frac{1}{2} \times 16 \times 6^2 = 288 \text{ Nm}$$
$$(KE)_f = 0$$
$$\text{Work done} = -mgh = -16 \times 9.81 \times h$$

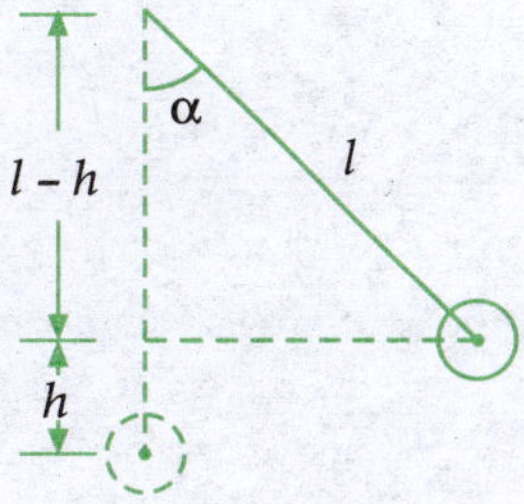

Fig. 17.27 a

By the work-energy principle, the work done equals the change in kinetic energy. That is

$$16 \times 9.81 \times h = 288 - 0$$
$$\therefore \quad h = \frac{288}{16 \times 9.81} = 1.83 \text{ m}$$
$$\alpha = \cos^{-1}\left(\frac{l - h}{l}\right) = \cos^{-1}\left(\frac{2.5 - 1.83}{2.5}\right) = \cos^{-1}(0.268) = 74.45°$$

Thus the swing of the sphere will be through an angle of 74.45° with the vertical.

(*b*) Let *F* be the impulsive force during impact. Then

$$F \times t = \text{change in momentum} = m(v_2 - v_1)$$
$$F \times 0.01 = 16(6 - 0)$$

$$\therefore \qquad F = \frac{16 \times 6}{0.01} = 9600 \text{ N}$$

(*c*) The distance travelled by the block can be worked out by applying work-energy principle to its motion.

work done by frictional force= change in kinetic energy

$$\mu m_1 g \times x = \frac{1}{2} m_2 v_2^2 - \frac{1}{2} m_1 v_1^2$$

$$\text{or} \qquad \mu m_1 g \times x = \frac{1}{2} m_1 v_1^2$$

$$\text{or} \qquad 0.18 \times 8 \times 9.81 \times x = -\frac{1}{2} \times 8 \times 2^2 = -16$$

$$\therefore \qquad x = \frac{-16}{0.18 \times 8 \times 9.81} = -1.132 \text{ m}$$

The negative sign implies that two block will move to left for a distance x = 1.132 m

(*d*) The maximum and minimum tensions in the string can be calculated by considering free body diagrams of the sphere just after the impact (Fig. 17.27 *b*) and at the highest position of the sphere (Fig. 17.27 *c*).

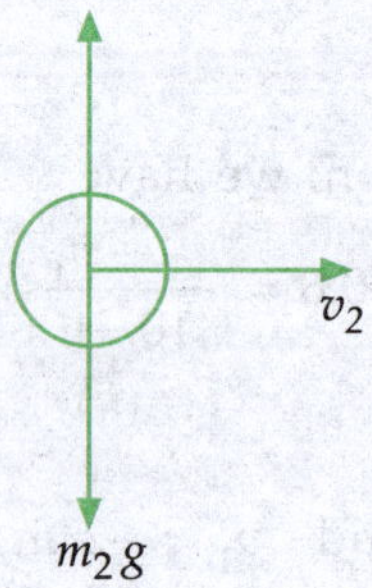

Fig. 17.27 b

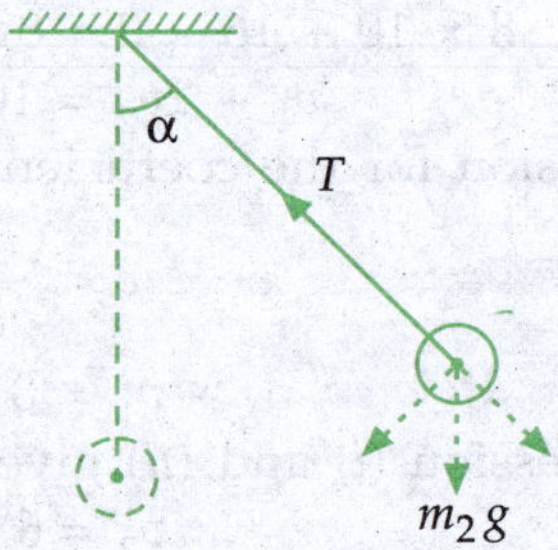

Fig. 17.27 c

(i) By D'Alembert's principle

$$T - m_2 g = m_2 a_2 = m_2 \frac{v_2^2}{l}$$

$$\therefore \qquad T = m_2 g + m_2 \frac{v_2^2}{l}$$

$$= 16 \times 9.81 + 16 \times \frac{6^2}{2.5} = 387.36 \text{ N} \text{ at the lowest position}$$

(*ii*) At the highest position, the velocity and hence the normal component of acceleration is zero. Hence

$$T = m_2 g \cos\theta = 16 \times 9.81 \times \cos 74.45°$$

$$= 42.08 \text{ N at the highest position.}$$

REVIEW AND SUMMARY

1. The term impact refers to the phenomenon of collision between two bodies which occurs in a very short interval of time, and during which the bodies exert very large force on each other.
2. The impact is elastic if the body rebounds after impact. Greater the elasticity of the body, greater will be the rebound.
   The impact is inelastic if the body does not rebound at all.

3. When the two bodies collide with one another, the resulting impact may be
   (*i*) central and eccentric impact
   (*ii*) direct and oblique impact
4. Momentum of the system before collision is equal to the momentum after collision. That is
$$m_1u_1 + m_2u_2 = m_1v_1 + m_2v_2$$
5. When two moving bodies collide with each other, their velocity of separation bears a constant ratio to their velocity of approach.
$$v_2 - v_1 = e(u_1 - u_2)$$
The constant of proportionality e is called the coefficient of restitution. The value of e lies between 0 and 1
   (*i*) For perfectly elastic bodies ; $e = 1$
   (*ii*) For inelsatic bodies : $e = 0$
6. The coefficient of restitution is a measure of the loss of energy during an impact.
$$\Delta E \text{ (energy loss)} = \frac{m_1 m_2}{2(m_1 + m_2)} (u_1 - u_2)^2 (1 - e^2)$$
When the colliding bodies are perfectly elastic, *e.g.*, impact between two hardened and polished balls ($e - 1$), energy loss is zero. That is
kinetic energy before impact = kinetic energy after impact
7. In case of perfectly elastic impact, both the momentum and kinetic energy are conserved.

REVIEW QUESTIONS

A. Conceptual and conventional questions :

1. When is a collision called an impact ?
2. Define line of impact.
3. Name the different types of impacts and illustrate those through appropriate sketches.
4. Distinguish between: (*i*) central impact and eccentric impact
   (*ii*) direct impact and oblique impact.
5. State the different stages in the phenomenon of collision.
6. What is meant by period of deformation and period of restitution.
7. Define coefficient of restitution and distinguish between perfectly elastic impact and perfectly plastic impact.
8. Set up an equation concerning principle of conservation of momentum for two bodies in direct central impact.
9. Set up the following expression for the loss of kinetic energy during impact:
$$\Delta E = \frac{m_1 m_2}{2(m_1 + m_2)} (u_1 - u_2)^2 (1 - e^2)$$
where the symbols have their usual meanings.
10. A sphere of mass 1 kg moving at 3 m/s overtakes another sphere of mass 5 kg moving along the same line at 0.6 m/s. Show that, due to impact, the direction of motion of the first sphere is reversed. Also calculate the loss of kinetic energy during impact.
(– 0.5 m/s, 10.5 Nm)

11. A ball is thrown vertically downwards with a velocity V m/s from a height of 1 m so that it hits the ground and just touches the ceiling after impact; the ceiling is at a height of 3.5 m from ground. Taking coefficient of restitution equal to 0.75, make calculations for the velocity V. **(6.62 m/s)**
12. Three perfectly elastic balls A, B and C of masses 2 kg, 6 kg and 12 kg move in the same direction with velocities of 12 m/s, 4 m/s and 2 m/s respectively. If the ball A impinges with ball B, which in turn, impinges with ball C, prove that the balls A and B will be brought to rest after the impact.
13. A smooth spherical ball A of mass 120 gms is moving from left to right with a velocity of 2 m/s in the horizontal plane. Another identical ball B travelling in a perpendicular direction with a velocity of 6 m/s collides with A as shown in in Fig. 17.28. What will be the velocities of balls A and B after the impact ?
14. A sphere of mass m_1 and initial velocity v_1 strikes another sphere of mass m_2 travelling in the same direction with velocity v_2. If the impact is perfectly elastic, calculate the final velocity of ball. Further, discuss the following situations :
    (a) $m_1 = m_2$ and $v_2 = 0$
    (b) $m_1 = m_2 = m$
    (c) $m_2 >>> m_1$ and $v_2 = 0$.
15. Two identical balls B and C are at rest when ball B is stuck by a ball A of the same mass moving with a velocity of 6 m/s. This causes a series of collisions between various balls. Presuming that coefficient of restitution is 0.5, determine the velocity of each ball after all collisions have taken place.
    (1.22 m/s, 1.41 m/s, 3.38 m/s)

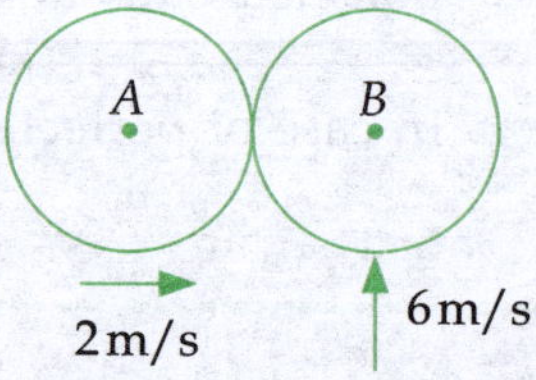

Fig. 17.28

16. A glass marble weighing 0.2 N falls from 10 m height onto a horizontal floor and rebounds to a height of 8 m. Neglecting air resistance, calculate
    (a) coefficient of restitution,
    (b) impulse and the average force between the marble and floor if the time of their contact is 0.1 second. **(0.894, 0.54 Ns, 5.406 N)**

B. Complete the following statements with most appropriate word/words :

1. The line of impact is the line .... to the surfaces of contact during impact.
2. In .... impact, the mass centres of the two colliding bodies lie on the line of contact.
3. The velocities of the two colliding bodies act along lines other than the line of contant. The impact is then known as .... impact.
4. The period during which the colliding bodies remain deformed is the ....
5. The two colliding bodies would join together and travel with the same velocity when the impact is ....
6. In a .... impact, the relative velocity before impact is equal to the relative velocity after the impact.
7. When two bodies collide with each other, then the ratio of their velocity of separation to their velocity of approach bears a constant ratio called the ....
8. The value of coefficient of restitution is .... for perfectly elastic bodies.

9. During elastic impact, both .... and .... are conserved.

10. A body of mass 3 kg moving with a velocity of 1.5 m/s impinges directly on a body of mass 6 kg at rest. After impinging, the first body comes to rest and the second body moves with a velocity of 0.75 m/s. The coefficient of restitution would be ....

Answers :

1. perpendicular 2. central 3. oblique 4. period of deformation 5. purely elastic 6. pure elastic 7. coefficient of restitution 8. one 9. kinetic energy, momentum 10. 0.5

C. Multiple choice questions :

1. During inelastic collision of two particles, which one of the following is conserved?
   (*a*) total linear momentum only
   (*b*) total kinetic energy only
   (*c*) both linear momentum and kinetic energy
   (*d*) neither linear momentum nor kinetic energy
2. The coefficient of restitution is defined on the basis of
   (*a*) velocity components along the line of impact only
   (*b*) velocity components normal to the line of impact
   (*c*) energies of bodies before and after the impact
   (*d*) impulses of recovery and deformation for either body
3. Two bodies moving in the same direction with velocity V_1 and V_2 undergo elastic collision and upon separation the corresponding velocities are V_1' and V_2' respectively. The coefficient of restitution is defined as
   (*a*) $\dfrac{V_2' - V_1'}{V_1 - V_2}$ (*b*) $\dfrac{V_2 - V_1}{V_1' - V_2'}$ (*c*) $\dfrac{V_1' - V_1}{V_2' - V_2}$ (*d*) $\dfrac{V_1 - V_2}{V_1' - V_2'}$
4. For perfectly elastic bodies, the value of coefficient of restitution is
   (*a*) 1 (*b*) 0.5 to 1 (*c*) 0 to 0.5 (*d*) zero
5. A tennis ball is dropped onto a plane surface from height 1.0 metre. After rebound, the ball rises to 0.64 metre height. The coefficient of restitution is
   (*a*) 0.64 (*b*) 0.8 (*c*) 0.97 (*d*) 0.51
6. A marble ball is rolled on smooth floor of a room and hits the wall after 1 second. Subsequent upon rebound, it takes twice the time to reach the initial position from where it was rolled. The coefficient of restitution between the wall and ball is
   (*a*) 0.25 (*b*) 0.50 (*c*) 0.75 (*d*) 1.0
7. A ball impinges directly upon another similar ball at rest and itself comes to rest due to impact. During collision, half of the kinetic energy gets dissipated. The coefficient of restitution is
   (*a*) 0.25 (*b*) 0.5 (*c*) 0.707 (*d*) 0.84
8. A sphere *P* impinges directly onto another identical sphere *Q* at rest. If the coefficient of restitution is 0.5, the ratio of velocities V_q/V_p after the impact would be
   (*a*) 1 : 1 (*b*) 2 : 1 (*c*) 3 : 1 (*d*) 2 : 3

Answers :

1. (*a*) 2. (*a*) 3. (*a*) 4. (*a*) 5. (*b*) 6. (*b*)
7. (*c*) 8. (*c*)

Kinetics: Impulse, Momentum, Work and Energy 18

Learning objectives : After a study of the subject matter included in this chapter, the readers will become familiar with the following aspects of kinetics

- *impulse, momentum, work and energy*
- *theory of motion of a lift and motion of connected bodies*
- *dynamic equilibrium and D'Alembert's principle*
- *conservative force and principle of conservation of energy ; work-energy principle*
- *work of force by gravity and by a spring*

18.1. FORCE, IMPULSE AND MOMENTUM

Force is an external agent which tends to change the state of rest or of uniform motion of a system. A force is applied whenever the system needs to be accelerated or decelerated. Force is a vector quantity specified completely by its

— magnitude
— point of application
— line of action, and
— direction.

The relationship between motion and force is provided by laws of dynamics; the most prominent being the Newton's second law. Newton's second law of motion states that force is proportional to rate of change of momentum. That is

$$\text{force} \propto \text{rate of change of momentum}$$

Momentum is the product of mass and velocity of a body and represents the energy of motion stored in a moving body.

$$\therefore \text{force} \propto \text{rate of change of (mass} \times \text{velocity)}$$
$$\propto \text{mass} \times \text{rate of change of velocity}$$
$$\propto \text{mass} \times \text{acceleration}$$

Thus $$F \propto ma\,; \quad F = \frac{ma}{g_c}$$

where m is the mass, a is acceleration and g_c is proportionality constant.

The SI unit of force is newton (N) which represents the force required to accelerate 1 kg mass with an acceleration of $1\ \text{m/s}^2$.

Thus $\quad 1\text{ N} = \dfrac{1\text{ kg} \times 1\text{ m/s}^2}{g_c}$ and $g_c = \dfrac{1\text{ kg m}}{\text{Ns}^2}$

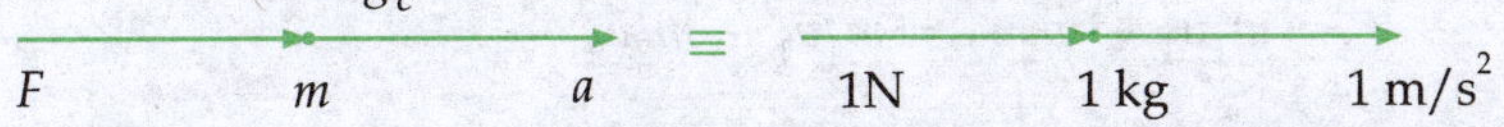

Fig. 18.1

The constant of proportionality g_c becomes unity for defining unit of force as Newton and accordingly is dropped. Such a practice corresponds to writing Newton's second law as

$$\text{Force} = \text{mass} \times \text{acceleration}\,; \quad F = ma$$

$$\therefore \quad 1\text{ N} = 1\text{kg}\frac{\text{m}}{s^2}$$

Other units of force are:

1 kilo newton (kN) = 10^3 N
1 mega newton (mN) = 10^6 N
1 giga newton (gN) = 10^9 N

Force is also measured in gravitational units expressed as kilogram force (kg_f). A kg_f is defined as that force which when applied to a body of 1 kg mass produces an acceleration equal to 9.81 m/s² (gravitational acceleration near the earth's surface).

$$1\text{ kg}_f = 1\text{ kg} \times 9.81\text{ m/s}^2 = 9.81\text{ kg m/s}^2$$

The newton and kilogram force are connected by the relation:

$$1\text{ kg}_f = 9.81\text{ N}$$

The identity $F = ma$ may be written as

$$F = m\frac{dV}{dt} = \frac{d}{dt}(mV) = \frac{mV_f - mV_i}{t} \qquad \text{...(18.1)}$$

where V_f and V_i are the final and initial velocities respectively.

The product of mass and velocity (mv) is called momentum of the body. Therefore

mV_i = momentum of the body at the beginning
mV_f = momentum of the body at the end of time t.

Thus *"the change in linear momentum per unit time is proportional to the impressed force and takes place in the direction of the force."*

Further, $\quad F \times t = mV_f - mV_i = m\,(V_f - V_i) \qquad \text{...(18.2)}$

The product of force and the time during which it acts ($F \times t$) is called the **impulse** of force. Thus the impulse equals the change in momentum.

Equation of 18.1 is referred to as the impulse-momentum relation.

Momentum and impulse are vector quantities with unit kg m/s or Ns.

The concept of impulse is useful in solving problems when a large force acts for a very short time, e.g., sudden blow of hammer striking a nail.

The *law of conservation of momentum* may be stated as:

"Total momentum of any group of objects always remains the same if no external force acts on it."

Consider that a body A of mass m_1 moving with velocity u_1 collides with another body B of mass m_2 and moving with velocity u_2. Let v_1 and v_2 be their velocities after the collision. Then

$$\text{Momentum of masses before collision} = m_1u_1 + m_2u_2$$
$$\text{Momentum of masses after collision} = m_1v_1 + m_2v_2$$

In accordance with law of conservation of momentum,

momentum before collision = momentum after collision

$$m_1u_1 + m_2u_2 = m_1v_1 + m_2v_2$$

EXAMPLE 18.1.

Determine the force necessary to produce an acceleration of 4 m/s² in a mass of 250 kg.

Solution: From Newton's second law, $F = ma$

$$\therefore \quad F = 250 \times 4 = \mathbf{1000\ N} \text{ (SI units)}$$

$$\text{or} \quad F = \frac{250 \times 4}{9.81} = \mathbf{101.94\ kg_f} \text{ (Gravitational units)}$$

EXAMPLE 18.2.

A body of mass 160 kg is made to move with a velocity of 20 m/s when a force of 80 N acts on it for 60 seconds. Make calculations for the velocity attained by the body when the

(a) force acts in the direction of motion

(b) force acts in the direction opposite to that of motion.

Solution: Using the relation $F = ma$,

$$\text{Acceleration/retardation of body } a = \frac{F}{m} = \frac{80}{160} = 0.5 \text{ m/s}^2$$

(*a*) When the force acts in the direction of motion, the body will have acceleration. The final velocity v is then given by

$$v = u + at = 20 + 0.5 \times 60 = \mathbf{50\ m/s}$$

(*b*) When the force acts in a direction opposite to that of motion, the body will have retardation. The final velocity v is then given by

$$v = u - at \quad \text{(– ve sign shows retardation)}$$

$$= 20 - 0.5 \times 60 = \mathbf{-10\ m/s}$$

The – ve sign implies that the body will be moving in the opposite (reversed) direction.

EXAMPLE 18.3.

A body of mass 50 kg rests on a rough horizontal surface ($\mu = 0.4$) and is acted upon by a push applied at an angle of 45° to the horizontal. Determine the magnitude of push if it causes the body to move with an acceleration of 2 m/s².

Solution: Refer Fig. 18.2 for the various force acting on the body

Normal reaction $R = W + P \sin 45°$

$= (50 \times 9.81) + P \sin 45°$

$= 0.707\,P + 490.5$

Force of friction $F = \mu R = 0.4\,(0.707\,P + 490.5)$

$= 0.2828\,P + 196.2$

Unbalanced force causing motion

$= P \cos 45° - (0.2828\,P + 196.2)$

$= 0.707\,P - 0.2828\,P - 196.2 = 0.4242\,P - 196.2$

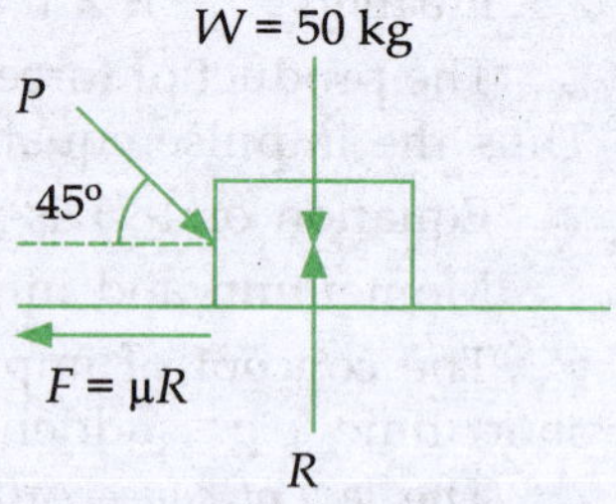

Fig. 18.2

Applying Newton's second law of motion $F = ma$, we have

$$0.4242\,P - 196.2 = 50 \times 2 \quad \therefore \quad \text{Push } P = \frac{100 + 1962.2}{0.4242} = \mathbf{698.25\ N}$$

EXAMPLE 18.4.

A cricket ball of mass 175 gm is moving with a speed of 36 km/hr. What average force will be required to stop the ball in 0.2 second.

Solution: Initial speed of the ball $u = 36 \text{ km/hr} = \dfrac{36 \times 1000}{3600} = 10 \text{ m/s}$

From the impulse-momentum relation,

$$F \times 5 = m\,(v - u)$$

$$F \times 0.2 = \frac{175}{1000}(0 - 10) \quad \therefore \quad F = -8.75 \text{ N}$$

The negative sign indicates that the force F is to be applied in a direction opposite to that in which the ball was moving.

EXAMPLE 18.5.

A machine gun fires 180 bullets per minute. Each bullet has a mass of 50 gm and the muzzle velocity of the bullet is 600 m/s. Presuming the firing to be continuous, make calculations for the average reaction of the gun against the support.

Solution: Time taken in firing one bullet $t = \dfrac{60}{180} = 0.333 \text{ s}$

From the impulse–momentum relation,

$$F \times t = m\,(v - u)$$

$$F \times 0.333 = \frac{50}{1000}\,(0 - 600) \quad \therefore \quad F = -90.1 \text{ N}$$

The negative sign indicates that this is the reaction of the gun on the bullet. The average reaction of the gun against the support will be numerically equal to it.

EXAMPLE 18.6.

A pile hammer of 250 kg mass is made to fall freely on a pile from a height of 6 m. If the hammer comes to rest in 0.012 second, determine the change in momentum, impulse and average force.

Solution: Since the initial velocity $u = 0$, the velocity with which the hammer falls on the pile is

$$v = \sqrt{2gh} = \sqrt{2 \times 9.81 \times 6} = 10.85 \text{ m/s}$$

Change in momentum = mass × change in velocity = 250 × (10.85 − 0) = 2712.5 Ns

Impulse equals the change in momentum

$\therefore$ Impulse = 2712.5 Ns

Further, impulse represents the product of force and time during which it acts. Then

$$F \times t = 2712.5$$

$$\therefore \quad \text{Average force } F = \frac{2712.5}{0.012} = 226042 \text{ N} \simeq 226.04 \text{ kN}$$

EXAMPLE 18.7.

A glass marble whose weight is 0.2 N falls from a height of 12 m and rebounds to a height of 9 m. The marble and the floor remain in contact for 0.1 second. Determine the impulse and the average force between the marble and the floor.

Solution: Velocity with which the marble strikes the floor

$$= \sqrt{2gh} = \sqrt{2 \times 9.81 \times 12} = 15.34 \text{ m/s (downward)}$$

Velocity of rebound $= \sqrt{2 \times 9.81 \times 9} = 13.29 \text{ m/s (upward)}$

$$\text{Impulse} = \text{change of momentum}$$
$$= \text{mass} \times \text{change of velocity}$$
$$= \frac{0.2}{9.81} \times [13.29 - (-15.34)] = \mathbf{0.5837\ Ns}$$

Impulse represents the product of force and the time during which it acts. Then

$$F \times t = 0.5837$$
$$\therefore \text{ Average force } F = \frac{0.5837}{0.1} = \mathbf{5.837\ N}$$

EXAMPLE 18.8.

A man throws a 10 kg suitcase with a horizontal velocity of 4 m/s into a 25 kg platform trolly. Determine the velocity of trolly after the suitcase has slid to stop on trolly.

Solution: Let the suffix s and t refer to suitcase and trolly respectively. Applying the principle of conservation of momentum to the trolly-suitcase system, we have

$$m_s (V_s)_1 + m_t (V_t)_1 = m_s (V_s)_2 + m_t (V_t)_2$$
$$= (m_s + m_t)V$$

where V is the common velocity with which the trolly and suitcase move together after the impact.

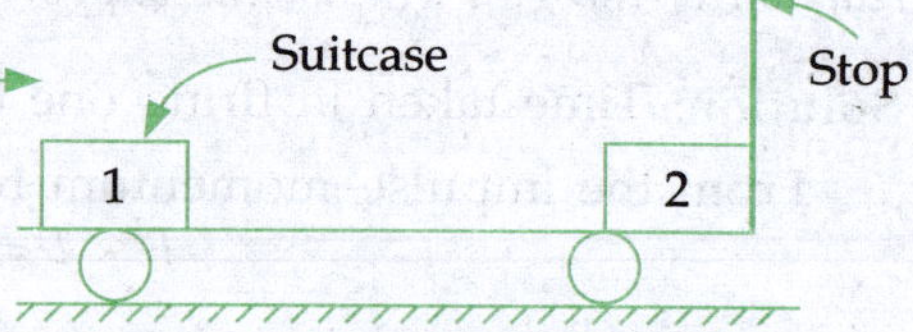

Fig. 18.3

Substituting the given values, we obtain

$$10 \times 4 + 25 \times 0 = (10 + 25)V$$
$$\therefore \quad V = \frac{40}{35} = \mathbf{1.143\ m/s}$$

EXAMPLE 18.9.

A 150 kg scootorist is travelling with a speed of 36 km/hr on a road that makes an angle of 30° with another road upon which a 60 kg cyclist is travelling at 8 km/hr. When they approach the crossing, they collide and move as one mass. Determine the final velocity; both in magnitude and direction.

Solution: Let suffix 1 and 2 refer to scootorist and cyclist respectively.

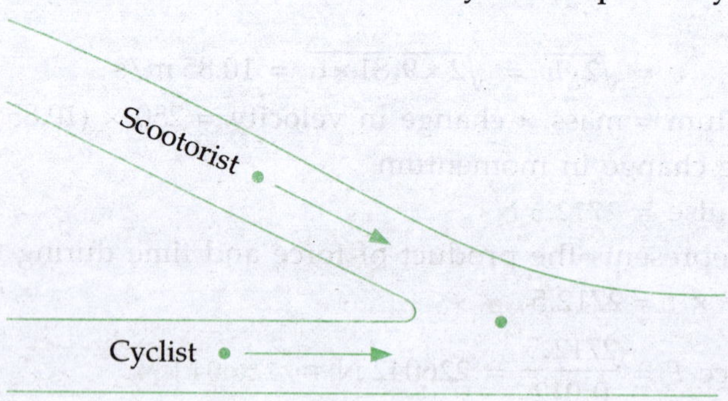

Fig. 18.4

Impact in x-direction :

$$(u_1)_x = \text{initial velocity of scootorist in x-direction before impact}$$
$$= \frac{36 \times 1000}{3600} \cos 30^\circ = 8.66\ \text{m/s}$$
$$(u_2)_x = \text{velocity of cyclist in } x\text{-direction before impact}$$
$$= \frac{8 \times 1000}{3600} = 2.22\ \text{m/s}$$

After the impact both will move with the same velocity. Let V_x be the final velocity in x-direction after the impact. Then from the principle of conservation of momentum,

$$m_1 u_1 + m_2 u_2 = (m_1 + m_2)\,V_x$$

$$150 \times 8.66 + 60 \times 2.22 = (150 + 60)\,V_x\,; \quad V_x = 6.82 \text{ m/s}$$

Impact in y-direction:

$(u_1)_y$ = velocity of scootorist in y-direction before impact

$$= \frac{36 \times 1000}{3600} \times \sin 30^\circ = 5 \text{ m/s } (\downarrow)$$

$(u_2)_y$ = velocity of cyclist in y-direction before impact = 0

Let V_y be the velocity of both vehicles after the impact. Then from the principle of momentum conservation,

$$150 \times 5 + 60 \times 0 = (150 + 60)\,V_y$$

$$V_y = 3.57 \text{ m/s } (\downarrow)$$

Resultant velocity after impact V

$$= \sqrt{V_x^2 + V_y^2} = \sqrt{6.82^2 + 3.57^2} = 7.69 \text{ m/s}$$

and its inclination with the horizontal is

$$\theta = \tan^{-1}\frac{3.57}{6.82} = 27.7^\circ$$

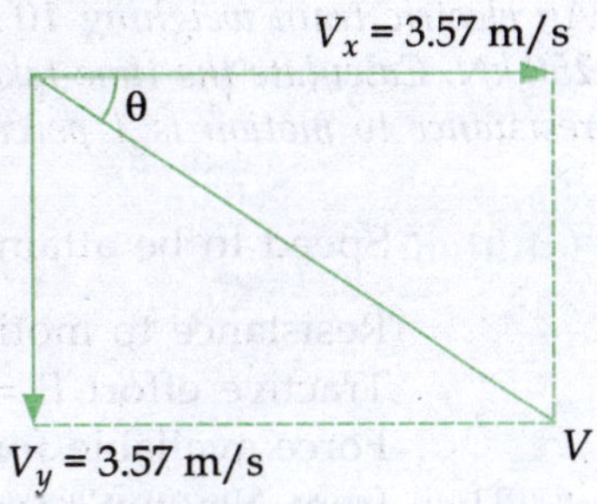

Fig. 18.5

EXAMPLE 18.10.

A block weighing 1.5 kN is initially at rest on a rough horizontal surface (μ = 0.2). It is acted upon by a force P which varies with time as shown in Fig. 18.6. Determine the velocity at the end of (i) 3 seconds (ii) 6 seconds.

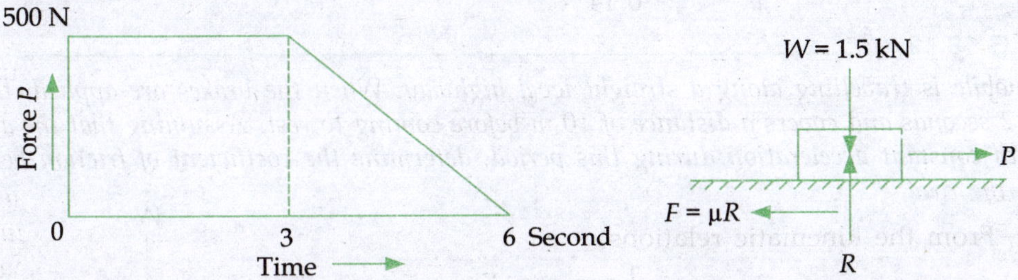

Fig. 18.6

Solution: Frictional force $F = \mu R = \mu W = 0.2 \times 1.5 = 0.3 \text{ kN} = 300 \text{ N}$

Case (*i*) From the impulse-momentum relation,

impulse = change in momentum

or $\quad F \times t = m \times dV$

or $\quad (500 - 300) \times 3 = \dfrac{1.5 \times 10^3}{9.81}(V_3 - 0)\,; \quad V_3 = 3.924 \text{ m/s}$

Thus the block will have a velocity of 3.924 m/s at the end of 3 seconds.

Case (*ii*) Impulse during 3rd and 6th second

= area of triangular portion of force-time diagram

$$= \frac{1}{2} \times 500 \times (6 - 3) = 750 \text{ Ns}$$

Impulse due to friction = 300 × (6 − 3) = 900 Ns

$$\text{Net impulse} = 750 - 900 = -150 \text{ Ns}$$

Again using the relation,

$$\text{impulse} = \text{change in momentum}$$

$$-150 = \frac{1.5 \times 10^3}{9.81}(V_6 - V_3); \quad V_6 - V_3 = -0.981$$

∴ Velocity of block at the end of 6 seconds,

$$V_6 = V_3 - 0.981 = 3.924 - 0.981 = \mathbf{2.943 \text{ m/s}}$$

EXAMPLE 18.11.

An electric train weighing 10 MN is driven by electric motors which provide a tractive effort equal to 250 kN. Calculate the time taken to accelerate the train to a speed of 80 km/hr. It may be assumed that resistance to motion is 1 percent of the weight of train.

Solution: Speed to be attained $v = 80 \text{ km/hr} = \frac{80 \times 1000}{3600} = 22.22 \text{ m/s}$

Resistance to motion $R = 0.01 \times (10 \times 10^6) = 100 \times 10^3$ N

Tractive effort $P = 250 \text{ kN} = 250 \times 10^3$ N

Force available for acceleration $F = P - R = 150 \times 10^3$ N

Then from Newton's second law $F = ma$, we have

$$150 \times 10^3 = \frac{10 \times 10^6}{9.81} a \qquad \therefore \quad \text{acceleration } a = 0.14 \text{ m/s}^2$$

Using kinematic relation $v = u + at$,

Time taken to accelerate the train from rest to 22.22 m/s velocity,

$$t = \frac{v - u}{a} = \frac{22.22 - 0}{0.14} = \mathbf{158.71 \text{ second}}$$

EXAMPLE 18.12.

An automobile is travelling along a straight level highway. When the brakes are applied, the vehicle slides for 2 seconds and covers a distance of 10 m before coming to rest. Assuming that the automobile moves with constant deceleration during this period, determine the coefficient of friction between the tyres and the road.

Solution: From the kinematic relations

$$v = u + at \quad \text{and} \quad s = ut + \frac{1}{2}at^2, \text{ we have}$$

$$0 = u + 2a \quad \text{or} \quad u = -2a \qquad \ldots(i)$$

and $$10 = u \times 2 + \frac{1}{2} a \times 2^2 \quad \text{or} \quad 24 + 2a = 10 \qquad \ldots(ii)$$

From expressions (*i*) and (*ii*),

$$a = -5 \text{ m/s}^2 \quad \text{(retardation)}$$

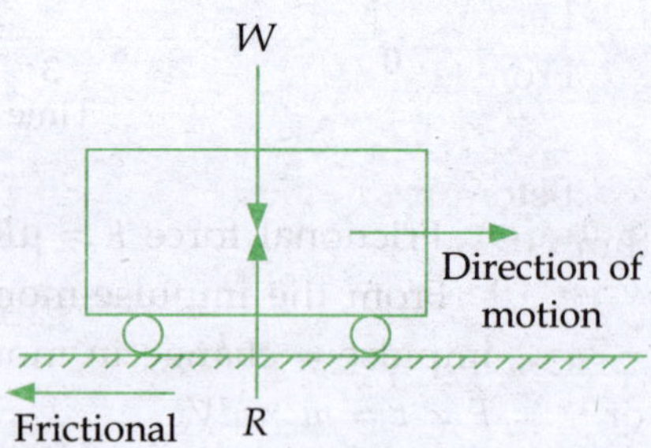

Fig. 18.7

Let μ be the coefficient of friction between the tyres and the road. Then

$$\text{Retarding force} = \mu W$$

where W is the weight of automobile.

Applying Newton's second law of motion,

Retarding force = mass × deceleration

$$\mu W = \frac{W}{g} \times 5 \qquad \therefore \quad \mu = \frac{5}{9.81} = \mathbf{0.51}$$

EXAMPLE 18.13.

A bullet of mass 0.1 kg and travelling at a speed of 180 m/s penetrated 10 cm when fired into a wooden log. Determine the velocity with which this bullet would emerge when fired with the same velocity into a similar 5 cm thick wooden plank. Also determine the force of resistance assuming it to be uniform.

Solution: From the kinematic relation $v^2 - u^2 = 2as$, we have

$$0 - 180^2 = 2a \times 0.1 \qquad \therefore \quad a = -162000 \text{ m/s}^2 \text{ (retardation)}$$

Let the bullet emerge with velocity V from the 5 cm thick plank of wood,

$$V^2 \quad 180^2 = 2 \times (\;162000) \times 0.05$$

The retardation has been taken same as the wooden planks are similar and the resistance offered by wood is uniform

$$V^2 = 32400 - 16200 = 16200 \qquad \therefore \quad V = \mathbf{127.28 \text{ m/s}}$$

Then from Newton's second law $F = ma$, we have

Force resistance $F_r = 0.1 \times 162000 =$ **16200 N**

EXAMPLE 18.14.

A shell of mass 60 kg is fired horizontally with a velocity of 250 m/s by a gun of 3000 kg mass. Make calculations for (a) the velocity with which the gun recoils, (b) the uniform force required to stop the gun in 0.5 m distance, and (c) the time required to stop the gun.

It may be presumed that momentum of the system comprising the gun and the shell is conserved.

Solution: Refer Fig. 18.8 for the given system comprising the gun and the shell.

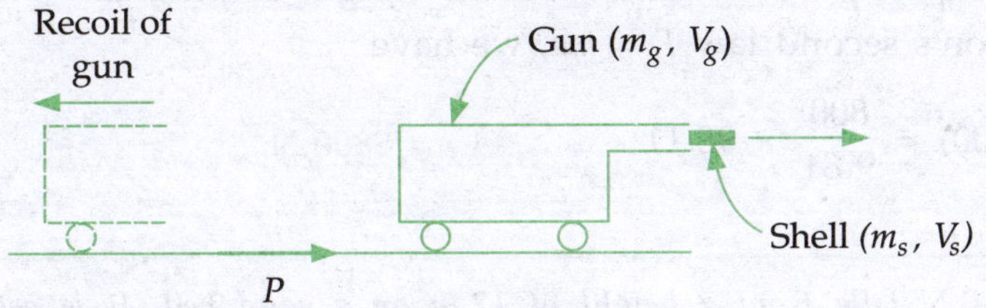

Fig. 18.8

Let suffix s and g denote the shell and gun respectively.

From the principle of conservation of momentum, we have

$$(m_sV_s + m_gV_g)_{\text{Before firing}} = (m_sV_s + m_gV_g)_{\text{After firing}}$$

Before firing, both the shell and the gun have zero velocity. As such initially the momentum is zero and therefore

$$m_sV_s + m_gV_g = 0 \quad \text{or} \quad m_sV_s = -m_gV_g$$

Obviously the velocities of the shell and the gun are in the opposite directions

$$60 \times 250 = -3000 \times V_g; \quad V_g = \mathbf{-5 \text{ m/s}^2}$$

(*b*) The gun moving with a velocity of 5 m/s² is brought to rest in a distance of 0.5 m. From the kinematic relation $v^2 - u^2 = 2as$, we have

$$a = \frac{v^2 - u^2}{2s} = \frac{0^2 - (-5^2)}{2 \times 0.5} = -25 \text{ m/s}^2 \text{ (retardation)}$$

From Newton's second law; $F = ma$, we have

Force required to stop the gun $= 3000 \times 25$

$$P = 75000 \text{ N} = \mathbf{75 \text{ kN}}$$

(c) From the kinematic relation $v = u + at$, we have

Time required to stop the gun $t = \frac{v-u}{a} = \frac{0-(-5)}{25} = 0.2$ **second**

EXAMPLE 18.15.

A man weighing 800 N dives into a swimming pool from a height of 20 m. It is estimated that he goes down in water by 2 m and then starts rising. Neglecting air resistance, determine the average resistance offered by water to the motion of the man.

Solution: Let v be the velocity of the man when he reaches the water surface. Using the relation $v^2 - u^2 = 2gh$ and noting that $u = 0$, we have

$$v = \sqrt{2gh} = \sqrt{2 \times 9.81 \times 20} = 19.81 \text{ m/s}$$

When the man rises upward through a distance of 2 m inside the water, his velocity changes from 19.81 m/s to zero. This occurs because of retardation due to water resistance. From the kinematic relation $v^2 - u^2 = 2gh$, we get

$$0^2 - 19.81^2 = 2a \times 2$$

$$a = -\frac{19.81^2}{2 \times 2} = -98.11 \text{ m/s}^2 \quad \text{(retardation)}$$

Let F_r be the average resistance of water acting on the man in the upward direction. Then Net force acting on the man in the upward direction

$$F_n = F_r - W \quad \text{where } W \text{ is the weight of the man}$$

Applying Newton's second law $F = ma$, we have

$$(F_r - 800) = \frac{800}{9.81} \times 98.11 \qquad \therefore \quad F_r = 8800 \text{ N}$$

EXAMPLE 18.16.

A body of weight 100 N falls from a height of 12 m on a sand bed. It is estimated that the body penetrates 1.2 m into sand before coming to rest. Make calculations for the average thrust exerted by the sand on the body.

Solution: Velocity of body after falling from 12 m height,

$$v = \sqrt{2gh} = \sqrt{2 \times 9.81 \times 12} = 15.34 \text{ m/s}$$

After penetration in the sand, the velocity of the body becomes zero. Using the kinematic relation $v^2 - u^2 = 2as$, we have

$$0^2 - 15.34^2 = 2a \times 1.2$$

$$a = -\frac{15.34^2}{2 \times 1.2} = -98.05 \text{ m/s}^2 \quad \text{(retardation)}$$

Let F_r be the average resistance exerted by the sand on the body. Then net force acting on the body in the upward direction,

$$F_n = F_r - W \quad \text{where } W \text{ is the weight of the body}$$

Applying Newton's second law $F = ma$, we have

$$(F_r - 100) = \frac{100}{9.81} \times 98.05 \qquad \therefore \quad F_r = 1100 \text{ N}$$

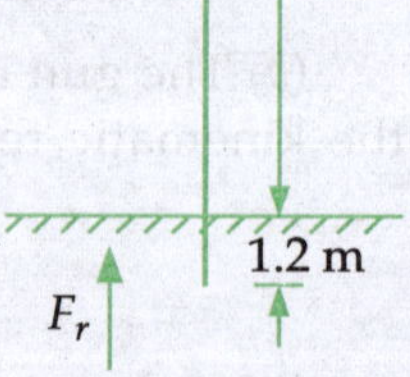

Fig. 18.9

EXAMPLE 18.17.

A train weighing 4000 kN has a frictional resistance of 5 N/kN of weight. Determine the steady pull which the locomotive must exert if the speed of the train is to be increased from 30 km/hr to 60 km/hr within a period of 1.5 minutes.

Solution: Initial velocity of train $u = 30\ \text{km/hr} = \dfrac{30 \times 1000}{3600} = 8.33\ \text{m/s}$

Final velocity $v = 60\ \text{km/hr} = \dfrac{60 \times 1000}{3600} = 16.66\ \text{m/s}$

Using the relation $v = u + at$, the acceleration of train is

$$\text{acceleration } a = \frac{v-u}{t} = \frac{16.66 - 8.33}{(2.5 \times 60)} = 0.092\ \text{m/s}$$

Frictional resistance $F_f = 5$ N per kN of train weight $= 5 \times 4000 = 20000$ N

The steady pull, *i.e.*, the tractive force F_t acts in the direction of motion, while the frictional resistance acts opposite to the direction of motion.

∴ Net force in the direction of motion

$$F_n = F_t - F_f = (F_t - 20000)\ \text{N}$$

Then from Newton's second law $F = ma$, we have

$$(F_t - 20000) = \frac{4000 \times 10^3}{9.81} \times 0.092 = 37512$$

∴ Steady (average) pull to be exerted by the locomotive,

$$F_t = 37512 + 20000 = 57512\ \text{N} = 57.512\ \text{kN}$$

EXAMPLE 18.18

An automobile weighing 12 kN is moving with a uniform speed of 72 km/hr over a vertical curve ABC of parabolic shape as indicated in the figure given below:

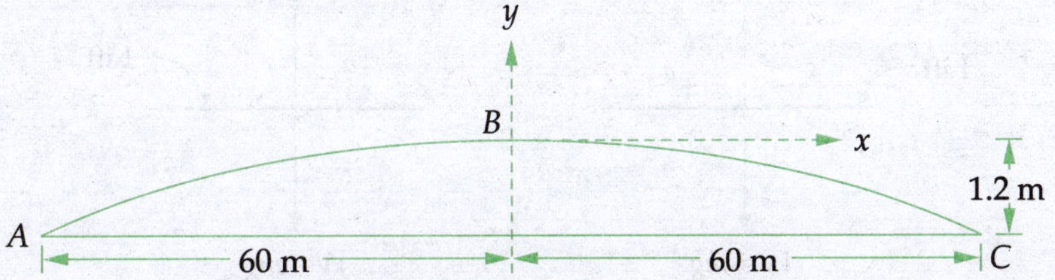

Fig. 18.10

Determine the total pressure exerted by the wheels of the automobile as it passes the topmost point B.

Solution: Let R be the reaction exerted by the wheels. Then

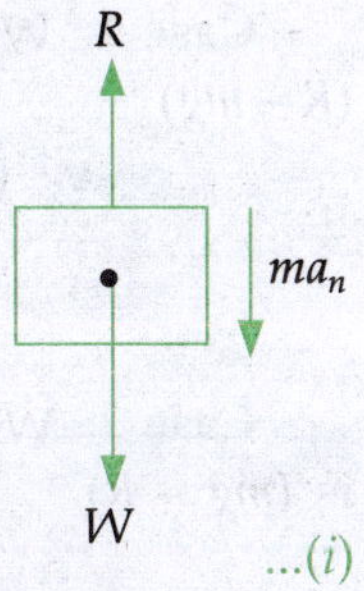

$$\text{Net force} = \text{mass} \times \text{acceleration}$$

$$W - R = ma_n$$

where a_n is the normal acceleration.

$$\therefore \quad R = W - ma_n = mg - m\frac{V^2}{\rho}$$

$$= mg\left[1 - \frac{V^2}{\rho g}\right] \quad \text{...}(i)$$

$$\text{Velocity of automobile } V = \frac{72 \times 1000}{3600} = 20\ \text{m/s}$$

The radius of curvature ρ of the road can be worked out by assuming that the equation of parabola is

$$x^2 = ky; \quad 60^2 = k \times 1.2 \quad \text{and therefore} \quad k = \frac{60^2}{1.2} = 3000$$

$$\therefore \quad x^2 = 3000\,y \quad \text{and} \quad \frac{dy}{dx} = \frac{2x}{3000}$$

$$\frac{d^2y}{dx^2} = \frac{2}{3000}$$

$\frac{dy}{dx} = 0$ at point B

$$\text{Now,} \quad \rho = \frac{\left[1 + \left(\frac{dy}{dx}\right)^2\right]^{3/2}}{\frac{d^2y}{dx^2}} = \frac{(1+0)^{3/2}}{2/3000} = 1500 \text{ m}$$

Substituting the relevant values in identity (*i*), we get

$$\text{Reaction exerted on wheels, } R = 12\left(1 - \frac{20^2}{1500 \times 9.81}\right)$$

$$= \mathbf{11.67\ kN}$$

18.2. MOTION OF LIFT

Consider a person of mass m inside a lift. The weight mg of the person acts vertically downwards while the reaction R acts vertically upwards

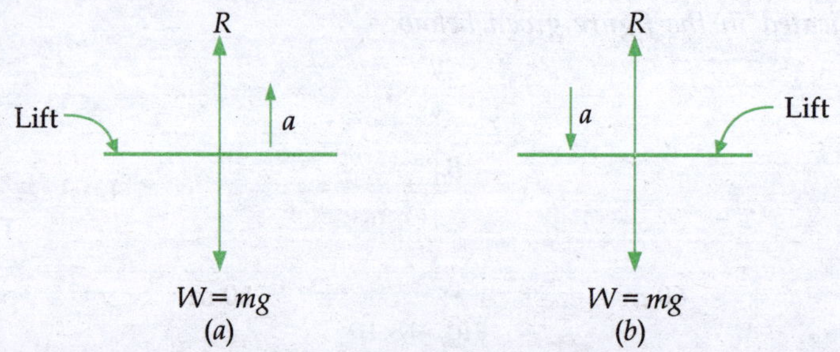

Fig. 18.11

Case 1: When the lift moves **upwards** with an acceleration a, the force causing motion is $(R - mg)$

$$\therefore \quad (R - mg) = ma$$

$$R = m(g + a) = mg\left(1 + \frac{a}{g}\right) \qquad \text{...}(i)$$

Case 2: When the lift moves **downwards** with an acceleration a, the force causing motion is $(mg - R)$

$$\therefore \quad mg - R = ma$$

$$R = m(g - a) = mg\left(1 - \frac{a}{g}\right) \qquad \text{...}(ii)$$

It is evident from identities (*i*) and (*ii*) that weight of the person apparently increases by the factor (a/g) when the platform moves up and decreases by the same factor when the platform moves down. This also implies that the person exerts more force on the floor of lift when moving upwards, and exerts less force when moving in downward direction.

Further, it may be noted that tension T in the cable supporting the load equals the reaction R. That is

$$T = W\left(1 + \frac{a}{g}\right) \quad \text{when lift moves upward}$$

and $$T = W\left(1 - \frac{a}{g}\right) \quad \text{when lift moves downwared} \qquad ...(18.3)$$

EXAMPLE 18.19

A lift carrying a load of 1000 N is moving with uniform acceleration of 2.5 m/s^2. Calculate the tension in the cable supporting the load when (a) lift is moving upward, (b) lift is moving downward.

Solution: Let T be the tension in the cables supporting the load

(*a*) *When lift moves upwards*

$$T - W = ma = \frac{W}{g}a \qquad \left(\because m = \frac{W}{g}\right)$$

$$T = W\left(1 + \frac{a}{g}\right) = 1000\left(1 + \frac{2.5}{9.81}\right) = \mathbf{1254.84\ N}$$

(*b*) *When lift moves downward*

$$W - T = ma = \frac{W}{g}a$$

$$T = W\left(1 - \frac{a}{g}\right) = 1000\left(1 - \frac{2.5}{9.81}\right) = \mathbf{745.16\ N}$$

EXAMPLE 18.20

An elevator of weight 25 kN is moving vertically downwards with a constant acceleration. Determine the limits of cable tension.

Starting from rest, the elevator is stated to travel a distance of 25 m during an interval of 8 seconds. What will be the cable tension during this time ?

Solution: Let T be the tension in the cables supporting the elevator. When the elevator moves downward,

$$W - T = ma = \frac{W}{g}a$$

$$T = W\left(1 - \frac{a}{g}\right)$$

The cable tension has two limits: when acceleration a is zero and when it has a maximum value equal to g = 9.81 m/s^2.

Cable tension when lift moves with zero acceleration,

$$T = 25 \times 10^3\left(1 - \frac{0}{g}\right) = \mathbf{25 \times 10^3\ N}$$

Cable tension when lift moves with acceleration equal to g,

$$T = 25 \times 10^3 \left(1 - \frac{g}{g}\right) = 0$$

(*b*) Applying kinematic equation, $s = ut + \frac{1}{2}at^2$, to the motion of elevator

$$25 = 0 \times t + \frac{1}{2} a \times 8^2 = 32a$$

$\therefore$ Constant acceleration of elevator, $a = \frac{25}{32} = 0.781 \text{ m/s}^2$

Again, using the relation

$$T = W\left(1 - \frac{a}{g}\right) = 25 \times 10^3 \left(1 - \frac{0.781}{9.81}\right) = 23 \times 10^3 \text{ N}$$

EXAMPLE 18.21

An elevator cage of a mine shaft, weighing 6 kN, is lifted or lowered by means of a wire rope. Starting from rest, it moves upwards with constant acceleration and acquires a velocity of 2.5 m/s within a distance of 3 m. Calculate the tensile force in the cable during the accelerated motion.

When a man weighing 800 N enters this elevator during its upward motion with a velocity of 2.5 m/s, the elevator is uniformly declerated to halt in 2 seconds. Determine the force exerted by the man on the floor of elevator.

Solution: Let T be the tensile force in the cable. Then considering upward motion of elevator,

$$T = mg\left(1 + \frac{a}{g}\right) = W\left(1 + \frac{a}{g}\right)$$

The acceleration acquired by the elevator after travelling 2.5 m is obtained by using the kinematic equation $v^2 - u^2 = 2as$. Then

$$2.5^2 - 0 = 2a \times 3 \qquad (u = 0 \text{ as elevator starts from rest})$$

$$a = \frac{2.5 \times 2.5}{6} = 1.042 \text{ m/s}^2$$

$$\therefore \quad T = 6 \times 1000 \left(1 + \frac{1.042}{9.81}\right) = 6637.3 \text{ N}$$

Considering motion of man,

$$v = u + at\,; \quad u = 2.5 \text{ m/s} \quad \text{and} \quad v = 0$$

$$0 = 2.5 + a \times 2\,; \quad a = -1.25 \text{ m/s}^2$$

The negative sign with acceleration suggests that it is opposite to the direction of velocity or is downward.

$$\text{Reaction } R = W\left(1 + \frac{a}{g}\right) = 800\left[1 + \frac{(-1.25)}{9.81}\right] = 523.55 \text{ N}$$

Thus the pressure at the floor of elevator under the feet of the man is 523.55 N.

EXAMPLE 18.22

An elevator of mass 600 kg is ascending with an acceleration of 2.5 m/s^2. During this ascent, its operator who weighs 700 N is standing on the scales placed on the floor of the elevator. Calculate the scale reading and total tension in the cables of elevator during this motion.

Solution: Weight of elevator = 600 × 9.81 = 5886 N. When the operator stands on the scale placed on the floor of the elevator, the reaction offered by the floor equals the reading on scale.

Considering motion of man,

$$R - W = ma = \frac{W}{g}\,a$$

$$R = W\left(1 + \frac{a}{g}\right) = 700\left(1 + \frac{2.5}{9.81}\right) = 878.4 \text{ N}$$

Considering motion of elevator and man taken together,

$$T - W = ma = \frac{W}{g}\,a$$

$$T = W\left(1 + \frac{a}{g}\right) = (5886 + 700)\left(1 + \frac{2.5}{9.81}\right) = 8264.4 \text{ N}$$

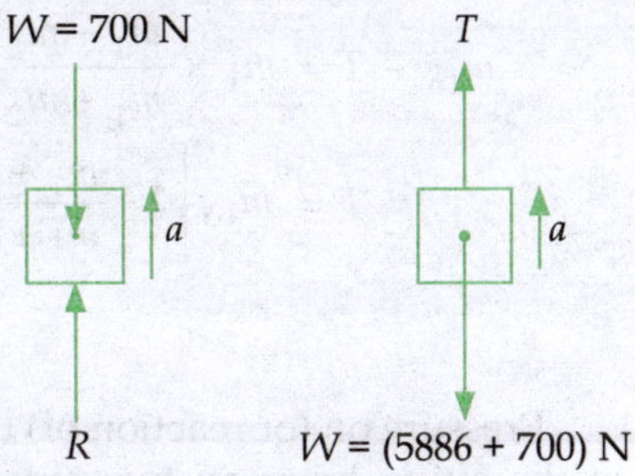

Fig. 18.12

18.3 MOTION OF CONNECTED BODIES

Attention has been focussed in the following section to examine the motion of a body when it is connected by a string to another body which is subjected to some external force. Both the bodies would be moving with the same velocity and acceleration.

18.3.1 Motion of two bodies connected by a string passing over a smooth pulley

Consider two bodies of masses m_1 and m_2 connected at the ends of a light inextensible string that passes over a smooth frictionless pulley (Fig. 18.13). If $m_1 > m_2$, then the mass m_1 moves downwards and the mass m_2 moves upwards.

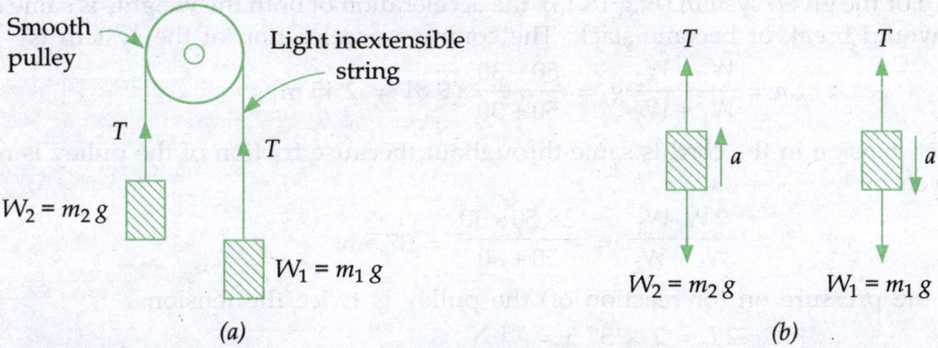

Fig. 18.13

It is to be noted that.

(*i*) the acceleration of both the weights is same otherwise the cord would break or become slack.

(*ii*) for a smooth frictionless pulley and continuous cord, the tension induced in the string will be same throughout.

With reference to the free body diagram [Fig 18.13 (*b*)], the equations of motion for the two masses are as follows:

Downward motion of mass m_1 : $\quad m_1 g - T = m_1 a \qquad ...(i)$

Upward motion of mass m_2 : $\quad T - m_2 g = m_2 a \qquad ...(ii)$

Adding expressions (*i*) and (*ii*)

$$m_1 g - m_2 g = m_1 a + m_2 a$$

$$\therefore \text{ Acceleration of the system, } a = \frac{m_1 - m_2}{m_1 + m_2}\,g = \frac{W_1/g - W_2/g}{W_1/g + W_2/g}\,g = \frac{W_1 - W_2}{W_1 + W_2}\,g \qquad ...(18.4)$$

Upon substitution of this value of acceleration *a* in expression (*i*), we get

$$m_1 g - T = m_1 \times \frac{m_1 - m_2}{m_1 + m_2} g$$

$$\therefore \quad T = m_1 g \left[1 - \frac{m_1 - m_2}{m_1 + m_2} \right] = m_1 g \times \frac{2m_2}{m_1 + m_2}$$

$$= \frac{2m_1 m_2}{m_1 + m_2} g = 2 \frac{W_1/g \times W_2/g}{W_1/g + W_2/g} g = \frac{2W_1 W_2}{W_1 + W_2} \quad \text{...(18.5)}$$

Pressure on (or reaction of) the pulley equals the vector sum of the tensions T_1 and T_2 induced in the string lying on two sides of the pulley. Here $T_1 = T_2 = T$ and they act along the same direction, *i.e.*, angle θ between then is zero. Then from the parallelogram law of forces

$$R = \sqrt{P^2 + Q^2 + 2PQ\cos\theta} = \sqrt{T^2 + T^2 + 2TT\cos 0^\circ} = \sqrt{4T^2} = 2T$$

Thus the pressure on the pulley is twice the tension in the string.

$$\therefore \quad R = 2T = \frac{4W_1 W_2}{W_1 + W_2} \quad \text{...(18.6)}$$

EXAMPLE 18.23

Two objects of weights 50 N and 30 N respectively are connected by a light cord passing over a smooth frictionless pulley. Determine (a) the acceleration with which the weights move, (b) the tension in the cord, and (c) the pressure on the pulley.

Solution: For the given system (Fi.g 18.13), the acceleration of both the weights is same otherwise the cord would break or become slack. The common acceleration of the system is

$$a = \frac{W_1 - W_2}{W_1 + W_2} g = \frac{50 - 30}{50 + 30} \times 9.81 = 2.45 \text{ m/s}^2$$

Further, tension in the cord is same throughout (because friction of the pulley is negligible) and is given by

$$T = \frac{2W_1 W_2}{W_1 + W_2} = \frac{2 \times 50 \times 30}{50 + 30} = 37.5 \text{ N}$$

Also, the pressure on (or reaction of) the pulley is twice the tension.

$$R = 2T = 2 \times 37.5 = 75 \text{ N}$$

EXAMPLE 18.24

Two blocks A and B of weights 600 N and 400 N respectively are connected to the two ends of cord which passes over pulley. Neglecting inertia of the pulley and assuming that the cord is inextensible, find the acceleration of block A. How much weight should be added to or taken away from block A, if the acceleration of the block is required to be g/3 downwards?

Solution: Let $W_1 = 600$ N and $W_2 = 400$ N.

Then for the given arrangement, the acceleration of block *A* or that of the system is

$$a = \frac{W_1 - W_2}{W_1 + W_2} g = \frac{600 - 400}{600 + 400} g = \frac{200}{1000} g = \frac{g}{5}$$

Let *W* be the weight of block *A* if its acceleration is to be $g/3$ downwards. Then

$$\frac{g}{3} = \frac{W - W_2}{W + W_2} g = \frac{W - 400}{W + 400} g$$

$$\text{or} \quad W + 400 = 3(W - 400) = 3W - 1200 \qquad \therefore \quad W = 800 \text{ N}$$

Hence a weight (800 – 600), *i.e.*, 200 N must be added to block *A*.

EXAMPLE 18.25

A system of frictionless pulleys carries two weights hung by inextensible cords as shown is Fig. 18.14. Make calculations for the tension in the cords and acceleration of the weights

Proceed to workout the displacement and velocity of weight W_1 after 4 seconds from start. Presume that the system is released from rest.

Solution: Since the pulleys are frictionless and the cord is continuous, the tension T induced in the cord will be same throughout.

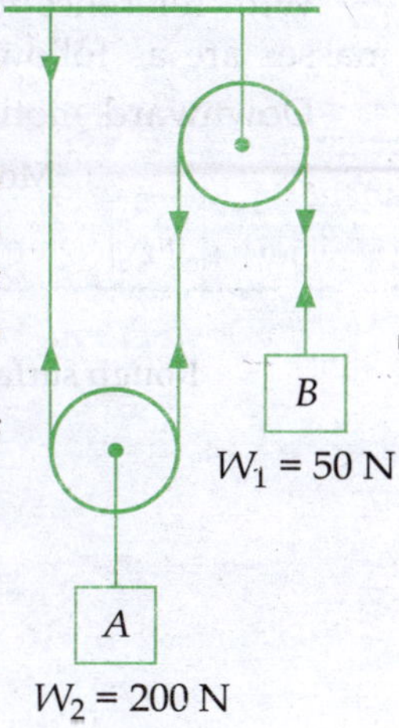

Fig. 18.14

It is evident from the given arrangement that when weight W_1 travels a unit distance, the weight W_2 travels half the distance. Acceleration is proportional to distance travelled, and therefore if a is acceleration of weight W_1, the weight W_2 will have acceleration equal to $a/2$. The weight W_1 moves upward and the weight W_2 moves downward.

With reference to the free body diagrams (Fig. 18.15), the equations of motion for the two weights are as follows:

$$T - 50 = \frac{50}{9.81} \times a \quad \text{(upward motion of block } B\text{)}$$

$$\text{or} \quad T = 50 + \frac{50}{9.81} a \qquad \ldots(i)$$

$$\text{Also} \quad 200 - T = \frac{200}{9.81} \times \frac{a}{2}$$

(downward motion of block A)

$$\text{or} \quad 2T = 200 - \frac{100}{9.81} a$$

$$\text{or} \quad T = 100 - \frac{50}{9.81} a \qquad \ldots(ii)$$

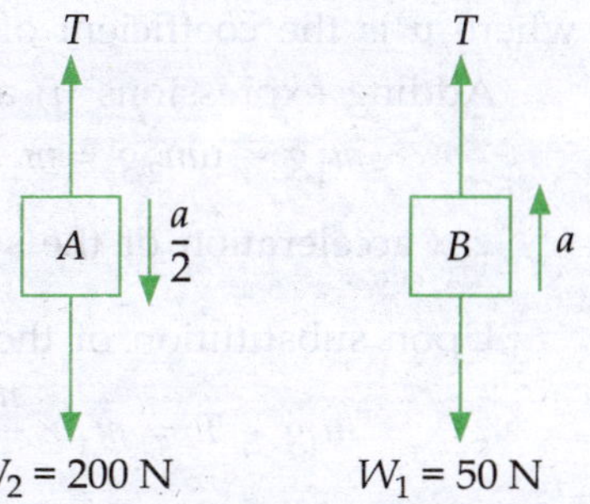

Fig. 18.15

From expression (i) and (ii), we obtain

$$50 + \frac{50}{9.81} a = 100 - \frac{50}{9.81} a$$

$$\text{or} \quad \frac{100}{9.81} a = 50 \qquad \therefore \; a = \frac{50 \times 9.81}{100} = 4.905 \text{ m/s}^2$$

Thus the acceleration of blocks A and B are 2.452 m/s^2 and 4.905 m/s^2 respectively.

From expression (i)

$$T = 50 + \frac{50}{9.81} \times 4.905 = 75 \text{ N}$$

(b) Applying kinematic equations to the motion of weight B (W_1 = 50 N)

$$v = u + at = 0 + 4.905 \times 4 = 19.62 \text{ m/s}$$

$$s = ut + \frac{1}{2} at^2 = 0 \times 4 + \frac{1}{2} \times 4.905 \times 4^2 = 39.24 \text{ m}$$

18.3.2 Motion of two bodies connected at the edge of a horizontal surface

Consider a block of mass m_2 resting on a rough horizontal surface. The block is attached to a light inextensible string which passes over a smooth frictionless pulley at the end of the surface. The portion of the string, beyond the pulley, hangs vertically down and carries a mass m_1 at the end.

Let the mass m_1 move downwards and the mass m_2 move along the surface of the plane. It is to be noted that :

(*i*) the acceleration of both the masses is same otherwise the string will break or become loose

(*ii*) for a smooth friction-less pulley and continuous string, the tension induced in the string is same throughout

With reference to the free body diagrams [18.16 (*b*)], the equations of motion for the two masses are as follows:

Downward motion of mass m_1: $m_1 g - T = m_1 a$ ...(*i*)

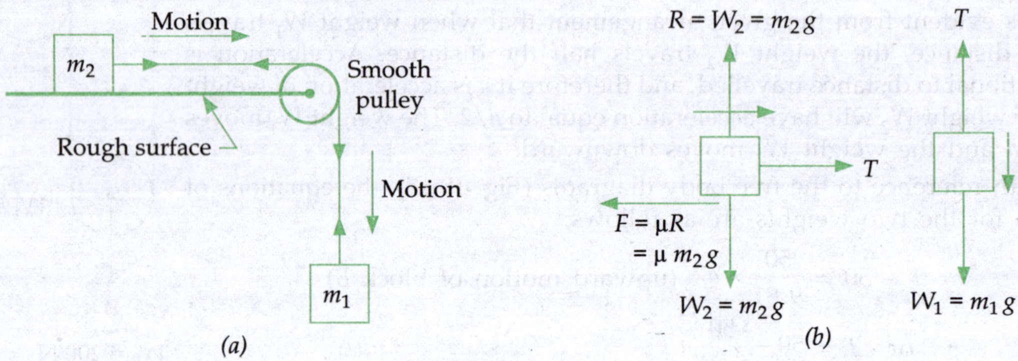

Fig. 18.16

Horizontal motion of mass m_2 : $T - \mu m_2 g = m_2 a$ ...(*ii*)

where μ is the coefficient of friction between the block and the rough surface.

Adding expressions (*i*) and (*ii*),

$$m_1 g - \mu m_2 g = m_1 a + m_2 a$$

∴ acceleration of the system, $a = \dfrac{m_1 g - \mu m_2 g}{m_1 + m_2} = \dfrac{W_1 - \mu W_2}{W_1 + W_2} g$ ...(18.7)

Upon substitution of the value of acceleration a in expression (*i*), we get

$$m_1 g - T = m_1 \times \frac{m_1 g - \mu m_2 g}{m_1 + m_2}$$

$$T = m_1 g - m_1 \frac{m_1 g - \mu m_2 g}{m_1 + m_2} = m_1 g \left[1 - \frac{m_1 - \mu m_2}{m_1 + m_2}\right] = m_1 g \left[\frac{m_2 (1 + \mu)}{m_1 + m_2}\right]$$

$$= W_1 \left[\frac{W_2 (1 + \mu)}{W_1 + W_2}\right] = \frac{W_1 W_2 (1 + \mu)}{W_1 + W_2} \quad ...(18.8)$$

For a smooth horizontal surface, μ = 0. In that case the equations 18.7 and 8.38 get modified to

$$a = \frac{W_1}{W_1 + W_2} g \quad \text{and} \quad T = \frac{W_1 W_2}{W_1 + W_2} \quad ...(18.9)$$

Pressure on (or reaction of) the pulley equals the vector sum of tensions T_1 and T_2 induced in the string lying on two sides of the pulley. Here $T_1 = T_2 = T$ and they act perpendicular to each other, *i.e.*, angle θ between them is 90º. Then from the parallelogram law of forces,

$$R = \sqrt{P^2 + Q^2 + 2PQ\cos\theta}$$

$$= \sqrt{T^2 + T^2 + 2TT\cos 90^\circ} = \sqrt{2T^2} = \sqrt{2}T \quad ...(18.10)$$

Thus the pressure on the pulley is $\sqrt{2}$ times the tension induced in the string.

EXAMPLE 18.26

A block of weight 10 N is resting on a rough horizontal table 1.5 m from its edge. This block A is attached to a string which passes over a pulley at the end of the table. The portion of the string beyond the pulley hangs vertically down and carries a block B of weight 8 N at the end. If the coefficient of friction between the block A and table is 0.2, make calculations for the acceleration of the system, tension induced in the string and pressure on the pulley. Neglect friction in the pulley and extension of the string.

If the system is released from rest, what time block A will take to fall over the edge ?

Solution: For the given system (Fig. 18.16), the acceleration of both the weights is same otherwise the string would break or become loose. The common acceleration of the system is

$$a = \frac{W_1 - \mu W_2}{W_1 + W_2} g = \frac{8 - 0.2 \times 10}{8 + 10} \times 9.81 = 3.27 \text{ m/s}^2$$

Further, tension in the cord is same throughout (because friction of pulley is negligible) and is given by

$$T = \frac{W_1 \; W_2 (1 + \mu)}{W_1 + W_2} = \frac{8 \times 10(1 + 0.2)}{8 + 10} = 5.33 \text{ N}$$

Pressure on the pulley $= \sqrt{2}T = \sqrt{2} \times 5.33 = 7.538$ N

(*b*) Applying kinematic equations to the motion of weight *A*,

$$v^2 - u^2 = 2as \; ; \quad u = 0 \text{ as the system is released from rest.}$$

$$\therefore \quad v = \sqrt{2as} = \sqrt{2 \times 3.27 \times 1.5} = 3.13 \text{ m/s}$$

Also, $$v = u + at \; ; \quad t = \frac{v}{a} = \frac{3.13}{3.27} = 0.957 \text{ s}$$

18.3.3 Motion of two bodies connected by a string; one hangs free and the other on a rough inclined plane.

Consider two bodies of masses m_1 and m_2 connected at the ends of a light inextensible string that passes over a smooth frictionless pulley. Further, mass m_1 hangs free and the mass m_2 lies on a rough surface inclined at an angle α with the horizontal.

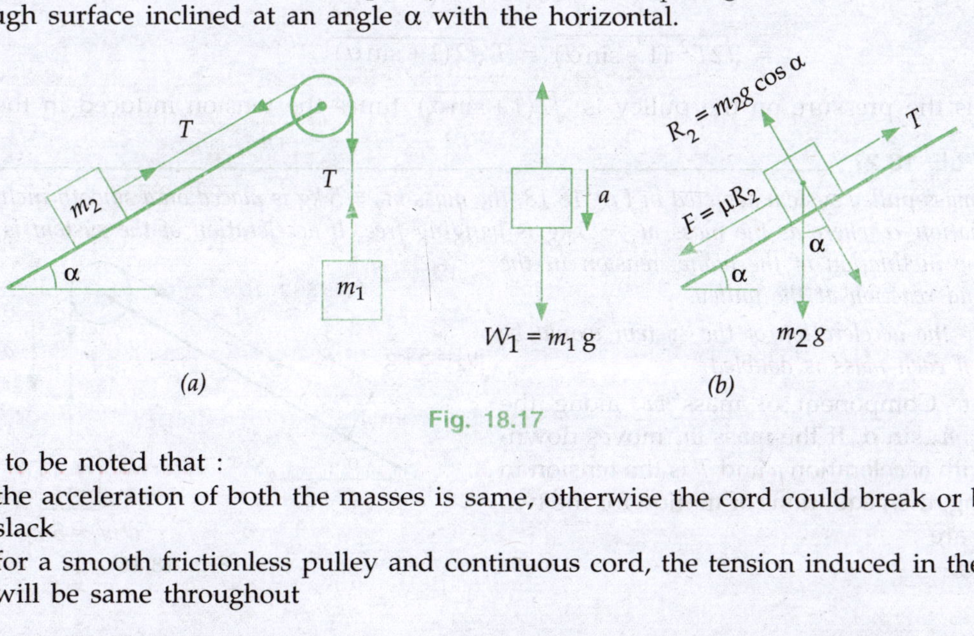

Fig. 18.17

It is to be noted that :

(*i*) the acceleration of both the masses is same, otherwise the cord could break or become slack

(*ii*) for a smooth frictionless pulley and continuous cord, the tension induced in the string will be same throughout

Let mass m_1 move downwards. Then with reference to free body diagrams [(Fig. 18.16 (*b*)], the equations of motion for the two masses are as follows:

$$m_1 g - T = m_1 a \qquad \text{...(i)}$$

$$T - m_2 g \sin\alpha - \mu\, m_2 g \cos\alpha = m_2 a \qquad \text{...(ii)}$$

Adding expressions (*i*) and (*ii*),

$$m_1 g - m_2 g \sin\alpha - \mu\, m_2 g \cos\alpha = (m_1 + m_2)\, a$$

$\therefore$ Acceleration of the system,

$$a = \frac{m_1 g - m_2 g \sin\alpha - \mu\, m_2 g \cos\alpha}{m_1 + m_2} = \frac{W_1 - W_2 \sin\alpha - \mu\, W_2 \cos\alpha}{W_1 + W_2} \times g \qquad \text{...(18.11)}$$

From expression (*i*), the tension T in the string is

$$T = m_1 (g - a) = m_1 \left[g - \frac{m_1 g - m_2 g \sin\alpha - \mu\, m_2\, g \cos\alpha}{m_1 + m_2} \right]$$

$$= m_1 \left[\frac{m_1 g + m_2 g - m_1 g + m_2 g \sin\alpha + \mu\, m_2\, g \cos\alpha}{m_1 + m_2} \right]$$

$$= m_1 \left[\frac{m_2 g (1 + \sin\alpha + \mu \cos\alpha)}{m_1 + m_2} \right] = \frac{W_1 W_2 (1 + \sin\alpha + \mu \cos\alpha)}{W_1 + W_2} \qquad \text{...(18.12)}$$

For a smooth inclined plane, $\mu = 0$. In that case the equations 18.11 and 18.12 get modified to

$$a = \frac{W_1 - W_2 \sin\alpha}{W_1 + W_2}\, g \quad \text{and} \quad T = \frac{W_1 W_2 (1 + \sin\alpha)}{W_1 + W_2} \qquad \text{...(8.13)}$$

Pressure on (or reaction of) the pulley equals the vector sum of tensions T_1 and T_2 induced in the string lying on two sides of the pulley. Here ($T_1 = T_2$) and the angle θ between them is $(90 - \alpha)$. Then from the parallelogram law of forces

$$R = \sqrt{P^2 + Q^2 + 2PQ\cos\theta} = \sqrt{T^2 + T^2 + 2TT\cos(90 - \alpha)}$$

$$= \sqrt{2T^2 (1 + \sin\alpha)} = T\sqrt{2(1 + \sin\alpha)} \qquad \text{...(8.14)}$$

Thus the pressure on the pulley is $\sqrt{2(1 + \sin\alpha)}$ times the tension induced in the string.

EXAMPLE 18.27

For the mass-pulley system depicted in Fig. 18.18, the mass $m_2 = 5$ kg is placed on a smooth inclined plane of inclination α where as the mass $m_1 = 5$ kg is hanging free. If acceleration of the system is 1.5 m/s², determine inclination of the plane, tension in the string and reaction at the pulley.

How the acceleration of the system would be affected if each mass is doubled?

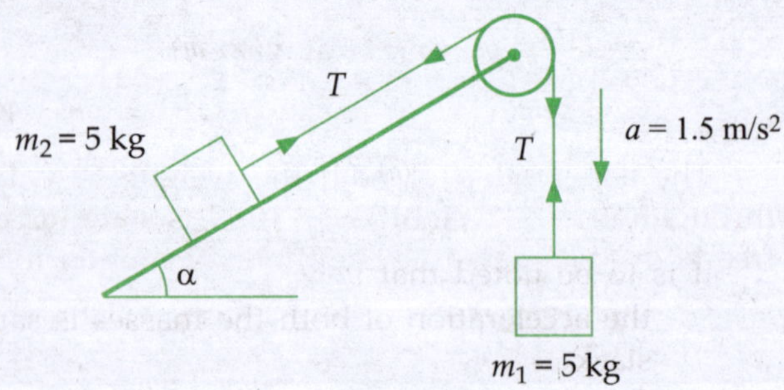

Fig. 18.18

Solution: Component of mass m_2 along the plane $= m_2 \sin\alpha$. If the mass m_1 moves downward with acceleration a and T is the tension in the string, then equations of motion for the two masses are :

$$m_1 g - T = m_1 a \quad ...(i)$$

and $$T - m_2 g \sin\alpha = m_2 a \quad ...(ii)$$

Upon adding these identities and solving, we get

$$\text{Acceleration } a = \frac{m_1 - m_2 \sin\alpha}{m_1 + m_2} \times g$$

$$1.5 = \frac{5 - 5\sin\alpha}{5+5} \times 9.81$$

$$\sin\alpha = \left(5 - \frac{1.5 \times 10}{9.81}\right) \times \frac{1}{5} = 0.692 \quad \text{and therefore} \quad \alpha = 43.96^\circ$$

From expression (*i*), the tension T in the string is

$$T = m_1(g - a) = 5(9.81 - 1.5) = 41.55 \text{ N}$$

$$\text{Reaction at the pulley } R = T\sqrt{2(1+\sin\alpha)} = 41.55\sqrt{2(1+\sin 43.96^\circ)}$$
$$= 76.48 \text{ N}$$

(*b*) $$\text{Acceleration } a = \frac{m_1 - m_2 \sin\alpha}{m_1 + m_2} \times g$$

An examination of the above expression shows that acceleration of the system would remain unaffected when each mass is doubled.

$$\therefore \quad a = 1.5 \text{ m/s}^2$$

EXAMPLE 18.28

Two bodies of weights 50 N and 10 N are connected to the two ends of a light inextensible string which passes over a smooth pulley. The weight 10 N is placed on a rough inclined plane of angle of inclination 20° while the weight of 50 N hangs vertically downward. If the coefficient of friction between the body and the plane is 0.2, make calculations for the acceleration of the system, tension induced in the string, reaction at the pulley and the distance moved by the body in 3 seconds starting from rest.

Solution: Refer Fig. 18.19 $W_1 = 50$ N, $W_2 = 10$ N and $\mu = 0.2$

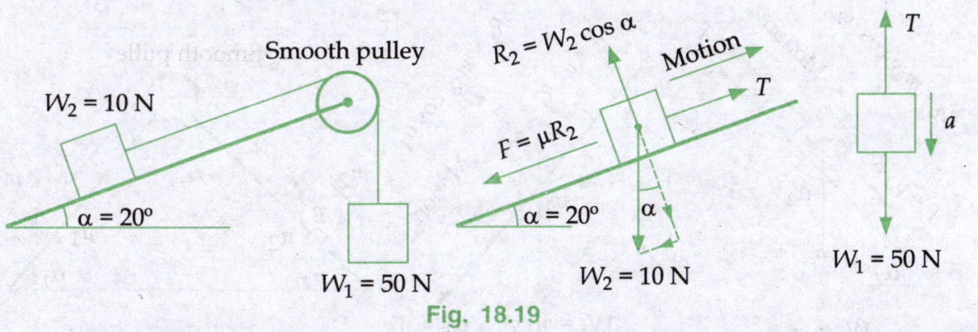

Fig. 18.19

The acceleration of both the weights is equal and tension induced in the string is same throughtout. The weight $W_2 = 10$ N is stated to move up the plane. Then with reference to the free body diagrams, the equations of motion for the two weights are as follows:

$$W_1 - T = \frac{W_1}{g} a \quad ...(i)$$

$$T - W_2 \sin\alpha - \mu W_2 \cos\alpha = \frac{W_2}{g} a \quad ...(ii)$$

Adding expressions (*i*) and (*ii*)

$$W_1 - W_2 \sin\alpha - \mu W_2 \cos\alpha = \left(\frac{W_1}{g} + \frac{W_2}{g}\right) a$$

$$a = \frac{W_1 - W_2 \sin\alpha - \mu W_2 \cos\alpha}{W_1 + W_2} \times g$$

$$= \frac{50 - 10\sin 20^\circ - 0.2 \times 10 \cos 20^\circ}{50 + 10} \times 9.81$$

$$= \frac{50 - 10 \times 0.342 - 0.2 \times 10 \times 0.939}{60} \times 9.81$$

$$= \frac{50 - 3.42 - 1.878}{60} \times 9.81 = 7.31 \text{ m/s}^2$$

(*b*) From expression (*i*), tension *T* in the string is

$$T = W_1 - \frac{W_1}{g} a = 50 - \frac{50}{9.81} \times 7.31 = 12.74 \text{ N}$$

(*c*) Reaction at the pulley,

$$R = T\sqrt{2(1 + \sin\alpha)} = 12.74\sqrt{2(1 + \sin 20^\circ)} = 20.87 \text{ N}$$

(*d*) Applying kinematic equation to the motion of weight $W_2 = 10$ N,

$$s = ut + \frac{1}{2}at^2 ; \quad u = 0 \text{ as the system starts from rest}$$

$$s = 0 + \frac{1}{2} \times 7.31 \times 3^2 = 32.89 \text{ m}$$

18.3.4 Motion of two bodies connected over inclined planes

Consider two blocks *A* and *B* of masses m_1 and m_2 respectively connected by a weightless inextensible rope passing over a frictionless pulley and placed on inclined planes making α_1 and α_2 with the horizontal as shown in Fig. 18.20.

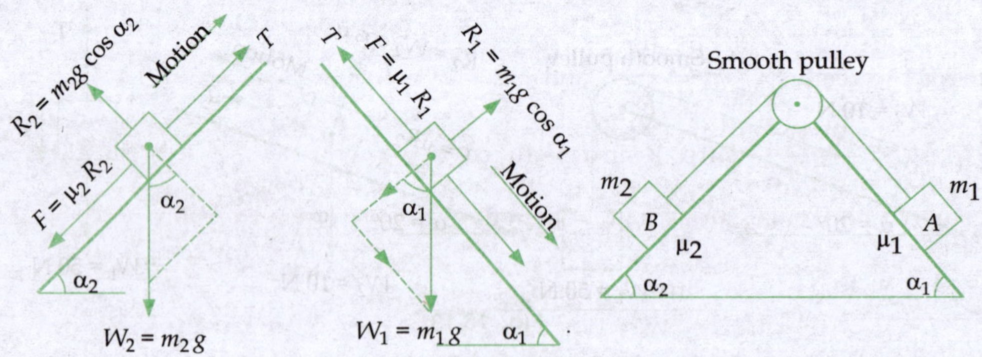

Fig. 18.20

It is to be noted that

(*i*) the acceleration of both the weights is same otherwise the rope would break or become slack

(*ii*) for a smooth frictionless pulley and continuous rope, the tension induced in the rope is same throughout

Let mass m_1 move downwards. Then with reference to free body diagrams, the equations of motion for the two masses are as follows:

$$m_1 g \sin \alpha_1 - T - \mu_1 m_1 g \cos \alpha_1 = m_1 a \quad ...(i)$$

$$T - m_2 g \sin \alpha_2 - \mu_2 m_2 g \cos \alpha_2 = m_2 a \quad ...(ii)$$

Adding expressions (*i*) and (*ii*),

$$m_1 g \sin \alpha_1 - \mu_1 m_1 g \cos \alpha_1 - m_2 g \sin \alpha_2 - \mu_2 m_2 g \cos \alpha_2 = (m_1 + m_2)\, a$$

∴ Acceleration of the system,

$$a = \frac{m_1 g \sin \alpha_1 - m_2 g \sin \alpha_2 - \mu_1 m_1 g \cos \alpha_1 - \mu_2 m_2 g \cos \alpha_2}{m_1 + m_2}$$

$$= \frac{W_1 \sin \alpha_1 - W_2 \sin \alpha_2 - \mu_1 W_1 \cos \alpha_1 - \mu_2 W_2 \cos \alpha_2}{W_1 + W_2} \times g \quad ...(18.15)$$

From expression (*i*), the tension T in the string is

$$T = m_1 g \sin \alpha_1 - \mu_1 m_1 g \cos \alpha_1 - m_1 a$$

$$= m_1 g \sin \alpha_1 - \mu_1 m_1 g \cos \alpha_1 - m_1 \left[\frac{m_1 g \sin \alpha_1 - m_2 g \sin \alpha_2 - \mu_1 m_1 g \cos \alpha_1 - \mu_2 m_2 g \cos \alpha_2}{m_1 + m_2} \right]$$

$$= \frac{1}{m_1 + m_2} \begin{bmatrix} (m_1 + m_2)(m_1 g \sin \alpha_1 - \mu_1 m_1 g \cos \alpha_1) \\ - m_1 (m_1 g \sin \alpha_1 - m_2 g \sin \alpha_2 - \mu_1 m_1 g \cos \alpha_1 - \mu_2 m_2 g \cos \alpha_2) \end{bmatrix}$$

$$= \frac{1}{m_1 + m_2} [m_1^2 g \sin \alpha_1 - \mu_1 m_1^2 g \cos \alpha_1 + m_1 m_2 g \sin \alpha_1 - \mu_1 m_1 m_2 g \cos \alpha_1$$

$$- m_1^2 g \sin \alpha_1 + m_1 m_2 g \sin \alpha_2 + \mu_1 m_1^2 g \cos \alpha_1 + \mu_2 m_1 m_2 \cos \alpha_2]$$

$$= \frac{1}{m_1 + m_2} [m_1 m_2 g \sin \alpha_1 + m_1 m_2 g \sin \alpha_2 - \mu\, m_1 m_2 g \cos \alpha_1 + \mu_2 m_1 m_2 \cos \alpha_2]$$

$$= \frac{1}{m_1 + m_2} [g m_1 m_2 (\sin \alpha_1 + \sin \alpha_2) - g m_1 m_2 (\mu_1 \cos \alpha_1 - \mu_2 \cos \alpha_2)]$$

$$= \frac{g m_1 m_2}{m_1 + m_2} [(\sin \alpha_1 + \sin \alpha_2) - (\mu_1 \cos \alpha_1 - \mu_2 \cos \alpha_2)]$$

In terms of weights : $m_1 = \dfrac{W_1}{g}$ and $m_2 = \dfrac{W_2}{g}$

$$T = \frac{W_1 W_2}{W_1 + W_2} [(\sin \alpha_1 + \sin \alpha_2) - (\mu_1 \cos \alpha_1 - \mu_2 \cos \alpha_2)] \quad ...(18.16)$$

For smooth horizontal surfaces, $\mu_1 = \mu_2 = 0$. In that case, the equations 18.15 and 18.16 get modified to

$$a = \frac{W_1 \sin \alpha_1 - W_2 \sin \alpha_2}{W_1 + W_2}\, g$$

$$T = \frac{W_1 W_2}{W_1 + W_2} (\sin \alpha_1 + \sin \alpha_2) \quad ...(18.17)$$

Pressure on (or reaction of) the pulley equals the vector sum of tensions T_1 and T_2 induced in the string lying on two sides of the pulley. Here $T_1 = T_2 = T$ and the angle θ between them is $180 - (\alpha_1 + \alpha_2)$. Then from the parallelogram law of forces

$$R = \sqrt{P^2 + Q^2 + 2PQ \cos \theta} = \sqrt{T^2 + T^2 + 2TT \cos\{180 - (\alpha_1 + \alpha_2)\}}$$

$$= \sqrt{2T^2 - 2T^2\cos(\alpha_1 + \alpha_2)} = \sqrt{2T^2[1 - \cos(\alpha_1 + \alpha_2)]}$$

$$= \sqrt{2T^2 \times 2\sin^2\frac{\alpha_1 + \alpha_2}{2}} = 2T\sin\frac{\alpha_1 + \alpha_2}{2} \quad ...(18.18)$$

Thus the pressure on the pulley is $2\sin\frac{\alpha_1 + \alpha_2}{2}$ times the tension induced in the string.

EXAMPLE 18.29

Two smooth inclined planes whose inclinations with the horizontal are 20° and 30°, are placed back to back. Two bodies of weight 100 N and 60 N are placed on them and are connected by a light inextensible string passing over a smooth pulley as shown in Fig. 18.21. Calculate the common acceleration of the system, tension in the string and pressure on the pulley.

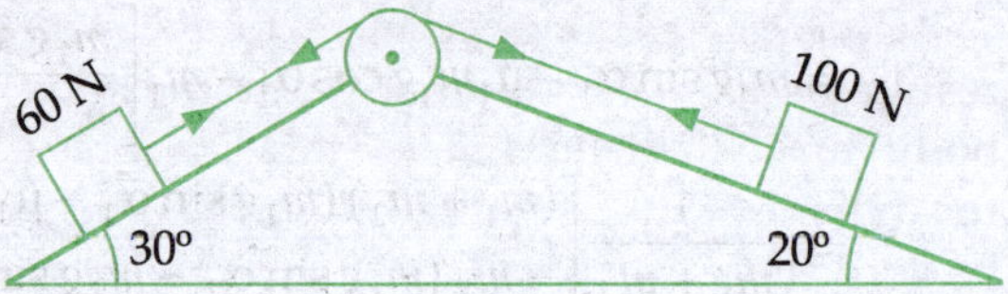

Fig. 18.21

Solution: With reference to Fig. 18.21

$$W_1 = 100 \text{ N} \quad \text{and} \quad \alpha_1 = 20°$$

$$W_2 = 60 \text{ N} \quad \text{and} \quad \alpha_2 = 30°$$

Components of weights W_1 and W_2 along respective planes are

$$W_1\sin\alpha_1 \quad \text{and} \quad W_2\sin\alpha_2$$

If weight W_1 moves downward with acceleration a and T is the tension in the string, then equations of motion for the two weights are

$$W_1\sin\alpha_1 - T = m_1 a = \frac{W_1}{g}a \quad ...(i)$$

and
$$T - W_2\sin\alpha_2 = m_2 a = \frac{W_2}{g}a \quad ...(ii)$$

Adding expressions (*i*) and (*ii*)

$$W_1\sin\alpha_1 - W_2\sin\alpha_2 = \frac{a}{g}(W_1 + W_2)$$

∴ Acceleration of the system,

$$a = \frac{g(W_1\sin\alpha_1 - W_2\sin\alpha_2)}{W_1 + W_2} = \frac{9.81(100\sin 20° - 60\sin 30°)}{100 + 60}$$

$$= \frac{9.81(100 \times 0.342 - 60 \times 0.5)}{160} = 0.2575 \text{ m/s}^2$$

Then from expression (*i*),

$$\text{Tension in the strings } T = W_1\sin\alpha_1 - \frac{W_1}{g}a = 100\sin 20° - \frac{100}{9.81} \times 0.2575$$

$$= 34.20 - 2.62 = 31.58 \text{ N}$$

$$\text{Pressure on the pulley } P = 2T\sin\frac{\alpha_1 + \alpha_2}{2} = 2 \times 31.58\sin\frac{20 + 30}{2} = 26.69 \text{ N}$$

18.4. D'ALEMBERT'S PRINCIPLE

The following section deals with the motion of a body when it is connected by a string to another body which is subjected to some external force. Solution to the system is worked out by applying D' Alembert's principle which states that "the system of forces acting on a body in motion is in dynamic equilibrium with the inertia force of the body"

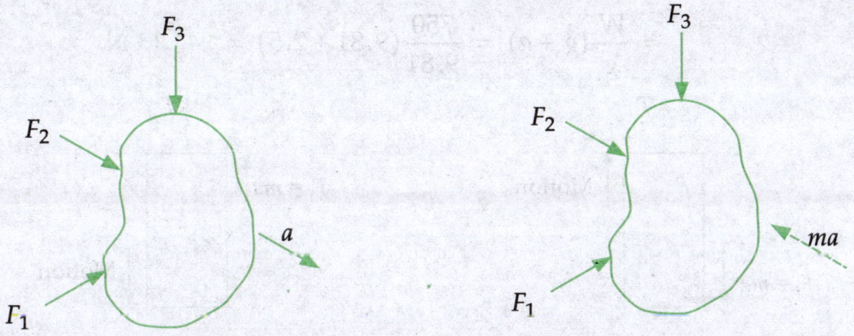

(a) (b)

Fig. 18.22

Consider a body subjected to a system of forces F_1, F_2 and F_3 as shown in Fig. 18.21 (*a*). The resultant of these forces will cause the body to move along its direction with an acceleration *a*. The body can be made to remain static if a force equal to *ma* is made to act in the reversed direction. Then for dynamic equilibrium of the body, the sum of the resultant force and the reversed force should be zero. That is

$$\Sigma F = 0; \quad R - ma = 0 \quad \text{or} \quad R = ma$$

The force $(-\ ma)$ is termed as the **inertia force** or **reverse effective force**.

EXAMPLE 18.30

Determine the acceleration produced in a body of 50 kg mass when it is acted upon by a force of 125 N. Use D'Alembert's principle.

Solution:

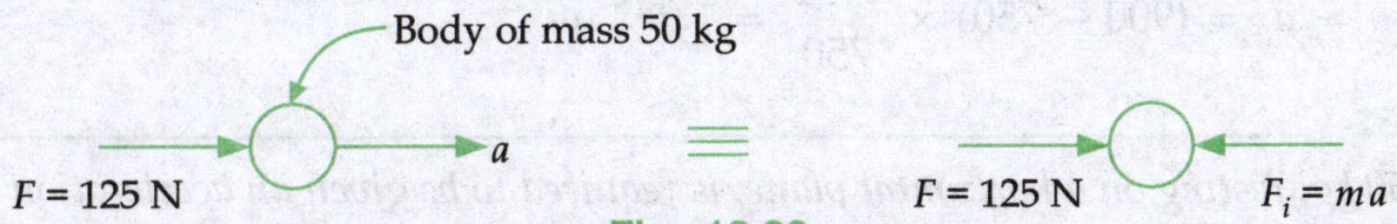

Fig. 18.23

The problem is solved using D'Alembert's principle by applying the inertial force $F_i = ma$ in a direction opposite to that of motion. The body is then in dynamic equilibrium and accordingly

$$\Sigma F_x = 0; \qquad 125 - ma = 0$$

$$125 - 50 \times a = 0 \qquad \therefore \quad a = \frac{125}{50} = 2.5 \text{ m/s}^2$$

EXAMPLE 18.31

A man weighing 750 N stands on the floor of a lift. Determine the pressure exerted on the floor when (a) the lift moves upward with an acceleration of 2.5 m/s^2 (b) the lift moves downward with an acceleration of 2.5 m/s^2.

If 900 N pressure is to be exerted on the floor, then with what acceleration the lift should move upward? Work-out your solution by using D'Alembert's principle

Solution: (*i*) *When the lift moves upwards* : The lift can be brought into dynamic equilibrium by applying an inertia force $F_i = ma$ in the downward direction [Fig. 18.23 (*a*)]. Accordingly,

$$\Sigma F_y = 0; \qquad T - mg - ma = 0$$

where *T* is the tension in the cable supporting the lift or the reaction of the lift or the pressure exerted by the man on the floor of the lift.

$$T = mg + ma = \frac{W}{g}g + \frac{W}{g}a$$

$$= \frac{W}{g}(g + a) = \frac{750}{9.81}(9.81 + 2.5) = \mathbf{941.13\ N}$$

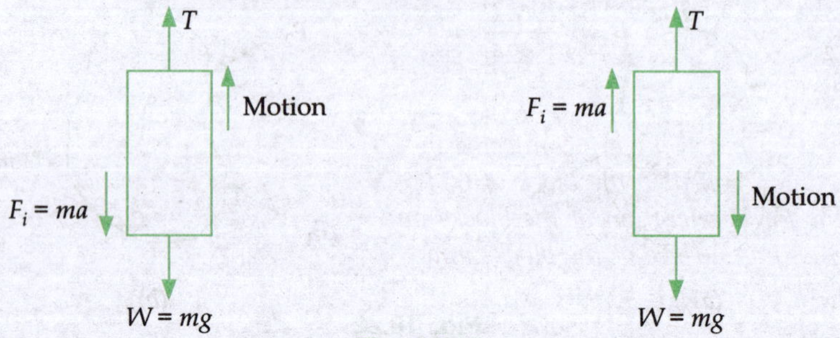

Fig. 18.24

(*ii*) *When the lift moves downwards* : The lift is brought into dynamic equilibrium by applying the inertia force $F_i = ma$ in the upward direction [Fig. 18.24] Accordingly

$\Sigma F_y = 0; \quad T + ma - mg = 0$

$$T = m(g - a) = \frac{W}{g}(g - a) = \frac{750}{9.81}(9.81 - 2.5) = \mathbf{558.87\ N}$$

(*iii*) Let a_u be the upward acceleration required to cause a pressure of 900 N on the floor. Then with reference to Fig. 18.24.

$$T - mg - ma_u = 0$$

$$900 - \frac{750}{9.81} \times 9.81 - \frac{750}{9.81}a_u = 0$$

$$\therefore \quad a_u = (900 - 750) \times \frac{9.81}{750} = \mathbf{1.962\ m/s^2}$$

EXAMPLE 18.32

A block of mass 50 kg, resting on a horizontal plane, is required to be given an acceleration of 2 m/s towards right by applying a push P at an angle of 45 degree with the horizontal. Assuming that coefficient of friction between the block and plane is 0.4, work-out the magnitude of push P. Obtain your solution by applying D'Alembert's principle.

Solution: Refer Fig. 18.25

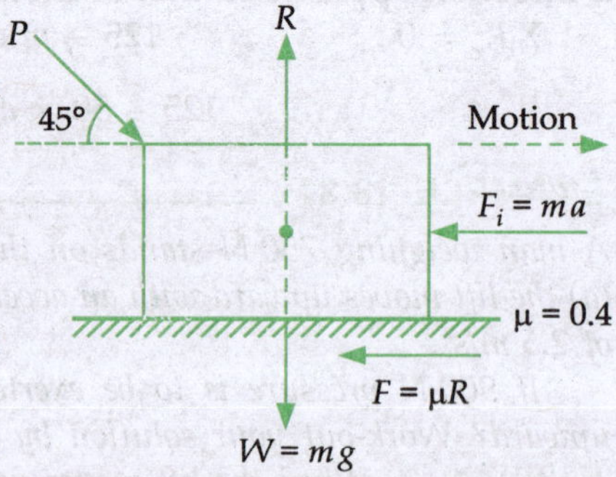

Fig. 18.25

Let an inertia force $F_i = ma$ be applied in a direction opposite to that of motion. The block is then in dynamic equilibrium under the action of following set of forces :

(*i*) Weight $W = mg$ of the block acting downwards

(*ii*) Inertia force $F_i = ma$ acting towards left

(*iii*) Frictional force $F = \mu R = \mu W$ acting opposite to direction of motion

(*iv*) Push P applied at angle of 45 degree with the horizontal.

Writing the equations for equilibrium, we get

$\Sigma F_x = 0; \quad P\cos 45° - \mu R - ma = 0$

$P\cos 45° - 0.4R - 50 \times 2 = 0 \quad \ldots(i)$

$\Sigma F_y = 0; \quad -P\sin 45° - mg + R = 0$

$R = P\sin 45° + 50 \times 9.81 \quad \ldots(ii)$

From expressions (*i*) and (*ii*),

$$P\cos 45° - 0.4\,(P\sin 45° + 50 \times 9.81) - 50 \times 2 = 0$$

or $\quad P\,(\cos 45° - 0.4\sin 45°) - 0.4 \times 50 \times 9.81 - 100 = 0$

or $\quad 0.424P = 196.2 + 100 = 296.2$

$\therefore \quad$ Necessary push $P = \dfrac{296.2}{0.424} = 698.58$ N

EXAMPLE 18.33

Two blocks of mass 60 kg and 15 kg are connected by a string and move along a rough horizontal surface when a force of 300 N is applied to the block of 60 kg mass as shown in Fig. 18.26. Apply D'Alembert's principle to determine the acceleration of the blocks and tension in the string. Presume that coefficient of friction between the sliding surface of the blocks and the plane is 0.25.

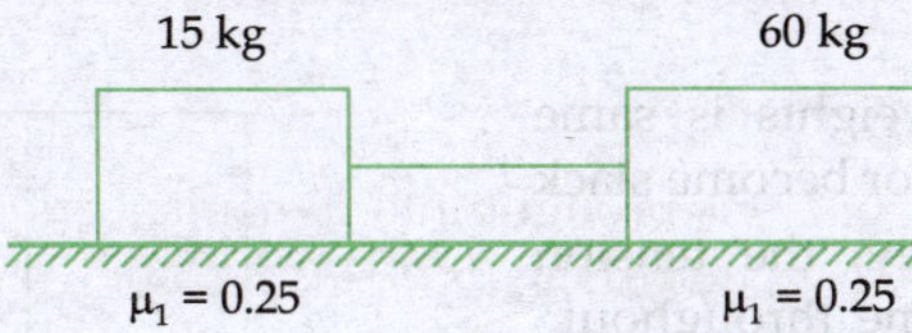

Fig. 18.26

Solution: Since the system moves towards right

(*i*) the friction force would act towards left, and

(*ii*) the inertia force needs to be applied towards left to bring the system in a state of dynamic equilibrium

$$F = \mu_1 R_1 + \mu_2 R_2 = \mu(W_1 + W_2) = \mu g(m_1 + m_2)$$
$$= 0.25 \times 9.81 \times (15 + 60) = 183.94 \text{ N } (\leftarrow)$$
$$F_i = ma = (m_1 + m_2)\,a = (15 + 60)a = 75\,a\ (\leftarrow)$$

Writing the equation for dynamic equilibrium (D'Alembert's principle), we get

$$300 - 183.94 - 75\,a = 0$$

$\therefore \quad$ Acceleration of the system $a = \dfrac{300 - 183.94}{75} = 1.547 \text{ m/s}^2$

(*iii*) Considering dynamic equilibrium of the block of mass 60 kg.

$\Sigma F_x = 0; \quad 300 - T - F_i - F = 0$

or $\quad 300 - T - 60 \times 1.547 - 0.25 \times (60 \times 9.81) = 0$

or $\quad 300 - T - 92.82 - 147.15 = 0$

$\therefore \quad$ Tension in the string $T = 300 - 92.82 - 147.15 = 60.30$ N

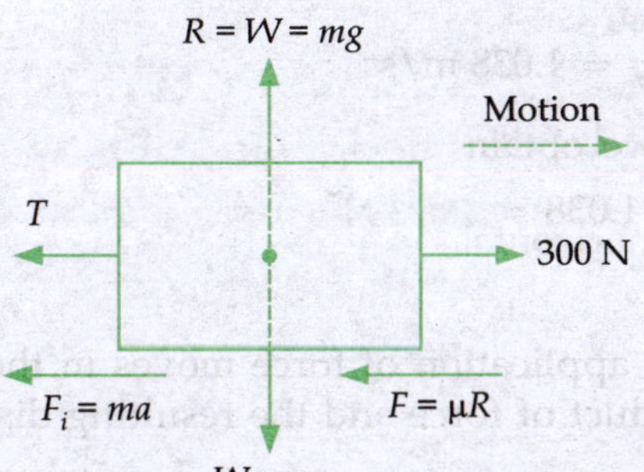

Fig. 18.27

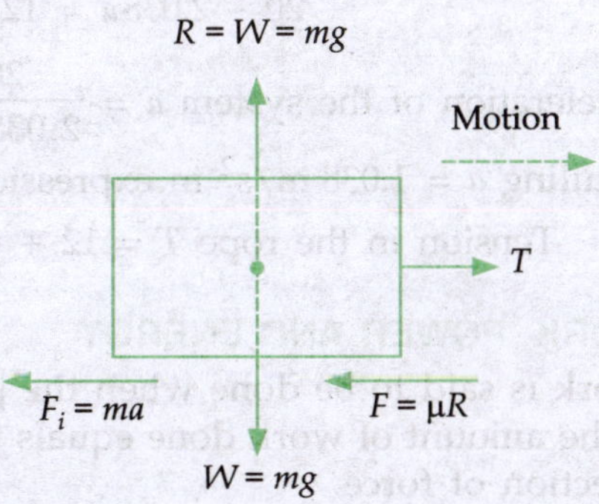

Fig. 18.28

The tension T in the string could also be worked out by considering dynamic equilibrium of the block of 15 kg mass.

$\Sigma F_x = 0$;

$$T - F_i - F = 0$$

or $\quad T - 15 \times 1.547 - 0.25 \times (15 \times 9.81)$

$\therefore \quad T = 15 \times 1.547 + 0.25 \times (15 \times 9.81) = 59.98 \text{ N}$

EXAMPLE 18.34

A block weighing 40 N is accelerated along the horizontal plane by means of a 20 N weight attached to it through flexible, inextensible and weightless rope passing over a smooth pulley as shown in Fig. 18.29. Using D'Alembert's principle, determine the acceleration of the system and the tension in the rope. Take coefficient of friction between the 40 N weight block and the horizontal plane as 0.3.

Solution : It is to be noted that

(*i*) the acceleration of both the weights is same otherwise the rope would break or become slack

(*ii*) for a smooth frictionless pulley, the tension induced in the rope will be same throughout.

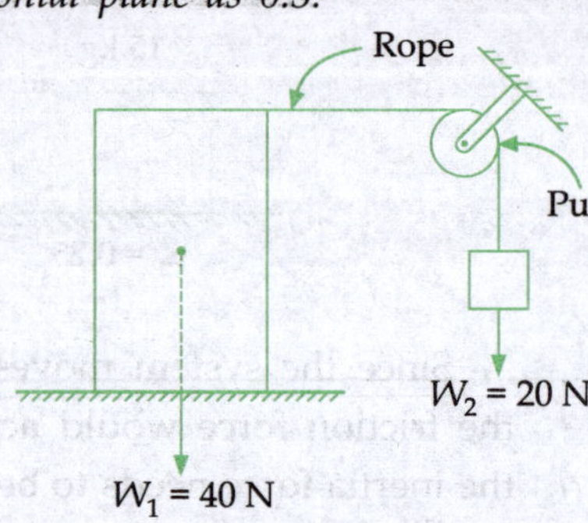

Fig. 18.29

Considering motion of 20 N *weight:* The weight is brought into dynamic equilibrium by applying inertia force F_i in a direction opposite to that of its motion which is downwards. Then

$$20 - T - F_i = 0$$

where $\quad F_i = ma = \dfrac{20}{9.81}a = 2.038\,a$

or $\quad T = 20 - F_i = 20 - 2.038\,a \quad \ldots(i)$

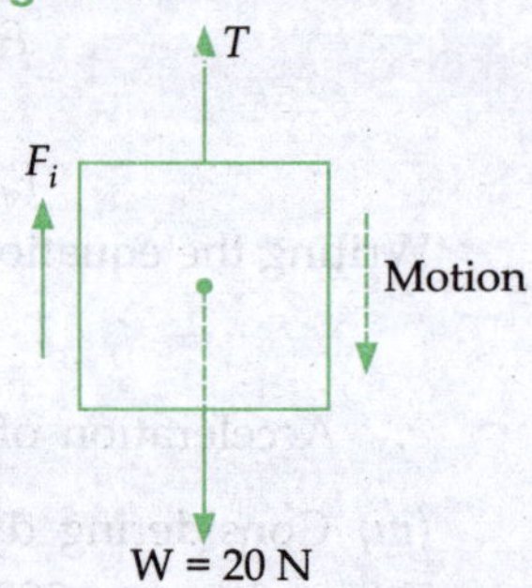

Fig. 18.30

Considering motion of block of 40 N *weight*

Frictional force $F = \mu R = \mu W = 0.3 \times 40 = 12 \text{ N} \ (\leftarrow)$

Inertia force $F_i = ma = \dfrac{W}{g}a = \dfrac{40}{9.81}a = 4.077\,a \ (\leftarrow)$

Then by D' Alembert's principle,

$$T - F - F_i = 0$$

or $\quad T = F + F_i = 12 + 4.077\,a \quad \ldots(ii)$

From expressions (*i*) and (*ii*),

$$20 - 2.038a = 12 + 4.077a$$

$\therefore$ Acceleration of the system $a = \dfrac{20-12}{2.038+4.077} = 1.038 \text{ m/s}^2$

Substituting $a = 1.038 \text{ m/s}^2$ in expression (*ii*), we obtain

Tension in the rope $T = 12 + 4.077 \times 1.038 = 16.23 \text{ N}$

18.5. WORK, POWER AND ENERGY

Work: Work is said to be done when the point of application of force moves in the direction of force. The amount of work done equals the product of force and the resulting displacement in the direction of force.

With reference to Fig. 18.31, F is the force acting an a body and the body moves a distance x in the direction of force. Then the work done on the body is

$$\text{Work done} = \text{force} \times \text{distance}\,; \quad W = Fx$$

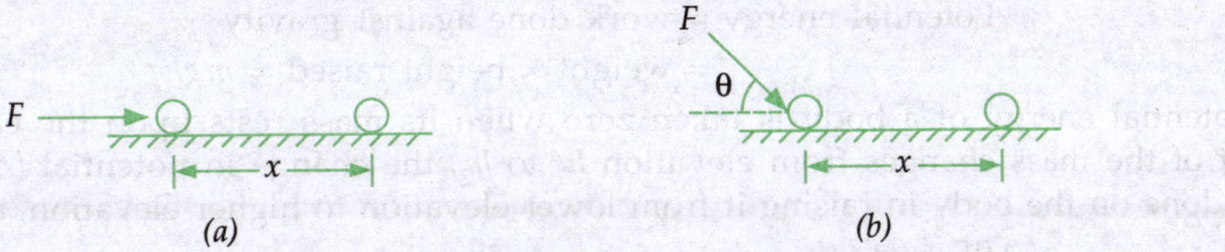

Fig. 18.31

Most of the time, the force and the displacement are not collinear. Work done by the force is then expressed as the product of the component of force in the direction of motion and the distance moved. With reference to Fig. 18.31, consider the force to be inclined at an angle θ with the direction of motion. Then

$$\text{Component of force in the direction of motion} = F\cos\theta$$

$$\text{Work done} = F\cos\theta \times x$$

When $\theta = 90^\circ$, $\cos\theta = 0$ and accordingly no work will be done. Hence if the force and displacement are at right angles, then the work done will be zero.

In case the force changes with displacement, then the work done can be obtained by the expression

$$W = \int_1^2 F(x)\,dx$$

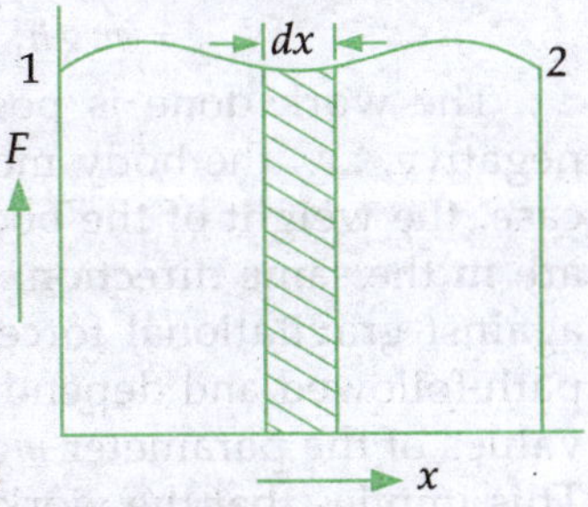

Fig. 18.32

If the force versus displacement curve is drawn (Fig. 18.32), then area under the curve gives the work done.

Note:

(*i*) work done is a scalar quantity; it has a magnitude and sense but no direction.

(*ii*) work done by a force is zero if either the displacement is zero or the force acts normal to displacement.

(*iii*) work done is taken to be positive if the point of application of force moves in the direction of the force. Work done is considered negative when the direction of force is opposite to that of displacement.

The unit of work is joule (J) which is defined as the work done by a constant force of one newton acting on a body and moving it through a distance of one metre in its direction.

$$1\text{ J} = 1\text{ Nm}$$

- **Power:** Power is the rate of doing work and is obtained by dividing the work done by time. The unit of power is watt (W), kilowatt (kW) or megawatt (MW). Watt represents a work equivalent of one joule done per second

$$\text{Power} = \text{work done per unit time} = \text{J/s}$$

Thus

$$1\text{ W} = 1\text{ J/s} = 1\text{ Nm/s}$$

$$1\text{ kW} = 10^3\text{ W} \quad \text{and} \quad 1\text{ MW} = 10^6\text{ W}$$

- **Potential energy and kinetic energy:** Energy is defined as the capacity to produce a change from existing conditions, *i.e.*, capacity to exert force through a distance and do work.

Potential energy is the energy possessed by a body due to its position or elevation; it represents the ability to do work against the body weight in lifting it from the surface of earth.

When a body of mass m kilogram is elevated through a distance h meters above the surface of earth. Then

$$\text{Weight of the body} = mg$$
$$\text{Potential energy} = \text{work done against gravity}$$
$$= \text{weight} \times \text{height raised} = mgh$$

Potential energy of a body is taken zero when its mass rests upon the earth. When the height of the mass changes from elevation h_1 to h_2, the change in potential (ΔPE) equals the work done on the body in raising it from lower elevation to higher elevation, and is given by

$$\Delta PE = mg\,(h_2 - h_1)$$

Consider a body of mass m shifting from position 1 (elevation h_1) to position 2 (elevation h_2) along path A.

Gravitational force on the body at position 1 = $-mg$

The negative sign is due to the fact that the weight of the body is acting downwards whilst the y-axis is chosen upwards.

Work done as the body moves from position 1 to position 2

$$W_{1\text{-}2} = \int F\,dy = \int_1^2 (-mg)\,dy = -mg\,(h_2 - h_1)$$

The work done against the force of gravity is stored as potential energy of the body, that is

$$W_{1\text{-}2} = mgh_1 - mgh_2 = (PE)_1 - (PE)_2 = \text{change in potential energy}$$

The work done is positive when $(h_2 - h_1)$ is negative, *i.e.*, the body moves downward. In that case, the weight of the body and its displacement are in the same direction. Further, the work done against gravitational force is independent of the path followed and depends on the initial and final values of the parameter mgh, *i.e.*, potential energy. This implies that the work done will be same and equal to $mg\,(h_1 - h_2)$ even when the body follows a path B which is different from the previous path.

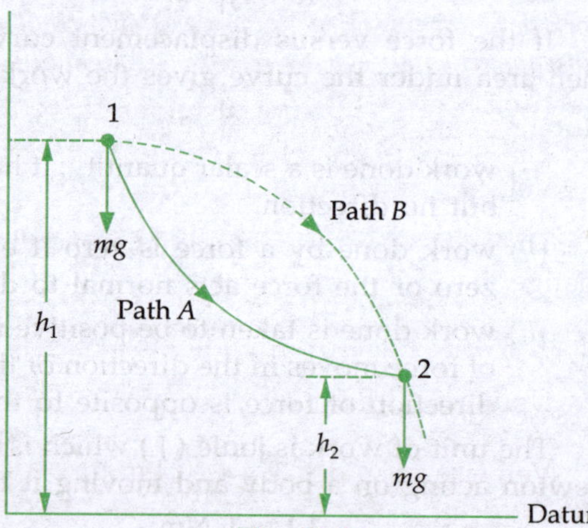

Fig. 18.33

Conservative force : If the work of force in moving an object between two positions is independent of the path followed by the object and can be expressed as a change in its potential energy, then such a force is called a *conservative force*. The notable examples of conservative force are the gravity force, elastic force, spring force etc.

Kinetic energy is the energy possessed by a body by virtue of its motion.

Consider a body of constant mass m which moves through a distance dx in the direction of force acting upon it.

Elementary work done = force × elemental distance moved

$$= \left(m\frac{dV}{dt}\right) \times dx = mV\,dV \qquad \left(\because \frac{dx}{dt} = V\right)$$

where V is the instantaneous velocity in the direction of distance moved. Then

Kinetic energy gained by the body = total work done = $m\int V\,dV$

If the body experiences a change in velocity from V_1 to V_2, then

$$\text{change in kinetic energy, } \Delta KE = m\int_1^2 V\,dV = \frac{1}{2}m\left(V_2^2 - V_1^2\right)$$

If the kinetic energy of the body at rest with respect to the earth is arbitrarily assigned a zero value, then

$$\text{Kinetic energy, } KE = \frac{1}{2}mV^2$$

where V is the velocity of the body relative to earth.

Since energy represents the capacity to do work, its units are same as those of work: Joule (J) or kilo joule (k J).

EXAMPLE 18.35

A box of mass 80 kg is moving at a speed of 10 m/s on a ropeway. If the box is 40 m above the ground level, estimate the potential energy and kinetic energy of the box.

Solution:

Potential energy $= mgh = 80 \times 9.81 \times 40 = 41392 \text{ J} = \mathbf{41.39 \text{ kJ}}$

Kinetic energy $= \frac{1}{2}mV^2 = \frac{1}{2} \times 80 \times 10^2 = 4000 \text{ J} = \mathbf{4 \text{ kJ}}$

EXAMPLE 18.36

A vehicle accelerates a glider of 125 kg mass from rest to a speed of 50 km/hr. Make calculations for the work done on the glider by the vehicle. What change would occur in the kinetic energy of the glider if subsequently its velocity reduces to 20 km/hr on the application of brakes?

Solution: Initial velocity of glider $V_1 = 0$ m/s

Velocity of glider after acceleration $V_2 = 50 \text{ km/hr} = \frac{50 \times 10^3}{3600} = 13.89 \text{ m/s}$

The work done on the glider in accelerating it from velocity V_1 to velocity V_2 equals the change in kinetic energy of the glider.

$\therefore$ Work done on the glider $W = \frac{1}{2}m\left(V_2^2 - V_1^2\right) = \frac{1}{2} \times 125 \times (13.89^2 - 0^2) = \mathbf{12058 \text{ J}}$

Upon application of brakes,

Velocity of glider $V_3 = 20 \text{ km/hr} = \frac{20 \times 10^3}{3600} = 5.56 \text{ m/s}$

Change in kinetic energy $\Delta KE = \frac{1}{2}m\left(V_3^2 - V_2^2\right) = \frac{1}{2} \times 125(5.56^2 - 13.89^2) = \mathbf{-10126 \text{ J}}$

EXAMPLE 18.37

What will be the kinetic energy in kWh of an aeroplane which has a mass of 30 tons and is travelling at 1000 km/hr speed? If this plane is made to nose vertically upwards at this speed with power off, calculate the vertical distance through which the plane will move. Take $g = 9.80 \text{ m/s}^2$.

Solution: Speed of the plane $V = 1000 \text{ km/hr} = \frac{1000 \times 10^3}{3600} = 277.78 \text{ m/s}$

Kinetic energy of the plane $= \frac{1}{2}mV^2 = \frac{1}{2} \times (30 \times 10^3) \times (277.78)^2 = 77161743 \text{ J} = 77162 \text{ kJ}$

1 kWh equals 3600 kJ

$\therefore$ Kinetic energy of the plane $= \dfrac{77162}{3600} = 21.43$ kWh

When the power is off and the plane is made to move vertically upwards, the kinetic energy gets converted to potential energy. Thus

$$77161743 = mgh$$

$\therefore$ Vertical distance through which the plane moves,

$$h = \frac{77161743}{(1000 \times 10^3) \times 9.80} = 7.87 \text{ m}$$

EXAMPLE 18.38

An artificial satellite of mass 500 kg is moving towards the moon. Make calculations for the kinetic and potential energies

(i) relative to the earth when 50 km from launching and travelling at 2500 km/hr. Take earth's gravitational field equal to 7.9 m/s².

(ii) relative to the moon when travelling at the same velocity and 50 km from its destination where 1 kg mass has a weight of 3N.

Solution: Speed of the satellite, $V = 2500 \text{ km/hr} = \dfrac{2500 \times 10^3}{3600} = 694.45 \text{ m/s}$

(i) Kinetic energy $= \dfrac{1}{2} \times 500 \times (694.45)^2 = 120.6 \times 10^6 \text{ J} = 120.6$ MJ

Potential energy $= mgh = 500 \times 7.9 \times (50 \times 10^3) = 197.5 \times 10^6 \text{J} =$ 197.5 MJ

(ii) 1 kg mass weighs 3 N, and hence 500 kg will weigh (relative to the moon) as 1500 N

Potential energy $= mgh = 1500 \times (50 \times 10^3) = 75 \times 10^6 \text{ J} =$ 75 MJ

$$\text{Kinetic energy} = \frac{1}{2} mV^2 = \frac{1}{2} \times 500 \times (694.45)^2 = 120.6 \times 10^6 = 120.6 \text{ MJ}$$

It is to be noted that kinetic energy is not dependent on gravitional field of the moon.

EXAMPLE 18.39

A block weighing 400 N is resting on a smooth horizontal surface. What work will be done if the block is to be moved through 2 m distance by applying

(a) a horizontal force of 150 N

(b) a force of 150 N whose line of action makes an angle of 30° with the horizontal.

Solution:

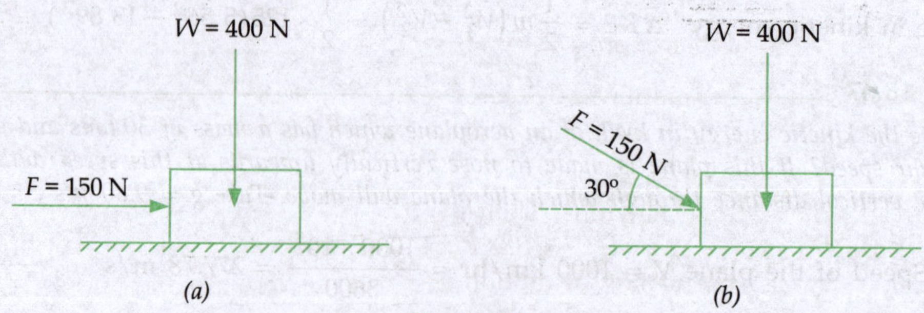

Fig. 18.34

Case (i): Work done = force × distance moved = 150 × 2 = 300 Nm

Case (*ii*): Work done = component of force in the direction of motion × distance moved

$$= F\cos\theta \times x = 150\cos 30^{\circ} \times 2 = 259.81 \text{ Nm}$$

Note: The weight of the block does not contribute towards work done as its line of action is at right angles to that of displacement.

EXAMPLE 18.40

A fire engine lifts 30 × 10³ litres of water to a height of 5 m and discharges it with a velocity of 8 m/s. What is the amount of energy spent during this process?

If the task is accomplished in quarter of an hour, determine input power of the fire engine which has an overall efficiency of 75 percent.

Solution: Quantity of water lifted = 30×10^3 litres = 30 m^3

Since 1 m^3 of water has a mass of 1000 kg,

mass of water lifted = 30×10^3 kg

Potential energy = $mgh = (30 \times 10^3) \times 9.81 \times 5 = 1471.5 \times 30^3$ Nm

Kinetic energy at delivery = $\frac{1}{2}mV^2 = \frac{1}{2} \times (30 \times 10^3) \times 8^2 = 960 \times 10^3$ Nm

∴ Total energy required to be spent by the fire engine

$$= 1471.5 \times 10^3 + 960 \times 10^3 = 2431.5 \times 10^3 \text{ Nm}$$

This energy is spent by the fire engine in quarter an hour, *i.e.*, 900 seconds,

output power of fire engine = energy spent per second

$$= \frac{2431.5 \times 10^3}{900} = 2.70 \times 10^3 \text{ W} = 2.70 \text{ kW}$$

$$\text{Input power} = \frac{\text{output power}}{\text{overall efficiency}} = \frac{2.70}{0.75} = 3.6 \text{ kW}$$

EXAMPLE 18.41

A train weighing 5000 kN starts from rest and accelerates uniformly to 75 km/hr in 40 seconds. If the frictional resistance is estimated to 3 kN per 1000 kN of weight of trains, work out the maximum power required and the power required to maintain the speed of 75 km/hr.

Solution: $u = 0$; $v = \frac{75 \times 1000}{3600} = 20.83 \text{ m/s}$; and $t = 40$ seconds

Then from the kinematic relation $v = u + at$,

$$\text{acceleration } a = \frac{v - u}{t} = \frac{20.83 - 0}{40} = 0.521 \text{ m/s}^2$$

frictional resistance to motion R = 3 kN per 1000 kN of weight of train

$$= 3 \times \frac{5000}{1000} = 15 \text{ kN}$$

Let P be tractive effort in newtons. Then the force available for traction,

$$= P - R = P - 15000$$

Applying Newton's second law of motion $F = ma$, we have

$$(P - 15000) = \frac{5000 \times 10^3}{9.81} \times 0.521 = 265545$$

$$P = 280545 \text{ N}$$

Maximum power required = $P \times v = 280545 \times 20.83$ Nm/s

$= 5843752$ W = **5.844 MW**

(*b*) At any time after 40 seconds, the force is required only to overcome frictional resistance of 15 kN

Power required = $(15 \times 10^3) \times 20.83$ Nm/s = 312450 W = **312.45 kW**

EXAMPLE 18.42

A wooden block of 750 N weight is placed on an inclined plane which makes an angle of 30° with the horizontal. What work will be done if the block is to be moved upwards the plane for 2.5 m distance and the (a) plane is smooth, (b) plane is rough with coefficient of friction 0.2.

Solution:

Case (*a*): When the plane is smooth, the force of friction will be zero. Then the force P on the block when it is on the verge of moving upwards is given by

$$P = W \sin\theta = 750 \times \sin 30° = 375 \text{ N}$$

Work done in pulling the block up

$= P \times s$ where s is the distance moved

$= 375 \times 2.5 =$ **937.5 Nm**

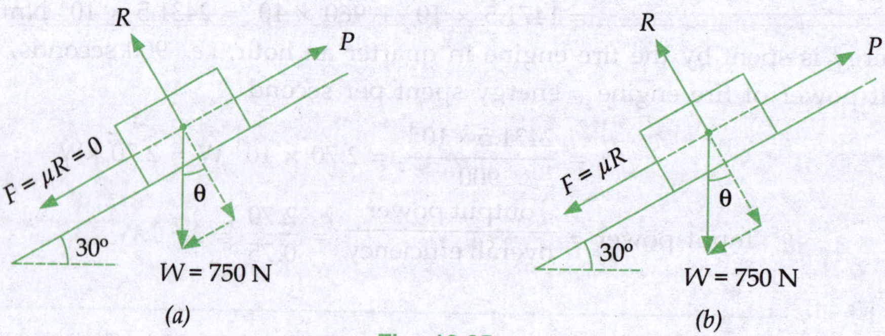

Fig. 18.35

Case (*b*): Refer Fig. 18.35 for the various forces acting on the block. For equilibrium of block,

Normal reaction $R = W \cos\theta = 750 \times \cos 30° = 649.5$ N

Force of friction $F = \mu R = 0.2 \times 649.5 = 129.9$ N

Then the force P on the block when it is on the verge of moving upwards is given by

$$P = W \sin\theta + F = 750 \sin 30° + 129.9 = 504.9 \text{ N}$$

Work done in pulling the block up

$= P \times s$ where s is the distance moved

$= 504.9 \times 2.5 =$ **1262.25 Nm**

EXAMPLE 18.43

A 750 kW power engine, working at full power propells a 2500 kN train up an incline of 1 in 100 at a speed of 60 km/hr. If the track resistance is 5 N per kN weight of train, determine the acceleration with which the train is moving.

Solution: Speed of train = 60 km/hr = $\dfrac{60 \times 1000}{3600} = 16.67$ m/s

Frictional resistance of track = $5 \times 2500 = 12.5 \times 10^3$ N

Slope of the track = 1 in 100

$$\therefore \quad \sin\theta = \tan\theta = \frac{1}{100}$$

Refer Fig. 18.36 for the various forces acting on the train. The component of weight $W\sin\theta$ acts down the plane and so does the frictional resistance F of the track (it has to be in a direction opposite to that of the motion of train).

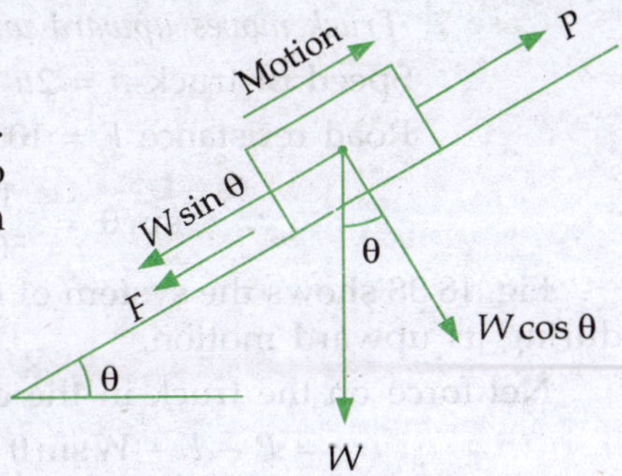

Fig. 18.36

$\therefore$ Net force on the train in the direction of motion

$$F_n = P - F - W\sin\theta$$

where P is the force exerted by the engine up the plane.

Power of the engine = 750 kW = 750×10^3 Nm/s

$$\therefore \quad P \times 16.67 = 750 \times 10^3; \quad P = 44.99 \times 10^3 \text{ N}$$

Thus $\quad F_n = 44.99 \times 10^3 - 12.5 \times 10^3 - 2500 \times 10^3 \times \frac{1}{100} = 7.49 \times 10^3$ N

Applying Newton's second law of motion $F = ma$, we have

$$7.49 \times 10^3 = \frac{2500 \times 10^3}{9.81} a$$

$$\therefore \quad \text{Acceleration of train } a = \frac{7.49 \times 10^3 \times 9.81}{2500 \times 10^3} = \mathbf{0.0294 \text{ m/s}}$$

EXAMPLE 18.44

A truck weighing 5 kN just moves freely (engine is not running) at 30 km/hr down a slope of 1 in 50, and the road resistance at this speed is just sufficient to prevent any acceleration. Estimate the road resistance per kN weight of truck.

What power will the engine have to exert to run up the same slope at double the speed when the road resistance remains the same? Take efficiency of engine as 40 percent.

Solution:

Case 1: *Truck moves freely down the plane.*

Refer Fig. 18.37 for the various forces acting on the truck.

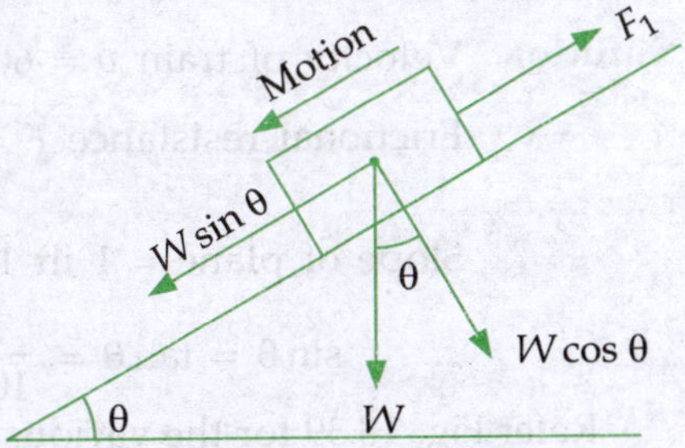

Fig. 18.37

Speed of truck = 30 km/hr

$$u = \frac{30 \times 1000}{3600} = 8.33 \text{ m/s}$$

slope of the truck = 1 in 50

$$\therefore \quad \sin\theta = \tan\theta = \frac{1}{50}$$

The component of weight $W \sin\theta$ acts down the plane and the road resistance F_1 acts up the plane (opposite to the direction of motion of truck).

Net force on the truck in the direction of motion,

$$= W\sin\theta - F_1$$

As the truck moves freely with uniform velocity, its acceleration is zero and so will be the net force ($F = ma$).

$$\therefore \quad W\sin\theta - F_1 = 0 \quad \text{or} \quad F_1 = W\sin\theta = (5 \times 10^3) \times \frac{1}{50} = 100 \text{ N}$$

$$\therefore \quad \text{Road resistance per kN weight of truck} = \frac{100}{5} = \mathbf{20 \text{ N}}$$

Case 2: *Truck moves upward with double the speed.*

Speed of truck $v = 2u = 2 \times 8.33 = 16.66$ m/s

Road resistance $F = 100$ N (same as before)

$$\sin\theta = \frac{1}{50}$$

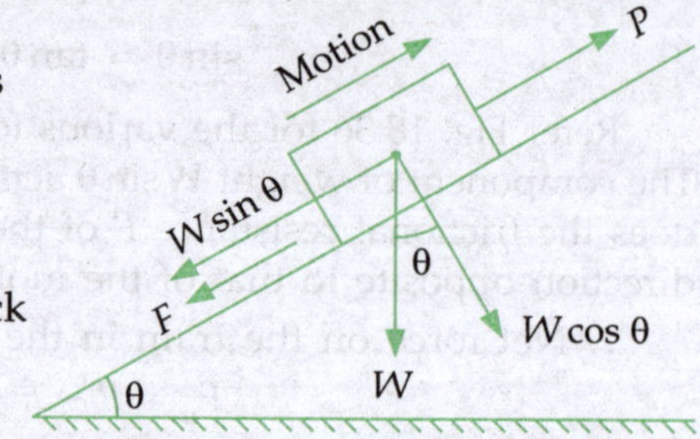

Fig. 18.38

Fig. 18.38 shows the system of forces acting on the truck during its upward motion.

Net force on the truck in the direction of motion

$$= P - F - W\sin\theta$$

where P is the force exerted by engine up the plane. Again the truck moves upward with uniform speed, *i.e.*, no acceleration and accordingly

$$P - F - W\sin\theta = 0$$

$$\therefore \quad P = F + W\sin\theta = 100 + (5 \times 10^3) \times \frac{1}{50} = 200 \text{ N}$$

Output power required $= P \times v$

$= 200 \times 16.66 = 3332$ W $= 3.332$ kW

$$\therefore \quad \text{Input power of engine} = \frac{\text{output power}}{\text{efficiency of engine}} = \frac{3.332}{0.4} = \mathbf{8.33\ kW}$$

EXAMPLE 18.45

Calculate the power required to pull a 400 kN train up an incline of 1 in 100 at a steady speed of 60 km/hr. Take frictional resistance equal to 5 N per kN weight of train.

At a particular instant while ascending the incline, the supply of steam to the locomotive pulling the train is shut off. How far the train will move before comming to rest? Assume that track resistance remains unchanged.

Solution: Velocity of train $v = 60 \text{ km/hr} = \frac{60 \times 1000}{3600} = 16.67$ m/s

Frictional resistance $F = 5$ N per kN weight of train

$= 5 \times 4000 = 2000$ N

Slope of plane = 1 in 100

$$\therefore \quad \sin\theta = \tan\theta = \frac{1}{100}$$

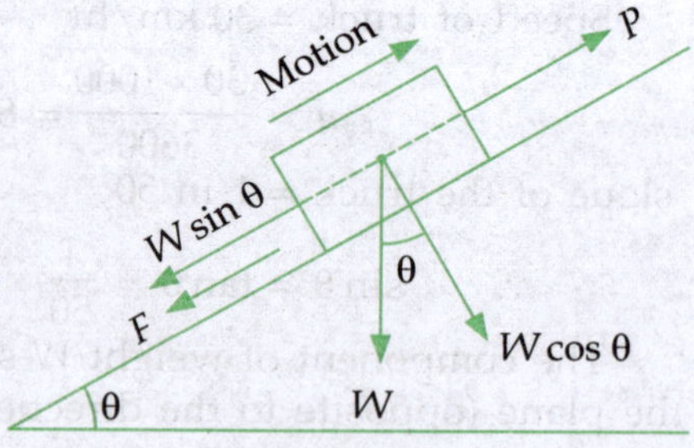

Fig. 18.39

Refer Fig. 18.39 for the various forces acting on the train. The component $W \sin\theta$ of the weight of train acts down the plane and so does the frictional resistance (it has to be in a direction opposite to that of the motion of the train).

Net force on the train in the direction of motion,

$$F_n = P - F - W\sin\theta$$

where P is the force exerted by the engine up the plane. As the train moves with uniform speed, its acceleration is zero and so will be the net force ($F = ma$)

$$\therefore \quad P - F - W\sin\theta = 0$$

$$P = F + W\sin\theta = 2000 + (400 \times 10^3) \times \frac{1}{100} = 6000 \text{ N}$$

Power required to pull the engine = $P \times v = 6000 \times 16.67$

$= 100.02 \times 10^3$ Nm/s = **100.02 kW**

(*b*) When the steam supply is shut off, there will be no pull exerted by the engine. The force acting down the plane is provided by frictional resistance and the component $W \sin\theta$ of the weight of train.

Total retarding force = $F + W\sin\theta$

$$F_r = 2000 + (400 \times 10^3) \times \frac{1}{100} = 6000 \text{ N}$$

Let the train travel a distance *s* before coming to rest.

initial velocity = 16.67 m/s and final velocity = 0

Then from the work – energy equation,

$$-F_r \times s = \frac{1}{2} m (v^2 - u^2)$$

The negative sign with F_r stems from the fact that retarding force is acting opposite to the direction of motion.

$$-6000 \times s = \frac{1}{2} \times \frac{400 \times 10^3}{9.81} \times (0^2 - 16.67^2)$$

$$\therefore \text{ Distance moved } s = \frac{400 \times 10^3}{2 \times 9.81} \times \frac{16.67^2}{6000} = \textbf{944.24 m}$$

EXAMPLE 18.46

A force of 500 N applied initially and horizontal is just sufficient to move a body of 1250 N weight placed on a smooth horizontal surface. The force applied does not remain constant but varies linearly with distance moved by the body, the linear variation is prescribed by the relation

$$P = P_0 + 8x$$

where P_0 is the initial force and x is the distance moved by the body in metres. Depict the force-distance diagram and determine the work done when the body has moved through 20 m distance.

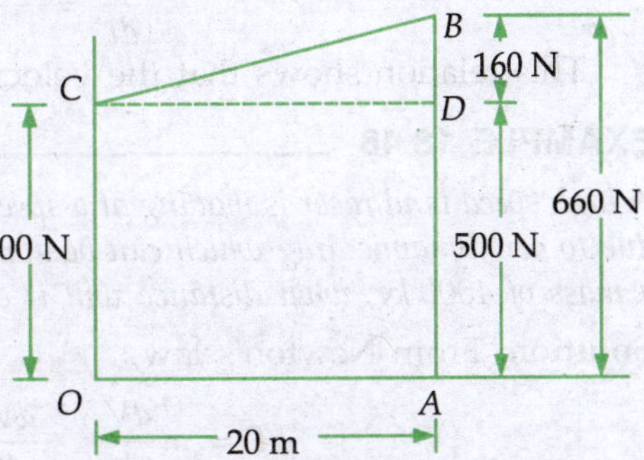

Fig. 18.40

Solution: Initial force P_0 = 500 N

Force after the body has moved a distance of 20 m

$$P = P_0 + 8 \times 20$$

$$= 500 + (8 \times 20) = 660 \text{ N}$$

The force-distance variation has been depicted in Fig. 18.40.

The area of this diagram gives the work done.

Work done = area of trapezium *OABC*

= area of rectangle *OADC* + area of triangle *BCD*

$$= (500 \times 20) + \frac{1}{2} \times 20 \times 160 = \textbf{11600 Nm}$$

Alternatively:

$$\text{average force} = \frac{500 + 660}{2} = 580 \text{ N}$$

work done = average force × distance moved

= 580 × 20 = **11600 Nm**

EXAMPLE 18.47

When a body of mass m moves through a liquid at low velocity, the force of resistance due to viscosity is given by F = kv where k is the resisting force at unit velocity. Show that velocity would decrease exponentially with time and linearly with displacement.

Solution: Retarding force $F = -kv$

$\therefore$ Retardation $\frac{dv}{dt} = -\frac{kv}{m}$

The negative sign stems from the fact that it is negative acceleration.
Separating the variables and upon integration, one obtains

$$\int \frac{dv}{v} = -\frac{k}{m}\int dt\,; \quad \log_e v = -\frac{kt}{m}$$

That gives: $v = e^{-kt/m}$

The above relation shows that the velocity decreases exponentially with time.

(*ii*) Again $\frac{dv}{dt} = -\frac{kv}{m}$

Since $v = \frac{ds}{dt}$, we can write

$$\frac{d}{dt}\left(\frac{ds}{dt}\right) = -\frac{k}{m}\left(\frac{ds}{dt}\right)$$

or

$$\frac{d^2s}{dt^2} = -\frac{k}{m}\frac{ds}{dt}$$

Upon integration:

$$\frac{ds}{dt} = -\frac{k}{m}s\,; \quad v = -\frac{k}{m}s$$

This relation shows that the velocity decreases linearly with displacement.

EXAMPLE 18.48

A high speed land racer is moving at a speed of 100 m/s. The resistance to motion of the vehicle is primarily due to aerodynamic drag which can be approximated as 0.2 V^2 newton with V in m/sec. If the vehicle has a mass of 4000 kg, what distance will it coast before its speed is reduced to 70 m/sec?

Solution: From Newton's law,

$$F = m\frac{dV}{dt}\,; \quad \frac{dV}{dt} = \frac{F}{m} = \frac{-0.2V^2}{4000} = -5 \times 10^{-5}V^2$$

The negative sign is because of drag.

Separating the variables and integrating, we have

$$\int \frac{dV}{V^2} = -5 \times 10^{-5}dt \quad \text{or} \quad -\frac{1}{V} = -5 \times 10^{-5}t + C_1$$

The constant of integration C_1 can be evaluating by noting that at $t = 0$, $V = 100$ m/s. That gives $C_1 = 0.01$ and accordingly

$$-\frac{1}{V} = -5 \times 10^{-5}t + 0.01 \qquad \ldots(a)$$

Replacing $V = dx/dt$ and then separating the variables

$$\frac{dt}{5\times 10^{-5} + 0.01} = dx$$

For integration of above expression, let

$$\eta = 5 \times 10^{-5}t + 0.01 \qquad \therefore \quad d\eta = 5 \times 10^{-5}dt$$

Making these substitutions

$$\frac{d\eta}{\eta} = 5 \times 10^{-5}dx$$

or $$\log_e \eta = 5 \times 10^{-5}x + C_2$$

or $$\log_e (5 \times 10^{-5}t + 0.01) = 5 \times 10^{-5}x + C_2$$

For evaluating the constant for integration, we use the condition $x = 0$ at $t = 0$. That gives $C_2 = \log_e 0.01$.

Accordingly $$\log_e (5 \times 10^{-5}t + 0.01) = 5 \times 10^{-5} + \log_e 0.01$$

or $$\log_e 0.01 + \log_e (5 \times 10^{-3}t + 1) = 5 \times 10^{-5}x + \log_e 0.01$$

$$\therefore \quad \log_e (5 \times 10^{-3}t + 1) = 5 \times 10^{-5}x \qquad \text{...(b)}$$

Substituting $V = 70$ m/s in expression (*a*),

$$\frac{1}{70} = 5 \times 10^{-5}t + 0.01 \;; \quad t = \left(\frac{1}{70} - 0.01\right) \div 5 \times 10^{-5}$$

$$= 85.71 \text{ sec}$$

The distance travelled by the racer is then obtained by substituting $t = 85.71$ sec in expression (*b*)

$$x = \frac{\log_e (5 \times 10^{-3} \times 85.71 + 1)}{5 \times 10^{-5}} = 7132.5 \text{ m}$$

$$= \mathbf{7.132 \text{ km}}$$

EXAMPLE 18.49

A particle of mass 1 kg moves in a straight line under the influence of a force which increases linearly with time at the rate of 60 N per second. At time t = 0, initial force may be taken as 50 N. Determine the acceleration and velocity of a particle 4 sec after it started from rest at the origin.

Solution: Let the variation of force with time be prescribed by the relation

$$F = at + b, \quad \text{where } a \text{ and } b \text{ are constants.}$$

$$\frac{dF}{dt} = a \;; \qquad 60 = a$$

Since $F = 50$ N at $t = 0$, we get

$$50 = 60 \times 0 + b \;; \qquad b = 50$$

Hence the force-time relation may be written as

$$F = 60t + 50$$

Putting $t = 4$ in the above relation, we get the value of force acting on the particle 4 sec after it started from rest. That is

$$F = 60 \times 4 + 50 = 290 \text{ N}$$

From the relation $F = ma$,

Acceleration after 4 sec, $a = \dfrac{290}{1} = \mathbf{290 \text{ m/s}^2}$

At any time t, the acceleration of the particle will be

$$a = \frac{F}{m} = \frac{60t + 50}{1} = 60t + 50$$

$$\frac{dv}{dt} = 60t + 50 \qquad \left(\because a = \frac{dv}{dt}\right)$$

Upon integration : $\int_0^v dv = (60t + 50)\,dt$

$$v = \frac{60t^2}{2} + 50t = 30t^2 + 50t$$

$\therefore$ velocity of particle 4 sec after start

$$v = 30 \times (4)^2 + 50 \times 4 = \textbf{680 m/s}$$

EXAMPLE 18.50

A hammer of mass 200 kg is made to fall freely from 2 m height on the head of a pile of 1500 kg mass. The pile is driven 5 cm into the ground in one blow. Determine (a) the common velocity of the pile and hammer after the impact, (b) the energy lost in the impact, and (c) the average resistance of the ground to penetration.

Solution: Let v be the velocity of the hammer when it reaches the pile. Then using the kinematic relation

$$v^2 - u^2 = 2\,gh\,;$$
$$v^2 - 0^2 = 2 \times 9.81 \times 2$$
$$v = \sqrt{2 \times 9.81 \times 2} = 6.26 \text{ m/s}$$

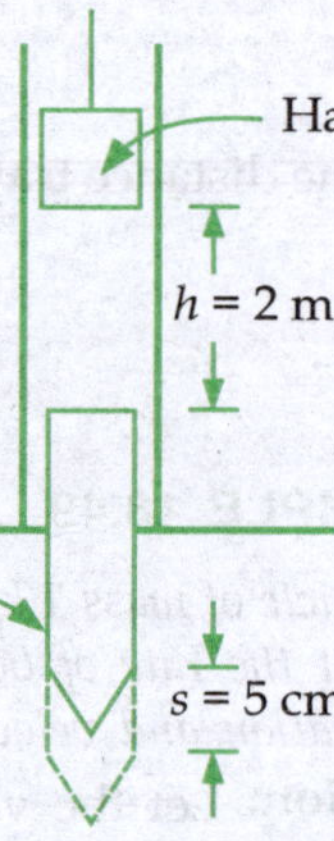

Fig. 18.41

Let V be the common velocity of the total mass $(M + m)$ in motion after the impact. Then from the principle of conservation of momentum,

momentum before impact = momentum after impact

$$mv = (M + m)V$$
$$200 \times 6.26 = (1500 + 200)V \quad \text{or} \quad V = 0.736 \text{ m/s}$$

(b) Kinetic energy before impact $= \frac{1}{2}\,mV^2 = \frac{1}{2} \times 200 \times (6.26)^2$
$= 3918.76$ Nm

Kinetic energy after impact $= \frac{1}{2}\,(M + m)\,V^2$

$= \frac{1}{2}\,(1500 + 200) \times (0.736)^2 = 460.44$ Nm

$\therefore$ Loss of kinetic energy at impact $= 3918.76 - 460.44 =$ **3458.32 Nm**

(c) After penetration in the ground, the velocity of the system (pile + hammer) becomes zero. Using the kinematic relatio $v^2 - u^2 = 2as$, we have

$$0^2 - 0.736^2 = 2a \times (0.05)$$

$$a = -\frac{0.736^2}{2 \times 0.05} = -5.417 \text{ m/s}^2 \quad \text{(retardation)}$$

Let F_r be the average resistance exerted by the ground. Then net force acting on the system in the upward direction,

$$F_n = F_r - W$$

where W is the weight of hammer and pile.

Applying Newton's second law of motion $F = ma$, we have

$$F_r - (1500 + 200) \times 9.81 = (1500 + 200) \times 5.417$$
$$F_r - 16677 = 9208.8 \qquad \therefore \quad F_r = \textbf{25885.8 N}$$

EXAMPLE 18.51

A hammer of mass 0.6 kg and moving with a velocity of 6 m/s drives a nail of 20 gm mass horizontally into a fixed wooden block. The nail is stated to penetrate 25 mm per blow of hammer. Workout the resistance offered by the block to the penetration of nail and the energy lost in driving the nail 75 mm inside the block.

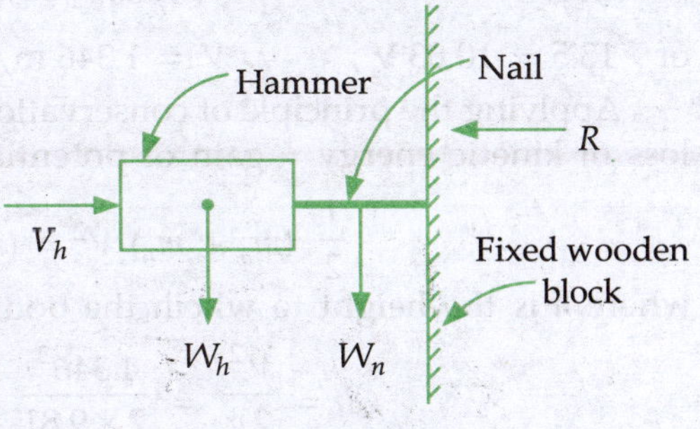

Fig. 18.42

Solution: Refer Fig. 18.42 which shows the system for the various parameters inserted.

Applying the principle of conservation of momentum momentum before impact

= momentum after impact

$$m_h V_h + m_n V_n = (m_h + m_n)\, V$$

where suffix h and n refer to hammer and nail respectively and V is the common velocity after impact.

$0.6 \times 5 + 0.02 \times 0 = (0.6 + 0.02)\, V \quad \therefore\ V = 4.838$ m/s

The resistance offered by the block will retard the motion and bring the nail and hammer to rest after blow has been given.

Then from the work energy relation,

$$-R \times s = \frac{1}{2} m (v^2 - u^2)$$

The negative sign with R stems from the fact that resistance force acts opposite to the direction of motion of nail

$$-R \times 0.025 = \frac{1}{2} \times (0.6 + 0.02)\,(0 - 4.838^2)$$

Solution gives: $R =$ **290.24 N**

The weight of the hammer and the nail do not contribute towards work done because their line of action is normal to that of displacement.

(*b*) There is 25 mm of penetration in 1 blow

$\therefore$ 75 mm of peneration would take place in 3 blows.

$$\text{Kinetic energy before impact} = \frac{1}{2} \times 0.6 \times 6^2 = 10.8 \text{ Nm}$$

$$\text{Kinetic energy after impact} = \frac{1}{2}\,(0.6 + 0.02) \times 4.838^2 = 7.26 \text{ Nm}$$

Loss in kinetic energy at each blow,

$$= 10.8 - 7.26 = 3.54 \text{ Nm}$$

$\therefore$ Total loss in energy $= 3.54 \times 3 =$ **10.62 Nm**

EXAMPLE 18.52

A body of mass 10 kg is suspended by a string of length 1 m. It is struck by a bullet travelling horizontally with a velocity of 450 m/s. The bullet weighs 30 gm and gets embedded into the body after striking it. Determine the maximum angle through which the body swings.

Solution: Refer Fig. 18.43 which shows the given system with the various parameters inserted.

Applying the principle of conservation of momentum,

momentum before impact

= momentum after impact

$$m_a V_a + m_b V_b = (m_a + m_b) V$$

where V is the common velocity with which the body and the bullet together after impact.

$$\frac{30}{1000}\times 450 + 10 \times 0 = \left(\frac{30}{1000} + 10\right) V$$

or $13.5 = 10.03\text{ V}$; $\therefore V = 1.346$ m/s

Applying the principle of conservation of energy, we have

loss of kinetic energy = gain of potential energy

$$\frac{1}{2}(m_a + m_b) V^2 = (m_a + m_b) gh$$

where h is the height to whcih the body rises

$$h = \frac{V^2}{2g} = \frac{1.346^2}{2 \times 9.81} = 0.0923 \text{ m}$$

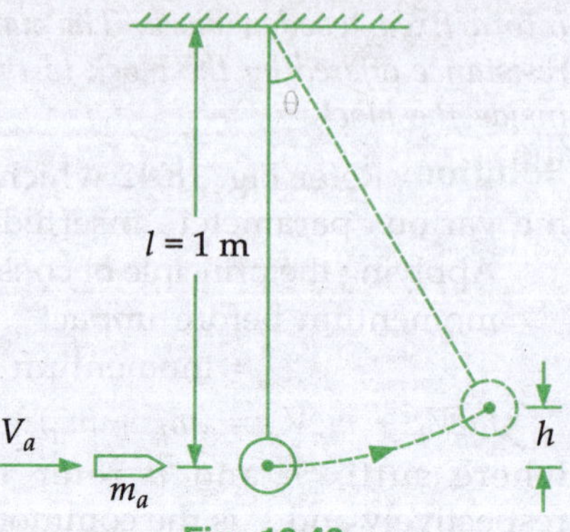

Fig. 18.43

From the geometry of Fig. 8.95,

$$\cos\theta = \frac{l-h}{l} = \frac{1-0.0923}{1} = 0.9077$$

$$\therefore \quad \theta = \cos^{-1}(0.9077) = \mathbf{24.81°}$$

18.6 WORK-ENERGY PRINCIPLE

Consider a force F acting on an object which may be displaced from position 1 to position 2 over the course of action covering distance s as shown in Fig. 18.44. When the object is subjected to more than one force, then F represents the net force or the resultant force acting on the object.

For an elementary distance ds travelled by the objected in time dt, the work done by the force would be

$$dW = F\,ds \quad \text{...(i)}$$

Newtons second law of motion provides that

$$F = \text{mass} \times \text{acceleration} = ma$$

Further, $a = \frac{dv}{dt} = \frac{dv}{ds} \times \frac{ds}{dt} = v\frac{dv}{ds}$

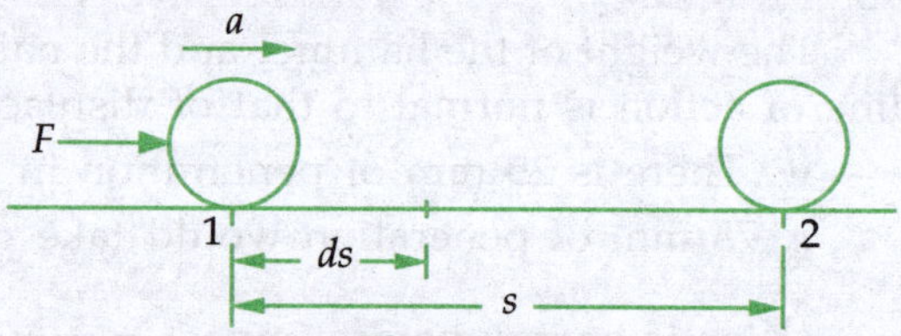

Fig. 18.44

Accordingly expression (i) may be rewritten as

$$dW = \left(m \times v\frac{dv}{ds}\right) \times ds = mv\,dv \quad \text{...(ii)}$$

Integrating the above identity with in appropriate limits

$$\int_1^2 dW = m\int_1^2 v\,dv$$

or $$W_{1-2} = m\left|\frac{v^2}{2}\right|_1^2 = \frac{1}{2}mv_2^2 - \frac{1}{2}mv_1^2 = (KE)_2 - (KE)_1 \quad \text{...(18.19)}$$

Thus the work done on the object equals the change in kinetic energy of the object Equation 18.19 represents the well-known **work-energy principle** in mechanics.

18.7 CONSERVATION OF MECHANICAL ENERGY

In the conservative force field, the work done in moving an object from position 1 to position 2 is independent of the path followed and depends on a change in the potential energy of the end states only. That is

$$W_{1-2} = (PE)_1 - (PE)_2$$

Equating this expression for work done with that provided by the work-energy equation, we get

$$(PE)_1 - (PE)_2 = (KE)_2 - (KE)_1$$

or $$(PE + KE)_1 = (PE + KE)_2 = \text{constant} \quad ...(18.20)$$

The above relation implies that the sum of *PE* and *KE* remains constant for all positions of the object during its motion in the conservative force field.

The function (*PE* + *KE*), *i.e.*, the sum of potential energy and kinetic energy of an object is called mechanical energy.

Equation 18.20 is referred to as the principle of conservation of mechanical energy.

In general, the law of conservation of energy may be stated as

"the total energy possessed by an object remains constant provided no energy is added to or subtracted from it"

Alternatively:

"the energy can neither be created nor destroyed though it can be transformed form one form to another".

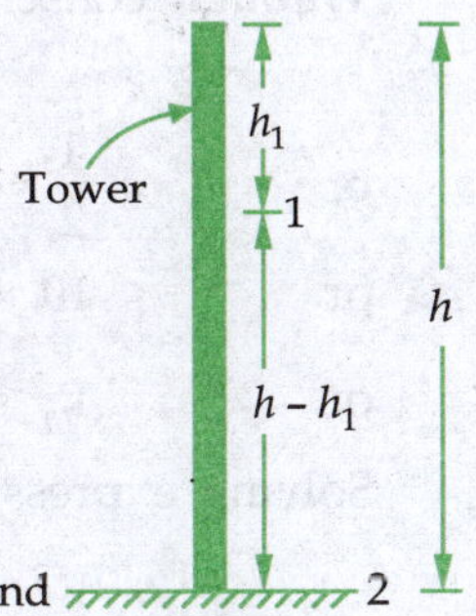

Fig. 18.45

Consider a body of mass *m* resting on the top of a tower of height *h*. At this position

$$KE \text{ of body} = \frac{1}{2}mv^2 = 0 \qquad (\because \; v = 0)$$

PE of body with respect to ground = *mgh*

$\therefore$ Total energy of the body at the top of tower is

$$= KE + PE = 0 + mgh = mgh \quad ...(i)$$

Energy at position 1:

Let the body fall to position 1 which is at a distance h_1 from the top of the tower. The velocity of the body at this location can be worked out from the kinematic equation $v^2 - u^2 = 2as$. That gives

$$v_1^2 - 0^2 = 2gh_1; \quad v_1^2 = 2gh_1$$

$$KE \text{ of the body at position } 1 = \frac{1}{2}mv_1^2 = \frac{1}{2}m \times 2gh_1 = mgh_1$$

PE of the body at position 1 = $mg(h - h_1)$

$\therefore$ Total energy of the body at position 1

$$= mgh_1 + mg(h - h_1) = mgh \quad ...(ii)$$

Energy at position 2:

Finally when the body has the complete fall from the top of tower to ground level (position 2), we have

$$v_2^2 - 0^2 = 2gh \; ; \quad v_2^2 = 2gh$$

$$KE \text{ at ground level} = \frac{1}{2}mv_2^2 = \frac{1}{2}m \times 2gh = mgh$$

PE at ground level = $mg \times 0 = 0$

$\therefore$ Total energy of the body at ground level

$$= mgh + 0 = mgh \quad ...(iii)$$

The above analysis does illustrate that the sum of *KE* and *PE* of a body remains constant under the action of gravity force (a conservative force). Thus, though the *KE* and *PE* individually change throughout the motion, their sum always remains constant.

EXAMPLE 18.53

A shell of 80 kg mass in travelling with a velocity of 500 m/s. During its motion, it splits into two equal portions which continue to travel in the same direction. If an energy equivalent to 1.5×10^6 Nm is released at the instant of explosion, determine the subsequent velocities of the two portions.

Solution: Let V_1 and V_2 be the velocities of the two portions. By conservation of momentum,

momentum before explosion = momentum after explosion

$$80 \times 500 = (m_1 V_1 + m_2 V_2) = 40\, V_1 + 40\, V_2$$

$$V_1 + V_2 = 1000 \quad \text{...(i)}$$

Through conservation of energy,

initial kinetic energy + kinetic energy of explosion = final kinetic energy

$$\text{or} \quad \frac{1}{2} \times 80 \times 500^2 + 1.5 \times 10^6 = \frac{1}{2} \times 40 \times V_1^2 + \frac{1}{2} \times 40 \times V_2^2$$

$$\text{or} \quad 10 \times 10^6 + 1.5 \times 10^6 = 20\, V_1^2 + 20\, V_2^2$$

$$\text{or} \quad V_1^2 + V_2^2 = 0.575 \times 10^6 \quad \text{...(ii)}$$

Solving expression (*i*) and (*ii*), we get

$$V_1^2 + (1000 - V_1)^2 = 0.575 \times 10^6$$

$$\text{or} \quad V_1^2 + 1 \times 10^6 + V_1^2 - 2000\, V_1 = 0.575 \times 10^6$$

$$\text{or} \quad 2V_1^2 - 2000\, V_1 + 0.425 \times 10^6 = 0$$

$$\therefore \quad V_1 = \frac{-(-2000) \pm \sqrt{(-2000)^2 - 4 \times 2 \times 0.425 \times 10^6}}{2 \times 2}$$

$$= \frac{2000 \pm 774.6}{4} = 693.65 \quad \text{or} \quad 306.35$$

Thus V_1 = **693.65 m/s** and V_2 = **306.35 m/s**

EXAMPLE 18.54

A body of 5 kg mass is initially at rest on a rough horizontal surface ($\mu = 0.2$) and is acted upon by a 20 N pull applied horizontally. Calculate:

(a) *the work done by the net force on the body in 5 seconds*

(b) *change in kinetic energy of the body in 5 seconds. Comment on the results.*

Solution: Refer Fig. 18.46 for the various forces acting on the body.

Normal reaction R = weight of the body

$$= 5 \times 9.81 = 49.05 \text{ N}$$

Force of friction $F_f = \mu R = 0.2 \times 49.05 = 9.81$

Net force causing motion = 20 − 9.81 = 10.19 N

Applying Newton's second law of motion $F = ma$, we have

$$10.19 = 5 \times a$$

$\therefore$ Acceleration of the body $a = \dfrac{10.19}{5} = 2.038 \text{ m/s}^2$

From kinetmatic equation $s = ut + \dfrac{1}{2} at^2$, the distance moved by the body in 5 seconds is

$$s = 0 \times 5 + \frac{1}{2} \times 2.038 \times 5^2 = 25.475 \text{ m}$$

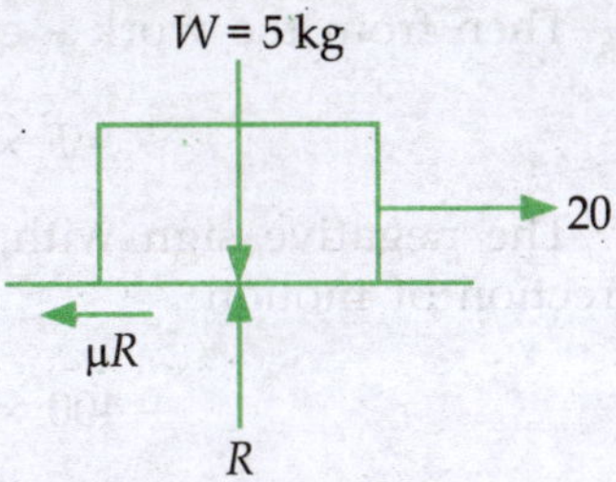

Fig. 18.46

$\therefore$ Work done by the net force = 10.19 × 25.475 = **259.59 Nm**

(*b*) The body starts from rest (u = 0) and accordingly initial kinetic energy of the body is zero (KE_1 = 0)

Final kinetic energy $KE_2 = \frac{1}{2} mV^2$

The velocity v can be worked out from the relation $v = u + at$,

$$v = 0 + 2.038 \times 5 = 10.19 \text{ m/s}$$

$$KE_2 = \frac{1}{2} \times 5 \times (10.19)^2 = 259.59$$

$\therefore$ Change in kinetic energy = $KE_2 - KE_1$ = 259.59 − 0 = **259.59 Nm**

Comment: Work done by net force on the body is equal to the change in kinetic energy of the body. This is in accordance with the work-energy principle.

EXAMPLE 18.55

A block of weight 2000 N rests on a rough horizontal surface (μ = 0.2) and is pulled by a force of 800 N applied at an angle of 30° to the horizontal. Determine the velocity attained by the block after it has moved 20 m starting from rest. Proceed to calculate the further distance moved by the body if the pull is removed. Use work–energy relation.

Solution: Refer Fig. 18.47 for the various forces acting on the block.

Normal reaction

$$R = W - P\sin\theta$$
$$= 2000 - 800\sin 30^\circ = 1600 \text{ N}$$

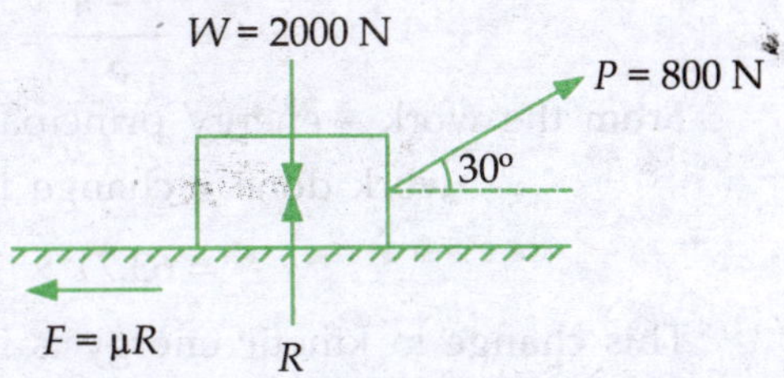

Fig. 18.47

Force of friction

$$F = \mu R = 0.2 \times 1600 = 320 \text{ N}$$

Unbalanced force causing motion

$$= P\cos 30^\circ - F$$
$$= 800 \times 0.866 - 320 = 372.82 \text{ N}$$

Applying the work-energy relation,

work done = change in kinetic energy

force × distance moved = change in kinetic energy

$$372.82 \times 20 = \frac{1}{2} m (v^2 - u^2) = \frac{1}{2} \times \frac{2000}{9.81} (v^2 - 0^2)$$

$$v^2 = \frac{372.82 \times 20 \times 2 \times 9.81}{2000} = 73.15 \text{ ; } v = \textbf{8.55 m/s}$$

(*b*) When the pull is removed, there will be only force of friction which retards the motion and brings the body to rest.

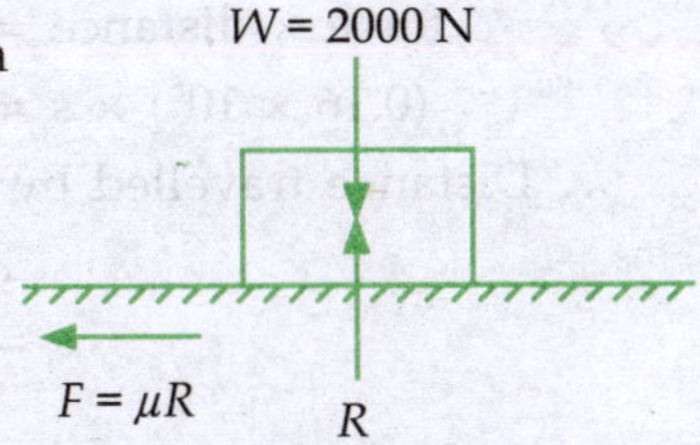

Fig. 18.48

$$F = \mu R = \mu W$$
$$= 0.2 \times 2000 = 400 \text{ N}$$

initial velocity = 8.55 m/s

final velocity = 0

Then from the work – energy equation,

$$-F \times s = \frac{1}{2}\, m\,(v^2 - u^2)$$

The negative sign with F stems from the fact that frictional force acts opposite to the direction of motion.

$$-400 \times s = \frac{1}{2} \times \frac{2000}{9.81}\,(0^2 - 8.55^2)$$

$\therefore$ Distance moved $s = \frac{2000}{2 \times 9.81} \times \frac{8.55^2}{400} = $ **18.63 m**

EXAMPLE 18.56

A train weighing 2×10^6 N starts from rest with an acceleration of 0.8 m/s^2 and acquires a speed of 90 km/hr. Determine the kinetic energy corresponding to final speed and the average power required.

Subsequently the power is shut off and the train is subjected to a retarding force equal to 8% of the weight of train. Calculate the distance the train will travel before coming to rest.

Solution: Final speed of train $V = 90$ km/hr $= \frac{90 \times 1000}{3600} = 25$ m/s

Kinetic energy of the train$= \frac{1}{2} mV^2 = \frac{1}{2} \times \frac{2 \times 10^6}{9.81} \times 25^2 =$ **63.71 × 10⁶ Nm**

From the kinematic relation $v = u + at$, the time taken to acquire the speed of 25 m/s is given by

$$t = \frac{v - u}{a} = \frac{25 - 0}{0.8} = 31.25 \text{ sec}$$

From the work – energy principal,

work done = change in kinetic energy

$= 63.71 \times 10^6$ Nm (∵ initial $KE = 0$)

This change in kinetic energy is in a period of 31.25 second.

Work done per second$= \frac{63.71 \times 10^6}{31.25} = 2.039 \times 10^6$ Nm/s

$\therefore$ Average power $= 2.039 \times 10^6$ J/s $= 2.039 \times 10^6$ W = **2.039 MW**

(*b*) Retarding force = 8% of weight of train

$$= \frac{8}{100} \times (2 \times 10^6) = 0.16 \times 10^6 \text{ N}$$

Again from the work – energy principle,

force × distance = change of kinetic energy

$(0.16 \times 10^6) \times s = 63.71 \times 10^6$

$\therefore$ Distance travelled by the train before coming to rest,

$$s = \frac{63.71 \times 10^6}{0.16 \times 10^6} = \textbf{398.19 m}$$

EXAMPLE 18.57

A wooden block of weight 40 N rests on a rough horizontal plane having friction coefficient μ = 0.35. The block is struck by a bullet travelling horizontally with a velocity of 750 m/s and weighing 0.25 N. Workout the distance by which the block is displaced from its initial position.

It may be presumed that the bullet after striking the block gets embedded in it.

Solution: Refer Fig. 18.49 which shows the block with various parameters inserted.

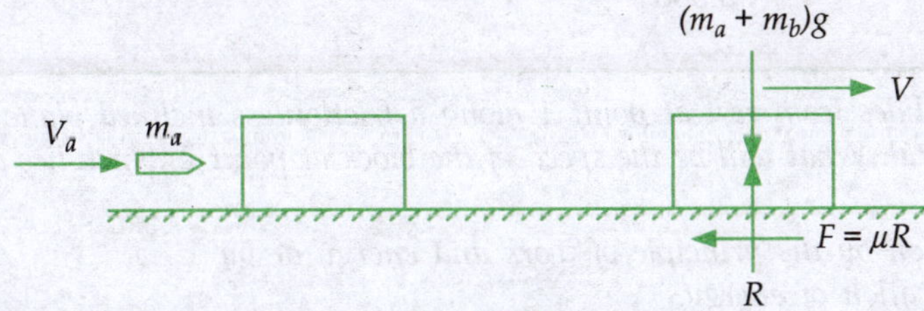

Fig. 18.49

Applying the principle of conservation of momentum,

momentum before impact = momentum after impact

$$m_aV_a + m_bV_b = (m_a + m_b)\, V$$

or $$\frac{0.25}{9.81} \times 750 + \frac{40}{9.81} \times 0 = \frac{0.25 + 40}{9.81} \times V$$

or $$19.11 = 4.103\, V\,; \quad V = \mathbf{4.657\ m/s}$$

$$\text{Normal reaction } R = (m_a + m_b)\, g = \frac{0.25 + 40}{g} \times g = 40.25 \text{ N}$$

$$\text{Force of friction } F = \mu R = 0.35 \times 40.25 = 14.087 \text{ N}$$

Applying the work – energy correlation,

work done to overcome the frictional force = kinetic energy lost by the block with bullet embedded

$$F \times s = \frac{1}{2}\,(m_a + m_b)\, V^2$$

$$14.087 \times s = \frac{1}{2}\left(\frac{0.25 + 40}{9.81}\right) \times 4.657^2 = 43.632$$

$$\therefore \quad \text{Displacement of block } s = \frac{43.632}{14.087} = \mathbf{3.097\ m}$$

EXAMPLE 18.58

A box of 60 N weight is pulled up along a 30° plane by applying a 50 N force as shown in Fig. 18.50. The box starts from rest and the surface of the plane is smooth. What speed the box will attain after travelling 3 m up the plane?

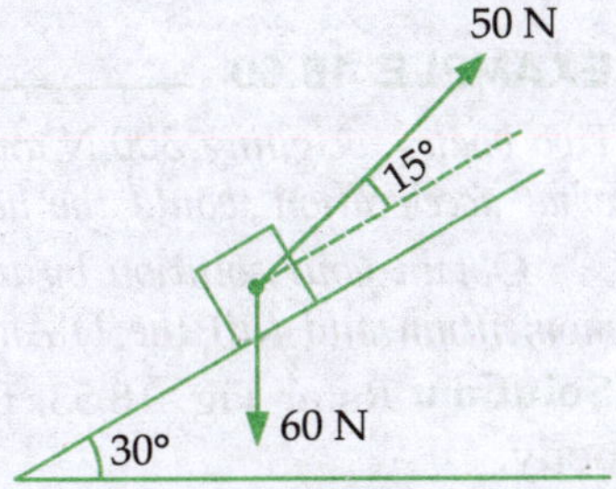

Fig. 18.50

Solution : Using the principle of work and energy

work done = change in kinetic energy

Component of weight parallel to plane

$$= 60 \sin 30° = 30 \text{ N } (\leftarrow)$$

$$\text{Component of pull parallel to plane} = 50 \sin 15° = 48.3 \text{ N } (\rightarrow)$$

$$\text{Effective force} = 48.3 - 30 = 18.3 \text{ N}$$

$$\text{Work done} = 18.3 \times 4 = 732.2 \text{ Nm}$$

Hence $\quad 73.2 = \frac{1}{2} mV_2^2 - \frac{1}{2} mV_1^2 = \frac{1}{2} mV_2^2 \qquad (\because \quad V_1 = 0)$

$$= \frac{1}{2} \times \frac{60}{9.81} V_2^2 = 3.058 V_2^2$$

$\therefore$ Speed of box $V_2 = \sqrt{\frac{73.2}{3.058}} = $ **4.892 m/s**

EXAMPLE 18.59

A block of 4 kg mass slides from rest at point 1 along a frictionless inclined plane that makes an angle of 30° with the horizontal. What will be the speed of the block at point 2 which lies at a distance of 2.5 m from point?

Obtain your solution by the principle of work and energy or by the principle of conservation of energy.

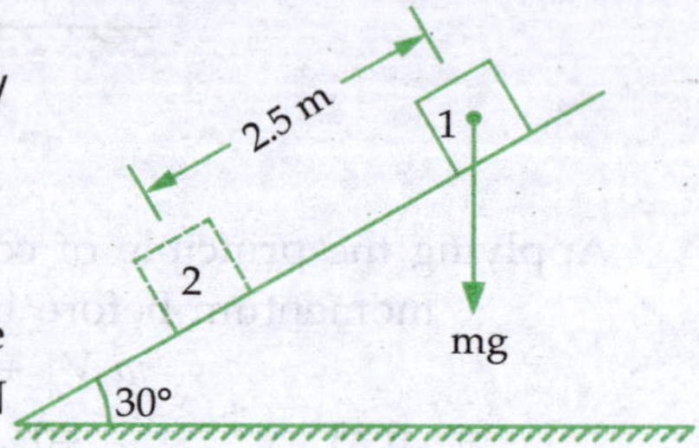

Fig. 18.51

Solution : Using the principle of work and energy

$$W_{12} = (KE)_2 - (KE)_1$$

Component of weight of block acting parallel to the plane

$$= mg \sin\theta = (4 \times 9.81) \times \sin 30^\circ = 19.62 \text{ N}$$

Work done during sliding from position 1 to position 2

$$W_{12} = 19.62 \times 2.5 = 49.05 \text{ Nm}$$

Kinetic energy at position 1 $= \frac{1}{2} mV_1^2 = 0 \qquad (\because \quad V_1 = 0)$

Kinetic energy at position 2 $= \frac{1}{2} mV_2^2 = \frac{1}{2} \times 4 \times V_2^2 = 2V_2^2$

$$\therefore \quad 49.05 = 2V_2^2; \quad V_2 = \sqrt{\frac{49.05}{2}} = \textbf{4.95 m/s}$$

(*b*) Using the principle of conservation of mechanical energy

$$(KE + PE)_1 = (KE + PE)_2$$

Taking the datum at the level of position 2,

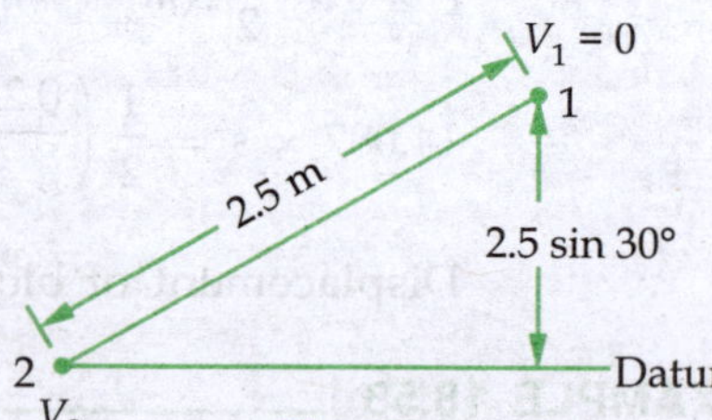

Fig. 18.52

$$0 + mg \times 2.5 \sin 30^\circ = \frac{1}{2} mV_2^2 + 0$$

or $\quad (4 \times 9.81) \times 2.5 \times 0.5 = \frac{1}{2} \times 4 \times V_2^2$

$$\therefore \quad V_2 = \sqrt{\frac{(4 \times 9.81) \times 2.5 \times 0.5}{2}} = \textbf{4.95 m/s}$$

EXAMPLE 18.60

Two bodies weighing 300 N and 450 N are hung to the ends of a rope passing over an ideal pulley. With what acceleration would the heavier body come down? What is the tension in the rope ?

Obtain your solution by using (i) the principle of work and energy (ii) the principle of impulse and momentum and (iii) the D'Alembert's principle.

Solution: Refer Fig. 18.53, for the arrangement of the system and free body diagram for each body.

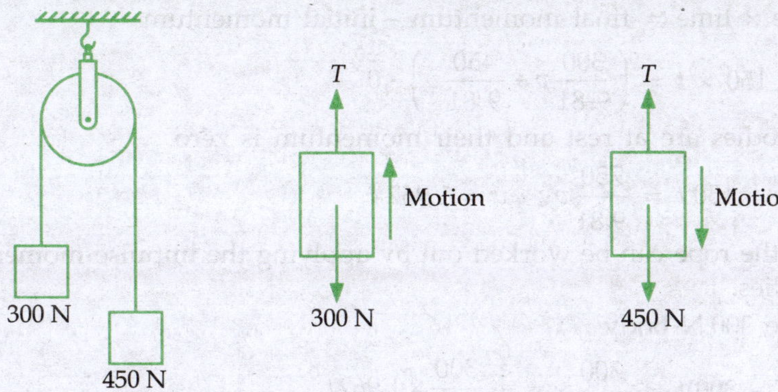

Fig. 18.53

It is to be noted that

(*i*) the acceleration of both the weights is same otherwise the rope would break or become slack.

(*ii*) for a smooth frictionless pulley, the tension induced in the rope will be same throughout.

- **Work-Energy Principle :** Let s be the distance moved by each body. Then

work done by 300 N body = $(T - 300)\, s$

work done by 450 N body = $-(T - 450)\, s$

The $-ve$ sign is due to the fact that the direction of motion of 450 N body is opposite to that of 300 N body.

Total work done by the system = $(T - 300)\, s + [-(T - 450)\, s] = 150\, s$

Change in kinetic energy of :

$$300 \text{ N body} = \frac{1}{2} \times \frac{300}{9.81}(v^2 - 0^2) = 15.29\, v^2$$

$$450 \text{ N body} = \frac{1}{2} \times \frac{450}{9.81}(v^2 - 0^2) = 22.93\, v^2$$

Total change in kinetic energy of the system = $15.29\, v^2 + 22.93\, v^2 = 38.22\, v^2$

From the principle of work and energy,

work done = change in kinetic energy

$$150\, s = 38.22\, v^2 \; ; \quad v^2 = 3.925\, s$$

Using the kinematic equation, $v^2 - u^2 = 2as$, we have

$$3.925\, s = 2as \qquad (\because u = 0)$$

$\therefore$ Acceleration of the bodies $a = \dfrac{3.925}{2} =$ **1.962 m/s²**

The tension in the rope can be worked out by applying the work-energy principle to any one of the bodies. Considering the 300 N body:

$$(T - 300)\, s = 15.29\, v^2$$

$$= 15.29 \times 3.925\, s$$

$\therefore$ Tension in the rope = $300 + (15.29 \times 3.925) =$ **360 N**

- **Impulse-momentum principle:** Net force in the direction of motion

$= (T - 300) - (T - 450) = 150$ N

Let this force act for time t seconds. Then from the impulse momentum principle,

impulse = change of momentum of the system

$$\text{net force} \times \text{time} = \text{final momentum} - \text{initial momentum}$$

$$150 \times t = \left(\frac{300}{9.81}v + \frac{450}{9.81}v\right) - 0$$

Initially both bodies are at rest and their momentum is zero

$$150\,t = \frac{750}{9.81}v\,; \quad v = 1.962\,t$$

The tension in the rope can be worked out by applying the impulse-momentum principle to any one of the bodies.

Considering the 300 N body

$$(T - 300)t = \frac{300}{9.81} \times v = \frac{300}{9.81} \times 1.962t$$

$\therefore$ Tension in the rope $T = 300 + \frac{300}{9.81} \times 1.962 =$ **360 N**

Using the kinematic equation $v = u + at$, we have

$$1.962t = 0 + at \qquad (\because u = 0)$$

$\therefore$ Acceleration of the bodies $a =$ **1.962 m/s²**

- **D'Alembert's Principle**

Considering motion of 300 N: The weight is brought into dynamic equilibrium by applying inertia force F_i in a direction opposite to that of its motion, which is upwards. Then by D'Alembert's principle

$$T - 300 - F_i = 0 \quad \text{where } F_i = ma$$

$$\therefore \quad T = 300 + F_i = 300 + \frac{300}{9.81}a \qquad ...(i)$$

Considering motion of 450 N: The weight moves down wards and accordingly the inertia force will act upwards. Then

$$T - 450 + F_i = 0$$

$$\therefore \quad T = 450 - F_i = 450 - \frac{450}{9.18}a \qquad ...(ii)$$

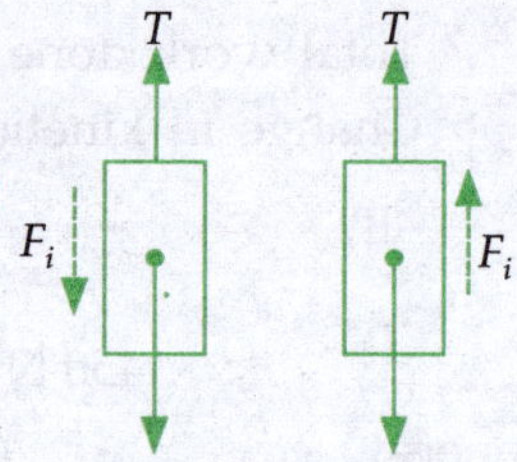

Fig. 18.54

From expressions (*i*) and (*ii*),

$$300 + \frac{300}{9.81}a = 450 - \frac{450}{9.81}a$$

or
$$\frac{300 + 450}{9.81}a = 450 - 300 = 150$$

$\therefore$ Acceleration of the system $a = \frac{150 \times 9.81}{750} =$ **1.962 m/s²**

Substituting $a = 1.962\ \text{m/s}^2$ in expression (*i*), we get

$$\text{Tension in the rope } T = 300 + \frac{300}{9.81} \times 1.962 = \mathbf{360\ N}$$

18.8 HAMILTON'S PRINCIPLE

The Hamilton's principle is an important mathematical concept which states that

"the time integral of the difference between the kinetic and potential energies of dynamical system is stationary"

Consider that a force F acts on a particle of mass m and that under the influence of this force, the particle moves from the fixed origin O along the path OBA. If R is the position vector at any time t, then from Newton's second law (force = mass × acceleration)

$$F = m\frac{d^2R}{dt^2}$$

Let the particle now follow path OCA with end position remaining same at A. The variation in path, called virtual displacement, is represented by δR. The work done during virtual displacement δR is

$$\delta W = F \times \delta R = m\frac{d^2R}{dt^2}\delta R \qquad ...(i)$$

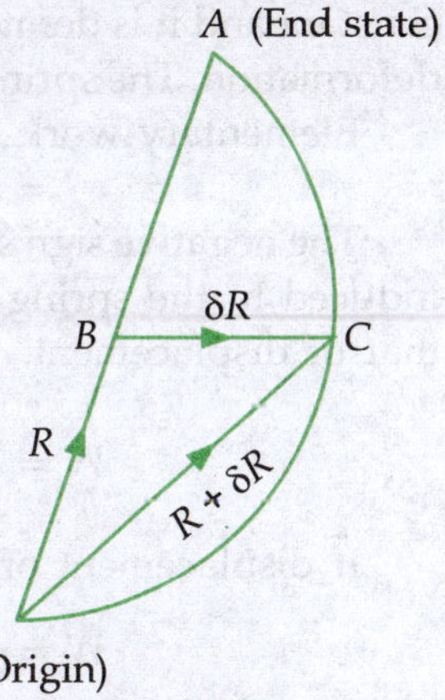

Fig. 18.55

The kinetic energy of the particle is

$$KE = \frac{1}{2}m\left(\frac{dR}{dt}\right)^2$$

Upon differentiation

$$\delta(KE) = m\left(\frac{dR}{dt}\right)\delta\left(\frac{dR}{dt}\right) = m\frac{dR}{dt}\frac{d}{dt}(\delta R) \qquad ...(ii)$$

Adding expressions (*i*) and (*ii*), we get

$$\delta W + \delta(KE) = m\frac{d^2R}{dt^2}\delta R + m\frac{dR}{dt}\frac{d}{dt}(\delta R) = m\frac{d}{dt}\left(\frac{dR}{dt}\delta R\right)$$

Integrating both sides within the time limits from t_0 to t_1

$$\int[\delta W + \delta(KE)]dt = \left| m\frac{dR}{dt}\delta R \right|_{t_0}^{t_1}$$

Since $\delta R = 0$ at t_0 and t_1, the above identity may be written as

$$\int[\delta W + \delta(KE)]dt = 0$$

It may recalled that when the force field is conservative, there exists a potential PE (potential energy) such that $W = -(PE)$

The above identity then takes the form

$$\int[\delta(KE) - \delta(PE)]dt = 0 \quad \text{or} \quad \delta\int[(KE - PE)]dt = 0$$

$$\text{or} \qquad \int(KE - PE)dt \text{ is stationary}$$

The Hamilton's principle is valid for a particle. However, its derivation can be extended to a system of particles by summation and to a rigid body by integration. As such the Hamilton's principle is true for any dynamical system.

The integral $\int(KE - PE)dt$ is minimum along the natural path in comparison with that along any other path that joins O to A.

The energy difference $(KE - PE)$ is called the kinetic potential or the Lagrangian function.

18.9 WORK DONE BY A SPRING

Consider a body attached to fixed support by a spring. Initially when the body is at position A_0, the spring is undeformed. When the body is acted upon by a force, the spring extends and the body shifts to a new position which depends upon the magnitude of applied force

Experimental results have shown that when a spring is deformed (compressed or elongated), a corresponding force is developed in the spring. The magnitude of force developed is directly proportional to the displacement of spring from the undeformed position.

$$F \propto x \text{ or } F = kx$$

The constant of proportionality is called the spring constant and it is defined as the force required for unit deformation. The spring constant is expressed in N/m.

Elementary work done by the spring force.

$$= F\,dx = -kx\,dx$$

The negative sign stems from the fact that the force induced in the spring acts in a direction opposite to that of displacement.

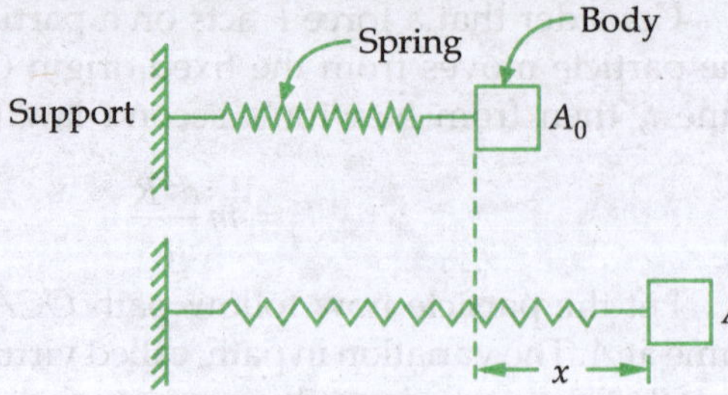

Fig. 18.56

$$W = -\int_0^x kx\,dx = -\frac{1}{2}kx^2$$

If displacement of body is from x_1 to x_2, then

$$W = -\frac{1}{2}k(x_2^2 - x_1^2) = \frac{1}{2}k\left(x_1^2 - x_2^2\right)$$

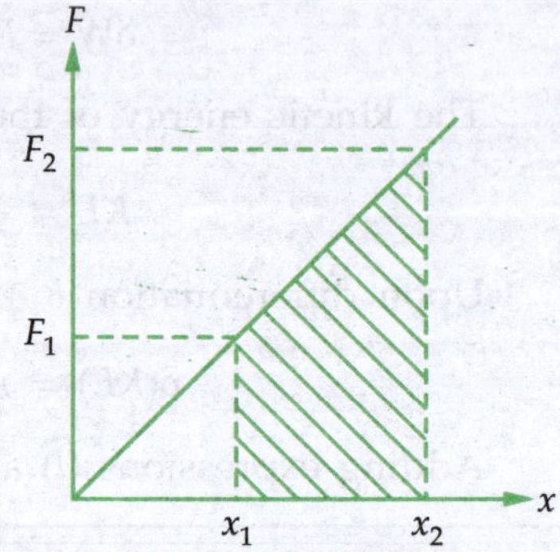

Fig. 18.57

The work done by the spring force is represented by the shaded area as shown in Fig. 18.57.

$$W = -\frac{F_2 + F_1}{2} \times (x_2 - x_1)$$

$$W = -\frac{kx_2 + kx_1}{2} \times \left(x_2 - x_1\right) = \frac{1}{2}k\left(x_1^2 - x_2^2\right)$$

The work done by spring force is positive when $x_2 < x_1$, *i.e.*, when the spring is returning to its original position.

EXAMPLE 18.61

Determine the length of cord AC in Fig. 18.58 so that 10 kg lamp is suspended in the position shown. The undeformed length of the spring AB is 0.6 m and the spring stiffness is 325 N/m.

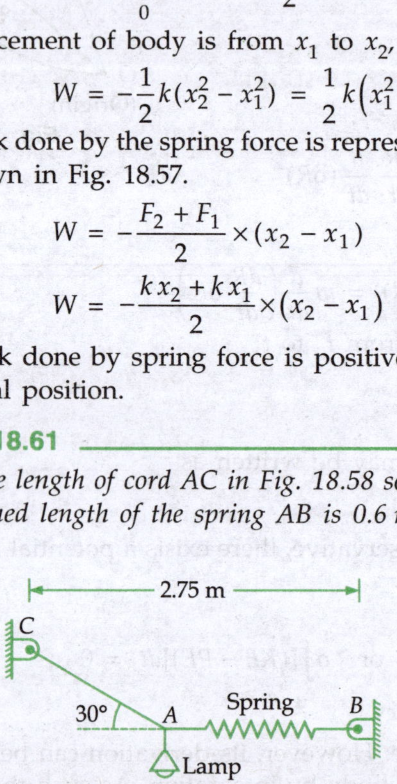

Fig. 18.58

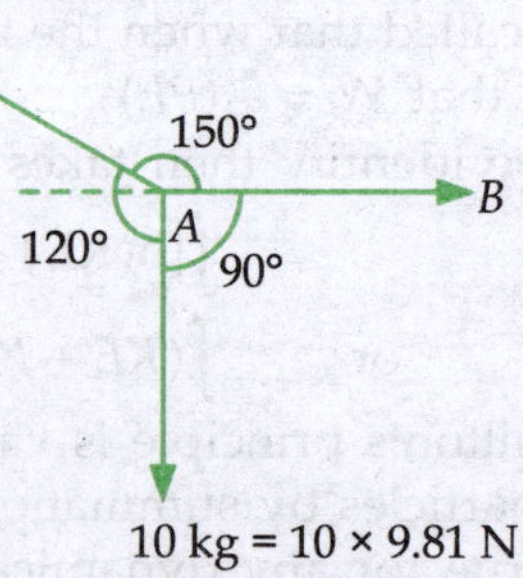

Fig. 18.59

Solution: With reference to free body diagram of junction A, the Lami's theorem gives

$$\frac{F_{ab}}{\sin 120^\circ} = \frac{(10 \times 9.81)}{\sin 150^\circ}$$

$$F_{ab} = (10 \times 9.81)\frac{\sin 120^\circ}{\sin 150^\circ} = (10 \times 9.81) \times \frac{0.866}{0.5} = 169.9 \text{ N}$$

Corresponding extension of spring $= \dfrac{169.9}{325} = 0.523$ m

$\therefore$ Deformed length of the spring = undeformed length + extension of spring

$$= 0.6 + 0.523 = 1.123 \text{ m}$$

From the given geometry : $AC \cos 30° = 2.75 - 1.123 = 1.627$

$$\therefore \quad \text{length of cord } AC = \frac{1.627}{\cos 30°} = \mathbf{1.879\ m}$$

EXAMPLE 18.62

The spring in a gun is held in vertical position and compressed by 0.25 m. A ball of 40 N weight is placed on the compressed spring which is subsequently released. Calculate the height to which the ball will rise and the velocity it will attain at a height of 2 m. It may be presumed that the spring has a stiffness of 5000 N/mm.

Solution: When the spring is compressed, it has a potential energy stored in it which is given by

$$= \frac{1}{2} k x^2 = \frac{1}{2} \times 5000 \times (0.25)^2 = 200 \text{ Nm}$$

The potential energy of the spring gets transformed into potential energy and kinetic energy of the ball.

(*a*) Let the ball rise to a maximum height of ***h*** m. At maximum height, the velocity of the ball will be zero and so will be its kinetic energy. Therefore

$$200 = 40h + 0 ; \quad h = \mathbf{5\ m}$$

(*b*) The law of conservation of energy can again be used to find the velocity of ball at a height of 2 m.

$$200 = 40 \times 2 + \frac{1}{2} \times \left(\frac{40}{9.81}\right) V^2 = 80 + 2.039\, V^2 ; \quad V = \mathbf{7.67\ m/s}$$

EXAMPLE 18.63

A body of mass 10 kg is made to fall 3 cm height on a spring of stiffness 120 N/cm. Find the displacement of spring. Use the concept that total energy of the mass-spring system remains constant.

Solution : Refer Fig. 18.60 for the system and let the energies be evaluated with respect to undeformed position of the spring as datum.

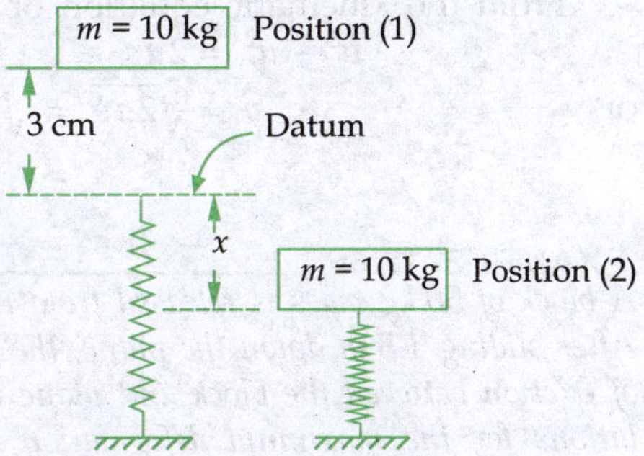

Fig. 18.60

At position (1), the body has only potential energy and the spring has no energy as it is not deformed. Accordingly

$$E_1 = mgh = 10 \times 9.81 \times 3 = 294.3 \text{ N cm}$$

At position (2), the body as well as the spring have potential energy.

$$E_2 = -mgx + \frac{1}{2} k x^2$$

where x is the compression of the spring in cm

$$E_2 = -10 \times 9.81 \times x + \frac{1}{2} \times 120 \times x^2 = -98.1x + 60x^2$$

Thus using the law of conservation of energy, we can write : $E_1 = E_2$

$$294.3 = -98.1x = 60x^2 \quad \text{or} \quad x^2 - 1.635x - 4.905 = 0$$

$$\therefore \quad x = \frac{-(-1.635) \pm \sqrt{(-1.635)^2 - 4 \times 1 \times (-4.905)}}{2 \times 1} = \frac{1.635 \pm 4.721}{2}$$

Neglecting the negative value which is not possible, we get

$$x = \frac{1.635 + 4.721}{2} = \mathbf{3.178\ cm}$$

EXAMPLE 18.64

A collar of 5 kg mass slides without friction along a vertical rod as shown in Fig. 18.61. The spring attached to the collar has a spring constant of 250 N/m and its undeformed length is 15 cm. The collar is released from rest at position A and slides 20 cm downward to position B. What will be the velocity of the collar at position B?

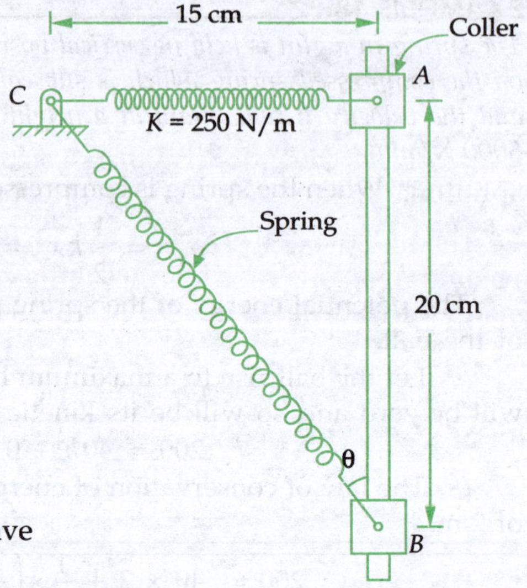

Fig. 18.61

Solution : Extension of spring

$$x = BC - AC = \sqrt{(AC)^2 + (AB)^2} - AC$$

$$= \sqrt{15^2 + 20^2} - 15 = 10 \text{ cm}$$

∴ Forced exerted by spring

$$F_s = kx = 250 \times 0.1 = 25 \text{ N}$$

Refer Fig. 18.62 for the free body diagram of the collar when it is at position B

$$\tan\theta = \frac{AC}{AB} = \frac{15}{20} = 0.667\,; \quad \theta = 33.70^\circ$$

From the equation of equilibrium : $\Sigma F_y = ma$, we have

$$mg - F_s \cos\theta = ma$$

or

$$5 \times 9.81 - 25\cos 33.70^\circ = 5 \times a$$

∴ Acceleration of collar motion, $a = \dfrac{5 \times 9.81 - 25 \times 0.832}{5} = 5.65 \text{ m/s}^2$

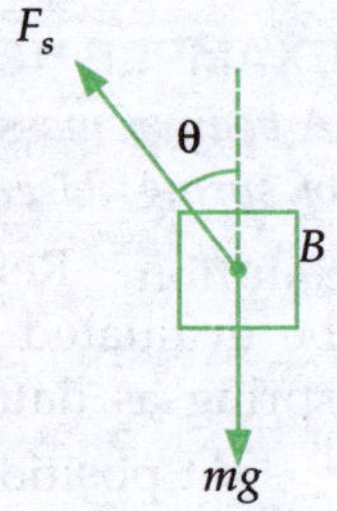

Fig. 18.62

From the kinematic equation of uniformly accelerated motion,

$$v^2 - u^2 = 2as$$

or

$$v = \sqrt{2as} = \sqrt{2 \times 5.65 \times 0.2} \qquad (\because \; u = 0)$$

$$= \textbf{1.50 m/s}$$

EXAMPLE 18.65

A block of 50 kg mass is released from rest and slides down a 30° inclined plane as shown in Fig. 18.63. After sliding 1.5 m down the plane, the block hits a spring of spring constant 25 N/mm. If the coefficient of friction between the block and plane is 0.2, make calculations for the maximum deformation of the spring and maximum velocity of the block. Also find the distance the block will move up the plane due to rebound.

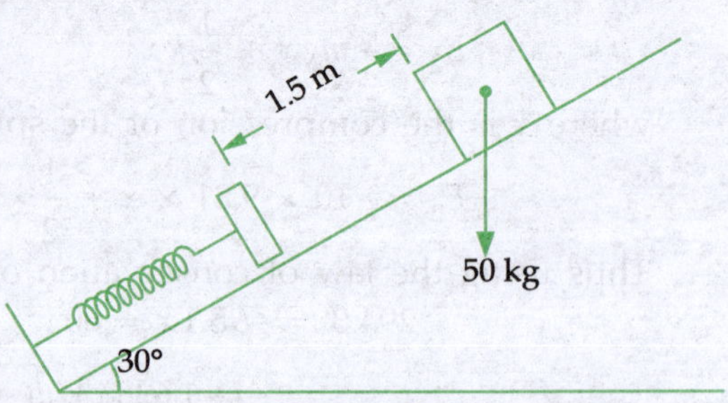

Fig. 18.63

Solution : Refer Fig. 18.64 for the various forces acting on the block.

Normal reaction $R = W\cos\theta = (50 \times 9.81)\cos 30^\circ$

$$= 424.78 \text{ N}$$

Force of friction $F = \mu R = 0.2 \times 424.78 = 84.96$ N

This frictional force acts in a direction opposite to that of motion.

Then the net force on the block when it starts moving downwards the plane is

$$= W\sin\theta - F = 50 \times 9.81 \times \sin 30^\circ - 84.96$$

$$= 160.29 \text{ N}$$

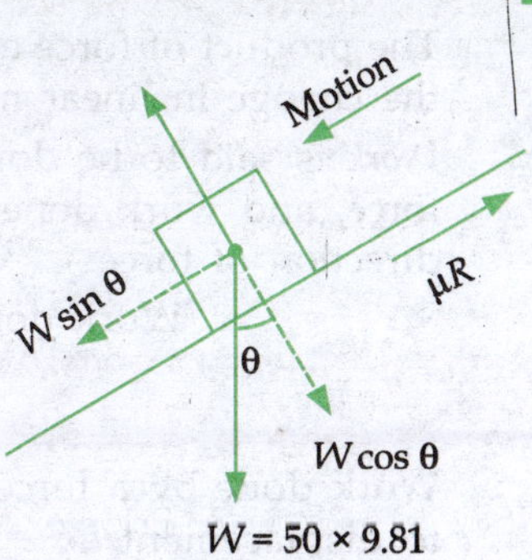

Fig. 18.64

If x is the deformation (compression) of the spring, then distance moved by the block

$$= (1.5 + x)$$

Work done by the block $= 160.29 \times (1.5 + x)$

$$\text{Work done by the spring} = -\frac{1}{2}kx^2 = -\frac{1}{2} \times (25 \times 10^3)x^2 = -12.5 \times 10^3 x^2$$

$\therefore$ Total work done by the system $= 160.29 \times (1.5 + x) - 12.5 \times 10^3 x^2$

The block starts from rest and finally comes to rest when there is maximum deformation of the spring. Obviously change in kinetic energy of the system is zero. Accordingly from the work-energy principle,

work done = change in kinetic energy

$$160.29 \times (1.5 + x) - 12.5 \times 10^3 x^2 = 0$$

or $\quad 12.5x^2 - 0.16x - 0.24 = 0$

$$x = \frac{-(-0.16) \pm \sqrt{(-0.16)^2 - 4(12.5)(-0.24)}}{2 \times 12.5} = \frac{0.16 \pm 3.47}{25} = 0.145 \text{ m} \quad \text{or} \quad 0.132 \text{ m}$$

The value $x = 0.145$ m is positive and hence acceptable

(*b*) According to Newton's second law $F = ma$

$$160.29 = 50 \times a; \quad a = 3.21 \text{ m/s}^2$$

From the kinematic relation $v^2 - u^2 = 2as$, we have

$$v^2 = 2 \times 3.21 \times 1.5 \qquad (\because u = 0)$$

$\therefore$ Maximum velocity of block $v = \sqrt{2 \times 3.21 \times 1.5} = 3.10$ m/s

(*c*) When the block moves up the plane due to rebound, both the components of weight of block along the plane and the frictional force act in the same direction. Therefore

$$\text{force on the block} = W \sin\theta + \text{frictional force} = (50 \times 9.81)\sin 30^\circ + 84.96 = 330.21 \text{ N}$$

If s is the distance moved up by the block, then

work done by forces acting on the block $= 330.21 \times s$

The work done equals the energy stored in the spring. Accordingly

$$\frac{1}{2}kx^2 = 330.21 \times s$$

or $\quad \frac{1}{2} \times (25 \times 10^3) \times (0.145)^2 = 330.21 \times s\,;\; s =$ **0.796 m**

Thus the distance of rebound of block on the plane is 0.796 m.

REVIEW AND SUMMARY

1. From Newton's second law,

$$F = m\frac{dV}{dt} = \frac{d}{dt}(mV) = \frac{mV_2 - mV_1}{t}$$

The product of mass and velociy (mV) is called *momentum*. Apparently the change in linear momentum per unit time is proportional to impressed force and takes place in the direction of force.

2. The product of force and time during which it acts ($F \times t$) is called *impulse*. Impulse equals the change in linear momentum.
3. *Work* is said to be done when the point of application of force moves in the direction of force, and work done equals the product of force and the resulting displacement in the direction of force.

   Work done $= F \times x$ when force and displacement are in the same direction
   $= F \cos\theta \times x$ when force acts at an angle θ with the direction of displacement

   Work done by a force is zero if either the displacement is zero or the force acts normal to displacement.
4. *Power* is the rate of doing work and is obtained by dividing the work by time taken to do that work. *Energy* is defined as the capacity to do work.
5. The energy possessed by a body by virtue of its velocity is known as kinetic energy. Potential energy is the energy by virtue of position of a body with respect to any given datum.

   $$KE = \frac{1}{2} mV^2 \quad \text{and} \quad PE = mgh$$

   The kinetic energy and potential energy individually change during the motion of body but their sum always remains the same. This is the essence of *law* of *conservation* of *energy*. The sum of *KE* and *PE* of a body is called the *mechanical energy*.
6. From work-energy principle, the work done on a body equals the change in its kinetic energy. Further, if the work of force in moving an object between two positions is independent of tha path followed by the object and can be expressed as change in its potential energy, then such a force is called *conservative force*.
7. Consider a lift (elevator or cage) carrying a load *W* and moving with uniform acceleration *a*. There will be reaction of lift which will equal the tension *T* in the cable supporting the lift. When the lift moves upwards,

   Net force = mass × acceleration

   $$T - W = \frac{W}{g} a$$

   $$\text{That gives: } T = W\left(1 + \frac{a}{g}\right)$$

   Likewise when the lift moves downwards

   $$T = W\left(1 + \frac{a}{g}\right)$$
8. D'Alembert's principle states that "the system of forces acting on a body in motion is in dynamic equilibrium with inertia force of the body".

   Let *R* be the resultant of a system of forces acting on the body which makes the body move along its direction with an acceleration *a*. Then for dynamic equilibrium of the body.

   $$\Sigma F = 0\,; \quad R - ma = 0 \quad \text{or} \quad R = ma$$

   The force ($-ma$) is termed as the inertia force or the reverse effective force.
9. The sum of kinetic energy and potential energy of a body remains constant under the action of gravity forces

   $$(KE + PE)_1 = (KE + PE)_2$$

10. When a spring is deformed (compressed or elongated), a force is developed the magnitude of which is directly proportional to displacement x of spring from the undeformed position

$$W = -\int_0^x kx\,dx = -\frac{1}{2}kx^2$$

The –ve sign stems from the fact that the force induced in the spring acts in a direction opposite to that of displacement.

If displacement of spring is from x_1 to x_2, then

$$W = -\frac{1}{2}k\left(x_2^2 - x_1^2\right) = \frac{1}{2}k\left(x_1^2 - x_2^2\right)$$

REVIEW QUESTIONS

A. Conceptual and conventional questions :

1. State Newton's law that governs the force acting on a body.
2. Define force, momentum and impulse
3. State the principle of conservation of linear momentum
4. Write the impulse-momentum equation and mention its applications
5. What is dynamic equilibrium?
6. State D'Alembert's principle. How it is applied in solving problems related to dynamics?
7. Define energy. What is its relationship with work?
8. Define kinetic energy, potential energy and total mechanical energy.
9. Explain the principle of conservation of mechanical energy.
10. Differentiate between work and power. Mention their units in SI system.
11. Define work. Is it a scalar or a vector quantity?
12. State whether any work is done by a man pressing against a vehicle which does not move. Explain.
13. What is the work done by the weight of a body if it is moved horizontally? Explain.
14. State the principle of work and energy.
15. What is a conservative force?
16. Write units for the following: impulse, momentum, work and power
17. A motorised boat of mass 25 tonnes drifts in towards a dock at a speed of 1.8m/s after its engine has been shut off. If the water offers a resistance of 40 N per tonne, what distance the boat will travel before coming to rest? **(4.5 m)**
18. A bullet of mass 10 gm leaves the barrel of a gun with a velocity of 350m/sec. Determine the impulse produced by the discharge of bullet. If the impulse lasts for 0.002 second, what will be average impulsive force? **(14 kgm/s; 7 kN)**
19. A body of mass M moves in outer space with velocity V. The body breaks into two parts so that mass of one part is $\frac{1}{10}$th of the total mass. After explosion, the heavier part comes to rest while the lighter part continues to move along the original direction of motion. Determine the velocity of smaller part in terms of velocity V. **(v = 10V)**
20. A body of 400 kg mass starts from rest and moves along a straight line under the action of a force which varies as the square of time from start. If the magnitude of force equals

400 N at 15 seconds from the start, make calculations for the velocity of body at 15 seconds from the start.

Hint: $F = kt^2 \; ; \; 400 = k\,(15)^2 \; ; \; k = \dfrac{16}{9}$

$$\therefore \quad 400\,a = \frac{16}{9}t^2 \; ; \quad a = \frac{1}{225}t^2$$

$$\int_0^v dv = \frac{1}{225}\int_0^{15} t^2\,dt \qquad \left(\because a = \frac{dv}{dt}\right)$$

$$v = \frac{1}{225} \times \frac{15^3}{3} = 5\text{m/s}$$

21. A running man has half the kinetic energy that a boy of half the mass has. The man has to increase his speed by 1 m/s if he wants to acquire the kinetic energy equal to that of the boy. Make calculations for the original speed of the man and the boy. **(2.4m/s, 4.8m/s)**
22. A man pulls a hand roller on a cricket pitch during 15 minutes recess for tea with an average speed of 0.2 m/s. What work will be done by the man if he applies a pull of 125 N at an upward angle of 30° to the horizontal? Neglect frictional effects, if any. **(19.5 kJ)**
23. A vehicle of 1000 kg mass moving with 48 km/hr collides with the back of another vehicle of 650 kg mass moving in the same direction at 24 km/hr. Find (*a*) their common speed after coupling, (*b*) loss in kinetic energy. **(10.71 m/s, 873 J)**
24. A pile driving hammer of mass 250 kg falls 3 metres from rest on a pile of mass 1200 kg. There is no rebound and the pile is driven 20 cm into the ground. Make calculations for the common velocity after impact and the average resistance offered by the ground. **(1.32 m/s, 20.55 kN)**
25. A spring of unstretched length 100 mm extends to a length of 125 mm when a load is applied to it. When the load is removed, the spring has to do 2 Joules of work in returning to its original position. Calculate the spring stiffness and the maximum load applied to the spring. **(6.4 N/mm, 160 N)**

B. Complete the following statements with most appropriate word/words :

1. Momentum and impulse are .... quantities with unit ....
2. Work done by a force is zero if the force and displacement are ....
3. Power is the .... of doing work and is obtained by dividing the work done by ....
4. The work done on a body equals the change in .... of the body.
5. The sum of kinetic energy and potential energy of a body is called the ....
6. If the work of force in moving an object between two positions is independent of the path followed by the object, and can be expressed as a change in its potential energy, then such a force is called ....
7. The system of forces acting on a body in motion is in dynamic equilibrium with the inertia force of the body. This is the essence of ....
8. When a bullet is fired from a gun, it recoils in the backward direction. It is due to ....
9. A lift carries a load W and moves up with uniform acceleration a. The tension T in the cables supporting the lift is ....

Answers :

1. vector, kg m/s 2. at right angles 3. rate, time 4. kinetic energy 5. mechanical energy

6. conservative force 7. D'Alembert's law 8. Conservation of momentum 9. $W\left(1+\frac{a}{g}\right)$

C. Multiple choice questions :

1. A man weighing 60 N jumps off a railway train running on horizontal rails at 20 km/hr with a packet weighing 100 N in his hand. The thrust of packet on his hand is
   (a) zero (b) 100 N (c) 500 N (d) 700 N
2. A truck of mass 3000 kg is moving at 10 m/s and is acted upon two forces: a forward force of 1000 N due to engine and a retarding force of 400 N due to friction. The truck then gains speed at the rate
   (a) 0.2 m/s^2 (b) 0.02 m/s^2 (c) 0.4 m/s^2 (d) 0.33 m/s^2
3. A bullet of mass 0.04 kg moving with a speed of 90 m/s enters a heavy wooden block and is stopped after a distance of 60 cm. The average resistive force exerted by the block on the bullet is
   (a) 180 N (b) 220 N (c) 270 N (d) 320 N
4. Two particles with masses in the ratio 1 : 4 are moving with equal kinetic energies. The magnitude of their linear momentums will conform to the ratio
   (a) 1 : 8 (b) 1 : 2 (c) $\sqrt{2}$: 1 (d) 4 : 1
5. Which of the following statements is wrong ?
   (a) impulse equals the change in momentum
   (b) action and reaction are equal and opposite and hence cancel each other
   (c) the momentum of a system of two bodies is conserved when there is no external force acting on either body
   (d) the work done on a particle must equal change in its kinetic energy.
6. A 150 gm cricket ball moving at 20 m/s is caught by a player and the catching process is completed in 0.1 second. The force of blow exerted by the ball on the hands of player is
   (a) 3 N (b) 8 N (c) 24 N (d) 30 N
7. When a bullet is fired from a gun, it is recoiled in the backward direction. It is due to
   (a) impulse (b) inertia
   (c) conservation of momentum (d) conservation of energy
8. A force f acts for one second on a body of mass 1 kg moving with an initial velocity u. Then which of the following statements is/are not true ?
   (a) body covers a distance $(u + f/2)$
   (b) final velocity of the body is $(u + f)$
   (c) change in kinetic energy of body is $\frac{1}{2}mf^2$
   (d) momentum of the body increases by f
9. A constant force $F = (20i + 30j + 10k)$ kN moves a particle from position
   $r_1 = (10i + 20j)$ m to $r_2 = (10i + 20j + 30k)$ m
   The work done by the force in kJ is
   (a) zero kJ (b) 300 kJ (c) 800 kJ (d) 1100 kJ

10. A bullet of mass 0.03 kg moving with a speed of 400 m/s penetrates 12 cm into a fixed block of wood. The average force exerted by the wood on the bullet is
(*a*) 10 kN (*b*) 20 kN (*c*) 25 kN (*d*) 30 kN

11. A man stands on a spring weighing scale in a lift which carries him upwards with an acceleration. The reading on the weighing scale will be
(*a*) true weight of the man (*b*) lower than the true weight
(*c*) greater than the true weight (*d*) zero

12. The tension in the cable supporting a lift moving upwards is twice the tension when the lift moves downwards. The acceleration of the lift is equal to
(*a*) g (*b*) $g/2$ (*c*) $g/3$ (*d*) $g/4$

13. A boy weighing 625 N stands on a spring weighing machine inside a lift. The lift commences its downward journey and the machine reads 500 N. At that instant, the lift is descending with an acceleration of
(*a*) 1.25 m/s^2 (*b*) 2 m/s^2 (*c*) 8 m/s^2 (*d*) 12 m/s^2
Take g = 10 m/s^2.

14. Two objects of masses 3 kg and 5 kg respectively are connected by a light cord passing over a smooth weighless pulley
(*i*) The acceleration with which the masses moves is
(*a*) $g/4$ (*b*) $g/2$ (*c*) $2g$ (*d*) $4g$
(*ii*) The tension in the string would be
(*a*) 37.5 N (*b*) 80 N (*c*) 150 N (*d*) 0.5 N
Take g = 10 m/s^2
(*iii*) The pressure on the pulley would be
(*a*) 37.5 N (*b*) 75 N (*c*) 50 N (*d*) 20 N

15. A horizontal force F pulls two masses m_1 and m_2 (lying on a smooth frictionless table) connected by a light inextensible string as shown in Fig. 18.65.

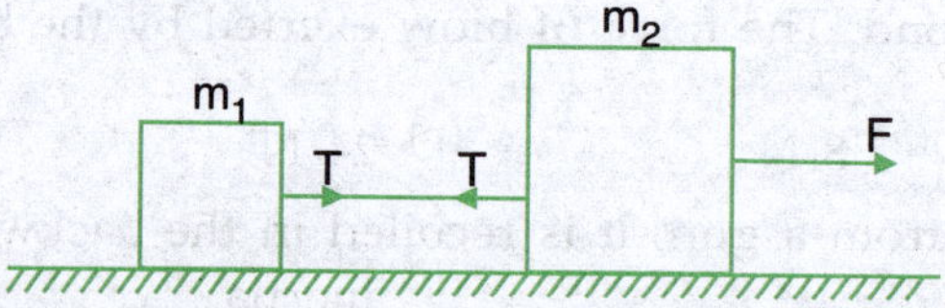

Fig. 18.65.

The tension in the string would be

(*a*) $\frac{1}{m_1 + m_2} F$ (*b*) $\frac{m_1}{m_1 + m_2} F$ (*c*) $\frac{m_2}{m_1 + m_2} F$ (*d*) $\frac{m_2}{m_1} F$

Answers :

1. (*a*)	2. (*a*)	3. (*c*)	4. (*b*)	5. (*b*)	6. (*d*)
7. (*c*)	8. (*c*)	9. (*b*)	10. (*b*)	11. (*c*)	12. (*c*)
13. (*b*)	14. *i*(*a*), *ii*(*a*), *iii*(*b*)	15. (*b*)			

Kinetics of Rigid Body (Rotary Motion) 19

Learning objectives : The contents of this chapter will make the readers familiar with the following aspects of kinetics of rotary motion of a rigid body

- *rotary motion and laws of rotary motion*
- *moment of momentum, torque and angular momentum*
- *kinetic energy of rolling motion and plane motion*
- *work-energy principle and principle of conservation of energy*
- *analogy between different parameters pertaining to linear and rotary motion*

The subject matter dealt with in this chapter pertains to kinetics of rigid bodies and relates the forces acting on the rigid body, the shape and mass of the body and the motion produced. The analysis, however, has been restricted to plane motion, *i.e*, to a motion in which each particle of the body remains at a constant distance from a fixed reference plane.

19.1 ROTARY MOTION AND TORQUE

• **Rotary motion:** motion of a rigid body in which the particles move in circular paths. The particles continue repeating the same path and have a fixed centre of rotation.

• **Torque:** turning moment of a force on a body on which it acts. Torque equals the product of force and perpendicular distance between the line of action of force and the point of rotation.

With reference to Fig. 19.1,

$$\text{Torque } T = F \times r$$

If F is in newtons (N) and r is in metres (m), then the unit of torque will be Nm.

When the body pivoted at point O rotates through a small angle θ (measured in radians), the point A shifts to B.

Work done by the tangential force F in rotating the body from A to B is

$$= \text{force} \times \text{distance moved}$$
$$= F \times \text{length of arc } AB$$
$$= F \times r\theta = (F \times r)\theta = T\theta$$

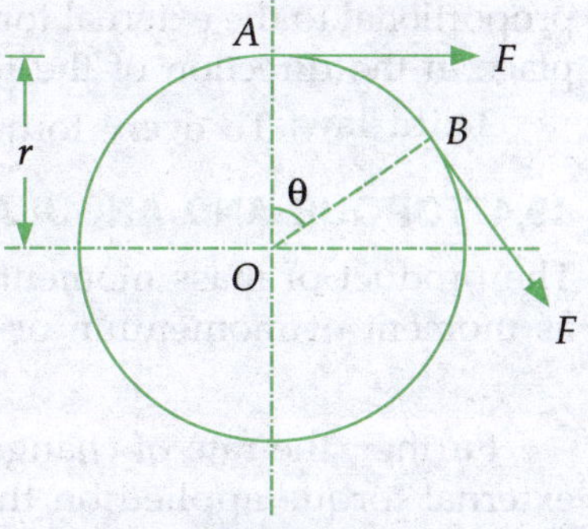

Fig. 19.1

Thus the work done by torque is given by the product of torque and the angular displacement.

In one rotation $\theta = 2\pi$. If the body turns N revolutions per minute, then the angular displacement per second is $\dfrac{2\pi N}{60}$

$$\text{Work done} = T \times \frac{2\pi N}{60} \text{ Nm/s}$$

$$\therefore \quad \text{Power of the system} = \frac{2\pi NT}{60} \text{ Watt}$$

19.2 MOMENT OF MOMENTUM

It may be recalled that linear momentum for a body moving along a straight line is the product of the mass of the body and its velocity

With reference of Fig. 19.2, a rigid body of mass m rotates about point O with constant angular velocity.

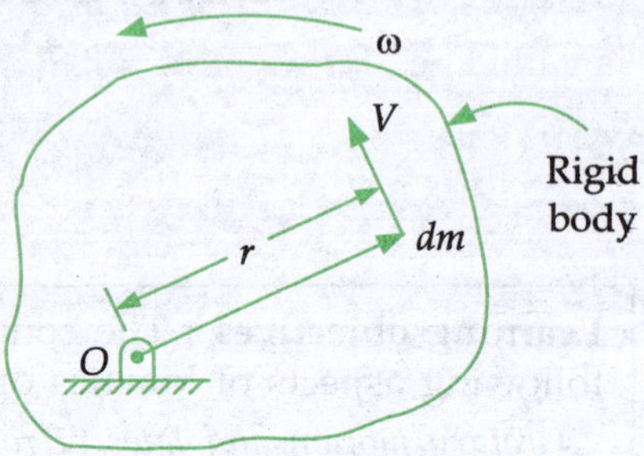

Fig. 19.2

Momentum of elementary mass about O

$$= dm \times V = dm \times \omega r \qquad (\because V = \omega r)$$

Moment of momentum of the elementary mass about O

$$= (dm \times \omega r) \times r = (dm \times r^2) \times \omega$$

For the entire mass, the moment of momentum about O

$$= \int (dm \times r^2) \times \omega = \omega \int (dm \times r^2)$$

The term $\int (dm \times r^2)$ represents the moment of inertia of the whole mass about O. Then

Moment of momentum of the whole body about O

$$= I\omega$$

Thus the moment of momentum may be defined as *"the product of mass moment of inertia and the angular velocity of a rotating body"*

Moment of momentum is also known as *angular momentum*

If the external torque acting on a system is zero, the angular momentum of the system remains constant

19.3 LAWS OF ROTARY MOTION

Similar to laws of linear motion, Newton's laws for rotary motion may be stated as follows:

First law: A body remains at rest or continues to rotate about an axis with constant angular velocity unless it is disturbed by an external torque that changes its state.

Second law: The rate of change of angular momentum of body that rotates about an axis is proportional to the external torque that acts on the body, and the change in angular moment takes place in the direction of the torque.

Third law: To every torque, there is always an equal and opposite torque.

19.4 TORQUE AND ANGULAR MOMENTUM

The product of mass moment of inertia and the angular velocity of a rotating body is known as moment of momentum or angular momentum.

$$\text{Angular momentum} = I\,\omega$$

Further, the rate of change of angular momentum of a rotating body is proportional to the external torque applied on the body and takes place in the direction of the torque.

$$T \propto \text{rate of change of angular momentum}$$

$$\propto \frac{d}{dt}(I\omega) \propto I\frac{d\omega}{dt} = k\,I\,\alpha$$

where $\alpha = \dfrac{d\omega}{dt}$ = angular acceleration and k is the constant of proportionality.

Take unit torque ($T = 1$) to produce unit angular acceleration ($\alpha = 1$) in a body of unit mass moment of inertia ($I = 1$), we get $k = 1$ and accordingly,

$$T = I\alpha \quad \text{...(19.1)}$$

The expression $T = I\,\alpha$ may be written as

$$T = I\frac{d\omega}{dt}$$

When the torque T is applied to the body for infinite small interval of time dt, the torque is called an *impulsive* or *angular torque*.

$\therefore$ Impulsive torque $T_i = I\, d\omega$ = change of angular momentum

19.5 MOMENT OF INERTIA AND RADIUS OF GYRATION

The moment of inertia I of a body is the second moment of all elemental areas (or masses) comprising the body.

For a circular lamina of mass m and radius r: $I = \dfrac{1}{2}\, mr^2$

For a solid sphere of mass m and radius r: $I = \dfrac{2}{5}\, mr^2$

For a uniform bar of mass m and length l: $I = \dfrac{1}{12} ml^2$

The radius of gyration k is defined as the distances of the point, from the axis of reference, where the whole mass of the body is assumed to be concentrated so as the moment of inertia about the given axis does not change.

The moment of inertia of a body in terms of radius of gyration is given by

$$I = mk^2 \quad \text{or} \quad k = \sqrt{\frac{I}{m}}$$

For a circular lamina : $k = \sqrt{\dfrac{1}{2}\dfrac{mr^2}{m}} = \dfrac{r}{\sqrt{2}} = 0.707\, r$

For a solid sphere : $k = \sqrt{\dfrac{2}{5}\dfrac{mr^2}{m}} = \sqrt{\dfrac{2}{5}}\, r = 0.632\, r$

19.6 KINETIC ENERGY DUE TO ROTATION

With reference to Fig. 19.2, a rigid body of mass m rotates about point O with constant angular velocity ω.

KE of an elementary mass

$$= \frac{1}{2}\, dm\, V^2 = \frac{1}{2}\, dm\, (\omega r)^2 = \frac{1}{2}\omega^2 r^2 dm$$

KE of the whole body,

$$= \int \frac{1}{2}\omega^2 r^2\, dm = \frac{1}{2}\omega^2 \int r^2 dm$$

The quantity $\int r^2\, dm$ represents the moment of inertia of the body about the axis of rotation.

$$\therefore \quad KE \text{ of a rotating body} = \frac{1}{2} I\,\omega^2 \quad \text{...(19.2)}$$

Work done equals the change in kinetic energy and therefore

$$\text{Work done} = T\theta = \frac{1}{2} I(\omega_2^2 - \omega_1^2) \quad ...(19.3)$$

The plane motion of a rigid body consists of rectilinear motion and a rolling motion without slipping. Then

Total kinetic energy of the body,

= KE in translation + KE in rotation

$$= \frac{1}{2} mV^2 + \frac{1}{2} I\omega^2 = \frac{1}{2} mV^2 + \frac{1}{2} (mk^2) \times \left(\frac{V}{r}\right)^2 = \frac{mV^2}{2}\left[1 + \frac{k^2}{r^2}\right] \quad ...(19.4)$$

The following analogy exists between different parameters pertaining to linear and rotary motion:

S.No.	Linear motion	Rotary motion
1.	Mass m	Moment of inertia I
2.	Force F	torque T
3.	Linear velocity v	Angular velocity ω
4.	Linear acceleration a	Angular acceleration α
5.	Linear momentum mv	Angular momentum $I\omega$
6.	Force equation $F = ma$	Torque equation $T = I\alpha$
7.	Linear kinetic energy $\frac{1}{2}mv^2$	Rotational kinetic energy $\frac{1}{2}I\omega^2$
8.	Work done $F s$	Work done $T\theta$

19.7 WORK-ENERGY PRINCIPLE AND PRINCIPLE OF CONSERVATION OF ENERGY

The work-energy principle connects the work done with change in kinetic energy and be stated as

Work done by a rigid body undergoing general plane motion

= change in KE of the rigid body due to linear motion

+ change in KE of the rigid body due to rotary motion

The work done results both from linear and angular displacement from the initial and final positions.

Then
$$Fs + T\theta = \frac{1}{2} m (V_2^2 - V_1^2) + \frac{1}{2} I (\omega_2^2 - \omega_1^2)$$

The principle of conservation of energy states that the sum of potential energy and kinetic energy of a rigid body moving under the influence of conservative forces is always a constant. That is

$$(PE + KE)_1 = (PE + KE)_2$$

19.8 EQUATION OF MOTION

For a rigid body in plane motion (combination of translation and rotation) the equations of motion are.

$$\left.\begin{aligned} \Sigma F_x &= m a_x \\ \Sigma F_y &= m a_y \end{aligned}\right\} \text{Translatory motion}$$

and
$$\Sigma M_G = I_G \alpha \quad \} \text{ Rotary motion}$$

where ΣM_G is resultant moment about the mass centre of the rigid body, α is angular acceleration of the rigid body, and I_G is the mass moment of inertia of the rigid body about its mass centre (axis of rotation).

EXAMPLE 19.1

A uniform homogeneous cylinder of 125 mm radius has a mass of 0.15 kg. This cylinder rolls without slipping along a level horizontal surface with a translational velocity of 20 cm/s. Determine its total kinetic energy.

Solution: Kinetic energy due to translation $(KE)_t$

$$= \frac{1}{2} mV^2 = \frac{1}{2} \times 0.15 \times 0.2^2 = 3 \times 10^{-3} \text{ Nm}$$

Kinetic energy due to rotation $(KE)_r = \frac{1}{2} I\omega^2$

$$\omega = \text{angular velocity} = \frac{V}{r} = \frac{0.2}{0.125} = 1.6 \text{ rad/s}$$

$$I = \text{mass moment of inertia } I = \frac{mr^2}{2}$$

$$= \frac{0.15 \times (0.125)^2}{2} = 1.172 \times 10^{-3} \text{ kgm}^2$$

That gives: $(KE)_r = \frac{1}{2} \times (1.172 \times 10^{-3}) \times 1.6^2 = 1.5 \times 10^{-3}$ Nm

$\therefore$ Total kinetic energy $= (KE)_t + (KE)_r = 3 \times 10^{-3} + 1.5 \times 10^{-3} =$ **4.5×10^{-3} Nm**

EXAMPLE 19.2

An angular acceleration $\alpha = (0.4\,t^2 + 0.6)$ rad/s² where t is in seconds is imparted to disc A by a motor. If initially the disc has an angular velocity of 5 rad/s, make calculations for the velocity and acceleration of block B at the time instant t = 1.5 seconds (Fig. 19.3).

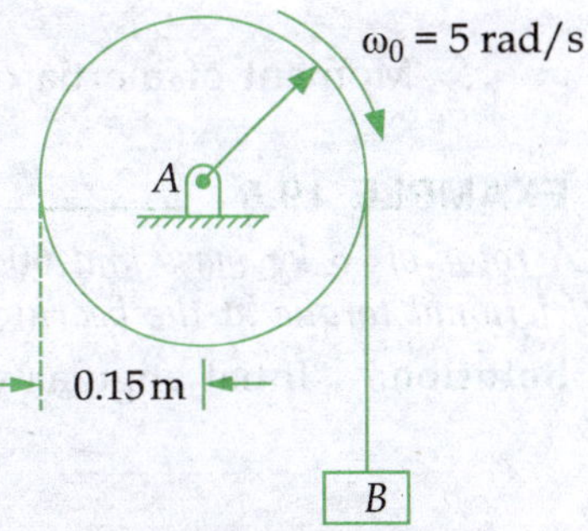

Fig. 19.3

Solution: Angular acceleration $\alpha = 0.4t^2 + 0.6$

$$\frac{d\omega}{dt} = 0.4t^2 + 0.6$$

Upon integration within appropriate limits, we get

$$\int_{\omega_0 = 5}^{\omega} d\omega = \int_0^t (0.4t^2 + 0.6)dt$$

$$\omega - 5 = 0.4 \times \frac{t^3}{3} + 0.6t$$

At t = 1.5 seconds: $\omega = 0.4 \times \frac{(1.5)^3}{3} + 0.6 \times 1.5 + 5 = 6.35$ rad/s

$$\alpha = 0.4 \times 1.5^2 + 0.6 = 1.5 \text{ rad/s}^2$$

Then velocity V and acceleration α of block B are:

$V = \omega r = 6.35 \times 0.15 =$ **0.9525 m/s² (↓)**

$a = \alpha r = 1.5 \times 0.15 =$ **0.225 m/s² (↓)**

EXAMPLE 19.3

A bar pivoted at one end and moving clockwise at 4 rad/s is subjected to a constant angular deceleration. After a certain time, the bar has angular displacement of 6 radians anticlockwise and has moved through total angle of 16 radians. Determine the angular velocity of the bar at the end of this time interval.

Solution: Refer Fig. 19.4. Considering motion between position OA and OA_1

$$\omega_1^2 - \omega_0^2 = 2\,\alpha\theta_1$$

At position A_1, the bar has travelled through 10 radians and comes to rest.

$$\therefore \quad 0 - (4)^2 = 2\alpha \times 10$$

$$\alpha = -0.8 \text{ rad/s}^2$$

Considering motion between position OA_1 and OA_2

$$\omega_2^2 - \omega_1^2 = 2\alpha\theta$$

$$\omega_2^2 - 0^2 = 2 \times 0.8 \times 16$$

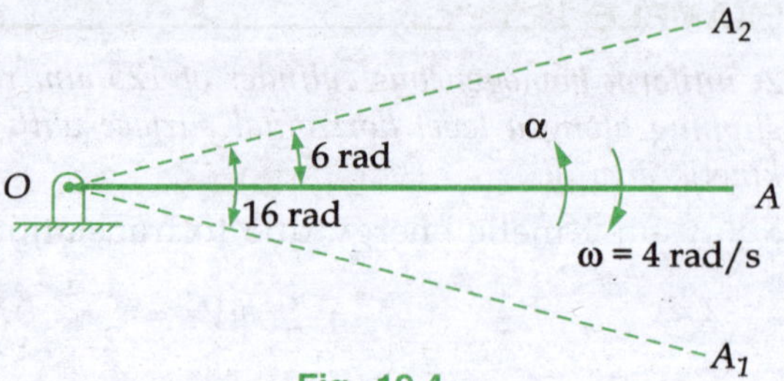

Fig. 19.4

$$\therefore \quad \omega_2 = \sqrt{2 \times 0.8 \times 16} = \mathbf{5.06} \text{ rad/s at the end of time interval.}$$

EXAMPLE 19.4

A string of 5 m length has been wrapped round the shaft of a wheel to determine its moment of inertia about its axis. A constant pull of 80 N is applied to the string and observations indicate that the wheel turns 180 rev/min at the instant the string leaves the axle. Calculate the moment of inertia of the wheel.

Solution: Work done in pulling the string

$$= \text{force applied} \times \text{distance moved} = 80 \times 5 = 400 \text{ Nm}$$

$$\text{Angular velocity of the wheel } \omega = \frac{2\pi N}{60} = \frac{2\pi \times 180}{60} = 18.84 \text{ rad/s}$$

Then from the work-energy correlation of a rotating system,

work done = kinetic energy of the wheel

$$400 = \frac{1}{2} I\omega^2 = \frac{1}{2} I \times (18.84)^2$$

$$\therefore \quad \text{Moment of inertia of the wheel } I = \frac{400 \times 2}{(18.84)^2} = 2.254 \text{ kg m}^2$$

EXAMPLE 19.5

A rotor of 8 kg mass and 60 cm radius of gyration is rotating at 1440 rev/min. What would be the frictional torque at the bearing, if the rotor comes to rest after 2 minutes?

Solution: Intial angular velocity

$$\omega_i = \frac{2\pi N}{60} = \frac{2 \times \pi \times 1440}{60} = 150.72 \text{ rad/s}$$

Final angular velocity $\omega_f = 0$

$$\therefore \quad \text{Angular acceleration} \quad \alpha = \frac{0 - 150.72}{2 \times 60} = -1.256 \text{ rad/s}^2$$

The negative sign implies that it is a retardation.

$$\text{Torque} = \text{mass moment of inertia} \times \text{angular retardation}$$

$$= (mk^2) \times \alpha$$

$$= (8 \times 0.6^2) \times 1.256 = 3.62 \text{ Nm}$$

The friction torque at the bearing would be in opposite direction.

EXAMPLE 19.6

A 40 Nm couple moment is applied to a wheel that rests on a plane and has 25 kg mass and 0.4 m radius as shown in Fig. 19.5. If the coefficient of friction between the wheel and plane at A is 0.25, determine:

(a) acceleration of the mass centre of wheel, and

(b) frictional force.

Presume that radius of gyration for the wheel is 0.20 cm

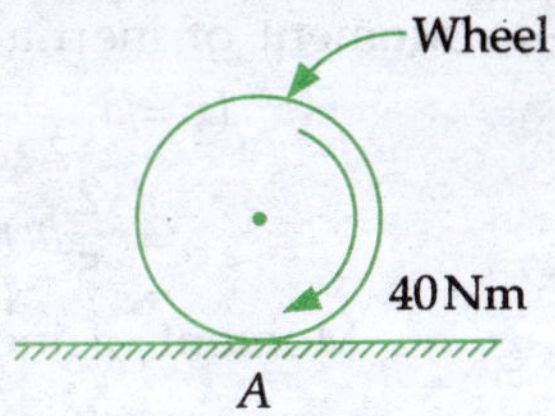

Fig. 19.5

Solution: $I_G = m K_G^2 = 25 \times (0.2)^2 = 1 \text{ kg m}^2$

Applying parallel axis theorem,

$$I_A = I_G + mh^2 = 1 + 25 \times (0.4)^2 = 5 \text{ kg m}^2$$

Recalling that for a rigid body in plane motion

$$\Sigma M_A = I_A \alpha \; ; \; 40 = 5\alpha$$

$\therefore$ Angular acceleration of the wheel, $\alpha = \dfrac{40}{5} = 8 \text{ rad/s}^2$

Acceleration of mass centre, $a = \alpha r = 8 \times 0.4 =$ **3.2 m/s²**

(*b*) Frictional force $F = \mu R$

The normal reaction equals the weight of the wheel

$$\therefore \quad F = \mu W = 0.25 \times (25 \times 9.81) = \textbf{61.31 N}$$

EXAMPLE 19.7

A slender rod AB of 12 kg mass and 1 m length is pivoted at point O which is 0.2 m from the end B (Fig. 19.6). The end A of the rod presses against a spring until the spring is compressed 20 mm and the rod lies horizontally. If the rod is released from this position, calculate the angular velocity when the rod has rotated through 60°. Take spring constant K = 250 kN/m

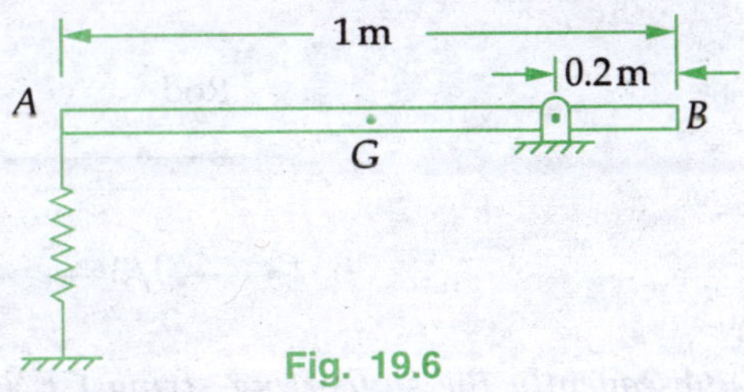

Fig. 19.6

Solution: Magnitude of force in the spring,

$$F = kx = 250 \times (0.02) = 5 \text{ kN} = 5000 \text{ N } (\downarrow)$$

Let I be the moment of inertia of the rod about the pivot point. Then from parallel axis theorem

$$I = I_G + mh^2 = \frac{ml^2}{12} + mh^2 = \frac{12 \times 1^2}{12} + 12\left(\frac{1}{2} - 0.2\right)^2 = 4.6 \text{ kg m}^2$$

For a rigid body in plane motion: $\Sigma M = I\alpha$

$$\therefore \quad 5000 \times 0.8 = 4.6\,\alpha \; ; \quad \alpha = 869.57 \text{ rad/s}^2$$

From the kinematic relation $\omega^2 - \omega_0^2 = 2\alpha\theta$ and noting that $\omega_0 = 0$ and $\theta = 60° = \pi/3$ radian, we have

$$\omega = \sqrt{2 \times 869.57 \times \pi / 3} = \textbf{42.66 rad/s}$$

EXAMPLE 19.8

Consider a composite system consisting of a rod AB and sphere BC fixed rigidly to the rod at B as shown in Fig. 19.7. Determine the angular acceleration of the system at the instant when it is released from the horizontal position.

Take mass of rod m_1 = 20 kg

mass of sphere m_2 = 5 kg

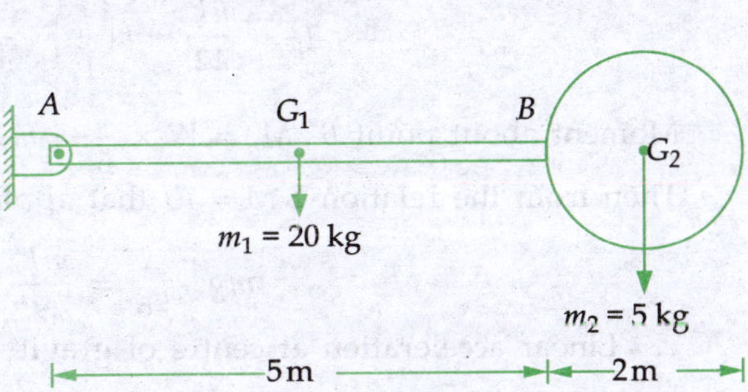

Fig. 19.7

Solution: Moment of inertia of the rod about A (centre of rotation),

$$I_1 = \frac{1}{3} ml^2$$

$$= \frac{1}{3} \times 20 \times 5^2 = 166.67 \text{ kg m}^2$$

Moment of inertia of sphere about A,

$$I_2 = I_{G_2} + mh^2 \quad \text{(parallel axes theorem)}$$

$$= \frac{2}{5}mr^2 + mh^2 = \frac{2}{5} \times 5 \times 1^2 + 5 \times (5+1)^2 = 2 + 180 = 182 \text{ kg m}^2$$

$\therefore$ Moment of inertia of the entire system

$$I_a = I_1 + I_2 = 166.67 + 182 = 348.67 \text{ kg m}^2$$

For rigid body undergoing rotary motion,

$$\Sigma M = I\alpha \; ; \quad M_a = I_a \times \alpha$$

$\therefore$ Angular acceleration of the system,

$$\alpha = \frac{M_a}{I_a} = \frac{(20 \times 9.81 \times 2.5) + (5 \times 9.81 \times 6)}{348.67} = \frac{490.5 + 294.3}{348.67} = \mathbf{2.251 \text{ rad/s}^2}$$

EXAMPLE 19.9

A slender rod ABCD of length l and mass m is supported in a horizontal position as shown in Fig. 19.8 given below:

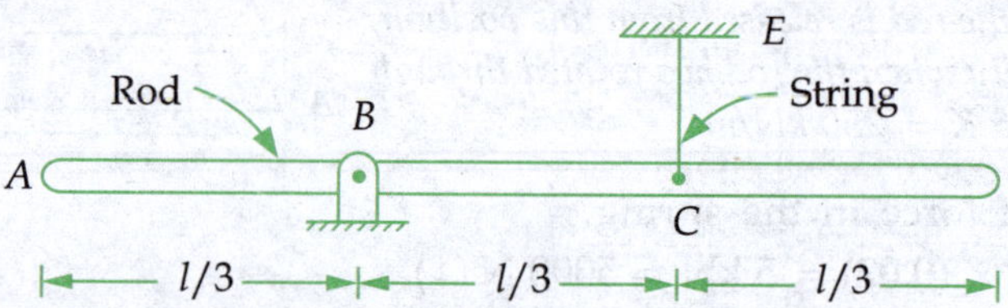

Fig. 19.8

Subsequently the supporting string CE is cut. Compute the instantaneous reaction at B immediately after string has been cut.

Solution: Refer Fig. 19.9 given below

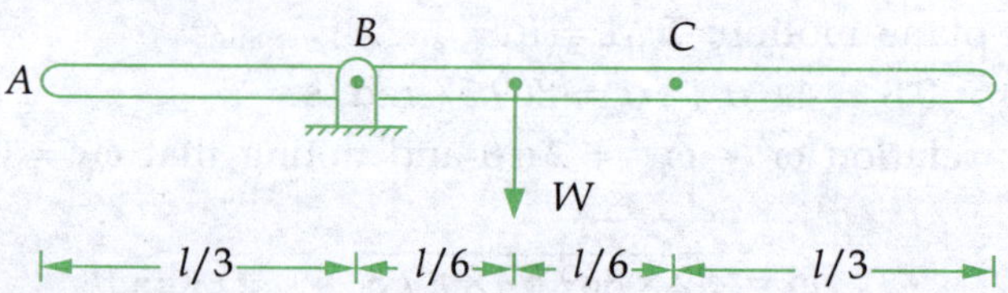

Fig. 19.9

Moment of inertia of the rod about the centre of rotation B,

$$I_b = I_G + mh^2 \quad \text{(parallel axes theorem)}$$

$$I_b = \frac{ml^2}{12} + m\left(\frac{l}{6}\right)^2 = \frac{ml^2}{9}$$

Moment about point $B : M_b = W \times \frac{l}{6} = mg \times \frac{l}{6}$

Then from the relation $\Sigma M = I\alpha$ that applies to rigid body rotation, we have

$$mg \times \frac{l}{6} = \frac{ml^2}{9} \times \alpha \; ; \quad \alpha = \frac{3}{2}\frac{g}{l}$$

$\therefore$ Linear acceleration at centre of gravity of the rod,

$$a = \alpha \times R = \frac{3}{2}\frac{g}{l} \times \frac{l}{6} = 2.45 \text{ m/s}^2$$

For a rigid body undergoing translatory motion,

$$\text{net force} = \text{mass} \times \text{acceleration}$$

$$mg - R_b = ma$$

∴ Vertical reaction at suspension point B,

$$R_b = m(g - a) = m\,(9.81 - 2.45) = 7.36\,m$$

Since there is no horizontal force acting on the system, the horizontal reaction at the suspension point B would be zero.

EXAMPLE 19.10

A bearing carries a shaft that supports a disc of 50 cm radius and 25 kg mass. The shaft is normal to the plane of the disc and is coaxial with the disc as shown in Fig. 19.10. The disc is made to rotate at 1500 rpm and then it decelerates because of bearing friction. If the disc comes to rest in 145 seconds, make calculations for the bearing torque and the number of revolutions the disc turns before it comes to rest. It may be assumed that the bearing friction is independent of the speed of the disc.

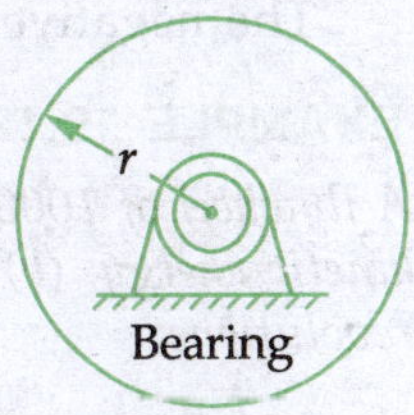

Fig. 19.10

Solution: $\omega_1 = \dfrac{2\pi \times 1500}{60} = 157 \text{ rad/s, and } \omega_2 = 0$

Since the bearing friction is independent of the speed of disc, the angular acceleration is constant and it is given by

$$\alpha = \frac{d\omega}{dt} = \frac{\omega_2 - \omega_1}{t} = \frac{0 - 157}{145} = -1.083 \text{ rad/s}$$

Mass moment of inertia $I = \dfrac{1}{2} mr^2 = \dfrac{1}{2} \times 25 \times 0.5^2 = 3.125 \text{ kg m}^2$

(*a*) Bearing torque $T = I\alpha = 3.125 \times (-1.083) = -3.384$ Nm

The negative sign implies that the frictional torque of the bearing resists the rotation of the disc.

(*b*) From the kinematic relation $\omega_2^2 - \omega_1^2 = 2\alpha\theta$ and noting that $\omega_2 = 0$, we have

Total angular displacement

$$\theta = -\frac{\omega_1^2}{2\alpha} = \frac{-157^2}{2 \times (-1.083)} = 11380 \text{ radians}$$

1 revolution = 2π radians

∴ Number of revolutions turned by the disc $= \dfrac{11380}{2\pi} = $ **1812**

EXAMPLE 19.11

A flywheel of mass 4000 kg and radius of gyration 0.8 m loses its speed from 360 rpm to 250 rpm in 2 minutes. Make calculations for (i) the torque acting on the wheel, (ii) change in the kinetic energy and angular momentum of the wheel during the given period.

Solution: Initial angular velocity $\omega_0 = \dfrac{2\pi N_0}{60} = \dfrac{2\pi \times 360}{60} = 37.68 \text{ rad/s}$

Final angular velocity $\omega = \dfrac{2\pi N}{60} = \dfrac{2\pi \times 250}{60} = 26.17 \text{ rad/s}$

Then from the correlation $\omega = \omega_0 + \alpha t$,

$$\text{angular acceleration } \alpha = \frac{\omega - \omega_0}{t} = \frac{26.17 - 37.68}{2 \times 60} = -0.0959 \text{ rad/s}^2$$

$$\text{Torque } T = I\alpha = mk^2\alpha = 400 \times (0.8)^2 \times (-0.0959) = -245.5 \text{ Nm}$$

The negative sign means that the torque is retarding torque.

(ii) Change in kinetic energy $= \frac{1}{2} I\omega^2 - \frac{1}{2} I\omega_0^2 = \frac{1}{2} I(\omega^2 - \omega_0^2) = \frac{1}{2} mk^2 (\omega^2 - \omega_0^2)$

$$= \frac{1}{2} \times (4000 \times 0.8^2)(26.17^2 - 37.68^2) = -9.407 \times 10^5 \text{ Nm}$$

The negative sign implies decrease in kinetic energy.

(iii) Change in angular momentum $= I\omega - I\omega_0 = I(\omega - \omega_0) = mk^2 (\omega - \omega_0)$

$$= 4000 \times 0.8^2 \times (26.17 - 37.68) = -29466 \text{ Nm/s}$$

The negative sign implies decrease in angular momentum.

EXAMPLE 19.12

A flywheel of 1000 kg mass and 0.8 m radius of gyration is rotating at 180 rev/min. Determine its (a) kinetic energy, (b) torque and average power required to give the wheel a speed of 180 rev/min in 75 revolutions.

Solution: Angular velocity of wheel $\omega = \frac{2\pi N}{60} = \frac{2\pi \times 80}{60} = 18.84 \text{ rad/s}$

Moment of inertia of flywheel $I = mk^2 = 1000 \times 0.8^2 = 640 \text{ kg m}^2$

(a) Kinetic energy of the flywheel $= \frac{1}{2} I\omega^2 = \frac{1}{2} \times 640 \times 18.84^2 = 113582 \text{ Nm}$

(b) Let T be the torque required to give a speed of 180 rev/min in 75 revolutions, *i.e.*, in $2\pi \times 75$ radians.

Then from the work-energy correlation of a rotating system,

change in kinetic energy = work done = torque × angular displacement

$$113582 = T \times (2\pi \times 75); \quad T = 241.15 \text{ Nm}$$

Average speed of the flywheel $= \frac{0 + 180}{2} = 90 \text{ rev/min}$

Time taken to make 75 revolutions $= \frac{75}{90} \times 60 = 50 \text{ sec}$

Work done/sec $= \frac{113582}{50} = 2271.64 \text{ Nm/s}$

Hence average power = 2271.64 W or **2.272 kW**

EXAMPLE 19.13

A mine cage of 500 kg mass is hung from the drum of hoisting engine which has a mass of 800 kg. The drum is of 150 cm diameter and its radius of gyration is 60 cm. Neglecting frictional effects, determine the velocity the cage will attain when allowed to fall for a period of 4 seconds.

Solution: For drum : Moment of inertia of the drum $I = mk^2 = 800 \times 0.6^2 = 288 \text{ kg m}^2$

angular impulse = change in angular momentum

$$(T \times 0.75) \times 4 = I\omega_2 - I\omega_1 \quad \text{or} \quad 3T = I\omega_2 \qquad (\because \omega_1 = 0)$$

where T is the tension in the hoisting rope.

Further, the linear velocity and the angular velocity are connected by the relation : $v = \omega r$. That gives

$$\omega_2 = \frac{v}{0.75}$$

$$\therefore \quad T = \frac{I}{3} \times \frac{v}{0.75} = \frac{288}{3} \times \frac{v}{0.75} = 128\,v \qquad ...(i)$$

For 500 kg mass: linear impulse = change in linear momentum

$$(500 \times 9.81 - T) \times 4 = 500\,v \quad \text{or} \quad 4905 - T = 125\,v \qquad ...(ii)$$

From expressions (*i*) ad (*ii*),

$$4905 - 128\,v = 125\,v$$

$$\therefore \text{ Velocity attained by the cage } v = \frac{4905}{253} = \mathbf{19.38\ m/s}$$

EXAMPLE 19.14

A cylindrical roller of 400 kg mass and 250 mm radius is at rest on a rough horizontal surface. The roller is pushed by a force of 500 N inclined at 25° to the horizontal and passing through its centre of gravity. The roller then starts rolling without slipping. Calculate the distance moved by the roller by the time it attains a speed of 1.8 m/s.

Solution : Refer Fig. 19.11 for the system of roller and a push being applied to it.

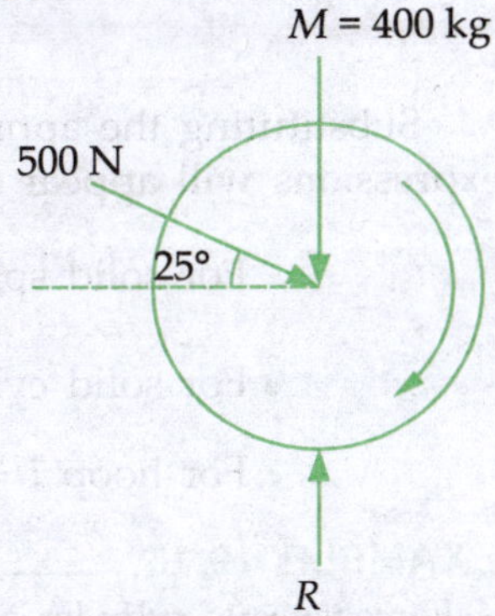

Fig. 19.11

It is to be noted that:

(*i*) the roller moves in the horizontal direction and as such the weight of the roller and the reaction at the rough surface donot contribute towards work.

(*ii*) the velocity at the point of contact is zero and accordingly no work will be done by the frictional force

(*iii*) the work will be done only by the horizontal component of 500 N push.

If S is the distance moved by the roller, then

$$\text{Work done} = (500 \times \cos 25^\circ) \times S = 453.15\,S$$

The roller starts from rest ($u = 0$) and accordingly initial kinetic energy for the roller is zero ($KE_1 = 0$).

Final kinetic energy = kinetic energy due to linear motion + kinetic energy due to rotation

$$KE_2 = \frac{1}{2}mV^2 + \frac{1}{2}I\omega^2 = \frac{1}{2}mV^2 + \frac{1}{2} \times \frac{mr^2}{2} \times \left(\frac{V}{r}\right)^2$$

$$= \frac{1}{2} \times 400 \times 1.8^2 + \frac{1}{2} \times \frac{400 \times 0.25^2}{2} \times \left(\frac{1.8}{0.25}\right)^2 = 648 + 324 = 972 \text{ Nm}$$

From the work-energy principle, the work done equals the change in kinetic energy. Thus

$$453.15\,S = 972 - 0 = 972$$

$$\therefore \quad \text{Distance moved by the roller } S = \frac{972}{453.15} = \mathbf{2.14\ m}$$

EXAMPLE 19.15

A solid sphere, a cylinder and a hoop are released from rest on an incline at the same elevation. In each case, the mass is m and the maximum radius is R. Determine the velocity of each body after it has rolled down the incline through a distance s as shown in Fig. 19.12.

Solution: It is to be noted that:

(*i*) motion takes place in the gravitational field of the earth

(*ii*) no work is done by the frictional forces in rolling motion

Accordingly the mechanical energy of the system is conserved. That is

$$(KE + PE)_1 = (KE + PE)_2 \quad ...(i)$$

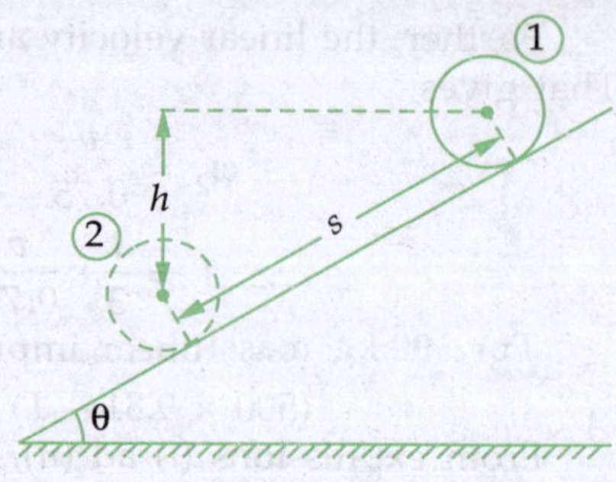

Fig. 19.12

Noting that:

$KE_1 = 0$ as the body starts from rest

$PE_1 = 0$ when the initial position is taken as datum

$$KE_2 = \frac{1}{2}mV_2^2 + \frac{1}{2}I\omega^2 = \frac{1}{2}mV_2^2 + \frac{1}{2}I\left(\frac{V_2}{R}\right)^2$$

$$PE_2 = -mgh = -mgs\sin\theta$$

Substitution of the above noted values in expression (*i*), we get

$$0 = \frac{1}{2}mV_2^2 + \frac{1}{2}I\left(\frac{V_2}{R}\right)^2 - mgs\sin\theta$$

Solution gives: $V_2 = \dfrac{\sqrt{2gs\sin\theta}}{\sqrt{1 + \dfrac{1}{mR^2}}}$

Substituting the appropriate values of moment of inertia for the given body, the following expressions will appear for the velocity of each body.

- For solid sphere $I = \frac{2}{5}mR^2$ and $V_2 = 0.845\sqrt{2gs\sin\theta}$
- For solid cylinder $I = \frac{1}{2}mR^2$ and $V_2 = 0.816\sqrt{2gs\sin\theta}$
- For hoop $I = mR^2$ and $V_2 = 0.707\sqrt{2gs\sin\theta}$

EXAMPLE 19.16

A homogeneous cylinder of mass 50 kg and 0.5 m in radius is having initial velocity of 6 m/s down a 30° inclined plane. Calculate the velocity of cylinder when it has reached 10 m down the plane from starting point. It may be presumed that the cylinder rolls without sliding.

Solution: Component of wight along the plane

$= mg\sin 30° = 50 \times 9.81 \times 0.5 = 245.25$ N

Work done in moving 10 m down the plane

$= 245.25 \times 10 = 2452.5$ Nm

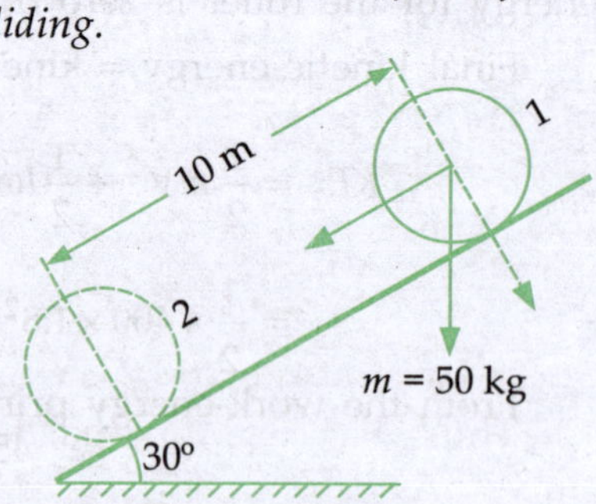

Fig. 19.13

Change in kinetic energy of the cylinder

$$\Delta KE = \frac{1}{2}m\left(V_2^2 - V_1^2\right) + \frac{1}{2}I\left(\omega_2^2 - \omega_1^2\right)$$

Noting that $I = \frac{1}{2}mr^2$, $\omega_1 = \frac{V_1}{r}$ and $\omega_2 = \frac{V_2}{r}$, we have

$$\Delta KE = \frac{1}{2} \times 50\left(V_2^2 - 6^2\right) + \frac{1}{2} \times \left(\frac{1}{2} \times 50 \times 0.5^2\right)\left[\left(\frac{V_2}{0.5}\right)^2 - \left(\frac{6}{0.5}\right)^2\right]$$

$$= 25\left(V_2^2 - 36\right) + 3.125\left(4V_2^2 - 144\right)$$

Since work done equals the change in kinetic energy, we have

$$2452.5 = 25\left(V_2^2 - 36\right) + 3.125\left(4V_2^2 - 144\right) = 25V_2^2 - 900 + 12.5V_2^2 - 450$$

$$\therefore \quad V_2 = \sqrt{\frac{2452.5 + 900 + 450}{37.5}} = \mathbf{10.07\ m/s}$$

EXAMPLE 19.17

A solid cylinder and a thin hoop of equal mass m and radius R are connected by a bar G_1G_2 as shown in Fig. 19.14. The system rolls down on the inclined plane without slipping. Determine the acceleration of the system and the tension in the bar.

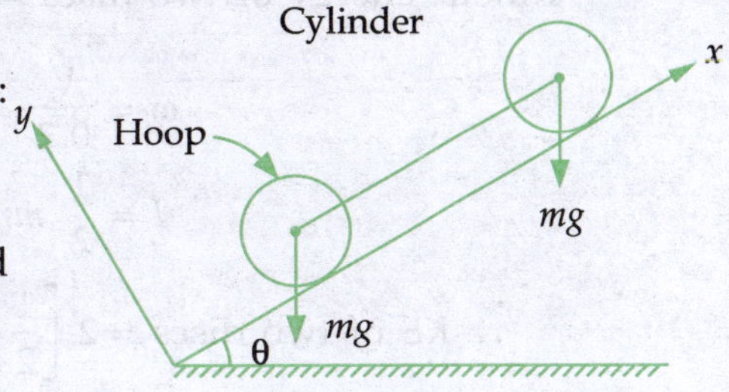

Fig. 19.14

Solution: For the cylinder, the equations of motion are:

$$\Sigma F_x = ma_x: \quad mg\sin\theta - F_1 + T = ma$$

$$F_1 = mg\sin\theta + T - ma \qquad \text{...(i)}$$

where F_1 is the force of Friction on the cylinder and T is the tension in the connecting bar G_1G_2

$$\Sigma M_{G_1} = T_{G_1}\alpha\ ; \quad F_1R = mR^2\alpha\ ; \quad F_1 = mR\alpha \qquad \text{...(ii)}$$

The corresponding expressions for hoop would be

$$F_2 = mg\sin\theta - T - ma \qquad \text{...(iii)}$$

$$F_2R = \frac{mR^2}{2}\alpha\ ; \quad F_2 = \frac{mR}{2}\alpha \qquad \text{...(iv)}$$

Adding expression (*i*) and (*iii*) and substituting for F_1 and F_2 as determined above vide expressions (*ii*) and (*iv*), we get

$$mR\alpha\left(1 + \frac{1}{2}\right) = 2m\,(g\sin\theta - a) \qquad (\alpha R = a)$$

$$\text{or} \quad \frac{3}{2}a = 2g\sin\theta - 2a \quad \therefore \quad a = \frac{4}{7}g\sin\theta$$

Thus the given system has a linear acceleration equal to $\frac{4}{7}g\sin\theta$

(*b*) From expression (*i*)

$$T = F_1 + ma - mg\sin\theta = mR\alpha + ma - mg\sin\theta$$

$$= mR\frac{a}{R} + ma - mg\sin\theta = 2ma - mg\sin\theta$$

$$= 2m \times \frac{4}{7}g\sin\theta - mg\sin\theta = \frac{mg}{7}\sin\theta$$

Thus the connecting bar will be subjected to tension equal to $\frac{mg}{7}\sin\theta$

EXAMPLE 19.18

A 12 kg plank is supported on two indentical uniform discs; each of mass 5 kg and radius 10 cm. The plank is pulled by 25 N force and its moves 6 m from the rest position. Assuming that the disc rolls without slipping, make calculations for the velocity with which the plank moves.

Solution: Work done = force × distance moved = 25 × 6 = 150 Nm

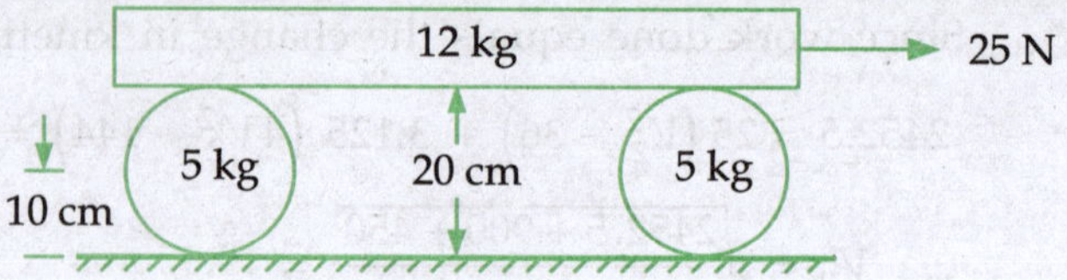

Fig. 19.15

Let the plank move with velocity V m/s towards right and the discs move clockwise with angular ω rad/s.

Initially: The kinetic energy of the plank and discs is zero since initially the system is at rest.

Finally: Kinetic energy of plank $= \frac{1}{2} mV^2 = \frac{1}{2} \times 12 \times V^2 = 6\, V^2$ Nm

$$\text{Kinetic energy of two discs} = 2\left[\frac{1}{2} m V_G^2 + \frac{1}{2} I_G \omega^2\right]$$

$$\omega = \frac{V}{0.2} = 5\,V; \quad V_G = \omega r = 5\,V \times 0.1 = 0.5\,V$$

$$I = \frac{1}{2} mr^2 = \frac{1}{2} \times 5 \times 0.1^2 = 0.025 \text{ kg m}^2$$

$$\therefore KE \text{ of two discs} = 2\left[\frac{1}{2} \times 5 \times (0.5V)^2 + \frac{1}{2} \times 0.025 \times (5V)^2\right]$$

$$= 2\,[0.625\,V^2 + 0.3125\,V^2] = 1.875\,V^2$$

Total kinetic energy of the system $= 6\,V^2 + 1.875\,V^2 = 7.875\,V^2$

This also equals the change in kinetic energy of the system as initial kinetic energy of the system is zero.

Since work done equals the change in kinetic energy, we have

$$150 = 7.875\,V^2; \quad V = \sqrt{19.05} = \mathbf{4.36 \text{ m/s}}$$

Thus the plank moves with 4.36 m/s velocity towards right.

EXAMPLE 19.19

A solid cylinder and a thin hoop are connected by a bar G_1G_2 as shown in Fig. 19.16. The system rolls down on an inclined plane without slipping. If both the cylinder and hoop are of equal mass and radius, determine the acceleration of the system and tension in the bar.

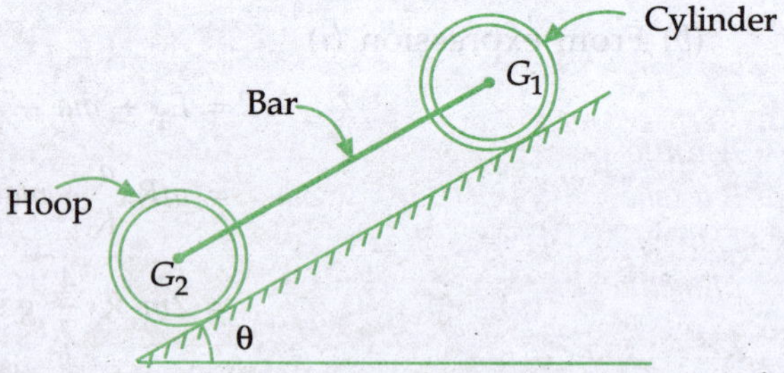

Fig. 19.16

Solution: Let m and r respectively designate the mass and radius of the cylinder.

With reference to the free body diagram as shown,

let T = tension in the connecting bar

F_1, F_2 = friction force on the cylinder and hoop

N_1, N_2 = Normal reaction on the cylinder and hoop

a = acceleration of the system

In accordance with Newton's law and D'Alemberts principle, the equations of motion are:

For the cylinder: Resolving the forces along and perpendicular to the incline.

$$mg\sin\theta - F_1 + T = ma \quad \text{or} \quad F_1 = mg\sin\theta + T - ma \qquad \ldots(i)$$

$$N_1 - mg\cos\theta = 0 \qquad \ldots(ii)$$

$$F_1 r = I_G \alpha \quad \text{or} \quad F_1 r = (m r^2)\,\alpha \qquad \ldots(iii)$$

where I_G is the mass moment of inertia about the centre of mass.

For the hoop :

$$mg\sin\theta - F_2 - T = ma \quad \text{or} \quad F_2 = mg\sin\theta - T - ma \qquad ...(iv)$$

$$N_2 - mg\cos\theta = 0 \qquad ...(v)$$

$$F_2 r = I_G\alpha \quad \text{or} \quad F_2 r = \left(\frac{1}{2}mr^2\right)\alpha \qquad ...(vi)$$

Adding the identities (*iii*) and (*iv*) for moments, we get

$$\alpha\left(mr^2 + \frac{1}{2}mr^2\right) = (F_1 + F_2)r$$

Substituting the values of friction forces F_1 and F_2 as obtained above in expressions (*i*) and (*iv*), and replacing α by $\frac{a}{r}$,

$$\frac{a}{2}\left(\frac{3}{r}mr^2\right) = (mg\sin\theta + T - ma + mg\sin\theta - T - ma)r$$

or
$$\frac{3}{2}mar = (mg\sin\theta) \times 2r - 2\,mar$$

Solution gives: $a = \dfrac{mg\sin\theta \times 2r}{\frac{3}{2}mr + 2mr} = \dfrac{4}{7}g\sin\theta$

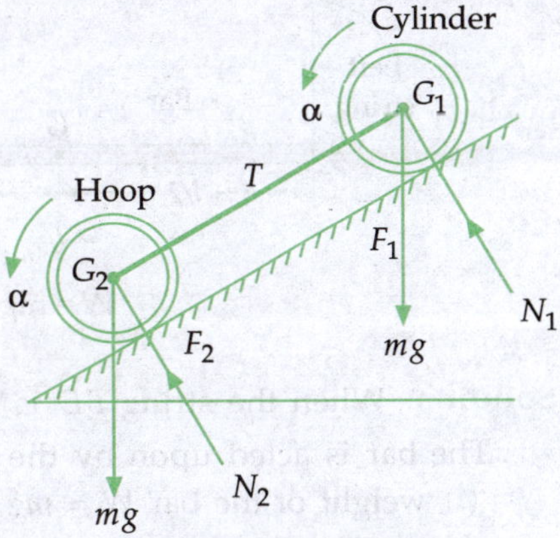

Free body diagram

(*b*) The tension T in the connected bar can be worked out from expression (*i*)

$$T = F_1\, mr\sin\theta + ma$$

From expression (*iii*), $F_1 = mr\alpha = mr \times \frac{a}{r} = ma$. Then

$$T = ma - mg\sin\theta + ma$$
$$= m(2a - g\sin\theta)$$
$$= m\left[2 \times \frac{4}{7}g\sin\theta - g\sin\theta\right] = \frac{1}{7}mg\sin\theta$$

EXAMPLE 19.20

A uniform disc of mass 20 kg and diameter 2 m is supported by two pins A and B as shown in Fig. 19.17. At a particular instant the pin B is suddenly removed. Determine the angular acceleration of the disc and the reaction at pin A immediately after the removal of pin B.

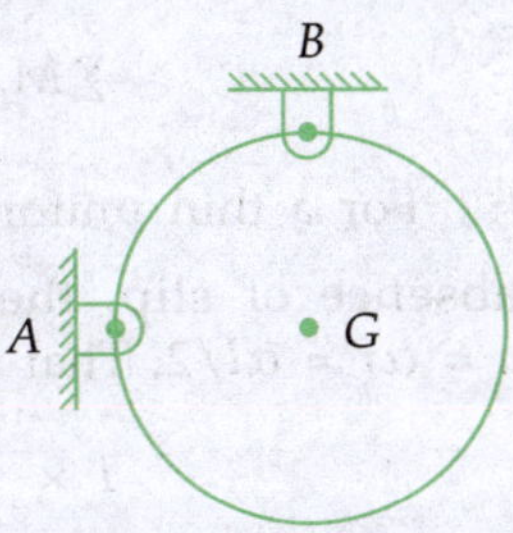

Fig. 19.17

Solution: Refer Fig. 19.17 *a*, which shows the various forces acting on the disc when the pin B is removed.

For a rigid body undergoing rotary motion,

$$\Sigma M_a = I_a\alpha$$

or
$$mg \times r = (I_G + mh^2)\,\alpha$$
$$= \left(\frac{1}{2}mr^2 + mr^2\right)\alpha = \frac{3}{2}mr^2 \times \alpha$$

$\therefore$ Angular acceleration $\alpha = \dfrac{2}{3}\dfrac{mg \times r}{mr^2} = \dfrac{2}{3}\dfrac{g}{r} = \dfrac{2}{3} \times \dfrac{9.81}{1}$ = **6.54 rad/s²**

Further, for translatory motion,

net force = mass × acceleration

$$mg - V_a = ma$$
$$= m\,(r\,\alpha)$$

∴ Vertical reaction at pin *A*,

$$V_a = mg - mr\alpha = 20 \times 9.81 - 20 \times 1 \times 6.54 = \mathbf{65.4\ N}$$

Horizontal reaction at pin *A* = 0

This is because there is no other horizontal component.

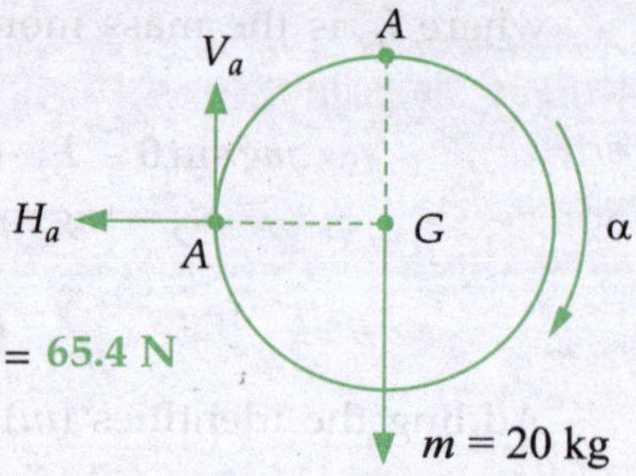

Fig. 19.17 a

EXAMPLE 19.21

A thin uniform bar of mass m and length l is suspended by two vertical inextensible strings (Fig. 19.18). If the right hand string BD is cut, set up expressions for the angular acceleration of the bar and the tension induced in the left string AC.

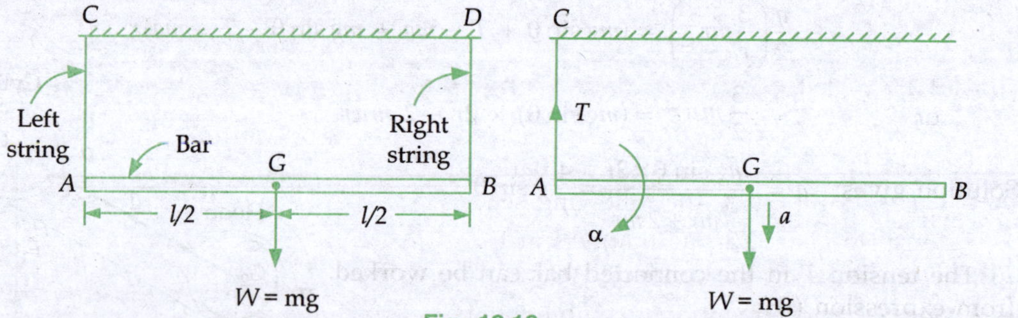

Fig. 19.18

Solution: When the string *BD* is cut, the bar *AB* rotates about its end *A*.

The bar is acted upon by the following set of forces:

(*i*) weight of the bar *W* = *mg* that acts at the point of its centre of gravity *G*.

(*ii*) tension *T* induced in the string *AC*.

For a rigid bar in plane motion (combination of translation and rotation), the equations of motion are

$$\Sigma F_x = ma_x \quad : \text{Here no force acts in the } x\text{-direction}$$

$$\therefore \quad a_x = 0$$

$$\Sigma F_y = ma_y \quad : mg - T = ma \qquad \ldots(i)$$

$$\Sigma M_G = I_G\alpha \quad : T \times \frac{l}{2} = I_G\alpha$$

For a thin uniform bar I_G (moment of inertia about the mass centre) = $\frac{ml^2}{12}$ and in the absence of slip, the angular acceleration α and linear acceleration *a* are connected as $a = \alpha r = \alpha l/2$. That gives

$$T \times \frac{l}{2} = \frac{ml^2}{12} \times \frac{2a}{l}; \quad T = \frac{1}{3}\,ma \qquad \ldots(ii)$$

Eliminating *T* from expressions (*i*) and (*ii*) we have

$$mg - \frac{1}{3}\,ma = ma \quad \text{or} \quad a = \frac{3}{4}\,g \quad \text{and} \quad \alpha = \frac{2a}{l} = \frac{3}{2}\,gl$$

Further :

$$T = m\,(g - a) = m \times \left(g - \frac{3}{4}g\right) = \frac{mg}{4}$$

EXAMPLE 19.22

A thin uniform bar of mass m and length l is held at an angle θ with the horizontal by means of two vertical inextensible strings, one at each end as shown in Fig. 19.19. Suddenly the string at the right end breaks, leaving the bar to swing. What will be the angular acceleration of the bar and tension induced in the string at the left end immediately after the string breaks?

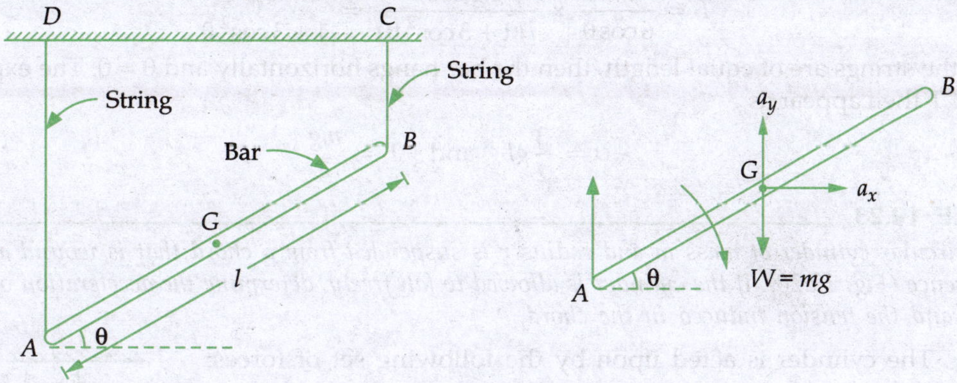

Fig. 19.19

Solution: When the string BC breaks, the bar AB rotates about its end A.

The bar is acted upon by the following set of forces:

(*i*) Weight of the bar $W = mg$ that acts at the point of its centre of gravity G.

(*ii*) Tension T induced in the string AD.

For a rigid bar in plane motion (combination of translation and rotation), the equations of motion are:

$$\Sigma F_x = ma_x \,; \quad \text{Here no force acts in the } x\text{-direction}$$

$$\therefore \quad a_x = 0$$

$$\Sigma F_y = mg_y \,; \quad mg - T = ma_y \qquad ...(i)$$

$$\Sigma M_G = I_G \alpha \,; \; T \times \frac{l}{2}\cos\theta = I_G \alpha \qquad ...(ii)$$

For a thin uniform bar I_G (moment of inertia about the mass centre = $ml^2/12$) and in the absence of slip, the angular acceleration α and linear acceleration a are connected as

$$a_y = \alpha r = \alpha \frac{l}{2}\cos\theta$$

The from expression (*i*) and (*ii*)

$$mg - T = m\alpha \frac{l}{2}\cos\theta \qquad ...(iii)$$

and

$$T\frac{l}{2}\cos\theta = \frac{ml^2}{12}\alpha \,; \quad T = \frac{ml}{6\cos\theta}\alpha \qquad ...(iv)$$

Substituting the value of T as defined by expression (*iv*) into expression (*iii*), we get

$$mg - \frac{ml}{6\cos\theta}\alpha = m\alpha\frac{l}{2}\cos\theta$$

or

$$mg = \alpha ml\left(\frac{\cos\theta}{2} + \frac{1}{6\cos\theta}\right) = \alpha ml\left(\frac{3\cos^2\theta + 1}{6\cos\theta}\right)$$

The expression for angular velocity α then appears as follows

$$\alpha = \frac{6g\cos\theta}{l(1+3\cos^2\theta)}$$

Then from expression (*iv*)

$$T = \frac{ml}{6\cos\theta} \times \frac{6g\cos\theta}{l(1+3\cos^2\theta)} = \frac{mg}{1+3\cos^2\theta}$$

Note: If the strings are of equal length, then the bar hangs horizontally and $\theta = 0$. The expressions for α and *T* then appear as

$$\alpha = \frac{3}{2}gl \quad \text{and} \quad T = \frac{mg}{4}$$

EXAMPLE 19.23

A right circular cylinder of mass m and radius r is suspended from a chord that is wound around its circumference (Fig. 19.20). If the cylinder is allowed to fall freely, determine the acceleration of its mass centre G and the tension induced in the chord.

Solution: The cylinder is acted upon by the following set of forces:

(*i*) weight of cylinder $W = mg$ that acts at the point of its centre of gravity *G*.

(*ii*) tension *T* induced in the chord.

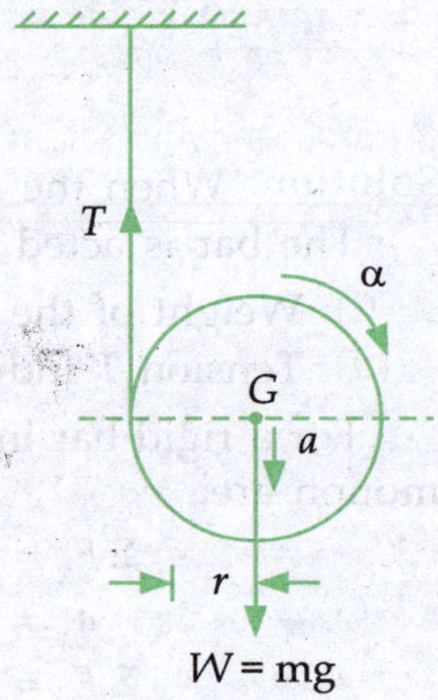

Fig. 19.20

For a rigid body in plane motion (combination of translation and rotation), the equation of motion are:

$\Sigma F_x = ma_x$: Here no force acts in the *x*-direction

$\therefore \quad a_x = 0$

$$\Sigma F_y = ma_y : \quad mg - T = ma \qquad \text{...(i)}$$

$$\Sigma M_G = I_G\alpha : \quad Tr = I_G\alpha$$

For a cylinder I_G (moment of inertia about the mass centre) $= \frac{1}{2}mr^2$ and in the absence of slip, the angular acceleration α and linear acceleration *a* are connected as $a = \alpha r$. That gives:

$$Tr = \frac{1}{2}mr^2 \times \frac{a}{r}; \quad T = \frac{1}{2}ma \qquad \text{...(ii)}$$

Eliminating *T* from expression (*i*) and (*ii*), we get

$$mg - \frac{1}{2}ma = ma; \quad mg = \frac{3}{2}ma$$

$$\therefore \quad a = \frac{2}{3}g$$

Further : $$T = m(g-a) = m\left(g - \frac{2}{3}g\right) = \frac{mg}{3}$$

EXAMPLE 19.24

A body of weight 8 N is suspended by a light rope wound round a pulley of weight 60 N and radius 30 cm. The other end of the rope is fixed to the periphery of the pulley. If the weight is moving downwards, make calculations for the acceleration of 8 N weight and tension in the string.

Solution: Refer Fig. 19.21 for the arrangement

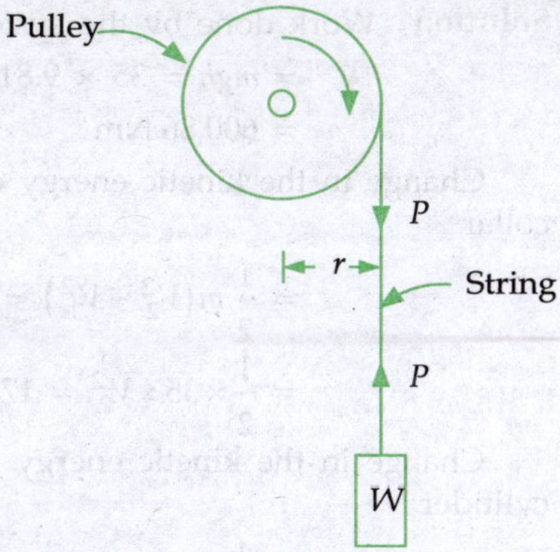

Fig. 19.21

Consider the motion of weight (coming downward):

Net downward force on the body

$$= W - P$$

If the body moves downward with an acceleration a, then from the equation $F = ma$, we have

$$W - P = \frac{W}{g} a \qquad \text{...(i)}$$

Consider the motion of pulley:

Torque on the pulley $T = I\alpha$

Noting that $T = P \times r$; $\alpha = \frac{a}{r}$, we get

$$P \times r = I \times \frac{a}{r}; \quad P = I \times \frac{a}{r^2}$$

If pulley is considered as a solid disc, then

$$I = \frac{W_0}{g}\frac{r^2}{2} \qquad \text{where } W_0 \text{ is the weight of pulley}$$

$$\therefore \quad P = \frac{W_0}{g}\frac{r^2}{2} \times \frac{a}{r^2} = \frac{W_0}{2g} a \qquad \text{...(ii)}$$

From expressions (*i*) and (*ii*),

$$W - \frac{W_0}{2g} a = \frac{W}{g} a; \quad W = \frac{a}{g}\left(W + \frac{W_0}{2}\right)$$

That gives: $$a = \frac{gW}{W + \frac{W_0}{2}}$$

Then from expression (*i*),

$$W - P = \frac{W}{g} \times \frac{gW}{W + \frac{W_0}{2}} = \frac{2W^2}{2W + W_0}$$

$$\text{or} \quad P = W - \frac{2W}{2W + W_0} = \frac{2W^2 + WW_0 - 2W^2}{2W + W_0} = \frac{WW_0}{2W + W_0}$$

Substituting the given data, we obtain

$$\text{Acceleration } a = \frac{9.81 \times 8}{8 + \frac{60}{2}} = \mathbf{2.065\ m/s^2}$$

$$\text{Tension induced in the string } P = \frac{8 \times 60}{2 \times 8 + 60} = \mathbf{6.31\ N}$$

EXAMPLE 19.25

An inextensible cord going around a homogeneous cylinder of mass 125 kg and radius 1 m holds a plate B of negligible mass.

A collar of 35 kg mass is released from rest in the position shown in Fig 19.22 and it drops on the plate. What will be the velocity of the collar after it has descended an additional distance of 0.5 m after striking the plate. It may be presumed that there is no rebound, i.e., the plate and collar move downwards locked together and the chord remains taut.

Solution : Work done by the system during fall of collar

$$= mgh = 35 \times 9.81 \times (1.25 + 0.5)$$
$$= 600.86 \text{ Nm}$$

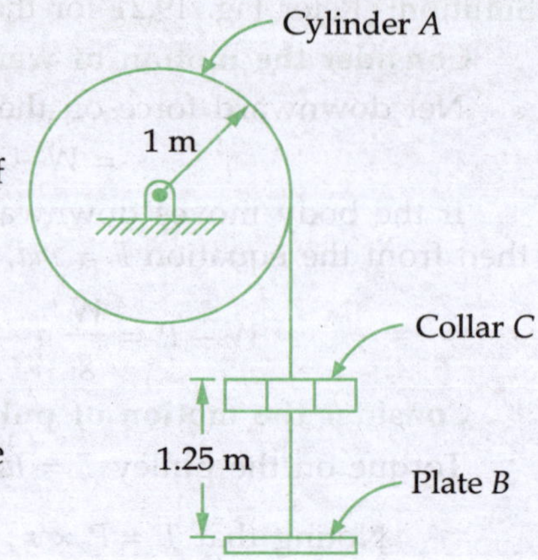

Fig. 19.22

Change in the kinetic energy due to translatory motion of collar

$$= \frac{1}{2} m\left(V_2^2 - V_1^2\right) = \frac{1}{2} mV_2^2 \qquad (\because V_1 = 0)$$
$$= \frac{1}{2} \times 35 \times V_2^2 = 17.5\, V_2^2$$

Change in the kinetic energy due to rotary motion of the cylinder

$$= \frac{1}{2} I(\omega_2^2 - \omega_1^2) = \frac{1}{2} I\omega_2^2 \qquad (\because \omega_1 = 0)$$
$$= \frac{1}{2}\left(\frac{1}{2} mr^2\right) \times \left(\frac{V_2}{r}\right)^2 = \frac{1}{2}\left(\frac{1}{2} \times 125 \times 1^2\right) \times \left(\frac{V_2}{1}\right)^2 = 31.25\, V_2^2$$

According to work – energy principle,

work done by the system = total change in kinetic energy of the system

$$600.86 = 17.5\, V_2^2 + 31.25\, V_2^2 = 48.75\, V_2^2$$

∴ Velocity attained by the collar

$$V_2 = \sqrt{\frac{600.86}{48.75}} = \mathbf{3.51 \text{ m/s}}$$

EXAMPLE 19.26

A homogeneous solid cylinder of weight 100 N rotates about its horizontal axis in frictionless bearings under the action of the weight of 10 N block which is carried by a rope wrapped round the cylinder as shown in Fig. 19.23. What will be the angular velocity of the cylinder two seconds after the motion starts? Assume diameter of cylinder as 100 cm.

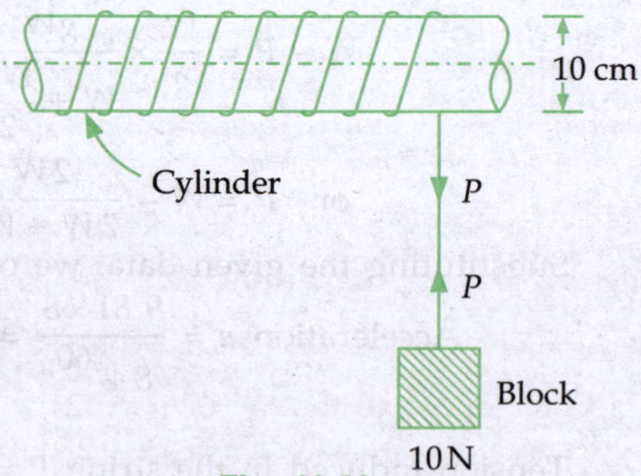

Fig. 19.23

Solution: When the weight moves downward, the cylinder starts rotating. The angular acceleration α of the cylinder and linear acceleration *a* of the system are correlated by the expression.

$$a = \alpha R \quad \text{where } R \text{ is the radius of the cylinder}$$
$$\therefore \quad a = 0.5\,\alpha$$

Considering downward motion of the block of 10 N weight and applying Newton's second law of motion, we have

$$\text{net force} = \text{mass} \times \text{acceleration}$$

$$10 - P = \frac{10}{9.81} \times a = \frac{10}{9.18} \times 0.5\alpha = 0.5097\,\alpha \qquad \ldots(i)$$

Considering the motion of cylinder,

$$\text{Mass moment of inertia } I = \frac{1}{2} mR^2 = \frac{1}{2} \times \frac{100}{9.81} \times (0.5)^2 = 1.274 \text{ kg m}^2$$

$$\text{Torque on the cylinder } T = P \times R = 0.5\,P$$

From Newton's second law: $T = I\alpha$

$$0.5\,P = 1.274 \times \alpha\,; \quad P = 2.548\,\alpha$$

Substituting this value of P in expression (*i*), we get

$$10 - 2.548\,\alpha = 0.5097\,\alpha$$

$\therefore$ Angular acceleration of rotation of cylinder,

$$\alpha = \frac{10}{2.548 + 0.5097} = 3.27 \text{ rad/s}^2$$

From the kinematic relation $\omega = \omega_0 + \alpha t$ and noting that initial angular velocity of cylinder $\omega_0 = 0$, we obtain

$$\omega = 0 + 3.27 \times 2 = \mathbf{6.54 \text{ rad/s}}$$

EXAMPLE 19.27.

A rope wrapped round a 1.25 m diameter drum is employed to lift a load of 2.5 kN as shown in Fig. 19.24. The drum is rotated by an electric motor that exerts a constant torque of 3.5 kN m. Presuming that the drum weighs 600 N and has 50 cm radius of gyration, make calculations for :

(*a*) *acceleration of the load*

(*b*) *tension in the rope, and*

(*c*) *time required to raise the load by a height of 15 m.*

The friction in the rope may be neglected.

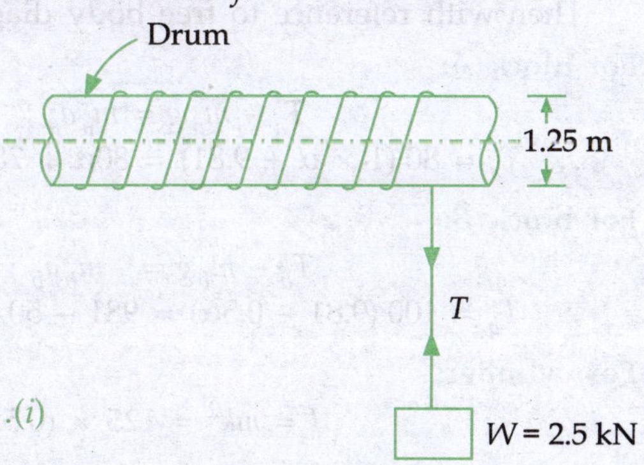

Fig. 19.24

Solution: Let T be the tension in the rope. Then consideration upward motion of load,

$$T - W = ma$$

or $$T - 2500 = \frac{2500}{9.81}a = 254.84\,a \qquad \ldots(i)$$

The net torque exerted on the drum is

= torque supplied by electric motor – torque due to rope tension

$= 3500 - T \times 0.625$

This torque equals: $I\alpha = I \times \frac{a}{r}$

$$= mk^2 \times \frac{a}{r}$$

$$= \frac{600}{9.81} \times (0.5)^2 \times \frac{a}{0.625} = 24.46\,a$$

$$\therefore \quad 3500 - 0.625\,T = 24.46\,a \qquad \ldots(ii)$$

From expression (*i*) and (*ii*)

$$3500 - 0.625\,(254.84\,a + 2500) = 24.46\,a$$

or $$3500 - 159.27\,a - 1562.5 = 24.46\,a$$

$\therefore$ Acceleration of load $a = \frac{3500 - 1562.5}{159.27 + 24.46} = \mathbf{10.54 \text{ m/s}^2}$

Substituting for acceleration $a = 10.54 \text{ m/s}^2$ in expression (*i*), we get

(*b*) Tension in the rope $T = 2500 + 254.84 \times 10.54$

$$= 2500 + 2686.01 = \mathbf{5186.01 \text{ N}}$$

(*c*) Using the equation, $s = ut + \frac{1}{2}at^2$, we have

$$15 = 0 + \frac{1}{2} \times 10.54\, t^2$$

Thus gives : $t = \left[\frac{15 \times 2}{10.54}\right]^{1/2}$ = **1.687 sec**

EXAMPLE 19.28.

A stepped cylinder shown in Fig. 19.25 has the dimensions R_1 = 50 cm and R_2 = 100 cm. The mass of the cylinder is 125 kg with radius of gyration 55 cm. Blocks A and B of masses 80 kg and 100 kg respectively are connected to the cylinder through the ropes. If the system is released from rest, determine the time required for block A to move by 2.5 meters.

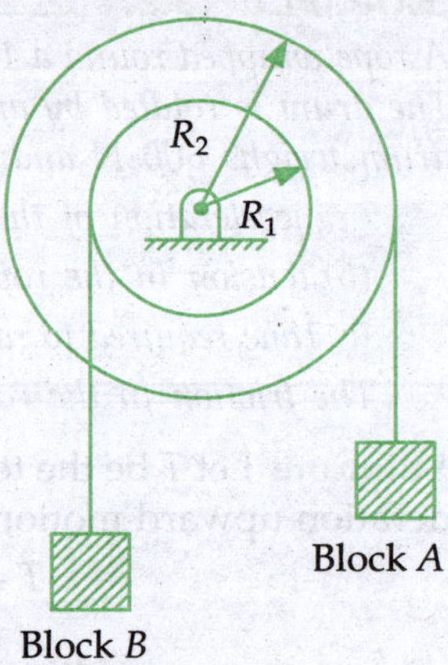

Fig. 19.25

Solution: Since block *B* is heavier than block *A*, the cylinder would rotate in the anti-clockwise direction and the block *A* would get lifted upwards

Then with reference to free body diagrams shown above,

For block *A*:

$$T_a - m_a g = m_a a_a \; ; \quad T_a = m_a (a_a + g)$$

$$\therefore \quad T_a = 80\,(1 \times \alpha + 9.81) = 80\alpha + 784.8 \quad (\because \alpha_a = \alpha_R)$$

For block *B*:

$$T_b - m_b g = -m_b a_b \; ; \quad T_b = m_b (g - a_b)$$

$$\therefore \quad T_b = 100\,(9.81 - 0.5\alpha) = 981 - 50\,\alpha$$

For cylinder:

$$I = mk^2 = 125 \times (0.55)^2 = 37.81 \text{ kg m}^2$$

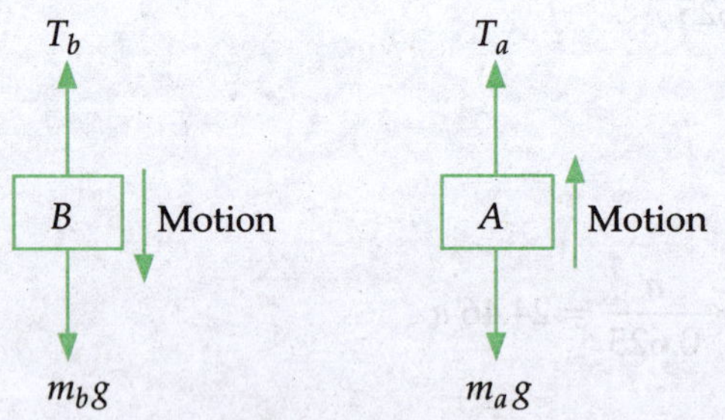

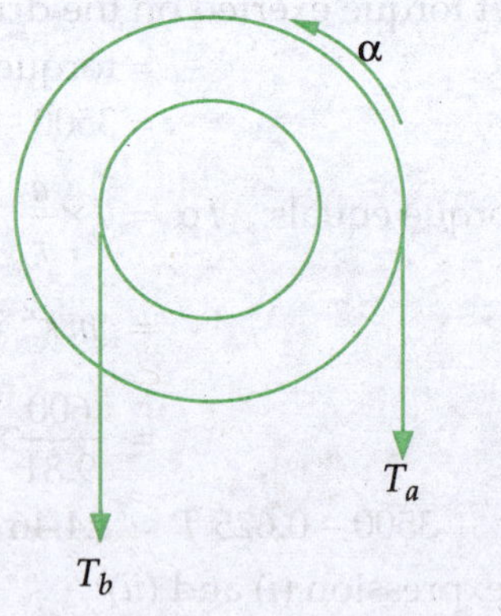

From Newton's second law; $T = I\alpha$

$$T_b R_1 - T_a R_2 = I\alpha$$

Substituting the appropriate values

$$(981 - 50\,\alpha) \times 0.5 - (80\,\alpha + 784.8) \times 1 = 37.81\,\alpha$$

$$(490.5 - 25\,\alpha) - (80\,\alpha + 784.8) = 37.81\,\alpha$$

$\therefore$ Angular acceleration of the cylinder $\alpha = -2.06$ rad/s^2

Corresponding to 2.5 m upward movement of block *A*, the angular displacement is $2.5/R_2$ = 2.5 radian

Then from the kinematic relation $\theta = \omega t + \frac{1}{2}\alpha t^2$, we have

$$2.5 = 0 \times t + \frac{1}{2} \times (2.06) \times t^2 \qquad (\because \omega = 0)$$

$\therefore$ Time required $t = \sqrt{\frac{2.5 \times 2}{2.06}}$ = **1.56 sec**

EXAMPLE 19.29.

When a compound disc with configuration shown in Fig. 19.26 is released from rest, it performs plane motion without slipping. Make calculations for :

(a) angular acceleration of the disc *(b) tension in the string*

(c) velocity of centre of disc when it has moved horizontally by 4 cm

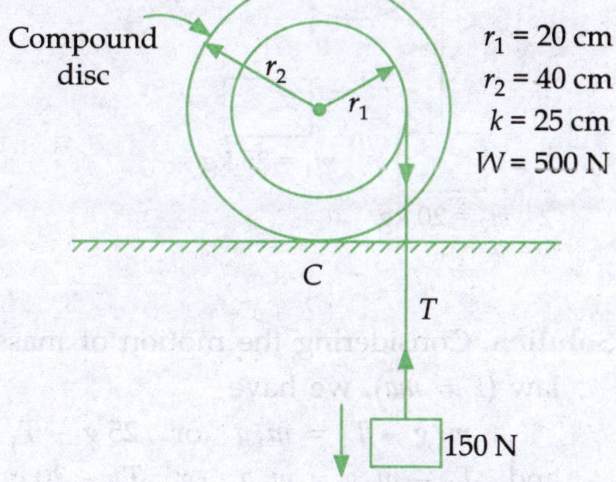

Fig. 19.26

Solution: Moment of inertia of the disc about the centre C of rotation

$$I = mk^2 + mh^2$$
$$= \frac{500}{9.81} \times 0.25^2 + \frac{500}{9.81} \times 0.4^2$$
$$= 3.185 + 8.155 = 11.34 \text{ kg m}^2$$

If α in rad/s^2 is the angular acceleration of the disc, than linear acceleration of block is,

$$a = r_1 \times \alpha = 0.2\,\alpha$$

Considering downward motion of block

$$W - T = ma$$

or $$150 - T = \frac{150}{9.81} \times 0.2\,\alpha$$

$\therefore$ $$T = 150 - \frac{150}{9.81} \times 0.2\,\alpha$$
$$= 150 - 3.058\,\alpha \qquad \ldots(i)$$

Taking moments about point C for the disc,

$$M = T \times r_1 = 0.2\,T$$

Since $M = I\alpha$, we get

$$0.2\,T = I\alpha = 11.34\,\alpha \qquad \ldots(ii)$$

From expressions (*i*) and (*ii*)

$$0.2\,(150 - 3.058\,\alpha) = 11.34\,\alpha$$

or $$30 - 0.6116\,\alpha = 11.34\,\alpha$$

That gives: Angular acceleation $\alpha = \frac{30}{11.34 \times 0.6116} = 2.518 \text{ rad/s}^2$

(*b*) Using expression (*ii*), tension in the string

$$T = \frac{11.34\,\alpha}{0.2} = \frac{11.34 \times 2.518}{0.2} = \textbf{142.77 N}$$

(*c*) Linear acceleration of the centre of disc

$$a = R_2 \times \alpha = 0.4 \times 2.518 = 1.0072 \text{ m/s}^2$$

Using the relation $v^2 - u^2 = 2as$ for the translation of disc,

$$v^2 - 0 = 2 \times 1.0072 \times 4 = 8.0576$$

$\therefore$ Velocity of centre of disc $v = \sqrt{8.0576} = \mathbf{2.84\ m/s}$

EXAMPLE 19.30

Two blocks of masses $m_1 = 25$ kg and $m_2 = 20$ kg are connected by a light inextensible string which passes over a 25 cm diameter pulley of 2.5 kg mass as shown in Fig. 19.27. Neglecting friction, work out the acceleration of the system and the tension in the string when the masses are released from rest. Assume radius of gyration of the pulley to be equal to its radius.

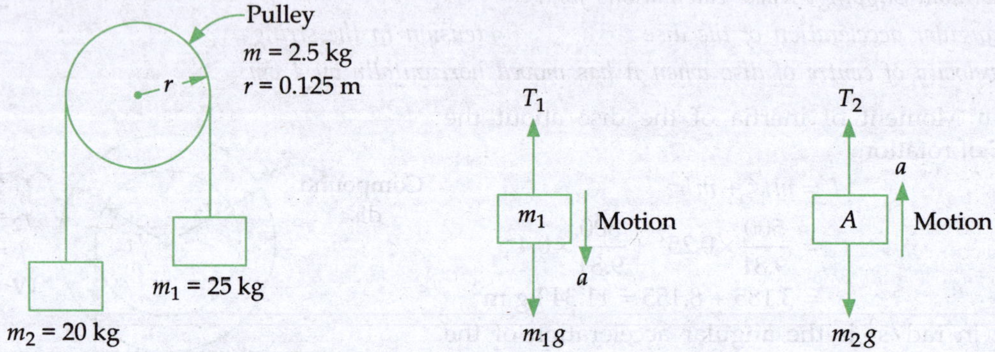

Fig. 19.27

Solution: Considering the motion of masses and applying Newton's second law ($F = ma$), we have

$$m_1 g - T_1 = m_1 a \quad \text{or} \quad 25g - T_1 = 25a \qquad ...(i)$$

and $$T_2 - m_2 g = m_2 a \quad \text{or} \quad T_2 - 20g = 20a \qquad ...(ii)$$

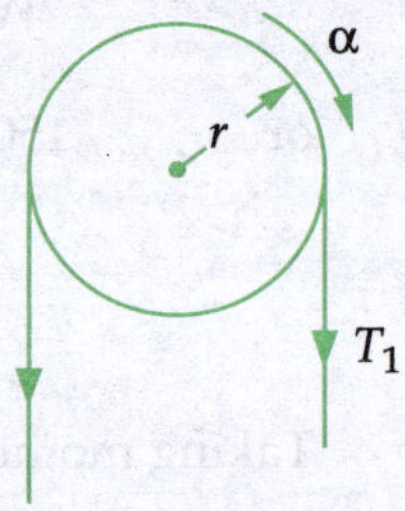

In the context of pulley

Resultant torque $T = (T_1 - T_2)\,r$

$= (T_1 - T_2) \times 0.125$

Mass moment of inertia $I = mk^2 = 2.5 \times (0.125)^2$

$= 0.0391$ kg m^2

Angular acceleration $\alpha = \dfrac{a}{r} = \dfrac{a}{0.125}$

From Newton's second law: $T = I\alpha$

$$(T_1 - T_2) \times 0.125 = 0.0391 \times \frac{a}{0.125}$$

or $$(T_1 - T_2) = 2.5a \qquad ...(iii)$$

Substituting the values of T_1 and T_2 from expression (*i*) and (*ii*) into expression (*iii*),

$$25g - 25a - (20a + 20g) = 2.5a; \quad 47.5a = 5g$$

$\therefore$ Acceleration of the system $a = \dfrac{5 \times 9.81}{47.5} = \mathbf{1.033\ m/s^2}$

Then from expression (*i*), $T_1 = 25(g - a) = 25(9.81 - 1.033) = \mathbf{219.42\ N}$

and from expression (*ii*), $T_2 = 20(g + a) = 20(9.18 + 1.033) = \mathbf{216.86\ N}$

EXAMPLE 19.31

Two bodies of weight 600 N and 400 N respectively are connected to the two ends of a chord which passes over a pulley of 80 cm diameter. The pulley may be assumed as a solid disc having mass 100 kg and 40 cm radius of gyration about its axis of rotation.

Make calculations for the torque which need to be applied to the pulley to raise the 600 N body with an acceleration of 1.25 m/s².

Neglect friction in the bearings of the pulley.

Solution: Refer Fig. 19.28 for the arrangement.

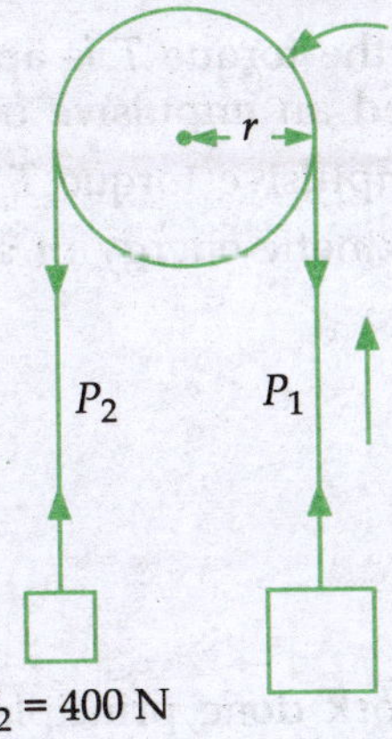

Fig. 19.28

Consider the motion of weight W_1:

The weight is being raised upwards with an acceleration a.

Net upward force on weight = $P_1 - W_1$

From the equation $F = ma$,

$$P_1 - W_1 = \frac{W_1}{g} \times a$$

$$\therefore \quad P_1 = \frac{W_1}{g}(g + a)$$

$$= \frac{600}{9.81}(9.81 + 1.25) = 676.45 \text{ N}$$

Consider the motion of weight W_2:

The weight is being lowered with an acceleration a.

Net downward force on weight = $W_2 - P_2$

From the equation $F = ma$

$$W_2 - P_2 = \frac{W_2}{g} \times a$$

$$\therefore \quad P_2 = \frac{W_2}{g}(g - a) = \frac{400}{9.81}(9.81 - 1.25) = 349.03 \text{ N}$$

Consider the motion of pulley:

Moment of inertia of the pulley $I = mk^2 = 100 \times (0.4)^2 = 16 \text{ kgm}^2$

Angular acceleration of pulley $\alpha = \frac{a}{r} = \frac{1.25}{0.4} = 3.125$ rad/s

Torque due to rotation of pulley $T_1 = I\alpha = 16 \times 3.125 = 50$ Nm

Torque acting on pulley due to difference in pulls in the two ropes,

$$T_2 = (P_1 - P_2)\, r = (6.76{:}45 - 349.03) \times 0.4 = 130.97 \text{ Nm}$$

$\therefore$ Torque which must be applied to the pulley

$$= 50 + 130.97 = \mathbf{180.97 \text{ Nm}}$$

REVIEW AND SUMMARY

1. Rotary motion is the motion of a rigid body in which the particles move in a circular path. The particles continue repeating the same path and have a fixed centre of rotation.
2. The product of mass moment of inertia and angular velocity of the body is called the moment of momentum or angular momentum

   Angular momentum = $I\omega$

3. Torque is the turning moment of force on a body on which it acts.
The rate of change of angular momentum of a rotating body equals the external torque applied on the body and takes place in the direction of the torque

$$T = I\frac{d\omega}{dt} = I\alpha$$

When the torque T is applied on the body for infinite small interval of time dt, the torque is called an impulsive or angular torque

Impulsive torque $T_i = I\,d\omega$ = change of angular momentum

4. Total kinetic energy of a rotating body

$$= KE \text{ in translation} + KE \text{ in rotation}$$

$$= \frac{1}{2}mV^2 + \frac{1}{2}I\omega^2$$

$$= \frac{1}{2}mV^2 + \frac{1}{2}(mk^2)\times\left(\frac{V}{r}\right)^2 = \frac{mV^2}{2}\left[1+\frac{k^2}{r^2}\right]$$

5. The work done principle connects the work done with change in kinetic energy. Further, work done results both from linear and angular displacement from the initial and final positions. Thus

$$Fs + T\theta = \frac{1}{2}m\left(V_2^2 - V_1^2\right) + \frac{1}{2}I\left(\omega_2^2 - \omega_1^2\right)$$

6. The principle of conservation of energy states that the sum of potential energy and kinetic energy of a rigid body moving under the influence of conservative forces is always a constant. That is

$$(PE + KE)_1 = (PK + KE)_2$$

REVIEW QUESTIONS

A. Conceptual and conventional questions :

1. Define rotary motion and explain the concept of torque.
2. Define moment of momentum (or angular momentum).
3. Set up an expression for the work done by torque.
4. State the law of conservation of angular momentum.
5. State Newton's laws of motion for angular motion
6. What is the kinetic energy of rotation?
7. Write the equations of motion for a rigid body in plane motion (combination of translation and rotation)
8. State the work-energy principle for a rigid body in plane motion
9. State the principle of conservation of energy for a rigid body.
10. A rotor of mass 20 kg and radius of gyration 1 m is rotating at a speed of 1440 revolutions per minute. If the rotor comes to rest in 2.5 minutes, determine the friction torque at the bearing. **(20.1 Nm)**

11. A uniform homogeneous cylinder of 0.01 kg mass and 8 cm radius rolls without slip along a horizontal level surface. If the translational velocity of the cylinder is 15 cm/s, calculate its total kinetic energy. **(0.000236 Nm)**

12. A flywheel of mass 5000 kg and having a radius of gyration 1 m loses its speed from 400 rpm to 280 rpm in 2 minutes. Calculate (a) the retarding torque acting on it (b) change in the kinetic energy during the above period and (c) change in its angular momentum during the period. **(523.6 Nm, 2236 kNm, 62835 Nm/s)**

13. Two blocks weighing 80 N and 30 N are supported at the ends of rope passing over the rough surface of a pulley mounted on a horizontal axle. The pulley may be assumed to be a solid disc with a weight of 50 N. Determine the tension on a two parts of the rope and the linear acceleration of the blocks. Neglect friction in the bearings of the pulley and the weight of the rope. **(50.37 N, 41.11 N, 3.63 m/s^2)**

B. Complete the following statements with most appropriate word/words :

1. The product of mass moment of inertia and the angular velocity of rolating body is known as ....
2. The .... of a rigid body consists of rectilinear motion and a rolling motion without slipping.
3. The torque applied to a body for infinite small interval of time is called ....
4. If the external torque acting on a system is zero, the .... of the system remains constant.
5. For a solid sphere of mass m and radius r, the radius of gyration is ....
6. The work-energy principle connects the .... with change in ....
7. The work done by torque is given by the product of .... and ....
8. Corresponding to force equation $F = ma$ in linear motion, the torque equation in rotary motion is ....
9. In the expression $\frac{1}{2} I\omega^2$ for the kinetic energy of a rigid body in rotation, I represents, the .... moment of inertia of the body about the ....
10. A flywheel has a mass moment of inertia of 10 kgm^2 about the axis of rotation. If it runs with angualr velocity of 100 rad/s, its kinetic energy in kNm would be ....

Answers :

1. angular momentum or moment of momentum 2. plane motions 3. impulsive torque 4. angular momentum 5. 0.632 r 6. Work done and kinetic energy 7. torque, angular displacement 8. $T = I\alpha$ 9. mass, axis of rotation 10. 50

Power Transmission : Methods and Devices 20

Learning objectives : A study of the subject matter included in this chapter will make the readers familiar with

- *open belt and crossed belt drive*
- *ratio of tensions for the flat belt and V-belt drives*
- *centrifugal tension and initial tension in belt drive system*
- *gear drive and commonly used forms of toothed gearing*
- *gear terminology, circular pitch, diametral pitch and module*
- *gear trains; their configuration and application*

Attention has been focused in this chapter on the methods and devices commonly used for power transmission from one shaft to another. Brief is given about belts and ropes, chains, gears and clutches.

- *Belts and ropes* are used when the distance between the axes of the two shafts to be connected is considerable. Such connectors are non-rigid and undergo strain while in motion. These devices are called non-positive drive because of the possibility of slip occurring between the belt and pulley.
- *Chain drive* is used when the distance between the shaft centres is short and no slip is required. These connectors are referred to as a positive or non-slip drive.
- *Gears* are used for transmitting motion and power when the distance between the driving and driven shafts is relatively small, and when a constant velocity ratio is desired.
- *Clutches* are used for power transmission between co-axial shafts.

20.1. BELT AND ROPE DRIVE

The transmission of power in factories from one rotating shaft to another that lies at a considerable distance is achieved through belts and ropes. The shafts are fitted with pulleys, the belt is wrapped round the pulleys and its ends are connected to form an endless connector. The belts and the pulley remain in contact by frictional grip.

With reference to Fig 20.1, the pulley A which is connected to the rotating shaft is called the *driver*. The pulley B that needs to be driven is termed as *follower*. When the driver rotates, it carries the belt because of friction that exists between the pulley and the belt. The frictional resistance develops all along the contact surfaces. That makes the belt carry the follower which too starts rotating. The driving pulley pulls the belt from one side (called tension side) and delivers it to the other side (called slack side). The tension T_1 in the belt on the tension side is more than tension T_2 on the slack side.

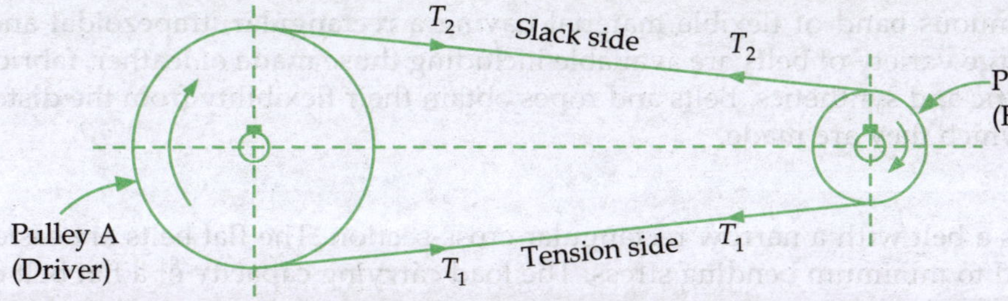

Fig. 20.1. Belt drive-open system

Two parallel shafts may be connected by *open belt* (Fig. 20.1) or by *cross belt* (Fig. 20.2). In the open belt system, the rotation of both the pulleys is in the same direction. If a crossed belt system is used, the rotation of pulleys will be in the opposite direction.

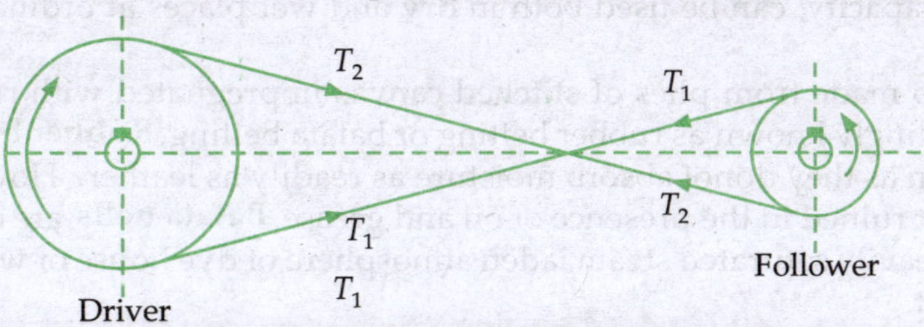

Fig. 20.2. Belt drive-cross system

The angle of contact in this system of drive is more and accordingly it can transmit more power than open belt drive system. However, the belt wears out fast at the places where crossing takes place in the crossed belt system. Further, for small centre distance, the belt is not fully utilised because of its larger slanted run off.

When a number of pulleys are used to transmit power from one shaft to another, then a *compound* drive is used.

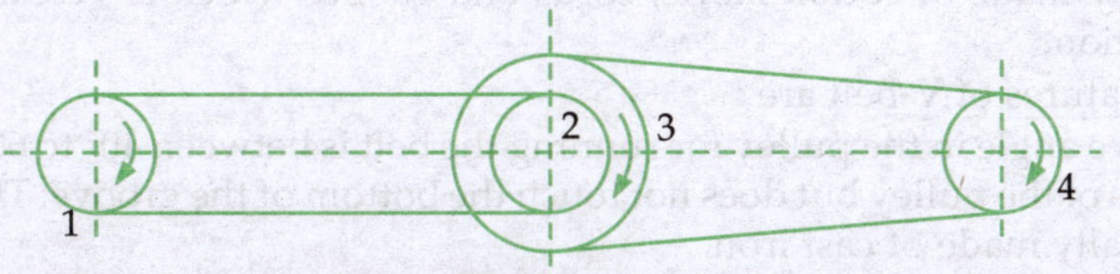

Fig. 20.3. Belt drive-compound system

With reference to Fig. 20.3, pulley 1 drives pulley 2. Since pulleys 2 and 3 are keyed to the same shaft, the pulley 3 is also driven by pulley 1. Subsequently, pulley 3 drives pulley 4. The arrangement conforms to extended open system and all the pulleys rotate in the same direction.

20.2. BELT MATERIAL AND SECTIONS

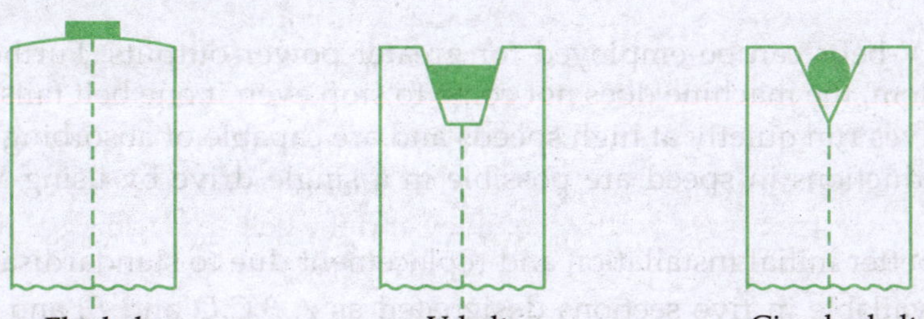

Fig. 20.4. Belt cross-section

A belt is a continuous band of flexible material having a rectangular, trapezoidal and round cross-section. Large variety of belts are available including those made of leather, fabric, rubber impregnated fabric and synthetics. Belts and ropes obtain their flexibility from the distortion of the material of which they are made.

20.2.1. Flat belt

A flat belt is a belt with a narrow rectangular cross-section. The flat belts are easier to use and are subjected to minimum bending stress. The load carrying capacity of a flat belt depends on its width.

Material used for belt is generally leather of various types having ultimate tensile strength between 4.5 to 7 N per cm width. For heavy duty, two or three plies of leather are cemented and pressed one above the other. Such belts are called double or triple ply belts. Leather belts have the best pulling capacity, can be used both in dry and wet places at ordinary temperature, but are costly.

Belting is also made from plies of stitched canvas impregnated with rubber or balata gum and is correspondingly known as rubber belting or balata belting. **Rubber belts** give best results in damp condition as they donot absorb moisture as readily as leather. However, they are quite expensive and get ruined in the presence of oil and grease. **Balata belts** are acid and water proof and are used in heavily saturated steam laden atmosphere of dye house or where chemical fumes are likely to effect.

Steel belts are immune from stretching and slipping, remain unaffected by dampness or heat, and transmit more power per cm width.

The mid section of flat pulleys is provided a slight dwell to prevent the belt from running off the pulley. This is referred to as **crowning** of pulleys and may be rounded or tapered.

20.2.2. V-belts

A V-belt is a belt of trapezoid section running on pulleys with grooves cut to match the belt. V-belts are usually made of cotton fabric, cords and rubber which is vulcanised in moulds of desired cross-section.

The salient features of V-belt are :

(i) The groove angle in the pulley for running the belt is between 40° to 60°. The belt rests an both sides of the pulley but does not touch the bottom of the groove. The grooved pulleys are generally made of cast iron.

(ii) The wedging action between the belt and sides of the groove increases the frictional grip and that reduces the chances of slipping ; there is no possibility of belt coming out of the groove. Due to reduced slipping, V-belts offer a more positive drive.

(iii) V-belts are used when the distance between the shafts is short, transmission members are large and large power is to be transmitted.
The ideal centre distance for *V*-belt drive is 1.25 to 1.5 times the diameter of the larger pulley.

(iv) Multiple V-belts can be employed for greater power outputs. Further in the multiple drive system, the machine does not come to stop even if one belt fails.

(v) V-belt drives run quietly at high speeds and are capable of absorbing high shock.

(vi) Larger reductions in speed are possible in a single drive by using V-belts over small pulleys.

(vii) There is better initial installation and replacement due to standardisation of V-belts.

V-belts are available in five sections designated as *A,B,C,D* and *E*, and these are used in order of increasing loads. That is, section *A* is used for light loads only, and section *E* is used for heavy duty machines. The angle of V-belt for all sections is about 40-degree.

20.2.3. Round belt

The round cross-section belts are employed when low power is to be transmitted such as in instruments, house hold appliances, table top machine tools and machinery of the clothing industry.

These belts are made of leather, canvas and rubber. Their diameter is usually within the range 4 mm to 8 mm, and the allowable ratio of the diameter of smaller pulley to the belt diameter is about 20.

20.3 VELOCITY RATIO (*VR*)

The velocity ratio of belt drives is the ratio of the speed of driven pulley to that of driving pulley.

Let d_1, d_2 = diameters of driver and driven pulleys

ω_1, ω_2 = angular velocities of driver and driven pulleys

N_1, N_2 = rotational speeds of driver and driven pulleys; expressed in revolutions per second (rps)

$$\text{Linear speed of driving pulley} = \omega_1 \times \frac{d_1}{2}$$

$$\text{Linear speed of driven pulley} = \omega_2 \times \frac{d_2}{2}$$

Presuming that the belt is inelastic and there is sufficient friction to prevent any slip between the belt and pulleys, both the pulleys will have the same linear speed. That is

$$\omega_1 \times \frac{d_1}{2} = \omega_2 \times \frac{d_2}{2}$$

$$\therefore \quad \frac{\omega_2}{\omega_1} = \frac{d_1}{d_2} \quad \text{or} \quad \frac{2\pi N_2}{2\pi N_1} = \frac{d_1}{d_2}$$

That gives :

$$\frac{N_2}{N_1} = \frac{d_1}{d_2} \qquad \text{...(20.1)}$$

The ratio N_2/N_1 is measure of the velocity ratio of the rotating pulleys. Further, it is apparent from equation 20.1 that *speed of a pulley is inversely proportional to its diameter*.

If thickness t of the belt is taken into account, then

$$v = \omega_1 \times \frac{d_1 + t}{2} = \omega_2 \times \frac{d_2 + t}{2}$$

$$\frac{\omega_2}{\omega_1} = \frac{d_1 + t}{d_2 + t} \quad \text{or} \quad \frac{2\pi N_2}{2\pi N_1} = \frac{d_1 + t}{d_2 + t}$$

$$\therefore \text{ Velocity ratio } \quad \frac{N_2}{N_1} = \frac{d_1 + t}{d_2 + t} \qquad \text{...(20.2)}$$

Slip and its effect on velocity ratio : When the frictional grip between the belt and pulley becomes insufficient, there occurs some forward motion of the driver without carrying the belt with it. The relative motion between the pulley and belt is called slip. The difference between the linear speeds of the pulley rim and belt is the measure of slip.

Let S_1 = percentage slip between driver and the belt

S_2 = percentage slip between belt and the follower (driven pulley)

Linear velocity of driving pulley

$$v_1 = \omega_1 \times \frac{d_1}{2}$$

Due to slip between the driving pulley and the belt, the velocity of belt will decrease

$$\text{Velocity of belt} = v_1 - v_1 \times \frac{S_1}{100} = v_1\left(1 - \frac{S_1}{100}\right)$$

This will also be the velocity of belt as it passes over the driven pulley. As there is slip at the driven pulley also, the velocity of the follower pulley will become less.

Linear speed of the driven pulley

$$= v_1\left(1 - \frac{S_1}{100}\right) - v_1\left(1 - \frac{S_1}{100}\right) \times \frac{S_2}{100} = v_1\left(1 - \frac{S_1}{100}\right)\left(1 - \frac{S_2}{100}\right)$$

$$= v_1\left(1 - \frac{S_1 + S_2 + 0.01 S_1 S_2}{100}\right) = v_1\left(1 - \frac{S}{100}\right) = \omega_1 \times \frac{d_1}{2}\left(1 - \frac{S}{100}\right)$$

where $S = S_1 + S_2 + 0.01\, S_1 S_2$ is the percentage of total effective slip.

The linear speed of the driven pulley is also given by :

$$v_2 = \omega_2 \times \frac{d_2}{2}$$

$$\therefore \quad \omega_2 \times \frac{d_2}{2} = \omega_1 \times \frac{d_1}{2} \times \left(1 - \frac{S}{100}\right)$$

or

$$\frac{\omega_2}{\omega_1} = \frac{d_1}{d_2}\left(1 - \frac{S}{100}\right)$$

or

$$\frac{2\pi N_2}{2\pi N_1} = \frac{d_1}{d_2}\left(1 - \frac{S}{100}\right)$$

$$\therefore \quad \text{Velocity ratio } \frac{N_2}{N_1} = \frac{d_1}{d_2}\left(1 - \frac{S}{100}\right) \qquad ...(20.3)$$

It is apparent from equation 20.3 that the velocity ratio decreases due to slipping of belt.

If thickness t of the belt is also taken into account, then

$$\frac{N_2}{N_1} = \frac{d_1 + t}{d_2 + t} \times \left(1 - \frac{S}{100}\right) \qquad ...(20.4)$$

Creep : When the belt passes from the slack side to the tight side a certain portion of the belt extends, and when the belt passes from the tight to slack side the belt contracts. Due to these changes in length, there is relative motion between the belt and pulley surfaces.

The relative motion is termed as **creep** of the belt. Like slip, creep also reduces the velocity of the belt drive system.

EXAMPLE 20.1

An engine shaft running at 240 rpm is required to drive a machine shaft by means of a belt. The pulley on the engine shaft is 600 mm in diameter. Determine diameter of the pulley on the machine shaft if it is to run at 360 rpm under the following conditions :

(a) *the belt thickness is negligible and there is no slip,*

(b) *the belt thickness is 5 mm and slip is neglected,*

(c) *the belt is 5 mm thick and a slip of 2 per cent is allowed between the belt and each pulley.*

Solution : (*a*) When the belt thickness is neglected and there is no slip, we have

$$\frac{N_2}{N_1} = \frac{d_1}{d_2} \quad \therefore \quad d_2 = d_1 \times \frac{N_1}{N_2} = 600 \times \frac{240}{360} = \mathbf{400\ mm}$$

(*b*) When belt thickness $t = 5$ mm is considered and slip is neglected, we have

$$\frac{N_2}{N_1} = \frac{d_1 + t}{d_2 + t}$$

$$d_2 + t = (d_1 + t) \times \frac{N_1}{N_2} = (600 + 5) \times \frac{240}{360} = 403.3 \text{ mm}$$

$$\therefore \quad d_2 = 403.3 - 5 = \mathbf{398.3\ mm}$$

(*c*) When both belt thickness and slip are considered

$$\frac{N_2}{N_1} = \frac{d_1 + t}{d_2 + t}\left(1 - \frac{S}{100}\right)$$

Effective slip, $S = S_1 + S_2 - 0.01\, S_1\, S_2 = 2 + 2 - 0.01 \times 2 \times 2 = 3.96\%$

$$d_2 + t = (d_1 + t) \times \frac{N_1}{N_2} \times \left(1 - \frac{S}{100}\right) = (600 + 5) \times \frac{240}{360} \times \left(1 - \frac{3.96}{100}\right) = 387.23 \text{ mm}$$

$$\therefore \quad d_2 = 387.23 - 5 = \mathbf{382.2\ mm}$$

EXAMPLE 20.2

A shaft of 48 cm diameter is driven with the help of belt by an engine turning at 200 rev/min. The diameter of the engine pulley is 30 cm, the belt is 5 mm thick and a slip of 3% between the belt and each pulley is allowed. Determine the percentage of total effective slip and the speed of driven shaft.

Solution : Effective slip $S = S_1 + S_2 - 0.01\, S_1\, S_2 = 3 + 3 - 0.01 \times 3 \times 3 = \mathbf{5.91\%}$

When both belt thickness and slip are considered,

$$\frac{N_2}{N_1} = \frac{d_1 + t}{d_2 + t}\left(1 - \frac{S}{100}\right)$$

∴ Speed of the driven shaft,

$$N_2 = N_1 \times \frac{d_1 + t}{d_2 + t} \times \left(1 - \frac{S}{100}\right) = 200 \times \frac{300 + 5}{480 + 5}\left(1 - \frac{5.91}{100}\right) = \mathbf{1183.4\ rev/min}$$

EXAMPLE 20.3

Two pulleys A and B, with the sum of their diameters 100 mm, are connected by a belt. The driving pulley A rotates at 120 rpm whilst the driven pulley B turns 2400 rev/min. Presuming that the belt is 10 mm thick and the slip in the whole drive system is 5 percent, make calculations for the diameter of each pulley.

Solution : When the belt thickness and slip are considered,

$$\frac{N_a}{N_b} = \frac{D_b + t}{D_a + t} \times \left(1 - \frac{S}{100}\right)$$

where t is the belt thickness and S is the percentage slip in the entire drive system.

$$\therefore \quad \frac{1200}{2400} = \frac{D_b + 10}{D_a + 10} \times \left(1 - \frac{5}{100}\right)$$

$$0.5 = \frac{D_b + 10}{D_a + 10} \times 0.95 \; ; \quad \frac{D_b + 10}{D_a + 10} = 0.5263 \qquad \text{...}(i)$$

Further $\quad D_a + D_b = 100 \quad$ or $\quad D_b = 100 - D_a \qquad$ ...(*ii*)

From identities (*i*) and (*ii*),

$$\frac{100 - D_a + 10}{D_a + 10} = 0.5263$$

or $$110 - D_a = 0.5263\ D_a + 5.263$$

$$D_a = \frac{110 - 5.263}{1.5263} = \mathbf{68.62\ mm}$$

and $$D_b = 100 - 68.62 = \mathbf{31.83\ mm}$$

20.4. LENGTH OF BELT

(*i*) *Open belt system* : Refer Fig 20.5 which shows the open belt system in which both the driving and driven pulleys rotate in the same direction.

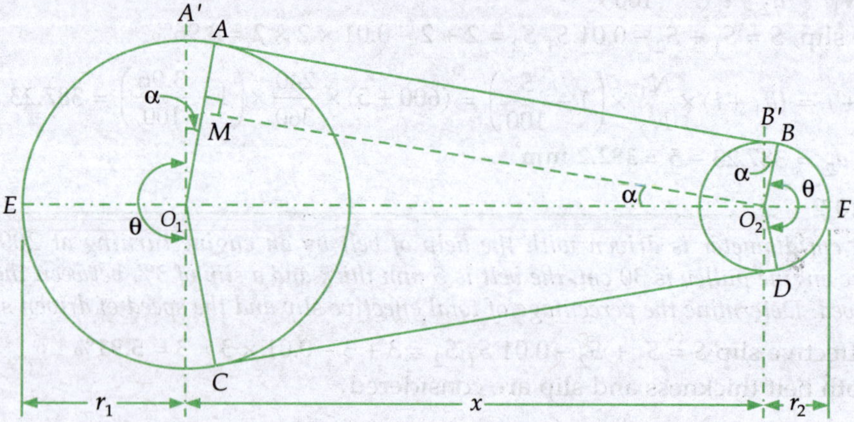

Fig. 20.5 Length of an open belt system

Let r_1, r_2 = radius of the two pulleys

x = distance between O_1 and O_2 ; the centres of the two pulleys.

The belt leaves the bigger pulley at A and C, and the smaller pulley at B and D.

A line O_2M drawn parallel to AB will be perpendicular to O_1A also.

$$\angle A'O_1A = \angle B'O_2B = \angle O_1O_2M = \alpha$$

$$\sin\alpha = \frac{r_1 - r_2}{x}$$

Since α is very small,

$$\sin\alpha = \alpha = \frac{r_1 - r_2}{x}$$

Length of the belt, $l = 2(\text{arc } EA + AB + \text{arc } BF)$

$$\text{arc } EA = r_1 \times \left(\frac{\pi}{2} + \alpha\right) \quad \text{and arc } BF = r_2 \times \left(\frac{\pi}{2} - \alpha\right)$$

$$AB = MO_2 = \sqrt{(O_1O_2)^2 - (O_1M)^2}$$

$$= \sqrt{x^2 - (r_1 - r_2)^2} = x\sqrt{1 - \left(\frac{r_1 - r_2}{x}\right)^2} = x\left[1 - \left(\frac{r_1 - r_2}{x}\right)^2\right]^{1/2}$$

Since $\left(\frac{r_1 - r_2}{x}\right)^2$ is very small, Binomial expansion would give

$$AB = x\left[1 - \frac{1}{2}\left(\frac{r_1 - r_2}{x}\right)^2\right] = x\left[1 - \frac{(r_1 - r_2)^2}{2x^2}\right]$$

$$\therefore \quad l = 2\left[r_1\left(\frac{\pi}{2} + \alpha\right) + x\left\{1 - \frac{(r_1 - r_2)^2}{2x^2}\right\} + r_2\left(\frac{\pi}{2} - \alpha\right)\right]$$

$$= 2\left[\frac{\pi}{2}(r_1 + r_2) + \alpha(r_1 - r_2) + \left\{x - \frac{(r_1 - r_2)^2}{2x}\right\}\right]$$

$$= \pi(r_1 + r_2) + 2\alpha(r_1 - r_2) + 2x - \frac{(r_1 - r_2)^2}{x}$$

Substituting the value $\alpha = \frac{r_1 - r_2}{x}$, we get

$$l = \pi(r_1 + r_2) + 2\frac{r_1 - r_2}{x} \times (r_1 - r_2) + 2x - \frac{(r_1 - r_2)^2}{x}$$

$$= \pi(r_1 + r_2) + \frac{(r_1 - r_2)^2}{x} + 2x \qquad \text{...(20.5)}$$

(ii) Crossed belt system : Refer Fig 20.6 which shows the crossed belt system in which the driving and the driven pulleys rotate in opposite directions.

The belt leaves the bigger pulley at A and C and the smaller pulley at B and D. A line O_2M drawn parallel at AB will be perpendicular to O_1A also.

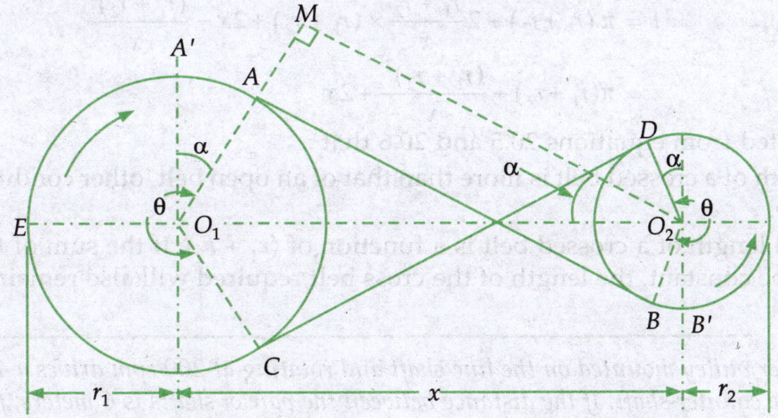

Fig. 20.6. Length of a crossed belt system

Let : r_1 and r_2 = radius of the two pulleys

x = distance between O_1 and O_2 ; the centres of the two pulleys

$\angle A'O_1A = \angle B'OB = \angle O_1O_2M = \alpha$

$$\sin\alpha = \frac{O_1M}{O_1O_2} = \frac{O_1A + AM}{O_1O_2} = \frac{r_1 + r_2}{x}$$

Since α is very small,

$$\sin\alpha = \alpha = \frac{r_1 + r_2}{x}$$

length of belt,

$$l = 2(\text{arc } EA + AB + \text{arc } BF)$$

$$\text{arc } EA = r_1\left(\frac{\pi}{2}+\alpha\right) \quad \text{and} \quad \text{arc } BF = r_2\left(\frac{\pi}{2}+\alpha\right)$$

$$AB = MO_2 = \sqrt{(O_1O_2)^2 - (O_1M)^2} = \sqrt{x^2 - (r_1 + r_2)^2}$$

$$= x\sqrt{1-\left(\frac{r_1 + r_2}{x}\right)^2} = x\left[1-\left(\frac{r_1 + r_2}{x}\right)^2\right]^{1/2}$$

Through Binomial expansion,

$$AB = x\left[1-\frac{1}{2}\left(\frac{r_1 + r_2}{x}\right)^2\right] = x\left[1-\frac{(r_1 + r_2)^2}{2x^2}\right]$$

$$\therefore \quad l = 2\left[r_1\left(\frac{\pi}{2}+\alpha\right)+x\left\{1-\frac{(r_1 + r_2)^2}{2x^2}\right\}+r_2\left(\frac{\pi}{2}+\alpha\right)\right]$$

$$= 2\left[\frac{\pi}{2}(r_1 + r_2)+\alpha(r_1 + r_2)+\left\{x-\frac{(r_1 + r_2)^2}{2x}\right\}\right]$$

$$= \pi(r_1 + r_2)+2\alpha(r_1 + r_2)+2x-\frac{(r_1 + r_2)^2}{x}$$

Substituting the value of $\alpha = \frac{r_1 + r_2}{x}$, we get

$$l = \pi(r_1 + r_2)+2\frac{r_1 + r_2}{x}\times(r_1 + r_2)+2x-\frac{(r_1 + r_2)^2}{x}$$

$$= \pi(r_1 + r_2)+\frac{(r_1 + r_2)^2}{x}+2x \qquad \text{...(20.6)}$$

It may be noted from equations 20.5 and 20.6 that :

(*i*) The length of a crossed belt is more than that of an open belt, other conditions remaining the same.

(*ii*) The total length of a crossed belt is a function of $(r_1 + r_2)$. If the sum of the radii of two pulleys be constant, the length of the cross belt required will also remain constant.

EXAMPLE 20.4

A 600 mm diameter pulley mounted on the line shaft and rotating at 200 rpm drives a 400 mm diameter pulley fitted on the counter shaft. If the distance between the pair of shafts is 6 meters, find the length of the belt connecting the shaft for the (a) open belt arrangement and (b) crossed belt arrangement.

Solution : r_1 = radius of larger pulley = 300 mm = 0.3 m

r_2 = radius of smaller pulley = 200 mm = 0.2 m

x = distance between centres of the pulley = 6 m

(*a*) Using open belt arrangement,

$$l_{open} = \pi(r_1 + r_2)+\frac{(r_1 - r_2)^2}{x}+2x = \pi(0.3+0.2)+\frac{(0.3-0.2)^2}{6}+2\times 6$$

$$= 1.57 + 0.0017 + 12 = \mathbf{13.572\ m}$$

(*b*) Using crossed belt arrangement,

$$l_{cross} = \pi(r_1 + r_2) + \frac{(r_1 + r_2)^2}{x} + 2x = \pi(0.3 + 0.2) + \frac{(0.3 + 0.2)^2}{6} + 2 \times 6$$

$$= 1.57 + 0.0416 + 12 = \mathbf{13.612\ m}$$

EXAMPLE 20.5

Two parallel shafts 5 m apart are provided with 600 mm and 400 mm diameter pulleys which are connected by a crossed belt. It is desired to reverse the direction of rotation of the driven pulley by changing over to an open belt drive system. Calculate the reduction in the length of belt required.

Solution : Let r_1 and r_2 be the radii of the larger and smaller pulleys, and x be the distance between the centres of the pulleys. Using crossed belt system,

$$l_{cross} = \pi(r_1 + r_2) + \frac{(r_1 + r_2)^2}{x} + 2x$$

Using open belt system,

$$l_{open} = \pi(r_1 + r_2) + \frac{(r_1 - r_2)^2}{x} + 2x$$

Then

$$l_{cross} - l_{open} = \frac{(r_1 + r_2)^2}{x} - \frac{(r_1 - r_2)^2}{x}$$

$$= \frac{(r_1^2 + r_2^2 + 2r_1 r_2) - (r_1^2 + r_2^2 - 2r_1 r_2)}{x}$$

$$= \frac{4 r_1 r_2}{x} = \frac{4 \times 300 \times 200}{5 \times 10^3} = \mathbf{48\ mm}$$

20.5 RATIO OF TENSIONS

Fig. 20.7 shows a flexible belt resting over the flat rim of a stationary pulley. The tensions T_1 and T_2 are such that the motion is impending (just to take place) between the belt and the pulley. Considering the impending motion to be clockwise relative to the drum, the tension T_1 is more than T_2. It is to be noted that only a part of the belt is in contact with the pulley. The angle subtended at the centre of the pulley by the position of belt in contact with it is called the *angle of contact* or the *angle of lap*

Angle of contact $\theta = \angle AOB$

Let attention be focused on small element *RS* of the belt which subtends an angle $\delta\theta$ at the centre. The segment *RS* is acted upon by the following set of forces :

- tension T in the belt acting tangentially at S,
- tension $(T + \delta T)$ in the belt acting tangentially at R
- normal reaction R exerted by the pulley rim, and
- friction force $F = \mu R$ which acts against the tendency to slip and is perpendicular to normal reaction R.

Considering equilibrium of forces in the radial (vertical) direction,

$$R = (T + \delta T)\sin\frac{\delta\theta}{2} + T\sin\frac{\delta\theta}{2}$$

For small values of $\delta\theta$; $\sin\frac{\delta\theta}{2} \rightarrow \frac{\delta\theta}{2}$

$$R = (T + \delta T)\frac{\delta\theta}{2} + T\frac{\delta\theta}{2} = T\frac{\delta\theta}{2} + \delta T\frac{\delta\theta}{2} + T\frac{\delta\theta}{2}$$

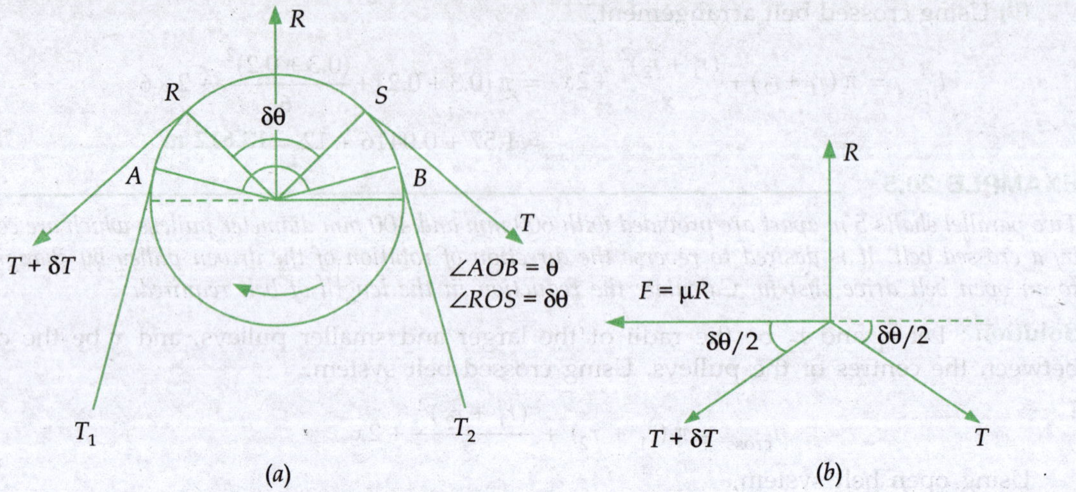

Fig. 20.7. Ratio of tension on tight and block side.

The term $\delta T \dfrac{\delta\theta}{2}$ is small in magnitude and can be neglected

$$R = T\frac{\delta\theta}{2} + T\frac{\delta\theta}{2} = T\,\delta\theta \qquad ...(i)$$

Considering equilibrium of forces in tangential (horizontal) direction,

$$\mu R = (T + \delta T)\cos\frac{\delta\theta}{2} - T\cos\frac{\delta\theta}{2}$$

For small values of $\delta\theta$; $\cos\dfrac{\delta\theta}{2} \to 1$

$$\mu R = (T + \delta T) - T = \delta T$$

$$R = \frac{\delta T}{\mu} \qquad ...(ii)$$

From expressions (*i*) and (*ii*)

$$T\,\delta\theta = \frac{\delta T}{\mu}$$

Separating the variables and integrating between the limit $T = T_2$ at $\theta = 0$ and $T = T_1$ at $\theta = \theta$, we get,

$$\int_{T_2}^{T_1} \frac{\delta T}{T} = \mu\int_{o}^{\theta} \delta\theta$$

$$\log_e \frac{T_1}{T_2} = \mu\theta \quad \therefore \quad \frac{T_1}{T_2} = e^{\mu\theta} \qquad ...(20.7)$$

When two pulley of unequal diameters are connected by open belt drive, the slip occurs first on the smaller pulley where the force of friction is less. Accordingly, the angle of contact on smaller pulley is taken into account while using the equation 20.7.

(*ii*) ***V-belt and rope drive :*** A V-belt or a rope runs into a grooved pulley.

Refer Fig. 20.8(*a*) which shows a V-belt of trapezoidal section resting in a grooved pulley.

Let 2α = angle of groove

N = normal reaction between belt and sides of the V-grooved pulley

R = total reaction in the plane of groove

μ = coefficient of friction between belt and groove

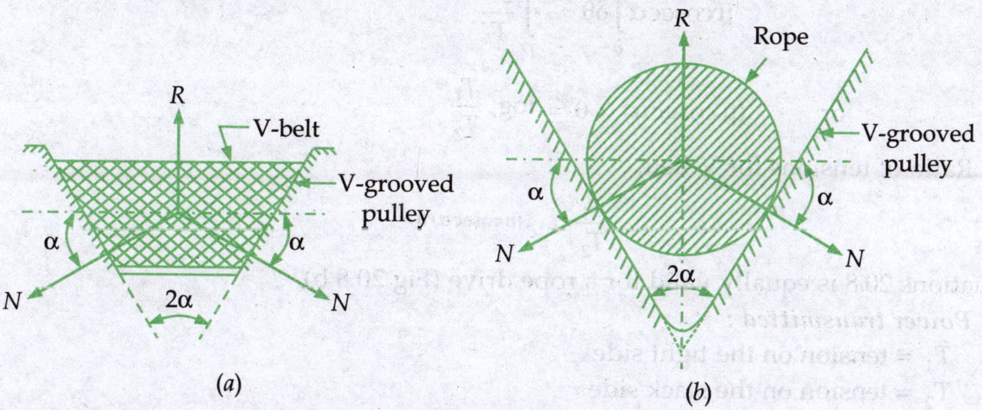

Fig. 20.8 V-belt and rope drive

Considering equilibrium between R and N, we have

$$R = N\sin\alpha + N\sin\alpha = 2N\sin\alpha$$

$$\therefore \quad N = \frac{R}{2\sin\alpha} = \frac{R}{2}\operatorname{cosec}\alpha$$

$$\text{Friction resistance} = \mu N + \mu N = 2\mu N = 2\mu \times \frac{R}{2}\operatorname{cosec}\alpha = \mu R \operatorname{cosec}\alpha$$

Then with reference to Fig. 20.7

$$R = (T + \delta T)\sin\frac{\delta\theta}{2} + T\sin\frac{\delta\theta}{2}$$

$$= (T + \delta T)\frac{\delta\theta}{2} + T\frac{\delta\theta}{2} \qquad (\because \delta\theta \text{ is small})$$

$$= T\frac{\delta\theta}{2} + \frac{\delta T\,\delta\theta}{2} + T\frac{\delta\theta}{2}$$

The term $\frac{\delta T\,\delta\theta}{2}$ is the product of two small quantities and hence neglected. That gives

$$R = T\frac{\delta\theta}{2} + T\frac{\delta\theta}{2} = T\,\delta\theta \qquad \ldots(a)$$

Also : $$\mu R \operatorname{cosec}\alpha = (T + \delta\theta)\cos\frac{\delta\theta}{2} - T\cos\frac{\delta\theta}{2}$$

$$= (T + \delta T) - T = \delta T \qquad \ldots(b)$$

This is because $\frac{\delta\theta}{2}$ is small and $\cos\frac{\delta\theta}{2} \to 1$

From expressions (a) and (b),

$$\mu\,(T\,\delta\theta)\operatorname{cosec}\alpha = \delta T$$

or $$(\mu \operatorname{cosec}\alpha)\,\delta\theta = \frac{\delta T}{T} \qquad \ldots(c)$$

Upon integration with in appropriate limits, we have

$$\mu \operatorname{cosec}\alpha \int_{0}^{\theta} \delta\theta = \int_{T_2}^{T_1} \frac{\delta T}{T}$$

or $$(\mu \operatorname{cosec}\alpha) \times \theta = \log_e \frac{T_1}{T_2}$$

∴ Ratio of tensions for V-belt,

$$\frac{T_1}{T_2} = e^{(\mu \operatorname{cosec}\alpha)\,\theta} \qquad \text{...(20.8)}$$

Equations 20.8 is equally valid for a rope drive (Fig 20.8 b).

(iii) ***Power transmitted :***

Let T_1 = tension on the tight side
T_2 = tension on the slack side
V = velocity of the belt

Then, effective turning force = $(T_1 - T_2)$

work done = force × distance moved = $(T_1 - T_2)\, V$ in Nm/s

and power = $(T_1 - T_2)\, V$ in watts ...(20.9)

where values of T_1 and T_2 are in N and V is the velocity of belt in m/s.

EXAMPLE 20.6

Calculate the force required to hold a weight of 10 kN suspended on a rope wrapped twice around a post. Take coefficient of friction $\mu = 0.2$.

Solution : $\mu = 0.2$ and $\theta = 2 \times 2\pi = 4\pi$ radian

Invoking the relation $\frac{T_1}{T_2} = e^{\mu\theta}$, we have

$$\frac{10}{T_2} = e^{0.2 \times 4\pi} = (2.718)^{2.512} = 12.326$$

$$\therefore \quad T_2 = \frac{10}{12.326} = 0.811 \text{ kN} = 811 \text{ N}$$

EXAMPLE 20.7

A belt is stretched over two identical pulleys of diameter D meter. The initial tension in the belt through out is 2.4 kN when the pulleys are at rest. When this belt-pulley system is used to transmit power, it is found that increase in tension on one side is equal to decrease in tension on the other side. Find the maximum torque that can be transmitted by the belt drive. Take coefficient of friction between belt and pulley as 0.3

Solution : Since both the pulleys are of the same diameter, angle of embrace (contact)

$\theta = 180° = \pi$ radian.

Let increase/decrease in tension be Δt. Then

$$T_1 = 2.4 + \Delta t \quad \text{and} \quad T_2 = 2.4 - \Delta t$$

Invoking the relation $\frac{T_1}{T_2} = e^{\mu\theta}$, we have

$$\frac{2.4 + \Delta T}{2.4 - \Delta T} = e^{0.3 \times \pi} = 2.57$$

or $\quad 2.4 + \Delta t = 2.57 \times 2.4 - 2.57\Delta T; \quad \Delta T = 1.06$ kN

$\therefore \quad T_1 = 2.4 + 1.06 = 3.46$ kN

and $\quad T_2 = 2.4 - 1.06 = 1.34$ kN

Maximum torque that can be transmitted by the belt

$$= (T_1 - T_2) \times \frac{D}{2} = (3.46 - 1.34)\frac{D}{2} = \mathbf{1.06}\ \boldsymbol{D}\ \mathbf{kNm}$$

EXAMPLE 20.8

Find out the number of turns a hauling rope must be wound round a rotating capstan in order to haul a load of 3 MN up to a gradient of 1 in 30. Presume the following data :

Resistance due to rolling = 0.00375 N per newton load

Friction coefficient between the rope and drum = 0.35

Pull on the free end of the rope = 250 N

Solution : The hauling is required up an inclined plane and for that T_1 has to overcome the frictional resistance as well as effort component of the incline. That is

$$T_1 = (3 \times 10^6 \times 0.00375) + \left(3 \times 10^6 \times \frac{1}{30}\right) = 111250 \text{ N}$$

$$T_2 = 250 \text{ N}$$

Invoking the relation $\dfrac{T_1}{T_2} = e^{\mu\theta}$, we have

$$e^{\mu\theta} = \frac{111250}{250} = 445$$

$$\mu\theta = \log_e 445 \,;\ \theta = \frac{\log_e 445}{0.35} = 17.42 \text{ radians}$$

$$\text{Number of turns} = \frac{17.42}{2\pi} = 2.77$$

Hence three turns need to be given.

EXAMPLE 20.9

A horizontal drum of belt drive carries the belt over a semicircle around it. The drum has a diameter of 1 m and when it rotates, 400 Nm of torque is transmitted. The drum has a mass of 25 kg and the coefficient of friction between the belt and drum is 0.25. Neglecting mass of belt, determine tensions induced in the limbs of the belt, and the vertical reaction on the bearings.

Solution: Since the drum carries the belt over a semi-circle around it,

$$\text{angle of contact} = 180° = \pi \text{ radians}$$

If T_1 and T_2 are the tensions on the tight and slack side of the belt, then

$$\frac{T_1}{T_2} = e^{\mu\theta} = e^{0.25 \times \pi} = 2.19 \qquad ...(i)$$

From the relation for torque transmitted: $T = (T_1 - T_2)\, r$

$$T_1 - T_2 = \frac{T}{r} = \frac{400}{0.5} = 800 \qquad ...(ii)$$

From expressions (i) and (ii)

$$2.19\,T_2 - T_2 = 800\,; \quad T_2 = \frac{800}{1.19} = 672.27\text{ N}$$

$$T_1 = 2.19 \times 672.27 = 1472.27\text{ N}$$

Vertical reaction on the drum bearings,

$$= (T_1 + T_2) + W \qquad \text{where } W \text{ is the weight of drum}$$

$$= (1472.27 + 672.27) + (25 \times 9.81) = 2389.79\text{ N}$$

EXAMPLE 20.10

Determine the minimum value of weight W required to cause motion of a block which rests on a horizontal plane. The block weighs 300 N and the coefficient of friction between the block and plane is 0.6. Angle of wrap over the pulley is 90° and the coefficient of friction between the pulley and rope is 0.3.

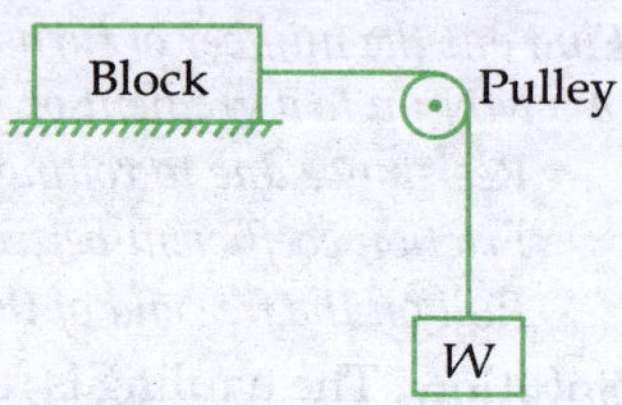

Fig. 20.9

Solution : Since the weight W impends vertical motion in the downward direction, the tension on the two sides of the pulley will be as shown in Fig. 20.10.

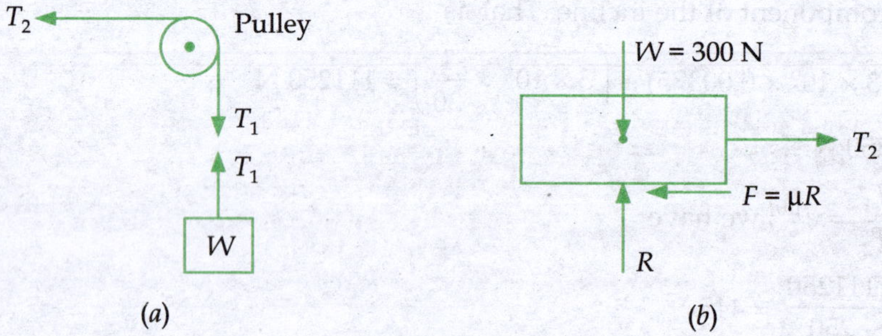

Fig. 20.10.

$$T_1 = W\,;\ \mu = 0.3\,;\ \theta = 90^\circ = \frac{\pi}{2}\text{ radians}$$

Invoking the relation

$$\frac{T_1}{T_2} = e^{\mu\theta}\,; \quad \frac{W}{T_2} = e^{0.3 \times \pi/2} = (2.718)^{0.471} = 1.60$$

Considering equilibrium of block,

$$\Sigma F_x = 0\,;\ T_2 = F = \mu R$$

$$\Sigma F_y = 0\,;\ R = 300\text{ N}$$

$\therefore \qquad T_2 = 0.6 \times 300 = 180$ N

Hence $\qquad W = 1.6 \times 180 =$ **288 N**

EXAMPLE 20.11

A flat belt running on a pulley 1m in diameter is to transmit 7.5 kW power at a speed of 200 rev/min. The angle of embrace of belt and pulley is 170 degree and the coefficient of friction between the belt and pulley is 0.25. If the pull in belt is not to exceed 200 N/cm, determine the width of belt.

Solution : Angle of embrace (lap)

$$= 170^\circ = \frac{170}{180} \times \pi = 2.965\text{ radians}$$

If T_1 and T_2 are the tensions on the tight and slack side of the belt, then

$$\frac{T_1}{T_2} = e^{\mu\theta} = e^{0.25 \times 2.965} = e^{0.741} = 2.1$$

$$\therefore \quad T_1 = 2.1\, T_2$$

Speed of belt, $V = \dfrac{\pi d_1 N}{60} = \dfrac{\pi \times 1 \times 200}{60} = 10.46$ m/s

Power transmitted by the belt

$$= (T_1 - T_2) \times V = (2.1\, T_2 - T_2) \times 10.46 = 11.506\, T_2 \text{ Nm/s}$$

$$\therefore \quad 7.5 \times 10^3 = 11.506\, T_2$$

$$T_2 = 651.83 \text{ N} \quad \text{and}$$

$$T_1 = 651.83 \times 2.1 = 1368.85 \text{ N}$$

Hence necessary width of belt $= \dfrac{1368.85}{200} = 6.84$ cm

EXAMPLE 20.12

A crossed belt running over two pulleys, 500 mm and 250 mm diameter, connects two parallel shafts 3 m apart. The larger pulley rotates at 300 rev/min, the coefficient of friction between the belt and pulley rim is 0.25 and the maximum permissible tension is 1.2 kN. Determine : (a) length of the belt required (b) angle of contact between the belt and pulley and (c) power transmitted by the belt.

Solution : Length of the belt,

$$L_{cross} = \pi(r_1 + r_2) + 2x + \frac{(r_1 + r_2)^2}{x}$$

$$= \pi(0.25 + 0.125) + 2 \times 3 + \frac{(0.25 + 0.125)^2}{3}$$

$$= 1.1775 + 6 + 0.0469 = 7.2244 \text{ m}$$

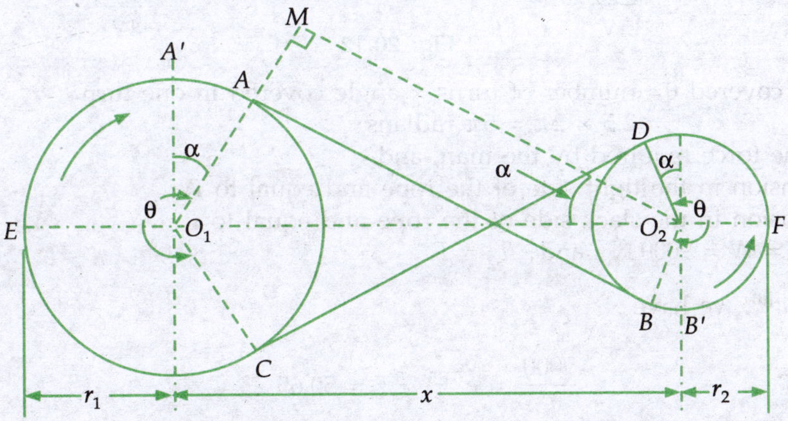

Fig. 20.11

(*b*) With reference to Fig. 20.11,

$$\sin\alpha = \frac{O_1M}{O_1O_2} = \frac{O_1A + AM}{O_1O_2} = \frac{r_1 + r_2}{x} = \frac{0.25 + 0.125}{3} = 0.125$$

$$\alpha = \sin^{-1}(0.125) = 7.18°$$

Angle of contact $\theta = 180° + 2\alpha = 180° + 2 \times 7.18 = 194.36°$

$$= \frac{\pi}{180} \times 194.36 = 3.39 \text{ radian}$$

For a crossed belt system, the angle of contact θ on both the pulleys is same.

(c) Speed of belt $V = \dfrac{\pi d_1 N}{60} = \dfrac{\pi \times 0.5 \times 300}{60} = 7.854$ m/s

Ratio of tensions $\dfrac{T_1}{T_2} = e^{\mu\theta} = e^{0.25 \times 3.39} = 2.334$

The tension T_1 equals the maximum permissible tension.

$$\therefore \quad T_1 = 1200 \text{ N} \quad \text{and} \quad T_2 = \frac{1200}{2.334} = 514.14 \text{ N}$$

Power transmitted $P = (T_1 - T_2)\,V = (1200 - 514.14) \times 7.854 = 5386.7$ W = **5.387 kW**

EXAMPLE 20.13

A casting weighing 9000 N hangs freely from a rope which makes 2.5 turns round a drum of 30 cm revolving at 20 rpm. The other end of the rope is pulled by a man. Presuming that coefficient of friction is 0.25, determine the force required by the man and the power expended in raising the casting

Solution: Refer the Fig. 20.12 for the arrangement of the given system.

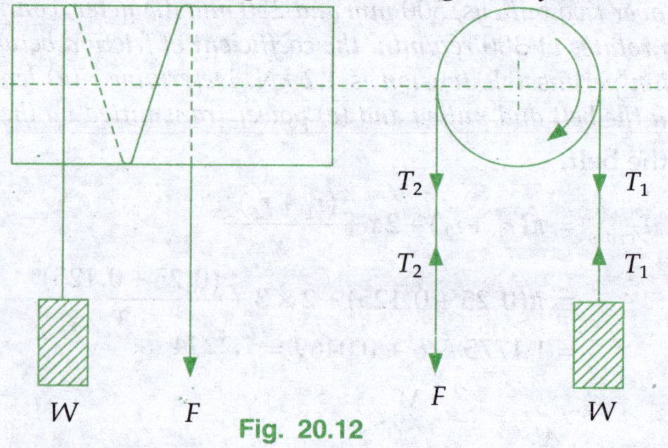

Fig. 20.12

Total angle covered θ= number of turns × angle covered in one turn
$= 2.5 \times 2\pi = 5\pi$ radians

Let F be the force required by the man, and

T_1 = tension in the tight side of the rope and equal to W

T_2 = tension in the slack side of the rope and equal to F

Then $\quad T_1 = W = 9000$ N $\quad$ and $\quad T_2 = F$

Since $\dfrac{T_1}{T_2} = e^{\mu\theta}$, we have

$$\frac{9000}{F} = e^{0.25 \times 5\pi} = 50.65$$

$\therefore \quad$ Force required by the man, $F = \dfrac{9000}{50.65} =$ **177.7 N**

(ii) Speed of belt, $V = \dfrac{\pi D N}{60} = \dfrac{\pi \times 0.3 \times 20}{60} = 0.314$ m/s

Power expended in raising the casting,

$$P = (T_1 - T_2)\,V = (9000 - 177.7) \times 0.314$$
$$= 2770 \text{ Nm/s} = \mathbf{2.77\ kW}$$

EXAMPLE 20.14

A flat belt open drive transmits 1.5 kW of power. The coefficient of friction between the belt and the drive pulley is 0.25 and lap angle is 159°. The drive pulley is rotating in the clockwise direction with linear velocity of 3.75 m/s. Determine the X and Y components of the reaction force on the drive pulley shaft.

Solution: Angle of embrace (lap) = $159° = \frac{159}{180} \times \pi = 2.77$ radians

If T_1 and T_2 are the tensions on the tight and slack side of the belt, then

$$\frac{T_1}{T_2} = e^{\mu\theta} = e^{0.25 \times 2.77} = 2$$

or $\quad T_1 = 2T_2 \qquad ...(i)$

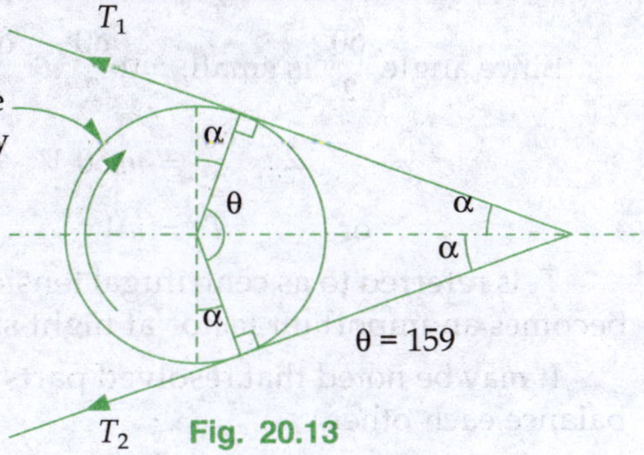

Fig. 20.13

From the relation for power transmitted:

$P = (T_1 - T_2)\,V$

$$1.5 \times 1000 = (T_1 - T_2) \times 3.75;$$

$$T_1 - T_2 = 400 \qquad ...(ii)$$

The solution of identities (*i*) and (*ii*) gives

$$T_1 = 800 \text{ N} \quad \text{and} \quad T_2 = 400 \text{ N}$$

Then X and Y components of the reaction force on the drive pulley shaft are:

$$X = T_1 \cos\alpha + T_2 \cos\alpha$$
$$Y = T_1 \sin\alpha - T_2 \sin\alpha$$

From the geometrical configuration;

$$2\alpha = 180 - \theta\ ; \quad \alpha = \frac{180-\theta}{2} = \frac{180-159}{2} = 10.5°$$

Therefore,

$$X = 800\cos 10.5° + 400\cos 10.5°$$
$$= 786.6 + 393.3 = \mathbf{1179.9\ N}$$

and $\quad Y = 800\sin 10.5° - 400\sin 10.5°$
$$= 145.79 - 72.89 = \mathbf{72.9\ N}$$

20.6 CENTRIFUGAL TENSION

When power is being transmitted with a belt drive system a part of the belt in contact with the pulley moves in a circular path of radius equal to the radius of pulley. Due to its mass and speed, the belt is subjected to centrifugal force that acts radially outwards. This force tends to lift the belt from the rim of the pulley. That reduces the normal reaction and hence the frictional resistance.

Consider an element AB of the belt and let it subtend an angle $\delta\theta$ at the centre of pulley. Further

Let r = radius of the pulley

V = speed of the belt

m = mass per metre length of the belt

Then, length of the elemental segment = $r\,\delta\theta$

mass of the elemental segment = $m\,r\,\delta\theta$

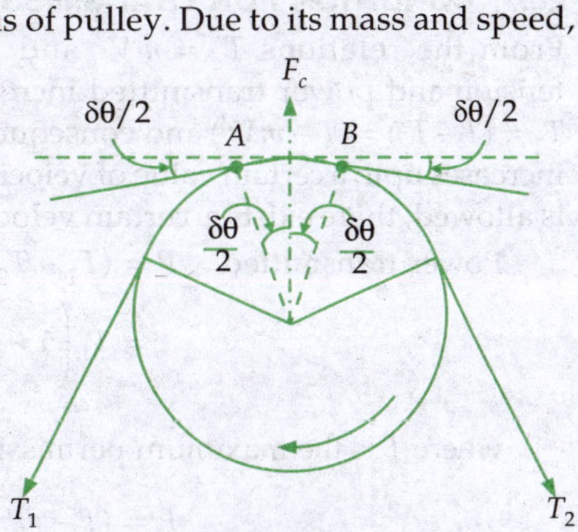

Fig. 20.14.

$$\text{centrifugal force } F_c = \frac{\text{mass} \times (\text{velocity})^2}{\text{radius}} = \frac{(m r \delta\theta) V^2}{r} = m \, \delta\theta \, V^2$$

This centrifugal force is counter - balanced by tensions at the ends A and B of the elemental segment.

Considering equilibrium of forces in the vertical direction,

$$2 T_c \sin \frac{\delta\theta}{2} = m \, \delta\theta \, V^2$$

Since angle $\frac{\delta\theta}{2}$ is small, $\sin \frac{\delta\theta}{2} \to \frac{\delta\theta}{2}$. That gives

$$2 T_c \frac{\delta\theta}{2} = m \, \delta\theta \, V^2$$

or $$T_c = m V^2$$

T_c is referred to as centrifugal tension. The centrifugal tension is negligible at low speeds but becomes an important factor at hight speed.

It may be noted that resolved parts of T_c at A and B acting along the horizontal direction will balance each other.

The centrifugal tension is independent of the angle of lap and accordingly will be same for the two ends of the entire belt. The total tension on the tight and slack sides of the belt would then be $(T_1 + T_c)$ and $(T_2 + T_c)$ respectively.

Power transmitted $P = [(T_1 + T_c) - (T_2 + T_c)] \, V$

$= (T_1 - T_2) \, V$ ...(same as per equation 20.9)

That is, the centrifugal tension has no effect on the power transmitted. However, with centrifugal tension, higher total tension is involved and that adversely affects the driving power capacity of the belt system.

The centrifugal tension in the belt drive system :

(*i*) increases in direction proportion to mass per unit length of the belt and to square of belt speed,

(*ii*) does not affect the power transmitted, and

(*iii*) reduces the driving power capacity.

20.7. CONDITION FOR TRANSMISSION OF MAXIMUM POWER

From the relations $T_c = mV^2$ and $P = (T_1 - T_2)V$, it is evident that both the centrifugal tension and power transmitted increase with the increase in belt speed. However the value of $T_1 = (T - T_c) = (T - mV^2)$ and consequently $(T_1 - T_2)$ diminishes. Evidently the power transmitted increases upto a certain value of velocity and thereafter it decreases. Thus when centrifugal tension is allowed, there exists a certain velocity of belt system at which maximum power is transmitted,

Power transmitted $P = (T_1 - T_2)V$

$$= T_1 \left(1 - \frac{T_2}{T_1}\right) V = T_1 \left(1 - \frac{1}{e^{\mu\theta}}\right) V = (T - T_c) \, kV$$

where T is the maximum permissible tension in the belt and k is a constant equal to $\left[1 - \frac{1}{e^{\mu\theta}}\right]$

$$P = (T - mV^2) \, kV$$
$$= (TV - mV^3) \, k$$

From strength consideration, value of T should not exceed the permissible limit. Keeping T constant, P is differentiated with respect to V and the differential equated to zero. That is

$$\frac{dP}{dV} = T - 3mV^2 = 0$$

or $$T = 3mV^2 = 3\,T_c$$

That gives : $$V = \sqrt{\frac{T}{3m}} \text{ and } T_c = \frac{T}{3}$$

Hence, maximum power will be transmitted when :

(*i*) the belt speed is $(T/3m)^{1/2}$, and

(*ii*) centrifugal tension is one- third of the belt strength,

(*iii*) the maximum permissible tension in the belt is thrice the centrifugal tension.

20.8. INITIAL TENSION

The belt is given an initial tension T_o when it is fitted an the driving and driven pulleys. This is to tighten the grip between the belt and the pulley rim. During motion and power transmission:

(*i*) The tight side of the belt stretches until the tension increases from T_o to T_1. The corresponding increase in length of belt on the tight side is

$$= \alpha\,(T_1 - T_o)$$

where α is the coefficient of belt length per unit force.

(*ii*) The slack side of the belt shortens in length until the tension decreases from T_0 to T_2. The resulting decrease in length of belt on the slack side is

$$= \alpha\,(T_o - T_2)$$

If the belt material is presumed to be inelastic, then length of the belt remains unchanged. For that increase in length on the tight side equals the decrease in length on the slack side. That is

$$\alpha\,(T_1 - T_0) = \alpha\,(T_0 - T_2)$$

or $$T_0 = \frac{T_1 + T_2}{2}$$

Thus the initial tension with which the belt is fitted onto the pulleys should be equal to the mean of the limiting values of tension on the tight and slack sides of the belt.

If centrifugal tension T_c is also taken into account, then

$$T_o = \frac{T_1 + T_2 + 2T_c}{2}$$

EXAMPLE 20.15

A belt connecting two pulleys has an initial tension of 750 N. From the data given below, determine power transmitted by the belt.

Angle of contact/lap = 150°

Coefficient of friction between the belt and pulley rim = 0.25

Speed of the belt = 250 meters/min

Solution : If T_1 and T_2 are the tensions on the tight and slack sides of the belt, then initial tension T_o is given by

$$T_o = \frac{T_1 + T_2}{2} \quad ; \quad 750 = \frac{T_1 + T_2}{2}$$

$$\therefore \quad T_1 + T_2 = 1500 \qquad \text{...}(i)$$

Angle of lap $\theta = 150° = \frac{150}{180} \times \pi = 2.617$ radians

The ratio of tension on the tight and slack sides of the belt is,

$$\frac{T_1}{T_2} = e^{\mu\theta} = e^{0.25 \times 2.617} = e^{0.654} = 1.92$$

$$\therefore \quad T_1 = 1.92\,T_2 \quad \text{...(ii)}$$

From expressions (*i*) and (*ii*)

$$1.92\,T_2 + T_2 = 1500$$

$$T_2 = \frac{1500}{2.92} = 513.7 \text{ N}$$

$$\text{and} \quad T_1 = 1.92 \times 513.17 = 986.3 \text{ N}$$

Power transmitted $P = (T_1 - T_2)\,V = (986.3 - 513.7) \times \frac{250}{60} = 328.2$ Nm/s

$= 328.2$ W $=$ **3.28 kW**

EXAMPLE 20.16

A belt transmits 10 kW through a pulley 1 m diameter running at 200 rev/min. The coefficient of friction between the belt and pulley is 0.25 and the allowable tension per mm width of belt is 18 N. Presuming an angle of lap on the pulley as 160 degrees, determine the necessary initial tension on the belt and the belt width.

Solution: Speed of the belt $V = \frac{\pi d N}{60} = \frac{\pi \times 1 \times 200}{60} = 10.47$ m/s

Angle of contact/lap $\theta = \frac{\pi}{180} \times 60 = 2.79$ radian

If T_1 and T_2 are the tensions an the tight and slack side respectively, then

$$\frac{T_1}{T_2} = e^{\mu\theta} = e^{0.25 \times 2.79} = 2.01 \quad \text{...(i)}$$

Also $\quad P = (T_1 - T_2)V\,; \quad 10 \times 10^3 = (T_1 - T_2) \times 10.47$

That gives:

$$T_1 - T_2 = \frac{10 \times 10^3}{10.47} = 955 \quad \text{...(ii)}$$

Solving identities (*i*) and (*ii*)

$$T_2 = 945.54 \quad \text{and} \quad T_1 = 1900.54 \text{ N}$$

Initial tension $T_0 = \frac{T_1 + T_2}{2} = \frac{945 \cdot 54 + 1900.54}{2} =$ **1423.04 N**

Maximum tension $T_{max} = T_1 = 1900.54$ N

$\therefore$ Necessary width of the belt,

$$b = \frac{1900.54}{18} = \textbf{105.58 mm}$$

EXAMPLE 20.17

Two pulleys of diameters 2 m and 1.5 m, mounted on two parallel shafts 5 m apart, are connected by an open belt drive. The initial tension in the belt connecting the pulleys is 2.5 kN, coefficient of friction between the belt and pulley rim is 0.3 and the mass of the belt material is 1.5 kg per metre length. What power would be transmitted when the smaller pulley turns 300 revolutions per minute ?

Solution : Speed of the belt $V = \frac{\pi d_2 N}{60} = \frac{\pi \times 1.5 \times 300}{60} = 23.55$ m/s

If T_1 and T_2 are the tensions on the tight and slack side of the belt, then initial tension T_o is give by

$$T_o = \frac{T_1 + T_2}{2} \quad ; \quad T_1 + T_2 = 2 \times 2.5$$

$$\therefore \quad T_1 + T_2 = 5 \qquad ...(i)$$

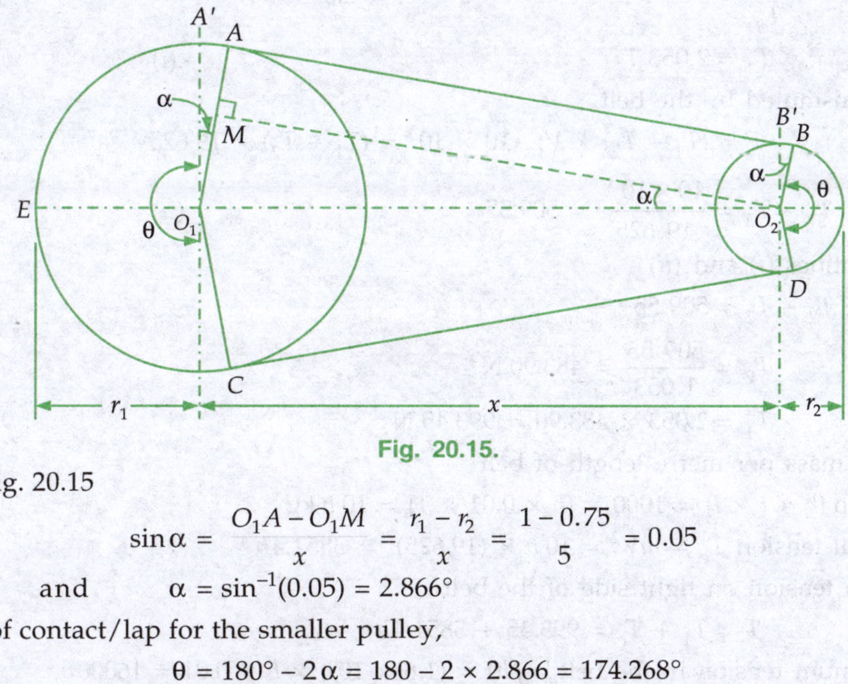

Fig. 20.15.

Refer Fig. 20.15

$$\sin\alpha = \frac{O_1A - O_1M}{x} = \frac{r_1 - r_2}{x} = \frac{1 - 0.75}{5} = 0.05$$

and $\quad \alpha = \sin^{-1}(0.05) = 2.866°$

Angle of contact/lap for the smaller pulley,

$$\theta = 180° - 2\alpha = 180 - 2 \times 2.866 = 174.268°$$

$$= \frac{\pi}{180} \times 174.268 = 3.04 \text{ radians}$$

The ratio of tensions on the tight and slack side is,

$$\frac{T_1}{T_2} = e^{\mu\theta} = e^{0.3 \times 3.04} = e^{0.912} = 2.489$$

$$\therefore \quad T_1 = 2.489\, T_2 \qquad ...(ii)$$

From identities (*i*) and (*ii*),

$$2.489\, T_2 + T_2 = 5 \; ; \quad T_2 = \frac{5}{3.489} = 1.433 \text{ kN}$$

and $\quad T_1 = 2.489 \times 1.433 = 3.567$ kN

Power transmitted $\quad P = (T_1 - T_2)V = (3.567 - 1.433) \times 23.55$

$= 50.25$ kNm/s $=$ **50.25 kW**

EXAMPLE 20.18

A 10 mm thick leather belt is used to transmit a power of 10 kW from a pulley of 1.25 m in diameter and turning 300 revolutions per minute. The angle of embrace of belt and pulley is 165° and the coefficient of friction between the belt and pulley is 0.25. The allowable stress and density for the leather belt are 1.6 MPa and 1000 kg/m³. Taking centrifugal tension into account, determine the width of belt.

Solution : Angle of embrace (lap)

$$\theta = 165° = \frac{\pi}{180} \times 165 = 2.878 \text{ radians}$$

$$\text{Speed of belt } V = \frac{\pi d N}{60} = \frac{\pi \times 1.25 \times 300}{60} = 19.625 \text{ m/s}$$

If T_1 and T_2 are the tensions on the tight and slack side of the belt, then

$$\frac{T_1}{T_2} = e^{\mu\theta} = e^{0.25 \times 2.878} = e^{0.7195} = 2.053$$

$$T_1 = 2.053\, T_2 \qquad \text{...}(i)$$

Power transmuted by the belt,

$$P = (T_1 - T_2) \times V; \quad 10 \times 10^3 = (T_1 - T_2) \times 19.625$$

$$T_1 - T_2 = \frac{10 \times 10^3}{19.625} = 509.55 \qquad \text{...}(ii)$$

From identities (*i*) and (*ii*)

$$2.053\, T_2 - T_2 = 509.55$$

$$T_2 = \frac{509.55}{1.053} = 483.90 \text{ N}$$

and $\quad T_1 = 2.053 \times 483.90 = 993.45$ N

Let $\quad m$ = mass per metre length of belt

$$= \rho\,(b \times t \times l) = 1000 \times (b \times 0.01 \times 1) = 10\,b \text{ kg}$$

Centrifugal tension $T_c = mV^2 = 10\,b \times (19.625)^2 = 3851.4b$

Maximum tension on tight side of the belt

$$T = T_1 + T_c = 993.45 + 3851.4b$$

But maximum tension in the belt $= fbt = (1.6 \times 10^6) \times b \times 0.01 = 16000b$

$$\therefore \quad 993.45 + 3851.4b = 16000b$$

That gives : Width of belt $b = \dfrac{993.45}{16000 - 3851.4} = 0.0818 \text{ m} =$ **81.8 mm**

EXAMPLE 20.19

A flat belt is required to transmit 35 kW from a pulley of 1.5 m effective diameter running at 300 rpm. The angle of contact is spread over 11/24 of the circumference and the coefficient of friction between belt and pulley is 0.3. The belt thickness is 9.5 mm and its density is 1100 kg/m³. If the permissible stress in belt is 2.5 N/mm², determine the width of belt required.

Solution : Since the belt covers 11/24 of circumference, the angle of embrace (lap)

$$\theta = \frac{11}{24} \times 2\pi = 2.878 \text{ radians}$$

If T_1 and T_2 are the tensions on tight and slack side of the belt, then

$$\frac{T_1}{T_2} = e^{\mu\theta} = e^{0.3 \times 2.878} = 2.372 \quad ...(i)$$

$$\text{Speed of the belt } V = \frac{\pi dN}{60} = \frac{\pi \times 1.5 \times 300}{60} = 23.55 \text{ m/s}$$

Mass per meter length of the belt $m = \rho \times (b \times t \times l)$

Taking b, t and l in metre units

$$= 1100 \times (b \times 9.5 \times 10^{-3} \times 1) = 10.45b \text{ kg}$$

Centrifugal tension induced on either side of the belt,

$$T_c = mV^2 = 10.45b \times 23.55^2 = 5796\, b \text{ N}$$

Maximum tension on tight side of the belt,

$$T_{max} = T_1 + T_c = T_1 + 5796b$$

Maximum tension is also given by,

$$T_{max} = f\,bt = (2.5 \times 1000^2) \times b \times (9.5 \times 10^{-3}) = 23750b$$

$$\therefore \quad 23750b = T_1 + 5796b$$

That gives:

$$T_1 = 17964\, b \text{ N}$$

Then from expression (*i*),

$$T_2 = \frac{T_1}{2.372} = \frac{17954b}{2.372} = 7569b\text{N}$$

Power transmitted by the belt $P = (T_1 - T_2)V$

or $\quad 35 \times 10^3 = (17954b - 7569b) \times 23.55$

or $$b = \frac{35 \times 10^3}{23.55(17954 - 7569)} = 0.143 \text{ m}$$

$\therefore$ Width of belt $b = 0.143$ m = **143 mm**

EXAMPLE 20.20

The smaller pulley in an open belt drive system is 30 cm in diameter, runs at 250 rev/min and has an angle of contact of 160°. The belt is on the point of slipping when 10 kW power is being transmitted. If the coefficient of friction between the belt and pulley rim is 0.25, which of the following alternatives would be more effective in increasing the power which could be transmitted:

(i) The initial tension in the belt is increased by 10 per cent

(ii) The coefficient of friction is increased by 10 per cent through suitable dressing to the belt ?

Solution : Speed of the belt $V = \frac{\pi dN}{60} = \frac{\pi \times 0.3 \times 250}{60} = 3.925 \text{ m/s}$

$$\text{Angle of contact/lap } \theta = \frac{\pi}{180} \times 160 = 2.79 \text{ radian}$$

If T_1 and T_2 are the tensions on the tight and slack side respectively, then

$$\frac{T_1}{T_2} = e^{\mu\theta} = e^{0.25 \times 2.79} = 2.01 \quad ...(i)$$

Also $P = (T_1 - T_2)\,V \quad \therefore \quad 10 \times 10^3 = (T_1 - T_2) \times 3.925$

That gives, $\quad T_1 - T_2 = 2548 \quad ...(ii)$

Solving identities (*i*) and (*ii*), we get

$$T_2 = 2523 \text{ N and } T_1 = 5071 \text{ N}$$

$$\text{Initial tension } T_0 = \frac{T_1 + T_2}{2} = \frac{5071 + 2523}{2} = 3797 \text{ N}$$

(*a*) Increased initial tension $T_o = 3797 \times 1.1 = 4177$ N

Let T_1 and T_2 be the corresponding tensions on the tight and slack side of the belt. Then

$$T_o = \frac{T_1 + T_2}{2}$$

or $\quad T_1 + T_2 = 2\,T_o = 2 \times 4177 = 8354 \quad \text{...(a)}$

$$\text{Ratio of tensions, } \frac{T_1}{T_2} = e^{\mu\theta} = e^{0.25 \times 2.79} = 2.01 \quad \text{...(b)}$$

From identities (*a*) and (*b*)

$$T_2 = 2775 \text{ N and } T_1 = 5578 \text{ N}$$

$\therefore$ Power transmitted $P = (T_1 - T_2)V$

$$= (5578 - 2775) \times 3.925 = 11002 \text{ W} = 11.002 \text{ kW}$$

(*b*) increased coefficient of friction $\mu = 0.25 \times 1.1 = 0.275$

$$\frac{T_1}{T_2} = e^{\mu\theta} = e^{0.275 \times 2.79} = 2.154 \quad \text{...(c)}$$

Initial tension $T_o = \frac{1}{2}(T_1 + T_2)$ remains the same and therefore

$$T_1 + T_2 = 2T_o = 2 \times 3797 = 7594 \quad \text{...(d)}$$

From identities (*c*) and (*d*)

$$T_2 = 2408 \text{ N and } T_1 = 5186 \text{ N}$$

$\therefore$ Power transmitted $P = (T_1 - T_2)V$

$$= (5186 - 2408) \times 3.925 = 10904 \text{ W} = 10.904 \text{ kW}$$

Comment : The increase in initial tension is more effective for getting an increased power transmission.

EXAMPLE 20.21

A rough rule for leather belting is that the difference between tight and slack side tensions should not exceed 100 N/cm of width for a belt of 5 mm thick. If this rule is applied under the following conditions, determine the maximum stress on the tight side of the belt.

Angle of contact = 170° ; coefficient of friction = 0.25

Belt speed = 100 m/min, and density of leather = 1×10^{-3} kg/cm^3

Solution : Angle of contact $= 170° = \frac{\pi}{180} \times 170 = 2.965$ radians

If T_1 and T_2 are the tensions on the tight and slack side of the belt, then

$$\frac{T_1}{T_2} = e^{\mu\theta} = e^{0.25 \times 2.965} = e^{0.741} = 2.1$$

or $\quad T_1 = 2.1\,T_2 \quad \text{...(i)}$

If *b* cm is the width of belt, then

$$T_1 - T_2 = 100 \times b \quad \text{...(ii)}$$

From expressions (*i*) and (*ii*)

$$2.1\,T_2 - T_2 = 100\,b$$

$$\therefore \quad T_2 = \frac{100\,b}{1.1} = 90.9\,b\,\text{N}$$

and $\quad T_1 = 2.1 \times 90.9\,b = 191\,b\,\text{N}$

Let m = mass per metre length of belt

$$= \rho \times (b \times t \times l) = 1 \times 10^{-3} \times (b \times 0.5 \times 100) = 0.05\,b \text{ kg}$$

Induced cetrifugal tension on either side of the belt,

$$T_c = mV^2 = 0.05\,b \times \left(\frac{1000}{60}\right)^2 = 13.89\,b\,\text{N}$$

Maximum tension on tight side of the belt,

$$T = T_1 + T_c - 191\,b + 13.89\,b = 204.89\,b\,\text{N}$$

$\therefore$ Maximum stress induced in the belt

$$= \frac{\text{maximum tension}}{\text{cross-sectional area of belt}} = \frac{204.89\,b}{b \times 0.5} = \mathbf{409.78\ N/cm^2}$$

EXAMPLE 20.22

A 8 mm thick × 100 mm wide, belt drives a 1.0 m diameter pulley at 200 revolutions per minute. The angle of contact/lap is 165° and mass of the belt material is 1000 kg/m³. If maximum permissible stress in the belt is 1.5 N/mm² and the coefficient of friction between the belt and pulley is 0.25, determine the power transmitted by the belt. Take the centrifugal tension into account.

Solution : Speed of belt $V = \frac{\pi d N}{60} = \frac{\pi \times 1.0 \times 200}{60} = 10.47$ m/s

Maximum tension in the belt $T = fbt = 1.5 \times 100 \times 8 = 1200$ N

Mass per meter length of belt $m = \rho \times (b \times t \times l) = 1000 \times (0.1 \times 0.008 \times 1) = 0.8$ kg

Induced centrifugal tension on each side of the belt,

$$T_c = mV^2 = 0.8 \times (10.47)^2 = 87.69 \text{ N}$$

$\therefore$ Tension on tight side of the belt,

$$T_1 = T - T_c = 1200 - 87.69 = 1112.31 \text{ N}$$

$$\text{Angle of contact } \theta = 165° = \frac{\pi}{180} \times 165 = 2.878$$

The ratio of tensions on the tight and slack side of the belt is

$$\frac{T_1}{T_2} = e^{\mu\theta} = e^{0.25 \times 2.878} = e^{0.7195} = 2.053$$

$$T_2 = \frac{T_1}{2.053} = \frac{1112.31}{2.053} = 541.79 \text{ N}$$

Power transmitted, $P = (T_1 - T_2)V = (1112.31 - 541.79) \times 10.47$

$= 5973.34$ Nm/s $= 5973.34$ W $=$ **5.97 kW**

EXAMPLE 20.23

A belt 10 mm thick and 125 mm wide is transmitting power at a speed of 20 m/s. The net driving tension is 1.5 times the tension on the slack side. Determine the power transmitted at this speed. It may be presumed that density of the belt material is 1000 kg/m³ and stress in the belt is not to exceed 1.5 N/mm².

Proceed to calculate the maximum power and the belt speed at which it can be transmitted.

Solution: Maximum tension in the belt,

$$T = f\,b\,t = 1.5 \times 125 \times 10 = 1875 \text{ N}$$

Mass per metre length of belt, $m = \rho \times (b \times t \times l) = 1000 \times (0.125 \times 0.01 \times 1) = 1.25$ N

Induced centrifugal tension on each side of the belt,

$$T_c = mV^2 = 1.25 \times (20)^2 = 500 \text{ N}$$

$$T_1 = T - T_c = 1875 - 500 = 1375 \text{ N}$$

Further, it is given that $(T_1 - T_2) = 1.5\,T_2$ and therefore

$$T_2 = \frac{T_1}{2.5} = \frac{1375}{2.5} = 550 \text{ N}$$

Power transmitted by the belt,

$$P = (T_1 - T_2)\,V = (1375 - 550) \times 20 = 16500 \text{ Nm/s}$$

$$= 16500 \text{ W} = \mathbf{16.5 \text{ kW}}$$

(*b*) For maximum power transmission,

$$\text{speed of belt } V = \sqrt{\frac{T}{3m}} = \sqrt{\frac{1875}{3 \times 1.25}} = \mathbf{22.36 \text{ m/s}}$$

$$\text{centrifugal tension } T_c = \frac{T}{3} = \frac{1875}{3} = 625 \text{ N}$$

$$\therefore \quad T_1 = T - T_c = 1875 - 625 = 1250 \text{ N}$$

$$\text{and} \quad T_2 = \frac{T_1}{2.5} = \frac{1250}{2.5} = 500 \text{ N}$$

$$\text{Power transmitted } P = (T_1 - T_2)V = (1250 - 500) \times 22.36$$

$$= 16770 \text{ Nm/s} = \mathbf{16.77 \text{ kW}}$$

EXAMPLE 20.24

The V-groove angle for a V-belt drive used for power transmission is 30°. The belt is 20 mm thick × 25 mm wide and the mass of the belt material is 0.5 kg per metre length. The angle of contact/lap is 160° and the coefficient of friction between the belt and pulley material is 0.20. If the maximum allowable stress is 1.8 MPa, determine the maximum power transmitted by the belt. Take the centrifugal tension into account.

Solution : Maximum tension in the belt,

$$T = f\,b\,t = (1.8 \times 10^6) \times 0.025 \times 0.02 = 900 \text{ N}$$

For maximum power transmission,

$$\text{speed of belt } V = \sqrt{\frac{T}{3m}} = \sqrt{\frac{900}{3 \times 0.5}} = 24.49 \text{ m/s}$$

$$\text{Centrifugal tension} \quad T_c = \frac{T}{3} = \frac{900}{3} = 300 \text{ N}$$

$$T_1 = T - T_c = 900 - 300 = 600 \text{ N}$$

The ratio of tension on the tight and slack side of the belt is

$$\frac{T_1}{T_2} = e^{\mu \operatorname{cosec} \alpha \times \theta} \quad \text{where} \quad \alpha = \text{half angle of groove}$$

$$\text{Angle of lap } \theta = 160° = \frac{\pi}{180} \times 160 = 2.791$$

$$\therefore \quad \frac{T_1}{T_2} = e^{0.20 \text{ cosec } 15° \times 2.791} = e^{2.157} = 8.644$$

$$T_2 = \frac{T_1}{8.644} = \frac{600}{8.644} = 69.41 \text{ N}$$

Maximum power transmitted $P = (T_1 - T_2)V = (600 - (69.41) \times 24.49$

$= 12994 \text{ Nm/s} =$ **12.994 kW**

EXAMPLE 20.25

Following data is given for a rope pulley transmitting 60 kW power :

dia of pulley = 1.25 m ; angle of groove = 45°

speed of the pulley = 180 rpm; coefficient of friction = 0.3

angle of contact/lap = 160°; mass of rope = 0.5 kg/m length

maximum permissible tension = 1000 N

Taking centrifugal tension into account, make calculations for the number of ropes required for the drive and initial tension in the rope.

Solution : Velocity of rope $= \frac{\pi d N}{60} = \frac{\pi \times 1.25 \times 180}{60} = 11.775 \text{ m/s}$

centrifugal tension induced on each side of the rope is

$$T_c = mV^2 = 0.5 \times (11.775)^2 = 69.32 \text{ N}$$

$$\therefore \quad T_1 = T - T_c = 750 - 69.32 = 660.68 \text{ N} \quad ...(i)$$

The ratio of tensions on the light and slack side of the rope is

$$\frac{T_1}{T_2} = e^{\mu \text{ cosec } \alpha \times \theta} \quad ; \quad \alpha \text{ is half angle of the groove}$$

The factor μ cosec $\alpha \times \theta$ works out as

$$= 0.3 \times \text{cosec } 22.5° \times \left(\frac{160}{180} \times \pi\right) = 2.188$$

$$\therefore \quad \frac{T_1}{T_2} = e^{2.188} = 8.17 \; ; \quad T_1 = 8.917\, T_2 \quad ...(ii)$$

From identities (*i*) and (*ii*),

$$T_1 = 660.68 \text{ N} \quad \text{and} \quad T_2 = \frac{660.68}{8.917} = 74.09 \text{ N}$$

Power transmitted by one rope $= (T_1 - T_2)V = (660.68 - 74.09) \times 11.775$

$= 6907.09 \text{ Nm/s} = 6.907 \text{ kW}$

$\therefore$ Number of ropes required,

$$n = \frac{\text{total power transmitted}}{\text{power transmitted by one rope}} = \frac{60}{6.907} = 8.686 \text{ say } \mathbf{9}$$

20.9. GEARS AND GEAR DRIVE

A toothed wheel or gear is essentially a wheel with teeth cut on its periphery.

Power or motion is transmitted from one shaft to another with gear drive when

- centre distances are relatively short
- speed of the shaft is low and the use of belt drive is not recommended

- positive drive is necessary, *i.e*, velocity ratio is fixed and known with certainty
- there is need to step up or step down the speed
- high torque is to be transmitted
- precise timing is required.

The gear drive has a compact layout and it provides a highly efficient and reliable service. However, the operation tends to become noisy, the whole set is affected when one tooth gets damaged, and the manufacture of gears requires special tools and equipment.

20.9.1. Types of gears

The commonly used forms of toothed gearing are :

- **Spur gear :** It is a cylindrical gear whose tooth traces are straight lines parallel to the gear axis. Further, the tooth profile is identical from one side of the face to the other.

The spur gears are used for transmitting motion between two shafts whose axis are parallel and coplanar.

These gears have a high (96 – 98%) efficiency of power transmission and are free from any axial thrust during tooth engagement. Accordingly they have been successfully employed as sliding gears for speed change mechanism in gear boxes of lathe. However, compared to gears of other types, spur gears are more noisy in operation, wear out readily and develop backlash.

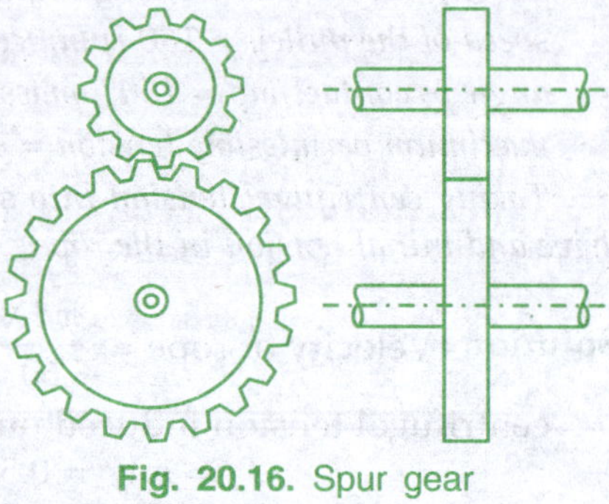

Fig. 20.16. Spur gear

- **Helical gear :** It is a cylindrical gear whose tooth traces are straight helices, teeth are inclined at an angle to the gear axis. The teeth are thus of helical or screw form. This ensures smooth action and accurate maintenance of velocity ratio. However, lateral thrust is set up due to teeth being inclined. This lateral thrust gets neutralised by using double helical gears called **Herringbone gears**. The double helical gears are the equivalent of two helical gears secured together ; but they are manufactured as one piece. The teeth may be continuous or separated by a small gap. Further, the two helics are of opposite hand and meet at a common axis.

The helical gears are used in automobile gear boxes, and in steam and gas turbines for speed reduction. The herringbone gears are used in machinery where large power is transmitted at low speeds.

- **Bevel gear :** The bevel gear wheels conform to the frusta of cones having a common vertex ; tooth traces are straight line generators of the cone.

The bevel gears are used to connect two shafts whose axis are coplanar but intersecting. When the shafts are at right angles and the wheels equal in size, the bevel gears are called *mitre gears*. When the bevel gears have their teeth inclined to the face of the bevel, they are known as *helical bevel gears*.

Fig. 20.17. Helical gear.

- **Spiral gear :** These are identical to helical gears with the difference that these gears have a point contact rather than a line contact. These gears are used when connection is to be made between intersecting and co-coplanar shafts.

- **Worm gear :** The system consists of a worm which is basically part of a screw. The worm meshes with the teeth on a gear wheel called worm wheel.

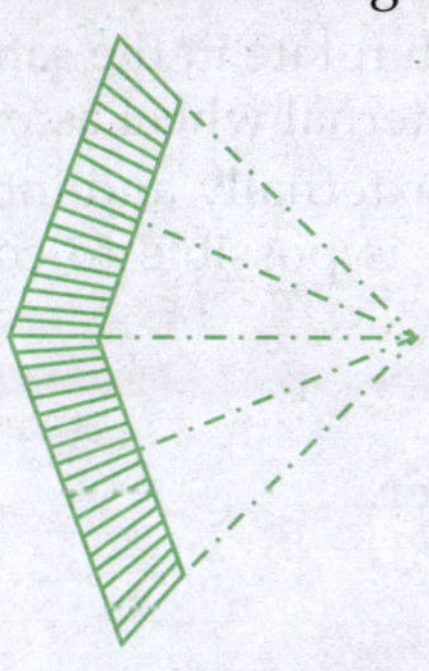

Fig. 20.18. Bevel gear

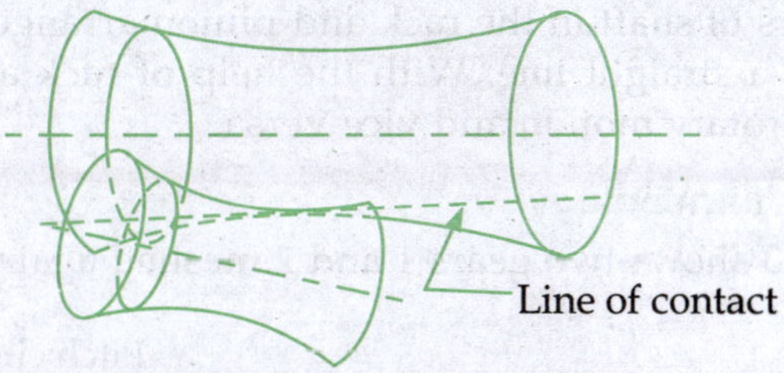

Fig. 20.19. Spiral gear

The worm gear is used for connecting two non-parallel, non-intersecting shafts which are usually at right angles. This gearing system is smooth and quiet in operation and provides a high gear ratio; rotational speed of the worm is quite high compared to that of wheel. Their use is recommended when high speed reduction (more than 10 : 1) is required.

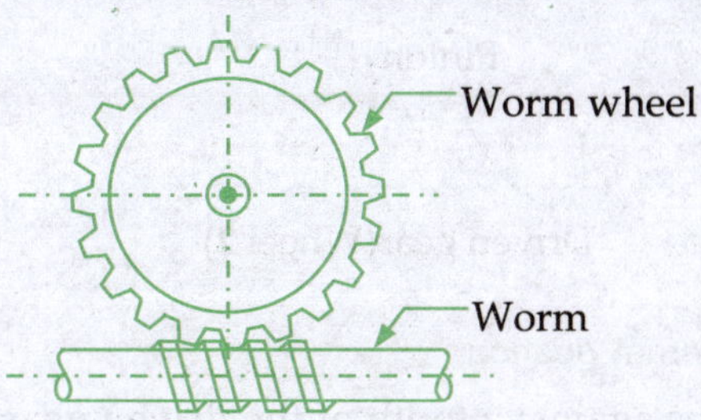

Fig. 20.20 Worm gear

Fig. 20.21 Rack and pinion

- **Rack and pinion :** Rack is a straight line spur gear of infinite diameter. It meshes, both internally and externally, with a circular wheel called pinion. The arrangement finds application where linear motion is to be converted into rotary motion and vice-versa
- **Internal and external gearing :** Two toothed wheels on parallel shafts may gear either externally or internally.

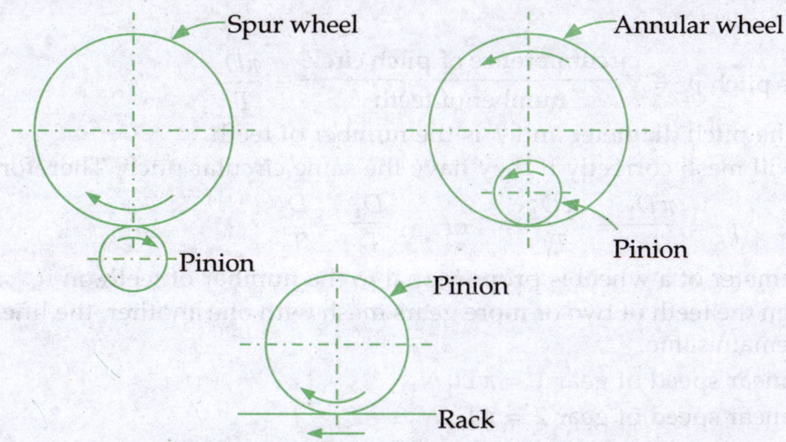

Fig. 20.22. Internal and external gearing.

In external gearing, the motion of the two wheels is unlike, *i.e.*, in the opposite direction. The larger wheel of the gear system is known as **spur** and the smaller wheel as **pinion**.

In internal gearing, the motion of the two wheels is alike ; both rotate in the same direction. The external large wheel is called **annular wheel** and the small internal wheel as **pinion**.

The gears of shaft in the rack and pinion arrangement mesh externally and internally with the gears in a straight line. With the help of rack and pinion, it is possible to convert linear motion into rotary motion and vice-versa.

20.9.2. Gear terminology

Fig. 20.23 shows two gears 1 and 2 meshed against one another.

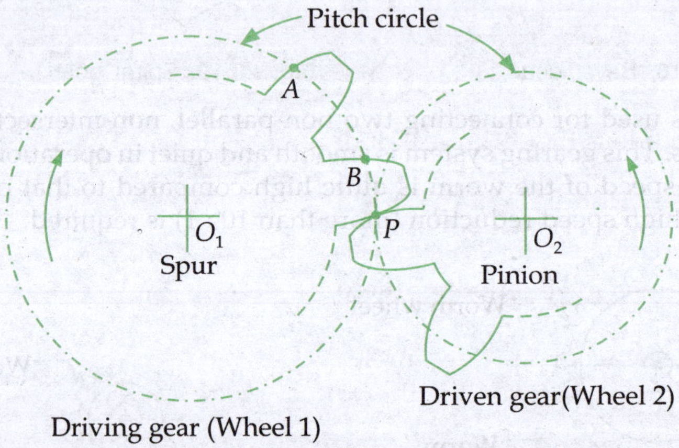

Fig. 20.23. External gearing.

A tooth of the driving gear (centre O_1) meshes against a tooth of the driven gear (centre O_2) at point P. The point P is called the *pitch point*. The circles whose centres are O_1 and O_2, and radii O_1P and O_2P are called *pitch circles* of wheels 1 and 2 respectively. A pitch circle is essentially an imaginary circle which by pure rolling action gives the same motion as the actual gear.

Further, on the pitch circle of wheel 1, let A be a point on one tooth and B be the corresponding point on the adjacent tooth. The distance AB (measured along the circular arc) is called *circular pitch* of wheel 1. Circular pitch may then be defined as

"distance measured on the circumference of the pitch circle from a point of one tooth to the corresponding point on the next tooth"

$$\text{Circular pitch } p_c = \frac{\text{circumference of pitch circle}}{\text{number of teeth}} = \frac{\pi D}{T} \qquad \text{...(20.10)}$$

where D is the pitch diameter and T is the number of teeth.

Two gears will mesh correctly if they have the same circular pitch. Therefore

$$p_c = \frac{\pi D_1}{T_1} = \frac{\pi D_2}{T_1} \quad \text{or} \quad \frac{D_1}{T_1} = \frac{D_2}{T_2}$$

Thus the diameter of a wheel is proportional to the number of teeth on it.

Further, when the teeth of two or more gears mesh with one another, the linear speeds of the two gears will remain same.

$$\text{Linear speed of gear 1} = \pi D_1 N_1$$

$$\text{Linear speed of gear 2} = \pi D_2 N_2$$

where D_1 and D_2 are the diameters, and N_1 and N_2 are the revolutions made by gears 1 and 2 per unit time.

Equating the linear speeds of the two gears,

$$\pi D_1 N_1 = \pi D_2 N_2 \; ; \quad \frac{N_2}{N_1} = \frac{D_1}{D_2}$$

Since the diameter of a wheel is proportional to the number of teeth on it, we get

$$\text{velocity ratio} = \frac{N_2}{N_1} = \frac{T_1}{T_2}$$

Diametral pitch : It represents the number of teeth on a wheel per unit of its diameter.

$$\text{Diametral pitch } p_d = \frac{\text{number of teeth}}{\text{pitch circle diameter}} = \frac{T}{D} \quad \text{...(20.11)}$$

From equations 20.10 and 20.11, we have

$$p_c \times p_d = \pi$$

Evidently if the circular pitches of two wheels are equal, their diametral pitches are also equal.

Obviously, the module is the reciprocal of diametral pitch.

$$\text{Module } \; m = \frac{D}{T} \quad \text{...(20.12)}$$

Refer Fig. 20.22 for some of the terms associated with profile of a gear tooth.

Addendum circle is the circle bounding the outer ends of the teeth and concentric with the pitch circle.

The radial distance between the pitch circle and addendum circle is called *addendum*.

Dedendum circle is the circle bounding the bottom of the tooth and concentric with the pitch circle.

The radial distance between the pitch circle and the dedendum circle is called *dedendum*.

The surface of tooth above the pitch surface is called the *face* of the tooth. *Flank* is the surface of tooth below the pitch surface. *Tooth space* is the width of tooth measured along the pitch circle. The length of arc between the sides of a gear tooth, measured on the pitch circle is called *thickness* of *tooth* or *circular thickness*.

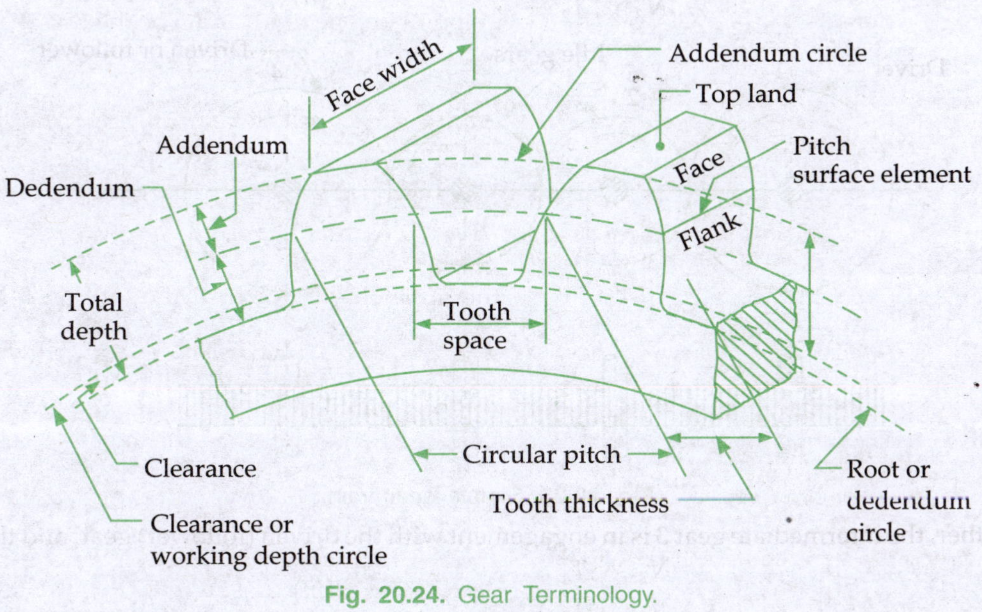

Fig. 20.24. Gear Terminology.

Module : Module represents the ratio of pitch circle diameter (in mm) to the number of teeth.

The radial distance between the addendum circle and dedendum circle is called the *total depth*. It equals the sum of the addendum and dedendum. *Working depth* equals the sum of the addenda of the two mating gears. The radial distance from the top of tooth to the bottom of tooth is called *clearance*. The standard value of clearance equals 0.157 m where m is the module. Then

$$\text{dedendum} = \text{addendum} + \text{clearance} = m + 0.157\,m = 1.157\,m$$

Addendum equals module and is always less than dedendum.

20.9.3. Gear Trains

Gear train is any combination of gear wheels by means of which power and motion is transmitted from one shaft to another.

Various types of gear trains are :

1. Simple gear train
2. Compound gear train
3. Reverted gear train
4. Epicyclic gear train

The nature of the train used depends upon the velocity ratio required and the relative position of the axes of the shafts.

• **Simple gear train :** Refer Fig. 20.25 for a simple gear train in which each shaft carries one wheel only. Between the driving gear 1 and the driven or follower gear 4, two intermediate gears 2 and 3 have been provided.

Let,

N_1, N_2, N_3 and N_4 be the rotational speeds of gears 1, 2, 3 and 4

T_1, T_2, T_3 and T_4 be the number of teeth on gears 1, 2, 3 and 4

The driving gear 1 is in mesh with the intermediate gear 2, and the speed ratio for these gears is

$$\frac{N_1}{N_2} = \frac{T_2}{T_1} \qquad \ldots(i)$$

Likewise the intermediate gears 2 and 3 are in mesh and accordingly

$$\frac{N_2}{N_3} = \frac{T_3}{T_2} \qquad \ldots(ii)$$

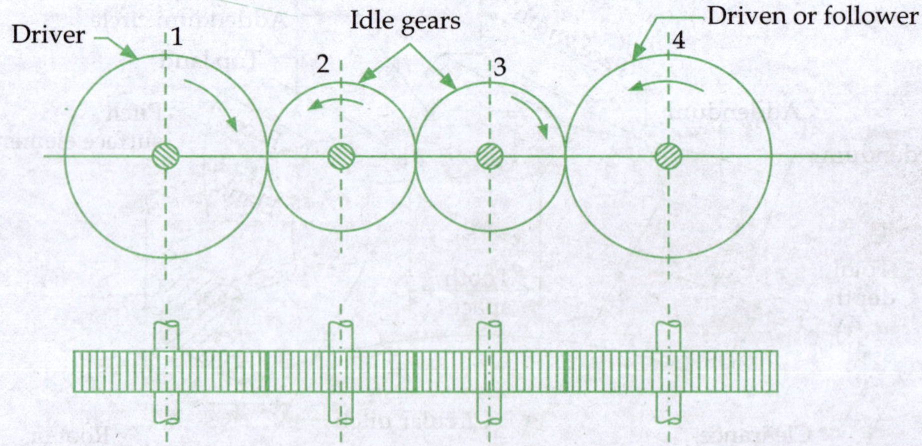

Fig. 20.25. Simple gear train.

Further, the intermediate gear 3 is in engagement with the driven (follower) gear, and therefore

$$\frac{N_3}{N_4} = \frac{T_4}{T_3} \qquad \ldots(iii)$$

The speed ratio N_1/N_4 for the gear assembly can be obtained by multiplying identities (*i*), (*ii*) and (*iii*).

$$\frac{N_1}{N_2} \times \frac{N_2}{N_3} \times \frac{N_3}{N_4} = \frac{T_2}{T_1} \times \frac{T_3}{T_2} \times \frac{T_4}{T_3} \quad \text{or} \quad \frac{N_1}{N_4} = \frac{T_4}{T_1}$$

$$\therefore \text{Speed or velocity ratio} = \frac{\text{speed of the driving wheel}}{\text{speed of the driven wheel}}$$

$$= \frac{\text{number of teeth on the driven wheel}}{\text{number of teeth on the driving wheel}}$$

Train value of a gear train is defined as the reciprocal of the speed ratio of gear train.

$$\text{Train value} = \frac{\text{speed of the driven wheel}}{\text{speed of the driving wheel}} = \frac{\text{number of teeth on the driving wheel}}{\text{number of teeth on the driven wheel}}$$

Comments :

(*i*) The speed ratio and the train value are independent of the size and number of intermediate gears.

(*ii*) The intermediate gears help to obtain the desired direction of motion of the driven gear. With odd number of idle gears, the direction of rotation of the driven wheel will be same as that of the driving wheel. Both the driving and driven wheels will have opposite direction of motion when the number of idle gears happens to be even.

The idle gears are also used when shafts with large centre distance are to be connected.

• **Compound gear train :** Refer Fig. 20.24 for a compound gear train in which all the gears donot mesh with one another. Between the driving gear 1 and the driven (follower) gears 6, two intermediate compound gears have been provided. A compound gear is the gear which carries two wheels mounted on the same shaft. The gears 2 and 3 constitute a compound gear mounted on shaft *M*, and the compound gear comprising gears 4 and 5 is mounted on shaft *P*.

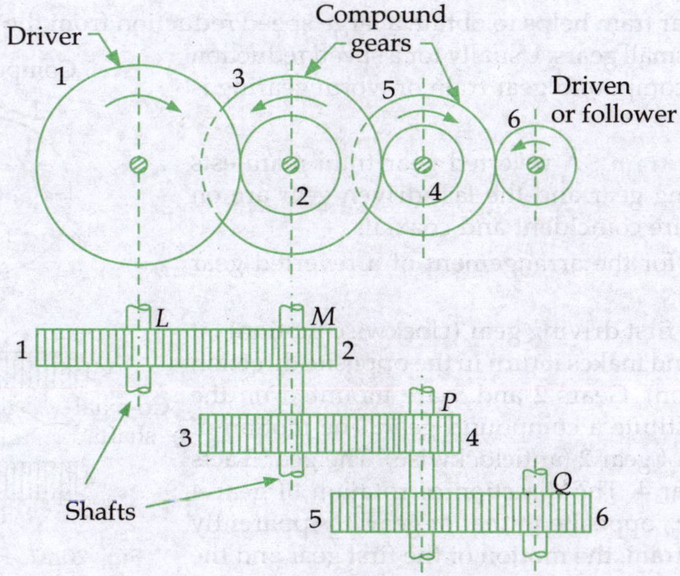

Fig. 20.26. Compound gear train

In the compound gear train, gear 2 is follower to the driving gear 1, and gear 3 is the driving wheel for gear 4. Obviously gears 1, 3 and 5 are the drivers, and gears 2, 4 and 6 are the drivens.

Let, N_1, N_2, N_3, N_4, N_5 and N_6 be the speeds of the respective gears

T_1, T_2, T_3, T_4, T_5 and T_6 be the number of teeth on respective gears.

Gears 1 is in mesh with gear 2 and therefore its speed ratio is

$$\frac{N_1}{N_2} = \frac{T_2}{T_1} \quad ...(a)$$

Gear 3 is in mesh with gear 4 and therefore the speed ratio is

$$\frac{N_3}{N_4} = \frac{T_4}{T_3} \quad ...(b)$$

Similarly for the gears 5 and 6 which engage each other, speed ratio is

$$\frac{N_5}{N_6} = \frac{T_6}{T_5} \quad ...(c)$$

The speed ratio N_1/N_6 for the gear assembly can be worked out by multiplying identities (*a*) (*b*) and (*c*)

$$\frac{N_1}{N_2} \times \frac{N_3}{N_4} \times \frac{N_5}{N_6} = \frac{T_2 \times T_4 \times T_6}{T_1 \times T_3 \times T_5} \quad ...(20.13)$$

Gears 2 and 3 are mounted on the same shaft *M* and therefore $N_2 = N_3$. Again gears 4 and 5 are mounted on the same shaft *P* and therefore $N_4 = N_5$. With these substitutions, equation 20.13 transforms to

$$\text{Speed or velocity ratio} = \frac{N_1}{N_6} = \frac{T_2 \times T_4 \times T_6}{T_1 \times T_3 \times T_5}$$

$$\text{That is} \quad \frac{N_1}{N_6} = \frac{\text{speed of the first driver}}{\text{speed of the last driven}}$$

$$= \frac{\text{product of the number of teeth on drivens}}{\text{product of the number of teeth on drivers}} \quad ...(20.14)$$

A compound gear train helps to obtain a large speed reduction from the first to the last shaft with an assembly of small gears. Usually for a speed reduction of the order of 8 : 1, compound gear train or worm gearing is employed.

• **Reverted gear train :** A reverted gear train manifests when the first driving gear and the last driven gear are on the same axis; axes are coincident and coaxial.

Refer Fig. 20.27 for the arrangement of a reverted gear train.

The gear 1 is the first driving gear (clockwise motion) ; it meshes with gear 2 and makes it turn in the opposite direction (anitclockwise motion). Gears 2 and 3 are mounted on the same shaft and constitute a compound gear. The motion of gear 3 is like to that of gear 2 (anticlockwise). The gear 3 acts as driver for the gear 4. The direction of rotation of gear 4 will be clockwise, *i.e.*, opposite to that of gear 3. Apparently in the reverted gear train, the motion of the first gear and the last gear is like.

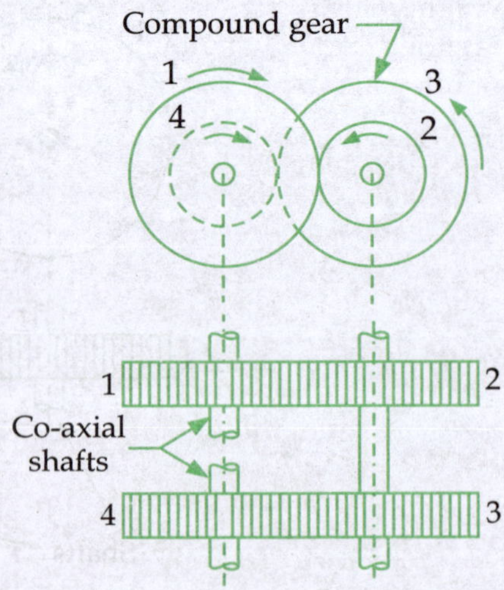

Fig. 20.27. Reverted gear train

Let D_1, D_2, D_3 and D_4 be the pitch circle diameters of the respective wheels, and their corresponding speeds be N_1, N_2, N_3 and N_4.

The distance between shaft centres of gears 1 and 2, and that between gears 3 and 4 is same. That gives

$$\frac{D_1 + D_2}{2} = \frac{D_3 + D_4}{2} \quad \text{or} \quad D_1 + D_2 = D_3 + D_4$$

Presuming that the circular pitch or module of all the gear wheels is same, the number of teeth on each gear is proportional to its circumference or diameter.

That gives

$$T_1 + T_2 = T_3 + T_4$$

Now, Speed ratio $\dfrac{N_1}{N_4} = \dfrac{\text{number of teeth on drivens}}{\text{number of teeth on drivers}} = \dfrac{T_2 \times T_4}{T_1 \times T_3}$

The reverted gear gear train is used to connect hour hand to minute hand in a clock mechanism. Other applications lie in automotive transmissions, lathe back gears and industrial speed reducers.

• **Epicyclic gear train :** In the simple and compound gear trains, the axis of the wheels remain fixed relative to one another. The epicyclic gear train is a special type of gear train in which axis of rotation of one or more of the wheels is carried on an arm and this arm is free to rotate about the axis of rotation of one or the other wheels in the train.

Refer Fig. 20.28 for the arrangement of epicyclic gear train. The gear A and the arm C can rotate about the axis at P, the gear B meshes with gear A and has its axis of rotation on the arm at Q. When the gear A is fixed and the arm C is made to rotate about P, the gear B would be forced to roll around the outside of the gear A.

With N_C revolutions of the arm, the gear wheel B rotates about its axis by speed N_B given by

$$\frac{N_B}{N_C} = 1 + \frac{T_A}{T_B}$$

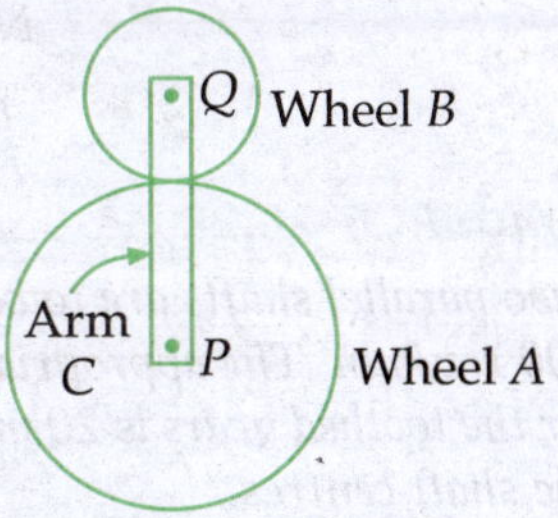

Fig. 20.28. Epicyclic gear train

The epicyclic gearing can bring about extensive speed reduction, and is capable of transmitting high velocity ratios with gears of moderate size in a comparatively short space. These aspects render it ideally suitable for use in differential gears of automobiles, wrist watches, hoists and pulley blocks etc.

EXAMPLE 20.26

A toothed gear is stated to have 48 teeth and circular pitch of 18. Determine the pitch diameter, diametral pitch and module of the gear.

Solution : Circular pitch $p_c = \dfrac{\pi D}{T}$

$\therefore$ pitch diameter $D = \dfrac{p_c \times T}{\pi} = \dfrac{18 \times 48}{\pi} = $ **275.16 mm**

(*b*) Invoking the relation : $p_c \times p_d = \pi$

$\therefore$ diametrical pitch $p_d = \dfrac{\pi}{p_c} = \dfrac{\pi}{18} = $ **0.174**

(*c*) Module, $m = \dfrac{D}{T} = \dfrac{275.16}{48} = $ **5.73 mm per tooth**

EXAMPLE 20.27

Two matching spur gears have 70 and 30 teeth respectively. Corresponding to a module of 5 mm, determine the centre to centre distance between the gears.

Solution : Centre to centre distance between two matching spur gears is

$$l = \frac{D_1 + D_2}{2}$$

From the relation $m = D/T$; $D = mT$ and therefore

$$l = \frac{m(T_1 + T_2)}{2} = \frac{5(70 + 30)}{2} = \mathbf{250\ mm}$$

EXAMPLE 20.28

A fixed gear having 200 teeth is in mesh with another gear having 50 teeth. The two gears are connected by an arm which makes one revolution about the centre of the bigger gear. Determine the number of turns made by the smaller gear.

Solution : Refer fig. 20.26. The arrangement corresponds to epicyclic gear train.

The ratio of the speed of wheel B to that of arm C is given by

$$\frac{N_B}{N_C} = 1 + \frac{T_A}{T_B} = 1 + \frac{200}{50} = 4$$

$$\therefore \quad N_B = 4N_C = 4 \times 1 = \mathbf{4}$$

EXAMPLE 20.29

Two parallel shafts are to be connected by spur gearing. One shaft runs at 100 rpm and other shaft turns 300 rev/min. The appropriate centre to centre distance between the shafts is 500 mm. If the circular pitch for the toothed gears is 20 mm, determine the number of teeth on each wheel and the exact distance between the shaft centres.

Solution : Centre to centre distance between two mating spur gears is

$$l = \frac{D_1 + D_2}{2} \ ; \ D_1 + D_2 = 2 \times 500 = 1000 \qquad ...(i)$$

For two meshing spur gears,

$$\frac{N_1}{N_2} = \frac{D_2}{D_1} \ ; \ \frac{100}{300} = \frac{D_2}{D_1} \ ; \ D_2 = 3D_1 \qquad ...(ii)$$

From the identities (*i*) and (*ii*)

$$D_1 + 3D_1 = 1000 \ ; \ D_1 = 250 \text{ mm}$$

and $\qquad D_2 = 3D_1 = 3 \times 250 = 750 \text{ mm}$

From the relation for circular pitch, $p_c = \frac{\pi D}{T}$, we get

$$\text{Number of teeth on the first gear } T_1 = \frac{\pi \times 250}{20} = 39.25$$

$$\text{Number of teeth on the second gear } T_2 = \frac{\pi \times 750}{20} = 117.75$$

However, the number of teeth on both the gears have to be an exact number. Accordingly, let it be presumed that the number of teeth on the first gear is 40. The second gear will then have 120 teeth. Then

Revised pitch circle diameter of first wheel $D_1' = \frac{T_1 \times p_c}{\pi} = \frac{40 \times 20}{\pi} = 254.78$ mm

Revised pitch circle diameter of second wheel $D_2' = \frac{T_2 \times p_c}{T_2} = \frac{120 \times 20}{\pi} = 764.34$ mm

∴ Correct distance between the shaft centres,

$$l_{correct} = \frac{D_1' + D_2'}{2} = \frac{254.78 + 764.34}{2} = \mathbf{509.65\ mm}$$

EXAMPLE 20.30

A simple gear train consists of two gears only, each gear mounted on separate parallel shafts. The number of teeth on the driving and driven gears are 28 and 70 respectively. If the driving gear turns 1200 revolutions per minute, determine:

(i) speed ratio and train value of the gear train,
(ii) speed of the driven gear, and
(iii) direction of rotation of the driven gear if the driving gear rotates clockwise

Solution : Refer Fig. 20.29 for the simple gear train consisting of two gears only. The gear 1 is driving the gear 2.

Speed ratio $\frac{N_1}{N_2} = \frac{T_2}{T_1} = \frac{70}{28} = \mathbf{2.5}$

Train value of a gear train is defined as the reciprocal of the speed ratio of gear train

Train value $= \frac{1}{\text{speed ratio}} = \frac{1}{2.5} = \mathbf{0.4}$

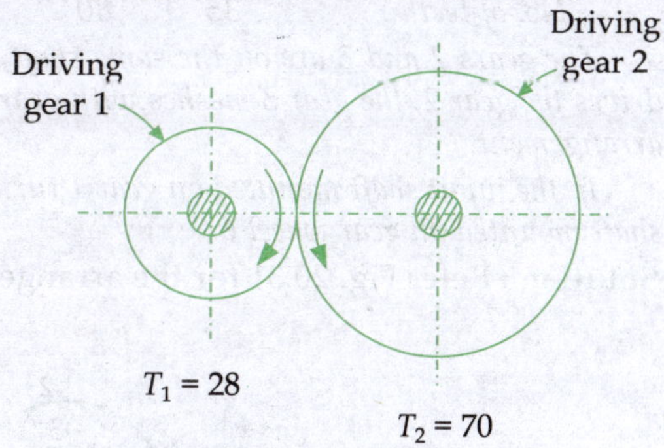

Fig. 20.29

(ii) Speed ratio $\frac{N_1}{N_2} = 2.5$ as calculated above

∴ Speed of driven gear, $N_2 = \frac{N_1}{2.5} = \frac{1200}{2.5} = \mathbf{480\ rev/min}$

(iii) In a simple gear train, the two meshing gears always move in opposite directions. As the driving gear is rotating clockwise, the direction of rotation of driven gear would be anti-clockwise.

EXAMPLE 20.31

A simple gear train consists of three gears; each mounted on separate parallel shaft as shown in Fig. 20.30.

The number of teeth on gears 1, 2 and 3 are 28, 42 and 70 respectively, and the driving gear 1 rotates clockwise at speed of 1000 rpm. Calculate (i) the speed ratio of the gear train; (ii) speed and the direction of rotation of the follower.

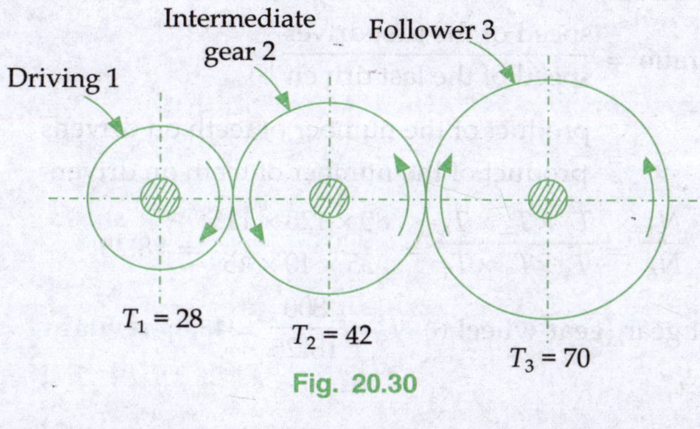

Fig. 20.30

Solution : Speed ratio $= \dfrac{\text{speed of driver}}{\text{speed of follower}} = \dfrac{\text{number of teeth on follower}}{\text{number of teeth on driver}}$

$$= \frac{70}{28} = 2.5$$

(*ii*) Speed ratio $= \dfrac{N_1}{N_3} = 2.5$ as calculated above

$\therefore$ Speed of follower $N_3 = \dfrac{N_1}{2.5} = \dfrac{1000}{2.5} = $ **400 rpm**

With odd number of idle (intermediate gears), the direction of rotation of the follower is same as that of driving wheel. Here the driving gear is rotating clockwise and as such the direction of rotation of the follower would be clockwise.

EXAMPLE 20.32

A compound gear train consists of six gears and the number of teeth on the gears are as follows:

Gear	*1*	*2*	*3*	*4*	*5*	*6*
No. of teeth	*35*	*80*	*40*	*125*	*45*	*115*

The gears 2 and 3 are on the same shaft, and so are the gears 4 and 5 on another shaft. The gear 1 drives the gear 2, the gear 3 meshes with gear 4, and the gear 5 is in engagement with gear 6. Sketch the arrangement.

If the input shaft mounted on gear 1 turns 800 rev/min, determine the rotational speed of the output shaft mounted on gear wheel 6.

Solution : Refer Fig. 20.31 for the arrangement of the given compound gear train

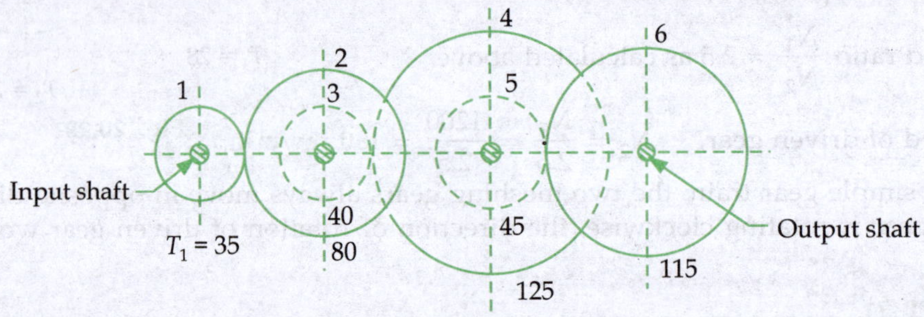

Fig. 20.31

Driving gears are : 1, 3 and 5

Driven gears are : 2, 4 and 6

For compound gear train,

$$\text{speed or velocity ratio} = \frac{\text{speed of the first driver}}{\text{speed of the last driven}}$$

$$= \frac{\text{product of the number of teeth on drivens}}{\text{product of the number of teeth on drivers}}$$

Thus, $$\frac{N_1}{N_6} = \frac{T_2 \times T_4 \times T_6}{T_1 \times T_3 \times T_5} = \frac{80 \times 125 \times 115}{35 \times 40 \times 45} = 18.25$$

$\therefore$ speed of the last gear (gear wheel 6) $N_6 = \dfrac{800}{18.25} = $ **43.83 rev/min**

REVIEW AND SUMMARY

1. Belt, rope and chain drives are used for transmission of power or rotary motion between shafts whose centre distance is too large and gear trains are not possible.
2. The angular velocities of two pulleys connected either by an open belt or a crossed belt are inversely proportional to their diameters. That is

$$\frac{N_2}{N_1} = \frac{d_1}{d_2}$$

3. The open belt drive is used with shafts arranged parallel and rotating in the same directions. The crossed belt drive is used with shafts arranged in parallel and rotating in opposite directions.
4. When frictional grip between the belt and shaft becomes insufficient, there occurs some forward motion of driver without carrying the belt with it. This phenomenon is called ***slip*** of the belt. The velocity ratio reduces due to belt slipping and is given by

$$\frac{N_2}{N_1} = \frac{d_1 + t}{d_2 + t}\left(1 - \frac{s}{100}\right)$$

where s is the total percentage of slip; sum of percentage slip between the driver and the belt and percentage of slip between the follower and the belt.
5. When the belt passes from the slack side to the tight side, a certain portion of the belt extends and when the belt passes from the tight to slack side the belt contracts. Due to these changes in length, there is relative motion between the belt and pulley surfaces. This relative motion is termed as ***creep*** of the belt.
6. The ratio of belt tensions on the tight and slack side is

$$\frac{T_1}{T_2} = e^{\mu\theta} \text{ for flat belt drive}$$

$$\frac{T_1}{T_2} = e^{\frac{\mu\theta}{\sin\alpha}} \text{ for V-belt drive}$$

Power transmitted $P = (T_1 - T_2)\ V$
7. A part of the belt in contact with the pulley moves in a circular path and is subjected to centrifugal force that acts radially outwards ($T_c = mV^2$). Maximum power is transmitted when

$$V = \left(\frac{T}{3m}\right)^{\frac{1}{2}}; \quad T_c = \frac{T}{3}$$

where T is the maximum permissible tension in the belt, V is the belt speed and m is mass per metre length of belt.
8. Gears or tooth wheels are used to transmit motion from one shaft to another when the distance between the two shafts (the driver and the follower) is very small. Gears use no intermediate link or connector and transmit the motion by direct contact. Hence, gears ensure definite velocity ratio and provide a positive drive.
9. The different types of gears with their application are :

   (*i*) ***Spur gears*** used for connecting two parallel and coplanar shafts with teeth parallel to the axes of gear wheel.

(ii) ***Helical gears*** used for connecting two parallel and coplanar shafts with teeth inclined to the axes of gear wheel.

(iii) ***Bevel gears*** used for connecting two non-parallel or intersecting but coplanar shafts.

10. For a spur gear of pitch circle diameter D and heaving T teeth,

$$\text{Circular pitch } p_c = \frac{\pi D}{T}$$

$$\text{Diameter pitch } p_d = \frac{T}{D} = \frac{\pi}{p_c}$$

$$\text{Module } m = \frac{D}{T} = \frac{1}{p_d}$$

11. A gear train is a combination of gears used to transmit motion from one shaft to another. Main types of gear trains are:

(i) *Simple gear train*: Each gear is on a separate shaft and all the gear axis remain fixed to the frame.

(ii) *Compound gear train*: Two or more gears are mounted on the same shaft and rotate about an axis with the same angular velocity.

(iii) *Reverted gear train*: The axis of the first gear and last gear are coaxial *i.e.*, coincide.

(iv) *Epicyclic gear train*: The axis of the shafts over which the gears are mounted may have relative motion.

REVIEW QUESTIONS

A. Conceptual and conventional questions :

1. Name the different driving elements used for the transmission of motion and power from one shaft to another.
2. (*a*) What is the material of which belts are made ?
   (*b*) What are the double and triple ply belts ?
   (*c*) How are the rubber belting and balata belting made ?
3. Give a brief description of the various types of belts used for power transmission.
4. Why is the *V*-belt drive preferred over other forms of mechanical drive ?
5. List the factors to be kept in view regarding maintenance of belts.
6. State the difference between open-belt and crossed-belt drive.
7. Setup expressions for the length of belt in (*i*) open belt system and (*ii*) crossed belt system.
8. What is slip ? How it affects the velocity ratio of belt drive ?
9. Define the following terms in relation to belt drives :
   (*a*) velocity ratio, (*b*) crowning of pulley, (*c*) creep
10. Setup the following expression for the flat belt drive

$$\frac{T_1}{T_2} = e^{\mu\theta}$$

where T_1 and T_2 are the tensions on the tight and slack sides of the belt; θ is the angle of contact and μ is the coefficient of friction between the belt and the pulley rim.

11. What leads to centrifugal tension in a belt ? With in permissible limits of stress (tension) in the belt, how does centrifugal tension affect the power transmitted by the belt ?
12. Setup an expression for the centrifugal tension induced in a belt transmitting power. Also determine the linear velocity of belt for maximum power transmission

(centrifugal tension taken into account).
13. Show that for maximum power transmission, the centrifugal tension should not exceed $^1/_3$rd of total tension.
14. Derive an expression for the ratio of driving tensions in rope drive assuming the angle of pulley to be 2β.
15. An engine turning 250 rev/min drives a 40 cm diameter shaft with the help of belt drive system. If the diameter of engine pulley is 60 cm, determine the speed of the shaft.

How would the shaft speed be affected if a belt thickness of 10 mm is accounted for ?

(375 rpm, 371.95 rpm)
16. Two parallel shafts 3 m apart are provided with 300 mm and 600 mm diameter pulleys and are connected by an open belt. The direction of rotation of the follower pulley is to be reversed by changing over to an open belt drive. How much additional length of the belt is required. **(6 cm)**
17. An open belt connects two flat pulleys. The smaller pulley is 600 mm in diameter and turns 200 revolutions per minute. The angle of lap on the pulley is 160° and the coefficient of friction between the belt and the pulley face is 0.25. If the maximum permissible tension in the belt is 2500 N, calculate the power transmitted by the belt. **(7.95 kW)**
18. A blower is driven by an electric motor through a belt drive. The speed of the belt is 75 m/s and 15 kW power is being transmitted. Calculate the difference of tensions on the two sides of the belt.

If the tension on the tight side is 1.25 times the tension on the slack side, determine the two tensions and the width of the belt necessary if the maximum permissible tension is 20 N per mm width of belt.
19. A leather belt of thickness 12 mm and width 100 mm connects two pulleys; their centres being 2.5 m apart. There is transmission of 11.25 kW power from a shaft running at 100 rpm to another shaft which turns 200 rev/min. If the diameter of smaller pulley is 500 mm and the coefficient of friction between the belt and pulley rims is 0.25, determine the stress induced in the belt material if the belt system is (*i*) open, (*ii*) crossed.

(3.45 N/mm^2, 2.95 N/mm^2)
20. A belt of thickness 100 mm and width 145 mm runs over two pulleys at a speed of 0.5 m/s. The other necessary specifications for the belt drive system are as follows :

density of belt material = 10 kN/m^3

safe working stress for the belt material = 2×10^6 N/m^2

coefficient of friction between belt and pulley rim = 0.3

The angle of contact is spread over 4/9 the of the circumference of the pulley.

Determine the centrifugal tension and the maximum power transmitted at this speed.

[Ans. **(332 N and 21085 kW)**
21. Explain briefly :
    (*i*) Why is the tension on tight side greater than that on the slack side ?
    (*ii*) Why a rope fitting in a V-shaped groove on a pulley rim does not slip as easily as a belt on cylindrical pulley ?
    (*iii*) Why is it that the velocity ratio of belt and ropes is variable ?

22. Enumerate the main types of toothed gears and point out their usual field of application.
23. List the advantages/disadvantages of gear drive.
24. Define circular pitch, diametrical pitch and module in relation to tooth gears.
25. Explain briefly the difference between simple, compound and epicyclic gear trains.
26. What is meant by gear train ? Name the different types of gear trains and state where each is used.
27. Define the term train value. How is it related to speed ratio ?

B. Complete the following statements with most appropriate word/words :

1. In a belt drive, the pulley connected to the rotating shaft is called .... and the pulley that needs to be driven is termed as ....
2. The relative motion between the pulley and belt is called .... and this .... the velocity ratio of the system.
3. The creep in a belt is due to .... on the tight and slack sides of the belt.
4. A slight dwell known as .... is provided on the pulley to prevent lateral slip of the belt.
5. The power transmitted by the belt is maximum when the maximum tension of the belt is .... of centrifugal tension.
6. The designation 6 × 19 of a wire rope means .... strands and .... wires.
7. Two parallel and coplanar shafts are connected by .... gears.
8. When two shafts are neither parallel nor intersecting, power can be transmitted by using a pair of .... gears
9. .... drive is preferred where high reduction in velocity ratio is required.
10. The ratio of number of teeth to the pitch circle diameter in millimeter is called ....
11. The train value of a gear train is reciprocal of ....
12. When the axes of the first and last wheels are co-axial, the gear train is known as .... gear train.

Answers :

1. driver, follower 2. slip, reduces 3. unequal tension 4. crowning 5. three lines 6. six, nineteen 7. spur 8. spiral 9. worm and wormwheel 10. diametral pitch 11. velocity ratio 12. reverted

C. Multiple choice questions :

1. Which one of the following statements relating to belt drives is correct?
   (*a*) the rotational speeds of the pulleys are directly proportional to their diameter
   (*b*) the length of the crossed belt increases as the sum of the diameters of the pulleys increases
   (*c*) the crowning of the pulleys is done to make the drive sturdy
   (*d*) the slip increases the velocity ratio
2. The ratio between tensions in the tight and slack side of a flat belt drive increases
   (*a*) in direct proportion to the angle of lap
   (*b*) exponentially to the product of lap angle and coefficient of friction

(*c*) in direct proportion to the coefficient of friction
(*d*) in direct proportion to the belt width

3. Which of the following statements is wrong about the power transmitted by a flat belt drive ?
(*a*) power transmitted becomes more when tension on the slack side is decreased
(*b*) power transmitted increases with an increase in tension on the tight side
(*c*) power transmitted becomes more when tensions on the two sides are increased by the same amount
(*d*) power transmitted remains unchanged when tensions on the two sides are increased by the same amount

4. The centrifugal tension in the flat belt drive system increases
(*a*) in direct proportion to mass per unit length of the belt
(*b*) in direct proportion to belt thickness
(*c*) in direct proportion to square of belt speed
(*d*) exponentially with maximum permissible tension in the belt

5. For maximum power transmitted by belt drive, the centrifugal tension is equal to
(*a*) half the effective tension on the tight side of the belt drive
(*b*) one-third of effective tension on the tight side of the belt drive
(*c*) two-third of effective tension on the tight side of the belt drive
(*d*) three-fourth of effective tension on the tight side of the belt drive

6. Considering centrifugal tension, the power transmitted by belt drive is maximum at velocity V equal to

(*a*) $V = \left(\dfrac{T}{m}\right)^{1/2}$ (*b*) $V = \left(\dfrac{T}{3m}\right)^{1/2}$

(*c*) $V = \left(\dfrac{3T}{m}\right)^{1/2}$ (*d*) $V = \left(\dfrac{2T}{3m}\right)^{1/2}$

where V is in m/s, T is the maximum permissible tension in newton and m is mass per unit length of the belt in kg.

7. The ratio of tension of tight side to slack side $\dfrac{T_1}{T_2}$ in case of a V-belt drive is

(*a*) $e^{\mu\theta/\cos\alpha}$ (*b*) $e^{\mu\theta\cos\alpha}$ (*c*) $e^{\mu\theta/\sin\alpha}$ (*d*) $e^{\mu\theta\sin\alpha}$

(μ = coefficient of friction, θ = angle of lap, and α = angle of groove)

8. Straight bevel gears are used for connecting
(*a*) two parallel coplanar shafts
(*b*) two intersecting coplanar shafts
(*c*) two non-parallel and non-intersecting shafts
(*d*) two non-intersecting and perpendicular shafts

9. A rack is a gear of infinite
(*a*) pitch (*b*) module (*c*) diameter (*d*) number of teeth

10. The product of circular pitch and diametral pitch equals
(*a*) 1 (*b*) 1.57 (*c*) π (*d*) infinity

11. Module of a gear is
(*a*) the ratio of pitch circle diameter to the number of teeth
(*b*) the ratio of number of teeth to pitch circle diameter
(*c*) the reciprocal of pitch circle diameter
(*d*) the product of pitch circle diameter and the number of teeth

12. Consider a simple gear train having odd number of idler gears. The direction of motion of the driven gear will
(*a*) be same as that of the driving gear
(*b*) be opposite to that of the driving gear
(*c*) depend upon the number of teath on the driving gear
(*d*) depend upon the total number of teeth on all gears of the train

13. Consider a simple gear train consisting of three gear wheels with 15, 10 and 30 teeth respectively. The velocity ratio of third wheel to first wheel would be
(*a*) 1/2 and same direction (*b*) 1/2 and opposite direction
(*c*) 2 and same direction (*d*) 2 and opposite direction

14. In a reverted gear train
(*a*) one gear is always fixed
(*b*) atleast one of the gear axis is in motion
(*c*) the axes of the first and the last gear are co-axial
(*d*) the axes of the first and the last gear are parallel

15. A gear train, in which atleast one of the gear is in motion relative to the frame, is known as
(*a*) compound gear train (*b*) reverted gear train
(*c*) bevel gear train (*d*) epicyclic gear train

16. The gear box in an automobile utilizes
(*a*) simple gear train (*b*) compound gear train
(*c*) epicyclic gear train (*d*) inverted gear train

Answers :

1. (*b*)	2. (*b*)	3. (*c*)	4. (*b* and *c*)	5. (*a*)	6. (*b*)
7. (*c*)	8. (*b*)	9. (*c*)	10. (*c*)	11. (*a*)	12. (*a*)
13. (*a*)	14. (*c*)	15. (*d*)	16. (*b*)		

INDEX

INDEX

Win Prizes

ATTENTION READERS

Win Prizes

We request for your frank assessment regarding various aspects of the book as given below :

ENGINEERING MECHANICS

(Statics & Dynamics)

By: Dr. D.S. Kumar

Please Fill up the Sheet in Neat and Clean Writing

(i) What topics of your syllabus, important for your examination point of view, are not covered in this book ?

....

....

....

....

(ii) What are the chapters wherein the treatment of the subject matter is not systematic or organised or updated ?

....

....

....

....

(iii) Have you come across misprints/mistakes/factual inaccuracies in the book? Please specify the chapters and the page numbers.

....

....

....

....

(iv) What other books on the same subject have you found/heard better than the present book? Please specify in terms of price and quality

....

....

....

....

(v) Further suggestions and comments for the improvement of this book.

....

....

....

....

(PS: If need be please attach a separate sheet to write your views)

OTHER DETAILS

(i) Who recommended you this book ? (Please tick (✓) in the box near the option relevant to you.

☐ Teacher ☐ Friends ☐ Bookseller

....

....

....

....

(ii) Name of the recommending teacher, his designation and address :

....

....

....

....

(iii) Name and address of the bookseller you purchased the book from :

....

....

....

....

(iv) Name and address of your institution (Please mention the University or Board as the case may be)

....

....

....

....

(v) Your name and complete postal address :

....

....

....

....

(vi) Course you are enrolled in :

....

....

....

....

The best assessment will be awarded each month: The award will be in the form of our publications as decided by the Editorial Board.

Please mail the filled up coupon at your earliest to :

S.K. KATARIA & SONS®

4885/109, Prakash Mahal, Dr. Subhash Bhargav Lane,
Opposite Delhi Medical Association, Daryaganj, New Delhi–110002 (INDIA)
Phone: +91-11-23243489, 43551243; *Telefax:* +91-11-23243489
e-mail: katariabooks@yahoo.com; katariabook@gmail.com
Website: www.skkatariaandsons.com